Chemistry and Toxicology of Pollution

Chemistry and Toxicology of Pollution

Ecological and Human Health

Second Edition

Des W. Connell
Griffith University
Nathan, Brisbane, Australia

Greg J. Miller
Griffith University
Nathan, Brisbane, Australia

This edition first published 2023
© 2023 John Wiley & Sons, Inc.

Edition History
First edition 1984 Wiley

All rights reserved. No part of this publication may be reproduced, stored in a retrieval system, or transmitted, in any form or by any means, electronic, mechanical, photocopying, recording or otherwise, except as permitted by law. Advice on how to obtain permission to reuse material from this title is available at http://www.wiley.com/go/permissions.

The right of Des W. Connell and Greg J. Miller to be identified as the authors of this work has been asserted in accordance with law.

Registered Office
John Wiley & Sons, Inc., 111 River Street, Hoboken, NJ 07030, USA

For details of our global editorial offices, customer services, and more information about Wiley products visit us at www.wiley.com.

Wiley also publishes its books in a variety of electronic formats and by print-on-demand. Some content that appears in standard print versions of this book may not be available in other formats.

Trademarks: Wiley and the Wiley logo are trademarks or registered trademarks of John Wiley & Sons, Inc. and/or its affiliates in the United States and other countries and may not be used without written permission. All other trademarks are the property of their respective owners. John Wiley & Sons, Inc. is not associated with any product or vendor mentioned in this book.

Limit of Liability/Disclaimer of Warranty
In view of ongoing research, equipment modifications, changes in governmental regulations, and the constant flow of information relating to the use of experimental reagents, equipment, and devices, the reader is urged to review and evaluate the information provided in the package insert or instructions for each chemical, piece of equipment, reagent, or device for, among other things, any changes in the instructions or indication of usage and for added warnings and precautions. While the publisher and authors have used their best efforts in preparing this work, they make no representations or warranties with respect to the accuracy or completeness of the contents of this work and specifically disclaim all warranties, including without limitation any implied warranties of merchantability or fitness for a particular purpose. No warranty may be created or extended by sales representatives, written sales materials or promotional statements for this work. The fact that an organization, website, or product is referred to in this work as a citation and/or potential source of further information does not mean that the publisher and authors endorse the information or services the organization, website, or product may provide or recommendations it may make. This work is sold with the understanding that the publisher is not engaged in rendering professional services. The advice and strategies contained herein may not be suitable for your situation. You should consult with a specialist where appropriate. Further, readers should be aware that websites listed in this work may have changed or disappeared between when this work was written and when it is read. Neither the publisher nor authors shall be liable for any loss of profit or any other commercial damages, including but not limited to special, incidental, consequential, or other damages.

Library of Congress Cataloging-in-Publication Data

Names: Connell, D. W., author. | Miller, Greg J., author.
Title: Chemistry and toxicology of pollution : ecological and human health /
  Des W. Connell, Griffith University, Nathan, Brisbane, Australia, Greg
  J. Miller, Griffith University, Nathan, Brisbane, Australia
Other titles: Chemistry and ecotoxicology of pollution
Description: Second edition. | Hoboken, NJ : Wiley, 2023. | Revised edition
  of: Chemistry and ecotoxicology of pollution / Des W. Connell, Gregory
  J. Miller. 1984. | Includes bibliographical references and index.
Identifiers: LCCN 2022017541 (print) | LCCN 2022017542 (ebook) | ISBN
  9781119377603 (cloth) | ISBN 9781119377627 (adobe pdf) | ISBN
  9781119377634 (epub)
Subjects: LCSH: Pollution. | Environmental toxicology.
Classification: LCC QH545.A1 C65 2023 (print) | LCC QH545.A1 (ebook) |
  DDC 571.9/5–dc23/eng/20220608
LC record available at https://lccn.loc.gov/2022017541
LC ebook record available at https://lccn.loc.gov/2022017542

Cover Design: Wiley
Cover Image: © Si-Gal/Getty Images

Set in 9.5/12.5pt STIXTwoText by Straive, Chennai, India

Contents

Preface *xi*
How to Use this Book *xiii*
About the Companion Website *xiv*

1 Introduction *1*
1.1 Background *1*
1.2 Environmental Pollution *2*
1.3 Drivers of Environmental Pollution *2*
1.4 Pollution and Environmental Health *3*
1.5 Chemicals and Pollution *4*
1.6 Health and Environmental Effects of Chemicals *10*
1.7 Science of Pollution: Assessment and Management *13*
References *18*

2 Environmental Pollutants *21*
2.1 Introduction *21*
2.2 Environmental Pollutants *21*
2.3 Types of Pollutants *26*
2.4 Some Special Classes of Pollutants *32*
2.5 Global Production, Emissions, and Releases of Chemicals and Hazardous Wastes *34*
2.6 Key Points *39*
References *40*

3 Pollutants, Health, and Environment *41*
3.1 Introduction *41*
3.2 Environmental Health *41*
3.3 Disease Causation *42*
3.4 Environmental Pollution and Burden of Disease *44*
3.5 Biodiversity and Ecosystem Services *51*
3.6 Human Impacts on Biodiversity and Ecosystem Services *52*
3.7 Environmental Assessment and Management of Pollution *56*
3.8 Key Points *59*
References *60*

4 Chemodynamics of Pollutants *63*
4.1 Introduction *63*
4.2 Some Fundamental Properties of Chemical Pollutants and Environmental Phases *63*
4.3 Fundamental Principles of Partition Behavior *66*
4.4 Relationships Describing Partition Behavior *67*

4.5	Partitioning Behavior in the Laboratory	*68*
4.6	Pollutant Behavior in Different Environmental Compartments	*69*
4.7	Models for Bioconcentration	*72*
4.8	The Persistent Organic Pollutants (POPs) – Long-Range Transport	*74*
4.9	Bioaccumulation of Metals and Metalloids	*76*
4.10	Fate of Chemicals in the Environment	*77*
4.11	Transport Processes: Advection, Diffusion, and Leaching	*81*
4.12	Transformation and Degradation Processes	*82*
4.13	Redox Conditions in the Environment	*83*
4.14	Kinetics of Transformation and Degradation	*84*
4.15	Acidification Processes in the Environment	*85*
4.16	Key Points	*87*
	References	*88*

5 Environmental Toxicology and Ecotoxicology *91*

5.1	Introduction	*91*
5.2	Exposure Pathways for Humans and Natural Biota	*92*
5.3	Toxicant Behavior in Organisms	*94*
5.4	Relationships Between Exposure to a Toxicant as Dose or Environmental Concentration and the Resultant Biological Effects	*95*
5.5	Models for the Distribution of Exposure Dose or Concentration	*102*
5.6	Relationships Between Exposure Level and Exposure Time Resulting in an Adverse Effect	*102*
5.7	Dose–Response and Dose Thresholds	*103*
5.8	Toxicity Due to Exposure to Multiple Toxicants	*104*
5.9	Quantitative Structure–Activity Relationships (QSARs)	*104*
5.10	Lethal and Sublethal Effects	*106*
5.11	Types and Classification of Toxicants	*108*
5.12	Effects on Populations and Ecosystems	*109*
5.13	Biomarkers	*109*
5.14	Key Points	*110*
	References	*111*

6 Genetic Toxicology and Endocrine Disruption: Environmental Chemicals *113*

6.1	Introduction	*113*
6.2	Genome	*113*
6.3	Genotoxicity and Mutagenicity	*114*
6.4	Reproductive and Developmental Toxicity	*118*
6.5	Endocrine Disruption	*118*
6.6	Key Points	*122*
	References	*124*

7 Some Principles of Pollution Ecology and Ecotoxicology *125*

7.1	Introduction	*125*
7.2	Controlling Factors in Natural and Human Systems	*126*
7.3	Classes and General Effects of Pollutants	*128*
7.4	Ecology of Deoxygenation and Nutrient Enrichment of Aquatic Systems	*129*
7.5	Ecotoxicology of Toxic Substances	*132*
7.6	Suspended Solids	*134*
7.7	Thermal Ecology	*134*
7.8	Pathogenic Microorganisms	*135*
7.9	Biotic Responses to Pollution at Different Environmental Temperatures and Latitudes	*135*
7.10	Biotic Indices	*136*

7.11	Indicator Species, Ecological Indicator Species, Chemical Monitor Species, and Biomarkers *139*
7.12	Key Points *139*
	References *140*

8 Pollutants in the Oceans, Estuaries, and Freshwater Systems *143*

8.1	Introduction *143*
8.2	Deoxygenating Substances *144*
8.3	Nutrient Enrichment and Eutrophication *160*
8.4	Key Points *174*
	References *177*

9 Pesticides *179*

9.1	Introduction *179*
9.2	Pesticides *179*
9.3	Nature and Properties of Various Pesticides *182*
9.4	Major Groups of Pesticides *183*
9.5	Sources and Emissions of Pesticides *189*
9.6	Environmental Behavior and Fate of Pesticides *190*
9.7	Environmental Exposures to Pesticides *198*
9.8	Absorption, Distribution, Biotransformation, and Excretion (ADBE) of Pesticides in Humans and Wildlife *200*
9.9	Toxic Effects of Pesticides on Biota *204*
9.10	Toxic Action of Pesticides *210*
9.11	Ecological Effects of Pesticides on Populations and Communities of Biota *211*
9.12	Human Health Effects of Pesticides *216*
9.13	Key Points *222*
	References *225*

10 Petroleum, Coal, and Biofuels *231*

10.1	Introduction *231*
10.2	Fossil Fuels *231*
10.3	Biofuels *235*
10.4	Physiochemical Properties of Fuels and Chemical Components *237*
10.5	Sources, Emissions, and Releases of Petroleum-Derived Hydrocarbons *237*
10.6	Environmental Behavior and Fate of Fossil Fuels and Biofuels *239*
10.7	Environmental Exposures *242*
10.8	Uptake, Metabolism, and Bioaccumulation of Petroleum Hydrocarbons *243*
10.9	Environmental Toxicities of Petroleum and Biofuels *244*
10.10	Ecosystem Effects of Petroleum Oils *246*
10.11	Effects on Human Health and Ecosystem Services *249*
10.12	Key Points *251*
	References *253*

11 Toxic Organic Pollutants *255*

11.1	Introduction *255*
11.2	Toxic Organic Chemicals *255*
11.3	Classification of Toxic Organic Chemicals *257*
11.4	Persistent, Bioaccumulative, and Toxic Chemicals *259*
11.5	Persistent Organic Pollutants (POPs) *261*
11.6	Less-Persistent and Emerging Toxic Organic Chemicals *267*
11.7	Sources and Emissions of Toxic Organic Chemicals *274*
11.8	Behavior and Fate in the Environment *277*
11.9	Environmental Exposures to Toxic Organic Chemicals *282*

- 11.10 Biological Processes Affecting Toxic Organic Chemicals in Wildlife and Humans *284*
- 11.11 Environmental Toxicity of Organic Chemicals *288*
- 11.12 Effects of Toxic Organic Chemicals on Wildlife Populations and Communities *290*
- 11.13 Human Health Effects of Toxic Organic Chemicals *292*
- 11.14 Key Points *296*
  - References *298*

12 Metals *303*
- 12.1 Introduction *303*
- 12.2 Metals *303*
- 12.3 Sources and Emissions of Metals *305*
- 12.4 Behavior and Fate of Metals in the Environment *306*
- 12.5 Environmental Exposures of Metals to Organisms *310*
- 12.6 Absorption, Distribution, Transformation, and Excretion in Organisms *311*
- 12.7 Toxic Effects of Metals on Organisms *314*
- 12.8 Ecological Effects of Metals *321*
- 12.9 Human Health Effects of Metals *322*
- 12.10 Key Points *324*
  - References *326*

13 Air Pollutants *331*
- 13.1 Introduction *331*
- 13.2 Earth's Atmosphere *331*
- 13.3 Air Pollution, Weather, and Climate *333*
- 13.4 Outdoor and Indoor Air Pollutants *333*
- 13.5 Sources and Emissions of Air Pollutants into the Atmosphere *338*
- 13.6 Behavior and Fate of Pollutants in the Atmospheric Environment *340*
- 13.7 Exposures from Pollutants in the Atmospheric Environment *347*
- 13.8 Effects of Air Pollutants on Wildlife and Natural Systems *349*
- 13.9 Human Health Effects of Air Pollutants *352*
- 13.10 Global Impacts and Risks of Air Pollutants *358*
- 13.11 Key Points *359*
  - References *360*

14 Greenhouse Gases, Global Warming, and Climate Change *365*
- 14.1 Introduction *365*
- 14.2 The Greenhouse Effect *366*
- 14.3 Greenhouse Gases and Atmospheric Particles *370*
- 14.4 Sources, Emissions, and Sinks of Greenhouse Gases *373*
- 14.5 Environmental Exposures from Atmospheric Greenhouse Gases *377*
- 14.6 Greenhouse Gases, Climate Change Processes, and Metrics *377*
- 14.7 Past Climate Changes *383*
- 14.8 Observed and Projected Climate Change Impacts for the Environment and Human Health *385*
- 14.9 Global Risks from Climate Warming and Climate Change *390*
- 14.10 Key Points *391*
  - References *393*

15 Soil and Goundwater Pollution *397*
- 15.1 Introduction *397*
- 15.2 Soil and Groundwater Systems and Key Environmental Properties *397*
- 15.3 Types and Properties of Soil and Groundwater Contaminants *400*
- 15.4 Sources and Releases of Soil and Groundwater Contaminants *400*

15.5 Environmental Behavior and Fate of Soil and Groundwater Contaminants *405*
15.6 Exposure Assessment of Soil and Groundwater Contaminants *411*
15.7 Biological Uptake and Bioaccumulation of Soil Contaminants *411*
15.8 Environmental Effects of Soil and Groundwater Contaminants *412*
15.9 Human Health Effects of Soil and Groundwater Contaminants *415*
15.10 Key Points *418*
References *419*

16 Solid, Liquid, and Hazardous Wastes *421*
16.1 Introduction *421*
16.2 Wastes: Types and Hazardous Characteristics *422*
16.3 Waste Sources, Generation, and Emissions *423*
16.4 Waste Management *424*
16.5 Environmental Behavior and Fate of Waste Contaminants *430*
16.6 Exposures to Contaminants from Wastes *430*
16.7 Environmental Effects and Risks from Wastes *432*
16.8 Human Health Effects from Waste Management *435*
16.9 Key Points *440*
References *441*

17 Pollution Monitoring, and Assessment *443*
17.1 Introduction *443*
17.2 Monitoring Objectives *443*
17.3 Monitoring Strategies *444*
17.4 Environmental Metrics of Pollution Monitoring *450*
17.5 Monitoring Programs and Methods *450*
17.6 Remote Sensing for Pollution Monitoring *457*
17.7 Monitoring and Assessment of Disease Impacts From Pollution *458*
17.8 Key Points *460*
References *461*

18 Human Health and Ecological Risk Assessment *465*
18.1 Introduction *465*
18.2 Risk Assesssment Processes and Principles *467*
18.3 Semiquantitative Risk Assessment *468*
18.4 Human Health Risk Assessment *470*
18.5 Ecological Risk Assessment *479*
18.6 Risk Assessment Using Probabilistic Techniques *482*
18.7 Key Points *484*
References *485*

19 Management of Hazardous Chemicals *487*
19.1 Introduction *487*
19.2 Goals and Strategies for Managing Chemicals *487*
19.3 Collection of Data on Chemicals *489*
19.4 Laboratory and Field Testing of Chemicals *490*
19.5 Regulation and Assessment of Chemicals *493*
19.6 Risk Characterization Example (19.1) *500*
19.7 Managing Chemicals to 2020 and Beyond *502*
19.8 Key Points *504*
References *506*

20 Pollution: Moving Toward a Healthy and Sustainable Future *509*

- 20.1 Introduction *509*
- 20.2 Sustainability *509*
- 20.3 Pollution and Planetary Health *517*
- 20.4 Pollution, Health, and Sustainable Solutions *524*
- 20.5 Key Points *531*
  - References *532*

Supplementary Chapters S21–S24 are available in the online website www.wiley.com/go/toxicologyofpollution2e *534*

S21 Thermal Pollution *535*

- S21.1 Introduction *535*
- S21.2 Heated Effluents *535*
- S21.3 Key Points *541*
  - References *542*

S22 Radionuclides *543*

- S22.1 Introduction to Radiation and Radionuclides *543*
- S22.2 Nuclear Energy and Reactions *544*
- S22.3 Sources and Exposures of Ionizing Radiation *546*
- S22.4 Environmental Behavior and Fate of Radionuclides *548*
- S22.5 Radiation *549*
- S22.6 Ecological Effects of Radionuclides *550*
- S22.7 Ionizing Radiation Exposures and Human Health Effects *550*
- S22.8 Key Points *553*
  - References *555*

S23 Sediment Pollution *557*

- S23.1 Introduction *557*
- S23.2 Sediment Types and Properties *557*
- S23.3 Sources and Releases to the Environment *558*
- S23.4 Environmental Behavior and Fate of Sediments *558*
- S23.5 Environmental Effects of Sediment Pollution *559*
- S23.6 Ecological Effects of Sediments *560*
  - References *561*

S24 Salinity *563*

- S24.1 Introduction *563*
- S24.2 Chemistry of Salinity *563*
- S24.3 Sources and Extent of Salinity *564*
- S24.4 Environmental Behavior and Fate of Salinity *565*
- S24.5 Environmental Exposures and Effects of Salinity *567*
- S24.6 Human Health Effects Due to Salinity *568*
- S24.7 Key Points *569*
  - References *569*

Index *571*

Preface

The First Edition of this book was published in 1984 and this is the Second Edition published in late 2022 – a hiatus of 38 years. This does not reflect a lack in the advance of science on the topic but rather the reverse. The mountain of data and information on pollution science has steadily grown over the years together with advances in data collection coupled with information processing techniques and modeling. Older experience and knowledge have been critically reviewed, then confirmed or discarded, allowing consolidation and the development of whole new areas of pollution science often with consequent social concerns. A major change has occurred in the way information is stored and accessed. Formerly, scientific information was stored as hard copy in journals kept in libraries accessed by on-site searches, but now it is stored in electronic databases and is readily accessed online through the Web.

The authors have been participants in these processes and have particularly noted the growth in knowledge and concerns regarding the adverse effects of pollution on **human health**. The First Edition was titled *Chemistry and Ecotoxicology of Pollution* and was focused on **natural ecosystem**s which remain at the core of pollution science. However, as a result of these developments, the authors have chosen to include the adverse effects of pollution on human health in the Second Edition causing a change in title to *Chemistry and Toxicology of Pollution*. This means that the adverse effects of pollution on natural ecosystems are still covered but the human health aspects are also evaluated. In fact, these two topics are complementary since similar techniques are used in both and the fields draw on knowledge and techniques from one another.

The world has become a much more crowded place with the global population increasing from about 5 billion in 1984 to almost 8 billion in 2021. This means the production and use of a vast, and increasing, assortment of industrial chemicals as fuels, fertilizers, agricultural pesticides, medications, detergents, plastics, and so on. As a result, there is increased pollution from human activities and wastes of the atmosphere, land, soil, oceans, lakes, and rivers as well as for the natural biota and humans that depend on them. These form the abiotic and biotic environment which supports all life on the planet. However, old pollutants continue to cause adverse effects, although in some cases we can claim victory by reduction or elimination of the pollutants. Our management strategies have been very successful in some cases and we continue to improve them.

Yet, the scientific evidence is clear that a coordinated, continual, and much greater effort is needed by human society to reduce and manage the magnitude and observed rate of pollution of the planet, its loss of biodiversity, increasing global warming from emissions of greenhouse gases, and burden of premature human disease and death from pollution and other environmental factors.

Socioeconomic and political forces play a major role in the control and management of pollution. Nevertheless, to allow pollution to be seen in its correct perspective and devise appropriate control measures, we need a clear understanding of its nature and effects. As with the First Edition, this Second Edition aims to provide an understanding of the chemical, toxicological, and ecological factors involved when the major classes of pollutants act on natural and human systems. It deals with the interactions and effects of pollutants on human health and natural systems. Even so it recognizes the critical role of socioeconomic, political, and cultural factors in achieving sustainable goals, strategies, and science-based solutions to pollution and health on a global scale as we move toward 2030 and beyond.

A framework for understanding the effects on ecosystems and human health was proposed in the First Edition and this has been preserved and extended in this Second Edition. This framework is contained in Figure 1.3 and it follows the pattern of change and consequent effects, from sources through to distribution in the environment governed by the physicochemical properties of the chemical pollutant. After distribution in the environment, then the biota, natural organisms, and humans are exposed to the contaminant and individuals respond according to the biological and biochemical properties of the pollutant.

These responses of the individuals lead to consequent changes to population, community, and ecosystem.

We believe this Second Edition, even though it is 38 years later, is also timely since a vast number of research papers and books have been published containing a huge array of scientific knowledge. Often the meaning, as a basis for management actions, cannot be derived although a great deal of data and information are available. We have attempted to collect and collate this knowledge and distil it into unifying theories and principles. This is urgently needed to properly interpret the information produced by the numerous international, national, and local agencies and lead the way forward in the effective control and management of pollution of the environment.

We gratefully acknowledge the enormous amount of research and number of publications made by dedicated people around the world who have advanced the scientific body of knowledge and processes we have relied upon in this book.

In the writing of this book, we have been assisted in a variety of ways by many people. In particular, special thanks to Michelle Baker for preparing most of the diagrams. These were of excellent quality although some were quite complex. Madison Mayfield also helped with the diagram work preparing a range of diagrams with great skill and excellent results.

Griffith University, School of Environment and Science provided a base for the writing of the book and access to the relevant databases on the Web. This has been greatly appreciated and the book could not have been completed without it. As well, the Centre for Environment and Population Health at Griffith University offered the opportunity for special insight into environmental health impacts on human populations in many parts of the world, especially Asia, and the urgent need for scientific understanding, evaluation, and management of pollution.

Our first contact with Wiley was with Bob Esposito who set up our contract and set us on this journey. Then, our early contact was with Beryl Mesiadhas as the Project Editor who was very helpful in organizing the guidelines for the writing of this book. Our later contact was the book's editors Andreas Sendtko and Summers Scholl who helped us to complete the submission of the manuscript and kindly through the publication process. We thank Ramya Vengaiyan (Wiley India) who assisted us in the final review and production phase. Our appreciation extends also to the work of other professionals and staff that contributed to the publication of our book.

Finally, we would like to thank our wives, Patricia and Gillian, who have supported us at all times in this journey. In particular, Gillian was very helpful in the editing of the final documents.

Des W. Connell and Greg J. Miller
Brisbane

How to Use this Book

Environmental pollution now occupies center stage in the management of this planet. This book has been written for those who are seeking a better understanding of the processes involved through the application of scientific principles of chemistry and toxicology and ecology, health sciences, and medicine. It clarifies and identifies the damage caused to the natural and human environment by pollutants and their mechanisms of action.

The strength of this book is that it uses a framework of the interaction of chemicals with the natural and human environment as its structure. This framework is outlined in Figure 1.3 and serves several purposes. First, it describes the basic scientific concepts of ecotoxicology and environmental chemistry and second, it is the format for this book.

The framework initially considers the global dimensions and adverse effects of chemical pollutants. Then, the distribution of chemicals to individuals of natural biota and human populations occurs resulting in toxic effects on these exposed individuals. Consequential effects on populations, communities, and ecosystems and global responses are then considered. Several chapters are devoted to a detailed consideration of important specific pollutant groups such as plant nutrients and deoxygenating substances, pesticides, petroleum, and metals. The great global environments of water, including oceans, estuaries, and freshwater systems and the atmosphere are considered together with the most significant pollutants. The techniques for assessment of the risk to human health and the natural environment are outlined.

This book can be used in many ways. The basic framework, as in Figure 1.3, can be used to understand and evaluate the interaction of chemicals with natural biota and humans to yield an ecotoxicological and human health evaluation. Also, the interaction of chemical pollutants with specific sectors of the environment, such as the water or atmosphere, can be evaluated. The collation of properties and data on specific chemical pollutant groups allows the consideration of pollutants from the perspective of the properties of the group.

To assist in consolidation of knowledge and evaluating progress, there are many case studies and worked examples included in the book as well as problems and worked answers with each chapter. These will assist both professionals and students in the use of the concepts and methods described. The risk assessment chapter should prove particularly helpful to those concerned with the possible adverse effects on the natural and human environment.

An Online Companion Book contains supplementary chapters, the questions and answers for chapters, as well as supplementary material, as shown in its Table of Contents.

The Companion Book is accessed via a Universal Password located in this Book.

About the Companion Website

A companion website with additional resources is available at:

www.wiley.com/go/toxicologyofpollution2e

The materials in the companion website includes supplementary Chapters 21 to 24 and answers to chapter questions.

1

Introduction

1.1 Background

Health and environmental impacts of worldwide pollution and related environmental changes are increasing to a pandemic level for human health. The World Health Organization (WHO) estimated that 12.6 million people died in 2012 as a result of living or working in unhealthy environments (WHO 2016). This number was nearly a quarter (1 in 4) of all global deaths. It is also likely that this number is an underestimate because the health impacts of exposures to many pollutants and environmental factors are poorly known. Environmental risk factors, such as air, water, and soil pollution, toxic chemical exposures, global warming and climate change, and ultraviolet radiation contribute to over 100 diseases and injuries. Deaths related to unhealthy environments were attributed mainly to diseases such as stroke, heart disease, cancers, and chronic respiratory diseases. Indoor and outdoor air pollution was estimated to result in 7 million people dying prematurely every year (Climate & Clean Air Coalition 2018).

At the same time, global biodiversity and ecosystems are under persistent to increasing pressures from habitat loss, degradation, overexploitation and unsustainable use, invasive species, climate change from emissions of greenhouse gases, and excessive nutrient loads and other sources of pollution (Secretariat of the Convention on Biological Diversity 2010, pp. 9–13) (also known as the Global Biodiversity Outlook 3 [GBO-3]). Human-induced activities, including pollution loads and interactions, continue to cause global biodiversity to be significantly reduced for many major groups of plants and animal species. As well, many endangered species face a future of accelerating extinction.

A major problem for evaluating biodiversity is predicting ecological thresholds or tipping points that result in large ecological impacts on biodiversity. These are difficult to model, control, and reverse their impacts (see Chapter 3 of this book). In the case of climate change, tipping point analyses of global warming projections indicate major biodiversity changes at temperatures near or below 2 °C for a wide range of global ecosystems and regions, such as degradation of coral reefs, shifts in Arctic phytoplankton communities, and widespread dieback of Amazon forests (Secretariat of the Convention on Biological Diversity 2010).

The great challenge is whether our planet can provide sustainable ecosystem services in the form of biologically productive land and water to satisfy our needs (e.g. food) and to absorb our wastes. The problem is that our ecological footprint, measured in terms of the area of biologically productive land and water needed to produce goods consumed and to assimilate the wastes generated, already exceeds the annual capacity of the Earth to regenerate itself by over 50% (equivalent to 1.7 Earths) (Global Footprint Network 2018).

This situation is due to factors such as population growth, mega-urbanization, excessive emissions and waste disposal, greenhouse gases and global warming, and major impacts on biodiversity, human health, and ecosystem services.

Despite international agreements, many national environmental regulations and pollution controls, there continues to be great scientific and societal demands to evaluate and reduce the effects of pollutants on the health of ecosystems and human populations using risk-based management solutions that meet the needs of sustainability principles and criteria.

To face the challenge due to pollution impacts on our planet, at strategic, policy, and practice levels, there are several things we can readily do, that continue to be an urgent priority. We can rapidly advance and disseminate our knowledge of the scientific principles that largely govern the effects of pollution on ecosystems and human health, improve and apply our ability to evaluate these effects and associated risks within this context, and enhance our capacity to achieve sustainable healthy environments, at all scales, and implement practical solutions.

In the Twenty-first Century, our future sustainable solutions and adaptation to environmental changes depend on environmental science, ecology, earth sciences,

Chemistry and Toxicology of Pollution: Ecological and Human Health, Second Edition. Des W. Connell and Greg J. Miller.
© 2023 John Wiley & Sons, Inc. Published 2023 by John Wiley & Sons, Inc.
Companion website: www.wiley.com/go/toxicologyofpollution2e

environmental engineering, environmental education, conservation, pollution control and regulation, risk communication, and sustainable environmental management. They are now major areas of scientific research, policy development, and practice in environmental management, greatly supported by the emergence and recent advances in pollution science, environmental chemistry, ecotoxicology, environmental health, and risk assessment.

1.2 Environmental Pollution

Pollution occurs when substances resulting from human activities are added to the environment, causing adverse alteration to its physical, chemical, biological, or aesthetic characteristics. All nonhuman organisms also produce wastes which are released to the environment, but these are generally considered part of the natural system, whether they have adverse effects or not. Pollution is usually considered to occur as a result of human actions such as from the discharge of poorly treated wastes. However, natural processes can result in situations in the natural environment where they can resemble those due to pollutants (Connell and Miller 1984, p. 1). *Fish kills*, for example, in water bodies can result from sudden loss of dissolved oxygen (deoxygenation) caused by microbial respiration during biodegradation of abnormal inputs of organic matter (e.g. sugars or carbohydrates) following intense rainfall events from their catchments.

Chemicals released to the environment may affect it in different ways. In the air, they can act as air pollutants, greenhouse gases, ozone-depleting chemicals, or may contribute to acid rain formation. Chemicals can contaminate water resources through direct discharges to bodies of water, stormwater runoff, or via deposition of air contaminants to water. As water pollutants, they can have adverse effects on aquatic organisms, including fish, and on the availability of water resources for drinking, bathing, and other activities. It is common for soil pollution to be a direct result of atmospheric deposition, dumping of waste, spills from industrial or waste facilities, mining activities, contaminated water, or pesticide applications. Soil contamination impacts include loss of agricultural productivity, contamination of food crops grown on polluted soil, adverse effects on soil microorganisms, and human uptake of contaminants, either through food or direct exposure to contaminated soil or dust.

While objective measures can be made about indicators of pollution or actual pollutants and biological effects, there is an element of subjective judgment about pollution as to whether an adverse effect has occurred or not. For example, the release of plant nutrients into a freshwater body may stimulate the growth of aquatic plants in the system such as phytoplankton (algae) and fish numbers. From a fishing perspective, this is likely to be viewed as a beneficial outcome and not pollution. In contrast, a water authority using this water body to supply household drinking water may find a significant increase in algal numbers in the raw water supply, as a detrimental effect, that requires treatment to ensure suitable water quality (Connell and Miller 1984, pp. 1–2).

It is evident that perceptions of pollution or subjective decisions about pollution are critical factors in the evaluation and management of pollution impacts and also risks. In fact, how to resolve conflicts between scientific and public judgments of technical and perceived risks about pollution have emerged as a major challenge for decision-makers to achieve consensus-based agreements and equitable solutions on management actions. This type of problem is clearly shown by the issue of global climate change where scientific consensus exists on the highly probable role between emissions of greenhouse gases from human activities and observed global warming events and modeling of risk-based scenarios for the Twenty-first Century (see Chapter 14).

1.3 Drivers of Environmental Pollution

Human activities cause changes to our environment and health through pressures involving *human driving forces*. These forces are seen as related to the following components (Landon 2006, p. 27):

- Population growth and urbanization
- Poverty and inequality
- Technological and scientific development
- Political and economic systems
- Cultural values

During the late 1960s and early 1970s, economic and ecological argument arose about rapid human population growth and the impacts of overuse of limited natural resources such as famine and starvation. Growth in GDP had increased steadily over a long time period, and was thought to drive an increase in human impacts on the environment, through increased consumption as many human populations became more affluent. In the early 1970s, the ecologists, Paul Ehrlich, John Holdren, and Barry Commoner proposed and developed the impact Eq. (1.1) that expressed human impacts (I) on the environment, as the product of population's size (P), its affluence (A), and the damage caused by the technologies used to supply each unit of consumption (T) (EJOLT 2012).

$$I = P \times A \times T \tag{1.1}$$

Population growth is the fundamental driver. Since the early period of the Industrial Revolution, the human population on Earth is estimated to have increased over seven times, from 1 billion in 1800 (McLamb 2011) to nearly 7.6 billion in mid-2017, and is projected by the United Nations to rise to 9.8 billion people by 2050 (United Nations Department of Economic and Social Affairs, Population Division 2017, p. 12). China (~1.4 billion) and India (~1.3 billion) had the largest populations in the world in 2015 (United Nations Department of Economic and Social Affairs, Population Division 2017, p. 1). According to the World Bank, extreme poverty remains unacceptably high, despite some reduction in recent decades, with 767 million people living on less than $US1.90 a day. The vast majority of the global poor live in rural areas; are poorly educated; work mostly in agriculture; and over half are under 18 years of age (World Bank 2016, pp. 3–4).

Human population growth, particularly in low- and middle-income countries, combined with large population migrations from rural to urban areas has resulted in rapid urbanization and formation of megacities such as Tokyo and Mexico City, with 50% of the human population projected to be living in urban areas by 2025 compared to 5% in 1800 (Landon 2006, p. 183). Already, 54% of the world's population live in urban areas, and is expected to increase to 66% by 2050 (United Nations 2014).

Rapid urbanization, high-density living, and lower incomes are commonly associated with unhealthy environments through placing pressures on available spaces, energy sources, biomass, wildlife and habitats, food, shelter, water resources, intensive industrial development, and exploitation of local natural resources. Pollutant emissions into the air, such as particulates and gases from burning fuels, waste heat, noise, solid, liquid and hazardous wastes, including sewage and waterborne pathogens, are readily concentrated in local environments and need integrated and sustainable land use planning, regulation and pollution controls such as drinking water and sewage treatment plants.

Affluence in a society generally leads to higher consumption of natural resources and other ecosystem services. It relates to the average consumption per person in the population while GDP measures production per capita. It is widely assumed that when consumption increases so does production. Generally, higher affluence drives consumption and leads to further exploitation of available natural resources, inefficient production, and greater generation of waste products disposed to the environment.

Another major challenge for the global environment is the emerging growth of affluence in developing and transition countries, such as China and India, where recent economic growth has improved incomes and lifestyles for large numbers of people. Increasing consumption among a large group of new consumers (over 1 billion) from such countries is matching and adding significantly to long-term environmental impacts caused by existing consumers (about 850 million) in rich or developed countries (Myers and Kent 2003).

We are readily aware of direct impacts of science and technology development on our environments, such as from motor vehicle emissions, uncontrolled mining and processing of heavy metals, and excessive use of synthetic fertilizers and pesticides. Nonetheless, there are positive and negative influences of science and technology development to consider. As populations grow and consumption of resources increases, developments in science and technology, mainly in high-income countries, have created more efficient agricultural production, e.g. *the green revolution*, advances in industrial production, health care, improvements in environmental and social conditions. Benefits also arise from the rapid spread of information and communication technology (e.g. worldwide web) and its use throughout high, medium, and low-income countries to improve quality of lifestyle and adaptation of sustainable technologies (Landon 2006, pp. 32–33).

In reality, the environmental impact of human activities is more complex and thought to be better described as a function of population, affluence, and technology, shown by Eq. (1.2).

$$I = f(P \times A \times T) \qquad (1.2)$$

Environmental impacts on natural resources, as expressed in Eq. (1.2), depend to different degrees on interactions involving human lifestyles, behavior, culture, and organization within political and socioeconomic systems. There are now many indications that we have exceeded the carrying capacity of the Earth as estimated. Pollution plays a major, and often silent and hidden role, in these impacts on our lives, health, and survival, with few exceptions.

1.4 Pollution and Environmental Health

Pollution results from the pressure of human activities on our environment (*human driving factors*). How humans change and use the resources of our environment on local, regional, and global scales influences the interaction between the environment and health. All human activity has environmental and health consequences (Landon 2006, pp. 26–28).

In the study of pollution, we are also focused on environmental health. We need to understand and act to identify environmental hazards, adverse impacts and risk factors, measure levels and adverse changes in the environment, evaluate, prevent, and reduce environmental health impacts and risks, within the context of sustainability. It covers the interaction between natural and human environments and how to sustain healthy environments.

Chapter 3 introduces the definition of environmental health according to the WHO, which also refers to *the theory and practice of assessing, correcting, controlling, and preventing these factors in the environment that potentially can adversely affect the health of present and future generations* (see Landon 2006, p. 5).

1.5 Chemicals and Pollution

Overview of Chemical Industry

The scope of the chemical industry in this book is generally treated as a broad and integrated industry rather than as multiple industries such as petroleum, minerals, metals, plastics, or polymers. Here we are concerned with the environmental and health impacts of chemicals as pollutants over their life cycles. This includes the extraction of natural resources, materials, processes, products, uses, emissions, and wastes.

Historical growth of the chemical industry occurred during the Industrial Revolution, from the mid-Eighteenth Century to the Nineteenth Century, initially involving basic inorganic chemicals, mineral processing, and base metals, and later the early synthesis of organic chemicals such as dyestuffs by the British chemist William Perkin, and later urea, as a synthetic nitrogen fertilizer, synthesized by the German chemist Fritz Haber, at the start of the Twentieth Century. Other major chemical industries such as large-scale petroleum extraction and refining, and the manufacture of plastics, synthetic pesticides, pharmaceuticals, and specialty chemicals developed in the Twentieth Century. After the Second World War, the chemical industry, including petrochemicals, rapidly expanded into a large complex and diverse industry worldwide, including the growth of nuclear power and related issue of radioactive wastes.

Chemical Industry

The chemical industry is a massive global industry, based on chemistry, that uses complex processes and operations to convert raw materials (from air, water, and minerals to petroleum, coal, and biomass) into a vast range of basic industrial chemicals, specialty and consumer products, numbering over 100 000 different commercial products

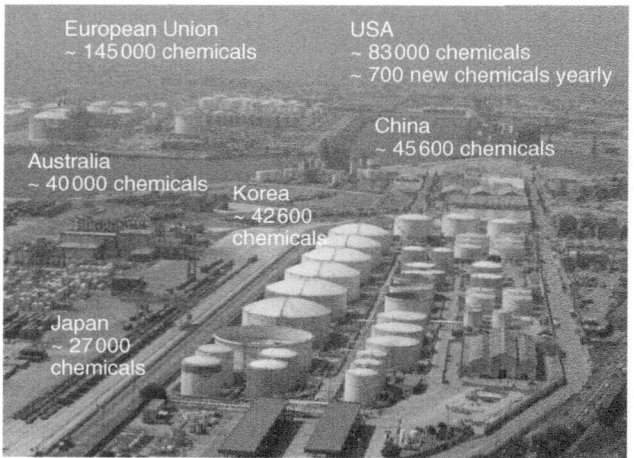

Figure 1.1 Numbers of different chemical products in commercial use in various countries and the European Union. Chemical data compiled from Chemical Inventories for selected countries and EU (Chemical SafetyPro 2016). Source: Photograph by G.J. Miller.

(see Figure 1.1). The products and applications form a vital part of our everyday lives, from personal and healthcare products, cleaning chemicals, clothing to building products, agrichemicals, and sources of energy such as gas and liquid fuels.

The nature of the chemical industry and its products is somewhat difficult to describe because it has evolved greatly since the era of the Industrial Revolution, through the rise of mass production in developed and developing countries, to combine with key suppliers of raw materials, such as the petroleum and minerals industries, and other industries, e.g. pharmaceutical manufacturing. A large proportion of chemicals made, such as bulk petrochemicals, are used within the chemical industry and by other industries to make a diverse range of end products. Table 1.1 outlines the major types of chemicals produced. There are three general categories: basic, specialty, and consumer products but life science products may also be included as a special category. Some degree of variance occurs in such classifications.

World production is measured more in the value of sales of chemicals (US dollars) rather than by mass of chemicals made (tonnes). Global output is increasing rapidly, from $171 billion in 1970 to over $4.1 trillion in 2013 (figures not adjusted for inflation or price changes) and sales are projected to grow at about 3% per year to 2050 (UNEP 2013, pp. 11–13). The industry is highly multinational and diversified. Production since the Second World War

Table 1.1 Major types of chemical categories and examples of chemical products from the chemical industry in 2014 based on the (%) values of sales per category in United States and European Markets.

Major chemical categories	Typical chemical products	United States (% value of sales for each category)	Europe (% value of sales for each category)
Basic chemicals		61	60
Polymers	Polyethylene (PE), polyvinyl chloride (PVC), polypropylene (PP), polystyrene (PS), and synthetic fibers (e.g. polyester, nylon, and acrylics)	18	27
Petrochemicals	Ethylene, propylene, benzene, toluene, xylenes, methanol, vinyl chloride monomer (VCM), styrene, butadiene, and ethylene oxide	28	20
Basic inorganics	Salt, chlorine, caustic soda, soda, soda ash, acids (e.g. sulfuric, nitric, and phosphoric), titanium dioxide, and hydrogen peroxide	15	13
Specialty chemicals	Electronic chemicals, industrial gases, catalysts, adhesive and sealants, paints, coatings and surface treatments, additives, pigments, dyes and inks, nanomaterials	25	28
Consumer chemicals	Soaps, surfactants, flavors and fragrances, cosmetics, and other personal care products	15	12

Source: Data from The Chemical Industry, Centre for Industry Education Collaboration (2016).

has been dominated by North America (US), Europe, and Japan but the percentage of production in China, other Asian countries, and India is rising rapidly. China was predicted to increase its chemical production from 2012 to 2020 by 66% compared to 25% for the United States over the same period (UNEP 2013, p. 13). An overview of the development of the modern chemical industry, including some of its key processes and manufacture of basic chemicals is given in Standen and Killheffer (2017) among many others. It reflects the progress of humankind in knowledge, science, advances in technology and communication, and use of natural resources leading to the synthesis of new chemicals, materials, consumer, personal care, and life science products.

Early History of Chemicals

The history of the chemical industry extends to ancient times when alkali and limestone were used by Middle Eastern artisans to make glass as early as 7000 BCE (Technofunc 2012). Soap making appears to be dated from around 2800 BCE. Early soap makers were Babylonians, Mesopotamians, Egyptians, ancient Greeks, and Romans who made soap by mixing fats, oils, and salts for cleaning and medicine purposes. The first reported use of natural chemicals as insecticides was about 4500 years ago by Sumerians who used sulfur compounds to control insects and mites. About 3200 years ago, the Chinese were using mercury and arsenical compounds for controlling body lice. Pyrethrum obtained from the dried flowers of *Chrysanthemum cinerariaefolium* (Pyrethrum daisies) has been used as an insecticide for over 2000 years (Unsworth 2010).

The early beginnings of chemistry originated among the great philosophers and thinkers of ancient Greece who developed theories on the nature of matter, and from the practical skills of Egyptians obtained from working with precious metals, embalming of the dead, and dyeing of clothing. In the Fifth Century BCE, the Greek philosopher Empedocles (c. 450 BCE) proposed that all matter is made up of four elemental substances: earth, air, fire, and water. In about 420 BCE, Democritus gave birth to the notion of atoms, stating matter consists of extremely small indivisible substances that join in different proportions. In the Third Century BCE, Alexandria became the center of the Greek world where early scientists started to explore the early Greek scientific theories and engage the practical skills of the Egyptians to experiment on and transform metals especially precious metals (Historyworld 2020).

These early experiments applied the ancient art of *alchemy* handed down from Egypt and Arabia, a practice of mysticism rather than a scientific approach. The main aims of the alchemists were to find the Stone of Knowledge (The Philosophers' Stone), to discover the medium of Eternal Youth and Health, and to discover the transmutation of metals (capacity to change base metals, such as lead, into gold) (Royal Society of Chemistry 2018). Alchemy is reported to have spread from Greece and Rome to western and central Europe and also developed further in Asia during the Eighth and Tenth Centuries CE, particularly in Baghdad and China. In the Eighth Century CE, while searching for the Elixir of eternal life, the Daoists (Taoists) in China invented black powder (early form of gunpowder) made by pounding a mixture of saltpeter, charcoal, and sulfur. Initially, it was used for skin treatments and as an insecticide before use as a powerful explosive was discovered.

In the late middle ages, alchemy declined as scientific methods and knowledge of metallurgy developed. The famous Swiss physician, Paracelsus, although an alchemist, believed that the processes that occur in the body are chemical in nature and that chemical medicines offer remedies to illness (known as the *Father of Toxicology*). In 1661, Robert Boyle, an Irish chemist of Boyle's gas law fame, published his book *The Sceptical Chemist*, attacking the ideas and approach of alchemists. He advocated a rational scientific approach and the use of the term chemistry to describe the science of materials (Connell 2005, p. 3).

Industrial Revolution and Nineteenth Century

The emergence of the chemical industry is associated with the Industrial Revolution that began in Great Britain during the mid-Eighteenth Century before spreading to Europe and other parts of the world such as the United States and Japan. The Industrial Revolution was characterized by technological, socioeconomic, and cultural change, shifting from an agrarian and handicraft economy to a largely industrial economy, based on factory systems, mass production, and division of labor. Technological change included new basic materials made in factories (e.g. iron and steel), the invention of new machines (e.g. steam engines and textile looms), new energy sources (e.g. steam, coal, petroleum, and electricity), new means of transport (e.g. steam locomotives and railways), and the application of science to industry (The Editors of Encyclopaedia Britannica 2018).

The transition to new manufacturing processes, from about 1760 to around 1840 (First Industrial Revolution), included new ways to manufacture chemicals on a large scale leading to the Age of Chemicals during the Twentieth Century (see Connell 2005, p. 5). Early chemical demand was for inorganic chemicals, acids, and alkalis, such as sulfuric acid and sodium carbonate (soda ash). Sulfuric acid was used for pickling iron and steel and for bleaching cloth in the textile industry. The use of sulfuric acid for bleaching was soon replaced by bleaching powder (calcium hypochlorite) developed in about 1800 by the Scottish chemist, Charles Tennant. The first sulfuric acid plants were built in Great Britain in 1740 (Richmond), France in 1766 (Rouen), Russia in 1805 (Moscow Province), and Germany in 1810 (near Leipzig) (Technofunc 2012). The production of sulfuric acid by the lead chamber process was invented early on by the Englishman John Roebuck in 1746 to enhance output for the British chemical industry (Rowe 1998).

The development of the textile, glass, soap, and paper industries created a large demand for sodium carbonate (soda ash) production. The first soda plants were built in France in 1793 (near Paris), Great Britain in 1823 (Liverpool), Germany in 1843 (Schönebeck), and Russia in 1864 (Barnaul). In 1823, British entrepreneur James Muspratt started mass-producing soda ash (needed for soap and glass) using the Leblanc process (using salt, limestone, sulfuric acid, and coal) developed by the French physician Nicolas Leblanc in 1790 (Technofunc 2012). The Solvay process or ammonia-soda process developed during the 1860s by the Belgian chemist Ernest Solvay later replaced the Leblanc process, and by the 1890s, this process produced most of the world's soda ash (Rowe 1998). Portland cement was another key material developed during the Nineteenth Century for the production of concrete.

The new industrial age of the Nineteenth Century saw major advances in metal industries, from early blast furnaces to cast iron and steel. By making coke from coal, mixing it with metal ores and heating in hot furnaces, metal ores could be reduced from the oxide to metals (e.g. iron) with less impurities, than was the case by using wood charcoal. More abundant coal was used to replace wood and other biofuels instead of cutting wood and converting it into charcoal.

The Nineteenth Century was also a pioneering era of modern chemistry, from the isolation of chemical elements, development of the early periodic table of elements, to the study of chemical reactions and organic chemistry, based on the chemistry of carbon atoms. This new field of organic chemistry led to the early synthesis of organic chemicals for commercial (e.g. synthetic dyes from coal tars) and medicinal use, such as the analgesic aspirin, derived from salicylic acid. It was the start of the mass production of synthetic organic chemicals that was to radically change the nature of the chemical industry, the health and lifestyles of human populations, and impact on the global environment and biodiversity of life.

> **Box 1.1 William Perkin – Founder of the Organic Chemical Industry**
>
> In the 1850s, the young British chemist, William Perkin (1838–1907), recognized the commercial potential of the synthesis of organic compounds. At 18 years of age, he tried to produce quinine for the treatment of malaria, from aniline, a chemical by-product of coal tar, but accidently ended up with an intense purple dye (named as *Mauveine, Mauve, or aniline purple*), that was stable when used to dye silk. He patented and commercially produced the first synthetic dye, achieving great success and personal wealth. Its success was driven by demand from the large textile industry for use on cotton and other natural fibers.
>
> Perkin laid the foundations of the organic chemical industry, continuing to discover and make other aniline dyes and synthetic perfumes until his retirement in the 1890s. By then, mass production of synthetic dyes such as indigo (1897) in Germany and other organic chemicals grew throughout Europe.
>
> Source: Adapted from Hart-Davis (2009) and Encyclopedia of World Biography (2004).

Development and Growth of the Modern Chemical Industry

The chemical industry expanded rapidly and prospered in the Twentieth Century, through periods of peace and the World Wars, driven by population growth, economic development, and technological innovations, underpinned by great discoveries in science and chemistry, in what is sometimes known as the Atomic Age (1890–1970), or perhaps, the Age of Chemicals. Major advances in the chemical industry included the manufacture of synthetic nitrogen fertilizers by the Haber–Bosch process, electrochemistry, the discovery and production of many types of organic chemical products such as artificial fibers, polymers and plastics, surfactants, pesticides, and petrochemicals derived from petroleum (crude oil and natural gas) to replace coal as a source of aromatic hydrocarbons, and pharmaceuticals. Table 1.1 indicates the major chemical types (basic chemicals, life science chemicals, specialty chemicals, and consumer products) made in the form of over 100 000 chemical products today.

> **Box 1.2 Fitz Haber – Synthesis of Ammonia by the Haber Process**
>
> From the mid-Nineteenth Century, artificial fertilizer plants had appeared in Britain, Germany, and Russia. In the early Twentieth Century, the German chemist Fritz Haber (1868–1934) invented the Haber process, one of the most important industrial processes of all time. In 1909, he demonstrated in a laboratory that at high pressure, nitrogen and hydrogen gases could combine, in the presence of an iron catalyst and heat, to form ammonia gas.
>
> In 1913, a research team lead by Carl Bosch from BASF, the giant German chemical manufacturer, developed the Haber (or Haber–Bosch) process on a commercial scale by fixing nitrogen gas from the atmosphere with hydrogen gas from natural gas to produce synthetic ammonia, as a feedstock for nitrogen fertilizers and the manufacture of explosives such as TNT. Its synthetic use for explosives enabled Germany to prolong the First World War when supplies of saltpeter (potassium nitrate) were unavailable to it.

The industrial synthesis of ammonia allows synthetic nitrogen fertilizers to be produced worldwide on a large scale for intensive crop production, replacing limited sources of natural nitrogen compounds. The Haber–Bosch process (and its analogues) continues to produce large quantities of nitrogen fertilizers (including synthetic urea and anhydrous ammonium nitrate) of over 100 million tonnes annually. It is sufficient to support agriculture that provides nourishment for at least 2 billion people (Flavell 2010; Gascoigne 2020).

Despite the food production benefits of synthetic fertilizers, adverse environmental impacts are a serious negative cost because a large proportion of nitrogen from synthetic nitrogen-containing fertilizers is not assimilated by plants and is lost, as water-soluble forms, e.g. ammonia or nitrates, from agricultural catchments into water bodies and the atmosphere, as volatile forms of nitrogen. Synthetic nitrogen fertilizers are major sources of nitrogen loadings to many water bodies and catchments suffering from eutrophication worldwide (see also Chapter 8 of this book).

Murmann (2002) in his review of *Chemical Industries after 1850* attributed three types of recurring innovations to the rise of the chemical industry: (i) industrial production of natural chemicals that are found in limited amounts and are simply cheaper to make, such as synthetic ammonia by the Haber–Bosch process, (ii) substitution of a more efficient process for an existing industrial process, such as the Solvay ammonia soda process for the Leblanc process, and (iii) the synthesis of chemical substances that are not found naturally, such as many synthetic dyes (e.g. Mauve invented by Perkin in 1856) and most modern drugs.

The early Twentieth Century to the Second World War saw the dominance of the United States, Europe, Great Britain, and Japan as chemical producers of new synthetic fertilizers, artificial fibers in the textile industry (rayon in 1914 and nylon by Dow Chemical in 1928), while early manufacture of plastics and petrochemicals from oil and natural gas in the 1920s and 1930s and synthetic rubber (in the 1940s) expanded outputs.

Post Second World War

After the Second World War, the production of petrochemicals, polymers and plastics, agrochemicals (e.g. pesticides), and pharmaceuticals (e.g. antibiotics) increased greatly. Major petroleum companies (e.g. BP, Shell, and Exxon) expanded into the production of petrochemicals using their feedstocks of crude oil and natural gas. Petrochemicals are the industry's largest sector.

Industrialization of the Nineteenth Century influenced the location of many of the world's major chemical companies, including siting of chemical complexes and refineries in the United Kingdom, Europe, and the United States through the Twentieth Century and into the Twenty-first Century. As demand for products spread worldwide, major companies invested in local chemical plants and refineries in other parts of the world, particularly in Asia (e.g. Shell in Daya Bay in south-east China), the Middle East (e.g. Dow Chemical at Al-Jubail in Saudi Arabia), and South America, to be close to these markets and sources of raw materials.

The United States became the world's largest producer of chemicals but its percentage of market share declined during the late Twentieth Century as other developed countries increased outputs, diversified and increased competition, followed by rapid expansion of chemical companies in developing regions of the world such as China and the rest of Asia. In the Twenty-first Century, the chemical industry is highly multinational. The three largest companies are BASF (Germany), Dow Chemical (USA) and Sinopec (China). China has the largest market share of chemical sales.

The crucial secret of success for the chemical industry results from applying science (e.g. chemistry) and technology (e.g. chemical engineering) to a cycle of funding research, development of discoveries, industrial process design and pilot plants, and efficient production of its products for profitable markets, based on vertical and horizontal integration.

Key factors for development of the industry were access to raw materials (e.g. gas and oil fields and minerals), water supplies, reliable energy supplies such as coal and petroleum, transport (road, rail, and ports for sea transport), availability of skilled labor and markets for intermediates and end products, including supply to other chemical processors. In particular, oil and natural gas were needed to feed the rapid growth of the petrochemical and polymer industries after 1945. Today, major chemical plants and refineries have become integrated into large complexes, usually built on or near coastlines for access to sea transport necessary for the import and export of raw materials and products (e.g. Gulf of Mexico, USA; Teesside, east coast England; Rotterdam [Netherlands], Singapore, and Daya Bay in south-east China) (Centre for Industry Education Collaboration 2016).

Future development of the chemical industry and its rate of adaption toward environmental sustainability are of vital interest for global pollution, mitigation, and health outcomes. The core of the chemical industry is shifting to Asia, especially China. By 2030, at least half of the top 10 chemical companies will be Asian or Middle Eastern. Production, market sales, and growth will be dominated by Asia with nearly half the world's population (consumers and workforce). As a result, developed countries such as in Europe should experience reduced growth and serious challenges from this increasing shift in industrial production and global competition toward Asia (Kearney 2017).

Hydrocarbon feedstocks (crude oil and natural gas) are projected to be sufficient until at least 2030 although subject to commodity supply and price volatility. Advances in the development of shale gas and biomass offer alternative sources of carbon feedstocks. Growth opportunities for established and agile chemical industries and suppliers are seen in highly innovative and *value-added* products (e.g. alternative feedstocks, alternative energy, environmental technology), energy storage (lithium batteries and fuel cells), and advanced or intelligent materials (e.g. lightweight materials and nanomaterials) based on existing strengths in R&D, efficiency, and commercialization (materials advantage, process excellence, patent control, application know-how, customer relationships, and superior brands) (Kearney 2017).

Chemicals and Environmental Challenges

While the human production and use of chemical products have grown enormously in volume and complexity through the synthesis of many thousands of new commercial chemicals, the scale of environmental and human health impacts from chemical pollution is now a serious global challenge and legacy for the future, despite the many benefits from chemical products such as new building materials, plastics, metals and alloys, transport, fuels, medicines, cleaning and sanitation chemicals, personal care products, and fertilizers.

In the last 50 years or more, chemical companies have been confronted by the rise and costs of environmental, health, and safety legislation and regulations due to widespread public concerns about potential and actual impacts of many chemical products, toxic emissions or releases to air, water and land, disposal of toxic wastes (e.g. Love Canal in the United States), and chemical accidents such as the world's worst at Bhopal in India. Environmental forms of pollution related to raw materials extraction, chemical production, use and release of chemical products and wastes are extensive. They include depletion of natural resources for raw materials, pesticide poisonings of wildlife, air, water, and land pollution, ozone depletion in the upper atmosphere, increased eutrophication of water bodies caused by inputs of synthetic fertilizers, acid rain, microplastic pollution of oceans, contamination of wildlife and humans with persistent synthetic chemicals and toxic metals, and excessive global emissions of greenhouse gases that are increasing the temperature of the Earth's atmosphere and surfaces.

The general response of the chemical industry has been driven by the emergence of environmental law and the enforcement of national regulations in many countries to reduce and control pollution emissions and wastes, improve health and safety, ensure acceptable land use risks from hazardous facilities, remediate contaminated chemical sites resulting from its activities, and require registration and toxicological evaluation of chemical products such as synthetic industrial chemicals, pesticides, personal care, and pharmaceuticals. Since the 1980s, the industry has pursued sound environmental management practices for chemicals based on a life cycle approach, continuous improvement and compliance in environmental, health, safety, and security, improved technology performance and use of nonrenewable resources, increased energy efficiency, reduction in carbon dioxide emissions, and sustainable development. Its worldwide initiative, in over 60 countries, is known as Responsible Care (Centre for Industry Education Collaboration 2016).

Air, water, and soil releases of hazardous chemicals from the extraction of raw materials, production and use of chemicals, and the disposal of toxic wastes are major sources of environmental exposure for wildlife and humans. If we consider the life cycle of chemicals, as illustrated in Figure 1.2, we can see how environmental exposures to chemicals from pollution may occur at each

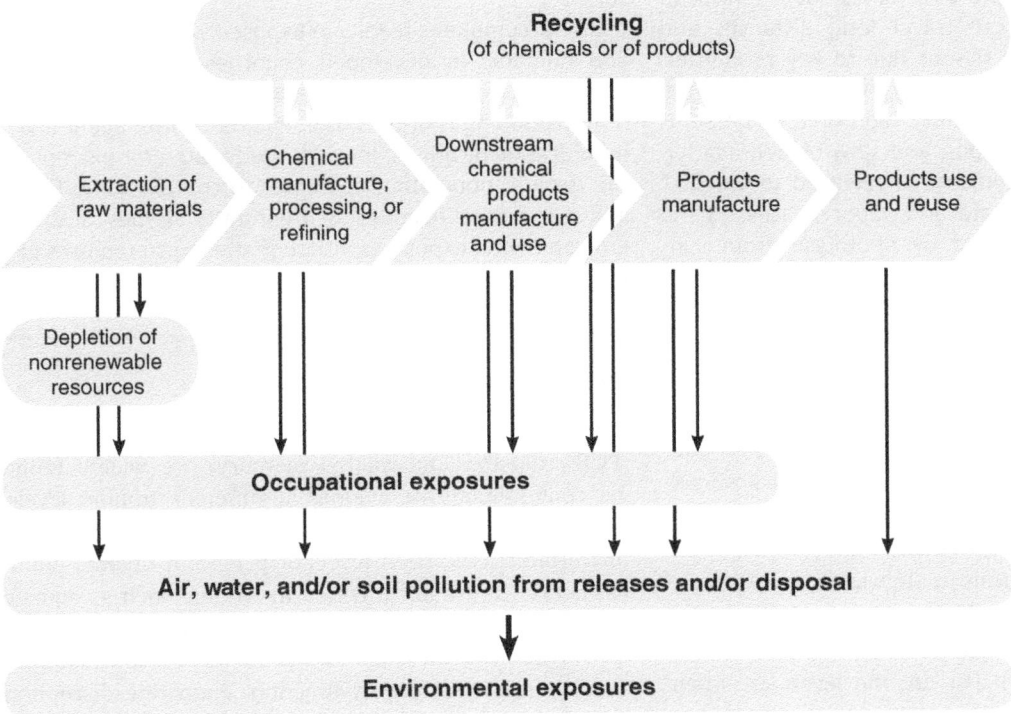

Figure 1.2 Life cycle of chemicals and pollutant exposures. Source: UNEP (2013).

stage of the process from raw materials, through processing, including recycling of materials or chemicals, to product use and reuse or final disposal to the environment.

The health and environmental effects of exposure to environmental chemicals are discussed below.

1.6 Health and Environmental Effects of Chemicals

As indicated in Figure 1.1, over 100 000 chemicals are used commercially worldwide. Chemical contamination and exposures in natural and human environments on a global scale result increasingly from the production, use, and disposal of these chemicals (UNEP 2013, p. 10). The scientific literature contains numerous observations and data on significant to serious effects on many natural and built ecosystems, the biodiversity of many known species, and human health on a worldwide magnitude, as indicated by recent WHO estimates of the global burden of disease from exposures to environmental pollutants. The following chapters in this book provide a scientific evaluation of pollutants and their effects on humans and other biota.

The release of chemicals into the environment from point sources, such as coal-fired power plant stacks, exhausts or wastewater effluents, and nonpoint sources, such as urban or agricultural stormwater runoff, permeates the atmosphere, waters, soils, sediments, wildlife, and food sources to different degrees or levels. They enter in many chemical forms into what we know as biogeochemical cycles, such as the water or hydrological cycle, the carbon, oxygen, nitrogen, and phosphorus cycles. These cycles usually involve interactions of chemicals between cycles and energy flows through ecosystems, on local to global scales. The inputs of a chemical, the quantity of a chemical contained in an environmental compartment, e.g. air, water, or sediment, the size of the compartment, and rate of mass movement of a chemical from one compartment to another (flux), are key quantitative measures (estimates) of chemical pollutants in the environment. Their physical and chemical properties together with transport (e.g. air movement, deposition, and sedimentation) and transformation processes (e.g. photolysis, oxidation, and biodegradation) play a major role in their distribution within and between environmental media, persistence and fate in both the physical environment and living organisms.

Box 1.3 DDT – Profile of the First Global Synthetic Pollutant

The chlorinated hydrocarbon (organochlorine) insecticide DDT is a classical example of a persistent chemical in the environment, that can travel long distances and will accumulate in fatty tissues due to key properties, such as its chemical structure (strong carbon–chlorine bonds), very low water solubility, high lipid or fat solubility, high adsorption to soils, and slow biodegradation in waters, soils, and sediments. If released to air, DDT will exist in the particulate and vapor phases. In the vapor phase, it is degraded by photolysis from sunlight while particulate DDT can be removed from air by wet and dry deposition. Vapor pressure is sufficient to allow progressive re-volatilization of DDT from various surfaces.

DDT accumulates readily in exposed wildlife, livestock, and humans. In developed countries, DDT has been phased out but continues to be used in many developing or transitioning countries as a malaria-control agent and in topical medications. In the United States, for example, the general population continues to contain some DDT and metabolite residues in their fatty tissues due to environmental exposures although the registered uses of DDT were cancelled in 1972 (US EPA 2016).

Exposures to chemical pollutants depend on the pathways of exposure, e.g. contact with air, water, or dusts, the levels of the pollutant, and rate of contact or exposure (dose per unit of time, such as mg day^{-1} or dose per unit of body weight, area or volume per unit of time, e.g. mg kg^{-1} body weight day^{-1}). Potentially toxic exposure levels for hazardous chemicals in air, waters, and soils for wildlife and humans are usually found in the range of parts per billion (ppb, µg L^{-1} or µg kg^{-1}) to parts per million (ppm, mg L^{-1} or mg kg^{-1}). For air, the levels are often converted to µg m^{-3} or mg m^{-3} (see also Chapter 2, Units of Measurement).

Environmental Effects

Persistent chemicals that bioaccumulate (e.g. DDT, PCBs, dioxins, and methyl mercury) are widely found as contaminants in wildlife at different trophic levels, and can concentrate up the food chain to top predators in certain cases. High levels of persistent organic pollutants (POPs) are documented in wildlife such as aquatic mammals, polar bears, and fish-eating birds (UNEP 2013, p. 47). Some of these chemicals are known to cause cancers, immune system dysfunction, endocrine disruption and reproductive disorders, and reduced biodiversity in wildlife, including near extinction in several species.

International action under a global treaty to restrict, control, and eliminate these types of chemicals from the environment commenced at the start of the Twenty-first Century. In 2004, an initial group of a dozen (12) halogenated chemicals, including dioxins and PCBs, known as POPs, were identified and listed under the Stockholm Convention of Persistent Organic Pollutants (see Chapters 2 and 11). Since then, it has been found that levels of dioxins and PCBs in wildlife from most areas are gradually decreasing due to control measures, but other emerging halogenated pollutants, including brominated flame retardants and perfluorinated compounds (PFOS) have increased in their place. Additional POPs have been added to the list of the Convention since, including these types of compounds (see Chapters 2 and 11).

Advances in ecotoxicology since the 1970s clearly show the widespread toxic effects and interactions of pollutants on populations and communities of nonhuman biota (aquatic, soil, and avian) and their physical environment (see Chapter 5). For example, chemicals such as many pesticides can affect aquatic microorganisms, invertebrates, and vertebrates. Adverse effects on aquatic animals include cancers, disrupted reproduction, immune dysfunction, damage to cellular structures and DNA, and gross deformities (UNEP 2013, p. 46). Ecotoxicology applies the same scientific principles of toxicology used in studies of individual species of mammals or human populations. It is concerned not only with the direct effect of pollutants such as a chemical on an organism but how that chemical may interact with an organism's environment (e.g. physical alteration of a habitat by a chemical spill; loss of a food source; biological or chemical degradation; increases in soil acidity or salinity; increases in air and water temperature due to global warming by greenhouse gases; and excess nutrient pollution of water bodies by human wastewaters and agricultural fertilizers) (e.g. Connell and Miller 1984).

Human Health Effects

Numerous human health studies and assessments have been made on health effects of exposure to environmental chemicals in the last 50 years although the body of information and data is still very limited for many of the over 100 000 chemicals used in commerce. Evidence of adverse human effects is derived from (i) experimental animal studies (e.g. bioassays), (ii) cellular or cytological testing, (iii) human clinical and epidemiological studies (occupational and environmental), (iv) biomarker and biomonitoring, (v) scientific assessments of carcinogenic and noncarcinogenic health risks, and (vi) global estimates of the burden of disease from various types of environmental exposures, such as airborne particulates in urban airsheds. Similar types of evidence-based approaches apply to effects from physical pollutants (e.g. heat stress, UV radiation, and noise) and biological pollutants such as air and waterborne microorganisms and toxins.

Acute and chronic exposure to a considerable number of toxic chemicals from air, water, soil, dust, and foodstuffs can cause or contribute to a broad range of human health effects, including death, as shown in Table 1.2. Adverse effects include eye, skin, and respiratory irritation; damage to organs such as the brain, lungs, liver, or kidneys; damage to the immune, respiratory, cardiovascular, nervous, reproductive, or endocrine systems; and birth defects or specific chronic diseases, such as cancer, asthma, or diabetes (UNEP 2013, p. 50).

Vulnerable Populations to Chemical Exposures

The effects of exposure to toxic chemicals are reported to be much greater for vulnerable populations described in the literature as workers in industries using chemicals; children, pregnant women, or persons living in poverty; the elderly and other vulnerable or susceptible groups. Children's responses to small doses of toxic chemicals are relatively large compared to adults because of differences in body size and immature metabolic pathways. Metabolism in children is slower to detoxify and excrete many environmental chemicals which may remain active and persist in their bodies for longer periods of time (UNEP 2013, pp. 49–50).

Biomonitoring of Chemical Pollutants

Many research studies and monitoring programs (e.g. United States Centers for Disease Control and Prevention's National Biomonitoring Program) have measured the levels of environmental toxicants in human specimens from toxic metals, organometallics (e.g. methyl mercury, radionuclides to synthetic organics such as pesticides, PCBs, dioxins, and other halogenated compounds). For instance, Patterson et al. (2009, p. 1211) reported human serum levels of selected POPs in the US population in 2003–2004. In developed countries, levels of toxic metals such as lead, organochlorine insecticides, and PCBs have greatly declined as exposures to these persistent and toxic substances were either banned or phased out during the 1970s to 1990s due to public health concerns, environmental and regulatory actions.

Recent biomonitoring programs in largely developed countries, especially in the United States, have revealed detailed evidence of large and complex numbers of environmental chemical contaminants in the human body that were not evident 50 or so years before. A 2009 study by

Table 1.2 Human health effects and examples of suspected or confirmed links to chemicals.

Health effect	No of chemicals confirmed or suspected	Examples of chemicals
Asthma (via non-respiratory sensitization)	NA	Chlorine, hydrochloric acid, sulfuric acid, diazinon, and malathion
Cancer	1070	Benzene, aromatic amines, TCDD, asbestos, PAHs, formaldehyde, arsenic, beryllium, chromium (VI), cadmium, lead
Diabetes	NA	Dioxins, arsenic, urea-based pesticides
Organ damage	466	Benzene, carbon tetrachloride, mercury, nickel, lead, cadmium, methanol, methyl butyl ketone, DDT, aldrin
Neurotoxicity	201	Benzene, toluene, trichloroethylene, vinyl chloride, chlorpyrifos, endosulfan, phthalates
Reproductive toxicity	261	1,2 Bromopropane, lead, mercury, phthalates, and 1,2-dibromo-3-chloropropane (DBCP)
Skin burns/irritation	837	Isocyanates, hydroxyl amine, and certain pesticides
Serious eye irritation/damage	892	
Respiratory irritation	224	
Skin sensitizers	997	Acetic anhydrides, amines and diisocyanates (MDI and TDI), formaldehyde and glutaraldehyde, chromium and nickel compounds
Respiratory sensitizers (e.g. allergic asthma)	114	

NA, not available.
Source: Modified from UNEP (2013).

the United States CDC found that all of the 212 chemicals studied were detected in some portion of the US population. Evaluation of results indicated widespread exposure to some industrial chemicals with detectable levels in 90–100% of samples assessed. These included perchlorate, mercury, bisphenol-A, acrylamide, multiple perfluorinated chemicals, and the flame retardant, polybrominated diphenyl ether-47 (BDE-47) (UNEP 2013, pp. 54–55). In particular, levels of organochlorine insecticides and PCBs indicated a significant decrease in human specimens along with blood lead levels in children.

Disease Burden due to Pollutants

WHO estimates of the total disease burden attributed to selected chemical exposures give 4.9 million deaths (8.35% of the total deaths) and 86 million DALYs[1] (5.7% of total) in 2004. Significantly, children suffered just over half of the global burden, mainly from lead. Chemical-based estimates are known to understate the real disease burden but clearly indicate the serious global health effects of many chemical exposures due to a multitude of diseases (UNEP 2013, p. 55). Chapter 3 of this book discusses the global burden of disease from pollutants, including toxic chemicals and DALYs, in more detail.

Evaluation of Chemicals

How to prevent or control the release of chemicals, with unwanted environmental or health properties, into the global environment is a major international and national challenge for regulatory and advisory authorities. According to UNEP (2013, p. 48), only a fraction of the tens of thousands of chemicals on the market have been thoroughly evaluated to determine their effects on human health and the environment. Progress is being made through the registration and submission of better information and data on the effects of individual chemicals (e.g. pesticides, pharmaceuticals, personal care products, and industrial chemicals) and under international and national agencies and legislation such as the European Union's Registration, Evaluation, Authorisation and Restriction of Chemicals (REACH) programme, United States Toxic Substances Control Act (TSCA), Canada's Chemicals Management Plan (CMP), and the Japanese Chemical Substances Control Law. Nonetheless, the limitations in experimental testing, scope of evaluations, and registration of chemicals, together with the difficulties of such international progress, persist as a core problem for the future of the chemical industry, regulatory agencies and community concerns

Effects of Mixtures

Another major challenge is how to deal with mixtures of toxic chemicals, especially at low levels or doses

---

[1] Disability-adjusted life years.

of environmental exposures, often below toxicological thresholds, environmental standards, or guideline levels based on single chemicals. For example, biomonitoring programs of human specimens reveal multiple exposures to hundreds of environmental chemicals (and many metabolites). In some cases, specific health effects of mixtures of some structurally related chemicals can be treated as additive (e.g. neurotoxic effects of low boiling benzenes but differ in their carcinogenic effects). Mixtures of organophosphate insecticides also similarly inhibit cholinesterase in humans. The toxicity of dioxin isomers is expressed as toxic equivalent oral doses (TEODs) of the isomer TCDD. In practice, very little information is available on the health and environmental effects of chemical mixtures, despite many of these chemicals causing well-known environmental and health risks (UNEP 2013, p. 48).

Exposome – New Approaches to Environmental Exposures and Health

Recent scientific advances are seeking to understand the role of how lifetime exposures to chemicals and other stressors in the environment, diet, lifestyle factors, behavior, etc., contribute to diseases that are making us ill. For many diseases such as cancer, cardiovascular disease, asthma, diabetes, and dementia, environmental exposures to pollution, for example, are increasingly seen as playing a major part in health risks for such diseases. By measuring multiple lifetime exposures and interactions with the genome, critical pathways of causation and risks for disease development have the potential to be evaluated and identified.

A new paradigm for environment and health known as the **exposome** has emerged to address the above issues that include the use of novel advances in exposure science, toxicology, and genomics. It can be defined basically as the measure of all the exposures of an individual in a lifetime, from before birth, and how those exposures relate to health. This involves the detailed study of how complex environmental exposures interact with the unique characteristics of individuals such as genetics, physiology, and epigenetics, and impact on their health. It is referred to as **exposomics** (e.g. NIOSH 2014**)**.

A key factor is the ability to accurately measure external and internal exposures and biological responses to exposures throughout lifespans. Many of the *omics* technologies such as genomics are being applied to further our understanding of disease causation and progression in human health (NIOSH 2014). Similarly, these approaches can be adapted to the study of plants and animals in ecotoxicology and eco-exposome. New approaches and methods are also offered for risk assessment of chemicals in evaluation of complex exposure–effect relationships (Barouki et al. 2021; Sillé et al. 2020).

The development of the exposome concept has been seen as one of the hallmarks in environmental and health research of the last decade and considered to be a major driver in this area for the future (Barouki et al. 2021).

1.7 Science of Pollution: Assessment and Management

The first edition of this book used an academic and professional approach to the study of pollution processes and ecotoxicology (Connell and Miller 1984). It provided a scientific understanding of how major classes of pollutants act on natural ecosystems and developed unifying theories on the chemical and ecological nature of pollution processes. It incorporated the relatively new field of ecotoxicology which seeks to examine and understand the effects of pollutants on populations, communities, and whole ecosystems using field and experimental studies, in addition to the use of laboratory bioassays involving effects on individual biological species. Ecotoxicology was considered as a sequence of exposure-related interactions and effects of pollutants, e.g. pesticides on natural ecosystems and their populations and communities of living organisms, influenced by the chemical and physical properties of the pollutants and the physical environment.

This second edition of the book advances the conceptual approach and basic format of the first edition, to incorporate pollutant impacts on not only natural ecosystems, but also human environments and human health. Worldwide, there is a rapidly growing need to recognize and act on how the health of both our natural and built environments impacts on human health at every scale, from local to global. The emphasis in this book is on explaining the ways pollutants interact with the environment and affect natural and human ecosystems together, and how we can use this knowledge to evaluate, predict, and achieve sustainable outcomes for humans and natural forms of life on Earth. For example, it describes the use of scientific risk assessment to achieve healthy environments as part of sustainable development and lifestyles for the present and future. New chemical pollutants are also emerging, some on a global scale. Human-derived emissions of greenhouse gases are driving unprecedented global warming and climate change effects in the history of humans.

Figure 1.3 presents the key conceptual model for this book. It integrates available scientific principles, processes, and practices in pollution, chemistry, toxicology, and ecology to evaluate environmental and human health effects

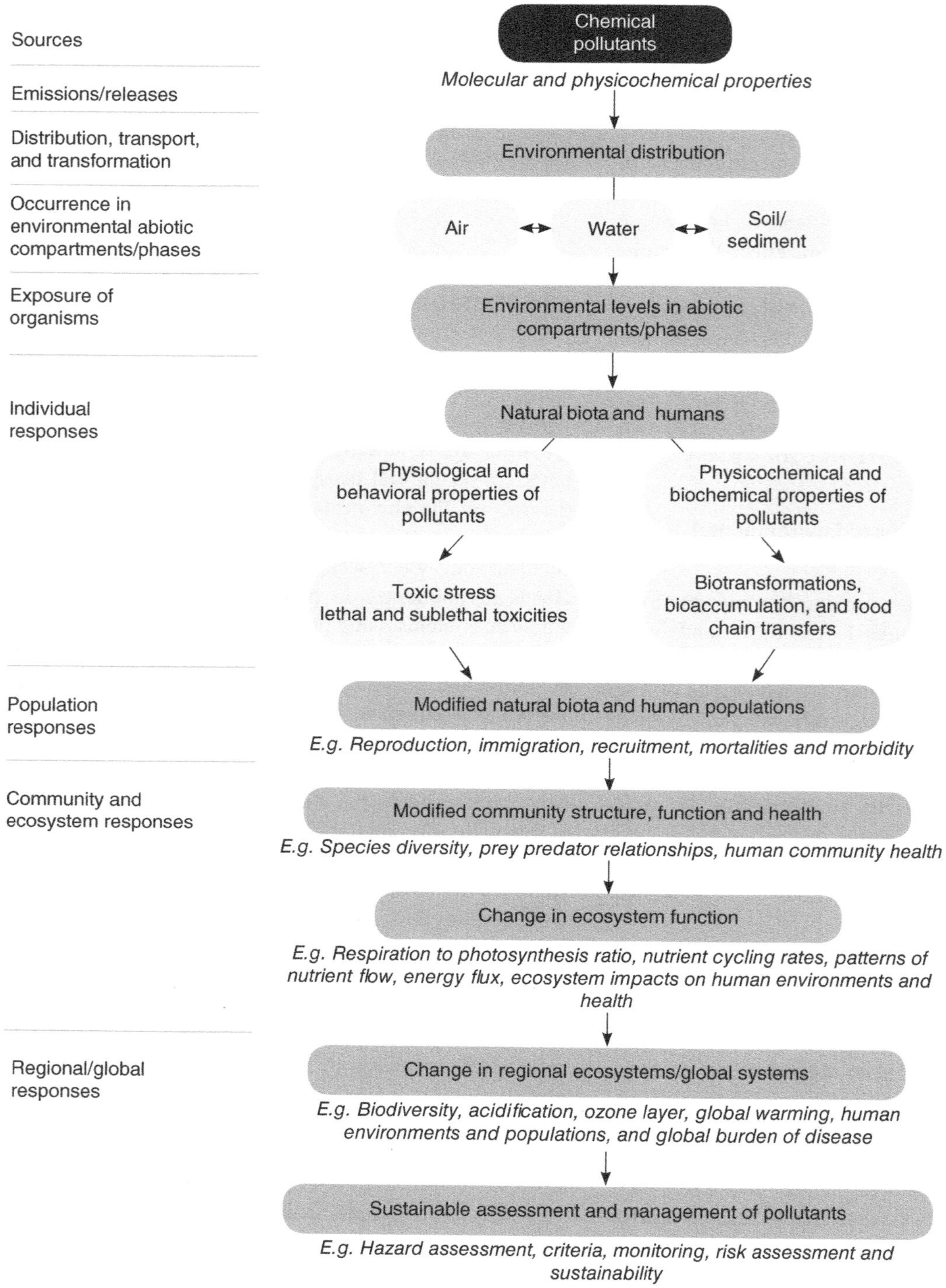

1 Introduction

Sources

Emissions/releases

Distribution, transport, and transformation

Occurrence in environmental abiotic compartments/phases

Exposure of organisms

Individual responses

Population responses

Community and ecosystem responses

Regional/global responses

```
Chemical pollutants
    │
    │ Molecular and physicochemical properties
    ▼
Environmental distribution
    │
    ▼
  Air ↔ Water ↔ Soil/sediment
    │
    ▼
Environmental levels in abiotic compartments/phases
    │
    ▼
Natural biota and humans
   ╱              ╲
Physiological and       Physicochemical and
behavioral properties   biochemical properties
of pollutants           of pollutants
   │                       │
   ▼                       ▼
Toxic stress            Biotransformations,
lethal and sublethal    bioaccumulation, and food
toxicities              chain transfers
        ╲              ╱
         ▼            ▼
Modified natural biota and human populations
E.g. Reproduction, immigration, recruitment, mortalities and morbidity
    │
    ▼
Modified community structure, function and health
E.g. Species diversity, prey predator relationships, human community health
    │
    ▼
Change in ecosystem function
E.g. Respiration to photosynthesis ratio, nutrient cycling rates, patterns of
nutrient flow, energy flux, ecosystem impacts on human environments and health
    │
    ▼
Change in regional ecosystems/global systems
E.g. Biodiversity, acidification, ozone layer, global warming, human
environments and populations, and global burden of disease
    │
    ▼
Sustainable assessment and management of pollutants
E.g. Hazard assessment, criteria, monitoring, risk assessment and sustainability
```

Figure 1.3 Integrated conceptual model of the impact of pollutants on the components and functions of human and natural ecosystems and their monitoring, assessment, and risk management.

and risks from exposures to pollutants, on local, regional, and global scales.

The approach in Figure 1.3 highlights the role of the physical, chemical, and biological properties of pollutants in understanding the environmental behavior, fate, and toxicological significance of pollutants (see Figure 2.1, Chapter 2). For instance, a pollutant released into the environment can be physically dispersed in air, water, soil, dust, or sediments depending upon its physicochemical properties. Simultaneously, it can be modified by abiotic processes (e.g. by oxidation or hydrolysis) or degraded by biotic processes (e.g. biodegradation by microorganisms).

Degradation products may be harmless or biologically inert, or sometimes form more hazardous products such as hydrogen sulfide or methane from anaerobic digestion or airborne particulates from burning of biomass. Pollutants and degradation products can also interact with their environment and change their intrinsic properties, such as in the case of deoxygenation of aerobic waters by microbial degradation of carbohydrates to produce anaerobic waters, acidification of soils by oxidation of iron sulfides, and increasing ocean acidification by absorption and hydrolysis of atmospheric carbon dioxide gas.

The basic sequence presented in Figure 1.3 is used as the framework in this book to evaluate the impacts and risks from individual or various types of pollutants such as chemicals that are increasingly released into our physical and biological environments from human and natural sources. It initially considers the key physicochemical and toxicity properties of pollutants, their sources, and the quantities and rates at which they are released into the physical environment (e.g. air, waters, soils, and sediments).

The behavior and fate of pollutants in the environment is largely determined by the intrinsic properties of the pollutants, environmental properties or factors, and processes that govern or influence their transport and distribution throughout the environmental compartments that make up ecosystems on local, regional, or global scales. These dynamic processes act via biogeochemical cycles and energy flows, as clearly shown by the emissions and dispersion of greenhouse gases, particularly carbon dioxide, throughout the Earth's atmosphere. The fate of pollutants involves interactions of physical, chemical, and biological processes that may remove or convert them into other forms or transfer them into other compartments (e.g. air, waters, particulates, soils, sediments, and biota). These processes interact within and between compartments, and are applicable to natural, modified, or built environments.

The exposure of living organisms, including humans, to pollutants depends on the degree of contact and levels contained in air, waters, soils, sediments, and food sources, such as plants and animals. The uptake of pollutants by living organisms then depends on the properties of the pollutants, the levels of exposure, environmental pathways (e.g. airborne lead or particulates in air), and biological routes of intake such as inhalation, ingestion, and dermal or skin absorption, leading to different degrees of absorption, metabolism, distribution in organs and/or tissues, storage or bioaccumulation, and excretion from the exposed organism or human body. The absorbed dose depends basically on the level or concentration of pollutant in the environmental medium, rate of intake with time (acute or chronic exposures), and the proportion absorbed by the organism via the route(s) of intake.

Organisms exhibit various and different degrees of reactions from absorbed doses of pollutants, from negligible or below an observed effect level to sublethal effects, such as biochemical, genetic, cellular alterations, immune system responses, reduced growth, reproduction and behavioral effects, to ultimately lethality. Biological effects or toxicities (lethal and sublethal) due to the doses absorbed are evaluated in exposed individuals and populations of living organisms according to various measures such as lethal doses, LD_{50}, derived from bioassays or probability distributions showing proportions of an exposed population affected by increasing doses, clinical observations, experimental studies (e.g. bioassays, outdoor experimental systems, mesocosms, and epidemiological methods) (e.g. Chapters 5, 9, 11, 12, and 18). Adverse biological or toxic effects are influenced by critical factors such as the intrinsic properties of pollutants, the physiological and behavioral properties of organisms, and environmental stressors acting on the exposed organisms and their habitats, including those of humans.

These adverse effects can result in significant mortalities, as well as physiological, genetic, reproductive, and behavioral modifications to exposed populations that can then further impact on and modify communities, and the structure and functions of ecosystems in acute and subtle ways. Complex ecosystems can respond in a variety of ways to the effects on their abiotic components and biota. Current scientific findings strongly suggest that humans and their health are similarly affected by their interactions with their environment. Ecosystem functions are driven by material cycles and energy flows, Thus, changes in ecosystem function can include disruption of material cycles such as nutrient cycling rates in accelerated eutrophication of water bodies, and energy flows or fluxes, such as through thermal pollution (e.g. heat stress), or increased exposure to ultraviolet radiation (e.g. skin cancer risks) due to depletion of the stratospheric ozone layer catalyzed by chlorofluorocarbon gases (CFCs). Indirect or secondary effects also arise due to human activities such as herbicide spray drift (e.g. phenoxyacid herbicides such as 2,4-D and 2,4,5-T) causing deforestation or food chain transfer of pollutants that readily bioaccumulate in tissues, such as DDT in the prey of predators or mercury from coal-fired power station emissions (Connell and Miller 1984; UNEP 2013).

Similarly, human populations are vulnerable to contamination of their foodstuffs, air, water supplies, soils, and dwellings by pollutants such as persistent and toxic pollutants that can act, for example, as endocrine disrupters and carcinogens. Overall, adverse effects can be identified, usefully described and expressed by various measures in terms of the health of individuals, populations, communities, and natural or human ecosystems (see also Chapters 3 and 5).

A key concept is that humans exist within the context of ecosystems and are subject to ecological functions and interactions, as for other living organisms. It follows that the ecology of natural biota and the environmental health of human populations are interwoven and need to be treated as interdependent in any evaluation of the pollution impacts or risks from human activities, rather than largely treated as separate entities.

Since the 1970s, there is now rigorous scientific evidence of the intensity and growing extent of impacts from human activities, such as urbanization, agriculture, industrial development, mining, forestry, burning of fossil fuels, and waste disposal and related pollutants on regional ecosystems and global systems. Remote sensing and monitoring from space confirm much of the global extent of environmental impacts.

The scope of evidence includes accelerated eutrophication and *dead zones* from excessive loads of nutrients and organics released into water bodies from worldwide wastes, agriculture runoff and atmospheric deposition, plastic pollution of the marine environments, excessive global mortality and morbidity among humans from outdoor and indoor pollution caused by fine and respirable particulates ($PM_{2.5}$ and PM_{10}), and combustion gases from the burning of fossil fuels and biomass. Recent assessments by the WHO suggest that the global burden of human disease due to environmental pollution factors has reached pandemic-like proportions for the majority of the Earth's population, mostly among children and the aged in developing countries and lower socioeconomic groups in developed countries (see Chapter 3). It is highly probable that a relatively rapid increase in emissions of greenhouse gases, primarily from the burning of fossil fuels, has produced global warming and climate change impacts at unprecedented rates over at least the last 800 000 years (IPCC 2014, p. 4). These types of human-induced impacts are being superimposed at abnormal rates of change upon observed natural variations due to normal environmental stressors. The natural resilience and capacity to adapt of many species and ecosystems is under threat or being exceeded. Observed recovery rates of populations or ecosystems modified or degraded by pollution are often long and uncertain.

There are many serious scientific challenges arising of global proportions for policy and decision-makers. These include what to do about emerging pollutants (new and known) and how to understand, measure, and predict the effects of mixtures or combinations of pollutants and environmental stressors acting simultaneously on ecosystems. Risk assessment and risk management are also confronted by the uncertainties of ecological thresholds or tipping points that can lead to abrupt changes from one ecological state to another one. Climate feedback systems may enhance (or reduce) global warming. Current projections show the net effect is to enhance global warming which also increases the risk of rapid change.

The approach outlined in Figure 1.3 is designed to meet the growing worldwide demand for scientific assessment and management of pollutant interactions, impacts, and risks to achieve sustainable biodiversity and human health. In essence, it forms an **integrated pollution model of ecological and human health** for such purposes.

The organization of this book is based on the development of scientific concepts indicated in Figure 1.3. It focuses on interactions and effects of pollutants on natural environments and human health. Figure 1.3 presents an integrated conceptual model of the impact of pollutants on the components and functions of human and natural ecosystems and a risk-based assessment and management approach to achieve sustainable development and health outcomes.

The book consists of four linked parts: (i) pollutants, environment, and health, (ii) scientific principles, (iii) chemical behavior and toxicology of pollutants, and (iv) applications in environmental assessment and management due to pollution impacts and risks from human activities:

- Part one: *Pollutants, Environment, and Health* introduces the conceptual basis of the book. It covers the increasing impact of pollutants on human health and biodiversity, and its scientific evaluation for environmental management purposes. The nature and key properties of pollutants (physical, chemical, and biological) are explained, along with relationships between pollutants, biodiversity, ecological, and human health and related measures, particularly the burden of disease.
- Part two: *Principles Governing the Interactions of Pollutants with Natural and Human Systems* provide the scientific foundations of how pollutants interact with and affect the environment, wildlife, and humans.
- Part three: *Chemical Behavior and Toxicology of Pollutants in Natural and Human Systems* evaluates the nature, exposures, and effects of major types of pollutants on environmental and human health using the basic principles, processes, and framework of the book from Parts one and two. An example is the emission of greenhouse gases, and how they accelerate global warming and the observed impacts and predicted risks of climate change for Earth and life on it.

1.7 Science of Pollution: Assessment and Management

- Part four: *Environmental Assessment and Management of Pollution* completes the book within the context and scope of the integrated pollution approach used throughout. Some new topics include environmental and health risk assessment, environmental sustainability and biodiversity, national pollutant inventories, and green chemistry. These are incorporated into updated chapters covering: pollution and biological monitoring and the toxicological assessment and management of chemicals. New chapters are presented on science-based approaches used in health and environmental risk assessment, management of hazardous chemicals, and management of pollutants and also pollution and environmental sustainability.

Planetary Health and Human Health

This book integrates Twenty-first Century advances in scientific understanding of the role of pollution due to human-induced activities (e.g. toxic chemical pollutants, air pollution, and global warming) with increasing evidence of observed impacts and risks for planetary health and global human health (e.g. The Lancet Commission on Pollution 2017; UNEP; Global Biodiversity Outlook GBO-6; The Stockholm Resilience Center; and Intergovernmental Panel on Climate Change).

The current nature and degree of these impacts and risks for the health and sustainability of life on Earth are indicated in Figure 1.4. It shows a conceptual framework

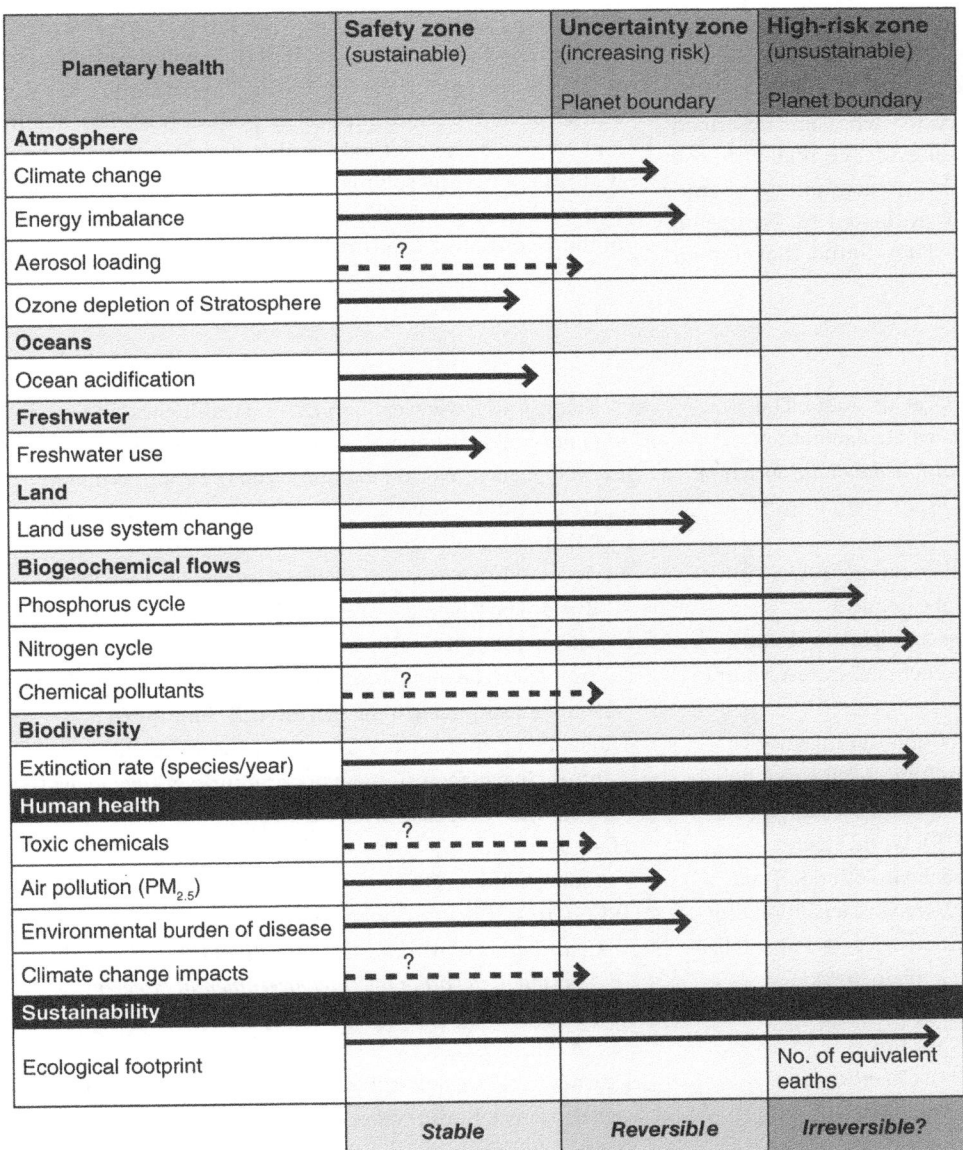

Figure 1.4 Conceptual status of key planetary and human health indicators and boundaries of sustainability on Earth. Source: Modified from Rockström et al. (2009) and Steffen et al. (2015).

of how environmental pollution and other human-induced stressors are affecting key indicators of planetary health and human health.

The basis of Figure 1.4 arises from the scientific development of (i) the concept of planetary boundaries for nine planetary life-support systems and their *safe operating spaces for humanity* and environmental impacts and uncertainties of changes on these systems (e.g. increases in greenhouse gases and global warming, losses of biodiversity, and increased loadings of phosphorus and nitrogen), (ii) global measures and estimates of impacts on human health (e.g. environmental burden of disease, and air pollution, as indicated by $PM_{2.5}$ levels), and environmental sustainability (e.g. ecological footprint). It is evident from Figure 1.4 that pollution plays a major role in the environmental impact and threat to the sustainability of humanity.

Critically, chemical pollution due to synthetic organic chemicals and plastics is now being recognized as exceeding the safe planetary boundary for what are described as "novel entities." The weight-of-evidence related to the global impacts and risks from this diverse mix of chemical pollutants has been recently evaluated by European, UK, and Canadian researchers. They found that annual production and releases of these types of chemicals are transgressing the physical and chemical capacity of the Earth system and our capacity to manage these environmental pollutants, including their assessment and monitoring, on a global scale (see Persson et al. 2022).

Our book presents the nature and role of pollution, through its chemistry and toxic effects at different levels of biological organization, and in the broader context of interactions with planetary life-support systems and the burden of disease in human populations attributed to pollution exposures. Key concepts, principles, and approaches used to evaluate, prevent, and manage emissions or releases of greenhouse gases, toxic chemicals, wastes, and other pollution problems, from local to global scales, are reviewed and discussed, including the challenges to achieve environmental sustainability such as defined by the 2030 Sustainable Development Goals of the United Nations.

Finally, the second edition of this book examines key scientific principles and impacts of pollution on our health and environment, methods, and various examples of how we are tackling and solving serious pollution issues, at local to global scales, and seeking sustainable healthy outcomes in the Twenty-first Century.

References

Barouki, R., Audouze, K., Becker, C. et al. (2021). The exposome and toxicology: a win–win collaboration. *Toxicological Sciences: An Official Journal of the Society of Toxicology* 186: 1–11. PMID 34878125. DOI: https://doi.org/10.1093/toxsci/kfab149.

Centre for Industry Education Collaboration (2016). The Chemical Industry. The Essential Chemical Industry-Online. http://www.essentialchemicalindustry.org/the-chemical-industry/the-chemical-industry.html (accessed 10 July 2018).

Chemical SafetyPro (2016). Global Chemical Inventories. http://www.chemsafetypro.com/Topics/Category/Global_Chemical_Inventories.html (accessed 11 July 2018).

Climate & Clean Air Coalition (2018). World Health Organization Releases New Global Air Pollution Data. CCAC Secretariat (2 May). http://www.ccacoalition.org/en/news/world-health-organization-releases-new-global-air-pollution-data (accessed 16 October 2018).

Connell, D.W. (2005). *Basic Concepts of Environmental Chemistry*, 2e. Boca Raton, FL: CRC Press.

Connell, D.W. and Miller, G.J. (1984). *Chemistry and Ecotoxicology of Pollution*. New York: Wiley.

EJOLT (2012). *Affluence and Environmental Impact*. Environmental Justice Organisations, Liabilities and Trade. http://www.ejolt.org/2012/11/affluence-and-environmental-impact-2/ (accessed 31 May 2018).

Encyclopedia of World Biography (2004). Perkin, William Henry. https://www.encyclopedia.com (accessed 10 July 2018).

Flavell, C. (2010). *Fritz Haber and Carl Bosch – Feed the World*. The Chemical Engineer (accessed 22 August 2020).

Gascoigne, B. (2020). "History of Chemistry" HistoryWorld. From 2001, ongoing. http://www.historyworld.net/about/sources.asp?gtrack=pthc (accessed 27 June 2020).

Global Footprint Network (2018). Ecological Footprint. https://www.footprintnetwork.org/our-work/ecological-footprint/ (accessed 10 July 2018).

Hart-Davis, A. (ed.) (2009). *Science – The Definitive Visual Guide*. London: DK.

IPCC (2014). *Climate Change 2014: Synthesis Report. Contribution of Working Groups I, II and III to the Fifth Assessment Report of the Intergovernmental Panel on Climate Change* (ed. Core Writing Team, R.K. Pachauri and L.A. Meyer). Geneva: IPCC.

Kearney (2017). Chemical Industry Vision 2030: A European Perspective. https://www.de.kearney.com/chemicals/article?/a/chemical-industry-vision-2030-a-european-perspective (accessed 24 June 2020).

Landon, M. (2006). *Environment, Health and Sustainable Development*. Maidenhead: Open University Press.

McLamb, E. (2011). *The Ecological Impact of the Industrial Revolution*. Ecology Global Network. http://www.ecology.com/2011/09/18/ecological-impact-industrial-revolution/ (accessed 11 July 2018).

Murmann, J.P. (2002). Chemical Industries After 1850. http://www.professor-murmann.net/murmann_oeeh.pdf (accessed 10 July 2018).

Myers, N. and Kent, T. (2003). New consumers: the influence of affluence on the environment. *Proceedings of National Academy of Sciences USA* 100 (8): 4963–4968.

NIOSH (2014). *Exposome and Exposomics*. Washington D.C.: National Institute for Occupational Health and Safety. Centers for Disease Control and Prevention. https://www.cdc.gov/NIOSH/ (accessed 3 February 2022).

Patterson, D.G. Jr., Wong, L.Y., Turner, W.E. et al. (2009). Levels in the U.S. population of those persistent organic pollutants (2003–2004) included in the Stockholm Convention or in other long range transboundary air pollution agreements. *Environmental Science & Technology* 43 (4): 1211–1218.

Persson, L., Carney Almroth, B.M., Collins, C.D. et al. (2022). Outside the safe operating space of the planetary boundary for novel entities. *Environmental Science & Technology* 56:1510–1521 doi: https://doi.org/10.1021/acs.est.1c04158

Rockström, J., Steffen, W., Noone, K. et al. (2009). Planetary boundaries: exploring the safe operating space for humanity. *Ecology and Society* 14 (2): 32.

Rowe, D.J.M. (1998). *History of the Chemical Industry 1750 to 1930 – An Outline*. Royal Society of Chemistry. http://www.rsc.org/learn-chemistry/resources/business-skills-and-commercial-awareness-for\ignorespaceschemists/docs/Rowe%20Chemical%20Industry.pdf (accessed 10 July 2018).

Royal Society of Chemistry (2018). What is Alchemy? https://www.rsc.org/periodic-table/alchemy/what-is-alchemy (accessed 10 July 2018).

Secretariat of the Convention on Biological Diversity (2010). Global Biodiversity Outlook 3. Montréal.

Sillé, F.C.M., Karakitsios, S., Kleensang, A. et al. (2020). The exposome – a new approach for risk assessment. *Altex* 37: 3–23. PMID 31960937. DOI:https://doi.org/10.14573/Altex.2001051.

Standen, A. and Killheffer, J.V. (2017). Chemical Industry. https://www.britannica.com/technology/chemical-industry (accessed 31 December 2016).

Steffen, W., Richardson, K., Rockström, J. et al. (2015). Planetary boundaries: guiding human development on a changing planet. *Science* 347 (736): 1259855.

Technofunc (2012). History of Chemicals Industry. http://www.technofunc.com/index.php/domain-knowledge/chemicals-industry/item/history-of-chemicals-industry (accessed 1 January 2017).

The Editors of Encyclopaedia Britannica (2018). Industrial Revolution. https://www.britannica.com/event/Industrial-Revolution (accessed 8 July 2018).

UNEP (2013). *Global Chemicals Outlook – Towards Sound Management of Chemicals*. Geneva: United Nations Environment Programme.

United Nations (2014). *World's Population Increasingly Urban with More Than Half Living in Urban Areas*. United Nations (10 July). http://www.un.org/en/development/desa/news/population/world-urbanization-prospects-2014.html (accessed 11 July 2018).

United Nations Department of Economic and Social Affairs, Population Division (2017). World population prospects: the 2017 revision, key findings and advance tables. In: *Working Paper No. ESA/P/WP/248*. New York: United Nations.

Unsworth, J. (2010). History of Pesticide Use. http://agrochemicals.iupac.org/index.php?option=com_sobi2&sobi2Task=sobi2Details&catid=3&sobi2Id=31 (accessed 1 November 2017).

US EPA (2016). DDT – A Brief History and Status. https://www.epa.gov/ingredients-used-pesticide-products/ddt-brief-history-and-status (accessed 22 November 2016).

WHO (2016). An estimated 12.6 million deaths each year are attributable to unhealthy environments. In: *Public Health, Environmental and Social Determinants of Health (PHE) Issue 82*. Geneva: World Health Organization. March 2016. http://www.who.int/phe/news/e-News-82.pdf (accessed 1 November 2018).

World Bank (2016). *Poverty and Shared Prosperity 2016: Taking on Inequality*. Washington, DC: World Bank. https://doi.org/10.1596/978-1-4648-0958-3. License: Creative Commons Attribution CC BY 3.0 IGO.

2

Environmental Pollutants

2.1 Introduction

This chapter describes the types of environmental pollutants that are rapidly changing the environment of the Earth, and the health of many of its inhabitants. This is essentially due to rapid expansion and growth of human populations, and their activities, since the Industrial Revolution of the Nineteenth Century. How we achieve a sustainable future in the Twenty-first Century greatly depends on how humans respond to and manage the environmental impacts and threats of human activities, pollutant emissions, and other waste streams on human health, biodiversity, and biomass of living organisms (see Landrigan et al. 2017).

How to understand, evaluate, and resolve the role of environmental pollutants in what is now a global crisis depends fundamentally on our level of scientific knowledge about these pollutants, their environmental interactions, and how they are affecting the diversity, capacity, and complexity of life forms on Earth. Initially, we need to know vital information about their nature, types, properties, sources, and the amounts and flows of hazardous or toxic substances that are generated and released into our living world. In this chapter, we start the journey, based on the conceptual framework for this book in Figure 1.3, with an introduction to the complex, often uncertain, and ubiquitous world of environmental pollutants.

Key aspects introduced in the chapter are:

- The nature, basic properties, and classification of pollutants
- The diverse types of pollutants: physical, chemical, and biological
- Sources of pollutants
- Emerging pollutants
- Production, use, and release of chemical pollutants into the environment

2.2 Environmental Pollutants

Nature of Pollutants

A pollutant is some form of matter or energy that causes pollution. It may be described as a waste material, substance, or energy form that pollutes the environment, or more specifically, air, water, soil, plants, animals, and humans. Pollutants can take the form of artificial substances such as synthetic pesticides and industrial chemicals (e.g. PFAS) or natural substances such as heavy metals, carbon dioxide, petroleum hydrocarbons, and ionizing radiation.

The definition of a pollutant may vary in scope from contamination of the environment, such as air or water, to adverse effects or damage to living organisms, including humans, components of the environments, environmental values, and use. As such, the effects of a pollutant are usually defined by the presence of harmful concentrations above normal, background, or measured *no effect* levels. Importantly, the nature, behavior, fate, and effects of a pollutant are influenced by its physical, chemical, and biological properties.

There are many different definitions of pollutants depending upon the context of pollution, such as a water or air pollutant, and popular, scientific, legal, or regulatory use of the term. Examples of a broad definition of a pollutant are given below:

Something that pollutes, especially a waste material that contaminates air, soil, or water – The American Heritage® Dictionary of the English Language (2020).

Based on a definition proposed by the European Environment Agency, The Lancet Commission on Pollution and Health defined pollution as *Any material introduced into the environment by human activity that endangers human health or harms living resources and ecosystem* (Fuller et al. 2018).

A regulatory example of an air pollutant is provided in The US Clean Air Act, as incorporated into the United States Code of Federal Regulations, Title 42, Chapter 85. Its Title III, Section 7602(g) defines an air pollutant:

> The term "air pollutant" means any air pollution agent or combination of such agents, including any physical, chemical, biological, radioactive (including source material, special nuclear material, and byproduct material) substance or matter which is emitted into or otherwise enters the ambient air. Such term includes any precursors to the formation of any air pollutant, to the extent the Administrator has identified such precursor or precursors for the particular purpose for which the term "air pollutant" is used
>
> (Legal Information Institute, Cornell University 2020).

Box 2.1 Definitions

Pollutant	a substance or form of energy that causes adverse effects on living organisms and/or their environments
Toxicant	a substance that causes adverse effects in living organisms
Toxin	a toxin is a toxicant produced by a living organism (e.g. microorganism, insect, spider, snake, and jelly fish)
Contaminant	a substance present in the environment or living organisms that causes impurity or potential adverse effects
Xenobiotic	an exogenous or foreign substance found in the body or an ecosystem
Hazardous substance	a hazardous substance may be toxic, corrosive, reactive, flammable, radioactive, or infectious or some combination of these

Characterization of Pollutants

Pollutants are generally characterized by their physical, chemical, or biological nature and related physicochemical, environmental, and toxic properties, or disease-causing properties (infectivity) in the case of pathogens. Key properties of pollutants to consider are:

- physical, chemical, and/or biological in nature: physical state(s) of matter (e.g. gases/vapors, liquids, and solids), form of energy (e.g. gamma radiation), mixture of substances (e.g. petrol, volatile organic chemicals [VOCs]), shape and size (e.g. asbestos fibers, $PM_{2.5}$, and nanoparticles), chemical structure (e.g. ionic forms, functional groups, monomers or polymers, and biomolecules), virus particle, unicellular or multicellular microorganism;
- volatility of substance (e.g. vapor pressure);
- solubility in water or lipids;
- electromagnetic radiation properties (e.g. absorption of solar radiation and re-emission of infrared radiation by greenhouse gases);
- mobility in different environmental compartments such as air and soils (e.g. water soluble, volatile, sorption capacity onto organic matter or carbon);
- chemical reactivity of a substance or mixture;
- persistence (e.g. biodegradability, half-life in air, water, soil, blood, and organs);
- bioaccumulation capacity in nonhuman organisms and humans, including fluids, tissues, and organs; and
- acute and chronic toxicity to different nonhuman organisms and humans.

The physical, chemical, and biological properties of pollutants, and interactions with their environment, influence their environmental behavior, fate, and biological effects on biota, and ecosystems, at all levels of organization. The fundamental Influence of physical, chemical, and biological properties of pollutants on environmental behavior and biological effects is illustrated in Figure 2.1.

How this set of key properties of a pollutant interacts with the physical environment, individuals, populations, communities, and their ecosystems forms the scientific basis for understanding pollution. Using this knowledge, we can act to predict, prevent, and manage pollution issues that affect our ecological and environmental health.

The significance of this approach is well shown by the study of chemical pollutants such as toxic metals, persistent organic pollutants (POPs), ozone-depleting chemicals, and greenhouse gases (e.g. carbon dioxide) in environmental chemistry and ecotoxicology. Effects may be direct such as for toxicity or indirect such as through changing the physical properties of the atmosphere in the case of the *greenhouse effect* leading to increased absorption and re-emission of infrared radiation by atmospheric carbon dioxide gas and global warming.

Connell (2005, pp. 10–11) has described the environmental properties of a chemical as an interdependent set of properties based on the characteristics of the molecule (e.g. surface area, molecular weight, functional groups, and chemical bonds). These characteristics are seen to govern the physical-chemical properties of the compound (e.g. water solubility, vapor pressure, and n-octanol–water coefficient) which then control transformation, distribution,

Figure 2.1 Influence of physical, chemical, and biological properties of pollutants on environmental behavior and biological effects.

Properties of pollutants

Physical characteristics of pollutant

(size, molecular weight, surface area, and so on)

↓

Chemical characteristics of the atomic or molecular structure

(reactive groups, chemical bonds, polarity, and so on)

↓

Physicochemical characteristics of the substance

(physical state, aqueous solubility, vapor pressure, melting point, octanol–water partition coefficient, and so on)

↓

Uptake by biota

(bioaccumulation, interaction with cellular components, activation of enzymatic systems, and so on)

↓

Biological effects

(toxicity and effects on biota and ecosystems at all levels of organization)

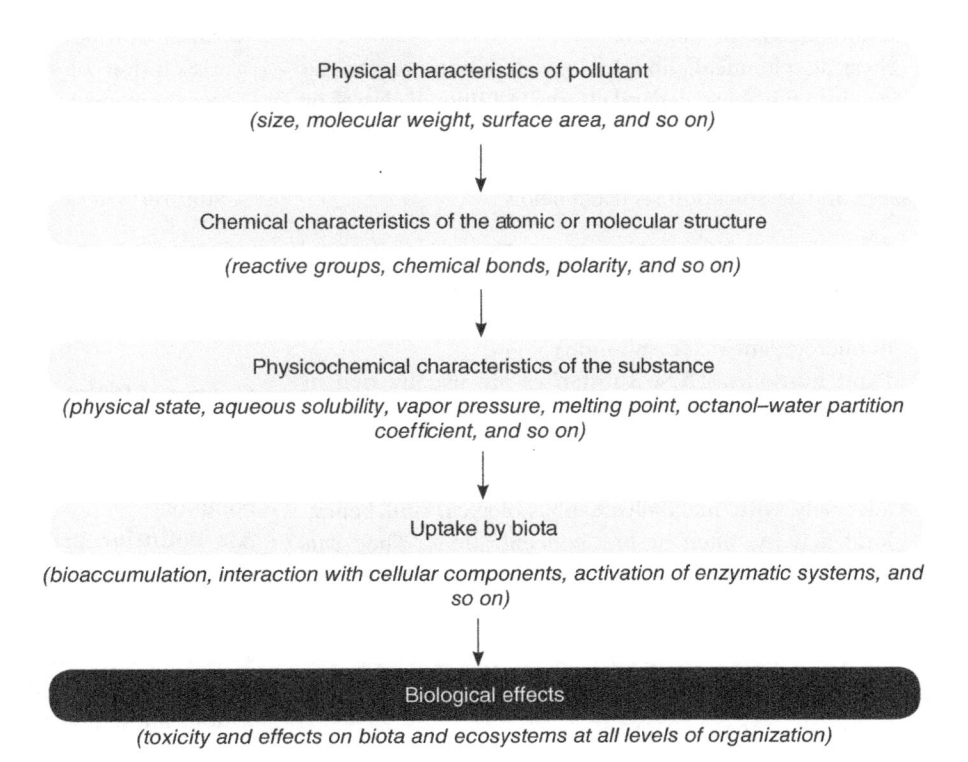

and biological effects in the environment. How physicochemical properties foreshadow their environmental toxicity is discussed in Chapter 5.

Disease-causing microorganisms or pathogens are infectious biological agents that are influenced by their intrinsic physical and chemical nature and related properties, as well as environmental factors (e.g. temperature, humidity, salinity, sunlight, available carbon, and nutrient sources). For example, waterborne pathogens outside of their hosts can persist to varying degrees in drinking water supplies and sometimes sediments but generally do not grow or proliferate in water. This includes viruses and cysts, oocysts, and ova forms of parasites. In contrast, some bacterial pathogens such as *Legionella, Vibrio cholerae, and Naegleria*, and nuisance organisms can grow under favorable conditions of high organic carbon, warm temperatures, and low residual chlorine in surface waters and distribution systems. In water, most pathogens gradually lose their viability and ability to infect. Nonetheless, the most common ones in water tend to have high infectivity due to persistence or ability to multiply. Survival of pathogens outside of their host is affected primarily by temperature. Other environmental factors include exposure to UV from sunlight and salinity. Die-off of pathogens usually follows an exponential decay (WHO 2011, Chapter 7).

Classifying Pollutants

Pollutants may be classified by various criteria, such as from the nature or origin of the pollutant (natural, synthetic, and genotoxic), effect on the environment, according to their physical, chemical, or biological properties (e.g. persistent, biodegradable, degradation product or metabolite, bioaccumulative, and toxicity), and treatability by pollution control processes.

The many different types of pollutants can condense into several broad groups that have similar properties and effects. For example, many thousands of chemical substances can be classified as toxic chemicals based on their adverse effects on nonhuman organisms and humans by acting as toxic substances at occupational and/or environmental levels. Some of these toxic chemicals will also be persistent in the environment and will be classified as persistent toxic chemicals (PTCs), POPs, or further as persistent, bioaccumulative, and toxic chemicals (PBTs).

In the case of water pollutants, the adverse effects can be related to three common environmental properties: (i) excess plant production, (ii) deoxygenation, and (iii) toxic or similar adverse physiological effects on organisms (Connell and Miller 1984, p. 74). Six general classes of water pollutants were described by Connell and Miller (1984, pp. 74–75): (i) organic matter (organic carbon),

(ii) plant nutrients, (iii) toxicants, (iv) thermal wastes (heat energy), (v) suspended solids, and (vi) disease-causing microorganisms (pathogens).

Physical, chemical, and biological pollutants can be placed into the above general classes. Additional classes or subclasses may be considered to cover atmosphere-altering pollutants (e.g. greenhouse gases and ozone-depleting gases), and acidification, as listed below.

1. Organic matter (organic carbon). This primarily consists of carbohydrates, proteins, and fats and leads to dissolved oxygen reduction by stimulating the growth of microorganisms (respiration).
2. Plant nutrients. These substances are usually rich in nitrogen and phosphorus compounds and stimulate excess plant growth (eutrophication).
3. Toxic substances. These are substances that interfere adversely with metabolism, physiological and behavioral activity, often at low concentrations. They can include saline (salinity) and radioactive substances.
4. Particulates. These substances have similar effects to toxic substances but act by physical interaction: suspended solids or particulates in water and airborne particulates (e.g. respirable particulates, $PM_{2.5}$, from combustion of fossil fuels).
5. Energy. This form of energy pollution is mainly due to thermal discharges (waste heat) or heat stress. Effects are similar to toxic effects but this activity is due to thermal or infrared energy inputs. Other forms include mechanical or kinetic energy such as sound energy causing noise pollution (e.g. nuisance and disruption to sleep) and ionizing (e.g. gamma radiation) and nonionizing electromagnetic radiation (e.g. UV, visible light, and microwaves).
6. Pathogenic microorganisms. These exhibit a toxic effect on organisms but the effects are due to infectious organisms rather than chemical substances.
7. Atmosphere-altering substances. These include (i) ozone-depleting gases (chlorofluorocarbons, CFCs) that remove UV-absorbing ozone gas from the stratosphere and allow increased UV radiation from sunlight to impact on living organisms, including humans; (ii) greenhouse gases, primarily carbon dioxide, nitrous oxide, and methane, that absorb reflected solar radiation in the atmosphere and re-emit longer wavelength infrared radiation back into the atmosphere causing the heat energy balance to shift toward a global warming effect; (iii) strong acid-forming gases (e.g. nitrogen dioxide and sulfur dioxide) leading to acid precipitation, and photochemical smog formation, including the production of toxic ozone gas in the lower atmosphere (see also Chapter 13).
8. Acidification-causing substances. Major examples include acid-forming gases (e.g. nitrogen dioxide and sulfur dioxide) leading to effects of acid precipitation, dissolution of carbon dioxide gas emissions in the oceans causing ocean acidification, and oxidation of exposed mineral sulfides in soils, sediments, and mining wastes to form acidic sulfates, leading to the release of sulfuric acid.

Section 2.2 classifies pollutants more specifically as types of physical, chemical, and biological pollutants.

Sources of Pollutants

The major sources of pollutants originate from the following forms of air, water, soil/sediment, and energy-related pollution:

Air Pollution caused by the release of hazardous and toxic chemicals, particulates, and bioaerosols into the atmosphere, usually in the form of mixtures:

- Common gaseous and vapor pollutants produced by industry, coal-fired power stations, oil and gas production, mining, agriculture and transport emissions (e.g. motor vehicles, shipping, and aircraft) include carbon monoxide, carbon dioxide, nitrogen oxides (NO_X), sulfur dioxide, sulfur trioxide, hydrogen sulfide, hydrocarbons or volatile organic compounds (VOCs), polyaromatic hydrocarbons (PAHs), and CFCs.
- Photochemical smog, including powerful pollutants – ozone and peroxyacyl nitrates (PAN), are formed in urban atmospheres as nitrogen oxides and hydrocarbons emitted from combustion sources (e.g. motor vehicles) react in the presence of sunlight (UV).
- Particulate matter or fine dust (e.g. PM_{10} and $PM_{2.5}$), diesel fumes, and airborne metals are also released into the air by combustion from coal-fired power stations, transport, construction, mining and extraction, metal refining, industrial, land clearing activities, agriculture, and wind erosion.
- Indoor air pollution is another form of air pollution, including emissions of VOCs, fine particulates, carbon monoxide, nitrogen oxides, tobacco smoke, and bioaerosols (e.g. airborne allergens and molds). Household combustion of fossil fuels and biomass for cooking and heating is a major source of respirable particulates, carbon monoxide, and other combustion by-products.

Water Pollution originating from the discharge or release of wastewaters, including contaminated runoff, from urban, commercial, industrial processes, mining, transport, and agriculture into surface waters or seepage into groundwaters:

- Various discharges of treated and untreated sewage, and chemical contaminants, such as nutrients, metals, endocrine-disrupting organic substances, chlorine, and disinfection by-products, from treated sewage.
- Releases of contaminated urban, agricultural, and rural runoff flowing to surface waters and leaching into groundwaters. Runoff may contain diverse chemical contaminants such as suspended sediments, toxic metals, petroleum residues, nutrients, salts, chemical fertilizers, antibiotics, and pesticides.
- Waste disposal on a global scale from urban development, industry, mining, quarrying, dredging, agriculture, and forestry and leaching into surface waters and groundwater.
- Waterborne pathogens from fecal wastes (e.g. sewage and warm-blooded animals) are also common contaminants from point and nonpoint sources in catchments, rivers, and along coastlines.
- The disposal of plastic wastes into the world's oceans which is also producing large areas of microplastic pollution at an unprecedented scale, including contamination of deep ocean waters and sediments.
- Natural events, such as flooding, or industrial accidents (e.g. oil spills and industrial fires). These events can also cause peak loads of waterborne pollutants.

Soil and Sediment Pollution resulting from erosion or releases of wastes and contaminants from discharges, spillages, leaks, leaching, burning wastes, applications of pesticides, nutrients, and biosolids, or landfilling involving multiple waste sources (e.g. urban, construction and demolition, industrial, mining, and agriculture):

- Significant soil contaminants include silt, salinity, acidification, excess nitrogen and phosphorus deposition, petroleum hydrocarbons, BTEX, PAHs, VOCs, arsenic and heavy metals, pesticide residues, chlorinated hydrocarbons, antibiotics, radioactive wastes, and asbestos residues or asbestos-containing fill.

Radioactive Contamination resulting from uranium mining, mineral sands processing (thorium in monazite), and waste sand fill, nuclear power generation, radioactive wastes (handling, reprocessing, storage, discharges, and accidents), and nuclear weapons research, manufacture, development, and testing (ionizing radiation [alpha, beta, and gamma] and radionuclides).

Energy-related Pollution involving global warming associated with greenhouse gas emissions, primarily from combustion of fossil fuels, thermal (heat), noise and nonionizing radiation (electromagnetic):

- Thermal pollution results from a temperature change in natural water bodies caused by the discharge of waste heat from industrial activities (e.g. coal-fired and nuclear power stations), such as use of water as a coolant in a power plant.
- Global warming and ocean acidification associated with greenhouse gas emissions, primarily from combustion of fossil fuels.
- Noise pollution is caused by abnormal sound energy intensity and frequency emissions such as roadway noise, construction noise, industrial noise and other urban noise (e.g. loudspeakers, air conditioners and pumps), high frequency and intensity sonar uses.
- Nonionizing radiation (e.g. UV radiation, infrared, microwave and radio frequency radiation) that may cause damage to tissues through heating or other possible effects on biological tissues.

Other Sources of Pollutants include litter, gross pollution (e.g. stormwater drains), shipping wastes, processing and disposal of electronic wastes (e-wastes).

Units of Measurement

Pollutants (or contaminants) in the environment (air, water, soils, and sediments) or in living organisms (e.g. body fluids and tissues) are commonly measured at low levels or concentrations. In many cases, toxic effects on exposed organisms are observed at these low levels. Concentrations are usually expressed in parts per million (ppm), or mass of pollutant per unit mass or volume of the solution, or solid substrate (e.g. milligrams per liter, $mg\,L^{-1}$ or milligrams per kilogram, $mg\,kg^{-1}$). The parts per million of a pollutant component is the number of grams of that component in 1 million grams of a solution or solid.

ppm by mass per unit mass:
Parts per million is the mass ratio between the pollutant component (p) and the solution or solid substrate (s):

$$ppm = 1\,000\,000 \text{ or } 10^6 \frac{m_p}{m_s} \qquad (2.1)$$

where m_p = mass of pollutant (kg) and m_s = mass of solution or solid substrate (kg).

One ppm can be usefully expressed in terms of a **milligram per kilogram**

i.e. $1\,mg\,kg^{-1} = 1$ part per million $= 1 \times 10^{-6}$

and

$$1\,mg\,kg^{-1} = 10^3\,\mu g\,L^{-1} = 10^{-3}\,g\,L^{-1}$$
$$= 10^{-6}\,kg\,L^{-1} = 1\,g\,m^{-3}$$

Lower concentrations than ppm can be expressed as:

ppb – parts per billion (1×10^{-9})

ppt – parts per trillion (1×10^{-12})

ppq – parts per quadrillion (1×10^{-15})

Larger concentrations of a pollutant can also be given in **weight percent**, i.e.

$$\text{weight percent} = 100 \frac{m_p}{m_s} \qquad (2.2)$$

One weight percent (1% w/w) of a pollutant component is equal to 10 000 ppm. For example, if a dry soil contains 0.2% (w/w) of lead, it has a lead concentration of 2000 ppm or 2000 mg kg^{-1} (dry weight).

ppm by mass per unit volume:
The concentration of a pollutant component can be measured as mass per unit volume (e.g. mg L^{-1}, mg cm^{-3}, mg m^{-3}, g m^{-3}, and kg m^{-3}). One ppm of a substance in a solution (or mixture) is equivalent to 1 mg L^{-1} of solution (homogeneous mixture). For example, if 100 mg of a substance are dissolved in one liter of water, the concentration of the substance in the aqueous solution would be 100 mg L^{-1} or 100 ppm. Other units equivalent to 1 ppm in solution include 1 mg L^{-1} and 1 g m^{-3}.

Note: For a pollutant component in air, such as particulates, or gases and vapors (e.g. carbon dioxide gas or benzene vapor), as a volume mixture, the mass per unit volume concentration of the component in the air mixture is usually given as mg m^{-3}, μg m^{-3}, or 1 g m^{-3}. However, where a gas or vapor concentration in parts per million by volume (ppm$_v$) is needed, Eq. (2.3) can be rearranged to convert mg m^{-3} into ppm$_v$.

$$\text{mg m}^{-3} = \left(\frac{[MW \times ppm]}{22.4}\right) \times \frac{T_1}{T_2} \qquad (2.3)$$

where MW is molecular weight; T_1 is 273 K and T_2 is the actual temperature of the air in Kelvin (K).

Worked Example 2.1 Convert 400 ppm of carbon dioxide gas in air to mg m^{-3} at 25 °C

$$\text{Using Eq. (2.3), mg m}^{-3} = \left(\frac{[MW \times ppm]}{22.4}\right) \times \frac{T_1}{T_2}$$
$$= \left(\frac{[44 \times 400]}{22.4}\right) \times \frac{273(K)}{298(K)}$$
$$= 720 \text{ mg m}^{-3}$$

ppm by volume per unit volume:
Parts per million can be expressed by volume of pollutant component per volume of mixture or solution using a modification of Eq. (2.1):

$$\text{ppm}_v = 10^6 \frac{V_p}{V_s} \qquad (2.4)$$

where V_p = volume of pollutant (m^3) and V_s = volume of mixture or solution (m^3). Here, ppm by volume is denoted by ppm$_v$.

Worked Example 2.2 If 10 L of toluene were well mixed in 100 m^3 of air, what would be the concentration of toluene (ppm) in the air mixture? (note: convert liters to cubic meters).

$$\text{Using Eq. (2.4), ppm}_v = 10^6 \frac{V_p}{V_s}$$
$$= 10^6 \times 0.01/100$$
$$= 100 \text{ ppm (v/v)}$$

2.3 Types of Pollutants

Physical Pollutants

Physical pollutants consist of particles of matter of various sizes and shapes (e.g. dusts, respirable particles, or suspended solids in water) and different forms of energy (e.g. heat and sound energy) that interact and impact on living organisms and their environments. Their impacts are characterized by (i) the mass transport of particles (aerosols to solids of different sizes, from ultrafine to coarse grains, gravel, rocks, and minerals), (ii) different types of energy forms (e.g. electromagnetic radiation spectrum, ionizing radiation, and thermal heat), and (iii) quantities and flows or fluxes through the physical environment and acting on biological systems (e.g. heat stress or ionizing radiation exposures).

For particles, their sizes, shapes, and particle size distributions determine many of their physical properties, e.g. aerodynamic diameter, mass, volume, density, surface area to mass ratios, optical properties (e.g. fluorescence), and chemical properties (e.g. reactivity). These properties and their chemical composition also influence how they will behave in the environment and cause adverse effects (e.g. respiratory and cardiovascular disease) on humans and natural biota, primarily through inhalation of excessive respirable particles by humans and air-breathing animals. The same applies to their transport or removal from the environment by natural processes (e.g. deposition from the atmosphere) or pollution control by filtration processes or sedimentation in treatment ponds. The biological effects of different energy forms such as UV, infrared radiation, and sound energy are largely influenced by the intensity, frequency, and wavelength of the energy, and exposure or dose.

Major types of physical pollutants are outlined below.

Airborne Particulates

Airborne particulates are derived from natural or human-made sources and exist as microscopic solid or liquid matter suspended in air or the atmosphere. They commonly form fine particulate–air mixtures of aerosols. Airborne particulate matter is generally classified according to their size (and shape) characteristics, from basically coarse to fine and ultrafine. Particle size depends on sampling method and aerodynamic behavior. It is described by aerodynamic diameter in terms of an idealized spherical particle. Particles with the same aerodynamic diameter may have different shapes and dimensions such as asbestos fibers.

The key types of airborne particles include: total suspended particulates (TSP) of fine to coarse airborne particles or aerosols that are less than 100 micrometers or microns (μm) in equivalent aerodynamic diameter, inhalable coarse particles, and respirable and ultrafine particles of less than 10 microns (see Chapter 13). Many common types of particles are distributed within these size ranges (e.g. dust fractions, mold spores, asbestos fibers, smoke, and nanoparticles) as indicated in Figure 2.2.

The particulates of serious health concern occur in the **inhalable or respirable fractions** (see Figure 2.2):

- PM_{10} – particulates with an equivalent aerodynamic diameter of less than approximately 10 μm. The largest of these are inhalable.

- $PM_{2.5}$ – particulates with an equivalent aerodynamic diameter of less than approximately 2.5 μm, known as the fine particle fraction. The particles in the fine fraction that are smaller than 0.1 μm are called ultrafine particles.

- $PM_{1.0}$ – particulates with an equivalent aerodynamic diameter of less than approximately 1 μm. These particles travel to the deepest areas of the lungs where they can be readily absorbed into the bloodstream.

Note: a human hair is about 60–100 μm in width.

Particles in the respirable size range make up a large proportion of dust that can be inhaled deep into the lungs. Larger inhalable particles tend to be trapped in the nose, mouth, or throat and can be swallowed. Most of the total mass of respirable particles consists of fine particles in the 10–2.5 μm range. However, while ultrafine particles may represent only a few percent of the total mass, they are usually the most numerous particles. Exposure to airborne particulate matter (as indicated by $PM_{2.5}$) is associated with most global deaths due to air pollution (Health Effects Institute 2019).

Several other types of respirable particles can cause specific and severe diseases in humans that are known to be associated with their size, shape, dimensions, and chemical compositions. The risk of developing these diseases tends to be dose-related.

- Crystalline silica – the three most common types of crystalline silica (SiO_2) are quartz, tridymite, and cristobalite. Respirable crystalline silica dust can occur in air due to natural sources, coal mining, industrial, and agricultural activities. Chronic exposure to relatively high concentrations of crystalline silica can cause a respiratory disease known as silicosis, primarily in occupations where particles of crystalline silica are released to the atmosphere (e.g. mining, quarrying, foundries, stone masonry, construction, and sandblasting).
- Asbestos fibers – the most common types are chrysotile, amosite, and crocidolite. Workplace and environmental exposures can lead to asbestosis, mesothelioma, and lung cancer.
- Smoke (tobacco, bushfire, and wood-fired cooking) is associated with respiratory and cardiac diseases and cancers.
- Diesel fumes and combustion soot particles are also associated with respiratory and cardiac diseases and cancers.
- Bioaerosols are fine suspensions of airborne particles that contain microbes and/or biological matter released from living organisms. They range in size from less than 1–100 μm (see Biological Pollutants).

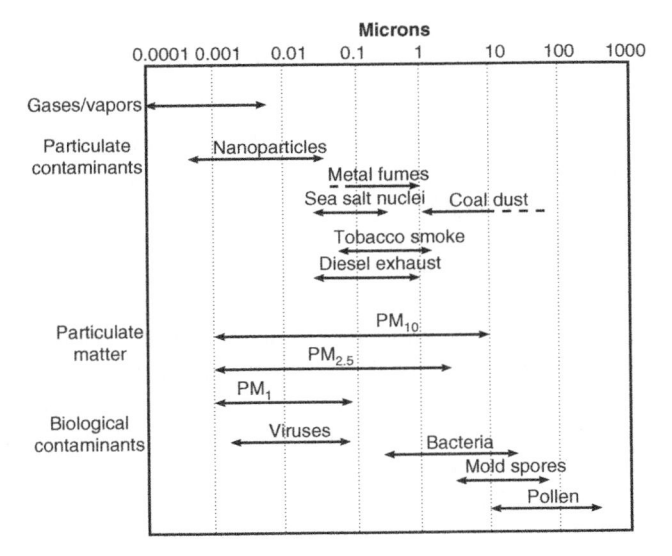

Figure 2.2 Aerodynamic particle size ranges for common airborne pollutants: gaseous molecules, specific particulate contaminants, defined ranges of particulate matter (PM), and biological contaminants.

Manufactured Particulates

Microbeads and particles: These particles consist of manufactured solid particles with dimensions that range from about 1 to 1000 μm depending upon applications such as in biomedical and health science research, cosmetics or other personal care products, and surface coatings. These types of particles are known to contaminate marine and fresh waters, some aquatic animals, human environments and more recently, humans, similar to other microplastics. Micro- and nanoplastic particles derived from plastics are also of current toxicological interest.

Nanoparticles: The emergence of nanotechnology has created a new type and range of ultrafine particles known as **nanoparticles** that are sized usually between 100 and 1 nm (0.1 and 0.001 μm) in diameter. Generally, these particles are engineered (e.g. carbon nanotubes, buckyballs, and quantum dots) for specific properties. Relatively little is known about the extent of their environmental or workplace exposures. Nanoparticle research and applications are an area of intense scientific and commercial interest due to a wide variety of potential applications in biomedical, optical, and electronic fields. At the nanometer scale, these particles exhibit a change in their properties (e.g. optical, semiconductor, or high surface area to volume ratios) compared to the bulk material. However, there is also current research and regulatory interest in the potential nanotoxicity of such particles because of their ultrafine size, shapes (e.g. respirable fibers and composition), and use of toxic elements, amalgams, and synthetic chemicals.

Waterborne Particles

Suspended solids are small solid particles of variable sizes, from coarse to colloidal (fine charged silt and clay particles), that remain in suspension in water or are transported by the mass flow of the water, such as in the case of stormwater runoff. The particle sizes of fine sediment (sand, silt, and clay) dispersed in the water medium are typically less than 2 mm (2000 microns). Silt particles are from 2 to 50 microns, and clay particles are very small, less than 2 microns.

Waterborne particles are commonly mixtures of inorganic and organic matter such as detritus that can also readily transport adsorbed pollutants (e.g. pesticide residues and metals) and pathogens on their surfaces. Higher potential loads in water flows are related to the greater surface area per unit mass on smaller particles. Depending upon particle size distributions (e.g. silt), concentrations, and environmental factors, these particles can impact on aquatic organisms through physical interference to movement or feeding, clogging, and smothering at levels elevated above background (see Chapter S23).

Energy Pollutants

Energy pollution commonly takes the form of heat, noise, or light. It is due to different types of energy, from mechanical or kinetic energy (e.g. vibrational sound energy) to heat energy (infrared) and other forms of electromagnetic radiation (e.g. infrared energy and visible light) and nuclear energy (e.g. radioactive particle emissions). Energy pollutants in this sense are components of energy pollution. Major examples are:

- *Thermal pollution (waste heat)* – the degradation of water quality by any process that changes ambient water temperature such as the use of river water as cooling water in thermal power stations (Chapter S21).
- *The greenhouse effect* – the process by which absorption and re-emission of infrared radiation by greenhouse gases (e.g. carbon dioxide and methane) in the atmosphere warm the Earth's lower atmosphere and surface. This is another form of physical pollution accelerated by emissions of greenhouse gases from burning of fossil fuels by human activities. It is also a competing process with fine particulate emissions to the atmosphere that may cause a cooling effect. Black soot or carbon particles from urbanization complicate these interactions further (Chapter 14).
- *Noise (sound energy)* – unwanted sound energy that results from mechanical vibrational energy. Excessive environmental noise can cause psychological effects (e.g. annoyance and aggression, and high stress levels) and physical health effects (e.g. hearing loss, sleep disturbances, hypertension, and other harmful effects). Sound energy pollution is not covered in this book.
- *Radiation* – two general forms of nonionizing and ionizing radiation (see also Chapter S22).

Nonionizing electromagnetic radiation (NR) covers near ultraviolet, visible light, infrared, microwave, radio waves, and low-frequency RF. It has insufficient energy to ionize matter or to directly damage DNA in tissues. However, near UV and visible light may induce photochemical reactions. Exposure to NR may result in thermal heating effects or tissue damage, depending on the frequency and field strength of the NR, its penetration depth, and the energy absorbed by the exposed body tissues. Potential cancer risks from magnetic field exposures associated with NR are also of current scientific and epidemiological interest.

Ionizing radiation consists of far ultraviolet light, X-rays, gamma-rays, and alpha and beta particles emitted from the radioactive decay of radionuclides (unstable atomic nuclei). This type of energy has sufficient energy to remove an electron from atoms or molecules and can alter chemical bonds and produce reactive ions. Nuclear fission of radioactive atoms also emits uncharged neutron particles. Exposure to ionizing radiation causes damage to living tissues, and can result in mutation, radiation sickness, cancer, and death (AGDH 2012).

There is also natural background radiation exposure to consider. It comes from cosmic rays and from natural radioactive materials found in the environment (e.g. soils, rocks, and living organisms).

Chemical Pollutants

Natural Pollutants

Many chemicals that occur naturally in the environment are potentially toxic to humans. Rocks, minerals, salts, and nutrients contained in the Earth's crust and living organisms release gases, metals, nonmetals, inorganic and organic compounds (e.g. crude oil from seeps in the oceans, PAHs from forest and bushfires, VOCs, mercury and sulfur dioxides from volcanoes) into the air, surface and groundwaters, soils, and sediments as part of biogeochemical cycles.

These chemicals form natural background levels that vary or change due to natural processes and human-induced pressures or a combination of both (e.g. greenhouse gas effect and global warming) over different time scales. Some plants and animals can metabolize toxins such as cyanobacteria (blue green algae).

Examples of naturally occurring chemical contaminants include:

- Excessive levels of arsenic in groundwaters from regions such as Bangladesh, Taiwan, China, and parts of the United States. Strong evidence links arsenic intake from groundwater with skin cancer and other cancers.
- Fluoride poisoning (e.g. fluorosis) in persons drinking water with elevated levels of natural fluoride, and also concentrated by human activities.
- Saltwater intrusion into groundwaters from overextraction of natural coastal aquifers for drinking water supply or irrigation.

Human-Caused Pollutants

These types of pollutants result from a multitude of human activities driven by complex global population and socioeconomic pressures despite limitations of available resources and capacity of the Earth's systems to absorb these pollutants. They readily occur from rapidly growing urbanization, agriculture, extraction, and processing of the Earth's natural resources including fossil fuels, development and production of synthetic chemicals, e.g. new pesticides, drugs, nanoparticles and engineered nanomaterials, waste disposal and treatment, sewage, landfills, and e-wastes.

Types of Chemical Pollutants

Chemical pollutants or contaminants can be basically divided into:

- Metals, metalloids, and nonmetals
- Inorganic compounds
- Organometallics
- Organic compounds

There are estimated to be well over 100 000 chemicals in commercial use and many more known chemicals (millions). Thousands of these commercial chemicals are well-known environmental contaminants or pollutants and may be used in many different products. Yet, for many of these chemicals, we have little or no knowledge of their environmental properties, fate, or effects on living organisms.

A general classification of the types of chemical pollutants and related contaminants known to be released into the environment by human activities (anthropogenic chemical contaminants and toxicants) is presented in Tables 2.1 and 2.2. Chemical classifications are usually based on related chemical structures and/or properties such as metals, nonmetals, petroleum hydrocarbons,

Table 2.1 Chemical pollutants: inorganics, radionuclides, and organometallics.

Chemical types	Examples
Metals and compounds (e.g. metal salts)	Alkali, alkaline earths, transition or heavy metals, lanthanides or rare earths, and actinides
Metalloids and compounds	Metal and nonmetal properties: As, Sb, Se
Nonmetals and compounds	Nonmetals: N, O, F, Cl, Br, C, P, S, Si
Inorganic gases	Ammonia, carbon monoxide, carbon dioxide, hydrogen fluoride, hydrogen sulfide, nitrogen oxides, sulfur dioxide
Acids	Hydrochloric acid, sulfuric acid, nitric acid, and phosphoric acid
Bases (alkalis)	Sodium hydroxide, potassium hydroxide, calcium hydroxide, magnesium hydroxide
Macronutrients (inorganic nitrogen and phosphorus compounds)	Inorganic nitrogen and phosphorus compounds: urea, calcium hydrogen phosphate, and potassium nitrate
Radionuclides	Radioactive isotopes, radon-222, uranium-238, plutonium-239
Organometallics	Organomercurials (methylmercury), organoarsenics, and organotins (tributyl tin)

Table 2.2 Common chemical pollutants: natural and synthetic organics.

Chemical types	Chemical classes
Petroleum	Gas (C_1–C_4), aliphatic and aromatic hydrocarbons, PAHs
Biofuels (biodiesel and ethanol)	Carbonyl compounds (acetaldehyde) and ethanol
Solvents (petroleum and water miscible)	Chlorinated hydrocarbons, benzene, toluene and xylenes, ethanol, butanol, ethylene glycol
Volatile organic chemicals (VOCs)[a]	Formaldehyde, hexane, decane, benzene, toluene, xylenes, ethanol, trimethylamine
Phenolics	Phenol, hydroxyl phenols (cresols), creosotes, chlorinated phenols, bisphenol A
Plastics	Monomers (ethylene, styrene, and methyl methacrylate), polymers and resins: polyethylene, polystyrene, polyvinyl chloride (PVC), fluoropolymers (Teflon) polyurethanes, acrylics (polyacrylates)
Plasticizers	Phthalate acid esters (or phthalates) (e.g. diethylhexyl phthalate, DEHP), tricreysl phosphate
Industrial chemicals and products	PCBs, CFCs, surfactants, disinfectants, dyestuffs, pigments, adhesives, surface coatings, flame retardants
Pesticides	Insecticides, herbicides, fungicides, and biocides
Veterinary chemicals	Antibiotics, parasiticides, anti-inflammatory drugs (nonsteroid)
Personal care products	Fragrances (perfumes): musk ketone, preservatives: parabens
Hormones and steroids	Estrogens (estriol, estrone), progesterone, androgens (androsterone, testosterone)
Pharmaceuticals	Antibiotics, anti-inflammatories, and beta adrenergic blockers
Disinfection by-products	Trihalomethanes (chloroform, bromoform, bromate, trichloroacetic acid), nitrosamines (NDEA, NDMA)
Combustion, processing or synthesis by-products	Polychlorinated dioxins (TCDD), PAHs, perfluorinated compounds (perfluorooctanoic acid, PFOA)

[a]The European Union defines a VOC as any organic compound having an initial boiling point less than or equal to 250 °C (482 °F) measured at a standard atmospheric pressure of 101.3 kPa.

alcohols, acids, chlorinated hydrocarbons, phenols, PAHs, and organometallics. This approach is useful for inorganic chemicals and families of many organic compounds but in some cases, the general type of chemical product (e.g. plastics and pharmaceuticals) or application (e.g. pesticides or surfactants) is a better descriptor for pollutants that consist of, or are derived from, mixtures of families of chemicals.

Biological Pollutants

Despite our concern about the role of pollutants in the growth of chronic noninfectious diseases in human populations, infectious diseases related to pollution sources such as human and livestock wastes remain ubiquitous in the global environment. Among pollutants, disease-causing microorganisms or pathogens, mainly ingested from fecal-contaminated water and foodstuffs due to poor sanitation, are a leading cause of disease and death among the world's human population, mainly affecting young children in developing countries. The World Health Organization reports that waterborne diarrheal diseases, for example, are responsible for 2 million deaths each year, with the majority occurring in children under 5 (WHO 2019).

Pathogens

Microorganisms are essential for all forms of life, building ecosystems, and sustaining food chains. There are millions of types of microorganisms, of which most have beneficial functions (antigenic). In contrast, microorganisms that induce harmful disease are called pathogens.

Biological pollutants are mainly pathogens or biological agents that produce disease in their host organisms (other microorganisms, plants, animals, and humans). Disease theories including the ecological nature of disease are introduced in Chapter 3. Pathogens exploit multiple pathways to infect unicellular and multicellular organisms. Depending on the organism, they may enter through pathways such as in tissues, cellular pores, disruption of cell walls or membranes, skin penetration, inhalation, and ingestion to form a site of infection and begin to multiply. Disease may result when infected cells are damaged or killed. The incidence of disease varies greatly with the type of pathogen and individual susceptibility. Infectious agents cover microorganisms or microbes such as bacteria, viruses, fungi, protozoa, prions, and multicellular parasites. Their characteristics and environmental behavior are summarized below.

Many wildlife species are reservoirs of pathogens that threaten domestic animals and human health. As well, emerging infectious diseases also threaten global biodiversity (Daszak et al. 2000). The main modes of transmission of infectious disease to human and nonhuman species are through *anthroponoses (diseases transmitted from humans to humans)* and *zoonoses* (diseases transmitted from animals such as insects, to humans). *Microbes that cause anthroponoses have adapted, through evolution, to the human species as their host, while nonhuman species are the natural reservoirs for those infectious agents that cause zoonoses.* Directly transmitted *anthroponoses* include TB, HIV/AIDS, and measles, and *zoonoses* such as rabies. There are also indirectly transmitted, vector-borne, anthroponoses such as malaria, dengue fever, and yellow fever, and *zoonoses* such as bubonic plague and Lyme disease (Patz et al. 2003).

Pathogens can be grouped according to their basic modes of transmission such as waterborne, airborne, soilborne, foodborne, and blood-borne. Environmental pathogens are commonly waterborne and airborne.

Waterborne Pathogens

The contamination of natural water bodies and drinking water supplies by microbial agents from animal and human wastes is a major global public health issue for morbidity and mortality. Microorganisms linked with waterborne disease are mainly enteric pathogens that have a fecal to oral route of infection and survive in water (WHO 2011). High densities of known pathogens occur in sewage in contrast to usually low levels found in water supplies for urban populations in developed countries. The types of common infectious agents that are found in raw sewage cover bacteria, viruses, protozoans, and helminths (intestinal worms, ova, and cysts). Pathogens found in sewage can vary considerably between countries and climatic regions.

Waterborne pathogens can occur from human and animal excretion and also from natural sources in the environment. The latter include free-living organisms that are opportunistic pathogens and toxic algae that produce toxins. Waterborne pathogens contaminate not only raw sewage, but some types can live, grow, and *bloom* in water supplies, including various bacteria, free-living protozoans, and toxic algae. Such pathogens that are of high to moderate health significance in water supplies are listed in Table 2.3.

Table 2.3 Important waterborne pathogens and their health significance in water supplies.

Pathogens	
Bacteria	**Protozoa**
Burkholderia pseudomallei	*Acanthamoeba* spp.
Campylobacter jejuni, C. coli	*Cryptosporidium parvum*
Escherichia coli – Pathogenic	*Cyclospora cayetanensis*
E. coli – Enterohaemorrhagic	*Entamoeba histolytica*
Legionella spp.	*Giardia intestinalis*
Pseudomonas aeruginosa	*Naegleria fowleri*
Salmonella spp.	*Toxoplasma gondii*
Shigella spp.	Helminths
Vibrio cholerae	*Dracunculus medinensis*
Yersinia enterocolitica	*Schistosoma* spp.
Viruses	Viruses
Adenoviruses	Sapoviruses
Enteroviruses	Noroviruses
Astroviruses	Rotavirus
Hepatitis A viruses	

Note: High health significance for all listed pathogens except moderate significance for *Pseudomonas aeruginosa* in drinking water supplies. Health significance relates to the severity of impact, including association with outbreaks.
Source: Data from WHO (2011).

Airborne Pathogens

Pathogens (and allergens) may be emitted, suspended, and spread in air (indoor and outdoor) in the form of biological aerosols (bioaerosols), dusts and liquid droplets from a wide range of environmental, body, and plant sources (e.g. infected animal or person, animal dander, fungi spores, pollen grains, endotoxins, excreta, and other biological wastes). Bioaerosols are fine suspensions of airborne particles that contain microbes and/or biological matter released from living organisms. They range in size from less than 1–100 microns.

As for waterborne pathogens, the spread and survival of airborne pathogens is affected by their physical properties (e.g. aerodynamic diameter) and environmental conditions (e.g. wind or air speed temperature, relative humidity, and exposure to UV light). Inhalation of airborne pathogens may cause inflammation in the nose, throat, sinuses, and the lungs, induce asthma and transmit infections such as influenza, measles, chickenpox, tuberculosis, cryptococcosis, and anthrax. Airborne diseases can also affect nonhumans. For example, Newcastle disease is an avian disease that affects many types of domestic poultry worldwide and is transmitted via airborne contamination (Prussin and Marr 2015).

2.4 Some Special Classes of Pollutants

Persistent Organic Pollutants (POPs)

POPs are chemical substances that possess certain toxic properties known to be particularly harmful for human health and the environment. POPs resist environmental degradation, accumulate in living organisms and their food chains, and in ecosystems. They are transported by air, water, and migratory species worldwide. Under a United Nations treaty known as the Stockholm Convention, the European community, the United States, China, and many other countries agreed in May 2001 in Stockholm, Sweden, to reduce or eliminate the production, use, and/or release of a dozen toxic chemicals, known as POPs. These priority POPs were: aldrin, chlordane, dichlorodiphenyltrichlorethane (DDT), dieldrin, endrin, heptachlor, mirex, toxaphene, polychlorobiphenyls (PCBs)[1], hexachlorobenzene (HCB)[1], and polychlorinated dioxins[1] and furans[1].

The initial POPs were informally called *the dirty dozen*. Note: They are *chlorinated compounds. Chlorine substitution tends to increase the potential persistence of an organic molecule.* Many of the POPs occur as isomeric mixtures. In particular, dioxins, furans, and PCBs occur in many different isomeric forms. Some isomers are more persistent than others, and more toxic to different organisms, e.g. the dioxin, tetrachlorodibenzodioxin (TCDD) is the most toxic form of the dioxin isomers.

In 2004, the United Nations (UNEP) – Stockholm Convention decided to limit pollution by POPs, define what are POPs, and also set the rules to control the production, importing, and exporting of POPs.

Following a Twelve Region review of the world, 26 chemicals or groups of chemicals were placed on the UNEP list of persistent toxic substances (PTS): the 12 priority POPs in the Stockholm Convention, three organometals (organotin, organomercury, and organolead compounds) and 11 chemicals (HCH, PAHs, endosulfan, pentachlorophenol, phthalates, PBDE, chlordecone (Kepone), octylphenols, nonylphenols, atrazine, and short-chained chlorinated paraffins).

After a scientific review process, the POPs list was expanded in May 2009 to include an additional set of persistent and halogenated chemicals made up of pesticides, industrial chemicals, and by-products: chlordecone (Kepone), alpha and beta hexachlorocyclohexane, hexabromobiphenyl, hexabromocyclododecane, hexabromodiphenyl ether and heptabromodiphenyl ether (commercial octabromodiphenyl ether), hexachlorobutadiene (HCBD), lindane, pentachlorobenzene (PeCB), pentachlorophenol and its salts and esters (PCP), perfluorooctane sulfonic acid and its salts, perfluorooctane sulfonyl fluoride, polychlorinated naphthalenes (PCNs), technical endosulfan and its related isomers, and tetrabromodiphenyl ether and pentabromodiphenyl ether (main components of commercial pentabromodiphenyl ether) (UNEP 2019a) (see also Chapter 11 on Toxic Organic Chemicals).

Air Pollutants

Major air pollutants are common throughout the world. Thus, countries have regulated air pollutants through developing, reviewing, or adopting sets of key or priority air pollutants from international organizations such as the WHO or regulatory agencies in other countries such the US EPA. Two well-known sets of regulatory criteria for air pollutants are described below.

Criteria Air Pollutants

The term *criteria air pollutants* originated in the 1970 Clean Air Act in the United States. This Act required the US EPA to set standards to protect human health and welfare from hazardous air pollutants in ambient air. Standards are set for six air pollutants: carbon monoxide (CO), ozone (O_3), sulfur dioxide (SO_2), nitrogen oxides (NO_x), lead (Pb), and particulate matter (PM_{10} and $PM_{2.5}$) (US EPA 2015). The WHO also adopted guidelines for the same set of ambient air pollutants that are now used widely in many countries. VOCs or specific components (e.g. benzenes) are sometimes included in major urban airsheds for monitoring purposes.

Hazardous Air Pollutants (HAPs)

Hazardous air pollutants (HAPs) or air toxics are those pollutants that are known or suspected to cause cancer, other serious health effects, or adverse environmental effects. There are 187 toxic air pollutants regulated by the EPA under the Clean Air Acts of 1970 and 1990 in the United States. Examples include organic chemicals such as acrylamide, benzene, toluene, perchloroethylene, methylene chloride, styrene, PAHs, metals, such as lead, cadmium, chromium, and mercury, and particulates, such as fibrous asbestos. Most of the listed air toxics are VOCs (US EPA 2017).

The major sources of hazardous air pollutants include motor vehicle emissions, products of burning fuels (fossil and biomass), and industrial emissions (e.g. solvents).

---

1 Listed as unintentional products/by-products by Stockholm Convention.

Water Pollutants

Toxic and Priority Pollutants

Under the Clean Water Act in the United States, pollutants in wastewater discharges to surface waters and from municipal sewage treatment plants are regulated by national effluent guidelines and measured with approved analytical methods. The US EPA issues these regulations for industrial categories, based on the performance of treatment and control technologies (US EPA 2019a). There are two lists of water pollutants that are regulated by the US EPA. These are the toxic pollutants and the priority pollutant lists. The toxic pollutants list consists of 65 chemicals or groups of chemicals that are a key starting point for developing national discharge standards such as effluent guidelines or in national permitting programs (US EPA 2019a).

Priority Pollutants are a set of chemical pollutants derived from the toxic pollutants list for which analytical test methods have been published (see US EPA 2019a). The current list of 126 Priority Pollutants can be found at 40 CFR Part 423, Appendix A (US EPA 2019a).

> (…The Priority Pollutant list is more practical for testing and for regulation in that chemicals are described by their individual chemical names. The list of toxic pollutants, in contrast, contains open-ended groups of pollutants, such as "chlorinated benzenes." That group contains hundreds of compounds; there is no test for the group as a whole, nor is it practical to regulate or test for all of these compounds
>
> (US EPA 2019b).

Carcinogens, Mutagens, Reproductive (CMR) Toxic Substances

Carcinogenic, mutagenic, and reproductive (CMR) toxic (reprotox) substances are chronically toxic and can cause serious adverse health effects or death. The abbreviation *CMR* is commonly used for this group of toxic substances (e.g. European Union). They are significant environmental pollutants. They are briefly described below and their toxicity is discussed further in Chapters 5 (Environmental Toxicology) and 6 (Genetic Toxicology and Endocrine Disruption: Environmental Chemicals). The following descriptions are based on Chapter 6.

- **Mutagens** increase the occurrence of mutations in the genetic material of the cells of organisms. They are physical, chemical, or biological agents (e.g. arsenic, aromatic amines, bromine, benzene, X-rays, gamma rays, alpha particles, and ultraviolet radiation) that cause direct or indirect damage to DNA that results in mutations (genetic alterations). These are changes in the DNA sequence that are retained in somatic cell divisions and passed onto progeny in germ cells.

- **Carcinogens** are broadly any substance or agent, or mixture of these that can induce or increase the incidence of cancer in living tissues. It can include exposures to UV radiation, fine air pollution particles, airborne asbestos fibers, benzene, arsenic, radon-222, and certain viruses and bacteria. Exposure to a carcinogen may damage directly or indirectly the DNA in cells causing genetic mutations (genotoxicity) or disrupt a wide variety of cellular processes that lead to abnormal cell growth and tumors, without damaging DNA.

- **Reproductive toxicity** is defined as adverse effects of a substance (e.g. chemical) on sexual function and fertility in adult males and females, and also **developmental toxicity** in the offspring. The latter refers to adverse toxic effects occurring in the developing embryo or fetus. Teratogens and endocrine-disrupting substances are two key groups of substances that are toxic for reproduction. (Teratogens are substances that affect the development of an embryo or fetus and cause birth defects or even death in offspring.) Examples of chemical teratogens include mercury (e.g. methyl mercury), PCBs, organochlorine pesticides (e.g. DDT), PBDEs (e.g. flame retardants), plasticizers (e.g. phthalates and bisphenol A [BPA]), PAHs, and perchlorate.

- **Endocrine-Disrupting Chemicals (EDCs)**
  Endocrine disruptors are generally described as natural and synthetic chemicals that may mimic or interfere with endocrine or hormone systems (endocrine glands and associated hormones) in the body and produce adverse developmental, reproductive, cancerous tumors, neurological, and immune effects in both humans and wildlife. There are many different types of EDCs that can cause endocrine disruption, including PCBs, dioxin and dioxin-like compounds, DDT and various other pesticides, drugs and plasticizers such as BPA.

Emerging Pollutants

Emerging pollutants are somewhat complex to detect, monitor, and evaluate their environmental behavior, exposure, and potential effects. They have been described as any synthetic or natural chemical or microorganism that is not routinely monitored in the environment but has the potential to enter, or is already detected in the environment, and cause known or suspected adverse effects on humans and wildlife. This description includes existing contaminants in the environment that have been detected or evaluated by advances in analytical technology, toxicology, scientific assessment, or review. It also applies to new sources of

Table 2.4 Some emerging pollutants of known or suspected environmental concern.

Emerging pollutants	Description
Chemicals of emerging concern in wastewater discharges, stormwaters, and recycled wastewaters	Residual mixtures of many synthetic organic chemicals: industrial chemicals (e.g. petrochemicals), drugs, personal care products, flame retardants, EDCs, pesticides, and food additives[a]
Pharmaceuticals and veterinary chemicals	Mainly in wastewaters from urban and agricultural sources[a]
Personal care products	Mainly in wastewaters from urban and agricultural sources[a]
Perfluoroalkylated substances (PFAS)	Surfactants: endocrine disrupters (e.g. PFOS and PFOA and derivatives)
Disinfection by-products (DBPs)	Contaminants in discharges from disinfection of wastewaters, recycled, and drinking waters[a]
Plastic pollution of the oceans and land	Large volume plastic wastes as litter, microplastic particles, and contaminants
Endocrine disruptive chemicals or compounds (EDCs)	Chemicals that block or mimic hormones in the body
Emerging pollutants in contaminated soils	Examples include: asbestos, 2,4,6-trinitrotoluene, perchlorate, 1,2,3 trichloropropane, N-nitroso-dimethylamine, and flame retardants
Electronic wastes (e-wastes)	Disposal and recycling: heavy metals, PCBs, flame retardants
Flame retardants[a]	Diverse group of chemicals: organophosphates and halogenated compounds, e.g. brominated compounds (PBDEs)
Nanomaterials (engineered NMs)	<100 nm particles: unique physical, chemical, and biological properties (e.g. carbon-based nanotubes and buckyballs, metal oxides, metals (zero-valent) and quantum dots (cadmium and zinc selenide), nanosilver, synthetic polymers, and resins
Shale and coal seam or coal bed gas releases	Methane gas (flammable and greenhouse gas) and fracturing fluid agents: e.g. biocides (glutaraldehyde), surfactants, scale inhibitors, oxygen scavengers, and corrosion inhibitors
Antibiotic-resistant bacterial pathogens	Multiple antibiotic-resistant bacteria: methicillin-resistant *staphylococcus aureus* (MRSA), *klebsiella oxytoca, and escherichia coli*

[a]Depending on sources and treatment processes.

emerging contaminants (new or existing substances) that are released or disposed to the environment.

Some emerging types of pollutants, including complex mixtures of pollutants from wastes, are listed in Table 2.4 and are discussed in other relevant chapters of this book such as Chapter 11 on Toxic Organic Chemicals. Many of these substances are organic pollutants that occur at residual or low levels due to disposal or degradation in unwanted wastes. There is usually a long delay period between their commercial use or application and recognition of their environmental impacts or potential risks.

2.5 Global Production, Emissions, and Releases of Chemicals and Hazardous Wastes

The degree of the potential environmental problem posed by pollutants, such as chemicals, can be evaluated from available statistics or estimates on the generation of pollutants from all sources (e.g. air emissions, wastewaters, stormwaters, and disposal to land) and their release into the environment. The situation is evident from available data on production trends and use of chemicals and estimates of the rates of emissions or release of hazardous chemicals and wastes into the air, water and onto land from pollution sources. Individual production of major chemical pollutants ranges from thousands to many millions of tonnes yearly. A multitude of these chemicals, such as greenhouse gases, toxic metals, nutrients, and known human and/or animal carcinogens, are released into the environment during their life cycles and are dispersed through biogeochemical cycles that interact on local to global scales.

Known and emerging scientific principles, properties, and environmental factors that apply to the behavior and fate of chemicals and other pollutants, and their toxic or adverse effects on humans and natural biota are evaluated in Parts 2 and 3 of this book.

Global production of chemicals is reported as chemical sales of $US5 trillion in 2017 according to the Global Chemical Outlook II update prepared by UNEP (2019b, p. 17). It is expected to double by 2030. Between 2000 and 2017, the global chemical industry's production capacity (excluding pharmaceuticals) increased from about 1.2 billion to 2.3 billion tonnes (UNEP 2019b, p. 17). The American

Chemical Council gives an amount of 455 million tonnes for the top 100 chemicals produced in the United States. China was the largest chemical manufacturer in 2010 with sales of $US754 billion. OECD countries as a group were the largest producers (63% of sales) in 2009 but BRIICS countries (Brazil, Russia, India, Indonesia, China, and South Africa) had increased their share to 28% (UNEP 2013, p. 11). China is expected to increase its market share of global sales to almost 50% by 2030 (UNEP 2019b, p. 18).

While sales values do not reflect the volume of chemical production, they indicate a strong future trend in chemical production and use among developing and economic transition countries. Consequently, there is a major shift in health and environmental impacts and risks from chemical production and use on a global scale.

Upstream Chemicals

Annual global production of key bulk chemicals (inorganics and organics), metals, and nonmetal halogens used in the manufacture of numerous downstream intermediates and chemical products are presented in Table 2.5. Petroleum (crude oil and natural gas) and natural minerals are the primary sources of these chemicals. Most of the bulk chemicals, heavy metals, and nonmetals (e.g. halogen gases) are produced in many millions of tonnes per year. Toxic lead compounds, for example, are produced at about 10 million tonnes per year. Several of the chemicals or related compounds in Table 2.5 are also human carcinogens (asbestos, benzene, 1,3 butadiene, cadmium, hexavalent chromium, lead, and nickel carbonyl). A few bulk inorganics make up the largest volume of chemicals, particularly for manufacture of agrichemicals and building products (limestone). Seven bulk organic chemicals start as the source of a large number of feedstock chemicals used to make end-products such as solvent plastics, resins, plasticizers, pesticides, latex, paints, dyes, coatings, adhesives, and surfactants. Halogen gases (chlorine, bromine, and fluorine) are used to derive many industrial chemicals from PVC to halogenated solvents and pesticides (see Table 2.5).

Downstream Chemicals

Major downstream chemical products of environmental concern and toxicity include synthetic fertilizers (nitrogen and phosphorus based), plastics, pesticides, and halogenated organic compounds listed in Table 2.6. Annual quantities produced range from thousands to tens of millions of tonnes, and 288 or more million tonnes for plastics. Toxic organic chemicals include millions of tonnes of synthetic pesticides and halogenated organics such as vinyl

Table 2.5 Global production (or consumption) and uses of upstream chemicals (bulk and feedstock).

Chemical	Principal uses	Global production (millions of metric tons year^{-1})
Bulk inorganic		
Lime/limestone (2008)	Metallurgy, building products, pulp and paper, environmental uses	285
Sulfuric acid (2010)	Production of phosphate fertilizers	198
Ammonia (2010)	Production of nitrogen fertilizers	134
Sulfur (2010)	Sulfuric acid production	77
Phosphoric acid (2009)	Production of phosphate fertilizers	46
Bulk organics		
Methanol (2010)	Phenol, formaldehyde, chloromethanes, methylmethacrylate	49.1
Ethylene (2010)	Production of organic chemicals^a	123.3
Propylene (2010)	Production of organic chemicals^a	74.9
Butadiene (2010)	Production of organic chemicals^a	10.2
Benzene (2010, 2012^b)	Production of organic chemicals^a	40.2 (42)^b
Toluene (2010)	Production of organic chemicals^a	19.8
Xylenes (2010)	Production of organic chemicals^a	42.5
Metals		
Arsenic (2012)		(38 655)
Cadmium (2012)		(21 861)
Chromium) (2002)^c		~13
Lead (2010)		~10
Mercury (2009)		(3 800)
Nickel (2012)		2.1
Rare earth oxides (2012)		(104 616)
Uranium (2010)		(53 663)
Zinc (2009)		11.2
Halogens		
Chlorine	Manufacture of ethylene dichloride, isocyanates, and propylene dioxide	56

(Continued)

Table 2.5 (Continued)

Chemical	Principal uses	Global production (millions of metric tons year^{-1})
Bromine	Production of brominated flame retardants, clear brine fluids, hydrogen bromide, and methyl bromide	0.53
Fluorine	Production of hydrofluoric acid, aluminum smelting, and steel manufacturing	5.6 (as fluorospar)
Asbestos (2010)	Asbestos cement building materials, insulation	~2

( ) Tonnes.
[a] Petrochemicals – primarily used in production of organic chemicals.
[b] IARC Monograph 120, p. 38.
[c] Including hexavalent chromium (Cr^{6+}).
Source: Data from UNEP (2013).

chloride, brominated flame retardants, solvents, and some POPs. Vinyl chloride is a very toxic chemical and a human carcinogen that has a large yearly production of about 47 million tonnes (2010).

Environmental Emissions and Releases of Chemical Pollutants

The total quantity of chemicals released to the global environment is simply unknown. Table 2.7 provides a compilation of annual estimates of emissions and releases of some major chemical pollutants to the global environment from production and use of large-volume chemical products and also data on international generation of hazardous wastes. The magnitude of emissions to the atmosphere is simply dominated by acid gases and particulates from primarily the combustion of fossil fuels (crude oil, gas, and coal), such as 50 120 million tonnes of carbon dioxide equivalents (CO_2-e) emitted in 2012 (UNEP 2013) (see Chapter 14 for recent emissions). Annual emissions of toxic metals are likely to be in the order of thousands of tonnes to over 100 000 tonnes for toxic lead.

Releases of chemical pollutants onto land are dominated by disposal, spills, and leaching of chemical-containing wastes (e.g. mining and hazardous wastes, solid waste in landfills), and application of synthetic fertilizers, pesticides, wastewaters, and increasingly, biosolids from sludge materials. Annually, releases of tens of millions of tonnes of chemical pollutants are indicated by production/use values in Table 2.7. Specific and recent estimates for many chemical pollutants are largely unavailable on a global scale.

By nature of their application, many tens of millions of tonnes of fertilizers and pesticides each year are widely

Table 2.6 Global production and uses of some downstream chemical products.

Chemical	Principal uses	Global production (millions of metric tons year^{-1})
Fertilizers (year 2009)		
Nitrogen (N)	Nitrogenous crop fertilizers	102
Phosphate (P_2O_5)	Phosphate crop fertilizers	37
Potassium (potash) (K_2O)	Potash crop fertilizers	23
Pesticides	Insecticides, herbicides, and fungicides	~2.6 (WHO)
Plastics (year 2012)	Polymers and resins	288
Halogenated organic compounds		
Vinyl chloride monomer (VCM)	Monomer for PVC polymer	47 (2010, Dow Chemical)
Trichloroethylene and perchloroethylene	Solvents	(>200 000)
DDT, lindane, endosulfan	Pesticides	(12 800) (endosulfan)
PCBs (obsolete)	Transformer and capacitor fluids	Banned
Polybrominated diphenyl ethers	Flame retardants	(50 000)
Fluoropolymers	Teflon, perfluorinated compounds	(100 000–180 000)
Fluorocarbons	CFCs (restricted)	NA

( ) Tonnes.
Source: Data from UNEP (2013).

Table 2.7 Global emissions and releases of some major chemical pollutants from production and use of chemical products and generation of hazardous wastes.

Chemical category	Global production use (million tonnes year^{-1})	Air (million tonnes year^{-1})	Water (million tonnes year^{-1})	Land (million tonnes year^{-1})
Fossil fuels	9 449 (as carbon) (2011)	~37 000 (CO_2)^c 50 120 (CO_2-e) (2012) (UNEP 2013)	~ 50 % of CO_2 emissions from fossil fuels absorbed by oceans ^{e} – acidification of seawater	^{b}
Petroleum^a	4 361.9	(CO_2-e in fossil fuels)^b	~ Up to 1 or more million (oceans)	^{b}
PAHs	^{b}	(520 000)^d (2004)	^{b}	^{b}
PCBs	1.32 million tonnes (1930–2000)	(7 710)^d	^{b}	^{b}
HCB		23 000 (kg year^{-1})^d (mid-1990s)		
Dioxins/furans (PCDD/Fs as kg TEQ)	^{b}	10.5 (kg year^{-1})^d (1995)	^{b}	^{b}
Pesticides	~2.6 (WHO) (4.6 million tons)	^{b}	^{b}	~2.6
Plastics	288	^{b}	~9 (2015) (oceans)	^{b}
Synthetic fertilizers				
Nitrogen	102	^{b}	^{b}	<102
Phosphate (P_2O_5)	37	^{b}	^{b}	<37
Metals				
Arsenic (2012)	(38 655)	(5 011)^d (1995)	^{b}	^{b}
Cadmium (2012)	(21 861)	(2 983)^d (1995)	^{b}	^{b}
Chromium (2002)	~13	(14 730)^d (1995)	^{b}	^{b}
Copper		(25 915)^d (1995)	^{b}	^{b}
Lead (2010)	~9.6	(119 259)^d (1995)	^{b}	^{b}
Mercury (2009)	(3 800)	(1 960)^d (2010)	^{b}	^{b}
Nickel (2012)	2.1	(95 287)^d (1995)	^{b}	^{b}
Rare earth oxides (2012)	(104 616)	^{b}	^{b}	^{b}
Selenium		(4 601)^d (1995)	^{b}	^{b}
Uranium (2010)	(53 663)	^{b}	^{b}	^{b}
Zinc (2009)	11.2	(57 010)^d (1995)	^{b}	^{b}
Hazardous wastes	Global generation			
Basel convention	>250 (2004–2006)	^{b}	^{b}	^{b}
Electronic wastes	~40 (2009)	^{b}	^{b}	<40 (2009)
Pesticides (obsolete)	0.54 (FAO inventory)	^{b}	^{b}	^{b}

(Tonnes per year)
a Crude oil, shale oil, oil sands, and natural gas liquids.
b Unavailable estimates.
c The Global Carbon Project (2020).
d MSCE Emissions for Global Modelling. Meteorological Synthesizing Centre-East (MSC-E) (MSCE 2020).
e Feely et al. (2006).

dispersed into air, deposited onto soils and plants, lost to water bodies, and absorbed by biomass, mainly nontarget. The United Nations Joint Group of Experts on the Scientific Aspects of Marine Pollution (GESAMP) estimated that land-based sources account for up to 80% of the world's marine pollution, 60–95% of the waste being plastics debris (Le Guern 2019). As global plastic production approaches 300 million tonnes yearly, an estimated 9 million tonnes of plastics end up in the world's oceans, much of it degrading into toxic and hazardous microplastic particles. Annual inputs of plastics into the oceans are equivalent to about 10% of the commercial harvest of marine fisheries (81.5 million tonnes in 2014) (FAO 2016). Up to a million or more tonnes of petroleum hydrocarbons are also discharged each year into the oceans from chronic sources and oil spills due to human activities (see Chapter 10). These estimates for oil inputs are reduced due to improved pollution controls but are likely to vary considerably.

Hazardous wastes are generated in large quantities by many countries based on available data, although this is likely to be greatly underreported. Data from UNEP (2013) indicate that 64 countries generated over 250 million tonnes of hazardous wastes, as reported in the years 2004, 2005, or 2006, under the Basel Convention for transboundary movement of wastes. It refers to the total amount of hazardous wastes generated per year in a country through industrial or other waste-generating activities, as defined by the Convention. There was a 12% increase in hazardous wastes over the period 2004 to 2006 as reported by 46 countries that provided yearly data. The generation of electronic wastes (e.g. PCs, laptops, mobile phones, and televisions) is increasing rapidly. E-waste generation was reported to be 40 million tonnes in 2009 (UNEP 2013, p. 42). These types of wastes contain large quantities of toxic metals, PCBs, and brominated flame retardants which are released during recycling operations and disposal. Globally, hundreds of millions of tonnes of hazardous wastes are released into the environment or managed at different stages of treatment or storage (see Chapter 16).

Hazardous Chemicals Released to Environment

Some data on the production of chemicals that are hazardous to human health and the environment exist for OECD countries through Pollutant Release and Transfer Registers (PRTR). In contrast, there is little available data for many non-OECD countries, despite the expansion of chemical production and rapid growth in the use of chemical products in many of these countries.

In Europe (EU-28), the production of chemicals hazardous to human health was 222.6 million tonnes in 2018, showing a decline from 233.5 million tonnes in 2004. Chemicals classified as carcinogenic, mutagenic, and showing reproductive or developmental toxicity (CMR) amounted to about 33 million tonnes in 2015 (European Environment Agency 2019). The European Environment Agency has reported that the production of hazardous chemicals to health has declined but it is unable to equate this to a reduction in environmental and health risks. It further highlighted the limitations of using chemical production volumes of hazardous chemicals as an indicator of exposure and potential health and environmental risks. (The EU is also a net importer of chemicals that are hazardous to health.)

PRTR data for North America (United States, Canada, and Mexico) reported in 2009, that 4.9 million tonnes of chemicals were released or disposed. Hazardous chemicals consisted of PBT chemicals (almost 1.5 million tonnes), known or suspected carcinogens (over 756 000 tonnes), and reproductive or developmental toxicants (nearly 667 000 tonnes). Chemicals released in large quantities as air pollutants were indicated to be inorganic chemicals such as ammonia, hydrogen sulfide, sulfuric acid, hydrochloric acid, and organic chemicals such as styrene, formaldehyde, toluene, and acetaldehyde. Chemicals discharged as water pollutants in large quantities were reported to include inorganic chemicals such as nitric acid/nitrate compounds, ammonia, manganese, and organic chemicals such as methanol, ethylene glycol, phenol, toluene, and formaldehyde (UNEP 2013, p. 38).

Pollutants discharged in large quantities to the environment in other countries with Toxic Release Inventory type data indicate some common inorganic and organic chemicals, yet some significant variations may occur. For example, in Mexico, the top chemicals released to surface water in 2009 were the carcinogens, vinyl chloride and acrylamide, and toxic pyridine (UNEP 2013, p. 38).

A serious problem identified in developing and economic transition countries is the lack of chemical management systems including inventories to collect chemical production data, track, monitor, and act on reducing health and environmental risks from known and suspect sources of chemical pollution and hazardous waste disposal (see Chapter 19).

Table 2.7 indicates there are large gaps in our knowledge of the global release of potential chemical pollutants into water and on land.

Future challenges to reduce and manage pollution of global environments from known and projected uses and releases of chemical pollutants are formidable.

Case Study S2.1 in the Companion Online Book looks at what is projected to happen to the global environment in the Twenty-first Century if we continue to increase our population and agricultural use of chemicals at similar rates to those in the Twentieth Century.

2.6 Key Points

1. A pollutant is a substance or form of energy that causes adverse effects on living organisms and/or their environments. Key properties of pollutants are their nature (physical, chemical, or biological), solubility in water or lipids, mobility, persistence, bioaccumulation capacity, and toxicity to humans and other living organisms (e.g. wildlife).
2. Pollutants can generally be divided into physical, chemical, and biological types. General classes may cover (i) organic matter (organic carbon), (ii) plant nutrients, (iii) toxicants, (iv) thermal wastes (heat energy), (v) suspended solids, (vi) disease-causing microorganisms (pathogens), (vii) atmosphere-altering substances (e.g. ozone-depleting gases [CFCs] and greenhouse gases), and (viii) acidification-causing substances (e.g. acid gases).
3. Pollutants can be of natural origin (e.g. silt, salts, nutrients, and heavy metals) or derived from human activities (e.g. petroleum extraction, refining of metals, synthetic pesticides, solvents, plastics, and radionuclides from nuclear energy processing).
4. Major sources of pollutants originate from air (outdoor and indoor), water, soil, waste disposal, and energy-related pollution. They cover point and non-point sources.
5. Physical pollutants consist of particles of matter of various characteristics: sizes, shapes, and other intrinsic properties (e.g. dusts and respirable particles: PM_{10}, $PM_{2.5}$, and $PM_{1.0}$), nanoparticles and suspended solids in water), and different forms of energy (e.g. heat, UV light, and sound energy) that interact and impact on living organisms and their environments.
6. Impacts or effects of these pollutants are characterized by (i) the mass transport of particles (from ultrafine to coarse grains, gravel, rocks, and minerals), (ii) different types of energy forms (e.g. electromagnetic radiation spectrum, thermal and sound energy), and (iii) quantity and rates of flows through the atmosphere, waters, soils, sediments, and biological systems.
7. Chemical pollutants can be basically divided into metals, metalloids and nonmetals, inorganics, organometallics, and organic compounds. Chemical types include petroleum hydrocarbons, synthetic pesticides, industrial chemicals, plastics and additives, surfactants, pharmaceuticals, and combustion, processing, or synthesis of by-products (e.g. dioxins and PAHs).
8. Biological pollutants cover infectious and disease-causing microorganisms or pathogens. They are mainly ingested from fecal-contaminated water and foodstuffs due to poor sanitation, although inhalation or contact can be critical routes of intake for some.
9. They include pathogenic bacteria, viruses, fungi and mold, parasitic protozoans and helminths (worms), cyanobacteria (toxic blue-green algae), and allergic pollen.
10. Special classes of environmental pollutants include POPs, and mutagenic, carcinogenic, reproductive, and developmental toxic substances that cause serious adverse health effects or death among human and wildlife populations.
11. Endocrine disruptors are a diverse class of natural and synthetic chemicals that may interfere with endocrine or hormone systems (endocrine glands and associated hormones) in the body and produce adverse developmental, reproductive, neurological, and immune effects in both humans and wildlife.
12. Regulatory legislation for management of air and water pollution also classifies groups of hazardous or toxic pollutants for these purposes (e.g. criteria air pollutants for protection of ambient air quality and HAPs and toxic air pollutants; toxic and priority water pollutants (chemicals) that apply in the United States). The WHO has published air quality guidelines for major ambient and selected indoor air pollutants.
13. Many environmental pollutants of current concern are described as emerging pollutants (e.g. drug and personal care product residues, flame retardants, and manufactured nanoparticles).
14. The degree of the potential environmental problem posed by pollutants, such as chemicals, can be evaluated from available statistics or estimates on the generation of pollutants from all sources (e.g. air emissions, wastewaters, stormwaters, and disposal to land) and their release into the environment.
15. A multitude of these chemicals, such as greenhouse gases, toxic metals, nutrients, and known and suspect carcinogens, are released into the environment during their life cycles and are dispersed through biogeochemical cycles that interact on local to global scales.
16. Known and emerging scientific principles, properties, and environmental factors that apply to the behavior and fate of pollutants, such as chemicals, and their toxic or adverse effects on humans and natural biota are evaluated in Parts 2 and 3 of this book.

References

AGDH (2012). *Ionising Radiation and Human Health.* Canberra: Australian Government Department of Health http://www.health.gov.au/internet/publications/publishing.nsf/Content/ohp-radiological-toc~ohp-radiological-05-ionising (accessed 19 January 2017).

Connell, D.W. (2005). *Basic Concepts of Environmental Chemistry*, 2e. Boca Raton, FL: CRC.

Connell, D.W. and Miller, G.J. (1984). *Chemistry and Ecotoxicology of Pollution.* New York: Wiley.

Daszak, P., Cunningham, A.A., and Hyatt, A.D. (2000). Emerging infectious diseases of wildlife-threats to biodiversity and human health. *Science* 287 (5452): 443–449.

European Environment Agency (2019). Chemicals Production and Consumption Statistics. https://ec.europa.eu/eurostat/statistics-explained/index.php/Chemicals_production_and_consumption_statistics#Total_production_of_chemicals (accessed 22 January 2020).

FAO (2016). *The State of World Fisheries and Aquaculture 2016.* Food and Agriculture Organization of the United Nations http://www.fao.org/3/a-i5555e.pdf (accessed 22 January 2020).

Feely, R.A., Sabine, C.L., Fabry, V. (2006). Carbon Dioxide and Our Ocean Legacy. https://www.pmel.noaa.gov/pubs/PDF/feel2899/feel2899.pdf (accessed 22 January 2020).

Fuller, R., Landrigan, P.J., and Preker, A.S. (2018). Pollution and global health – a time for action. *World Hospitals and Health Services Journal* 54 (4): 58–61.

Health Effects Institute (2019). *State of Global Air 2019. Special Report.* Boston, MA: Health Effects Institute https://www.stateofglobalair.org/sites/default/files/soga_2019_report.pdf (accessed 22 August 2020).

Landrigan, P.J., Fuller, R., Acosta, N.J.R. et al. (2017). The Lancet Commission on Pollution and Health. *Lancet* 391: 10119.

Le Guern, C. (2019). *When Mermaids Cry: The Great Plastic Tide, Coastal Care.* Santa Agulla Foundation http://plastic-pollution.org (accessed 22 January 2020).

Legal Information Institute, Cornell University (2020) 42 U.S. Code ξ 7602. Definitions. https://www.law.cornell.edu/uscode/text/42/7602 (accessed 20 January 2020).

MSCE (2020). Emissions for Global Modelling. Meteorological Synthesizing Centre-East (MSC-E), Moscow, Russian Federation. http://en.msceast.org/index.php/j-stuff/content/list-layout/global (accessed 23 January 2020).

Patz, J.A., Githeko, A.K., McCarty, J.P. et al. (2003). Climate change and infectious diseases, chapter 6. In: *Climate Change and Human Health – Risks and Responses* (ed. A.J. McMichael, D.H. Campbell-Lendrum, C.F. Corvalán, et al.). Geneva: World Health Organization, World Meteorological Organization, and United Nations Environment Programme.

Prussin, A.J. and Marr, L.C. (2015). Sources of airborne microorganisms in the built environment. *Microbiome* 3: 78.

The American Heritage Dictionary of the English Language (2020). Pollutant. https://ahdictionary.com/word/search.html?q=Pollutant&submit.x=0&submit.y=0mpany (accessed 20 January 2020).

The Global Carbon Project (2020). https://www.globalcarbonproject.org (accessed 22 January 2020).

Tilman, D., Fargione, J., Wolff, B. et al. (2001). Forecasting agriculturally driven global environmental change. *Science* 292 (5515): 281–284.

UNEP (2013). *Global Chemicals Outlook -Towards Sound Management of Chemicals.* Geneva: United Nations Environment Programme.

UNEP (2019a). Stockholm Convention. United Nations Environment Programme. http://chm.pops.int/TheConvention/ThePOPs/TheNewPOPs/tabid/2511/Default.aspx (accessed 21 January 2020).

UNEP (2019b). *Global Chemicals Outlook II – From Legacies to Innovative Solutions: Implementing the 2030 Agenda for Sustainable Development.* Geneva: United Nations Environment Programme.

US EPA (2015). *Criteria Air Pollutants.* United States Environmental Protection Agency https://www.epa.gov/sites/production/files/2015-10/documents/ace3_criteria_air_pollutants.pdf (accessed 21 January 2020).

US EPA (2017). *What are Hazardous Air Pollutants?* United States Environmental Protection Agency https://www.epa.gov/haps/what-are-hazardous-air-pollutants (accessed 21 January 2020).

US EPA (2019a). *Effluent Guidelines.* United States Environmental Protection Agency https://www.epa.gov/eg (accessed 21 January 2020).

US EPA (2019b). *Toxic and Priority Pollutants Under the Clean Water Act.* United States Environmental Protection Agency https://www.epa.gov/eg/toxic-and-priority-pollutants-under-clean-water-act (accessed 21 January 2020).

WHO (2011). *Guidelines for Drinking-water Quality*, 4e. Geneva: World Health Organization.

WHO (2019). *Waterborne Disease Related to Unsafe Water and Sanitation.* World Health Organization https://www.who.int/sustainable-development/housing/health-risks/waterborne-disease/en/ (accessed 3 April 2019).

3

Pollutants, Health, and Environment

3.1 Introduction

This chapter introduces the threats and impacts of pollution on human health, ecological health, and the environment, as measured by the burden of disease in humans and the loss of biodiversity and ecosystem services. Understanding the links between human activities, concepts of health, and ecosystems is essential for solving current and future pollution-related issues. The global challenge to meet United Nations Sustainable Development Goals that are affected by pollution and other human activities has become critical (see Landrigan et al. 2017).

The environmental state of the Earth is under serious threat. The Report of the Sixth Global Environment Outlook (GEO 6) (UNEP 2019) found that the overall condition of the global environment has continued to deteriorate since 1997 (GEO First edition), despite environmental policy efforts worldwide. Between 1970 and 2014, population abundances of global vertebrate species have decreased by on average 60%. The GEO 6 report *concludes that unsustainable human activities globally have degraded the Earth's ecosystems, endangering the ecological foundations of society.*

Here, we introduce concepts of health and theories of disease causation, and their progression from environmental to planetary health, and the role of human activities and pollution in its impacts and threats to human health and ecological health. This information forms the contextual basis for the scientific evaluation of pollution, pollutants, and their toxic effects in the following chapters.

3.2 Environmental Health

Concept of Health

What is health? Human health is basically seen as a person's physical or mental state or condition rather than an absence of disease. It can be viewed as a dynamic condition that results from a body's constant adaptation in response to stress and changes (e.g. environmental) to maintain an internal equilibrium condition, known as homeostasis. In a medical sense, homeostasis is described as a healthy state that is maintained by the constant adjustment of biochemical and physiological pathways, such as the need to keep body temperature within a narrow range (e.g. Stöppler 2020).

Importantly, the World Health Organization defined health in its 1948 Constitution, in a broader sense, as a "state of complete physical, mental, and social well-being, and not merely the absence of disease or infirmity." Being able to enjoy the highest attainable state of health was adopted as a fundamental human right, among other rights (WHO 2017a). The concept of *complete well-being* within the definition has been generally well received and resilient. Even so, various international health experts have expressed the need to revise the meaning of health to include the ability to adapt and self-manage, particularly from health challenges caused by increasing chronic diseases, rather than an overemphasis on medical and drug-based interventions, and healthcare systems, to achieve *complete well-being* (BMJ 2011).

Health and Environment

Environmental health looks at how human health and behavior are potentially affected by all the physical, chemical, and biological factors that are related to the environment. It then determines and acts to control or prevent external environmental factors that may or can affect human health, such as vector-borne diseases or ensure suitable and safe drinking water quality (risk-based approach).

Recent interpretation of environmental health by the World Health Organization targets (i) external environmental factors that impact on human health and disease, (ii) evaluation and control of environmental factors that potentially affect health, and (iii) preventative measures and health-supportive services.

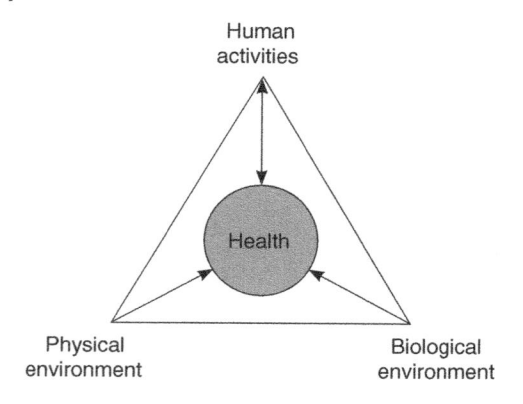

Figure 3.1 Interactions between human activities and the physical and biological environments (or ecosystems) and the health of human individuals and populations.

Environmental health addresses all the physical, chemical, and biological factors external to a person, and all the related factors impacting behaviors. It encompasses the assessment and control of those environmental factors that can potentially affect health. It is targeted towards preventing disease and creating health-supportive environments. This definition excludes behavior not related to environment, as well as behavior related to the social and cultural environment, as well as genetics

(Wikipedia 2022).

The WHO definition of environmental health shows the connection between the state of the environment and the health experiences of individuals and communities (Landon 2006, p. 6).

Environmental health refers to the interdependence between the health of individuals and communities and the health of the environment (Brown et al. 2005, p. xiv). It depends on interactions between human activities, such as extraction, production, and uses of resources, and the environment (e.g. ecosystems). These activities result in exposures to environmental agents and factors (human-derived and natural) and effects on humans and the environment, adaptation to changes, and management responses to sustain environmental health. The general relationship between human activities, health, and the environment is indicated in Figure 3.1.

Wilcox and Jessop (2010, pp. 3–47), for example, have introduced the general scientific principles of ecology and how human activities can disrupt the energy and material cycles of ecosystem functioning, and affect the populations and biodiversity of ecosystems, leading to the emergence of environmental health risks, including infectious diseases. They also introduce a key concept that humans are within, and are not separate from, ecosystems and ecological interactions.

Ecological Health and Sustainable Development

The second half of the Twentieth Century has seen the scientific links between the impacts of human activities on the state of the environment, and increasing adverse effects on the health of human populations, becoming well established. Human changes to the environment have disrupted the global cycles (and energy flows) and the rate of change is expected to increase.

In the Twenty-first Century, the challenge for environmental health is to develop, incorporate, and rapidly advance an ecological approach to protect and maintain the complex relationships between diversity of life on Earth and human needs, health, and social systems that are likely to meet sustainable development goals for existing and future generations (see Brown et al. 2005).

The following sections of this chapter introduce theories on the nature and cause of diseases, the concept and measures of the human burden of disease due to human activities (e.g. occupational exposures) and environmental factors. As well, the role of human-related pollutants in causing impacts on ecosystems, such as biological diversity and ecosystem services that are vital to human and ecological health, is outlined. Understanding cause–effect relationships and interactions with other factors in disease and toxicity from exposures to environmental agents underpins much of our approaches to solving environmental health challenges and how to achieve sustainable environmental goals.

3.3 Disease Causation

Disease

In contrast to the concept of well-being, disease is described as a dysfunction of the dynamic state of health. For humans, it is usually considered as a medical condition associated with specific symptoms or signs. If we apply it generally to living organisms, it is a disorder of a structure or function in an organism (human, animal, or plant) that affects part of or all of an organism. It can result from various causes such as an infection, genetic defect, environmental stress, or exposure to a toxic agent. Pathology is the science or study of the origin, nature, and course of disease.

Several types of disease are identified: infectious diseases, physiological diseases, deficiency diseases, genetic diseases (hereditary and nonhereditary), and mental illness or disorders. Generally, diseases may also be classified as communicable (infectious diseases resulting from an invasion of body tissues by microorganisms) or non-communicable (noninfectious or non-transmissible from an infected host to another). The latter type is commonly a chronic disease such as cancer or cardiovascular diseases.

The gold standard for classifying diseases is the International Classification for Diseases (ICD) and Related Health Problems run by the WHO. It is the standard diagnostic tool for epidemiology, health statistics (mortalities and morbidities from diseases), health management, and clinical purposes. Injuries are also included in the ICD (WHO 2020a).

Environmental diseases are of particular interest because they result from direct exposure to environmental factors (physical, chemical, biological, and genetic predisposition). They include all pollution-related diseases. In 2006, the WHO reported that almost a quarter of all disease is caused by environmental exposure and could be avoided or reduced through better environmental management and healthy environments (WHO 2006).

Causation of Disease

General theories or models of disease describe how exposures to determinants, risk factors, or possible causes of disease (e.g. a chemical or biological agent) are linked to a disease. Causal associations between exposure and disease usually involve a multitude of interacting factors/conditions, including behavioral, genetic, and environmental.

Analytical epidemiology is widely used to determine a relationship or an association, such as relative risk, between a specific exposure, among various interacting factors, and the frequency of disease in populations. It is then necessary to determine the strength of an association (unlikely to be caused by chance or bias) and a plausible biological mechanism between exposure and the disease to infer causation. The problem is how to establish causal association.

For causal inference, Bradford Hill's criteria (strength of association, consistency, specificity, temporality, biological gradient, plausibility, coherence, experiment, and analogy) are widely used to evaluate hypothesized relationships between occupational and environmental exposures in populations and observed disease outcomes (e.g. Wilkinson 2006, pp. 89–91).

The overall process is judgmental, based on a *weight of evidence* approach using a critical evaluation of systematic review of studies, causal models, data and explanation, such as used by the International Agency for Cancer Research (IARC) to classify potential carcinogens (see also Chapter 18).

Theories of Disease

Major theories of disease causation have developed following the discovery of microorganisms as infectious agents in the Nineteenth Century. These are briefly described below.

Germ Theory of Disease

The germ theory expounds the view that microorganisms (pathogens) are the cause of diseases. In the Nineteenth Century, the dominant belief was that diseases were related to *miasma* or unhealthy smells, poisonous vapors, and mists in the air from rotting matter. Convincing evidence of the germ theory arose during the Nineteenth Century due to the discovery of infectious microbes by Louis Pasteur, the French Chemist, in 1864. He proved that microbes caused infection by travelling through air, by physical contact, or from ingestion of contaminated food or water. His germ theory of disease proposed that specific microbes were the causal agent of each disease. In the 1880s, the Prussian doctor, Robert Koch and colleagues discovered that specific types of bacteria caused the dreaded diseases: anthrax, cholera, and consumption, leading to further discoveries of infectious bacteria and viruses (1890s) by other scientists. Medical treatment involved killing or inhibiting the specific pathogen.

Epidemiological Triad

A classical model of infectious disease causation is known as the Epidemiologic Triad. It consists of an agent, a host, and an environment in which host and agent interact causing the disease to occur in the host. It recognized that disease was caused by factors other than microorganisms because not everyone exposed to the disease-causing agent contracted the disease. Key components of the triad model are summarized below (e.g. Stone et al. 1996, pp. 22–27).

- Agents can be infectious organisms (pathogens), chemical substances (e.g. industrial chemicals and toxins), or physical agents (e.g. heat, ionizing radiation, sharp objects, and forces). Other agents have been described as psychological (mental and physical stressors), social agents (underlying causes such as poverty, war, and social violence), and genetic (hereditary factors).
- Hosts are organisms (e.g. humans) that come into contact (e.g. via skin, water, air, soil, or food) with an agent in an external environment and are susceptible to it. (For example, individual persons who develop an illness from exposure to an infectious virus; poisoned by inhaling a volatile toxic chemical; or develop mesothelioma from inhaling asbestos fibers at worksites).

- External environments contain a multitude of physical, chemical, biological, and socioeconomic factors that interact with the host and exposure from the agent. It provides suitable conditions for the agent and host to interact and induce disease.

Disease prevention and control can be achieved by modifying factors which influence the nature of exposure and susceptibility to the agent.

Multifactorial Causation of Disease

In contrast to the emphasis on the disease-causing agent, such as in the triad theory, the multifactorial theory focuses on multifactor interactions between the host and the environment that induce chronic noninfectious diseases or disorders such as heart disease, obesity, and diabetes. They are likely to be associated with effects of multiple genes in combination with lifestyle and environmental factors.

Web of Causation

A web of causation model links interacting causes of a disease by using a diagram to illustrate how they are interconnected (e.g. Stone et al. 1996, pp. 18–19). This epidemiological approach considers that the occurrence of the disease results from a complex web of interactions involving multiple factors, and disease is resolved by interventions that modify these risk factors. It applies more to chronic diseases such as cardiovascular disease (CVD) where the disease-causing agent is unknown or uncertain but many factors interact in the progression of the disease.

Disease Ecology

Infectious diseases can be caused by pathogens or parasites that can affect or kill hosts, such as wildlife, and pervade ecosystems. Diseases are not transmitted between hosts but by pathogens and parasites that cause disease. This has led to the emergence of disease ecology which is concerned with the ecological study of host–pathogen interactions and their populations within the context of their environment, climate, and evolution (Kilpatrick and Altizer 2010).

More broadly, disease ecology examines the underlying principles that influence spatial and temporal patterns of diseases within their environment rather than specific disease characteristics as in epidemiology (Ostfeld 2018).

3.4 Environmental Pollution and Burden of Disease

In this book, we are interested in understanding the impact of pollution and related environmental risk factors on humans and nonhuman biota. A summary metric for the health of human populations and communities is the burden of disease, injuries, and risk factors that cause them.

The focus here is on the proportion of the population affected by environmental risk factors attributed to pollution and pollutants, particularly chemicals. This type of information is critical for health risk management involving planning, policy development, intervention, and remediation to reduce the burden of disease due to pollution and other risk factors on human populations.

Burden of Disease

The burden of disease measures the impact of living with illness and injury and dying prematurely. The essential measure is described as *disability-adjusted life years* (or DALY) which measures the years of healthy life lost from disability and death. The Global Burden of Disease Study (GBD) is an independent and comprehensive global-based health research project that evaluates disease burden from disease, injuries, and risk factors. The strengths of the GBD approach is that consistent methods are applied to *critically analyze available information on each condition, make this information comparable and systematic, estimate results from countries with incomplete data, and produce results using standardized metrics* (Murray and Lopez 2013).

A method for assessing the burden of disease was developed in the 1980s by the WHO. In 1990, the World Bank commissioned the WHO to produce the initial GBD study that assessed the GBD using the DALY, as the unit of measure (Landrigan et al. 2017, pp. 8–9).

WHO (2017b) describes the DALY as follows:

> One DALY can be thought of as one lost year of "healthy" life. The sum of these DALYs across the population, or the burden of disease, can be thought of as a measurement of the gap between current health status and an ideal health situation where the entire population lives to an advanced age, free of disease and disability.

The use of the DALY is central to the GBD project initiated by the WHO in partnership with the World Bank, and the Harvard School of Public Health, and continued by the WHO and Institute of Health Metrics and Evaluation (Murray and Lopez 2013); and The Lancet Commission on Pollution and Health (Landrigan et al. 2017).

The calculation of the DALY is a somewhat complex procedure needing available clinical and population data for each risk factor or agent involved such as a waterborne pathogen or a toxic chemical (e.g. lead or arsenic). Examples using pathogens are shown in Box 3.1.

> **Box 3.1 Disability-Adjusted Life Years (DALYs)**
>
> DALYs for a disease or health condition are calculated as the sum of the Years of Life Lost (YLL) due to premature mortality in the population and the Years Lost due to Disability (YLD) for people living with the health condition or its consequences: Hence, DALYs = YLL (years of life lost) + YLD (years lived with a disability or illness).
>
> The basic principle of the DALY is to weight each health impact in terms of severity within the range of 0 for good health to 1 for death. The weighting is then multiplied by duration of the effect and the number of people affected by the effect. In the case of death, duration is regarded as the years lost in relation to normal life expectancy. In this context, disability refers to conditions such as illness or disease, physical or mental impairment that detract from good health.
>
> **Worked Examples using Australian Data for Water Recycling:**
>
> Using the DALY approach, mild diarrhea with a severity weighting of 0.1 and lasting for 7 days, results in a DALY of 0.002, whereas death of a 1 year old resulting in a loss of 80 years of life equates to a DALY of 80.
>
> In evaluating the acceptable level of treatment for water recycling, reference pathogens usually include a virus (rotavirus), a protozoan (*Cryptosporidium* spp.), and a bacterium (*Campylobacter* spp.). Using Australian examples, the DALY per case of infection for each of these pathogens is calculated below.
>
> Infection with rotavirus causes:
>
> - Mild diarrhea (severity rating of 0.1) lasting three days in 97.5% of cases
> - Severe diarrhea (severity rating of 0.23) lasting seven days in 2.5% of cases
> - Rare deaths of very young children in 0.015% of cases
>
> The DALY per case:
>
> $= (0.1 \times 3/365 \times 0.975) + (0.23 \times 7/365 \times 0.025)$
> $+ (1 \times 80 \times 0.00015)$
> $= 0.0008 + 0.0001 + 0.012$
> $= 0.013$
>
> Infection with *Cryptosporidium* can cause watery diarrhea (severity weighting of 0.067) lasting for seven days with extremely rare deaths in 0.0001% of cases. This equates to a DALY per case of 0.0015.
>
> *Campylobacter* can cause diarrhea of varying severity, Guillain-Barré syndrome of varying severity, reactive arthritis, and occasional deaths. The calculated DALY per case is 0.0046.
>
> Based on DALYs per case, the order of health impacts of the three pathogens are rotavirus > *Campylobacter* > *Cryptosporidium*.
>
> DALYs per case is based on the work of Havelaar and Melse (2003), with a modification using Australian data for rotavirus.
>
> Source: Modified from AWA (2006).

Basic Formula Used to Calculate DALYS for Mortality and Morbidity

The DALY is the sum of the Years of Life Lost (YLL) due to premature mortality in the population and the Years Lost due to Disability (YLD) for people living with the health condition or its consequences (Eq. 3.1) (WHO 2017b). The YLL basically correspond to the number of deaths multiplied by the standard life expectancy at the age at which death occurs.

$$DALY = YLL + YLD \qquad (3.1)$$

The basic formula for YLL (not including social weightings mentioned below) is shown as Eq. (3.2) for a given cause, age, and sex:

$$YLL = N \times L \qquad (3.2)$$

where N = number of deaths and L = standard life expectancy at age of death in years.

Because YLL measure the incident stream of lost years of life due to deaths, an incidence perspective[1] has also been taken for the calculation of YLD in the original GBD Study for the year 1990 and in subsequent WHO updates for years 2000–2004. Apart from disability weighting, social value weights (age-weighting and discounting) were also included for years 2000–2004.

The YLD value for a particular cause in a particular time period is estimated by multiplying the number of incident cases[1] in that period by the average duration of the disease and a weight factor that reflects the severity of the disease on a scale from 0 (perfect health) to 1 (dead). The basic formula for YLD is the following:

$$YLD = I \times DW \times L \qquad (3.3)$$

where I = number of incident cases; DW = disability weight; and L = average duration of the case until remission or death (years).

---
1 Incidence-based YLD.

The GBD study published by IHME in December 2012 used an updated life expectancy standard for the calculation of YLL, and based the YLD calculation on prevalence[2] rather than incidence:

$$YLD = P \times DW \quad (3.4)$$

where P = number of prevalent cases and DW is the disability weight.

Worked Examples for Exposed Populations using DALYS per Case

Quantitative examples of DALY calculations for pathogens and chemicals (arsenic and bromate) in drinking waters are described in Havelaar and Melese (2003).

A simplified burden of disease calculation for a case study of persons exposed to arsenic in drinking water is given below:

Using selected exposure and epidemiological data:

Mortality from skin cancer[a]	Severity of outcome	Duration in years	Burden of disease per case in DALYs
	1	54	54

[a] All cancers in adults.

Population-based estimate of number of DALYs per 1 000 skin cancer deaths

= Number × Severity weight × Duration

= 1 000 × 1 × 54 = 54 000 DALYS

Box 3.1 provides more detailed worked examples for DALY calculations per case using reference waterborne pathogens to evaluate the level of water treatment for potential exposure to recycled water.

Further GBD studies by the WHO and later, by the Institute for Health Metrics and Evaluation (IHME) based at the University of Washington in Seattle, USA, have continued to update and develop a robust framework to evaluate and compare how important are diseases, injuries, and risk factors in causing premature death, loss of health, and disability in different populations. The IHME has become the coordinator and database website for the GBD and related studies following the GBD 2010 study, published in 2012.

In particular, *GBD 2010 addressed a number of major limitations to previous analyses, including strengthening the statistical methods used for estimation and using disability weights derived from surveys of the general population.*

Metrics produced include leading causes of death, years of life lost, years lived with disability, and disability-adjusted life years (DALYs), which are the years of healthy life lost by a person due to death or disability (Murray and Lopez 2013).

Environmental Burden of Disease

What are the environmental risks to human health? From a pollution perspective, we are concerned with the contribution of pollutants to the burden of disease. While there is no readily available study on this aspect, except for selected major chemical pollutants, the WHO has examined the broader question on environmental risks and the GBD (e.g. Prüss-Üstün and Corvalán 2006 and 2007; Prüss-Üstün et al. 2016).

Prüss-Üstün et al. (2016) defined environmental health risks as ...*all the physical, chemical and biological factors external to a person, and all related behaviours, but excluding those natural environments that cannot reasonably be modified*. The concept of *modifiable* environmental factors includes those that can reasonably be managed or changed given current knowledge and technology (e.g. natural biological agents and the natural environments of vectors such as lakes and wetlands).

Methodology used by the 2016 study adopted a systematic review and meta-analysis of scientific literature and data (up to December 2014) that related diseases and injuries to the environment. Out of 133 diseases or injuries, or their groupings, 101 had significant links to the environment, and 92 were quantified, some partially. Results were used to make quantitative estimates of the disease burden attributed to the environment (PAF, see Box 3.2) using a combination of comparative risk assessment (CRA), epidemiological data, transmission pathways, and expert opinion approaches. Several pathways of reducing the burden of disease from unhealthy environments were reviewed and proposed (Prüss-Üstün et al. 2016).

Prüss-Üstün and coresearchers estimated in 2012 that 23% (95% CI: 13–34%) of all deaths globally (12.6 million deaths) were due to environmental factors included in this study. When accounting for both death and disability, the fraction of the GBD (deaths and disability) due to the environment was 22% (95% CI: 13–32%), similar to earlier estimates reported by the WHO (Prüss-Üstün et al. 2016). (Diseases are the main component of the environmental burden of disease.)

Previously, Prüss-Üstün and Corvalán (2007) concluded that the environmental disease burden could largely be prevented by available cost-effective measures such as clean water, clean air, and basic safety measures.

---

2 Prevalence YLD.

Box 3.2 Population Attributable Fraction (PAF)

The contribution of a risk factor to a disease or a death is calculated using the term population attributable fraction (PAF). It is defined as the proportional reduction in population disease or mortality that would occur if exposure to a risk factor were reduced to an alternative ideal exposure scenario (e.g. no tobacco). Many diseases are caused by multiple risk factors, and individual risk factors may interact in their impact on overall risk of disease. As a result, PAFs for individual risk factors often overlap and add up to more than 100%.

$$\text{PAF} = \frac{\sum_{i=1}^{n} P_i RR_i - \sum_{i=1}^{n} P'_i RR_i}{\sum_{i=1}^{n} P_i RR_i} \quad (3.5)$$

P_i = proportion of population at exposure level i, current exposure
P'_i = proportion of population at exposure level i, counterfactual or ideal level of exposure
RR = the relative risk at exposure level i
n = the number of exposure levels

Source: Based on WHO (2020b).

Key results from their study were:

- Environmental risks account for a large fraction of the GBD.
- Environmental impacts on health are uneven across life course and gender.
- Low- and middle-income countries bear the greatest share of environmental disease.
- Total environmental deaths are unchanged since 2002, but show a strong shift to noncommunicable disease.
- The evidence on quantitative links between health and environment has increased.

Children under five years and older adults were most affected by the environment for different reasons. In young children, it was mainly due to infectious and parasitic diseases, neonatal, and nutritional diseases, and injuries. In older adults, noncommunicable diseases caused by the environment were prevalent, although injuries remained significant.

This is an important study because it examines and provides an estimate of how much disease can be prevented by taking measures to reduce environmental risks to health. It also acts to some degree as a benchmark to assess the impact of pollutants such as chemicals on environmental health. The preventative approach of the study toward disease from unhealthy environments (including workplaces) is vital because it offers a sound basis to prevent environmental risks to human health, through setting priorities for targeted environmental actions (interventions) and measuring quantitative performance. In this way, it forms a key part of environmental risk management (see also Part 4 of this book).

Here, a conceptual model of the environmental burden of disease approach used by the WHO has been constructed and is presented as Figure 3.2. It illustrates the framework for estimation of environmental burden of disease and

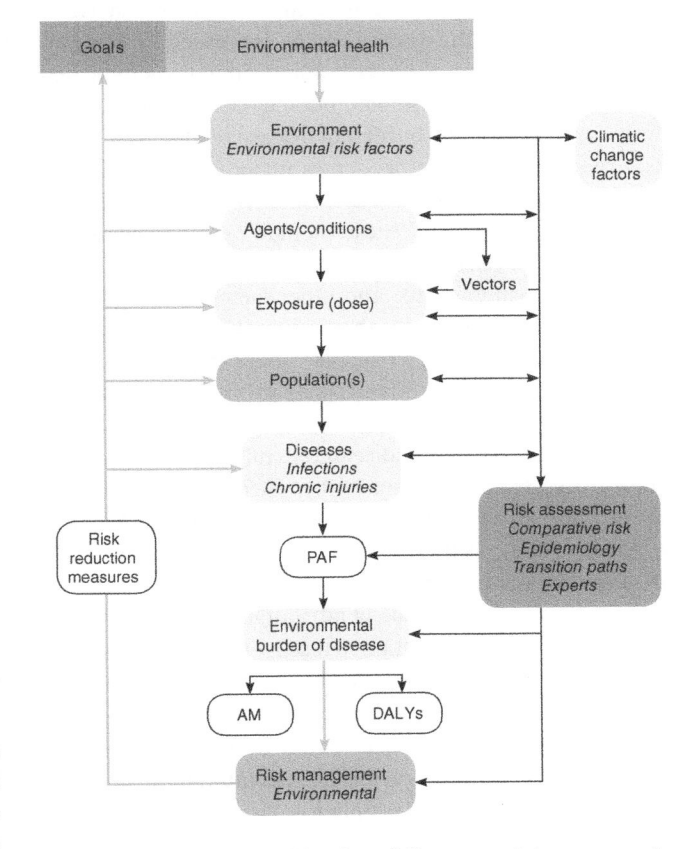

Figure 3.2 Environmental burden of disease model: assessment, management, and intervention.

key metrics (attributed mortality, AM, and DALYS) and environmental intervention, within a risk assessment and management approach to reduce risks in selected populations. Outcomes are measured as attributed mortalities and DALYS, while other metrics could be adopted within the framework (e.g. compliance with air pollution goals, relative risks at exposure levels, and prevalence of infectious disease or smoking rates in populations). Social and economic factors are not considered in this model.

The model follows a stepwise approach used by the WHO to estimate the burden of disease in populations due to exposures from risk factors and interactions with the external environment. It incorporates the WHO approach within a risk assessment and management framework designed to prevent or reduce diseases through environmental interventions (healthy environments) promoted by the WHO.

The assessment model is based on the assumption of the WHO (Prüss-Üstün and Corvalán 2007) that *Environmental health action can improve population health in a sustainable manner and improve equity*. This depends on defining what health and sustainability goals are to be achieved. The primary objective is to estimate the burden of disease to see how much of the disease can be prevented by environmental intervention using risk management. As indicated in Figure 3.2, a stepwise methodology determines quantitative estimates of the burden of disease related to selected environmental risk factors across major categories of diseases and injuries. A hazard assessment step identifies environmental risk factors. These include physical, chemical, and biological hazards that may act directly as disease or disability-causing agents, and other factors or conditions, such as physical inactivity, that indirectly contribute.

In what can be called a risk assessment step in Figure 3.2, the probability distribution of exposure to the agent and/or condition in a population is estimated and relative risk estimates for an associated disease(s) (infectious, chronic,or injury) are obtained (usually by selecting from the literature by meta-analysis or review) (Prüss-Üstün and Corvalán 2006). As indicated in the model, some infectious diseases are vector-borne.

Estimates are then made of attributable fractions (PAFs) of the proportional reduction in disease or death that would occur in an exposed population if exposure to the risk was reduced to a reasonably achievable baseline exposure, in the short- or medium-term (Prüss-Üstün and Corvalán 2006). As mentioned previously, a combination of CRA, epidemiological data, transmission pathways, and expert opinion techniques are used in WHO estimates.

The disease burden attributable to the environment can be calculated from the following equations:

$$AM = PAF \times M \qquad (3.6)$$

$$\text{and } AB \text{ (DALYs)} = PAF \times B \text{ (DALYs)} \qquad (3.7)$$

where AM = attributable mortality, PAF = population attributable fraction, M = mortality, B = disease burden calculated in DALYs, and AB = attributable disease burden in DALYs, stratified according to regional, sex, and age groups as necessary.

The disease burden estimates for the environment are characterized for evaluation of intervention options in the risk management step and decision-making on strategies, priorities, and management actions or measures to prevent or reduce disease burdens caused by individual or groups of environmental risk factors. As shown in Figure 3.2, interventions may occur at each step in the process, from goal setting, modifying, or eliminating risk factors (agents and conditions), and interactions with environmental factors, to reducing exposure levels, protecting sensitive populations, and prevention of diseases. Reductions in environmental disease burdens and performance of specific intervention and management measures, for different diseases in populations of interest, can then be evaluated and monitored using the burden of disease model approach.

Burden of Disease due to Chemical Pollution

Environmental risk factors include chemical hazards. Many chemicals and mixtures of chemicals are known or suspected environmental pollutants that can cause a multitude of diseases, poisonings, or injuries to humans due to exposures from air, water, soils, foodstuffs, other media, and products. How to reduce health impacts and risks from environmental exposures to these chemicals is a major global and national challenge for many countries, as indicated by the WHO estimates of the environmental burden of disease.

The WHO has investigated evidence on the burden of disease due to chemicals, to answer critical questions for health policy-makers: what fraction of the burden of disease do chemicals cause? Among these chemicals, what ones are of concern? It then follows, what are the policy actions and priorities for policy-makers to prevent or reduce health impacts or disease from exposures to chemicals?

Prüs-Üstün and colleagues from the WHO have conducted a systematic review of the published literature (1990–2009) for GBD estimates related to chemicals (Prüs-Üstün et al. 2011). Of these, 14 chemicals or their groups, or mixtures, were identified as relevant. Methods underlying total burden of disease estimates due to chemicals applied the standard methodology developed by the WHO for GBD estimates.

The total burden of disease due to environmental exposures and use of selected chemicals was estimated to be

Table 3.1 Global overview of available disease burden estimates attributable to chemical exposures (reference year 2004).

Chemical/groups of chemicals	Deaths	DALYs
Chemicals in acute poisonings[a]	526 000	9 666 000
Unintentional acute poisonings	240 000	5 246 000
Unintentional occupational	30 000	643 000
Self-inflicted pesticide injuries	186 000[c]	4 420 000[c]
Occupational chemical[a] (chronic effects)	581 000	6 763 000
Asbestos	107 000	1 523 000
Occupational lung carcinogens	111 000	1 011 000
Occupational leukaemogens	7 400	113 000
Occupational particulates (COPD)	375 000	3 804 000
Occupational particulates (other respiratory)	29 000	1 062 000
Air pollutants[a]	3 720 000	60 669 000
Outdoor air pollutants	1 152 000	8 747 000
Outdoor air pollutants (emissions from ships)	60 000[c]	NA
Indoor air pollutants (from solid fuels combustion)	1 965 000	41 009 000
Second hand smoke	603 000	10 913 000
Individual chemicals[a] (mostly chronic effects)	152 000	9 102 000
Lead	143 000	8 977 000
Arsenic (drinking water)[d]	9 100	125 000
Total[b]	4 879 00	86 200 000
Percentage of total deaths/DALYs (year 2004)	8.3%	5.7%

[a]Subtotals.
[b]Children <15 years: deaths 1 073 000 (22% of total); DALYs 46 627 000 (54% of total).
[c]2002 data year.
[d]2001 data year for arsenic in drinking water.
Source: Data from Prüss-Üstün et al. (2011, 2016) and (Landrigan et al. (2017).

a total of 4.9 million deaths (8.3% of total) and 86 million DALYs (5.7% of total) in 2004, the reference year of study. In comparison, children under 15 years suffered 54% of the total burden (DALYs) due to chemicals (see Table 3.1). The total burden of disease due to known chemicals also exceeds the burden of disease for all cancers estimated worldwide, indicating the size of its public health impact (Prüs-Üstün et al. 2011) (see also Landrigan et al. 2017, p. 10).

Available disease burden estimates attributable to chemicals are compiled in Table 3.1 from data given in Prüs-Üstün et al. (2011). They are grouped in four categories of chemicals:

- Chemicals in acute poisonings
- Chemicals in occupational exposures (long-term effects)
- Air pollutant mixtures (outdoor and indoor)
- Single chemicals with mostly longer term effects

Environmental exposures to air pollutant mixtures (outdoor and indoor) were related to most (70%) of the known burden of disease (DALYs) compared to the chronic disease burden estimated for occupational exposures to chemicals (8%). The latter are essentially air pollutants also. The predominance of air pollutants, mainly from combustion of fossil fuels and biomass, highlights these sources as major health problems.

Prüs-Üstün et al. (2011) concluded from their study that the disease burden from chemicals is considerable but is probably underestimated because a number of known toxic chemicals (e.g. dioxins, mercury, cadmium, and various pesticides) had inadequate data to be included. Even so, they advocated that effective intervention allows chemicals to be managed and to limit their public health impacts

Case Study 3.1 presents an example of GBD estimates for lead exposure and environmental intervention (phasing out of lead in gasoline) for the period 2000–2004.

> **Case Study 3.1**
>
> **Global Burden of Disease Estimates for Lead and Environmental Intervention**
>
> How much can the burden of disease due to a chemical, or another environmental risk factor, be reduced by environmental intervention?
>
> The extent that the disease burden can be prevented by targeted action is indicated by the burden of disease estimate and its preventable fraction (Prüs-Üstün et al. 2011). An example of how public health intervention reduced the GBD caused by lead exposure is shown in Figure 3.3, using the environmental burden of disease model presented as Figure 3.2. The total burden of lead-induced disease in 2004 is estimated to be 0.6% of the global disease burden for all diseases (and injuries) and 10.4% of the disease burden estimate attributed to selected chemicals.
>
> In this case, the GBD estimates (mild mental retardation, MMR, and CVD) induced by lead are given in Figure 3.3 for the years 2000 (Annex 4) and 2004 (Prüs-Üstün et al. 2011). During this period, the proportion of people with blood lead levels above $10\,\mu g\,dL^{-1}$ (WHO criterion at time) decreased from 20 to 14%, a reduction of 30%. This reduction over a short period was mainly attributed to the phase out of lead from gasoline in most countries (Prüs-Üstün et al. 2011). Over the same time period, the burden of disease estimates for lead reduced by about 37.5% for deaths (mostly attributed to CVD) and ~33% for DALYS, supporting the effectiveness of this preventive action.
>
> Note: Generally, comparisons between mean estimates of disease burdens made at different time periods may experience subtle changes in an evolving methodology and sometimes have relatively large confidence intervals.
>
> In 2016, a large global increase in lead-related deaths was reported relative to 2004. The IHME estimated that lead exposure accounted for 540 000 deaths and 13.9 million years of healthy life lost (disability-adjusted life years [DALYs]) worldwide due to long-term effects on health. Low- and middle-income countries suffered the highest burden. Overall, lead exposure accounted for 63.8% of the global burden of idiopathic developmental intellectual disability, 3% of the global burden of ischemic heart disease, and 3.1% of the global burden of stroke (WHO 2019). It is possible this result reflects an updated statistical methodology and reporting by global collaborators at the IHME.
>
> Lead is highlighted here as it remains a major global pollutant in many countries.
>
> Figure 3.3 Model of global burden of disease attributed to lead exposure and preventative intervention based on WHO estimates (2000–2004) derived from Prüss-Üstün and Corvalán (2007), Annex 4 Estimating the global disease burden of environmental lead exposure. Source: Prüss-Üstün and Corvalán (2007) and Prüs-Üstün et al. (2011).

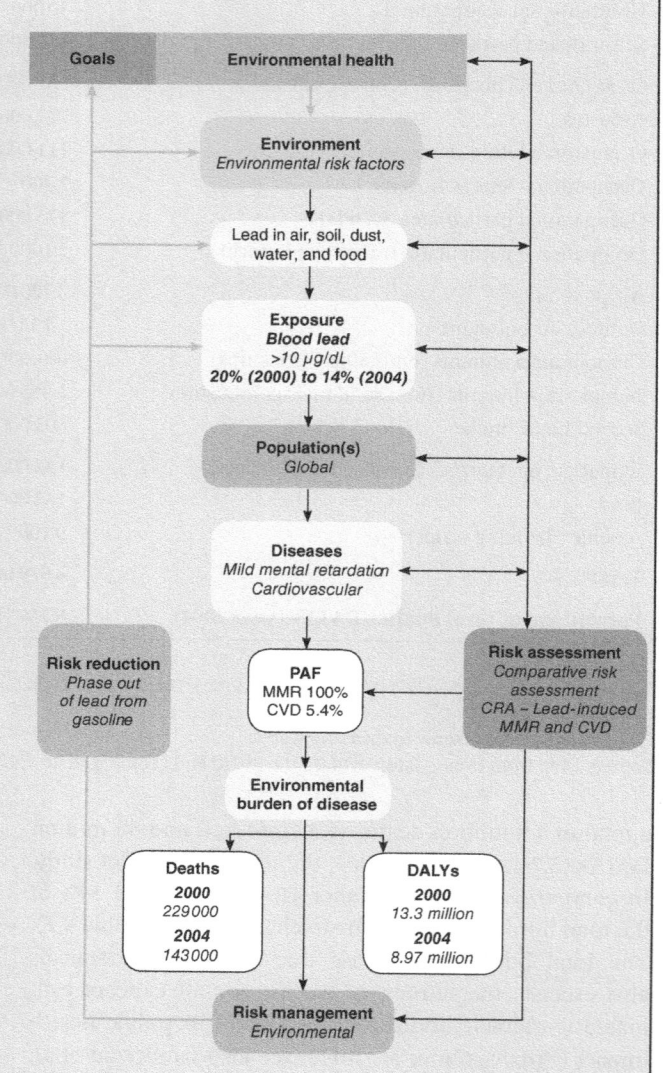

Note: MMR: mild mental retardation; CVD: cardiovascular disease

A comparative overview of the GBD and premature death due to pollution is available from The Lancet Commission on Pollution and Health (Landrigan et al. 2017), as given in Table 3.2. The GBD (2015) study reported 9 million premature deaths from pollution risk factors (air, water, soils, heavy metals, chemicals, and occupational), which is

Table 3.2 Global estimated deaths (millions) due to pollution risk factors from the Global Burden of Disease Study (GBD 2015)[a] versus WHO data 2012[b] is from Prüs-Üstün et al. (2016).

Pollution risk factors	GBD study best estimate (95%CI)[a]	WHO best estimate (95%CI)[b]
Air (total)	6.5 (5.7–7.3)	6.5 (5.4–7.4)
Household air	2.9 (2.2–3.6)	4.3 (3.7–4.8)
Ambient particulate	4.2 (3.7–4.8)	3.0 (3.7–4.8)
Ambient ozone	03 (0.1–0.4)	—
Water (total)	1.8 (1.4–2.2)	0.8 (0.7–1.0)
Unsafe sanitation	0.8 (0.7–0.9)	0.3 (0.1–0.4)
Unsafe source	1.3 (1.0–1.4)	0.5 (0.2–0.7)
Occupational	0.8 (0.8–0.9)	0.4 (03–0.4)
Carcinogens	0.5 (0.5–0.5)	0.1 (0.1–0.1)
Particulates	0.4 (0.3–0.4)	0.2 (0.2–0.3)
Soil, heavy metals, and chemicals	0.5 (0.2–0.8)	0.7 (0.2–0.8)
Lead	0.5 (0.2–0.8)	0.7 (0.2–0.8)
Total	9.0	8.4

Note: The totals for air pollution, water pollution, and all pollution are less than the arithmetic sum of the individual risk factors within each of these categories because these have overlapping contributions – e.g. household air pollution also contributes to ambient pollution and vice versa.
[a]GBD (2015).
[b]WHO 2012 data (Prüs-Üstün et al. 2016).
Source: Data from Global Burden of Disease Study (GBD) (2015) and Landrigan et al. (2017).

slightly higher than the earlier WHO 2012 study which was updated by Prüs-Üstün et al. (2016). Both studies are likely to be underestimates, as indicated by the lack of detailed data on deaths from metal and chemical exposures.

3.5 Biodiversity and Ecosystem Services

Biodiversity

Biological diversity (biodiversity) is ... *the variety of life on Earth – its genes, species, populations and ecosystems* (Pimm et al. 2008, p. 3). It can be seen as the variety of all the different plants, animals, and microorganisms, the genetic information they contain, and the ecosystems they form. As such, biodiversity is usually considered at three levels of organization – genetic diversity, species diversity, and ecosystem diversity – described below (Australian Museum 2018):

- **Genetic diversity** is the variety of genes within a species. Each species is made up of individuals that have their own particular genetic composition. This means a species may have different populations, each having different genetic compositions. To conserve genetic diversity, different populations of a species must be conserved.
- **Species diversity** is the variety of species within a habitat or a region. Some habitats, such as rainforests and coral reefs, have many species. Others, such as salt flats or a polluted stream, have fewer.
- **Ecosystem diversity** is the variety of ecosystems in a given place. An ecosystem is a community of organisms and their physical environment interacting together. An ecosystem can cover a large area, such as a whole forest, or a small area, such as a pond.

Biodiversity reflects the number and different types of plants and animals in a biological community and their ecological niches. The stability of ecosystems depends on the diversity of plants and animals, partly because ecosystem functions are more efficient at extracting energy and available nutrients, with different species occupying more niches. More complexity increases adaptability to environmental change. However, loss of biodiversity makes the system less stable and less adaptable (Landon 2006, p. 160).

Ecosystem Services

Ecosystems sustain life on Earth through their natural functions that freely produce goods and services such as food and timber, primary productivity (photosynthesis), breakdown of wastes, and nutrient cycling for the benefits of humans and other living organisms. The concept of ecosystem services explains the benefits from ecosystem functions generated by natural cycles (biogeochemical) that are complex and vary greatly in time and spatial scales.

There are four major categories of ecosystem services, as described by Melillo and Sala (2008, p. 76) (see Table 3.3):

- Provisioning services provide products obtained from ecosystems (e.g. food, wood, and medicine).
- Regulating services are the benefits people obtain from ecosystem controls including many environmental functions that regulate climate, plant pests, and pathogens, animal diseases, water quality, air quality, and soil erosion.
- Cultural services are nonmaterial benefits people experience from ecosystems such as recreational, aesthetic, spiritual, and intellectual.
- Supporting services, such as photosynthetic production of new organic matter by plants and nutrient cycling, are

Table 3.3 Outline of ecosystem services.

Services and description	Examples
Provisioning: Products obtained from ecosystems	• Food • Fuel wood • Fiber • Medicines
Regulating: Benefits obtained from environmental regulation of ecosystem processes	• Cleaning air • Purifying water • Mitigating floods • Controlling erosion • Detoxifying soils • Modifying climate
Cultural: Nonmaterial benefits obtained from ecosystems	• Aesthetics • Intellectual stimulation • A sense of place
Supporting: Services necessary for the production of all other ecosystems	• Primary productivity • Nutrient cycling • Pollination and seed dispersal

Source: Modified from Melillo and Sala (2008).

essential for living organisms and all other ecosystem services.

The current disruption of natural cycles and related ecosystem functions by human activities, such as large-scale forest clearing, overharvesting of fisheries, and pollution of the atmosphere, including excessive emissions of greenhouse gases resulting in global warming and climate change, is greatly impacting on ecosystem services available to humans, natural ecosystems, human health, and biodiversity (Chivian and Bernstein 2008a, Chapter 2).

3.6 Human Impacts on Biodiversity and Ecosystem Services

Biological diversity – the variety of life on Earth – is at the heart of our efforts to relieve suffering, raise standards of living, and achieve the U.N. Millennium Development Goals... We cannot do without the countless services provided by biodiversity...

(Kofi Annan, Former Secretary-General, United Nations).

Prologue to the major review *Sustaining Life: How Human Health Depends on Biodiversity* by Eric Chivian and Aaron Bernstein, Center for Health and Global Environment, Harvard Medical School (Chivian and Bernstein 2008b).

The book, *Sustaining Life*, explains and details the links between the decline in biodiversity and the consequences for human health. In Figure 3.4, the conceptual links between humans and their health and the living environment, as a function of biodiversity and ecosystem services, are illustrated.

Among the direct impacts of human activities on human health (e.g. injuries and air pollution exposures from urbanization), there is strong worldwide evidence that direct and indirect impacts (e.g. habitat destruction, eutrophication, endocrine disruption from toxicants in food chains, pesticide spraying, and global warming) are rapidly increasing the loss of biodiversity and ecosystem services which are essential for human health.

The global outlook for the future health of the planet from threats such as increasing habitat loss, overexploitation of resources, pollution, and climate change is indicated in Box 3.3.

Box 3.3 Twenty-first Century Biodiversity Outlook

- There is a high risk of dramatic biodiversity loss and accompanying degradation of a broad range of ecosystem services if the Earth system is pushed beyond certain threshold or tipping points.
- The loss of such services is likely to impact the poor first and most severely, as they tend to be most directly dependent on their immediate environments, but all societies will be impacted.
- However, there is greater potential than was recognized in earlier assessment to address both climate change and rising food demand without further widespread loss of habitats.

Source: Modified from Secretariat of the Convention on Biological Diversity (2010).

Major environmental risk factors such as chronic pollution and climate change can cause changes in biodiversity and the state of ecosystems, that include shifts in species distribution, and accelerated rates of change over time and space (e.g. species extinction). With increasing rates of change and environmental stressors, there are ecological thresholds at which different ecosystems or their important functions may rapidly change or even collapse. These situations are of great scientific concern but are difficult to predict.

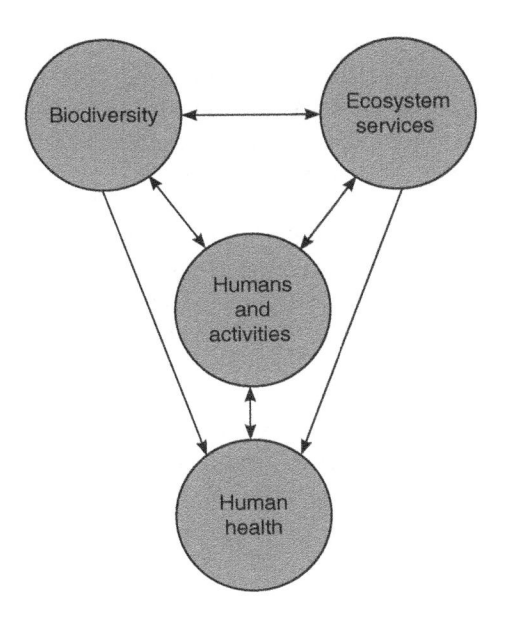

Figure 3.4 Interactions between impact of humans and their activities on biodiversity and ecosystems services and links to human health.

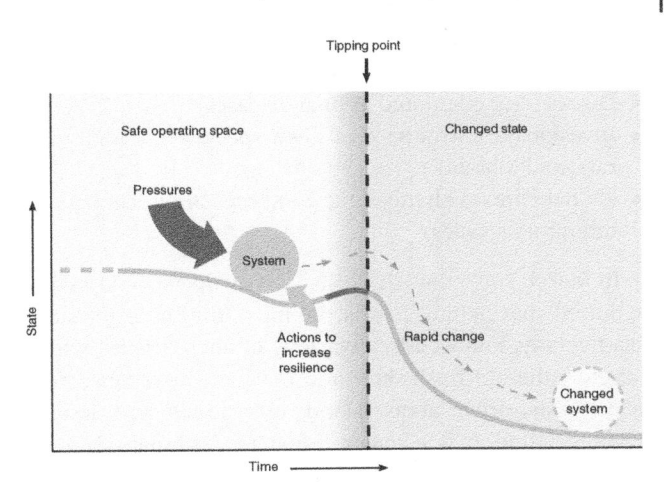

Figure 3.5 Concept of an ecological tipping point. Here, pressures or stresses on an existing ecosystem overcome resilience, and shift the system rapidly to a new state, causing significant changes to its biodiversity and ecosystem services. Source: Modified from Secretariat of the Convention on Biological Diversity (2010).

The concept of an ecological threshold, for instance, is defined as follows:

> An ecological threshold is the point at which there is an abrupt change in an ecosystem quality, property, or phenomenon, or where small changes in an environmental driver produce large responses in the ecosystem (Groffman et al. 2006).

Beyond certain thresholds, called tipping points, ecosystems may collapse or change into new or different states, such as illustrated and defined by the Secretariat of the Convention on Biological Diversity (2010) in Figure 3.5. Tipping points of ecosystems can involve potentially large, rapid, and irreversible impacts and it is difficult for many organisms and societies to adapt to such changes. They tend also to be complex to measure and predict with some degree of precision, with many interactive factors and nonlinear processes in natural systems.

An overview of threats from human activities, including pollution, on biodiversity and ecosystems is indicated in Figure 3.6, along with monitoring, conservation, and management responses to reduce these threats and existing impacts on biodiversity (and ecosystem services).

The major threats to biodiversity from human activities are recognized as:

- Habitat destruction and degradation (e.g. deforestation and wetland drainage)
- Overexploitation (e.g. extraction of resources, hunting, and fishing)

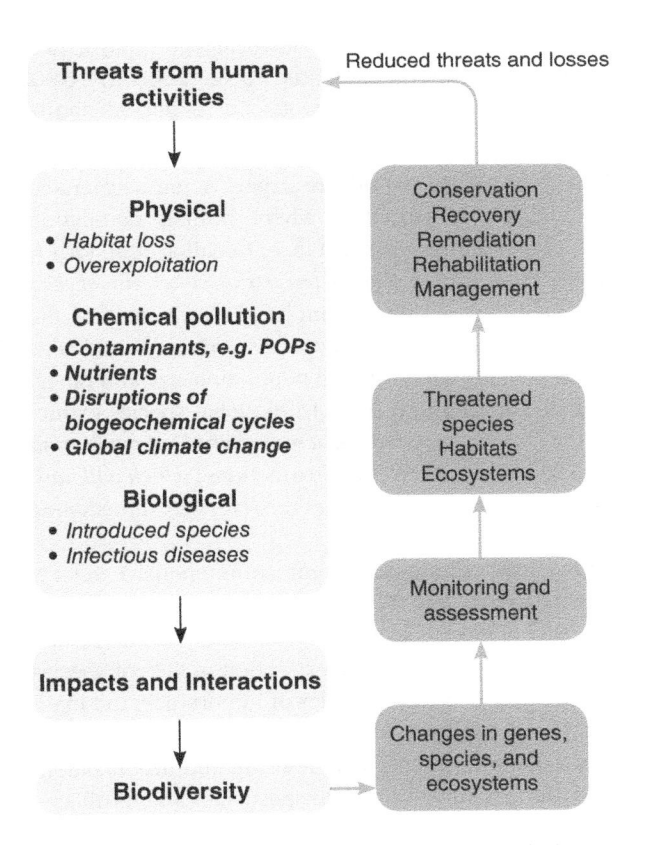

Figure 3.6 Physical, chemical pollution, and biological threats from human activities on biodiversity (changes in genes, species, habitats, and ecosystems) and general management responses to reduce threats and losses to biodiversity.

- Pollution (e.g. air and water pollution, persistent organic pollutants, and microplastics in oceans)
- Disease (e.g. plant and animal diseases)
- Invasions of introduced or alien species (e.g. feral cats, rats, and rabbits)
- Global climate change (e.g. coral bleaching, changes in migratory species)

In March 2007, the IUCN Red List on Species Extinction reported that habitat loss is the most dominant threat to biodiversity, followed by invasions of alien species which severely disrupt freshwater and marine ecosystems, tropical forests, urban areas, islands, grasslands, and deserts. Overexploitation is a major threat to mammals. Habitat loss is emerging as the main threat for birds, followed by invasive species and overexploitation. Amphibians are affected by pollution (including climate change) (29% of species) and 17% by disease (e.g. chytridiomycosis). The greatest threat for marine species is overexploitation, then habitat loss and incidental death in fisheries (e.g. marine animals). Freshwater species are mainly affected by habitat loss, followed by pollution and invasive alien species. Climate change is projected to increase the extent of biodiversity threats (and impacts) depending upon which of the mitigation of greenhouse gases and adaptation scenarios prevails (IUCN 2007).

Critically, many plant and animal species are increasingly threatened with extinction (see Box 3.4) but only a small percentage of described species have been assessed.

Box 3.4 Species Extinction – IUCN Red List

Biological extinction is a natural process due to environmental drivers or evolutionary changes. Long and dramatic periods of geological changes have made the vast majority of species that have ever lived vanish, long before the arrival of humans, according to the fossil record. There have been five known mass extinction events in geological history. However, the rapid loss of species seen today is estimated to be 1 000 to 10 000 times higher than the *background* or expected natural rate, such that it is often called the *sixth extinction crisis* (IUCN 2007; De Vos et al. 2015).

Biodiversity continues to decline. There were more than 120 000 species on the IUCN Red List, with more than 32 000 species threatened with extinction, including 41% of amphibians, 34% of conifers, 33% of reef building corals, 25% of mammals, 30% of sharks and rays, 14% of birds, and 28% of selected crustaceans (IUCN Red List 2019).

Many parts of the world are experiencing accelerated losses of species, especially where human population growth rates are high. For instance, Ceballos et al. (2017) found that *the rate of loss in terrestrial vertebrate species is extremely high*. From a sample of nearly half of the known vertebrate species, these researchers found 32% (8 851/27 600) are decreasing in population size and range. Their data indicate that aside from global species extinctions, the *Earth is experiencing a huge episode of population declines and extirpations* (local extinctions) *which will have negative cascading consequences on ecosystem functioning and services*.

Among invertebrates, a recent comprehensive review by Sanchez-Bayo and Wyckhuys (2019) found that the biodiversity of insect species is declining worldwide, to the extent that it may lead to the extinction of 40% of species in the next few decades. In order of importance, the main drivers of species decrease were related to (i) habitat loss and conversion to intensive agriculture and urbanization; (ii) pollution, mainly by synthetic pesticides and fertilizers; (iii) biological factors, including pathogens and introduced species; and (iv) climate change. The fourth driver is related to global warming caused by greenhouse gas emissions (Chapter 14).

It is increasingly evident that pollution-related impacts, including global warming, are playing a substantial role in the observed loss of biodiversity, and also ecosystem services, discussed further below.

The Sixth Global Environment Outlook (2019) report has concluded a major species extinction event is unfolding, on a scale that will affect planetary integrity and the Earth's capacity to meet human needs (UNEP 2019, p. 8).

Of major ecological concern for the future, quantitative scenarios for global terrestrial, freshwater, and marine biodiversity evaluated by Henrique Pereira and coauthors, using a range of measures (e.g. extinctions, changes in species abundance, habitat loss, and distribution shifts) found that these scenarios indicate that biodiversity will continue to decline over the Twenty-first Century (Pereira et al. 2010).

Pollution and Biodiversity

Pollution, including effects of greenhouse gas emissions on climate change, has a major impact on global biodiversity and ecosystems, although the magnitude and rate of chronic impacts are often poorly defined. Exposure–response relationships (ecotoxicity) are difficult to establish in field studies on ecosystem scales. Common types of physical, chemical, and biological pollutants affecting biodiversity are summarized in Table 3.4. Climate change effects are discussed in the next subsection.

Table 3.4 Some major pollutants causing impacts on global biodiversity.

Pollutants	Global impacts	Examples
Physical		
Ultraviolet radiation	Increased surface UV due to human-caused depletion of ozone layer. Damages DNA and affects stability of proteins and cell membranes.	Impairs photosynthesis and reproduction in phytoplankton and other marine plants. UVB radiation damages DNA in amphibians and can affect reproduction success (e.g. Western toad *Bufo boreas*).
Chemicals		
Acid deposition (wet and dry)	Increasing acidity (lower pH) from acid gas emissions: acid–base balance of atmosphere disrupted since industrial revolution.	Cascading effects of acidity on many terrestrial and freshwater organisms and ecosystems (e.g. North America and Europe). Acid gases downwind of coal-fired power plants cause acidification of temperate forests and soils, depleting available calcium uptake from soils, resulting in increased stress effects (e.g. disease and pest infestations) on trees.
Metals	Worldwide pollutants – toxic metals (Pb, Cd, Hg, and As) accumulate in and affect a diverse range of wildlife, including their fetal development and nervous system.	Mercury, especially methyl mercury, is a potent nerve poison in wild vertebrates worldwide and also affects some plants (e.g. disrupting photosynthesis by substituting for magnesium in chlorophyll).
Nutrients	Excessive nutrient releases, particularly nitrogen, from agriculture, sewage discharges, etc. Major worldwide impacts: overfertilization or accelerated eutrophication, algal blooms, toxins, and marine *dead zones* (anoxic), leading to deaths of fish and other aquatic organisms.	Increasing numbers of *dead zones* worldwide. Major marine dead zones are in the Baltic Sea (world's largest), Black Sea, northern Gulf of Mexico. and Bay of Bengal, near Eastern India, Bangladesh, and Myanmar (see also Chapter 8). Harmful algal blooms tend to be caused by phosphorus in freshwaters, at low levels, and nitrogen in marine waters. Algal toxins produced by blooms may transfer up the food web and cause toxic effects in higher aquatic organisms. Endangered Brown Pelicans (*Pelecanus accidentalis*) and *West Indian Manatees (Trichechus manatus)* are reported victims of harmful algal blooms.
Petroleum hydrocarbons	Petroleum hydrocarbons, particularly toxic soluble aromatic fractions, from oil spills and chronic discharges cause high mortalities of marine wildlife, long-term contamination damage to vulnerable ecosystems (e.g. wetlands, sea grasses, and coral reefs and fisheries).	Major oil spills and biodiversity effects from chronic inputs (e.g. Connell and Miller 1984; Volkman et al. 1994).
Persistent organic pollutants (POPs)	POPs mimic hormones, (endocrine disrupters) and substitute for other biologically active molecules, affect immune systems, act as carcinogens, interfere with development and nervous systems in animals.	DDT-caused thinning of eggshells in raptor populations (e.g. Peregrine Falcon and American Bald Eagle). Beluga Whales living in the mouth of St Lawrence River in Canada experience high rates of infections, cancer, and reproductive failure suspected to be due to their high POP burdens.
Pesticides	~2 600 million tonnes year^{-1} released globally (US EPA). Lethal and sublethal toxicity to many nontarget plants and animals.	Acute and subtle poisonings of many wildlife species; herbicide, atrazine, can cause sex change in Leopard frogs (*Rana pipiens*) and slow their gonadal development at low environmental levels (0.1 ppb).
Pharmaceuticals	Use of human medicines (e.g. hormones and antibiotics) in livestock and aquaculture; discharges of drug residues in wastewaters (sewage and some agricultural). Increased antibiotic resistance; endocrine disruption effects in wild animals.	Synthetic estrogens shown to kill trout and to reduce fertility of male trout at low environmental levels (1 ppb and less). In India, populations of three species of vultures declined by more than 90% following the eating of dead livestock treated with the veterinary drug, diclofenac. This drug was banned in India in 2005 and replaced with a safe and effective substitute.

(Continued)

Table 3.4 (Continued)

Pollutants	Global impacts	Examples
Plastics	~9 million tonnes enter oceans each year (80% from land-based sources) (GEO 6): marine litter of oceans, coastline, and sediments resulting in entanglement and choking; ingestion and toxicity, particularly from microparticles and adsorbed toxics.	Plastics are lethal to marine animals at different life stages. Among other marine animals, zooplankton can ingest and concentrate toxic microparticles of plastic. Plastics kill up to 1 million seabirds and 100 000 sea mammals and turtles each year.
Biological		
Infectious disease	Introduction of new pathogens due to human activities or increases in disease among wildlife due to environmental stresses from pollution.	Outbreaks of lethal infectious diseases among coral species after bleaching due to warm ocean temperatures (e.g. severe losses of the dominant coral *Acropora cervicornis* in Caribbean reefs).

Source: Adapted from Connell and Miller (1984), Chivian and Bernstein (2008a), UNEP (2019) and GESAMP (2020).

The mode of action for many pollutant impacts follows a series of biological interactions that may be described as an *ecological cascade* involving tipping points where pressures or forcings reach a threshold sufficient to shift the state of the system. This event (e.g. fish kills, eutrophication, and climate change effects) may occur rapidly, from local to regional scales, with increasing risks for worldwide effects (e.g. observed declines in the world's insects and mammal species).

Global Climate Change and Biodiversity

There is ample scientific evidence and consensus (e.g. Intergovernmental Programme on Climate Change [IPCC] Reports and UNEP GEO 6 Report) that global increases in anthropogenic greenhouse gas emissions are the major contributing factor in warming the surface of the Earth and causing ocean acidification. Global climate changes and mechanisms are strongly associated with increasing carbon dioxide and other greenhouse gas levels in the atmosphere (see Chapter 14).

If the current rates of burning fossil fuels continue, the IPCC has predicted that the Earth's surface will warm on average by 2–3°, and even possibly up to 6 °C or so by 2100. Fundamentally, climate change is a threat to survival for many species because they have evolved to live within certain temperature ranges, and may not adapt to changes in temperatures that exceed such ranges, or loss of food species that cannot adapt.

On global to local scales, climate change is a serious and increasing threat to biodiversity. It is expected to exceed even the major threat from habitat loss on land. By 2050, extinction is predicted to threaten about one quarter or more of all species on land. Marine and freshwater species are also at great risk from climate change, especially those that are highly sensitive to warming temperatures and live in vulnerable ecosystems such as coral reefs (see Chivian and Bernstein 2008a, chapter 2).

Pollution Threats to Ecosystem Services

Some of the major threats to ecosystem services involve massive impacts from human activities on a global scale from climate change, deforestation, desertification, urbanization, wetland drainage, pollution, dams, and water diversion, as reviewed by Melillo and Sala (2008, p. 107). Pollution impacts are usually more diffuse and difficult to evaluate compared to the physical impacts of habitat destruction, urbanization, and infrastructure projects on ecosystems. Table 3.5 emphasizes some known pollution-related threats, including climate change, on ecosystem services.

3.7 Environmental Assessment and Management of Pollution

The primary drivers of environmental change are seen as human population pressures and economic development, in association with recent rapid urbanization and accelerating technological innovation. These drivers are increasing in scale, global extent, and speed of change, posing urgent challenges for managing environmental and climate change problems (UNEP 2019, p. 7). In this context, scientific understanding of the role of pollution in environmental degradation, climate change, ecosystem, and human health effects becomes vital as an urgent basis for adopting socioecological solutions that are sustainable.

How can we evaluate and manage the threats and impacts of pollutants to meet sustainable development goals for human health and biodiversity? In the following Parts of this book, we explore how to answer this question

Table 3.5 Some major pollution threats to ecosystem services.

Pollution threats	Ecosystem services impacts	Examples
Climate change – significant to high	Terrestrial, freshwater, and marine ecosystems.	Increased changes in weather patterns and extreme weather impacts (droughts and floods) on local and regional ecosystem habitats, functions, species diversity, population abundance, and food webs.
Air pollution – significant to high	Air pollutants such as ozone (O_3), nitrogen oxides (NO_x), ammonia (NH_3), and sulfur dioxide (SO_2) all have major negative impacts on ecosystem services.	Effects range from marked reductions in food provisioning due to crop yield impacts (O_3) (e.g. China) to changes in ecosystem functioning driven by eutrophication and acidification (NO_x, NH_3, and SO_2).
Acid precipitation – significant	Air pollution of rain and snow with sulfur and nitrogen results in acid rain that damages plants, degrades soils, and acidifies surface waters killing aquatic plants and animals.	Nitrogen inputs in acid rain can cause harmful algal blooms in some estuaries including fish kills (e.g. Chesapeake Bay, USA).
Heavy metals – significant	Air emissions downwind from smelters and coal-fired power stations (Hg emission) have contaminated soils and waterways, impacting plants and aquatic species.	Heavy metals from smelters such as Sudbury, Ontario, have accumulated in soils downwind, killing plants in affected areas and removing vegetation leading to major erosion.
Energy generation – significant	Energy use unsustainable for poor; generation impacts on all four ecosystem services.	Lower crop yields, reduced water quality and fresh water volumes; increased pest problems, water purification, and waste treatment.
Pesticides – significant	Impacts on all four ecosystem services: pollination, natural pest control, nutrient cycling, and wild food supplies	Insecticides like fipronil, carbaryl, and cypermethrin used by farmers (e.g. in Laos, Cambodia, and Vietnam) impact on the supply of wild foods collected in or nearby the rice paddies (e.g. insects, frogs, crabs, fish, and snails).
		Widespread exposure of pollinating animals such as bees to multiple pesticides.
Nutrients – high	Global eutrophication of fresh and marine water from excessive nutrient inputs mainly from human activities.	Eutrophication and large anoxic *dead zones* in the shallow and stratified Baltic Sea are a serious problem for marine life and fisheries.

Source: Modified from Persson et al. (2010).

using the conceptual integrated pollution model or framework presented in Chapter 1 (Figure 1.3) as the basis of our approach. We examine key scientific, chemical, toxicological, and environmental principles that determine the behavior, exposure, and adverse or toxic effects of pollutants at individual, population, community, and ecosystem levels on local, regional, and global scales (Part 2), and apply these to major groups of pollutants (Part 3), followed by a critical review of major assessment and management methods and practices used to resolve environmental pollution problems (Part 4).

A concise Case Study 3.2 is given below, as an introduction or outline, to show stepwise how the main components of the integrated conceptual framework (Figure 1.3) can be used to evaluate and develop a basic pollution profile for a pollutant or group of pollutants. In this case, we are looking at an emerging chemical family of synthetic surfactants, known as perfluoroalkylated substances (PFAS) that are persistent organic pollutants like DDT. We are focused on two toxic chemicals of special concern in this family, referred to as PFOS and PFOA.

Note: At this stage, the overview is important and unfamiliar chemical or technical terms and details will be covered in following chapters of the book or readily obtained from literature searches.

Case Study 3.2

Integrated Pollution Profile of Perfluoroalkylated Substances (PFAS): PFOS and PFOA

The chemical structures of the two very persistent, bioaccumulative, and toxic PFAS compounds increasingly found in the environment, humans, and wildlife are shown in Figure 3.7.

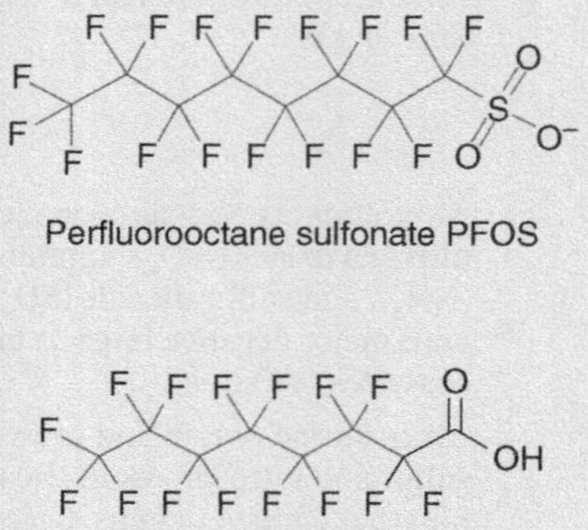

Figure 3.7 Chemical structures of perfluoroalkylated substances (PFAS): PFOS and PFOA.

Properties	• PFAS are synthetic fluorinated organic chemicals, widely used as surfactants; many compounds and derivatives: PFOS and PFOA are the main pollutants. • They are stable, have low surface tensions; most are insoluble in water, organic solvents, and lipids (PFOA soluble in water), and have low volatility. Hydrophobic, lipophobic but can be proteinphilic (e.g. PFAS in blood). • PFOS and PFOA are persistent, bioaccumulative, and toxic pollutants of air, water, soils, wildlife, and humans (see ATSDR 2019).
Sources	• Manufacturing, global uses, and disposal. Many diverse industrial and commercial products: surface coating and protectant products due to surfactant properties; manufacture of fluoropolymers (e.g. Teflon). • Industrial surfactants, emulsifiers, wetting agents, additives, coatings, waterproofing, fire-fighting foams, and nano-products.
Emissions	• Tens of thousands of tonnes of PFOS and PFOA are likely to have been produced from 1951 to 2004. In 2002, US production for PFOS (6–227 tonnes) and PFOA (227–454 tonnes) (ATSDR 2019); production phased out in the United States. China now main producer: 2008 production of PFOS estimated to be over 200 tons. • ~10% lost during production, mainly waste streams; unknown emissions from applications and disposal (e.g. fire-fighting foams).
Environmental transport and transformation	• Perfluoroalkyls are very stable and persist in air, water, soils, and sediments. • Air dispersion and deposition can lead to soil contamination and long-range transport. Mobile in water and can leach from soil into groundwater. • Resistant to photooxidation, photolysis, hydrolysis, and biodegradation; PFAS alcohols are volatile and can degrade into persistent PFOA.
Exposures	• PFOA and PFOS in outdoor air: 46–919 pg m^{-3} and indoor air, up to 28 μg m^{-3}. • Perfluoroalkyls vary greatly in waters, drinking waters, soils, sediments, indoor dusts, and foodstuffs (usually at low ppb or some ppm levels); elevated near industrial or contaminated sites); bioaccumulate in wildlife but not in fat; highest levels in top predators (blood, liver, kidneys, and eggs); half-lives of PFAS vary with carbon chain length and animal species. • Major exposure pathways for uptake in humans appear to be ingestion from food, water, dust, and hand to mouth contact, and inhalation of indoor air and dusts. • Perfluoroalkyls detected widely in human serum, breast milk, and cord blood. Accumulate mainly in blood, liver, and kidneys. Av. blood serum: 10–30 ng PFOS mL^{-1} (ppb), tending to decrease, but up to >100 ppb in elevated exposures. Long half-lives in blood serum of 3.8 years (PFOA) and 5.4 years (PFOS) (Jensen and Leffers 2008).

Case Study 3.2 (Continued)

Toxicities: individual and populations	• Moderately toxic. Oral animal toxicity studies report various reproductive and developmental effects, liver and kidney toxicity, suppressed immune effects, and cancer. • Genotoxic effects in human cell assays. • Human epidemiology data report associations between PFOS exposure and adverse effects on the thyroid, immune system, reproduction and development, and indications of cancer (bladder, colon, and prostate). PFOA exposure is associated with various developmental effects, liver and kidney toxicity, immune effects, and cancer (liver, testicular, and pancreatic). • Possibly carcinogenic to humans (2B) – IARC (Monograph 110) (IARC 2014).
Community and ecosystem responses	• Ubiquitous and persistent in marine and aquatic ecosystems. • Bioaccumulate, less so in fish, and potential to biomagnify in food webs to apex predators such as some marine mammals (e.g. in Arctic ecosystems). • Endocrine disruption chemicals of concern in wildlife (UNEP review).
Regional and global changes	• Volatile precursors of PFOS and PFOA subject to atmospheric transport; long-distance transport of PFOS and PFOA in oceans. • Many global regions contaminated with PFAS including remote areas; highest levels near urban and industrial sites. • Biomonitoring shows widespread levels in different human populations and higher levels in exposed workers or persons living near PFAS sites.
Assessment and management	• Listed as persistent organic pollutants under Stockholm Convention. • PFOS production in the United States and Europe phased out in 2001/2; reduction also in PFOA production and emissions. Production, however, increased in China. • Transition to short-chain replacements with low half-lives based on perfluorobutane sulfonamide and related alcohols. • Ongoing issues with carcinogenicity classification, emerging effects of replacement chemicals, including other PFAS compounds, continued persistence of PFAS (or PFCs) in environment and humans.

Source: Adapted from ATSDR (2019), Pubchem (2020a, b).

3.8 Key Points

1. The World Health Organization in 1948 originally defined health, as a "state of complete physical, mental, and social well-being, and not merely the absence of disease or infirmity." Being able to enjoy the highest attainable state of health was adopted as a fundamental human right, among other rights.
2. Environmental health addresses all the physical, chemical, and biological factors external to a person, and all the related factors impacting behaviors… WHO.
3. Concisely, environmental health refers to the interdependence between the health of individuals and communities and the health of the environment.
4. Disease is described as a dysfunction of the dynamic state of health. It is a disorder of a structure or function in an organism that affects part of or the whole organism. It can result from various causes such as an infection, genetic defect, environmental stress, or exposure to a toxic agent.
5. Four main types of disease are identified: infectious diseases, deficiency diseases, genetic diseases (hereditary and nonhereditary), and physiological diseases (communicable and noncommunicable).
6. Environmental diseases result from direct exposure to environmental factors (physical, chemical, biological, and genetic predisposition). They include all pollution-related diseases.
7. General theories or models of disease describe how exposures to determinants, risk factors, or possible causes of disease (e.g. a chemical or biological agent) are linked to a disease. Causal associations between exposure and disease usually involve a multitude of interacting factors or conditions, including behavioral, genetic, and environmental.
8. A key metric for the health of human populations and communities is the burden of disease, injuries, and risk factors that cause them.
9. The burden of disease (BoD) measures the impact of living with illness and injury and dying prematurely. The essential measure is described as "disability-adjusted life years" (or DALY) which measures the years of healthy life lost from disability and death. DALYs = YLL (years of life lost) + YLD (years lived with a disability or illness).
10. GBD estimates (WHO) for deaths due to environmental factors are almost a quarter of the total annual deaths. Diseases are the main component and low- and middle-income countries bear the greatest share of environmental disease.

11. The available GBD estimates due to selected chemical exposures (occupational and environmental) show that air pollution is a major factor in total deaths. Deaths from other chemical exposures appear to be underestimated because of inadequate or unavailable data for many toxic chemicals and their environmental exposures.
12. Case Study 3.1 is an example of GBD estimates for lead exposure and environmental intervention (mainly phasing out of lead in gasoline) for the period 2000–2004. GBD estimates are associated with substantially reduced blood levels in exposed human populations.
13. Biodiversity reflects the number and different types of plants and animals in a biological community and their ecological niches. It is usually considered at three levels of organization – genetic diversity, species diversity, and ecosystem diversity.
14. The concept of ecosystem services explains the benefits from ecosystem functions generated by natural cycles (biogeochemical) that are complex and vary greatly in time and spatial scales. Four major categories of ecosystem services are: provisioning, regulating, cultural, and supporting.
15. The major threats to biodiversity are habitat destruction and degradation, overexploitation of resources, hunting, fishing, etc., pollution (e.g. air and water pollution, persistent organic pollutants, and microplastics in oceans), diseases, invasions of introduced or pest species, and climate change due to global warming.
16. Many plant and animal species are threatened with extinction. In particular, mammals and birds face a high risk of extinction while one in three amphibians are threatened, especially tortoises and freshwater turtles.
17. Pollution, including effects of greenhouse gas emissions in climate change, has a major impact on global biodiversity and ecosystems, although the magnitude and rate of chronic impacts are often poorly defined.
18. The mode of action for many pollutant impacts follows a series of biological interactions that may be described as an *ecological cascade* involving tipping points where pressures or level of *forcings* reach a threshold sufficient to shift the state of the system.
19. Major threats to ecosystem services involve massive impacts from human activities on a global scale from climate change, deforestation, desertification, urbanization, wetland drainage, pollution, dams, and water diversion. Pollution impacts (e.g. ecotoxicity) are usually more diffuse and difficult to evaluate compared to the physical impacts of habitat destruction.
20. Important pollution principles, evaluation and management of the threats, and impacts of pollutants are examined in the following Parts 2, 3, and 4 of this book.

References

ATSDR (2019). Toxicological profile for perfluoroalkyls. In: *Agency for Toxic Substances and Disease Registry*. Washington, DC: US Department of Health and Human Services https://www.atsdr.cdc.gov/toxprofiles/tp.asp?id=1117&tid=237 (accessed 31 January 2020).

Australian Museum (2018). What is Biodiversity? https://australianmuseum.net.au/what-is-biodiversity/ (accessed 28 January 2020).

AWA (2006). *Australian Guidelines for Water Recycling: Managing Health and Environmental Risks (Phase 1)*. Australian Water Association https://www.awa.asn.au/Documents/water-recycling-guidelines-health-environmental-21.pdf (accessed 1 February 2020).

BMJ (2011). How should we define health? *British Medical Journal*. 343:d4163. http://www.bmj.com/content/343/bmj.d4163 (accessed 6 March 2017).

Brown, V.A., Grootjans, J., Ritchie, J. et al. (2005). *Sustainability and Health*. Crows Nest, Australia: Allen and Unwin.

Ceballos, G., Ehrlich, P.R., and Dirzo, R. (2017). Biological annihilation via the ongoing sixth mass extinction signaled by vertebrate population losses and declines. *PNAS* 114 (30): E6089–E6096.

Chivian, E. and Bernstein, A. (2008a). How is biodiversity threatened by human activity? In: *Sustaining Life: How Human Health Depends On Biodiversity*. (ed. E. Chivian and A. Bernstein). Center for Health and the Global Environment, Harvard Medical School, Oxford University Press. Oxford.

Chivian, E. and Bernstein, A. (ed.) (2008b). *Sustaining Life: How Human Health Depends on Biodiversity*. Oxford: Center for Health and the Global Environment, Harvard Medical School, Harvard Medical School, Oxford University Press.

Connell, D.W. and Miller, G.J. (1984). *Chemistry and Ecotoxicology of Pollution*. New York: Wiley.

De Vos, J.M., Joppa, L.N., Gittleman, J.L. et al. (2015). Estimating the normal background rate of species extinction. *Conservation Biology* 29 (2): 452–462.

GESAMP (2020). Joint Group of Experts on the Scientific Aspects of Marine Environmental Protection. http://www.gesamp.org (accessed 29 January 2020).

Global Burden of Disease Study (GBD) (2015). Risk Factors Collaborators. Global, regional, and national comparative risk assessment of 79 behavioral, environmental and occupational, and metabolic risks or clusters 64 of risks, 1990–2015: a systematic analysis for the Global Burden of Disease. *The Lancet* 388 (10053): 1659–1724.

Groffman, P.M., Baron, J.S., Blett, T. et al. (2006). Ecological thresholds: the key to successful environmental management or an important concept with no practical application. *Ecosystems* 9: 1–3.

Havelaar, A.H. and Melse, J.M. (2003). Quantifying public health risks in the WHO Guidelines for Drinking-Water Quality. A burden of disease approach. *RIVM Report 734301022/2003*. https://www.ircwash.org/sites/default/files/Havelaar-2003-Quantifying.pdf (accessed 1 February 2020).

IARC (2014). *IARC Monograph 110*. Lyon, France: International Agency for Cancer Research, United Nations, IARC.

IUCN (2019). *The IUCN Red List of Threatened Species. Version 2020-2*. International Union for Conservation of Nature http://www.iucnredlist.org (accessed 2 November 2020).

IUCN Red List (2007). *Species Extinction – The Facts*. International Union for Conservation of Nature (accessed 19 March 2017).

Jensen, A.A. and Leffers, H. (2008). Emerging endocrine disrupters: perfluoroalkylated substances. *International Journal of Andrology* 31 (20): 161–169.

Kilpatrick, A.M. and Altizer, S. (2010). Disease ecology. *Nature Education Knowledge* 3 (10): 55.

Landon, M. (2006). *Environment, Health and Sustainable Development*. Maidenhead: Open University Press.

Landrigan, P.J., Fuller, R., Acosta, N.J.R. et al. (2017). The Lancet Commission on Pollution and Health. *Lancet* 391: 10119.

Melillo, J. and Sala, O. (2008). Ecosystem services. In: *Sustaining Life: How Human Health Depends on Biodiversity* (ed. E. Chivian and A. Bernstein). Oxford, UK: Center for Health and the Global Environment, Harvard Medical School, Oxford University Press.

Murray, C.J.L. and Lopez, A.D. (2013). Measuring the global burden of disease. *New England Journal of Medicine* 369: 448–457.

Ostfeld, R.S. (2018). *Disease Ecology*. Oxford: Oxford Bibliographies, Oxford University Press https://www.oxfordbibliographies.com/view/document/obo-9780199830060/obo-9780199830060-0128.xml (accessed 26 January 2020).

Pereira, H.M., Leadley, P.W., Proença, V. et al. (2010). Scenarios for global biodiversity in the 21st century. *Science* 330: 1496–1501.

Persson, L., Arvidson, A., Lannerstad, M. et al. (2010). *Impacts of Pollution on Ecosystem Services for the Millennium Development Goals, SEI Project Report*. Stockholm: Stockholm Environment Institute.

Pimm, S.L., Alves, M.A.S., Chivian, E. et al. (2008). *What Is Biodiversity? In: Sustaining Life: How Human Health Depends on Biodiversity* (ed. E. Chivian and A. Bernstein). Oxford, UK: Center for Health and the Global Environment, Harvard Medical School, Oxford University Press.

Prüss-Üstün, A. and Corvalán, C. (2006). *Preventing disease through healthy environments. Towards an estimate of the environmental burden of disease*. World Health Organization: Geneva.

Prüss-Üstün, A. and Corvalán, C. (2007). How much disease burden can be prevented by environmental interventions? *Epidemiology* 18 (1): 167–178.

Prüss-Üstün, A., Vickers, C., Haefliger, P. et al. (2011). Knowns and unknowns on burden of disease due to chemicals: a systematic review. *Environmental Health* 10 (9): https://doi.org/10.1186/1476-069X-10-9.

Prüss-Üstün, A., Wolf, J., Corvalán, C.F. et al. (2016). *Preventing Disease Through Healthy Environments: A Global Assessment of the Burden of Disease from Environmental Risks*. Geneva: World Health Organization https://apps.who.int/iris/handle/10665/204585.

Pubchem (2020a). *Perfluorooctane Sulfonate-Pubchem*. US National Library of Medicine https://pubchem.ncbi.nlm.nih.gov/compound/74483 (accessed 29 January 2020).

Pubchem (2020b). *Perfluorooctanoic Acid-Pubchem*. US National Library of Medicine https://pubchem.ncbi.nlm.nih.gov/compound/9554 (accessed 29 January 2020).

Sanchez-Bayo, F. and Wyckhuys, A.G. (2019). Worldwide decline of the entomofauna: a review of its drivers. *Biological Conservation* 232: 8–27.

Secretariat of the Convention on Biological Diversity (2010). Global Biodiversity Outlook 3. Montréal, Canada. https://www.cbd.int/doc/publications/gbo/gbo3-final-en.pdf (accessed 14 July 2020).

Stone, D.B., Armstrong, R.W., Macrina, D.M. et al. (1996). *Introduction to Epidemiology*. Boston, MA: WCB McGraw-Hill.

Stöppler, M.C. (2020). *Medical Definition of Homeostasis*. MedicineNet https://www.medicinenet.com/script/main/art.asp?articlekey=88522 (accessed 1 February 2020).

UNEP (2019). *Global Environment Outlook GEO-6. Healthy Planet, Healthy People*. Cambridge, UK: United Nations Environment Programme, Cambridge University Press.

Volkman, J.K., Miller, G.J., Revill, A.T., and Connell, D.W. (1994). Environmental implications of offshore oil and gas development in Australia – oil spills. In: *Environmental Implications of Offshore Oil and Gas Development in*

Australia – the Findings of an Independent Scientific Review (ed. J.M. Swan, J.M. Neff and P.C. Young), 509–695. Sydney: Australian Petroleum Exploration Association.

WHO (2006). *Almost a Quarter of All Disease Caused by Environmental Exposure.* World Health Organization https://www.who.int/mediacentre/news/releases/2006/pr32/en/ (accessed 10 April 2019).

WHO (2017a). *Constitution.* Geneva: World Health Organization http://www.who.int/about/mission/en/ (accessed 6 March 2017).

WHO (2017b). *Metrics: Disability-Adjusted Life Year (DALY).* World Heallth Organization http://www.who.int/entity/healthinfo/global_burden_disease/metrics_daly/en/index.html (accessed 11 March 2017).

WHO (2019). *Lead Poisoning and Health.* World Health Organization https://www.who.int/health-topics/newsroom/fact-sheets/detail/lead-poisoning-and-health (accessed 29 January 2020).

WHO (2020a). *International Classification of Diseases. 11th Revision.* World Health Organization http://www.who.int/classifications/icd/en/ (accessed 26 January 2020).

WHO (2020b). *Metrics: Population Attributable Fraction (PAF).* World Health Organization https://www.who.int/healthinfo/global_burden_disease/metrics_paf/en/ (accessed 1 February 2020).

Wikipedia (2022). Environmental health. https://en.wikipedia.org/wiki/Environmental_health (accessed 23 February 2022).

Wilcox, B. and Jessop, H. (2010). Ecology and environmental health. In: *Environmental Health. From Local to Global*, 2e (ed. H. Frumpkin). New Jersey: Wiley.

Wilkinson, P. (ed.) (2006). *Environmental Epidemiology.* Maidenhead, England: Open University Press, McGraw-Hill.

4

Chemodynamics of Pollutants

4.1 Introduction

The basic diagram which describes **ecotoxicology**, as well as the human interaction with environmental toxicants, is illustrated as shown in Figure 4.1. This indicates how a chemical pollutant enters the environment in emissions and releases, and through other possible pathways. The physical state of the pollutant as a liquid, solid, or gas together with concurrent discharges, such as water and air emissions, influence the environmental fate of the chemical pollutant (Mackay and Patterson 1981, 1982; Connell 2005, pp. 17–48). The physical state and basic physicochemical properties of a chemical pollutant are determined by the inherent characteristics of the molecule such as molecular weight, surface area, and so on, as illustrated in Figure 4.1.

The physical state (solid, liquid, or gas) is controlled by inherent molecular characteristics including molecular weight, surface area, chemical structure, and so on. These molecular characteristics determine the physicochemical properties of the compound such as reactive groups, chemical bonds, polarity, and so on (Walker 2009, pp. 67–73). These characteristics then, in turn, govern important environmental physicochemical properties including aqueous solubility, octanol–water partition coefficient, vapor pressure, and so on. Thus, the first step in understanding the ecotoxicology and human toxicology of a chemical pollutant is to understand the inherent characteristics of the molecule and basic physicochemical properties of the compound (Connell 2005, pp. 17–48).

However, we also need to understand the properties of the environment itself and how the properties of the chemical interact with the properties of the environment. To develop an understanding of the properties of the environment, it is necessary to divide the environment into **compartments**. The environment can be seen to consist of different compartments which are sectors of the environment that are relatively homogenous and within which a chemical behaves in a uniform manner. Often these compartments are described as **phases** when considering partitioning processes. Commonly, the environment is seen to have several abiotic phases – an atmospheric phase (the air), an aquatic phase (the sea, rivers, lakes, etc.), and soils and sediments. In many situations, biota (humans, fish, invertebrates, etc.) can act as a phase whereas in other types of systems, the biota can act separately from this system. Some of these phases are complex and some are relatively simple, for example, the atmosphere is of relatively consistent composition globally whereas biota have complex interactions with chemicals. When a chemical enters the environment, it moves between the phases according to the properties of the chemical and the phase itself. It should be noted that biota, such as fish, invertebrates, humans, and so on, can be considered to be phases and take up the chemical according to the properties of the chemical and the biota involved. The environment involved is not necessarily the large external environment. It can be any environment including the often relatively small occupational environment.

Distribution within the phases in the environment leads to the presence of the chemical pollutant in various abiotic phases at different levels according to the properties of the chemical and the phase (Mackay et al. 2006; Walker 2009, pp. 67–98). Some phases may not have resident biota or distributed chemical and thus no biotic exposure results but distribution in other phases can result in biotic exposure. But to account for all the chemicals involved and correctly evaluate the biotic exposure, then all phases need to be taken into account. Nevertheless, it is the exposure of biota which is of prime concern in toxicology.

4.2 Some Fundamental Properties of Chemical Pollutants and Environmental Phases

Figure 4.1 traces out conceptually the fate of a chemical in the environment and indicates that the molecular

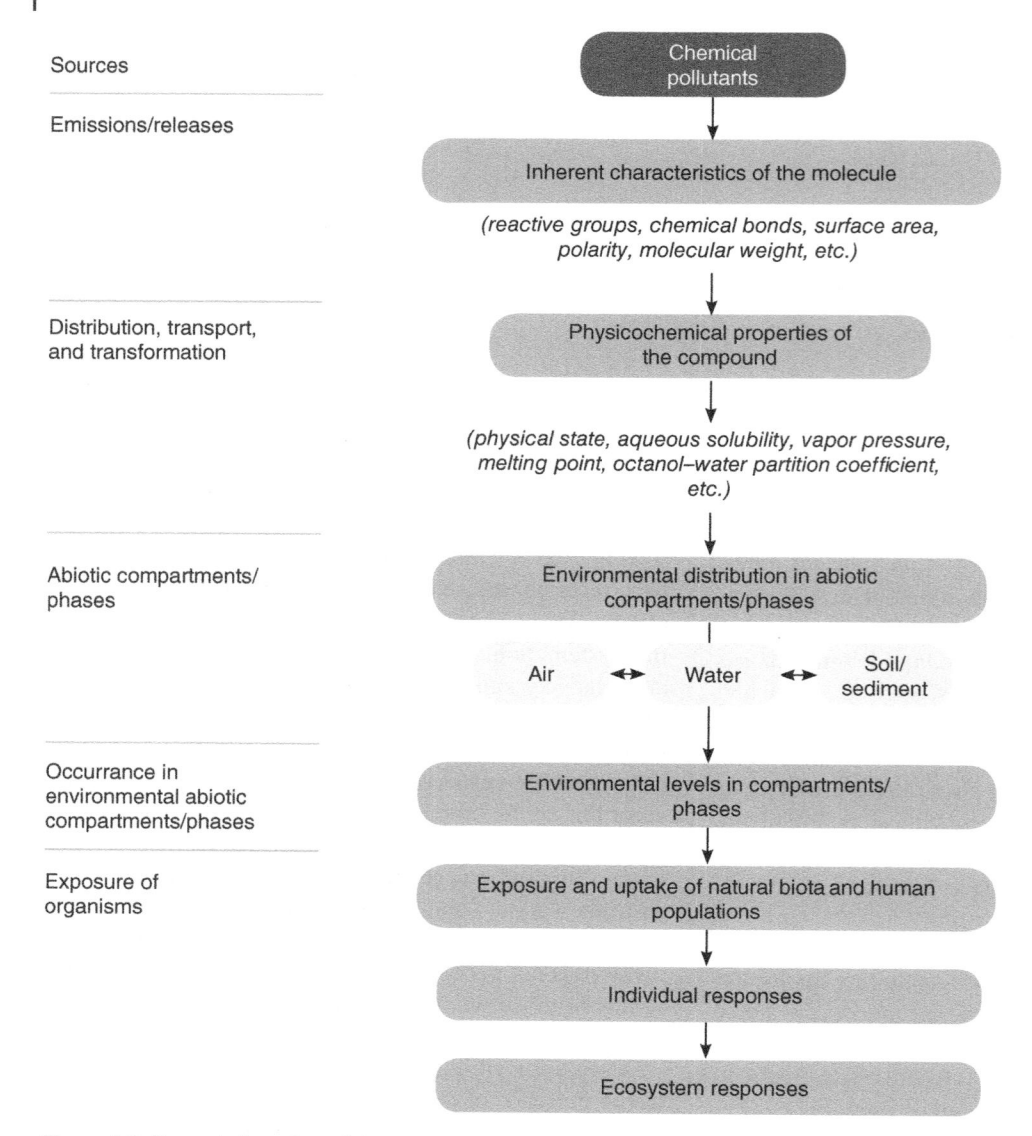

Figure 4.1 Expanded section of the general Figure 1.3 showing the initial interactions of a chemical pollutant with the environment where distribution of the pollutant between phases occurs as a result of its molecular characteristics and molecular properties and exposure of individuals leading to effects on other components of the ecosystem.

characteristics and physicochemical properties are a powerful influence on these processes. Figure 4.1 also illustrates the interdependencies of the fundamental properties of the molecule and physicochemical properties of pollutants and how these properties determine the nature of the environmental response from environmental distribution to biological effects (Connell 2005, pp. 17–48). However, metals and related compounds have a somewhat different pattern of behavior as described later in this chapter.

The values of these fundamental characteristics of the molecule itself and the related actual compound are the starting point for the evaluation of the effects on the natural ecosystem and human health. The characteristics of the molecule govern the initial physical state of the substance – whether it is a liquid, solid, or gas, and subsequent behavior and effects in the environment. Various physicochemical properties and molecular characteristics are listed in Table 4.1. Some of these characteristics are measured in the laboratory and some are calculated or evaluated by other methods. Generally, most of the physicochemical characteristics can be measured in the laboratory and where this is not possible, they are usually calculated. The measurement procedures, for example melting point and boiling point, can be seen as a quantitative controlled and idealized imitation of the corresponding

Table 4.1 Some important environmental properties of molecules and compounds.

Property	Symbol	Units	Measurement technique
Characteristics of the molecule			
Total molecular surface area	$Å^2$	Square Ångstroms	Calculation
Molecular weight	u or Da	Dalton	Laboratory measurement
Polarity	μ	Debye	Calculation
Chemical structure	Structure	None	Laboratory measurement
Physicochemical properties of the compound			
Melting point	°C or K	Degrees centigrade or Kelvins	Laboratory measurement
Aqueous solubility	Mass/volume	mg L^{-1}, M/L, etc.	Laboratory measurement
Persistence/half-life	$t_{1/2}$	Units of time	Laboratory measurement
Octanol–water partition coefficient	P or K_{OW}	Usually reported as unitless	Laboratory measurement

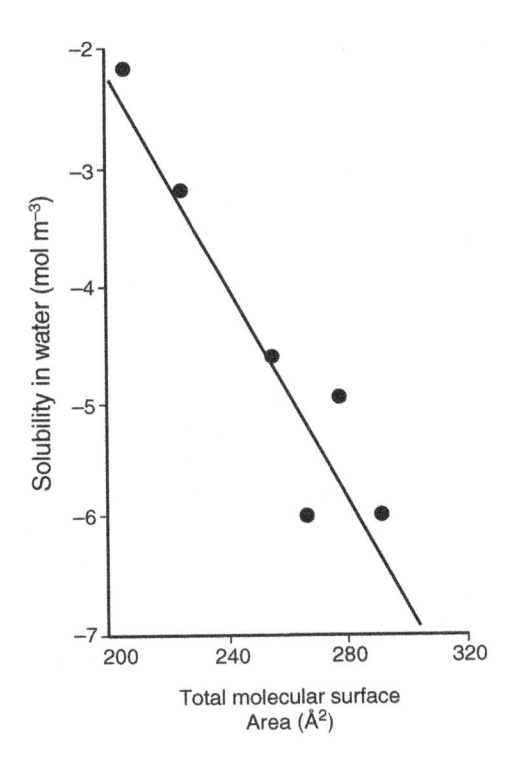

Figure 4.2 Relationship between total surface area of the molecule and aqueous solubility for polychlorinated biphenyls (PCBs). Source: Connell (2005).

environmental processes on a small scale in the laboratory. A chemical in the environment can be subject to melting and boiling processes as part of their initial behavior on release to the environment.

The molecular surface area (MSA) is a fundamental molecular characteristic that has a major influence on the aqueous solubility of nonpolar organic compounds, which include a wide range of environmentally important compounds including DDT, polychlorinated biphenyls (PCBs), and chlorobenzenes. This is illustrated by the plot in Figure 4.2 which demonstrates the relationship of total molecular surface area (TMSA) to aqueous solubility with PCBs.

Another important environmental characteristic of organic compounds is molecular polarity. The polarity of a compound relates to its chemical structure and results from a small (δ^+ and δ^-) charge on the ends of bonds and atomic groups within organic compounds. This occurs because the electron density in a bond is modified as a result of the differing electronegativity of the atoms forming the bond. The polarity of a compound has a profound impact on its solubility characteristics in solvents of differing polarity (Connell 2005, pp. 17–48). The solvents of major importance in investigations of environmental behavior are water, which is polar, and biotic lipid, which is nonpolar. For example, the polar compounds, such as the insecticides, imidacloprid, and cypermethrin, are soluble in water, while nonpolar compounds, such as DDT and polyaromatic hydrocarbons (PAHs), are soluble in biota lipid. Conversely, these chemicals are insoluble in the solvent of opposite polarity.

The presence of a chemical pollutant in an environmental phase is dependent on the properties of the pollutant as well as the phase itself. Some common environmental phases are shown in Figure 4.1, where air, water, and soil/sediment are categorized as environmental phases. There may be other phases depending on the particular environment being considered. The properties of each phase are critical factors influencing the presence of a pollutant in a specific phase. For example, the presence in air is influenced by the boiling point and vapor pressure and the presence in water is influenced by the aqueous solubility. The physical state of a phase, i.e. solid, liquid, or gas, has an impact on the uptake of pollutants. Phases such as air and water are fluids which have a prime effect on the kinetics of environmental distribution, with fluids having faster kinetics than solids. Other physical properties of significance with kinetics are permeability and porosity with soils. Chemical properties of soils, such as organic matter and carbon content, and lipid content with biota, are important characteristics controlling the uptake of chemical pollutants.

A group of pollutants which has many of the characteristics of the pollutant chemicals described above are the solid pollutants. These pollutants occur as suspended solids in aquatic areas, particulates in the atmosphere, soot from burning, particulate matter in foods, and so on. Their chemical nature is quite variable ranging from minerals and metals in soil, organic matter essentially from plant decay in the environment, to nanoparticles of synthetic plastics in the environment. These polluting substances have biological properties which relate to sorbed substances on surfaces and internally, or relate to the shape and size of the particles. With atmospheric pollutants, this is considered in more detail in Chapter 13.

4.3 Fundamental Principles of Partition Behavior

The behavior and effects of environmental pollutants are related to their dynamics in the four major non-biotic compartments or phases which constitute the Earth's ecosphere, that is, air, water, and soil/sediments (see Figure 4.1). Thus, the properties of chemicals and the environmental phases, as outlined above, and the mechanisms that control their forms and distribution within and between environmental compartments are of major importance (Mackay et al. 2006). Aspects of particular interest are those that control the rates of chemical and energy transfer within environmental compartments and across the major interfaces between compartments.

Environmental Interfaces and Chemical Equilibria

An environmental interface can be described as the junction where two different compartments meet and interact across a common boundary. Spontaneous transfer of chemical and thermal energy occurs across interfaces between compartments or phases until net movement of chemicals and thermal energy ceases at equilibrium. However, chemical species continue to cross the interface but the rates of transfer between the two phases are equal.

If we assume a chemical equilibrium results from reversible transfers between environmental phases, thermodynamic considerations require that the chemical potentials (μ) or fugacities ($\underline{f}°$) of a component, A, in both phases of a multicomponent system be equal for constant temperature and pressure. Thus, for component A, between two phases, air (a) and water (w):

$$f°_{A\text{ air}} = f°_{A\text{ water}} \quad (4.1)$$

In each phase, the fugacity of component, A, can be related to the activity coefficient (γ); mole fractions X or Y; concentration, C; phase molar volume V_A; and the reference fugacity f° as:

$$f°_A = X_A \gamma_A f°_A = C_A V_A \gamma_A f°_A \quad (4.2)$$

A more useful form of (4.1) can be derived from (4.2) as

$$[X_A \gamma_A f°_A]_a = [C_A V_A \gamma_A f°_A]_w \quad (4.3)$$

Thus, the distribution between phases depends on solubility in each phase which is independent of the quantity of chemical but dependent on temperature and pressure.

The Partition Coefficient

The equilibrium distribution of a chemical substance between two immiscible phases occurs as a fixed proportion, the **Partition Coefficient** (**K**) where K is the Concentration in Phase 1/Concentration in Phase 2, thus $K = C_1/C_2$ or in the case of air and water, C_a/C_w. Thus, the Partition Coefficient is independent of concentration and measured at a constant temperature and pressure (Connell 2005, pp. 38, 345).

There is an interface at the junction of the two phases in the system where transfer of mass and heat occurs. The transfer of mass and heat through environmental interfaces can be defined in terms of equilibrium and nonequilibrium concepts. Profiles of concentration and temperature in the region near the interface occur but equilibrium, both chemical and thermal, is assumed to exist at the interface itself. Thus, the chemical concentration profile is usually discontinuous at the interface since concentrations in the phases are usually different from that at equilibrium.

The bulk phases, that is, the regions far from the interface, are not at thermal or chemical equilibrium so interface movement of heat and mass spontaneously proceeds. The mass of material associated with the interface regions where temperature and concentration gradients exist is a diminishingly small fraction of the total mass, and, by definition, the equilibrium interface itself is assumed to be a hypothetical physical region two molecules thick, one monomolecular layer in each phase. In natural systems, although not usually truly reversible, it is realistic to assume an equilibrium between compartments in appropriate situations. This provides a basis for the theoretical evaluation of transport and transformation of chemicals and energy in the environment.

Influence of Solubility on Phase Equilibria

Solubility is the extent to which a chemical substance mixes with a liquid to form a homogeneous system. Solubility in water is a basic intrinsic property of a pollutant and is a major determinant of the transport of the substance in the aquatic environment. Solubility equilibria for pure chemicals in water can be represented by equal fugacities in both participating phases, water and another phase, given by Eq. (4.4). Thus, for a chemical pollutant A,

$$Y_A \gamma_{Aa} f^\circ_{Aa} = X_A \gamma_{Aw} f^\circ_{Aw} \qquad (4.4)$$

For a pure gas, $\gamma_A = 1$ and $\gamma_{Aa} = 1$ (ideal gas behavior assumed at 1 atm),

$f^\circ_{Aa} = 1$, and Eq. (4.4) becomes

$$1 = X_A \gamma_{Aw} f^\circ_{Aw} \qquad (4.5)$$

where γ_{Aw} is the activity coefficient of the chemical A in water, f°_{Aw} is the pure component fugacity of A in water, and X_A is the mole fraction solubility of A in water.

For a pure liquid, Eq. (4.5) applies also. For this system, since f°_A is approximately unity, then the solubility of a pure liquid can be represented by

$$X_A = \frac{1}{\gamma_{Aw}}$$

For a pure solid (s), Eq. (4.5) becomes:

$$f^\circ_{As} = X_A \gamma_{Aw} f^\circ_{Aw}$$

where f°_{As} is the fugacity of pure solid chemical A when a solid chemical is in equilibrium with an aqueous solution of concentration A, f°_{As} is estimated as the vapor pressure of the solid. With chemical pollutants of low solubility, it can be further shown that components for a pure chemical A (solid or liquid) in equilibrium with water give:

$$P^\circ_A = X^*_A \gamma_{Aw} f^\circ_{Aw} \qquad (4.6)$$

where P°_A is the vapor pressure of the pure solid or liquid and X^*_A is the solubility in water at equilibrium.

4.4 Relationships Describing Partition Behavior

The partition coefficient (K) at different concentrations has a constant value ($K = C_1/C_2$) but while the partition coefficient is constant, it is measured at different levels giving different values of C_1 and the corresponding C_2. The sequence of results of concentration in one phase and the corresponding concentration in the second phase can be plotted as a relationship usually described as an *isotherm*.

Two widely used relationships for partition processes giving isotherms, at equilibrium, are set out below.

1. *Langmuir Isotherm.* This isotherm is applicable to the adsorption of gases on solids, but not for the absorption of chemicals from solution, particularly with heterogeneous adsorbents such as soils (Hanaor et al. 2014). It is a plot of the chemical at different concentrations usually resulting from a set of experiments in the laboratory. The concentration of solute adsorbed per gram of solid is expressed as a function of the equilibrium concentration of solute in the gas according to the following equation:

$$X = \frac{(X_M \cdot b \cdot C)}{(1 + b \cdot C)} \qquad (4.7)$$

where X_M is the number of moles of solute adsorbed per gram of chemical; C, the equilibrium concentration of the chemical in the gas; b, a constant related to energy of adsorption or carbon content with sediment or soil.

2. *Freundlich Isotherm.* This is an empirical relationship (see Figure 4.3) and is expressed as follows:

$$C_1 = K C_2^{1/n} \qquad (4.8)$$

where C_1, the equilibrium concentration of the chemical in the liquid; C_2, equilibrium concentration of the chemical in the solid or second liquid; K, the equilibrium constant and $1/n$ is a constant describing the degree of nonlinearity.

A linear form of this relationship is often used in the analysis of experimental data. This is obtained by taking logarithms of both sides of Eq. (4.8). Thus,

$$\log C_1 = \log K + \left(\frac{1}{n}\right) \log C_2 \qquad (4.9)$$

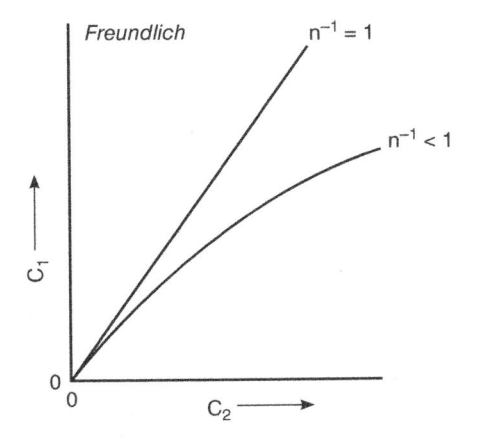

Figure 4.3 Plot of the Freundlich Isotherms which does not approach any limiting value and is linear when $1/n = 1$.

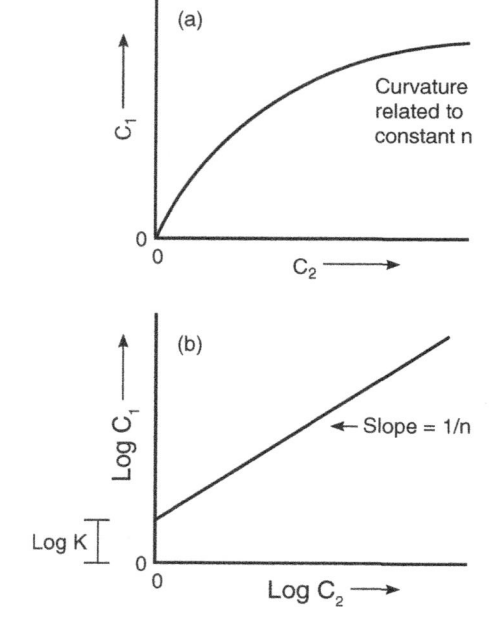

Figure 4.4 Hypothetical plot of the concentration in one phase (C_1) against the concentration in the other phase (C_2) at equilibrium (Plot A) and using the log form of the Freundlich Isotherm, Eq. (4.9) (Plot B).

A hypothetical plot of data using this equation is shown in Figure 4.4 where the intercept in the y axis at zero is the Partition Coefficient (K) of the partition between phase 1 and phase 2 and the slope is the 1/n value for nonlinearity in Eq. (4.9).

For many nonionic organic compounds in solution, the slope of the Freundlich Isotherm (1/n) approaches unity. Equation (4.8) then becomes:

$$C_1 = KC_2 \quad \text{or} \quad K = \frac{C_1}{C_2} \quad (4.10)$$

where K is the sorption or partition or distribution coefficient.

4.5 Partitioning Behavior in the Laboratory

Important environmental systems can be imitated in miniature in the laboratory and the partition coefficient measured. For example, the sediment and water in the natural environment can be representatively sampled and the samples taken to the laboratory and the partition coefficient measured. In some cases, this can be simply done by the **shake flask** method where the sediment is placed in a flask, water is added, and the system is shaken to equilibrium. Equilibrium takes a variable time to establish, but in some instances can take several weeks or more. However, there are many methods to measure or calculate partition coefficients including HPLC and electrochemical-based techniques as well as methods which involve calculation from basic properties.

Aqueous solubility is an important environmental property which can be estimated from the MSA with some nonpolar organic compounds. The MSA has a major influence on aqueous solubility as illustrated in Figure 4.2. The MSA has a direct effect on the aqueous solubility as a result of the polar properties of water and the nonpolar properties of some important environmental chemicals such as DDT, chlorobenzenes, and dioxins. An example of the relationship between MSA and aqueous solubility is shown in Figure 4.3 (Connell 2005, pp. 17–48). This relationship can be used to estimate the aqueous solubility of PCBs where this characteristic is unknown.

n-Octanol–Water Partition Coefficient

Probably the most widely used property in environmental chemistry and pharmaceutical sciences to predict environmental behavior and effects is the n-octanol–water partition coefficient usually represented by the symbols K_{ow}, P, or P_{ow} (OECD 2006). It is a measure of the **lipophilicity** of a chemical which evaluates the potential of a chemical to accumulate in lipids. It is the partition coefficient between n-octanol and water where octanol represents the lipid and water is the other phase. In biota, the chemicals with a high K_{ow} value would be expected to accumulate in the fatty tissues, and chemicals with a low K_{ow} would accumulate in the water-based tissues. Alternative descriptive terms can be used including *hydrophilicity* which designates water-loving chemicals and *hydrophobicity*, as an alternative to lipophilicity, designates water-hating chemicals. Table 4.2 contains some values for representative compounds. Since the values for K_{ow} can extend over a wide range, these values are expressed on a log basis. Lipophilic compounds are usually taken to have log K_{ow} from 2 to 6 and hydrophilic compounds are less than 2. Thus, ethanol, 2,4,5-T, and 2,4-D would be classified as hydrophilic, and benzene, 1,4-dichlorobenzene, and DDT would be classified as lipophilic.

There are many methods for the prediction of the K_{ow} value of compounds where this is unknown. These fall into a number of categories as mentioned above. One of these methods makes use of the basic relationship between properties and characteristics as outlined

Figure 4.5 Plot between the total molecular surface area (TMSA) and log K_{ow} for the chlorobenzenes with extrapolation and interpolation to predict the log K_{ow} of tetrachlorobenzene and hexachlorobenzene. Source: Connell 2005.

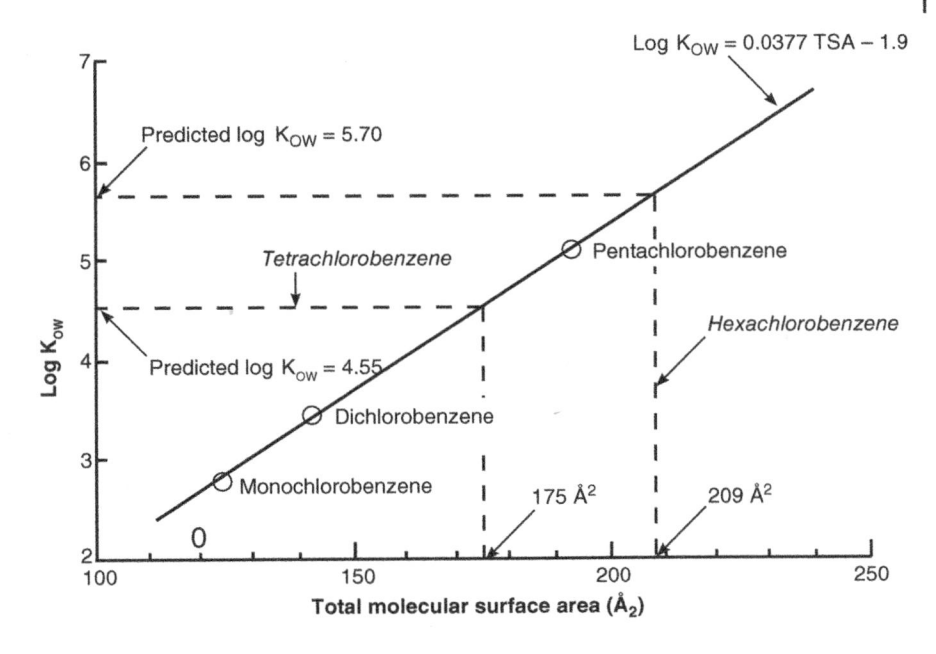

Table 4.2 Some octanol–water partition coefficient values for various compounds.

Compound	K_{ow}	Log K_{ow}
Ethanol	0.49	−0.31
2,4,5-T	3.98	0.60
2,4-D	37.00	1.57
Benzene	135.0	2.13
1,4-Dichlorobenzene	3310.00	3.52
DDT	2.39×10^6	6.36

Source: Connell (2005).

in Figures 4.1 and 4.2 where MSA is a major influence on the octanol–water partition coefficient (log K_{ow} value). A plot of a linear relationship is shown in Figure 4.5 of the MSA against the log K_{ow} value for the chlorobenzenes. In addition, the use of this plot to predict the log K_{ow} values for hexachlorobenzene and tetrachlorobenzene is illustrated.

4.6 Pollutant Behavior in Different Environmental Compartments

When a chemical pollutant is discharged to the environment, it distributes into the various phases present according to the properties of the chemical and the properties of the environmental phases. This is illustrated in Figure 4.1 where chemicals that are partitioned into the abiotic phases of air, water, soil/sediments, and biota are exposed as a result. However, the partitioning behavior of chemicals in some of the various phases in the environment has been investigated separately, usually in the laboratory. Some of these processes are of major importance, particularly those that result in elevated levels of environmental contaminants in biota. Some of the various environmental partition processes which have been investigated are outlined below.

Air–Water Partitioning – Henry's Law

Behavior of chemicals in the water–atmosphere system is historically one of the first partition processes ever investigated and is described as Henry's Law. It was formulated in 1803 by English scientist William Henry in interpreting his experiments on gases in water. It states that the amount of dissolved gas is proportional to its partial pressure in the gas phase with the proportionality factor being called the Henry's Law constant (H). Thus,

$$H = \frac{P}{C_W} \quad (4.11)$$

where P is the partial pressure of the compound in air and C_W, the corresponding concentration in water at equilibrium. There are many different units for the expression of the values of H which depend on the specific situation being considered.

Many books have been published by Boethling and Mackay (2000) and Sander (2015) on a range of important environmental properties and methods to calculate values for them. Henry's Law has been applied in many situations and it is valuable for the investigation of environmental distributions involving water in rivers, oceans, dams, and so on.

Soil and Sediment Partitioning with Water

As previously mentioned, the partition coefficient between sediment or soil and water (K_D) can be successfully investigated in the laboratory (OECD 2001; Gouin et al. 2014). Sediment/soil from the environment and water can be brought together in an experimental system in the laboratory and the concentration of chemical in the two phases measured at equilibrium. Thus,

$$K_D = \text{concentration in whole solid phase},$$

$$\frac{C_S}{\text{concentration in water } C_W},$$

$$\text{thus } K_D = \frac{C_S}{C_W} \text{ at equilibrium}$$

The soil, sediment, and solid particulates usually contain organic matter originating from the decomposition of plant and animal detritus. However, in some situations, the organic matter can be a very low proportion of the solid material present as with beach sand but usually ranges up to about 6% or more. Partitioning occurs between the soil organic matter (SOM) in the solid phase and water. The other solid-phase components, such as silica sand, clay, and calcium carbonate, are not considered to play a significant role in this process with lipophilic compounds. Thus, at equilibrium, the partition coefficient K_{OM} between organic matter and soil can be given as:

$$K_{OM} = \frac{C_{SOM}}{C_W} = \text{constant}$$

where C_{SOM} is the concentration in the solid organic matter in terms of organic matter. If f_{OM} is the fraction of organic matter in the solid, then:

$$C_S(\text{whole solid phase}) = f_{OM}.C_{SOM}$$

$$\text{and } K_{OM} = \frac{C_{SOM}}{C_W} = \frac{C_S}{(C_W f_{OM})}$$

$$\text{and } K_D = \frac{C_S}{C_W} = \frac{(f_{OM} C_{SOM})}{C_W} \quad (4.12)$$

If octanol is a good surrogate for organic matter in the partitioning process, but not as good as it is with the biota lipids, then C_{SOM} is equal to C_O (concentration in octanol in an equivalent octanol/water system) multiplied by constant, x, less than unity, a proportionality factor related to the efficiency of octanol as a surrogate for the organic matter in the solid phase. Thus,

$$K_D = \frac{(x f_{OM} C_O)}{C_W} = x f_{OM}.K_{OW}$$

$$\text{and } K_{OM} = x.K_{OW} \quad (4.13)$$

The constant, x, would be unity if octanol was a perfect surrogate but it generally has a value of about 0.66 with soils, although other values are used with specific sets of compounds and specific solid phases. Also, the assumption that octanol is a perfect surrogate for organic matter when the proportionality factor, x, is used, is not fully accurate and a power coefficient, a, is used in addition with K_{OW} to allow for this. The value of the power coefficient, a, is usually about unity (Connell 2005, pp. 323–344).

Thus, with soils,

$$K_D = 0.66.f_{OM}.K_{OW}^a \quad (4.14)$$

There is a lack of precision in the laboratory analyses for the value of the content of the organic matter in soils, f_{OC}, and the more precise value of organic carbon, f_{OC}, is often measured. This can be substituted for f_{OM} in the expressions above but constants x and a have different values. The logarithmic form of Eq. (4.14) is usually used to evaluate soil and sediment partitioning with water. Thus,

$$\log K_D = \log 0.66.f_{OM} + a \log K_{OW} \quad (4.15)$$

This linear relationship has a slope, a, with the intercept on the y axis having a value of log 0.66. f_{OM} as shown in Figure 4.6.

Many measurements have been made of pollutant chemicals in the sediment–water system at equilibrium and have been used in evaluating the distribution behavior in the environment.

Uptake and Retention of Chemicals by Organisms

A wide variety of terms have been used in an inconsistent and confusing manner to describe organism uptake and retention of pollutants by different paths and mechanisms. Terms often used are: bioamplification, bioaccumulation along food chains, cumulative transfer, trophic contamination, biological magnification, and trophic magnification. However, the following three terms can be applied to all of these processes according to the processes involved with the following definitions, which are now widely accepted:

- **Bioaccumulation** is the uptake and retention of pollutants from the environment by organisms via any mechanism or pathway.
- **Bioconcentration** is the uptake and retention of pollutants directly from the water mass by organisms through such tissues as the gills or epithelial tissues.

The n-octanol–water partition coefficient (K_{OW}) is a major predictor of bioconcentration (OECD 2012). This is particularly due to the significance of octanol which is a surrogate for biotic lipids. High K_{OW} compounds (log K_{OW} from approximately 2–3 and up to about 6–7) tend to partition strongly into these biotic lipid-rich environmental phases. Above and below these values, bioconcentration does not usually occur according to the same mechanisms. The lipophilic chemicals under consideration should be resistant to biodegradation, otherwise true equilibrium cannot be established. These characteristics fit many pollutant chemicals of environmental importance. With these compounds, the value of log K_{OW} can be used to estimate the partition coefficients for the partition processes involved.

The biota/water partition coefficient, K_B, is the bioconcentration factor at equilibrium. It is measured by placing fish, or other aquatic biota, into aquaria and measuring the concentration of the chemical in the biota after exposure to a fixed concentration in the water when an equilibrium is established. Thus,

$$K_B = \frac{\text{concentration of chemical in biota } (C_B)}{\text{concentration in water } (C_W)} = \text{constant} \tag{4.16}$$

Experimental values of K_B are known for many chemicals and species of aquatic biota. However, an approximate theoretical relationship between K_B, K_{OW}, and the lipid fraction of the biota can be derived to estimate K_B from K_{OW} as set out below.

With lipophilic compounds, the nonlipoid phases in the aquatic organism can be considered not to play any significant role in the partition process and only the biota lipids participate. After partitioning between biota lipid and water has occurred through the gills of the organism in the aquaria, and equilibrium is established, then:

$$\frac{C_L}{C_W} = \text{constant} = K_{BL}$$

where C_L is the concentration of the chemical in the biota lipid and K_{BL} is the bioconcentration factor in lipid terms. If f_{lipid} is the fraction of lipid present in the biota, then:

$$C_B(\text{whole biota weight}) = f_{lipid} C_L$$

$$\text{and } K_B = \frac{C_B}{C_W} = \frac{(f_{lipid} C_L)}{C_W}$$

If octanol is a perfect surrogate for biota lipid, then C_L is equal to C_O and

$$K_B = \frac{(f_{lipid} C_L)}{C_W} = \frac{(f_{lipid} C_O)}{C_W} = f_{lipid} K_{OW}$$

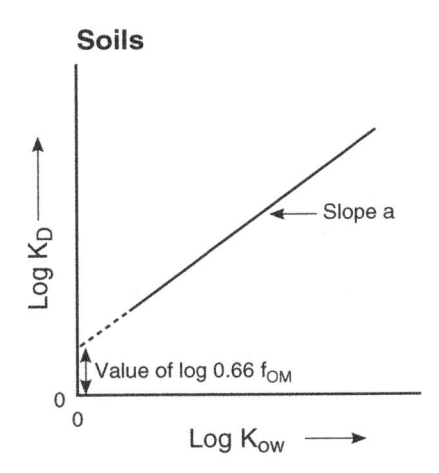

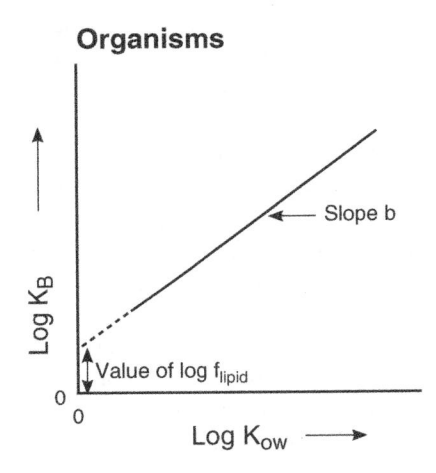

$$\text{so } K_B = f_{lipid} K_{OW}{}^b \tag{4.17}$$

Figure 4.6 Hypothetical plot of log K_D and log K_B against log K_{OW} for persistent lipophilic compounds with soils (Eq. 4.15) and organisms (Eq. 4.18) where the values of the intercept on the vertical axis are log 0.66 f_{OM} for soils and log f_{lipid} for organisms, and the slopes have values of a and b, respectively.

- **Biomagnification** is the process whereby pollutants are passed from one trophic level to another and exhibit increasing concentrations in organisms related to their trophic status.

Bioconcentration Processes

The partitioning process occurs between fully aquatic organisms and waterborne chemicals through the respiratory pathway whereby oxygen is taken up by organisms. Thus, the gills and other oxygen permeable surfaces are the entry surfaces and from there the chemicals are distributed throughout the organism by circulatory fluids. In addition, this is a return pathway resulting in the loss of chemicals allowing an equilibrium to be established. Aquatic mammals, such as whales, dolphins, and so on, do not take up chemicals from water by this mechanism.

where the constant, b, is introduced to account for differences between the biota lipids and octanol. Thus, taking logarithms,

$$\log K_B = \log f_{lipid} + b \log K_{OW} \quad (4.18)$$

This relationship has the slope of the linear relationship as b, and the intercept on the vertical axis at a value of log f_{lipid} as shown in Figure 4.6.

The power coefficient, b, is used to better express the relationship between biota lipids and octanol. The value of coefficient, b, would be unity if octanol was a perfect surrogate for the biota lipid and it often approaches unity since octanol is quite a good surrogate for lipid in most situations.

Often the lipid fraction, f_{lipid}, for aquatic biota is about 0.05 or 5% and the coefficient, b, is unity. Then, taking logarithms,

$$\log K_B = 1 \cdot \log K_{OW} - 1.30 \quad (4.19)$$

Relationships having these approximate values are found for fish and other aquatic organisms. It is important to remember that these relationships apply at equilibrium when there are constant concentrations present in the biota. But aquatic and other biota have a capacity to degrade the pollutant chemicals which leads to a declining chemical concentration. If this is sufficiently rapid, it will interfere with the relationships expected.

Biomagnification

The uptake and retention of persistent lipophilic pollutants from food is described as **biomagnification**. Connell (1990, p. 145) has described the significance of this mechanism in natural ecosystems. The mechanism of uptake effectively deals with two systems – aquatic ecosystems with **bioconcentration** and terrestrial ecosystems with **biomagnification**. With aquatic systems, the occurrence of chemicals in an aquatic organism, no matter what the route of uptake is, must be governed by equilibration with the ambient water through the respiratory pathways. On the other hand, terrestrial organisms do not have an ambient water phase as a source of chemicals and thus have no requirement for the chemicals in the organism to equilibrate with the water phase. The biomagnification mechanism has been used to explain the high concentrations of some pollutants, including DDT, in top carnivores, such as eagles. Eagles, and other carnivorous birds, are members of terrestrial ecosystems and not participants in the partitioning processes involved in aquatic ecosystems, but at the same time they utilize food derived from the aquatic systems such as fish.

The suggested mechanism involves uptake of persistent pollutants in the food but, on metabolism of the food, the pollutant, being resistant to degradation, is preferentially retained. This process results in a loss of energy and food matter as a result of biodegradation by respiration processes and consequent concentration of the pollutant. The energy content at some trophic levels is approximately 90% less than the level below. Thus, an approximately tenfold concentration of a persistent chemical could be expected for each trophic level through which a chemical is transferred in this system. This could lead to a sequential increase in concentration in a food chain, with organisms at the highest trophic levels exhibiting the highest concentrations.

The overall significance of biomagnification for a persistent pollutant in natural ecosystems is somewhat uncertain, particularly with birds, marine mammals, and other organisms which are not directly exposed to the pollutant in the aquatic system. These organisms lack a direct respiratory interface with the aquatic phase in an ecosystem and so there is no direct uptake of persistent chemicals or equilibrium processes. Another difficulty with terrestrial ecosystems is that the structure of the ecosystem is often uncertain and there are variable relationships between predator and prey which could lead to uncertain flows of persistent chemicals in relation to trophic structure. The evidence suggests that biomagnification occurs in an unpredictable manner but can lead to elevated concentrations of persistent chemicals in top carnivores. In addition, carnivorous birds often exhibit evidence that pollutants present in muscle tissue are due to biomagnification.

In fully aquatic systems, including such organisms as fish and aquatic invertebrates, direct uptake from water appears to be the dominant process for persistent pollutants according to the partition processes previously described. It follows that the levels present in organisms can be quantitatively related to partition processes. Even if the chemical has been taken up in food, the levels in all organisms (including food organisms) would be controlled by partition processes. The retention of pollutants by different organisms depends more on partition processes influenced by the properties of the chemicals involved rather than differing rates of metabolism and excretion of pollutant compared with food and position in the food web.

4.7 Models for Bioconcentration

Only a few theoretical models attempt to explain uptake and elimination processes for persistent pollutants in organisms. Of these, the compartmental model, as applied in pharmacokinetics, is a useful approach (Moriarty 1975). In this model, a compartment is defined as a phase in the environment or an organism that has uniform kinetics of transformation and transport, and whose kinetics are

Figure 4.7 Conceptual compartment models for the bioconcentration process with (a) representing a single compartment model and (b) representing a two-compartment model.

discrete from other compartments. The model can be based on one (single) or two compartments whichever fits the bioconcentration characteristics best (see Figure 4.7). However, the single compartment model is most commonly used, as it is usually the simplest and most widely applicable, and is outlined below.

Single Compartment Model

The bioconcentration process can be seen as a balance between two kinetic processes, uptake and depuration (see Figures 4.7a and 4.8) as characterized by the first-order rate constants, k_1 and k_2, respectively.

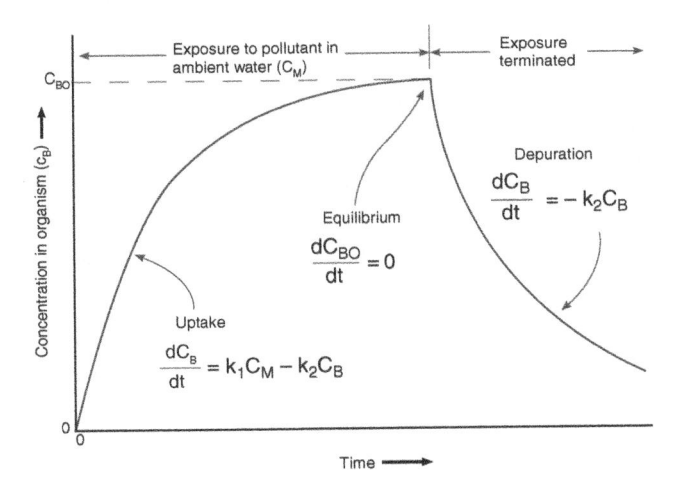

Figure 4.8 Changes in biotic concentration (C_B) over time with the uptake and depuration processes operating in an aquatic organism.

The rate of change of chemical pollutant in an organism using a single compartment model (Figure 4.7a) is given by

$$\frac{dC_B}{dt} = k_1 C_M - k_2 C_B \quad (4.20)$$

where C_B is the biotic concentration; C_M, the concentration in the ambient environment (water); k_1, the uptake rate constant; k_2, the depuration rate constant; and t, the elapsed time.

On integration of Eq. (4.20) from an initial $C_B = 0$ and $t = 0$, the biotic concentration, C_B, at time, t, is:

$$C_B = \left(\frac{k_1}{k_2}\right) C_M \left(1 - e^{k_2 t}\right) \quad (4.21)$$

When the chemical pollutant approaches a steady state, the uptake and depuration processes will be in equilibrium and

$$\frac{dC_B}{dt} = 0 = k_1 C_M - k_2 C_B$$
$$\text{and } k_1 C_M = k_2 C_B \quad (4.22)$$

Figure 4.8 illustrates these kinetic rate processes for changes in biotic concentration (C_B) over time for an aquatic organism.

If exposure to the pollutant is terminated for an aquatic organism, and it is placed in clean pollutant-free water, then depuration of the pollutant occurs.

$$K_1 C_M = 0$$

Thus for depuration

$$\frac{dC_B}{dt} = -k_2 C_B \quad (4.23)$$

and on integration

$$C_B = C_{BO} \left(1 - e^{k_2 t}\right) \quad (4.24)$$

where C_{BO} is the concentration in the organism at the start of the depuration process.

Then, taking natural logarithms, the following equation is obtained where at any time the biotic concentration, as $\ln C_B$, is

$$\ln C_B = \ln C_{BO} - (k_2 \cdot t) \quad (4.25)$$

Thus, a semilogarithmic plot of C_B against time (t) will be linear if first-order kinetics are followed. The **half-life** ($t_{1/2}$) can be calculated from the equation:

$$t_{1/2} = \frac{0.693}{k_2} \quad (4.26)$$

The kinetics of environmental movements and transformations are discussed further in Section 4.14 on Kinetics of Transformation and Degradation later in this chapter.

4.8 The Persistent Organic Pollutants (POPs) – Long-Range Transport

Pesticides are probably the most widely criticized environmental chemicals due mainly to their specific environmental properties and their broadcast distribution over large areas of crops as well as the natural and human environment. The use of pesticides in agriculture and other areas is not recent. Prior to the 1940s, insecticides such as lime, sulfur, nicotine, pyrethrum, kerosene, and rotenone were extensively used but they lacked potency with a wide range of insects, and lacked persistence to reduce repeated applications, and attracted little adverse comment.

A major development in insect control came in 1939 when the Swiss chemist Paul Müller working for the Geigy Company patented DDT as an insecticide. It found extensive use during World War II due to its relatively long persistence, cheap cost, and potency to a wide range of insect species. It achieved such a high level of success in helping control pests in food crops and pests bearing human diseases that Müller was awarded the Nobel Prize in 1948 for its development. However, DDT represents a significant milestone in the development of awareness of environmental chemicals and control of chemicals in the environment.

Usage of DDT and related chlorohydrocarbon insecticides rapidly accelerated during the 1940s and subsequent decades, and the organophosphate pesticides also became widely used as well. Little thought was given to ecological implications although there were a few reports of possible consequences due to the occasional "kills" of fish and other aquatic organisms associated with the use of these substances. In 1962, the book **Silent Spring** was published by Rachel Carson (1962). This book raised many possible ecological problems which could be associated with the usage of DDT. It has had a major influence in that it initiated a large research program in countries throughout the world. As a result, there have been many scientific and governmental enquiries into the usage of DDT and many other chemicals. These enquiries have led to the banning of DDT in many countries throughout the world. In addition, this research into trace chemicals in the environment continues to the present day with an increasing array of chemicals of concern with concurrent implications for human health and the natural environment.

During the late 1990s, investigations started to uncover the global presence of pesticides and other substances in the world's oceans, water, air, soil, sediments, and biota. In fact, this distribution was not directly related to the sources of the substances but, in some cases, was on a global basis. This aroused further concerns and in 2001 there was a further step taken in this evolving process. The European Union and 91 other countries agreed to a United Nations Treaty, the Stockholm Convention on Persistent Organic Pollutants (POPs), developed in Stockholm to address the control and management of toxic environmental chemicals with adverse effects on both the natural and human environment (see UNEP 2020). This was as a result of the following issues:

- Environmental transport of POPs through several different countries due to their persistence and other environmental properties, resulting in adverse effects on areas different from their area of release. Although the concentrations of POPs were at a low trace level in the air and water phases in receiving areas, their bioconcentration capacity caused elevated concentrations to occur in biota.
- Ability of POPs to bioaccumulate and pass from one species to another through natural food chains.

A summary of the distinctive properties of POPs is shown in Table 4.3.

The Stockholm Convention on POPs was formulated in 2001 and at that time it applied to 12 substances, known as the Dirty Dozen, including DDT and other chlorohydrocarbon pesticides, PCBs, dioxins, and furans. The current convention covers about 28 substances which include pesticides, industrial chemicals, or by-products, and there are several additional chemicals listed for consideration (US EPA 2017).

The Long-Range Transport (LRT) process has been described by Beyer et al. (2000) and results from the use of POPs in equatorial areas which are warmer on a global scale. The processes involved are illustrated in Figure 4.9.

Table 4.3 Some of the distinctive properties of persistent organic pollutants (POPs).

- Toxic and can have adverse effects on the human and natural environment
- Chemically stable and do not readily degrade in the environment, with a half-life in soil of 6 months (183 days) or more
- Low aqueous solubility and high solubility in lipids/fats
- Lipophilic with log K_{ow} between 2 and 6 and bioaccumulative
- Vapor pressure between 1.0 and 10^{-4} Pa at STP and can evaporate slowly
- Henry's Law constant of 10^{-5} to 10^{-2} atm-m^3 mol^{-1}
- Move over long distances in nature and can be found in regions far from their points of origin

Figure 4.9 Diagram illustrating the processes described and **Global Distillation** and the **Grasshopper Effect** showing how POPs move from equatorial areas toward polar regions.

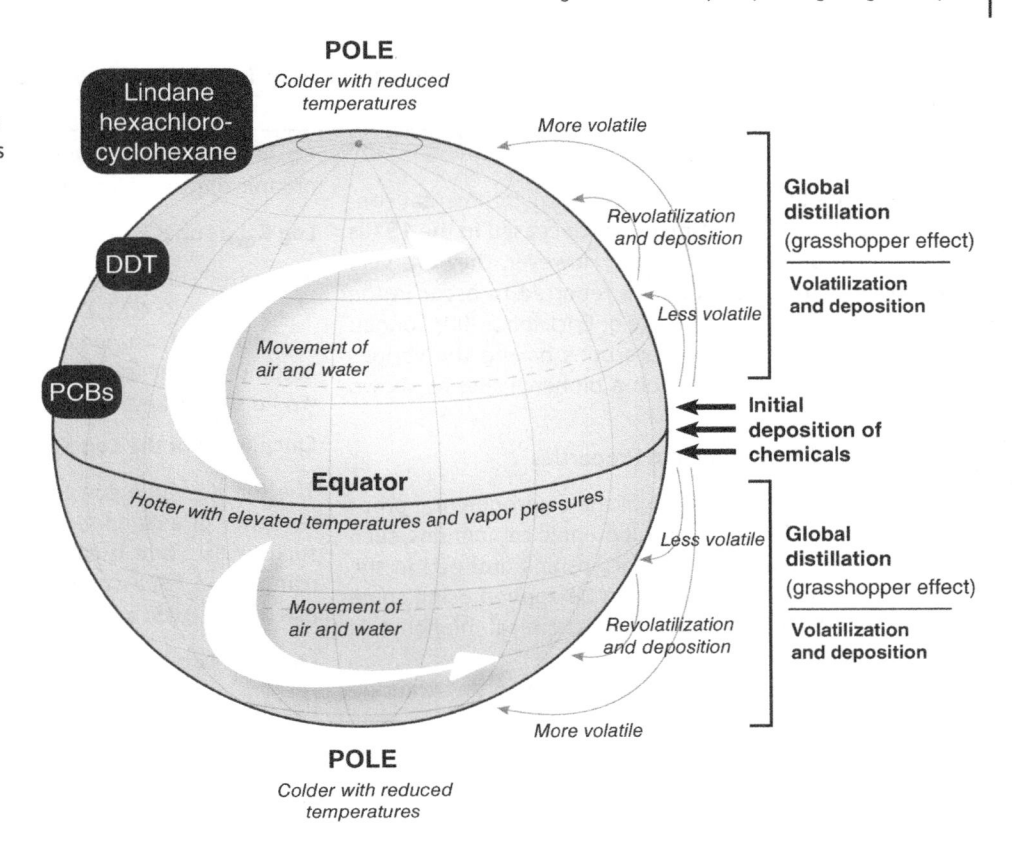

On deposition in these warmer areas, there are small proportions of these substances which volatilize into the atmosphere in relation to the ambient temperature and in accord with their vapor pressure and Henry's Law constant. Since the POPs are resistant to degradation by processes in the atmosphere, they can persist there for a relatively long time. In addition, the atmosphere and ocean currents move in accord with prevailing wind and water directions, some of which are in the direction of colder regions. On entering colder regions, some of the volatilized POP is deposited in soil and water in trace levels. However, if a POP has high bioaccumulation potential, it can be accumulated in biota and may reach levels which can result in adverse effects. This mechanism can be repeated on a global scale and has been described as the **Grasshopper effect** resulting from **Global Distillation**.

If sufficient physicochemical data on partition behavior and persistence in different environmental phases are available, then models, particularly the OECD Screening Tool for estimating overall persistence (POV) and long-range transport potential (LRTP) of organic chemicals (OECD 2009), can be used. This consists of software in a spreadsheet format containing multimedia chemical fate models at the Level III Multimedia level. This tool can be used for comparative assessment of environmental hazard properties of different chemicals and help identify potential POPs according to persistence and long-range transport metrics (OECD 2009).

Quantitative Structure–Activity Relationships – QSARs

Quantitative Structure–Activity Relationships (QSARs) are relationships between structure and properties of compounds and their biological activity. The principles involved are illustrated by the first section of Figure 4.1 where the inherent characteristics of the molecule control the physicochemical properties of the compound, which then guide the environmental distribution of the chemical. However, with QSARs, there are similar relationships usually involving **biological activity** which are discussed in Section 5.9. QSARs can use these characteristics and properties of a chemical in the development of the relationships (Connell 1988). Box 4.1 provides several examples of the estimation of key environmental properties from TMSA of PCBs.

> **Box 4.1 Estimation of Some Key Environmental Properties from Total Molecular Surface Area of the Polychlorinated Biphenyls (PCBs)**
>
> **The PCBs**
>
> The PCBs are an important group of environmental contaminants, the manufacture of which ceased in the 1970s due to environmental concerns. However, they are very persistent and occasionally are reported to occur in the environment in recent times (e.g. Erickson 2000; Connell 2005). They are chlorohydrocarbons having the various chlorine substituent atoms on a biphenyl ring as shown below.
>
> **Calculation of Environmental Properties**
>
> It is possible to deduce many of the environmental properties from the TMSA of neutral organic compounds, such as the PCBs, using various relationships outlined in the text. An example of how this can be applied using some PCBs and the corresponding base chemical, biphenyl, is outlined below.
>
> BP – biphenyl – structure above without any chlorine substituents.
>
> PCBs – 2-monochlorobiphenyl (2-MCB); 2′,3,4-trichlorobiphenyl (2′,3,4-TCB); decachlorobiphenyl (DB).
>
> **Calculation of Log K_{OW}**
>
> The log K_{OW} was calculated using the following equation derived by de Bruijn and Hermens (1990):
>
> $$\log K_{OW} = 0.0285 (TMSA \text{ Å}^2) - 1.87 \quad (4.27)$$
>
> TMSA: log K_{OW}. Results of calculations: BP 216: 4.2; 2-MCB: 234: 4.8; 2′,3,4-TCB: 267: 5.7; DB: 376: 8.8.
>
> **Calculation of the Log K_D Using the Log K_{OW}**
>
> The log K_D value can be used to evaluate the inherent tendency of a compound to accumulate in soils and sediments and it can be calculated from the log K_{OW}, using Eq. (4.15) assuming typical soil and sediment values for organic matter of 5% (0.05) and slope value, a, of unity.
>
> **Log K_{OW}: Log K_D**
>
> Results of calculations: BP: 4.2: 2.72; 2-MCB: 4.8: 3.32; 2′,3,4-TCB: 5.7: 4.22; DB: 8.8: 7.32.
>
> These results suggest that all of the compounds would accumulate in soils and sediments, but accumulation would be weakest with BP.
>
> **Calculation of the Log K_B Using the Log K_{OW}**
>
> The log K_B value can be used to evaluate the inherent tendency of a compound to accumulate in aquatic biota, such as fish, and it can be calculated from the log K_{OW} using Eq. (4.18) assuming typical lipid content values for fish of 5% (0.05) and slope value, b, of unity.
>
> **Log K_{OW}: Log K_B**
>
> Results of calculations: BP 4.2: 2.9; 2-MCB: 4.8:3.5; 2′,3,4-TCB: 5.7: 4.4; DB: 8.8: 7.5.
>
> These results suggest that the compounds 2-MCB and 2′,3,4-TCB would bioaccumulate in fish and other aquatic organisms since these compounds not only have an appropriate log K_{OW} but are resistant to degradation. Resistance to degradation by the fish arises from the presence of many C–Cl bonds in the PCB molecule which are very stable and resist attack by biotic agents. However, DB lies outside of the bioconcentration range which usually is at its highest at log K_{OW} of 6–7 and the compound has a log K_{OW} 8.8. BP has doubtful bioconcentration potential since it is at the low end of the bioconcentration range which is at log K_{OW} of 2–3. In addition, it does not have any C–Cl bonds, which means that it would not be very persistent in fish.

4.9 Bioaccumulation of Metals and Metalloids

The relationships described above apply to neutral organic compounds and not to metals, metalloids, and related chemical compounds. But metals and metalloids are common components of the environment having wide occurrence in soils, rocks, and seawater, including metals essential to biota such as sodium, potassium, iron, and magnesium. However, some nonessential **metals** are of considerable environmental management importance, such as mercury, lead, and cadmium, since they often present hazards to human health and the natural environment. This group of metals is usually of relatively high atomic weight and is referred to as the **heavy metals**. **Metalloids** are elements that have properties of both metals and nonmetals while some common metalloids are boron, silicon, arsenic, and selenium. Some metalloids are hazardous in the environment. They are considered in Chapter 12 – Metals. In this section, we are considering only the bioaccumulation of metals and metalloids rather than their broader environmental properties and effects.

The metals and metalloids exist in the environment in many different chemical forms ranging from the metal itself in ionic form to combinations with various chemical groups to form an ionic complex, sorbed to particulate matter and organometallic forms without charge

4.10 Fate of Chemicals in the Environment

When a chemical enters the environment, there is active exchange of the chemical between the phases, particularly from the phase or phases in which the chemical is initially discharged. The processes occurring are shown in Figure 4.10. Of course, a phase can only exchange chemicals with a phase in which it is in direct contact. So, exchange would be expected between air and water, water and sediment, and so on, as shown in Figure 4.10. Alternatively, there would be no direct exchange between phases that are not in contact such as fish and vegetation, atmosphere and sediment, and so on. But a chemical can move in a stepwise fashion from phase to phase so that all phases are effectively interacting. We can consider these exchange processes to occur by diffusive movement of the chemical backward and forward as shown by the arrows indicating these movements in Figure 4.10. All of the exchange processes between phases in environmental systems can be represented by sets of two phase processes. So, the processes which occur in the environment shown in Figure 4.10 can be represented by the following two phase exchange processes:

Vegetation–atmosphere
Soil–atmosphere
Atmosphere–water

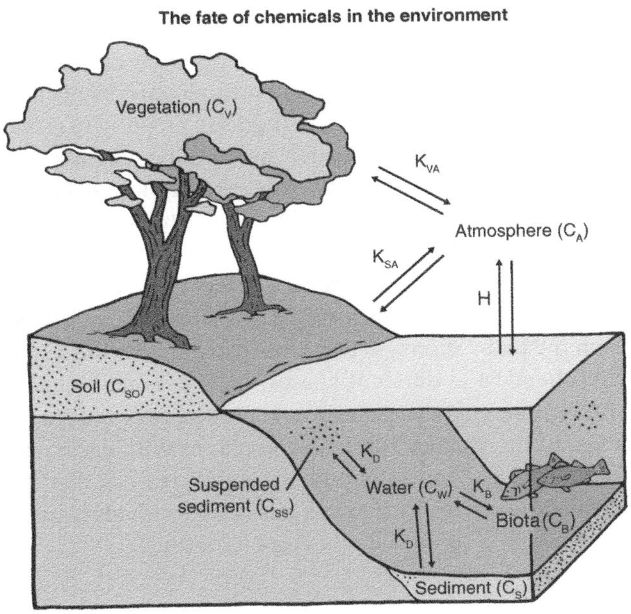

Table 4.4 The mean Bioaccumulation Factors for heavy metals with a wide range of aquatic organisms.

Metal	Mean Bioaccumulation Factor
Zinc	3394
Cadmium	1866
Copper	1144
Lead	598
Nickel	157
Silver	1233
Mercury	6830

Source: Data taken from McGeer et al. (2003).

(Egorova and Ananikov 2017). Organisms can take up metals and metalloids in food, particulate matter in air or water depending on the organisms as well as the chemical form of the metal and metalloid involved (Donati 2018). The chemical form also can vary with such environmental factors as pH and dissolved oxygen concentration in aqueous systems. After uptake, the metals and metalloids can be deposited in biota in the ionic state, sorbed or chemically bound to form many biologically important substances and participate in vital biological processes.

Many metals are indispensable to life since they are required to form enzymes, structural biotic components, or other essential biotic components. These are described as the **essential trace elements** and include such elements as iron, copper, zinc, and selenium. The toxic behavior of these elements is outlined in Chapter 12. However, when the biotic exposure concentration exceeds the essential level, they may exhibit adverse toxic effects. Some heavy metals have no known role in biotic processes and have adverse biotic effects at all levels.

McGeer et al. (2003) have reviewed the data on bioaccumulation of Zn, Cd, Cu, Pb, Ni, Ag, and Hg by a wide range of aquatic organisms and concluded that the values for the Bioaccumulation Factor are characterized by extreme variability. With most of the heavy metals, the Bioaccumulation Factor was inversely related to the exposure concentration, thus as the exposure concentration increases the Bioaccumulation Factor declines. The mean Bioaccumulation Factors derived by these authors from the review of the data in aquatic organisms are shown in Table 4.4.

It can be concluded that while some metals and metalloids, particularly the heavy metals, bioaccumulate in aquatic and other organisms, the mechanisms are variable. The values of the Bioaccumulation Factor reflect this variability and relate to chemical form and mechanisms of uptake of the particular organisms involved.

Figure 4.10 A conceptual diagram of the distribution of a neutral organic chemical in the environment consisting of several phases with the partition coefficients labeled (K) and the concentrations in the phases at equilibrium (C).

Sediment–water
Suspended sediment–water
Biota–water

These are the major processes occurring but some processes are not as well understood as others, for example the atmosphere–soil and atmosphere–vegetation processes. Also, less important processes could be included in the environment which is represented in Figure 4.10. For example, the water environment could be expanded to include the aquatic vegetation/water process. This could make our diagram represent the actual environment more accurately, depending on the levels of aquatic vegetation present, and thus give a better representation of the system. So, the set of two phase processes indicated should be representative of the actual environment in which we are interested. Often the aquatic section of the environment is modeled alone since it is the best understood and can be modeled alone.

In the sections above, the behavior of neutral organic compounds, such as DDT, PCBs, and PAHs, in various phases in the environment is described. However, these descriptions are limited to particular phase combinations such as water–sediment or the atmosphere–water. But a chemical discharged to the environment distributes into all of the various phases present such as water in the form of rivers, lakes, the oceans, air in the form of the atmosphere, soils, aquatic sediments, and so on. In fact, the chemical distributes according to its intrinsic partitioning properties and the chemical nature of the phases themselves. It continues to distribute until equilibrium is established and the concentrations in the phases are effectively constant. A method of combining all of the partition coefficients and phases together is based on the **fugacity** property to arrive at the equilibrium distribution as described by Mackay (2001). Fugacity is not a new concept but Mackay (1979) applied it to modeling the behavior of chemicals in the environment.

The fugacity of a chemical in a phase is its tendency to escape that phase. In this sense, the *fugacity* name is derived from the Latin, *fugere*, to flee. Clearly the fugacity of a specific chemical in different phases, such as water and soil, will be different depending on the properties of the chemical and the characteristics of the phase. With gases, the fugacity of a chemical in a phase is the pressure that the chemical exerts external to that phase, and with ideal gases, the pressure, P, is equal to the fugacity, f, thus

$$f = P$$

When a chemical enters the environment, it partitions between all the phases to a lesser or greater extent. More of it may go into certain phases than others and, in fact, some phases may take up very little. In simple terms, this means the chemical separates into parts which move into the different phases. For example, *fat loving* or *lipophilic* chemicals, because their preference is to dissolve in fat, will move into phases high in fat or lipid and will be reluctant to enter phases with little or no fat. For example, DDT and PAHs, lipophilic compounds, would be expected to move into lipid-rich phases, such as fish, but only relatively low amounts would enter lipid-deficient phases such as water and air.

Fugacity has a distinct advantage in that it is proportional to concentration at the relatively low levels of contaminants which are encountered in the environment. This relationship is described by

$$C \alpha f$$

By inserting a proportionality constant, Z, then

$$C = Zf$$

The proportionality constant, Z, is called the **fugacity capacity factor**. It has units of concentration and reciprocal pressure, e.g. mol m^{-3} Pa. By rearranging,

$$C = f.Z \tag{4.28}$$

where C is the concentration in the phase, Z is the fugacity capacity constant with units of mol m^{-3} Pa, and f is the fugacity. A single chemical in each phase has its own particular Z value and in terms of a partitioning model, at equilibrium, chemicals will tend to accumulate in phases with high fugacity constants. Fugacity has a further distinct advantage in that it is proportional to concentration at the relatively low levels of contaminants which are encountered in the environment.

The fugacity capacity constants can be estimated from basic physicochemical data. The fugacity capacity constant is a constant for a given substance in a given phase and is a physicochemical characteristic of that substance in that phase. The fugacity capacity constants can be calculated using the following relationships:

$$Z_{air} = \frac{1}{RT} \tag{4.29}$$

$$Z_{water} = \frac{1}{H} \tag{4.30}$$

$$Z_{sediments} = \frac{K_D}{H} \tag{4.31}$$

$$Z_{biota} = K_B H \tag{4.32}$$

$$Z_{soil} = K_{SA} H \tag{4.33}$$

where R is the universal gas constant; H, Henry's Law constant; T, the absolute temperature; K_D, the equilibrium constant between sediment and water; K_B, the equilibrium constant between biota and water; K_{SA}, the

Table 4.5 Some values of properties of environmental importance for various compounds.

Compound	Henry's Law constant		K_{OW}	Measured K_B (fish)	Calculated K_B	Measured K_{OC}	Calculated K_{OC}
	Dimensional (H_D)	Dimensionless (H)					
	atm-m³ mol⁻¹						
Lindane	4.8×10^{-7}	2.2×10^{-5}	5 250	470	263	1 080	2 153
DDT	3.8×10^{-5}	1.7×10^{-3}	2 290 000	1 100 000	1 145 000	243 000	940 000
Arochlor 1242	5.6×10^{-4}	2.4×10^{-2}	199 600	3 200	5 480	5 600	44 700
Naphthalene	1.15×10^{-3}	4.9×10^{-2}	3 900	430	199	1 300	1 600
Benzene	5.5×10^{-3}	2.4×10^{-1}	135	13	7	83	55
Mercury	1.1×10^{-2}	4.8×10^{-1}	—	—	—	—	—
Vinyl chloride	2.4	99	—	—	—	—	—

Source: Data from Lyman et al. (1990).

equilibrium constant between soil and the atmosphere. This means that the equilibrium partition coefficients such as K_B, H, and K_{SA} should be known. Examples of the physicochemical properties are shown in Table 4.5.

In most situations, these constants are not known but often can be calculated using QSARs. QSARs have been extensively used to predict the behavior of chemicals in the environment. These applications have been most successful with nonpolar lipophilic compounds that generally exhibit negligible biodegradation.

If all phases are in equilibrium in terms of a distributed chemical, then the fugacities of the chemical in each phase are equal and are given by

$$f = \frac{C_i}{Z_i} \quad (4.34)$$

If $C_I = M_i/V_i$ in any particular phase, then substituting in Eq. (4.34),

$$f = \frac{M_i}{V_i Z_i} \quad (4.35)$$

and if we take the system as a whole, then $f = \Sigma M_i / \Sigma(V_i Z_i)$ and

$$f = \frac{M_{total}}{\Sigma(V_i Z_i)} \quad (4.36)$$

where M_i is the number of moles of chemical in a phase of volume V_i; C_i, the concentration of the chemical; Z_i, the fugacity capacity constant of the chemical in that phase.

Using this equation, the prevailing equilibrium fugacity can be determined and from this the amount and concentration of the substance in each individual phase can be calculated. However, to calculate the overall fugacity, f, the total mass involved (M_{total}) expressed as moles, must be known, the volume of each phase (V_i) measured, and the fugacity capacity constant (Z_i) of the substance in each phase estimated. Examples of the calculation of environmental distributions of chemicals in the environment using the fugacity concept are shown in Table 4.6, with a detailed example using benzene in Box 4.2.

Table 4.6 Calculated distribution and concentration of some representative chemicals in a typical environment.

Compound Phase[a]	Penta-chlorophenol		Phenanthrene		Tetra-chloroethane		DDT		Hexa-chlorobiphenyl	
	µg kg⁻¹	% Moles[b]	µg kg⁻¹	% Moles[b]	µg kg⁻¹	% Moles[b]	µg kg⁻¹	% Moles[b]	µg kg⁻¹	% Moles[b]
Air	5.7×10^{-2}	5.8	3.0×10^{-1}	76.2	6.1×10^{-1}	99.9	4.6×10^{-3}	0.4	1.0×10^{-1}	7.5
Water	5.2×10^{-2}	0.6	2.4×10^{-2}	0.4	1.8×10^{-3}	4.0×10^{-2}	4.9×10^{-3}	5.0×10^{-2}	2.0×10^{-4}	2.0×10^{-3}
Soil	33.2	81.0	5.5	20.2	4.6×10^{-3}	2.0×10^{-2}	47.5	86.7	44.5	80.0
Suspended sediments	100	0.2	16.6	4.0×10^{-2}	1.4×10^{-2}	4.0×10^{-5}	142	0.2	134	0.2
Sediment	100	11.9	16.6	3.0	1.4×10^{-2}	3.0×10^{-3}	142	12.7	134	11.7
Aquatic biota	265	2.0×10^{-3}	44.0	4.0×10^{-4}	3.6×10^{-2}	4.0×10^{-7}	376	2.0×10^{-3}	354	2.0×10^{-3}
Vegetation	106	0.6	17.7	0.1	1.5×10^{-2}	1.0×10^{-4}	151	0.6	141	0.6

[a] Concentration in a single phase.
[b] Percentage distribution of the compound between all phases.

Box 4.2 Example Calculation of Distribution of Benzene in a Model Environment

The Model Environment

Now that we have methods to calculate Z values, estimations can be made of the distribution of some chemicals in the environment. To do this, we need further important information: the volumes of the environmental phases are needed as shown in Eqs. (4.35) and (4.36). However, these volumes can be made available by using a typical or model environment, where the size of the phases is an educated guess. We are not so concerned with the accuracy of the volume in our model environment, but can focus instead on using the properties of the chemical to evaluate the way it distributes.

A typical model environment is described by Mackay (1991) and Connell (2005, p. 345). The model has the six major phases: air, water, soil, sediments, suspended solids, and biota with volumes in m^3: air (10^{10}), water (7×10^6), soil (6×10^3), suspended solids (35), sediments (2.1×10^4), and fish (3.5). Each of the six phases is assumed to be homogeneous and at a temperature of 25 °C.

The Properties of Benzene

An arbitrary amount of benzene, 100 mols (78 kg), is added to the model environment. The benzene is assumed to distribute to equilibrium and disperse till eventually steady concentrations in each phase are reached.

The important physicochemical properties for benzene should be assembled. These are:

Calculation of the Distribution of Benzene

First, we need to calculate the Z values for each phase as follows using the expressions derived above and shown in Eqs. (4.29)–(4.33):

$$Z_{air} = \frac{1}{RT} = 4.04 \times 10^{-4} \text{ m}^3 \text{ mol}^{-1} \text{ Pa}$$

$$Z_{water} = \frac{1}{H} = 1.8 \times 10^{-3} \text{ m}^3 \text{ mol}^{-1} \text{ Pa}$$

$$Z_{soil} = \frac{K_{sorb}(\text{soil})}{H} \text{ or}$$

$$\frac{K_D}{H} = 2.0 \times 10^{-3} \text{ m}^3 \text{ mol}^{-1} \text{ Pa}$$

$$Z_{sediment, \text{ sus. solids}} = \frac{K_{sorb}(\text{sediment, sus. solids})}{H}$$
$$= 4.0 \times 10^{-3} \text{ m}^3 \text{ mol}^{-1} \text{ Pa}$$

$$Z_{biota} = \frac{K_B}{H} = 1.2 \times 10^{-2} \text{ m}^3 \text{ mol}^{-1} \text{ Pa}$$

In the next step, we calculate ZV for each phase and sum these products as $\Sigma Z_i V_i$ ($4.05/10^6$ mol Pa^{-1}) and calculate the prevailing equilibrium fugacity, recalling that $f = M_{total}/\Sigma (V_i Z_i)$ (Eq. (4.36)) as 2.47×10^{-5} Pa. We can then use this fugacity value to calculate the benzene mass in each phase, recalling that $M_i = f Z_i V_i$ as in Eq. (4.36).

Finally, we can calculate the benzene concentration in each phase using ($C_i = M_i/V_i$) and obtain the following distribution:

Phase	Air	Water	Soil	Suspended solid	Sediment	Fish	Total
Moles in each phase (%)	99.69	0.31	4.4×10^{-4}	3.5×10^{-6}	2.07×10^{-3}	1.0×10^{-6}	100
Concentration in each phase (C = M/V, mol m^{-3})	1×10^{-8}	4.4×10^{-8}	5×10^{-8}	1×10^{-7}	1×10^{-7}	3×10^{-7}	—

H, 557 Pa m^3 mol^{-1}; K_{OW} 135; K_{sorb} (soil, 2% organic carbon) 1.1; K_{sorb} (sediment, suspended solids, 4% organic carbon) 2.2; K_B (fish, 5% lipid) 6.7. The K values for soil, sediment, suspended solids and biota have been estimated from K_{OW} using equations in Section 0, Pollutant Behavior in Different Environmental Compartments.

These calculations indicate that greater than 99% of the benzene is dispersed into the air. This is expected, since benzene is a volatile compound and most of the model environment is taken up by air (>99.9% by volume).

Finally, a reminder that a simple, static view of chemical distribution has been illustrated. Other factors may influence a pollutant's fate. For example, environmental phases such as air and water move about, described as advection, carrying pollutants with them. The pollutant discharge may be continuous and the chemical can decompose.

The fugacity models are available on the Web at different levels with each level building on the previous. Level I is the basic level, which has been considered above with a closed environment at equilibrium; Level II is an open system in equilibrium with continuous emissions and degradation and transformation of the chemical taken into account; Level III builds further and is a system in a steady state, rather than at equilibrium, and takes into account transport of phases; Level IV is an open system in a nonsteady state. These factors and others need to be covered when distributions in specific environments are being considered. The Canadian Centre for Environmental Modeling and Chemistry provides a full set of models on the Web (CEMC 2020).

These models provide a valuable basis for comparison of chemicals and the identification of phases in the environment that are likely to be repositories of discharged chemicals. If estimations of actual concentrations likely to occur in an actual environment are required, then additional environmental-specific factors may need to be taken into account. The accuracy of the results obtained is related to how well the data reflect the actual environment.

4.11 Transport Processes: Advection, Diffusion, and Leaching

In the sections above, some of the major processes influencing the distribution of contaminants in the environment are described, particularly those involving partitioning processes which form a basis for modeling and understanding distribution processes. Even so, there are several other very important processes which occur in the environment in conjunction with the partition processes and can have a powerful influence on distribution.

Advection is the passive movement of a contaminant by the physical movement of the phase which contains it. Contaminants are transported by the winds in air and by river water flow and ocean currents in water.

Diffusion is a process of movement of compounds at the atomic, molecular, or ionic level by the random motion of these entities in the environment leading to an ongoing process of dispersal in the surrounding matrix. Diffusion is a major force in the movement of chemicals from one phase to another in the environment to establish equilibrium in the partitioning process. Thus, in the Level 1 fugacity model outlined above, the only movement of contaminant is by diffusion.

Leaching is an important process in the removal of contaminants from soil and other media and involves a combination of diffusion and advection processes as illustrated in Figure 4.11. Four phases are involved (air, pore water, soil particles, and groundwater) as a chemical in the soil equilibrates with the pore water by diffusion and moves by advection to the water surface or the groundwater. Both the air and groundwater are moved from the system by advection processes such as wind and groundwater movement with the possible removal of contaminants. The removal

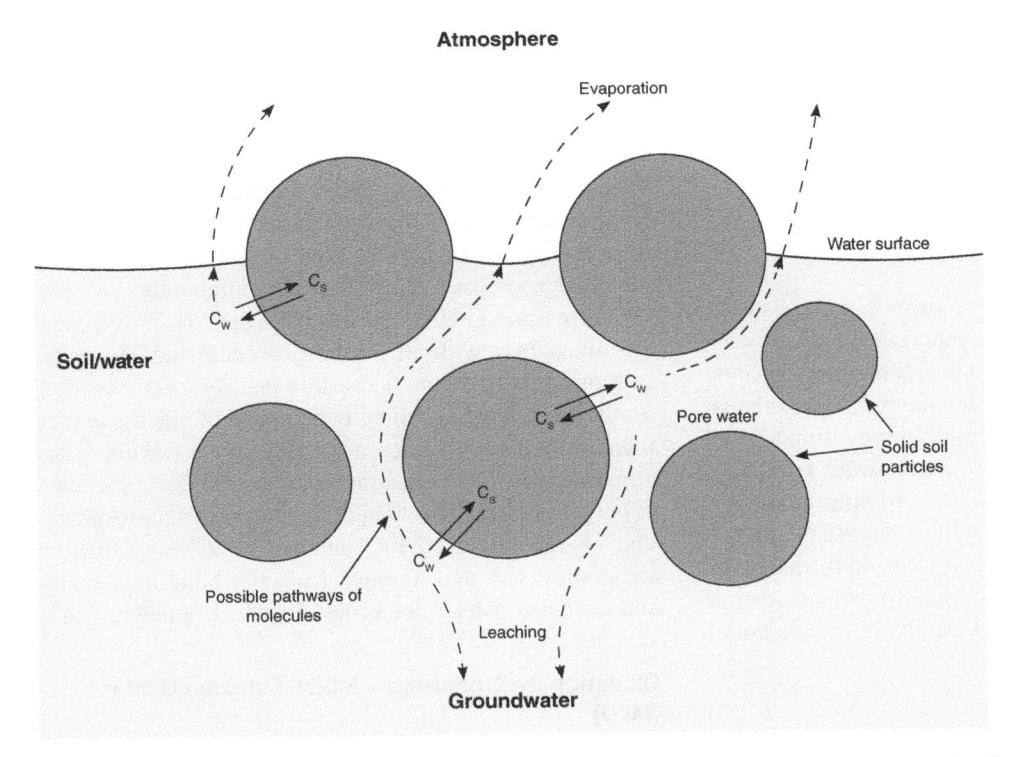

Figure 4.11 Illustration of the processes involved in leaching from a soil environment with both advection and diffusion processes involved.

of contaminants by evaporation is dependent on the vapor pressure of the contaminant while removal in the groundwater is dependent on solubility characteristics.

4.12 Transformation and Degradation Processes

Chemicals from decomposition and excretions of animals and plants, as well as contaminant chemicals added to the environment by human activities, occur in water, soil, and the atmosphere. By the actions of this complex array together with the physical conditions, including temperature and radiation, both natural and added chemicals may undergo **chemical transformation**. This change in chemical nature could be a rearrangement of a molecule into another form; alternatively, it could be the addition or loss of chemical groups by environmental processes. **Degradation** is a term commonly used which usually refers to the breakdown of the original molecule or chemical entity by the loss of various component parts or by fragmentation into smaller substances with a lower molecular weight (Connell 2005).

These can be facilitated by the input of energy in the form of radiation or heat. Alternatively, biota may be involved, leading to the transformation and degradation of compounds through biological processes. Both oxygen and water are substances that are reactive and available in large quantities in the environment for transformation and degradation of organic compounds. Oxygen comprises about 20% of the atmosphere, and water occurs in high proportions in biota, as well as existing in large quantities in the oceans, lakes, and rivers.

Oxidation

The addition of oxygen to the molecule can result in the formation of a molecule that is increased in size by the addition of oxygen or it may result in splitting the molecule into smaller oxygen-containing fragments. The oxidative degradation of an organic compound to the ultimate level results in the formation of carbon dioxide, water, ammonia, nitrate ion (NO_3^-), nitrite ion (NO_2^-), orthophosphate ion (PO_4^{3-}), hydrogen sulfide (H_2S), sulfate ion (SO_4^{2-}), and so on, depending on the conditions involved and the nature of the original compound. A simple example is the oxidation of methane by combustion which results in the release of energy:

$$CH_4 + 2O_2 \rightarrow CO_2 + 2H_2O \qquad (4.37)$$

Hydrolysis is a process that results in the addition of water to a molecule (Wolfe and Jeffers 2000). But this often

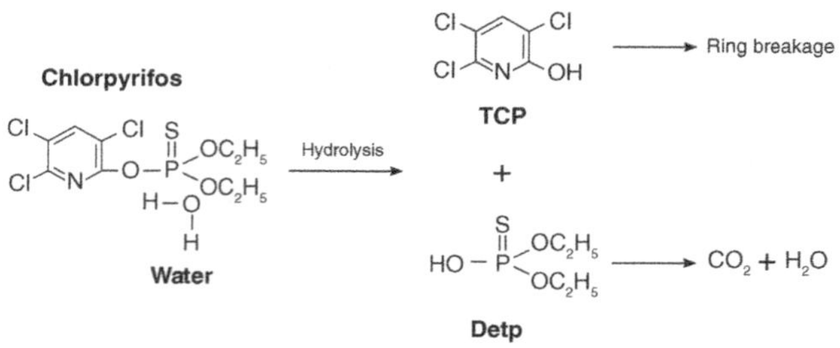

Figure 4.12 Hydrolysis of the insecticide chlorpyrifos in the natural environment or by metabolic processes within an organism.

results in the fragmentation of the molecule into smaller fragments that may contain additional hydrogen and oxygen. An example of this is the hydrolysis of the organophosphorus functional group in chlorpyrifos, as represented by the equation shown in Figure 4.12. Chlorpyrifos is one of the most widely used pesticides in the world and has many uses. However, this pesticide can be hydrolyzed in the environment or subject to metabolic transformation with organisms to yield mainly trichloropyridinol (TCP).

Both oxidation and hydrolysis can occur when substances come in contact with oxygen and/or water under the appropriate conditions in air, water, soil, and biota. These processes may occur with the intervention of biota or as a result of abiotic processes. Biotic processes mediate reactions facilitated by animals or plants. These reactions are influenced by environmental factors such as prevailing temperature, the presence of oxygen, water, and light.

Combustion

Often, spectacular oxidation of organic matter by atmospheric oxygen occurs by combustion. The burning of trees, grass, and petroleum is an example of oxidation through the combustion process. Many organic compounds can exist in the environment in the presence of the 20% of oxygen in the atmosphere without combustion occurring. However, ignition by a spark or flame initiates the occurrence of combustion. The burning or combustion of organic matter such as wood and petroleum is a major source of energy in human society. This has a major impact on the occurrence of oxygen and carbon dioxide in the Earth's atmosphere. The energy produced by combustion of many organic compounds can be estimated from the bond association and dissociation energies as described by Connell (2005).

Oxidation by Organisms – Mixed Function Oxidase (MFO)

Lipophilic organic compounds are particularly difficult substances for animals to remove from their internal metabolic and physiological systems and prevent their toxic actions. If the compound is insoluble in water, it is effectively protected from chemical attack. However, when an animal is exposed to lipophilic compounds, it induces a response within the animal which results in the production of the MFO enzyme system. This enzyme system converts the lipophilic compounds to oxidized products which are more soluble in water and thus can be removed from the animal in water-based excretion products.

Phototransformation

Many organic chemicals are introduced into the environment and can absorb radiation and, as a result, undergo chemical transformation. All chemical processes require the reacting substances to have attained a certain energy. This energy can be obtained either from thermal energy (heat) as with combustion, which was considered in the previous section, or by absorption of radiation. The internal energy of a molecule is in several different forms. Thermal energy in a molecule results from being *jostled* by its neighbors, and it is manifested as translation, rotation, and vibration of the molecules. Absorption of electromagnetic radiation with frequencies in the infrared (IR) range causes increased molecular rotation and vibration. A molecule exists in specific electronic energy level states, generally the lowest energy one or the ground state. Most molecules absorb IR radiation and thermal energy, but to absorb radiation to change the electronic energy state requires ultraviolet (UV) or visible (VIS) light and specific chemical groupings in an organic molecule (chromophores). If the light absorbed is of sufficient energy, the excited molecule produced can undergo various transformations that would essentially not occur under normal conditions. The basic characteristics of radiation can be described by relatively simple equations. Two important equations relate energy, frequency, and wavelength of radiation. First,

$$E = h\nu \quad (4.38)$$

where E is energy in joules, h is Plank's constant (6.63×10^{-34} J), and ν is frequency (Hertz, or cycles per second, s^{-1}).

Second,

$$C = \lambda\nu \quad (4.39)$$

where C is the velocity of light (assumed constant at 3×10^8 m s^{-1}) and λ is the wavelength in meters. The velocity of light is dependent on the medium that it is passing through and reaches a maximum in a vacuum. The equations above indicate that in general terms, as the wavelength (λ) increases, the frequency (ν) will correspondingly decline and the energy of a photon of radiation will also decline since energy (E) is directly dependent on frequency. Accordingly, a photon of UV radiation has greater energy than one of VIS radiation, which has greater energy than an IR photon.

Reactions induced by UV/VIS include fragmentation, oxidation, and polymerization. It is also possible that an excited molecule can return to its original state and in the process emit radiation at a different wavelength. The following are examples of photochemical reactions.

Phototransformation in the environment can only occur if the UV/VIS absorption spectrum of the compound and the solar admission spectrum overlap. Solar radiation contains a substantial amount of UV/VIS radiation when it reaches the Earth's surface.

Chemical groups (chromophores) which absorb UV/VIS radiation in solar radiation all contain double bonds, in some form, aromatic rings or series of double bonds separated by a single bond. Some structures or functional groups are poor absorbers of solar radiation and these include alcohols (–OH), ethers (R–O–R), and amines (R–NH$_2$). For such groups, or chemicals containing only these functional groups, phototransformation is likely to be unimportant. In addition, phototransformation can only occur when the chemical is likely to be exposed to solar radiation. For example, the atmosphere, upper layers of water bodies, and the surface of soil are likely locations in which chemicals would be exposed to solar radiation.

4.13 Redox Conditions in the Environment

As outlined above, **oxidation** is a key process in the environment, particularly the aquatic environment. In addition, the removal of oxygen is an important process and is described as **reduction**. The redox potential, the potential to accept or remove oxygen, can be measured in millivolts and is a continuum from reducing at 100 mV or less, to oxidizing at 300–500 mV or more. In natural surface waters, the dissolved oxygen content of the water can be fully saturated at about 12–14 mg L^{-1} registering a redox of 300–500 mV while the organic bottom waters, enriched with organic matter, of a stratified water body can have an oxygen content of close to zero with a redox of 100 mV. These redox conditions have a strong influence on the chemical processes that can occur. For example, low redox potential can result in the release of iron and phosphorus from the bottom sediments enriched with organic matter, such as plant debris, and result in the formation of H$_2$S, NH$_3$, and other substances. These substances are not generated in the surface waters with high redox potential and high dissolved oxygen content. The redox characteristics

4.14 Kinetics of Transformation and Degradation

It is frequently important to determine how rapidly transformation and degradation occur in the environment so that the extent and period of contamination can be evaluated. As a result, numerous methods have been developed to simulate transformation of organic compounds in the various compartments of the environment, such as the degradation of environmental contaminants in soil as conducted in laboratory experiments to evaluate persistence. Environmental transformation and degradation involve reaction of an organic compound with another substance usually either water or oxygen or, more often, both of these substances. The kinetics of bioconcentration have been described above where a *first-order model* was used to describe the kinetics of loss of a contaminant from an organism. A similar kinetic expression can be developed for loss of a substance by any process. In this case, the rate of reaction is declining and has a negative sign, as follows:

$$-\frac{dC}{dt} \alpha\, CX$$

where C is the concentration of organic compound present; t, the elapsed time period; and X, the concentration of oxygen or water.

This is a *second-order kinetic* expression where the rate of degradation is proportional to the concentration of two reactants. In most environmental situations, the amount of oxygen or water available is very large and effectively constant. This means that the concentration of the environmental contaminant is relatively low and the following expression applies:

$$-\frac{dC}{dt} \alpha\, C$$

This expression describes first-order kinetics, and the rate of reaction is proportional to the concentration of the environmental contaminant present. Assuming this to be the normal situation in the environment, kinetic expressions for degradation of growth substances by microorganisms can be relatively complex, but the following expression for degradation is often used:

$$-\frac{dC}{dt} \alpha\, BC$$

where B is the size of the bacterial population.

This expression is a second-order rate expression, similar to the expression above, in which the population of bacteria was not considered. In this case, the second-order rate of reaction is proportional to the population of microorganisms but this population tends to keep relatively constant in environments such as open water, soils, air, and sediments. This means that the first-order expression indicated previously can generally be applied to most environmental degradation processes; that is,

$$-\frac{dC}{dt} \alpha\, C \qquad (4.40)$$

and

$$-\frac{dC}{dt} = kC \qquad (4.41)$$

where k is the first-order transformation or degradation rate constant with units of time^{-1}. On integration and rearrangement, the following equation can be derived:

$$\ln\left(\frac{C_o}{C_t}\right) = kt \quad \text{or} \quad C_t = C_o e^{-kt} \qquad (4.42)$$

where C_o is the concentration of the organic substance at time zero and C_t is the concentration of the substance after a time period t.

This can be plotted as shown in Figure 4.13a with $\ln(C_o/C_t)$ against the elapsed time period, t. The slope of

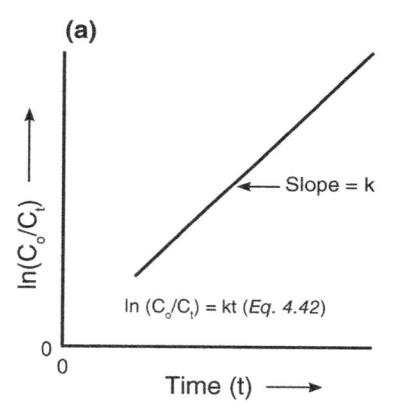

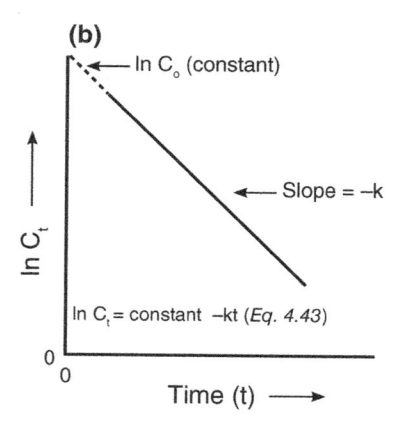

Figure 4.13 Plots of data for the loss of an environmental contaminant from an environmental phase according to first-order kinetics with Eqs. (4.42) (a) and (4.43) (b).

this plot represents the rate constant, k. From the equation above, it can be shown that:

$$\ln C_t = \ln C_o - kt \quad (4.43)$$

Thus,

$$\ln C_t = \text{constant} - kt \quad (4.44)$$

This relationship can be plotted as shown in Figure 4.13b and a measure of persistence of the compound is the rate of loss that can be determined as the rate loss constant, k, which is the reverse slope of the plot in Figure 4.13a.

A more convenient measure of the persistence of a chemical in the environment is the **Half-Life**, $t_{1/2}$. An expression for this can be derived as follows:

$$t_{1/2} = \ln\left(\frac{C_o}{C_t}\right) / k$$

Starting at any point in time when half of the original substance has been degraded, then

$$\frac{C_o}{C_t} = \frac{1}{0.5} = 2$$

Thus,

$$t_{1/2} = \ln\frac{2}{k} = \frac{0.693}{k} = \text{constant} \quad (4.45)$$

This means that the half-life ($t_{1/2}$) is constant for a given compound and a given environmental degradation process that occurs under specific conditions. This has found to be generally true for most environmental processes since these obey approximate pseudo-first-order kinetics. The actual degradation rate constants measured in the environment show a high level of variability. The pH, temperature, availability of water, availability of oxygen, and other factors all influence persistence in a particular situation. These factors vary within different phases of the environment, causing this high degree of variation. Examples of some half-lives that have been measured are shown in Table 4.7.

Table 4.7 Half-lives of some compounds in soil and water.

Compound	Environment	Half-life (days)
2,4-D	Soil	1 to 14
2,4-D	Water	1 to >20
Pyrethrins	Soil	2 to 10
Glyphosate	Soil	3 to 130
Glyphosate	Water	35 to 63
Aldrin	Soil	20 to 100
Atrazine	Soil	1 to 240
Benzo(a)pyrene	Soil	2 to 299
Benzo(a)pyrene	Water	57 to 825
DDT	Soil	31 to 12 775

4.15 Acidification Processes in the Environment

Acid Drainage Water

There are processes which occur within the environment that change the chemical conditions in the ambient water, soils, lakes, and the oceans. In drainage basins within the environment, there are igneous and metamorphic rocks and other materials which contain trace amounts of metals, particularly iron and other metals in chemical combination with sulfur. Natural input to the oceans originates from erosion and infiltration and enters through rivers, streams, and groundwater.

However, these processes can be supplemented and augmented by human activities. Mining, drinking water supply using bores, drainage works, and related activities can expose areas of previously covered chemically inaccessible rock and subsoil to water and oxygen from the atmosphere. There are trace levels of microorganisms within the environment which can utilize the iron and sulfur containing minerals, such as pyrites, as a source of energy leading to an increase in their populations. These processes can take a lengthy period, up to decades, even centuries, to develop and can result in major alterations to the ambient chemical conditions in natural waters and soils (see also Chapter 12 on Metals).

A typical sequence of reactions that can result from iron pyrites (FeS_2) in an area is shown in Figure 4.14 (Akcil and Koldas 2006). The iron pyrites in rocks and soil is freshly exposed to oxygen and water and is utilized by *Thiobacillus* and *Ferrobacillus* bacteria as a source

Iron pyrites (FeS_2)

$FeS_2 + 2H_2O + 7O_2 \longrightarrow 4H^+ + 4SO_4^{2-} + 2Fe^{2+}$

$4Fe^{2+} + O_2 + 4H^+ \longrightarrow 4Fe^{3+} + 2H_2O$

$FeS_2 + 14Fe^{3+} + 8H_2O \longrightarrow 15Fe^{2+} + 2SO_4^{2-} + 16H^+$

$Fe^{3+} + 3H_2O \longrightarrow Fe(OH)_3 + 3H^+$

Acid conditions

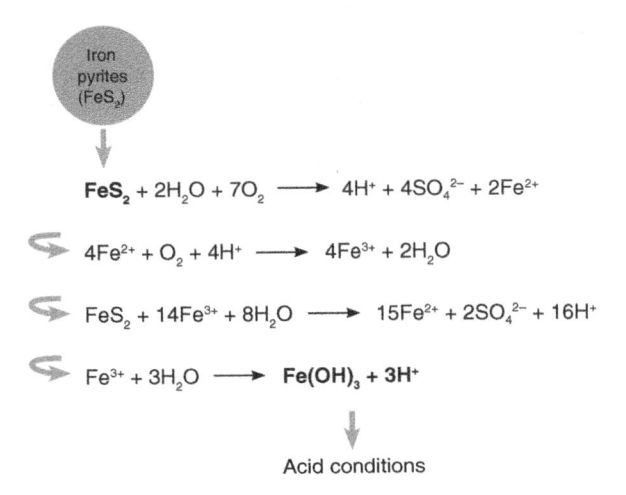

Figure 4.14 Sequence of reactions mediated by *Thiobacillus* and *Ferrobacillus* in the environment starting with iron pyrites and finishing with ferric hydroxide and many hydrogen ions leading to acidic conditions.

of energy. The chemical reaction sequence in Figure 4.14 is set in motion with the slow buildup of microorganism numbers and the alteration of ambient chemical conditions. Finally, there is the formation of ferric hydroxide together with many other metallic components, such as arsenic, copper, and nickel, depending on the levels of these metals in the environment. However, the amounts of ferric hydroxide can be very high leading to large solid deposits in waterways having an orange or yellow color. In addition, a major reaction product is the hydrogen ion (H^+) leading to strongly acid conditions.

The early history of this **acid drainage** process is principally concerned with mining activities and related operations such as tailings dams and mine rock waste dumps. In these situations, evidence of contamination by metals and low measures of pH reflecting acidic conditions have been reported in drainage water and adjacent streams. Thus, there are many instances described in the scientific literature of **acid mine drainage water** as a variant of acid drainage. In more recent times, there are many instances of this process occurring in other environments and waters including drinking water, agricultural runoff, drainage in swampland, and so on.

Possibly the most serious of these incidences of **acid drainage** is that which occurs in Bangladesh. Here, drinking water is provided by an extensive set of bores penetrating into the substrata. The bores have exposed the mineral elements in the substrata to oxygen from the atmosphere together with water for the first time. In this environment, the acid drainage process has developed and become prevalent in the bores and adjacent substrata with the subsequent contamination of the drinking water from the bores. Millions of people have consumed drinking water contaminated mainly with arsenic and continue to do so with major adverse effects on public health. Nevertheless, acid drainage is a major problem manifesting itself in various ways in a range of environments contaminating waterways and waters throughout the world inflicting extensive damage on the natural and human environment.

Acidification of the Oceans

Currently, carbon dioxide occurs in the Earth's atmosphere at a level of about 408 parts per million (~0.04%) and on a geological scale has seen considerable fluctuations over time and global conditions. It dissolves in water to form carbonic acid which then dissociates to form carbonate and bicarbonate ions and hydrogen ions as shown in Figure 4.15. For pure water, as in rainfall, the pH, due to the hydrogen ions formed by the dissolved carbon dioxide, would be about 5.7. However, the composition of the oceans is not pure water but water with many dissolved components including sodium, chloride ions, and, importantly, carbonate ions.

Calcium carbonate ($CaCO_3$) is a component of seawater which is of major biological importance. This substance is needed for the formation of structural skeletons or shells in corals, molluscs, and many other organisms. The world's ocean waters contain large amounts of calcium carbonate in the ionic forms of Ca^{2+} and CO_3^{2-} derived largely from limestone rock formations. These ions are taken up by organisms under precise conditions of pH and other factors and converted into skeletons and shells.

An increase in the concentration of carbon dioxide in the atmosphere has resulted from the discharge of carbon dioxide during the industrial era leading to a concentration increase from about 300 ppm in 1900 to over 400 ppm. This process is continuing and is predicted to lead to further increases in atmospheric carbon dioxide (see Chapter 14). This has produced a corresponding increase in dissolved carbon dioxide and related ionic products. The overall effect is an increase in the acidification of the world's oceans, an increase in the hydrogen ion concentration, and an average decrease in the pH of the oceans, of about 0.3 of a pH unit from about an ocean pH of 8.2–7.9.

Diurnal and seasonal variations in pH and related factors occur without the influence of carbon dioxide particularly in coastal regions. However, most organisms established in these environments for long periods of time are thought

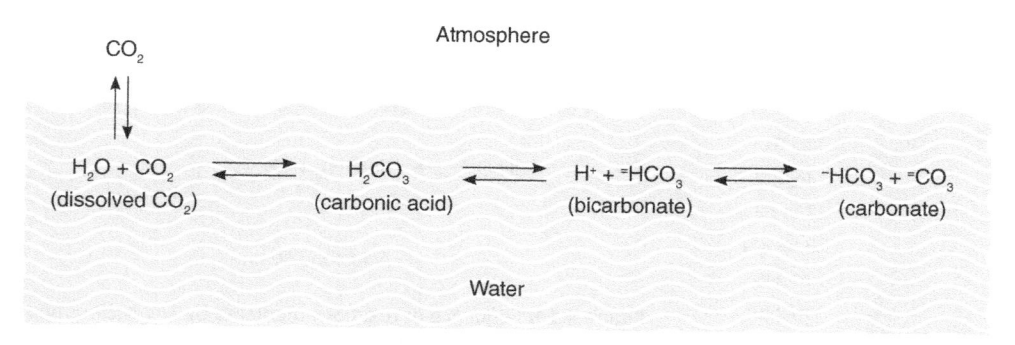

Figure 4.15 Interactions of carbon dioxide in the atmosphere with water in natural water bodies such as the oceans, lakes, rivers, and other water bodies.

Acid Rain

It was mentioned in the previous section that carbon dioxide in the atmosphere dissolves in rainwater and forms carbonic acid by the processes shown in Figure 4.15. If the rainwater is unaffected by any other contaminants, this process results in a lowering of the pH to about 5.7. However, there may be other atmospheric contaminants present such as sulfur dioxide (SO_2) and nitrogen oxides which also dissolve in the rainwater resulting in lower pH levels below 5.7. Acid rain has an adverse effect on aquatic and terrestrial ecosystems. Fish and other aquatic organisms are adversely affected which increases with declining pH. In addition, trees and terrestrial plants can be badly affected sometimes resulting in death (see also Chapter 13).

4.16 Key Points

1. Pollutants can enter the environment in the form of liquids, gases, solids, and other forms and be distributed according to physicochemical and physical processes of the pollutant and the environment itself.
2. The physical state and fundamental physicochemical properties of a chemical pollutant are determined by the inherent characteristics of the molecule such as molecular weight, surface area, and so on, as illustrated in Figure 4.1.
3. The environment can be seen to consist of different **compartments** or **phases.** Commonly, the environment is seen to have several abiotic phases – an atmospheric phase (the air), an aquatic phase (the sea, rivers, lakes, etc.), soils, and sediments. In many situations, biota (humans, fish, invertebrates, etc.) can act as a phase.
4. Distribution within the phases in the environment leads to the presence of the chemical pollutant in various abiotic phases at different levels according to the properties of the chemical and the phase and the exposure of biota present.
5. The basic properties of the molecule and the physicochemical properties of pollutants are interdependent, and these properties determine the nature of the environmental response from environmental distribution to biological effects.
6. Generally, most of the physicochemical characteristics can be measured in the laboratory and where this is not possible, they are usually calculated as is done with the basic molecular characteristics.
7. The properties of chemicals and the environmental phases govern the partition mechanisms that control the distribution of pollutants between environmental compartments.
8. **The Partition Coefficient.** The equilibrium distribution of a chemical substance between two immiscible phases occurs as a fixed proportion, the **Partition Coefficient (K)** where K is the Concentration in Phase 1/Concentration in Phase 2, i.e. $K = C_1/C_2$. Thus, the Partition Coefficient is independent of concentration and measured at a constant temperature and pressure.
9. **Freundlich Isotherm.** This is an empirical relationship (see Figure 4.3) and is expressed as follows:

$$C_1 = KC_2^{1/n} \tag{4.8}$$

where C_1 is the equilibrium concentration of the chemical in the liquid; C_2, equilibrium concentration of the chemical in the solid or second liquid; K, the equilibrium constant indicative; and $1/n$ is a constant describing the degree of nonlinearity.

A linear form of this relationship is log

$$C_1 = \log K + (1/n) \log C_2 \tag{4.9}$$

10. **The n-Octanol–Water Partition Coefficient.** Probably the most widely used property in environmental chemistry and pharmaceutical sciences to predict environmental behavior and effects is usually represented by the symbols K_{ow}, P, or P_{ow}. It is a measure of the **lipophilicity** of a chemical which evaluates the potential of a chemical to accumulate in lipids. It is the partition coefficient between n-octanol and water.
11. **Air–Water Partitioning – Henry's Law.** Describes the behavior of chemicals in the water–atmosphere system and states that the amount of dissolved gas is proportional to its partial pressure in the gas phase with the proportionality factor being called the Henry's Law constant (H). Thus,

$$H = P/C_W \tag{4.11}$$

where P is the partial pressure of the compound in air and C_W the corresponding concentration in water at equilibrium.

12. **Soil and Sediment Partitioning with Water.** Sediment/soil from the environment and water can be brought together in an experimental system in the laboratory and the concentration in the two phases measured at equilibrium. Thus, the soil–water partition coefficient (K_D) is

$$K_D = \frac{C_S}{C_W} \quad \text{at equilibrium}$$

$$\text{With soils,} \quad K_D = 0.66 \cdot f_{OM} \cdot K^a_{OW} \tag{4.14}$$

where C_S and C_W are soil and water concentrations at equilibrium; f_{OM}, fraction of organic matter; and K_{OW}, the octanol–water coefficient.

13. **Uptake and Retention of Chemicals by Organisms.** These processes can be described by the following terms:

    Bioaccumulation is the uptake and retention of pollutants from the environment by organisms via any mechanism or pathway.

    Bioconcentration is uptake and retention of pollutants directly from the water mass by organisms through such tissues as the gills or epithelial tissues.

    Biomagnification is the process whereby pollutants are passed from one trophic level to another and exhibit increasing concentrations in organisms related to their trophic status.

14. **Single Compartment Model.** The bioconcentration process can be seen as a balance between two kinetic processes, uptake and depuration (see Figure 4.7), as characterized by the first-order rate constants, k_1 and k_2, respectively. The rate of change of chemical pollutant in an organism is given by

    $$\frac{dC_B}{dt} = k_1 C_M - k_2 C_B \qquad (4.20)$$

    where C_B is the biotic concentration; C_M, the concentration in the ambient environment (water); k_1, the uptake rate constant; k_2, the depuration rate constant; and t, the elapsed time. Thus,

    $$C_B = \left(\frac{k_1}{k_2}\right) C_M \left(1 - e^{k_2 t}\right)$$

    Thus, for depuration: $C_B = C_{BO}\left(1 - e^{k_2 t}\right)$

    (4.24)

15. The Stockholm Convention on PoPs was formulated in 2001 and at that time it applied to 12 substances, known as the Dirty Dozen, including DDT and other chlorohydrocarbon pesticides, PCBs, dioxins, and furans. The current convention covers about 24 substances which includes pesticides and industrial chemicals, or by-products, and there are several additional chemicals listed for consideration.

16. The LRT process results from the use of POPs in warm equatorial areas subsequently leading to deposition in cooler regions through a series of volatilizations and depositions. This mechanism can be repeated on a global scale and has been described as the **Grasshopper effect** resulting from **Global distillation**.

17. Some **heavy metals,** such as mercury, lead, and cadmium, and some metalloids, including arsenic, are of considerable environmental management importance, since they often present hazards to human health and the natural environment. Often, these substances accumulate in biological tissues and organs.

18. **Quantitative Structure–Activity Relationships.** QSARs are relationships between structure and properties of compounds and their biological activity. The inherent characteristics of the molecule control the physicochemical properties of the compound which then guide the environmental distribution and other characteristics of the chemical.

19. The environment can be considered to be made up of compartments which can also be described as phases, for example, water, air, soil, sediments, and when a chemical is discharged, it distributes between these environmental phases according to partition processes.

20. A method of combining all of the partition coefficients and phases together is based on the **fugacity** property to arrive at the equilibrium distribution.

21. The fugacity models are available on the Web at different levels of complexity with each level building on the previous level. Level I is the basic level and there are levels up to Level IV which is an open system taking into account many of the factors which exist in natural environments.

22. Advection, diffusion, and leaching are important transport processes resulting in the movement of chemicals in the environment.

23. Chemical transformation and degradation occur in the environment principally by oxidation and hydrolysis. The kinetics of these reactions can often be described by first-order kinetics described in Key Point 14.

References

Akcil, A. and Koldas, S. (2006). Acid Mine Drainage (AMD): causes, treatment and case studies. *Journal of Cleaner Production* 14: 1139–1145.

Beyer, A., Mackay, D., Matthies, M. et al. (2000). Assessing long-range transport potential of persistent organic pollutants. *Environmental Sciences and Technology* 34 (4): 699–703.

Boethling, R.S. and Mackay, D. (2000). *Handbook of Property Estimation Methods for Chemicals: Environmental and Health Sciences*. Boca Raton, FL: Lewis.

de Bruijn, J. and Hermens, J. (1990). Relationships between octanol/water partition coefficients and total molecular surface area and total molecular volume of hydrophobic organic chemicals. *Quantitative Structure-Activity Relationships* 9: 11–21.

Carson, R. (1962). *Silent Spring*. Boston, MA: Houghton Mifflin, First Printing.

CEMC (2020). *Canadian Centre for Environmental Modeling and Chemistry*. Canada: Trent University https://www.trentu.ca/cemc/resources-and-models (accessed 15 September 2020).

Connell, D.W. (1988). Quantitative structure activity and relationships and the ecotoxicology of chemicals in aquatic systems. *Atlas of Science* 1: 221–225.

Connell, D.W. (1990). *Bioaccumulation of Xenobiotic Compounds. Biomagnification of lipophilic compounds in terrestrial and aquatic systems*. Boca Raton, FL: CRC Press.

Connell, D.W. (2005). *Basic Concepts of Environmental Chemistry*, 2e. Boca Raton, FL: CRC Press.

Donati, E. (ed.) (2018). *Heavy Metals in the Environment*. Boca Raton, FL: CRC Press.

Egorova, K.S. and Ananikov, V.P. (2017). Toxicity of metal compounds: knowledge and myths. *Organometallics* 36: 4071–4090.

Erickson, M.D. (2000). *PCB Properties, Uses, Occurrence, and Regulatory History*. New York: Wordpress.

Gouin, T., Coombes, V., Ericson, J. et al. (2014). Adsorption–Desorption Distribution (Kd) and Organic Carbon–Water Partition (KOC) Coefficients. Technical Report 123. European Centre for Ecotoxicology and Toxicology of Chemicals (Ecetoc), Brussels. http://www.ecetoc.org/report/measured-partitioning-property-data/adsorption-desorption-distribution-kd-and-organic-carbon-water-partition-koc-coefficients/ (accessed 15 September 2020)

Hanaor, D.A.H., Ghadiri, M., Chrzanowski, W., and Gan, Y. (2014). Scalable surface area characterization by electrokinetic analysis of complex anion adsorption. *Langmuir* 30 (50): 15143–15152.

Lyman, W.J., Reehl, W.F., and Rosenblat, D.H. (1990). *Handbook of Chemical Property Estimation Methods – Environmental Behaviour of Organic Compounds*. Washington, DC: American Chemical Society.

Mackay, D. (1979). *Finding fugacity feasible. Environmental Science and Technology* 13 (10): 1218–1223.

Mackay, D. (1991). *Multimedia Environmental Models: The Fugacity Approach*. Boca Raton, FL: Lewis Publisher.

Mackay, D. (2001). *Multimedia Environmental Models: The Fugacity Approach*, 2e. Boca Raton, FL: CRC Press.

Mackay, D. and Patterson, S. (1981). Calculating fugacity. *Environmental Science and Technology* 15: 1006–1014.

Mackay, D. and Patterson, S. (1982). Fugacity revisited. *Environmental Science and Technology* 16: 654–660.

Mackay, D., Shiu, W.-Y., Ma, K.-C., and Lee, S.C. (2006). *Handbook of Physical–Chemical Properties and Environmental Fate for Organic Chemicals*, 2e. Boca Raton, FL: CRC Press.

McGeer, K.V., Brix, J.M., Skeaff, D.K. et al. (2003). Inverse relationships between bioconcentration factor and exposure concentration for metals: Implications for hazard assessment of metals in the aquatic environment. *Environmental Toxicology and Chemistry* 22 (5): 1017–1037.

Moriarty, F. (1975). *Organochlorine Pesticides: Persistent Organic Pollutants*. London: Academic Press.

OECD (2001). *Test No. 121: Estimation of the Adsorption Coefficient (K_{oc}) on Soil and on Sewage Sludge using High Performance Liquid Chromatography (HPLC)*. Paris: Organisation for Economic Cooperation and Development https://www.oecd-ilibrary.org/environment/test-no-121-estimation-of-the-adsorption-coefficient-koc-on-soil-and-on-sewage-sludge-using-high-performance-liquid-chromatography-hplc_9789264069909-en (accessed 15 September 2020).

OECD (2006). Guidelines for the testing of chemicals. In: *Test No. 123: Partition Coefficient (1-Octanol/Water): Slow-Stirring Method*. Paris: Organisation for Economic Cooperation and Development https://www.oecd-ilibrary.org/environment/test-no-123-partition-coefficient-1-octanol-water-slow-stirring-method_9789264015845-en (accessed 15 September 2020).

OECD (2009). *The OECD Pov and LRTP Screening Tool (Version 2.2)*. Paris: Organisation for Economic Cooperation and Development http://www.oecd.org/chemicalsafety/risk-assessment/oecdpovandlrtpscreeningtool.htm (accessed 15 September 2020).

OECD (2012). *Test No. 305. Bioaccumulation in Fish: Aqueous and Dietary Exposure*. Paris: Organisation for Economic Cooperation and Development https://www.oecd-ilibrary.org/environment/test-no-305-bioaccumulation-in-fish-aqueous-and-dietary-exposure_9789264185296-en (accessed 15 September 2020).

Sander, R. (2015). Compilation of Henry's law constants (version 4.0) for water as solvent. *Atmospheric Chemistry and Physics* 15: 4399–4981.

UNEP (2020). *All POPs Listed in the Stockholm Convention*. United Nations Environment Programme http://www.pops.int/TheConvention/ThePOPs/AllPOPs/tabid/2509/Default.aspx (accessed 15 September 2020).

US EPA (2017). Persistent Organic Pollutants: A Global Issue, A Global Response. https://www.epa.gov/.../persistent-organic-pollutants-global-issue-global-response (accessed 15 September 2020).

Walker, C.H. (2009). *Organic Pollutants - An Ecological Perspective*, 2e. Boca Raton, FL: CRC Press.

Wolfe, N.L. and Jeffers, P.M. (2000). Hydrolysis. In: *Handbook of Property Estimation Methods for Chemicals* (ed. R.S. Boethling and D. Mackay), 311–333. Boca Raton, FL: Lewis Publishers.

Note: Other Chapter 4 Figures are in the Chapter 4 Tests/Answers in Companion Book.

5

Environmental Toxicology and Ecotoxicology

5.1 Introduction

Environmental toxicology is broadly concerned with the harmful interaction of physical, chemical, and biological agents from pollution of the environment, with living organisms and systems. Thus, it encompasses interactions with living organisms and systems at all levels of organization from molecular and cellular to global. Environmental toxicology can be seen as consisting of two major areas, one area covering the interaction of pollutants with humans, including such fields as epidemiology, occupational health, and so on, as well as another area covering interactions with natural organisms and systems. This latter area is often described as **Ecotoxicology** and is specifically concerned with the harmful interactions of environmental agents with ecosystems.

The conceptual diagram in Figure 1.3 illustrates the sequence of processes and interactions of a toxic chemical pollutant with the environment. The sources and overall impacts of pollutants on the global environmental system are described in Chapters 1, 2, and 3. Chapter 4 outlines the initial processes after discharge or releases when a toxicant distributes within the phases in the environment according to its physiochemical properties and the properties of the phases themselves. As a result of this process, the chemical becomes distributed in air, water, soil, aquatic sediments, and possibly other phases, according to the environment being considered. These processes can be seen as preceding the actual exposure of individual organisms in both the natural and human environments and occur as depicted in Figure 5.1.

Environmental toxicology and ecotoxicology are concerned with the evaluation of harmful adverse effects and these disciplines form the basis of scientific efforts to control and manage adverse effects in the environment. An important management procedure involves the testing of chemicals in the laboratory as a basis for the registration of chemicals. These key laboratory tests are now an accepted international procedure for the control and management of environmental chemicals and pollutants.

The OECD (2020a) has developed a collection of the most relevant internationally agreed testing methods to be used by governments, industry, and independent laboratories. The Guidelines seek to evaluate chemicals on a broad basis following the concept, as illustrated in Figure 5.1, from distribution characteristics through to assimilation of toxicants, to consequent toxicological evaluations.

The assimilation of pollutant chemicals within organisms at sublethal or lethal levels may induce a sequence of adverse biological effects. These range from molecular interference with biochemical mechanisms and interactions with cellular organelles (e.g. DNA and RNA molecules), through to pathological changes at the cellular, tissue, and organ levels. Finally, these result in an integrated functional or behavioral response, experienced at the whole organism level, which may be reversible or irreversible.

There is a need to extend our understanding of the significance of such toxic effects from the individual level to higher levels of biological organization and complexity to determine the effects at the population, community, and ecosystem levels. On the other hand, new insights into biochemistry, cellular biology, cytogenetics, and pharmacology have focused increased attention on the ability of toxic agents to induce responses and pathological changes at the cellular level through interaction with cellular organelles and enzymes. Concurrent with this trend, an extensive array of sublethal effects in organisms subjected to environmental stresses or physical and chemical agents has been revealed.

The problem confronting ecotoxicologists is the complex task of identifying (i) the consequences to individuals in the natural environment of effects generally demonstrated in laboratory studies and (ii) the ecological significance of these often subtle effects experienced by an individual. Approaches to this task are based on the structural and functional interrelationships existing between each succeeding level of biological organization. Thus, it is important to determine the relationships between effects demonstrated at the macromolecular or cellular level and

Chemistry and Toxicology of Pollution: Ecological and Human Health, Second Edition. Des W. Connell and Greg J. Miller.
© 2023 John Wiley & Sons, Inc. Published 2023 by John Wiley & Sons, Inc.
Companion website: www.wiley.com/go/toxicologyofpollution2e

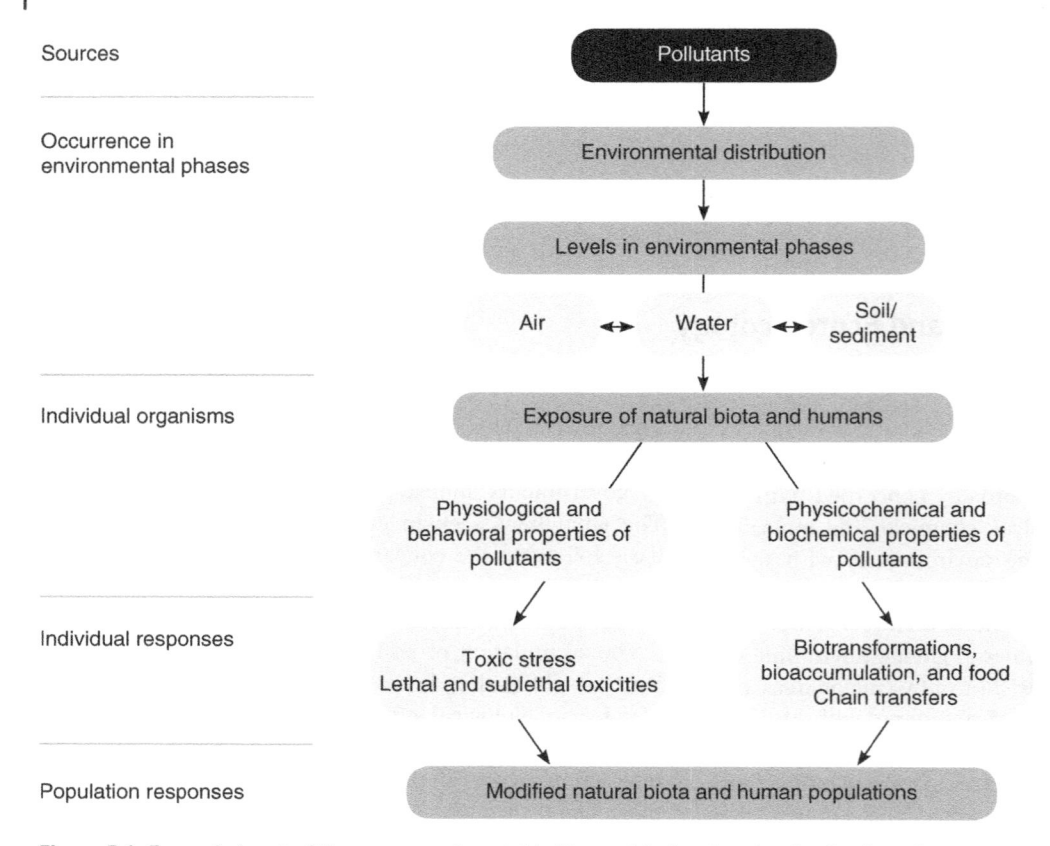

Figure 5.1 Expanded part of the conceptual model in Figure 1.3 showing the distribution of pollutants in the environment resulting in the exposure of individuals in systems of natural biota and humans.

the ultimate response of the organism, and the response at the population, community, and ecosystem level.

Human health evaluations are also subject to similar complexity, particularly due to the fact that toxicological investigations are usually carried out on surrogate animals such as rats and mice. With these laboratory investigations, extrapolations from the surrogate animals to humans are required. In addition, while the human population involves only one species, it has great complexity due to the presence of different age groups from babies to the aged, different exposure settings, and different health status of individuals. It is very rare for these subgroups in the human population to be the specific subject of either experimental or epidemiological investigation. Of course, experimental investigations of humans in the laboratory are ethically forbidden.

A thorough knowledge of basic mechanisms by which toxic agents impair organisms is fundamental to any ecological assessment. The relevant properties of the chemical contaminant can be divided into two groups – first, *the physiological and behavioral properties* which lead to toxic stress and lethality and second, *physicochemical and biochemical properties* which lead to biotransformations, bioaccumulation, and food chain transfer as shown in Figure 5.1. But, in addition to toxicant–effect relationships, it is important to evaluate the interaction of environmental factors such as ambient temperature, light, and so on, which may modify responses observed in organisms.

5.2 Exposure Pathways for Humans and Natural Biota

The processes from discharge of the contaminant to distribution in the environment all lead to exposure of individual organisms in the natural or the human environment and consequent effects at different levels of biological organization as shown in Figure 1.3. The responses at the individual level are illustrated in Figure 5.1. The levels of a contaminant which occur in air, water, soil/sediments, and other phases result in uptake of the contaminant through exposure pathways, and resultant physiological, biochemical as well as behavioral responses occur consequently. These responses depend on the physiological and behavioral properties as well as the physicochemical and biochemical properties of the contaminant and the biological characteristics of the individual organism.

5.2 Exposure Pathways for Humans and Natural Biota

Of course, this quantification will be specific for human individuals, or groups of individuals, or types of biota in the natural environment. For example, children will usually differ from adults, occupational exposure will differ from residential exposure, so these have to be treated separately. In the natural environment, mammals will differ from fish, fish will differ from molluscs, and so on. The US EPA (1992, 2017) has developed tools for the estimation of exposure through different routes.

When the exposure pathways have been identified with a specific population of individuals, these can be quantified using data on measured or estimated levels in the appropriate media. Thus, with mammals

Total Daily Dose or Intake (TDD or TDI)

$$= \Sigma(\text{Intake through the individual pathways}) \quad (5.1)$$

The Average Daily Dose or Intake (ADD or ADI) of a pathway can be calculated from the general equation below:

$$\text{ADD or ADI} = \frac{(C_i A_i BA_i)}{\text{bw}} \text{ mg kg}^{-1} \text{ (body weight) day}^{-1} \quad (5.2)$$

where C_i is concentration in the medium involved in the pathway; A_i, the amount of medium involved per day; and BA_i, the bioavailability of the contaminant to the organism which is a unitless proportion.

This calculation of ADD introduces several important factors into the estimation of exposure of mammals. First, there is a time factor where the *day* is used as a basic time unit for exposure, and second, there is the introduction of the weight of the organism. The size of the organism allows the distribution and dilution of the contaminant to occur; thus, this is accounted for by *body weight* (bw) in the equation.

However in many situations, the environment is the occupational environment and exposure occurs on a routine basis such as daily, weekly, monthly, or seasonally. Thus, in the human occupational environment, the estimation of exposure must allow for this repeated exposure over time. In this situation, the following equation for Lifetime Average Daily Dose (LADD) can be derived based on the ADD:

Lifetime Average Daily Dose (LADD)

$$= \text{ADD} \times \left(\frac{\text{EF.ED}}{\text{AT}}\right) \quad (5.3)$$

where (EF. ED/AT) is an adjustment to the ADD and EF is Exposure Frequency (days year^{-1}), ED is the Exposure Duration (years), and AT is the Averaging Time (this is often taken over a lifetime of 70 years × 365 days).

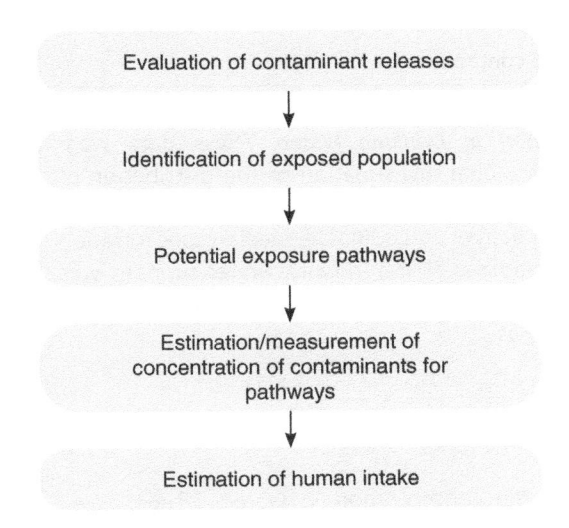

Figure 5.2 Steps involved in quantifying the exposure of individual organisms to a contaminant.

In the evaluation of the significance and biological effects, there is a need to identify the exposure pathways to the individual organisms. Following the concept outlined in Figures 1.3 and 5.1, the quantitative evaluation of exposure can be visualized as involving a number of steps as shown in Figure 5.2. The possible exposure pathways with human uptake of contaminants are shown in Figure 5.3.

The pathways or routes of exposure have been described by the International Program on Chemical Safety (IPCS 2004) for human exposure and the pathways involved with most biota can be identified within the following general classes.

- Inhalation through vapors or dust with mammals.
- Uptake through the respiratory pathways with aquatic organisms.
- Ingestion with food and drinking water.
- Direct ingestion of contaminated materials.
- Dermal sorption.

Inhalation
- Air vapors
- Airborne dust/soil

Ingestion
- Dust/soil
- Drinking water
- Food (crops, meat, dairy, seafood)

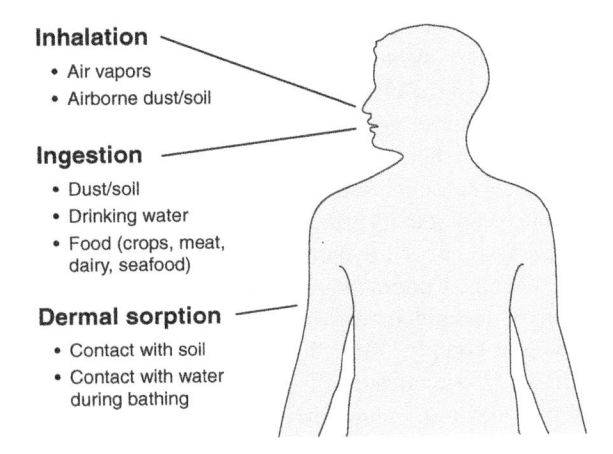

Dermal sorption
- Contact with soil
- Contact with water during bathing

Figure 5.3 Possible exposure pathways with humans.

> **Box 5.1 Some examples of the calculation of the exposure of humans to contaminants**
>
> **Worked Example 5.1** *Dermal Exposure of Humans to Chlorpyrifos in an Occupational Environment* Data used in this example were obtained from farmers in Ghana spraying chlorpyrifos onto rice crops (Atabila et al. 2017). The Total Dermal Exposure (TDE) was obtained from the farmers wearing absorptive suits during a spraying event and the analysis of the suits for sorbed chlorpyrifos. Other information was obtained by questionnaire. The data are for a typical farmer where the Dermal Average Daily Dose (DADD) can be calculated using Eq. (5.4).
>
> $$\text{DADD} = \frac{(\text{TDE} \times \text{DAF})}{\text{bw}} = \frac{(26\,000 \times 0.043)}{70}$$
> $$= 16\,\mu\text{g kg}^{-1}\,\text{bw day}^{-1} \qquad (5.4)$$
>
> where TDE is 26 000 µg; DAF (Dermal Absorption Factor), 0.043 unitless; and bw, 70 kg.
>
> The Lifetime Dermal Average Daily Dose (LDADD) is calculated from Eq. (5.5).
>
> $$\text{LDADD} = \text{DADD} \times \left(\frac{\text{EF.ED}}{\text{AT}}\right) = 16\,\frac{(6 \times 48)}{(62 \times 365)}$$
> $$= 0.20\,\mu\text{g kg}^{-1}\,\text{bw day}^{-1} \qquad (5.5)$$
>
> where EF is Exposure Frequency in days year^{-1}, i.e. 2 spraying events at 3 days per event; ED, the exposure duration (years) 48 working years; and AT, time over which exposure is calculated. The life expectancy in Ghana is 62 years; thus, AT is 62×365 days.
>
> **Worked Example 5.2** *Human Exposure to Trichloromethane (THM) in Drinking Water* These data were taken from a global investigation of the distribution of chlorination by-products in drinking water in world water supplies (Hamidin et al. 2008). The median concentration of trichloromethane (THM) in world water supplies was reported as $38\,\mu\text{g L}^{-1}$ and the ADD can be calculated from Eq. (5.2) as
>
> $$\text{ADD} = \frac{(C_i A_i \text{BA}_i)}{\text{bw}} = \frac{(38 \times 2 \times 1)}{70}$$
> $$= 1.1\,\mu\text{g kg}^{-1}\,\text{bw day}^{-1} \qquad (5.6)$$
>
> where C is the concentration in water; $38\,\mu\text{g L}^{-1}$; A_i, amount of water consumed per day (L day^{-1}); BA_i, bioavailability is assumed to be unitless and unity; and bw is 70 kg.
>
> This can be converted to a lifetime basis using Eq. (5.3); thus, the LADD is calculated as
>
> $$\text{LADD} = \text{ADD} \times \left(\frac{\text{EF.ED}}{\text{AT}}\right) = 1.1 \times \frac{(365 \times 70)}{(70 \times 365)}$$
> $$= 1.1\,\mu\text{g kg}^{-1}\,\text{bw day}^{-1} \qquad (5.7)$$
>
> It should be noted that the ADD and the LADD are the same. This occurs because the exposure is on a daily basis and it is continuous for a lifetime.

This calculation adjusts the ADD for the number of times exposure occurs in a year and for the number of years over which this occurs, and then averages this over a specific time, often a human lifetime. Of course, these factors can be changed to fit the circumstances involved.

The above set of calculations can be applied to mammals, particularly humans, but a different approach is used with aquatic organisms. In this case, the levels in the water environment can be used as a measure of exposure.

Some examples of calculation of the exposure of humans to toxicants in the environment are contained in Box 5.1.

5.3 Toxicant Behavior in Organisms

Background

The relationships between chemical structures and the biological properties of compounds allow mechanisms to be developed explaining the interaction of certain chemical structures with specific cellular processes. This approach also permits the development of Quantitative Structure–Activity Relationships (QSARs) between structure and toxic action which is considered later in this chapter. Current knowledge of toxic mechanisms offers a basis to formulate generalized concepts of toxicity shown in Figure 5.4. This illustrates a basic toxicological framework which involves a succession of processes from uptake or absorption of the toxicant to the presence of the toxicant at the site of action within the organism leading to the observed toxic effects exhibited by the exposed organism. The initial phase covers those biological processes that control the entry of the chemical into the organism which then lead to processes governed by the distribution and metabolism of the chemical agent. It is this kinetic phase that determines the chemical form and transport of the active chemical agent to its primary site of action.

At this point, there is a dynamic phase where the proximal toxic chemical interacts with the primary target or receptor (enzyme, lipid, membrane, nucleic acid, etc.). The ligand–receptor complex formed in the primary reaction, or, alternatively, any reactive free radicals (e.g. peroxides and hydroxyl ions) may initiate a complex sequence of biological effects that transform into a lethal or sublethal response. Table 5.1 summarizes some of the possible

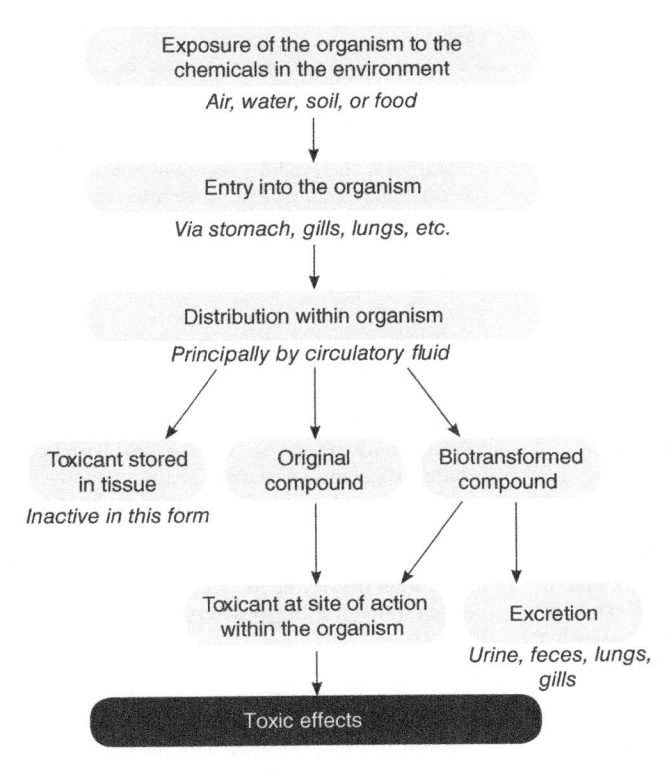

Figure 5.4 Uptake of a toxicant from the environment and distribution within an organism leading to the presence of the toxicant, or its biotransformed product, at the site of action leading to the observed toxic effects.

mechanisms through which toxic agents can impair important biochemical processes and physiological functions in living organisms leading to observed toxic effects. The degree of response will depend on the actual concentration, or dose, that reaches the receptor or target tissues in the dynamic phase.

Toxicological Reactions

Two basic types of toxic reactions can be distinguished which involve the parent compound, or a metabolite, as the active agent.

1. Chemical lesions which are caused by irreversible covalent binding between the chemical agent and the biological substrate or receptor, for instance, the covalent binding of toxic metals such as lead, cadmium, and mercury, to SH groups in enzymes and other important proteins, as well as the binding of hydrogen cyanide (HCN) and hydrogen sulfide (H_2S) to the iron groups in cytochromes. Other types of lesions include carcinogenic and mutagenic reactions involving DNA molecules and reactive intermediates such as biological alkylating agents and microsomal metabolites (e.g. epoxides).

2. Reversible interactions which occur between the exogenous substance and biological substrate or receptor. These substances undergo reversible interactions with specific receptors, usually leaving the substance and receptor unaltered. In some instances, the receptor may experience temporary alterations in its conformations during the formation of the ligand–receptor complex.

Further resolution of reaction mechanisms associated with toxic reactions, such as cellular necrosis, has resulted in the postulation of several possible molecular mechanisms of toxicity (e.g. Gillette 1980; Farber 1980).

5.4 Relationships Between Exposure to a Toxicant as Dose or Environmental Concentration and the Resultant Biological Effects

Introduction

In considering the relationship between dose or environmental concentration of an environmental toxicant and the resultant reaction in biological systems, it is useful to distinguish between different types of responses and effects. In general terms, the dose–response can be seen as the relation between a measurable stimulus of a physical, chemical, or biological nature and the response of living matter in measurable terms. The responses to one single stimulus can be multiple with each observed response being of a specific type and quantity. A specific response can be evaluated in several different ways, for example, the magnitude of the effect, whether the effect is produced or not, or the time taken to produce the effect. The influence of the exposure time on toxicity is extremely important when considering all biological effects. However, it is common to use the terms **acute** to describe short-term readily apparent adverse effects, and **chronic** to describe long-term effects. Lethal effects are usually relatively easy to observe and record but sublethal effects can be relatively difficult to observe and record quantitatively. The OECD (2020a) has produced a set of internationally accepted guidelines for the experimental conduct of tests on chemicals.

Unless exact mechanisms or modes of toxic action within organisms are known, it is common to measure the relation between different dose levels of toxicants and induced effects on organisms or biological systems, that can be either readily observed in nature or under experimental test conditions in the laboratory. Relationships are generally expressed in the form of dose–response or effect plots which take the usual curved form as described generally in Figure 5.5.

Table 5.1 Summary of some important biochemical and physiological effects of toxic agents.

Biochemical/physiological effect	Function	Toxic response
Disruption of cellular membrane function	Membranes form semipermeable barriers with specific properties which control the rate and transfer of substances into and out of cells and cell organelles.	Disruption and modification of membrane permeability leading to adverse interference with transfer systems.
Modification of enzyme action	Enzymes perform highly specialized catalytic activities with many complex reactions which support the living process with cells and whole organisms.	Reversible or irreversible modification of catalytic properties of the enzyme.
Disturbance of lipid metabolism	Lipids are one of the basic chemical components of cells and organisms. They have structural functions as well as being involved with storage and transfer of energy.	Disturbance may result in impaired liver function, including pathological lipid accumulation in the liver. The capacity of the liver to produce cholesterol may be impaired.
Interference with protein biosynthesis	The biosynthesis of specific cellular proteins, including enzymes, involves assembly of the constituent amino acids into a specific sequence in the polypeptide chain. The sequence required is determined by the structure of the nuclear DNA.	Protein biosynthesis can be interfered with by many exogenous substances including many toxic agents which modify the nuclear DNA causing the formation of proteins which have adverse biological effects.
Microsomal enzyme system modification	There is a multifunction enzyme system located in the hepatic microsomal section of the cell. This is responsible for the oxidation of substances which are hydrophobic and lipophilic (water insoluble) such as DDT and similar compounds converting them to water-soluble compounds suitable for excretion.	Alteration in enzyme function by stimulation or inhibition leading to adverse biological effects.
Interference with carbohydrate metabolism	Carbohydrates and related compounds are major components of cells which have important structural functions as well as being involved with the transfer of energy.	Toxic agents may interfere with the normal processes of carbohydrate biosynthesis and metabolism.
Interference with respiration	Respiration facilitates the process of introduction of atmospheric oxygen to organisms allowing biological oxidation in an organism to occur. This is vital to living organisms since it releases energy for normal life functions. This respiratory chain involves the transfer of energy utilizing ATP and ADP.	The respiratory chain can be disrupted at various sites by different chemical agents causing adverse biological effects with the organism involved.

Hypothetical relationships between observed effects and toxicant exposure dose or concentration, described as a dose–response curve, are represented in Figure 5.5. Two basic types of dose–response curves and biological effects are represented by Substance 1 and 2. Substance 1 is essential for life at trace concentrations and toxic at higher concentrations, whereas Substance 2 is a nonessential toxic substance which is toxic at all concentrations. This latter plot is typical of the dose–response plots for a toxic substance.

With Substance 1 below the concentration C_1, a deficiency in the essential substance exists which increases as the concentration becomes lower until there is no essential substance present and the organism cannot survive and death results; between concentration C_1 and concentration C_2 an adequate amount of the essential substance is present, and there are no harmful effects, in fact beneficial effects result. At excessive concentrations above C_2, the Substance 2 acts as a toxicant, and irreversible damage and ultimately death may occur. Important examples of this type of substance are the essential trace elements, including Na, K, P, and Mg. These substances are all essential in trace quantities to support the growth of living organisms but become toxic at higher concentrations than the essential requirement.

With Substance 2 below the concentration C_1, the substance performs no essential biological function and a very low level, effectively no harmful effect occurs as a result of exposure. This plot follows a normal toxicity plot with adverse effects continuing to occur up to concentration C_2. From this point onward, Substance 1 and 2 follow essentially the same general plot of dose–response against toxicant concentration.

It is noteworthy that the plot for Substance 2 is the usual shape for dose–response curves observed with most toxicants. At very low doses or concentrations, a small number of adverse effects are observed with the most sensitive individuals. But with increasing dose or concentration, the

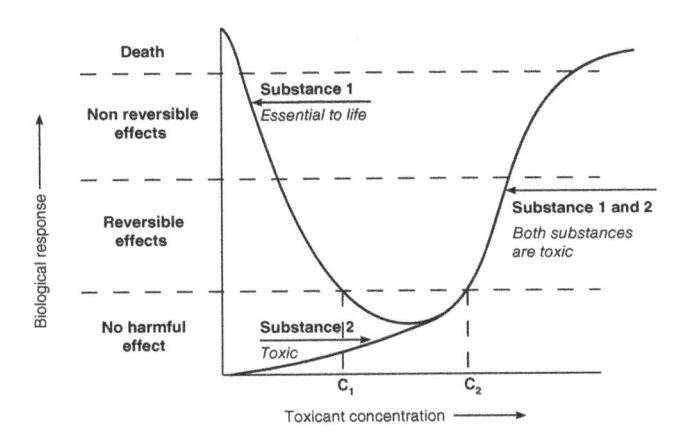

Figure 5.5 Two basic types of dose–response curves and biological effects are represented by Substance 1 and 2. The Substance 1 plot is of the effects on an organism exposed to a substance essential for life at trace concentrations but toxic at higher concentrations. Substance 2 is a plot of the effects induced by a nonessential toxic substance at all concentrations. This plot is typical of the dose–response plots for a toxic substance.

number of adverse effects increases with most individuals exhibiting adverse effects. At the highest dose or concentration, even the most resistant individuals exhibit adverse effects.

Relationships Between Dose or Environmental Concentration and Biological Response

When a toxicant interacts with the receptors at a site of action within the tissues of an organism, it alters the function of the receptor and a toxic response occurs (Mailman 2008). These interactions can be very complex but often can be understood by assuming that they occur according to the Law of Mass Action and Fractional Occupancy (FO). Thus, the interaction of a toxicant and a receptor can be described by a simple equilibrium reaction:

$$\text{Toxicant} + \text{Receptor} \underset{k_{OFF}}{\overset{k_{ON}}{\rightleftharpoons}} \text{Toxicant} - \text{Receptor Complex}$$

The induced effect, E, is a function of the quantity of ligand–receptor complexes formed or the FO where RA is the number of receptors occupied.

$$E = f[RA]$$

or a fraction of the total number [r] of receptors occupied, [RA]/[r]. When this quotient is unity, that is, all receptors are occupied, a substance induces its maximal effect, Emax (100% effect).

The principle behind the dose–effect concept rests with the affinity, expressed as the equilibrium constant, of a substance for specific receptor sites and its intrinsic ability to cause changes in the receptor. These changes result in adverse changes in the physiological, biochemical, or behavioral characteristics of the organism. The magnitude of the effect will depend on the dose of the active substance, for example, reactive metabolite, and its affinity and intrinsic activity in competition with other substances, including essential endogenous substances or antagonists for receptor sites. From dose–effect curves, the dose can be determined at which a certain effect (e.g. 50% of the maximal effect) is produced and the magnitude of the maximum effect that can be reached with the particular substance.

With individual organisms, there is a dose-response relationship which exhibits somewhat different characteristics to that obtained with a population. A hypothetical relationship with individual humans exposed to increasing acute levels of a toxic agent in the environment and the consequent adverse biological effects is shown in Figure 5.6. At low levels, the exposure causes no adverse effects and the human is able to cope with trace toxicants in the environment. At increasing levels of exposure, the adverse effects are expressed as headache, dizziness, tiredness, and so on, which are reversible, but with further increased levels of exposure, the adverse effects become irreversible and of a more serious nature, finally resulting in death. This pattern of dose–response with individuals would be expected to occur with individual organisms of most species. It follows a pattern of increasingly serious adverse responses corresponding to increasing dose with the ultimate response of death being registered.

Dose–Response Models with Populations

Dose–(or environmental concentration) response models refer to the relationships between the quantitative characteristics of exposure to a toxic agent and a measured or observed adverse biological response for the individuals which constitute a population. The adverse biological response is within the spectrum of consequent effects which result from the exposure. The relationships imply a reasonable presumption that the observed effects are induced by the toxic agent which has known biological effects. Numerical and graphical expressions of the dose-response relationship are then based on assumptions that the response and dose are causally related. Mathematical models for dose-response relationships have been reviewed recently by Lutz et al. (2014) and Sussman et al. (2011).

In toxicological investigations of responses, a sensitive and unequivocal measure or index of toxicity is necessary. This may be related, directly or indirectly, to a biochemical, physiological, or behavioral function such that a causal

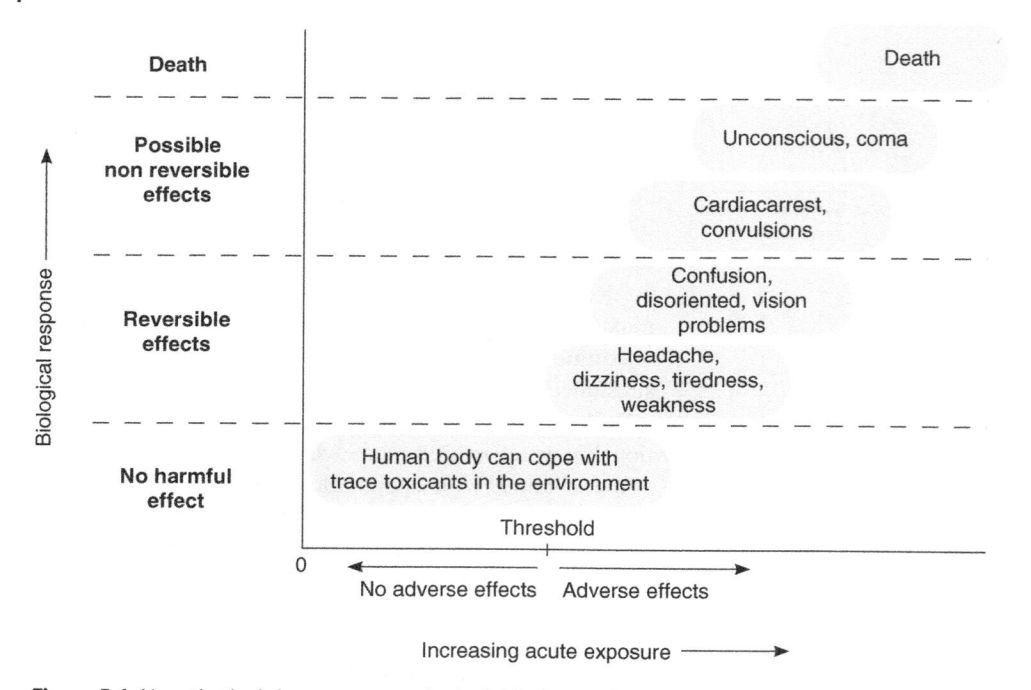

Figure 5.6 Hypothetical dose–response for individual organisms, for example, humans, exposed at acute levels to a toxic chemical in the environment.

relationship can be established or reasonably presumed. One type of response commonly measured is the quantal (*all or nothing*) response, for example, death, particularly where quantification of a response is difficult or impossible. In practice, lethality represents a precise and unequivocal toxicity measure that is widely used to evaluate response in dose–response relationships within an exposed test population of organisms.

When the response is quantal, with an individual there will be some level of stimulus, dose, or concentration, at which, under constant environmental conditions, the response will not occur and above which it will. This level is referred to as the tolerance of the individual organism within the population. However, it is more useful to consider the distribution of tolerances over the whole population of organisms since whole populations are of principal concern. This situation can be represented in the form of a tolerance distribution, f(D)dD, along with its corresponding cumulative distribution P(D), as:

$$P(D_1) = \int_0^D f(D)dD$$

when a population is exposed to a dose of D_1. The function P(D) represents the dose–response relationship for the population when the response is quantal in nature and the parameter values are assumed to be P(O) = 0 (no responders for zero dose) and P(∞) = 1 (all respond to some high dose).

Quantal responses, such as lethality, exhibit a range of differences in tolerance reflecting the susceptibility to a toxic agent among individuals within a population. Usually, these responses follow a normal Gaussian distribution in relationship to log (dose) with a Bell shape as represented by experimental data on the percentage of individuals in the population responding to the toxicant, as shown in Figure 5.7a with the line of best fit. The distribution of the data in general terms of the response of individuals can be seen more clearly by using the line of best fit, as shown in the corresponding Figure 5.7b. In a normally distributed population, the mean ± standard deviation (SD) represents 68% of the individuals in a population, the mean ± 2 × SD represents 95% of the population, and the mean ± 3 × SD equals 99.7% of the population. The sensitive individuals in the population occur in low proportions and are represented on the lower dose or left hand side of the plot, the average or median individuals, the major proportion in the population, are in the center and the tolerant individuals, also in low proportions, at the high dose or right hand side of the plot.

More conveniently, the data in Figure 5.7a can be replotted in the form of a sigmoidal cumulative probability dose–response curve, as in Figure 5.7c. It is important to note that the center section of this plot is approximately linear. This allows linear regression lines and equations to be fitted which permits this section to be easily described quantitatively. A Probit transformation is frequently used to plot mortality, or other quantal data, against dose since with a normal population this results in a straight line for most of the data, as shown in Figure 5.7d. Alternatively, the

5.4 Relationships Between Exposure to a Toxicant as Dose or Environmental Concentration and the Resultant Biological Effects

Table 5.2 Some typical LD_{50} values with rats for oral doses of some common substances.

Substance	Lethal dose (LD_{50} mg kg^{-1} body weight)
Sucrose	30 000
Cadmium sulfide	7 080
Ethanol	7 060
Table salt (sodium chloride)	3 000
Chlorine	850
Aspirin (acetyl salicylic acid)	200
DDT	100–800
Arsenic (arsenic trioxide)	15

vertical scale can be converted to a probability scale which is nonlinear and somewhat unconventional. However, plots of normally distributed data using this scale are often linear over an extended range.

Since quantal dose–response phenomena are usually normally distributed, the percent response can be converted to units of deviation from the mean or Normal Equivalent Deviations (NED 34%). The NED for a 50% response of individuals in a population is zero. A NED of +1 is equated with 84% response, and −1 represents a 16% response. Units of NED are usually converted by the addition of 5 to the value to avoid negative numbers and these converted units are called *probit* units. The probit, then, is a NED +5. Thus, a set of relationships between percent response, normal equivalent deviation, and probit units can be established. By using probit units, normal distributions can then be plotted as straight lines for most of the data, as shown for mortality data in Figure 5.7d. The dose–response relationship can be simply described by the LD_{50} (or LC_{50}) value, which is the dose determined from the line of best fit at the 50% mortality value (see Figure 5.7c), or equivalent probit unit for 50% mortality (i.e. 5) (see Figure 5.7d).

With aquatic organisms, the units of the LC_{50} are mass/volume (e.g. mg L^{-1}), whereas with other organisms, including mammals, the units of the LD_{50} are mass/body mass (e.g. mg kg^{-1} body weight) or mass/body weight/day (e.g. mg kg^{-1} body weight day^{-1}). Some values for the LD_{50} for oral doses with rats are shown in Table 5.2. These data show that even common, apparently nontoxic, chemicals have measurable toxicity and inherent toxicity of different chemical ranges over large values of the LD_{50}. The LD_{50} values are used to classify chemicals into toxic classes which are then used in management and control of these substances.

For example, the WHO uses the lowest published rat oral LD_{50}, as a basis for its classification of acute hazard rankings of pesticides shown in Table 5.3. However, the

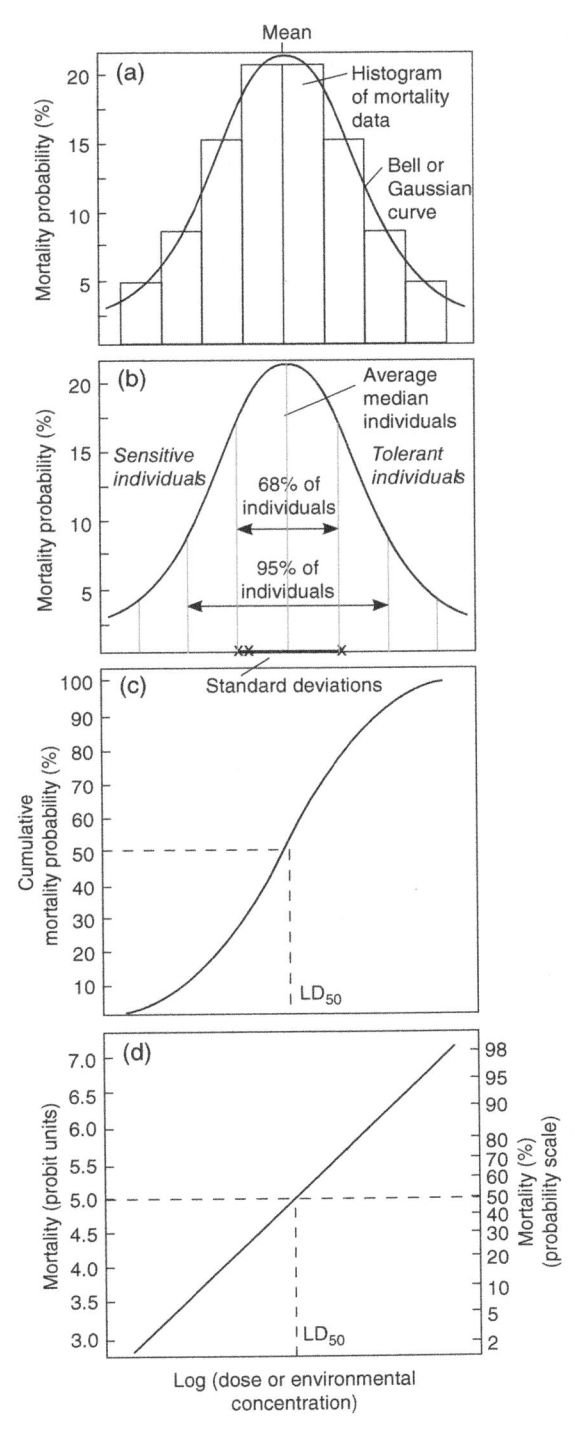

Figure 5.7 Hypothetical dose–response relationships for a homogeneous population of individuals being tested to mortality using a toxicant at different doses, or environmental concentrations, with the primary data shown in (a). (a) Data as the percentages of the population responding at different levels of toxicant. These data can be fitted with a line of best fit which is usually the Gaussian or Bell Curve. A generalized interpretation of the significance of this curve with respect to individuals within a population is given in (b). These data can be replotted as the Cumulative Mortality Probability (%) as shown in (c) and with Probit Units and the Cumulative Probability Scale in (d).

Table 5.3 The WHO classification of pesticide hazard.

Hazard class	Description	Solids (oral)	Liquids (oral)	Solids (dermal)	Liquids (dermal)
Ia	Extremely hazardous	≤5	≤20	≤10	≤40
Ib	Highly hazardous	5–50	20–200	10–100	40–400
II	Moderately hazardous	50–500	200–2000	100–1000	400–4000
III	Slightly hazardous	>500	>2000	>1000	>4000

Based on the lowest published rat oral LD_{50}.
Source: Based on WHO (2019).

LD_{50} or LC_{50} values can vary with the experimental conditions and types of animals used and do not reflect the toxicity of the whole population which includes organisms at many different life stages. For example, this occurs with human populations from babies to elderly people and with aquatic species from eggs to adult organisms. There are several other factors not taken into account with the LD_{50} including sublethal effects of longer term exposure, such as carcinogenesis, reduction of breeding success, and so on.

The dose–response models described above are based on tolerance distributions. However, there are other similar models which have been used in diverse applications including the Weibull model. In addition, there are stochastic biological models derived from what has been called "hit theory." These models do not assume a dose–tolerance distribution to produce a dose–response curve, but rather general mechanistic dose–response assumptions are made.

An example of a typical plot of data (Figure 5.8) is from a laboratory experiment on fish which were subject to different concentrations of toxicant for a fixed period of time. The mortalities were plotted against the log of the concentration to give a characteristic plot with a sigmoidal shape. But it is important to note that the line is approximately linear in the cumulative probability range of about 10% to about 90%. The toxicity at the median level for 50% of the fish is indicated which is the approximate toxic level to most of the fish. However, of prime concern in risk assessment and setting of guidelines is the most sensitive individuals rather than the median sensitivity of exposed fish. So, the No Observable Adverse Effect Level (NOAEL) and the Lowest Observable Adverse Effect Level (LOAEL) are usually used in these applications. The NOAEL is the highest level at which no adverse effect is observed and the LOAEL is the lowest level an adverse effect is observed (Figure 5.8).

The data used in most dose–response relationships are obtained from experiments conducted in the laboratory where the organisms are selected from a particular species with size, age, condition, and other characteristics controlled. Similarly, the exposure is tightly controlled for the chemical nature of toxicant, its purity, and the conditions of exposure such as ambient temperature, and so on. The value and relevance of these data to ecosystems can be shown by using it in what are called Species Sensitivity Distribution (SSD) plots. These SSD plots describe the variation in sensitivity to a toxicant within a set of species. The species set can be from within a biological group, for example, a taxon, species from a specific area, a particular ecosystem, a natural community, or some other relevant situation. These data on a specific toxicity measure can be plotted as a Cumulative Probability Distribution (CPD) as shown in Figure 5.9. A corresponding set of toxicity measures can then be obtained from laboratory experiments on species in a relevant biological system such as the LD_{50}, NOAEL, or LOAEL and can be plotted as CPD to give a SSD. Figure 5.9 demonstrates the variation in toxicity to dissolved cadmium, as the LD_{50}, within an arthropod group. The most sensitive species are on the left hand side reflected by the low LD_{50} and the most tolerant are on the right hand side reflected by the relatively high LD_{50}.

Alternatively, in the natural and human environment, many of these conditions, such as the same species and condition, fixed exposure times, and so on, used to derive the LD_{50} or the LC_{50} data do not apply. In natural environments, whole ecosystems involving many species at different life stages and under different environmental conditions are involved. With the human situation while there is only one species involved there are many different factors including different life stages from babies to elderly people, different exposure routes from dietary to dermal, different sexes, and so on. These circumstances pose many additional difficulties requiring novel approaches.

A somewhat similar CPD plot can be made for human data which extends the value of the available data. Accurate data for dose–response with humans are unavailable and are unlikely to become available in the near future. Even so, there is a wealth of epidemiological data on adverse human health outcomes for human exposure to toxicants with certain human populations, such as workers subject to occupational exposure. These data have been produced from different investigations with different populations and different study designs and methods to evaluate the health outcomes. Nevertheless, it can be collated as the lowest exposure level to a toxicant where an adverse effect has been observed, and then the data are plotted as a CPD to give a Toxicant Sensitivity Distribution (TSD) for humans. A TSD developed by Phung et al. (2015) is shown for chlorpyrifos, an organophosphorus insecticide, commonly used with rice crops in Figure 5.10. We can expect that the adverse effects are cumulative with increasing

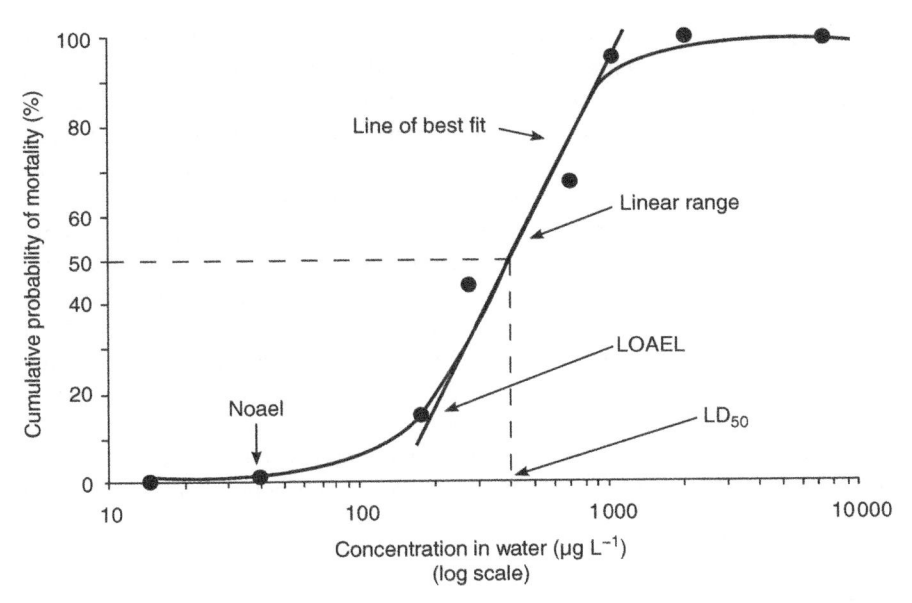

Figure 5.8 Typical plot of the cumulative probability data from an experiment on exposure of fish to a toxicant in water where the percentage mortality is recorded after a fixed time and is plotted against the concentration in water on a log scale.

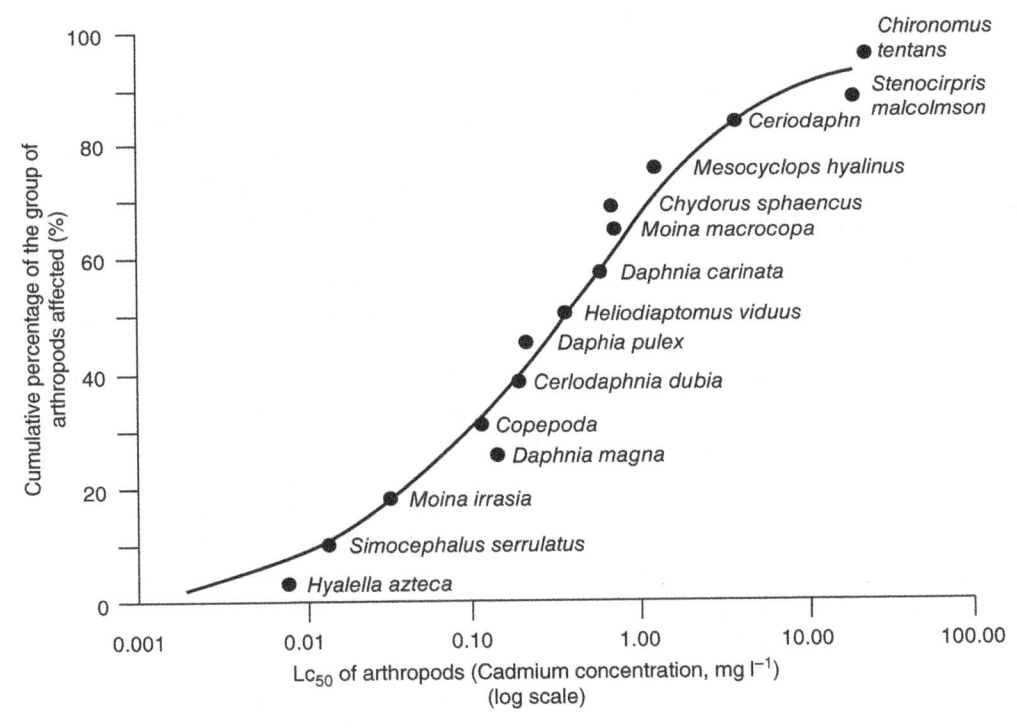

Figure 5.9 Species Sensitivity Distribution (SSD) of the toxicity of dissolved cadmium to arthropods represented by the experimental data on the LC_{50}. Source: US EPA (2019a).

dose, and would increase in severity at the same time. At the low doses observed in this evaluation, the doses are quite low and the adverse effects are very diverse and sublethal in nature.

The SSD relationships, previously described, are not derived from a single designed laboratory experiment where the various species are subject to the same experimental procedure and conditions, and a consistent set of adverse effects observations made. Similarly, the TSD relationships are not the result of a consistent set of laboratory experiments. Thus, these relationships lack the accuracy of the laboratory experiments but have qualities

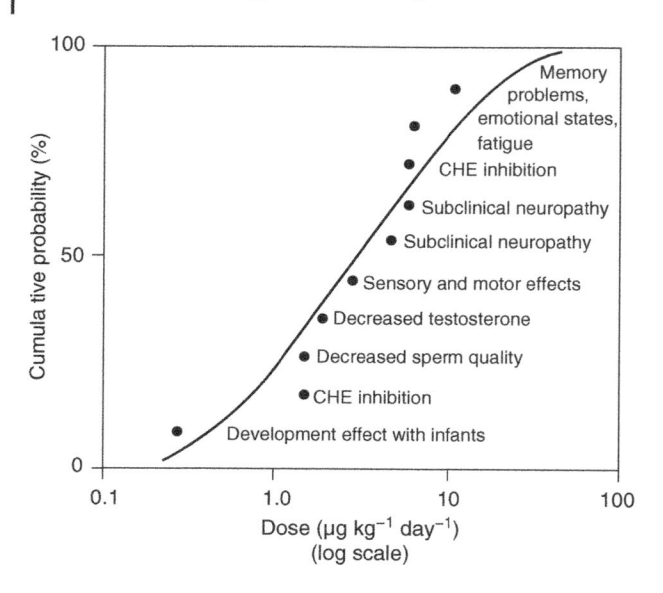

Figure 5.10 Toxicant Sensitivity Distribution (TSD) for the organophosphate insecticide, chlorpyrifos, based on epidemiological data with humans. Source: Modified from Phung et al. (2015).

of improved relevance to the natural environment and human population that make them useful (Phung et al. 2015).

5.5 Models for the Distribution of Exposure Dose or Concentration

The data used to develop dose–response relationships, discussed above in Dose-Response Models for Populations, are derived principally from laboratory experiments where the exposure dose is known with a reasonable level of accuracy. The dose is usually controlled at set values and the adverse response observed at set time intervals which allows the relationships in Figures 5.7 and 5.8 to be plotted. Most of the exposure data relate to organisms from the natural environment, such as fish, or surrogate animals, such as rats. However, in the natural and human environments, the exposure to toxicants is not controlled but is often continuous and variable. These exposure data can be modeled giving descriptions of the data which can be used to understand patterns of occurrence and behavior in space and time. Various models have been used including the normal, log-normal, exponential, and Weibull as described by Theodore (2016). The most common model for exposure data is the log-normal where the data are plotted as a probabilistic distribution of the dose or concentration; effectively the same model that is used with most toxicity dose–response data. These plots take the usual sigmoidal shape with a central approximately linear section, representing most of the data, and curved sections representing the high and low doses or concentrations.

Examples of these CPD plots are shown in Figure 5.11 where the global occurrence of disinfection by-products of chlorination of drinking water (trichloromethane, bromodichloromethane, dibromochloromethane, and tribromomethane) is plotted. Linear regression lines are fitted to the linear sections of the plots which can exhibit different ranges of linearity depending on the characteristics of the specific data set. Using these plots, the level of exposure of different segments of the population being investigated can be evaluated. Most of the population is exposed to the dose at about the CP50% level but the highly exposed group in the population occurs at the CP95% level which represents the 5% of the population which is the most highly exposed.

5.6 Relationships Between Exposure Level and Exposure Time Resulting in an Adverse Effect

Apart from dose–response relationships, the relationships between the exposure level and the time period to cause the occurrence of an adverse effect, such as lethal toxicity, are also important in understanding toxic effects. Figure 5.12 gives a hypothetical example of the relationships between exposure time, dose, and response (lethality). The concentration at which the plot becomes asymptotic to the exposure time axis is referred to as the "asymptotic," "threshold," or "incipient" LC_{50}. Concentrations below this will not cause lethality at the LC_{50} in the long term. This measure can be useful in establishing risk criteria for natural populations of organisms exposed to chronic levels of toxicants. Thus, the exposure time or duration of a bioassay experiment is a critical factor in deriving "safe" levels of exposure. However, it should be remembered that although long exposure times do not result in an LC_{50} with a mortality of 50% of the test organisms, a lower percentage of mortalities can occur.

Various mathematical models have been developed which relate dose levels or environmental concentrations to the adverse effect. Perhaps the most widely used of these relationships is "Habers Rule." This relationship states that the product of the effective concentration to cause lethality (LC_{50}) and the Exposure Time is constant for a given situation:

$$LC_{50} \times \text{Exposure Time} = \text{Constant}$$

Once this constant is calculated from known data relating to a specific toxicant under specific conditions, which

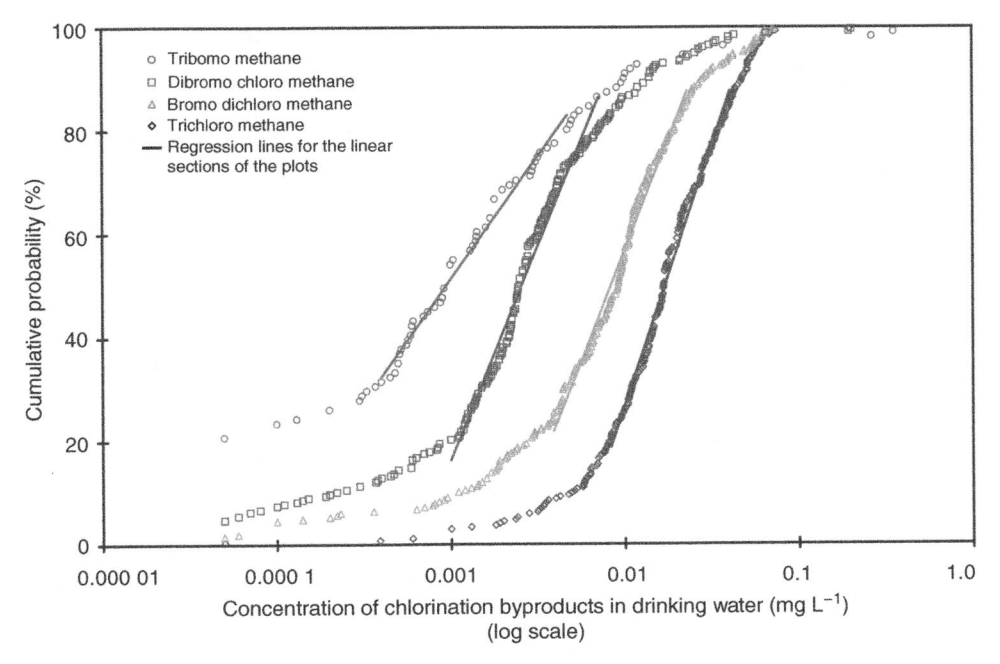

Figure 5.11 Cumulative Probability Distribution (CPD) for the distribution of chlorination by-products in drinking water taken from published results of drinking water composition reported in countries throughout the globe. Source: Based on Hamidin et al. (2008).

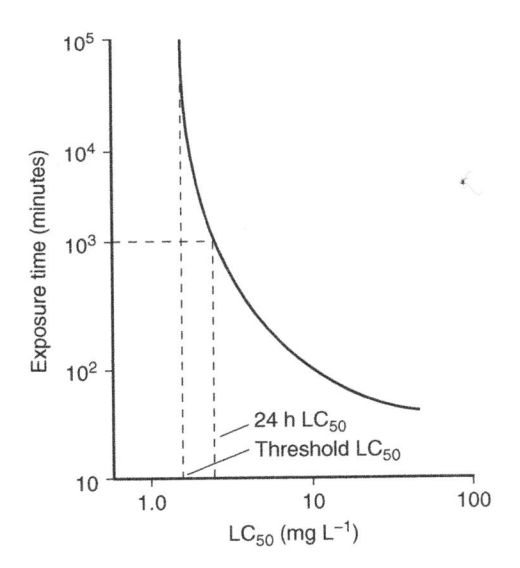

Figure 5.12 Toxicity curve of the exposure time against the log of the LC_{50} with the "incipient" or "threshold" LC_{50} indicated when the curve becomes asymptotic to the time axis.

includes the LC_{50} and corresponding Exposure Time, then the LC_{50} can be calculated for unknown Exposure Times.

Another approach has been suggested by Connell et al. (2016) by using the Reduced Life Expectancy (RLE) model. As a general rule, the exposure time required to cause lethality increases as the lethal concentration declines as illustrated in Figure 5.13. This decline follows a regular pattern and is terminated when the exposure time reaches the Normal Life Expectancy (NLE) of the organism. At this point, the lethal concentration reaches zero since at the NLE there is no exposure to any toxicant. This model utilizes existing data on the organism of interest to establish the equation for the relationship between LT_{50} and LC_{50} as below.

$$\ln(LT_{50}) = -a(LC_{50})^v + b$$

where v is an empirical nonlinearity constant related to the organism and toxicant; LC_{50}, the lethal concentration to the average organism; LT_{50}, the average exposure time for the expression of lethality LC_{50}; b, constant; and a, empirical constant related to the toxicity of the chemical to the test organism over extended times. The NLE is used as a fixed data point when the LC_{50} is zero in fitting the model equation to the available data with all data sets. This model has been evaluated with existing data on fish, invertebrates, and mammals involving 115 data sets and with a wide range of organic and inorganic toxicants (Connell et al. 2016). Once the equation has been developed, it can be used to calculate the LC_{50} at any exposure time.

5.7 Dose–Response and Dose Thresholds

The **threshold level** can be defined as the level above which there is a response to a toxicant and below which

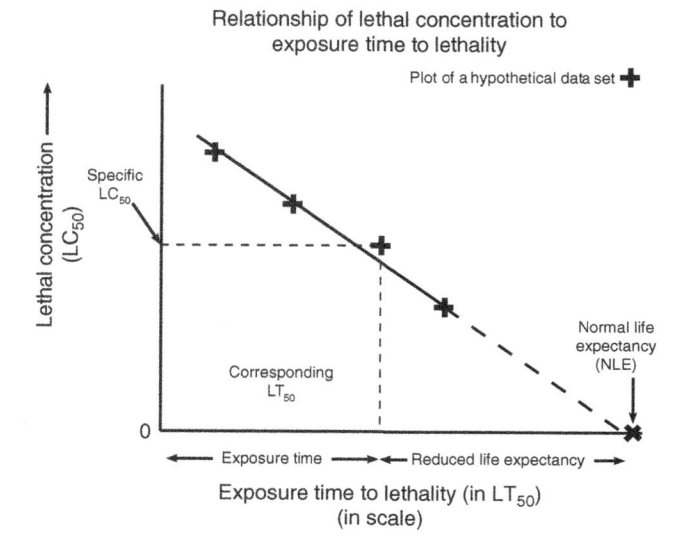

Figure 5.13 Conceptual diagram of the relationship of lethal toxicity to exposure time as the basis of the Reduced Life Expectancy (RLE) model. Source: Based on Connell et al. (2016).

there is no response. It can be observed as the highest level at which there appears to be no added response over controls. Threshold levels are very important in setting guidelines and in risk assessment since these values define when risk is essentially zero and when it becomes apparent.

With this work, it is important to determine whether the dose–response curve displays a threshold or quasi-threshold, or is linear. However, statistical prediction of the existence of threshold levels appears to be inconclusive. Brown (1976) has shown that statistical analysis of bioassay results cannot discriminate between mathematical models which assume the existence or nonexistence of an actual threshold. Furthermore, experimental observations of no effect levels, such as the NOAEL, do not validate that the probability of a response is equal to zero (Brown 1978).

It is a common toxicological practice to predict risks associated with low levels by extrapolation from high-level response curves. However, at low levels, the shape of the level–response relationship cannot be established with statistical confidence; hence, it is usual to assume a linear-level response relationship without threshold. This assumption is a useful approximation in the absence of a knowledge of exact toxicity mechanisms. In terms of risk assessment, the linearity assumptions can be extended to cover small increments in doses above already existing or background doses such as, for example, natural sources of radiation. However, for large increments, particularly at high doses, nonlinear dose–response relations are more likely.

5.8 Toxicity Due to Exposure to Multiple Toxicants

In most polluted systems in the natural and human environments, the individuals are usually exposed to a diverse range of toxicants rather than a single specific agent. Interactions between these pollutants may mutually be additive, enhance, or alternatively, inhibit toxic responses in exposed organisms. *Additive effects* are simply where the effects of the individual toxicants are the same as the equivalent mixture. Where the mixture exhibits enhanced toxicity over the individual toxicants, this is described as *synergy* and, on the other hand, where there is inhibition of toxicity, this is described as *antagony*. Mechanisms of interaction between constituents of pollutant mixtures and models for assessing multiple toxicity responses can occur (i) outside the organism in the ambient environment affecting bioavailability and uptake, and (ii) internally within the organism by influencing transport to the active site, metabolic processes, binding at the target site, and excretion.

Important environmental toxicants can be classified into related groups such as metals, pesticides, and antifoulants. Cedergreen (2014) has extensively reviewed the literature for many different species of organism and evaluated the data for pesticides (195), metals (20), and antifoulants (103) in binary mixtures using a calculated Model Deviation Ratio (MDR) for synergy to calculate the interactions of the toxicants. The results are summarized in Figure 5.14 which indicates the Cumulative Frequency for the calculated MDR for the set of three binary mixtures evaluated – metals, pesticides, and antifoulants. It was found that synergy occurred with 7, 3, and 26% of these groups, respectively, as illustrated in Figure 5.14, and that the most common effect was additivity with a much lower proportion exhibiting antagony. Cedergreen (2014) concluded that interactions with metabolic processes was the most common mechanism of synergy although interactions with uptake and availability could be important with metals. Also, in considering the cumulative toxic effect of environmental toxicants, chemical additivity was an appropriate approach in most cases.

5.9 Quantitative Structure–Activity Relationships (QSARs)

In many countries, there are restrictions on the use of surrogate animals to test for the adverse effects of chemicals. The results of such testing are then extrapolated to humans or other organisms by appropriate methods. There are many tests using surrogate animals described in

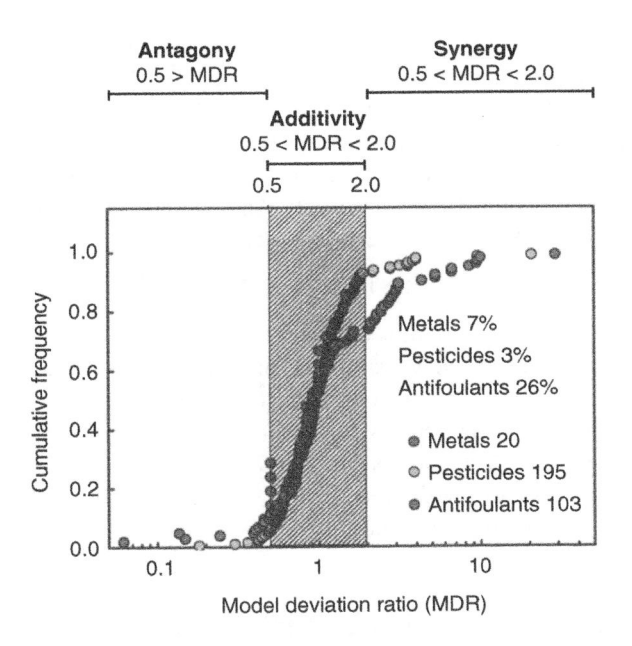

Figure 5.14 Plots of the Model Deviation Ratio (MDR) calculated from literature toxicity data on binary mixtures of common environmental contaminants – metals (20), pesticides (195), and antifoulants (100) against the cumulative frequency for each value. Source: Cedergreen (2014). Licenced under CC BY-4.0.

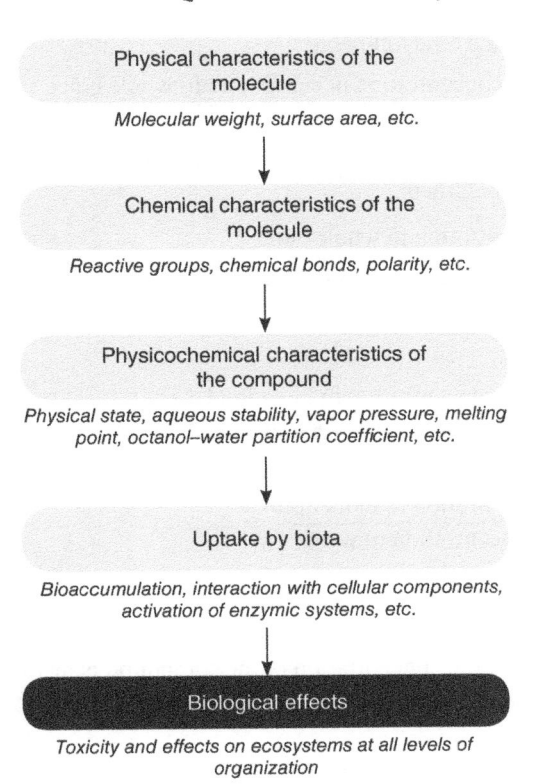

Figure 5.15 Characteristics of the molecule and how these relate to the physicochemical characteristics of the compound and its biological effects.

the OECD Guidelines for the Testing of Chemicals which is regarded as the standard reference for the testing of environmental chemicals. However, many organizations, including International Council on Animal Protection in OECD Programs (ICAPO), People for the Ethical Treatment of Animals (PETA), Australian Society of Cosmetic Chemists (ASCC), are concerned about cruel and inhuman practices in the use of surrogate animals and the need for the development of nonanimal methods to obtain data on the possible human response to chemicals.

The dimensions of this problem are indicated by the official records in the European Union which show that over 11 million animals are used in scientific procedures in the EU every year. However, Ferdowsian and Beck (2011) have reported a figure of 100 million for the number of animals used in scientific experiments, with the types of animals used including fish, mice, rats, birds, and nonhuman primates.

A response to this concern has been an increase in the use of QSARs often described as Quantitative Structure–Property Relationships (QSPR), Ecological Structure–Activity Relationships (ECOSAR) or Structure–Activity Relationships (SAR) which are based on the principle described in Figure 5.15. Here, the physical characteristics of a molecule, such as molecular weight, surface area, and so on, govern the chemical characteristics, such as reactive groups, chemical bonds, polarity, etc., which in turn control the physicochemical properties such as aqueous solubility, octanol–water partition coefficient (K_{ow}) with these properties finally governing the uptake by biota and the subsequent biological effects.

This approach is simplest at the beginning of the sequence in Figure 5.15 where only the characteristics of molecules and compounds are considered, and becomes more complex when biota and biological effects are involved. However, relationships can be expected between many of the characteristics, for example, with aromatic and aliphatic hydrocarbons the total surface area of the molecule and molar volume are directly related, and these properties are directly related to the octanol–water partition coefficient and aqueous solubility of the compounds.

Important widely used QSARs have been established for the bioconcentration of neutral nonpolar organic compounds which are resistant to biodegradation. A measure of the bioconcentration property of chemicals with aquatic organisms is the K_B or Bioconcentration Factor (BCF) value, which is the ratio of the concentration in the biota to the concentration in the ambient water at equilibrium. This characteristic can be measured with biota in an aquarium under laboratory conditions. A QSAR for bioconcentration can be derived by the following:

$$K_B = \frac{\text{concentration of compound in whole biota}}{\text{concentration in water at equilibrium}}$$

For lipophilic compounds which are effectively only soluble in biota lipid, then

Concentration in whole biota
= concentration in biota lipid, L

where L is the lipid fraction.

Thus,

$$K_B = \frac{(\text{concentration in biota lipid, L})}{\text{concentration in water}}$$

If octanol perfectly represents biota lipid, then:

$$\frac{\text{Concentration in biota lipid}}{\text{concentration in water}} = K_{OW}$$

and $K_B = K_{OW}.L$

where K_{OW} is obtained from a laboratory partition experiment in which the concentration of the compound is measured in octanol and water at equilibrium.

Taking logs:

$$\log K_B = 1 \log K_{OW} + \log L$$

Thus, the bioconcentration factor of a neutral non-biodegradable compound can be predicted using this relationship from the K_{ow} value measured in the laboratory. This QSAR predicts that as the log K_{ow} values of a set of compounds increase, the bioconcentration increases accordingly. Deviations from this relationship could be expected since octanol is not a perfect representative for biota lipid and compounds can exhibit deviations from neutral and nonbiodegradable properties. Many QSARs have been formulated for the prediction of different classes of compounds with different organisms utilizing a variety of molecular descriptors and methods.

Similarly, many QSARs have been developed for toxicity to aquatic organisms. For example, the OECD, EChA QSAR Toolbox (OECD 2020b), and ECOSAR US EPA (US EPA 2019b) are software used to predict toxicological properties for assessing the hazards due to exposure to chemicals.

Many toxicants act by narcosis which means these compounds do not have chemical reactions with the biotic processes of an organism to produce toxic effects but dissolve in the biotic lipids and change their physical nature. This is often referred to as *baseline* toxicity. QSARs for this type of toxicity usually take the form:

$$\log\left(\frac{1}{LC_{50}}\right) = A \log K_{ow} + B$$

where A is the slope of the linear relationship and B is a constant. These values are dependent on the chemicals involved, the organisms, units used, and test conditions. This QSAR shows that as the log K_{ow} increases, the toxicity value falls. In other words, the toxicity increases since numerically lower values are predicted.

For example, the K_{ow} value can be used as a predictor of general adult fish acute toxicity (96 h LC_{50}). The QSAR reported by Kluver et al. (2016) takes the form:

$$\log LC_{50} = -0.898 \log K_{ow} + 1.711$$

where the LC_{50} is in units of mM.

QSARs take a variety of different forms and are applicable to many chemical processes leading to toxic adverse effects in biota. This approach has proved effective in extrapolating existing knowledge on toxic effects to provide new knowledge on toxicants without the use of laboratory testing and the use of surrogate animals.

5.10 Lethal and Sublethal Effects

Pollution stresses extend from effects on individual species through to successively higher levels of biological organization incorporating populations, communities, and ecosystems. Analyses or predictions of any systematic effects or changes depend on an adequate database on the response of the individual organisms.

Under increasing degrees of environmental pollution, these effects may transcend the normal range of adaptation and tolerance exhibited by living organisms. There usually exists a range of sublethal responses, where long-term survival potential of the organism is reduced and ultimately a level of pollution is reached which results in comparatively rapid death. Effects on organisms can be categorized into those-causing (i) direct lethal toxicity and (ii) sublethal disruption of physiological or behavioral activities.

In identifying these effects, it is pertinent to recognize not only the direct effects of a pollutant on an organism, but the effects due to interactions of pollutants with the organism's environment (e.g. physical alteration of habitat by oil spills, loss of food organisms, and increased competition), and modification to the pollutant caused by environmental factors (e.g. photochemical decomposition) or metabolism within the organism.

Of particular interest is the distinction between lethal and sublethal effects of pollutants on organisms. Qualitatively, lethal effects can be defined as those responses that occur when physical or chemical agents interfere with cellular and subcellular processes in the organism to such an extent that death follows directly. In severe cases, this may take the form of smothering and suffocation, interference with movements to obtain food or escape predators, or destruction of habitat (e.g. for sedentary organisms).

In comparison, sublethal effects are those that disrupt physiological or behavioral activities but do not cause immediate mortality, although death may follow because of interference with feeding, abnormal growth or behavior, greater susceptibility to predation, less ability to colonize, or other indirect causes. These effects may not only lead to changes in populations of individual species, but may also result in shifts in species composition and diversity (see GESAMP 1977, p. 27).

A critical aspect of these investigations is the distinction between an adverse effect and an adaptive response. Thus, interpretation may be confused by adaptive changes, such as acclimation, in the short term (less than one generation) or by genetic changes in populations in the long term. Furthermore, sublethal changes in organisms must be evaluated in terms of several levels of organization of the organism, which integrate stepwise from within a cell to, eventually, the success or otherwise of groups of organisms within a system.

Assessment of sublethal effects depends on two important factors: (i) the selection of physiological and behavioral parameters which predict ecologically significant responses and (ii) the experimental measurement of sublethal responses. These factors may include the recognition of critical effects which occur at different life stages (see Table 5.4). The task of extrapolating significant responses to the ecosystem level is complex and currently uncertain. For most organisms, there is an inadequate understanding of natural variations in physiological responses and interactions with environmental parameters. Also, natural perturbations may quite easily mask any physiological changes induced by long-term exposure to low levels of exogenous agents.

Relationships Between Lethal and Sublethal Levels of Toxicity

Measurements of lethality are frequently used to derive "safe" levels of exposure to toxicants. This includes the use of "application factors," "safety factors," or "uncertainty factors." For example, a safety factor of 1000 can be applied as 0.001 of the NOAEL or the 96 h LC_{50} to fish to calculate "safe" levels for fish in natural ecosystems. These values may also function as water quality guidelines for certain toxicants under specific conditions. A similar approach is used with setting guidelines for protection of human health by use of results on surrogate animals with the application of safety factors. The assumptions adopted in the safety factor approach are not well supported empirically, and as an alternative, the use of chronic, sublethal tests may be more appropriate. Sublethal measurements can be considered more suitable for predicting "safe" or "ecologically insignificant" levels of toxicants if responses are quantified and statistically significant relationships are derived. Conceptually, this approach appears feasible but selection of sublethal tests and criteria, test organisms, and experimental conditions are major obstacles.

Relationships between sublethal and lethal toxicities continue to be of major importance in setting guidelines and developing policies for protection of both human health and the natural environment. In the natural environment, the range of species in many aquatic ecosystems can be extensive as is illustrated by the SSD shown in Figure 5.9. While all these species may not necessarily occur in the same ecosystem, the data illustrate the wide range of sensitivities that can occur with any toxicant. In addition, an aquatic species can exist at many life stages, for example, egg, larva, juvenile, and adult which all have differing sensitivities to a toxicant. With human health, we

Table 5.4 Selected parameters for sublethal studies.

Uptake, Accumulation, and Excretion

Complexation and storage, distribution within tissues and organs, kinetics of uptake and release, bioconcentration, and bioaccumulation.

Physiological Studies

Metabolism, photosynthesis and respiration, osmoregulation, feeding and nutrition, heartbeat rate, blood circulation, body temperature, and water balance.

Biochemical Studies

Carbohydrate, lipid and protein metabolism, pigmentation, enzyme activities, blood characteristics, and hormonal functions.

Behavioral Studies (Individual Responses)

Sensory capacity, rhythmic activities, motor activity, and motivation and learning phenomena.

Behavioral Studies (Interindividual Responses)

Migration, intraspecific attraction, aggregation, aggression, predation, vulnerability, and mating.

Reproduction

Viability of eggs and sperm, breeding/mating behavior, fertilization and fertility, survival, life stages, and development.

Genetic

Chromosome damage, mutagenetic and teratogenic effects.

Growth Alterations and Delays

Cell production, body and organ weights, and developmental stages, e.g. larval and juvenile stages.

Histopathological Studies

Abnormal growths, respiratory and sensory membranes, tissues, and organs, e.g. reproductive organs.

Interactions

Environmental or ecospecific factors with pollutants, multiple pollutant combinations.

are dealing with a single species but at many different life stages. Also, many different adverse responses are significant, some of which are recorded in Figure 5.10 for a TSD at sublethal levels in a human population. The complexity of these processes indicates that the use of probabilistic distributions can assist greatly in interpretation of the data on exposure and responses.

5.11 Types and Classification of Toxicants

In the use and evaluation of toxic substances, it is helpful to place a substance into a class. The properties of the class can then be used to help understand the properties of individual toxicants. The general classification

Table 5.5 Common terms and characteristics of classes of environmental toxicants.

Term	Characteristics	Examples
Biological property-based class		
Pesticide	Toxic to pests	DDT, 2,4-D
Insecticide	Toxic to insects	Chlorpyrifos, DDT, parathion
Herbicides	Toxic to plants	2,4-D, glyphosate
Fungicides	Toxic to fungi	Phenyl mercury acetate
Rodenticides	Toxic to rodents	Hydrogen cyanide
Carcinogens	Induce cancer	Benzene, benzo(a)pyrene
Chemical structure-based class		
Chlorohydrocarbons (chlorinated hydrocarbons)	Compounds based on Cl, C, and H alone	DDT
Hydrocarbons	Compounds based on C and H alone	Hexane
Polycyclic aromatic hydrocarbons (PAHs)	Polycyclic aromatic hydrocarbons containing two or more aromatic rings	Benzo(a)pyrene
Organochlorines (OCs)	Organic compounds containing Cl	DDT, 2,4-D
Persistent organochlorines	Organic compounds containing Cl with environmental persistence	DDT
Dioxins	Combustion and industrial products having polychlorodibenzodioxin structure	Tetrachlorodibenzo-dioxin
Furans	Combustion and industrial products having polychlorodibenzofuran structure	Tetrachlorodibenzo-furan
Heavy metals	Toxic metals of higher specific gravity	Mercury
Organometallics	Organic compounds containing metals	Tributyltin
Pyrethroids	Usually synthetic pesticides related to pyrethrum	Fenvalerate
Organophosphates	Compounds based on the organophosphate structure	Parathion, chlorpyrifos
Phenoxy acetic acids	Compounds based on the phenoxy acetic acid structure	2,4-D
Physicochemical property-based class		
Lipophilic (or hydrophobic)	Fat soluble and water insoluble	DDT
Hydrophilic	Water soluble and fat insoluble	Phenol
Neutral organic compounds	Organic compounds without ionic charges	DDT
Radionuclides	Substances having radioactivity	Uranium-238
Surfactants	Compounds which act at interfaces by altering surface tension	Alkylbenzene sulphonates
Physical property-based class		
Nanotoxicants	Ultrafine particles usually about 1–100 nm of wide-ranging chemical composition	Dust, soot, fibers, plastic particles in the oceans, and engineered nanoparticles

of toxicants does not follow a systematic pattern but is influenced by factors such as usage, chemical structure, adverse effects on target organisms, and physiochemical properties. The common terms used to describe the properties used in the classification of toxicants, which can be environmental pollutants, are shown in Table 5.5. Classes based on the observed biological effects are very common; thus, a toxicant can be a pesticide, insecticide, and so on. The chemical structure-based class includes chlorohydrocarbons, organophosphates, pyrethroids, and so on. The physicochemical property-based group includes radionuclides and surfactants. The classes are not mutually exclusive and there is considerable overlap in classes. In fact, almost all the biologically based groups also belong to another, or several other, classes. For example, parathion can be described by any one or more of the following terms: pesticide, insecticide, organophosphate, and so on, and DDT can be described as a pesticide, insecticide, chlorohydrocarbon, persistent organochlorine, lipophilic, and neutral organic compound.

Nanotoxicants and nanoparticles are ultrafine particles (1–100 nm) that can exist in many different chemical forms where the density and surface area of the particles are major properties leading to adverse effects on biota. Nowack and Bucheli (2007) have reviewed the overall occurrence in the natural environment as well as sources and effects of these particles as pollutants. The particles are sufficiently small as to be able to be taken up by mammals and transferred internally. Some particles such as soot can migrate to the lungs and can be carcinogenic partly due to the presence of PAHs. Interactions with plants have been shown to have adverse biological effects.

5.12 Effects on Populations and Ecosystems

The general format for the interpretation of the effects of toxicants on ecosystems is contained in Figure 5.1. It is possible to generalize on the adverse effects on natural and human populations as shown in Figure 5.16. If a population is exposed to zero levels of toxicant, there will be no effect even over extended periods of exposure. However, if the exposure is increased to a low level, then over an extended period there will be mortalities of sensitive individuals such as the very young or very old organisms, diseased organisms, and the very sensitive individuals within the population. When this occurs, then sublethal effects would be expected in biological systems such as those shown in Figure 5.10 which are applicable to a human population. Somewhat similar adverse effects would be expected in populations in the natural environment as well although

Dose	Exposure time period	Dose/concentration	Exposure period	Expected response
		Very low	Very long (many years)	• No detectable effects
		Low	Long (months/years)	• Death of sensitive individuals and species • Sublethal effects in survivors
		Intermediate	Intermediate (days)	• Equal numbers of deaths and survivors • Severe effects in some survivors
		High	Short (hours/days)	• Few resistant individuals and species survive
Increasing dose	Increasing exposure period	Very high	Very short (hours)	• Death to all members of the population and ecosystem

Figure 5.16 Range of effects of changes in dose and exposure time of a toxicant on biological systems.

these are difficult to measure. At intermediate and high doses, the number of mortalities and sublethal effects would be expected to increase even though the exposure period has decreased. Finally, at very high doses although the exposure period is relatively short, the complete mortality of all members of the population will occur. With natural ecosystems, these effects on a component population would be expected to result in adverse effects on the ecosystem as a whole.

There has been considerable effort to reproduce simplified ecosystems in the laboratory and in the natural environment. These can then serve to investigate the effects of natural variants such as temperature on a system which exhibits some of the characteristics of natural ecosystems. In addition, the behavior and effects of toxicants in simple ecosystems may be extrapolated to a complex natural ecosystem. Ferard and Blaise (2013) have collated the many aspects of the use of mesocosms in their Encyclopedia of Ecotoxicology, while Caquet (2013) has described some of these applications with aquatic systems.

5.13 Biomarkers

Biomarkers are extensively used in ecotoxicology, medicine, and other fields where the presence of specific compounds are used as indicators of exposure to a substance or the occurrence of an important biological process. This could be presence of a particular disease, for example, the PSA (prostate specific antigen) test for prostate size in men, use of radioactive isotopes to evaluate cardiac function, presence of cancers, or many other effects. Biomarkers have had extensive use in ecotoxicology particularly with the use of mussels, *Mytilus edulis* (e.g. Gupta 2014), as indicators of exposure to pollutants, histological changes reflecting morphological alterations, and stress due to contaminants. Extensive analyses of mussels have been carried out, particularly along coastlines, to evaluate exposure and adverse changes of environmental significance.

5.14 Key Points

1. Environmental toxicology is concerned with the adverse effects of physical, chemical, and toxicants in the environment on living organisms at all levels of organization from cellular to global.
2. Ecotoxicology is specifically concerned with the harmful effects of environmental toxicants on ecosystems at all levels of organization from individual organisms to whole ecosystems.
3. Figure 5.1 illustrates the sequence of processes and interactions of a toxic chemical pollutant with the environment. The sources and overall impacts of pollutants on the global environmental system are described in Chapters 1, 2, and 3. Chapter 4 outlines the initial processes after discharge when a toxicant distributes within the phases in the environment according to its physiochemical properties and the phases themselves. As a result, levels of the chemical become distributed in air, water, soil, food, and possibly other phases, according to the environment being considered, and exposure of individual organisms in both the natural and human environments occurs as shown in Figure 5.1.
4. The testing of chemicals in the laboratory as a basis for the registration of chemicals is now an accepted international procedure for the control and management of environmental chemicals. The OECD has developed a collection of the most relevant internationally agreed testing methods used by governments, industry, and independent laboratories. The Guidelines seek to evaluate chemicals on a broad basis, as illustrated in Figure 5.1, from distribution characteristics through to assimilation of toxicants to consequent toxicological evaluations.
5. The assimilation of pollutant chemicals within organisms at sublethal or lethal levels may induce a sequence of adverse biological effects. These range from molecular interference with biochemical mechanisms and interactions with cellular organelles (e.g. DNA and RNA molecules), through to pathological changes at the cellular, tissue, and organ levels. Finally, these result in an integrated functional or behavioral response, experienced at the whole organism level, which may be reversible or irreversible.
6. The consequences to individuals in the natural environment of toxic effects are generally demonstrated in laboratory studies, but these effects need to be translated to the population, community, and ecosystem levels to evaluate the ecological significance.
7. Human health evaluations are also complex, particularly due to the fact that toxicological investigations are usually carried out on surrogate animals, such as rats. Thus, extrapolations from the surrogate animals to humans are required. In addition, the human population is complex due to the presence of different age groups from babies to the aged at different exposure settings and different health status.
8. The pathways or routes of exposure with humans have been described by the International Program on Chemical Safety (IPCS 2004) for human exposure. However, the pathways involved with most biota can be identified within the following general classes:
    - inhalation through vapors or dust with mammals,
    - uptake through the respiratory pathways with aquatic organisms,
    - ingestion with food and drinking water,
    - direct ingestion of contaminated materials, and
    - dermal sorption.
9. The ADD or ADI of a pathway can be calculated from the general equation below:

$$\text{ADD or ADI} = \frac{(C_i A_i BA_i)}{bw} \quad (5.2)$$

    where C_i is the concentration in the medium involved in the pathway; A_i, the amount of medium involved per day;, BA_i, the bioavailability of the contaminant to the organism which is a unitless proportion; and bw, the body weight in mg kg^{-1} (body weight) day^{-1}.
10. The following equation for LADD with humans can be derived based on the ADD:

$$\text{Lifetime Average Daily Dose (LADD)} = \text{ADD} \times \left(\frac{EF.ED}{AT}\right) \quad (5.3)$$

    where (EF. ED/AT) is an adjustment to the ADD and EF is Exposure Frequency (days year^{-1}), ED the Exposure Duration (years) and AT the Averaging Time (days – a lifetime is 70 years × 365 days).
11. Uptake of a toxicant from the environment through the routes outlined above leads to distribution within an organism and a proportion of the toxicant becomes located at the site of action. This causes an alteration to the characteristics of the site and changes to its biochemical and physiological functions which result in the observed toxic effects.
12. The relation between different dose levels of toxicants and induced effects on organisms or biological systems

13. can be either readily observed in nature or under experimental test conditions in the laboratory. Relationships are generally expressed in the form of dose–response or effect plots which usually take a sigmoid form.
13. Most dose–response data are derived from experiments in the laboratory and are normally distributed; thus, plots of log dose versus response have a sigmoidal cumulative probability dose–response curve. It is important to note that the center section of these plots can be approximately linear which allows linear regression lines and equations to be fitted permitting this section to be described quantitatively. Also, the lethal toxicity, for example, at any level of response can be calculated as the LD_X where x is the percentage response.
14. The NOAEL (no observed adverse effects level) and the LOAEL (lowest observed adverse effect level) can also be delineated from the dose–response plots.
15. Epidemiological data on adverse human health outcomes for exposure to a specific toxicant have been produced from different investigations with different populations with different study designs and methods to evaluate the health outcomes. Nevertheless, it can be collated and plotted as a CPD for the toxicant and thus can be described as a TSD.
16. The influence of exposure time on toxicity can be calculated using the Haber Rule or the RLE model.
17. Exposure to multiple toxicants can cause additive, antagonistic, or synergistic effects but with environmental toxicants, the effects are usually additive.
18. QSARs often described as QSPR or ECOSAR or SAR relate toxicological properties, such as LC_{50} and LD_{50}, to physicochemical properties, such as K_{OW}, which can be obtained from experiments in the laboratory.
19. Sublethal effects can involve a wide range of factors including uptake, accumulation and excretion, physiological factors, and biochemical factors. The relationship of the levels of exposure to induce sublethal responses and lethal responses is difficult to define.
20. Toxicants can be classified into a number of classes depending on their biological effects, chemical structure, and so on.
21. Biomarkers are important indicators of the occurrence of an important biological process such as the presence of disease or presence of pollutants.

References

Atabila, A., Phung, D.T., Hogarh, J.N. et al. (2017). Dermal exposure of applicators to chlorpyrifos on rice farms in Ghana. *Chemosphere* 178: 350–358.

Brown, C.C. (1976). Mathematical aspects of dose–response studies in carcinogenesis – the concept of thresholds. *Oncology* 33: 62.

Brown, C.C. (1978). The statistical analysis of dose–effect relationships. In: *Principles of Ecotoxicology, SCOPE 12* (ed. G.C. Butler). New York: Wiley.

Caquet, T. (2013). Aquatic mesocosms in ecotoxicology. In: *Encyclopedia of Aquatic Ecotoxicology* (ed. J.-F. Férard and C. Blaise), 99–108. Dordrecht: Springer.

Cedergreen, N. (2014). Quantifying synergy: a systematic review of mixture toxicity studies within environmental toxicology. *PLoS ONE* 9 (5): e96580. https://doi.org/10.1371/journal.pone.0096580.

Connell, D.W., Yu, Q.J., and Vibha, V. (2016). Influence of exposure time on toxicity – an overview. *Toxicology* 355: 49–53.

Farber, J.L. (1980). Molecular mechanisms of toxic cell death. In: *The Scientific Basis of Toxicity Assessment* (ed. H.R. Witschi), 201–210. Amsterdam: Elsevier/North Holland Biomedical Press.

Ferard, J.-F. and Blaise, C. (2013). *Encyclopedia of Aquatic Ecotoxicology* (ed. J.-F. Ferard and C. Blaise), 99–108. Dordrecht: Springer.

Ferdowsian, H.R. and Beck, N. (2011). Ethical and scientific considerations regarding animal testing and research. *PLoS ONE* 6 (9): e24059. https://doi.org/10.1371/journal.pone.0024059.

GESAMP (1977). *Impact of Oil on Marine Environment. UN Joint Group of Experts on the Scientific Aspects of Marine Pollution (GESAMP). Reports and Studies No. 6*. Rome: Food and Agricultural Organization of the United Nations.

Gillette, J.R. (1980). Pharmacokinetic factors governing the steady-state concentrations of foreign chemicals and their metabolites. In: *Symposium 76, Environmental Chemicals, Enzyme Function, and Human Disease* (ed. Ciba Foundation), 191–217. Amsterdam: Excerpta Medica.

Gupta, R.C. (2014). *Biomarkers in Toxicology*, 1e. Amsterdam: Academic Press/Elsevier.

Hamidin, N., Yu, Q.J., and Connell, D.W. (2008). Human health risk assessment of chlorinated disinfection byproducts in drinking water using a probabilistic approach. *Water Research* 42: 3263–3274.

IPCS (2004). *IPCS Risk Assessment Terminology: Part 2: IPCS Glossary of Key Exposure Assessment Terminology*. Geneva: International Programme on Chemical Safety, World Health Organization.

Kluver, N., Vogs, C., Altenberger, R. et al. (2016). Development of a general baseline toxicity QSAR model for the fish embryo acute toxicity test. *Chemosphere* 164: 164–173.

Lutz, W., Lutz, R., and Gaylor, D. (2014). Dose–response relationships and extrapolation in toxicology – mechanistic and statistical considerations. In: *Regulatory Toxicology* (ed. F.-X. Reichl and M.K. Schwenk), 547–568. Heidelberg: Springer.

Mailman, R.B. (2008). Toxicant – receptor interactions: fundamental principles. In: *Molecular and Biochemical Toxicology*, 4e (ed. R.C. Smart and E. Hodgson), 359–388. Hoboken, NJ: Wiley.

Nowack, B. and Bucheli, T.D. (2007). Occurrence, behavior and effects of nanoparticles in the environment. *Environmental Pollution* 150: 5–22.

OECD (2020a). *Guidelines for the Testing of Chemicals*. The Organisation for Economic Co-operation and Development https://www.oecd.org/chemicalsafety/testing/oecdguidelinesforthetestingofchemicals.htm (accessed 18 September 2020).

OECD (2020b). *The OECD QSAR Toolbox*. Organisation for Economic Co-operation and Development http://www.oecd.org/chemicalsafety/oecd-qsar-toolbox.htm (accessed 3 August 2020).

Phung, D.T., Connell, D., and Chu, C. (2015). New method to set guidelines to protect human health from agricultural exposure using chlorpyrifos as an example. *Annals of Agricultural and Environmental Medicine* 22: 274–227.

Sussman, R.G., Sargent, E.V., and Davidson, T.L. (2011). Dose–response analysis in experimental toxicology and risk assessment. In: *General, Applied and Systems Toxicology*. Wiley Online Library https://doi.org/10.1002/9780470744307.gat122.

Theodore, L. (2016). *Environmental Risk Analysis – Probability Distribution Calculations*. Boca Raton, FL: CRC Press.

US EPA (1992). *Guidelines for Exposure Assessment. Report EPA/600/Z-92/001*. Washington, DC: United States Environmental Protection Agency.

US EPA (2017). *Exposure Assessment Tools by Routes, EPA ExpoBox*. United States Environmental Protection Agency https://www.epa.gov/expobox/exposure-assessment-tools-routes (accessed 15 November 2020).

US EPA (2019a). *Causal Analysis/Diagnosis Decision Information System (CADDIS). Data Analysis, Species Sensitivity Distributions, SSDs*, vol. 4. Washington, DC: United States Environmental Protection Agency https://www.epa.gov/caddis-vol4 (accessed 18 September 2020).

US EPA (2019b). *Ecological Structure Activity Relationships (ECOSAR) Predictive Model*. United States Environmental Protection Agency https://www.epa.gov/tsca-screening-tools/ecological-structure-activity-relationships-ecosar-predictive-model (accessed 3 August 2020).

WHO (2019). *Classification of Pesticides by Hazard and Guidelines to Classification 2019*. Geneva: World Health Organization https://apps.who.int/iris/bitstream/handle/10665/332193/9789240005662-eng.pdf?ua=1 (accessed 18 September 2020).

6

Genetic Toxicology and Endocrine Disruption: Environmental Chemicals

6.1 Introduction

This chapter introduces several critical toxic effects of environmental chemicals on humans and wildlife that are of major importance for current and future global environmental health. These effects relate to genetic toxicity, mutagenicity, carcinogenicity, reproductive and developmental toxicity, including teratogenicity, and the growing impacts of endocrine-disrupting chemicals on humans and many wildlife populations.

Genetic toxicology involves the study of the effects of physical, chemical, and biological substances or agents on hereditary material (DNA or deoxyribonucleic acid) and on the genetic processes of living cells. Genotoxicity is seen as the process by which an agent produces an adverse effect on DNA and other cellular targets that control the integrity of genetic material.

Many environmental chemicals are also becoming well known as endocrine disruptors because they can mimic or interfere with the function of endogenous hormones leading to reproductive, developmental, neural, immune, and other dysfunctions observed in laboratory animals, humans, and wildlife.

6.2 Genome

All the genetic information needed for a living organism to develop, grow, and reproduce is contained in the **genome**. It contains the complete set of genes or genetic material present in a cell or organism. This information is chemically encoded in DNA, a double helix molecule made up of nucleotides, contained in cells, or in the case of many types of viruses, in RNA (ribonucleic acid) – a single helix molecule of nucleotides. In DNA, each nucleotide contains a phosphate group, a sugar group, and a nitrogen base (nucleic acid). The four types of nitrogen bases are adenine (A), thymine (T), guanine (G), and cytosine (C). RNA has a single strand of bases, but instead of thymine (T), it has the base, uracil (U).

Within the genome, sequences of nucleotides make up functional groups known as **genes**. They occur along a single long DNA chain or molecule called a **chromosome**. In eukaryotes (mainly animals and plants), the genome of each cell is contained within the nucleus (inner membrane bound structure). Prokaryotes (bacteria and archaea) store their genome in a region of the cytoplasm called the nucleoid, in the absence of a cell nucleus.

The role of genes is to code for the synthesis of RNA and protein molecules needed by an organism to function. This process is known as gene expression in which instructions encoded within a gene are converted into a functional product (RNA or proteins). RNA is primarily encoded to make proteins, according to the information encoded within DNA. Genes control and regulate many aspects of protein synthesis in eukaryotes (organisms whose cells have a nucleus enclosed within a membrane).

In protein synthesis within the cell, DNA copies the coding sequence for a protein (genetic code) into a messenger RNA (mRNA) by a process called transcription. The mRNA then travels to the ribosome (site of protein synthesis) which reads or decodes the sequence and makes a protein (specific amino acid chain or polypeptide) which is coded to that sequence from amino acids transported by transfer RNA (tRNA) molecules. Proteins consist of hundreds to thousands of amino acids bonded to each other in long chains. They are essential for cellular function, structure, and regulation of the body's tissues and organs.

The entire human genome has over 3 billion DNA base pairs which are present in all cells that have a nucleus. However, they are found in most cells in duplicated pairs. In cell division by mitosis, DNA molecules are replicated by breaking the weak hydrogen bonding between the two chains and then forming bases, according to rules, through bonding with new nucleotides. Two new cells result, with each containing paired chromosomes of the same genetic material (genetic code) as the parent cell (Fischer-Cripps 2012, p. 167). There are normally 23 pairs of chromosomes in the human genome (22 pairs called autosomes and 1 pair of sex chromosomes, X and Y).

Chemistry and Toxicology of Pollution: Ecological and Human Health, Second Edition. Des W. Connell and Greg J. Miller.
© 2023 John Wiley & Sons, Inc. Published 2023 by John Wiley & Sons, Inc.
Companion website: www.wiley.com/go/toxicologyofpollution2e

> **Box 6.1 Cell Communication and Signaling**
>
> In genetic and cellular toxicity, we are concerned about interference or disruption to cellular processes that control normal cell functions. These cell functions and responses to changes in their surrounding environment are controlled in organisms by chemical messages or signals that cells send, receive, process, and trigger specific responses. In multicellular organisms, these chemical messages stimulate cell processes and can coordinate the actions of organs, tissues, and cells.
>
> In other words, cells communicate through their own form of language made up of chemical signals (specific molecules).
>
> There are three stages of cell signaling: reception, signal transduction, and response.
>
> - Cells receive messages via signaling molecules, often many at once, such as neurotransmitters or hormones, which bind with selective receptors (e.g. membrane proteins) on the cell surface, or sometimes inside the cell by passing through cell membranes.
> - The binding of the signal molecule triggers the receptor protein of the target cell which initiates signal transduction through a chain of biochemical or molecular events.
>
> This process transfers and amplifies the signal within the cell where it integrates this information and triggers a specific cellular response. For example, it may activate specific genes in the nucleus or send messages to nearby or cells located elsewhere in the organism.

6.3 Genotoxicity and Mutagenicity

Some chemical agents, known as genotoxins, can damage cellular DNA (or impair DNA repair mechanisms) causing cell death or mutations that may lead to cancer. DNA damage can occur in a number of forms: single- and double-strand breaks, loss of excision repair, cross-linking of DNA strands, alkali-labile sites, point mutations, and structural and numerical chromosomal aberrations. Some chemicals, for example, are known to cause chromosomal mutations by cross-linking chromosomes or by changing the number of chromosomes in cells. These include alkylating agents such as epoxides, aldehydes, alkaline halides, alkyl sulfonates, nitrosoureas, and triazines (Connell 2005, pp. 376–377).

A genotoxin will cause damage to a DNA sequence but cells may prevent expression of genotoxic mutations by either DNA repair mechanisms or cell death (apoptosis). If the damage is not repaired, permanent genetic changes or mutations in the genetic material of cells or organisms (mutations) can result (mutagenicity). The process is known as mutagenesis and the agent causing the change is called a mutagen. It follows that all mutagens are genotoxic, but not all genotoxins are associated with mutations (i.e. mutagenic).

Mutagenesis is a well-known natural process in which mutation is recognized as the source of variation and change in the theory of natural selection and evolution. Such mutations can result from mechanisms such as errors in DNA replication, repair and recombination, or hydrolysis. Natural mutagens include ultraviolet light, ionizing radiation from the decay of natural radioactive materials, alkaloids and flavonoids from plants, and mycotoxins from fungi (Connell 2005, p. 375). Overall, mutagens act to increase the occurrence of mutations in the genetic material of cells and organisms.

Mutations can either occur in **germ cells** or **somatic cells**. If the mutation occurs in a **germ cell**, the effect is heritable. The exposed person is unaffected but the effect is passed on to future generations. If the mutation occurs in a **somatic cell**, it can cause altered cell growth (e.g. cancer) or cell death (e.g. teratogenesis) in the exposed person.

- **Germ cells** are those cells that are involved in the reproductive process and can give rise to a new organism. Male germ cells give rise to sperm and female germ cells develop into ova. Toxicity to germ cells can cause effects on the developing fetus (such as birth defects and abortions).
- **Somatic cells** are all body cells except the reproductive germ cells. They have two sets (or pairs) of chromosomes. Toxicity to somatic cells causes a variety of toxic effects to the exposed individual (such as dermatitis, cancer, or death).

Considerable evidence of a positive correlation between the mutagenicity of substances *in vivo* and their carcinogenicity has been observed in long-term animal studies (Griffiths et al. 2000).

Mutagens may be of physical, chemical, or biological origin. Many mutagens have been identified by experimental bioassays such as the Ames test, Comet assays, and a battery of other tests. Examples of mutagens include: arsenic, aromatic amines, bromine, benzene, dioxane, ethanol (via metabolite acetaldehyde), X-rays, gamma rays, alpha particles, and ultraviolet radiation.

Genetic toxicity assays are widely used to identify mutagens for purposes of hazard identification and risk assessment (e.g. dose–response relationship between mutagens and cancer-causing agents and mutagenic mechanisms).

Table 6.1 Some key genotoxicity tests.

In vitro testing	Biomarkers	Type of test
Ames test/assay Bacterial Reverse Mutation Assay (OECD 471)	Mutagenicity in *Salmonella typhimurium* or *Escherichia coli* (*E. coli*) tester strains.	Sensitive test that uses amino-acid requiring strains of these bacteria to detect point mutations, which involve substitution, addition, or deletion of one or a few DNA base pairs.
In vivo testing		
Mammalian erythrocyte micronucleus test (OECD 474)	An increase in the frequency of micronucleated polychromatic erythrocytes in treated animals is an indication of induced chromosome damage.	Uses analysis of erythrocytes as sampled in bone marrow and/or peripheral blood cells of animals (usually mice or rats) to detect damage induced by the test substance to the chromosomes or the mitotic apparatus of erythroblasts.
In vitro and *in vivo* testing		
Comet assay	Assay detects DNA strand breaks in single cells from both *in vitro* and *in vivo* sources. The latter test applies to tissues that can be dispersed to a single cell suspension.	The comet assay uses single-cell gel electrophoresis and microscopy as a sensitive and rapid tool for measuring DNA strand breaks in eukaryotic cells.

Source: Based on Bajpayee et al. (2013) and Chemsafetypro (2016).

There are both *in vitro* and *in vivo* mutagenicity/genotoxicity studies for germ cells and somatic cells. The basic screening test for mutagenicity is the Ames test – an *in vitro* gene mutation study in bacteria (bacterial reverse mutation test, OECD 471). If this test is positive, *in vivo* mutagenicity studies are needed for confirmation (see also Chapter 19 in this book). Table 6.1 provides an outline of some key genetic toxicity tests.

A carcinogen is broadly any substance or agent, or mixture of these that can induce or increase the incidence of cancer in living tissues. It can include exposures to UV radiation, fine air pollution particles, airborne asbestos fibers, benzene, arsenic, and certain viruses and bacteria.

Exposure to a carcinogen may damage directly or indirectly the DNA in cells causing genetic mutations (genotoxicity) or disrupt a wide variety of cellular processes that lead to abnormal cell growth and tumors, without damaging DNA.

The current multistage theory of cancer considers three critical stages in the development and progression of carcinogenesis: (i) initiation – alteration of the normal cell to a cancer cell through mutation, (ii) promotion – multiplication and selection of initiated cells, and (iii) progression – expansion of the cancer cell line by invasion of local tissue and metastasis (development of secondary malignant growths at distant organ sites) (e.g. LaGrega et al. 2001, p. 281). Figure 6.1 shows the key steps and stages involved in the development of cancer.

Carcinogenic chemicals can be divided into two types of carcinogens that are defined according to their general modes of action in carcinogenesis:

- Genotoxic carcinogen (GTXs): "Chemical substances or agents causing tumors by interaction with the genetic

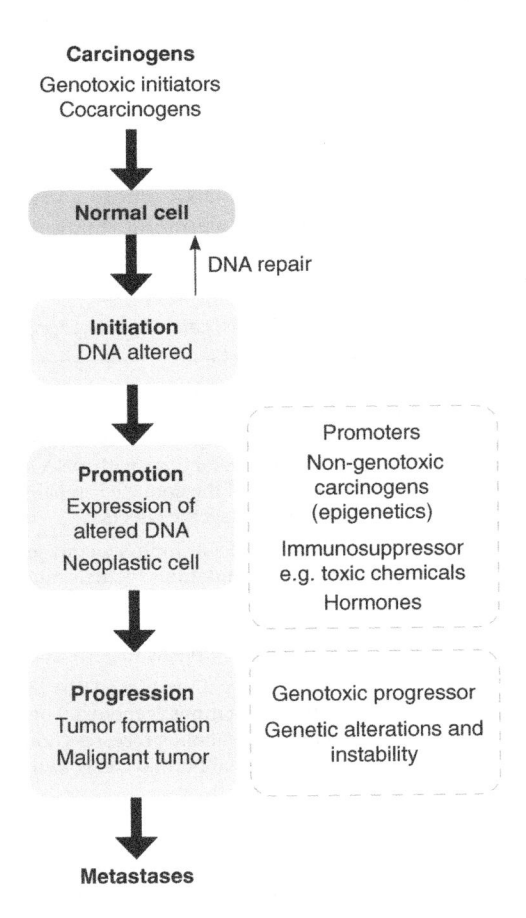

Figure 6.1 The general stages and the steps involved in the genetic mutation by carcinogens, development, and progression of cancer. Source: Modified from LaGrega et al. (2001) and Connell (2005).

material (e.g. DNA, chromosomes). Such substances are usually both genotoxic and carcinogenic, and can induce gene mutations, structural chromosome mutations, and genome mutations."

These chemicals are active in carcinogenic initiation and progression. Examples include aflatoxin, arsenic, benzo(a)pyrene, hexavalent chromium, ethylene oxide, formaldehyde, N-nitrosodimethylamine, and radon-222.

- Non-genotoxic carcinogen (NGTXs): "Chemical substances or agents causing tumors by non-genotoxic mechanism (e.g. peroxisome proliferators, hormones, and local irritants). Such substances do not have genotoxicity as a primary biological activity."

Discrimination of genotoxic from non-genotoxic carcinogens has been reported, for example, by van Delft et al. (2004).

The modes of action implicated in the case of chemical carcinogenesis are given as:

Genotoxicity – direct or indirect damage to DNA or chromosomes through several mechanisms:

- Direct mechanisms involve interaction of the chemical or metabolite (GTX) with DNA by binding covalently to form DNA adducts, that damage the DNA within a cell causing mutations, which may lead to tumor formation. In response, the cell may initiate repair of the damage, arrest of the cell cycle, or induction of normal apoptosis to eliminate unwanted or healthy cells. In many cases, cells may not be able to repair damage or prevent propagation of cell mutations resulting in cancer.
- Indirect mechanisms by GTXs can cause damage to DNA or chromosomes by toxic interactions with non-DNA targets such as proteins (e.g. aneugens), or overloading the system/metabolism and exceeding natural protective mechanisms. The latter mechanism can involve substances such as reactive oxygen species (ROS).

Non-Genotoxicity – non-primary biological activity

- It involves chemical substances or agents (e.g. peroxisome proliferators, hormones, and local irritants), also known as epigenetic carcinogens, that cause tumors without direct modification or damage to DNA. Their mechanisms may involve a wide variety of cellular processes. In effect, non-genotoxic agents can cause carcinogenic promotion. Examples of these chemicals include carbon tetrachloride, chloroform, DDT, PCBs, and dioxins (TCDD).

An overview of genetic toxicants and chemical examples is given in Table 6.2. It includes reproductive and developmental toxicants (e.g. teratogens) discussed in Section 6.3. Various endocrine-disrupting chemicals can also cause adverse reproductive effects (see Section 6.4).

Table 6.2 Overview of genotoxic, mutagenic, carcinogenic, reproductive, and developmental toxicants.

Toxicant classes	Toxicity description	Examples
Genotoxicant	Induces damage to the genetic material in cells through interactions with DNA sequence and structure. If the damage is repaired, it may not lead to mutagenic effects.	Note: not all genotoxins are mutagens.
Mutagen	Causes direct or indirect damage to DNA that results in mutations (genetic alterations). These are changes in the DNA sequence that are retained in somatic cell divisions and passed onto progeny in germ cells.	Acrylamides, aflatoxins, benzo(a)pyrene, bisphenol A, dichloromethane, nitrosamines, phthalates, PCBs, PFOA, PBDE, and trichloroethylene.
Carcinogen	May damage directly the DNA in cells causing genetic mutations (genotoxic) or disrupt cellular metabolic processes (non-genotoxic) that lead to abnormal cell growth and tumors.	**Genotoxic:** Aflatoxins, benzene, 1,3 butadiene, benzo(a)pyrene, formaldehyde, vinyl chloride, 1,2 dichloropropane, pentachlorophenol, and N-nitrosodiethylamine. **Non-genotoxic:** 1,4-dichlorobenzene, 17 β-estradiol, PCBs, TCDD, and 1,4 dioxane.
Reproductive and developmental toxicant (reprotoxic)	Interferes with normal reproduction and may also cause developmental toxicity in the offspring. (Teratogens affect the development of an embryo or fetus and cause birth defects or even death in offspring.)	Bisphenol A, PCBs, benzo(a)pyrene, cytotoxic drugs. **Teratogens:** Methyl mercury, PCBs, DDT, PBDEs, phthalates, bisphenol A (BPA), PAHs, and perchlorate.

Source: Modified from LaGrega et al. (2001), and Stepa et al. (2017).

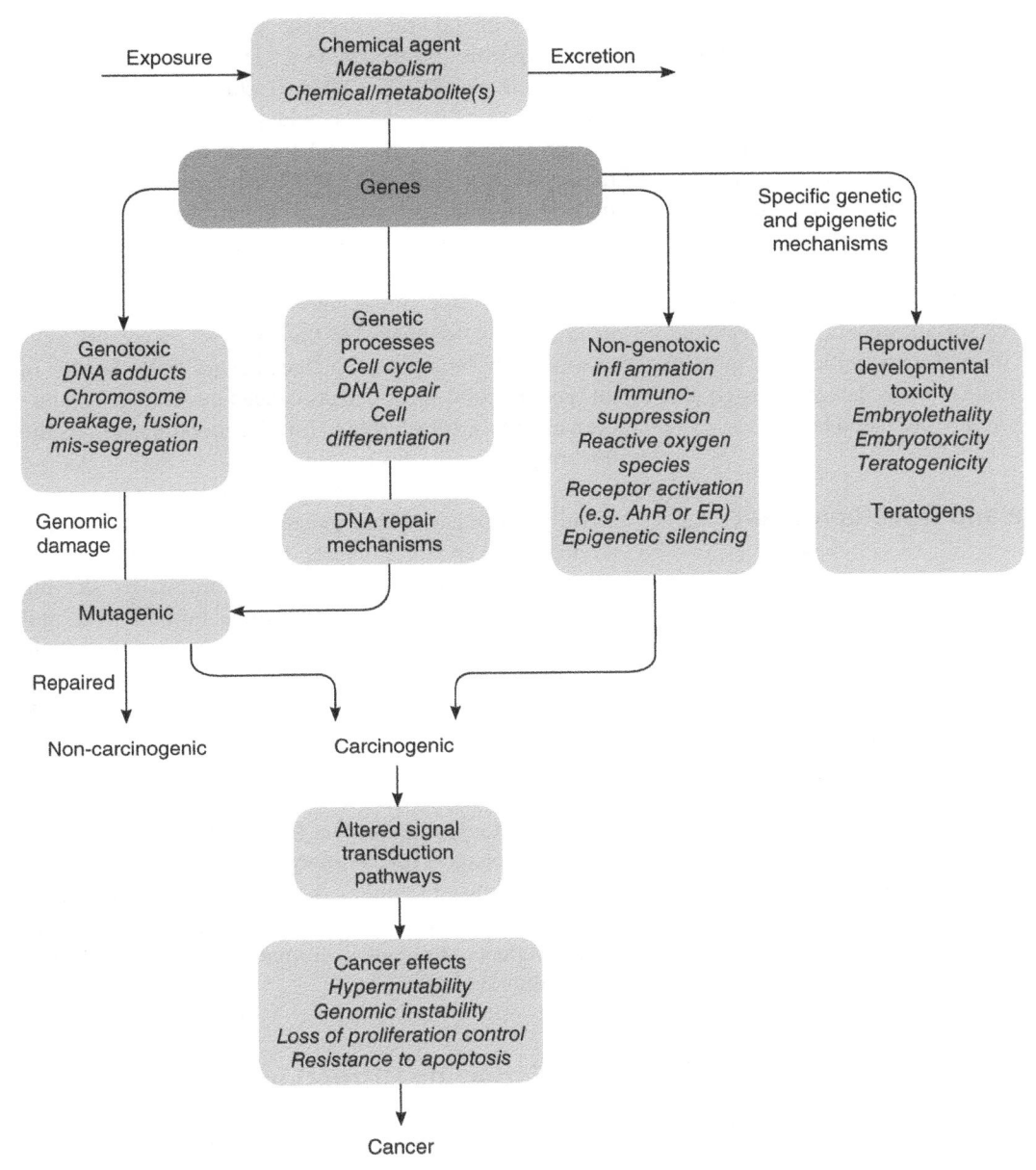

Figure 6.2 Outline of genetic and non-genetic toxicity mechanisms of chemical agents related to genotoxicity, mutagenicity, carcinogenicity, and reproductive/developmental effects.

As indicated in Figure 6.2, chemical agents may be absorbed by cells, retained, or metabolized and excreted from cells. Inside the cell, genotoxic chemicals or their metabolites can cause direct or indirect effects on the regulation and expression of genes involved in the cell-cycle control, DNA repair, cell differentiation, or apoptosis. Mechanisms involve genomic damage (mutations) such as forming DNA adducts or inducing chromosome breakage, fusion, deletion, mis-segregation, and nondisjunction. Two types of chromosome damage result from chemical agents known as **clastogens**, that cause breaks or disruption in chromosomes, and **aneugens**, that cause daughter cells to have an abnormal number of chromosomes. In response, a cell may initiate repair of the damage, arrest of the cell cycle, or induction of normal apoptosis to eliminate unwanted or healthy cells. In this case, the genotoxic chemicals become noncarcinogens. If cells are unable to repair damage or prevent propagation of cell mutations resulting in cancer, then the genotoxic chemicals act as carcinogens.

Importantly, some other chemicals may act as epigenetic carcinogens by **non-genotoxic mechanisms** that do not cause direct modification or damage to DNA, such as induction of inflammation, immunosuppression, formation of ROS, activation of receptors, e.g. the arylhydrocarbon receptor (AhR) or estrogen receptor (ER), and epigenetic silencing. These genotoxic and non-genotoxic mechanisms can act together to alter signal-transduction pathways that result in hypermutability, genomic instability, loss of proliferation control, and resistance to apoptosis.

Reproductive/developmental toxic effects can be caused by various chemicals (and other physical and biological agents) and are induced by specific genetic and epigenetic mechanisms separate from carcinogenicity. Examples include teratogens that induce births defects. This type of toxicity and its mechanisms are discussed below.

6.4 Reproductive and Developmental Toxicity

Reproductive toxicity is defined as adverse effects of a chemical substance on sexual function and fertility in adult males and females, as well as developmental toxicity in the offspring. Developmental toxicity refers to adverse toxic effects in the developing embryo or fetus. Chemicals cause developmental toxicity by two ways. They can act directly on cells of the embryo or fetus causing cell death or cell damage leading to abnormal organ development. A chemical may also induce a mutation in the germ cell of a parent which is transmitted to the fertilized ovum. Some mutated fertilized ova develop into abnormal embryos.

There are three basic types of developmental toxicity:

- Embryolethality – failure to conceive, spontaneous abortion, or stillbirth.
- Embryotoxicity – growth retardation or delayed growth of specific organ system.
- Teratogenicity – irreversible conditions that leave permanent birth defects in live offspring.

Effects via lactation are also recognized where a toxicant may interfere with lactation, or occur in breast milk and may cause harm to breast-fed children.

Teratogens cover a select and diverse group of physical, chemical, and biological agents, in combination with environmental factors that interfere with the normal reproduction process and cause either a reduction of successful births, or offspring to be born with physical, mental, developmental, or behavioral defects. Exposure to teratogens during the prenatal stage in humans and animals can significantly increase the risk of birth defects. Reported effects range from infertility, restricted prenatal onset of growth, structural defects, and functional central nervous system abnormalities, to miscarriage or death. Chemical teratogens include mercury (e.g. methyl mercury), PCBs, organochlorine pesticides (e.g. DDT), polybrominated diethyl ethers (PBDEs) (flame retardants), phthalates, bisphenol A (BPA), PAHs, and perchlorate. Significant uptake of lead and placental transfer can cause spontaneous abortion and stillbirth. Other forms of teratogens include ionizing radiation, certain viruses, drugs, stressors (e.g. hyperthermia, alcohol, and tobacco), and even malnutrition (Gilbert-Barnes 2010).

Teratogens may cause defects by a variety of means, including DNA damage, but they only act on somatic cells in contrast to mutagens, which can cause DNA damage in all types of cells. Birth defects caused by teratogens, however, cannot be transferred to future generations unless the offspring are similarly exposed.

The specific toxicity mechanisms of teratogens can be broadly classed as either genetic or epigenetic. Genetic teratogens exert their effects by gene mutation, chromosomal abnormality, or inhibiting mitosis by slowing DNA synthesis or preventing spindle formation, a vital step in mitosis. Epigenetic mechanisms of action include affecting metabolic processes, inhibition of enzymes, and altering the permeability of cell membranes (Connell 2005, pp. 371–372).

Many environmental teratogens are similar in physical or chemical nature to other developmental toxins such as endocrine disruptors (Gilbert-Barnes 2010). The major toxicological role of many environmental chemicals is as endocrine disruptors that can cause reproductive, developmental, carcinogenic, neurological, and other adverse effects observed in humans, laboratory animals, and wildlife (see Section 6.5).

6.5 Endocrine Disruption

Among the many global pollutants that are known or suspected to cause adverse toxic effects and induce chronic (noninfectious) diseases in humans and wildlife are a diverse number of environmental chemicals that can affect the hormone system, by interfering with the actions of hormones, leading to a multitude of adverse effects, including reduced fertility, birth defects, obesity, and some types of cancers. These chemicals are generally known as endocrine-disrupting chemicals (EDCs).

Thousands of manufactured chemicals may be EDCs. They exert a special type of toxicity that is relatively complex and is not fully understood. In the Twenty-first Century, a great deal of research is providing new insights into their toxicity mechanisms, the extent of environmental contamination from these chemicals, and the relationship between chemical exposures and health

Endocrine System

The endocrine system is responsible for controlling and coordinating numerous normal body functions. It consists of a series of glands that produce hormones which act as chemical messengers and regulate essential body functions such as reproduction, sexual function, metabolism, growth and development, tissue function, sleep, and mood. The endocrine system includes major glands: pituitary gland, thyroid gland, parathyroid gland, adrenal glands, pancreas, ovaries, testicles, and others (see Figure 6.3).

Endogenous hormones are synthesized in cells within a gland, secreted into the circulatory system, and transported through the bloodstream until they reach a target tissue or organ and bind to specific receptors within target cells. The process of binding is determined by the specific chemical nature and shape of the hormone and its complementary receptor(s). Specific activation of a receptor by a hormone triggers a response via complex interactions of signaling pathways that may produce a protein, another hormone, a change in metabolism, a behavioral response, or other responses (see Figure 6.4) (WHO 2013a, p. 6; Gore et al. 2014, pp. 12–13). Receptor activation, however, depends upon various factors such as the quantity of hormone reaching the receptor, its potency, and time of exposure and activation.

There are over 50 different hormones and hormone-related molecules (cytokines and neurotransmitters) in

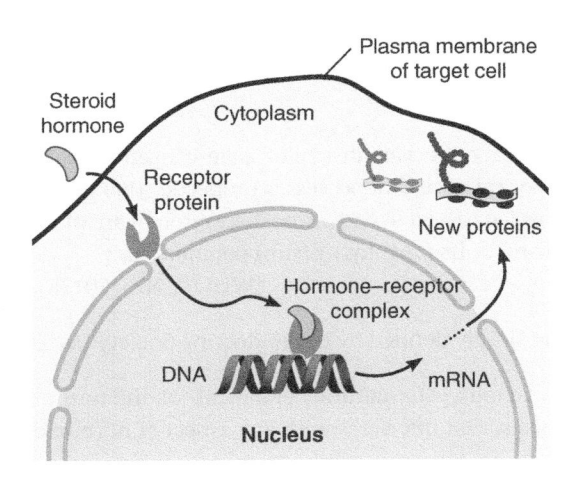

Figure 6.4 Example of hormone action. Many hormones (i) act via binding to specific receptors and (ii) stimulate the synthesis of new proteins, which then control tissue function. Some hormones also act via receptors on the membrane. In that case, the actions are more immediate in nature. Source: Adapted from WHO/UNEP (2013a).

humans that integrate and control normal body functions across and between tissues and organs over the lifespan. This is also the case in wildlife.

Hormones and their signaling pathways are essential to the normal functioning of every tissue (WHO 2013a, p. 4).

Estrogens are an example of natural or exogenous hormones that play fundamental roles in female reproduction, and are also involved in male reproduction, neurobiological functions, bone development and maintenance, cardiovascular functions, and many other functions. Natural estrogens exert these actions, after being released from the gonad (ovary-female or testis-male), by binding to ERs in the target tissues (Gore et al. 2014, pp. 12–13).

What are Endocrine Disruptors?

Many environmental chemicals are increasingly known to act as endocrine disruptors by mechanisms that mimic or interfere with the function of endogenous hormones through affecting the signals carried by these hormones, leading to reproductive, developmental, neural, immune, and other dysfunctions observed in laboratory animals, humans, and wildlife (Gore et al. 2015; Yang et al. 2015).

Various definitions of endocrine disruptors exist. The WHO Report on State of the Science of Endocrine Disrupting Chemicals – 2012 adopted the IPCC (2002) definition of an endocrine disruptor as,

> …an exogenous substance or mixture that alters function(s) of the endocrine system and consequently

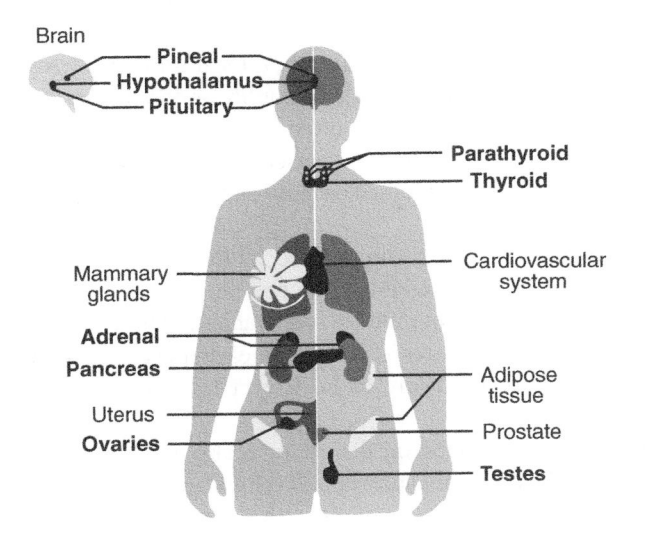

Figure 6.3 Diagram of major endocrine glands in the human body, shown in a female (left) and male (right). Source: Modified from Gore et al. (2014).

causes adverse health effects in an intact organism, or its progeny, or (sub) populations,

and,

A potential endocrine disruptor is an exogenous substance or mixture that possesses properties that might be expected to lead to endocrine disruption in an intact organism, or its progeny, or (sub) populations
(WHO/UNEP 2013a).

EDCs are simply defined by the Endocrine Society[1] as:

an exogenous [non-natural] chemical, or mixture of chemicals, that interferes with any aspect of hormone action
(Gore et al. 2014, p. 1).

Numerous EDCs are found widely in industrial, consumer, and personal care products, and wastes from human activities. Hundreds of potential EDCs are dispersed widely in air, waters, soils, sediments, foodstuffs, and many may accumulate in natural biota, human tissues, and organs.

EDCs are usually found at low concentrations in the environment and biological tissues, but may act as endocrine disruptors at these levels. They range from persistent organic pollutants (POPs), various industrial chemicals, pesticides, and toxic metals to emerging pollutants such as residues from drug and personal care products. Examples of EDCs of concern include BPA (plasticizer), nonyl phenol (detergents, insecticides), DEHP plasticizer (phthalate used in PVC production), endosulfan (insecticide), PCBs (industrial chemical), dioxins (by-products of combustion and certain chemical synthesis), PBDEs (flame retardants), and cadmium (toxic metal).

The Endocrine Society's 2015 review of the scientific literature on EDCs found:

… strong mechanistic, experimental, animal, and epidemiological evidence for endocrine disruption, namely: obesity and diabetes, female reproduction, male reproduction, hormone-sensitive cancers in females, prostate cancer, thyroid, and neurodevelopment and neuroendocrine systems.

Prevalent EDCs identified in the reviewed literature included bisphenol A, phthalates, pesticides, and POPs such as polychlorinated biphenyls, PBDEs, and dioxins (Gore et al. 2015). Exposures to EDCs are also critical during early life stages (fetus and infant) because perturbations of hormones can increase the probability of a disease or dysfunction later in life (Gore et al. 2015).

Toxicity Mechanisms and Effects of Endocrine-Disrupting Chemicals

Some EDCs can act directly on hormone receptors as hormone mimics or antagonists, and others can act directly on any number of proteins that control the delivery of a hormone to its normal target cell or tissue (WHO/UNEP 2013b). Hormone mimics can activate a hormone's receptor and trigger processes normally activated only by a natural hormone. In the case of hormone blockers, an EDC can bind to a hormone's receptor, and prevent activation, even if the natural hormone is present (Gore et al. 2014, pp. 12–13).

The best-known example of endocrine disruption involves estrogenic hormones, which act upon the body's estrogen receptors (ERs). In both males and females, ERs are present in many cells in the brain, in bone, in vascular tissues, and in reproductive tissues (Gore et al. 2014, pp. 12–13). Not only are ERs targeted by EDCs, but receptors for androgens (testosterone), progesterone, thyroid hormones, and many others, are also disrupted in their functioning by EDCs. Some EDCs can disrupt multiple receptors and hormone signaling pathways at the same time (WHO/UNEP 2013b; Gore et al. 2014, pp. 12–13).

The WHO Report on State of the Science of EDCs – 2012 (see WHO/UNEP 2013a, b) presented several key messages about the observed dose–responses of EDCs on hormone receptors:

- The affinity of an endocrine disruptor to a hormone receptor is not equivalent to its potency. Chemical potency on a hormone system is dependent upon many factors including receptor abundance.
- They bind to receptors at very low concentrations and have the ability to be active at low concentrations, many in the range of current human and wildlife exposures.
- Endocrine disruptors produce nonlinear dose responses both *in vitro* and *in vivo*; these nonlinear dose responses can be quite complex and often include non-monotonic dose responses[2], due to a variety of mechanisms. No threshold can be assumed because endogenous hormone levels fluctuate.
- Mixtures of EDCs can induce additive or synergistic effects, even when combined at low doses that individually do not produce observable effects.
- Timing of exposures is critical as early developmental effects are also likely to be irreversible unlike adult effects.
- Sensitivity to endocrine disruption is also highest during tissue development such that developmental effects will occur at lower doses than those observed in adults.

---

1 World's oldest, largest, and most active organization devoted to research on hormones and the clinical practice of endocrinology.

2 Non-monotonic dose–response curves are mathematically defined as a change in the sign (positive/negative) of the slope of a dose–response relationship over the range of doses tested.

Toxicological Mechanisms

The toxicity mechanisms underlying the disruptive actions of EDCs on the endocrine system are not well understood. Various researchers and reviewers have proposed general molecular mechanisms to explain how EDCs bind to hormone receptors and interact with genomic and non-genomic signaling pathways to induce adverse effects that include abnormal reproduction, hormone-related cancers, obesity, and dysfunction of the neural system and immune system. Many EDCs bind to steroid receptors, for example, Lee et al. (2013) have described potential mechanisms for the interference of EDCs with the steroid hormone, estrogen (E2), and binding to ERs that lead to abnormal estrogen-related signaling responses in diverse cells and tissues. These are given in Figure 6.5.

Toxicity mechanisms of EDCs may follow two general pathways: (i) *genomic pathway* of EDC interference with estrogen (E2) binding to ERs. EDCs bind to ERs instead of E2, and can thus affect the transcription of target genes in the nucleus by binding to the estrogen response element (ERE) of target gene, and (ii) *non-genomic pathway* of EDC action may occur through the ER such as G protein-coupled receptor (GPR30) located in the cytoplasmic membrane. Activation of GPR30 by EDCs leads to rapid downstream cellular signaling. This induces subsequent stimulation of protein kinase activation and phosphorylation, which in turn may affect the transcription of target genes. The resulting changes by interaction between ERs and GPR30 in gene expression and intracellular signaling can cause cellular response without regulation, which may produce adverse effects of EDCs on organs (Lee et al. 2013).

Common molecular mechanisms for endocrine disruption are also proposed by Yang et al. (2015), developed from using a bioinformatics tool, known as molecular pathway analysis, which is widely used to predict the mechanism or biological network of certain chemicals. These researchers investigated molecular mechanisms of three EDCs (BPA, phthalate, and nonyl phenol) related to hormone disruption (e.g. estrogenic), obesity, and cancer. They found the following:

1. Hormone disruption: inhibition of endocrine receptors by endocrine disruptors in target tissues is most common mode of action and is widely studied, although EDCs can disrupt enzymes involved in the production of steroids and metabolism of estrogens, and certain transport proteins of endogenous hormones. Two types of ERs (ERα and ERβ) are considered to mediate distinct biological effects in different tissues which suggest endocrine disruptors could show selective agonist and antagonist activity in different tissues or during development.

2. Obesity mechanism: endocrine disruptors can also act as *obesogens* when their target cells are adipocytes (specialized cells for fat storage in connective tissue), resulting in a change to the regulation of lipid metabolism

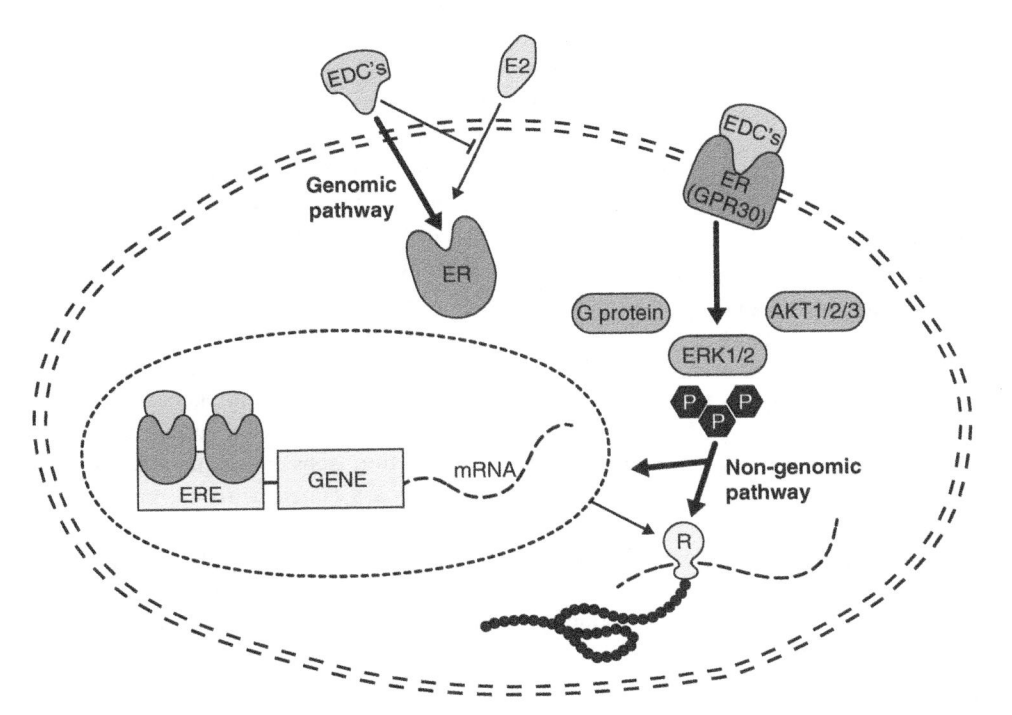

Figure 6.5 Potential mechanism(s) of endocrine-disrupting chemicals (EDCs) action. Source: Modified from Lee et al. (2013).

and adipogenesis which can promote obesity. In this case, the endocrine disruptors induce lipogenesis and inhibit lipolysis in the adipocyte cells, increasing fat storage.

3. Cancer mechanisms: endocrine disruptors are considered to act on the cell cycle, via endocrine receptor (ERα), where cyclin protein and p21 protein are known to regulate cancer cells when exposed to endocrine receptors. Breast cancer and prostate cancer are typical cancers associated with endocrine disruptors. Even so, the direct mode of action of endocrine disruptors in cancer is complex and largely unclear.

Epigenome and Trans-Generational Effects

There is a growing concern that maternal, fetal, and childhood exposure to EDCs could play a larger role in the causation of many endocrine diseases and disorders than previously believed. Studies of wildlife populations and of laboratory animals show associations between exposure to EDCs and adverse health effects. These observations are supported by evidence of increased incidence and prevalence of several endocrine disorders that cannot be explained by genetic factors (WHO 2013b).

Environmental exposure to EDCs during early development and pregnancy may modify chemical compounds of the epigenome that are not part of the DNA sequence but can attach to it, and regulate the activity of all genes in the genome. Examples of epigenetic modifications or changes include DNA methylation of part of the DNA molecule which inhibits certain genes being expressed, and modification of histone proteins that DNA wraps around. Epigenetic modifications or changes can remain as cells divide, and may be transferred or inherited across multiple generations[3] (US National Human Genome Research Institute 2016).

There is increasing evidence that epigenetic gene alterations can induce trans-generational diseases such as asthma, autism, cancer, cardiovascular dysfunctions, diabetes, obesity, schizophrenia, infertility, reproductive diseases, and dysfunction later in life (Shahidehnia 2016).

How exposures to pollutants in air, water, soil, food, and other stressors in our environment can harm human health in specific ways or mechanisms, such as epigenetic gene alterations and endocrine disruption, over a lifetime are being investigated by advances in exposome methodologies. Annette Peters and coauthors in their paper "Hallmarks of environmental insults" published in the Journal *Cell* have proposed a set of key toxicological mechanisms that include *oxidative stress and inflammation, genomic alterations and mutations, epigenetic alterations, mitochondrial dysfunction, endocrine disruption, altered intercellular communication, altered microbiome communities, and impaired nervous system function (Peters et al. 2021)*.

Importantly, as these authors stated *They provide a framework to understand why complex mixtures of environmental exposures induce severe health effects even at relatively modest concentrations.*

6.6 Key Points

1. Genotoxicity is the study of the adverse effects of substances or agents on the genetic material of cells (DNA) leading to expression of these changes as biological effects (mutagenicity, carcinogenicity, and reproductive or developmental).
2. All the genetic information needed for a living organism is chemically encoded in DNA, a long double helix molecule made up of nucleotides, which is contained in the genome.
3. Within the genome, sequences of nucleotides along the DNA molecule make up functional groups known as genes. A single long DNA chain or molecule is called a chromosome. The role of genes is to code for the synthesis of RNA and protein molecules needed by an organism to function.
4. Various natural and synthetic agents, including environmental chemicals, are known as genotoxins, and can damage cellular DNA (or impair DNA repair mechanisms) or chromosomes causing cell death or mutations that may lead to cancer.
5. Cells may prevent expression of genotoxic mutations by either DNA repair mechanisms or cell death (apoptosis). If unrepaired, mutations in the genetic material of cells or organisms can result (mutagenicity). The process is known as mutagenesis and the agent causing the change is called a mutagen.
6. It follows that all mutagens are genotoxic, but not all genotoxins are associated with mutations (i.e. mutagenic).
7. Spontaneous mutations occur naturally due to errors in natural biological processes, while induced mutations are due to agents in the environment that cause changes in DNA structure.
8. Mutations can either occur in germ cells or somatic cells. If the mutation occurs in a germ cell, the effect is heritable.

---

3 Transgenerational **epigenetic** inheritance is the **transmission** of information from one generation of an organism to the next (i.e. parent–child **transmission**) that affects the traits of offspring without alteration of the primary structure of DNA (i.e. the sequence of nucleotides).

9. Genetic toxicology assays are widely used to identify mutagens for purposes of hazard identification and risk assessment (e.g. dose–response relationship between mutagens and cancer-causing agents and mutagenic mechanisms).
10. A carcinogen is broadly any substance or agent, or mixture of these that can induce or increase the incidence of cancer in living tissues. It can include exposures to UV radiation, fine air pollution particles, airborne asbestos fibers, benzene, arsenic, and certain viruses and bacteria.
11. It may (i) directly or indirectly damage the DNA in cells causing genetic mutations (genotoxicity), or (ii) disrupt a wide variety of cellular processes that lead to abnormal cell growth and tumors, without damaging DNA.
12. Genotoxic carcinogens cause direct or indirect damage to DNA or chromosomes through several mechanisms that are indicated in Figure 6.2.
13. Non-genotoxic carcinogens involve chemical substances or agents (e.g. peroxisome proliferators, hormones, and local irritants), also known as epigenetic carcinogens, that cause tumors by carcinogenic promotion without direct modification or damage to DNA. Their mechanisms can involve a wide variety of cellular processes (see Table 6.2).
14. The current multistage theory of cancer considers three critical stages in the development and progression of carcinogenesis: (i) initiation – alteration of the normal cell to a cancer cell through mutation, (ii) promotion – multiplication and selection of initiated cells, and (iii) progression – expansion of the cancer cell line by invasion of local tissue and metastasis to distant organ sites.
15. Certain environmental chemicals can cause reproductive/developmental toxicity in exposed persons, laboratory animals, and wildlife.
16. Reproductive toxicity is defined as adverse effects of a chemical substance on sexual function and fertility in adult males and females as well as developmental toxicity in the offspring.
17. Developmental toxicity refers to adverse toxic effects in the developing embryo or fetus. Chemicals cause developmental toxicity by two ways. They can act directly on cells of the embryo or fetus causing cell death, or damage leading to abnormal organ development.
18. A chemical may also induce a mutation in the germ cell of a parent which is transmitted to the fertilized ovum. Some mutated fertilized ova develop into abnormal embryos.
19. Teratogens are a diverse group of physical, chemical, and biological agents, and environmental factors that interfere with the normal reproduction process and cause either a reduction of successful births, or offspring to be born with physical, mental, developmental, or behavioral defects.
20. Many environmental teratogens are similar in physical or chemical nature to other developmental toxins such as endocrine disruptors.
21. A major toxicological role of many environmental chemicals is as endocrine disruptors (EDCs) that can cause reproductive, developmental, carcinogenic, neurological, and other adverse effects observed in humans, laboratory animals, and wildlife.
22. The endocrine system consists of a series of glands that produce hormones which act as chemical messengers and regulate essential body functions such as reproduction, sexual function, metabolism, growth and development, tissue function, sleep, and mood.
23. Many environmental chemicals can affect the hormone system (e.g. estrogenic hormones), by interfering with the actions of hormones, leading to a multitude of adverse effects, including reduced fertility, birth defects, obesity, and some types of cancers.
24. Toxicity mechanisms are relatively complex but they basically mimic or interfere with the function of endogenous hormones through affecting the signals carried by these hormones, leading to reproductive, developmental, neural, immune, and other dysfunctions observed in laboratory animals, humans, and wildlife.
25. Some EDCs can disrupt multiple receptors and hormone signaling pathways at the same time.
26. EDCs bind to receptors at very low concentrations and have the ability to be active at low concentrations, many in the range of current human and wildlife exposures. They produce variable nonlinear dose–responses.
27. Mixtures of EDCs can induce additive or synergistic effects, even when combined at low doses that individually do not produce observable effects.
28. Timing of exposures is critical as early developmental effects are also likely to be irreversible unlike adult effects.
29. Sensitivity to endocrine disruption is also highest during tissue development such that developmental effects will occur at lower doses than those observed in adults.
30. Environmental exposure to EDCs during early development and pregnancy may also modify epigenomes and induce diseases or disorders that are transgenerational later in life. There is evidence showing that some EDCs can induce epigenetic gene alterations that can be transferred into subsequent generations.

References

Bajpayee, M., Kumar, A., and Dhawan, A. (2013). The Comet Assay: Assessment of In Vitro and In Vivo DNA Damage. In: *Genotoxicity Assessment. Methods in Molecular Biology (Methods and Protocols)*, vol. 1044 (ed. A. Dhawan and M. Bajpayee). Totowa, NJ: Humana Press.

Chemsafetypro (2016). Mutagenicity and Genotoxicity. https://www.chemsafetypro.com/Topics/CRA/Mutagenicity_and_Genotoxicity.html (accessed 3 February 2019).

Connell, D.W. (2005). *Basic Concepts of Environmental Chemistry*, 2e. Boca Raton, FL: CRC Press.

van Delft, J.H.M., van Agen, E., van Breda, S.G.J. et al. (2004). Discrimination of genotoxic from non-genotoxic carcinogens by gene expression profiling. *Carcinogenesis* 25 (7): 1265–1276.

Fischer-Cripps, A.C. (2012). *The Chemistry Companion*. New York: CRC Press.

Gilbert-Barnes, E. (2010). Teratogenic causes of malformations. *Annals of Clinical and Laboratory Science* 40 (2): 99–114.

Gore, A.C., Crews, D., Doan, L.L. et al. (2014). *Introduction to Endocrine Disrupting Chemicals (EDCs): A Guide for Public Interest Organizations and Policy-Makers*. Endocrine Society/IPEN. https://www.endocrine.org/-/media/endosociety/files/advocacy-and-outreach/important-documents/introduction-to-endocrine-disrupting-chemicals.pdf (accessed 4 February 2020).

Gore, A.C., Chappell, V.A., Fenton, S.E. et al. (2015). Executive summary to EDC-2: the endocrine society's second scientific statement on endocrine-disrupting chemicals. *Endocrine Reviews* 36 (6): 593–602.

Griffiths, A.J.F., Miller, J.H., Suzuki, D.T. et al. (2000). *An Introduction to Genetic Analysis*, 7e. New York: W.H. Freeman.

LaGrega, M.D., Buckingham, P.L., Evans, J.C. et al. (2001). *Hazardous Waste Management*, 2e. New York: McGraw Hill.

Lee, H.-R., Jeung, E.-B., Cho, M.-H. et al. (2013). Molecular mechanism(s) of endocrine-disrupting chemicals and their potent oestrogenicity in diverse cells and tissues that express oestrogen receptors. *Journal of Cellular and Molecular Medicine* 17 (1): 1–11.

Peters, A., Nawrot, T.S., and Baccarelli, A.A. (2021). Hallmarks of environmental insults. *Cell* 184 (6): 1455–1468.

Shahidehnia, M. (2016). Epigenetic effects of endocrine disrupting chemicals. *Journal of Environmental and Analytical Toxicology* 6 (4): 381.

Stepa, R.A., Schmitz-Felten, E., and Brenzel, S. (2017). *Carcinogenic, Mutagenic, Reprotoxic (CMR) Substances*. Hamburg: The Cooperation Centre (Kooperationsstelle). https://oshwiki.eu/wiki/Carcinogenic,_mutagenic,_reprotoxic_(CMR)_substances (accessed 4 February 2020).

United States National Human Genome Research Institute (2016). *Epigenomics Fact Sheet*. National Institutes of Health (NIH). https://www.genome.gov/27532724/epigenomics-fact-sheet/ (accessed 6 March 2019).

WHO/UNEP (2013a). *State of the Science of Endocrine Disrupting Chemicals – 2012. Summary for Decision-Makers*. World Health Organization/United Nations Environment Programme. https://apps.who.int/iris/bitstream/handle/10665/78102/WHO_HSE_PHE_IHE_2013.1_eng.pdf?sequence=1 (accessed 4 February 2020).

WHO/UNEP (2013b). *State of the Science of Endocrine Disrupting Chemicals – 2012*. Geneva: World Health Organization/United Nations Environment Programme. https://www.who.int/ceh/publications/endocrine/en/ (accessed 4 February 2020).

Yang, O., Kim, H.L., Weon, J.-I. et al. (2015). Endocrine-disrupting chemicals: review of toxicological mechanisms using molecular pathway analysis. *Journal of Cancer Prevention* 20 (1): 12–24.

7

Some Principles of Pollution Ecology and Ecotoxicology

7.1 Introduction

A concept of the way in which pollutants interact with living systems, both natural and human, is presented in Figure 1.3. This concept can be seen as describing the basis of **Pollution Ecology** and **Ecotoxicology** and providing a way in which these branches of ecology and relevant aspects of chemistry can be integrated and understood. This concept is preserved in Figure 7.1 where chemical pollutants originating from emissions and discharges enter the environment. The interrelationships of the content of a chapter to one another and to the content of this chapter are indicated. These follow a logical sequence of steps with the processes, which occur in each step being dependent on the processes occurring in the preceding step. After release, the chemical pollutant undergoes this sequence dependent on its physicochemical and biological properties. As a result, these chemicals are transported, distributed, and chemically modified resulting in their occurrence in the various phases/compartments present in the environment.

Thus, the original chemical and metabolites can occur in phases in which animals, plants, and humans can cohabit leading to exposure of individual organisms, which generates a biological response from them. The properties of the chemical contaminant can be seen as being expressed as *physiological and behavioral* leading to *toxic stresses*, and as *physiochemical and biochemical* being expressed as *biotransformations, bioaccumulations, and food chain transfers*. Thus, these responses can range from lethality to a wide array of sublethal effects, which can be adverse. These adverse effects on individuals then consequently have an initial impact on the populations of biota present as shown in Figure 7.1. With natural ecosystems, the populations of more sensitive larval and juvenile stages of some species can be reduced in size by lethal effects, and similarly with human populations the very young and very old individuals can be reduced in size. The structure of the populations can be altered leading to modified community structure, function, and health. These changes in the population lead to alterations to the ambient ecosystems, which can be measured by such factors as overall respiration to photosynthesis ratio, nutrient cycling rates, and energy flux as shown in Figure 7.1. These changes are then reflected at different spatial scales, which can indicate that the pollution is of local, regional, or global significance.

Interestingly, the pollutant and its behavior are depicted as most influential in the initial steps of this concept of pollutant interactions with living systems. The chemical influence on ecosystems is principally through the adverse effects on individuals, while its direct impact on ecosystems is lower.

The concept of Pollution Ecology and Ecotoxicology shown in Figure 7.1 does not include the dimension of time, and considerations of this factor lead us to a clearer understanding of this concept. We only have a general idea of the rates at which the processes in Figure 7.1 occur. But it is clear that some parts in the concept occur at greatly different rates than others. For example, the time it takes for a chemical to distribute in a system is much shorter than the time it takes for changes to occur in the structure of a natural or human population and the ecosystem. Also the rate of change may alter depending on the stage the system has reached and perhaps it may slow up or increase. Often when the pollution reaches a certain threshold level, then other changes may occur and there can be feedback to other biological components at a different level. These feedback mechanisms can ameliorate the rate of occurrence of the process.

Another concept, which is commonly used in the behavior of natural systems, is the **tipping point**. This refers to a point in time when a natural system has been subject to many small changes over time such that it has reached a critical point. At this point, the changes cannot reverse in any predictable manner and future changes may take the form of significant, perhaps destructive, unstoppable effects. This concept has been described by The Economist (2009) as based on the epidemiological concept of the spread of diseases from a normal outbreak to an epidemic or pandemic. Gladwell (2001) has applied this concept to

Chemistry and Toxicology of Pollution: Ecological and Human Health, Second Edition. Des W. Connell and Greg J. Miller.
© 2023 John Wiley & Sons, Inc. Published 2023 by John Wiley & Sons, Inc.
Companion website: www.wiley.com/go/toxicologyofpollution2e

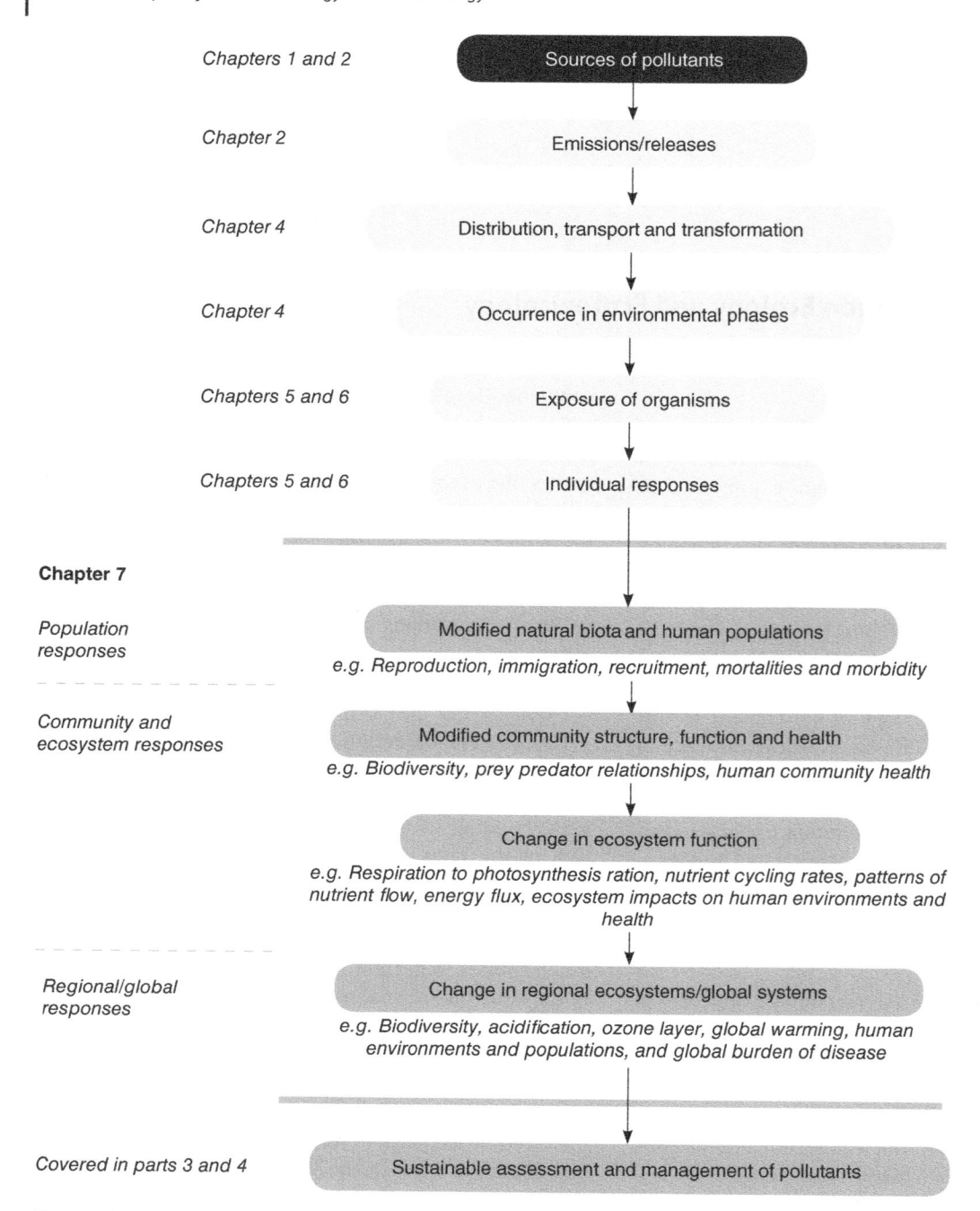

Figure 7.1 Integrated conceptual model of pollutants on the components and functions of human and natural systems based on Figure 1.3 with the processes involved with population responses, community and ecosystems, and regional/global responses as covered in this emphasized. The relationships of the topics covered in Chapters 1, 2, 3, 4, 5, and 6 to this chapter are shown.

many systems in our society but epidemiology provides perhaps the best examples. For example, the spread of influenza can occur as sporadic outbreaks of the disease, but at some point it can become an epidemic. This concept has been applied to some ecosystems where change is occurring at a rapid rate and involves many relatively small changes at different levels.

7.2 Controlling Factors in Natural and Human Systems

The natural environment of an organism contains a wide variety of physical, chemical, and biological factors controlling its development. Somewhat similarly, the human environment also has a set of factors, but these are subject to human-related control. While only one species is involved, it is subdivided into many groups such as adults, juveniles, and so on. The set of organisms comprising a natural ecosystem are genetically adapted to these factors as part of their natural environment. In addition, the organisms are adapted to the natural changes, which occur throughout the seasons, diurnally, and in other ways. Pollution introduces a change into the set of factors operating and controlling a natural ecosystem and the human environment.

For example, an ecosystem is subjected to the natural ranges of temperature, which occur in its environment. However, thermal pollution can lead to excessively low or excessively high environmental temperatures. Also, the pH of water may be altered by acid rain or discharges to give acid or alkaline conditions beyond the normal range. Some substances or environmental factors may cause pollution only when they extend beyond the normal range in one direction and not the other. For example, plant nutrients in excess lead to pollution but if plant nutrient concentrations are too low, this is usually not due to a pollution situation.

Each organism including humans in the system has a natural ability to tolerate these changes and exhibits a characteristic optimal activity and range of changing activities in relation to the concentration or factor. Thus, many environmental changes, which result in a shift toward the optimum for some organisms will be beneficial to those particular organisms in the system but may be detrimental to others.

Figure 7.2a shows how hypothetical populations of two organisms could respond in individual and characteristic ways to changes in concentrations or environmental factors of the general type discussed above. Thus, at level x, organism a will be favored by the conditions which exist, and at level y, organism b will be favored. Thus, there will be a change in the composition in the community of organism a and organism b related to the concentration or factor operating in the system.

Another general type of change is that due to concentrations or factors to which the organisms are not specifically adapted, generally the introduction of exogenous substances to the system such as synthetic pesticides, and some petrochemicals, and heavy metals. In this situation, the organisms show a characteristic response related to metabolic, physiological, and other characteristics. Usually, there are no observable effects on organisms or the ecosystem if the concentrations or factors are sufficiently low, but at higher levels, characteristic effects on the population are observed.

Figure 7.2 Generalized patterns of change in a population of organisms induced by pollution. (a) Population response produced by naturally occurring concentrations of substances and factors essential to the survival of organisms, but which are detrimental in excess of the range encountered in natural ecosystems. (b) Response induced by substances or factors which are not part of the natural ecosystem and are detrimental at some level.

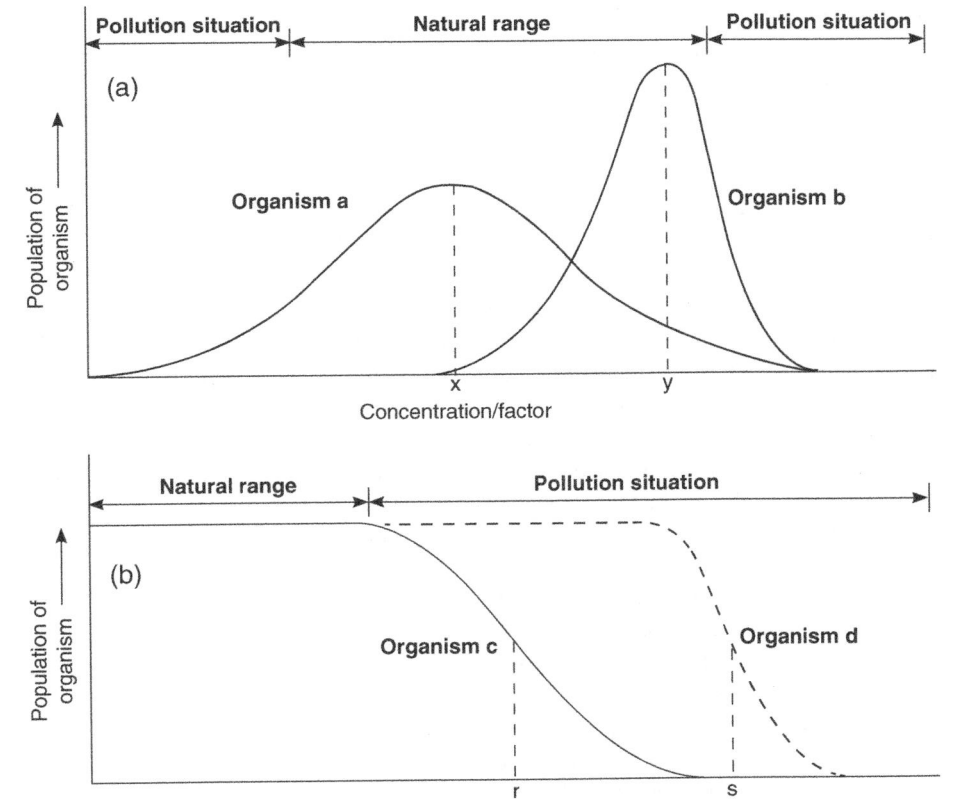

Figure 7.2b shows the hypothetical response of the populations of two organisms c and d to different levels of pollutant substances that have a natural range of zero. Thus, in a pollution situation, conditions at r will favor c and conditions at s will favor d. So, there will be a set of changes in the community structure related to the level of the concentration or factor involved.

This concept is also applicable to human populations as well. Some sectors of the human population such as the very young or very old are often more susceptible to adverse effects of pollution than the rest of the population. This can lead to changes in the population structure.

Furthermore, factors such as competition, food, and a variety of other ecological factors have an impact on organisms in the system and will also be affected by environmental conditions. In a pollution situation, the magnitude of the concentration or environmental factor change is often sufficient to give a set of related and comparatively easily identified effects. Nevertheless, a clear understanding of more subtle and long-term effects in pollution situations is usually unavailable with both natural and human populations.

In polluted systems, organisms are naturally selected for tolerance, which can lead to genetic adaptation over time with successive generations becoming more tolerant to the pollution conditions. The occurrence of this depends on a variety of factors, including the proportion of interacting and reproducing population affected by the pollution, reproduction rate, and genetic variability. If only a small proportion of the reproducing population is affected, as often occurs in pollution situations, little adaptation can be expected.

In natural ecosystems, there are sets of animals and plants adapted to the particular environmental conditions which exist, for example, temperature, availability of oxygen, and availability of nutrients. Usually, there are a wide range of interacting organisms present including microorganisms, plants, invertebrates, vertebrates, and so on. But the interrelationships within the groups are not particularly well understood, which can lead to difficulties in the understanding of pollution effects. However, in many situations, the pollution effect is large compared with variations experienced in the natural system, and this leads to a larger and more obvious biological effect.

Some pollutants such as toxicants may act directly on individual organisms, while others may alter the environment to give an effect on the ecosystem. But since all organisms are affected in different ways by pollution, it can be expected that there will be an alteration to the structure of a food web resulting from pollution (see Figure 7.3).

With some exceptions, the general effects of pollution on an area can be summarized as follows:

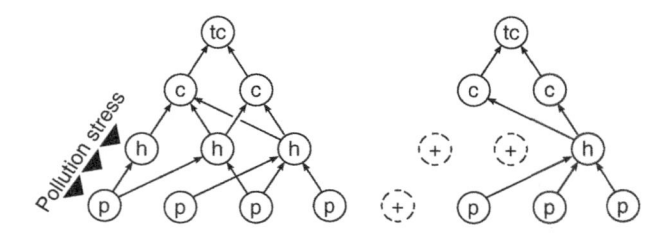

Figure 7.3 A hypothetical food web illustrating modification as a result of pollution where tc = top carnivore, c = carnivore, h = herbivore, and p = producer.

1. There is a decrease in the suitability of the area as a human habitat or habitat for the living components of the ecosystem, which have been naturally established and adapted to the area.
2. There is a detrimental impact on humans or certain species and groups related to the intensity and type of pollution.
3. An alteration to the community structure or natural system occurs and, as a general rule, the biodiversity present declines.
4. The flows of energy and matter in the natural ecosystems are changed.

Patterns of change may be apparent in pollution situations since pollutants released to the environment are acted on by a variety of physical, chemical, and biological forces, which generally lead to systematic loss and dilution. Thus, for a point discharge there is a sequence of physiochemical and related biological changes. These patterns can occur in space or in time. Similarly, nonpoint emissions or discharges of pollutants may also result in subtle patterns of change on receiving environments (e.g. watersheds or airsheds).

7.3 Classes and General Effects of Pollutants

The deleterious effects of pollutants can be generally related to three environmental factors:

1. Excess plant production
2. Deoxygenation
3. Toxic or similar deleterious effects on organisms

Factor 2 mentioned above applies exclusively to aquatic systems and factor 1 almost so, whereas factor 3 applies to all systems. A limited number of pollutants do not fit this scheme particularly well, for example thermal pollution. Nevertheless, it provides a reasonable basis for comparison and contrast of pollution effects. For example, pathogenic microorganisms lead to direct toxic or similar deleterious physiological effects, but this group does not have an impact on dissolved oxygen or stimulate excess plant production. Suspended solids can be classified as having toxic or similar deleterious physiological effects and causing reduced plant production with a resultant impact on dissolved oxygen. Other examples are detergents that contain phosphates, causing stimulation of plant growth, and surfactants with a toxic impact. Petroleum can lead to dissolved oxygen reduction due to microbial degradation and also toxic effects due to the presence of aromatic hydrocarbons.

Using this concept, it can be seen that there are related groups of pollutants (physical, chemical, and biological). These are set out below:

1. *Organic Matter*. This consists principally of carbohydrates, proteins, and fats and leads to dissolved oxygen reduction by stimulating the growth of microorganisms in aquatic systems.
2. *Plant Nutrients*. These substances are usually rich in nitrogen and phosphorus and stimulate excess plant growth in aquatic systems.
3. *Toxic Substances*. These are substances that interfere with metabolism and physiological activity of organisms in a detrimental manner in low concentrations.
4. *Suspended Solids in Water and the Atmosphere*. These substances have similar effects to toxic substances but act by physical interaction at comparatively high concentrations.
5. *Energy*. Energy pollution is mainly due to thermal discharges. The effects are similar to toxic effects, but this activity is due to thermal energy inputs.
6. *Pathogenic Microorganisms*. These exhibit a toxic effect on organisms, but the effects are due to organisms rather than chemical substances.

These classes provide the basis for discussion of the overall interaction of pollutants with ecosystems.

7.4 Ecology of Deoxygenation and Nutrient Enrichment of Aquatic Systems

On a worldwide scale, these two processes are the most important pollution effects in aquatic areas. Deoxygenation and nutrient enrichment are due to the two classes, organic matter and plant nutrients, respectively, and can be discussed together since many interrelated processes occur in both situations. Deoxygenation is caused by the addition of organic matter, rich in chemical energy (carbohydrate, protein, and fat), to a water body. If these substances are in a form whereby they can be readily attacked by microorganisms, they are often referred to as putrescible matter. A wide variety of microorganisms are present in all water bodies, and some of these can readily utilize organic matter and are capable of rapid population expansion. Thus, respiration occurs and dissolved oxygen in the water mass is consumed in this process. Dissolved oxygen is essential for the survival of most aerobic aquatic organisms. Reduction in dissolved oxygen leads to a change in the ecological structure of the affected community.

Figure 7.4 shows how increasing respiration and bacterial numbers are accompanied by a decrease in primary production and dissolved oxygen in an aquatic system, leading to a set of changes in the associated biota. Fish, which require reasonably high dissolved oxygen levels, survive best in the range indicated. Benthic invertebrates, such as oligochaetes and chironomids, need less dissolved oxygen and so occur in a lower concentration range.

The addition of organic matter therefore causes a corresponding set of changes in dissolved oxygen, primary production, respiration, and biota, which are reasonably consistent and related to a particular level of pollution. These changes can be classified into a sequence of states, zones, or classes containing a particular range of chemical and related biological characteristics. This can be done in a variety of ways, but one of the simplest is clean, poor, polluted, seriously polluted (Figure 7.4). Other classifications use terms such as clean, active decomposition, septic, degradation, and so on. The classification with the longest history is the **Saprobic System**, which uses a saprobic level or index (see Figure 7.3). All the schemes are similar in principle but differ in nomenclature and methods of zone classification. Considerable controversy surrounds the validity of these classifications and their application to pollution situations. Nevertheless, the basic principle of a related sequence of chemical and related biological changes is sound and has been used as a framework for the following discussion.

The initial increase in primary production, which occurs with increasing organic pollution, is due to the release of plant nutrients (e.g. nitrogen and phosphorus) by the decomposition of organic matter containing these substances. The absorption of solar radiation initiates photosynthesis during daylight hours, producing an input in photosynthetic oxygen. In comparison, respiration occurs throughout the daily cycle, and this leads to a diurnal cycle of dissolved oxygen with a maximum during the day and a minimum at night. Overall, if the rate of respiration exceeds primary production, this leads to a deficit in dissolved oxygen and a reduction in the dissolved oxygen present in the water mass, which increases with organic pollution (see Figure 7.4). When the dissolved oxygen falls to low levels, hydrogen sulfide and other toxic substances

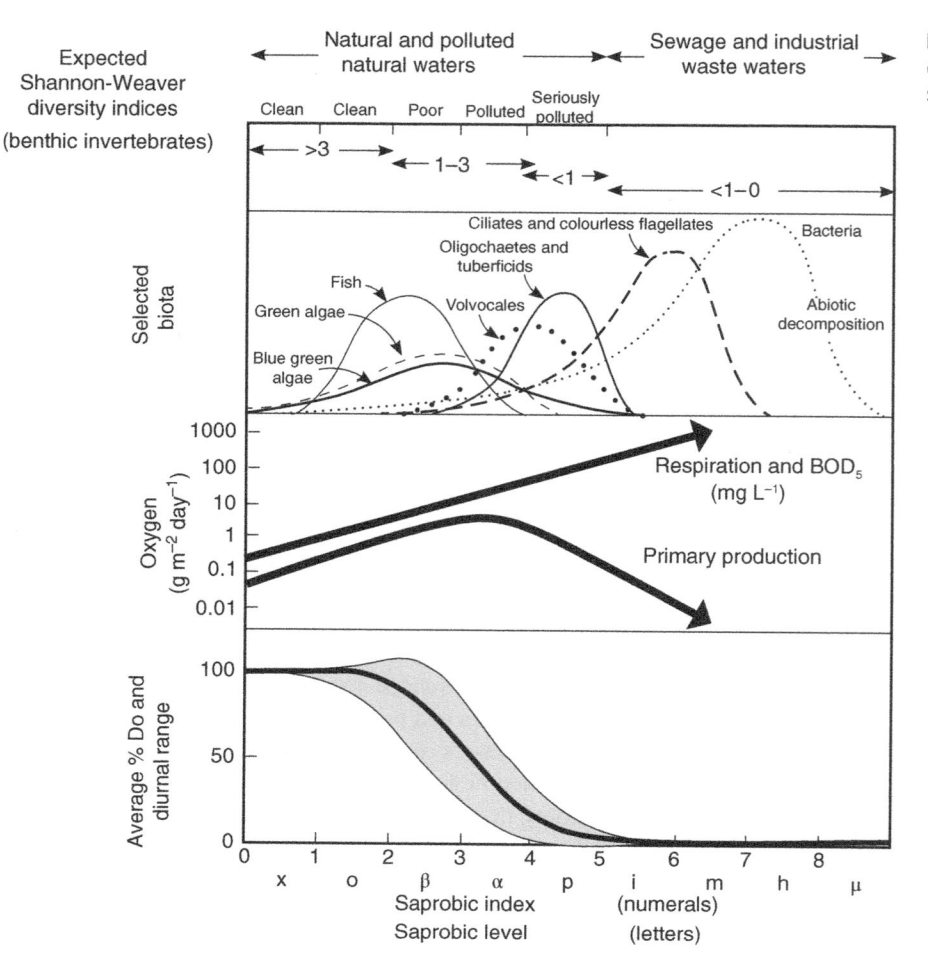

Figure 7.4 Some generalized characteristics of waters at different saprobic levels.

are produced, which inhibit plant growth, and thus there is a decrease in primary production at higher saprobic levels. In addition, there is a steady fall in the species diversity of benthic invertebrates due to increasing divergence in environmental conditions from the natural situation to which the system has adjusted (see Figure 7.4).

The higher saprobic levels do not occur in natural systems, which have been polluted. These are confined to wastewaters but can be placed in this saprobic sequence.

When organic matter is discharged to a stream, a sequence of characteristic changes occurs. The dissolved oxygen content of the water decreases due to increased microbial respiration and replenishment occurs by solution of atmospheric oxygen at the air–water interface. This leads to a characteristic *dissolved oxygen sag* shown in Figure 7.5. The pollution level rises and falls in relation to the dissolved oxygen decrease and subsequent increase. Other related changes in biota, BOD, dissolved salts, saprobic index, and diversity follow. This leads to a characteristic set of changes related to the discharge point (see Figure 7.5).

Some of the features of nutrient enrichment are related to the factors outlined above. A discharge can consist of water containing high concentrations of plant nutrients but low in organic matter, and thus little direct demand on dissolved oxygen occurs. The receiving waters are enriched with plant nutrients, usually nitrogen and phosphorus compounds, and there is an increase in photosynthesis and primary production. Changes in phytoplankton biomass can be detected by measuring the concentration of chlorophyll-a, the green pigment in plants. Also, an increase in respiration occurs due to the increase in organic matter present from plant matter production. A resultant set of changes in environmental conditions and biota occurs, including such factors as water turbidity, type and density of plants, and type and density of animals (see Figure 7.6). The available data suggest that, as a general rule, the number of species present decreases with increasing enrichment in natural systems. However, it would be expected that a maximum in species would occur at a particular enrichment level. Lower species numbers than this would occur with water extremely low in nutrients, for example ultraoligotrophic, while a decrease in species would occur at higher levels.

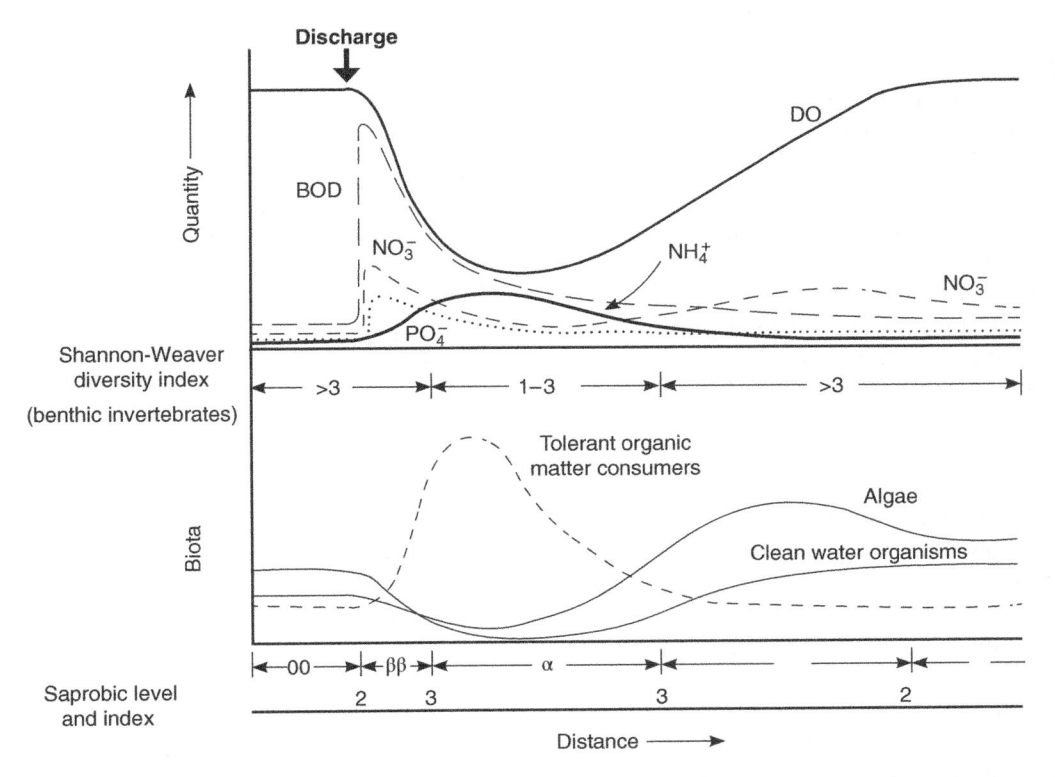

Figure 7.5 Chemical and biological changes resulting from a discharge of organic matter into a stream.

Many aquatic areas, from small pools to the continental shelves, exhibit summer stratification due to heating of the surface water. This can increase plant production in the epilimnion and, upon death, the plant matter falls to the bottom increasing microbial respiration, dissolved oxygen consumption, and nutrient concentrations. In oligotrophic waters, there is little change from surface to bottom in dissolved oxygen and plant nutrients, but in eutrophic conditions, distinct vertical profiles are developed (see Figure 7.7). Thus, the depletion of hypolimnetic dissolved oxygen increases with increasing nutrient enrichment.

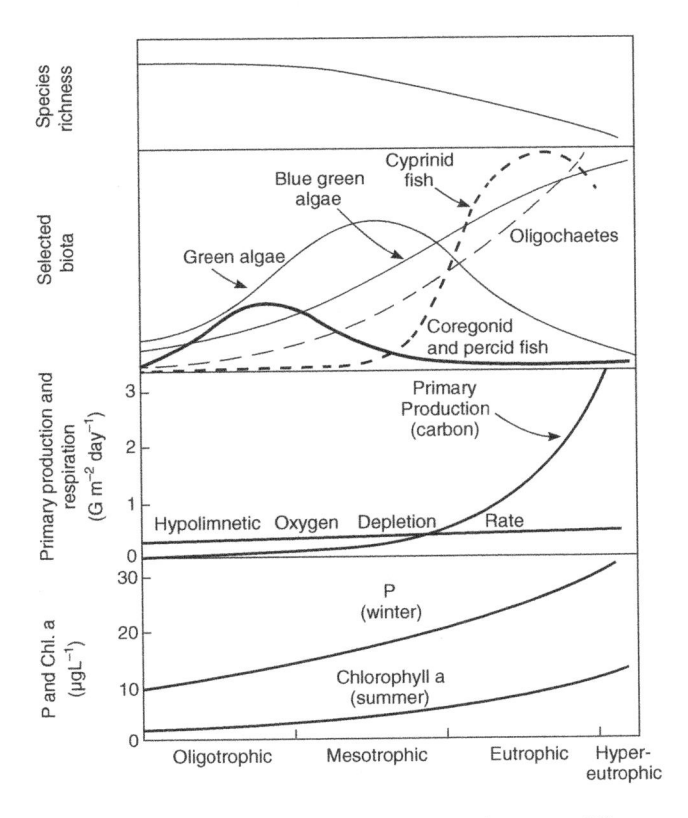

Figure 7.6 Some generalized characteristics of water at different trophic levels.

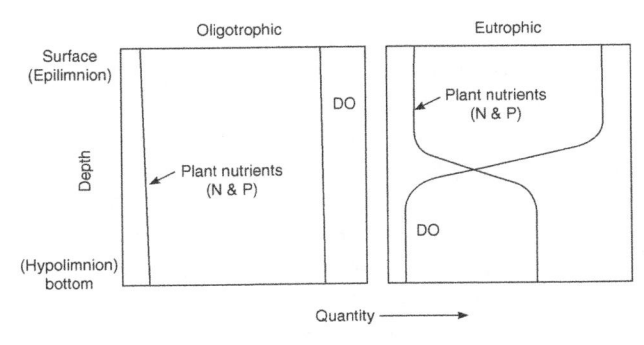

Figure 7.7 Generalized vertical distribution patterns of some water constituents in oligotrophic and eutrophic conditions with stratification.

In addition, there are a related set of seasonal changes in the epilimnion of many larger water bodies. In winter, the surface and bottom waters are mixed, but the conditions of temperature and solar radiation are unfavorable for the growth of phytoplankton. With the onset of summer, there is the development of good growth conditions and also the stratification of the water column. Rapid growth of phytoplankton proceeds with the consequent consumption of waterborne plant nutrients, and, due to the isolation of the nutrient-rich bottom waters, there is depletion of plant nutrients in the epilimnion, ultimately leading to reduced plant growth. Mixing of the water column can occur in autumn leading to a small growth of phytoplankton, while the temperature conditions and solar radiation are still favorable. With the onset of unfavorable winter conditions, low phytoplankton numbers once again are established. Generally, zooplankton growth follows the phytoplankton pattern but is displaced in time.

There have been many attempts to relate saprobity to trophic states. These are usually based on the two major processes, which occur in both situations, primary production, and respiration. There is disagreement on the primary production and respiration characteristics of different trophic and saprobic classes. However, Figure 7.8 represents an evaluation of the position of the various classes of saprobity and trophy on a primary production versus respiration graph derived from the available data.

The two basic processes of nutrient enrichment (trophic status or eutrophication) and organic enrichment (deoxygenation and saprobity) are distinctly different but related. In the nutrient-enrichment process, additions of plant nutrient into the water mass lead to acceleration of the primary production process. Accordingly, the nutrient enrichment process is controlled by primary production, which would be expected to exceed respiration at all stages (see Figure 7.8). On the other hand, saprobity is due to the addition of decomposable organic matter (putrescible matter), which leads to deoxygenation. Consequently, this process and its impact are controlled by respiration, which would be expected to exceed primary production at all stages (see Figure 7.8).

The development of the various classes of saprobity and trophy has proceeded independently and according to different criteria. Eutrophication classes have been developed according to plant growth and related characteristics, whereas the saprobic classes have been developed according to deoxygenation effects. Thus, it would be expected that the various classes of saprobity and trophic status having the same prefix would not necessarily be equivalent. The relationships between the various classes are shown approximately in Figure 7.8. Even so, there are

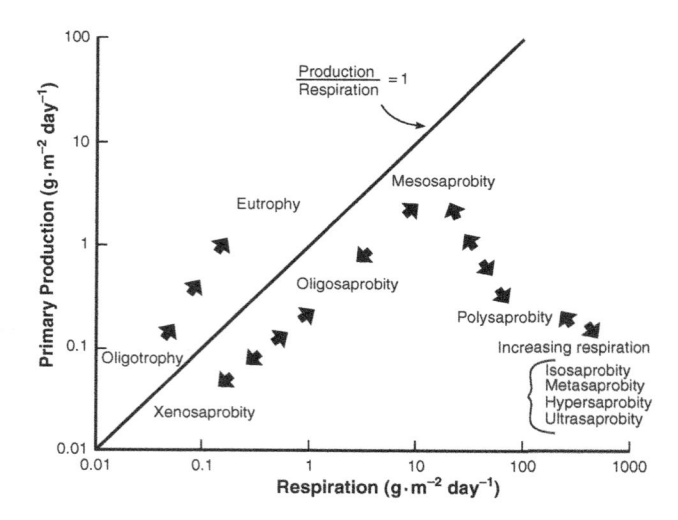

Figure 7.8 Plot of primary production vs. respiration for a variety of saprobic and trophic conditions in aquatic areas and the sequence of interrelationships.

differences between the characteristics used by various authors to define the various classes. It can be seen from Figure 7.8 that the physicochemical and biological criteria are different for all saprobic and trophic classes.

An important difference between nutrient enrichment and organic enrichment is the reversibility of the process. Organic enrichment is a reversible process, which can be seen in a forward and reverse sequence in a stream receiving an organic-rich discharge. In this way, the sequence xenosaprobic through to polysaprobic and the reverse can be seen in stream situations. On the other hand, nutrient enrichment is a process which generally proceeds in one direction – toward increasing eutrophication.

The closest relationship between the classes of saprobity and trophy occurs when the levels of plant nutrients and organic matter, that is, primary production and respiration, are very low. Hence, oligotrophic and xenosaprobic conditions are somewhat similar.

7.5 Ecotoxicology of Toxic Substances

Toxic substances can affect ecosystems in a number of different ways, but in its simplest form, two basic types of effect are possible:

1. Acute lethal toxicity over a short time period due to discharge of a toxic substance, or treatment of an area with a toxic material on a single occasion.
2. Chronic sublethal effects can occur in an area due to exposure to sublethal concentrations over a longer time period on a continuous or intermittent basis.

Figure 7.9 Some generalized characteristics of the effects of a toxicant on an ecosystem.

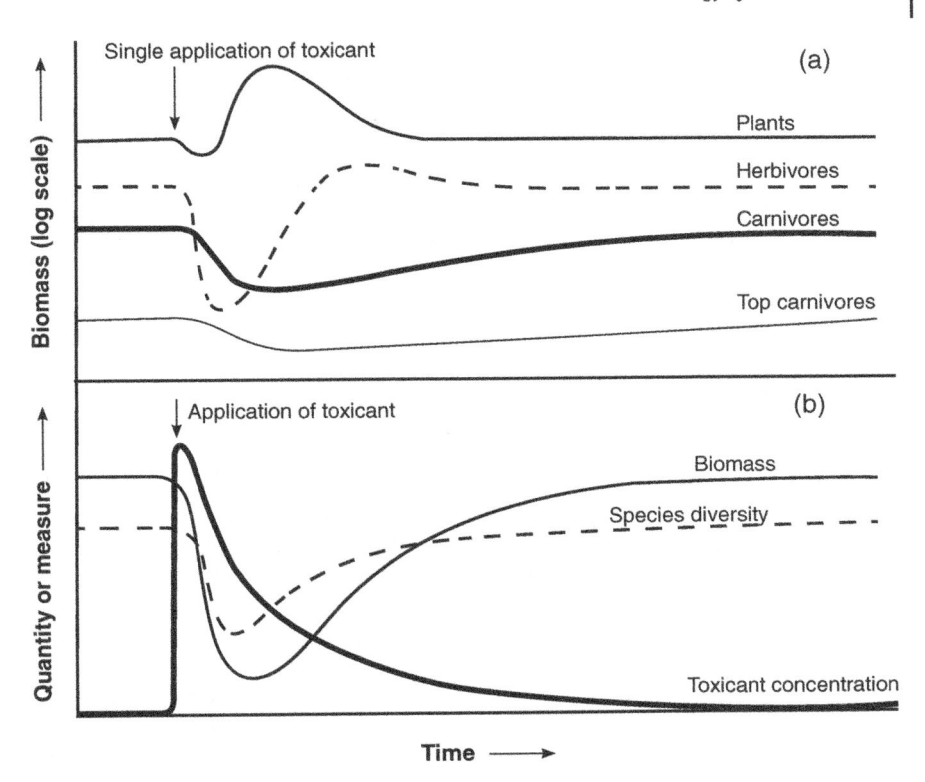

Acute lethal toxicity with limited time exposure can occur in many pollution situations. However, the concentration of toxicant would be expected to decrease with time due to processes such as dispersal, dilution, and degradation (see Figure 7.9). Generally, animals are more sensitive to environmental toxicants than plants. But many carnivores are highly mobile and able to avoid the initial toxic discharge or treatment. Good examples of organisms exhibiting this behavior are eagles, hawks, carnivorous fish, and mammals. Consequently, toxic effects are often most acute with herbivores and omnivores.

The general pattern of ecosystem response to a toxicant causing herbivore lethality is shown in Figure 7.9. Although the initial impact on carnivores and top carnivores is delayed, there is, nevertheless, a longer-term impact and recovery lags in comparison with the herbivores. Initially, both species diversity and biomass fall and then recover as the toxicant concentration deceases.

This pattern of ecosystem response is often observed in streams receiving a toxic discharge where the generalized pattern shown in Figure 7.9 occurs with distance downstream from the discharge point. Similarly, following oil spills in marine ecosystems, this general pattern occurs although changes in the carnivores and top carnivores can be difficult to quantify. However, phytoplankton blooms following oil spills are quite common. Also, this general type of sequence has been observed in the application of insecticides. With this situation, predator organisms are often preferentially removed from the population leading to an outbreak of prey species rather than plants.

The chronic sublethal effects of persistent toxicants are difficult to quantify to enable establishment of a general pattern. Figure 7.10 shows a simulation of the concentration of a persistent toxicant in organisms within an

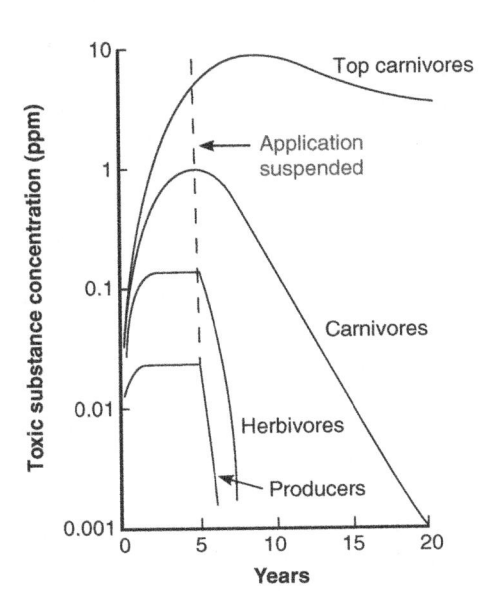

Figure 7.10 Simulation of uptake and loss of a persistent toxicant in an ecosystem with five years of initial constant application. Source: Eschenroeder et al. (1980).

ecosystem over time. The organisms at high trophic levels show a slow build-up in concentrations but with much higher ultimate concentrations and slower rate of loss than other organisms in the ecosystem. In this case, the greatest impact of the toxicant would be expected at the highest trophic levels, and there are many examples of population reductions with top carnivorous birds as a result of insecticide usage.

7.6 Suspended Solids

Suspended solids can be expected to have broadly similar effects to toxicants in aquatic ecosystems. Many organisms show tolerance to comparatively high concentrations of suspended solids. Nevertheless, plants would be expected to show a population decrease due to the reduction in light penetration. A major impact on the ecosystem is the elimination of food organisms. Also, a reduction in the numbers of fish and other organisms using visual means to seek prey occurs as well.

7.7 Thermal Ecology

There is little information of the type required to allow the formulation of general principles for the effects of thermal discharges on aquatic ecosystems. With thermal pollution, aquatic ecosystems are mainly affected because the heat input to a specific system in this situation is much more substantial than with terrestrial organisms. In thermal ecology, we are more concerned with the localized effects of thermal discharges on ecosystems rather than the global effects of temperature rises due to carbon dioxide.

Heat represents an input of energy into an ecosystem, which is not readily assimilated by organisms. However, it can be utilized by organisms in providing a suitable environmental temperature for metabolic processes.

Thermal pollution usually causes similar changes in organisms and ecosystems to those noted previously with other forms of pollution. Generally, there can be expected to be a systematic distribution of numbers of organisms, number of species, or some other measure of ecosystem success against temperature. Thus, there will be an optimum temperature, T_O, where maximum success would be expected and, on either side of this optimum, a range of temperatures where the system was less successful (see Figure 7.11). In natural ecosystems, there will be a range of temperatures due principally to diurnal and seasonal changes with consequent changes in ecosystem success. Here, T_U represents the upper limit of the normal range of

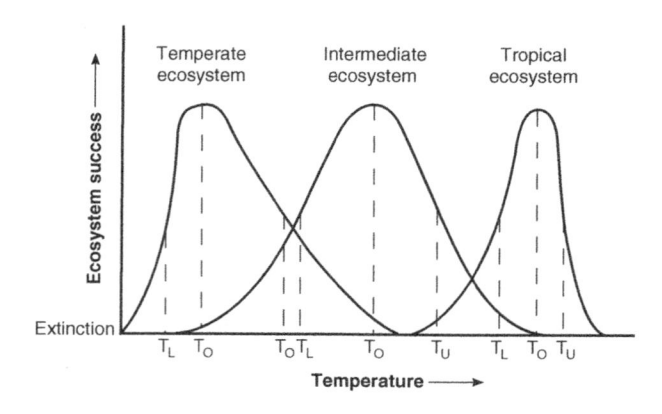

Figure 7.11 Hypothetical plot of ecosystem success against temperature with T_L and T_U representing the natural variation in the ecosystem and T_O as the optimum.

temperature and T_L represents the lower temperature limit in a normal ecosystem (see Figure 7.11).

Any ecosystem contains a diversity of organisms, which are naturally adapted to these thermal conditions. The seasonal patterns of change may result in a corresponding change of species related to temperature or other seasonal characteristics. Also, there is an absolute upper and lower limit for the survival of living organisms. So, irrespective of normal environmental temperature, we would expect that all ecosystems will approach extinction at these limits.

Tropical systems have a relatively stable ambient temperature regime with little annual variation between summer and winter, and the ecosystems present have adjusted to these conditions. There is a natural temperature regime, which is close to the upper limit for the organisms present. A skewed ecosystem success curve would be expected with a limited range between the maximum and the upper extinction limit (Miller and Stillman 2012). This is illustrated in Figure 7.11, which shows that constraints on temperatures lower than the optimum are not as acute as the upper temperatures, so the ecosystem success curve would extend further in this direction. It would be expected that there would be a limited seasonal range of temperatures and thus the tropical ecosystems would be adapted to a narrow range of temperature tolerance. Somewhat similar arguments apply with temperate ecosystems, but in this situation the temperature range of adaptation would be larger and a lower level of susceptibility to increased temperatures would prevail.

Changes in environmental temperatures due to thermal pollution or climate change (e.g. Settele et al. 2014) will give a change in ecosystem success. This change will depend on the position in the seasonal and diurnal cycle when discharge occurs or the magnitude of change due to climate. If it results in temperature movement toward the maximum or optimum, it will be beneficial to the ecosystem. On the other hand, if it results in a movement away from the maximum, a detrimental impact will occur. As a general rule, temperatures outside of the natural range (i.e. greater than T_U and less than T_L) would be expected to be detrimental to the particular ecosystem involved. In general terms, on this basis, tropical ecosystems would be expected to be the most susceptible to detrimental effects resulting from thermal additions and temperature elevation. This general conclusion is substantiated by results described by Khaliq et al. (2014), as well as Stevens (1989).

Also an extensive study by Nguyen et al. (2011) investigated tropical marine animals across different rates of temperature change and compared that with published data for temperate and polar species. These authors suggest that (i) animals living in thermally stable environments have reduced acclimatory ability, and (ii) animals living constantly close to their upper limits are particularly susceptible to increases in temperature. This combined with the fact that regions nearer to the equator have relatively faster global warming rates suggests that tropical regions are possibly among the most vulnerable to climate change conditions.

Generally, thermal stress leads to a somewhat similar pattern of ecosystem change as that caused by the discharge of toxicants. But in this case, it is not clear which groups of organisms will be affected and which will benefit from changes in thermal conditions. Nevertheless, the overall changes indicated in Figure 7.9b would be expected.

Organisms can adapt to thermal stress. An ecosystem subject to thermal stress results in a process of selection and the development of a temperature-tolerant breeding stock. Thus, long-term changes in ecosystems may differ from the short-term effects.

Radionuclides also result in an energy input to organisms and ecosystems, but this energy cannot be utilized in any significant way. In fact, radionuclides act as toxicants by interfering with metabolism and physiological activity in a detrimental manner at low concentrations. Thus, the effects of exposure to radionuclides are of a similar nature to that described for toxicants previously.

7.8 Pathogenic Microorganisms

Pathogenic microorganisms are a very important water pollution factor, particularly regarding human health. Their impact on natural populations is much less important. Nevertheless, microorganism-borne disease from pollution can possibly be expected to affect natural populations. The effects are probably similar to the general pattern of toxicant impact illustrated in Figure 7.9b.

7.9 Biotic Responses to Pollution at Different Environmental Temperatures and Latitudes

There are basic differences between temperate and tropical regions, which lead to difficulties in applying knowledge developed in temperate regions to the tropics. Most of the background knowledge we have on these differences relates to aquatic systems. The differences are caused fundamentally by different levels of incoming solar radiation leading to a variety of different physical, chemical, and biotic characteristics, which stimulate different responses of these systems in a pollution situation.

A basic characteristic, which influences many differences between tropical and temperate regions, is water temperature. The temperature of water governs the rate of environmental transformations, solubility of natural substances and pollutants, stability of pollutants, and the metabolic rate of organisms. So, the behavior and impact of pollutants would be expected to vary with temperature and, consequently, with latitude. Table 7.1 summarizes some of the differences between tropical and temperate regions.

Previously, it was suggested that tropical ecosystems operate closer to the upper thermal limit and thus would be expected to be more susceptible to thermal pollution. In laboratory experiments, many toxicants have been found to increase in toxicity with increasing temperature. Conversely, respiration rates increase with temperature leading to a higher rate of metabolism and excretion. In this situation, it would be expected that the Q_{10} rule would apply and thus the rates of metabolism and excretion would double for every 10 °C increase in temperature.

Table 7.1 How the behavior and impact of water pollutants may be expected to differ quantitatively in the tropics from temperate systems[a].

Ambient temperature	Higher
Ambient water temperature rise to upper thermal limit	Lower
Aqueous solubility of oxygen	Lower
Biological uptake rates of toxicants	Higher
Rates of physiochemical degradation	Higher
Rates of biological degradation	Higher
Toxicity	Higher
Toxicity thresholds	Lower
Rates of biological oxygen depletion	Higher
Biological impact of nutrients	Higher

[a] There are some exceptions to these generalizations.
Source: Based on Johannes and Betzer (1975).

Therefore, toxicants would tend to be more toxic, but their toxic effects would decrease more rapidly with time.

An important aspect of pollution in the tropics concerns the interaction of dissolved oxygen concentrations with the respiration rates of organisms. With increasing temperature, the respiration rate of organisms increases while the solubility of oxygen in water decreases. In addition, as the respiration of microorganisms increases with temperature, an increasing rate of dissolved oxygen consumption occurs. The effects have been summarized by Johannes and Betzer (1975) as follows:

1. The dissolved oxygen levels in tropical waters are generally lower than temperate waters due to the higher ambient temperatures.
2. The rate of oxygen depletion of water due to discharges of organic matter will be more rapid due to the higher respiration rates of microorganisms.
3. Oxygen demand will decrease more rapidly with distance from a discharge point and with time.
4. Due to higher respiration rates and lower dissolved oxygen levels, tropical aquatic ecosystems will generally be closer to the limiting levels of dissolved oxygen.

Many tropical marine systems have developed over geological time in very stable conditions of temperature, salinity, solar radiation, and other physiochemical factors allowing the development of a rich and diverse ecosystem. Therefore, it is suggested that the tolerance of many of these organisms to environmental fluctuations would be relatively limited. In this way, susceptibility to pollution could be elevated. The concept that increasing trophic complexity confers stability on ecosystems due to the multiplicity of checks and balances does not seem to hold with these communities (Johannes and Betzer 1975).

7.10 Biotic Indices

The previous sections have considered the essential interrelationships between pollution factors and the resultant change in biological characteristics. In most pollution investigations of natural environments, concern is mainly centered on the health of the living system. Thus, direct measurements of biotic characteristics are of particular importance. The response of biota to pollution in a natural ecosystem is an integral part of a pollution evaluation.

There are a number of different measures of the functions of the ecosystems, which are modified by pollution. One of the most fundamental is the change in the respiration to photosynthesis ratio. This has been discussed previously and measures of the respiration to photosynthesis ratio provide an estimation of ecosystem function related to enrichment by organic matter and also eutrophication (see Figure 7.8).

Changes in community structure in an area affected by pollution involves identification, classification, and quantification of biota in the area. This can be applied to a wide range of biota, but the most commonly used group are the benthic macroinvertebrates. This group consists of relatively sedentary organisms and therefore would be expected to reflect conditions in the affected area. They are relatively easily sampled, and the species show a wide range of tolerance to pollution conditions. However, algae, fish, and other organisms have also been used to determine community structure in areas affected by pollution.

A simple structural biotic index is the number of higher taxa, families, genera, or species in an area. Figure 7.12 shows the variation in numbers of taxa with distance downstream from an organic discharge, such as from a secondary sewage treatment plant. The dissolved oxygen profile shows a sag with a minimum at the third site below the discharge. This is due to suppression of microbiological activity by chlorine in the discharge, which evaporates rapidly from the water, allowing major microbiological activity by the third site downstream. The data show improved sensitivity of this index by identification to the species level. The number of species and species diversity drop when the dissolved oxygen plus chlorine have a major impact in the area. A minimum number of species can be seen at site 2, the first sample point below the discharge, where pollution conditions would be expected to be at their worst. The same set of species would be expected in similar conditions in other situations. Thus, a limited number of species could be selected as indicators of the presence of this type of pollution condition. Similarly, sets of indicator species could be selected to represent different levels of pollution. There are a number of limitations with this approach to the biotic evaluation of pollution situations. These are mainly due to the presence of indicator species in unpolluted areas, the need to identify animals to the species level due to the different pollution tolerance of closely related taxonomic groups, and the nonquantitative nature of results obtained.

The use of the number of species as a numerical biotic index for pollution is an expression of species diversity in the area. However, this technique also has several limitations. One is that there is no weighting for the different numbers of individuals represented by the different species present. All species are given an equal weighting, irrespective of their numerical representation in an area. This can be overcome by using equations, which provide a weighting for the number of individuals of a species and the different relationships between number of individuals, number of species, and number of individuals representing

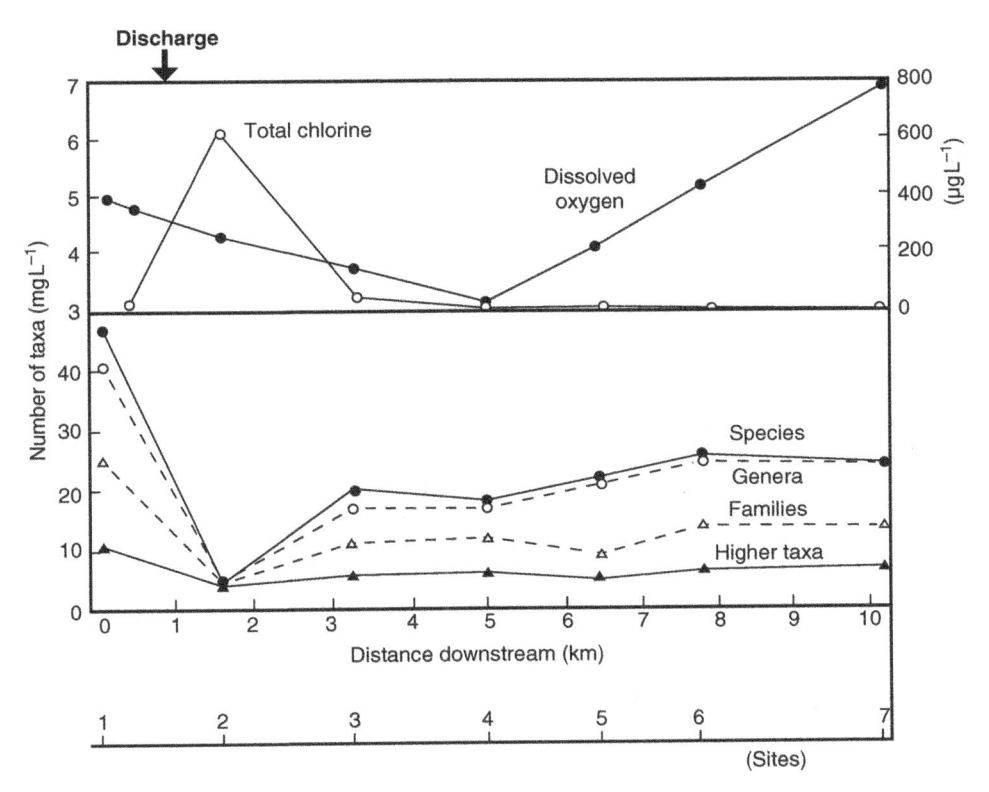

Figure 7.12 Number of species, genera, families, and higher taxa of benthic invertebrates in a stream receiving a discharge from a secondary sewage treatment plant. Source: From Arthington et al. (1982).

Table 7.2 Some equations for various indices used in assessment of biota in polluted areas.

Index	Equation[a]
Shannon–Weaver	$H' = -\Sigma \dfrac{n_i}{N} \log_2 \dfrac{(n_i)}{N}$
Margalef	$I = \dfrac{S-1}{\ln N}$
Brillouin	$H = \dfrac{1}{N} \ln \dfrac{N!}{n_1! n_2! \ldots n_s!}$
Pielou	$E = \dfrac{H}{H_{max}} = \dfrac{H}{\log_2 S}$

[a] N, total number of individuals; S, number of species; n_i, number of individuals of the ith species.

each species. The indices in Table 7.2 present a variety of different ways in which the data can be combined to a single biotic index. Figure 7.13 illustrates the results obtained by applying these indices to a pollution situation resulting from organic enrichment. Arthington et al. (1982) have suggested that Margalef's species richness index and the number of species provided the most satisfactory measure of pollution conditions in this situation. These indices have a close relationship to the chemical parameters of dissolved oxygen and total chlorine and also distance downstream from the discharge. A range of biotic indices have been developed for different purposes and different areas of the world (Abbasi and Abbasi 2011). Some of these are shown in Table 7.3. These indices are focused on different biotic groups as shown in the biotypes column, utilize different sampling methods, as well as being coalesced into a biotic index using different indices.

However, species diversity may be affected by factors other than pollution. For example, stream headwaters with high water velocity may have low diversity but high water quality. In addition, the species composition factors used in the species diversity index take no account of the physiological stress caused by different pollution conditions on the differently susceptible species present.

Jackson et al. (2016) in a global overview of biotic indices have made recommendations for future needs. These researchers point out that existing biotic indices are not useful at a global level and have been developed to assess changes in biota at local levels as reflected in the regions in which current indices are being used (see Table 7.3). They suggest that:

1. Other taxonomic groups than those currently used should be evaluated;

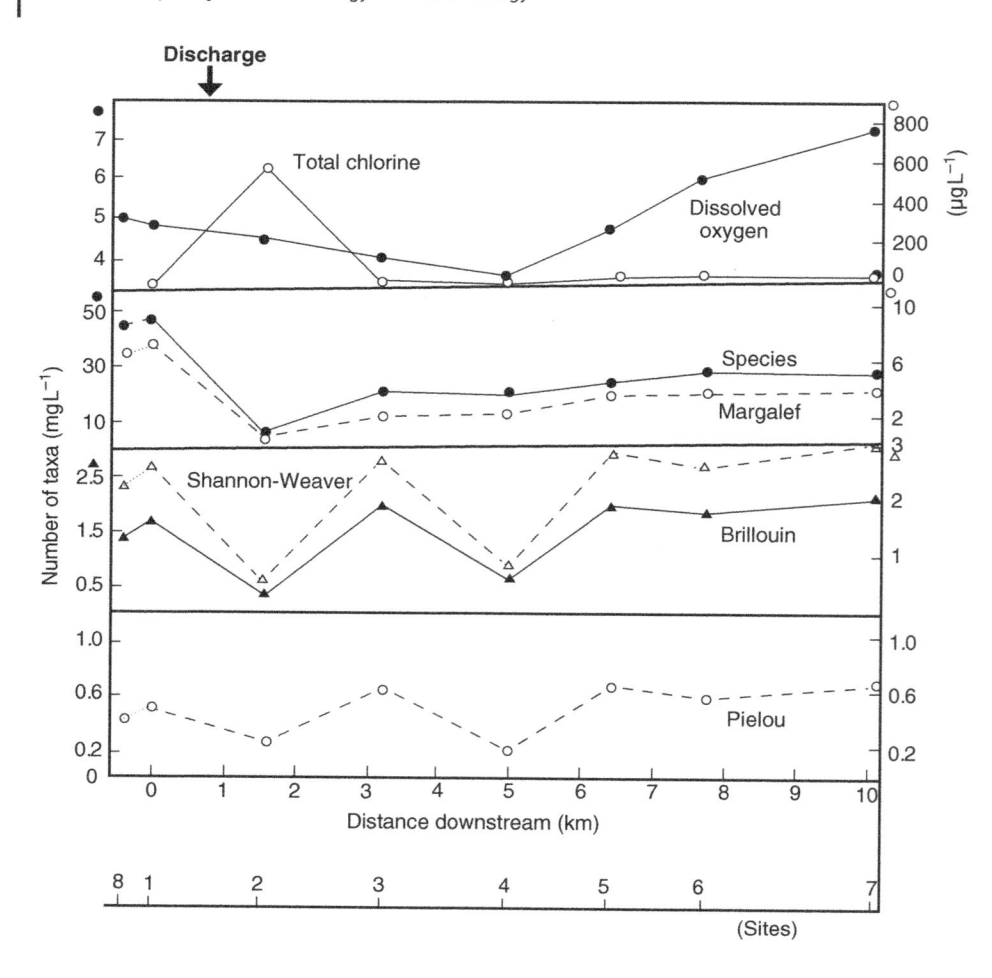

Figure 7.13 Diversity indices of benthic invertebrates, dissolved oxygen, and chlorine in a stream receiving a discharge from a secondary sewage treatment plant. Source: Arthington et al. (1982).

Table 7.3 An overview of the major biotic indices based on aquatic macroinvertebrates

Biotic index	Biotypes sampled	Sampling equipment	Sampling protocol	Taxonomic level	Regions in which currently used
Beck's biotic index	All, combined	Not stipulated	Nonquantitative	Species	—
Trent biotic index	All, combined	Hand net	Nonquantitative	Family + genus + species	—
Hilsenhoff's biotic index	Stones-in-current	Hand net	Quantitative >100	Genus + species	USA
Belgian biotic index	All, combined	Hand net	Nonquantitative	Family + genus	Belgium and surrounding countries
Macroinvertebrate community index	Stones-in-current	Hand net/surber	Nonquantitative	Genus	New Zealand
Iberian BMWP	Lotic + lentic combined/separate	Hand net	Nonquantitative	Family	Spain, Italy
Stream invertebrate index	Six per-defined	Hand net	Nonquantitative, 100 organisms	Family	Australia
Danish stream fauna index	All, combined	Hand net (500 μm)	Semi-quantitative, 12 samples	Family + genus	Denmark, Sweden
Balkan biotic index	All, combined	Benthos net	Quantitative	Family + sub-family + genus	Serbia

Source: Based on Abbasi and Abbasi (2011).

2. Biotic indices are developed to apply across regions and allow global comparisons;
3. Measure biological change at multiple time points. They also suggest that in addressing these objectives, remote sensing, molecular tools, and community involvement to carry out biological surveys be used (see Chapter 8.1).

Another approach has been suggested by Pawlowski et al. (2018) who have conducted pilot metabarcoding studies using environmental DNA to infer biotic indices. This overcomes some of the problems with the use of morphological identification of bioindicator taxa.

Indices alone or in combination with other information are a valuable technique in the investigation of pollution situations. But it should be kept in mind that an index represents a simplification of a complex set of data and there is a loss of information in the process. This may introduce errors and inaccuracies into the final result.

7.11 Indicator Species, Ecological Indicator Species, Chemical Monitor Species, and Biomarkers

Indicator species have been used in two distinctly different ways:

1. To describe particular species that are selectively adapted to certain pollution conditions, for example heavily polluted or clean. The presence of these particular species can be used to indicate pollution conditions as described in the previous section. Also the presence of certain species can be used to identify climatic and microclimatic types and zones. Overall, indicator species have wide applications. Perhaps one of the best known examples is the use of the presence of lichens as indicators of atmospheric pollution.
2. Organisms that bioaccumulate toxic substances present in trace amounts in the environment can be used as chemical monitor species. Chemical analysis of these species then indicates the presence of toxicants in the environment more effectively than direct analysis of an environmental sample, such as water.

The bioaccumulation of chemicals by aquatic organisms is described in Chapter 4 – Chemodynamics of Pollutants – where the bioconcentration of persistent lipophilic components can be explained by partition processes. In many environments, the presence in many sectors or phases in the environment can be estimated from the concentration in the chemical monitor species. However, there are many chemicals, which do not behave in this way, for example, hydrophilic or non-lipophilic chemicals, nonpersistent chemicals, and non-bioaccumulative metals. Nevertheless, available results for environmental management can be obtained. For example, mussels (*Mytilus species*) have a high capacity to accumulate a range of contaminants in water. These organisms have been used to biomonitor occurrence of trace metals and organic contaminants in coastal waters particularly in the United States. A major project commenced in 1986 and is ongoing under the name, "Mussel Watch." Information obtained has been used to monitor the effectiveness of pollution control measures, contamination from natural disasters, and effectiveness of coastal remediation (Kimbrough et al. 2008).

Biomarkers are used in human health as an observation or measurement of an outcome of the incidence of a disease. This may reflect adverse health effects or exposure to environmental chemicals, or treatments and interventions. The biomarker can be physical, chemical or biological and is expected to describe a measurable relationship with human clinical endpoints (Califf 2018).

7.12 Key Points

1. The basic concept of the interaction of pollutant chemicals with both natural and human systems can be described as **Pollution Ecology** or **Ecotoxicology** and involves the following steps:
   (1) Discharge and distribution of the chemical
   (2) Exposure of individual organisms to the chemical
   (3) Individual responses to the exposure
   (4) Modified natural biota and human populations
   (5) Modified community structure, function and health
   (6) Change in ecosystem function
   (7) Change in regional and global systems
   The first three of these steps result in the exposure of individuals, and the final four steps are the biological outcomes of this exposure of individuals, which has resultant effects on populations and ecosystems.
2. The **tipping point** refers to a point in time when a natural system has been subject to many small changes over time such that it has reached a critical point. At this point, the changes cannot reverse in any predictable manner, and future changes may take the form of significant, perhaps destructive, unstoppable effects.
3. Organisms in the natural environment, as well as humans, are adapted to the ambient environment with the changes, which occur throughout the daily and annual cycle and in other ways. Changes to this set of factors extending them beyond the normal range can result from pollution and can have adverse effects.

4. There are many species present in most natural ecosystems, but the human population consists of only one species, however, that has many subgroups such as adults, juveniles, and so on. Generally, the adverse effects of pollution have their greatest impact on the very young and old individuals in the human population.
5. The effects of pollution result in a decline in the suitability of a natural or human environment as a habitat for animals, plants, and humans usually resulting in a decline in biodiversity.
6. The deleterious effects of pollution are generally related to at least one of three environmental factors – excess plant production, deoxygenation (these two factors relate to aquatic systems only), and toxic or similar deleterious effects on organisms.
7. Generally pollutants can be placed into one or more of the following groups: organic matter (aquatic systems), plant nutrients (e.g. aquatic systems), toxic substances, suspended solids in water and the atmosphere, energy and pathogenic microorganisms.
8. The discharge of organic matter to aquatic systems results in a characteristic set of changes in the receiving system. Overall, the dissolved oxygen content of the water falls together with primary production, while respiration increases and hydrogen sulfide is produced. The species diversity of benthic invertebrates and fish falls, while the numbers and species of bacteria increases. The status of these water bodies can be classified according to the **Saprobic System.**
9. The discharge of plant nutrients, usually N and P, to a water body low in nutrients and thus of oligotrophic status results in a characteristic set of changes. Initially, there is an increase in primary production leading to mesotrophic status and with further discharges eutrophic status.
10. The relationship of enrichment of water bodies with organic matter and plant nutrients can be understood by plotting respiration against primary production for a set of water bodies. Nutrient-enriched bodies have primary production > respiration and bodies enriched with organic matter have primary production < respiration.
11. The generalized biological effects of a toxic substance on a natural ecosystem relates to the specific group, which is targeted by the toxicant. The adverse effects will initially impact this group and cause a population decline, which may have adverse effects on the populations dependent on this group, for example food organisms. But overall, the levels of the toxicant will decline over time as a result of environmental processes, and the populations and ecosystems will recover.
12. All natural ecosystems experience variations in the ambient temperature experienced throughout the daily and yearly cycle. In tropical systems, this cycle is limited and thus the ecosystem has adapted to these conditions. However, in temperate ecosystems, the daily and yearly cycle has variations in temperature, which are much greater. Thus, the adverse biological effects of thermal discharges, as well as temperature changes resulting from climate change, would be expected to be greater in tropical ecosystems.
13. Biotic Indices, such as the Saprobic and the Trophic Systems, are a convenient way to summarize and describe any aquatic ecosystem based on several of the physical, chemical, and principally biological properties of an aquatic system. However, there are many specific biotic indices developed for use in specific areas such as the Trent Biotic Index and Becks Biotic Index. Some biotic indices are based on mathematical principles reflecting the occurrence and number of individuals and species such as Shannon-Weaver and Margalef Indices.
14. The occurrence of **Indicator Species** can be used to identify specific climatic zones or ecosystems and, since they are sensitive to different levels of pollutants, the levels of pollution which occur in an area. Another application of Indicator Species is as **Chemical Monitor Species**, which are species having the capacity to accumulate pollutants to higher levels than the ambient environment and can thus serve as biomonitors of pollutants.

References

Abbasi, T. and Abbasi, S.A. (2011). Water quality indices based on bioassessment: the biotic indices. *Journal of Water and Health* 9 (2): 330–348.

Arthington, A.H., Conrick, D.L., Connell, D.W. et al. (1982). The ecology of a polluted urban creek. Australian Water Resources Council. Technical Paper No. 68. Canberra: Australian Government Publishing Service.

Califf, R.M. (2018). Biomarker definitions and their applications. *Experimental Biology and Medicine* 243 (3): 213–221.

Eschenroeder, A., Irvine, E., Lloyd, A. et al. (1980). Computer simulation models for assessment of toxic substances. In: *Dynamics, Exposure and Hazard Assessment of Toxic Chemicals* (ed. R. Haque), 323–368. Ann Arbor, MI: Ann Arbor Sciences.

Gladwell, M. (2001). *The Tipping Point – How Little Things Can Make a Big Difference*. New York: Little Brown and Company.

Jackson, M.C., Weyl, O.L.F., Altermatt, F. et al. (2016). Recommendations for the next generation of global freshwater biological monitoring tools. *Advances in Ecological Research* 55 (12): 615–636.

Johannes, R.E. and Betzer, S.B. (1975). Introduction: marine communities respond differently to pollution in the tropics than at higher latitudes. In: *Tropical Marine Pollution* (ed. E.J. Ferguson-Wood and R.E. Johannes), 1–12. Amsterdam: Elsevier.

Khaliq, I., Hof, C., Prinzinger, R. et al. (2014). Global variation in thermal tolerances and vulnerability of endotherms to climate change. *Proceedings of the Royal Society B: Biological Sciences* 281 (1789): 20141097. DOI: https://doi.org/10.1098/rspb.2014.1097.

Kimbrough, K.L., Johnson, W.E., and Lauenstein, G.G. (2008). An assessment of two decades of contaminant monitoring in the nation's coastal zone. *NOAA Technical Memorandum NOS NCCOS 74*. National Oceanic and Atmospheric Administration, Silver Springs, MD.

Miller, N.A. and Stillman, J.H. (2012). Physiological optima and critical limits. *Nature Education Knowledge* 3 (10): 1.

Nguyen, K.D., Morley, S.A., Lai, C.-H. et al. (2011). Upper temperature limits of tropical marine ectotherms: global warming implications. *PLoS ONE* 6 (12): e29340. https://doi.org/10.1371/journal.pone.0029340.

Pawlowski, J., Kelly-Quinn, M., Altermatt, F. et al. (2018). The future of biotic indices in the ecogenomic era: Integrating (e)DNA metabarcoding in biological assessment of aquatic ecosystems. *Science of the Total Environment* 637–638: 1295–1310.

Settele, J., Scholes, R., Betts, R. et al. (2014). Terrestrial and inland water systems. In: *Climate Change 2014: Impacts, Adaptation and Vulnerability. Part A: Global and Sectoral Aspects. Contribution of Working Group II to the Fourth Assessment Report of the Intergovernmental Panel on Climate Change* (ed. C.B. Field, V.R. Barros, D.J. Dokken, et al.). Cambridge and New York: Cambridge University Press.

Stevens, G.C. (1989). The latitudinal gradient in geographical range: how so many species coexist in the tropics. *American Naturalist* 133: 240–256.

The Economist (2009). Tipping point – idea, April 20, https://www.economist.com›news›2009/04/20›tipping-point (accessed 9 August 2020).

8

Pollutants in the Oceans, Estuaries, and Freshwater Systems

8.1 Introduction

Water covers about 71% of the global surface, so the water environment comprises a major part of the human and natural environment. About 97% of this water environment is in the oceans and the remainder in lakes, rivers, aquifers, groundwater, glaciers, and other water bodies. The water environment sustains an important component of the global biota such as coral reefs, commercial fish stocks, and cetacean communities.

The study of water bodies has been a major area of scientific activity since the beginning of scientific endeavors. This includes oceanographic voyages, such as that of the research ship the **Beagle** with **Charles Darwin** on board. Darwin was later to use the scientific information gathered to formulate the biological evolution concept. Research on the oceans has been stimulated by the use of the oceans for ship-borne trade with oceans currents being a major factor. Physical and biological research on water bodies and their inhabitants has been a major focus for scientific research and remains so in recent times.

Water pollution is the contamination of water bodies by addition of substances that alter the properties of the water making it unsuitable for use by human society or the natural biological system. Pollution is usually considered to be as a result of human activities. Concern regarding water pollution was evident in Ancient Rome with the construction of the Cloaca Maxima commencing in 600 BCE. The great sewer was constructed to receive storm and wastewater and is still in use today protecting the quality of water in Rome.

In the 1800s, the pollution of the Thames River in England by untreated sewage resulted in extensive deaths due to cholera and other waterborne diseases. As a result, sewerage systems were introduced to collect the sewage, and special plants for its treatment before discharge. Sources of water pollution can be classified as point or nonpoint sources. Point sources have a single source or discharge, such as a storm drain or a wastewater discharge. Nonpoint sources are more diffuse, such as urban and agricultural runoff.

The contaminants that pollute water and make it unsuitable for human use or to sustain the natural environment are principally chemical and microbiological in nature. Principal human use is as drinking water and with the natural environment, the protection and the survival of natural ecosystems is usually the use. The microbiological agents in drinking water, which cause waterborne disease to spread, are many and varied and each has its own particular characteristics. However, the principal chemical agents that are of importance in the protection of natural ecosystems can be generally reduced to two classes.

Deoxygenating Substances and Plant Nutrients

Surprisingly these major water pollutants produced by human society are substances that have comparatively little directly harmful or toxic effect. They are relatively nontoxic animal and vegetable wastes, as well as agricultural and urban runoff, which are often discharged into streams and inshore marine areas. In certain limited quantities, these pollutants can be beneficial in some bodies of water by releasing nutrients in the form of nitrogen and phosphorus salts, which stimulate plant and animal growth. However, in excessive quantities, their secondary effects are usually extremely harmful. These wastewaters can be rich in organic carbon and can be discharged into waterways in the form of human sewage, food processing wastes from the fruit, meat, and dairy industries, as well as wastewaters from paper manufacturing and a variety of other industries. However, in many areas they originate from point sources and can be collected and treated in specially designed plants. Sewage treatment proceeds through a sequence of treatment steps. Initially, the sewage is subject to **primary** treatment, which is the removal of solids and other materials, then **secondary** treatment, which is a microorganism-mediated oxidation of the organic carbon present to carbon dioxide and water. This can be followed

---

Chemistry and Toxicology of Pollution: Ecological and Human Health, Second Edition. Des W. Connell and Greg J. Miller.
© 2023 John Wiley & Sons, Inc. Published 2023 by John Wiley & Sons, Inc.
Companion website: www.wiley.com/go/toxicologyofpollution2e

by **tertiary** treatment, which is the removal of plant nutrients. Agricultural and urban runoff are usually not rich in carbon but have excess amounts of the nutrient chemicals, principally nitrogen and phosphorus, originating from the fertilizers used to stimulate crop growth. The pollutants are in a diffuse nonpoint form, which is not amenable to relatively simple treatment.

These polluting substances in the global aquatic environment are considered in the following sections: Section 8.2, and plant nutrients are considered under the title of Section 8.3.

8.2 Deoxygenating Substances

Photosynthesis and respiration are fundamental to life on Earth since these processes involve the harvesting and consumption of solar energy by living systems. The chemical processes involved can be simply expressed with glucose as set out in equation (8.1), but a highly complex series of reactions are involved in proceeding from one side of this reaction to the other. Also, the formation and consumption of proteins and fats involve different reaction sequences.

$$6CO_2 + 6H_2O \underset{\text{Respiration}}{\overset{\text{Photosynthesis}}{\rightleftharpoons}} 6CH_2O + 6O_2 \quad (8.1)$$

These processes involve carbon dioxide and oxygen in the Earth's atmosphere, which, in aquatic ecosystems, are dissolved in the water mass. The involvement of the water mass is a marked difference between aquatic and terrestrial systems, which is an important aspect of pollution ecology. So terrestrial systems are mediated by the atmosphere, and aquatic systems are mediated by water in which the atmospheric components are dissolved.

The basic transformations of carbon in natural aquatic ecosystems are shown in Figure 8.1. Carbon from dissolved carbon dioxide is incorporated by photosynthesis into the autotrophic biomass of the ecosystem consisting mainly of photosynthetic plants such as algae. Respiration of the heterotrophic biota involves the consumption of autotrophic biomass, algae, and other plants as food, and finally the decomposition of body tissues after death. These processes

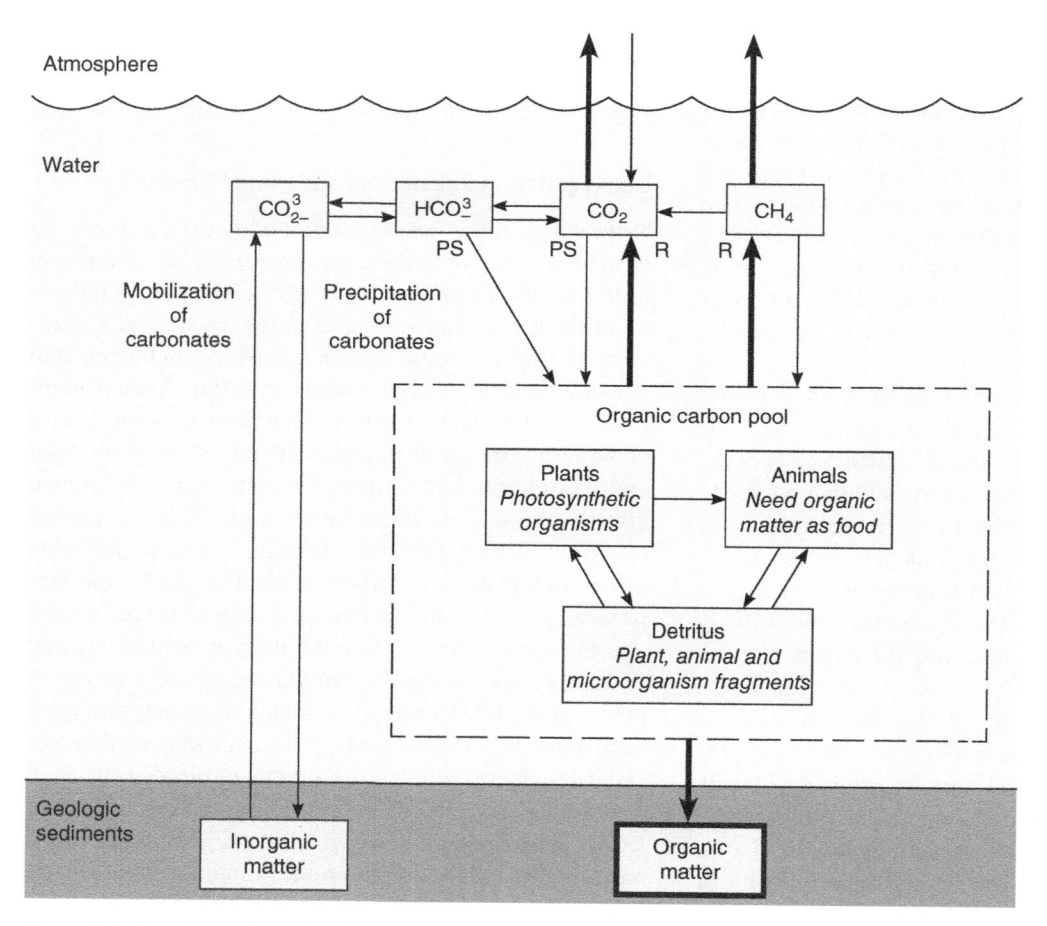

Figure 8.1 Transformations of carbon in aquatic systems with processes accelerated or increased by added organic matter shown in heavy lines with PS indicating *photosynthesis* and R indicating *respiration*.

produce residual-resistant fragmental matter described as detritus. Both of these processes result in the formation of carbon dioxide and methane, which are returned to the water and, ultimately, the atmosphere. Some carbon, in the form of detritus, is semipermanently incorporated into the sediments and thus is ultimately removed from the system by incorporation in geological strata sediments.

If substances rich in organic carbon are added to the system, some of the pathways shown in Figure 8.1 are increased in magnitude and also some of the pools of organic carbon are increased in size. This results in an increase in respiration, mainly through the respiration of microorganisms, giving rise to increased amounts of carbon dioxide and methane (see Figure 8.1).

The transformations of carbon involve oxygen and the related transformations of oxygen are shown in Figure 8.2. The oxygen needed for aerobic respiration is obtained from the dissolved oxygen (DO) in the water mass. This oxygen is usually substantially derived from the atmosphere and converted into carbon dioxide, which is discharged into the water mass and ultimately to the atmosphere. Increased respiration, due to an increased amount of organic carbon in the water, results in changes in the magnitude of the pathways and changes in the pools of oxygen involved (see Figure 8.2). Most importantly, there is a demand on the reservoir of dissolved oxygen in the water mass, which is comparatively small since oxygen has limited solubility in water, usually ranging from about 6 to 14 mg L^{-1}. Therefore, substantial reductions in dissolved oxygen can occur, which have significant implications for aquatic organisms.

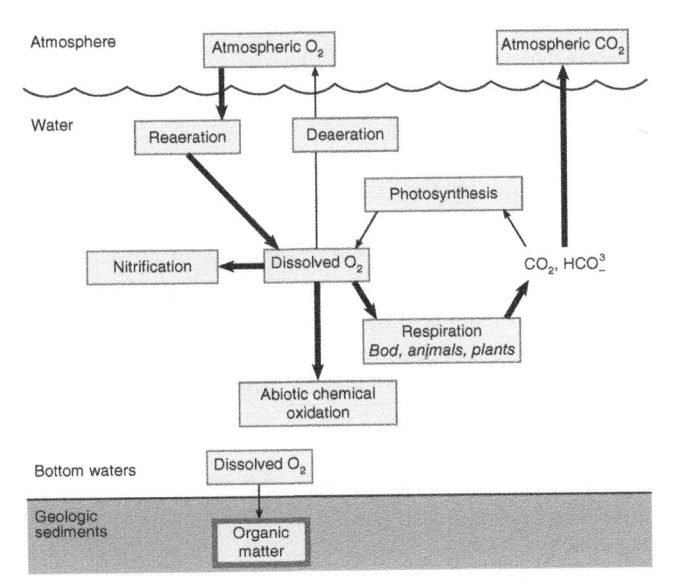

Figure 8.2 Oxygen dynamics in aquatic systems with processes increased or accelerated by organic discharges indicated in heavy lines.

When the DO reaches levels of about 2 mg L^{-1} or less, the water cannot support normal aquatic life and is described as **hypoxic** and the condition is known as **hypoxia**. If the DO falls to levels at or near 0 mg L^{-1}, **anoxia** occurs, and the conditions are described as **anoxic**. This results in severe damage to the normal aquatic system at a level more intense than hypoxia.

Additionally, the detritus-organic matter in the geologic sediments can play an important role in the dynamics of dissolved oxygen in an aquatic system. If the waters are turbulent, it can influence the overall amount of dissolved oxygen present. Alternatively, if the waters are not turbulent but relatively still, the bottom waters then represent a separate and discrete layer where respiration, mainly by microorganisms and invertebrates, can deplete the dissolved oxygen present as shown in Figure 8.2. In this manner, the bottom waters in all bodies of water ranging from the oceans to ponds, as well as swamps in a natural condition, can exhibit hypoxia and anoxia. Also flowing waters such as rivers can exhibit hypoxia on a periodic or permanent basis. Well-known urban rivers such as the Thames River in London, the Hudson River in the City of New York, and the Rhine River in Germany have had hypoxia problems, as well as many lesser known streams.

The most spectacular examples of this phenomena are in the world's large water bodies. In these bodies, there has been the natural accumulation of detritus-organic matter originating from both the catchment discharges and growth of phytoplankton within the water mass over geological time. This has resulted in the accumulation of detritus-organic matter in the bottom sediments and waters resulting in hypoxia and anoxia occurs. These areas in the oceans are often described as **dead zones (DZ)** since there is an absence of normal aquatic life. The area of total **oxygen minimum zones (OMZ)** has been estimated by Paulmier and Ruiz-Pino (2009) as 30.4×10^6 km^2 in the global ocean. However, Diaz and Rosenberg (2008) have reported that the *dead zone*s in these OMZ have expanded over recent years with the expansion of the world's population and subsequent increase in discharges of organic wastes and plant nutrients in wastewater. They are usually variable in size with seasonal, environmental, and global factors having an influence. Some examples are shown in Table 8.1.

Factors Affecting Dissolved Oxygen in Water Bodies

Respiration and Nitrification

In water bodies, microorganisms are usually the most important heterotrophic organisms, in terms of respiration. In well-oxygenated waters, aerobic respiration occurs mediated by aerobic microorganisms. The chemical

Table 8.1 Some examples of oxygen minimum zones (OMZ) and dead zones (DZ) in the world's oceans.

Name	Location	Approximate area (km^2)
Global oceans	Total globe	30 400 000 (OMZ)
Baltic Sea	Northern Europe	5000 (DZ)
Gulf of Mexico	Central America	23 000 (DZ)
Black Sea	Eastern Europe	2000 (DZ)
Bay of Bengal	North Indian Ocean	1 600 000 (OMZ)

reaction involved can be simply expressed, for glucose and carbohydrates, as:

$$6CH_2O + 6O_2 \rightarrow 6H_2O + 6CO_2 \quad (8.2)$$

However, while this is appropriate for carbohydrates, other important substances are present in natural systems, such as proteins and fats, which are expressed by different equations.

Environmental temperature is a strong influence on this process since different species of microorganisms have a narrow range of temperatures for their optimal activity. Even so, in most aquatic bodies, communities of microorganisms exist which cover a wide range of optimal temperatures.

Equation (8.2) outlines the reaction of carbon during aerobic respiration, but organic matter in discharges also contains various other elements, particularly nitrogen and sulfur. Nitrogen is present mainly in the amino group of the peptide link in proteins. Complete oxidation of the organic nitrogen present leads to the formation of the nitrate ion. Sulfur is present in organic matter in a variety of different forms but principally as the sulfhydryl group, the disulfide group, and the sulfide group. Complete oxidation yields the sulfate ion. In the absence of oxygen, under anoxic or hypoxic conditions, the anaerobic microorganisms take over the degradation and decomposition of organic matter. Some are facultative organisms, which can function as either aerobic or anaerobic organisms, while others, particularly the methane-producing group, are obligate anaerobes. In fact, molecular oxygen is toxic to this group. With glucose, the anaerobic respiration reaction can be simply expressed as:

$$6CH_2O \rightarrow 3CH_4 + 3CO_2 \quad (8.3)$$

This process makes no demand on the oxygen present in the water mass, but it is very important for the removal of oxygen-demanding substances in waterways. The methane and carbon dioxide produced are released into the water mass and then to the atmosphere, resulting in the removal of organic carbon and oxygen demand from the system. This process occurs in natural swamps, organic-rich bottom muds, and water bodies suffering from pollution by organic wastes.

Anaerobic respiration with sulfur- and nitrogen-containing organic substances gives rise to hydrogen sulfide and ammonia, respectively. Ammonia in the presence of oxygen is readily oxidized to nitrate. Nitrite is formed under suitable oxidation and reduction conditions but is less commonly the end product of nitrogen metabolism. The formation of nitrate from organic matter is described as the **nitrification reaction** and is mediated by a variety of microorganisms. A series of reactions occurs, which can be simply expressed as follows:

$$2O_2 + NH_4^+ \rightarrow NO_3^- + 2H^+ + H_2O \quad (8.4)$$

This process can make an important demand on dissolved oxygen in natural water bodies, especially where sewage contamination, containing suitable microorganisms, occurs in the temperature ranges between 25 and 30 °C.

These oxygen-demanding processes can be measured by the **biological oxygen demand** (**BOD test**) (Clesceri et al. 2005). This laboratory test measures the consumption of oxygen by water samples incubated over a period of five days under standard conditions, often with seed microorganisms added. In general, clean natural water has a BOD of 1 mg L^{-1} or less and seriously polluted water contains greater than 10 mg L^{-1}. The nitrification reaction occurs in what is described as the second stage of the BOD test at time periods greater than five days. The precision of the BOD test is about ±17% and may not satisfactorily reflect the actual conditions existing in a natural water body. The test conditions may differ from those existing in the environment in terms of temperature, microorganisms present, nutrient status, and so on.

The kinetics of BOD changes in waterways have been subject to intense study from the earliest days of scientific interest in water pollution. Velz (1970) has described one of the principles governing BOD in natural waterways first formulated by Phelps as "The rate of biochemical oxidation of organic matter is proportional to the remaining concentrations of unoxidized substance measured in terms of oxidizability." Thus, the oxidation of organic matter follows first-order decay reaction kinetics, which are described in Chapter 4.

From this, the following series of mathematical expressions can be derived:

$$-dL/dt = K_1 L$$

which can be integrated to

$$\ln(L/L_0) = -K_1 t$$

or

$$\log(L/L_0) = -0.434k_1 t = -k_1 t \quad (8.5)$$

$$L/L_0 = 10^{-k_1 t}$$

Where L_0 is the total BOD debt at time zero; L the BOD debt at time t; t the time period since time zero; K_1 and k_1 the BOD decay rate coefficients, that is, empirical constants ($0.434 K_1 = k_1$). Logarithms to the base 10 were commonly used in the past, but it is more correct to use natural logarithms, so both have been included here.

By measurements of BOD in waterways over long periods, Velz (1970) reported that at 20 °C, $k_1 = 0.1$ day^{-1}, if the time period is measured in days. Since first-order kinetics apply, it can be shown that $t_{1/2} = 0.301/k_1$ (see Chapter 4) and also that the half-life normally exhibited by BOD demanding substances in natural waterways is equal to about three days. Also, from the mathematical treatment outlined above, a plot of log percentage of remaining BOD versus time should give a straight line (see Figure 8.3). Thus, after five days, $\log(L/L_0) = -0.5$ and from this it can be shown that the BOD_5 is 0.32 (32%) of the total BOD debt or ultimate BOD, L_0.

The BOD decay rate coefficient (k) varies with temperature. It can be shown that the deviation of this constant from the 20 °C value can be derived from the equation:

$$K_T = k_{20\,°C} \times 1.047^{(T-20)} \quad (8.6)$$

Since the BOD measure has some disadvantages associated with it, such as lack of precision and application to real aquatic systems, other measures have been developed for measuring oxygen demand. One of these is the **chemical oxygen demand** (COD), which is performed by subjecting the test sample to strong oxidizing agents so that extensive and vigorous oxidation proceeds. While this is more rapid than the BOD test, the oxygen taken up by strong oxidizing agents may not reflect the amount of oxygen consumed under natural conditions. Another method measures the **total organic carbon** (TOC) content of the water, from which the oxygen demand can be estimated. This procedure is rapid and comparatively precise compared with the BOD test, However, once again its relevance to natural systems is not as clear as that of the BOD test.

Temperature and Salinity

The solubility of molecular oxygen in water is affected by (i) the partial pressure of atmospheric oxygen gas in contact with the water, (ii) the temperature, and (iii) the salinity. The influence of the partial pressure of the gas on solubility is given by Henry's Law:

$$\text{Solubility (molar concentration)} = H_C P_x \quad (8.7)$$

Where H_C is Henry's Law constant and P_x is the partial pressure.

Ambient temperature has a strong influence on the solubility of oxygen in water as shown by the Clausius-Clapeyron equation:

$$\log(C_1/C_2) = \Delta H/2303 R \,(1/T_1 - 1/T_2) \quad (8.8)$$

Where C_1 and C_2 are gas concentrations in water; T the absolute temperature in Kelvins; H the heat of solution (cal mol^{-1}); and R the gas constant (1.987 cal K^{-1} mol^{-1}).

Thus, the solubility of oxygen in water decreases with temperature. Actual figures are shown in Table 8.2. In fact, water at 0 °C contains almost twice the concentration of water at 30 °C. Dissolved oxygen also decreases with increasing salinity (see Table 8.2).

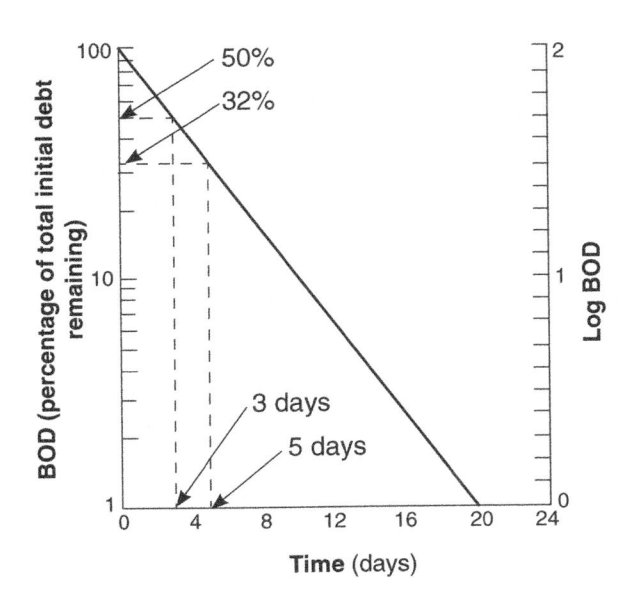

Figure 8.3 Normal decline of organic BOD debt at 20 °C (k = 0.1) in an aquatic area.

Table 8.2 Solubility of oxygen in water exposed to water-saturated air.[a]

	Chloride concentrations (mg L^{-1})				
	0	5000	10 000	15 000	20 000
Temperature (°C)	Dissolved oxygen (mg L^{-1})				
0	14.6	13.7	12.9	12.1	11.4
10	11.3	10.7	10.1	9.5	9.0
20	9.1	8.6	8.2	7.7	7.3
25	8.2	8.0	7.4	7.1	6.7
30	7.5	7.2	6.8	6.5	6.2

[a] At 760 mm Hg.

The Dissolved Oxygen "Sag"

The modelling of the dynamics of dissolved oxygen in rivers receiving discharges of wastewater containing significant amounts of BOD commenced in 1925 with Harold Streeter and Harold Phelps (Schnoor 1996). It probably represents the first widespread and highly successful application of mathematics to an environmental pollution problem.

One of the most important sources of dissolved oxygen in waterways is oxygen in the atmosphere, which dissolves in the water mass at the water surface. The rate of reaeration is proportional to the oxygen deficit in the water mass. Thus,

$$dC/dt = k_2 D \quad (8.9)$$

Where C is the concentration of oxygen; k_2 the reaeration coefficient; and D the oxygen deficit (i.e. $C_{saturation} - C_{actual}$).

Losses of dissolved oxygen are often associated with BOD due to organic discharges. The rate of loss of BOD follows first-order kinetics. Thus, the rate of loss of oxygen is proportional to BOD present at the time:

$$dD/dt = -k_1 L \quad (8.10)$$

Where D is the oxygen deficit; L the BOD debt or concentration of BOD; and k_1 the BOD decay rate coefficient.

The deoxygenation of water containing organic wastes follows a first-order rate law, then this can be integrated to give the following equations to describe this process:

$$\ln(L_1/L_0) = -k_1 t$$
and $\quad (8.11)$
$$\ln L_t = \ln L_o - k_1 t$$

Where L_0 is complete oxidation capacity (ultimate BOD), L_t oxidation capacity at time t (ultimate BOD − BOD at time = t), and k_1 decay rate constant (using logs to base e).

By combining equations (8.9, 8.10), the actual oxygen deficit can be obtained by subtracting the uptake due to reaeration from the deficit due to oxygen consumption due to the BOD. This is commonly referred to as the Streeter-Phelps equation. Thus,

$$dD/dt = k_1 L - k_2 D \quad (8.12)$$

These two processes are illustrated diagrammatically in Figure 8.4. Curve "e" is the cumulative deoxygenation curve, which would represent dissolved oxygen values if no reaeration was to occur. However, as soon as an oxygen deficit occurs, soon after zero, reaeration commences (curve "d") and increases as the deficit increases. Curve "d" represents the cumulative oxygen input to the water as a result of deoxygenation and reaeration. The summation of these curves gives the actual dissolved oxygen profile, the dissolved oxygen **"sag"** (curve "a"). An expression

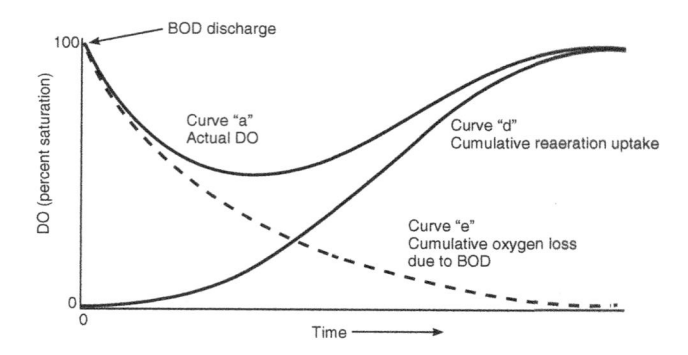

Figure 8.4 Resultant downstream effects in a flowing stream of an increase in BOD at time zero causing deoxygenation, reaeration, and a consequent oxygen *sag* curve.

for this curve is mathematically obtained by integrating equation (8.12):

$$D = (k_2 L_0 / k_2 - k_1)\left(e^{-k_1 t} - e^{-k_2 t}\right) + D_0 e^{-k_2 t} \quad (8.13)$$

Where L_0 and D_0 are values at time zero.

Critical factors in this expression are k_1 and k_2. The BOD decay coefficient (k_1) can be calculated from experimental results and is usually 0.1 day^{-1} at 20 °C with time in days. The reaeration coefficient (K_2 and k_2) is more difficult to measure and several methods have been used. Many methods take into account mean velocity, depth of the water, and turbulence, which can have an important influence.

The equation below has been used for streams showing isotropic flow pattern with little turbulence, that is, the velocity in different vertical layers is similar. Thus, for streams with isotropic flow, the following expression can be applied:

$$k_2 = \left(127 (D_L U)^{1/2}\right) / h^{3/2} \quad (8.14)$$

Where D_L is the coefficient of molecular diffusion (liquid film ft^2 day^{-1}); h the average depth (ft); U the mean velocity of flow (ft sec^{-1}); and k_2 reaeration (per day). The equation for nonisotropic flow includes a value for S, the slope of river channel (ft ft^{-1}) (see Connell and Miller 1984, p. 104).

Several other methods of calculation are described by Schnoor (1996).

In the discussion above, only oxygen demand due to BOD is considered. This is due to organic matter present as suspended or dissolved matter in the water mass. However, organic matter can be sedimented out of the water mass to form bottom sediments, which are rich in organic matter as shown in Figure 8.1. These organic-rich layers may induce anaerobic degradation in the bottom sediments and waters but above this aerobic respiration can occur. This leads to an oxygen demand not measured by BOD of the water mass, which can exceed the BOD in some cases

Figure 8.5 Stream water receiving wastewater rich in organic matter causing a reduction of the stream water DO to zero resulting in anoxic conditions, which convert plant tannins and related substances to black-colored tannins. Source: Photograph by Des Connell (Author).

(e.g. Connell et al. 1982). Additionally, when anoxic conditions develop, the tannins and related substances dissolved in the water may change color to black giving the water a distinctive black color. An example of this is shown in Figure 8.5.

Treatment of Sewage

Sewage can be treated to remove the BOD, firstly by **primary treatment,** which removes the solids, described as **sludge**. This sludge can be used directly as a conditioner and source of nutrients for soils. However, it can be subject to anaerobic digestion, according to equation (8.3), yielding methane, which can be used as a fuel to generate electricity. The liquids produced in primary treatment are usually subject to aerobic digestion, according to equation (8.2), in the **secondary treatment** process. This treatment effectively removes the BOD from the sewage, limiting the amount of deoxygenation, which the treated sewage can cause when discharged to a natural water body.

Photosynthesis and Diurnal Variations in Dissolved Oxygen

Many polluted waterways have very little plant biomass but, in some cases, conditions can be suitable for the growth of aquatic plants. Large rivers and estuaries can have significant populations of phytoplankton, but many small and shallow streams are dominated by rooted aquatic plants and attached algae. Plants can have a very important influence on dissolved oxygen content through photosynthesis and respiration.

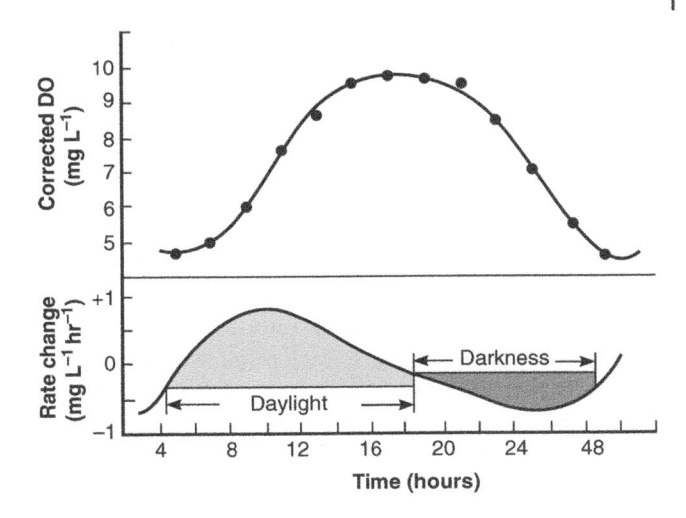

Figure 8.6 Plot of average variation of corrected dissolved oxygen concentrations and rate of change of dissolved oxygen concentration in a 24-hour cycle in an estuary. Source: Connell et al. (1982)

Photosynthesis occurs during the daylight hours but respiration by plants occurs throughout the diurnal cycle. Thus, if significant plant growths are present, this will lead to an input of photosynthetic oxygen during the daylight hours but a continuous consumption of dissolved oxygen by respiration. Figure 8.6 shows these processes cause a rise in dissolved oxygen during the daylight hours with a maximum in the afternoon. At sunset, production of photosynthetic oxygen ceases, and so the dissolved oxygen content of the water starts to drop due to respiration by plants and other aquatic organisms. This continues overnight reaching a minimum before dawn when sunlight once again initiates photosynthesis.

These diurnal curves can be used to calculate the total oxygen produced by photosynthesis and the total oxygen consumed by respiration throughout the diurnal cycle. This is done by plotting the rate of change of dissolved oxygen with time and measuring the area under or over the curves (see Figure 8.6).

Dissolved oxygen production and consumption is influenced by light intensity, plant biomass, as well as ambient water temperatures. During periods of overcast weather, photosynthetic oxygen production can be very low and oxygen consumption due to plant respiration can exceed the photosynthetic oxygen production leading to a decrease in dissolved oxygen in the water mass. In addition, large growths of plants upon death deposit on the bottom and decay, enriching the bottom sediments with organic matter. However, this can lead to anaerobic respiration in bottom waters and sediments, but on the surface, aerobic respiration can occur. A similar situation can occur after the use of aquatic herbicides, which cause extensive death and decay of aquatic plants leading to substantial reductions in dissolved oxygen.

Diurnal variations in aquatic areas, particularly estuaries, can result from tidal flow. The tides may bring oceanic water rich in dissolved oxygen into an area that has lowered dissolved oxygen levels due to high respiratory activities in the sediments. This results in a cycle of dissolved oxygen concentrations related to tidal flows. Also, changes in the concentrations of BOD in estuarine waters are often due to the influx of oceanic water and often low in BOD. Tidal flows follow a time cycle out of phase with the periods of daylight and darkness, so they will introduce patterns out of phase with those produced by photosynthetic activity. These factors can complicate the analysis of dissolved oxygen variation in estuaries.

Seasonal Variations and Vertical Profiles of Dissolved Oxygen

Many environmental factors vary seasonally. For example, the incidence of sunlight has an important impact on photosynthesis and resultant dissolved oxygen concentrations in many aquatic areas. In this case, the production of photosynthetic oxygen is related to the incidence of sunlight, with a maximum occurring in summer and a minimum in winter. In addition, ambient temperatures affect algal photosynthesis, with extreme temperatures giving the lowest photosynthetic activity.

Turbulent water conditions give well-mixed waters in which vertical profiles of dissolved oxygen are constant from the surface to the bottom waters. However, if the physical conditions are not turbulent and comparatively still, then **thermal stratification**, due to solar heating of the surface waters, can lead to isolation of the bottom waters. If the bottom sediments are enriched with organic matter, the bottom waters may become depleted in dissolved oxygen while the surface waters remain unaffected. In this manner, a vertical profile showing considerable variation in dissolved oxygen concentration can be the result.

In temperate areas, **vertical stratification** usually occurs in a seasonal pattern with stratification in the summer, and **mixing** or **overturn** in the winter. Corresponding to this, there is usually a seasonal pattern of dissolved oxygen variation in bottom waters, which is common in lakes and dams, as well as relatively small water bodies, but can also occur on the continental shelves of the oceans.

Oxygen Balance

In investigating the dissolved oxygen levels and final oxygen balance in a water body, it is necessary to identify all the factors contributing to losses and additions of oxygen. These factors have been identified by Connell et al. (1982) as:

Additions

1. *Reaeration.* Solubilization of atmospheric oxygen at the water surface from the atmosphere.
2. *Photosynthesis.* Production of photosynthetic oxygen by aquatic plants.
3. *Accrual.* The addition of dissolved oxygen to the body by discharge of tributaries.

Losses

1. *Aerobic Respiration.* The aerobic respiration activities of all organisms present in the water body, including microorganisms in the water mass (measured as BOD), benthic organisms (particularly microorganisms), plants, fish, and other large organisms.
2. *Export.* The discharge of water containing dissolved oxygen from the water body.
3. *Deaeration.* If saturation exceeds 100%, there is a loss of dissolved oxygen from the water mass to the atmosphere.

The amounts in these various categories would be expected to vary from area to area. However, Connell et al. (1982) have quantified the amounts for a tidal sub-tropical creek as shown in Table 8.3.

Table 8.3 Oxygen budget for an urban estuary.

	kg day^{-1}	Standard deviation	Percent
	Additions		
Accrual	76	48	13
Aeration	225	24	45
Photosynthesis	214	—	42
Total	515	—	100
	Losses		
Export	30	6	5
Deaeration	89	24	14
BOD	91	35	14
Plant respiration	97	—	15
Benthic respiration	335	~45	52
Total	642	—	100

Source: Connell et al. (1982).

Box 8.1 Major DO Depletion Event in the New York Bight in 1976

One of the most spectacular environmental disasters resulting from oxygen depletion occurred in 1976 in the New York Bight located on the continental shelf of North Eastern United States. Figure 8.7 shows the normal seasonal effect of stratification on dissolved oxygen concentration in subsurface waters in the Bight. Armstrong (1977) has suggested that the bottom waters and sediments were enriched by phytoplankton decay, river discharge containing oxygen-demanding substances and nutrients, as well as dumping of sewage sludge transported in barges from New York City. In 1976, there was an early arrival of spring leading to elevated surface temperatures, which coincided with high discharge of water from the adjacent Hudson and Delaware Rivers (see Figure 8.7). This led to a prolonged period of stratification, in conjunction with the presence of relatively large quantities of organic matter, causing severe depletion of dissolved oxygen in the bottom waters (see Figure 8.7). These conditions resulted in extensive "**kills**" of marine organisms in the New York Bight during 1976, which are considered later in this chapter – Overall Impact on Aquatic Systems.

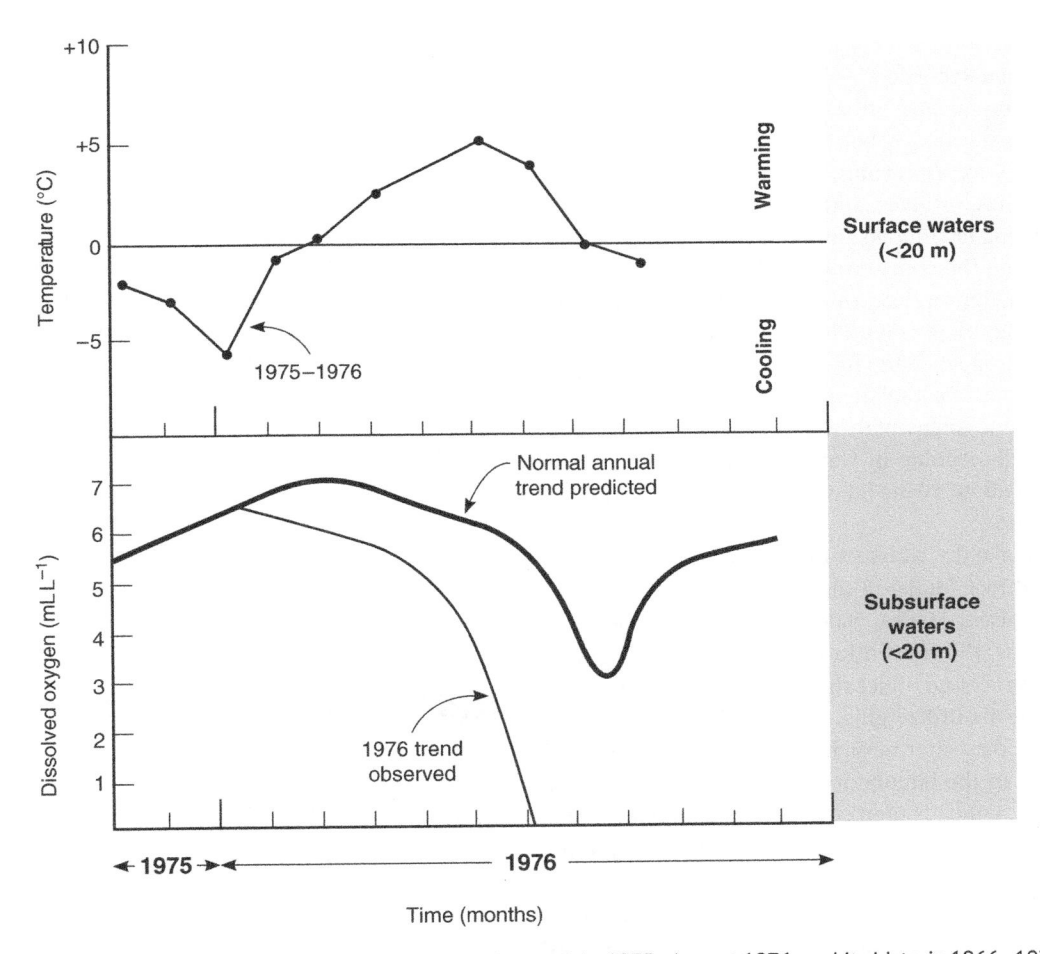

Figure 8.7 New York Bight monthly sea surface temperature change, July 1975–August 1976, and its historic 1966–1975 range in surface waters. Subsurface waters (>20 m) dissolved oxygen as predicted and observed in 1976 and historical range and mean at the same location. (Values from gulf stream, National Weather Service, NOAA, January 1975–August 1976, and The Gulf Stream Monthly Summary, U.S. Naval Oceanographic Office, January 1966 – December 1974.) Source: Armstrong (1977).

Response of Individuals to Reduced Dissolved Oxygen and Associated Contaminants

In natural systems, low dissolved oxygen concentrations (OMZ, hypoxia and anoxia) in natural waters are usually associated with **anaerobic respiration** and the production of substances that are toxic to aquatic organisms. Hydrogen sulfide is probably the most important substance in this category. In this case, it can be difficult to distinguish between the effects on organisms caused by low dissolved oxygen concentrations and hydrogen sulfide. Thus, in considering the response of individuals to reduced oxygen, it is also important to consider the possible effects of other substances, particularly hydrogen sulfide. The effects of both these factors on organisms in aquatic systems are outlined below.

Dissolved Oxygen Reduction

Most aquatic organisms utilize the **aerobic respiration** process to obtain energy, but some invertebrates have a limited capacity for **anaerobic respiration**. For example, some polychaetes, bivalves, and even mesopelagic fish are reported as being facultative anaerobes.

In considering the concentration of dissolved oxygen in water, the most appropriate units to use are concentration, usually as milligram per liter. This is more appropriate than percent saturation since the metabolic demand for oxygen by fish is in terms of actual quantity of oxygen and the percent saturation changes with water temperature. Oxygen is only sparingly soluble in water and dissolves to maximum levels of about $10\,\text{mg}\,\text{L}^{-1}$ under the usual conditions in water bodies.

Oxygen uptake by aquatic mammals and reptiles is achieved by direct intake of air at the water surface, for example porpoises, whales, water snakes, and turtles. For these organisms, the concentration of dissolved oxygen in the water mass is of no direct significance. However, many aquatic organisms utilize gills, whereby dissolved oxygen is passed from the water mass into the circulatory fluid of the organism. In the circulatory fluid, oxygen is attached to hemoglobin, which is then circulated by the heart to the muscles where oxygen is consumed together with carbohydrates to produce carbon dioxide in the aerobic respiration process. Carbon dioxide is respired to the external environment by the reverse path. Smaller organisms, such as microorganisms and some invertebrates, utilize diffusion through the external body surfaces as a source of oxygen. Circulatory fluids are not necessary since the distances involved are short and uptake can occur relatively rapidly. This process is referred to as **cutaneous respiration**.

Lowered dissolved oxygen concentrations in the water mass lead to low oxygen uptake by organisms and, consequently, muscles are not fed with sufficient oxygen for aerobic respiration to continue at an optimum rate. This can be compensated for in fish and other organisms by more rapid pumping of water over the gills. In some situations, the fish may be able to come to the surface and gulp air containing oxygen at the surface, as shown in Figure 8.8. But if oxygen uptake is inadequate, insufficient muscle activity will occur and eventually death of the organism will result (Erichsen Jones 1964). These conditions can result in **"kills"** of aquatic organisms including fish as shown in Figure 8.9. Moderately reduced dissolved oxygen levels decrease physiological activity of aquatic organisms. For example, with fish there is a decrease in food consumption, growth, and swimming velocity at dissolved oxygen

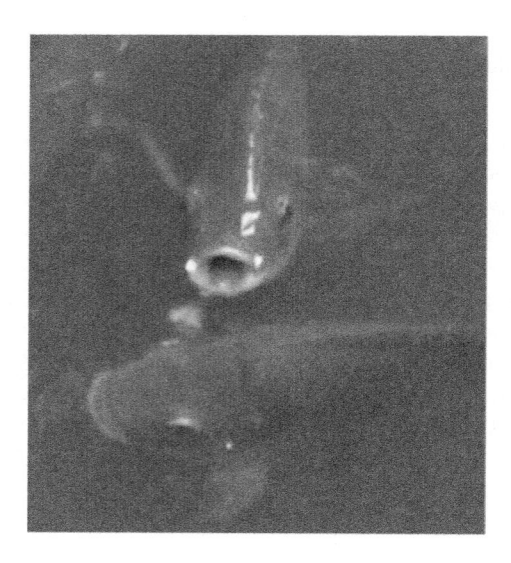

Figure 8.8 Fish in water, which is depleted in DO at the surface gulping oxygen directly at the water surface. Source: Photograph by Des Connell (Author).

Figure 8.9 "Kill" of marine fish on a beach in Australia due to deoxygenation by discharge of wastewater rich in organic matter. Source: Photograph by Des Connell (Author).

concentrations less than 8–10 mg L^{-1}. Thus, adverse effects, although not lethal, can be detected in many fish species when concentrations fall below 100% saturation.

Mobile species of aquatic organisms can exhibit altered schooling and avoidance behavior to low dissolved oxygen concentrations in water. Domenici et al. (2017) have reported that several species of fish avoid low concentrations of dissolved oxygen produced by increased hypoxia in oceanic waters. Similarly, invertebrate species have been shown to avoid water with dissolved oxygen concentrations less than full saturation. It has been suggested that trout in freshwater can detect an oxygen gradient and thus move into areas with more favorable oxygen content.

Some aquatic organisms have adapted to environmental conditions of low dissolved oxygen concentrations. For example, the respiratory pigments of annelid worms are adapted to take up comparatively high quantities of oxygen by changes in their molecular structures. The proportion of oxygen taken up is directly correlated with the molecular weight of the pigment. In addition, some benthic organisms have developed high concentrations of respiratory pigments in body fluids to compensate for low dissolved oxygen in the water mass, for example the chironomid larvae which have a highly visible red coloration. Furthermore, there are physical adaptations to low dissolved oxygen concentrations. The larvae of *Eristalis* (the rat-tailed maggot) have a long telescopic tail with an air tube through which it can breathe.

Different organisms exhibit different rates of oxygen consumption dependent upon size, metabolic characteristics, activity, and many other factors. For example, oxygen consumption rates in terms of cubic centimeters of O_2 per gram of body per hour are 0.0034–0.005 for the jelly fish, 0.055 for the mussel, 0.04 for the eel, and 0.22 for the rainbow trout (Erichsen Jones 1964).

Similarly, organisms show different minimum levels of dissolved oxygen for survival (see Table 8.4). It is interesting to note that often closely related species (e.g. amphipods in Table 8.4) can show widely different minimum dissolved oxygen levels. Also, many organisms show a different range of responses to lowered dissolved oxygen concentrations in the water mass.

Another important aspect of organism response to reduced dissolved oxygen is the different oxygen requirements of the different life stages of aquatic organisms. For example, the development of the embryo of many fish species is dependent on an increasing minimum dissolved oxygen as it increases in size (Ziober et al. 2012).

Dissolved Oxygen Criteria

The maintenance of DO in water bodies is probably the single most important objective of water quality management authorities throughout the world. Criteria for its control and management have been set by most authorities to meet the objectives of the use the water is to be put to. Some objectives may be related to drinking water, recreation, aesthetics, and industry but usually the most important uses relate to the maintenance of a healthy aquatic ecosystem. DO standards and criteria differ from those with most other substances in that the DO criteria are set at a minimum, whereas with other substances the level is set at a maximum (US EPA 2019).

Until recently, 5 ppm (5 mg L^{-1}) of dissolved oxygen was considered an acceptable minimum for normal growth and reproduction of fish (Erichsen Jones 1964). However, there are a number of complicating environmental factors, which can have an impact on the minimal dissolved oxygen level required for fish to complete a normal life cycle. Other aquatic organisms need to be taken into account, so the whole ecosystem can be protected. Seasonal variations in the dissolved oxygen requirement of aquatic organisms are known to occur. Also, there is a different susceptibility of aquatic organisms to lowered dissolved oxygen levels at differing latitudes. Much of this variation in susceptibility to lowered dissolved oxygen levels is related to water temperature. As water temperature increases, the metabolic

Table 8.4 Examples of limiting oxygen concentrations for aquatic organisms.[a,b]

Organism	Temperature (°C)	Test	Oxygen (mg L^{-1})
Brown trout (*Salmo trutta*)	6.4–24	Limiting concentration	1.28–2.9
Coho salmon (*Oncorhynchus kisutch*)	16–24	Limiting concentration	1.3–2.0
Rainbow trout (*Salmo gairdnerii*)	11.1–20	Limiting concentration	1.05–3.7
Worm (*Nereis grubei*)	21.7–26.3	28 days LC$_{50}$	2.95
Worm (*Capitella capitata*)	21.7–26.3	28 days LC$_{50}$	1.50
Amphipod (*Hyalella azteca*)	—	Threshold	0.7
Amphipod (*Gammarus fasciatus*)	—	Threshold	4.3

[a]Compiled from Erichsen Jones (1964).
[b]Limiting values for existence.

Table 8.5 Examples of recommended minimum concentrations of dissolved oxygen (DO).

Estimated natural seasonal minimum DO	Recommended minimum DO for selected levels of fish protection (mg L^{-1})			
	Yearly maximum	High	Moderate	Low
5	5	4.7	4.2	4.0
6	6	5.6	4.8	4.0
7	7	6.4	5.3	4.0
8	8	7.1	5.8	4.3
9	9	7.7	6.2	4.5
10	10	8.2	6.5	4.6
12	12	8.9	6.8	4.8
14	14	9.3	6.8	4.9

Source: Based on US EPA (1973); US EPA (2019).

Table 8.7 Water quality criteria for minimum ambient dissolved oxygen concentration (mg L^{-1}) to protect freshwater aquatic life.

Parameter	Cold water criteria		Warm water criteria	
	Early life stages[a]	Other life stages	Early life stages	Other life stages
30 day mean	NA	6.5	NA	5.5
7 day mean	9.5	NA	6.0	NA
1 day minimum[b]	8.0	4.0	5.0	3.0

[a] This includes all embryonic and larval stages and all juvenile forms to 30 days following hatching
[b] All minima should be considered as instantons concentrations to be achieved at all times
Source: US EPA (1986).

rate, and thus oxygen demands, of organisms increases, but the solubility of oxygen in the water mass decreases.

The setting of criteria or standards can be approached in a number of different ways. With one method, the criterion can be based on the seasonal minimum dissolved oxygen concentrations (see Table 8.5). Thus, this method assumes that organisms in the area are adjusted to these seasonal variations, and criteria can be accordingly related to the seasonal minima. Another method is to use the solubility at different temperatures together with the metabolic needs of organisms at those temperatures (see Table 8.6). An example of the water quality criteria for dissolved oxygen to protect freshwater organisms in the United States as set by the US EPA is shown in Table 8.7.

Hydrogen Sulfide (H$_2$S)

Hydrogen sulfide can occur as a significant pollutant in some wastewater discharges, as well as a natural

Table 8.6 Example of minimum acceptable concentrations of dissolved oxygen in freshwaters.

Temperature (°C)	DO complete saturation (mg L^{-1})	Minimal levels for protection of aquatic life	
		mg L^{-1}	Saturation (%)
36.0	7	5.8	82.9
27.5	8	5.8	72.5
21.0	9	6.2	68.9
16.0	10	6.5	65.0
7.7	12	6.8	56.7
1.5	14	6.8	48.6

Source: Based on US EPA (1973); US EPA (2019).

component of swamps, eutrophic lakes, and dead zones in the oceans. However, in our considerations of deoxygenation of natural waters by pollution, hydrogen sulfide is a product of anaerobic respiration by natural microorganisms in the degradation of polluting organic matter. It is produced at reduced dissolved oxygen levels in aquatic areas by organic discharges, discharge of plant nutrients, and the accumulation of plant and animal detritus as shown in Figures 8.1 and 8.2. Anoxia, in the absence of oxygen, will produce hydrogen sulfide, but in natural systems hypoxic systems usually contain some anoxic areas that produce hydrogen sulfide as well. It follows that natural ecosystems affected by reduced dissolved oxygen will often also be influenced by hydrogen sulfide.

Hydrogen sulfide in aqueous solution undergoes the following dissociation:

$$H_2S \rightarrow H^+ + HS^- \rightarrow H^+ + S^{2-} \quad (8.15)$$

At temperatures of 25 °C and pH <6, most of the hydrogen sulfide is dissolved as the undissociated form. But at pH >7.8, the bisulfite ion predominates.

The toxicity of hydrogen sulfide at intermediate pH values is shown in Table 8.8. It can be seen that hydrogen sulfide is very toxic to a wide range of aquatic organisms. In most situations where reduced dissolved oxygen occurs, there will be the two effects acting on organisms in the area, that is, reduced dissolved oxygen and toxic effects due to hydrogen sulfide.

Long-term tests have shown that many adverse effects can occur at very low concentrations of hydrogen sulfide (Poole et al. 1978). For example, *Gammarus pseudolimnaeus*, a freshwater amphipod, shows effects on reproduction and growth of young at concentrations as low as 2 μg L^{-1}. It is noteworthy that this value is 10 times lower than the 96 hours LC$_{50}$. Similarly, other adverse effects have been found with fish and eggs at low concentrations. Poole et al. (1978) have reported that generally

Table 8.8 Lethal effects of hydrogen sulfide on freshwater aquatic fauna.

Species	96 hours LC$_{50}$ (μg L^{-1})	pH	Temperature (°C)
Gammarus pseudolimnaeus	22	7.7–7.9	17.8–18.1
Asellus militaria	1070	—	—
Crangonyx richmondensis laurentianus	840	—	—
Gammarus pseudolimnaeus	59	7.5	15.0
Baetis vagans	20	—	—
Ephemera simulans	316	—	—
Hexagenia limbata	11	—	—
Notropis cornutus	278[a]	6.7–7.9	13.4–14.1
Carassius auratus	110	7.8	15.0

[a]Recalculated from total sulfide concentration.
Source: Poole et al. (1978).

the toxicity resulting from hypoxia and hydrogen sulfide together is less than additive. The toxic effect of hydrogen sulfide is believed to result from inhibition of metalloenzymes by reaction of the substance with the metals present.

Effects of Reduced Dissolved Oxygen and Associated Factors on Communities and Ecosystems

There are several factors associated with dissolved oxygen reduction, which have an ecological impact. Dissolved oxygen reduction usually results from the addition of organic matter to a water body. However, organic matter itself, apart from its secondary effect of reducing dissolved oxygen, has an ecological impact. It results in an additional supply of food and energy for aquatic organisms, and hydrogen sulfide, as well as other substances, are produced. So the ecological effects usually associated with organic discharges are due to the combined effect of reduced dissolved oxygen, organic enrichment, and the toxic effect of hydrogen sulfide and other substances. Although laboratory experiments can be conducted to demonstrate the different effects of these factors, similar distinctions cannot be made in affected natural ecosystems.

It should be noted that in the longer term, the degradation of organic matter releases nitrogen and phosphorus salts. These substances play an important role in the eutrophication process, and thus organic enrichment is connected with eutrophication. This aspect is considered in this chapter (see Section 8.3).

Overall Impact on Aquatic Systems

In broad terms, freshwater, estuarine, and marine systems exhibit similar responses to organic enrichment and dissolved oxygen reduction. The basic metabolism of an aquatic system is altered by the addition of organic matter. As discussed previously, photosynthesis and respiration are the fundamental chemical processes occurring in aquatic ecosystems, and these processes regulate and control the living components. The discharge of organic matter to an aquatic area results in an increase in respiration and, in addition, may reduce photosynthesis. There will be an alteration in the basic metabolism in the aquatic area as measured by the ratio of photosynthesis to respiration.

Odum (1956) has classified a wide variety of aquatic systems according to their photosynthesis to respiration ratio as shown in Figure 8.10. Water receiving a high level of organic pollution, often described as polysaprobic, is located in the bottom right-hand area of the diagram where respiration greatly exceeds photosynthesis. Respiration in this zone is due to large numbers of bacteria and protozoa and counts in excess of one million per milliliter (mL) have been obtained. As the quantity of organic matter decreases and the availability of plant nutrients (nitrogen and phosphorus salts) increases, photosynthesis and plant growth are stimulated leading to the development of a photosynthesis to respiration ratio approaching 1 (see Figure 8.10). When the organic matter has been removed by respiration, the conditions are described as oligosaprobic.

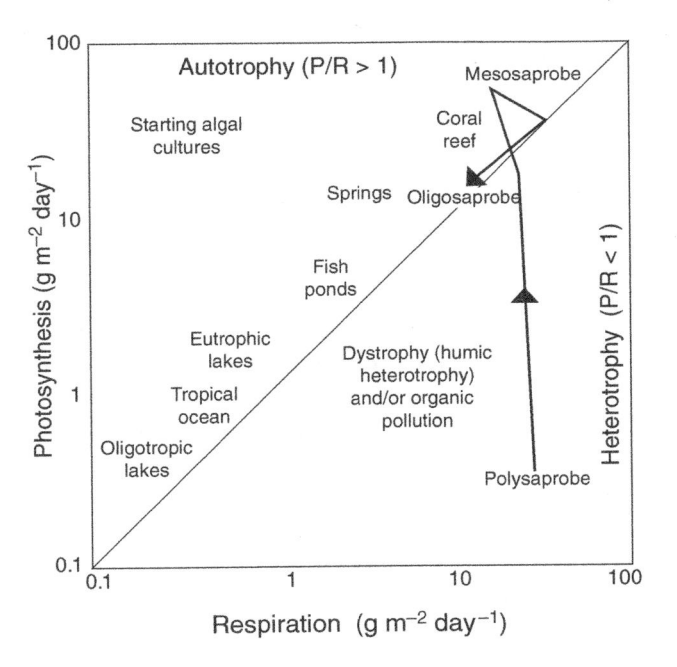

Figure 8.10 A diagram showing a functional classification of aquatic communities according to the total metabolism and relative dominance of photosynthesis and respiration
Source: Odum (1956).

The increase in respiration resulting from additions of organic matter mentioned above is mainly due to bacteria and protozoa, but there is a large population increase in many other heterotrophic organisms. The major larger heterotrophs include the polychaetes (e.g. *Capitella* sp.) and tuberficids. Overall, there is an increase in the biomass even when the dissolved oxygen levels drop to zero (Poole et al. 1978).

When the dissolved oxygen reduction extends to the lethal levels of aquatic organisms in an aquatic system then "**kills**" occur. These are more common in rivers and streams due to the limited volume of water that is present compared with the discharge volume. In oceanic areas, the volume of water is usually very large compared with the discharge volume and thus dissolved oxygen reductions are less frequent. Nevertheless, stratification can occur isolating the bottom layers, which may then become anoxic (see Figures 8.1 and 8.2, and Box 8.1).

Changes in the structure of aquatic communities are related, to a certain extent, to the mobility of the organisms in the area. Fish can undertake avoidance movement to remain clear of areas where there are low dissolved oxygen and concentrations of hydrogen sulfide. But large *kills* of fish often occur in rivers where there may be dissolved oxygen reduction and fish have restricted movement. Fish *kills* in oceanic areas are less common due to comparatively unrestricted movement, the lengthy period it usually takes to develop anoxic conditions, and the restriction of anoxic conditions to bottom waters in many areas. Usually large bodies of water exhibit *kills* of immobile benthic animals (Garlo et al. 1979) and, under extreme conditions, whole groups of organisms (e.g. fish) may be removed from an ecosystem. Also, anoxic conditions on the continental shelf in North Eastern United States, see Box 8.1, have caused avoidance movements, which have altered the migration patterns of finfish and lobsters (Steimle and Sindermann 1978; Garlo et al. 1979).

Different organisms exhibit different ranges of oxygen consumption related to the availability of dissolved oxygen in the water mass. Similarly, different species exhibit different lower lethal limits with reduced dissolved oxygen concentrations. Therefore, in an area where oxygen reduction occurs, we would expect to see a differentiation of species and a reduction in species diversity related to the different response to reduced oxygen levels.

The presence of organic matter from a discharge alters the type and availability of food and the community will also respond to this type of change. Welch (1980, p. 337) has described broad food preferences and indicated how changes in food supply could cause shifts in community structure. The community members can be classified as follows: (i) Detrital feeders, e.g. the net spinning caddis flies, aquatic sow bug, chironomids, clams, snails, some mayflies, and blackflies (largely collectors of fine particulate matter); (ii) Grazers, e.g. most stone flies, some mayflies, case-building caddis flies, and snails; and (iii) Predators as represented by the dragonflies, leeches, a few stone flies, beetles, some midges, and a few caddis flies.

An increase in the amount of detritus present through the discharge of organic matter would result in an increase in the detrital feeders, especially at the expense of the grazers. In addition, predators are often reduced or eliminated by low dissolved oxygen concentrations in the water mass due to their generally greater sensitivity.

Hawkes (1979) has summarized the tolerance of a wide variety of freshwater riffle organisms to organic enrichment. Three major groupings have been described – the **tolerant** group which includes sludge worms, certain midges, leeches, and certain snails; the **mildly tolerant** group which includes most snails, sow bugs, skuds, blackflies, crane flies, fingernail clams, dragonflies, and some midges; and the **intolerant** group which includes mayflies, stone flies, caddis flies, riffle beetles, and hellgrammites.

The reproduction rate of organisms is an important factor affecting their abundance in polluted conditions. Welch (1980, p. 337) has reported that although oligochaete worms are adversely affected by dissolved oxygen reduction, large numbers occur because of the rapid reproduction rate in a wide range of differing environmental conditions.

Biotic Indices

There are a number of quantitative measures of the biotic changes that occur in an aquatic ecosystem from the effects described above. Predictably, a relationship should exist between water composition and measures of changes in the aquatic ecosystem. As a result of this, biotic indices are often used together with physicochemical measures to characterize the quality of water in an area (Sladecek et al. 1982).

The measure with the longest history is the **saprobic** or **saprobien** system, initially developed for continental Europe. Sladecek (1979) has defined **saprobity** as the state of water quality with respect to the content of putrescible organic material as reflected by species composition of the community. This system has been subject to many modifications over time and some of these are described by Sladecek (1979). The method basically consists of dividing the various stages of recovery of a stream polluted by an organic discharge into zones related to the organisms present. There are basically three zones: **polysaprobic**, **mesosaprobic**, and **oligosaprobic**, but further subdivision of these zones can be used.

The **polysaprobic** zone is characterized by high concentrations of decomposable organic matter, the absence of oxygen, and the presence of hydrogen sulfide. The biotic community is restricted to a few groups – primarily large numbers of bacteria and protozoa. The **mesosaprobic** zone has a well-established oxidation process occurring and is often subdivided into two zones. The **α-mesosaprobic** zone may contain substantial quantities of oxygen, that is often in excess of 50% saturation, and hydrogen sulfide is not present. Biologically, it is rich in bacteria and protozoa. The **ß-mesosaprobic** zone has continued oxidation or mineralization with the oxygen content never being less than 50% saturation. However, in this zone there is a decrease in the number of bacteria and protozoa and an increase in the diversity of plants and animals. In the **oligosaprobic** zone, the oxidation or mineralization processes are complete and the organic content is low. It is biologically characterized by low numbers of bacteria and a high species diversity including fish.

Sladecek (1979) has described more advanced applications of the saprobien system. The system can be placed on a more quantitative basis by the use of the **saprobic valency**. This takes account of the fact that single species rarely are represented in only one zone or one saprobic level. Each species is considered to be distributed in an approximately normal fashion in the saprobic scale and is assigned numerical values according to this. Thus, each species has a maximum valency in a particular area of the saprobic scale and a distribution on either side of this.

To apply this system requires a detailed knowledge of the characteristics of individual species. But it should be noted that there are a number of criticisms of the Saprobic System. For example, the major criticisms have been listed by Persoone and De Pauw (1979) as:

(1) necessity to identify organisms to species level;
(2) paucity of basic knowledge of the ecological characteristics and requirements of individual species and communities;
(3) failure to properly quantitate species relationships;
(4) failure to accommodate the uniqueness of each stream, each pollutant, and each problem; and
(5) restriction of the use of saprobic systems to pollution caused by municipal sewage or similar acting organic wastes.

These factors limit the application of the system and indicate precautions necessary when it is used.

Somewhat similar systems have also been developed elsewhere to evaluate stream water quality to support natural ecosystems. While these indices are usually based on the aquatic macroinvertebrates, they are generally applicable to evaluation of deoxygenation effects. A summation is shown in Table 8.9. (It is also included as Table 7.3.) A range of biotic indices has been developed for different purposes and different areas of the world (Abbasi and Abbasi 2012). These indices are focused on different biotic groups, as shown in the biotypes column, utilize different

Table 8.9 An overview of the major biotic indices based on aquatic macroinvertebrates.

Biotic index	Biotypes sampled	Sampling equipment	Sampling protocol	Taxonomic level	Regions in which currently used
Beck's biotic index	All, combined	Not stipulated	Nonquantitative	Species	—
Trent biotic index	All, combined	Hand net	Nonquantitative	Family + genus + species	—
Hilsenhoff's biotic index	Stones-in-current	Hand net	Quantitative >100	Genus + species	USA
Belgian biotic index	All, combined	Hand net	Nonquantitative	Family + genus	Belgium and surrounding countries
Macroinvertebrate community index	Stones-in-current	Hand net/surber	Nonquantitative	Genus	New Zealand
Iberian BMWP	Lotic + lentic combined/separate	Hand net	Nonquantitative	Family	Spain, Italy
Stream invertebrate index	6 per-defined	Hand net	Nonquantitative, 100 organisms	Family	Australia
Danish stream fauna index	All, combined	Hand net (500 μm)	Semiquantitative, 12 samples	Family + genus	Denmark, Sweden
Balkan biotic index	All, combined	Benthos net	Quantitative	Family + sub-family + genus	Serbia

Source: Based on Abbasi and Abbasi (2011).

Ecological Effects of Deoxygenation Related to Spatial and Temporal Patterns of Change

Reduction in dissolved oxygen content of water and the other effects associated with organic enrichment often lead to a systematic sequence of physiochemical and biological changes related to the source of discharge, time, and environmental conditions. These are best considered by discussion of related groups of situations as set out below.

Streams

With streams, there is a unidirectional flow of water, which leads to the formation of the dissolved oxygen sag described previously. This gives a set of zones grading into one another but having different dissolved oxygen content, organic matter, and biota. In Figure 8.11, sections a and b show variations of chemical components as related to a discharge of organic matter and consequent changes in dissolved oxygen. Related and consequent biological changes are shown in Figure 8.11, sections c and d. The stream section below the discharge point can be classified into various zones using the indices previously described.

Lakes and the Open Sea

The classification and systematic investigation of ecological changes resulting from organic discharges have been most successful in flowing streams. The unidirectional flow of water gives a clear sequence of events, which simplifies study. Such events are not so clearly systematic in lakes and the sea and thus investigations of these areas have not yielded clear results. In lakes and the sea, there is a multidirectional flow of water, which, in the sea, is related to tides and, to a lesser extent, ocean currents. Changes in water level in the intertidal zone, as well as turbulence due to wave action and salinity gradients, can also cause variations in the ecological patterns in these areas. Also, most marine organisms have planktonic larvae. It follows that factors which influence the planktonic larval stage may have an important impact on the distribution of organisms. These factors add to the complexity of interpreting changes in biota in marine areas, which are more subtle than in freshwater areas.

The oceans usually contain large reservoirs of dissolved oxygen due to the comparatively large volume of water as compared to the volume of freshwater streams. Thus, comparatively large volumes of organic matter are generally needed to cause reductions in dissolved oxygen. Many of the effects due to discharges of organic matter into the sea occur in areas of relatively shallow and confined water.

Nevertheless, somewhat similar ecological changes occur in lakes and the sea as compared to rivers. These are summarized in Table 8.10. There is a reduction in the number of species and diversity in heavily polluted areas with an increase in the biomass of some tolerant organisms. With increasing distance from the discharge point, this zone leads gradually through to clean conditions with the intervening situation showing intermediate circumstances. Pearson and Stanley (1979) have related this situation to Eh values in sediment, which decrease as the organic content in marine areas increases. There are various macroinvertebrate species involved in the different zones of pollution resulting from discharges of organic matter. In addition, a decrease in the size of animals has been observed by Pearson and Rosenberg (1978) with increasing organic pollution. Increasing organic matter was also found to cause a change in the trophic structure of the affected communities with the proportion of deposit feeders increasing while carnivores and omnivores decrease.

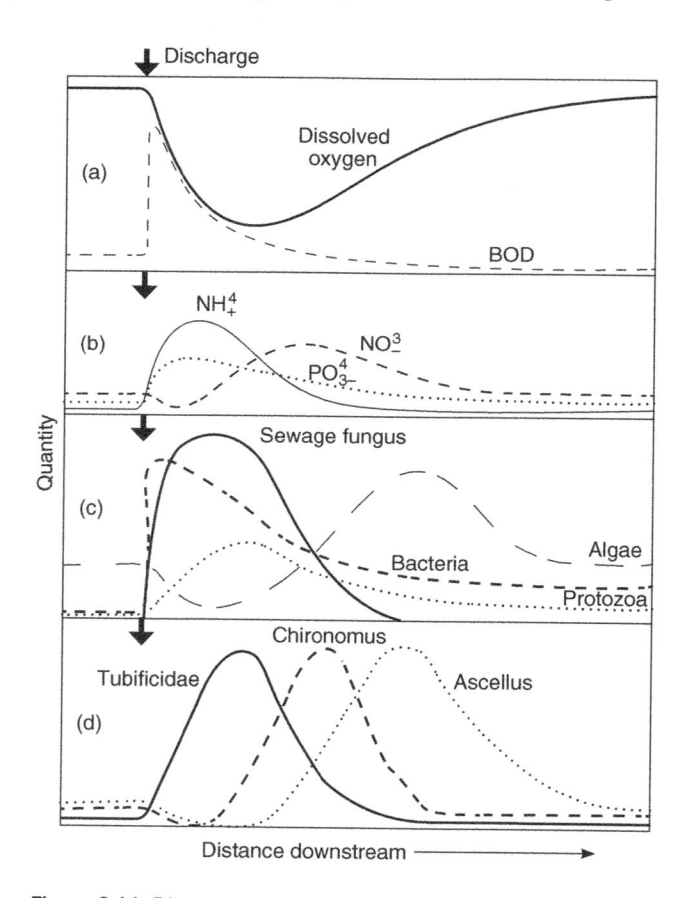

Figure 8.11 Diagrammatic representation of the effects of an organic discharge on the downstream section of a stream: (a) and (b) chemical changes; (c) microorganisms; and (d) other organisms. Source: Compiled from Hynes (1960)

Table 8.10 Changes in dominant macroinvertebrate fauna with increasing organic enrichment in some marine areas.[a]

Area/effluent type	Normal	Transitory	Polluted	Grossly polluted
Lochs in, Scotland, and Fjords in Sweden/paper mill effluent	*Nucula Amphiura Terebellides Rhodine Echinocardium Nephrops*	*Lapidoplax Corbula Goniada Thyasira Pholoe Chaetozone Anaitides Pectinaria Myriochele Ophiodromus*	*Capitella Scolelepis*	No fauna. Surface covered by a fiber blanket
Off Marseilles/sewage effluent	Not described	*Nereis caudata Staurocephalus rudolphii Cirriformia tentaculate Corbula gibba Thyaira flexuosa* Rich polychaete fauna	*Capitella capitata Scolelepis fuliginosa*	No fauna

[a]Based on Bellan (1970) and Pearson and Rosenberg (1978).

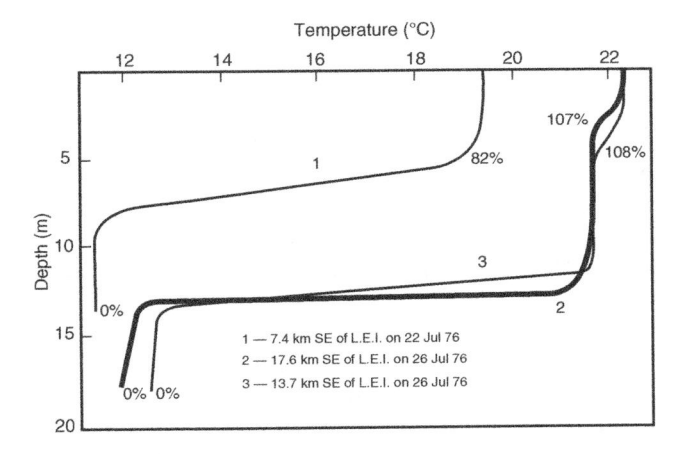

Figure 8.12 Temperature profiles and associated dissolved oxygen saturation values at ocean stations in the vicinity of Little Egg Inlet, New Jersey. (1) 7.4 km SE of L.E.L on 22 July 1976; (2) 17.6 km SE of L.E.L. on July 1976 and (3) 13.7 km SE of L.E.L. on 26 July 1976. Source: Garlo et al. (1979).

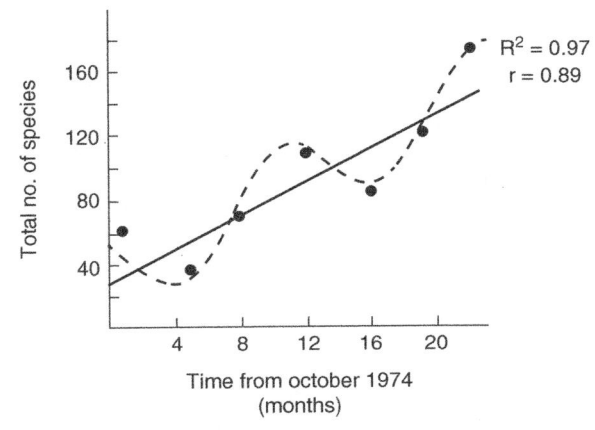

Figure 8.13 Changes in number of species over time in an estuarine area in tropical Australia. Source: Saenger et al. (1980).

Vertical Profiles (See also Seasonal Variations and Vertical Profiles of Dissolved Oxygen)

Vertical stratification of water bodies, often due to solar radiation at the surface as shown in Figure 8.12, leads to the isolation of bottom waters causing dissolved oxygen reduction as shown diagrammatically in Figures 8.1 and 8.2. This can occur in water bodies ranging in size from the ocean itself to small pools. For example, Figure 8.12 shows vertical profiles of temperature and dissolved oxygen in New York Bight in 1976, and Box 8.1 describes the general factors involved. The vertical distribution of fish and of other mobile organisms in these areas is related to these factors.

Temporal Patterns

Temporal patterns of physical changes in a water body can take a wide variety of forms particularly in lakes in relation to seasons as described in Section 8.3 on Physical Factors Affecting Nutrient Enrichment and Eutrophication. Biological changes are strongly influenced by these changes, as well as the influence of pollution. Firstly, the population in many aquatic areas including those affected by organic matter exhibit seasonal patterns of change. Figure 8.13 shows an annual cycle in the number of species or macrobenthos in an estuarine area. The species diversity also shows a steady increase over approximately two years of investigation. This is believed to be part of a longer time sequence of change in which the species-number increase is due to recovery from freshwater flooding of the estuary, which occurs on a periodic basis. These changes provide a basic natural sequence of change in aquatic areas on which changes due to dissolved oxygen and organic matter enrichment are superimposed.

Seasonal effects are particularly marked where rainfall shows distinct seasonal variations. Reduced rainfall leads to a lower stream flow and therefore lower dilution of discharges and a greater impact on dissolved oxygen and the organic matter content of the area. This pattern of change leads to a consequent change in the community structure of biota in the area (see Figure 8.14). Diurnal changes in

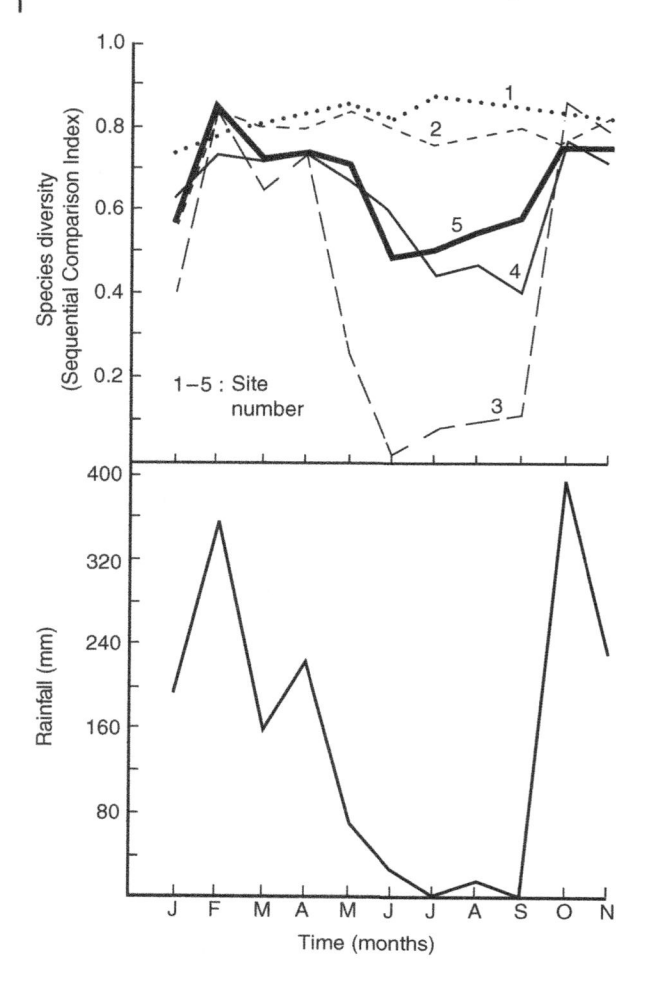

Figure 8.14 Variation of species diversity and rainfall in a stream subject to a seasonal pattern of rainfall (Sites 1 and 2 are above the discharge and sites 3, 4, and 5 are in numerical order below the discharge). Source: McIvor (1976).

dissolved oxygen in some areas require biota able to tolerate the overnight minimums.

The initial discharge of organic matter into an area gives a temporal sequence of changes in dissolved oxygen, and associated factors, with consequent effects on biota. These are generally related to changes observed with distance from a source. Recovery of an area after discharge has ceased shows a somewhat similar reverse sequence, which has been described by Pearson (1980).

8.3 Nutrient Enrichment and Eutrophication

Introduction

The enrichment of aquatic areas with plant nutrients is an important process in aquatic pollution and the major significant adverse effect of this is described as **eutrophication**. Eutrophication was described by Weber in 1907 when he introduced the descriptive terms **oligotrophic, mesotrophic, and eutrophic** (Hutchinson 1969). These terms describe the eutrophication process as a sequence from a clear lake to a bog by enrichment with plant nutrients and increased plant growth. Since that time there have been many descriptions and criteria for these terms, and the introduction of new terms has proliferated (Farley 2012). Hutchinson (1969) has suggested that the trophic state of an aquatic area should be considered in terms of the whole water system including the catchment and bottom sediments rather than the water alone. The nutrient status should be related to total nutrients present, and potentially available, rather than simply the concentration in water at any particular time.

The OECD has defined eutrophication as *the nutrient enrichment of waters which results in stimulation of an array of symptomatic changes among which increased production of algae and macrophytes, deterioration of fisheries, deterioration of water quality and other symptomatic changes are found to be undesirable and interfere with water uses* (Wood 1975, p. 238).

In the natural eutrophication process, plant detritus, salts, silt, and so on from a catchment are entrained in runoff water and deposited in the water body over geological time. This leads to nutrient enrichment, sedimentation, infilling, and increased biomass. Figure 8.15 illustrates in general terms how eutrophication is related to aging. The final stage of the process results in the formation of bogs, swamps, and the extinction of the water body. It is believed the process slows with increasing time due to increased turbidity causing limited light penetration and a consequent fall in primary production.

Water bodies with little flushing, such as lakes, dams, and enclosed seas, become eutrophic through nutrient enrichment over an extended time scale as described above. This follows the generally accepted eutrophication pattern. Nutrient enrichment also occurs in situations where infilling and increased sedimentation leading to the formation of swamps, bogs, and wetlands is less likely due to comparatively rapid water movement and flushing. This situation arises in streams, estuaries, the continental shelf, and the open seas. Nevertheless, these water bodies may show many of the characteristics of eutrophication and are often referred to in eutrophic terms. However, it is noteworthy that nutrient-enrichment processes can occur in the oceans as well leading to the enrichment of bottom waters with plant matter and organic carbon. This is described earlier in Section 8.2 and leads to the formation of **"dead zones"** in the bottom waters of the oceans.

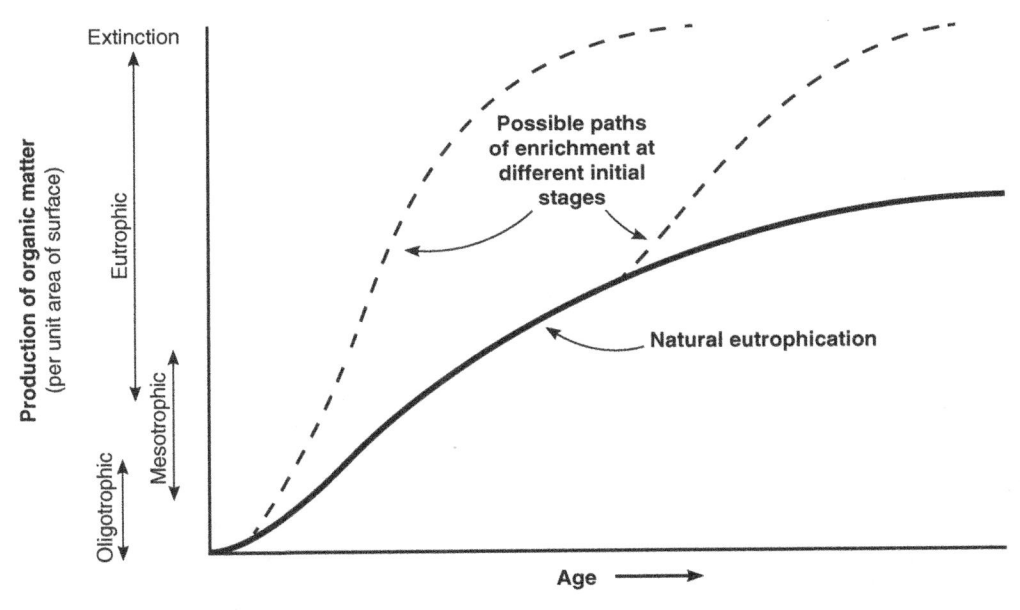

Figure 8.15 Hypothetical curve of the course of eutrophication in a water body. The broken lines show the possible course of accelerated eutrophication when enrichment from pollution occurs.

Nutrient enrichment and eutrophication have been greatly accelerated by human activities. Perhaps the greatest chemical innovation ever made, with enormous implications for human society and the global environment, was made by Fritz Haber and Carl Bosch in the first decade of the 1900s as mentioned in Chapter 1 of this book. Haber was a controversial scientist who was awarded the Nobel Prize for Chemistry in 1918 for his studies of chemical reactions. Derived from this research was the Haber-Bosch process for the industrial fixation of nitrogen from the atmosphere using hydrogen and relatively large amounts of energy (see Figure 8.16)

$$N_2 + 3H_2 \rightarrow 2NH_3 \tag{8.16}$$

It is surprising that nitrogen is in short supply for food crops since it comprises about 78% of the atmosphere. But most plants cannot readily utilize this form of the nutrient since it requires a large amount of energy to be converted into organic nitrogen in plant tissue by the photosynthetic process. Some plants have this capacity, but these are usually in low numbers in the natural environment.

In fact, many lakes have been rapidly enriched with nutrients over the last 100 years due to pollution (Vollenweider 1971; Allen and Kramer 1972, p. 457; Farley 2012). Discharges, such as domestic sewage, septic tank runoff, some industrial wastes, urban runoff, runoff from agricultural and managed forests, and animal wastes, contain plant nutrients that often lead to nutrient enrichment and accelerated eutrophication.

Figure 8.16 Industrial plant for the fixation of atmospheric nitrogen by the Haber-Bosch process to produce ammonia for use as a crop fertilizer. Source: Photograph by Des Connell (Author).

Eutrophication can cause quite a number of important problems in water use (Istvánovics 2009; Lemley and Adams 2019). An increase in the populations of plants can lead to a decrease in the dissolved oxygen content of the water on plant death and decomposition by microorganisms. This decreases the suitability of the area as a habitat for many species of fish and other organisms. The increase in turbidity and color, which occurs during eutrophication, renders the water unsuitable for domestic use or

Figure 8.17 A shallow lake which is eutrophic and exhibits the excessive growth of aquatic plants. Source: Photograph by Des Connell (Author).

Figure 8.18 Aquatic plant growth of blue-green algae on the surface of a water body used to supply drinking water. Source: Photograph by Des Connell (Author).

difficult to treat to a suitable standard for this purpose (see Figure 8.17). Odors are also produced by many of the algal growths, which create problems in domestic use. Blooms, pulses, and so on of aquatic plants become more frequent and with advanced nutrient pollution can lead to the production of toxic blooms of **blue-green algae** also described as **Cyanobacteria** (see Figure 8.18). If the blooms are toxic, they can lead to the death of fish and other aquatic organisms and also of terrestrial wild animals and livestock using the water. In addition, the use of the water as drinking water can become a health hazard for human society. Floating macrophytes and algal scums can render a water body unsuitable for recreation and water sports and also cause navigation problems as shown in Figures 8.17 and 8.18.

Nutrients and Plant Growth

If the growth of algal cells is not limited by any environmental or nutrient factor, then population growth occurs according to an exponential function. Thus, the number of cells at any time regulates the rate of growth:

$$dN/dt = \beta N \qquad (8.17)$$

Where N is the number of cells, t the time, and β the growth rate constant. This expression can be integrated to give:

$$N_t = N_0 e\beta^t \qquad (8.18)$$

Where N_0 is the number of cells at time zero and N_t the number of cells at time t.

The growth rate is often expressed as the doubling time (t_d), which is constant under fixed conditions and can be derived from the equation (8.18) as

$$t_d = \ln 2/\beta \qquad (8.19)$$

Golterman (1975) has reported that the growth rate constants (β) for a wide variety of algae range from 0.12 to 3.55 ($\log_{10}$, day units) with doubling times from 2.0 to 97.7 days.

Exponential growth of the kind outlined above cannot be maintained for very long by algae due to limitations of various kinds. For example, elements, such as carbon, nitrogen, sulfur, hydrogen, and oxygen, are needed to construct plant tissue, particularly proteins and carbohydrates. In addition to these major elements, phosphorus, iron, magnesium, sodium, and a variety of other elements are needed to construct various vital components. Table 8.11 lists the relative quantities of essential elements which occur in plant tissue. While the composition of plant tissue is variable the general pattern is apparent from this data. The function of all of these elements in plant processes is unclear at the present time, but the complexity of plant growth processes suggests that there is a role for them.

In most aquatic areas, carbon and oxygen are readily available from carbon dioxide in the atmosphere, and hydrogen and oxygen can also be readily obtained from water. The other elements mentioned above are usually obtained from dissolved salts in the water or sediments. However, these substances are not always available in the quantities required to maintain maximum growth. For example, Table 8.11 shows a comparison of the relative quantities of the elements required for plant growth with their occurrence in river water. River water shows considerable variability, but these data indicate the general pattern encountered. This suggests that phosphorus and nitrogen are in comparatively short supply compared with all the other elements and that phosphorus is likely to be less available than nitrogen. This situation is generally applicable to aquatic areas. These elements are often

Table 8.11 Relative quantities of essential elements in plant tissue (demand) and their supply in river water.

Element	Demand plants (%)	Supply water (%)	Demand plants/supply water (approx.)
Oxygen	80.5	89	1
Hydrogen	9.7	11	1
Carbon	6.5	0.0012	5000
Silicon	1.3	0.00065	2000
Nitrogen	0.7	0.000023	30 000
Calcium	0.4	0.0015	<1000
Potassium	0.3	0.00023	1300
Phosphorus	0.08	0.000001	80 000
Magnesium	0.07	0.0004	<1000
Sulfur	0.06	0.0004	<1000
Chlorine	0.06	0.0008	<1000
Sodium	0.04	0.0006	<1000
Iron	0.02	0.00007	<1000
Manganese	0.0007	0.0000015	<1000
Boron	0.001	0.00001	<1000
Zinc	0.0003	0.000001	<1000
Copper	0.0001	0.000001	<1000
Molybdenum	0.00005	0.0000003	<1000
Cobalt	0.000002	0.000000005	<1000

Source: Vallentyne (1973).

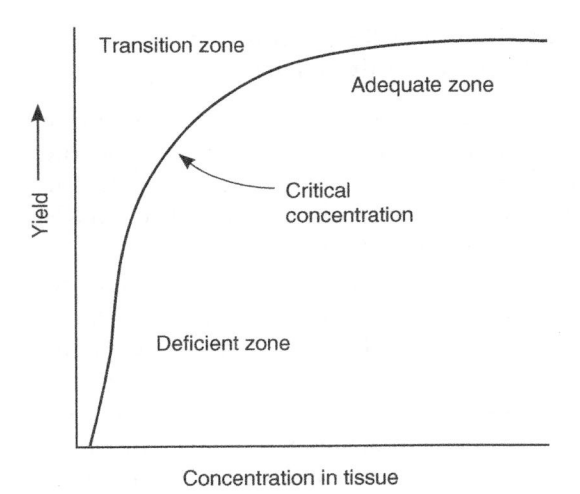

Figure 8.19 Schematic illustration of the relationship between plant growth and the concentration of an essential element in specific plant parts of a definite physiological age.

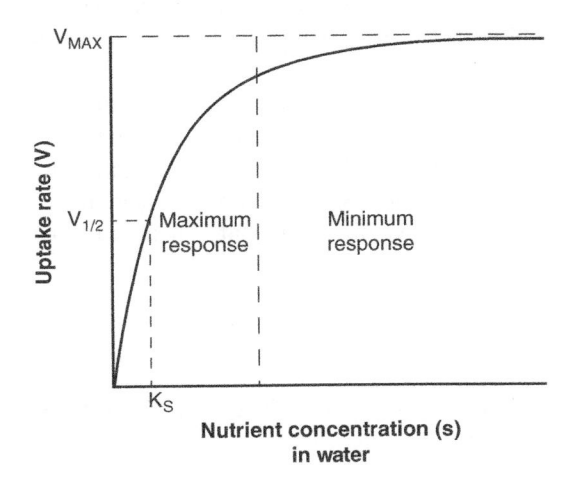

Figure 8.20 Relationship between uptake rate of nutrient and nutrient concentration in medium for aquatic plants as described by the Michaelis–Menten equation.

growth limiting and an addition of them to a water body will stimulate plant growth. Of course, these data are generalized and, in individual cases, other elements or combinations of elements may be limiting.

Figure 8.19 shows the pattern of uptake of an essential element, when there is a deficiency of that element and addition leads to a rapid increase in plant yield. The vertical section of the curve in Figure 8.19 shows rapid growth with little change in the plant tissue concentration of the element. When all needs are satisfied, the plant continues to take up nutrient, but this leads to no increase in yield. This is the adequate zone or the zone of luxury consumption.

The uptake of nutrient by phytoplankton usually follows Michaelis–Menten kinetics as illustrated in Figure 8.20. This can be expressed mathematically as:

$$V = (V_{max}\, S/(K_S + S)) \qquad (8.20)$$

Where V is the rate of nutrient assimilation; V_{max} the maximum rate of nutrient assimilation; S the nutrient concentration, and K_S the nutrient concentration at $V_{1/2}$ (see Figure 8.20).

Phytoplankton shows a maximum growth response at low concentrations and a minimum response at somewhat higher concentrations (see Figure 8.20). Vollenweider (1971) has compiled information on the concentrations of nutrients required by various algal species and concluded that the phosphorus requirements increased in order from *Asterionella formosa, Tabellaria, Fragilaria, Scenedesmus,* and *Oscillatoria rubescens*.

In assessing the effects of nutrient discharges on a water body, it is necessary to establish which are the limiting nutrients and their relationships to plant growth. This can be done by bioassay of cultures of the appropriate plant, enriched with an essential element singly or in combinations with others. Even so, there are difficulties in applying the results of bioassay experiments to actual situations since the test water may not accurately represent the water body. Water composition in natural bodies can show large variations with environmental conditions.

Sources and Transformations of Nitrogen and Phosphorus in Aquatic Systems

The availability of nitrogen and phosphorus to growing plants depends on a complex set of biologically mediated reactions. Nitrogen occurs in the aquatic environment in a wide variety of forms and chemical combinations involving different oxidation states. Organic nitrogen is bound into cellular constituents of living organisms, for example proteins, purines, peptides, and amino acids, while inorganic nitrogen (e.g. ammonia, nitrite, and nitrate) and nitrogen gas is dissolved in the water mass. These components are linked as illustrated in Figure 8.21. The transformation in the water mass of inorganic nitrogen in the form of NH_3 (ammonia) and NO_3^- (nitrate), as well as other N forms into organic nitrogen, occurs by photosynthetic growth of aquatic plants (see Figure 8.21). The reverse of this process results in the formation of ammonia from organic matter by a number of mechanisms involving cell autolysis, microorganisms, and excretion from large organisms. Ammonia can be lost from water by volatilization, but oxidation results in nitrification mainly by microorganisms and produces nitrate, which is nonvolatile and very soluble. Nitrate may undergo the denitrification reaction resulting in a loss of nitrogen gas to the atmosphere. Nitrogen fixation can be carried out by various organisms but is believed to be an adaptive process, which is only important to the overall process when nitrogen is in limited supply.

Phosphorus exists in a single oxidation state as inorganic phosphorus or organic phosphorus. Inorganic forms are mainly orthophosphate (PO_4^{3-}) and polyphosphates. Organic forms are usually associated with complex cellular substances and most phosphorus in natural waters is in the organic form. The inorganic forms, particularly orthophosphate, are readily assimilated during photosynthesis. Figure 8.22 summarizes the main phosphorus transformations in aquatic areas.

Sediments play a major role in the availability of phosphorus in many aquatic areas. A high proportion of phosphorus is removed from the water mass by sorption onto sediment minerals. One fraction adsorbs onto anionic sites and another into the crystal lattice structure by substitution of hydroxyl ions. The principal factors controlling this process are oxidation–reduction potential, pH values, and concentrations of other substances. A large proportion of phosphorus is adsorbed onto ferric hydroxide and oxides which dissolve, releasing the phosphorus at low redox potentials. These conditions result when the dissolved oxygen is less than $2\,mg\,L^{-1}$. Conversely, phosphorus is adsorbed at dissolved oxygen concentrations greater than $2\,mg\,L^{-1}$. Dissolved phosphate in natural waters is usually at a minimum at pH values of 5–7 (Brezonik 1972).

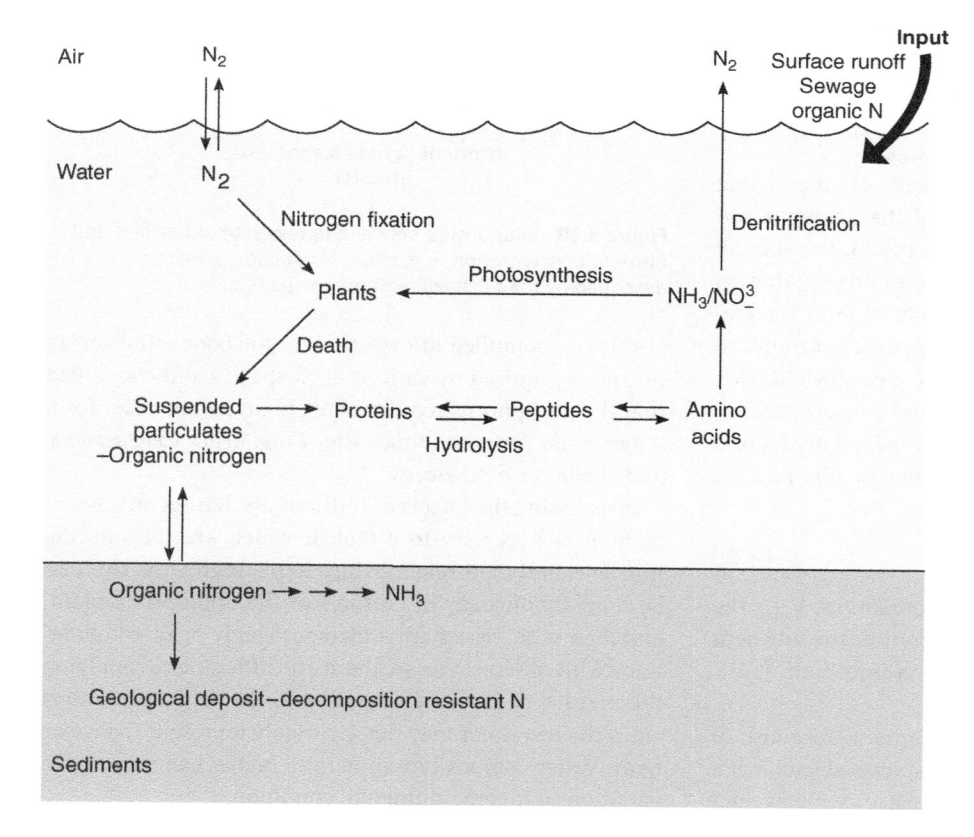

Figure 8.21 Simplified nitrogen cycle showing main molecular transformations in a water body.

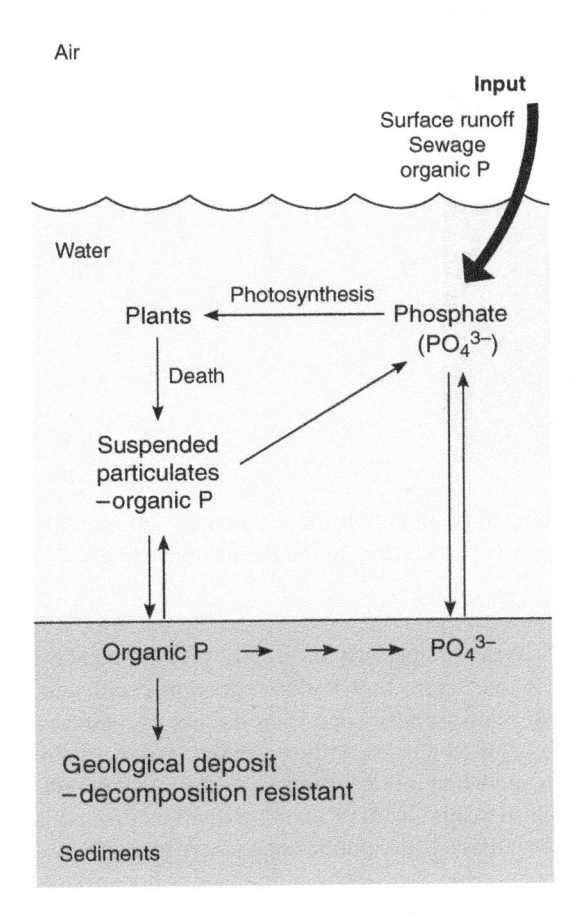

Figure 8.22 Simplified phosphorus cycle in aquatic areas illustrating the main molecular transformations.

Table 8.12 Summary of estimated nitrogen and phosphorus reaching a lake from the catchment.

Source	N	P	N	P
	(kg year^{-1} × 10^3)		(% of total)	
Municipal treatment facilities	9000	3200	24.5	55.7
Private sewage systems	2200	130	5.9	2.2
Industrial wastes[a]	680	45	1.8	0.8
Rural sources				
Manured lands	3670	1200	9.9	21.5
Other cropland	261	174	0.7	3.1
Forest land	197	20	0.5	0.3
Pasture, woodlot, and other lands	245	163	0.7	2.9
Groundwater	15 600	129	42.0	2.3
Urban runoff	2020	1570	5.5	10.0
Precipitation on water areas	3150	70	8.5	1.2
Total	35 205	5701	100.0	100.0

[a]Excludes industrial wastes that discharge to municipal systems.
Table does not include contributions from aquatic nitrogen fixation, waterfowl, and wetland drainage.
Source: Hasler (1968).

The metal content of water has also been shown to influence phosphorus dynamics since other metals apart from iron can complex with phosphorus.

In any aquatic system, there is a set of nutrient sources and processes for their removal. Removal can be in such processes as effluent loss, ground water recharge, harvesting of fish and plants, chemical volatilization, permanent loss in geological sediments, and so on. Major pollution sources of nutrients are surface and subsurface agricultural and urban drainage, animal waste runoff, as well as domestic and industrial waste effluents including sewage (see Table 8.12).

These wastes contain a variety of nitrogen- and phosphorus-containing substances. For example, nitrogen can occur as organic nitrogen, ammonia, nitrite, or nitrate, which are derived from protein, nucleic acids, urea, and other substances. Phosphorus compounds result from degradation of compounds such as nucleic acids and phospholipids and occur as inorganic phosphates (orthophosphate and polyphosphate), or organic phosphorus. In addition, phosphorus can originate from phosphate builders in detergents. These can be readily hydrolyzed to yield orthophosphate, which is readily assimilated by plants giving increased growth. However, improvements in wastewater treatment and reduced detergent phosphorus have resulted in a reduction in the load of phosphorus entering water bodies globally and an improvement in water quality.

Most sources of nitrogen and phosphorus in aquatic areas result from food production or waste in the form of sewage (e.g. see Table 8.12). Nitrogen and phosphorus are being substantially mobilized from new sources in the global environment, which are used to fertilize crops and increase food production. In fact, Ringerval et al. (2014) estimate that the quantities of new nitrogen and phosphorus now being used is 7 and 3.5 times, respectively, that being used about 60 years ago. New anthropogenic nitrogen originates from the industrial conversion of atmospheric nitrogen gas by fixation with hydrogen gas into nitrogenous fertilizers. New anthropogenic phosphorus is derived from the mining of phosphorus-rich minerals, which are converted into crop fertilizers and then taken up by food crops during large-scale production. The consumption of food leads to increased sewage discharges. Wastes originating from these sources enter aquatic areas causing eutrophication and enrichment of water bodies on a global scale as shown in Figure 8.23.

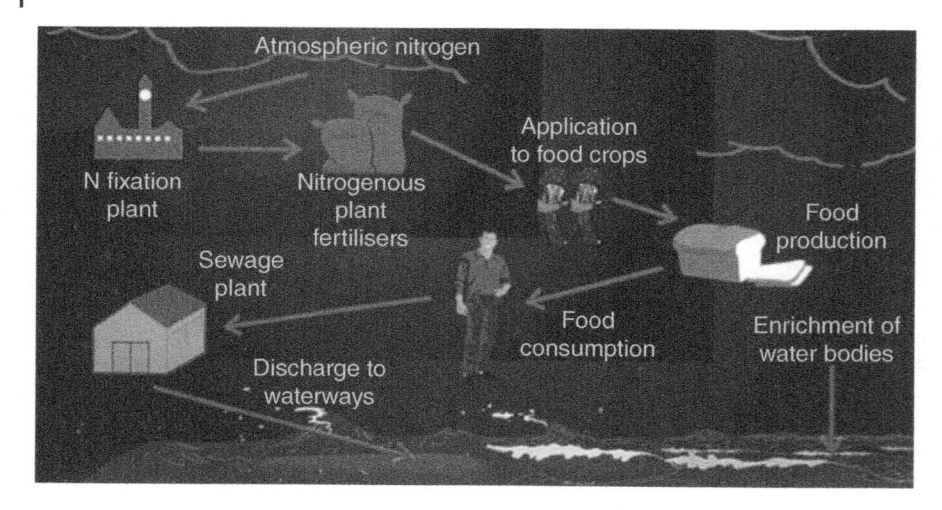

Figure 8.23 Pathway of new anthropogenic nitrogen fixed by the Haber-Bosch process through use as a crop fertilizer to food and human consumption to treatment in a sewage plant and discharge to a water body.

Global nitrogen fixation from all sources, natural and industrial, have been quantified by Fowler et al. (2013) who found that $413\,g \times 10^{12}$ of nitrogen were contributed annually to terrestrial and marine ecosystems of which industrial fixation is responsible for about half, $210\,g \times 10^{12}$ N. The majority of the chemical transformations of anthropogenic N are on land within soils and vegetation with the applications of nitrogenous fertilizers on crops.

Overall, there has been a significant alteration to the Earth's nitrogen cycle. Similarly, the Earth's phosphorus cycle has been significantly changed. In fact, nitrogen and phosphorus are being utilized from these new anthropogenic sources, and these elements are being transferred from the atmosphere and mineral deposits to the oceans and other water bodies. Here, they provide the basis for increased plant production and the eutrophication of waterways on a global scale (see also Case Study 2.1).

Treatment of nutrients in sewage is an important water pollution control process. Sewage can be treated to remove the BOD by **primary treatment** (anaerobic process), which removes most of the solids (e.g. primary sludges), and wastewaters separated in the primary treatment are treated further in the **secondary treatment** process (aerobic) as simplified by equation (8.2). Even so, this leaves a substantial amount of the nutrients in the wastewater produced, which require **tertiary treatment** to reduce to low environmental loadings of N and P. This level of treatment is relatively expensive but can be achieved by a number of treatment processes.

Physical Factors Affecting Nutrient Enrichment and Eutrophication

Solar Radiation and Primary Production

Primary production by photosynthesis depends on solar radiation reaching the Earth's surface. In the water environment, this radiation is attenuated on passing through the water according to the Beer-Lambert Law:

$$I_Z = I_0\, e^{-KZ} \qquad (8.21)$$

Where I_Z is the intensity at any depth; I_0 is the incident radiation; Z the depth; and K the extinction coefficient, which varies with turbidity and the presence of dissolved substances. Thus, light intensity decreases exponentially with depth and when it is less than 1% of the incident radiation results in insignificant primary production. From this depth to the surface is the photic zone where primary production occurs. However, at the immediate water surface, there is usually inhibition of growth due to excess light. The compensation depth occurs where photosynthesis is equal to respiration, but this concept is only applicable to well-mixed photic zones.

The incidence of light and related primary production vary with latitude with a minimum production at high latitudes. Water temperature also has a strong influence on the growth of aquatic plants. Phytoplankton species have an optimal growth temperature, but with mixed species in natural areas an increase in growth rate with temperature occurs for the whole community. The increase factor for each 10 °C is about double in accord with the Q10 Law (Welch 1980).

Thermal Stratification

Thermal stratification is a major factor in influencing water quality and it can occur due to solar heating of the surface waters in the photic zone forming an upper layer, the **epilimnion,** of a water body, which is then separated from the underlying waters, the **hypolimnion.** In this situation, the epilimnion and hypolimnion can exhibit different physicochemical characteristics (see Figure 8.24). The center of gravity of a stratified water body is lower than that of the same body in an unstratified state. For mixing to occur, the

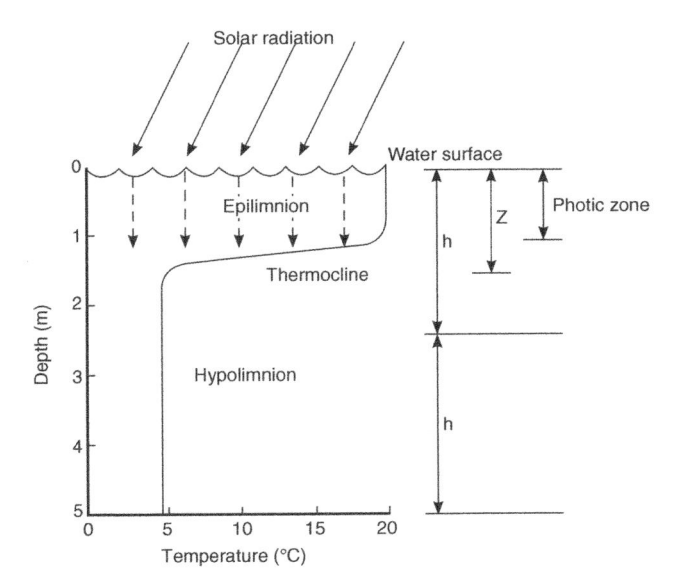

Figure 8.24 Vertical depth profile of temperature and other characteristics of a thermally stratified water body.

center of gravity must be raised and the work required to do this provides a measure of the stability of stratification. A simple expression can be obtained by viewing the water body as having perpendicular sides and, in the stratified state, a horizontal thermocline (see Figure 8.24) at depth Z. In the unstratified state, uniform temperatures in the epilimnion and hypolimnion lead to correspondingly uniform water densities of D1 and D2, respectively, and the center of gravity lies at depth h. The stability per unit (S) can be derived as (Ruttner 1974, p. 307):

$$S = (D2 - D1)(2h - Z)(Z/2) \quad (8.22)$$

The principal agent providing the work to mix the different layers in natural water bodies is the wind. In some cases, water in-flow from a catchment may play a role in such mixing. Some important generalizations regarding stratification can be made from the expression above. But it is important to note that water density is not directly related to temperature. There is a maximum density at about 4 °C, and above this temperature, the density decreases at a more rapid rate than the temperature increases. Thus, small differences in temperature at elevated temperatures (e.g. 25–30 °C) produce larger differences in density than the same changes at lower temperatures (e.g. 1–10 °C). Hence, for the same temperature differential, water bodies with warmer temperatures show greater stability than those with cooler temperatures. If all other factors are unchanged, as the thermocline descends, the stability increases until a maximum results at Z = h. At further depths, the stability decreases.

Seasonal patterns of vertical water movement are a common characteristic of water bodies. In temperate areas, stratification in the summer leads to summer stagnation and in the winter stratification does not occur due to the lack of a temperature differential produced by summer solar radiation. Mixing of the epilimnion and hypolimnion can occur when stability is at its lowest. Bodies that are ice-covered, although unstable, are protected from wind-generated movement but may exhibit a turnover in spring when the ice melts and produces water at a maximum density of 4 °C before surface heating generates high stability. Similarly, mixing may occur in the autumn and winter when cooling of the surface waters leads to instability. According to these patterns of movement, lakes can be classified into a number of different types, for example dimictic, which means exhibiting two turnovers usually in autumn and spring; monomictic, with one turnover usually in autumn; and polymictic, with no stratification pattern.

Due to turbulence, thermal stratification is uncommon in flowing streams and estuaries. Estuaries can exhibit stratification due to the formation of a salt wedge by incoming seawater flowing along the estuary floor. Seasonal thermal stratification has been noted on the continental shelves, such as the New York Bight (Steimle and Sindermann 1978; Box 8.1). In fact, the deep oceans exhibit a high level of stable thermal stratification unaffected by the seasons. In all oceans, water well below the photic zone at about 500 m (i.e. not on the continental shelf) has a temperature of about 4 °C. Movement of bottom waters to the photic zone near the surface occurs in areas of **upwelling**, principally due to movement of ocean currents and turbulence caused by the interaction of ocean and wind-induced currents.

Stratification has a number of important implications for nutrient enrichment and eutrophication. Figure 8.25 indicates the nitrogen transformations that would be expected to occur in an idealized stratified body of water. In the epilimnion, photosynthesis produces organic matter, principally phytoplankton, containing carbon, nitrogen, phosphorus, and a variety of other elements. Death followed by sedimentation transfers this organic matter to the hypolimnion where microbiological degradation occurs, resulting in the formation of carbon dioxide, orthophosphate, ammonia, and nitrite (under different conditions) with consumption of dissolved oxygen. In an oligotrophic situation, this may not result in any marked differences between the hypolimnion and the epilimnion since all processes occur at a low rate. However, in eutrophic conditions the hypolimnion is usually depleted in dissolved oxygen since direct reaeration from the atmosphere cannot occur. In addition, there is usually an accumulation of the

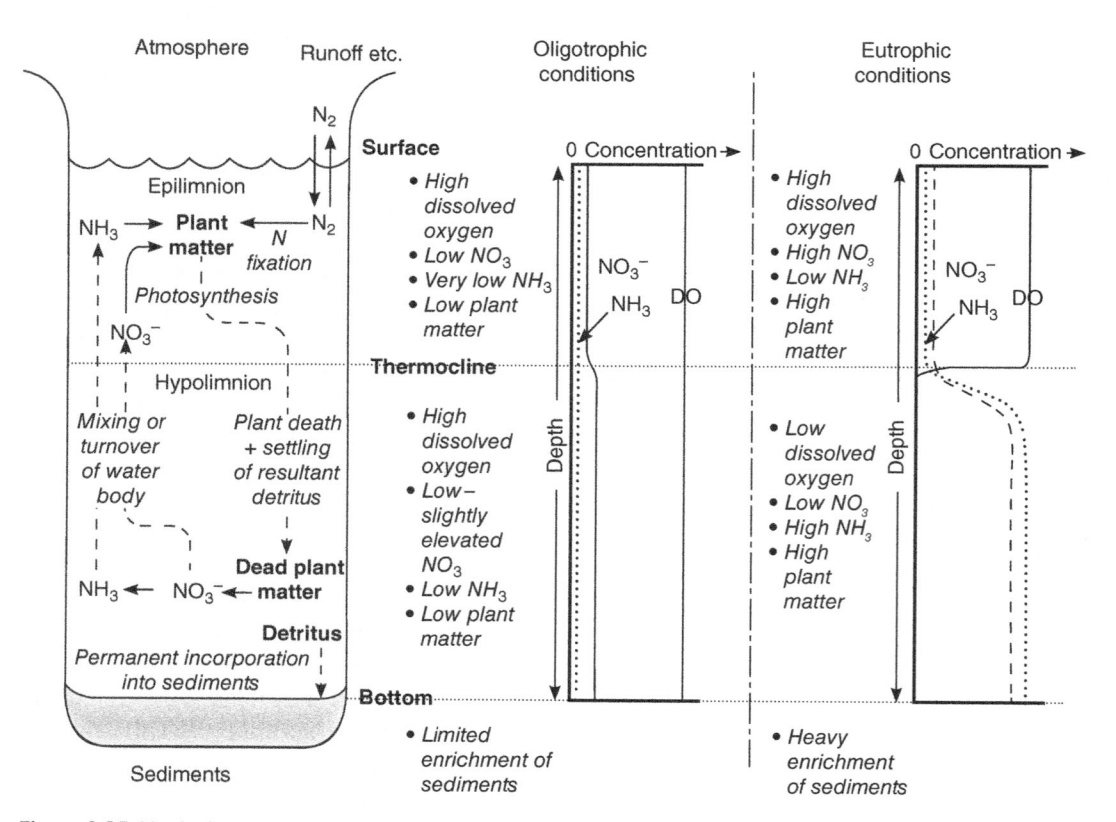

Figure 8.25 Vertical water column profiles of chemical components in an idealized stratified water body with related transformations of nitrogen, oxygen, and other component reactions. Note that both aerobic and anaerobic transformations are shown in the hypolimnion, although in a real lake they would not occur simultaneously.

inorganic materials formed during anaerobic respiration, such as ammonia and hydrogen sulfide.

The vertical profiles of nitrogen, oxygen, phosphorus, and other chemical components in the water column provide valuable information, which can be used in the management evaluations of water bodies as shown in Figure 8.25. The DO levels are high in both eutrophic and oligotrophic conditions in the epilimnion, but under eutrophic conditions, this zone is usually high in plant matter. With eutrophic conditions, the hypolimnion has high levels of the forms of nitrogen and phosphorus, but the DO is very low. Oxygen depletion of the hypolimnion occurs in many water bodies throughout the world due to the introduction of plant nutrients yielding excess plant growth. An extreme example occurred on the northeast continental shelf of the United States, in the New York Bight as described in Box 8.1.

If dissolved oxygen is zero in the bottom waters, the nitrification reaction leading to nitrate cannot proceed and organic nitrogen yields ammonia as the principal end product. This ammonia is not usually formed in surface waters since these are well aerated by atmospheric and photosynthetic oxygen. Here, the principal nitrogen form is the fully oxidized form – nitrate. Eutrophic conditions produce characteristic vertical profiles of nitrate and ammonia, which correspond with the above observations as shown in Figure 8.25

Dissolved oxygen depletion also results in the release of adsorbed orthophosphate as discussed previously. Therefore, while the epilimnion may be depleted by photosynthesis and subsequent sedimentation of the primary production, the hypolimnion becomes enriched. During periods of mixing of bottom and surface waters, there can be an increase in plant growth in the photic zone if other environmental conditions are satisfactory. Thus, surface waters may exhibit a seasonal cycle of nutrient concentrations and phytoplankton growth related to the factors outlined above, as well as the pattern of vertical mixing previously outlined (see Figure 8.25).

The surface area of a lake, its catchment, and depth have a strong influence on primary production. If the inputs from the catchment and precipitation have reached a

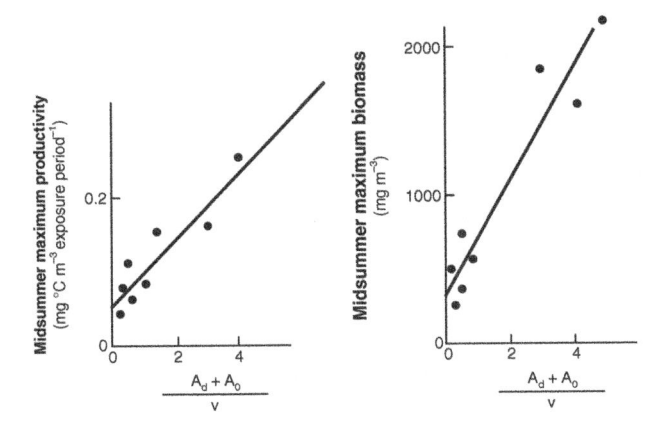

Figure 8.26 Relationships of lake primary production response to watershed area (A_d) plus lake area (A_o) divided by volume (V). Source: Schindler (2011).

steady state, nutrient input will be proportional to the total area and its influence on the nutrient concentration in the water will be related to the lake volume, as demonstrated by Schindler (2011) (see Figure **8.26**).

Flushing of a water body also influences the nutrient concentration and therefore affects primary production. In a stream or estuary, flushing occurs at a relatively rapid rate and somewhat different conditions prevail to those in water bodies where water movement is at a lower level. In streams, nutrient levels usually decrease downstream from a discharge leading to a set of related chemical and biological changes.

Somewhat similar processes of phytoplankton growth in the upper layers of the photic zone have been occurring in the oceans over geological time. On the death of the phytoplankton, the dead plant matter falls to the bottom and many areas in the bottom waters of the oceans have dissolved oxygen levels of zero due to the accumulation of organic matter from primary production in the photic zone. These often form OMZ as described in Section 8.2. These OMZ have increased in area in the bottom waters of the oceans over recent years.

In the oceans, it is well known that oceanic primary production enables fisheries to be developed in the most productive areas of upwelling where nutrient-rich bottom waters enter the photic zone. On the other hand, these processes in surface waters where upwelling does not occur cause depletion in nutrients, without corresponding replenishment processes, giving oligotrophic conditions in the photic zone. This continual nutrient uptake and removal from the photic zone results in there being no significant level of plant nutrients in the oceans in its more distant and remote sections. This situation and the nutrient depletion, which results in about 40% of the ocean surface.

Nutrient Loadings, Primary Production, and Trophic State

The trophic state of a water body depends on a variety of physical, chemical, and biological properties. Clearly, the supply of any chemical factor, which promotes plant growth, is of importance. The nutrient loading approach has been most successful with lakes in which productivity is controlled by the phosphorus loading (Vollenweider and Kerekes 1980). The nutrient loading must take into account external inputs, as well as the availability of nutrients already within the body. However, it has been found that the availability of nutrients already within the body is not of primary significance in many lake systems. The models relating nutrient loading to primary productivity consist of two submodels:

1. The relationships between nutrient input and nutrient concentration in the water body.
2. The relation between nutrient concentrations in the water body and primary productivity.

The establishment of these relationships is only applicable if it is assumed that complete mixing of the lake occurs and that steady-state conditions operate.

The relationship between nutrient concentration in the water (or load per unit area) and nutrient input has been reported in a series of papers by Vollenweider and co-workers (e.g. Vollenweider 1976; Vollenweider and Kerekes 1980). One relationship that has found wide use is:

$$L_P = P_w (Z/\tau_w + Z\sigma_P) \quad (8.23)$$

Where L_P is the surface loading of phosphorus per unit area per year; P_w is the average total phosphorus concentration of dissolved and particulate phosphorus in water at the spring overturn; Z the mean depth; τ_w the average residence time of water; and $Z\sigma_P$ relates to the sedimentation rate of phosphorus. As a general rule, $Z\sigma_P$ is approximately 10/Z. Thus, substituting in equation (8.23),

$$L_P = P_w (Z/\tau_w + 10) \quad (8.24)$$

If oligotrophic lakes are considered to have $P_w < 10\,\text{mg}\,\text{m}^{-3}$, mesotrophic $10-20\,\text{mg}\,\text{m}^{-3}$, and eutrophic $>20\,\text{mg}\,\text{m}^{-3}$ with a depth to residence time ratio of 0.1, tolerance lines are obtained for the limiting positions of trophic state as shown in Figure 8.27. In addition, Figure 8.27 indicates plots for 60 lakes in North America with expert classification into trophic class. These points are in good agreement with the tolerance lines.

Vollenweider and Kerekes (1980) has also reported the relationship:

$$P_w = (\sigma_P/\tau_w)\, P_i \quad (8.25)$$

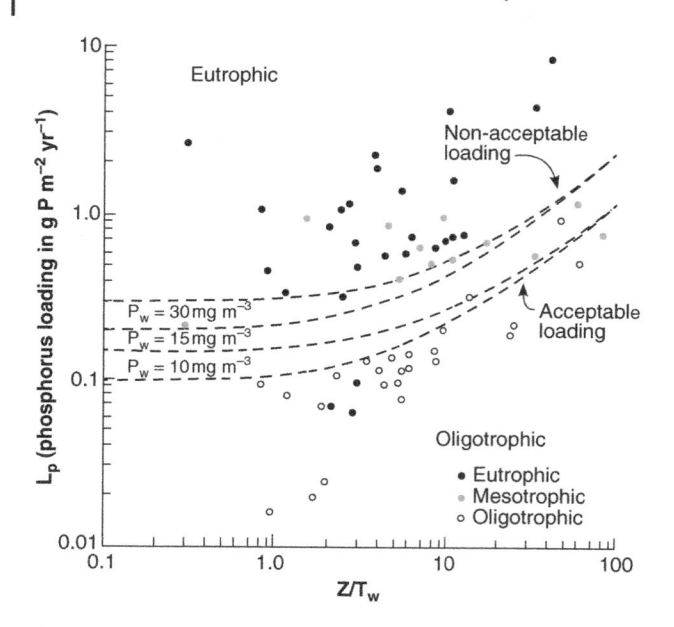

Figure 8.27 Plots of phosphorus loading against depth (Z) divided by water flushing time (τ_w) for 60 lakes classified as oligotrophic, mesotrophic, or eutrophic and also the phosphorus loading tolerance according to the equation $L_p = P_w(Z/\tau_w + 10)$. Source: Vollenweider (1976).

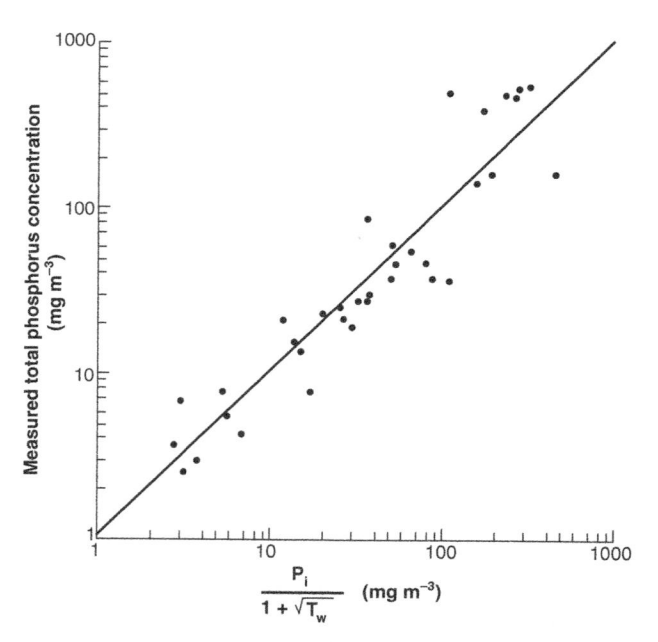

Figure 8.28 Relationship between total phosphorus concentration at spring overturn and total phosphorus predicted from Vollenweider's equation, $P_w = P_i/(1 + \sqrt{\tau_w})$. Source: Vollenweider and Kerekes (1980).

Where σ_P/τ_w is the average residence time of phosphorus relative to the average residence time of water, and P_i the average inflow concentration of total phosphorus. In addition, Vollenweider (1976) found the following approximate relationships:

$$\sigma_P/\tau_w = 1/(1 + \sqrt{\tau_w}) \quad (8.26)$$

Thus, substituting in equation (8.25),

$$P_w = P_i/(1 + \sqrt{\tau_w}) \quad (8.27)$$

A plot of predicted phosphorus concentrations in water (P_w) versus actual phosphorus at spring turnover for selected OECD lakes is shown in Figure 8.28.

$$P_w = P_i(1 - R_D) \quad (8.28)$$

Where R_D is the phosphorus retention coefficient.

Vollenweider (1976) has also shown that his relationship above for P_w can be expressed as:

$$P_w = L_P/q_s (1/1 + \sqrt{Z/q_s}) \quad (8.29)$$

Where q_s is the hydraulic load.

The loading model is also concerned with the relationships between primary production and phosphorus concentrations in the water. Primary production in terms of a biomass parameter (e.g. chlorophyll) would be expected to be related to phosphorus concentration in water (P_w) and hence to phosphorus loadings (P_i). Vollenweider and Kerekes (1980) have found significant relationships of the form:

$$\text{Average chlorophyll} - a = 0.37 (P_i/1 + \sqrt{\tau_w})^{0.91} \quad (8.30)$$

This relationship has been found to hold for 60 lakes classified subjectively as oligotrophic, mesotrophic, or eutrophic.

These models have been updated, modified, and reorganized with applications to water bodies in several countries in many different parts of the world. For example, Chang et al. (2019) have reviewed the models and classified them into three key approaches:

1. the empirical approach that employs data from field surveys,
2. the theoretical approach in which models based on first principles are tested against the results of laboratory experiments, and
3. the process-based approach that uses parameters and functions representing detailed biogeochemical processes.

These researchers found that a comparison and integration of these different approaches were hampered by their large differences in scale and complexity. They proposed a generically parameterized lake eutrophication model that links field-, lab-, and model-based knowledge based on nutrients and light. Utilizing only the parameters for lake depth, residence time, and nutrient loading, a first-order assessment of water quality can be made. Using this model, measures such as reducing nutrient load, decreasing residence time, or changing depth can be evaluated.

Ecosystem Changes Resulting from Nutrient Enrichment and Eutrophication

Changes in Community Metabolism

Elevated concentrations of nutrients are related to increases in chlorophyll-a concentrations and primary production leading to increasing eutrophication (see equation (8.30)). A substantial proportion of this increased primary production would be expected to be transferred to the hypolimnion where deoxygenation can occur as described previously. Thus, oxygen depletion rates in the hypolimnion would be expected to be related to trophic state.

The ratio of primary production to respiration in different water bodies can be used as the basis of a scheme of classification and comparison (see Figure 8.10). In this scheme, oligotrophic and eutrophic water bodies have approximately equal rates of primary production (P) and respiration (R), that is, P/R is approximately equal to 1. But the rates of both photosynthesis and respiration are much higher in eutrophic, compared to oligotrophic waters. In contrast, organically polluted waters have a P/R ratio of much less than 1.

Community and Population Changes with Nutrient Enrichment

Physicochemical conditions change with trophic state stimulating changes in biological composition. With phytoplankton in lakes, there is a seasonal change in community composition related to temperature, light, and other seasonal factors. Welch (1980) has reported that in temperate regions, diatoms generally dominate in the spring, green algae in the summer, blue-green algae in late summer, and possibly diatoms in late autumn. There can be considerable variation in this pattern, since different phytoplankton also have different dynamics and requirements for nitrogen, phosphorus, carbon dioxide, and other factors, which produce changes in the community composition with increasing eutrophication. The most prominent change with increasing nutrient enrichment is that the blue-green algae (Cyanophyceae) become increasingly dominant (see Figure 8.29).

Toxic algal blooms generally increase in occurrence with increasing eutrophication. This can cause mortalities of large numbers of aquatic organisms and terrestrial animals utilizing the water. Also, toxic algal blooms known as *red tides* can occur in enriched marine areas resulting in the occurrence of toxic components in edible shellfish.

In flowing waters, the periphyton (attached biota) include the most important primary producers (Welch 1980). The physical form and method of attachment are very important particularly with regard to the ability to withstand erosion and scouring by water movement. Under these conditions, periphyton usually take the form of slick coatings on rocks or strands of filamentous formation. The most common species are filamentous blue-green algae (e.g. *Oscillortoria*), filamentous green algae (e.g. Cladophora), and filamentous bacteria (e.g. *Sphaerotilum*).

Rooted macrophytes may obtain nutrients from sediments or the water mass, but the water is probably of secondary importance. However, growth is generally

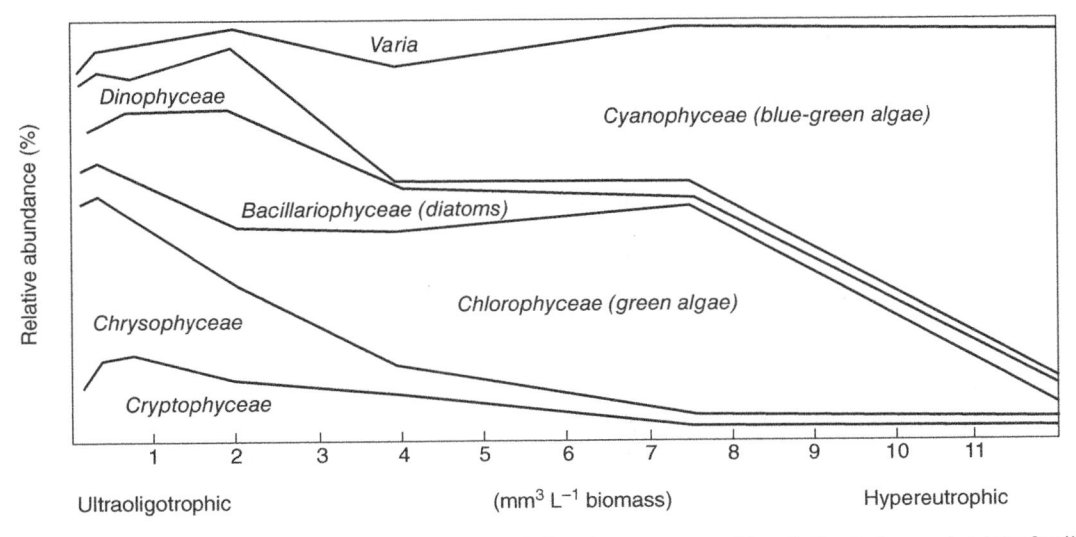

Figure 8.29 Composition of phytoplankton and algal abundance expressed in relation to increasing lake fertility. Source: Skulberg (1980).

greatest in organically enriched sediments, and this group of plants can display large growths in restricted and shallow waterways.

The zooplankton communities are probably not directly modified by nutrient enrichment. Even so, indirect effects are of considerable importance. For example, the rate of primary production and hypolimnetic oxygen concentrations have a significant impact on zooplankton. As a general rule, if the diversity of the phytoplankton is high, the diversity of zooplankton will also be high. Zooplankton grazing also has an impact on phytoplankton community composition and biomass. Increasing eutrophication leads to decreasing efficiency of utilization of energy in the phytoplankton-based food web but a decreasing biomass of zooplankton.

The benthos generally reflect conditions in the water mass. Benthic animals are influenced by material produced in the water mass or discharged into the water body. Macrozoobenthos are often used as these are most suitable for investigation (usually >0.5–0.6 mm mesh size sieve). Early spring or autumn are commonly used as standard sampling periods to minimize seasonal variations. The soft bottom communities are the richest and usually selected for investigation. These communities are dominated by chironomids and polychaetes with lesser numbers of molluscs, crustacea, and insects.

Saether (1979, 1980) has delineated 15 chironomid communities as indicative of trophic classes or types, designated from α to ° with increasing eutrophication. Figure 8.30 shows the chlorophyll-a content of a variety of lakes in North America and Europe plotted against trophic type, as designated from the chironomid communities

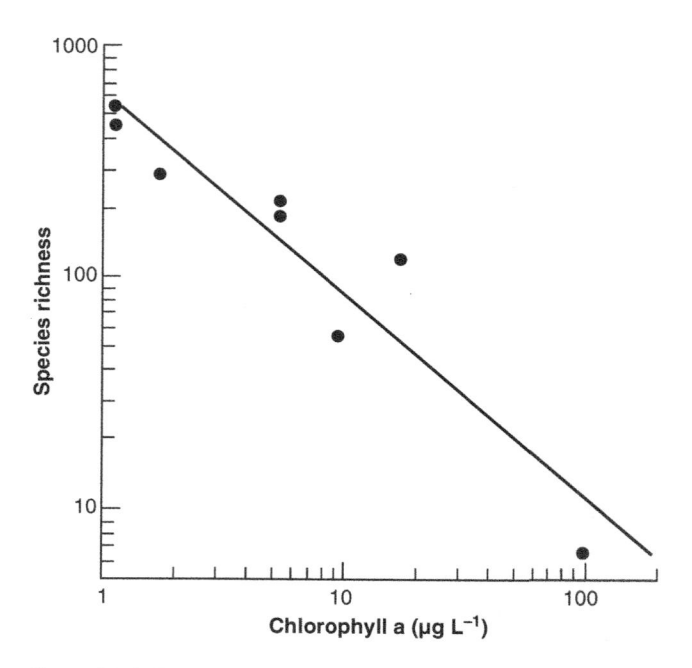

Figure 8.31 Chlorophyll-a in relation to species richness adjusted for depth in the Swedish Great Lakes. Source: Weiderholm (1980).

present. A similar relationship was found when total phosphorus divided by depth was plotted against type for the same set of lakes.

Weiderholm (1980) found that species richness decreased with increasing eutrophication, measured as chlorophyll-a, in Swedish lakes (see Figure 8.31). However, he has suggested that species richness would be expected to reach a maximum at some intermediate lake fertility. Also, oligochaete numbers increase in relationship to sedentary chironomid numbers with increasing eutrophication, and this ratio can be used as a diagnostic characteristic.

The principal effect of increasing eutrophication on fish is due to dissolved oxygen depletion. However, moderate increases in primary production may be beneficial to fish numbers and biomass. Summer stratification of the water mass causes a similar vertical stratification of fish species according to their tolerance to temperature and dissolved oxygen reduction. But benthic areas depleted in dissolved oxygen, which expand with increasing eutrophication, decrease the habitat suitable for fish and can lead to an overall decline in fish numbers. Figure 8.32 shows the general trends in species tolerance to increasing eutrophication. The cyprinids (e.g. carps and minnows) are comparatively tolerant to environmental stress and become increasingly dominant with increasing trophic state.

Characteristics and Criteria for Trophic State

A clear assessment of the degree of eutrophication is needed for the development of procedures for water

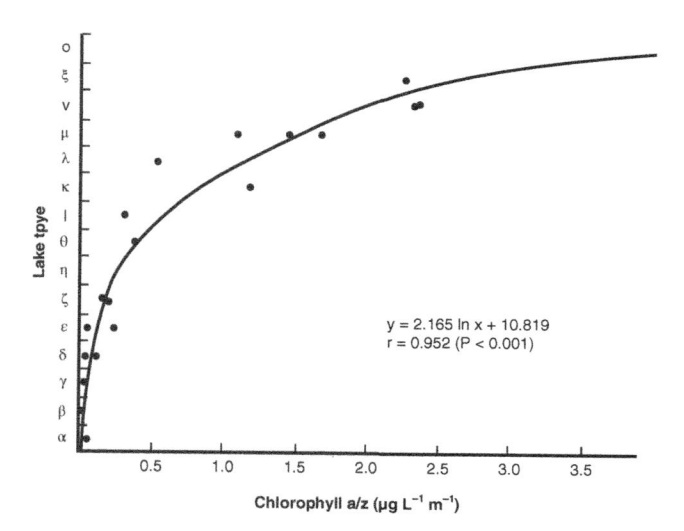

Figure 8.30 Chlorophyll-a/z (mean lake depth) in relation to 15 lake types based on chironomid communities (α to °). Source: Saether (1979).

$y = 2.165 \ln x + 10.819$
$r = 0.952$ ($P < 0.001$)

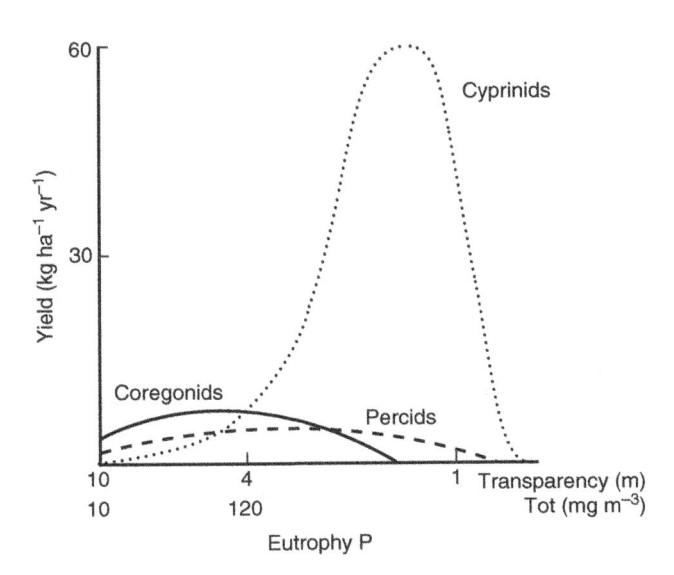

Figure 8.32 Summary of relations between increasing eutrophy and yields of coregonids, percids, and cyprinids for 17 European lakes. Source: Hartmann (1986).

Table 8.13 Some characteristics of oligotrophic and eutrophic lakes.

Parameter	Oligotrophic	Eutrophic
Depth	Deep	Shallow
Primary production	Low	High
Rooted macrophytes	Few	Many
Secchi disc evaluation	High	Low
Algal species richness	Many species	Few species
Distribution	To great depths	Photic layer
Algal blooms	Very rare	Frequent
Hypolimnetic depletion	Rare	Common
Characteristic algal groups or genera	Chlorophyceae Desmids *Staurastrum* Diatomaceae *Tabellaria Cyclotella* Chrysophyceae *Dinobryon*	Cyanophyceae *Anabaena Aphanizomenon Microcystis* Diatomaceae *Melosira, Stephanodiscus Asterionella*

Table 8.14 OECD key quantitative characteristics of oligotrophic mesotrophic and eutrophic lakes.

Trophic category	Mean total, P (μg L^{-1})	Mean (μg chl-a L^{-1})	Max. (μg chl-a L^{-1})	Mean Secchi depth (m)
Oligotrophic	<10	<2.5	<8	>6
Mesotrophic	10–35	2.5–8	8–25	6–3
Eutrophic	>35	>8	>25	<3

Source: Istvánovics (2009).

quality management. Most of the characteristics of trophic state presently available have been based on freshwater lakes with a set of primary characteristics as shown in Table 8.13.

The OECD has developed a set of criteria as shown in Table 8.14 for freshwater bodies, which can be used to classify where these water bodies lie in terms of trophic state (Istvánovics 2009). These criteria interrelate reasonably well, but some water bodies fall into indeterminate classes. Phosphorus has been identified as the limiting nutrient, which is the most critical growth factor in most lakes. Chlorophyll-a, the concentration of chlorophyll in the water, is used as a measure of the primary production, and the Secchi disc evaluation has been widely used to measure the turbidity of natural waters. It can be seen from Table 8.14 that increasing trophic state results from increasing levels of phosphorus and chlorophyll, but over that change of trophic state, the Secchi disc evaluation declines.

Biotic indices have also been suggested and a large number have been developed for eutrophication. For example, Weiderholm (1980) has developed a benthic quality index (BQI) with the following characteristics:

$$BQI = \sum_{i=0}^{5} k_i n_i / N \qquad (8.31)$$

Where k_i represents a value constant for each species; n_i the number of individuals in the various groups; and N the total number of indicator species. This expression combines the principal indicator species into a single index. Similarly, other indices, for example species richness and oligochaete to chironomid ratio, have been shown to be useful.

In the oceans and on the continental shelves, many of the characteristics in Tables 8.13 and 8.14 can apply. Officer and Ryther (1980) have reported that excessive marine phytoplankton growth leading to adverse eutrophication effects have been related to flagellate blooms, which are not utilized by most grazing organisms. Nutrient elements usually stimulating the growth of marine phytoplankton have been identified principally as nitrogen and silicon (Officer and Ryther 1980) although phosphorus has been implicated as well. Primary production and nitrate levels in the ocean surface waters generally range from 70 mg C m^{-2} day^{-1} and 1 μg N L^{-1} in oligotrophic waters of subtropical halistatic areas, to 1000 mg C m^{-2} day^{-1} and about 300 μg N L^{-1} in open coastal waters.

8.4 Key Points

Section 8.2

1. Water covers about 71% of the global surface with about 97% of this water environment in the oceans and the remainder in lakes, rivers, aquifers, groundwater, glaciers, and other water bodies.
2. The water environment sustains an important component of the global biota such as coral reef organisms, commercial fish stocks, and cetacean communities.
3. **Water pollution** is the contamination of water bodies by addition of substances that alter the properties of the water making it unsuitable for use by human society or by natural ecosystems. Pollution is usually considered to be as a result of human activities.
4. The principal chemical agents, which are of importance in the protection of natural ecosystems can be generally reduced to two classes – **Deoxygenating Substances** and **Plant Nutrients**.
5. Photosynthesis and respiration are fundamental to life on Earth and the chemical processes involved can be simply expressed with glucose as set out below

$$6CO_2 + 6H_2O \underset{\text{Respiration}}{\overset{\text{Photosynthesis}}{\rightleftharpoons}} 6CH_2O + 6O_2$$

These processes involve carbon dioxide and oxygen in the Earth's atmosphere, which, in aquatic ecosystems, are dissolved in the water mass.

6. In a water body, carbon from dissolved carbon dioxide is incorporated by photosynthesis into photosynthetic plants such as algae. Respiration of the heterotrophic biota involves the consumption of autotrophic biomass, algae, and other plants as food, and finally the decomposition of body tissues after death.
7. The oxygen needed for aerobic respiration is obtained from the dissolved oxygen in the water mass. Most importantly, the reservoir of dissolved oxygen in the water mass is comparatively small since oxygen has limited solubility in water, usually ranging from about 6 to 14 mg L^{-1}.
8. When the DO reaches levels of about 2 mg L^{-1} or less, the water cannot support normal aquatic life and is described as **hypoxic** and the condition is known as **hypoxia**. If the DO falls to levels at or near 0 mg L^{-1}, **anoxia** occurs and the conditions are described as **anoxic**. This results in severe damage to the normal aquatic system at a level more intense than hypoxia.
9. The bottom waters in all bodies of water ranging from the oceans to ponds, as well as swamps in a natural condition, can exhibit hypoxia and anoxia. Also flowing waters such as rivers can exhibit hypoxia on a periodic or permanent basis. Well-known urban rivers such as the Thames River in London, the Hudson River in the City of New York, and the Rhine River in Germany have had hypoxia problems.
10. In the oceans and other large water bodies, there has been the natural accumulation of detritus-organic matter originating from both the catchment discharges and growth of phytoplankton within the water mass over geological time. This has resulted in the accumulation of detritus-organic matter in the bottom sediments and waters resulting in hypoxia and anoxia occurring. These areas in the oceans are often described as "**dead zones**" since there is an absence of normal aquatic life.
11. The total OMZ occupy 30.4×10^6 km^2 in the global ocean. However, these OMZ have expanded over recent years with the expansion of the world's population and subsequent increase in discharges of organic wastes and plant nutrients in wastewater.
12. Usually, the most important heterotrophic aquatic organisms, in terms of respiration, are the microorganisms. In well-oxygenated waters, aerobic respiration occurs mediated by aerobic microorganisms. The chemical reaction involved can be simply expressed, for glucose, as

$$6CH_2O + 6O_2 \rightarrow 6H_2O + 6CO_2$$

13. In the absence of oxygen, anoxic or hypoxic conditions develop and anaerobic microorganisms take over the degradation and decomposition of organic matter. In fact, molecular oxygen is toxic to this group. With glucose the anaerobic respiration reaction can be simply expressed as

$$6CH_2O \rightarrow 3CH_4 + 3CO_2$$

14. Anaerobic respiration with sulfur- and nitrogen-containing organic substances gives rise to hydrogen sulfide and ammonia, respectively. Ammonia in the presence of oxygen is readily oxidized to nitrate. Nitrite is formed under suitable oxidation and reduction conditions but is less commonly the end product of nitrogen metabolism. The formation of nitrate from organic matter is described as the nitrification reaction and is mediated by a variety of microorganisms.
15. These oxygen-demanding processes can be measured by the BOD test. This test measures the consumption of oxygen by water samples incubated over a period of five days under standard conditions. In general, clean water has a BOD of 1 mg L^{-1} or less and seriously polluted water contains greater than 10 mg L^{-1}. The nitrification reaction occurs in what is described as the second stage of the BOD test at time periods greater than five days.

16. The kinetics of BOD changes in waterways have been subject to intense study, thus the oxidation of organic matter follows first-order decay reaction and the following series of mathematical expressions can be derived:

$$\ln (L/L_0) = -K_1 t$$

or

$$\log (L/L_0) = -0.434 k_1 t = -k_1 t$$

17. The modeling of the dynamics of dissolved oxygen in rivers receiving discharges of wastewater containing significant amounts of BOD commenced in 1925 with Harold Streeter and Harold Phelps. It probably represents the first widespread and highly successful application of mathematics to an environmental problem.

18. The actual dissolved oxygen profile in a stream receiving wastewater rich in organic matter produces a dissolved oxygen "sag." An expression for this curve is mathematically obtained by utilizing the following equation

$$D = (k_2 L_0 / k_2 - k_1) \left(e^{-k_1 t} - e^{-k_2 t} \right) + D_0 e^{-k_2 t}$$

19. Photosynthesis occurs during the daylight hours but respiration by aquatic plants occurs throughout the diurnal cycle. Thus, these processes cause a rise in dissolved oxygen in water during the daylight hours, and at sunset, the dissolved oxygen starts to drop. This continues overnight reaching a minimum before dawn when sunlight once again initiates photosynthesis.

20. One of the most spectacular environmental disasters resulting from oxygen depletion occurred in 1976 in the New York Bight located on the continental shelf of northeastern United States. Here, the bottom waters and sediments were enriched by phytoplankton decay, river discharge containing oxygen-demanding substances and nutrients, as well as dumping of sewage sludge transported in barges from New York City. This caused extensive mortalities of marine organisms over a large area.

21. Hydrogen sulfide is a product of anaerobic respiration by natural microorganisms in the degradation of polluting and natural organic matter. It is produced at reduced dissolved oxygen levels in aquatic areas. Thus, natural ecosystems affected by reduced dissolved oxygen will often also be influenced by hydrogen sulfide.

22. The increase in respiration resulting from additions of organic matter is principally due to bacteria and protozoa, but there is a large population increase in many other heterotrophic organisms. The major larger heterotrophs include the polychaetes (e.g. *Capitella* sp.) and tuberficids. Overall, there is an increase in the biomass even when the dissolved oxygen levels drop to zero.

23. The biotic index with the longest history is the **saprobic index** or **saprobien** system, initially developed for continental Europe. **Saprobity** has been defined as the state of water quality with respect to the content of putrescible organic material as reflected by species composition of the community. There are basically three levels – **polysaprobic**, **mesosaprobic**, and **oligosaprobic**, but further subdivision of these zones can be used.

24. Vertical stratification of water bodies leads to the isolation of bottom waters causing dissolved oxygen reduction. This can occur in water bodies ranging in size from the ocean itself to small pools. For example, vertical profiles of temperature and dissolved oxygen were observed in New York Bight in 1976.

Section 8.3

1. The enrichment of aquatic areas with plant nutrients is an important process in aquatic pollution and is described as **eutrophication**. It is described by the terms **oligotrophic, mesotrophic, and eutrophic** for increasing levels of nutrient enrichment of principally freshwater water bodies.

2. The OECD has defined eutrophication as *the nutrient enrichment of waters which results in stimulation of an array of symptomatic changes among which increased production of algae and macrophytes, deterioration of fisheries, deterioration of water quality and other symptomatic changes are found to be undesirable and interfere with water uses.*

3. Increased populations of plants can lead to a decrease in the dissolved oxygen content of the water on plant death and decomposition. This decreases the suitability of the area as a habitat for many species of fish and other organisms. The increase in turbidity, color, and odors renders the water unsuitable for domestic use.

4. Blooms, pulses, and so on of aquatic plants become more frequent and in eutrophic water bodies, toxic blooms of **blue–green algae, cyanobacteria** often occur.

5. Nutrient enrichment processes can occur in the oceans, as well as freshwater bodies, leading to the enrichment of bottom waters with plant matter and organic carbon producing deoxygenation leading to the formation of "**dead zones.**"

6. The growth of algae is inhibited due to limitations in the supply of elements needed for the growth of plant

tissue particularly proteins and carbohydrates. Elements such as phosphorus, nitrogen, iron, magnesium, sodium, and a variety of other elements are needed to construct various vital components. So addition of these elements, particularly nitrogen and phosphorus, can stimulate the growth of plants.

7. The uptake of nutrient by phytoplankton usually follows Michaelis–Menten kinetics, which can be expressed mathematically as:

$$V = (V_{max} S/(K_S + S)) \quad (8.20)$$

8. Wastewaters from sewage treatment, as well as from food processing and farm runoff, contain a variety of nitrogen- and phosphorus-containing substances derived from proteins, nucleic acids, phospholipids, urea, and other substances. In addition, phosphorus can originate from phosphate builders in detergents that can be readily hydrolyzed to yield orthophosphate, which is readily assimilated by plants giving increased growth.

9. New anthropogenic nitrogen originates from the industrial conversion of atmospheric nitrogen by fixation with hydrogen into nitrogenous fertilizers. New anthropogenic phosphorus is derived from the mining of phosphorus-rich minerals that are converted into crop fertilizers then consequent food crop production. The consumption of food leads to increased sewage discharges and eutrophication on a global scale.

10. Global nitrogen fixation from all sources, natural and industrial, has been quantified at $413\,g \times 10^{12}$ of nitrogen with industrial fixation responsible for about half, $210\,g \times 10^{12}$ N.

11. **Thermal stratification** can occur due to solar heating of the surface waters in the **photic zone** forming an upper layer, the **epilimnion**, of a water body, which is then separated from the underlying waters, the **hypolimnion**.

12. The stability of stratification can be expressed as

$$S = (D2 - D1)(2h - Z)(Z/2) \quad (8.22)$$

13. Lakes can be classified into a number of different types, for example **dimictic**, which means exhibiting two turnovers usually in autumn and spring; **monomictic**, with one turnover usually in autumn; and **polymictic**, with no stratification pattern.

14. The vertical profiles of nitrogen, oxygen, phosphorus, and other chemical components show that the DO levels are high in both eutrophic and oligotrophic conditions in the epilimnion, but under eutrophic conditions, this zone is usually high in plant matter. With eutrophic conditions, the hypolimnion has high levels of the forms of nitrogen and phosphorus, but the DO is very low.

15. Phytoplankton growth in the upper layers of the oceans has been occurring over geological time. On death of the phytoplankton, the dead plant matter sinks to the bottom, and in many areas, the bottom waters of the oceans have dissolved oxygen levels of zero due to the accumulation of organic matter. These areas often form OMZ as described in Section 8.2. These OMZ have increased in area in the bottom waters of the oceans over recent years.

16. Over geological time, phytoplanktonic growth in the photic zone of the oceans has taken up the available nutrients and transferred them to the bottom waters after death without any corresponding replacement process. This has led to large areas of the more remote and distant ocean water being in an oligotrophic condition. However, upwelling processes in certain areas can bring these nutrient-rich waters to the surface, giving high phytoplanktonic growth and the basis for fisheries.

17. The relationship between nutrient concentration in the water (or load per unit area) and nutrient input has been reported in a series of papers by Vollenweider and co-workers. One relationship that has found wide use is

$$L_P = P_W (Z/\tau_w + Z\sigma_P) \quad (8.23)$$

18. Oligotrophic and eutrophic water bodies have approximately equal rates of primary production (P) and respiration (R), that is P/R is approximately equal to 1. But the rates of both photosynthesis and respiration are much higher in eutrophic waters compared to oligotrophic waters. In contrast, organically polluted waters have a P/R ratio of much less than 1.

19. In temperate regions, diatoms generally dominate in the spring, green algae in the summer, blue-green algae in late summer, and possibly diatoms in late autumn. The most prominent change with increasing nutrient enrichment is that the blue-green algae (*Cyanophyceae*) become increasingly dominant (see Figure 8.29).

20. Toxic algal blooms, with **"Red Tides"** in marine areas, generally increase in occurrence with increasing eutrophication. This can cause mortalities of large numbers of aquatic organisms and terrestrial animals utilizing the water.

21. The OECD has developed a set of criteria as shown in Table 8.14 for freshwater bodies, which can be used to classify where these water bodies lie in terms of trophic state.

References

Abbasi, T. and Abbasi, S.A. (2011). Water quality indices based on bioassessment: the biotic indices. *Journal of Water and Health* 9 (2): 330–348.

Abbasi, T. and Abbasi, S.A. (ed.) (2012). *Water Quality Indices*. London: Elsevier.

Allen, H.E. and Kramer, J.R. (1972). *Nutrients in Natural Waters*. New York: Wiley.

Armstrong, R.S. (1977). Climatic conditions related to the occurrence of anoxia in the waters off New Jersey during the summer of 1976. In: *Oxygen Depletion and Associated Environmental Disturbance in the Middle Atlantic Bight in 1976.* (ed. Northeast Fisheries Center), 17–35. Technical Series Report No. 3, Northeast Fisheries Center, National Marine Fisheries Service, National Oceanic and Atmospheric Administration, Sandy Hook, NJ.

Bellan, G. (1970). Pollution by sewage in Marseilles. *Marine Pollution Bulletin* 1: 59–60.

Brezonik, P.L. (1972). Nitrogen: sources and transformations in natural waters. In: *Nutrients in Natural Waters* (ed. H.E. Allen and J.R. Kramer), 1–50. New York: Wiley.

Chang, M., Teurlincx, S., DeAngelis, D.L. et al. (2019). A generically parameterized model of lake eutrophication (GPLake) that links field-, lab- and model-based knowledge. *Science of the Total Environment* 695: https://doi.org/10.1016/j.scitotenv.2019.133887.

Clesceri, L.S., Eaton, A.D., and Rice, E.W. (2005). *Standard Methods for Examination of Water & Wastewater. Method 5210B Biochemical Oxygen Demand (BOD)*. Washington, DC: American Public Health Association, American Water Works Association, and the Water Environment Association.

Connell, D.W. and Miller, G.J. (1984). *Chemistry and Ecotoxicology of Pollution*. New York: Wiley.

Connell, D.W., Morton, H.C., and Bycroft, B.M. (1982). Oxygen budget for an urban estuary. *Australian Journal of Marine and Freshwater Research* 33: 607–616.

Diaz, R.J. and Rosenberg, R. (2008). Spreading dead zones and consequences for marine ecosystems. *Science* 321 (5891): 926–929.

Domenici, P., Steffensen, J.F., and Marras, S. (2017). The effect of hypoxia on fish schooling. *Philosophical Transactions of the Royal Society of London B: Biological Sciences* 372 (1727): 20160236.

Erichsen Jones, J.R. (1964). *Fish and River Pollution*. London: Butterworth.

Farley, M. (2012). Eutrophication in fresh waters: an international review. In: *Encyclopedia of Lakes and Reservoirs*. Encyclopedia of Earth Sciences Series (ed. L. Bengtsson, R.W. Herschy and R.W. Fairbridge). Dordrecht: Springer.

Fowler, D., Coyle, M., Skiba, U. et al. (2013). The global nitrogen cycle in the twenty-first century. *Philosophical Transactions of the Royal Society of London B: Biological Sciences* 368 (1621): http://dx.doi.org/10.1098/rstb.2013.0164.

Garlo, E.V., Milstein, C.B., and Jahn, A.E. (1979). Impact of hypoxic conditions in the vicinity of Little Egg Inlet, New Jersey in Summer, 1976. *Estuarine Coastal Marine Science* 8: 421.

Golterman, H.L. (1975). *Physiological Limnology – An Approach to the Physiology of Lake Ecosystems*, 9e. Amsterdam: Elsevier.

Hartmann, J. (1986). Biological nutrient recycling and problems of eutrophication in lakes. PhD Dissertation, Washington State Water Research Center, University of Washington.

Hasler, A.D. (1968). Man-induced eutrophication of lakes. In: *Global Effects of Environmental Pollution* (ed. S.F. Sanger), 111–125. Dordrecht: D. Reidel!

Hawkes, H.A. (1979). Invertebrates as indicators of river water quality. In: *Biological Indicators of Water Quality* (ed. A. James and L. Evison), 2-1-2-45. Chichester: Wiley.

Hutchinson, G.E. (1969). Eutrophication, past and present. In: *Eutrophication: Causes, Consequences, Correctives.* (ed. National Academy of Sciences), 17–28. Washington, DC: National Academy of Sciences.

Hynes, H.B.N. (1960). *The Biology of Polluted Waters*. Liverpool: Liverpool University Press.

Istvánovics, V. (2009). Eutrophication of lakes and reservoirs. In: *Encyclopedia of Inland Waters*. 157–165. (ed. G.E. Likens) Amsterdam: Elsevier.

Lemley, D.A. and Adams, J.B. (2019). *Eutrophication in Encyclopedia of Ecology*, 2ee. Amsterdam: Elsevier.

McIvor, C.C. (1976). The effects of organic and nutrient enrichments on the benthic macroinvertebrate community of Moggill Creek, Queensland. *Water* 3: 16–21.

Odum, H.T. (1956). Primary production in flowing waters. *Limnology and Oceanography* 1 (2): 102–117.

Officer, C.B. and Ryther, J.H. (1980). The possible importance of silicon in marine eutrophication. *Marine Ecology – Progress Series* 3: 83–91.

Paulmier, A. and Ruiz-Pino, D. (2009). Oxygen minimum zones (OMZs) in the modern ocean. *Progress in Oceanography* 80: 113–128.

Pearson, T.H. (1980). Marine pollution effects of pulp and paper industry wastes. *Helgolander Meeresunters* 33: 340–365.

Pearson, T.H. and Rosenberg, R. (1978). Macrobenthic succession in relation to organic enrichment and pollution of the marine environment. *Oceanography and Marine Biology* 16: 229–311.

Pearson, T.H. and Stanley, S.O. (1979). Comparative measurement of the redox potential of marine sediments as a rapid means of assessing the effect of organic pollution. *Marine Biology* 53: 371–379.

Persoone, G. and DePauw, N. (1979). Systems of biological indicators for water quality assessment. In: *Biological Aspects of Freshwater Pollution* (ed. O. Ravera), 39. Oxford: Pergamon Press.

Poole, N.J., Wildish, D.J., and Kristmanson, D.D. (1978). The effects of the pulp and paper industry on the aquatic environment. *Critical Reviews in Environmental Control* 8: 153–195.

Ringeval, B., Nowak, B., Nesme, T. et al. (2014). Contribution of anthropogenic phosphorus to agricultural soil fertility and food production. *Global Biogeochemical Cycles* 28: 743–756.

Ruttner, F. (1974). *Fundamentals of Limnology*, 3ee. Toronto, Canada: University of Toronto.

Saenger, P., Stephenson, W., and Moverley, J. (1980). The estuarine macrobenthos of the Calliope River and Auckland Creek, Queensland. *Memoirs of the Queensland Museum* 20 (1): 143–161.

Saether, O.A. (1979). Chironomid communities as water quality indicators. *Holarctic Ecology* 2: 65–74.

Saether, O.A. (1980). The influence of eutrophication on deep lake benthic invertebrate communities. *Progress in Water Technology* 12: 161–180.

Schindler, D.W. (2011). A hypothesis to explain the differences and similarities among lakes in the experimental lakes area, Northwestern Ontario. *Journal of Fisheries Research Board of Canada* 28 (2): 295–301.

Schnoor, J. (1996). *Environmental Modeling: Fate and Transport of Pollutants in Water, Air and Soil*. New York: Wiley-Interscience.

Skulberg, O.M. (1980). Blue-green algae in Lake Mjøsa and other Norwegian lakes. In: *Eutrophication of Deep Lakes. Proceedings of Seminar on Eutrophication of Deep Lakes, 19–20 June 1978 at Gjøvik, Norway*. (ed. S.H. Jenkins), 121–140. Amsterdam: Elsevier.

Sladecek, V. (1979). Continental systems for the assessment of river water quality. In: *Biological Indicators of Water Quality* (ed. A. James and L. Evison), 3-1–3-32. Chichester: Wiley.

Sladecek, V., Hawkes, H.A., Alabaster, J.S. et al. (1982). Biological examination. In: *Examination of Water for Pollution Control – A Reference Handbook*, vol. 3 (ed. M.J. Suess), 1. World Health Organization – Regional Office for Europe. Oxford: Pergamon Press.

Steimle, F.W. and Sindermann, C.J. (1978). Review of oxygen depletion and associated mass mortalities of shellfish in the middle Atlantic Bight in 1976. *Marine Fisheries Review* 40 (12): 17–26.

US EPA (1973). Water quality criteria 72. Report EPA-R3-73-033. United States Environmental Protection Agency, Washington, DC.

US EPA (1986). Ambient water criteria for dissolved oxygen. EPA 44D/S-86-003, AR1326, Office of Water, Regulations and Standards Division, United States Environmental Protection Agency. Washington, DC.

US EPA (2019). National recommended water quality criteria – aquatic life criteria table. United States Environmental Protection Agency. https://www.epa.gov/wqc/national-recommended-water-quality-criteria-aquatic-life-criteria-table (accessed 1 September 2020).

Vallentyne, J.R. (1973). The algae bowl – a Faustian view of eutrophication. *Proceedings of Federation of American Societies for Experimental Biology* 32 (7): 1754–1757.

Velz, C.J. (1970). *Applied Stream Sanitation*. New York: Wiley-Interscience.

Vollenweider, R.A. (1971). Scientific fundamentals of the eutrophication of lakes and flowing waters, with particular reference to phosphorus and nitrogen as factors in eutrophication. OECD Technical Report DAS/CSI/68.27 (Revised 1971).

Vollenweider, R.A. (1976). Advances in defining critical loading levels for phosphorus in lake eutrophication. *Memorie dell'Istituto italiano di idrobiologia* 33: 53–83.

Vollenweider, R.A. and Kerekes, J. (1980). The loading concept as basis for controlling eutrophication philosophy and preliminary results of the OECD programme on eutrophication. *Progress in Water Technology* 12: 5–38.

Weiderholm, T. (1980). Use of benthos in lake monitoring. *Journal of Water Pollution Control Federation* 52: 537–547.

Welch, E.B. (1980). *Ecological Effects of Waste Water*. Cambridge: Cambridge University Press.

Wood, G. (1975). *An Assessment of Eutrophication in Australian Inland Waters*. Australian Water Resources Council Technical Paper No. 15. Canberra: Australian Government Publishing Service.

Ziober, S.R., Bialetzki, A., and de Fátima Mateus, L.A. (2012). Effect of abiotic variables on fish eggs and larvae distribution in headwaters of Cuiabá River, Mato Grosso State, Brazil. *Neotropical Ichthyology* 10 (1): https://doi.org/10.1590/S1679-62252012000100012 (accessed 6 September 2020).

9

Pesticides

9.1 Introduction

Pesticides are biologically active substances used worldwide that are designed for the control of pests or target organisms, such as insects, rodents, weeds, fungi, nematodes, and disease vectors in agriculture, forestry, and public health. They are also used in veterinary and human medicine to control parasites (WHO 2010).

The environmental impacts of pesticide use on non-target organisms, natural and agricultural ecosystems, and human health risks from exposure to toxic pesticide residues are a major global challenge for plant and animal protection in agriculture and pest control in human environments, including disease vectors (e.g. malaria). There is a basic conflict in design and use for many pesticides because the properties that enhance their efficacy and effectiveness can also result in adverse impacts and risks. These can include selective or nonselective toxicity to a diverse taxonomic range of biota, persistence, or mobility in the environment depending upon dispersion and environmental degradation, bioaccumulation, and increased pest resistance. Various conventional pesticides, including DDT and other POPs, have been *banned*, phased out of use, or restricted from use in many countries.

Historically, natural pesticides of inorganic mineral and plant extract origin (e.g. arsenic, copper, derris dust, sulfur, nicotine, and pyrethrins) have been used but suffer from lack of potency and specificity, and also high cost, to meet the demands of rapid human population growth for food production and better public health protection during the Twentieth Century and now. The development of synthetic chemicals also produced synthetic pesticides, such as DDT and other organochlorine insecticides, as well as organophosphate, and carbamate insecticides, that offered greater efficacy and economic benefits over the limited inorganic and botanical pesticides available for pest and vector disease control.

Since the Second World War era, the use of synthetic pesticides in agriculture, vector disease control, and many other applications, has resulted in a dramatic growth in worldwide pesticide production from an estimated 0.2 million tonnes in the 1950s to more than 5 million tonnes of pesticide formulations (including active ingredients) by the year 2000. Pesticides include a number of groups of compounds that have been progressively developed, from organochlorines, such as DDT, to organophosphates, carbamates, pyrethroids, growth regulators, neonicotinoids, and more recently, biopesticides. Herbicides such as glyphosate are now the main type of pesticides sold, followed by insecticides and fungicides (Carvalho 2017).

Global trends are toward less-persistent and low-toxicity pesticides, including biopesticides, and the development of pest-control strategies that reduce the use of chemical pesticides such as integrated pest management (IPM) and sustainable pest controls to combat pest resistance, and enhanced biological controls in agricultural ecosystems. Organic farming without pesticides has also increased.

9.2 Pesticides

Pesticides are defined as substances or mixtures of substances intended to control, prevent, kill, repel, or attract any biological organism defined as a pest. There are many kinds of pesticides of which insecticides, herbicides, and fungicides are major groups, while others include avicides, defoliants, desiccants, fumigants, and nematicides.

Classification

Pesticides consist of biologically active ingredients, derived from natural organic and inorganic materials, certain microorganisms, and synthetic organic chemicals, that are used to control specific or a broad range of pest species. The active ingredients are formulated into different types of commercial products, usually according to a variety of application needs (e.g. various concentrations, liquid or solid concentrates, sprays, wettable powders, or baits).

```
                          ┌─────────────┐
                          │  Pesticides │
                          └──────┬──────┘
                    ┌────────────┴────────────┐
              ┌─────┴─────┐              ┌────┴────┐
              │ Synthetic │              │ Natural │
              └─────┬─────┘              └────┬────┘
```

Synthetic

- **Insecticides**
  - Chlorinated hydrocarbons
  - Organophosphates
  - Carbamates
  - Pyrethroids
  - Neonicotinoids
  - Benzoylureas

- **Herbicides**
  - Triazines
  - Phenoxyacids
  - Ureas
  - Glyphosate
  - Bipyridiliums
  - Uracils
  - Dinitroanilines
  - Sulfonylureas

- **Fungicides**
  - Dithiocarbamates
  - Benzimidazoles
  - Phenylamides
  - Pyrimidines
  - Strobilurins

- **Fumigants**
  - Some growth regulators
  - Rodenticides

Natural

- **Biopesticides**
  - Derived from natural materials and certain minerals

- **Biochemical pesticides**
  - Control of pests by nontoxic mechanisms (e.g. insect sex pheromones and attractants)

- **Microbial pesticides**
  - Bacteria, viruses, fungi, and protozoa

- **Plant-incorporated protectants (PIPs)**
  - Pesticide substances produced by plants from added genetic material

- **Mineral derived**
  - e.g. copper compounds

- **Antimicrobial pesticides**

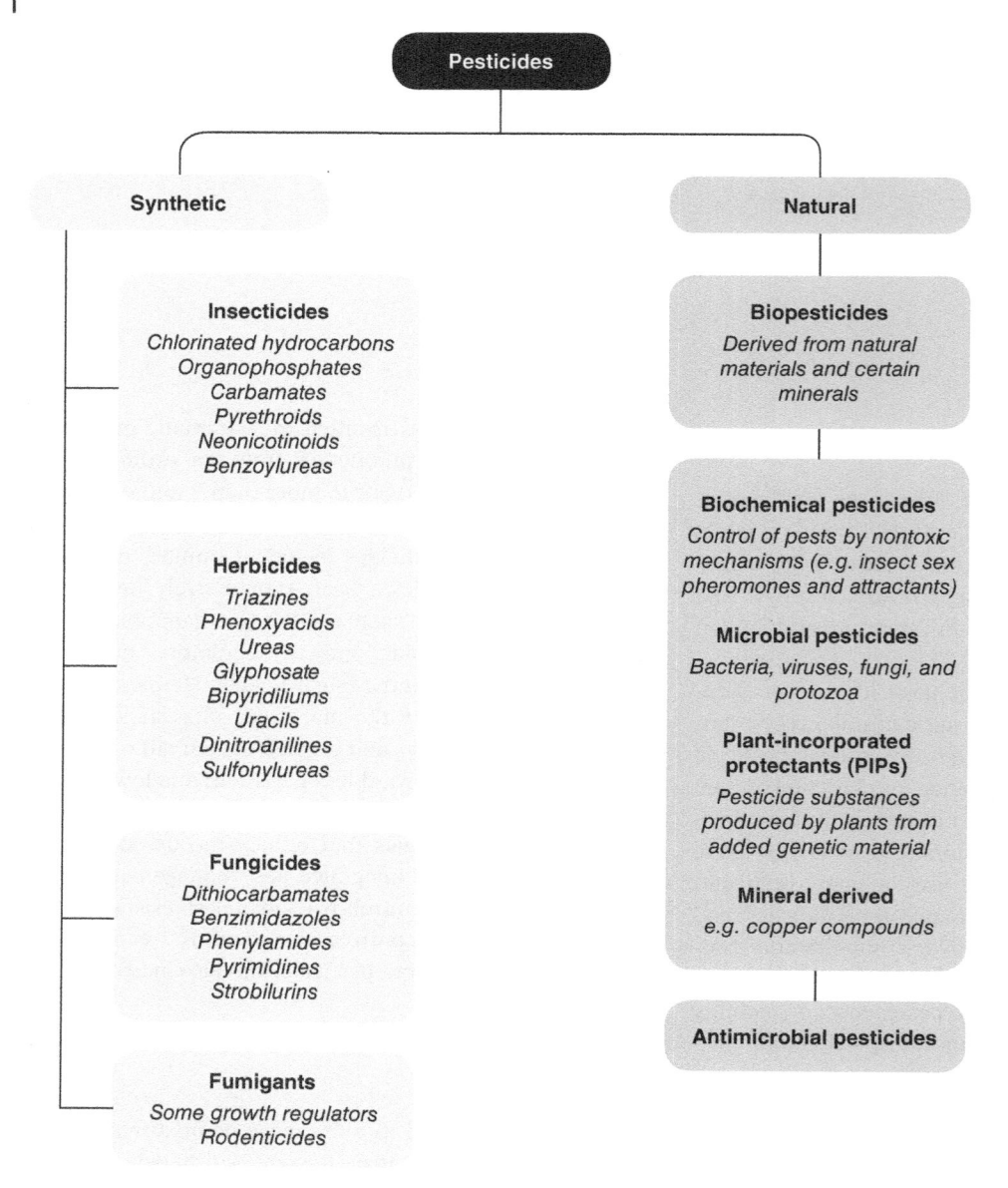

Figure 9.1 General classification of pesticides.

They can be broadly separated into synthetic or natural pesticides as shown in Figure 9.1.

Synthetic chemicals used primarily as insecticides, herbicides, fungicides, and fumigants are the dominant pesticides. Many occur in structurally related chemical families or classes, often with similar biological activities (e.g. selective or nonselective herbicides) against target pest organisms. Well-known chemical classes include chlorinated hydrocarbon (or organochlorine) insecticides, organophosphorus insecticides, synthetic pyrethroid insecticides, triazines, and phenoxyacid herbicides. New or more recent classes of insecticides, for example, include the neonicotinoids (e.g. imidacloprid), benzoylureas (e.g. hexaflumuron), and the phenylpyrazole family (e.g. fipronil). Types of synthetic pesticides are discussed in Section 9.3.

Natural pesticides are extracted or derived from natural plants, animals, microorganisms (e.g. live bacteria and toxins) and certain minerals. Among the earliest pesticides were inorganic chemicals such as sulfur, lime, arsenic trioxide, and copper compounds that are less chemically complex than natural organic pesticides such as pyrethrum extracted from the *Chrysanthemum* flower. The growth of biological pest control and management has created a major category of biopesticides that are derived from natural materials. For pesticide registration purposes in the United States, the US EPA has divided biopesticides into three classes (US EPA 2016):

1. Biochemical pesticides consisting of natural substances (e.g. insect sex hormones or insect attractants) that control pests by nontoxic mechanisms.
2. Microbial pesticides that use a microorganism (e.g. a bacterium, fungus, virus, or protozoan) as the active ingredient to specifically control a target pest(s). For example, subspecies and strains of *Bacillus thuringiensis* (Bt) are widely used to control specific species of insect larvae (e.g. flies and mosquitoes).
3. Plant incorporated protectants (PIPs) that are pesticide substances produced by plants from genetic material that has been added to the plant.

Biopesticides may also include predatory insects used for crop protection, such as ladybugs that predate on aphids.

Antimicrobial pesticides are another special and valuable category of substances or mixtures of substances used to control the growth of infectious or other harmful microorganisms for a wide range of public health, agricultural, industrial, commercial, and domestic applications. These include disinfectants, sanitizers, germicides, chemical sterilants, and biocides. In the United States, about 275 different active ingredients for pesticides were registered with the EPA in 2017 (US EPA 2017a).

Pesticides are usually classified according to the type of target organisms they control or kill, as shown in Table 9.1. The major types of pesticides are synthetic chemicals in the form of insecticides, herbicides, and fungicides. Biopesticides represent an emerging market (~5% of crop protection

Table 9.1 Classification of pesticides according to types of target organisms.

Pesticides	Examples	Target organisms
Acaricides	Permethrin, carbamate, and organophosphate insecticides	Mites
Algicides	Copper sulfate, terbutryn, endothal	Algae
Antifouling agents	tributyl tin (TBT), copper compounds	Biofouling organisms, e.g. molluscs, barnacles, polychaetes, and seaweed
Avicides	CPTH, fenthion	Birds
biopesticides	Microbial, biochemical (e.g. neem oil), and plant incorporated protectants (GMOs)	Diverse pests and diseases – animals, plants, and microbes
Biocides	Pesticides and antimicrobial agents	Harmful organisms to human and animal health
Defoliants	2,4-D; 2,4,5-T (agent orange)	Stripping leaves off plants
Dessicants	Silica gel, calcium chloride, zeolites, and boric acid	Drying out of living tissues, e.g. insects and plants
Disinfectants	Glutaraldehyde, hexachlorophene	Microorganisms
Fumigants	Methyl bromide, chloropicrin	Insects
Fungicides	Tea tree oil, vinclozolin, mancozeb, and chlorothalonil	Fungi (e.g. mildews, molds, and rusts)
Herbicides	Atrazine, simazine, glyphosate, 2,4-D, diuron, triclopyr, metolachlor, and diquat	Plants
Insecticides	DDT, dieldrin, endosulfan, chlorpyrifos, imidacloprid, bifenthrin, aldicarb, fipronil, and pymetrozine	Insects
Insect growth regulators	Methoprene, pyriproxyfen, and triflumuron	Insects – mimic juvenile growth in hormones
Larvicides	*Bacillus thuringiensis israelensis*, Bt or Bti, temephos, and methoprene	Insect larvae
Molluscicides	Metaldehyde, methiocarb	Snails, slugs
Nematicides	Aldicarb, carbofuran, methyl bromide, and terbufos	Nematodes
Plant growth regulators	2,4-D, 2,4,5-T, cytokinins (e.g. thidiazuron)	Plants
Ovicides	Malathion, spinosad	Eggs of insects and mites
Pheromones	Methyl eugenol, beta-farnesene	Disrupt insect behavior
Piscicide	Rotenone, saponins, and antimycin	Fish
Repellents	N,N-diethyl-*meta*-toluamide (DEET) and picaridin	Repel insects (e.g. mosquitoes) and birds
Rodenticides	Difethialone, sodium fluoroacetate, and bromadiolone	Rodents (e.g. rats and mice)
Termiticides	Fipronil, bifenthrin, hexaflumuron, triflumuron, and pyriproxyfen	Termites

market) that is projected to challenge the more toxic and less specific synthetic pesticides such as organophosphates.

Alternatively, pesticides may be classified according to chemical structures, which is useful for many families of synthetic chemicals that act as major insecticides, herbicides, or common fungicides. Yet the diverse chemical structures and biological activities inherent in active ingredients of pesticides can affect a broad-based classification. Even so, our understanding of the environmental behavior, fate, and toxicity of pesticides depends largely on their physicochemical nature and properties and critically applies to chemical families with known environmental and health hazards.

9.3 Nature and Properties of Various Pesticides

Pesticides include a vast array of natural and synthetic substances of widely different chemical nature. Organic pesticides consist of individual or combinations of carbon skeletons in the form of chains, aromatic, and heterocyclic rings, with various substituent elements, such as hydrogen and halogens (e.g. chlorine and fluorine), and/or functional groups (e.g. hydroxyl or phosphates) that confer different physical, chemical, and toxicological properties on their structures.

Their only common property is their ability to destroy or control pest organisms. Environmental toxicology and human health effects are principally concerned with the synthetic organic chemicals although others can have a significant environmental and health effect in some situations. The pesticides of major ecological and health significance are the insecticides, herbicides, and fungicides. Biopesticides are generally less toxic for nontarget species than synthetic chemical pesticides. They are an emerging group that needs more environmental evaluation because of their increasing use and potential as a substitute for many synthetic chemical pesticides on a worldwide scale.

The synthesis and properties of organic pesticides are generally related to the nature of their chemical structures and functional groups needed to achieve their desired environmental behavior and toxicity to target organisms. For instance, the type and range of bonds present in a chemical family such as chlorinated hydrocarbon insecticides (CHCs) (e.g. C—C, C—H, and C—Cl) influence their polarity and dipole moments and the solubility of the overall molecule in water or lipids (fats). The structures and bond types in CHCs tend to have low polarities and dipole moments resulting in low water solubility and high lipid solubility (lipophilic) (Connell 2005, p. 171).

Key physical and chemical properties used to evaluate the behavior and fate of chemical pesticides include the following: molecular weight, density, solubility in water, vapor pressure, n-octanol – water partition coefficient (K_{OW}), Henry's Law constant, and dissociation constant in water (pK_a and pK_b). Other useful properties include air–water partition coefficient (K_{AW}) and n-octanol–air partition coefficient and UV/visible light absorption, although these values may be unavailable (US EPA 2019).

Table 9.2 (and a more detailed Table S9.1, Section S9.1) lists some of these properties for a set of common synthetic organic pesticides. For example, the aqueous solubility of the pesticides ranges from very high (paraquat) in $g\,L^{-1}$ to about $5\,\mu g\,L^{-1}$ for DDT (5.50×10^{-3} $mg\,L^{-1}$) depending upon chemical family, temperature. and pH (hydrolysis or ionization). Conversely, the lipophilicity of these pesticides is indicated by the octanol–water partition coefficient (K_{OW}). Log K_{OW} values range from between <1 for negligible solubility (e.g. glyphosate) to 6 or more for very high lipid solubility (e.g. DDT). In Section 9.6, pesticide fate in the environment is described using some of these types of properties and pesticide examples.

The behavior and fate of pesticides in the environment are also influenced by their environmental properties such as their partitioning between different environmental phases (e.g. water and soil or water and lipids in biological tissues). The distribution and mobility of pesticides in soils, for instance, can be affected by sorption between the chemical pesticide and solids such as soils, sediments, plants, and other surfaces.

Bioconcentration properties are also a characteristic of the lipophilic nature of a particular pesticide if it occurs by partitioning between the lipid in an organism and the surrounding water at equilibrium and without biological degradation and removal. The bioconcentration factor K_B (organism-water partition coefficient) is defined as follows:

$$K_B = C_{organism}/C_{water} \quad (9.1)$$

Pesticides such as CHCs with high log K_{OW} values (e.g. 5 or 6) would be expected to have high K_B values. Bioconcentration is considered further in Section 9.8.

Other key environmental properties of pesticides that influence their persistence include their environmental degradation rates (often measured by half-lives) in different media due to combinations of physical, chemical and biological processes (see Chapter 4). Again, physical and chemical properties of pesticides, including their structures and bond types, are key factors in influencing environmental processes such as biodegradation. The half-life of a pesticide in an environmental medium (e.g. air, water, or organism) is a relevant environmental property for persistence.

Table 9.2 Some physicochemical properties of various pesticides.

Pesticide	Molecular weight (g mol^{-1})	Water solubility (mg L^{-1}) 25 °C	Log K_{OW}	Vapor pressure (mm Hg) 25 °C	Henry's law constant (atm-m^3 mol^{-1}) 25 °C
Insecticides					
DDT	354.49	0.0055	6.91	1.6×10^{-7}	8.32×10^{-6}
Endosulfan	406.93	0.53 α	3.83 α	1.73×10^{-7}	6.5×10^{-5} 20 °C
		0.28 β	3.62 β		
Chlorpyrifos	350.586	1.4	4.96	2.02×10^{-5}	3.55×10^{-5}
Diazinon	304.35	40	3.81	9.01×10^{-5}	6.25×10^{-8}
Aldicarb	190.26	4930 (20 °C)	1.13	2.9×10^{-5} (20 °C)	1.5×10^{-9}
Bifenthrin	422.868	$<1 \times 10^{-3}$ (20 °C)	6.00	1.335×10^{-8} (20 °C)	1.0×10^{-6}
Imidacloprid	255.69	6.1×10^2 (20 °C)	0.57 (21 °C)	3.0×10^{-12}	1.65×10^{-15}
Fipronil	437.15	1.9 (pH 5), 2.4 (pH 9) (20 °C)	4.0	2.78×10^{-9}	8.42×10^{-10}
Herbicides					
Atrazine	215.684	34.7 (26 °C)	2.61	2.89×10^{-7}	2.6×10^{-9}
Glyphosate	169.07	Soluble	−3.40	9.8×10^{-8}	5.82×10^{-11}
2,4-D	221.03	677 ppm	2.81	1.40×10^{-7} (20 °C)	9.75×10^{-8} (20 °C)
Diuron	233.09	42	2.68	6.9×10^{-8}	5.0×10^{-8}
Metolachlor	283.79	530 (20 °C)	3.13	3.14×10^{-5}	9×10^{-9} (20 °C)
Fungicides					
Azoxystrobin	403.39	6.0	2.50	8.25×10^{-13}	7.3×10^{-14}
Chlorothalonil	265.9	0.81	3.05	5.7×10^{-7}	2.5×10^{-7}

Source: Data compiled from PubChem database, National Institutes of Health, US; Chemspider database, The Royal Society of Chemistry, London.

9.4 Major Groups of Pesticides

Insecticides

Insecticides cover a number of families of synthetic chemicals, including termiticides, ovicides, larvicides, and insect growth regulators, and a broad range of natural or biopesticides of chemical or biological origin. Among the major synthetic chemical insecticides are the chlorinated hydrocarbons, organophosphates, carbamates, pyrethroids, nicotinoids, and benzoylureas. Natural insecticides include inorganic insecticides, such as sulfur, boric acid, borax, arsenic trioxide, and some biopesticides, such as pyrethrum, nicotine, neem extracts, and spinosyns (e.g. spinosad) derived by fermentation from soil bacteria or bacterial toxins (e.g. *Bacillus thuringiensis*).

Chlorinated Hydrocarbon Compounds (CHCs)

The chlorinated hydrocarbons or organochlorine compounds are the most controversial group of pesticides. This group includes DDT and related substances and were extensively used in agriculture and to varying degrees in public health (e.g. DDT for malaria control and lindane [gamma BHC] as a miticide), since the Second World War until the 1990s, due to their *effectiveness* and low cost to applicators. (Note: DDT continues to be used for malaria control in certain countries.) Maximum usage occurred in the early 1970s, but CHCs have been largely phased out from use due to insect resistance and their ecological and health impacts on a global scale. Chlorinated hydrocarbon insecticides are a major sub-group of persistent organic pollutants (POPs) listed under the Stockholm Convention (see Chapter 2).

The chemical structures of some common CHC insecticides are shown in Figure 9.2. Chemical sub-classes of these insecticides consist of the diphenyl aliphatics (e.g. DDT, dicofol, and methoxychlor), hexachlorocyclohexanes (HCH) (also known as benzene hexachloride, BHC) – these are alpha, beta, gamma, delta, and epsilon isomers of HCH (the gamma isomer is also called lindane or γ BHC); cyclodienes (aldrin, dieldrin, chlordane,

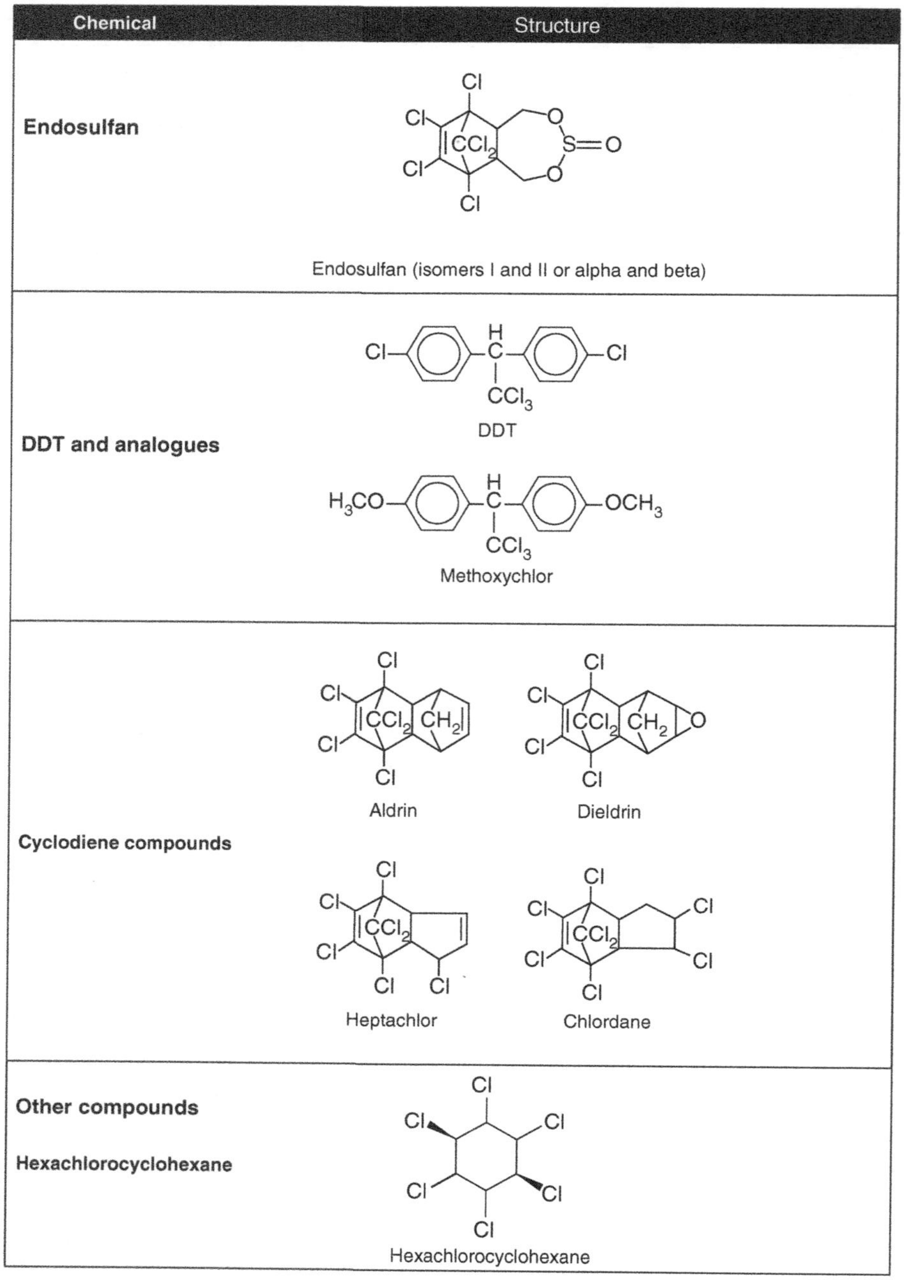

Chemical	Structure
Endosulfan	Endosulfan (isomers I and II or alpha and beta)
DDT and analogues	DDT; Methoxychlor
Cyclodiene compounds	Aldrin; Dieldrin; Heptachlor; Chlordane
Other compounds: Hexachlorocyclohexane	Hexachlorocyclohexane

Figure 9.2 Chemical structures of common organochlorine or chlorinated hydrocarbon insecticides.

heptachlor, endrin, mirex, endosulfan, and chlordecone); and polychloroterpenes (e.g. toxaphene).

CHC insecticides show some general similarities in properties. They have low water solubility, high lipid or fat solubility and are particularly persistent in the natural environment, although endosulfan is less so. The types of bonds in CHCs are generally resistant to attack by abiotic or biotic agents, so degradation rates proceed slowly. In soils, half-lives of these insecticides can be up to years depending on soil conditions (e.g. microbial populations, temperatures, and moisture). They are expected to strongly bioconcentrate based on log K_{OW} values, ranging from usually 2 to 6, and with increasing K_B values (see Connell 2005, pp. 171–173). Field and monitoring studies show they can bioaccumulate in individual organisms and may biomagnify in some food chains, such as for DDT/DDE in top air-breathing predators.

Historically, DDT is commonly used as the benchmark chemical for persistent organic pesticides and pollutants.

Organophosphates

The organophosphate insecticides (OPs) are esters that are derivatives of phosphoric acid with varying combinations of oxygen, sulfur, carbon and nitrogen atoms. They can be generally divided into three groups: aliphatics (e.g. TEPP, malathion, dimethoate, dichlorvos, and methamidophos), phenyl derivatives (e.g. parathion, sulfrofos, and fenitrothion) and heterocyclic derivatives (azinphos-ethyl, chlorpyrifos, and phosmet).

The general formula shows the phosphorus ester contains one P=O or P=S and three other P bonds as shown in Figure 9.3.

The two R groups are usually methyl or ethyl groups. The oxygen atom in the P=O can be replaced by sulfur S in some compounds, and the X can be aliphatic carbon chains, phenyl or heterocyclic structures attached to the O or S atom. Examples of the chemical structures of organophosphate insecticides are shown in Figure 9.4. Chlorpyrifos is the most widely used of these insecticides.

Their general properties differ markedly from the CHC insecticides. They are chemically reactive, most being unstable (e.g.undergo hydrolysis in water), and have limited persistence in the natural environment. The P—O bonds in the phosphorus ester are likely to be polar similar to the O—H bond. In contrast, the alkyl R—O bonds would have low polarity, while other attached bond types on the molecule may vary in polarity. Overall, these compounds are expected to have greater water solubility than CHC insecticides and have lower lipophilicity (Connell 2005, pp. 176–178).

Chemical	Structure
General formula for organophosphorus insecticides	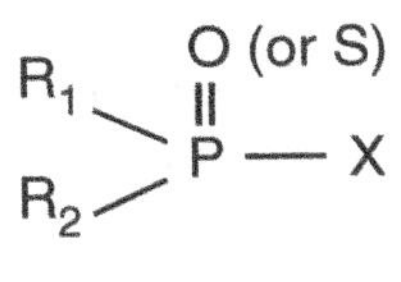

Figure 9.3 General structural formula for organophosphate insecticides.

Chemical	Structure
Chlorpyrifos	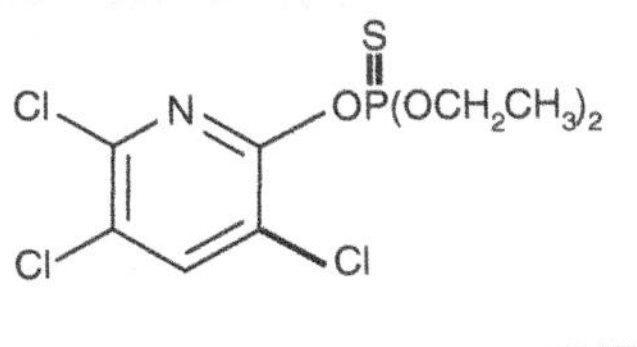
Fenitrothion	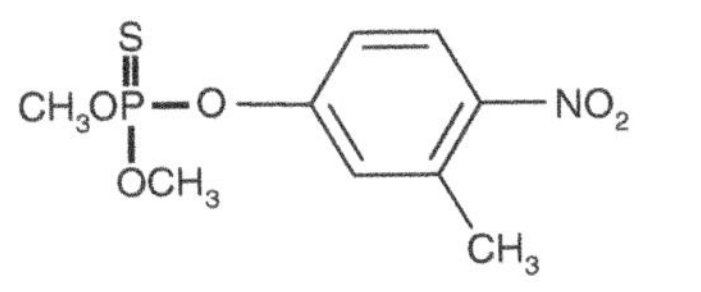

Figure 9.4 Examples of common organophosphate insecticides.

As indicated in Table 9.2 and from the pesticide literature, these pesticides tend to be moderately water soluble, practically non-bioaccumulative, and do not biomagnify in food chains. These properties were originally useful as substitutes for the persistent CHC insecticides, but they suffer with the disadvantage of being generally more toxic to vertebrates than other insecticides, through inhibiting key enzymes (cholinesterase) in the nervous system. The binding of OPs to the enzyme tends to be irreversible.

Carbamates

The carbamate insecticides are derivatives of carbamic acid and have the following general structural formula shown in Figure 9.5. R can be a variety of groups, often containing aromatic rings, and R_2 can be hydrogen or other groups.

The general properties of carbamates are similar to those of organophosphates. The carbamate function group is polar leading to a relatively polar molecule like OPs of moderate water solubility. The ester group in carbamates is readily hydrolyzed in contact with water and contributes to their low persistence in water and soils. Their mode of action also inhibits vital cholinesterase enzymes but binding to the enzyme is reversible. Common carbamates include carbaryl, aldicarb, propoxur, bendiocarb, and primicarb (Connell 2005, pp. 179–180).

Pyrethroids

The synthesis of pyrethroids is derived from the chemical nature of natural insecticides known as pyrethrins that are

Chemical	Structure
General structure for carbamate insecticides	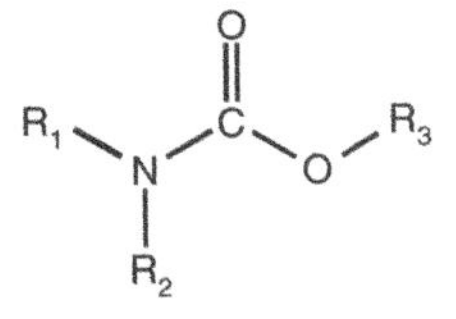
Carbaryl	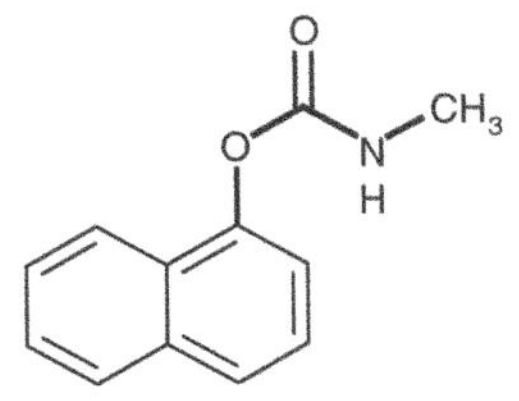
Methomyl	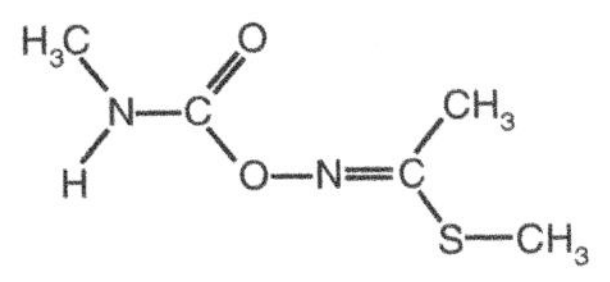

Figure 9.5 Examples of common carbamate insecticides.

found in flowers belonging to the genus *Chrysanthemum*. The general approach has been to retain the molecular geometry of their natural insecticidal activity while increasing their environmental persistence. Common pyrethroids such as permethrin, bifenthrin, cypermethrin, cyfluthrin, and deltamethrin are stable in sunlight, act as broad-spectrum insecticides, and are effective at low application rates. Pyrethroids are neurotoxic substances that have a similar toxic action as DDT. While mammalian toxicity is relatively low, pyrethroids are toxic to fish. They contain some polar bonds but have a large number of nonpolar groups, resulting in low water solubility and a lipophilic nature, in contrast to OP and CHC insecticides. Most have a low volatility due to their higher molecular weight (see Table 9.2) (Connell 2005, pp. 180–183). Examples of chemical structures of synthetic pyrethroids are given in Figure 9.6, along with other insecticides discussed below.

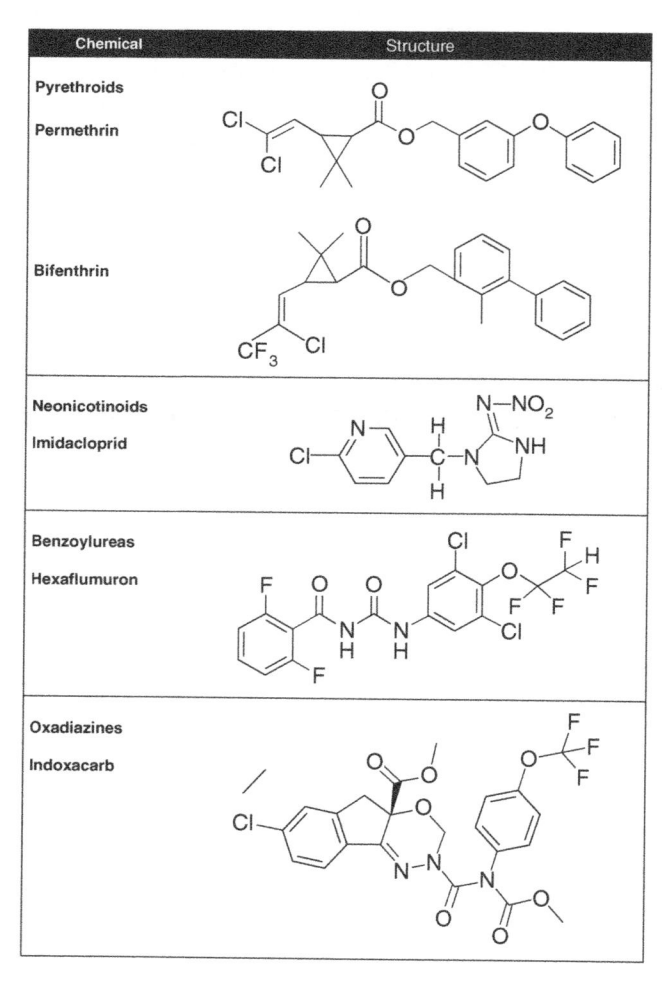

Figure 9.6 Examples of other common insecticides including synthetic pyrethroids and neonicotinoids that are widely used.

Neonicotinoids

Neonicotinoids or simply nicotinoids are a more recent class of neuro-active insecticides modeled after natural nicotine. Neonicotinoids and fipronil (a phenylpyrazole insecticide) are among the most widely used pesticides in the world, mainly due to their high toxicity to invertebrates, ease and flexibility of application, their long persistence, and their systematic nature, which ensures they transfer to all parts of the crop (Bonmatin et al. 2015). The nicotinoid family includes imidacloprid, one of the most widely used insecticides in the world, thiacloprid and thiamethoxam. The chemical structure of imidacloprid is included in Figure 9.6. They act on the central nervous system of insects by causing irreversible blockage of postsynaptic nicotinergic acetylcholine receptors. These insecticides are highly toxic to many classes of insects and show low toxicity to vertebrates such as mammals. Most nicotinoids are soluble in water, low in lipid solubility, and also vapor pressure but degrade slowly in the environment, especially in the absence of sunlight and microbial activity. Neonicotinoids, fipronil, and metabolites are implicated in widespread exposure of nontarget organisms, including pollinators such as honey bees and their colonies, and environmental contamination, particularly from dispersion of contaminated dusts during large-scale treated seed planting (Bonmatin et al. 2015).

Benzoylureas

These insecticides selectively control insect pests. They are insect growth inhibitors that act by preventing the production of chitin that is used by an insect to build its exoskeleton. Diflubenzuron is a common member used as an insecticide on cattle, field crops, in forest management, and as a larvicide (mosquitoes and flies). It has a low acute toxicity to mammals but can have high acute and chronic toxicity to some aquatic invertebrates (US EPA 1997a).

Hexaflumuron is another benzoylurea being used as an insect growth regulator (see Figure 9.6). It was the first pesticide to be classified as a low-risk pesticide by the US EPA (Baird and Cann 2012, p. 605). It is effectively used in bait stations as a termiticide. It has very low water solubility, a moderate vapor pressure, binds strongly to soil (and suspended solids), and is practically immobile. With a high log K_{OW} value of 5.68, the estimated bioconcentration factor (BCF) suggests the potential for bioconcentration in aquatic species is very high (estimated BCF of 4700, PubChem website).

Bioinsecticides

These types of insecticides are designed as alternatives to higher risk synthetic insecticides, especially from a mammalian toxicity perspective. S9.2 text provides a concise description of several bioinsecticides (on-line website).

Herbicides

Herbicides make up the largest group of pesticides in commercial use. Many chemical classes of herbicides exist (see Ware and Whitacre 2004; Tu et al. 2001). A wide variety of essentially organic substances are used as herbicides, primarily synthetic organic chemicals. Herbicides in the same class or family usually have the same mode of action due to their chemical structure (see Ware and Whitacre 2004; Tu et al. 2001).

There are five general types of herbicides:

1. Broad spectrum – affect a wide variety of plants.
2. Selective – act on a narrow range of plants.
3. Contact – kill plant tissue at or near the point of contact with it (require even coverage in their application).
4. Systemic – move through the plant tissues via the circulation system of the plant.
5. Residual – can be applied to the soil in order to kill weeds by root uptake. They remain active in the ground for a certain length of time and can control germinating seedlings.

Major herbicide classes and chemical structures of some common examples are shown in Figures 9.7 and 9.8, and can be divided broadly into mainly nitrogen-based herbicides with some non-nitrogen containing herbicides. Glyphosate and atrazine are examples of major global herbicides.

Similar to other pesticides, the physical and chemical properties of herbicides affect their environmental properties, mobility, and persistence. Herbicides are usually soluble in water, which increases their mobility in water and potential leaching from soils, depending upon other environmental factors such as sorption or binding capacity with soil particles. Trifluralin, for example, has a low solubility in alkaline water (pH = 9) (Table 9.2). Some herbicides, such as glyphosate and the bipyridiniums, paraquat, and diquat, can bind strongly to soils despite their high water solubility.

Most herbicides have an intermediate to low vapor pressure, as indicated in Table 9.2, and are relatively nonvolatile under field conditions. Volatile herbicides include members of the thiocarbamate family and dinitroanilines (e.g. trifluralin). The chemical structure of a herbicide influences how it degrades by chemical reaction (e.g. hydrolysis and oxidation) and/or microbiological activity and hence its persistence in water and soil. The chemical structure of 2,4-D, for example, allows microbes to rapidly detoxify the molecule into inactive metabolites, whereas atrazine is not as prone to microbial attack; hence degradation is slower. The pH of water and soil can also be a critical factor for chemical reactions (Penn State Extension 1999).

The K_{OW} values of herbicides tend to be low, indicating low bioaccumulation potential in invertebrates and vertebrates. Despite this, atrazine (one of the most widely used herbicides) can bioaccumulate in fish species.

The development of selective herbicides, with plant growth control or auxin-like properties, has resulted in extensive use of these substances for weed control in agriculture and forestry. Plants are destroyed by disruption of their growth processes. Herbicides in this group are the phenoxy acids including 2,4-dichlorophenoxy acetic acid (2,4-D) and 2,4,5 trichlorophenoxy acetic acid (2,4,5-T) (see Figure 9.8). Generally, the phenoxy acids have limited persistence in the natural environment, are moderately water soluble, and do not bioaccumulate. 2,4-D is a major herbicide, for example, in the United States. In contrast, 2,4,5-T, which contains residues of the toxic dioxin, TCDD, has practically been phased out of use (see Ware and Whitacre 2004).

An example of a low-toxicity synthetic herbicide is imazaquin (classed as an imidazolinone). It has a water solubility of 60 mg L^{-1} and its half-life in soil is about 60 days. It is therefore categorized as a moderately persistent pesticide.

Bioherbicides are commonly compounds and secondary metabolites derived from microbes, such as fungi, bacteria, or protozoa, or substances extracted or derived from other plant species, parasitic plants, and certain insects such as parasitic wasps. For example, glycopeptides can be produced by a number of bacteria to induce disease in target plant species or various commercial suspensions of fungal spores can be applied to cause, for instance, root rot or stem and leaf blight in susceptible plants.

Since the 1980s, a new strategy for weed control in major crop production has been the use of genetic engineering to produce transgenic plants that have enhanced resistance to herbicides such as the broad-spectrum herbicide, glyphosate (Roundup®). The pesticide manufacturer, Monsanto, developed genetically altered seeds for soy, corn, alfalfa, sorghum, canola, and cotton to grow into glyphosate-tolerant plants (genetically modified organisms, GMOs), which allowed greater quantities of glyphosate to be applied against resistant weeds (Baird and Cann 2012, p. 612).

Fungicides

Fungicides are basically chemical compounds or biological organisms that can kill fungi or fungal spores by contact or systemic absorption. They can be of natural or synthetic origin. Natural substances include inorganic minerals and natural plant extracts (e.g. tea tree, citronella, clove,

Chemical	Structure
Pyridine acids Picloram	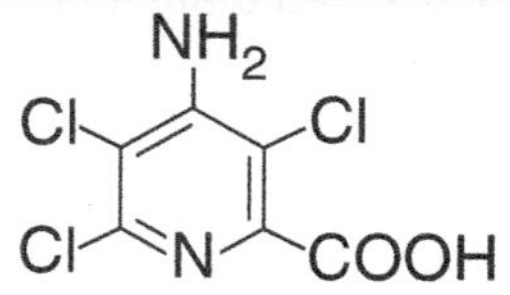 4-amino-3,5,6-trichloropyridine-2-carboxylic acid structure
Triazines Atrazine	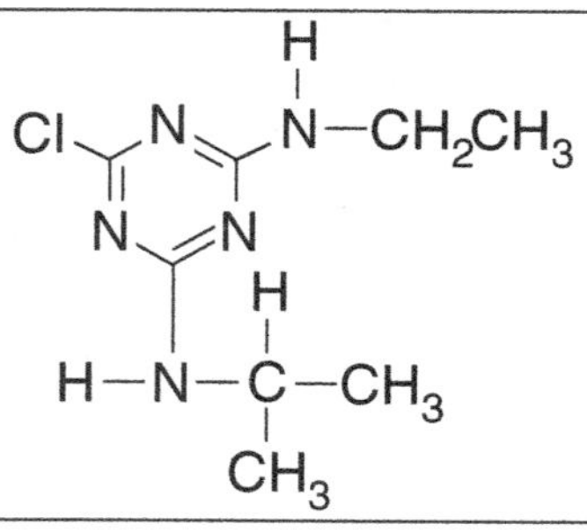 2-chloro-4-(ethylamino)-6-(isopropylamino)-1,3,5-triazine structure
Ureas Diuron	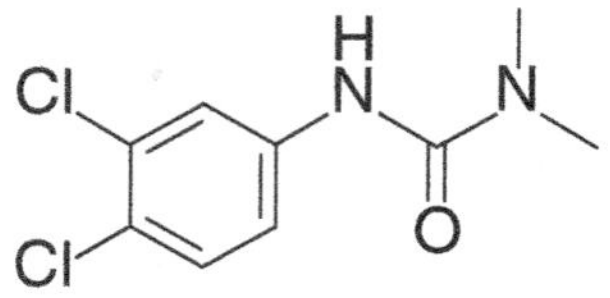 3-(3,4-dichlorophenyl)-1,1-dimethylurea structure
Dinitroanilines Trifluralin	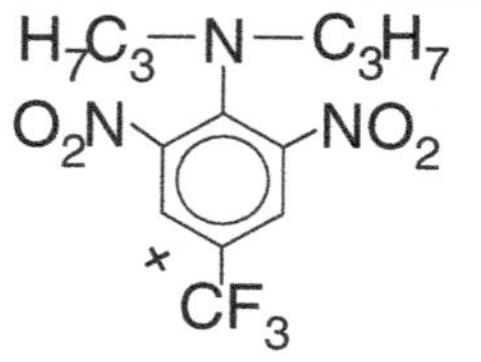 N,N-dipropyl-2,6-dinitro-4-(trifluoromethyl)aniline structure
Bipyridyliums Diquat	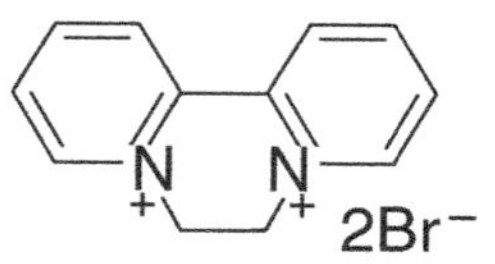 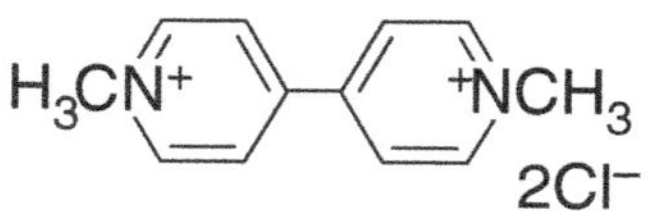 Diquat dibromide and paraquat dichloride structures
Phosphonic acid Glyphosate	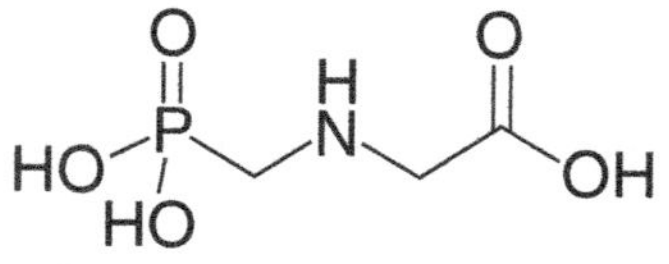 N-(phosphonomethyl)glycine structure

Figure 9.7 Chemical structures of some common herbicides.

or rosemary oils) or microbial organisms (e.g. *Bacillus subtilis*). These are generally classified as biopesticides.

Synthetic fungicides are chemically diverse: organic, inorganic and organometallic chemicals (e.g. mancozeb, benomyl, carbendazim, imazaquin, trifloxystrobin, vinclozolin, copper oxychloride, and copper hydroxide) and it is very difficult to generalize on their properties. Some early synthetic fungicides such as hexachlorobenzene (HCB) (a persistent organic pollutant), phenylmercuric acetate (mercury-based), and pentachlorophenol (PCP) are toxic to many nontarget organisms.

A new highly effective and major class of fungicides is the strobilurins, which are derived from the fungus, *Strobilurus tenacellus*. For instance, they are used in controlling downy mildew diseases. Strobilurins include one of the world's biggest selling fungicides, azoxystrobin. This fungicide was considered to be of low toxicity for humans and a reduced risk pesticide by the US EPA. It is highly toxic to aquatic

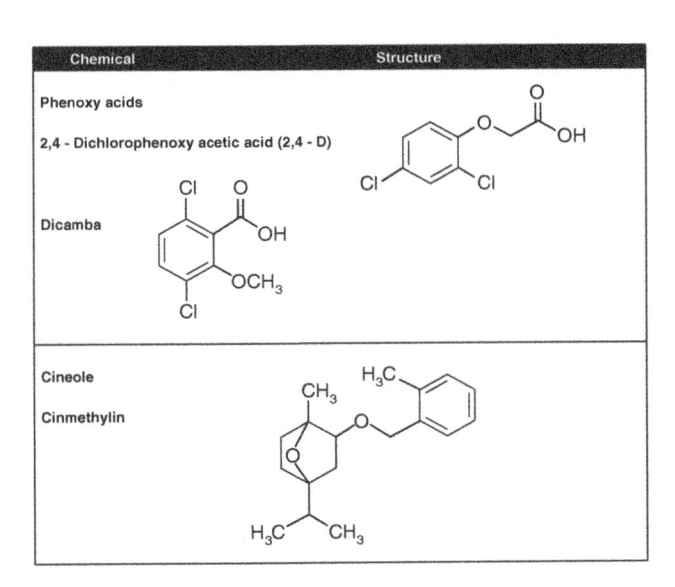

Figure 9.8 Chemical structures of some other herbicides such as 2,4-D – a common chlorinated phenoxy acid herbicide, and cinmethylin, a novel derivative of the cineole family found in essential oils (e.g. Eucalyptus).

fish and invertebrates although environmental risks were evaluated to be low at time of registration by the US EPA (1997b).

Many biofungicidal products are being used in seed treatments to control soil-borne pathogens and also in crop treatments to stimulate the plant host defense and other physiological processes, which can make treated crops more resistant to various environmental stresses.

Fungicides are used in lower quantities than either the insecticides or herbicides but are used in a wide range of valuable applications such as inhibition of fungal growth in paper manufacture, mildew-proofing of fabrics, control of post-harvest rot in fruits, vegetables, stored grains, and so on. They are chemically very diverse, as indicated, for example, in Figure 9.9, and their environmental behavior and toxicity are difficult to generalize. Fungicides may persist for varying periods in the environment due to their chemical structures, properties, climatic, and environmental factors, like herbicides. Vinclozolin is an example of a fungicide that can exhibit prolonged persistence of residues, including metabolites.

Wightwick and colleagues in 2010 evaluated environmental risks of fungicides used widely in horticulture production and concluded that it was not currently possible to characterize environmental risks due to fungicides with any degree of certainty, except possibly for risks to soil organisms from copper accumulation (copper-based fungicides). They also compiled a set of values (ranges or individual) for physical and chemical properties of fungicides registered for use in Australian viticulture and

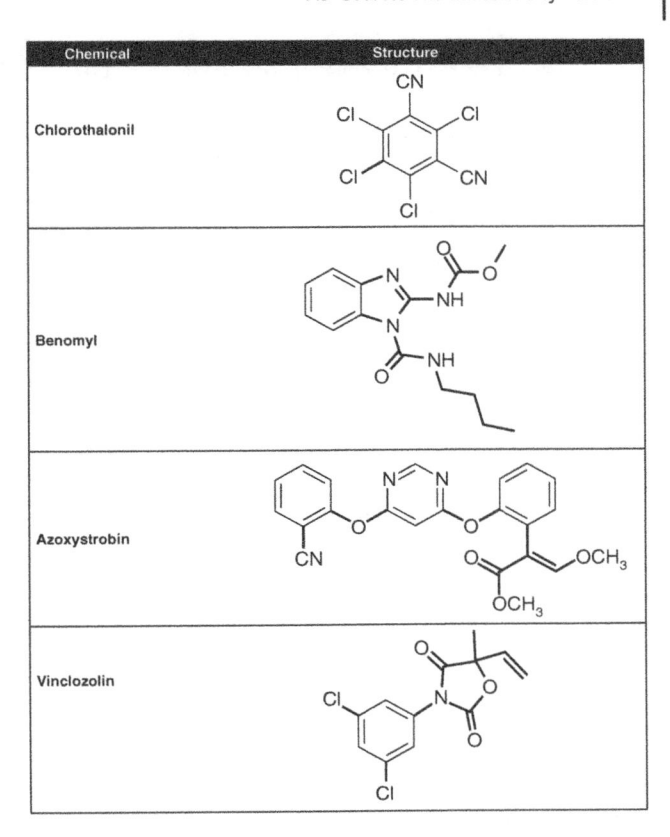

Figure 9.9 Examples of chemical structures of common fungicides.

their chemical classes. These include triazoles, carboxamides, dithiocarbamates, strobilurins, amides, anilide, benzimidazole, and phthalimide (Wightwick et al. 2010).

9.5 Sources and Emissions of Pesticides

Sources of pesticide residues from production, formulation, and multi-use applications occur almost globally and are increasing. World use of pesticides was estimated to be 5.2 billion pounds (2.36 million metric tonnes) in 2007, as shown in Table 9.3. The world's largest market for pesticides in 2007 was the United States (32% of the almost US$40 billion world market in 2007). It used 1.1 billion pounds (0.5 million metric tonnes) of pesticides, equivalent to 22% world usage (Grube et al. 2011; Table 9.3).

The extent of worldwide use and distribution of pesticides is indicated by global pesticide sales given in the Global Chemicals Outlook Report in 2013 (UNEP 2013, pp. 33–35). In 2009, North America accounted for 27% of the world market, Asia/Pacific (24%), Central and South America (19%), and Western Europe (17%). Africa and the Middle East have relatively small expenditures on pesticides but rapid agricultural growth and pesticide

Table 9.3 World and United States estimates of quantities and market share of pesticides (active ingredients) used in 2007.

Pesticide type	World market		US market		US share of world market %
	Million pounds	%	Million pounds	%	
Herbicides[b]	2096	40	531	47	25
Insecticides	892	17	93	8	10
Fungicides	518	10	70	6	14
Other[b]	1705	33	439	39	26
Total	5211	100	1133	100	22

Notes: Table does not include wood preservatives, speciality biocides, and chlorine/hypochlorites.
[a]Includes plant growth regulators.
[b]Includes nematicides, fumigants, other miscellaneous conventional pesticides, and other chemicals used as pesticides (e.g. sulfur, petroleum oil, and sulfuric acid).
Source: Grube et al. (2011).

use, particularly insecticides, are occurring in these two regions. South Africa, the largest consumer in Africa, accounts for 2% of global pesticide consumption.

Pesticides are mainly used in agriculture throughout the world. Section S9.3 in the Companion Book Online provides further insight into pesticide uses, mainly in the United States and the EU.

Little is known about the extent or amounts of emissions from worldwide applications of pesticides inside homes and other buildings, gardens, weed control by local authorities, and large-scale vector-borne pest control. United States data on conventional pesticide use shows that 12% of these pesticides were used by industrial, commercial, and government activities, mainly as herbicides, fungicides, and fumigants, while home and garden use accounted for 8% in 2007, mostly as herbicides and insecticides. Significant amounts of pesticides could be expected to accumulate on surfaces and in dusts in human built environments (e.g. inside buildings from pest control treatments). Other countries with pesticide use patterns similar to the United States may be expected to release comparable quantities into built environments.

Additional sources of pesticides released into the environment include during manufacture and formulation, spillages (e.g. handling, storage, and transport), waste disposal and leaching from landfills, contaminated lands, or storages of obsolete pesticides. An inventory maintained by FAO was reported to contain 537 000 tonnes of obsolete pesticides located in Africa, Asia, Eastern Europe, Latin America, and the Middle East (UNEP 2013, p. 42). This quantity indicates wastes or unwanted pesticides are a major potential source and hazardous waste management issue.

Estimates of pesticide losses or emissions to local environments and nontarget organisms from agricultural uses by spray drift, wind erosion, runoff to surface waters, and leaching to groundwater are highly variable due to a range of factors. Tyler Miller (2004, pp. 211–216) reported that over 98% of sprayed insecticides and 95% of applied herbicides are dispersed beyond their target species when sprayed or spread across cultivated farmlands.

Pesticide loss by volatilization from farmlands may also be substantial, up to 16% or more, depending on properties of pesticide, soil types, site environment, and cropping. However, in a major field and analytical modeling study of pesticide leaching and runoff from farm fields (involving 12 crops) in the United States, covering a time series from the 1960s to 1990s, losses from surface runoff were up to 5% of the amount applied, while losses to sub-surface drainage were typically less than 0.5%. The sum of loss estimates for both leaching and runoff ranged from about 4 to 5% of the amount applied, representing the 95 percentile of total loss estimates. Key determinants of pesticide loss, via leaching or runoff from farm fields, were the chemical properties of the pesticides, annual rainfall and its relationship to leaching and runoff, and changes in cropping patterns (Kellogg et al. 2000).

9.6 Environmental Behavior and Fate of Pesticides

Environmental Properties, Distribution, and Processes

The environmental distribution and fate of pesticides are determined largely by their physical and chemical properties (e.g. water solubility and vapor pressure) and environmental properties (e.g. soil sorption coefficient, Henry's Law constant, and half-life), controlled by several key environmental processes under the influence of a number of environmental factors:

- Types and methods of pesticide application (e.g. aerial spraying) or environmental release (e.g. fugitive emissions from manufacture or waste disposal).
- Environmental processes, such as the dissolution of pesticides in water, soil sorption, deposition from air and water, volatility, leaching, and degradation, interact to influence pesticide distribution and fate within and between environmental media (e.g. air, soils, and water).
- Environmental factors include temperature, sunlight intensity and spectra (UV), wind, water flows, and properties of environmental media such as moisture, organic matter, particle size, pH, and organic matter content.

Chapter 4 describes the nature of key properties and environmental processes that apply to the environmental fate of chemical pollutants in general. The overall process that affects the potential fate of pesticides released into the environment is illustrated in Figure 9.10.

The environmental fate of a pesticide can be evaluated from knowledge of its persistence and mobility due to its key properties and the processes that act upon it. Persistence of a pesticide is particularly affected by photo-, chemical and microbial degradation, mobility by sorption, volatilization, biological uptake, wind erosion, runoff and leaching.

An important measure of persistence of a chemical is its half-life ($t_{1/2}$) in an environmental medium or compartment due to its rate of loss or transfer from one medium to another, or breakdown by an environmental process (e.g. volatilization, hydrolysis, microbial activity, or biological uptake) (see Chapter 4). The same applies to pesticides. The persistence of some common insecticides, herbicides, and fungicides as measured by their half-lives in soils ($t_{1/2}$) is indicated later in Table 9.4. The mobility of some pesticides in soils is also discussed in this section, as indicated by groundwater ubiquity scores (GUS) or other leaching indices.

Generally, persistence in the environmental media increases in the following order: atmosphere (air) < surface waters < soils < aquatic sediments. It reflects differences in environmental processes and decreasing rates of degradation from air to water (slower chemical reactions); water to solid phases of soils and sediments (decreasing photodegradation, with less sunlight, and increasing sorption) (Barbash 1993; Mackay et al. 1997).

As shown in Chapter 4 and Connell (2005, pp. 333–339), the distribution pattern of chemicals in the environment can be evaluated by simplifying the environment into sets of two-phase partition processes between phases in contact: atmosphere-water, soil-atmosphere, soil-vegetation, soil-pore water, water-biota, and sediment-water. The partitioning process of molecules between phases depends on the physical-chemical properties of the chemical and the two-phase equilibrium involved. The two-phase processes are described by a set of partition coefficients and equations, which can also be applied to pesticides to predict or estimate their concentrations in air, water, soil, sediments, and biota under equilibrium conditions. Key partition coefficients and equations (phase equilibria) that can be used to describe the distribution of pesticides are given in Chapters 4 and 15 of this book.

The quantity of pesticide released to the environment and its local distribution in air, soils, water, plants and animals are determined by its properties, source (e.g. formulation, application method and rate, kg

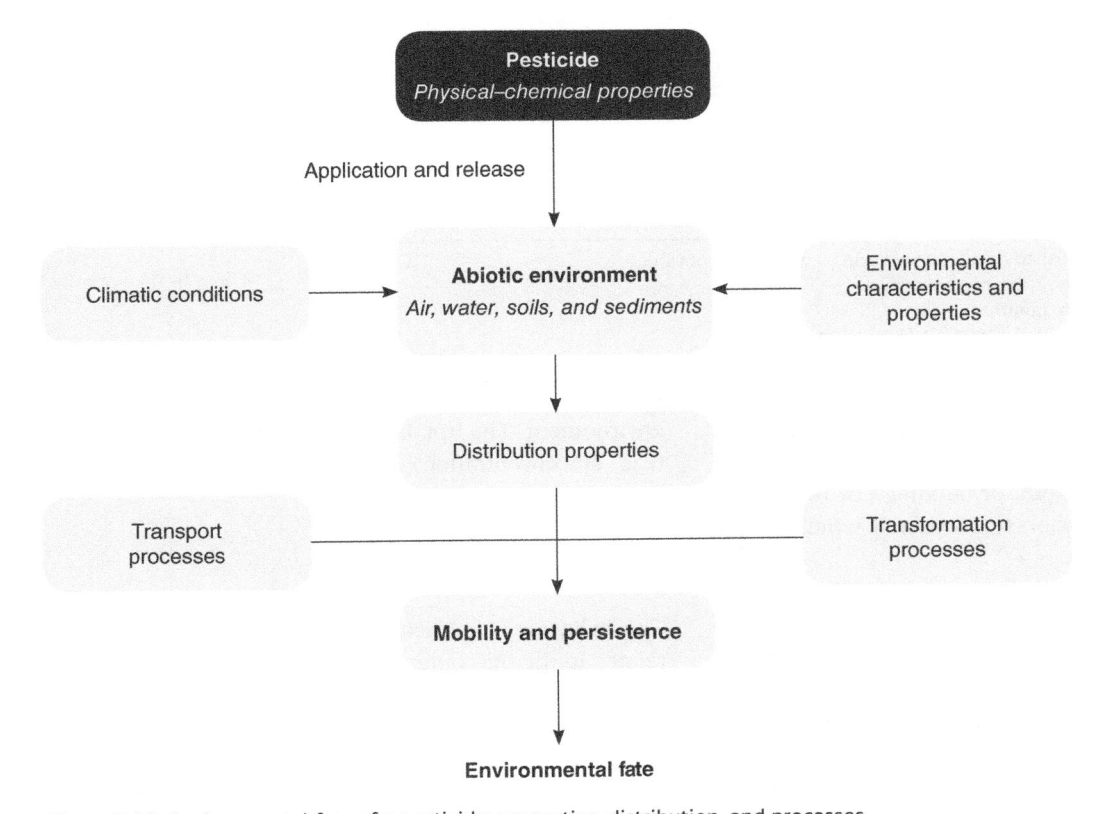

Figure 9.10 Environmental fate of a pesticide: properties, distribution, and processes.

Table 9.4 Generalized mean half-lives and ranges ($t_{1/2}$) of some common pesticides in air, surface waters, soil, and aquatic sediments.

Chemical class	Pesticide examples	Air	Water	Soil	Aquatic sediment
		\multicolumn{4}{c}{Mean half-lives ($t_{1/2}$)^{a,b}}			
Herbicides					
Acetanilides	Metolachlor	1(w)	2(m)	2(m)	8(m)
Phosphonate	Glyphosate	1(w)	2(m)	2(m)	8(m)
Chlorophenoxy acids	2,4-D	1(d)	2(d)	3(w)	2(m)
	2,4,5-T	1(d)	3(w)	3(w)	2(m)
Dinitroanilines	Isopropalin	1(d)	3(w)	2(m)	8(m)
	Trifluralin	1(w)	2(m)	2(m)	8(m)
Triazines	Atrazine	5(h)	2(y)	2(m)	2(m)
	Simazine	2(d)	3(w)	2(m)	8(m)
Ureas	Diuron	1(d)	3(w)	2(m)	8(m)
	Linuron	1(d)	3(w)	2(m)	8(m)
Insecticides					
Organochlorines	p,p′ DDT	1(w)	8(m)	2(y)	6(y)
	Lindane	1(w)	2(y)	2(y)	6(y)
	Chlordane	2(d)	2(y)	2(y)	6(y)
Organophosphates	Chlorpyrifos	1(d)	1(w)	1(w)	2(m)
	Diazinon	3(w)	2(m)	2(m)	8(m)
	Malathion	1(d)	2(d)	2(d)	3(w)
Carbamates	Aldicarb	5(h)	3(w)	2(m)	2((y)
	Carbaryl	2(d)	1(w)	3(w)	2(m)
	Carbofuran	5(h)	1(w)	3(w)	2(m)
Fungicides					
Imides	Captan	1(d)	1(d)	3(w)	3(w)
Organochlorines	Chlorothalonil	1(w)	1(w)	5(w)	2(m)
Fumigants					
Organochlorines	Chloropicrin	1(w)	2(d)	2(d)	1(w)

Compiled from extensive published data, often covering a wide range of values.
^{a}1 (w), one week; 1(d), one day; 2(m), two months etc.
^{b}h, hours; d, days; w, weeks; m, months; y, years
Source: Modified from Mackay et al. (1997); Barbash (1993).

(active ingredients, a.i.) ha^{-1}, and area covered), site characteristics (e.g. soil properties such as particle sizes, moisture, pH, and organic matter), vegetation cover, drainage, groundwater depth, or building environments), and prevailing climatic factors (e.g. seasonal, wind, rainfall, runoff, and temperature).

Pesticides are commonly dispersed beyond the target area into the local environment. They are then subject to transport and degradation processes that may redistribute residues on local, regional, and sometimes long-range global environments, depending upon mobility and persistence properties. General principles influencing the behavior and fate of pesticides, primarily in soils (and sediments), water, and air are considered below. In this section, we are focusing on the fate of pesticides in the abiotic environment. The uptake and concentration of pesticides (e.g. bioconcentration) by organisms are discussed in Section 9.8.

Terrestrial

The behavior of pesticides in the lithosphere is mainly related to the thin outer layer, several meters deep, on the Earth's surface (Robinson 1973). The soil environment contains a relatively immobile matrix of solid, liquid, and gaseous phases. The solid phase contains clay minerals, organic matter, aluminum and silicon oxides, and hydroxides, whereas the liquid phase consists of water and dissolved salts, termed the soil solution. Air forms most of the gaseous phase. Pesticides enter the soil environment via direct application, dust deposition, rainout or precipitation, from runoff, and waste disposal. Interactions between pesticides and the soil environment are controlled by three key factors: (i) sorption processes; (ii) leaching-diffusion; and (iii) degradation (e.g. Connell and Miller 1984, pp. 173–83).

Distribution of Soil Pesticides

Pesticide residues can redistribute from soil solids, depending upon their properties, into other environmental phases (abiotic and biotic), such as the atmosphere and pore water. They undergo two-phase partitioning between these phases:

$$\text{air } (C_A) : \text{soil solids } (C_S) : \text{pore water } (C_W).$$

Where C_A, C_S, and C_W are the pesticide concentrations in air, soil solids, and water, respectively (Connell 2005, p. 333).

Soil–Water Partition Process

The solid sorption factor K_D is the ratio between the concentration in the whole solid phase (soil, sediments and suspended sediments) and water at equilibrium:

$$K_D = C_S/C_W \qquad (9.2)$$

It can also be expressed in terms of the fraction of organic matter, f_{OM}, present in the solid phase and the octanol–water partition coefficient of the pesticide, as $K_D = 0.66\ f_{OM}\ K_{OW}$ (see Chapter 4 in this book).

Mobility and Persistence

Mobility of pesticides in soil can occur by leaching, runoff from sloping surfaces, and volatilization. These processes are affected by a pesticide's water solubility, sorption, and vapor pressure properties and are discussed below. The loss of pesticides in soils is also caused by the interactions of degradation processes: chemical degradation, photodecomposition, and microbial action, as measured by their half-lives ($t_{1/2}$) in various soils.

Aqueous Solubility

Water solubility of a pesticide is measured in $mg\,L^{-1}$, the mass of the pesticide (in milligrams) that will dissolve in one liter of water (L). General guidelines for water solubility are taken from NIPC (2016):

Low water solubility	<10 mg L^{-1} or 10 ppm
Moderate water solubility	10–1000 mg L^{-1} or 10–1000 ppm
High water solubility	>1000 mg L^{-1} or 1000 ppm

From Table 9.2, pesticide solubility in water varies greatly. Some chemical families have similar solubilities in the water phase, for example, chlorinated hydrocarbon, and synthetic pyrethroid insecticides have a very low solubility, while herbicides such as triazines have a moderate water solubility, as indicated by atrazine, and glyphosate, diquat, and paraquat have a very high water solubility.

Sorption Process

Sorption involves the attraction between a chemical and solids such as soils, sediments, plants, and other surfaces. Major factors that influence sorption or adsorption/desorption of pesticides on soil colloids include soil characteristics and composition (e.g. water, clay and organic matter content, pH, and Eh), and pesticide characteristics (e.g. chemical character, shape and configuration of the pesticides, pK_A or pK_B water solubility, charge distribution on cations, polarity, and ionization). Organic matter content of soils exerts an important influence on adsorption. In soils with an organic matter content of up to about 6%, adsorption involves both mineral and organic surfaces, but at high organic contents, adsorption will occur mostly on organic surfaces (Khan 1980).

Bailey and White (1970) identified four basic structural factors in a pesticide molecule that relate to soil adsorption potentials:

1. Nature of functional groups present particularly acidic, carboxyl, alcoholic hydroxyl, and amine groups.
2. Nature of substituent groups that may alter the behavior of functional groups.
3. Position of substituent groups related to functional groups that may enhance or hinder intramolecular bonding or permit coordination with transition metal ions.
4. Presence and magnitude of unsaturation in the molecule that may affect the balance between water and lipid solubility.

Khan (1980) suggested that the charge characteristics of a pesticide are probably the most important properties governing its adsorption. This may range from weakly polar to a relatively strong ionic charge. It follows that the pH of the soil system will also be an important factor.

Some pesticides exhibit a relationship between water solubility and adsorption. This has been shown for some acidic herbicides on muck soil, and for some nonionic pesticides on organic matter and several substituted ureas. Highly water-soluble pesticides such as diquat and paraquat readily adsorb onto clay surfaces, the moderately soluble 2,4-D adsorbs to most soil surfaces, and the hydrophobic chlorinated hydrocarbons (e.g. DDT) strongly adsorb onto organic surfaces. In other cases, the relationship is either direct or nonexistent (Khan 1980). Generally, the sorption of pesticides onto soil is influenced by pesticide characteristics and soil properties of moisture, organic matter, clay, and sand content. For pesticides that are weak acids or weak bases (e.g. dicamba salt and prometon), it is dependent on the pH of the soil.

The sorption of a pesticide on soil is experimentally measured by the solid–water distribution coefficient (K_D) and the sorption coefficient (K_{OC}), as described in Chapter 4. Pesticide sorption to soil increases in strength from low to high K_D values. However, sorption values in laboratory tests vary according to the ratio of water to soil used and properties of the pesticide and soil. As a result, the sorption coefficient (K_{OC}) is commonly used because soil organic carbon is directly proportional to soil organic matter content, which strongly influences soil sorption properties. Soil sorption increases with higher K_{OC} values and related mobility decreases. Measured K_{OW} and K_{OC} for some organic pesticides in soils and sediments are given in Table 9.5.

Importantly, the **organic carbon–water partition co-efficient, K_{OC}**[1] is widely used as a key transport and environmental fate parameter and is defined as:

$$K_{OC} = (K_D \cdot 100) / \% \text{ organic carbon} \quad (9.3)$$

The sorption process on solid surfaces can also be represented by adsorption isotherms (see Chapter 4). In general, the Freundlich and Langmuir isotherms have been used to develop mathematical relationships for pesticide adsorption on soil materials. Freundlich exponents can indicate whether sorption is concentration dependent when measured over a sufficient range of concentrations.

Table 9.5 Measured partition coefficients K_{OW} and K_{OC} for some organic pesticides found in soils and sediments.

Pollutant	Log K_{OW}	Log K_{OC}
γ BHC (lindane)	3.72	3.30
p,p′ DDT	6.19	5.38
Atrazine	2.33	2.17
Carbaryl	2.81	2.36
Malathion	2.89	3.25
Chlorpyrifos	3.31–5.11	4.13
Diuron	1.97, 2.81	2.60, 2.58

Source: Data compiled from Karickhoff (1981).

---

1 K_{OC} is a measure of the tendency of a chemical to bind to soils, corrected for soil organic carbon content. K_{OC} values can vary substantially, depending on soil type, soil pH, the acid–base properties of the pesticide, and the type of organic matter in the soil (Weber et al. 2004).

Leaching and Diffusion

The movement of pesticides in the soil environment involves the interaction of several processes: diffusion and dispersion, mass transfer, leaching, and volatilization. Diffusion, leaching, and vapor loss of pesticides in soil are inhibited as adsorption increases. Physical transport processes are discussed in more detail in Chapter 4. Various authors have reviewed the movement of chemicals including pesticides in soils and their physical transfer mechanisms (e.g. Khan 1980; Connell and Miller 1984; and Connell 2005).

Diffusion of pesticides in the soil environment can occur in solution, at the soil–water interface or air–solid interface, and in the air. Generally, diffusion coefficients (D) of pesticides, as derived from Fick's Law of Diffusion, are $1–3 \times 10^4$ times greater in air than in water (Khan 1980). Thus, pesticides with a water to air ratio under 1×10^4 should diffuse mainly through air and those with ratio above 3×10^4 should diffuse mainly through water.

The general diffusion of pesticides in soil, however, is complex and influenced by environmental variables such as solubility, vapor density, adsorption, bulk density, soil water content and porosity, and uptake by plants. Increasing temperature tends to increase diffusion, which may be significant in summer and in tropical regions, with implications for global heating of air, soil, and water phases.

Leaching of soluble pesticides by water passing through soil is dependent mainly upon soil adsorption although diffusion and water dispersion, adsorption dynamics, and evapotranspiration play a role (Hamaker 1975; Khan 1980). Pesticides that are likely to readily leach are very water soluble, relatively persistent, and have a low sorption potential.

Techniques used in predicting the distribution of pesticides through soil profiles do not reliably predict leaching behavior under field conditions, particularly the slow moving or trailing fraction. There is less movement of chemical than indicated by laboratory leaching tests (Hamaker 1975). For example, a simple measure of the leaching potential of a chemical such as a pesticide in soil, R, is given by,

$$R = 1/K_D \left(1 - \Phi^{2/3}\right) d_s \quad (9.4)$$

Where K_D is the soil–water partition coefficient, Φ is the pore water fraction of the soil, and d_s is the density of soil solids. Thus, the leachability of a pesticide decreases as K_D increases, and increases as the pore water fraction increases (Connell 2005, pp. 332–333).

Volatilization

Volatilization of pesticides can be a major route of loss to the atmosphere. The rate of volatilization is related to the vapor pressure of the pesticide within the soil and its rate of movement to and away from the evaporating surface. Adsorption on soils and other surfaces, as well as solution in soil water, significantly alters the vapor pressure of the pesticide and its rate of movement within soils. The magnitude of these effects is dependent mainly on the nature of the chemical, soil chemical concentration, soil water content, and soil characteristics, for example, organic matter and clay content (Spencer and Cliath 1975). Generally, pesticides evaporate faster from wet soils than dry soils because of mechanisms such as water competition for adsorption sites and upward capillary movement of the soil solution, as water evaporates from the soil surface.

For many pesticides (e.g. dieldrin and lindane), an inverse relationship between the rate of pesticide volatilization and soil organic matter content has been widely reported. Temperature also affects the volatilization of pesticides from soils primarily by a direct influence on vapor pressure and the physical and chemical properties of the soil (e.g. Khan 1980).

The volatility of pesticides from soil is a measure of their tendency to go into the air phase and is used to help predict their potential movement from where they are applied. It can be simply indicated from the following general classification of their vapor pressure (mm Hg at 25 °C) (see Table 9.2):

Low volatility $<1 \times 10^{-8}$ – intermediate volatility 1×10^{-8} to 1×10^{-3} – high volatility $>1 \times 10^{-3}$.

A US EPA classification scheme for volatility of organic pesticides is given in S9.4.

The potential for pesticide movement through soil can be influenced by several factors that include pesticide persistence ($t_{1/2}$) in soil, solubility in water, sorption potential (K_{OC}), and vapor pressure. S9.5 shows an example of a screening approach for the movement of groundwater through soil (Groundwater Ubiquity Score).

Pesticide Degradation Processes

The degradation of pesticide residues is a major route for the loss of pesticides from soils. Key aspects are:

- The chemical nature of the pesticide is particularly important in determining pathways and rates of degradation.
- Transformation products may also exhibit significant biological activity.
- Many conversion products persist in soil, water, and sediment, and may have ecological effects in the same way as the original compound (e.g. the formation of photoproducts from chlorinated cyclodienes by the actions of sunlight and also microorganisms in soils).

The degradation of pathways of pesticides are extensively reviewed in the scientific literature. Three main types of pesticide transformations in soils are identified: photochemical, chemical, and microbiological. These are briefly considered below (Connell and Miller 1984, p. 178).

1. *Photodecomposition:* Although photodecomposition of pesticides in air and water occurs widely, Goring et al. (1975) suggest that photodecomposition of pesticides in soil is of doubtful significance. Radiant energy is strongly absorbed by soil and consequently little photodecomposition of pesticides would be expected, except on or very near to the surface.

2. *Chemical Transformations:* These types of transformations are important processes for their removal from soils. Khan (1980) has concluded that reactions are mediated by water functioning as a reaction medium, a reactant, or both.

Common reactions involve hydrolysis such as for many organophosphorus compounds and s-triazines – and oxidation in the case of many sulfur-containing pesticides – but reduction and isomerization are important with certain compounds. N-nitrosation of certain nitrogenous pesticides, for instance, atrazine in soils, has received attention because of the formation of potentially carcinogenic nitrosamines. Nucleophilic substitution reactions, other than hydrolysis, take place with reactants dissolved in water or with groups in soil organic matter. Chemical degradation reactions in soils may be mediated by free radicals or catalyzed by soil constituents such as clay surfaces, metal oxides, metal ions, and organic matter. Extracellular soil enzymes play a significant role in the degradation of many pesticides and represent the transition between chemical and intracellular microbiological breakdown (Goring et al. 1975).

3. *Microbiological Degradation.* The major groups of soil microorganisms (actinomycetes, fungi, and bacteria) can readily adapt to and degrade pesticides through oxidation, ether-cleavage, ester and amide hydrolysis, oxidation of alcohols and aldehydes, dealkylation, hydroxylation, dehydrohalogenation, epoxidation, reductive dehalogenation, and N-dealkylation (Matsumura 1973).

Dehydrohalogenation is a major process since a large proportion of pesticides contains halogens. Microbiological processes affecting pesticides in soil, including specific structural characteristics of pesticides associated with microbial degradation are described (Khan 1980). The processes of pesticide degradation in soil include chemical reaction, microbial enrichment, and co-metabolism. Simultaneous chemical and microbiological transformations, however, in soil are difficult to distinguish. The complex major pathways of degradation that are known include dehydrohalogenation and isomerization reactions.

While the end products of pesticide transformations in soil are carbon dioxide, water, mineral salts, metabolites naturally occurring in soil and humic substances, the fate and toxicity of many pesticide degradation products in soils remain relatively unknown.

Pesticide Degradation Products

The environmental fate of pesticide degradation products has received less attention although many potential metabolites (free and conjugated) can be formed from parent compounds. The production of N-nitrosamines from the oxidation of amine groups and reduction of nitro groups is a potential toxicity issue for nitrogen-containing pesticides in soils. Generally, soil pH levels of about 3–4 and excess nitrite favor the N-nitrosation reaction and yield only trace amounts of nitrosamines. As well, the high levels of pesticides and sodium nitrite used in some experimental studies are unlikely to occur in agricultural practice. Even so, the persistence of N-nitroso derivatives, such as N-nitrosoglyphosate, in soils has been shown. N-nitroso compounds are also found in technical and formulated products used in agricultural and domestic applications (Fine et al. 1980).

Environmental changes and conversions of primary amines and related compounds appear to occur more rapidly than mineralization reactions (Parris 1980). Consequently, it appears the related compounds can be regarded as latent forms of aromatic amines in the environment. One interesting set of condensation products of aromatic amines is the chlorinated azobenzenes. For example, 3,3′, 4,4′-tetrachloroazobenzene has received attention because of its steric similarity to 2,3,7,8-tetrachlorodibenzodioxin (TCDD).

Persistence

Pesticide persistence in a soil is dependent on the effect of many interacting factors between soils and plants (see Connell and Miller 1984, p. 180). Persistence is interpreted as the residence time of a pesticide in the soil environment, or as a half-life (see Chapter 4) assuming first-order kinetics. As indicated in Table 9.4, the most persistent group is the chlorinated hydrocarbon insecticides. Herbicides can show a wide range of persistence, from days to weeks to over a year in certain cases, although mean half-lives are given as two months. Organophosphate and carbamate insecticides are generally short-lived in soils, from a few days to several months.

Structure-activity and degradation relationships have been studied in pesticides. For instance, the relative degradation reactions of various herbicide classes show that herbicides that initially undergo ester hydrolysis are relatively short-lived in comparison to those that experience an initial dealkylation reaction. In this latter case, persistence is variable but tends to increase with complexity of the molecule.

Variations in herbicide persistence can be considerable even within closely related classes. For example, some methoxy-s-triazines and their metabolites are found to be persistent in soils for up to several years but chloro- and methylthio-s-triazines are much less persistent. Numerous studies show soil conditions have a strong influence on persistence as shown, for example, by the half-lives of 2,4,5-T, which varied from 4 days at 35 °C and 34% moisture to 60 days at 10 °C and 20% moisture content. The persistence of bipyridylium herbicides, paraquat, and diquat appears to depend on the organic matter and clay contents of soil. In a mineral soil, about 8% of the total paraquat applied accumulated as a residue over a period of nine years (Khan 1980). Overall, certain herbicide residues, for example phenoxy acids, uracils, substituted ureas, s-triazines (Connell and Miller 1984, pp. 181–183), and glyphosate, may persist in agricultural soils for a period of over one year, especially soils receiving repeated applications.

Atmosphere

Worldwide use of pesticides has resulted in the distribution of toxic and persistent pesticide residues throughout the Earth's ecosphere. The atmosphere is considered a major route for this chemical pollution phenomenon (see Connell and Miller 1984, p. 167). Pesticides escape into the atmosphere either in particulate form or as a vapor from usage zones. In general, atmospheric pesticides are partitioned between both vapor and particulate forms. Particulates may be either liquids (aerosols) or solids and may range in composition from active pesticide molecules to mostly an inert carrier with absorbed or adsorbed pesticide molecules. Combustion particulates can act as carrier substrates for adsorbed pesticide residues (Wheatley 1973).

They are usually emitted into the atmosphere in intermittent patterns by agricultural applications such as through misting, spraying and spray drift, or by fugitive emissions due to volatilization and wind erosion from surfaces (e.g. soils, plants, water, and treated building environments). Dynamic exchange and redistribution of pesticides can occur between the atmosphere and solid or water surfaces, which may result in widespread distribution, including long-range transport for persistent pesticides. Long-distance transport of pesticides via the atmosphere is discussed in Chapter 4.

The rate of entry and the distance over which pesticides move are dependent upon their vapor pressure and the meteorological conditions. Most of the pesticides entering the atmosphere are deposited locally, due to scavenging processes such as dust deposition, photochemical decomposition, and rain washout. The dominant processes are (i) vapor phase and particulate behavior, (ii) photochemical reactions, and (iii) dry and wet deposition (Connell and Miller 1984, p. 168).

Evaporation and soil erosion losses from application sites appear to be more significant sources of atmospheric pesticides than spray drift since this involves particulate matter which settles out in a relatively short time (Seiber et al. 1975). However, pesticide molecules will interchange between vapor and particulate states as the system approaches equilibrium. Wheatley (1973) suggests that the highly dynamic natural systems equilibrate rapidly tending to damp-out perturbations in pesticide concentrations caused by the intermittent escape of relatively large quantities of pesticide as air moves from a pesticide source to a sink. Oceans are an example of a large sink since the net exchange is from air to water.

Vaporization and vapor phase movements are important in the dissipation of pesticides in air, water, and from plant surfaces. Evaporation of chemicals from surface deposits is dependent on vapor pressure of the chemical and its rate of movement away from the evaporating surface. Many variables interact under field conditions, making development of predictive approaches difficult. K_{WA} may be calculated from the reciprocal of K_{AW} when available.

Photochemical transformations occur readily in the atmosphere. The rate and nature of the transformation products depend largely upon the chemical and spectral characteristic of the parent pesticides. The wavelength range of solar radiation is generally >290 nm, which yields sufficient energy to disrupt the bonds of organic molecules (Tinsley 1979). However, photochemical reactions are mediated by the presence of many naturally occurring sensitizers, oxidants, and catalytic surfaces, resulting in photoisomerization and intramolecular condensation.

Significant photochemical transformations under atmospheric conditions can be summarized as (i) dechlorination reactions, which often result in the loss of pesticide toxicity; (ii) photoisomerization reactions, which may form relatively stable and more toxic photoisomers; (iii) photooxidation reactions in which oxidants such as O, O_2, O_3, and NO_2 may react strongly with olefinic and aromatic groups; and (iv) photomineralization processes, where chlorinated hydrocarbon pesticides can be converted into CO_2 and HCl (Korte 1978). However, some pesticides (e.g. parathion) when exposed to UV light do not produce photoproducts in the vapor phase in contrast to photooxidation effects when adsorbed on fine dust particulates. Airborne particulate surfaces may play an important role in the environmental photochemistry of pesticides. In certain cases, photochemical transformations produce more toxic and stable compounds, for example, the epoxidation of aldrin to dieldrin and heptachlor to heptachlor epoxide (Connell and Miller 1984, p. 171).

The fate of pesticides in the atmosphere is relatively short compared to water, soils, and sediment. Mean half-lives tend to be about one week or less, as indicated in Table 9.4.

Waters

The hydrosphere acts as a major reservoir for persistent pesticide residues. Pesticides enter into the hydrosphere via many pathways including: (i) direct application for pest and disease vector control; (ii) urban and industrial wastewater discharges; (iii) runoff from nonpoint sources including agricultural soils; (iv) leaching through soil; (v) aerosol and particulate deposition, and precipitation, and (v) absorption from the vapor phase at the air–water interface. The relative inputs from these sources remain difficult to assess. Generally, water bodies associated with urban regions receive substantial pesticide inputs from industrial and domestic effluents, and stormwater runoff. The major input is likely to originate from agricultural and forestry practices in conjunction with control programs for human disease vectors (see Kerr and Vass 1973).

The major routes of pesticide transport into the hydrosphere are widely accepted as surface runoff, and aerial transport and deposition. Pesticide runoff varies according to the rate of surface water flow and soil types, while leaching depends primarily on adsorption/desorption between the soil constituents and water percolating through it. Significant amounts of pesticide residues appear to be related to sources such as industrial effluents and waste disposal activities (e.g. disposal of large quantities of obsolete or unwanted pesticides, cattle and sheep dips or sprays, and washing of spray equipment).

Pesticides vary greatly in water solubility despite a more recent trend toward more water soluble and biodegradable pesticides. Many pesticides are still lipophilic and therefore relatively insoluble in water (see Table 9.2). For example, the solubility of persistent CHC insecticides is generally less than $10\,mg\,L^{-1}$, similar to synthetic pyrethroids. In natural waters, the major proportion of the lipophilic pesticides is either adsorbed onto suspended and settled particles or partitioned into organic substrates. These pesticides show strong affinity for the lipoid components of living tissues and dead organic matter. The quantities involved depend on the chemical characteristics and solubility of the pesticide, as well as characteristics of the sediment such as organic content, clay content, and pH of pore water.

Significantly, pesticide residues tend to be associated with the fine particulates in the water column and sediments. pH has a marked effect on the sorption of pesticides containing acidic functional groups. For example, at any pH the amount of 4-amine-3,5,6-trichloropicolinic acid adsorbed is correlated with the amount of unionized acid present (Hamaker et al. 1966). In addition, temperature affects solubility and volatility of pesticides in solution. Pesticide persistence in waters is commonly measured by their half-lives.

Removal of pesticides from waters may occur by (i) volatilization, (ii) absorption by aquatic organisms, and (iii) settling of particles to which pesticides are adsorbed (Robinson 1973). The distribution of pesticides between air and water phases under equilibrium conditions can be represented by K_{AW}, the air–water distribution ratio (C_{air}/C_{water}), which is unitless. The potential volatility of pesticides from water can be predicted from the value of their Henry's Law constant (H). As the value of H increases, the pesticide is more likely to transfer from the water phase into air.

For neutral compounds in a dilute water solution, the air–water distribution ratio (K_{AW}) can be approximated as Henry's Law constant, which is the relation between the vapor pressure of the compound and its solubility in water at equilibrium. Essentially it is the partition coefficient of the compound in air (gas phase) and water (aqueous phase).

$$H = P_i/C_W \ (\text{atm L mol}^{-1}) \quad (9.5)$$

Where P_i is the partial pressure of the gas phase of the pesticide and C_W is the molar concentration in water. In comparison, K_{AW} is defined as a unitless ratio of concentrations.

Volatility classes for pesticides from water are given in Section S9.6.

Available evidence suggests absorption of pesticides by biota is only a very minor route for removal from the hydrosphere. Pesticides in the water column may be removed by (i) sedimentary sorption, (ii) degradation by microorganisms, (iii) uptake by organisms, or (iv) diluted further, especially upon translocation in the oceans (Hassett and Lee 1975). Photochemical decomposition of pesticides may also occur via a series of photolysis reactions – for example, photooxidation, photonucleophilic hydrolysis, and reductive dechlorination – influenced by factors such as natural photosensitizers, pH conditions, and availability of dissolved oxygen. For example, the aquatic herbicide, trifluralin, undergoes photodecomposition rapidly under acidic and aerobic conditions in water compared to basic anaerobic conditions (Connell and Miller 1984, pp. 183–184).

In the oceans, persistent DDT residues are mainly associated with particulates and circulate in the upper mixed layer, which frequently extends to a depth of 75–100 m. Sedimentation slowly transfers the DDT below the thermocline into the much larger volume of the abyss with a transfer time of about 14 years (Woodwell et al. 1971). Within the abyss, the transfer rate to sediment may be extended from hundreds to thousands of years. Once in the sediment, a pesticide may be (i) re-released in the water, (ii) absorbed by organisms, (iii) transformed or degraded by microbial activity, or (iv) fixed in the sediment. In certain cases, degradation may proceed more favorably under anaerobic conditions (e.g. DDT and lindane). Ultimate removal will result by mineralization (see Connell and Miller 1984, p. 185).

9.7 Environmental Exposures to Pesticides

Human production, distribution, storage and uses of pesticides have resulted in local, regional, and global exposures of human populations and wildlife to many of these mainly synthetic substances, at acute and chronic toxicity levels. Persistent pesticide residues have widely contaminated the abiotic environment, foodstuffs, ecosystem food webs and natural biota, and human populations from the Tropics to the Polar Regions of the Earth.

Human exposures to a wide range of pesticides are complex and involve (i) accidental and/or suicide poisonings; (ii) occupational exposures from activities such as manufacturing, mixing/loading, application, harvesting, and handling of crops; (iii) bystanders, individuals or communities exposed to spray drift and residues; (iv) the general public consuming foodstuffs contaminated with pesticide residues from many different sources, including fisheries and other wildlife; (v) drinking water contamination, from surface waters, groundwaters to water storages; (vi) inhalation of contaminated particulates, aerosols, and vapors in rural and outdoor urban air; (vii) inhalation of indoor air and contact with dusts in homes, dwellings, and other buildings from pest control and hygiene uses of pesticides; (viii) landscaping and garden uses; and (ix) contact with contaminated soils and dusts resulting from agriculture, industrial, commercial, and government authorities uses or activities such as weed control or land contamination and remediation.

High-risk human populations for acute or chronic exposures include those exposed in workplaces, from manufacturing to application and remediation of pesticides. The majority of pesticide use covers agriculture in which over one billion persons are engaged worldwide, mainly among low-income earners, in developing and transition countries (Ecobichon 2001). For example, Phung et al. (2012) evaluated chlorpyrifos exposures to rice farmers in Vietnam. Post application levels ranged from 0.35 to 94 $\mu g\,kg^{-1}\,day^{-1}$, which exceeded most of the acute guidelines at the 95th percentile level.

Environmental pathways for human exposures to pesticides are primarily from airborne contamination, drinking water sources, contact with soils, dusts and residues on surfaces such as in indoor environments, and dietary intake of foodstuffs. There are numerous published studies and monitoring reports on environmental levels from many countries although key gaps remain for estimates of actual exposures or doses, especially for multiple residue exposures indicated by the results of human biomonitoring data.

Potential exposure ranges of measured pesticide levels found in the abiotic environment are indicated in Table 9.6. Types of pesticides of concern tend to be relatively persistent insecticides, herbicides, and fungicides. Values are highly variable. They are activity or event dependent and influenced by pesticide properties, temporal and spatial distributions, environmental releases, site specific, and climate characteristics. High levels in non-target areas are generally in the low parts per million (e.g $mg\,kg^{-1}$ of soil/sediment) or $mg\,m^{-3}$ in workplace air. Excessive levels may occur from spillages, excessive applications, spraying drift, or in contaminated sites from waste pesticide residues. Most significant levels in water are usually in the parts per billion such as $\mu g\,L^{-1}$, or sediment at $\mu g\,kg^{-1}$, or $\mu g\,m^{-3}$ in air.

Water levels of pesticides are commonly reported as undetected in ambient monitoring studies. Levels tend to be <0.1 $\mu g\,L^{-1}$ in surface waters. Factors such as their aqueous solubility, mobility from soils, dispersion, dilution, often short persistence periods in water, and capacity to adsorb onto available suspended solids or settled sediments, influence their concentration and fate in surface and groundwaters.

Stehle and Schulz (2015) in a major meta-analysis of peer-reviewed studies (>2500 sites in 73 countries) evaluated the significance of exposure of surface waters (and sediments) to toxic agricultural insecticides on a global scale, using quantifiable levels rather than excessive numbers of nondetectable levels. Their risk analysis revealed, among other findings, that (i) no measurable insecticide concentrations (MICs) were found in an estimated 94.7% of tests, and (ii) there was no scientific monitoring data for about 90% of global croplands, that make up the world's largest terrestrial biome. Yet, significantly, they estimated that over half of the detected insecticide levels (52.4%) out of 11 300 values exceeded accepted regulatory threshold levels (RTLs) for either surface waters or sediments (5915 cases at 68.5% of the sites).

Dietary intake of pesticide residues from environmental contamination of foodstuffs is often the major source of exposure for non-occupationally exposed humans. Risk assessments of international long-term dietary intake of some common pesticides conducted by the WHO/FAO (2004) indicate that the intake range (% of maximum acceptable daily intake, ADI) can vary from <1% to up to 110–330% (e.g. fenitrothion). The percentage intake of the ADI value for most pesticides in the small data set was 30% or less (e.g. chlorpyrifos, 3-30%).

Exposures of wildlife or non-target organisms from human use of pesticides are a major ecological impact and global concern for biodiversity, driven largely by agriculture that affects about 40% of the Earth's land surface. Numerous scientific monitoring studies of persistent and common pesticides exist worldwide. Pesticides such as herbicides and insecticides are strongly implicated in habitat and species loss on a worldwide scale in terrestrial and aquatic ecosystems. Persistent and bioaccumulative pesticides continue to be widely used (e.g. atrazine and synthetic pyrethroids), including CHC insecticides in some developing countries.

Persistent residues can remain a problem in many countries. For example, large scale residual DDT and hexachlorocyclohexane (HCH) contamination of aquatic environments in China has been reviewed by Grung et al. (2015), covering English and international peer-reviewed papers in the last two decades until December 2013. The use of both pesticides was banned in China in 1983. Mean and maximum levels of these pesticides in waters (marine and freshwater), sediments, and aquatic biota are summarized in Table 9.7. Maximum levels in water are <10 $\mu g\,L^{-1}$ (ppb) but maximum aquatic biota levels for DDT are in the low $mg\,kg^{-1}$ or ppm (1000 $ng\,g^{-1}$, wet weight) or about hundreds of times higher. Residues indicated a mix of aged and recent sources from urban, agriculture, industry, and also shipping use of DDT.

Table 9.6 Typical ranges of measured environmental levels of pesticides.

Media	Units	Pesticides
Air		
Workplace	$\mu g\,m^{-3}$	1–10,000
Ambient	$\mu g\,m^{-3}$	0.0001–10
Surface and groundwaters	$\mu g\,L^{-1}$	0.001–10
Sediments	$\mu g\,kg^{-1}$ (dw)	0.1–10,000
Soils	$\mu g\,kg^{-1}$ (dw)	1–10,000

dw, dry weight.

Table 9.7 Total DDT and total HCH levels in aquatic environments of China.

Media	Units	Total DDT		Total HCH	
		Ranges of means (maximum level)			
Waters	$ng\,L^{-1}$	<1–180	(9300)	0.16–300	(1200)
Sediments	$ng\,g^{-1}$ (dw)	0.2–340	(7400)	0.2–550	(1600)
Aquatic biota					
Fish	$ng\,g^{-1}$ (ww)	1.1–130	(1600)	0.26–2.6	(130)
Crustaceans	$ng\,g^{-1}$ (ww)	0.23–111	(2400)	0.16–14	(38)
Bivalves	$ng\,g^{-1}$ (ww)	6–650	(2400)	0.13–3.5	(38)
Birds	$ng\,g^{-1}$ (ww)	45–1200	(1700)	5.1–65	(130)
Bird eggs	$ng\,g^{-1}$ (ww)	32	(5800)	11	(440)

dw, dry weight; ww, wet weight.
Source: Data compiled from Grung et al. (2015).

Dose calculations of pesticide intakes from environmental exposures via inhalation, ingestion, and dermal absorption for humans are described in Chapter 5. Ecotoxicity studies involving environmental exposures to chemicals and dose–response relationships are also discussed in Chapter 5, and specifically on pesticides in Section 9.9.

9.8 Absorption, Distribution, Biotransformation, and Excretion (ADBE) of Pesticides in Humans and Wildlife

Pesticides are distributed in the environment by flow and transport systems (air and water) to biotic surfaces where they are adsorbed and from which they may be absorbed by living organisms through cell membranes usually by passive diffusion. They are then partitioned or actively transported (transporter proteins) through surfaces, organs, and other cell barriers, via aqueous flow systems (e.g. water, blood, and sap), and returned to external flow systems either unchanged or transformed by metabolic processes (Connell and Miller 1984, p. 185).

General interrelated processes of pollutant absorption, biotransformation, storage, and elimination in organisms are described in Chapter 4. Factors influencing the absorption and distribution of pesticides in biological systems are related to (i) inherent physical and chemical properties of the pesticide (e.g. volatility, water and fat solubility, and sorption characteristics); (ii) physiological characteristics of various species (e.g. feeding behavior, routes of uptake, and habitat); and (iii) ecosystem specific properties (e.g. types of flow systems, temperature, pH, organic matter, food web structure, and so on) (Connell and Miller 1984, p. 185).

Humans and Animals

The major routes of exposure to pesticides and the general toxicokinetics of their uptake in humans are summarized in Figure 9.11. Pesticide absorption by humans or animals may occur either directly from the physical environment or from gastrointestinal absorption. For aquatic organisms, intake can result from (i) ingestion of contaminated food, (ii) uptake from water passing over gill membranes, (iii) cuticular diffusion, and (iv) direct absorption from sediments (Livingston 1977).

Terrestrial species may absorb pesticides by the following processes: (i) through contaminated food and water, and sometimes ingestion of soil, via the gut; (ii) directly via percutaneous absorption; and (iii) inhalation of airborne pesticides. For example, insects absorb pesticides via the cuticle and trachea, whereas with soil fauna it occurs through ingestion and also direct contact with pesticides adsorbed to soil minerals or dissolved in soil water (Walker 1975).

Factors influencing the absorption and distribution of pesticides in biological systems are related to (i) inherent physical and chemical properties of the pesticide (e.g. volatility, water and fat solubility, and sorption characteristics); (ii) physiological characteristics of various species (e.g. feeding behavior, routes of uptake, and habitat); and (iii) ecosystem-specific properties (e.g. types of flow systems, temperature, pH, organic matter, food web structure, and so on).

Following absorption, pesticides are distributed to the organs and tissues of the body by the circulatory systems (e.g. blood and lymph in vertebrates and the hemolymph in insects), extracellular fluids, and by transport across internal cellular barriers. The distribution pattern is dependent upon the nature of the pesticide, the route of uptake, its metabolism, and also characteristics of exposed

Multiple routes of exposure to pesticides	
Ingestion	Air, water, soils, dusts, dietary sources, accidental, and self-ingestion
Inhalation	Outdoor air and indoor air Occupational exposure
Dermal absorption	Contact residues on body surfaces from different exposure sources
Transplacental	Residue transfer from mother via placenta to fetus

Toxicokinetics for different types of pesticide(s)	
Routes of absorption	Dermal, ocular, ingestion, inhalation, injection
Distribution and storage (bioaccumulation)	Fat soluble pesticides stored in adipose tissues Other tissues (e.g. bone and hair)
Biotransformations (metabolism)	Formation of inactive or bioactive metabolites (e.g. carcinogenic)
Elimination	Urinary excretion Bilary / fecal excretion Excretion in milk

Figure 9.11 Pesticide routes of exposure and toxicokinetics for humans (absorption, distribution, bioaccumulation, transformations, and elimination).

species. Lipophilic compounds, such as organochlorine insecticides, tend to bind reversibly to plasma proteins as part of an internal transport mechanism and are readily stored in body fat stores. Rapid depletion of body fat, however, may redistribute stored pesticide residues and has the potential to increase exposure of target organs or tissues. Pesticides usually undergo biotransformation (metabolism), and then its metabolites are eliminated or excreted from the body. Some metabolites may also be biologically active or toxic to target organs and to external organisms in the environment when excreted.

Plants

Biological processes that influence the uptake, distribution, storage, and transformations of pesticides in plants are broadly described in Connell and Miller (1984, pp. 186–187). Pesticides penetrate the outer layers of plants through foliage, the epidermis of stems, bark, and the roots. The most common pathways are considered to be (i) the walls of root hairs or epidermal cells of roots, (ii) stomata and the cuticle of cells in the spongy mesophyll, and (iii) lenticles or cracks in the cuticle and periderm (Finlayson and MacCarthy 1973). The degree and routes of pesticide absorption through plant surfaces are strongly influenced by factors such as properties of the pesticides, type of formulation, environmental conditions, and biological properties of the plant. In penetrating internal structures, pesticides have to pass through the cuticle, which covers leaf surfaces exposed to air.

Potential routes of absorption of pesticides involve the lipoid and aqueous pathways. Lipophilic pesticides enter comparatively rapidly via the lipoid components of the cuticle compared to diffusion of polar molecules and ionic forms via the aqueous route, although this can be enhanced by surfactants. Stomata penetration by herbicides may also be an effective route if open at the time of application. Residues of insecticides in plants have also been attributed to penetration of leaves by the vapor phase of compounds with comparatively low vapor pressures. Root absorption of pesticides is a major site of absorption, particularly in the zone of root hairs, and tends to favor water-soluble herbicides.

Translocation or transport of pesticides in plants may be upward (acropetally), downward (basipetally), or laterally.

There are two systems by which herbicide molecules may move rapidly in plants, the phloem and xylem:

1. Pesticides absorbed through the leaves, for example 2,4-D, tend to move in the assimilate stream, that is via the phloem, and accumulate in developing parts of the plant such as the root and shoot tips.
2. Translocation of root-absorbed herbicides involves movement via the sap stream of the xylem in the direction of the transpiration stream. This is the normal movement pattern for the systemic pesticides.

Many of the herbicides, however, that enter the phloem may also move into the xylem and circulate in the plant.

Considerable variations in uptake and movement occur between species. Selectivity results from differential rates of penetration through the plant surfaces and different rates of translocation to sites of biological action. There are many mechanisms and factors that influence the local storage, binding, and persistence of pesticides, such as herbicides, in plants. For example, immobilization of active molecules may result from adsorption during uptake or translocation. Generally, evidence of pesticide persistence in plants is confined to field and experimental studies with agricultural crops (see Connell and Miller 1984, pp. 186–187).

Bioaccumulation and Bioconcentration

Bioaccumulation of chlorinated hydrocarbon insecticides by biota has been widely investigated revealing significant factors influencing the process. As well, mathematical models have been developed and attempts made to predict bioaccumulation potential of pesticides and other organic chemicals from experimental tests. Various tests are designed to measure the physicochemical characteristics of these substances, as well as uptake and elimination kinetics for selected species (see Connell and Miller 1984; Chapters 4, 11 and 19 of this book).

The chemical structure of a pesticide exerts a major influence on bioaccumulation processes and ultimately its elimination from an organism. Recently, structural features have received increased attention as a predictive measure for bioaccumulation potential. For example, some significant differences in environmental behavior of cyclodienes, particularly in metabolism, have been related to substituents (epoxy groups) and stereochemistry. BHC and PCB isomers also exhibit significant differences in environmental behavior (e.g. Shaw and Connell 1983).

The bioaccumulation of pesticides from aqueous solution can be related to a simple partitioning or sorption process between animal and water. Experimental studies such as aquatic bioassays allow the two-phase partition coefficient of a pesticide between water and biota, K_B, known as the bioconcentration factor (BCF), to be measured under equilibrium conditions:

$$K_B = \text{Concentration of pesticide in biota } (C_B) / \text{Concentration in water } (C_W) \quad (9.6)$$

Some bioconcentration values for common pesticides are given in Table 9.8.

Table 9.8 Bioconcentration factors and biological half-lives for some common pesticides.

Pesticide	Aqueous solubility (mg L^{-1}) 25 °C	Log K_{OW}	BCF (K_B) Fish	Biological half-lives ($t_{1/2}$)		
				Fish	Rats	Humans
Atrazine	34.7	2.61	<0.27–12	—	—	31.3 h
Bifenthrin	0.001	6.00	6100	—	19–51 d (tissues/organs)	—
Chlorpyrifos	1.4	4.96	100–4667	—	12–23 h	27 h
Chlorothalonil	0.81	3.05	16–264	~425^a	–	—
2,4-D	677 ppm	2.81	3	1.32 d	2–3 h	17.7 h
DDT	0.0055	6.90	600–100 000	—	~5 wk	—
Dieldrin	1.95	5.40	3.300–12 500	—	—	100–369 d
Glyphosate	Soluble	−3.40	0.52	—	Biphasic: α = ~6 h, β = 79 h	—
Imidacloprid	610 (20 °C)	0.57	3	—	26–118 h	—
Metolachlor	530 (20 °C)	3.13	<1–74	<1	—	—

Data on BCF and biological half-lives of individual pesticides extracted from US National Library of Medicine, TOXNET: Hazardous Substances Data Bank (HSDB) (accessed 17 May 2017). HSDB data is now integrated into PubChem database.
^{a}10 mg kg^{-1} (body weight, bw) dose.

Empirical relationships have been observed between distribution phenomena, as represented by BF, and sorption constants, K_S, and n-octanol/water partition coefficients for various organisms. A useful theoretical relationship can be derived between K_B, the octanol–water partition coefficient K_{OW}, and the lipid fraction of the biota to estimate K_B from K_{OW} (see Chapter 4).

Ellgehausen et al. (1980) measured the bioaccumulation factor K_B for water for ten pesticides using algae, daphnids, and catfish. The bioaccumulation of stable nonionic pesticides can be described as sorption and was correlated with the n-octanol/water partition coefficient. Linear relationships are also observed between the partition coefficients and aqueous solubility of various nonionic chemicals. Generally, as water solubility increases, the values of the partition coefficient and bioaccumulation potential of a pesticide in an organism decrease. Hamelink et al. (1971) indicated the inverse relationship of the solubility of several organochlorine pesticides in water to their accumulation in whole fish.

With unstable pesticides, predicted bioaccumulation factors and depuration rates can vary greatly from experimental estimates. For example, Ellgehausen et al. (1980) found that the pesticide, profenofos, was readily degraded in catfish and the calculated bioaccumulation after 24 hours exposure was less than half the value that would be expected from the n-octanol/water partition coefficient.

These researchers concluded from their studies with three different organisms that the final degree of bioaccumulation was related to the total biomass and not the surface area to mass ratio. Bioaccumulation factors decreased with increasing biomass at the given pesticide concentration.

In the case of soil organisms (e.g. invertebrates such as earthworms or rooted plants), a simple three phase partition model can be used to estimate the transfer of lipophilic pesticides and other chemical contaminants from soils (or sediments) to the organism, via the water phase, as shown by equation (9.7). The bioaccumulation factor (BF) is the ratio of the concentration in the biota and the concentration in the soil, so that $BF = C_B/C_S$.

$$BF = (C_B/C_W)(C_W/C_S) = K_B/K_D \quad (9.7)$$

Equation (9.7) can be simplified further to the expression, as shown by Connell (2005, pp. 338–339), which gives an approximate value of the bioaccumulation of lipophilic pesticides by soil (or sediment) biota, largely independent of their K_{OW} value or chemical properties (see also Chapter 4):

$$BF = f_{lipid}/0.66 f_{OM} \quad (9.8)$$

Bioaccumulation Models

The theoretical basis and limitations of bioaccumulation models used to estimate the uptake and elimination kinetics of pollutants such as pesticides are discussed in Chapter 4. Compartmental models have been applied to the kinetics of organochlorine insecticide uptake and elimination. Much of these data indicate that a single-compartment model, following first-order kinetics, can provide a reasonable approximation to an aquatic organism. But in many cases, a better model is the two-compartment model, both compartments following first-order kinetics (see Connell and Miller 1984, pp. 31–35, 189) or Chapter 4 in this book.

Moriarty (1975) found that the half-lives of the pesticides, as well as the reciprocal of the rate constants of depuration, were highly correlated with the lipophilicity of the pesticides evaluated. However, he concluded that bioaccumulation of organochlorines does not depend solely on passive diffusion, but many active processes of transport, metabolism, and excretion. Therefore, the rates of transfer between compartments are considered more relevant than partition coefficients.

Biomagnification

The process of biomagnification is discussed in Section S9.7, of the on-line website.

Biotransformations

Pesticides absorbed by microorganisms, plants, animals and humans undergo biotransformations, which include biodegradation, detoxification, and metabolism. These processes have been investigated in detail. Vertebrates including humans have similar processes. Most pesticides along with other pollutants are transformed in organisms by several major reaction pathways in which the pesticide is oxidized, hydrolyzed, or reduced (Phase 1) or conjugated (Phase II), or both Phase I and Phase II reactions occur (Figure 9.12). Generally, lipophilic pesticides are metabolized by Phase I enzymes into polar or more water-soluble and less toxic metabolites, and also provide sites for conjugation reactions. Phase II enzymes are conjugating enzymes that can interact directly with pesticides or more commonly interact with metabolites produced by the Phase I enzymes (e.g. Croom 2012, pp. 31–88).

Chemical reactions that detoxify pesticides and conjugate metabolites are catalyzed by enzyme systems that occur in the plasma or various organs, such as the liver. These enzymes may be distributed in the microsomal

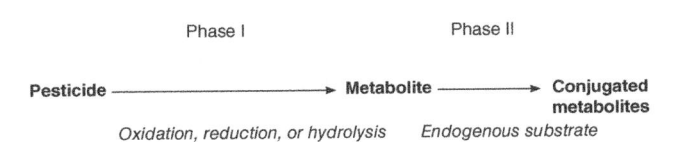

Phase I | Phase II
Pesticide ⟶ Metabolite ⟶ Conjugated metabolites
Oxidation, reduction, or hydrolysis | *Endogenous substrate*

Figure 9.12 Generalized biotransformation pathway for pesticides.

subcellular fraction, which is derived from the endoplasmic reticulum, the mitochondria (or are present in cell sap). Enzyme systems that metabolize pesticides in Phase I reactions include mixed function oxidases, located in the microsomes, notably in the vertebrate liver and in insect body fat, phosphatases, carboxyesterases, epoxide hydrase, DDT dehydrochlorinase, hydrolases (e.g. esterases and amidases), and nitroreductases (Connell and Miller, pp. 194–197).

Major detoxification pathways mediated by these enzyme systems include initial metabolic alterations such as oxidations and hydrolyses (Khan et al. 1975). Some examples of major detoxification reactions for pesticides in biota are given in Table 9.9. In many cases, detoxification may involve more than one reaction pathway.

The polar metabolite resulting from the initial detoxification of a pesticide may be conjugated before excretion. Common enzyme systems in Phase II reactions are glucuronyl transferases, which are widespread in vertebrates, other than fish, but not in insects, and glutathione-S-transferases, also found in insects (Walker 1975). Several pathways for the conjugation and elimination of pesticides and their metabolites from biota exist. Microorganisms do not perform conjugation reactions because excretion takes place through the cell surface. However, plants do not excrete pesticides and their metabolites. Instead, they conjugate these substances with endogenous compounds and deposit them in metabolically inactive sites in the cell (e.g. vacuoles). Animals have efficient excretory systems to eliminate conjugates or pesticides primarily in urine, bile, and feces, but other means include eggs, milk, and sweat (Walker 1975; Menn 1978).

Rates of these various reactions determine selective toxicity of a pesticide to different organisms and are also related to differences in the activities of detoxifying enzyme systems (Khan et al. 1975; Menn 1978). The presence and activity of these enzyme systems in major phylogenetic groups of nontarget organisms is a critical factor in the susceptibility or resistance of these organisms to xenobiotic exposures. Pesticides may reside in plants for longer periods because of their contact times and less efficient circulation and excretory systems than animals (Menn 1978). Aside from pesticide persistence and development of pest resistance, the formation of biologically active metabolites of pesticides, such as in the case of DDT, aldrin, heptachlor, parathion, and malathion, in various natural biota and humans is a critical factor in the evaluation of the effects of residues (Connell and Miller 1984, p. 197).

9.9 Toxic Effects of Pesticides on Biota

Pesticides can induce a broad spectrum of selective and nonselective toxic effects on target and non-target organisms due to their design, formulation, application, exposure levels, routes of intake, metabolism, and toxicity mechanisms. Toxic effects can impact upon almost all taxonomic groups. Many environmental toxicological studies show that considerable variation is observed (i) between species in sensitivity and tolerance to exposure from a particular pesticide, and (ii) in the toxicity of different pesticides to a particular species. Other factors such as sex, age, nutritional state, stress, and habitat or microenvironment influence individual sensitivity (Connell and Miller 1984, p. 203).

Direct toxic effects may originate from several different exposure pathways such as from eating of pesticide-treated seeds or contaminated prey, uptake of residues from water and/or via the skin following a spray event. Alternatively, indirect toxic effects may result from habitat modification or loss of prey or food through pesticide use, such as spraying of nontarget biota.

Acute and chronic toxicity tests are designed to determine the short-term and long-term effects of pesticide

Table 9.9 Some major types of biotransformation reactions in biota.

Reaction	Examples
Oxidation	
Aromatic C—H hydroxylation	Carbaryl, 24-D
Aliphatic C—H hydroxylation	Cyclodienes, alkyl-groups, pyrethroids, DDT
Dealkylation	Nicotine, malathion, parathion abate
N-dealkylation	Phenylurea, carbamates, trifluralin
O-desulfuration	Phosphorothioates
Sulfoxidation	Disulfoton, fenitrothion, abate, phorate, thiocarbamate herbicides
Thioether oxidation	Carbophenothion, demeton
Epoxidation	Cyclodienes, IGRs
Epoxidation hydration	Cyclodienes
Ether cleavage	2,4-D, piperonyl butoxide
Hydrolysis	
Ester hydrolysis	Organophosphate and carbamate insecticides
Amide hydrolysis	Dimethoate
Reduction	
Dechlorination	DDT, γ HCH, atrazine, 2,4-D
Dehydrochlorination	DDT, trifluralin, dinitrothion
Ring cleavage	2,4-D, DDT, simazine

Source: Modified from Menn (1978); Connell and Miller (1984).

exposure on a variety of **toxicity endpoints**, including survival, physiological, biochemical, and reproduction responses. The term, biomarker, may also refer to some toxicity endpoints. Results derived from these tests (e.g. no observed effect levels, NOELs) can be used (i) to determine pathological effects of exposures to pesticides, in doses or concentrations in the diet, water, air or sediment, (ii) to analyze observed field effects, (iii) to identify potential effects, and (iv) to provide dose–response data for comparison to exposure levels in the field or environment (Kendall et al. 2001).

In ecotoxicology, studies of effects on natural biota and their populations have relied primarily on laboratory test systems or bioassays of experimental organisms, biological systems or processes, field trials or monitoring of selected toxic endpoints, which can be measured. Ecosystem level effects on structure and functions are also investigated using microcosm bioassays with multispecies on a laboratory scale and sometimes in large-scale mesocosms in outdoor experimental settings to examine effects on natural environments under controlled conditions (LaGrega et al. 2001, pp. 304–305). Field studies usually relate dosage rates (e.g. pounds/acre) to differences in the number of species and individuals between treated and control areas, commonly involving plant communities, soil organisms, and arthropod fauna of plant communities (Connell and Miller 1984, p. 203).

The evaluation of the toxic effects of pesticides on humans involves studying observations of accidental exposures and poisoning, in vivo exposures on experimental animals, in vitro studies using short-term cellular assays, clinical studies, and epidemiological studies of exposed populations. Toxic effects of pesticide exposures on human health are discussed in Section 9.12.

Limitations of Animal Studies

While animal-based (in vivo) studies have been traditionally used in regulatory toxicity testing of chemicals for risk assessment and risk management of chemical use and exposure, there has been a recent international shift toward integrating alternative testing strategies into chemical safety management. Animal toxicity testing for humans is a complex interspecies process that extrapolates experimental dose–responses of observed effects (endpoints) based on animal biology to predicted human physiological and behavioral responses, depending upon certain assumptions, development of animal models, use of safety factors, etc. Limitations occur at each step in the process, from chemical characterization to dose–response and extrapolation modeling.

The global use of large numbers of animals in toxicity testing has become ethically contentious. Aside from scientific issues, practical limitations include relatively high costs, time-consuming methods, urgent need for high-throughput screening, and use of high doses to overcome relatively small numbers of test animals for statistical assessment of low doses. These aspects apply equally to pesticide testing. Emerging and proposed approaches for integrated testing strategies (e.g. nonanimal testing) are discussed in Chapter 19.

Advances in Toxicity Testing and Assessment

OECD guidelines for the testing and assessment of chemicals such as pesticides for environmental and human health are widely used internationally (OECD 2019) (see Chapter 19 of this book). S9.8 also summarizes a set of US EPA toxicity testing guidelines.

Acute and Chronic Toxicities

Acute Toxicity to Aquatic and Terrestrial Species

Acute toxicities of pesticides to organisms (e.g. LD_{50} and LC_{50}) are derived from field observations and a vast array of bioassay data on experimental test species, particularly rodents and other mammals, aquatic organisms such as fish, crustaceans (e.g. mysid shrimp and daphnids), duckweed and algae, and more recently, beneficial insects, pollinators (e.g. honey bee), amphibians, and earthworms.

Acute oral, dermal, and inhalation toxicities of pesticides (and formulations) are commonly measured by bioassays using experimental rodents (e.g. rats and mice), and other mammals (e.g. rabbits or hamsters) to evaluate potential human toxicity, and sometimes, other mammals. Acute toxicities of a range of selected pesticides for classification of human health hazards are given in Section 9.12.

Comparisons of available toxicity data between major taxonomic groups or classes for families of and individual pesticides are somewhat difficult because of variations and key gaps in test species used and non-standardized bioassays. Even so, available toxicity data suggest the general pattern of lethal toxicities to nontarget organisms from chemical pesticides reported by Connell and Miller (1984, p. 205) appears to remain: insecticides > herbicides > fungicides, although considerable variations in species tolerances and sensitivities occur for individual pesticides.

The development of standardized or harmonized testing protocols (e.g. OECD toxicity testing guidelines) for regulation and registration of chemicals has improved the availability of comparable acute toxicity data, although many older pesticides were evaluated using test species selected from a limited number of taxonomic groups.

Aquatic Species

Acute effects of pesticides on nontarget organisms, especially fish, have been caused by broad-scale applications of insecticides resulting in large mortalities or *kills*. Organochlorine insecticides are highly toxic to many aquatic organisms. For example, with fish species, most 96 hours LC_{50} values range from 1 to 200 ppb, while other pesticides tend to exhibit 96 hours LC_{50} values above this range.

Organophosphorus and carbamate insecticides, however, are toxic to many aquatic invertebrates with 96 hours LC_{50} values often in the low ppb range. Acute toxicities of herbicides to aquatic species are generally in the ppm range, with exceptions including diuron, dinitrocresol, and endothal. Butler (1963) demonstrated that organochlorine insecticides suppress photosynthesis in natural estuarine phytoplankton assemblages at the relatively high concentration of 1000 ppb (four hour exposure). Amphibia such as frogs, toads, and their tadpoles are considerably less susceptible to insecticide poisoning than fish (Brown 1978). Many pesticides, including widely used neonicotinoid insecticides (e.g. Gibbons et al. 2015; Sanchez-Bayo et al. 2016), are also known to be toxic to various aquatic species including many fish, crustaceans, and benthic species. Toxicity varies between species and life stages. Early life stages are generally more sensitive.

For example, Section S9.9 in the Companion Book presents a set of acute toxicities (96 hours LC_{50}) of selected freshwater fish and crustaceans exposed to pesticides based on formulations used in the United States (see also Section S9.11 for a general classification of acute toxicities for aquatic species).

Terrestrial Species

Generally, insecticides are more toxic to warm-blooded animals than herbicides or fungicides as suggested by LD_{50} values for the laboratory rat, *Rattus norvegicus* (see also Table 9.10). Brown (1978) suggests that nearly all the compounds developed as herbicides have a low toxicity to warm-blooded animals and are unlikely to have a direct toxic effect on animals in natural systems. With insecticides, the cyclodienes are particularly toxic (LD_{50} values, <100 mg kg^{-1}) to mammals, along with organophosphorus insecticides, such as parathion, azinphosmethyl, phosphonidon, and methyl parathion and the carbamate, mexacarbate.

Feeding studies using game birds indicate the sensitivity of many avian species to many insecticides, often having acute LD_{50} values <100 mg kg^{-1} (e.g. Connell and Miller 1984, p. 204 and in Table 9.10). Less sensitivity is usually exhibited with herbicides and fungicides. Critically, wild populations of birds can suffer considerable mortalities from insecticide exposures.

Comparative acute and chronic toxicity data for mammals, birds, terrestrial, and aquatic species are also given in Table 9.10 for six major synthetic pesticides, representing herbicides (atrazine and glyphosate), insecticides (chlorpyrifos, imidacloprid, and permethrin), and a fungicide (chlorothalonil). Results are derived from standardized toxicity tests. However, despite the economic value and widespread use of these pesticides, there are still toxicity data gaps for evaluation purposes.

The insecticides, especially chlorpyrifos, are more toxic to the test species than the herbicides and the fungicides. These insecticides are generally less persistent than the organochlorine insecticides that they have replaced but also show considerable toxicity to aquatic test species, such as fish and crustaceans, and pollinators, such as bees, and can be acutely toxicity to birds and mammals (chlorpyrifos and imidacloprid). The fungicide, chlorothalonil, is similarly highly toxic to aquatic test species. The herbicides exhibit less toxicity although showing moderate toxicity to some aquatic and terrestrial test species.

Chronic Toxicity

In many cases, environmental exposures to pesticides are at levels well below those likely to cause acute toxicities. Measuring and evaluating sublethal effects of contaminant exposures in aquatic and terrestrial systems and human environments are critical factors in environmental toxicology. Sublethal effects of pesticides may indirectly lead to lower chances of survival or reproduction in natural populations, at chronic concentrations most likely to be experienced by nontarget organisms. However, it can be difficult to distinguish between an adverse effect induced by the presence of an individual or mixtures of pesticides at sublethal levels and an adaptive response by an organism or system. This aspect becomes more apparent when effects are measured within natural populations. Hence, there is a continuing need to establish sublethal criteria that allow evaluation of significant adverse changes in populations with chronic exposure.

Specific sublethal effects are numerous and relate to a broad spectrum of physiological and behavioral responses, such as alterations in enzyme production, growth rate, reproduction, behavior and activity, production of tumors, and teratogenic effects (see Connell and Miller 1984, p. 209).

Among others, Kendall et al. (2001, pp. 1022–1023) have identified monitoring approaches for a number of key sublethal measures or biomarkers and toxicity endpoints that can indicate short-term and delayed adverse effects on individual organisms, with the potential to cause adverse

Table 9.10 Acute and chronic toxicity of selected pesticides to mammals, birds, terrestrial, and aquatic species.

Ecotoxicity test system	Units	Organism	Pesticide						
			Atrazine	Glyphosate	Chlorpyrifos	Imidacloprid	Permethrin	Chlorothalonil
Mammals Acute oral LD$_{50}$	mg kg^{-1}	Mouse	—	—	—	131	—	—
		Rat	1869	>2000	64	424	>430	>5000
Short-term dietary NOEL	mg kg^{-1} ppm	Rat	—	150	1 (2 yr)	>13	100	3
			200			>120		10
Birds Acute LD$_{50}$	mg kg^{-1}	Quail						
		-Japanese	4237	—	—	31	—	>2000
		-Bobwhite	783	>2250	13.3	—	>9800	—
Short-term dietary (LC$_{50}$/LD$_{50}$)	mg kg^{-1} feed^{-1}	Mallard duck	—	—	203	—	—	>5200
		Japanese quail	—	>4640 U sp	—	—	—	
Fish Acute 96 h LC$_{50}$	mg L^{-1}	Rainbow trout	5.3	38	0.0013	>83	0.032	0.017; 0.047
Chronic 21 d NOEC	mg L^{-1}	Rainbow trout	2	25	0.00014	9.02	0.00012 U sp.	0.003
Aquatic invertebrates Acute 48 h EC$_{50}$	mg L^{-1}	Daphnid	85	40	0.0001	85	0.00032	0.054
Chronic 21 d NOEC	mg L^{-1}	Daphnid	0.25	30	0.0046	1.8	—	0.009
Aquatic crustacean Acute 96 h LC$_{50}$	mg L^{-1}	Mysid shrimp	1.0	40	0.00004	0.034	0.00002	—
Benthic organisms Acute 96 h LC$_{50}$	mg L^{-1}	Chironomid midge	1.0	—	0.00024	0.055	0.0029	0.061
Chronic 28 d NOEC, static water	mg L^{-1}		—	—	0.0001	>0.0021	—	0.04
Aquatic plants Acute 7 d EC$_{50}$, biomass	mg L^{-1}	Duckweed	(0.037) 14 d 0.019	12	—	—	—	0.29

(Continued)

Table 9.10 (Continued)

Ecotoxicity test system	Units	Organism	Pesticide					
			Atrazine	Glyphosate	Chlorpyrifos	Imidacloprid	Permethrin	Chlorothalonil
Algae								
Acute 72 h EC$_{50}$, growth	mg L^{-1}	Green algae	0.059 Raphidocelis Subcapitata	4.4	0.48 U sp	>10	0.0125 U sp	0.21 Raphidocelis subcapitata
Chronic 96 h NOEC, growth	mg L^{-1}	Green algae	0.1	2 U sp	0.043 U sp	10	0.0009 U sp	0.033
Honey bees								
Contact acute 48 h LD$_{50}$	µg/bee	—	—	>100	0.059	0.081	0.29	101
Oral acute 48 h LD$_{50}$	µg/bee	—	100	100	0.025	0.0037	—	>63
Earthworms								
Acute 14 d LC$_{50}$	mg kg^{-1}	Eisenea foetida Unknown sp	79	>5600	129	10.7	1440	268.5
Chronic 14 d NOEC, reproduction	mg kg^{-1}	Eisenea foetida	—	>28.8	127	≥0.178	—	50

Source: Data compiled from PubChem database; Hazardous Substances Data Bank (HSDB); OPP Pesticide Ecotoxicity Database, US EPA.

impacts upon structure and function of populations and communities at the ecosystem level. These approaches cover:

1. Biochemical and physiological measures of endpoints in sentinel plants and animals to assess exposure and effects in many different species. For example, inhibition of cholinesterases (ChEs) is known to be sensitive and a diagnostic biomarker for organophosphate and carbamate exposures. Induction of enzyme systems, such as mixed function oxidases, is another well-known sublethal biomarker of pollutant exposures. Other important sublethal endpoints are related to immune function, genotoxicity, fecundity and reproductive success, and growth rates and body size of organisms.
2. Behavioral responses and traits of individual organisms exposed to sublethal levels of pesticides that indirectly affect their survival, growth and reproduction success, populations, and the structure and functions of communities within ecosystems. Examples include reduced capacity to avoid predators or success in prey capture. Foraging efficiency, feeding behavior, migration, and homing can also be disrupted, reducing, for example, the fitness or potential survival of individuals. Changes in fish behavior patterns, such as avoidance, are observed in polluted waters (e.g. chronic oil spills or effluent plumes). Reproductive success can also be affected by decreased fecundity through impaired nest-building and courtship behavior, territorial defense, and parental care of young.

The widespread use of organochlorine insecticides has resulted in population declines in some bird populations due primarily to behavioral and physiological changes induced by chronic sublethal exposure to these substances in their diet. Generally, the most important effect has been the death of embryos due primarily to premature breakage of thin eggshells and secondly to lethal pesticide levels in the embryos. DDT and its breakdown products DDE and DDD have caused eggshell thinning and population declines in birds of prey species such as the bald eagle, brown pelican, peregrine falcon, and osprey. Sensitivity to these insecticides is observed to vary between bird species. DDE-related eggshell thinning is considered a major reason for the decline of exposed populations of these species (e.g. Connell and Miller 1984, pp. 209–211).

Biological mechanisms are complex with multiple toxicity mechanisms suggested for different species. In particular, p,p'-DDE is considered to inhibit calcium ATPase in the membrane of the shell gland and reduces the transport of calcium carbonate from blood into the eggshell gland. This results in a dose-dependent reduction in thickness or what is known as the *thin eggshell effect*.

Chronic toxicities of selected pesticides are shown in Table 9.10. Many of the chronic toxicity levels occur at parts per billion in sensitive aquatic organisms, while moderate toxicity is indicated for some test species of nontarget terrestrial invertebrates, such as honey bees and other arthropods, and earthworms.

Acute toxicities and various sublethal effects (e.g. physiological, biochemical, neurotoxic, reproductive, or behavioral) of some pesticides (e.g. imidacloprid, synthetic pyrethroids, fenoxycarb, and spinosad) on beneficial insects, such as bees or pollinators and natural enemies to insect pest species, are now well documented in the literature. Sublethal effects include altered mobility, developmental rate, malformation rates (natural enemies), disruption of foraging patterns in parasitoids and honey bees, and changes in feeding behavior and learning processes (honey bees). There is a need to develop experimental approaches for sublethal effects in integrated pest management and pesticide registration procedures (see Desneux et al. 2007).

Pesticides that are listed as endocrine-disrupting chemicals (EDCs) are associated with disruption of reproductive function and development in certain wildlife animals (mammals, birds, fish, amphibians, and mollusks) and certain diseases in the endocrine system of humans, although in many cases for EDCs, causal links between exposures and adverse effects are difficult to establish.

Diverse effects of EDCs are reported on the thyroid, retinoid, androgen, estrogen, and corticosteroid systems in many different animals (Jobling and Tyler 2006). Few studies have examined endocrine disruption effects of pesticides at very low levels or as mixtures applied to crops such as corn in the United States. Using this approach, Hayes et al. (2006) demonstrated endocrine disruption effects (inhibition of larval growth and development) of individual and mixtures of nine pesticides (e.g. atrazine and S-metolachlor) at very low concentrations (0.1 ppb each) on the larvae of leopard frogs.

Pesticide mixtures significantly affected metamorphosis compared to controls. In a follow-up study with adult male African clawed frogs, the nine-pesticide mixture (0.1 ppb of each) also induced damage to the thymus, causing immunosuppression and contraction of bacterial meningitis. These researchers proposed that these adverse effects may be due to an increase in measured plasma levels of the stress hormone, corticosterone. The contributions of individual pesticides in the mixtures to adverse effects could not be determined. Nonetheless, Hayes et al. (2006) questioned whether existing ecotoxicity testing using individual pesticides at higher concentrations is underestimating the ecological risk to amphibians.

9.10 Toxic Action of Pesticides

Knowledge of the modes of toxic action (MoA) of toxicants, or likewise pesticides, helps us classify (e.g. into chemical groups) and understand their toxicity (e.g. cause-effect mechanisms) and predict their potential effects on target and nontarget organisms, including complex issues such as management of pesticide resistance. Much of the toxicity information is obtained during research, development, registration, or regulatory phases for new pesticides and their formulations. The process is complex, long, and costly, but lack of or improper understanding can also result in severe health and environmental costs.

The chemical structure of a pesticide usually defines its mode of action, which is the way it causes physiological disruption at the target or receptor site within an organism. Lushchak et al. (2018), among others, have described general modes of action of pesticides.

Many insecticides target the insect nervous system through several modes of action and are known as neurotoxins. These include the most widely used insecticide family or subgroup, neonicotinoids, and the well-known chlorinated hydrocarbons, organophosphates, carbamates, synthetic pyrethroids, and spinosyns. For example, neonicotinoids work by interfering with neural transmission in the central nervous system, by binding to the nicotinic acetylcholine receptors (*nAChR*) in the postsynaptic neuron and acting as *false neurotransmitters* (agonists) (Gibbons et al. 2015). Nervous systems in many nontarget invertebrates and vertebrates, including humans, are similarly affected. Broadly, insecticides can affect nerves and muscle, growth and development, and energy production.

Knowledge of MoA and target-site resistance is vital in managing pest resistance to individual pesticides and cross-resistance to other pesticides within a related chemical group, and also across multi-groups. For example, The Insecticide Resistance Action Committee (IRAC 2020), an industry-based expert authority, has classified insecticides into 29 groups and an unknown or uncertain group reflecting common modes of action (see also Sparks and Nauen 2015).

Herbicides are known as growth regulators, seedling growth inhibitors, photosynthesis inhibitors, and disrupters of amino acid and lipid biosynthesis, cell membranes, and pigment biosynthesis in plants. They act by a large specific number of modes of action. Herbicides are generally classified according to their site of action, because herbicides within the same site of action class will act similarly on susceptible plants, with some exceptions. Weeds can also exhibit multiple resistance to herbicides across several classes or groups.

The Herbicide Resistance Action Committee (HRAC) (https://www.hracglobal.com) and the Weed Science Society of America (WSSA) (http://wssa.net) have classified herbicides into groups by site of action to enhance herbicide resistance management in practice.

Fungicides control fungal disease by specifically killing (fungicidal) or inhibiting the disease-causing fungus (fungistatic). Fungistats must be applied continually over the lifetime of the plant to suppress the growth of the fungal disease. The general modes of action of fungicides against fungi involve damaging cell membranes, inactivating critical enzymes or proteins, or by interfering with key processes (e.g. energy production and respiration, or specific metabolic pathways (e.g. biosynthesis of sterols or chitin). Biochemical mode of action is the primary basis used to classify fungicides into chemical groups. The Fungicide Resistance Action Committee (FRAC) (https://www.frac.info) maintains a list of fungicides organized by its chemical group.

Some recently developed products are unique in that they do not directly affect the pathogen itself. Many of these elicit a response from the host plant known as *systemic acquired resistance* (SAR). These SAR inducers basically mimic chemical signals in plants that activate plant defense mechanisms such as the production of thicker cell walls and antifungal proteins. The utility of SAR inducers has been limited so far since many pathogens are capable of over-powering such defenses.

Pesticide Interactions

Some pesticide formulations, containing pyrethrins, carbamates, DDT, cyclodienes, may also contain synergists, for example, piperonylbutoxide, sesamin, isothiocyanates, chloroethylethers, and dicarboximides, to potentiate their activity toward target insects, although this induces increased mammalian susceptibility. Methylenedioxyphenyl synergists (e.g. piperonylbutoxide) act by blocking the enzyme system responsible for detoxification of the insecticide (De Bruin 1976). Thus, the basic mechanism of potentiation relates to the inhibition of detoxifying enzymes (Seume and O'Brien 1960).

Antagonistic interactions involve the inhibition by one pesticide of the enzyme responsible for the activation of another. These interactions of pesticides include combinations such as parathion and malathion and dipterex plus EPN (De Bruin 1976). Microsomal enzyme inducers (e.g. organochlorine insecticides) usually interact with organophosphorus insecticides to exhibit accelerated deactivation. Substituted urea herbicides also confer protection against parathion toxicity, probably due to increased MFO detoxication of parathion (De Bruin 1976).

For nontarget organisms, an array of synergistic and antagonistic pesticide combinations has been described particularly for fish, birds, and mammals. The majority of these results are derived from laboratory bioassay studies, which are limited in their application to potential effects on natural populations. Global evidence of multiple pesticide residues in tissues and organs from nontarget species indicates the need to understand the biological implications of pesticide interactions.

Liver Enzyme Induction

Some pesticides, especially the organochlorine insecticides, are capable of inducing increased activity of hepatic enzyme systems, for example mixed-function oxidase (MFO) induction in mammals, birds, and fish at extremely low levels of intake. This phenomenon has the potential for influencing several environmental effects including (i) synergistic or antagonistic effects through stimulation of enzyme systems responsible for metabolizing other pesticides; (ii) antagonistic storage of lipophilic insecticides in animal tissues due to accelerated metabolic detoxication, which may result in lowered chronic toxicity for certain insecticide combinations; and (iii) increased hormone turnover by induced enzymes, causing disturbances to endocrine relationships, which may lead to physiological aberrations in certain bird species, for example thin-eggshell phenomenon (Connell and Miller 1984, p. 202).

Endocrine Disruption

Endocrine disrupters and their basic mechanisms are discussed in Chapter 6. They are a highly heterogeneous group of synthetic and natural chemicals that include insecticides such as DDT, methoxychlor, chlorpyrifos, and vinclozolin. Current research suggests EDCs act via nuclear hormone receptors, nonnuclear steroid hormone receptors (e.g. membrane ERs), nonsteroid receptors (e.g. neurotransmitter receptors such as serotonin, dopamine, and norepinephrine receptors), orphan receptors (e.g. aryl hydrocarbon receptor [AhR]), enzymatic pathways within steroid biosynthesis and/or metabolism, and many other mechanisms that impact upon endocrine and reproductive systems (Diamanti-Kandarakis et al. 2009).

9.11 Ecological Effects of Pesticides on Populations and Communities of Biota

Adverse ecological effects of different pesticides affect all levels of biological organization, from local to global scales, over the short term to the long term. The spectrum of these effects can cover changes in growth, reproduction, and development, and/or behavior of individuals and their populations, resulting in alterations to community structure, decreases in biodiversity, and changes to ecological functions (e.g. loss of productivity and changes to nutrient cycling) and ecosystem services that benefit human health and well-being. Individual species and organisms in the natural environment differ widely in sensitivity to any pesticide. This variation in response means that a pesticide can eliminate susceptible individuals from an exposed population or an entire susceptible species from a community of organisms (Pimentel and Goodman 1974).

Most ecotoxicity data apply to bioassay tests of individual pesticides or formulations on single species of biota. The effects of pesticides on populations and communities of organisms, however, have been studied considerably under field and experimental conditions, and in many terrestrial and aquatic ecosystems, especially on nontarget species since the 1960s.

Generally, the findings from these studies show that major effects of pesticides on populations and communities of animals and plants in ecosystems are (i) their selective or broad spectrum of toxic effects on target and nontarget species, and (ii) their capacity to cause significant changes in species abundance and associated shifts in population dynamics that may severely impact upon communities in agro-ecosystems, and related terrestrial and aquatic ecosystems.

Evidence of adverse ecological effects on the biodiversity of organisms in agricultural ecosystems continues. Recently, Zaller and Bruhl (2019) have provided an editorial overview of research and knowledge gaps on the effects of pesticides on nontarget organisms living in agro-ecosystems.

The focus in this section is on the major types of ecological effects of pesticides observed on populations and communities of non-target species within agricultural, terrestrial, and aquatic ecosystems. The basis of these effects is described in Connell and Miller (1984, pp. 212–214):

Reductions in Populations Caused by Direct Toxic Effects, Secondary Poisoning, and Indirect Effects

Pesticides are usually selected for their direct toxicity toward soil and/or plant organisms. Nontarget arthropods (e.g. insects, earthworms, slugs, and gastropods) are readily susceptible to insecticides (e.g. organophosphates) because of their similar physiological and habitat relationships to target organisms. Many pollinators such as bees are also affected through contact and ingestion. Neonicotinoid insecticides that are now widely used show acute and cumulative toxicity to honey bees and other types of bees, leading to adverse effects on colonies and hives, including colony collapse. Secondary effects may occur through complex interactions between pesticides, pathogens, and parasites. The magnitude and duration of effects may vary greatly.

Relatively insignificant changes in density levels may occur due to factors such as only a localized effect on the total population or a capacity for rapid recolonization and reproduction in response to a density change. Alternatively, elimination of a particularly sensitive species can result. For example, nontarget birds and small terrestrial vertebrates are known to suffer direct mortality from exposure to toxic insecticides although their populations may recover in the longterm due to compensatory effects (Sanchez-Bayo 2011).

Sanchez-Bayo et al. (2016) have identified negative impacts of neonicotinoids on populations of aquatic invertebrates. These reviewers found that the decline of many invertebrate populations, attributed mostly to extensive residue contamination and extreme chronic toxicity (low ppb levels), is affecting the structure and function of aquatic ecosystems. Indirect effects include the starvation of insectivores and invertebrate feeders leading to declines in populations of vertebrates (e.g. insectivorous birds) that depend on insects and other invertebrates.

In animal populations, secondary poisoning of consumer and predatory birds, reptiles, and mammals through ingestion of contaminated prey has adversely affected some populations of vertebrate predators and scavengers (e.g. Sanchez-Bayo 2011; Gibbons et al. 2015).

Severe reductions in prey organisms, such as by insecticides, may also result in a reduction of specific parasites or predators who are unable to adapt to alternative feeding strategies following loss of food. Indirect impacts of herbicide applications, in particular, include declining population densities and biodiversity of birds and possibly amphibians. Herbicides can affect populations through various pathways, including (i) the direct removal of the food base of granivorous species, and (ii) reduction of invertebrate abundance by removing plants that invertebrates depend on for food sources or habitats (e.g. Sanchez-Bayo 2011). Ecological effects of herbicides on plant communities are outlined in Box 9.1.

Pest Resurgence and Secondary Outbreaks

The phenomenon of pest resurgence or rebound is well documented and invariably results from the elimination or suppression of the natural enemies or competitors of the pest and survival of the pest. In some cases, pest populations may increase to similar numbers or significantly greater after spraying, or new pests may emerge. For example, insecticides used to control mosquitoes may reduce populations in the short term but may rebound to produce larger populations through suppression of natural pest controls. Loss of one species from a habitat may also allow another pest or non-target species to invade.

The loss of predator or parasite species can also cause a related phenomenon known as secondary pest outbreaks in which major pest problems can be greater after the use of the pesticides because the natural enemies were more susceptible than the actual pests. Application of insecticides in agriculture and forestry may result in the resurgence of pest species, followed by the emergence of minor pests as major or secondary pests. For example, the use of DDT to control the codling moth resulted in increased numbers and then outbreaks of aphids, scale insects, mealy bugs, and mites (Brown 1978).

Sublethal Effects

Sublethal physiological and behavioral effects of pesticides on animals can affect the survival and reproduction in populations and communities by causing (i) accumulation of sublethal doses leading to lethal mobilization of lipid-stored residues; (ii) shifts in prey–predator relationships and differential survival; (iii) changes in reproduction success (e.g. thin eggshell effects in predatory birds); and (iv) adverse behavior changes such as neonicotinoids affecting foraging, learning, and memory in worker bees. For example, Wu-Smart and Spivak (2016) found adverse effects of imidacloprid on queen bees (egg-laying and locomotor activity), worker bees (foraging and hygienic activities), and colony development (brood production and pollen stores) in all treated colonies.

Desneux et al. (2007) have reviewed sublethal effects of pesticides on the physiology, development, and behavior of beneficial arthropods (pollinators and natural enemies). These reviewers found (i) the link between sublethal effects of pesticides and consequences at the population and community levels is still not well understood in either pollinators or natural enemies, (ii) this also applies to how sublethal effects are taken into account for the development of integrated pest management (IPM) programs, and (iii) although many studies have documented sublethal effects of pesticides on natural enemies, only mortality tests are considered when a choice between several pesticides must be made. Importantly, they concluded that risk assessment needed to establish a link between the toxicity of a given product in laboratory assays and the risk associated with exposure under field conditions.

Pesticide Interactions and Natural Stressors

Research into interactive effects on organisms of exposure to mixtures of pesticide residues and other contaminants, combined with natural or inherent stressors (e.g. nutritional, disease, predation, climate, and water quality) on terrestrial and aquatic organisms is an expanding area of ecotoxicology. The degree of toxic effects of chemicals is commonly dependent upon the nature of the environmental stress (Kendall et al. 2001). These researchers suggest that the main inherent stressors are nutritional restrictions and climate extremes.

As well, a major scientific challenge involves a better understanding of the toxic mechanisms and potential ecological effects or risks for terrestrial and aquatic organisms found to contain mixtures of low levels of pesticides and other chemicals in their environment and tissues.

Pesticide Resistance

The development of resistance to pesticide toxicity in animal and plant populations emerges due to differential survival of individuals within an exposed population as a result of genetic variability. The survivors are thus genotypes selected for toxicity resistance. With continued exposure, the proportion of resistant individuals may significantly increase in succeeding generations. The application of pesticides to agricultural, forest, grassland, and other ecosystems to control populations of specific animal and plant species may also markedly affect populations and communities of nontarget organisms, either directly or indirectly.

Resistance is common in populations of terrestrial invertebrates. During the 1970s, hundreds of species of arthropods developed resistance to organochlorine and organophosphate insecticides. In arthropods, the rate of resistance appears to depend on (i) the intensity of selection of resistant genotypes in successive populations, (ii) the number of generations each year, and (iii) the degree of isolation of the population from dilution by immigration from untreated populations. Basically, resistance is species dependent and variations in its onset between interacting species can cause disruption to an ecosystem (e.g. Brown 1978).

Managing pesticide resistance is now a complex global challenge for agriculture and agro-ecosystems. The scope, advances, and principles in prevention and management of pesticide resistance include agriculture (transgenic crops, insects, weeds, fungal diseases, and rodents), forestry, public health pests, and vector diseases. Integrated Pest Management, reduced use of chemical pesticides, and alternative pest management are key strategies (see FAO 2012).

Loss of Biodiversity

The loss of biodiversity caused by human activities is described as unprecedented in human history (Millennium Ecosystem Assessment 2005). Serious large-scale declines in aquatic and terrestrial biodiversity of insects, birds, invertebrates, and other organisms continue to be reported (e.g. Sanchez-Bayo et al. 2016; Zaller and Bruhl 2019).

In a Europe-wide study of agro-ecosystems, Geiger and co-researchers found that the use of insecticides and fungicides had consistent negative effects on biodiversity while insecticides reduced the biological control potential. They concluded "... that despite decades of European policy to ban harmful pesticides, the negative effects of pesticides on wild plant and animal species persist, at the same time reducing the opportunities for biological pest control" (Geiger et al. 2010).

Beketov et al. (2013) analyzed the effects of pesticides on regional taxa richness of stream invertebrates in Europe (France and Germany) and Australia (southern Victoria). Their analysis showed pesticides significantly reduced regional biodiversity in the stream invertebrates (species and family richness). Importantly, their results indicated that current ecological risk assessment of pesticides (e.g. no effects level below 1/100 of the EC_{50} of *Daphnia* spp.) was insufficient to protect biodiversity at spatial, time, and regional scales.

Global decline of wildlife populations and biodiversity are a critical challenge for sustainability of ecosystem services. Among human activities, pesticides have a major effect on biological diversity, alongside habitat loss and climate change. The large quantities and intensity of pesticides released and dispersed globally each year are well known to have acute and chronic toxic effects on exposed organisms, which can cause major population declines that threaten endangered and rare species. This situation occurs although toxicity testing, registration, and regulation of pesticides are growing in many countries (see Chapter 19 in this book).

Effects of Endocrine Disrupting Chemicals on Animal Populations

Many chemicals identified as endocrine disrupters are pesticides that alter the normal functioning of the endocrine system of both wildlife and humans (see Chapter 11 in this book). McKinlay et al. (2008) identified 127 pesticides from the literature and reported toxic modes of action.

Wildlife populations have been affected by endocrine disruption, with widespread negative impacts on growth and reproduction, due mainly to POPs. Bans on these chemicals have reduced exposures and led to recovery of some populations. Wildlife populations affected by other environmental stressors are particularly vulnerable to EDC exposures. Impacts appear to be underestimated. The future situation for wildlife (and human populations) due to EDCs has developed into a global ecological challenge for more sophisticated scientific studies and advanced testing, policy, and risk management actions (see WHO/UNEP 2013a, pp. IX–XVII).

In many cases, exposures to EDCs involve mixtures of industrial chemicals and pesticides. Exposures to EDCs affect the reproductive health of wildlife species, but few studies have translated these effects into impacts at the population level (WHO/UNEP 2013b, pp. 10–11). WHO/UNEP (2013a) further considers that the best evidence of EDCs such as the pesticides, DDT and tributyl tin, affecting wildlife populations, comes from long-term monitoring of numbers of birds and molluscs, for example, in regions where exposure to these substances has been reduced. Nicotinoids are another major group of insecticides implicated in EDC effects (Gibbons et al. 2015).

Worldwide, a long list of adverse effects on wildlife (birds, fish, shellfish, turtles, mammals, and gastropods) associated with exposure to EDCs, including various pesticides, has been documented in the scientific literature. Effects include abnormal thyroid dysfunction in birds and fish; decreased fertility in birds, fish, shellfish, and mammals; decreased hatching success in fish, birds, and turtles; demasculinization and feminization of male fish, birds, and mammals, the opposite for female fish, gastropods, and birds; and alteration of immune function in birds and mammals.

Although many publications on EDCs report effects on endocrine systems of individual species of wild animals such as fish, amphibians, birds and mammals in field and laboratory studies, only a limited number of studies have demonstrated causal relationships between exposure, and reproduction or population structure effects on wildlife vertebrates. There is sufficient evidence of EDC effects by organochlorines on the reproductive system in some bird species, that are linked to population declines (e.g. DDT/DDE through eggshell thinning). EDCs also affect the endocrine system in fish and may alter sexual development and fertility leading to population effects in a few studies. Similarly, mammal population declines have been correlated with organochlorine pollution (Bernanke and Kohler 2009).

DDT metabolites and two nematocides (DBCP and ethylene dibromide, EDB) have been implicated in reproductive failures and dramatic declines in juvenile alligator populations (1980–1987) at Lake Apopka in Florida, based largely on elevated levels of DDT metabolites in alligator eggs (Semenza et al. 1997). Lake Apopka experienced a pesticide spill in 1980 from the Tower Chemical Company site. Milnes and Guillette (2008) have reported that since the initial decline, egg viability and the number of juvenile alligators gradually increased. Long-term reproductive success remains to be determined for this species that takes 10–15 years to reach sexual maturity.

Amphibian populations are declining in many parts of the world. Yet the degree to which chemicals such as EDCs contribute to the decline of amphibian populations, either alone or through interactions with other environmental stressors (e.g. habitat loss, climate change, or UV-B) is uncertain (Bernanke and Kohler 2009).

A critical issue for EDCs is whether adverse endocrine effects can occur at low ppb levels measured in environmental waters. As a case study, the role of the major herbicide, atrazine, as an EDC, has been controversial in USA-based studies and adversarial in scientific and registration reviews of its ecological effects. The European Union banned the use of atrazine in 2004 because of water contamination. Various investigations and reviews have focused on the relationship between exposure to atrazine and adverse effects on amphibians and other wildlife. Published literature reviews on the ecotoxicity of atrazine to fish, amphibians, and reptiles have used epidemiological criteria (e.g. Hill's criteria) and developed weight of evidence analyses to derive conclusions (see Farruggia et al. 2016, pp. 182–184). Findings vary from evidence of adverse endocrine effects at low levels to no weight of supporting evidence or observed effects that would translate to adverse outcomes in typical apical endpoints. Solomon et al. (2008), for example, concluded that the vast majority of observations did not support the theory that atrazine affected reproduction and/or reproductive development in fish, amphibians, and reptiles at relevant environmental levels.

In 2016, a preliminary draft ecological risk assessment produced for atrazine by the US EPA, based on hundreds of toxicity studies found "… that aquatic plant communities are impacted in many areas where atrazine use is heaviest, and there is potential chronic risk to fish, amphibians, and aquatic invertebrates in these same locations. In the terrestrial environment, there are risk concerns for mammals, birds, reptiles, plants and plant communities across the country for many atrazine uses. EPA levels of concern for chronic risk are exceeded by as much as 22, 198, and 62 times for birds, mammals, and fish, respectively" (Farruggia et al. 2016).

Long-term studies to monitor the effects and toxicity mechanisms of various environmental chemicals such as EDCs on wildlife populations are strongly advocated by ecotoxicologists (e.g. Bernanke and Kohler 2009) for scientific policy development, ecological risk assessment, and management. McKinlay et al. (2008) have proposed

> **Box 9.1 Herbicide Effects on Plant Communities**
>
> Herbicides exhibit high biological activity with plants (Way and Chancellor 1976). The effects vary from stimulation of growth at low application rates to lethality at higher rates. They can be selective or nonselective in their activities because of their varied modes of biological action.
>
> The capacity of these compounds to selectively modify, revert, and suppress plant growth has led to many applications such as in agriculture and forestry. For example, certain compounds are effective against grassy weeds, while others are effective against broadleaf weeds in cereal crops and grasses.
>
> Selective activity is dependent on (i) the time of application (e.g. pre-planting, pre-emergence, or post-emergence), (ii) the method and rate of application, and (iii) the actual amount that is transported to the site of action.
>
> Most of the information on the effects of herbicides on plant communities refers to the weed flora of cultivated or agricultural land where there is a constant recycling of the initial stages of a plant succession that rarely exceeds one year and generally exists for less than six months.
>
> Basic ecological effects in these systems are outlined from Way and Chancellor (1976):
>
> - Pre-existing natural tolerances and differences in susceptibilities between plant species
> - Changes in the density of plant populations
> - Development of resistant populations
> - Changes in the number and composition of plant species
>
> Herbicides and other pesticides can contaminate waterways directly from use in aquatic weed control or in spray drift and water runoff.
>
> For example, herbicides from agricultural sources in river catchments are discharged mainly by flood plumes into the Great Barrier Reef Region on the eastern tropical coast of Australia. Photosystem II inhibiting herbicides (e.g. diuron, atrazine, hexazinone, and tebuthiuron) are detected in reef waters at ng L^{-1} to low µg L^{-1} levels that can adversely affect photosynthesis in seagrass and corals and growth in tropical algae (eAtlas n.d.).

a more precautionary approach for the use of endocrine disrupting pesticides.

Trophic-Level Transfer of Pesticides

Sublethal levels of pesticides may accumulate in the tissues of prey species and transfer through the food web to higher trophic level species and top predators like mammals or raptors and induce direct or indirect toxic effects in their populations (e.g. *thin eggshell effect*, as demonstrated by lipid-soluble chlorinated hydrocarbon insecticides such as DDT and its metabolite DDE). Nontargeted predatory mammals (e.g. dogs and foxes) and raptors often suffer *secondary poisoning* by eating contaminated prey such as, for instance, mice, which have been poisoned by rodenticides.

Overall Ecosystem Effects of Pesticides

A diverse range of ecological effects may be generated following a sequence of lethal or sublethal effects of pesticides on the growth, reproduction, or survival of individual organisms (Connell and Miller 1984, pp. 219–221).

Because the dynamics of species populations with each other and their abiotic environment are extremely complex and difficult to measure and understand, relatively less emphasis has been placed on investigating the effects of pesticides on natural ecosystems than estimates of direct toxicities to individual organisms.

It is the interrelations of species and abiotic factors within an ecosystem that must also be considered in any assessment of pesticide impact on individuals or groups of species. The complete or partial removal of a species from an ecosystem by use of a pesticide will be followed by changes in the prey, predator, competitor, and other species populations with which that species interacts. Again, each of these species interacts with others (Hurlbert 1975). Thus, the actual extent of pesticide effects on nontarget organisms can be seen as ecosystem dependent.

In Japan, for example, the use of toxic and persistent neonicotinoid insecticides in agricultural watersheds (rice paddies) has been linked to cascading effects on aquatic food webs and annual fishery yields of smelt and eel in nearby Lake Sinji, Shimane Prefecture. Since the use of these insecticides in 1993, Yamamuro and co-researchers observed an 83% decrease in average zooplankton biomass in spring, which caused the smelt harvest to collapse, from 240 to 22 tonnes (Yamamuro et al. 2019).

Ecosystem response to pesticide exposures is mainly measured in terms of changes in species composition and population numbers. Typically, these changes follow a sequence of dynamic events as outlined below (e.g. Pimentel and Goodman 1974; Brown 1978; Connell and Miller 1984, pp. 219–221).

1. If lethal or sublethal concentrations of pesticides are dispersed in an ecosystem, the number of species in the ecosystem becomes reduced.
2. If the reduction in number of species is sufficient, this may lead to instability within the ecosystem and subsequently to population outbreaks in some nontarget species. Outbreaks result from a breakdown in the normal check-balance structure of the system.
3. After a pesticide disappears from the affected ecosystem, species in the lower trophic levels (e.g. herbivores) usually increase to outbreak levels.
4. Predators and parasites existing at the higher trophic levels become susceptible to loss of a species or large-scale fluctuations in numbers of species in the lower parts of the food chain upon which they are dependent.

Use of herbicides such as atrazine can impact plant communities and their biodiversity via primary productivity, structure and function.

9.12 Human Health Effects of Pesticides

Pesticide Poisonings

Humans may experience pesticide poisonings through (i) accidental or suicidal poisonings, (ii) occupational exposures (manufacturing, mixing/loading, application, harvesting, and handling of crops), (iii) by-stander exposure to off-target drift from spraying, and (iv) environmental or dietary contamination in the case of the general public (Ecobichon 2001, p. 767). Prenatal exposure of parents is also a special risk for developmental effects in children.

Acute and chronic pesticide poisonings are a major problem worldwide, especially in low- to middle-income countries. Over one billion people are engaged in pesticide use mostly in agriculture (see Section 9.7). Children are facing high risks from pesticide poisonings according to major UN agencies. A joint report published by FAO/WHO/UNEP in 2004 identified pesticide poisoning as a major public health problem (WHO 2004). It estimated one million to five million cases of pesticide poisonings occurred each year. There are no reliable estimates of pesticide-related poisoning in many countries. Pesticide self-poisoning accounts for about one-third of the world's suicides, estimated to be conservatively about 258 000 deaths (Gunnell et al. 2007).

Typical symptoms of acute poisoning in humans are fatigue, headaches and body aches, skin discomfort, skin rashes, poor concentration, feelings of weakness, circulatory problems, dizziness, nausea, and vomiting, but progression to serious poisoning can lead to muscle weakness, heart rate changes, coma, and death (e.g. Dowdall and Klotz 2016).

Pesticide poisonings tend to be difficult to diagnose and classify. They possess diverse structure–activity relations and exhibit a wide range of toxicities, signs and symptoms of poisoning, and observed health effects. Thundiyil and co-researchers have proposed a set of guidelines for evaluating acute pesticide poisonings by classifying (i) adverse effects for selected classes of pesticides, and (ii) signs and symptoms by organ system and severity (Thundiyil et al. 2008).

Acute Toxicity of Pesticides

Acute toxicity of pesticides (e.g. active ingredient) to humans is usually expressed and classified on the basis of acute oral and dermal LD_{50} values in test mammal populations, particularly rats. If the rat is not the most suitable test animal, then another more sensitive or suitable species as a human toxicity model may be used for a particular compound. Hazard classification can also take into account special cases: (i) toxicity risks where dermal LD_{50} values exceed oral LD_{50} values; (ii) use of other applicable toxicity criteria such as inhalation LC_{50} values for volatile fumigants, solvents, and so on; and (iii) available technical or proportionate calculation of toxicological data for non-inert ingredients in formulations (WHO/IPCS 2010).

An example of acute toxicities to mammals of various types of pesticides in commercial formulations and indications of toxicity classifications are given in Table S9.10, in Section S9.10.

A general classification of acute toxicity (oral and dermal) of pesticides to mammals is given in Table 9.11. This type of table is sometimes used to indicate the general hazard classification of pesticides to humans.

The toxicity results in Table 9.11 can also be compared with Table S9.10 in Section S9.10.

Recent international changes to guidelines for hazard classifications and labeling of chemicals through the Global Harmonization System (GHS) have included the acute toxicity hazards of pesticides and classification

Table 9.11 General classification of acute toxicity of pesticides to mammals.

Mammals toxicity category	Rat oral LD_{50} (mg kg^{-1})	Rabbit dermal LD_{50} (mg kg^{-1})
Super toxic	<5	<20
Extremely toxic	5–50	20–200
Highly toxic	>50–500	>200–1000
Moderately toxic	>500–5000	>1000–2000
Slightly toxic	>5000–15 000	>2000–20 000
Practically nontoxic	>15 000	>20 000

Table 9.12 Acute toxicity hazard of selected pesticides, according to GHS classification, using oral LD$_{50}$ values for rats.

Pesticide (active ingredients, technical grade)	Type	GHS	LD$_{50}$ (oral rat) mg kg^{-1} bw	Pesticide (active ingredients, technical grade)	Type	GHS	LD$_{50}$ (oral rat) mg kg^{-1} bw
Extremely hazardous class 1a				**Moderately hazardous class II continued**			
Aldicarb (I)	C	1	0.93	Deltamethrin (I)	PY	3	135
Disulfoton (I)	OP	1	2.6	Diquat (H)	BP	3	231
Parathion (I)	OP	2	13	Endosulfan (I)	OC	3	80
Phorate (I)	OP	1	2	Fenitrothion (I)	OP	4	503
Highly hazardous class 1b				Fipronil (I)		3	92
Azinphos-ethyl (I)	OP	2	12	Lindane (I)	OC	3	88
Cyfluthrin (I)	PY	2	11	Hexazinone (H)		4	1690
Dichlorvos (I)	OP	3	56	Imidacloprid (I)		4	450
Methomyl (I)	C	2	17	Metaldehyde (M)		3	227
Moncrotophos (I)	OP	2	14	Metam-sodium (F-S)		3	285
Pentachlorophenol (I, F, H)		2	80^a	Paraquat (H)	BP	3	150
Moderately hazardous class II				Permethrin (I)	PY	4	500
Allethrin (I)	PY	4	685	Pirimicarb (AP)	C	3	147
Bendiocarb (I)	C	3	55	Propoxur (I)	C	3	95
Bifenthrin (I)	PY	3	55	Pyrethrins (I)		4	500–1000
Carbaryl (I)	C	3	300	Rotenone (I)		3	132–1500
Chlorpyrifos (I)	OP	3	135	Sodium chlorate (H)		4	1200
Copper hydroxide (F)	CU	4	1000	Spiroxamine (F)		4	500
Cypermethrin (I)	PY	3	250	Thiram (F)		4	560
2,4-D (H)	PAA	4	375	Triclopyr (H)		4	710
2,4-DP (H)		4	700	Triflumizole (F)		4	695
DDT (I)	OC	3	113	Ziram (F)		4	1,400

(I) insecticide, (H) herbicide, (F) fungicide; (C) carbamate, (CU) copper compound, (OC) organochlorine compound, (OP) organophosphorus compound, (PY) pyrethroid, (BP) bipyridylium derivative, (PAA) phenoxyacetic acid.
^{a}Dermal toxicity value > oral toxicity value.
Source: Adapted from WHO/IPCS (2010).

criteria, as indicated in Tables 9.12 and 9.13 (WHO/IPCS 2010). Earlier WHO hazard classifications for pesticides are also included in WHO/IPCS (2010).

Under this classification scheme, the final classification of any product is based on the actual formulation. In the case of mixtures of pesticides in a formulation, three possible approaches are given in the guidelines (e.g. formulator should supply reliable toxicity data for rats on the actual mixture or apply a recommended formula to mixture of constituents).

Significantly, many older pesticides have become obsolete or discontinued, some of these because of human health or environmental concerns (e.g. POPs or long-term health effects) although a number of extremely hazardous pesticides remain. Classifications of chronic hazards derived from toxicity testing protocols, clinical, and epidemiological studies are more complex, depending increasingly upon weight of evidence assessments. Many pesticides also lack adequate toxicological evaluation (e.g. endocrine disrupters, epigenetic toxicity) and many human studies of pesticide exposures face many confounding factors leading to non-specific, conflicting or inconclusive evidence.

Yet the strength of the body of evidence on chronic and serious human health effects for many pesticides on exposed individuals and populations has grown markedly this Century, covering neurotoxic diseases, disorders and deficits in development, multiple cancers including endocrine-related, reproductive defects and developmental diseases and disorders, especially for children.

Chronic Toxic Effects of Pesticides

Chronic exposures to pesticides are associated with a wide range of non-specific symptoms such as headache, dizzi-

Table 9.13 GHS hazard classification criteria for pesticides.

	GHS classification criteria			
GHS category	LD_{50} (mg kg^{-1} bw)	Oral[a] Hazard Statement	LD_{50} (mg kg^{-1} bw)	Dermal[b] Hazard statement
Category 1	<5	Fatal if swallowed	<5	Fatal in contact with skin
Category 2	5–50	Fatal if swallowed	5–50	Fatal in contact with skin
Category 3	50–300	Toxic if swallowed	50–300	Toxic in contact with skin
Category 4	300–2000	Harmful if swallowed	300–2000	Harmful in contact with skin
Category 5	2000–5000	May be harmful if swallowed	2000–5000	May be harmful in contact with skin

[a]For oral data, the rat is the preferred species, though data from other species may be appropriate when scientifically justified.
[b]For dermal data, the rat or rabbit are the preferred species, though data from other species may be appropriate when scientifically justified.
Source: WHO/IPCS (2010).

ness, fatigue, weakness, nausea, chest tightness, difficulty breathing, insomnia, confusion, and difficulty with concentration (Alavanja et al. 2004). Exposures are often complex, variable, and may involve multiple pesticide use patterns with time, and other chemicals or agents. Symptoms are difficult to evaluate at low levels of exposure, usually self-reported, and prevalence varies. Evaluation of any related adverse or toxic effects largely depends upon the effectiveness of clinical and epidemiological assessment in occupational, environmental, and indoor settings.

The following assessment of reported chronic effects based on pesticide exposures focuses on neurological, carcinogenic, reproductive and developmental effects, including EDCs.

Neurological Effects

Neurotoxicity Pesticide exposure is known to affect the nervous system, depending upon (i) acute exposures causing short-term signs and symptoms of poisoning, sometimes with delayed impairments long after the exposure episode, or (ii) chronic/long-term exposures. It is also associated with changes in mood and affect, including depression among some exposed workers and pesticide-related illness.

Neurotoxic effects may cover from deficits in neurobehavioral performance and abnormalities in nerve function to neurological disease (e.g. Parkinson's disease and Alzheimer's disease) and neurodevelopmental disorders, including learning disabilities in children. An overview of neurotoxic effects related to pesticide exposures of adults and children is given below, derived primarily from reviews of human epidemiology studies. Many animal studies support key neurotoxic effects of pesticides observed in humans.

Neurobehavioral Performance Most studies of workers spraying or exposed to pesticides such as DDT, organophosphates, and fumigants indicate one or more deficits in cognitive function (reasoning, memory, attention, and language used to obtain knowledge) and psychomotor function (skills of movement, coordination, manipulation, dexterity, grace, strength, speed – related to fine motor skills). Specific tests of sensory or motor function and nerve conduction suggest mostly negative or inconclusive findings or inadequate sensitivity for individuals exposed to low levels of pesticides (Alavanja et al. 2004).

Neurologic Disease Pesticide exposure and some other environmental agents are increasingly linked to neurodegenerative diseases and disorders. Alavanja et al. (2004) also reported that pesticide exposure is linked to an increase in the risk of Parkinson's disease (PD). Many studies have related PD risk to occupational exposures such as farming, living in rural areas, and well water. Some insecticides, herbicides, and fungicides are implicated in several studies, including higher levels of organochlorine insecticides in postmortem brains of PD sufferers compared to patients with other neurological diseases. Occupational exposures to pesticides were associated with Alzheimer's disease (AD) and dementia (D) but not so with environmental exposure to the general public.

Several studies linked the risk of amyotrophic lateral sclerosis (ALS) to pesticides and farming. For example, in a case–control study, McGuire et al. (1997) found more than a twofold increase in ALS from workplace pesticide exposures in men in an industrial hygiene assessment using detailed lifetime job histories.

Neurodevelopmental Disorders Neurodevelopmental disorders are disabilities associated primarily with the functioning of the neurological system and brain. Examples of neurodevelopmental disorders in children include attention-deficit/hyperactivity disorder (ADHD), autism, learning disabilities, intellectual disability (also known as mental retardation), conduct disorders, cerebral palsy, and impairments in vision and hearing (US EPA 2015, p. 1).

Developmental disorders in children are observed to be increasing. About 15% of all US children have one or more developmental disabilities (Boyle et al. 2011). Neurodevelopmental disorders in children and in wildlife studies have been linked to EDCs. Since 2002, increased evidence supports the action of thyroid hormone mechanisms in these disorders in humans and wildlife, and the sensitivity of embryonic and postnatal development to EDCs when compared with adulthood (WHO/UNEP 2013a). Pesticides (e.g. organophosphates such as chlorpyrifos) can also interfere with brain function and development through neurotransmitter control, developing brain cells at low levels (e.g. interference with neural cell replication, differentiation, and survival) (Rauh et al. 2012), and disruption of sodium flows into nerve cells (e.g. pyrethroids) (Soderlund 2012).

Pesticides, such as organophosphates and organochlorines, and prenatal exposures are linked to higher risks of autism or related disorders among children of exposed parents (e.g. Roberts et al. 2007). Prenatal exposures to organophosphates and children with higher levels of organophosphate metabolites in their urine or bodies have been linked to increased risk of attention deficit/hyperactivity disorders (ADHD) (e.g. Bouchard et al. 2010). Similarly, in other studies, prenatal exposures to organophosphate insecticides can lead to cognitive effects, such as deficits in IQ levels and delays in neurodevelopment of children (e.g. Bouchard et al. 2010; Engel et al. 2011).

Pesticide Exposures and Cancers

Overview

There are hundreds of pesticides (active ingredients) used in many thousands of formulated products for commercial use. Among these pesticides, the International Agency for Cancer Research (IARC) has classified three chemicals used as pesticides – arsenic, ethylene oxide, and lindane – and the dioxin contaminant (TCDD) found in certain pesticides such as 2,4,5-T, as human carcinogens (Group 1). For example, arsenic compounds are a known cause of lung cancer. Significantly, the IARC has also classified the "spraying and application of non-arsenical insecticides as a probable cause of cancer" (IARC 1991).

Even so, only six specific pesticides – captafol (widely banned), ethylene dibromide, glyphosate, malathion, diazinon, and dichlorodiphenyltrichloroethane (DDT) –are classed as a probable cause of cancer (Group 2A). A limited number of other pesticides, including tetrachlorvinphos, parathion, metolachlor, pendimethalin, permethrin, trifluralin, and 2,4-dichlorophenoxyacetic acid (2,4-D) have been classified as possible causes of cancer (Group 2B), based primarily on animal evidence and insufficient human data. IARC classifications of pesticides are derived from expert evaluations of scientific studies, including human epidemiological investigations, but are not exposure-based assessments of cancer risks for specific populations.

Numerous international studies (epidemiological and meta-analyses) have investigated the relationship between pesticide exposures and cancers in the workplace and the general environment. Our overview of various human studies, meta-analyses, and reviews related to pesticide uses, levels of human exposures and statistical evaluation of cancer associations and risks, observes the increasing strength of available evidence that many pesticides show significant associations with specific human cancers, such as leukemia, lymphomas, cancers of the brain, and cancers of the endocrine system (e.g. testicular, prostate, and breast). Some childhood cancers are also increasingly implicated.

The interpretation of scientific studies on human health effects and risks associated with pesticide exposures is a major and conflicting issue for public health decision-makers. Many epidemiological studies of occupational or environmental exposures to pesticides are known to suffer limitations due to inadequate analytical or statistical design (e.g. methods used, small study numbers of exposed and non-exposed persons, and lack of exposure or specific biomarker measurements), confounding factors (e.g. multiple risk factors or exposures to multiple pesticides or other potential carcinogens or toxic substances), and bias (e.g. problems with self-reporting, perceptions of bias, or conflicting interests). In many cases, findings are inconclusive or human health studies are lacking for specific pesticides (e.g. less persistent) and potential cancers.

The adequacy of toxicity assessments used for registration of pesticides is another scientific challenge. Alavanja and co-authors reported that epidemiological studies indicate that existing exposures are associated with risks to human health, despite premarket animal testing (Alavanja et al. 2004). They advocate the need to improve the quality of epidemiological studies and to integrate this information with toxicology data that will better characterize the human health risks of pesticide exposures for public health policymakers.

Workplace exposure studies including meta-analyses indicate higher levels and risks of cancers among farmers, pesticide applicators, and pesticide manufacturing workers. Alavanja and Bonner (2012) in a major worldwide review of occupational pesticide exposures and cancer risk found that "… Chemicals in every major functional class of pesticides including insecticides, herbicide, fungicides, and fumigants have been observed to have significant associations with an array of cancer sites." Their review identified twenty-one pesticides since the last IARC review in 1991 that showed significant exposure-response associations in studies on specific cancers. Importantly though, they observed that most pesticides in epidemiological studies were not carcinogenic to humans. This finding is complex because of problems or limitations with many epidemiological studies involving chemical exposures.

An integrative literature review by Pluth and co-researchers focusing on farmers, pesticide applicators, rural workers, and rural populations found significantly higher cancer risks for farmers, living near crops, or in high agricultural areas. Among 73 selected studies, 64 showed a significant association between pesticide exposures and one or more types of cancers. Overall, 53 different types of pesticides were significantly associated (Pluth et al. 2019).

Generally, cancers showing significant association with exposures to various pesticides include non-Hodgkin lymphoma, leukemia, multiple myeloma, brain cancer, prostate cancer, bladder cancer, colon cancer, pancreatic cancer, and lung cancer (e.g. Blair and Freeman 2009; Alavanja and Bonner 2012; Alavanja et al. 2013; Pluth et al. 2019).

The prospective US Agricultural Healthy Study on pesticide applicators and spouses found 19 out of 32 pesticides to be associated with one or more cancers and identified 6 for further investigation, based on related animal toxicity data: alachlor, carbaryl, metolachlor, pendimethalin, permethrin, and trifluralin (Weichenthal et al. 2010).

Analyses of pooled data from case–control studies of occupational pesticide exposures during pregnancy and at time of conception indicate significant increased risks for leukemia but sub-types (AML and ALL) were different between paternal (at conception) and maternal (pregnancy) exposures. For example, a study by the Childhood Leukemia International Consortium of pooled data from 13 case–control studies found an increased risk of AML in the offspring, following maternal occupational pesticide exposure during pregnancy. It also found that the risk of ALL increased slightly with paternal occupational pesticide exposure around conception (Bailey et al. 2014).

Other evidence of pesticide exposure links to human cancers involves residential, environmental, and childhood exposures (see below).

Residential and Environmental Exposures to Pesticides
A number of studies have assessed the risk of various cancers among both adults and children following residential pesticide exposure from pest control services, home and garden use of retail pesticides, chemicals, and imported contamination from workplaces. Some pesticide exposures are associated with prostate cancer, neuroblastoma, and childhood brain tumors. Leukemia and brain cancers are the most common cancers among children. Recent studies link childhood cancers (leukemia, brain, and neuroblastoma) to fetal development and parental exposure before conception. Residential exposure to pesticides during pregnancy substantially increased the risk of childhood leukemia, particularly during pregnancy, and with insecticide exposure (Van Maele-Fabry et al. 2010). Other studies have suggested that parental exposures to pesticides, preconception, during pregnancy or after birth, may also be associated with increased risks of childhood brain cancer in the next generation.

Examples include: a meta-analysis of 20 studies (1974 to 2010) that supported an association between parental occupational exposure to pesticides and brain tumors in children and also young adults (Van Maele-Fabry et al. 2013); and an Australian case–control study also published in 2013 that indicated preconception exposure to pesticides, and possibly exposure during pregnancy, are associated with an increased risk of childhood brain tumor (Greenop et al. 2013).

Widespread evidence of human intake of persistent and endocrine-disrupting pesticide residues such as organochlorines in foodstuffs and the environment (outdoor and indoor air, drinking water, soil, and dusts) is indicated by biomonitoring of body fluids (blood, urine, and breast milk) and tissues for parent compounds and/or metabolites, often found in low levels with other environmental contaminants. Dietary intakes of pesticides in developed countries are generally well below acceptable daily intakes for individual pesticides as determined by the Joint FAO/WHO Meeting on Pesticide Residues (JMPR). Globally, pesticide residues are reported in many types of foodstuffs, such as meats, fish, cereals, fruits, and vegetables, but there appears to be no clear epidemiological evidence that dietary intake of synthetic pesticide residues in foodstuffs is associated with cancer effects in exposed human populations. Despite strong evidence of recent increases in human hormonal cancers (e.g. breast and testicular), there are relatively few statistical associations with specific pesticides that are endocrine disrupters (e.g. DDT). The limited carcinogenicity of DDT (IARC Group 2A) is supported by studies on non-Hodgkin lymphoma, liver cancer, and testicular cancer (Loomis et al. 2015).

There is also some evidence that exposure to DDT in early life or adolescence could increase the longer-term risk of breast cancer (Cohn et al. 2015). In a large, prospective case-control study, Cohn and co-researchers linked measured DDT exposure in utero to risk of breast cancer.

They concluded that early lifetime exposure to p,p′-DDT may increase the risk of breast cancer. These findings suggest DDT is an endocrine disrupter and a predictor of breast cancer, which appears to be underestimated as a risk factor in the prevalence of breast cancer among female populations exposed to the peak period of DDT use.

Pesticides Exposures and Reproduction and Developmental Effects

Endocrine Disrupters

Many pesticides are known or suspect endocrine disrupters, among about 800 chemicals known or suspected to be endocrine disrupters. According to the World Health Organization, there have been significant increases in reproduction disorders and related diseases over the last few decades in many parts of the World linked to the endocrine system (e.g. testicular and breast cancers). The mounting evidence related to EDCs is given as three strands (WHO/UNEP 2013a, p. 2):

- The high incidence and increasing trends of many endocrine disorders in humans;
- Observations of endocrine-related effects in wildlife populations;
- The identification of chemicals with endocrine disrupting properties linked to disease outcomes in laboratory studies.

Certain human studies have also shown stronger associations with pesticide and other EDC exposures, particularly through the use of animal model data and human evidence (Lauretta et al. 2019). Children are among the most vulnerable humans. Exposures to EDCs during fetal development and puberty are considered to play a role in increased incidences of endocrine-related diseases and dysfunctions including infertility, endocrine-related cancers, behavioral and learning problems, for instance, ADHD, infections and asthma, and possibly obesity and diabetes in humans (WHO/UNEP 2013b, p. 7). In adults, EDCs have been linked with obesity, cardiovascular disease, diabetes, and metabolic syndrome. Table 9.14 outlines developmental diseases in humans related to exposures by various pesticides and other EDCs.

Fertility

Increasing infertility is a major human reproduction issue. Low sperm quality is reported for many young men in some countries. Many pesticides and other EDCs are reported to affect the male reproductive system by a number of mechanisms such as reduction of sperm density and motility, inhibition of spermatogenesis, reduction of testis weights, reduction of sperm counts, motility, viability and density,

Table 9.14 Diseases induced by exposure to various pesticides and other EDCs during development in animal model and human studies.

Reproductive/endocrine	Cardiopulmonary
Breast/prostate cancer	Asthma
Endometriosis	Heart disease/hypertension
Infertility	Stroke
Diabetes/metabolic syndrome	
Early puberty	*Brain/nervous system*
Obesity	Alzheimer's disease
Immune system	Parkinson's disease
Susceptibility to infections	ADHD
Autoimmune disease	Learning disabilities

Source: Modified from WHO/UNEP (2013a).

inducing sperm DNA damage, and increasing abnormal sperm morphology. Results in occupational exposures indicate that semen changes are related to multifactorial factors (Mehrpour et al. 2014). Nonetheless, a systematic review by Martenies and Perry (2013) supported the hypothesis that exposures to pesticides at relevant occupational and environmental levels may be associated with decreased sperm health.

Birth Defects and Early Puberty

Birth defects are increasing with trends varying according to the type of defect. They are a major cause of infant mortality in the United States. Some studies show that pesticide exposures (occupational, community, or residential) of parents may increase the risk of birth defects in newborn, particularly during the period or season of conception (see Schafer and Marquez 2012. In Denmark, Wohlfahrt-Veje et al. (2012) found that prenatal exposure to currently used pesticides (non-persistent) in greenhouses may be related to earlier breast development in girls.

A few studies link exposures to pesticides, e.g. organochlorines, during fetal development or early childhood to puberty effects (e.g. onset of early puberty). Further endocrine-related studies of puberty effects from exposures to recent pesticides are advocated. "Mechanisms by which EDC exposure during development can alter the development of specific tissues, leading to increased susceptibility to diseases later in life, are just beginning to be understood" (WHO/UNEP 2013b, p. 12).

Immune System Disorders

Susceptibility to allergic disease is one of many adverse outcomes that could be caused by developmental immunotoxicity. Allergic responses result from the hypersensitivity of the immune system to an allergen in the environment.

> **Box 9.2 Pesticides and Children**
>
> - Children are continually exposed to pesticides via different routes of uptake in urban and rural settings from pesticide applications and residues from multiple outdoor and indoor sources (e.g. air, water, soils, dusts, crop surfaces, playgrounds, toys, furniture, pets, foodstuffs, and child labor).
> - For many children, diet is the main source of intake. Routes of perinatal exposure: mother's intake and body burden transferred across placenta. Breast milk can also be contaminated.*
> - Children may receive acute and high-level exposure, including *accidental* ingestion, leading to poisoning or chronic, low-level exposures linked to more subtle, developmental, and other effects.
> - Acute effects can cover dermal, ocular, and respiratory irritations or allergic responses, gastrointestinal and neurological symptoms, and specific pesticide syndromes (e.g. cholinergic crisis from organophosphates).
> - Chronic toxicity effects may include adverse birth outcomes (e.g. preterm birth, low birth weight, and congenital abnormalities), pediatric cancers (acute lymphocytic leukemia, non-Hodgkin lymphomas, and brain cancers), neurobehavioral and cognitive defects (IQ, ADHD, and autism), and asthma.
> - Pesticide exposure before or during pregnancy has been associated with increased health risks (infertility, perinatal death, congenital malformations, and early childhood cancer).
> - Mechanisms of toxicity include endocrine disruption for some pesticides.
> - More critical actions, measures, and their implementation are greatly needed for prevention of exposure and poisoning by pesticides.
>
> Based on WHO (2008). Children's Health and the Environment WHO Training Package for the Health Sector World Health Organization. World Health Organization, Geneva. https://www.who.int/ceh/capacity/Pesticides.pdf (accessed 27 August 2019).; WHO (2020). Children's Environmental Health. World Health Organization, Geneva. https://www.who.int/health-topics/children-environmental-health#tab=tab_1 (accessed 8 May 2020).
>
> Council on Environmental Health (2012). Policy Statement: Pesticide Exposure in Children; Robert, J.R. and Karr C.J. Council on Environmental Health (2012). Technical Report.
>
> *Well demonstrated for persistent organochlorine insecticides, PCBs and dioxins.

Epidemiological studies suggest that exposure to chemicals may be involved in the etiology of childhood asthma or cancers such as childhood leukemia (Dietert 2009; WHO 2013b, p. 26). Among respiratory irritants, pesticides have been linked to inducing asthma attacks and increasing the risk of developing asthma (e.g. Hernández et al. 2011).

Neurodevelopmental

These types of disorders of the brain and neurological system (e.g. ADHD, autism, and learning problems) are considered in the sub-section on Neurological Effects.

Pesticides and Children

Children are a critical and vulnerable sub-group affected by exposures to many pesticides. A summary of potential health-related effects from pesticide sources and exposures is given in Box 9.2.

9.13 Key Points

1. Pesticides are defined as substances or mixtures of substances intended to control, prevent, kill, repel, or attract any biological organism defined as a pest. They are usually classified according to the type of target organisms they control or kill. Generally, they are chemical or biological agents.
2. The main types or groups of pesticides are usually synthetic organic insecticides, herbicides, and fungicides. Each group consists of many different chemical families. Pesticides within a family are structurally related with similar modes of toxic action.
3. An emerging class of pesticides is known as biopesticides, derived from certain natural substances, various microorganisms, and plants, including protectants produced by plants from added genetic material (PIPs).
4. Biopesticides are generally less toxic for nontarget species than synthetic chemical pesticides. Even so, there is a need for more environmental evaluation because of their increasing use and potential as a substitute for many synthetic chemical pesticides on a worldwide scale.
5. The synthesis and properties of pesticides are generally related to their chemical structures, covalent bond types formed between carbon atoms and also with and between atoms of other elements, functional groups

present, and any ionization. The physical and chemical properties of a pesticide affect its environmental properties, mobility, and persistence.
6. Key physical and chemical properties used to evaluate the behavior and fate of chemical pesticides include molecular weight, density, solubility in water, vapor pressure, n-octanol–water partition coefficient (K_{OW}), Henry's Law constant, and the dissociation constant in water, given by pK values.
7. Important environmental properties of pesticides include their bioconcentration factor (K_B) in different organisms and half-lives in different environmental media, e.g.

$$K_B = C_{organism}/C_{water}$$

8. Physical and chemical properties of pesticides, including their chemical structures and bond types, are key factors in influencing environmental processes such as biodegradation. The half-life of a pesticide in an environmental medium (e.g. air, water, or organism) is a relevant environmental property for persistence.
9. Pesticides are very diverse substances with a number of different groups (e.g. insecticides, herbicides, fungicides, biopesticides). It is important to be familiar with the nature and properties of common pesticides within major groups (e.g. insecticides), and their families (see Section 9.4).
10. Pesticides are mainly used in agriculture, and also for public health protection, throughout the world. In the United States, agriculture accounts for 80% of pesticide use. Herbicides, such as glyphosate and atrazine, are usually the most widely used pesticides, especially in developed countries, such as the United States.
11. Emissions or releases of pesticides into the global environment continue to grow in most countries through worldwide production, distribution, and use in agriculture and human environments. In developed countries, the production and use of pesticides is regulated and managed to some degree. The intensity of application (kg ha^{-1}) on agricultural lands can vary by 10 times or more.
12. The environmental distribution and fate of pesticides are determined largely by their physical and chemical properties (e.g. water solubility and vapor pressure) and environmental properties (e.g. soil sorption coefficient, Henry's Law constant, and half-life), controlled by several key environmental processes (e.g. sorption, volatilization, leaching and diffusion, and degradation) under the influence of a number of environmental factors (e.g. wind, particle size, moisture, pH, UV light intensity, organic matter).
13. Human exposures to pesticides are primarily from airborne contamination, drinking water sources, contact with soils, dusts, and residues on surfaces such as in indoor environments, and dietary intake of foodstuffs. Dietary intake is often the major source of exposure for nonoccupationally exposed humans.
14. Human populations at high risk are those exposed in workplaces, from manufacturing to application, and also disposal and remediation of pesticide wastes. The majority of pesticide use covers agriculture in which over 1 billion persons are engaged worldwide, mainly in developing and transition countries. Children are also identified as a special vulnerable group for pesticide exposure and adverse effects.
15. Exposures of wildlife or nontarget organisms to toxic pesticides from human use of pesticides are associated with major ecological impacts and global concern for biodiversity, driven largely by agriculture that affects about 40% of the Earth's land surface.
16. Pesticide residues may be absorbed from the environment and/or diet by microorganisms, animals, plants, and humans via different routes of exposure (e.g. inhalation, surface or dermal absorption, ingestion and gastrointestinal absorption, gill surfaces, or root systems) depending upon the exposed organism.
17. Absorbed pesticides are transferred and may be stored within a living organism before undergoing biotransformations by Phase I and/or Phase II reactions, which include biodegradation, detoxification, and metabolism, to eliminate usually more water-soluble and less-toxic metabolites from the organism.
18. Bioaccumulation of persistent pesticides is well known to occur in the lipids of tissues of living organisms when the rate of accumulation in tissues is faster than their metabolism or elimination. Biomagnification may occur if a persistent pesticide tends to concentrate when transferred up the food chain from lower to higher trophic levels such as observed in top predators that are air-breathing animals.
19. Pesticides can induce a broad spectrum of selective and nonselective toxic effects on target and nontarget organisms due to their design, formulation, application, exposure levels, routes of intake, metabolism, and toxicity mechanisms. Toxic effects can impact upon almost all taxonomic groups.
20. The general pattern of lethal toxicities to nontarget organisms from chemical pesticides appears to remain: insecticides > herbicides > fungicides, although considerable variations in species tolerances and sensitivities occur for individual pesticides.
21. The development of standardized or harmonized testing protocols (e.g. OECD toxicity testing guidelines)

for regulation and registration of chemicals has improved the availability of comparable acute toxicity data, although many older pesticides were evaluated using test species selected from a limited number of taxonomic groups.
22. Sublethal effects at chronic exposures to low pesticide levels are associated with a broad spectrum of physiological and behavioral responses, such as alterations in enzyme production, growth rate, reproduction, behavior and activity, production of tumors, and teratogenic effects. Disruption of endocrine systems in diverse organisms by a number of pesticides is increasingly recognized in terrestrial and aquatic wildlife populations.
23. Modes of action (MoA) of pesticides vary according to their target organisms (e.g. insects, weeds, and fungi) and are usually complex. Many insecticides are neurotoxins; herbicides usually inhibit plant growth and photosynthetic processes; and fungicides kill or inhibit disease-causing fungus. They damage cell membranes, inactivate critical enzymes or proteins, or interfere with key physiological processes (e.g. respiration and metabolism).
24. The general mode of pesticide action is by targeting biological systems or enzymes in the pest species. These systems or enzymes, however, may be comparable to those in humans and other nontarget organisms in the environment.
25. Knowledge of MoA and target-site resistance is vital in managing pest resistance to individual pesticides and cross-resistance to other pesticides within a related chemical group and also across multi-groups.
26. Major effects of pesticides on populations and communities of animals and plants in ecosystems are (i) their selective or broad spectrum of toxic effects on target and nontarget species, and (ii) their capacity to cause significant changes in species abundance and associated shifts in population dynamics that may severely impact upon communities and related terrestrial and aquatic ecosystems. These effects also apply to agricultural ecosystems.
27. Adverse ecological effects of different pesticides affect all levels of biological organization, from local to global scales, over the shortterm to the longterm.
28. The spectrum of these effects can cover changes in growth, reproduction and development, and/or behavior of individuals and their populations, resulting in alterations to community structure, decreases in biodiversity, and changes to ecological functions (e.g. loss of productivity and changes to nutrient cycling),and ecosystem services that benefit human health and well-being.
29. Individual species and organisms in the natural environment differ widely in sensitivity to any pesticide. This variation in response means that a pesticide can eliminate susceptible individuals from an exposed population or an entire susceptible species from a community of organisms.
30. Humans may experience pesticide poisonings through (i) accidental or suicidal poisonings, (ii) occupational exposures, (iii) by-stander exposure to off-target drift from spraying, and (iv) environmental and/or dietary contamination in the case of the general public. Preconception and prenatal exposures of parents are also a special risk for developmental effects and also cancers (e.g. leukemia) in children.
31. Acute pesticide poisonings for humans are a major problem worldwide, especially in low- to middle-income countries. Children are facing high risks from pesticide poisonings according to major UN agencies.
32. Typical symptoms of acute poisoning in humans are fatigue, headaches and body aches, skin discomfort, skin rashes, poor concentration, feelings of weakness, circulatory problems, dizziness, nausea, and vomiting but progression to serious poisoning can lead to muscle weakness, heart rate changes, coma, and death.
33. Chronic exposures to pesticides are associated with a wide range of nonspecific symptoms such as headache, dizziness, fatigue, weakness, nausea, chest tightness, difficulty breathing, insomnia, confusion, and difficulty concentrating.
34. Exposures are often complex, variable, and may involve multiple pesticide use patterns with time, and other chemicals or agents. Symptoms are difficult to evaluate at low levels of exposure, usually self-reported, and prevalence varies.
35. Chronic exposure to pesticides is well known to affect the nervous system. Neurotoxic effects may cover from deficits in neurobehavioral performance and abnormalities in nerve function to neurological disease (e.g. Parkinson's disease and Alzheimer's disease) and neurodevelopmental disorders, including learning disabilities in children.
36. Some pesticides are listed as human carcinogens or probable carcinogens (IARC). A number of studies have found possible to significant associations with leukemia, lymphomas, brain cancers, and cancers of the endocrine system (e.g. testicular, prostate, and breast). Some childhood cancers are also increasingly implicated.
37. Many pesticides in use are known or suspect endocrine disrupters. The World Health Organization has identified mounting evidence related to EDCs including various pesticides:

- The high incidence and increasing trends of many endocrine disorders in humans;
- Observations of endocrine-related effects in wildlife populations;
- The identification of chemicals with endocrine disrupting properties linked to disease outcomes in experimental animal studies.

References

Alavanja, M.C.R., Hoppin, J.A., and Kamel, F. (2004). Health effects of chronic pesticide exposure: cancer and neurotoxicity. *Annual Review of Public Health* 25: 155–197.

Alavanja, M.C.R. and Bonner, M.R. (2012). Occupational exposures and cancer risk: a review. *Journal of Toxicology and Environmental Health Part B Critical Reviews* 15 (4): 238–263.

Alavanja, M.C.R., Ross, M.K., and Bonner, M.R. (2013). Increased cancer burden among pesticide applicators and others due to pesticide exposure. *CA: A Cancer Journal for Clinicians* 63: 120–142.

Bailey, G.W. and White, J.L. (1970). Factors influencing the adsorption, desorption, and movement of pesticides in soils. *Residues Reviews* 32: 29–92.

Bailey, H.D., Fritschi, L., Infante-Rivard, C. et al. (2014). Parental occupational pesticide exposure and the risk of childhood leukemia in the offspring: Findings from the Childhood Leukemia International Consortium. *International Journal of Cancer* 135 (9): 2157–2172.

Baird, C. and Cann, M. (2012). *Environmental Chemistry*, 5e. New York: W.H. Freeman and Company.

Barbash, J.E. (1993). The geochemistry of pesticides. In: *Treatise on Geochemistry*, vol. 9 (ed. B.S. Lollar), 541–577. Oxford: Elsevier.

Beketov, M.A., Kefford, B.J., Schäfer, R.B., and Liess, M. (2013). Pesticides reduce regional biodiversity of stream invertebrates. *Proceedings of the National Academy of Sciences United States of America* 110 (27): 11039–11043. https://doi.org/10.1073/pnas.1305618110.

Bernanke, J. and Kohler, H.-R. (2009). The impact of environmental chemicals on wildlife vertebrates. *Reviews in Environmental Contamination and Toxicology* 198: 1–47.

Blair, A. and Freeman, L.B. (2009). Epidemiologic studies of cancer in agricultural populations: observations and future directions. *Journal of Agromedicine* 14 (2): 125–131.

Bonmatin, J.-M., Giorio, C., Girolami, V. et al. (2015). Environmental fate and exposure; neonicotinoids and fipronil. *Environmental Science and Pollution Research (International)* 22 (1): 35–67.

Bouchard, M.F., Bellinger, D.C., Wright, R.O., and Weisskopf, M.G. (2010). Attention-deficit/hyperactivity disorder and urinary metabolites of organophosphate pesticides in U.S. children 8–15 years. *Pediatrics* 125 (6): e1270–e1277.

Boyle, C.A., Boulet, S., Schieve, L.A. et al. (2011). Trends in the prevalence of developmental disabilities in US children, 1997–2008. *Pediatrics* 127 (6): 1034–1042.

Brown, A.W.A. (1978). *Ecology of Pesticides*. New York: Wiley.

Bunch, T.R., Bond, C., Buhl, K., and Stone, D. (2014). Spinosad general fact sheet. National Pesticide Information Center, Oregon State University Extension Services. http://npic.orst.edu/factsheets/spinosadgen.html (accessed 28 April 2020).

Butler, P.A. (1963). Commercial fisheries investigations. In: *Pesticide-Wildlife Studies: A Review of Fisheries and Wildlife Service Investigations during 1961 and 1962*, Fisheries and Wildlife Service Circular, vol. 167, 11–25. United States Department of Interior.

Canton, J.W., Greve, P.A., Slooff, W., and van Esch, G.J. (1975). Toxicity, accumulation and elimination studies of alpha-hexachlorocyclohexane (alpha-HCH) with freshwater organisms of different trophic levels. *Water Research* 9: 1163–1169.

Carlson, G.R., Dhadialla, T.S., Hunter, R. et al. (2001). The chemical and biological properties of methoxyfenozide, a new insecticidal ecdysteroid agonist. *Pesticide Science Management* 57 (2): 115–119.

Carvalho, F.P. (2017). Pesticides, environment, and food safety. *Food and Energy Security* 6 (2): 48–60. https://doi.org/10.1002/fes3.108 (accessed 12 May 2020).

Cohn, B.A., Wolff, M.S., Cirillo, P.M., and Sholtz, R.I. (2015). DDT and breast cancer in young women: new data on the significance of age at exposure. *Environmental Health Perspectives* 115 (10): 1406–1414.

Connell, D.W. (2005). *Basic Concepts of Environmental Chemistry*, 2e. Boca Raton, FL: CRC Press.

Connell, D.W. and Miller, G.J. (1984). *Chemistry and Ecotoxicology of Pollution*. New York: Wiley.

Council on Environmental Health (2012). Policy statement: pesticide exposure in children. *Pediatrics* 130 (6): e1757–e1763.

Croom, E. (2012). Metabolism of xenobiotics of human environments. In: *Progress in Molecular Biology and Translational Science*, vol. 112 (ed. E. Hodgson), 31–88. Oxford: Academic Press.

De, A., Bose, R., Kumar, A., and Mozumdar, S. (ed.) (2014). Worldwide pesticide use. In: *Targeted Delivery of Pesticides Using Biodegradable Polymeric Nanoparticles*, Springer Briefs in Molecular Science, 1–99. New Delhi: Springer.

De Bruin, A. (1976). *Biochemical Toxicology of Environmental Agents*. Amsterdam: Elsevier.

Desneux, N., Decourtye, A., and Delpuech, J.-M. (2007). The sublethal effects of pesticides on beneficial arthropods. *Annual Review of Entomology* 52 (1): 81–106.

Diamanti-Kandarakis, E., Bourguignon, J.-P., and Giudice, L.C. (2009). Endocrine-disrupting chemicals: an endocrine society scientific statement. *Endocrine Reviews* 30 (4): 293–342.

Dietert, R.R. (2009). Distinguishing environmental causes of immune dysfunction from pediatric triggers of disease. *The Open Pediatric Medicine Journal* 3: 38–44.

Dowdall, C.M. and Klotz, R.J. (2016). *Pesticides and Global Health: Understanding Agrochemical Dependence and Investing in Sustainable Solutions*. eBook New York: Routledge.

eAtlas (n.d.). Toxicology testing of herbicides entering the Great Barrier Reef. Australian Institute of Marine Science/Australian Government's National Environmental Science Programme. https://eatlas.org.au/content/about-e-atlas (accessed 9 September 2019).

Ecobichon, D.J. (2001). Toxic effects of pesticides. In: *Casarett and Doull's Toxicology. The Basic Science of Poisons*, 6e (ed. C.D. Klaasen), 763–810. New York: McGraw-Hill.

Ellgehausen, H., Guth, J.A., and Esser, H.O. (1980). Factors determining the bioaccumulation potential of pesticides in the individual compartments of aquatic food chains. *Ecotoxicology and Environmental Safety* 4: 134–157.

Engel, S.M., Wetmur, J., Chen, J. et al. (2011). Prenatal exposure to organophosphates, paraoxonase 1, and cognitive development in childhood. *Environmental Health Perspectives* 119: 1182–1188.

Farruggia, F.T., Rossmeisl, C., Hetrick, J.A. et al. (2016). Refined ecological risk assessment for atrazine. Environmental Fate and Effects Division, Office of Pesticide Programs, United States Environmental Protection Agency, Washington, DC. https://www.biologicaldiversity.org/campaigns/pesticides_reduction/pdfs/AtrazinePreliminaryERA.pdf (accessed 26 October 2020).

FAO (2012). *International Code of Conduct on the Distribution and Use of Pesticides: Guidelines on Prevention and Management of Pesticide Resistance*. Rome: Food and Agriculture Organization of the United Nations http://www.fao.org/3/a-bt561e.pdf (accessed 24 May 2020).

Fine, D.H., Krull, I.S., Rounbhler, D.P., and Edwards, G.S. (1980). N-nitroso compound impurities in consumer and commercial products. In: *Dynamics Exposure, and Hazard Assessment of Toxic Chemicals* (ed. R. Haque), 417. Ann Arbor, MI: Ann Arbor Science.

Finlayson, D.G. and MacCarthy, H.R. (1973). Pesticides residues in plants. In: *Environmental Pollution by Pesticides* (ed. C.A. Edwards), 57. London: Plenum Press.

Geiger, F., Bengtsson, J., Berendse, F. et al. (2010). Persistent negative effects of pesticides on biodiversity and biological control potential on European farmland. *Basic and Applied Ecology* 11: 97–105.

Gibbons, D., Morrissey, C., and Mineau, P. (2015). A review of the direct and indirect effects of neonicotinoids and fipronil on vertebrate wildlife. *Environmental Science Research International* 22: 103–118.

Goring, C.A.I., Laskowski, D.A., Hamaker, J.W., and Meikle, R.W. (1975). Principles of pesticide degradation in soil. In: *Environmental Dynamics of Pesticides*, Environmental Science Research, vol. 6 (ed. R. Haque and V.H. Freed), 135. Boston, MA: Springer https://doi.org/10.1007/978-1-4684-2862-9_9.

Greenop, K., Peters, S., Bailey, H. et al. (2013). Exposure to pesticides and the risk of childhood brain tumors. *Cancer Causes & Control* 24 (7): 1269–1278. https://doi.org/10.1007/s10552-013-0205-1.

Grube, A., Donaldson, D., Kiely, T., and Wu, T. (2011). *Pesticides Industry Sales and Usage: 2006 and 2007 Market Estimates*. Washington, DC: United States Environmental Protection Agency https://www.epa.gov/sites/production/files/2015-10/documents/market_estimates2007.pdf (accessed 29 April 2020).

Grung, M., Lin, Y., Zhang, H. et al. (2015). Pesticide levels and environmental risk in aquatic environments in China – a review. *Environment International* 81: 87–97.

Gunnell, D., Eddleston, M., Phillips, M.R. et al. (2007). The global distribution of fatal pesticide self-poisoning: systematic review. *BMC Public Health* 7: 357.

Gustafson, D.I. (1989). Groundwater ubiquity score: a simple method for assessing pesticide leachability. *Environmental Toxicology and Chemistry* 8: 339–357.

Hamaker, J.W. (1975). The interpretation of soil leaching experiments. In: *Environmental Dynamics of Pesticides* (ed. R. Haque and V.H. Freed), 115. New York: Plenum Press.

Hamaker, J.W., Goring, C.A.I., and Youngson, C.R. (1966). Sorption and leaching of 4-amino-3,5,6-picolinic acid in soils. In: *Organic Pesticides in the Environment*, Advances in Chemistry Series, vol. 60 (ed. R.F. Gould), 23–37. Washington, DC: American Chemical Society.

Hamelink, J.L., Waybrant, R.C., and Ball, R.C. (1971). A proposal: exchange equilibria control the degree chlorinated hydrocarbons are magnified in benthic environments. *Transactions of the American Fisheries Society* 100: 207–214.

Hassett, J.P. and Lee, G.F. (1975). Modeling of pesticides in the aqueous environment. In: *Environmental Pollution by Pesticides* (ed. C.A. Edwards), 173. London: Plenum Press.

Hayes, T.B., Case, P., Chui, S. et al. (2006). Pesticide mixtures, endocrine disruption, and amphibian declines: are we underestimating the impact? *Environmental Health Perspectives* 114 (Suppl 1): 40–50.

Hernández, A.F., Parrón, T., and Alarcón, R. (2011). Pesticides and asthma. *Current Opinion in Allergy and Clinical Immunology* 11 (1): 90–96.

Hurlbert, S.H. (1975). Secondary effects of pesticides on aquatic ecosystems. *Residue Reviews* 57: 81–148.

IARC (1991). Occupational exposures in insecticide application and some pesticides. In: *IARC Monographs Volume 53: Evaluation of Carcinogenic Risks to Humans* (ed. International Agency for Research on Cancer). Lyon, France: IARC.

IRAC (2020). The IRAC mode of action classification online. The Insecticide Resistance Action Committee. https://irac-online.org/modes-of-action/ (accessed 2 May 2020).

Jobling, S. and Tyler, C.R. (2006). Introduction: the ecological relevance of chemically induced endocrine disruption in wildlife. *Environmental Health Perspectives* 114 (Suppl 1): 7–8.

Karickhoff, S.W. (1981). Semi-empirical estimation of sorption of hydrophobic pollutants on natural sediments and soils. *Chemosphere* 10 (8): 833–846.

Kellogg, R.L., Nehring, R., Grube, A. et al. (2000). Environmental indicators of pesticide leaching and runoff from farm fields. Presented at a *Conference on Agricultural Productivity: Data, Methods, and Measures*. Washington, DC. (March 9–10 2000). https://www.nrcs.usda.gov/wps/portal/nrcs/detail/national/technical/?cid=nrcs143_014053(accessed 29 April 2020).

Kendall, R.J., Anderson, T.A., Baker, R.J. et al. (2001). Environmental toxicology. In: *Casarett and Doull's Toxicology: The Basic Science of Poisons*, 6e (ed. C.D. Klaasen), 1013–1045. New York: McGraw-Hill.

Kerr, S.R. and Vass, W.P. (1973). Pesticide residues in aquatic invertebrates. In: *Environmental Pollution by Pesticides* (ed. C.A. Edwards), 134. London: Plenum Press.

Khan, S.U. (1980). *Pesticides in the Soil Environment*. Amsterdam: Elsevier.

Khan, M.A.Q., Gassman, M.L., and Ashrafi, S.H. (1975). Detoxification of pesticides by biota. In: *Environmental Dynamics of Pesticides* (ed. R. Haque and V.H. Freed), 289. New York: Plenum Press.

Korte, F. (1978). Abiotic processes. In: *Principles of Ecotoxicology* (ed. G.C. Butler) SCOPE 12, 11. New York: Wiley.

LaGrega, M.D., Buckingham, P.L., Evans, J.C. et al. (2001). *Hazardous Waste Management*, 2e. New York: McGraw-Hill.

Lamichhane, J.H., Dachbrodt-Saaydeh, S., Kudsk, K., and Messéan, A. (2016). Toward a reduced reliance on conventional pesticides in european agriculture. *Plant Disease* 100 (1): 10–24. http://apsjournals.apsnet.org/doi/pdf/10.1094/PDIS-05-15-0574-FE/ (accessed 14 May 2017).

Lauretta, R., Sansone, A., and Sansone, M. (2019). Endocrine disrupting chemicals: effects on endocrine glands. *Frontiers in Endocrinology* 10: 178.

Livingston, R.J. (1977). Review of current literature concerning the acute and chronic effects of pesticides on aquatic organisms. *CRC Critical Reviews in Environmental Control* 7 (4): 235–351.

Loomis, D., Guyton, K., Grosse, Y. et al. (2015). Carcinogenicity of lindane, DDT, and 2,4-dichlorophenoxyacetic acid. *Lancet Oncology* 16 (8): 891–892.

Lushchak, V., Matviishyn, T.M., Husak, V.V. et al. (2018). Pesticide toxicity: a mechanistic approach. *EXCLI Journal* 17: 1101–1136.

Macek, K.J., Petrocelli, S.R., and Sleight, B.H. III, (1979). Considerations in assessing the potential for, and significance of, biomagnification of chemical residues in aquatic food chains. In: *Aquatic Toxicology* (ed. L.L. Marking and R.A. Kimerle), 251–268. Philadelphia, PA: American Society for Testing and Materials, STP 657.

Mackay, D., Shiu, W.-Y., and Ma, K.-C. (1997). *Illustrated Handbook of Physical-Chemical Properties and Environmental Fate for Organic Chemicals. Pesticide Chemicals*, vol. V. Boca Raton, FL: Lewis Publishers.

Martenies, S.E. and Perry, M.J. (2013). Environmental and occupational pesticide exposure and human sperm parameters: a systematic review. *Toxicology* 307: 66–73.

Matsumura, F. (1973). Degradation of pesticide residues in the environment. In: *Environmental Pollution by Pesticides* (ed. C.A. Edwards), 494. London: Plenum Press.

McGuire, V., Longstreth, W.T. Jr., Nelson, L.M. et al. (1997). Occupational exposures and amyotrophic lateral sclerosis. a population-based case-control study. *American Journal of Epidemiology* 145 (12): 1076–1088.

McKinlay, R., Plant, J.A., Bell, J.N.B., and Voulvoulis, N. (2008). Endocrine disrupting pesticides: implications for risk assessment. *Environment International* 34 (2): 168–183.

Mehrpour, O., Karrari, P., and Zamani, N. (2014). Occupational exposure to pesticides and consequences on male semen and fertility: a review. *Toxicology Letters* 230 (2): 146–156.

Menn, J.J. (1978). Comparative aspects of pesticide metabolism in plants and animals. *Environmental Health Perspectives* 27: 113–124.

Millennium Ecosystem Assessment (2005). *Ecosystems and Human Well-Being: Biodiversity Synthesis*. Washington, DC: World Resources Institute.

Milnes, M.R. and Guillette, L.J. Jr., (2008). Alligator tales: new lessons about environmental contaminants from a sentinel species. *Bioscience* 58 (11): 1027–1036.

Mischke, C. and Avery, J. (2013). Toxicities of agricultural pesticides to selected aquatic organisms. SRAC Publication No. 4600. Southern Regional Aquaculture Center. Mississippi University, Mississippi State, MS. grilife.org/fisheries2/files/2013/09/SRAC-Publication-No.-4600-Toxicities-of-Agricultural-Pesticides-to-Selected-Aquatic-Organisms1.pdf (accessed 19 May 2020).

Morgan, E.R. and Brunson, M.R. (2002). Toxicities of agricultural pesticides to selected aquatic organisms. SRAC Publication No. 4600. Southern Regional Aquaculture Center. Mississippi University, Mississippi State, MS. http://fisheries.tamu.edu/files/2013/09/SRAC-Publication-No.-4600-Toxicities-of-Agricultural-Pesticides-to-Selected-Aquatic-Organisms.pdf (accessed 19 May 2020).

Moriarty, F. (ed.) (1975). *Organochlorine Insecticides: Persistent Organic Pollutants*. London: Academic Press.

Moriarty, F. (1978). Terrestrial animals. In: *Principles of Toxicology* (ed. G.C. Butler), 169. SCOPE 12. New York: Wiley.

NPIC (2015). Bacillus thuringiensis. General fact sheet. National Pesticide Information Center, Oregon State University.http://npic.orst.edu/factsheets/BTgen.pdf (accessed 26 April 2017).

NPIC (2016). Water solubility. National Pesticide Information Center, Oregon State University. http://npic.orst.edu/envir/watersol.html (accessed 30 April 2017).

NPIC (2018). Groundwater ubiquity score (GUS). National Pesticide Information Center, Oregon State University. http://npic.orst.edu/envir/gus.html(accessed 27 October 2020).

OECD (2019). OECD test guidelines programme. Organisation for Economic Co-operation and Development. http://www.oecd.org/chemicalsafety/testing/oecd-guidelines-testing-chemicals-related-documents.htm (accessed 2 May 2020).

Parris, G.E. (1980). Environmental and metabolic transformations of primary aromatic amines and related compounds. In: *Residue Reviews. Residues of Pesticides and Other contaminants in the Total Environment*, vol. 76 (ed. F.A. Gunther and J.D. Gunther), 1–30. New York: Springer-Verlag.

Penn State Extension (1999). Persistence of herbicides in soil. Agronomy Facts 36.

Penn State Extension (2016). Toxicity of pesticides. Pennsylvania State University. https://extension.psu.edu/toxicity-of-pesticides (accessed 7 July 2017).

Phung, D.T., Connell, D., Miller, G., and Chu, C. (2012). Probabilistic assessment of chlorpyrifos exposure to rice farmers in Viet Nam. *Journal of Exposure Science and Environmental Epidemiology* 22 (4): 417–423.

Pimentel, D. and Goodman, N. (1974). Environmental Impact of Pesticides. In: *Survival in Toxic Environments* (ed. M.A.Q. Khan and J.P. Bederka Jr.), 25–52. New York: Academic Press.

Pluth, T.B., Zanini, L.A.G., and Battisti, I.D.E. (2019). Pesticide exposure and cancer: an integrative literature review. *Saúde Debate* 43 (122): 906–924. https://doi.org/10.1590/0103-1104201912220.

Rauh, V.A., Perera, F.P., Horton, M.K. et al. (2012). Brain anomalies in children exposed prenatally to a common organophosphate pesticide. *Proceedings of the National Academy of Sciences United States of America* 109 (2): 7871–7876.

Reinert, R.E. (1972). Accumulation of dieldrin in an alga (*Scenedesmus obliquus*), *Daphnia magna,* and the Guppy (*Poecilia reticulata*). *Journal of Fisheries Research Board of Canada* 29 (10): 1413–1418.

Robert, J.R., Karr, C.J., and American Academy of Pediatrics, Council on Environmental Health (2012). Technical report – pesticide exposure in children. *Pediatrics* 130 (6): e1765–e1788.

Roberts, E.M., English, P.B., and Grether, J.K. (2007). Maternal residence near agricultural pesticide applications and autism spectrum disorders among children in the California central valley. *Environmental Health Perspectives* 115: 1482–1489.

Robinson, J. (1973). Dynamics of pesticide residues in the environment. In: *Environmental Pollution by Pesticides* (ed. C.A. Edwards), 459. London: Plenum Press.

Sanchez-Bayo, F. (2011). Impacts of agricultural pesticides on terrestrial ecosystems. In: *Ecological Impacts of Toxic Chemicals* (ed. F. Sánchez-Bayo, P.J. van den Brink and R.M. Mann), 63–87. Bussum: Bentham Science Publishers.

Sanchez-Bayo, F., Goka, K., and Havasaka, D. (2016). Contamination of the aquatic environment with neonicotinoids and its implication for ecosystems. *Frontiers in Environmental Science* 4: 71.

Schafer, K.S. and Marquez, E.C. (2012). A Generation in Jeopardy. How Pesticides are undermining our children's health and intelligence. Pesticide Action Network North America. https://www.panna.org/sites/default/files/KidsHealthReportOct2012.pdf (accessed 8 May 2020).

Seiber, J.N., Woodrow, J.D., Shafik, T.A., and Enos, H.F. (1975). Determination of pesticides and their transformation products in air. In: *Environmental Dynamics of Pesticides* (ed. R. Haque and V.H. Freed), 17. New York: Plenum Press.

Semenza, J.C., Tolbert, T.E., Rubin, C.H. et al. (1997). Reproductive toxins and alligator abnormalities at Lake

Apopka, Florida. *Environmental Health Perspectives* 105 (10): 1030–1032.

Seume, F.W. and O'Brien, R.D. (1960). Potentiation of the toxicity to insects and mice of phosphorothionates containing carboxyester and carboxyamide groups. *Toxicology and Applied Pharmacology* 2: 495–503.

Shaw, G.R. and Connell, D.W. (1983). Factors influencing concentrations of polychlorinated biphenyls in organisms from an estuarine ecosystem. *Australian Journal of Marine and Freshwater Research* 33 (6): 1057–1070.

Soderlund, D.M. (2012). Molecular mechanisms of pyrethroid insecticide neurotoxicity: recent advances. *Archives of Toxicology* 86 (2): 165–181.

Solomon, K.R., Carr, J.A., Du Preez, L., and Giesy, J.P. (2008). Effects of atrazine on fish, amphibians, and aquatic reptiles: a critical review. *Critical Reviews in Toxicology* 38 (9): 721–772.

Sparks, T.C. and Nauen, R. (2015). IRAC: mode of action classification and insecticide resistance management. *Pesticide Biochemistry and Physiology* 121: 122–128.

Spencer, W.F. and Cliath, M.M. (1975). Vaporization of chemicals. In: *Environmental Dynamics of Pesticides* (ed. R. Haque and V.H. Freed), 61–78. New York: Plenum Press.

Stehle, S. and Schulz, R. (2015). Agricultural insecticides threaten surface waters at the global scale. *Proceeding of the National Academy of Sciences United States of America* 112 (18): 5750–5755.

Stokstad, E. and Grullon, G. (2016). Infographic: pesticide use on arable land between 2005 – 2009. Genetic Literacy Project. *Science/AAAS*. November 28, 2016. https://geneticliteracyproject.org/2016/11/28/infographic-pesticide-use-on-arable-land-between-2005-2009/ (accessed 29 April 2020).

Thundiyil, J.G., Stober, J., Besbelli, N., and Pronczuk, J. (2008). Acute pesticide poisoning: a proposed classification tool. *Bulletin of the World Health Organization* 86 (3): 205–209.

Tinsley, I.J. (1979). *Chemical Concepts in Pollutant Behavior*. New York: Wiley.

Tu, M., Hurd, C., and Randall, J. (2001). *Weed Control Methods Handbook: Tools and Techniques for Use in Natural Areas*. Arlington, VA: The Nature Conservancy https://www.invasive.org/gist/products/handbook/methods-handbook.pdf (accessed 28 April 2020).

Tyler Miller, G. Jr., (2004). *Sustaining the Earth: An Integrated Approach*, 211–216. Pacific Grove, CA: Brooks/Cole.

UNEP (2013). *Global Chemicals Outlook – Towards Sound Management of Chemicals*. Geneva: United Nations Environmental Programme.

US EPA (1997a). Diflubenzuron. R.E.D. Facts. United States Environmental Protection Agency. https://www3.epa.gov/pesticides/chem_search/reg_actions/reregistration/fs_PC-108201_1-Aug-97.pdf (accessed 26 April 2017).

US EPA (1997b). Azoxystrobin. Pesticide Fact Sheet. United States Environmental Protection Agency. https://www3.epa.gov/pesticides/chem_search/reg_actions/registration/fs_PC-128810_07-Feb-97.pdf (accessed 28 April 2017).

US EPA (ed.) (2015). Health: neurodevelopmental disorders. In: *America's Children and the Environment*. 3. United States Environmental Protection Agency, Washington, DC. https://www.epa.gov/sites/production/files/2015-10/documents/ace3_neurodevelopmental.pdf (accessed 4 July 2017).

US EPA (2016). What are biopesticides?. United States Environmental Protection Agency. https://www.epa.gov/ingredients-used-pesticide-products/what-are-biopesticides (accessed 12 May 2020).

US EPA (2017a). What are antimicrobial pesticides?. United States Environmental Protection Agency. https://www.epa.gov/ingredients-used-pesticide-products/what-are-biopesticides (accessed 23 April 2017).

US EPA (2017b). Pesticides industry sales and usage 2008 – 2012 market estimates. Office of Chemical Safety and Pollution Prevention, United States Environmental Protection Agency. Washington, DC. https://www.epa.gov/sites/production/files/2017-01/documents/pesticides-industry-sales-usage-2016_0.pdf (accessed 29 April 2020).

US EPA (2017c). Test guidelines for pesticides and toxic substances United States Environmental Protection Agency. Washington, DC. https://www.epa.gov/test-guidelines-pesticides-and-toxic-substances (accessed 11 June 2017).

US EPA (2019). Guidance for reporting on the environmental fate and transport of the stressors of concern. United States Environmental Protection Agency. https://www.epa.gov/pesticide-science-and-assessing-pesticide-risks/guidance-reporting-environmental-fate-and-transport#III (accessed 23 April 2017).

Van Maele-Fabry, G., Hoet, P., Lantin, A.-C., and Lison, D. (2010). Residential exposure to pesticides and childhood leukaemia: a systematic review and meta-analysis. *Environment International* 37 (1): 280–291.

Van Maele-Fabry, G., Hoet, P., and Lison, D. (2013). Parental occupational exposure to pesticides as risk factor for brain tumors in children and young adults: a systematic review and meta-analysis. *Environment International* 5 (56C): 19–31.

Walker, C.H. (1975). Variations in the intake and elimination of pollutants. In: *Organochlorine Insecticides: Persistent Organic Pollutants* (ed. F. Moriarty), 73. London: Academic Press.

Ware, G.W. and Whitacre, D.M. (2004). An introduction to herbicides. In: *Radcliffe's IPM World Textbook*, 2e. College of Food, Agricultural and Natural Resource Sciences, University of Minnesota, St Paul Minnesota, MN. Willoughby, OH: MeisterPro Information Resources.

Way, J.M. and Chancellor, R.J. (1976). Herbicides and higher plant ecology. In: *Herbicides, Physiology, Biochemistry, Ecology*, vol. 2 (ed. L.J. Audus), 345–372. London: Academic Press.

Weber, J.B., Wilkerson, G.G., and Reinhardt, C.F. (2004). Calculating pesticide sorption coefficients (K_d) using selected soil properties. *Chemosphere* 55: 157–166.

Weichenthal, S., Moase, C., and Chan, P. (2010). A review of pesticide exposure and cancer incidence in the Agricultural Health Study cohort. *Environmental Health Perspectives* 118 (8): 1117–1125.

Wheatley, G.A. (1973). Pesticides in the atmosphere. In: *Environmental Pollution by Pesticides* (ed. C.A. Edwards), 365. London: Plenum Press.

WHO (2004). Children are facing high risks from pesticide poisoning. World Health Organization, Geneva. https://www.who.int/mediacentre/news/notes/2004/np19/en/ (accessed 4 May 2020).

WHO (2008). Children's health and the environment WHO training package for the health sector world health organization. World Health Organization, Geneva. https://www.who.int/ceh/capacity/Pesticides.pdf (accessed 27 August 2019).

WHO (2010). Exposure to highly hazardous pesticides: A major public health concern. World Health Organization, Geneva. https://www.who.int/ipcs/features/hazardous_pesticides.pdf?ua=1 (accessed 16 August 2019).

WHO (2020). Children's environmental health. World Health Organization, Geneva. https://www.who.int/health-topics/children-environmental-health#tab=tab_1 (accessed 8 May 2020).

WHO/FAO (2004). Pesticide residues in food – 2004. FAO Plant Production and Protection Paper 178. World Health Organization, Geneva, and Food and Agricultural Organization, Rome, United Nations.

WHO/IPCS (2010). The WHO recommended classification of pesticides by hazard and guidelines to classification 2009. World Health Organization/International Programme on Chemical Safety. https://apps.who.int/iris/handle/10665/44271 (accessed 5 May 2020).

WHO/UNEP (2013a). State of the science of endocrine disrupting chemicals – 2012. World Health Organization/United Nations Environment Programme. https://www.who.int/ceh/publications/endocrine/en/ (accessed 4 February 2020).

WHO/UNEP (2013b). State of the science of endocrine disrupting chemicals – 2012. Summary for decision-makers. World Health Organization/United Nations Environment Programme. https://apps.who.int/iris/bitstream/handle/10665/78102/WHO_HSE_PHE_IHE_2013.1_eng.pdf?sequence=1 (accessed 4 February 2020).

Wightwick, A., Walters, R., Allinson, G.et al. (2010). Environmental risks of fungicides used in horticultural production systems. In: *Fungicides*. (ed. O. Carisse), InTech. http://www.intechopen.com/books/fungicides/environmental-risks-of-fungicides-used-in-horticultural-production-systems (accessed 29 April 2020).

Wohlfahrt-Veje, C., Andersen, H.R., Schmidt, I.M. et al. (2012). Early breast development in girls after prenatal exposure to non-persistent pesticides. *International Journal of Andrology* 35: 273–282.

Woodwell, G.M., Craig, P.P., and Johnson, H.A. (1971). DDT in the biosphere: Where does it go? *Science* 174: 1101.

Wu-Smart, J. and Spivak, M. (2016). Sub-lethal effects of dietary neonicotinoid insecticide exposure on honey bee queen fecundity and colony development. *Scientific Reports* 6: 32108. https://doi.org/10.1038/srep32108 (accessed 2 July 2020).

Yamamuro, M., Komuro, T., Kamiya, H. et al. (2019). Neonicotinoids disrupt aquatic food webs and decrease fishery yields. *Science* 366 (6465): 620–623.

Zaller, J.G. and Bruhl, C.A. (2019). Editorial: non-target effects of pesticides on organisms inhabiting agroecosystems. *Frontiers in Environmental Science* 7: 75. https://doi.org/10.3389/fenvs.2019.00075 (accessed 2 July 2020).

10

Petroleum, Coal, and Biofuels

10.1 Introduction

Most of the world's energy use by human populations results from the burning of fossil fuels, mainly for electricity generation, transport, industrial use, and heating. Fossil fuels are derived from sedimentary deposits of combustible organic materials formed from decayed plants and animals by exposure to heat and pressure in the Earth's crust over hundreds of millions of years. Most carbonaceous material formed before the Devonian Period (419–359 million years ago) was derived from algae and bacteria, while most formed during and after this Period was derived from plants (Kopp n.d.).

The major forms of fossil fuels are coal, oil, and natural gas (see Table 10.1). The term, petroleum, is used to describe crude oil, refined oil products, and natural gas. Fossil fuels have been the major energy driver and source of organic chemicals for human society since the start of the Industrial Age. The burning of fossil fuels by humans, however, is the major source of air pollution and emissions of atmospheric carbon dioxide (key greenhouse gas) that contributes to global warming (Chapter 14). Fossil fuels are projected to decrease as the dominant energy source during the twenty-first Century as their environmental health impacts and economic uses are challenged by renewable sources of energy, including bioenergy.

Bioenergy is derived from biological materials (biomass) that are not embedded in geological sources such as fossilized organic matter (fossil fuels). It supplies about 10% of total global energy, mostly as unprocessed biomass in developing countries. Processed forms of bioenergy such as liquid biofuels (ethanol and biodiesel) are increasing as substitutes for liquid petroleum (FAO 2008, Chapter 2). The use of hydrocarbon-based biofuels is not expected to increase the net amount of carbon dioxide in the atmosphere.

Biofuels are currently only a small fraction of petroleum oil. Globally, transportation accounts for most of petroleum production. The equivalent production of biofuels represents only a small percentage of petroleum used for transportation, based on the values in Table 10.1. Even so, we need to evaluate any potential pollution impacts from new fuel sources because they are expected to increase market share and reduce petroleum-related emissions.

10.2 Fossil Fuels

Types of fossil fuels include coal, petroleum (crude oils and refined petroleum products), natural gas (including coal seam and shale gas), oil shales, bitumens (natural or refined from crude oil), heavy oils, and tar sands. Peat is another type of fuel formed over thousands of years from the partial decay and accumulation of organic matter (e.g. peat bogs) and considered an early geological form of coal and other fossil fuels.

Petroleum

Natural Gas

Petroleum is a complex mixture of hydrocarbons and other components that occurs in natural sedimentary reservoirs in gaseous, liquid, or solid-like forms. The gaseous form is **natural gas**, a mixture of flammable lightweight alkanes, consisting primarily of methane, CH_4, small amounts of ethane and propane and traces of other C_3, C_4, and C_5 alkane hydrocarbons, nitrogen, carbon dioxide, hydrogen sulfide, oxygen, and metals. Natural gas is found in several different types of rocks, including coal seams, sandstones, and shales. Another very large source of natural gas, in the form of **methane hydrate** (crystalline solid lattice, $4CH_4 \cdot 23H_2O$), occurs naturally in sub-surface deposits where it is formed by suitable pressure and temperature conditions. It is found below the Arctic permafrost, under the Antarctic ice, in sedimentary deposits along continental margins, and in deep-water sediments of inland lakes and seas. The release of methane from the Arctic region is an important contributor to global warming.

Commercial-grade natural gas is essentially a mixture of methane and some ethane. Wet natural gas is processed,

Table 10.1 Global primary energy consumption by fuel (million tonnes oil equivalent, Mtoe).

Fuel type	2018
Oil (petroleum)	4662.1
Natural gas	3309.4
Coal	3772.1
Nuclear energy	611.3
Hydroelectricity	948.8
Renewables[a]	561.3
Total	13 864.9

[a] Wind, solar, geothermal, biomass, and biofuels.
Source: Data compiled from BP (2019).

or petroleum (crude oil) is refined to remove propane and butane under pressure, and produced as liquefied petroleum gases (LPG), in the form of a mixture of these gases, or separately as propane, and butane.

Crude and Refined Petroleum Oils

Liquid petroleum is found as **crude oil** (50–95%, by weight). The physical nature and chemical composition of crude oils from different deposits or oil fields can vary considerably, from light to heavy oils. It contains mainly complex mixtures of liquid to waxy hydrocarbons, together with small amounts of more polar organic compounds containing sulfur, nitrogen or oxygen, asphaltenes, resins, and various trace elements (e.g. iron, nickel, copper, and vanadium) in complex and unbound forms, see Figure 10.1a and b. A viscous black liquid or semisolid form is known as asphalt or bitumen (pitch), which can occur as natural asphalt, such as in the massive tar sand deposits found in Alberta, Canada, or as a refined residue produced during the distillation process of certain crude oils.

Crude oil is converted into commercial products such as LPG, feedstock for petrochemical production, liquid fuels (kerosene, gasoline, and diesel), refined oils, lubricants, and asphalt, by **petroleum refining** (see Table 10.2). There are several processes involved: (i) separation of crude oil into fractions by distillation, (ii) treatment of fractions by various methods (e.g. cracking, reforming, alkylation, polymerization, and isomerization) to produce more refined mixtures and new hydrocarbons, (iii) further separation of treated mixtures (e.g. fractionation and solvent extraction), and (iv) removal of impurities by various methods (e.g. dehydration, desalting, sulfur removal, and hydrotreating) to obtain commercial products (The Essential Chemical Industry-online 2014).

The chemical components in liquid petroleum can be separated into four groups: saturates, aromatics, resins, and asphaltenes. In crude oil, there are two major chemical classes of hydrocarbons termed aliphatic and aromatic.

Aliphatic hydrocarbons consist of saturated hydrocarbons (saturates) or single-bonded carbon atoms. These are alkanes or paraffins – n-normal or straight chains, branched (isoalkanes) or cycloalkanes (naphthenes). Soft paraffin or petroleum wax forms from complex mixtures of alkane hydrocarbon molecules with 18–65 carbon atoms. Alkenes with a nonconjugated double bond are rarely found in natural crude oil although formed during catalytic *cracking* during refining of crude oils.

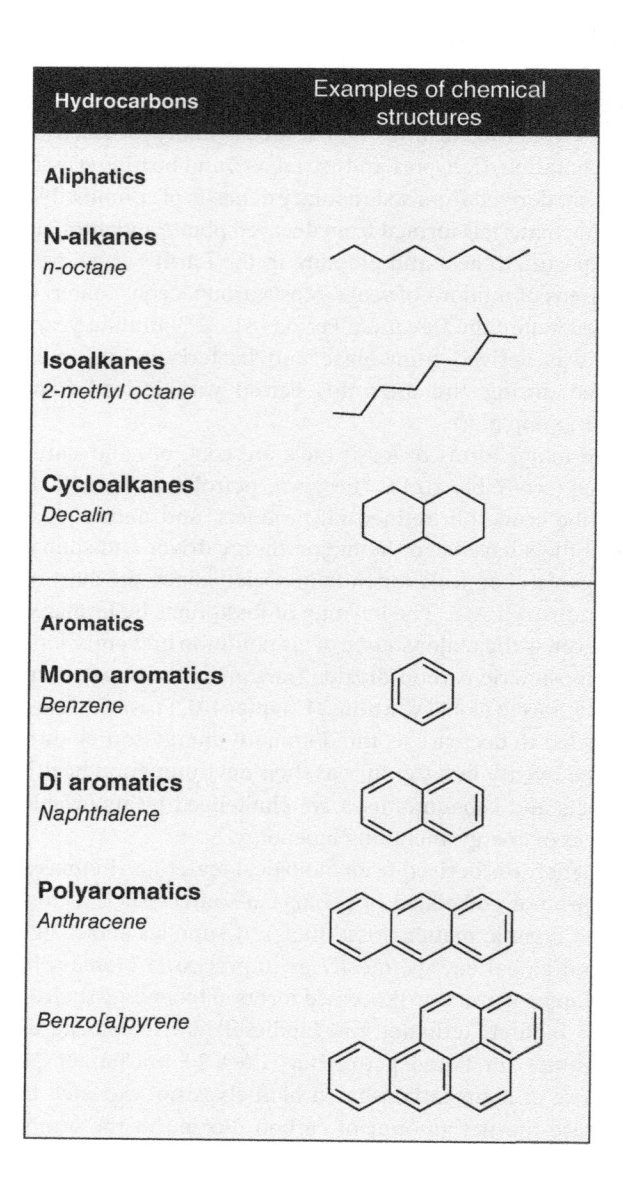

Figure 10.1 (a) Examples of the chemical structures of some common types of hydrocarbons found in crude petroleum. (b) Examples of the chemical structures of some common types of non-hydrocarbons found in crude petroleum.

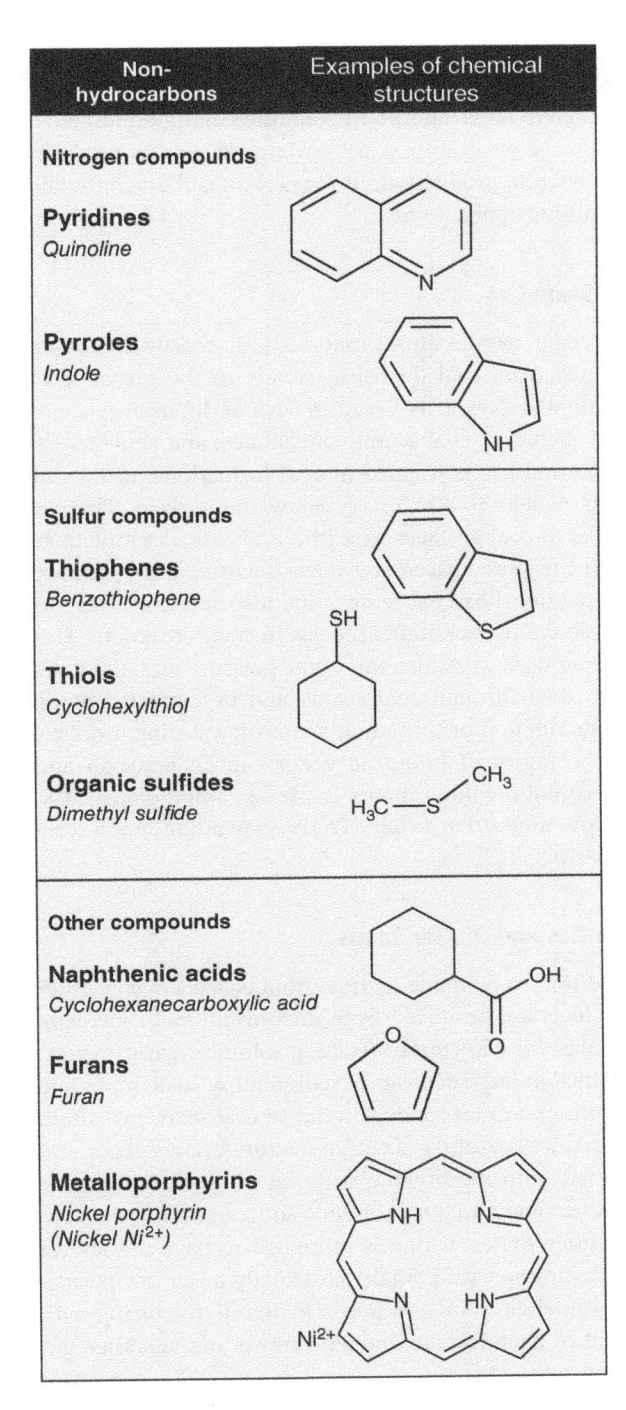

Non-hydrocarbons	Examples of chemical structures
Nitrogen compounds **Pyridines** *Quinoline*	
Pyrroles *Indole*	
Sulfur compounds **Thiophenes** *Benzothiophene*	
Thiols *Cyclohexylthiol*	
Organic sulfides *Dimethyl sulfide*	
Other compounds **Naphthenic acids** *Cyclohexanecarboxylic acid*	
Furans *Furan*	
Metalloporphyrins *Nickel porphyrin* (Nickel Ni^{2+})	

Figure 10.1 (*Continued*)

Table 10.2 Petroleum fractions obtained from refined petroleum oil.

Petroleum fraction	Boiling range (°C)	Number of carbon atoms
Natural gas	<20	C_1–C_4
Petroleum ether	20–60	C_5–C_6
Gasoline	40–200	C_5–C_{12}
Kerosene	150–260	~C_{12}–C_{13}
Fuel oils	>260	≥C_{14}
Lubricants	>400	≥C_{20}
Asphalt	Residue	Polycyclic

in crude oil. They are relatively high molecular weight, complex polar, polycyclic, aromatic ring compounds with side chains (alkanes). Resins have a highly polar end group (aromatic and naphthenic rings, often with heteroatoms of O, S, and N) and long alkane tails. Figure 10.1 shows some examples of the types of chemical structures for (a) hydrocarbons and (b) non-hydrocarbons found in crude petroleum. Hundreds to thousands of individual compounds are reported to exist in complex mixtures of petroleum.

Aromatic Hydrocarbons

Aromatic hydrocarbons are major constituents of crude oils, usually present as 20–45% of total hydrocarbons. The low-boiling or volatile benzenes (benzene, toluene or methyl benzene, ethyl benzene, and xylenes or dimethyl benzene isomers) are commonly known as BTEX. These are major toxic components of crude and many refined petroleum oils. For example, benzene is also a known human carcinogen (IARC and US EPA).

Polynuclear aromatic hydrocarbons (PAHs) contain two or more fused benzene rings (see Figure 10.1a and include the naphthalenes (two fused rings). These rings can also be substituted with alkyl groups. They are semi-volatile toxic components of crude oils and many refined oils. The main routes of formation in the environment are (i) burning or pyrolysis of organic matter such as plants, fossil fuels, and biomass, and (ii) generation in sedimentary organic matter and fossil fuels. Pyrolysis and incomplete combustion of organic matter yield PAHs in the breakdown products. Essentially, complex molecules such as proteins and carbohydrates in the organic matter degrade to form lower molecular weight free radicals, which rearrange and combine to form PAHs. Soot that contains a large number of linked aromatic rings is often formed, including PAHs. The conditions of pyrolysis or combustion strongly influence the type and quantity of PAHs produced. Soot is classified as a human carcinogen by the IARC.

Aromatic hydrocarbons have one to six or more benzene rings, which may also have alkyl substituents. These range from benzenes (single aromatic ring), to naphthalenes (two aromatic rings) and polynuclear or polycyclic aromatic hydrocarbons (PAHs) (three or more fused aromatic rings). The term, PAHs, can also include the two-ring naphthalenes. Resins and asphaltenes may also be present

Crude oil and refined petroleum products show a large variation in aromatic and PAH content. Synthetic crude petroleum produced by coal liquefying plants and shale oil pyrolysis generally has a higher PAH content.

Biogenic Hydrocarbons

The analytical investigation of petroleum-derived hydrocarbons in the environment can also be complicated by the presence of hydrocarbons of recent biogenic origin (alkanes, isoprenoids, alkenes, and aromatics). These hydrocarbons occur widely in terrestrial and marine organisms and may be formed by biosynthesis, transferred from foods, or formed following ingestion of precursors from food and abiotic sources. Alkanes (straight and branched) and isoprenoids (e.g. pristine, phytane, and squalene) tend to predominate in organisms although unsaturated hydrocarbons may be significant in bacteria and algae (see Connell and Miller 1981, pp. 40–49; 1984, p. 232).

Coal

Coal is a combustible and carbon-rich type of sedimentary rock found mainly as anthracite, bituminous, sub-bituminous, and lignite forms. *Anthracite coal* (86–98% C w/w) is a dense, hard rock with a jet-black color and a metallic luster. *Bituminous coal*, or soft or black coal (~60–80% C w/w) is the most abundant form of coal. *Sub-bituminous coal* contains less carbon and more water while *lignite coal*, or brown coal, is a very soft coal that contains up to 70% water by weight. Elemental analysis gives empirical formulas such as $C_{137}H_{97}O_9NS$ for bituminous coal and $C_{240}H_{90}O_4NS$ for high-grade anthracite.

Coal is primarily a complex structure of layered forms of fused aromatic and cyclic forms of hydrocarbons combined or joined with oxygen, nitrogen, and sulfur containing rings and hydrogen, oxygen, nitrogen, and sulfur atoms in various structural forms (substituents, functional groups, or chains). It also contains a number of minerals with other common and trace elements. Upon combustion, such as in coal-fired power stations, coal also releases some toxic elements, including mercury, arsenic, selenium, and chromium, in gaseous and/or particulate forms such as in fly ash or in bottom ash as wastes.

The carbonization of coal to produce coke and/or natural gas also forms coal tars (dark viscous liquids or semisolids) as by-products. These tars contain complex combinations of PAHs, phenols, and heterocyclic O, N, and S compounds. Distillation of coal tars produces oily liquids of creosote coal tars (aromatic hydrocarbons and PAHs) and a residue of coal tar pitch (methylated PAHs and heteronuclear compounds) (ATSDR 2002). Coal combustion by-products or residues such as large quantities of fly and bottom ash, flue gas desulfurization residues, and boiler slag are usually deposited in large landfills or contained in impoundments, while some residues (e.g. fly ash) are re-used as products (e.g. concrete production, or for certain soil amendments and mining applications).

Coal Seam Gas

Coal seam gas is simply natural gas, mostly methane, extracted from coal deposits, usually in the presence of formation waters. It is found in several different types of rocks, including coal seams, sandstones, and shale rocks. The natural gas is trapped in coal formations, usually at depths of 400–1000 m or so below the surface. The gas adsorbs to coal surfaces as a film and can accumulate as free gas in pore spaces and open fractures. The pressure of water that fills coal seams, and also saturates free gas in these seams, maintains the gas in coal formations. Gas wells are used to reduce the water pressure and allow the gas to flow through coal seams and be extracted along with production or formation water (dewatering process). Large volumes of produced water can be brackish and may contain contaminants such as ammonia, BTEX, and low-molecular-weight PAHs depending upon coal sub-basin.

Shale Gas and Oil/Tar Sands

In addition to methane hydrates and coal seam gas, other fossil fuels are produced from unconventional sources of natural gas and **kerogens** (solid, insoluble organic matter) contained in large deposits of sedimentary shale rocks and also bitumen in tar sands. Similar to coal seam gas, **shale gas** occurs as natural gas within fractures, pore spaces, and adsorbed onto the organic material in shale formations. It has become a major economic source of natural gas in the United States. China is estimated to have the largest reserves in the world. Shale gas usually has a low permeability in shale rock and needs hydraulic fracturing with the aid of chemicals to extract trapped and adsorbed gas via multiple wells.

Shale oil is produced from pulverized oil shale rocks by pyrolysis, hydrogenation, or thermal dissolution. Raw shale oil is a complex hydrocarbon mixture containing large to substantial proportions of alkenes, aromatics, PAHs, heteroatomic-organics (N, S and O), and residual minerals. It can be processed to obtain refined products similar to crude oil. Tar sands, such as found on a massive scale in Alberta, Canada, contain bituminous hydrocarbons that can also be processed to produce synthetic crude oil (see Baird and Cann 2012, pp. 241–242).

Fracking Chemicals

During oil, shale gas, and coal seam gas drilling, hydraulic fracturing (*fracking*) is widely used. This process involves injecting large volumes of a fluid and ceramic materials (*proppants*) such as sand or resin-coated sand, which may also include various chemical additives (fracking chemicals), under high pressure into shale rock or coal seams to widen existing fractures and make new ones to improve economic recovery of gases. The chemical additives or fracking chemicals may include mixtures of low boiling aromatic hydrocarbons (BTEX), acids (e.g. hydrochloric acid), gelling agents, stabilizers, pH buffers, chemical breakers, corrosion scale inhibitors, biocides, foaming agents, preservatives, and surfactants. Environmental issues associated with hydraulic fracturing and use of fracking chemicals relate to surface water and groundwater contamination through spills at the well site, discharge of wastewaters, or seepage of hydraulic fluids into aquifers.

Syngas

Fossil fuels, particularly coal, can be decomposed at high temperatures (700–1000 °C) to produce a mixture of carbon monoxide and hydrogen gases, known as **synthesis gas or syngas**, and residual tar. It has been used as a fuel source (e.g. town gas) and in the organic synthesis of alkane-like hydrocarbons (synthetic gasoline) in variants of the *Fischer-Tropsch* process.

Catalytic conversion of syngas may also form methanol (Baird and Cann 2012, pp. 314–315). The production of syngas has been greatly limited by its need for intensive energy use, hazardous and toxic components, production emissions, residues, and readily available alternative sources of natural gas and crude oil for fuel and petrochemicals. Syngas, however, may also be produced from biomass (e.g. wood or crop wastes) instead of coal, as a form of biogas.

10.3 Biofuels

The emergence of biofuels as a renewable energy source is projected to challenge the future use of fossil fuels, especially liquid petroleum, and reduce emissions of toxic air pollutants and greenhouse gases, along with other renewable energy sources. Their environmental and health risks are under current research investigation, including life cycle assessments.

Bioenergy is an alternative and renewable source of energy obtained from recent biomass. Primary production of biomass results via photosynthesis from sunlight at an efficiency of only about 5%. Nonetheless, potential bioenergy from this annual biomass amounts to about five to six times total global energy consumption of 560 EJ (exajoules, EJ = 1×10^{-18} J).

The conversion of solid biomass into bioenergy, mostly by burning for heating and cooking, has long been practiced by humans. Today, around 3 billion people cook and heat their homes using open fireplaces and inefficient stoves burning biomass (wood, animal dung, and crop waste) and also coal. Most of these people are poor, and live in developing countries.

Biomass can be used as a solid fuel for combustion or converted to liquid fuels (ethanol, methanol, butanol, and biodiesel) and gas biofuels, such as methane and syngas, via several processes, for heating, electricity, and transportation end uses. To a minor extent, it can also be used as a raw material to make plastics, fertilizers, and other chemicals, similar to fossil fuels (Diesendorf 2014, p. 49). Biofuels can be classified into two types: primary (unprocessed), such as firewood, wood chips and pellets, and secondary (processed), such as solids (e.g. charcoal), liquids (e.g. ethanol and biodiesel), or gases (e.g. biogas) (FAO 2008, pp. 10–22). These sources are discussed below.

Solid Biomass

The combustion of solid biomass, usually wood feedstock and other dry plant matter, is the oldest source of heat energy for human use and remains widely used. It involves biomass and oxygen gas (air) burning in a high temperature environment to form carbon dioxide and water vapor and to release heat energy (about 20 MJ kg^{-1} of biomass). Ash containing inorganic and carbon substances is also produced. Combustion conditions, such as the nature and amount of feedstock, air supply, moisture content of the feedstock, combustion temperature, and time, control the energy efficiency and emissions of particulates, gases, volatile organics, and ash content.

Liquid Biofuels

The emergence of biofuels as a renewable energy source is projected to challenge the future use of fossil fuels, such as liquid petroleum, and reduce emissions of toxic air pollutants and greenhouse gases, along with other renewable energy sources. Their environmental and health risks are under current research investigation, including life cycle assessments.

Biofuels are derived from recently formed biomass in contrast to the ancient biomass of fossil fuels. The two main types are bioethanol and biodiesel.

Bioalcohols

Bioalcohols (methanol, ethanol, isopropanol, and butanol) are produced by the fermentation of carbohydrates such as sugars (e.g. sugar cane, sugar beet, and molasses), starches (e.g. corn, wheat, or *cassava*) or cellulose and hemicellulose in plants, through the actions of microorganisms and enzymes.

Bioethanol is the most commonly used biofuel. First-generation bioethanol is produced from edible food crops by enzyme digestion of stored starches to release sugars, followed by fermentation of the sugars by zymase (produced by yeasts) into ethanol (12–15% v/v) and carbon dioxide, and then fractional distillation to high ethanol distillate.

Brazil has the most advanced biofuel economy in the world based on the production of anhydrous ethanol from sugar cane and blending with gasoline (e.g. 25% ethanol/gasoline) for regular use in motor vehicles. Residual cane waste, *bagasse*, is used to produce heat and power, including sugar mill cogeneration. Large life-cycle reduction in greenhouse gas emissions has been reported due to ethanol production and use.

Bioethanol produced from lignocellulosic biomass such as waste plant material or crop residues (stems, husks, and leaves) is considered more sustainable than using edible food crops because of factors such as reduced impacts on food crops and arable land, less use of agrichemicals and irrigation, and reduced greenhouse gas emissions during processing. However, the process is more difficult than for using starch. The United States plans to produce 16 million gallons of biofuels from cellulosic materials by 2022 although no commercial biofuel of this type was produced in 2010 (Baird and Cann 2012, pp. 301–302).

Methanol was historically made as *wood alcohol* by the destructive distillation of wood but is now produced mainly from natural gas (fossil fuel) using the synthesis gas process. However, the production of **biomethanol** from biomass via biogas (see Biogas below) is being developed as an alternative fuel source. It can be blended with gasoline (e.g. 15% methanol blends in gasoline in China) to produce a fuel mixture with less emissions than gasoline (Baird and Cann 2012, p. 318).

Biodiesel

Biodiesel fuel (also known as B100) refers to mono-alkyl esters of long-chain fatty acids derived from vegetable oils and animal fats. It is usually blended with traditional diesel fuel or can be burnt as a pure fuel in compression ignition engines. Biodiesel is produced by the esterification of vegetable and animal oils and fats, by reacting the triglycerides in extracted oil or fat with an alcohol (ethanol or methanol), in the presence of a strong base catalyst (sodium hydroxide), to form esters and glycerol (transesterification reaction). In this reaction, the triglyceride consists of three fatty acid chains bonded by individual ester linkages to a glycerine (trialcohol) molecule. The alcohol reacts with the fatty acids to form the mono-alkyl ester(s) – biodiesel – and crude glycerol mixture from which the biodiesel is separated. The mono-alkyl esters of biodiesel have similar fatty acid carbon chains to the hydrocarbons in traditional diesel.

The properties and quality of commercial biodiesel are influenced by the reaction, chemical nature of the fatty acids in triglyceride feedstock, and removal of residual reactants (e.g. methanol) and by-products. First-generation biodiesel is commonly made from food crops and second generation uses nonfood crops, cellulosic wastes, and algae (FAO 2008, Chapter 2).

Biogas

Biomass can also be converted into biogas by thermochemical or anaerobic digestion processes, mainly for use in heating and to generate electricity. As a biofuel for vehicles, it needs further cleanup or refining to remove by-products: carbon dioxide, hydrogen sulfide, trace elements, and organic contaminants. Major biogas products are:

(a) Syngas or fuel gas from gasification or pyrolysis of lignocellulose-containing biomass (e.g. wood, crop, and forest residues).
The organic material is heated at high temperatures (>700 °C) in controlled conditions of oxygen and/or steam to produce the syngas, composed of carbon monoxide, carbon dioxide, and hydrogen gases. It can be burnt directly in gas engines, used to produce methanol and hydrogen gas, or refined further into a synthetic liquid fuel (carbon monoxide and hydrogen).
(b) Methane gas from anaerobic digestion of animal manures, green crops, and wet wastes.
Anaerobic digestion is a series of biological processes in which microorganisms break down biomass material, in the absence of oxygen, to produce biogas, mainly methane, for combustion to generate electricity and heat. It can also be further processed into higher-grade natural gas and transportation fuels.

There are a number of anaerobic digestion technologies used to convert biomass wastes: animal manure, food wastes, fats, oils and grease, municipal wastewater solids, and various other organic waste streams, into biogas. In the United States, for example, there were about 250 anaerobic digester systems operating at commercial livestock farms in August 2017. Most of these facilities use biogas for electricity generation, while several farms use biogas to produce biofuel for transportation. The process also produces waste streams of digested solids and liquids. Uncontaminated waste streams can be separated and solids composted for reuse as biosolids and nutrients in liquids applied as a fertilizer in agriculture.

Biogas or landfill gas (LFG) in the form of methane emissions from anaerobic digestion of organic wastes in landfills is also a significant source of this greenhouse gas. LFG projects are increasingly being used to generate electricity from this biogas source. For example, 619 LFG projects were reported to be operational in the United States as of February 2019 (AFDC 2020).

Gasoline additives can also be significant or serious environmental contaminants. They are usually added to liquid petroleum fuels and biofuels to enhance the octane rating of the fuel, engine performance, power, and efficiency. Types of additives cover antiknock agents (e.g. tetraethyl lead, toluene, and isooctane), antioxidants and stabilizers (e.g. butylate hydroxytoluene, BHT), lead scavengers, (e.g. tricresyl phosphate and 1,2 bromoethane), and oxygenates or oxyfuels (e.g. alcohols and ethers). Some of these additives and combustion breakdown products are known to be very toxic to humans and persistent contaminants of air, waters, soils, and biota. For example, emissions of lead compounds from the combustion of leaded gasoline, containing tetraethyl lead, have contaminated urban air, major roads, and highways, throughout the world.

The phasing out of lead in gasoline has resulted in increased use of oxygenated additives such as alcohols and ethers that increase the oxygen content of gasoline, promoting more complete combustion and reducing carbon monoxide emissions. Even so, oxygenated additives produce toxic and reactive aldehyde combustion and other breakdown products. Ethers such as MTBE are also persistent contaminants, such as in groundwaters, through emissions, leakages from fuel storages or spills. There are several types of ethers. In particular, MTBE (methyl tertiary-butyl ether) has a higher octane number than ethanol and a lower evaporation rate but suffers with several environmental limitations: an objectionable odor, higher water solubility than hydrocarbons in gasoline, environmental persistence, and combustion emissions of aldehydes and other oxygenated by-products like alcohols. Its use in unleaded gasoline in North America and Europe is controversial and subject to environmental restrictions, controls, and replacement by alternative fuel additives (Baird and Cann 2012, pp. 319–320).

10.4 Physiochemical Properties of Fuels and Chemical Components

The nature, fate, and toxicity of fuels and chemical components in the environment are influenced by a number of key physical and chemical properties that apply to the fuels as mixtures, specific chemical components, or combustion by-products. Petroleum products and biofuels are produced according to international standards such as the ASTM that specify and allow testing of fuels for many physical and chemical properties, including their composition, boiling point ranges, purity, density, evaporation or volatility, viscosity, miscibility, compatibility with other fluids and materials, sulfur content, flash point, thermal stability, and heat of combustion. Emissions of toxic combustion by-products and chemicals in fuels are being reduced through global harmonization of fuel quality standards, which includes lead additive free, low sulfur and benzene content in fuels, and strict limits on emission standards for toxic pollutants such as nitrogen oxides, fine particulates ($PM_{2.5}$), and carbon dioxide.

Key properties of hydrocarbons and oxygenated compounds in fuels are related to their chemical structures (e.g. saturated and unsaturated carbon bonding, carbon chain lengths, rings and functional groups, melting and boiling points, and polarity). These factors influence their movement and distribution through volatility, water solubility, and partitioning between water and lipid phases, as indicated by K_{OW} values (Sangster 1989). In Table 10.3, water solubility and log K_{OW} values are given for common compounds (hydrocarbons and oxygenated derivatives) found in or derived from fossil fuels (alkanes, benzenes, PAHs, and phenolics), biofuels (ethanol and methyl esters of fatty acids), and fuel additives (ethers). For toxic aromatics, K_{OW} values tend to increase with the number of aromatic rings, from single ring benzenes to five-ring benzo(a)pyrene. The importance of these properties, among others, is considered in Section 10.6 on behavior and fate of fuels in the environment and Section 10.8 on uptake by aquatic biota.

10.5 Sources, Emissions, and Releases of Petroleum-Derived Hydrocarbons

A diverse and complex range of hydrocarbons is widely distributed in terrestrial, aquatic, and atmospheric environments. They can originate readily from petroleum or biogenic sources. Petroleum-derived hydrocarbons are major pollutants in the world's coastal waters and oceans. Global sources of petroleum hydrocarbons occur from natural seeps in many oil and gas-rich regions of the world

Table 10.3 Water solubility and octanol–water partition coefficients of some common chemical compounds found in fossil fuels and biofuels.

Compound	Aqueous solubility (mg L^{-1})[a]	Log K_{ow}[a]
n-Hexane	9.5	3.90
n-Octane	0.66	5.18
n-Tetradecane	0.0022	8.306
Benzenes (BTEX)		
Benzene	1790	2.13
Toluene	526	2.73
p-Xylene	162	3.15
PAHS		
Naphthalene	31	3.3
Phenanthrene	1.15	4.46
Pyrene	0.135	4.88
Benzo(a)pyrene	0.174	5.956
Phenolics		
Phenol	82 800	1.46
p-Cresol	9246	1.94
Naphthol	866[b]	2.85[b]
Biofuels		
Ethanol	Miscible[b]	−0.31[b]
Fatty acid esters (biodiesel)	Insol. in water	
Hexadecanoic acid, methyl ester, C16 : 0	Insol. in water[b]	7.38[b]
Octadecadienoic acid, methyl ester, C18 : 2	Insol. in water[b]	8.35[b]
Fuel Additives		
Methyl tert-butyl ether (MTBE)	51 000 (20 °C)[b]	0.94[b]
2-Methoxy-2 methylbutane (TAME)	10 710 (20 °C)[b]	1.55[b]

[a]Source: Data compiled from Chemspider database, The Royal Society of Chemistry, London.
[b]Pubchem database, National Institutes of Health, Maryland. Chemspider: Water solubility data predicted or calculated.

Table 10.4 Annual average estimates of petroleum releases: worldwide and from North America by source in kilotonnes from 1990 to 1999.

Sources	North America (best estimate)	Worldwide (best estimate)
Natural seeps	160	600
Oil and gas operations		
Extraction of petroleum	3.0	38
Platforms	0.16	0.86
Atmospheric deposition	0.12	1.3
Produced waters	2.7	36
Transportation of petroleum	9.1	150
Pipeline spills	1.9	12
Tank vessel spills	5.3	100
Operational discharges (cargo washings)	na	36
Coastal facility spills	1.9	4.9
Atmospheric deposition	0.01	0.4
Consumption of petroleum	84	480
Land-based (river and runoff)	54	140
Recreational marine vessel	5.6	nd
Spills (non-tank vessels)	1.2	7.1
Operational discharges (vessels ≥100 GT)	0.10	270
Operational discharges (vessels <100 GT)	0.12	nd
Atmospheric deposition	21	52
Jettisoned aircraft fuel	1.5	7.5
Total	**260**	**1300**
Minimum to maximum	110–2300	470–8300

nd, not determined; na, not available
Source: Modified from TRB/NRC (2003).

and through losses (evaporation, spills, discharges, and combustion) during anthropogenic production, transport, storage, and use. Atmospheric emissions of non-methane hydrocarbons from evaporation and combustion of petroleum contribute to urban air pollution (e.g. photochemical smog) and deposition (Chapter 13). Large-scale disposal, deposition, and releases of petroleum hydrocarbons to terrestrial environments contribute to site contamination, evaporation, and land-based runoff to the oceans.

Petroleum-derived hydrocarbons remain major pollutants in the world's coastal waters and oceans.

Estimated average inputs of petroleum hydrocarbons into the world's oceans over the period (1990–1999) are available from a detailed report by the Transportation Research Board and National Research Council (TRB/NRC) of the United States in 2003. Table 10.4 summarizes their best estimate and range for worldwide inputs of petroleum hydrocarbons from all sources into the oceans and also in North American waters.

Considerable uncertainty exists for estimates of ranges. The best estimate is 1 300 000 tons year^{-1}, with a range of 470 000–8 400 000 tons year^{-1}. Natural seeps (600 000 tons year^{-1} or 46% of the total) are the largest input. Human-related inputs are mainly from consumption of petroleum (480 000 tons year^{-1}, 37%), followed by transport spills and operational discharges (160 000 tons year^{-1}, 12%), and extraction processes (38 000 tons year^{-1}, 3%). In comparison, an extreme event such as the wreck of the *Amoco Cadiz* off the coast of Brittany, France, resulted in the release of 230 000 tons of crude oil into coastal waters (TRB/NRC 2003, p. 122).

The majority of petroleum hydrocarbon inputs (61%) into North American waters are estimated to originate from diffuse sources (natural seeps and land-based runoff), while average discharges from extraction and marine transportation of petroleum are estimated to be less than 3% of the hydrocarbon inputs (TRB/NRC 2003, p. 53). However, infrequent major oil spill events, such as the *Exxon Valdez* in Alaska (1989) (260 000 barrels; 41 000 m^3; 35 000 metric tons) and the Deepwater Horizon oil spill (2010) (~650 000–780 000 m^3; ~4.1–4.9 million barrels) in the Gulf of Mexico, show the severity of inputs from acute events, which can impact over long time periods on wildlife, fisheries, tourism, and other resource uses.

Aside from petroleum inputs into the oceans, polycyclic aromatic hydrocarbons (PAHs) are generated primarily during the incomplete combustion of organic materials that include fossil fuels and biofuels (e.g. coal, oil, petrol, and wood). Emissions from anthropogenic activities predominate, although some PAHs in the environment originate from natural sources such as open burning, natural losses or seepage of petroleum or coal deposits, and volcanic activities (Abdel-Shafy and Mansour 2016). Global atmospheric estimates of PAHs for 2004 have been reported as 520 Giga grams per year (Gg year^{-1}) or 520 000 tons year^{-1} by Zhang and Tao (2009), with biofuels (56.7%), wildfires (17.0%), and consumer product usage (6.9%) being the main sources. Among countries, the highest emitters were China (114 Gg year^{-1}), India (90 Gg year^{-1}), and the United States (32 Gg year^{-1}).

10.6 Environmental Behavior and Fate of Fossil Fuels and Biofuels

The behavior and fate of fossil fuels or biofuels and their constituents in the environment are determined essentially by their sources, composition, chemical structures, physicochemical properties, and interactions with key environmental transport and transformation processes over time. The persistence of individual chemicals and degraded mixtures may extend from hours to many years, and even decades (residual oil spills) or much longer (e.g. particulates, coal ash, or heavy metal residues). Bioalcohols (ethanol) and ethers (e.g. MTBE) used as fuel additives are water soluble, while alkyl esters of fatty acids in biodiesel have very low water solubilities.

For example, the fate of petroleum (e.g. crude oil) discharged into an aquatic environment (e.g. sea) is illustrated in Figure 10.2, which shows the sequence of environmental processes acting on petroleum, initially at the sea-atmosphere surface. At the same time, transport and transformation processes disperse, partition, and degrade petroleum components between the atmosphere, water mass, and sediments (or soils in the case of land) depending upon their properties (e.g. structures, mixtures, water solubility, volatility, molecular weight, viscosity, polarity, photochemical, and sorption capacities), and prevailing environmental conditions.

When crude oil spills into the sea, a variety of physical, chemical, and biological processes act upon the oil mixture. These effects cause initial spreading of the oil, dissolution and evaporation of volatile hydrocarbons, and emulsification of oil in seawater within the first 10 or so days. The horizontal transport or movement of crude oil occurs through spreading, advection, dispersion, and entrainment, whereas the vertical transport of oil involves dispersion, entrainment, Langmuir circulation, sinking, and sedimentation, over-washing or submergence of oil below the water surface, and partitioning (see TRB/NRC 2003, Chapter 4). Evaporation may remove about one third of the spilled oil in a few days, depending on its composition and weather conditions (Volkman et al. 1994, p. 641).

Other processes, such as sedimentation of insoluble particles and biodegradation, act over much longer periods, depending upon composition and persistence of higher molecular components, as microorganism populations grow and adapt. The oil composition undergoes weathering due to evaporation (volatilization), emulsification, dissolution, and oxidation (chemical, photolysis and microbiological). Its composition changes, as short-chain alkanes and low-molecular-weight aromatic hydrocarbons are lost or degraded over a few months or so, but cyclic alkanes and PAHs may persist for years. Typically, gas chromatograms of heavily degraded oil show a large "hump" or unresolved complex mixture (UCM) of branched hydrocarbon constituents (Volkman et al. 1994, pp. 528–531). Persistent oil residues from seeps, spills, and discharges often form tarballs in offshore waters or become stranded along shorelines for spills near to shore (TRB/NRC 2003, p. 103).

Transport Processes

Petroleum hydrocarbons exhibit low solubility in water, decreasing logarithmically as molecular weight increases (McAuliffe 1966). They are strongly lipophilic as indicated by K_{OW} values (see Table 10.3). For example, the simplest PAH (naphthalene) has a water solubility of about 30 mg L^{-1}, and this decreases with increasing molecular weight of PAHs. Low-boiling aromatic hydrocarbons (benzenes) are generally orders of magnitude more soluble in water than aliphatic hydrocarbons and PAHs.

Oxygenated derivatives of hydrocarbons (e.g. alcohols, esters, aldehydes, and phenols) are more water soluble, depending mainly upon carbon chain lengths, cyclic,

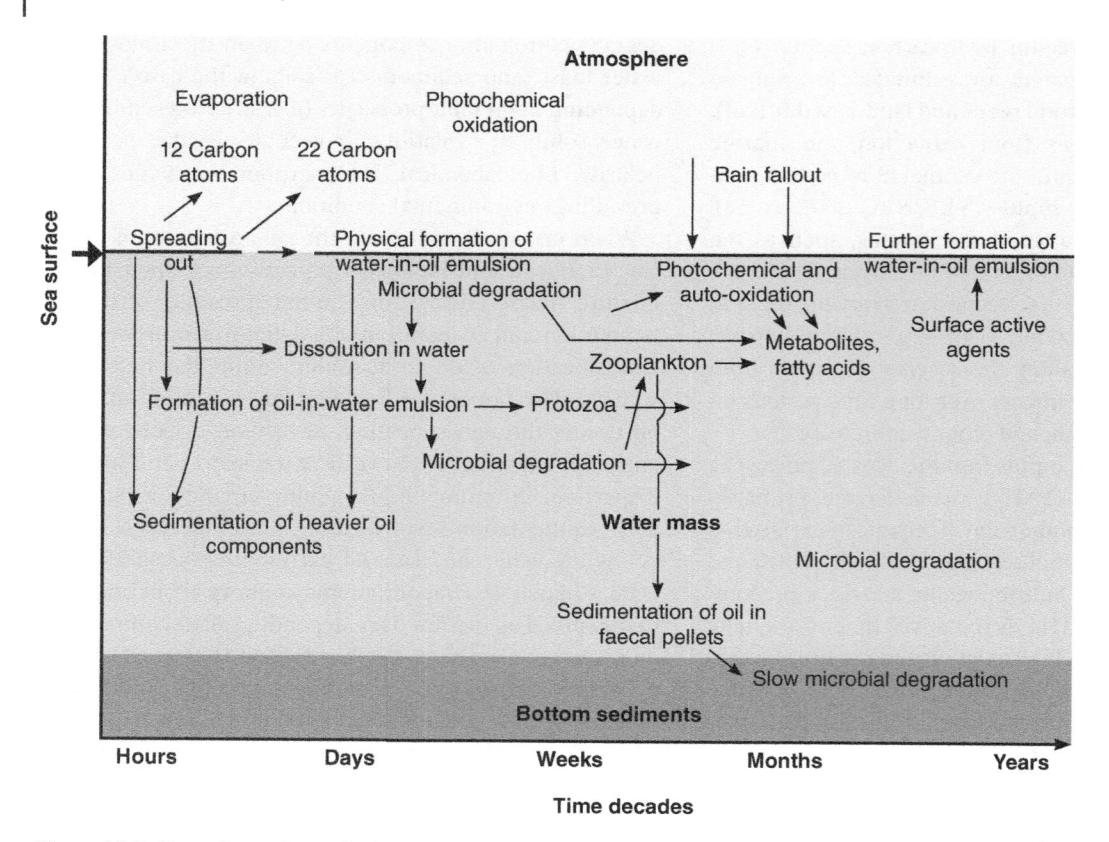

Figure 10.2 Fate of petroleum discharged to the aquatic environment (e.g. sea) (Source: GESAMP (1977).

and aromatic rings. Low-molecular-weight phenolic compounds formed during oxidation of liquid fossil fuels and degradation of coal (e.g. coal tars) are moderately water soluble, decreasing with alkyl substitution and number of rings. Bioalcohol fuels (ethanol) and ethers (e.g. MTBE) used as fuel additives are water soluble, while alkyl esters of long chain fatty acids in biodiesel have very low water solubilities (see Table 10.3). Where used as a fuel additive, MTBE is widely distributed as a surface water and groundwater contaminant (e.g. United States).

Partition processes affect the distribution of petroleum hydrocarbons between the water-sediment and water-air phases. These processes and fugacity modeling used to estimate distributions and predict concentrations of persistent lipophilic hydrocarbons are discussed in Volkman et al. (1994, pp. 550–554).

Petroleum hydrocarbons sorb strongly to the organic matter in sediments with K_{OC} values up to 10^6. For example, PAH concentrations (2–6 rings) in sediment samples collected from Kyeonggi Bay, Incheon, Korea, in December 1995, were positively correlated with organic carbon content (OC) and negatively correlated with mean sediment grain size. Generally, silt would be expected to contain relatively high concentrations compared with coarse sands/gravels (Kim et al. 1999).

As indicated by log K_{OW} values in Table 10.3, the partitioning of petroleum hydrocarbons into lipid phases from water increases with K_{OW} value and decreasing water solubility. Bioethanol does not accumulate in lipids and low-molecular-weight phenols show little accumulation. As expected, fatty acid esters in biodiesel are very soluble in lipids but are readily biodegraded in the environment and metabolized in biota.

PAHs are widely detected in freshwater and marine sediments. As described by Neff et al. (2005), sediment PAHs are derived from combustion of organic matter, fossil fuels, and biosynthesis by microbes. Pyrogenic PAHs, particularly those associated with combustion particles (soot), have a low accessibility and bioavailability in sediments. Polycyclic aromatic hydrocarbons associated with petroleum, creosote, or coal tar in sediments may have a moderate accessibility/bioavailability, particularly if the PAHs are part of a nonaqueous phase liquid (NAPL) that is in contact with sediment pore water. Strong sorption of PAHs (and PCBs) onto some types of common plastic pellets in an urban bay study has demonstrated implications for contamination of marine debris and microplastics and ingestion by marine animals (Rochman et al. 2013).

Petroleum hydrocarbons also enter the coastal ocean from the atmosphere by wet and dry aerosol deposition and gas exchange. However, volatilization is the dominant fate process for petroleum hydrocarbons because hydrocarbon loadings from land-based and near-shore sources support dissolved hydrocarbon loadings in coastal waters that greatly exceed the loadings in equilibrium with the atmosphere (see TRB/NRC 2003, p. 115). Atmospheric partitioning of PAH compounds between the particulate and the gaseous phases strongly influences their transport in the atmosphere, persistence, and removal from the atmosphere by dry and wet deposition. Atmospheric deposition is a major source for PAHs in soil.

Transformation Processes

Organic fuels and their components are transformed at different rates in the environment, primarily by oxidation, photolysis, and microbiological degradation, into many other substances, particularly carbon dioxide, that may act as environmental pollutants (e.g. greenhouse gases, air, water, and soil contaminants) depending upon prevailing conditions, exposure, and exposed organisms. In a simple case, pure methane and ethanol are converted to carbon dioxide and water only. However, commercial fuels, including blends, can contain more complex molecules, mixtures, additives, and/or contaminants leading to a variety of by-products or emissions.

Oxidation of fossils and biofuels covers combustion, photo-oxidation, or microbial oxidation of carbon-based compounds, which releases energy and produces carbon dioxide, water, and a potential range of other breakdown products, such as fine particulates, carbon monoxide, other toxic gases (e.g. nitrogen oxides and sulfur oxides), aldehydes, alcohols, organic acids, esters, and PAHs, depending on chemical composition, structures, properties, and oxidation conditions.

Photochemical transformations of crude and fuel oils involve formation of (i) reactive radicals, (ii) potentially toxic intermediate peroxides and hydro-peroxides, and (iii) oxidation products such as carboxylic acids, esters, oxygenated aromatics, carbonyl compounds, and carbon dioxide (e.g. GESAMP 1977). Preferential oxidation of aromatic hydrocarbons and sulfur compounds in surface films of crude oils on seawater irradiated at 250 nm has been observed (see Connell and Miller 1984, p. 236).

Many microorganisms (e.g. bacteria and fungi) can metabolize the hydrocarbons from petroleum oils or fractions, either completely to carbon dioxide or partially to more polar compounds, usually oxygenated forms.

Microbial oxidation is dominated by bacterial action, which appears to be species dependent. These organisms are capable of oxidizing aromatic hydrocarbons from monocyclic to polycyclic compounds. Biodegradation of highly condensed ring structures is much slower (Connell and Miller 1984, p. 236).

The general mechanisms for the bacterial oxidation of aromatics involve the incorporation of both atoms of molecular oxygen into aromatic hydrocarbons and the formation of *cis*-dihydro-diols, followed by further oxidation to catechol-type compounds (Gibson 1977). Catechols can undergo enzymatic cleavage of the benzene ring between or adjacent to the hydroxyl groups. Oxidation can proceed to or end at various oxygenated metabolites (e.g. salicylic acids, naphthenic acids, and naphthanols) in the presence of an alternative substrate. In most cases the higher molecular weight PAH cannot be used as the sole carbon source (Varanasi and Malins 1977). An example of the microbial oxidation of benzene is shown in Figure 10.3.

The distribution of hydrocarbon-utilizing bacteria and fungi is worldwide, but populations in marine waters appear to be extremely low except in oil-contaminated areas. Within aerobic sediments, higher levels of activity are observed, possibly facilitated by the presence of organic carbon substrates. Oxidation of oil in the natural environment is dependent on factors such as temperature, salinity, concentrations of inorganic nutrients, extent of dispersion in water, abundance and types of microorganisms, and chemical composition of oil.

The rate of biodegradation of hydrocarbons in the environment generally decreases in the order of n-alkanes > branched or cyclic alkanes > aromatic compounds, depending on the composition of petroleum, structure of molecules, and environmental characteristics. Under aerobic conditions, the rate of biodegradation of aromatic compounds decreases with increasing numbers of aromatic rings, degree of alkylation, and position of alkyl substituents (Volkman et al. 1994, p. 545).

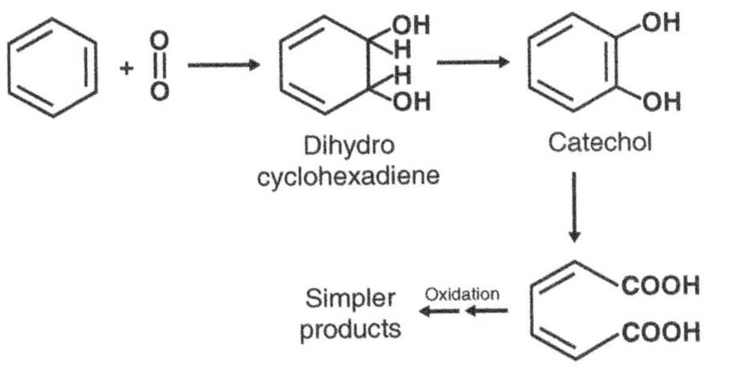

Figure 10.3 Microbial oxidation of benzene. Source: Connell and Miller (1984).

Extreme Environmental Effects

The distribution and persistence of petroleum in the environment is subject to change by climatic or environmental extremes such as low temperature and ice conditions in the Arctic, as described in the Oil in the Sea III Report (TRB/NRC 2003). The general effect of low temperatures and ice on oil spills is to slow the spreading and contain the oil when compared to open water conditions. Evaporation is reduced, viscosity is increased, and oil tends to adhere to snow and ice. Oil trapped in ice and snow over winter is generally released in spring when the ice melts, resulting in widespread distribution of the oil during spring or summer (TRB/NRC 2003, p. 105).

Another critical consideration applies to the fate of oil in sub-surface water releases, especially at depth, in events such as exploration well blowouts, pipeline ruptures, and shipwrecks. The main issues include enhanced dissolution in the water column, potential emulsification, presence of soluble natural gas and possible methane hydrate formation, and effects on the nature of the underwater jet and plume dispersion and slick spread at surface. Dissolution of most natural gas is expected due to the high solubility of methane in seawater at the high pressures and cold temperatures found in deeper water. In deepwater releases, most oil is predicted to rise to the surface but may take up to several hours. Hydrate formation can reduce the buoyancy of the plume and delay time for hydrocarbons to reach the surface (TRB/NRC 2003, p. 108).

10.7 Environmental Exposures

Human extraction, processing, combustion, and other uses of fossil fuels and biomass, to a lesser extent, on a global scale result in millions of tonnes of oil residues and emission products (black carbon, fine particulates, diesel fumes, toxic gases, greenhouse gases, volatile organic compounds, PAHs, and toxic metals) into the air, waters, sediments, and soils of natural and human environments.

The distribution and concentration of these pollutants in waters, sediments, soils and also biota tend to be uneven or heterogeneous, related more to locations of urbanized population clusters, coal mining, oil and gas production fields, coal-fired power generation, industrial uses, oil refining, aluminum smelting, chemical processing activities, ports, storage of fuels, major highways, shipping transport routes around the globe, runoff discharges from major urbanized and petroleum production catchments and incineration.

Petroleum-like hydrocarbons, mainly alkanes, are ubiquitous throughout the world's oceans, but the background concentrations in water are very low except in the vicinity of natural seeps or chronic oil discharges. At low levels, biogenic hydrocarbons are also prevalent but their components and mixtures are less complex than petroleum hydrocarbon mixtures. Low levels of PAHs are also present in the environment from combustion sources and petrogenic (oil-derived) sources. Relatively heavy contamination of water, sediments, and biota can occur in the vicinity of petrochemical activities and urban areas, with variable levels of up to 1–10 mg L^{-1} in waters, and $\geq$1000 mg kg^{-1} in sediments, soils, and infauna.

Contaminated land sites for fossil fuel remediation may experience much higher soil and groundwater levels of petroleum hydrocarbons, ranging from free-form refined fuels to mixtures of BTEX, PAHs, metals (e.g. Pb, Ni, and Zn), coal tars, creosotes, fly ash, petro-solvents, and other petrochemical residues.

BTEX and PAH fractions are of particular concern. Many types of PAHs are found in the environment but usually those with three to six benzene rings, some with alkyl side chains, are most common. BTEX and PAHs can occur in wastewaters at levels of up to mg L^{-1}, such as in produced waters from oil-producing areas, sewage and industrial effluents, and urban runoff. Other PAHs may be derived from combustion of solid biomass as fuels, by burning of forests, peat lands and agriculture wastes, and forest, bush, or wildfires. A distinction is generally made between petrogenic and pyrogenic (combustion-derived) PAHs (Hylland 2006).

Emissions of volatile hydrocarbons (e.g. benzene and toluene), respirable particulates, and other combustion products (e.g. NO$_x$, and SO$_2$) from the use of primarily fossil fuels contaminate many urban airsheds while long-lasting greenhouse gases (CO$_2$-e) are more ubiquitously mixed in the Earth's atmosphere (CO$_2$, average ~408 ppm). As mentioned, global atmospheric estimates of PAHs of 520 000 tons year^{-1} by Zhang and Tao (2009) implicate biofuel, wildfire, and consumer product usage as the main sources. These emissions were positively correlated to the country's gross domestic product and negatively correlated with average income.

Exposures to emissions from fossil fuel combustion are the main source of global air pollution, affecting mainly children. Outdoor and indoor airborne concentrations of multiple air toxics range from μg m^{-3} (e.g. respirable particulates, benzenes, Pb, and Hg) to mg m^{-3} (NO$_X$, CO, SO$_2$, non-methane hydrocarbons). In occupational and polluted environments, levels can increase to 10 or more times higher. Exposures to diesel fumes are now a worldwide health concern, while air lead exposures have greatly decreased in urban environments in developed countries since 2000. Environmental exposures and health effects of air emissions from fossil fuels are evaluated also in Chapter 13 on Air Pollutants and Chapter 14 on Greenhouse Gases, Global Warming, and Climate Change.

Wildlife and their habitats are impacted acutely by crude and fuel oil spills, intensive operations, and uncontrolled discharges (see Section 10.10) involving contact with oils, oil droplets, oil–water emulsions, and toxic aromatic fractions such as benzenes, naphthalenes, and other PAHs. Chronic exposures in waters to toxic fuel oil fractions are widely documented for many aquatic invertebrates, vertebrates, plankton, and benthic plants. Adverse exposures to soluble aromatic fractions and derivatives of crude and fuel oils from spills, discharges, and runoff usually vary from sublethal levels of 0.1–1 mg L^{-1} to acute levels of 1–100 mg L^{-1}, depending on sensitivity or avoidance mobility of exposed biota. Oiled-sediment levels can range from <100 to 10 000 mg kg^{-1} (dry weight) and include exposures to toxic aromatics and PAHs (e.g. carcinogenic benzo(a)pyrene). The degree of weathering of oil affects exposures to toxic fractions. Potential wildlife exposures to toxic hydrocarbons in marine tar residues (pelagic and benthic), derived by weathering of natural petroleum seeps and oil spills, appear to remain unresolved despite studies of their widespread occurrences and accumulations in the oceans (Warnock et al. 2015).

10.8 Uptake, Metabolism, and Bioaccumulation of Petroleum Hydrocarbons

Petroleum hydrocarbons are absorbed from the environment into organisms largely by partitioning processes, as for other lipophilic substances. Low water solubilities and high log K_{ow} values favor the transfer of hydrocarbons from water into the lipid phases in organisms (Connell and Miller 1984, p. 237). Absorption of hydrocarbons into lipid-rich tissues occurs through respiratory surfaces, the gastrointestinal tract, and external surfaces. For instance, dissolved hydrocarbons can diffuse across gill and cell membrane surfaces, and those on particles can be ingested during feeding. Marine filter feeders (e.g. oysters and mussels) can also absorb PAHs by direct uptake of the oil droplets from water. Single-cell organisms, such as phytoplankton, are exposed to hydrocarbons mainly through partitioning of *dissolved* hydrocarbons, including colloid and small particles (TRB/NRC 2003, pp. 102–103). Uptake rates are a function of the concentration in water and are influenced by chemical structure, physicochemical properties, and environmental parameters, such as temperature and salinity.

Partitioning strongly affects the mechanisms and magnitude of exposure of aquatic organisms to hydrocarbons.

Table 10.5 Bioconcentration factors (BCF) for some aromatic hydrocarbons found in liquid petroleum fuels.

Aromatic hydrocarbons	Aquatic test organisms	BCF values
Benzene	Fathead minnow	19
	Pacific herrings	4.4
Naphthalene	Sheephead minnow	692 and 774
	Amphipods (*Diporeia* spp.)	490–736
Pyrene	Rainbow trout	72
	Fathead minnow	600–970
	Amphipods (*Diporeia* spp.)	12 316–36 329
Benzo(a)pyrene	Amphipod (*Daphnia magna*)	800–3400
	Clam (*Macoma inquinata*)	861

Source: Compiled from PubChem database. Published by National Institutes of Health, US (accessed 13 March 2020).

Bioconcentration of some toxic aromatic hydrocarbons from water and/or sediments into aquatic organisms is indicated by BCF values in Table 10.5.

In Table 10.5, BCF values range from low (benzene) to moderate or high for PAHs in fish and amphipods. Values were derived from experimental studies with various exposures in µg L^{-1} ranges. Humic acids and dissolved organic carbon tended to reduce BCF values for freshwater organisms. In contrast, reported BCF values for ethanol (3), and the fuel additive, MTBE, (1.5) are very low (see PubChem database).

Many marine organisms (e.g. vertebrates) possess metabolic systems that can readily metabolize hydrocarbons into more hydrophilic conjugates. These metabolites can be excreted or eliminated. Metabolites are usually formed by cytochrome P-450-mediated mixed function oxidase systems (MFO) and transformed into hydrophilic conjugates by microsomal epoxide hydrase and glutathione-S-transferases (e.g. Connell and Miller 1984, pp. 237–238) (see Figure 10.4). Alkanes are readily metabolized despite their lipophilic properties.

In contrast, PAHs are also highly lipid soluble and readily absorbed, such as from the gastrointestinal tract of mammals, distributed in lipid tissues, and metabolized by the MFO system. Induction of arylhydrocarbon hydroxylase can result in the formation of intermediate oxidized metabolites that have the capacity to interact with cellular macromolecules such as DNA and form primary carcinogens, mutagens, and cytotoxins.

Food Web Transfer and Bioaccumulation

The fate of petroleum hydrocarbons in biota also includes transfer of petroleum hydrocarbons up the food chain

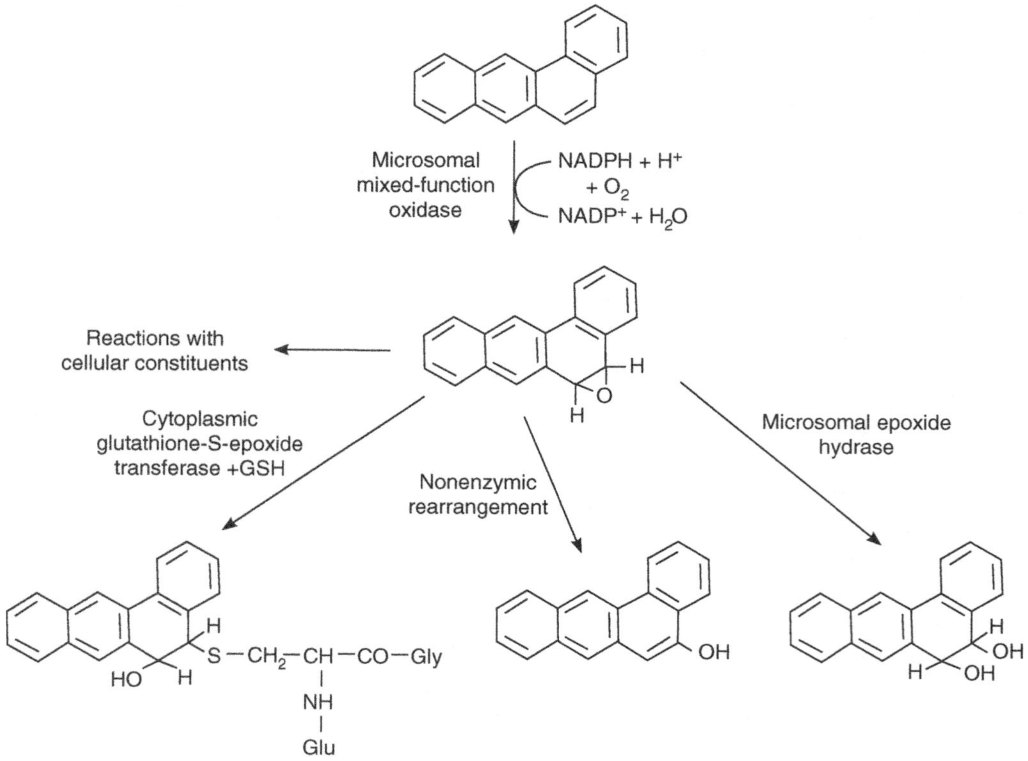

Figure 10.4 Pathways involved in the metabolism of benzo(a)anthracene or 1,2 benzanthracene. Source: Connell and Miller (1984).

through predator–prey relationships, aside from metabolism and excretion of metabolites and physical transfer by biota of ingested or sorbed hydrocarbons and metabolites to other habitats or environments. Food web transfer occurs through partitioning of hydrocarbons onto organic-rich particles, including plankton and detritus, and the transfer of hydrocarbons to higher trophic levels, including filter feeders (e.g. mussels, oysters), fish, and mammals. PAHs may accumulate in the food chain depending on the metabolic rate of the organism. The extent of bioaccumulation of hydrocarbons in organisms is controlled by the desorption rate of hydrocarbons from particles in the gut of higher trophic organisms and metabolic rates that degrade or transfer hydrocarbons outside the cell (TRB/NRC 2003, pp. 102–103). Although there is some evidence of the concentration of certain natural or biogenic hydrocarbons in the food chain, there appears to be no convincing evidence of food chain magnification of petroleum hydrocarbons.

10.9 Environmental Toxicities of Petroleum and Biofuels

Studies on the toxic effects of petroleum fuels in the environment have mostly focused on observations and field studies of oil spills and experimental bioassays of crude and fuel oils on aquatic organisms, usually marine species. Crude and refined oils are found to be toxic to a wide phylogenetic range of cold and warm water aquatic organisms, with considerable variation in tolerance and sensitivity to oils between species within classes. Broadly, adults of animals and mature plants tend to be more tolerant than earlier life stages.

Importantly, the toxic fraction of petroleum oils in the water column is attributed to soluble aromatic-type hydrocarbons (e.g. benzenes and low-molecular-weight naphthalenes, and other PAHs). Refined oils tend to be more toxic than crude oils but the content of water-soluble aromatic fractions (WSF) varies markedly between oils (at ppm levels). Produced waters commonly have a higher proportion of total aromatics (especially benzenes or BTEX) relative to alkanes than are found in crude oils. From a global perspective, toxicity studies have focused more on cold-water species than tropical and sub-tropical species (e.g. Connell and Miller 1984, pp. 240–243; Volkman et al. 1994).

Lethal Toxicities

Laboratory bioassay comparisons between estimated lethal toxicities of soluble aromatics to types of marine organisms

Table 10.6 Toxicities of soluble aromatics to types of marine organisms.

Types of organisms	Estimated ranges of lethal concentrations of soluble aromatics[a] (ppm)		
	Moore and Dwyer (1974)	Crude oils (96 hours LC_{50})	No 2 fuel oil (96 hours LC_{50})
Flora	10–100	ID[b]	ID
Finfish	5–50	1–>10	1–>10
Crustaceans	1–10	1–>10	0.1–10
Bivalves	5–50	1–>10	0.5[c]–>5
Gastropods	1–100	1–>10	1–>5
Other benthic invertebrates	1–10	>5–10[b]	>1[b]
Larvae (all species)	0.1–1	0.1–5	ID
Juveniles	NR[d]	5–>10	1–>10

[a]Soluble aromatic derivatives.
[b]ID, inadequate data.
[c]tentative result.
[d]NR, not reported.
Source: Miller (1982).

are given in Table 10.6. Toxicities within classes vary by one or two orders of magnitude. For exposure periods of up to 96 hours, lethal responses for adults can be expected in the 1–100 ppm (mg L^{-1}) range of soluble aromatic derivatives (SAD), with some fish, crustaceans, and molluscs susceptible at levels of <1–10 ppm. Lethal toxicity to oil at larval stages can extend from 0.1 to 10 ppm, and for juveniles, from 1 to 10 ppm or greater (Connell and Miller 1984, pp. 240–243).

Further toxicity studies on the environmental effects of biodiesel and bioalcohols, as substitutes or alternatives for liquid petroleum fuels, are needed. Khan et al. (2007) compared the acute toxicity of biodiesel, biodiesel blends, and diesel on aquatic organisms (*Daphnia magna* juveniles) and rainbow trout (*Oncorhynchus mykiss* frys). These researchers found that biodiesel and biodiesel blends were less toxic than conventional diesel but *their risk to aquatic organisms is still quite substantial.*

Sublethal Effects

Sublethal effects or responses of marine organisms exposed to petroleum hydrocarbons (crude and fuel oils) are generally observed at water-soluble aromatic concentrations from under 0.01 ppm (mg L^{-1}) or 10 ppb (μg L^{-1}) to over 1000 ppb, as indicated in Figure 10.5. Bioaccumulation and behavioral responses are detectable at the lowest concentrations and growth and reproduction effects tend to occur at higher levels. Larval and juvenile forms appear to be the most sensitive at levels of <1000 ppb of soluble aromatic derivatives (SAD). Birds can be affected by doses of 0.5 g fresh oil and embryos may die by adding as little as 5 μg fresh oil to eggs. Lethal toxicities vary greatly in the range of 0.1 (larvae) to over 10 ppm (10 000 ppb) for more tolerant adults.

PAHs in low concentrations have been shown to decrease growth, development, and feeding rates in aquatic organisms. Various PAHs are also carcinogenic, mutagenic, and teratogenic. There is also extensive evidence linking sediment-associated PAHs to induction of phase-1 enzymes, development of DNA adducts, and eventually neoplastic lesions in fish (Connell and Miller 1984, p. 243).

Chronic toxicity effects of biofuels on aquatic organisms are also under investigation as their use grows and potential environmental risks from discharges and spills

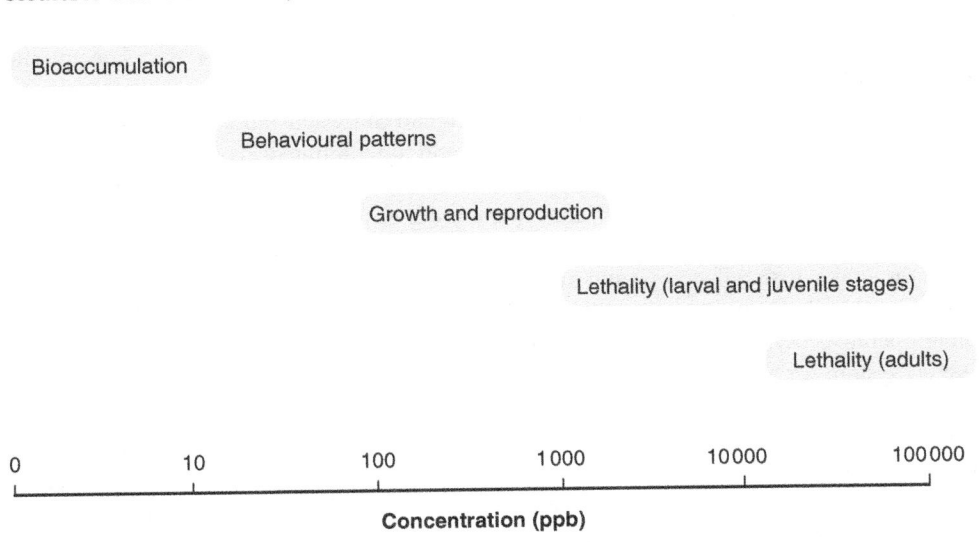

Figure 10.5 Range of effects in marine organisms for exposures to soluble aromatic hydrocarbons. Source: Miller (1982).

increase. Chronic toxicity studies (sea urchin embryonic development test and microalgae growth-inhibition test) by Brazilian researchers (Leite et al. 2011) showed significant toxic responses (inhibition concentrations, EC_{50} values) compared to controls for two local marine organisms (sea urchin and microalgae) exposed to water-soluble fractions of biodiesel fuels (methyl esters of fatty acids derived from castor oil, palm oil, and waste cooking oil). Methanol was identified by gas chromatography as the … *most prominent contaminant*.

10.10 Ecosystem Effects of Petroleum Oils

Individuals, Populations, and Communities

The biophysical impacts of oil spills depend generally on a complex set of factors and interactions, which include the nature, toxicity, and dosage of oil type, weather conditions, hydrological regime, ecological characteristics, sensitivities of marine systems, or habitats exposed to the oil spill. Oil affects aquatic organisms via several interrelated mechanisms: (i) direct mortalities; (ii) sublethal disruption of physiological functions and behavioral activities; (iii) direct coating of animal or plant surfaces/roots by oil; (iv) ingestion, uptake, and accumulation of hydrocarbons in tissues, including possible tainting of seafoods; and (v) changes in biological habitats, especially substrate, through oiling and sub-surface penetration (e.g. burrows), or by clean-up techniques (Volkman et al. 1994, pp. 641–642).

The ecological impacts of petroleum oils on individuals, populations, and communities of species in ecosystems (aquatic and terrestrial) are evaluated from two sets of extensive literature: (i) field surveys (observations and measurements), often in the absence of baseline studies, which generally apply to effects caused by oil spills, and (ii) field trials, mesocosms, and laboratory dose–response studies (bioassays) of experimental organisms under controlled conditions designed to measure potential effects on populations or communities, and a link to field surveys. Despite these limitations and those of ecological understanding, the scientific body of evidence on the major toxic effects of petroleum oils is relatively strong, especially for marine populations and communities in temperate ecosystems.

The effects of oil on a marine ecosystem are generally the same as for other disturbances by pollution or natural stressors. That is, after an oil spill, some or all of the exposed organisms suffer mortalities, and replacement occurs through a succession of opportunistic species until diversity and stability are regained (Southward 1982). Essentially, the ecosystem is modified, destabilized, and

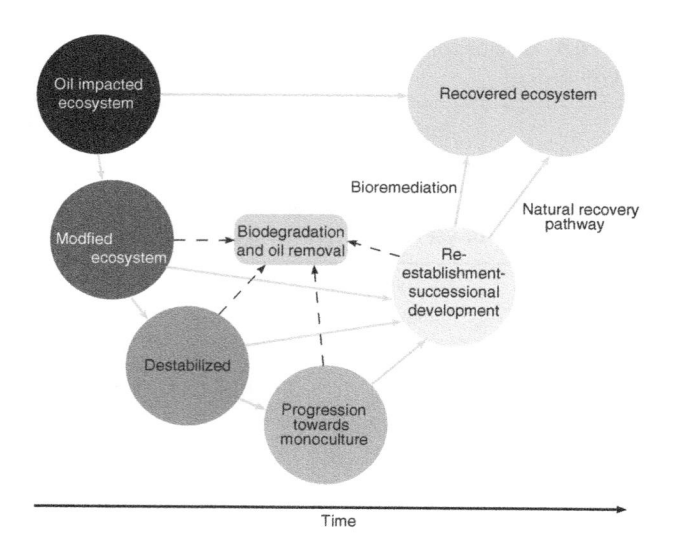

Figure 10.6 Stages of oil-impacted ecosystem, depending upon severity, with pathways to recovery via biodegradation and oil removal followed by bioremediation or natural processes.

reverts toward a monoculture in severe events. Adverse temporal changes may range from short term (days to months) to decades. Once the oil toxicity has declined to a tolerable level for robust colonizing organisms, natural recovery processes will respond (see Figure 10.6).

In acute exposures, the ecosystem structure generally reverts back at least one successional stage. In some oil spills, successional changes in marine communities may be more substantial in cases of high mortalities. Chronic exposures result in more gradual changes in community structure (e.g. diversity and influxes of opportunistic species) and basic processes (e.g. nutrient fluxes and production) (Connell and Miller 1984, pp. 244–246). Major factors that influence ecotoxicity are (i) direct and indirect effects on critical species and changes in community structure, such as the loss of species that control the populations of many other species through interspecific interactions. This can cause substantial changes in the rest of the structure and functioning of ecosystem (e.g. losses of oiled herbivores or mangrove trees or roots); and (ii) effects on ecosystem processes, especially impacts on primary production, biodegradation rates of oil or organic matter, and nutrient cycling (e.g. Harwell and Harwell 1989).

An oil-impacted ecosystem may be modified, destabilized, or progress toward a monoculture (in a severe event). Natural recovery will occur when oil is degraded and removed below a threshold level that allows re-colonization to commence and successional changes of ecosystem to continue, until a recovered state is achieved over time. Intervention by clean-up/bioremediation to degrade/remove oil is also indicated, with recovery time expected to be shorter than by natural processes only.

Petroleum oil effects on wildlife in marine environments are widely described but considerably less so for terrestrial and freshwater environments, although many similar toxic effects are evident. Overall, the extent of scientific knowledge and observations of adverse effects of oil spills on marine systems tend to decrease from intertidal systems to pelagic environments and from temperate to tropical and polar regions. Scientific areas of ecological uncertainty on specific adverse effects still exist despite the scale of our current global production, use and emissions of petroleum oils and products.

General effects of petroleum oils on major marine communities and environments are summarized briefly in Table 10.7. Qualitative vulnerability ratings for oil-exposed environments are given on an increasing scale of 1–10. Vulnerability within a defined environment can vary greatly, emphasizing the need for careful evaluation, mapping, and protection of sensitive marine resources and communities.

In the Arctic, there is uncertainty in attributing long-term ecosystem and species changes to an oil spill event due to increased rates of warming from climate change and other possible environmental stressors. Another related phenomenon is the predicted increased emissions of methane from permafrost in Arctic regions. Climate change factors also are likely to affect the behavior, fate, and effects of hydrocarbons in ecosystems within other climatic systems.

Ecological Recovery and Remediation of Petroleum Oil Spills

The ecological impacts of petroleum oils on individuals, populations, and communities of species in ecosystems (aquatic and terrestrial) are evaluated from two sets of extensive sources of literature:

- Field observations and measurements, often in the absence of baseline studies, which generally apply to effects caused by oil spills, and
- Dose–response studies of experimental organisms under controlled conditions designed to measure potential effects within complex communities and their habitats.

Despite some limitations and gaps in ecological understanding, the scientific body of evidence on the adverse ecological effects of petroleum oils and toxic water-soluble fractions is generally sufficient to undertake preventative and remediation measures.

Recovery

Recovery periods for oiled environments appear to be relatively short-term for some communities, but substantial for some others. A few long-term studies of acute oil spills indicate recovery periods of up to 10 or more years for sedimentary environments in estuarine and intertidal regions (see Volkman et al. 1994, pp. 587–588). Baker et al. (1990) reported that recovery times from oil spills vary from days to months for the vast majority of organisms, not seabirds, in pelagic environments; 1–10 years in littoral zones; and usually 1–5 years in sub-tidal environments. Recoveries of oiled-seabird populations ranged from one year for Eider colonies to a predicted 20–70 years for Guillemot populations.

Predicted recovery rates for certain sensitive systems (e.g. coral reefs or polar regions) are expected to be much slower, although considerable uncertainties exist (Volkman et al. 1994, pp. 643–644; TRB/NRC 2003; 2014). Most long-term effects relate to low energy environments, such as lagoons, estuaries, and salt marshes, and may extend into decades for some perturbations (Volkman et al. 1994, pp. 643–644; TRB/NRC 2003).

The recovery process for an oiled ecosystem is illustrated in Figure 10.6. It relies on biodegradation and removal of oil to tolerable levels followed by natural or facilitated succession until a new base state is reached. It may involve some shift from the pre-impact state or development trend. The rate of the process may be enhanced by dispersion of oil, clean-up, removal, and bioremediation techniques, particularly to achieve tolerable levels for re-establishment of the ecosystem.

Factors controlling recovery include (e.g. Harwell and Harwell 1989):

1. The degree of oil spill disturbance on a spatial scale and key components and processes of the system;
2. The rates and effectiveness of removal of oil contamination such as biodegradation, physical and mechanical removal of oil spills, and export or dispersion of oil residues out of the system;
3. Re-establishment of physical and biological habitats that link to re-colonization of biota and further successional development of biota; and
4. Availability of significant sources of dispersed and relatively large areas of intact populations of biota to enhance propagation, re-colonization, and regeneration of impacted areas.

Remediation

Major responses to oil spills in ecosystems may involve the use of containment devices such as bunds or booms and skimmers in aquatic environments, free phase oil recovery, use of dispersants at sea and coastal waters, shoreline clean-up, e.g. beaches, use of bioremediation techniques to enhance biodegradation, in situ and remote monitoring (e.g. biomarkers and recovery). Use of dispersants (e.g. nonionic surfactants) is equivocal in pelagic environments, depending on dispersion and dilution in water column, and usually avoided in sensitive shorelines

Table 10.7 Expected effects of petroleum oils on major marine environments.

Marine environment	Vulnerability rating[a]	Effects on habitat, populations/communities and ecosystem functions	Recovery
Oceanic			
Open to coastal shelf (outer)	1–8	Low to high impact on phyto- and zooplankton organisms, severe effects on spawning populations of fish larvae; light soluble oil fractions may cause tainting in fisheries and/or affect behavior of benthos in shallow areas (e.g. banks). Oil slicks and droplets may affect surface dwellers	Relatively fast dispersion and degradation of oils; fast to moderate recovery for most pelagic organisms and any exposed benthos; tainting of seafoods may persist
		Severe seabird mortalities can occur among surface feeders (e.g. auks and diving sea ducks)	Slow recovery is expected for local and migratory seabirds with low reproduction rates
Coastal and estuarine			
Bays, channels, harbors, rivers, estuaries	5–10	Moderate to heavy effects from oil spills and chronic oil inputs	Recovery varies from fast to slow, depending upon flushing, route to benthos, and shoreline characteristics. Long-term effects are indicated for sub-tidal habitats and larval fauna
		Oil spill effects depend on time of year (spawning, migration, etc.) and persistence. Many bird species vulnerable	
		Effects include alteration of photosynthesis by algae, reduced populations of fish and benthos, tainting of estuarine seafoods, and individual losses among mammals	Recovery of bird species depends on range, population numbers, and reproduction rates, with breeding and migratory birds at high risk.
Rocky coasts	1–8	Varying degrees from less affected and high wave energy coasts along exposed shorelines (e.g. rocky platforms or cliffs) to heavy on sheltered shorelines. Tidal pools are also affected.	Recovery periods vary according to morphology and energy. Slow recovery in sheltered tidal flats/pools depending on oil penetration.
Beaches and tidal flats	3–9		
		Severe to short-term losses of algae and fauna to significant infauna losses in oil sediments	High risks for infauna. Oil can persist for over 10 years.
Wetlands			
Marshes and mangroves	5–10	Heavy vulnerability from oil spills and chronic inputs due to extent of impacts on key ecosystem functions (high productivity area, basis of detrital food chain, nursery and breeding grounds, species abundance, and diversity)	Moderate to slow recovery depending upon persistence of oil in sediments. Upon removal of oil, biological succession may be fairly rapid (e.g. re-establishment of mangrove seedlings) although mangroves are complex. Oil spill recovery in low energy intertidal communities can take 10–15 years. Persistence of oil in sediments can be from 5 to over 20 years
		Low to severe damage to or losses of marsh plants and mangroves, productivity, epi- and infauna. Local and migratory waders and wildfowl may suffer severe population losses.	
		Low to heavy impacts: Intertidal seagrasses are vulnerable, especially associated fauna	
Seagrasses	3–10	Low to heavy impacts: intertidal and sub-tidal seagrasses and associated fauna exhibit variable responses. Oiling of intertidal seagrasses may produce long-term effects as for mangroves. Sub-tidal bed may be less affected and have shorter recovery periods.	Recovery of oiled seagrasses related to survival of root rhizomes in sediments. Severity of oiling of sediments and loss of seagrasses are critical factors in recovery period.

Table 10.7 (Continued)

Marine environment	Vulnerability rating[a]	Effects on habitat, populations/communities and ecosystem functions	Recovery
Coral reefs	3–10	Low to heavy: severe mortalities among intertidal and sub-tidal reef communities on oiled areas of reefs; losses of larval forms of reef fauna expected; extensive sublethal effects include tissue damage, growth, and behavioral effects	Recovery can be slow due to structural complexity of coral fauna. Observed rates of recovery (natural and otherwise) vary widely. Uncertainty about recovery of coral reefs suffering from climate change effects (e.g. heat bleaching of corals)
Polar ecosystems	5–10	Acute and chronic impacts observed as for other sensitive environments (e.g. Exxon Valdez oil spill). Negative effects observed on plankton, sensitive larval stages, and benthic organisms (deposit and filter feeders). Marine bird populations are likely to be vulnerable to oil spills. Some marine mammals (e.g. polar bears, sea otters, killer whales, and bowhead whales) affected by oil spills through direct and indirect contact. Uncertainty over causes of observed deaths and disease among ringed seals in north and west Alaska in 2011	Recovery is uncertain. Spilled oil may persist longer at cold temperatures, and polar organisms have slower growth rates, extended life cycles, longer reproduction periods, and narrow dispersal stages. Adaptation to harsh and variable conditions, however, may allow some coastal communities to recover rapidly

[a]Vulnerability scale: 1–10: low 1–3; medium 4–7; high 8–10.
Source: Compiled from Connell and Miller (1984); Volkman et al. (1994).

and habitats, coral reefs, lagoons, and other water bodies, because of field observations and experimental field studies of impacts. Dispersants (and their formulations) can cause similar toxic effects to oil and dispersed oil (e.g. Volkman et al. 1994; TRB/NRC 2003).

10.11 Effects on Human Health and Ecosystem Services

Aside from major environmental and economic effects (e.g. fisheries, tourism, and conservation), oil spills can also affect human health (e.g. skin and eye irritation, neurological, breathing problems, and stress), with persons who clean up the oil spill more at risk. In a review of the human health effects of exposure to a number of major oil spills (~38 accidents involving supertankers over five decades), most of the studies showed evidence of a relationship between exposure to spilled oils and the appearance of acute physical, psychological, genotoxic, and endocrine effects in the exposed individuals(Aquilera et al. 2010). In an updated review, Laffron et al. (2016) found additional evidence to support these adverse health effects, including several symptoms that may persist for some years after exposure. As an outcome of this review, health risk assessment was recommended to extend from the initial exposure to prolonged periods of post exposure to detect any potential harmful effects.

The main human exposures to fossil fuels relate to inhalation of their air emissions. Almost everyone around the world is directly exposed to some outdoor air pollutants mostly from combustion of fossil fuels and solid biomass. About 40% of the population is also exposed to indoor pollution caused by burning solid fuels (Smith et al. 2013). Emissions from the use of fossil fuels, and traditional solid biomass (wood, charcoal, coal, crop wastes, and dung) to some degree, contribute largely to the global burden of environmental disease and premature deaths (see Chapter 13 of this book).

The toxic effects of pollutants (respirable particulates, toxic gases, organic chemicals, and metals) on humans, mainly from exposure to combustion by-products of fossil fuels, are summarized in Table 10.8. Critical chronic effects are generally respiratory, cardiovascular, carcinogenic, or neurotoxic. Airborne lead pollution from use of tetraethyl lead is still significant in many countries but has decreased generally in developed countries. Coal combustion is a major source of global mercury emissions, which are likely to increase until 2050. Mercury is readily bioaccumulative in food webs (e.g. seafoods) and humans (Sundseth et al. 2017). Pregnant women are vulnerable to mercury exposures.

Primary health-related impacts and risks for occupational and community exposures to major types of fossil fuel and biofuel sources are also concisely described in Smith et al. (2013), along with an estimate of

Table 10.8 Toxic effects of pollutants on exposed humans from emissions of fossil fuels and biofuels.[a]

Pollutant	Known human toxic effects
Particulates PM_{10} and $PM_{2.5}$[b]	Bronchial irritation, inflammation, increased reactivity, reduced mucociliary clearance, reduced macrophage response, cardiopulmonary disease, increased cardiovascular mortality, carcinogenicity
Diesel exhaust particles/fumes (gases)	
Respirable coal dusts/silica	Silicosis, coal miner's pneumoconiosis, diffuse fibrosis, chronic obstructive pulmonary disease
Carbon monoxide	Reduced oxygen transfer to tissues due to formation of carboxyhemoglobin, acute exposures can be fatal
Nitrogen dioxide	Bronchial reactivity, increased susceptibility to bacterial and viral lung infections
Sulfur dioxide	Bronchial reactivity and other toxic effects as for particulates
Hydrogen sulfide	Odorous irritant at low levels; toxic chemical and asphyxiant
Methane	Elevated to high levels: loss of oxygen effects – asphyxiant
Hexane, benzenes (BTEX)	Neurotoxicity.
Formaldehyde	Carcinogenicity
Acetaldehyde	Co-carcinogenicity
1,3 Butadiene	Mucus coagulation, cilia toxicity
Benzene	Increased allergic sensitization
Phenols/cresols	Increased airway reactivity
Pyrene	
Benzo(a)pyrene	
Dibenzo(a)pyrenes	
Benzo(a)anthracene	
Dibenzocarbazoles	
Ethanol	Acute, chronic, and delayed effects of intoxication on central nervous system can be fatal at acute doses; main metabolite, acetaldehyde, is a human carcinogen
Metals (arsenic, lead, mercury, cadmium, nickel, vanadium)	Systemic toxicants, cellular damage by initiating oxidative stress leading to multiple organ damage; induction of carcinogenicity (e.g. As, Cd, Cr, and Ni). Adverse effects on child development (Pb and Hg) and neurotoxic effects (Hg)

[a]By volatilization, coal dust generation and combustion.
[b]Includes respirable particles from combustion of solid biomass fuels.
Sources: Compiled from Aquilera, F., Méndez, J., Pásaro, E., Laffon, B. (2010). Review on the effects of exposure to spilled oils on human health. Journal of Applied Toxicology 30: 291–301.; Smith, K.R., Frumkin, H., Balakrishnan, K. et al. (2013). Energy and Human Health. Annual Reviews of Public Health 34: 159–188.

climate-related risks due to carbon dioxide emissions from each fuel type.

Increased health and environmental impacts and risks from air pollutants and CO_2 emissions, due to major contributions from fossil fuel combustions, are already evident on a global scale. Future environmental health of humans and wildlife is likely to depend largely on the rate of global conversion from a fossil fuel-based economy to a renewable energy one to reduce current gross inputs of petroleum hydrocarbons, toxic air pollutants, and greenhouse gases into our environments.

Even so, large-scale use of resources such as fertilizer, fossil fuels, water, and land that are needed to produce sufficient biofuel to reduce fossil fuel consumption can also lead to loss of food and nutrition, increased pollution, health, and ecological impacts or risks. Alternative fuel sources such as biofuels, hydrogen, hydropower, nuclear energy, solar, and wind also need robust environmental and health risk evaluation. Sustainable approaches, including use of life-cycle assessments and carbon footprints, are essential for current and future transition to renewable energy, and development and management of future energy resources.

10.12 Key Points

1. Types of fossil fuels include coal, petroleum (crude oils and refined petroleum products), natural gas (including coal seam and shale gas), oil shales, bitumens (natural or refined from crude oil), heavy oils, and tar sands.
2. Natural gas is a mixture of flammable lightweight alkanes, consisting primarily of methane, CH_4, small amounts of ethane and propane, and traces of other C_3, C_4, and C_5 alkane hydrocarbons, gases: nitrogen, carbon dioxide, hydrogen sulfide, and oxygen, and metals. It can also occur in the form of methane hydrate (crystalline solid lattice, $4CH_4 \cdot 23H_2O$).
3. Liquid petroleum can be separated into four groups: saturates or aliphatics, aromatics, resins, and asphaltenes. In crude oil and refined oil products (e.g. diesel oil), there are two major chemical classes of hydrocarbons, termed aliphatic and aromatic, of varying chemical compositions.
4. Biogenic hydrocarbons also occur widely in terrestrial and marine organisms and may be formed by biosynthesis, transferred from foods, or formed following ingestion of precursors from food and abiotic sources.
5. Coal is primarily a complex structure of layered forms of fused aromatic and cyclic forms of hydrocarbons combined or joined with oxygen, nitrogen, and sulfur-containing rings and hydrogen, oxygen, nitrogen, and sulfur atoms in various structural forms (substituents, functional groups, or chains). It also contains a number of minerals with other common and trace elements (e.g. arsenic and mercury).
6. Coal tars containing complex combinations of PAHs, phenols, and heterocyclic O, N, and S compounds can also be formed from heating of coal (carbonization).
7. Coal seam gas is a natural gas, mostly methane, extracted from coal deposits where it adsorbs to coal surfaces as a film and can accumulate as free gas in pore spaces and open fractures. Shale gas also occurs as natural gas within fractures, pore spaces, and adsorbed onto the organic material in shale formations.
8. Hydraulic fracturing (*fracking*) is widely used during oil, shale gas and coal seam gas drilling. This process involves injecting large volumes of a fluid and particles and may also include potentially hazardous chemical additives.
9. Fossil fuels, particularly coal, can be decomposed at high temperatures (700–1000 °C) to produce a mixture of carbon monoxide and hydrogen gases, known as synthesis gas or syngas, and residual tar.
10. Bioenergy is an alternative and renewable source of energy obtained from recent biomass. Biomass can be used as a solid fuel for combustion or converted to liquid fuels (ethanol, methanol, butanol, and biodiesel) and gas biofuels, such as methane and syngas, via several chemical processes, for heating, electricity, and transportation end uses.
11. Gasoline additives can also be significant or serious environmental contaminants. Types of additives cover antiknock agents (e.g. tetraethyl lead, toluene, and isooctane), antioxidants and stabilizers (e.g. butylate hydroxytoluene, BHT), lead scavengers, (e.g. tricresyl phosphate and 1,2 bromoethane), and oxygenates or oxyfuels (e.g. alcohols and ethers).
12. The behavior and fate of fossil fuels or biofuels and their constituents in the environment are determined essentially by their sources, composition, chemical structures, physicochemical properties, and interactions of key environmental transport and transformation processes over time.
13. Bioalcohols (ethanol) and ethers (e.g. MTBE) used as fuel additives are water-soluble, while alkyl esters of fatty acids in biodiesel have very low water solubilities. Where used in gasoline, MTBE is known to contaminate groundwaters.
14. The persistence of individual chemicals and degraded mixtures may extend from hours to many years, and even decades (residual oil spills) or much longer (e.g. particulates, coal ash, or heavy metal residues).
15. The partitioning of petroleum hydrocarbons into lipid phases from water increases with K_{OW} value and decreasing water solubility.
16. Bioethanol and low-molecular-weight phenols do not accumulate in lipids. However, fatty acid esters in biodiesel are very soluble in lipids but are readily biodegraded in the environment and metabolized in biota.
17. The rate of biodegradation of hydrocarbons in the environment generally decreases in the order of n-alkanes > branched or cyclic alkanes > aromatic compounds, depending on the composition of petroleum, structure of molecules, and environmental characteristics.
18. Under aerobic conditions, the rate of biodegradation of aromatic compounds decreases with increasing numbers of aromatic rings, degree of alkylation, and position of alkyl substituents.
19. Petroleum-like hydrocarbons, mainly alkanes, are ubiquitous throughout the world's oceans, but the background concentrations in water are very low except in the vicinity of natural seeps.
20. At low levels, biogenic hydrocarbons are also prevalent, but their components and mixtures are less complex than petroleum hydrocarbon mixtures.

21. Low levels of polyaromatic hydrocarbons (PAHs) are also present in the environment from combustion sources and petrogenic (oil-derived) sources.
22. Relatively heavy petroleum contamination of water, sediments, and biota can occur in the vicinity of petrochemical activities and urban areas, with variable levels of petroleum hydrocarbons of up to 1–10 mg L^{-1} in waters and $\geq$1000 mg kg^{-1} in sediments, soils, and infauna.
23. Contaminated land sites for fossil fuel remediation may experience much higher soil and groundwater levels of petroleum hydrocarbons. Contaminants can range from free-form fuels to BTEX, PAHs, metals (e.g. Pb, Ni, V, and Zn), coal tars, creosotes, fly ash, petro-solvents, and other petrochemical residues.
24. Exposures to emissions from fossil fuel combustion are the main source of global air pollution. Outdoor and indoor airborne levels of multiple air toxics range from μg m^{-3} (e.g. respirable particulates, benzenes, Pb, and Hg) to mg m^{-3} (NO$_X$, CO, SO$_2$, non-methane hydrocarbons).
25. Wildlife and their habitats are impacted acutely by crude and fuel oil spills, intensive petroleum operations, and uncontrolled discharges, involving contact with oils, oil droplets, oil-water emulsions, and toxic aromatic fractions such as benzenes, naphthalenes, and PAHs.
26. Oiled-sediment levels can range from <100 to 10 000 mg kg^{-1} and include exposures to toxic aromatics and PAHs.
27. Petroleum hydrocarbons are absorbed from the environment into organisms largely by partitioning processes, as for other lipophilic substances. Low water solubilities and high log K$_{OW}$ values favor the transfer of hydrocarbons from water into the lipid phases in organisms. Bioconcentration of some toxic aromatic hydrocarbons and PAHs from water and/or sediments into aquatic organisms is indicated by BCF values.
28. Many marine organisms, particularly vertebrates, possess metabolic systems that can readily metabolize hydrocarbons into more hydrophilic conjugates that can be excreted or eliminated. Metabolites are usually formed by cytochrome P-450-mediated mixed function oxidase systems (MFO) and transformed into hydrophilic conjugates by microsomal epoxide hydrase and glutathione-S-transferases.
29. The toxic fraction of petroleum oils in the water column is attributed to soluble aromatic-type hydrocarbons (e.g. benzenes, and low-molecular-weight naphthalenes and PAHs). Refined oils tend to be more toxic than crude oils, but the content of water-soluble aromatic fractions (WSF) varies markedly between oils (at ppm levels). Produced waters commonly have a higher proportion of total aromatics (especially benzenes or BTEX) relative to alkanes than are found in crude oils.
30. Crude and refined oils are found to be toxic to a wide phylogenetic range of cold and warm water aquatic organisms with considerable variation in tolerance and sensitivity to oils between species within classes. Broadly, adults and mature plants tend to be more tolerant than earlier life stages.
31. Sublethal effects or responses of marine organisms exposed to petroleum oils and fuels are generally observed at water-soluble aromatic concentrations from under 0.01 ppm (mg L^{-1}) or 10 ppb (μg L^{-1}) to over 1000 ppb. Larval and juvenile forms appear to be the most sensitive.
32. Generally, oil spills cause some or all of the exposed organisms to suffer mortalities, followed by replacement through a succession of opportunistic species until diversity and stability are regained. Natural recovery processes occur once the oil toxicity has declined to a tolerable level for robust colonizing organisms. Adverse temporal changes may range from short term (days to months) to decades.
33. Overall, the extent of scientific knowledge and observations of adverse effects of oil spills on marine systems tends to decrease from intertidal systems to pelagic environments and from temperate to tropical and polar regions.
34. During oil spill remediation in marine environments, use of dispersants (e.g. nonionic surfactants) is equivocal in pelagic environments, depending on dispersion and dilution in water column, and usually avoided in sensitive habitats. Dispersants (and their formulations) can cause similar toxic effects to oil and dispersed oil.
35. The main human health effects from exposures to fossil fuels relate to uncontrolled inhalation of their air emissions (outdoor and ambient) and combustion by-products such as benzene, PM$_{2.5}$, and NO$_X$.
36. The use and burning of fossil fuels, and traditional solid biomass to some degree, contribute largely to the global burden of environmental disease and death, including effects of global warming and climate change (Chapter 14).
37. Benefits of using biofuels to reduce greenhouse gas emissions may be negated in certain cases by fossil fuel inputs, land use changes, and other factors such as air pollution effects from combustion of biomass, wastewaters, clearing and burning of forests for oil palms and other crops.

References

Aquilera, F., Méndez, J., Pásaro, E., and Laffon, B. (2010). Review on the effects of exposure to spilled oils on human health. *Journal of Applied Toxicology* 30: 291–301.

Abdel-Shafy, H.I. and Mansour, M.S.M. (2016). A review on polycyclic aromatic hydrocarbons: Source, environmental impact, effect on human health and remediation. *Egyptian Journal of Petroleum* 25 (1): 107–123.

AFDC (2020). Alternative fuels data center. US Department of Energy. http://www.afdc.energy.gov/fuels/natural_gas_renewable.html (accessed 10 March 2020).

ATSDR (2002). Toxicological profile for creosote. Agency for Toxic Substances and Disease Registry. https://www.atsdr.cdc.gov/ToxProfiles/tp.asp?id=66&tid=18 (accessed 7 August 2017).

Baird, C. and Cann, M. (2012). *Environmental Chemistry*, 5e. New York: W.H. Freeman.

Baker, J.M., Clark, R.B., Kingston, P.F., and Jenkins, R. (1990). Natural recovery of cold water marine environments after an oil spill. *Thirteenth Annual Arctic Marine Oilspill Program Technical Seminar*, Edmonton, Alberta. 1–111.

BP (2019). BP statistical review of world energy – 2019. 68e. https://www.bp.com/content/dam/bp/business-sites/en/global/corporate/pdfs/energy-economics/statistical-review/bp-stats-review-2019-primary-energy.pdf (accessed 11 March 2020)

Connell, D.W. and Miller, G.J. (1981). Petroleum hydrocarbons in aquatic ecosystems –behavior and effects of sublethal concentrations: part 1. *Critical Reviews in Environmental Control* 11: 37–104.

Connell, D.W. and Miller, G.J. (1984). *Chemistry and Ecotoxicology of Pollution*. New York: Wiley.

Diesendorf, M. (2014). *Sustainable Energy Solutions for Climate Change*. Sydney: UNSW Press.

FAO (2008). *The State of Food and Agriculture 2008 – Biofuels: Prospects, Risks and Opportunities*. Rome: Food and Agriculture Organization of the United Nations http://www.fao.org/3/a-i0100e.pdf (accessed 10 March 2020).

GESAMP (1977). *Impact of Oil on the Marine Environment*. Rome: UN Joint Group of Experts on the Scientific Aspects of Marine Pollution (GESAMP). Reports and Studies No. 6. Food and Agriculture Organization of the United Nations.

Gibson, D.T. (1977). Biodegradation of Aromatic Petroleum Hydrocarbons. In: *Fate and Effects of Petroleum Hydrocarbons in Marine Organisms and Ecosystems* (ed. D.A. Wolfe), 36. New York: Pergamon Press.

Harwell, M.A. and Harwell, C.C. (1989). Environmental decision making in the presence of uncertainty. In: *Ecotoxicology: Problems and Approaches* (ed. A. Levin, M.A. Harwell, J.R. Kelly and K.D. Kimball), 517–540. New York: Springer-Verlag.

Hylland, K. (2006). Polycyclic aromatic hydrocarbon (PAH) ecotoxicology in marine ecosystems. *Journal of Toxicology and Environmental Health Part A* 69 (1-2): 109–123.

Khan, N., Warith, M.A., and Luk, G. (2007). A comparison of acute toxicity of biodiesel, biodiesel blends, and diesel on aquatic organisms. *Journal of the Air & Waste Management Association* 57 (3): 286–296.

Kim, G.B., Maruya, K.A., Lee, R.F. et al. (1999). Distribution and sources of polycyclic aromatic hydrocarbons from Kyeonggi Bay, Korea. *Marine Pollution Bulletin* 38 (1): 7–15.

Kopp, O.C. (n.d.). Fossil fuel. Britannica. https://www.britannica.com/science/fossil-fuel (accessed 5 May 2019).

Laffron, B., Pásaro, E., and Valdiglesias, V. (2016). Effects of exposure to oil spills on human health: updated review. *Journal of Toxicology and Environmental Health Part B* 19 (3-4): 105–128.

Leite, M.B., de Araújo, M.M., Nascimento, I.A. et al. (2011). Toxicity of water-soluble fractions of biodiesel fuels derived from castor oil, palm oil, and waste cooking oil. *Environmental Toxicology and Chemistry* 30 (4): 893–897.

McAuliffe, C. (1966). Solubility in water of paraffin, cycloparaffin, olefin, acetylene, cycloolefin, and aromatic hydrocarbons. *Journal of Physical Chemistry* 70 (4): 1267–1275.

Miller, G.J. (1982). Ecotoxicology of petroleum hydrocarbons in the marine environment. *Journal of Applied Toxicology* 2 (2): 88–98.

Moore, S.F. and Dwyer, R.L. (1974). Effects of oil on marine organisms: a critical assessment of published data. *Water Research* 8: 819–827.

Neff, J., Stout, S., and Gunster, D. (2005). Ecological risk assessment of polycyclic aromatic hydrocarbons in sediments: identifying sources and ecological hazard. *Integrated Environmental Assessment and Management* 1 (1): 22–33.

Rochman, C.M., Hoh, E., Hentschel, B.T., and Kaye, S. (2013). Long-term field measurement of sorption of organic contaminants to five types of plastic pellets: implications for plastic marine debris. *Environmental Science and Technology* 47 (3): 1646–1654.

Sangster, J. (1989). Octanol-water partition coefficients of simple organic compounds. *Journal of Physical Chemistry Reference Data* 18: 1111–1228. https://doi.org/10.1063/1.555833.

Smith, K.R., Frumkin, H., Balakrishnan, K. et al. (2013). Energy and human health. *Annual Reviews of Public Health* 34: 159–188.

Southward, A.J. (1982). An ecologist's view of the implications of the observed physiological and biochemical effects of petroleum compounds on marine organisms and ecosystems. In: *The Long-Term Effects of Oil Pollution on Marine Populations, Communities and Ecosystems* (ed. R.B. Clark), 57–71. London: The Royal Society.

Sundseth, K., Pacyna, J.M., Pacyna, E.G. et al. (2017). Global sources and pathways of mercury in the context of human health. *International Journal of Environmental Research and Public Health* 14: 105. doi:https://doi.org/10.3390/ijerph14010105.

TRB/NRC (2003). *Oil in the Sea III: Inputs, Fates, and Effects.* Transportation Research Board and National Research Council. Washington, DC: The National Academies Press.

TRB/NRC (2014). *Responding to Oil Spills in the U.S. Arctic Marine Environment.* Transportation Research Board and National Research Council. Washington, DC: The National Academies Press.

The Essential Chemical Industry-online (2014). Cracking and related refinery processes. Centre for Industry Education Collaboration. York UK: York University. http://www.essentialchemicalindustry.org/processes/cracking-isomerisation-and-reforming.html (accessed 7 August 2017).

Varanasi, V. and Malins, D.C. (1977). Metabolism of petroleum hydrocarbons: accumulation and biotransformation in marine organisms. In: *Effects of Petroleum on Arctic and Sub-arctic Marine Environments and Organisms*, vol. 2 (ed. D.C. Malins), 172. New York: Academic Press.

Volkman, J.K., Miller, G.J., Revill, A.T., and Connell, D.W. (1994). Environmental implications of offshore oil and gas development in Australia – oil spills. In: *Environmental Implications of Offshore Oil and Gas Development in Australia – The Findings of an Independent Scientific Review* (ed. J.M. Swan, J.M. Neff and P.C. Young), 509–695. Sydney: Australian Petroleum Exploration Association.

Warnock, A.M., Hagen, S., and Passeri, D.L. (2015). Marine tar residues: a review. *Water Air and Soil Pollution* 226 (3): 68–72.

Zhang, Y. and Tao, S. (2009). Global atmospheric emission inventory of polycyclic aromatic hydrocarbons (PAHs) for 2004. *Atmospheric Environment* 43 (4): 812–819.

11

Toxic Organic Pollutants

11.1 Introduction

Life on Earth has evolved from the chemistry of carbon and its complex interactions between carbon atoms and various other elements to form a multitude of organic molecules, from methane to the genetic code of life, deoxyribonucleic acid, DNA. Millions of organic chemicals are known to exist. Increasingly, new and synthetic chemical compounds are being produced for commercial use. How living organisms interact with exposures to toxic organic chemicals in their environments is a matter of life, survival, and death. For some species, it can also cause or contribute to extinction.

Organic chemicals are major building blocks and functional components of modern society, our lifestyles, health, and state of well-being. They include fuels, plastics, pharmaceuticals, personal care products, surfactants, pesticides, disinfectants, dyestuffs, surface coatings, adhesives, solvents, explosives, and many other industrial and specialty chemical products. Despite their known or perceived benefits, many of these organic chemicals possess the potential to cause adverse effects on humans and natural biota, usually unintended, when dispersed into the environment. At the same time, the environmental effects of many organic chemicals in use remain either unknown or understudied.

In this chapter, we apply the basic principles and conceptual model of this book to a major and special group of organic pollutants – **toxic organic chemicals (TOCs)**–known to pervade, persist in many cases, and contaminate our environments, and affect living organisms, including humans. This pervasive mix of chemicals of concern consists primarily of synthetic organic chemicals released into the environment by human activities during industrial production, uses of chemical products containing these substances, or formed as by-products of combustion of organic substances, degradation, metabolism, or during chemical synthesis. Part 1 of this book introduces the sources, nature, and effects of such pollutants.

The scale of production and continual use of TOCs, their physical and chemical forms and properties enhance their capacity to persist and induce a diverse range of toxic effects in exposed humans and natural biota. Exposure is often through multiple pathways (air, water, soils, and food), and at chronic doses. In many cases, multiple chemical exposures occur from mixtures of related chemicals and also other pollutants. Of utmost concern, known or potential ecological and human health impacts for many of these organic pollutants have rapidly evolved and progressed from local to regional or global scales since the 1950s (e.g. POPs).

> The term synthetic chemical is used to describe substances that generally do not occur in nature but have been synthesized by chemists from simpler substances
> – Baird and Cann (2012, p. 575).

Here, we refer to commercial organic chemicals that are generally derived from the production of fossil fuels, mostly petroleum oils or natural gas (see Chapter 10). Table 11.1 lists the main types of organic chemicals that occur as organic pollutants in the environment. Hundreds of these chemicals exhibit endocrine-disrupting properties.

A large number of the TOCs such as the POPs are increasingly regulated in many countries, including their use, emissions and residual levels in foods, water, air, soil, and wastes. Scientific research on TOCs in the Twenty-first Century continues to discover the extent and effects of known organic contaminants and reveal the emergence of new organic contaminants.

11.2 Toxic Organic Chemicals

Overview

There are many chemical sub-groups or families that make up toxic organic chemicals (TOCs), ranging from persistent organic pollutants (POPs) to contaminants of

Table 11.1 Organic chemical pollutants.

Chemical types	Chemical examples
Solvents (petroleum and water miscible)	Chlorinated hydrocarbons, benzene, toluene, and xylenes, ethanol and butanol, ethylene glycol
Volatile organic chemicals (VOCs)[a]	Formaldehyde, hexane, decane, benzene, toluene, xylenes, ethanol, trimethyl amine
Phenolics	Phenol, hydroxyl phenols (cresols), creosotes, chlorinated phenols, bisphenol A
Plastics	Monomers (ethylene, styrene, methyl methacrylate, vinyl chloride), polymers and resins: polyethylene, polystyrene, polyvinyl chloride (PVC), fluoropolymers (Teflon™), polyurethanes, acrylics (polyacrylates), phenol-formaldehyde resins
Plasticizers	Phthalate acid esters (or phthalates) (e.g.diethylhexyl phthalate, DEHP), tricresyl phosphate
Industrial chemicals and products	PCBs, CFCs, surfactants, disinfectants, dyestuffs, pigments, adhesives, surface coatings, flame retardants
Pesticides	Insecticides, herbicides, fungicides, and biocides
Veterinary chemicals	Antibiotics, parasiticides, anti-inflammatory drugs (non-steroid)
Personal care products	Fragrances (perfumes): musk ketone, preservatives: parabens
Hormones and steroids	Estrogens, (estriol, estrone), progesterone, androgens (androsterone, testosterone)
Pharmaceuticals	Antibiotics, anti-inflammatories, and beta-adrenergic blockers
Disinfection by-products	Trihalomethanes (chloroform, bromoform), bromate, trichloroacetic acid, nitrosoamines (NDEA, NDMA)
Combustion, processing or synthesis by-products	Polychlorinated dioxins and furans (e.g. TCDD), PAHs, perfluorinated compounds (perfluorooctanoic acid, PFOA)

[a] The European Union defines a VOC as *any organic compound having an initial boiling point less than or equal to 250 °C (482 °F) measured at a standard atmospheric pressure of 101.3 kPa.*

emerging concern (CECs), such as brominated flame retardants (e.g. PBDE), perfluorinated surfactants (e.g. PFOS), disinfection by-products (DBPs), and microplastic particles. Well-known TOCs include the industrial chemical isomers or congeners of polychlorinated biphenyls (PCBs), chlorinated organic pesticides (Chapter 9), the polynuclear aromatic hydrocarbons (PAHs), generated mainly by incomplete combustion of biogenic matter or fossil fuels, and chlorinated dioxins (e.g. TCDD) formed as trace by-products during synthesis of certain chlorinated chemicals or during combustion of organic substances.

In this chapter, major sub-groups or families of many TOCs are identified but evaluated as a group of organic pollutants similar to synthetic organic chemicals and biopesticides in Chapter 9, petroleum hydrocarbons in Chapter 10, and PAHs in this chapter and others.

Major groups of organic pollutants are released into the environment during the production, storage, distribution, handling, use, recycling, and waste disposal of refined petrochemicals, bulk chemicals, industrial or specialty chemicals, and their chemical products (Chapters 1 and 2). Synthetic polymers, pesticides, and the emerging group of pharmaceuticals, personal care, and consumer products (PPCPs) are examples of global chemical products based on organic chemicals. Potential organic pollutants can also be released from recent development and applications of biogenic organic compounds such as in biofuels and biopesticides. Another broad group covers natural or recent biogenic sources of organic pollutants such as terpenes from plants, phytoestrogens (e.g. isoflavones), mycotoxins, PAHs from combustion of organic matter, and algal toxins (e.g. microcystin).

Origin of Toxic Organic Chemicals

Toxic organic chemicals of concern are usually derived from the processing of natural gases and petroleum oils, and synthesis of petrochemicals into industrial, commercial, speciality and consumer chemical products, as indicated in Figure 11.1. For example, benzene derived from petroleum oil or by conversion from toluene is used as a precursor for the production of benzene derivatives, solvents, ABS detergents, monomers, polymers (polystyrenes, polycarbonates, and polyurethanes), and epoxy resins as illustrated in Figure 11.2.

These types of chemicals are hazardous and exhibit different degrees of toxic effects depending upon physical/chemical form, use, and exposure to humans or natural biota in the environment.

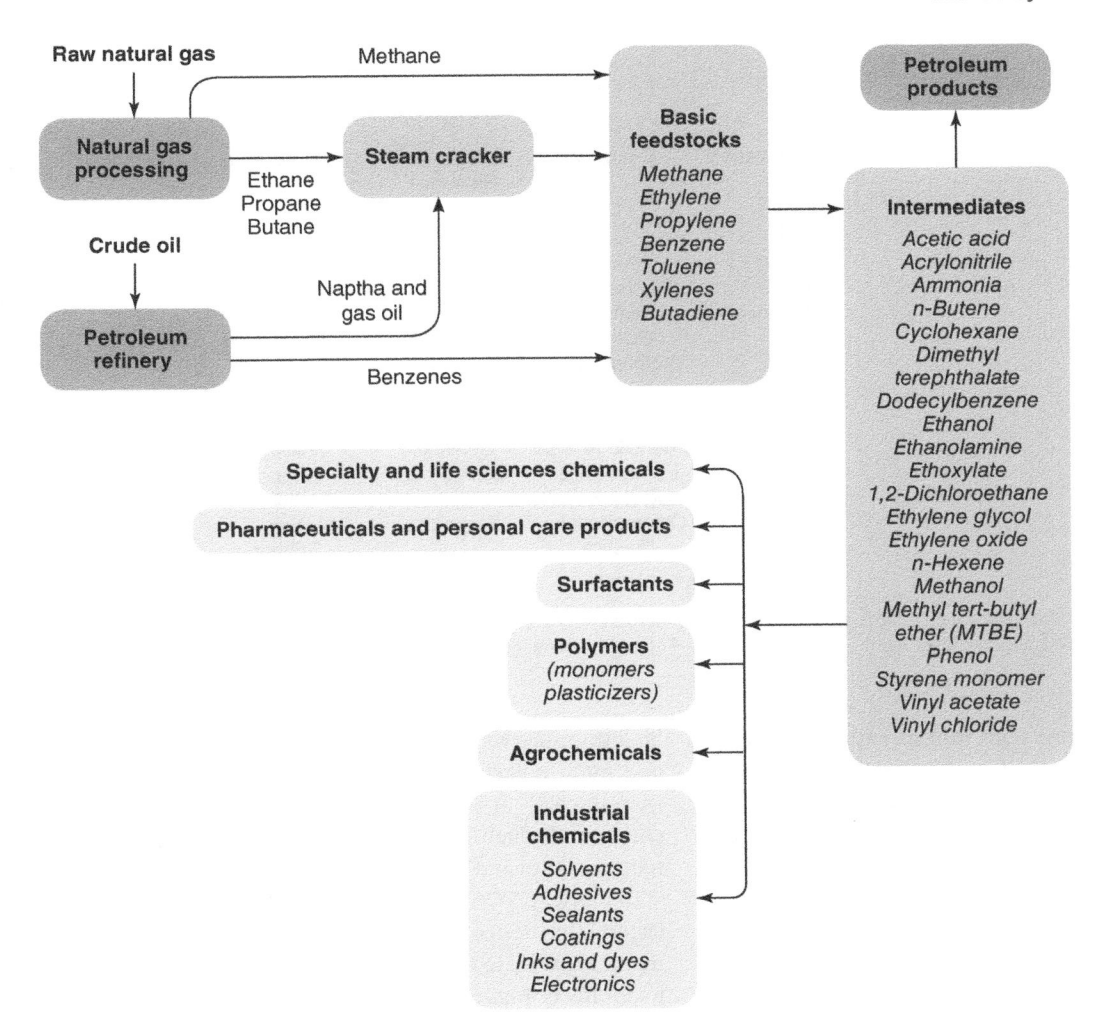

Figure 11.1 Flow chart showing production of industrial and commercial organic chemicals from petroleum that are usually converted on a large scale by integrated industrial processes: from bulk chemical feedstocks to intermediates, and then industrial, specialty, life science, and consumer chemicals. Coal was once the principal feedstock of chemicals (e.g. from coal tar) prior to the 1950s but has been largely replaced by more economical production from natural gas and petroleum oils.

The major types of organic chemicals produced are petrochemicals and derivatives, agrochemicals (including pesticides), polymers, surfactants, many other industrial, specialty and life science chemicals, pharmaceuticals and veterinary chemicals, and personal care products. As discussed in Chapters 1 and 2, tens of thousands of commercial chemicals are used worldwide; the vast majority being organic chemicals. Even so, bulk inorganic chemicals are also prevalent in the top 100 chemicals used.

The emissions and releases of many of these organic chemicals, their by-products and wastes, into the environment are widely reported in the scientific literature. Individual chemical residues and complex mixtures of known and emerging chemical pollutants are found widely in the global environment. Concentrations are usually detected at low levels that can be difficult to identify, measure, and evaluate for any harmful effects.

11.3 Classification of Toxic Organic Chemicals

In practice, the scientific classification of known or potential toxic organic chemicals depends largely on evaluation of their chemical structures, functional groups, physical, chemical, and environmental properties (e.g. persistence, mobility, and bioaccumulation), and scientific studies of their toxicity. Toxic effects can include endocrine-disrupting properties, genetic toxicology, and ecotoxicity. For many chemicals, this type of information and data are limited, inadequate, or under regulatory review (see Chapter 19).

Many synthetic organic contaminants found at parts per billion levels in environmental and biomonitoring samples tend to consist of numerous different structural

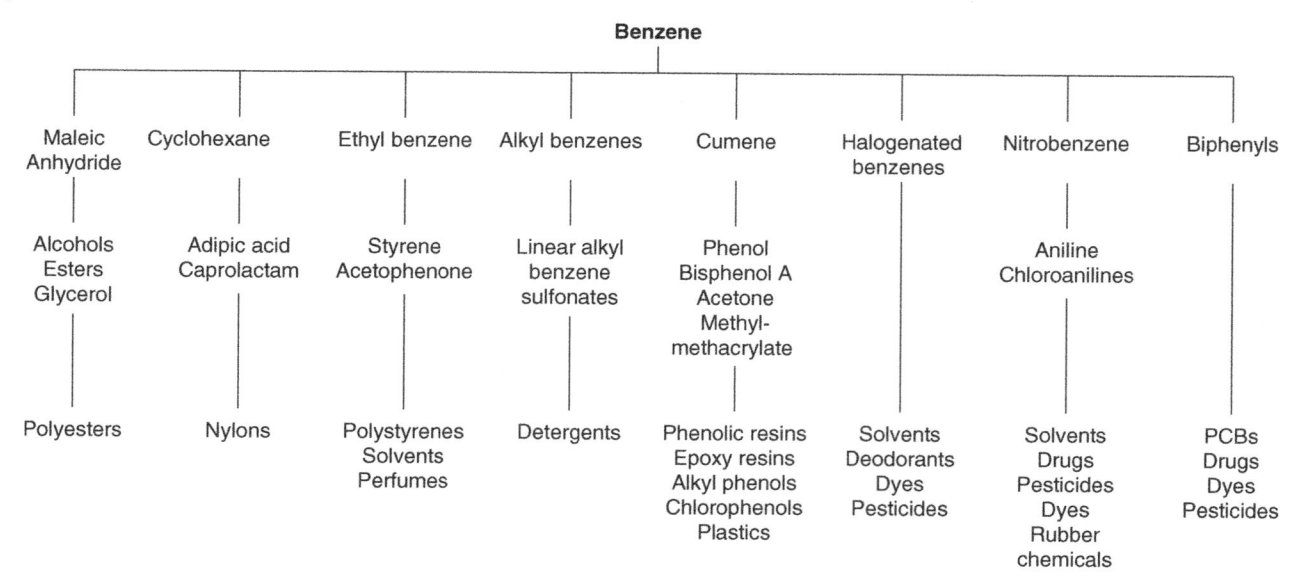

Figure 11.2 Flowchart of some industrial organic chemicals derived from benzene as an example of a basic chemical feedstock. (Note: PCBs are no longer produced.) Source: Data from Stepa, R., Schmitz-Felton, E., Brenzel, S. (2017). Carcinogenic, mutagenic, reprotoxic (CMR) substances. OSHWiki.

types of organic chemicals, with multiple types of potential toxicities, and originate from various categories of chemical production and use (e.g. pesticides, industrial chemicals, and pharmaceuticals). Due to the complex nature of organic chemicals, they are generally categorized according to their industrial group, use, or function (e.g. pesticides, solvents, flame retardants, polymers, and surfactants). Subcategories are commonly based on chemical families with related structures, isomers, functional groups, properties, and uses, such as for halogenated hydrocarbon solvents or organophosphate insecticides.

Major groups of organic chemicals that occur as chemical pollutants are described in the Companion Book online in Section S11.1.

Types of TOC Pollutants

In practice, the classification of organic pollutants (e.g. persistent, bioaccumulative and/or toxic, PBTs) has evolved more through a progressive mix of state-of-the-art methods of sampling and analyses, scientific evaluation of pollutant properties, regulatory needs, and international agreements on control of hazardous chemicals and risk assessment approaches. Assessment criteria for PBT chemicals and related POPs are not defined purely on the basis of science but are also subject to the aims of policy and regulation (Matthies et al. 2016).

Emerging pollutants have been broadly characterized according to physical/chemical properties into (i) organic substances that can be subdivided into PBTs (persistent, bioaccumulative, and toxic substances including POPs) and more polar substances (e.g. various pesticides, pharmaceuticals, and industrial chemicals), (ii) inorganic compounds such as metals, and (iii) particulates such as nanoparticles and microplastics (Geissen et al. 2015).

In this way, TOCs can generally be separated according to criteria that define their persistence and fate in the environment, capacity to bioaccumulate in organisms, and induce toxic effects. Figure 11.3 presents a scheme for classification of toxic organic chemicals that reflects (i) their synthetic or natural origin, (ii) PBT properties, and (iii) current groupings (e.g. POPS or emerging chemicals of concern) for regulatory, analytical or scientific purposes.

There are five main groups of TOCs discussed in this chapter that cover tens of thousands of known and potential organic pollutants derived mainly from industrial and commercial chemicals, including specialty chemicals, pharmaceutical and personal care products that are produced or used in a modern industrial society. Some of these chemicals are generated unintentionally as contaminants during various manufacturing processes. The five groups are:

(1) PBTs (Persistent, bioaccumulative and toxic substances),
(2) POPs (Persistent halogenated and non-halogenated organic chemicals),
(3) Less persistent and less bioaccumulative TOCs, i.e. less to more polar,
(4) Contaminants of emerging concern (CECs), and
(5) PPCPs (Pharmaceutical and personal care products).

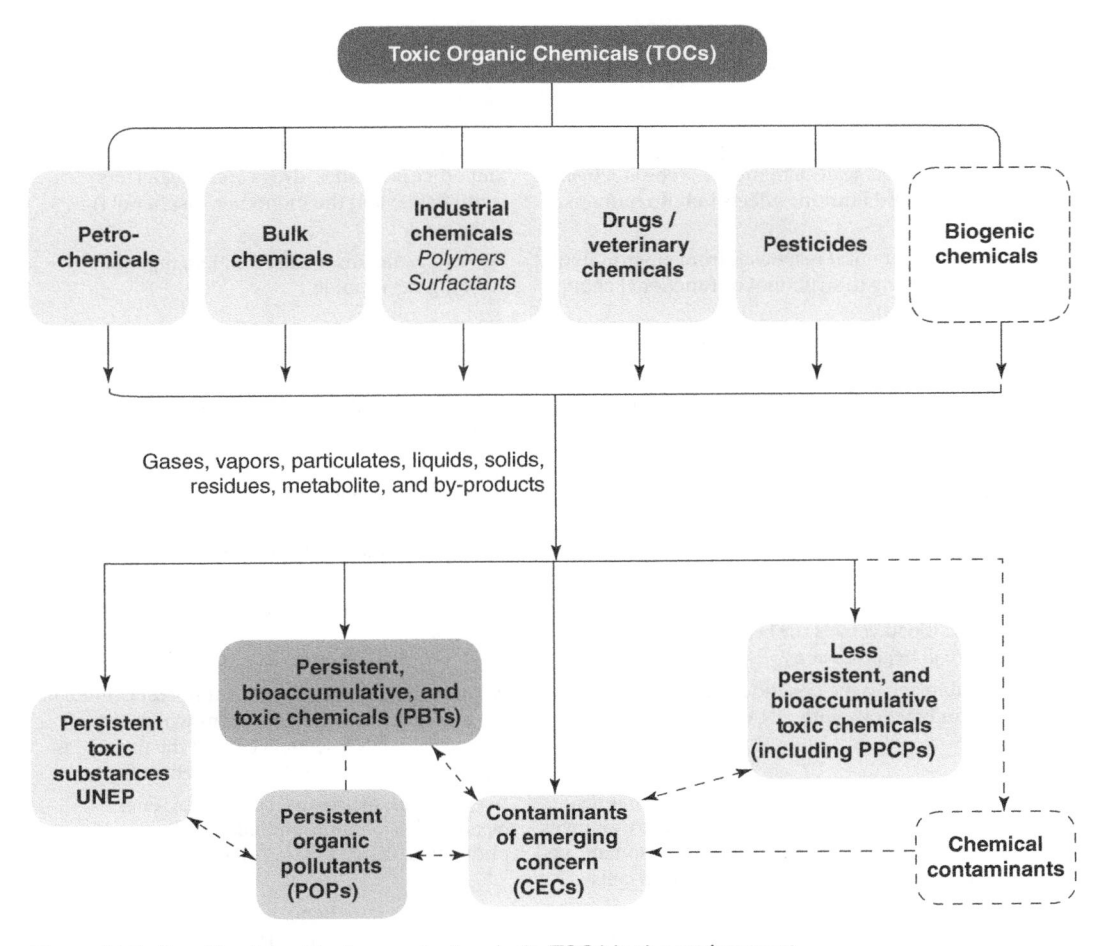

Figure 11.3 Classification of toxic organic chemicals (TOCs) in the environment.

The PPCPs form a special and emerging group of known and potentially toxic organic chemicals that can also be considered a sub-group of group 4, and similar in multiple toxicity to current-use pesticides.

Another potential group consisting of unknown or unresolved organic chemicals is evident at low levels in many environmental samples, especially among chemical mixtures of variable composition, or predicted to theoretically exist in the environment and be toxic to living organisms.

These groups of organic pollutants are discussed further in Section 11.3.

A classification of toxic organic chemicals (TOCs) in the environment is outlined in Figure 11.3.

Toxicity Classification and Priority Pollutants

As described in Chapter 2, organic pollutants may also be classified by their toxicity or as priority pollutants, usually for regulatory and analytical purposes in various countries, such as developed under the Clean Water Act in the United States of America (US EPA 2017a).

Toxicity classifications are also commonly applied to groups of toxic chemicals, including organic pollutants. Broad groupings include non-carcinogenic and carcinogenic toxicants, or more specifically, by body organ or system affected, such as neurotoxicant or respiratory toxicant. Common types of toxicity groups or classes that include TOCs are listed in Table 11.2.

11.4 Persistent, Bioaccumulative, and Toxic Chemicals

Persistent, bioaccumulative, and toxic chemicals are a special category that includes thousands of toxic organic chemicals found in our environment, and many as residues (chemical and/or its metabolites) in the tissues, organs and fluids of wildlife, livestock, and humans. However, the vast majority of toxic organic chemicals that need to be evaluated as known or potential pollutants, and managed, are not classed as PTBs or POPs. The discovery and emergence of synthetic organic chemicals as environmental pollutants,

Table 11.2 Common toxicity classifications for organic pollutants (TOCs).

Common toxicant classes	Toxicity description	Examples
Endocrine-disrupting chemical (xenoestrogen)	Disrupts endocrine or hormone systems in the body and produce adverse developmental, reproductive, neurological, and immune effects in both humans and wildlife	PCBs, dioxin and dioxin-like compounds, DDT and other pesticides, drugs and plasticizers (phthalates) and the monomer, bisphenol A
Neurotoxicant	Causes adverse neural responses from environmental exposures leading to structural or functional change in nervous system	POPs (e.g. dioxins, PCBs, PBDEs), organic solvents, acrylamide
Hematotoxicant	Causes adverse effects on blood and blood forming agents	Ethanol, benzene, halogenated hydrocarbons
Immunotoxicant	Disrupts functioning of immune system, may lead to diseases (e.g. cancers, autoimmune diseases), hypersensitivity, and allergic reactions	PCBs, organochlorines, benzene, toluene, TDI, halogenated aromatic hydrocarbons and PAHs
Pulmonary or respiratory system toxicants	Can trigger effects ranging from mild irritation to pulmonary edema and death by asphyxiation	Respirable particles, isocyanates, organic solvents, plasticizers, acetaldehyde
Hepatotoxicant	Induces liver cell injury or damage	Ethanol, halogenated hydrocarbons, vinyl chloride
Cardiovascular toxicants	Can cause adverse effects on the cardiovascular (heart and blood vessels) or hematopoietic (blood) systems (e.g. hypertension)	Benzene, respirable particles, phthalate esters, solvents
Genotoxicant	Induces damage to the genetic material in cells through interactions with DNA sequence and structure	All mutagens are genotoxic, but not all genotoxins are mutagens. A genotoxic agent will cause damage to a DNA sequence but if the damage is repaired it may not lead to mutagenic effects
Mutagen	Causes direct or indirect damage to DNA that results in mutations (genetic alterations). These are changes in the DNA sequence that are retained in somatic cell divisions and passed onto progeny in germ cells	Acrylamides, aflatoxins, benzo(a)pyrene, bisphenol A, dichloromethane, nitrosoamines, phthalates, PCBs, PFOA, PBDE, trichloroethylene
Carcinogen	May damage directly the DNA in cells causing genetic mutations (genotoxic) or disrupt cellular metabolic processes (non-genotoxic) that lead to abnormal cell growth and tumors	Benzene, 1,3 butadiene, benzo(a)pyrene, formaldehyde, vinyl chloride, 1,2 dichloropropane, PCBs, pentachlorophenol, N-nitrosodiethylamine **Non-genotoxic:** 1,4-dichlorobenzene, 17 β-estradiol, TCDD, 1,4 dioxane
Reproductive and developmental toxicant (reprotoxic)	Interferes with normal reproduction. Includes adverse effects on sexual function and fertility in adult males and females, and also developmental toxicity in the offspring. Teratogens and endocrine-disrupting substances affect reproduction. (Teratogens affect the development of an embryo or fetus and cause birth defects or even death in offspring.)	Bisphenol A, PCBs, benzo(a)pyrene, cytotoxic drugs

Source: Based on Compiled from various sources including Chapter 6, and Stepa, R., Schmitz-Felton, E., Brenzel, S. (2017). Carcinogenic, mutagenic, reprotoxic (CMR) substances. OSHWiki (accessed 14 February 2017).

particularly persistent organochlorine chemicals and pesticides, can be seen as an outcome of the post Second World War growth of the chemical industry in the Twentieth Century to meet global demand for food, economic development, security, health and personal care, and lifestyle.

The early environmental focus was on chemicals, such as organochlorine pesticides (e.g. DDT) and the industrial chemicals, PCBs. These chemicals have high mobility in the environment, readily resist degradation, and persist for long periods in environmental phases, exhibit high bioaccumulation in wildlife and humans, can transfer in food webs, and in certain instances may biomagnify in concentration, with increasing trophic levels to top predators.

The PBT classification is attributed to the Great Lakes Binational Toxic Strategy (GLBNS) developed in 1997 between the United States and Canada. The Great Lakes – Superior, Michigan, Huron, Erie, and Ontario – and their connecting channels form the largest system of fresh surface water in the world. Top priority PBTs listed in this Strategy covered mercury, PCBs, dioxins/furans, benzo(a)pyrene, hexachlorobenzene (HCB), alkyl lead, organochlorine insecticides (mirex, dieldrin/aldrin, chlordane, and toxaphene), and ocatachlorostyrene. In the United States, the US EPA evaluates and controls new and existing PBT chemicals under the Toxics Substances Control Act (TSCA) and through chemical rules and reporting of the Toxics Release Inventory (TRI). Previously, the United States phased out the use and registration of DDT such that it is no longer legally sold or distributed in the United States (US EPA 2009a).

Environmental criteria used by the European Union (EU) and the United States of America to define PBT chemicals are compared in Table S11.1. As well, persistent organic pollutants (POPs) under the UN Stockholm Convention are discussed and compared with classification criteria in Table S11.1. Types of POPs are listed in Table 11.3.

The development of criteria for PBTs in the European Community and United States reflects consideration of POPs criteria and subtle differences in regulatory approaches for management of toxic chemicals, similar to other countries such as Japan, Canada, and Australia.

Half-lives criteria for PBTs in waters, sediments, and soils vary between the EU and United States. In waters, the criteria range from >40 and >60 days (EU) to >60 and >180 days (United States). For sediments, the US criterion for persistence is >60 days, in contrast to >120 and >180 days for the EU. The soil persistence criterion in the EU is >120 days compared to >60 days in the United States. Very persistent is >180 days in both cases. In the United States, a persistent criterion for air is given as greater than two days, the same as for POPs.

Criteria for bioaccumulation properties in the EU and United States are based on BCF values in aquatic species. These are ≥2000 (bioaccumulative) and ≥5000 (very bioaccumulative) in the EU, while the United States has a lower BCF value of ≥1000 for bioaccumulative.

Toxicity criteria for PBTs are given as chronic toxicity endpoints in fish in the EU and United States. However, the EU also includes other chronic toxicity criteria relevant to human health. In the United States, when the EPA reviews a chemical for its PBT characteristics, the Agency also considers potential human health effects due to environmental exposure.

11.5 Persistent Organic Pollutants (POPs)

Overview

On a global scale, the problem of persistent organic pollutants (POPs) in the environment, wildlife, and humans has been addressed through the United Nations. POPs are chemical substances that possess certain toxic properties and, unlike other pollutants, resist degradation. POPs are particularly harmful for human health and the environment. They accumulate in living organisms, are transported by air, water, and migratory species, and accumulate in terrestrial and aquatic ecosystems around the world (UNEP 2017b).

Under the United Nations treaty known as the Stockholm Convention, the EU, the United States, and 90 other countries agreed in May 2001 in Stockholm, Sweden, to reduce or eliminate the production, use and/or release of a dozen toxic chemicals, known as POPs. The 12 priority POPs were listed: aldrin, chlordane, dichlorodiphenyltrichloroethane (DDT), dieldrin, endrin, heptachlor, mirex, toxaphene, polychlorobiphenyls (PCBs),[1] hexachlorobenzene (HCB),[1] and polychlorinated dioxins[1] and furans[1] (US EPA 2009a).

They are informally called the *dirty dozen*. Note: *These POPs are chlorinated compounds. Chlorine substitution tends to increase the potential persistence of an organic molecule.* Various POPs occur as isomeric mixtures (dioxins, furans, and PCBs exist in many different isomeric forms). Some isomers are more persistent than others and more toxic to different organisms, e.g. the dioxin, tetrachlorodibenzodioxin, TCDD, is the most toxic form of the dioxin isomers.

Following a 12 region review of the world, 26 chemicals or groups of chemicals were placed on the UNEP list of persistent toxic substances (PTS): the 12 priority POPs in the Stockholm Convention, 3 organometals (organotin, organomercury, and organolead compounds) and 11 chemicals (HCH, PAHs, endosulfan, pentachlorophenol, phthalates, PBDE, chlordecone (Kepone), octylphenols, nonyl phenols, atrazine, and short-chained chlorinated paraffins) (see UNEP 2003).

In 2004, the United Nations (UNEP) – Stockholm Convention decided to limit pollution by POPs, define what are POPs, and also set the rules to control the production, importing, and exporting of POPs. In May, 2009, and since, new POPs have been added to the List. POPS consist primarily of persistent and halogenated chemicals made up of pesticides, industrial chemicals, and by-products. Table 11.3 shows POPs listed under the Stockholm Convention and their general use. Some are no longer produced.

The criteria for defining persistent organics pollutants (POPs) are outlined in Table S11.1, along with examples of criteria for PBTs in the EU and United States. Specific scientific criteria adopted for POPs are based primarily on their persistence and bioaccumulative properties. Environmental persistence of POPs is indicated by half-lives

---
1 Unintentionally formed by some industrial processes and combustion.

Table 11.3 Persistent organic pollutants (POPs).

POPs	Uses
May 2001	
Aldrin	Pesticide
Chlordane	Pesticide
DDT	Pesticide
Dieldrin	Pesticide
Endrin	Pesticide
Heptachlor	Pesticide
Hexachlorobenzene	Pesticide
Mirex	Pesticide
Toxaphene	Pesticide
Polychlorobiphenyls (PCBs)	Industrial (heat exchange fluids, transformer and capacitor fluids, additives in paint, plastics, and carbonless paper)
Polychlorinated dibenzo-p-dioxins (PCDD)	Unintentional production
Polychlorinated dibenzofurans (PCDF)	Unintentional production
May 2009 to 2020	
Alpha hexachlorocyclohexane	Pesticide
Beta hexachlorocyclohexane	Pesticide
Chlordecone	Pesticide
Dicofol	None
Decabromodiphenyl ether (c-decaBDE)	Industrial (flame-retardant additive)
Hexabromobiphenyl	None
Hexabromocyclododecane	Industrial – flame-retardant additive)
Hexabromodiphenyl ether and heptabromodiphenyl ether (commercial octabromodiphenyl ether)	Industrial
Hexabromobiphenyl	Industrial (flame retardant)
Hexachlorobutadiene	None
Lindane (gamma-BHC)	Control of head lice and scabies (former pesticide)
Pentachlorobenzene (PeCB)	None (pesticide, industrial, unintentional production)
Pentachlorophenol and its salts and esters	Pesticide
Perfluorooctane sulfonic acid, its salts, and perfluorooctane sulfonyl fluoride	Industrial (electronic parts, photo imaging, hydraulic fluids, and textiles). Former use – fire-fighting foam)
Polychlorinated naphthalenes	Industrial
Short-chain chlorinated paraffins (SCCPs)	Industrial (plasticizer, paints, adhesives, flame retardants, extreme pressure lubricant)
Technical endosulfan and its related isomers	Pesticide
Tetrabromodiphenyl ether and pentabromodiphenyl ether (commercial)	Industrial (flame-retardant additive)

Source: Based on UNEP (2017c). The New POPs under the Stockholm Convention. United Nations Environment Programme. http://chm.pops.int/TheConvention/ThePOPs/TheNewPOPs/tabid/2511/Default.aspx (accessed 18 February 2017); UNEP (2019). All POPs listed in the Stockholm Convention – United Nations Environment Programme. http://www.pops.int/TheConvention/ThePOPs/AllPOPs/tabid/2509/Default.aspx (accessed 26 July 2020).

of >60 days (waters), >120 days (sediments and soils), and >2 days (air). Bioaccumulation is indicated by a bioconcentration factor (BCF) in aquatic species of >5000 or a $K_{OW} > 5$ (in absence of BCF data). Other criteria for POPs depend on the evaluation and weighting of evidence of long-range environmental transport, monitoring and modeling data of transport, transfer and fate, ecotoxicity effects, and adverse toxic effects on human health.

Some Examples of Persistent Organic Pollutants

Polychlorinated Biphenyls (PCBs)

PCBs are synthetic industrial compounds that possess low volatility, low water solubility, and high resistance to heat, chemical, and biological degradation. They also have high dielectric constants making them suitable for use in electrical equipment. Commercially, they have been used widely as heat exchange fluids, in electric transformers and capacitors, and as additives in paint, carbonless copy paper, and plastics.

The industrial synthesis of PCBs involves the progressive chlorination of biphenyl to produce a variety of products with differing chlorine content. Chlorine can be substituted in any of the biphenyl ring positions, from 2 to 6 or 2′ to 6′ to give compounds with 1–10 chlorine atoms (see Figure 11.4) (Connell and Miller 1984, p. 252). PCBs are mixtures of various isomers based on biphenyl and are known to consist of 209 congeners.[2] They were marketed in large quantities in various commercial mixtures of congeners.

PCBs can also contain a variety of chlorinated impurities, mostly polychlorinated dibenzofurans and chlorinated naphthalenes, which can influence the toxicities of commercial PCB mixtures (Connell and Miller 1984, p. 254). Related products that have been produced in limited quantities are polychlorinated terphenyls (PCTs), and chlorinated naphthalenes, a listed POP.

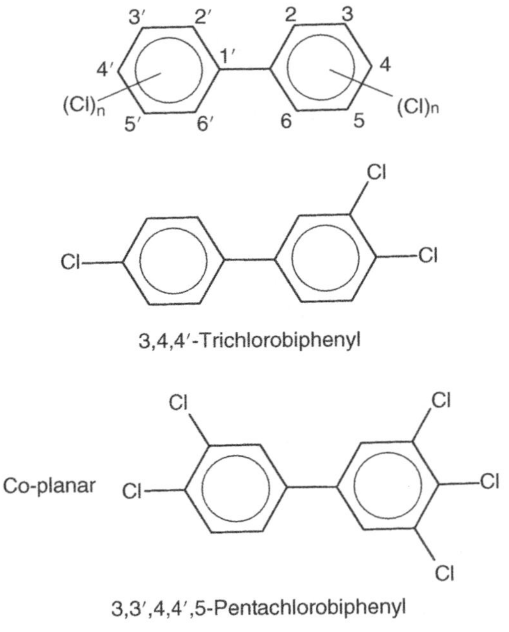

Figure 11.4 PCB structure showing substitution positions on biphenyl.

---
2 Congeners are different members of a chemical family that differ only in the number and position of the same substituent.

The persistence of PCBs in the environment corresponds to the degree of chlorination, and half-lives can vary from 10 days to 1.5 years. PCBs are soluble in lipids, resulting in bioaccumulation in fatty tissues of biota to relatively high levels in higher tropic levels, especially for higher chlorinated biphenyls.

Significantly, 13 of the 209 congener forms of PCBs exhibit a dioxin-like toxicity. PCBs are acutely toxic to fish at higher doses and cause spawning failures at lower doses. As endocrine disrupters, they are also linked to reproductive failure and suppression of the immune system in various wild animals, such as seals and mink.

Episodes of human exposures to PCBs are reported from food contamination. These include consumption of PCB-contaminated rice oil in Japan in 1968 and in Taiwan in 1979 that caused pigmentation of nails and mucous membranes and swelling of the eyelids, along with fatigue, nausea, and vomiting. Due to the persistence of PCBs in their mothers' bodies, children born up to seven years after the Taiwan incident showed developmental delays and behavioral problems. Similarly, children of mothers who ate large amounts of contaminated fish from Lake Michigan in North America showed impaired short-term memory function. PCBs are also known to suppress the human immune system. Dioxin-like PCBs are listed as human carcinogens (UNEP 2017d).

Polychlorinated Dibenzo-p-Dioxins (PCDDs)

The term, *dioxins*, is commonly used to describe a family of chlorinated dibenzo-p-dioxins. It may also be applied to a group of toxic POPs that share closely related chemical structures and biological properties:

- Chlorinated dibenzo-p-dioxins
- Chlorinated dibenzofurans
- Certain polychlorinated biphenyls (PCBs)

(see basic chemical structure of dibenzo-p-dioxin and furans in Figure 11.5).

The first chlorinated dioxin was prepared in 1872 by two German chemists, Merz and Weith (Rappe 1978). These chemicals are produced unintentionally due to incomplete combustion, as well as during the manufacture of pesticides such as the herbicide 2,4,5-T and some other chlorinated substances. They are emitted mostly from the burning of wastes (hospital, municipal, and hazardous) and also from automobile emissions, peat, coal, and wood.

An example of the formation of dioxins is shown in Figure 11.6. There are 75 different types of dioxins, of which seven are considered to be of concern, particularly 2,3,7,8-tetrachlorodibenzo-p-dioxin or 2,3,7,8-TCDD. The basic structure of these substances is shown in Figure 11.5 as dibenzo-p-dioxin, in which substituent positions 1–4 or

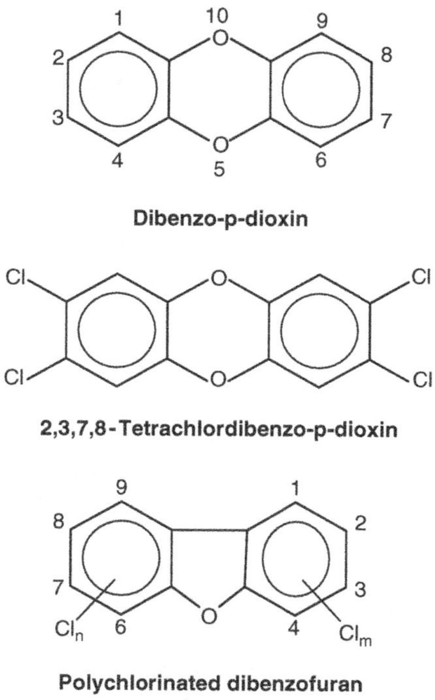

Figure 11.5 Showing basic chemical structure of dibenzo-p-dioxin and polychlorinated dibenzofurans.

6–9 can hold chlorine or hydrogen atoms or an organic group. All dioxin congeners are planar.

Dioxins are very persistent in soils but may be transported long distances in air by volatilization and cold condensation. Half-life estimates for 2,3,7,8-tetrachlorodibenzo-*p*-dioxin (TCDD) on surface soils range from 9 to 15 years, whereas the half-life in subsurface soil may range from 25 to 100 years (ATSDR 1998). They have been associated with a number of adverse effects in humans, including chloracne, immune, and enzyme disorders. TCDD is now classified as a human carcinogen (IARC). Laboratory animals given dioxins suffered a variety of effects, including an increase in birth defects and stillbirths. Fish exposed to these substances died shortly after the exposure ended. Food is the major source of exposure for humans. Dioxins accumulate in the fatty tissue of living organisms including humans and are found at higher concentrations at higher trophic levels in the food chain (UNEP 2017e).

Polychlorinated Dibenzofurans (PCDFs)
These compounds are also produced unintentionally from many of the same processes that produce dioxins and also during the production of PCBs. They have been detected in emissions from waste incinerators and automobiles. Furans are structurally similar to dioxins and share many of their toxic effects (see Figure 11.5). There are 135 different types, and their toxicity varies. Furans persist in the environment for long periods and are classified as probable human carcinogens. Food is the major source of exposure for humans. Furans have also been detected in breast-fed infants (UNEP 2017e).

Perfluorinated Alkyl Substances
Per- and polyfluorinated alkyl substances (PFAS) or chemicals (PFCs) and their derivatives are stable synthetic chemicals that can uniquely repel oil, grease, and water. There are two groups of PFCs used in industry: (i) perfluorosulfonic acid (PFSA) group, including perfluorooctane sulfonic acid (PFOS), and (ii) perfluorocarboxylic acid (PFCA) group, including perfluorooctanoic acid (PFOA), which is currently under assessment as a potential POP. PFAS are widely used in industrial and consumer products (e.g. electronic parts, fire-fighting foam, photo imaging, hydraulic fluids, and textiles).

Perfluorooctane sulfonic acid (PFOS), its salts, and perfluorooctane sulfonyl fluoride (PFOS-F)

PFOS is a fully fluorinated anion, which is commonly used as a salt or incorporated into larger polymers (see Figure 11.7). PFOS and its closely related compounds, which may contain PFOS impurities or substances that can result in PFOS, are members of the large family of perfluoroalkyl sulfonate substances.

PFOS is both intentionally produced and an unintended degradation product of related anthropogenic chemicals. It is highly persistent and readily bioaccumulates by binding to proteins in the blood and the liver, in contrast to normal partitioning into fatty tissues by other POPs. It also has a capacity to undergo long-range transport and meet the toxicity criteria of the Stockholm Convention.

> PFOS levels that have been detected in wildlife are considered high enough to affect health parameters, and recently higher serum levels of PFOS were found to be associated with increased risk of chronic kidney disease in the general United States population, consistent with earlier animal studies
> (US National Center for Biotechnology (2020) – Information – Perfluorooctanesulfonic acid).

PFOS is being phased out of some applications but many remain acceptable or exempt, particularly where alternatives are unavailable.

Perfluorooctane sulfonyl fluoride (PFOS-F) is a neutral persistent perfluorinated chemical used to make PFOS and related compounds (see Figure 11.7). It can degrade to form PFOS from PFOS-F-based polymers and in the environment. In 2000–2002, the main US producer, 3M, phased out

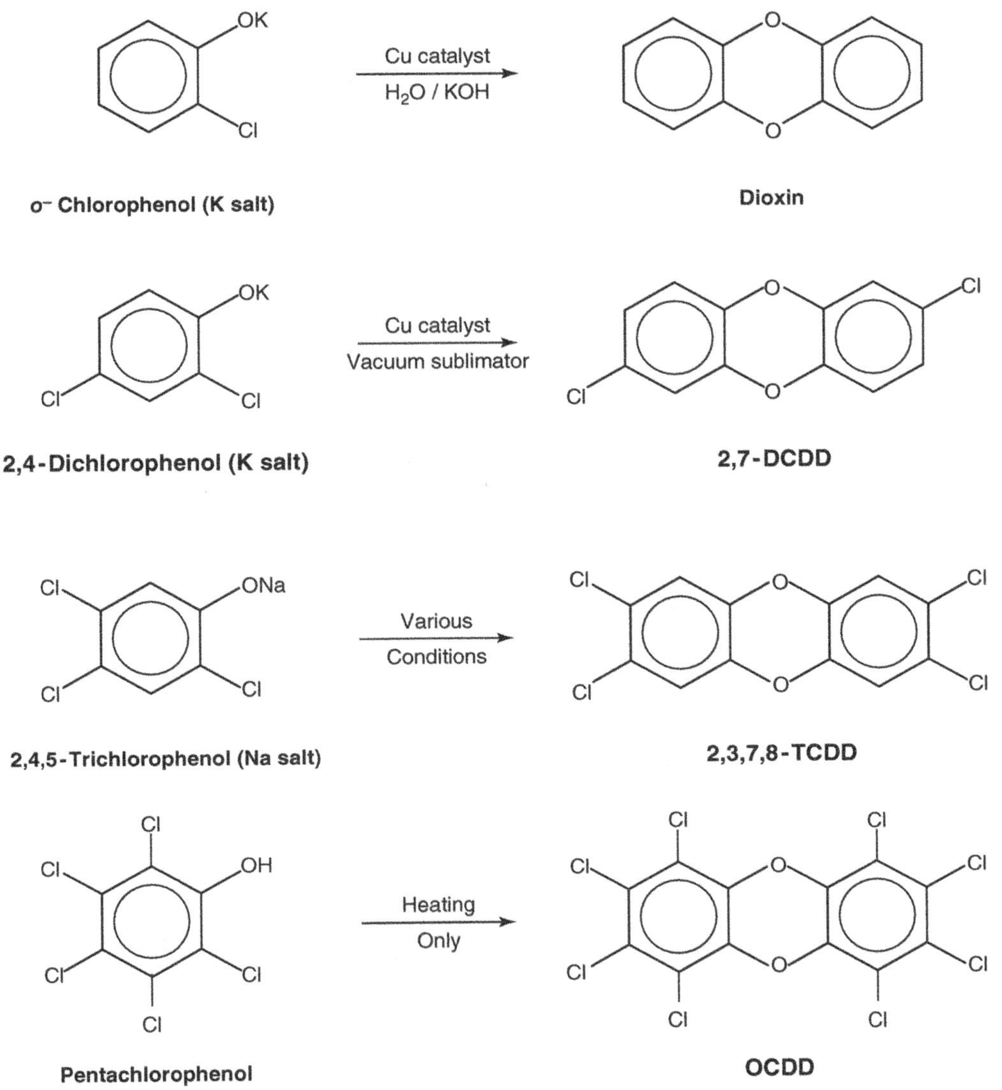

Figure 11.6 Example of formation of dioxins under different conditions from chlorophenols. Sources: Based on Esposito, N.P., Tiernan, T.O., Dryden, F.E. (1980). Dioxins, Industrial Environmental Research Laboratory, Office of Research and Development, United States Environmental Protection Agency, Cincinnati; Connell, D.W. and Miller, G.J. (1984). Chemistry and Ecotoxicology of Pollution. New York: Wiley.

production of PFOS, PFOA, and related PFOS-F-type products but production has shifted to China as the dominant producer.

Perfluorooctanoic acid (PFOA), also known as C8, is used to make Teflon™ (polytetrafluoroethylene) and other fluorotelemers. Low residues of PFOA are found widely in the environment and in human bodies, but higher levels are found in persons exposed to PFOA sources. Flu-like symptoms are observed in some persons inhaling fumes from overheated Teflon-coated surfaces. It is a possible human carcinogen based on animal studies and limited evidence in humans (IARC Group 2B) (The American Cancer Society 2016).

Brominated Organic Compounds (Flame Retardants)

Polybrominated diphenyl ethers (PBDEs) and polybrominated biphenyls (PBBs) are brominated organic compounds used to inhibit or suppress combustion in organic material as additive flame retardants in many types of products, from building materials, furnishings, electronics to plastics, foams, and textiles. They are structurally related to PCBs and other polyhalogenated compounds consisting of two halogenated rings (see Figure 11.8). PBDEs are classified according to the average number of bromine atoms in the molecule. PBDEs were listed as POPs in 2009 under the Stockholm Convention (UNEP 2017e).

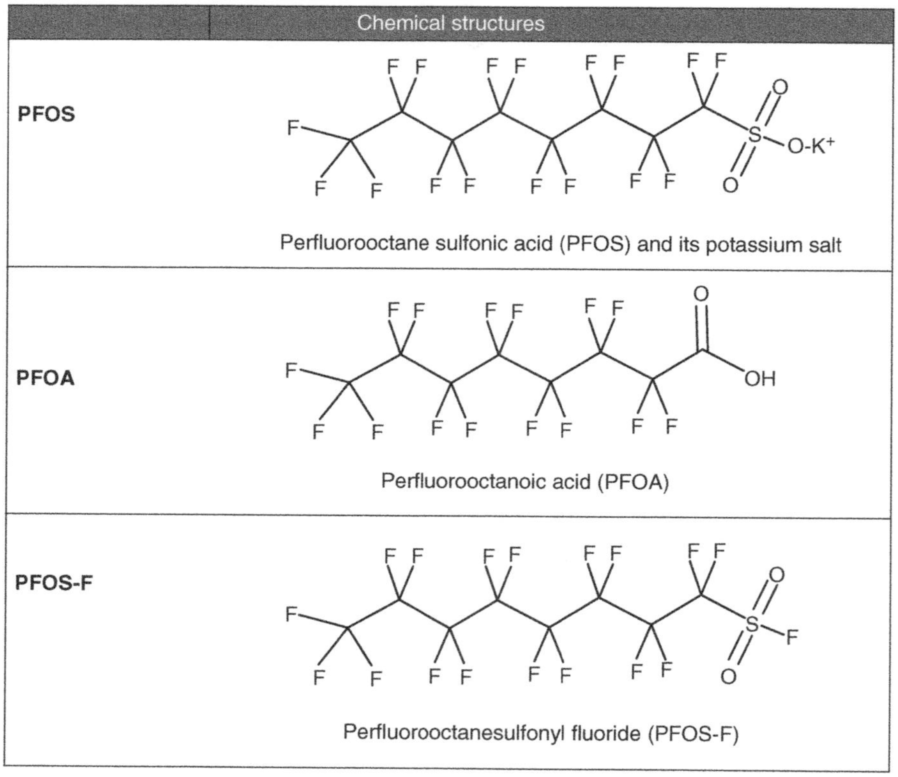

Chemical structures	
PFOS	Perfluorooctane sulfonic acid (PFOS) and its potassium salt
PFOA	Perfluorooctanoic acid (PFOA)
PFOS-F	Perfluorooctanesulfonyl fluoride (PFOS-F)

Figure 11.7 Chemical structures of PFOS, PFOA, and PFOS-F.

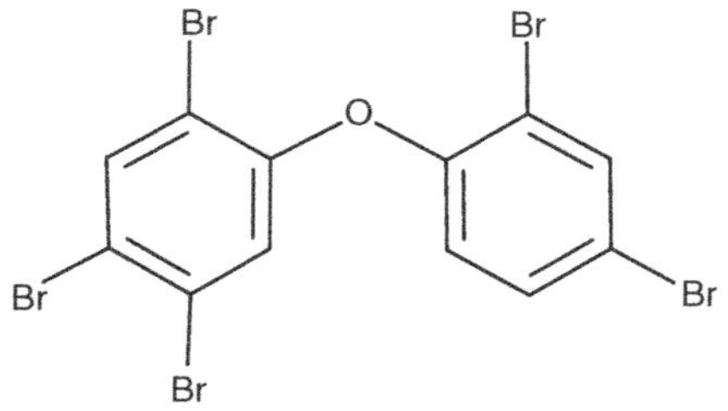

2,2′,4,4′,5 - Pentabromodiphenyl ether

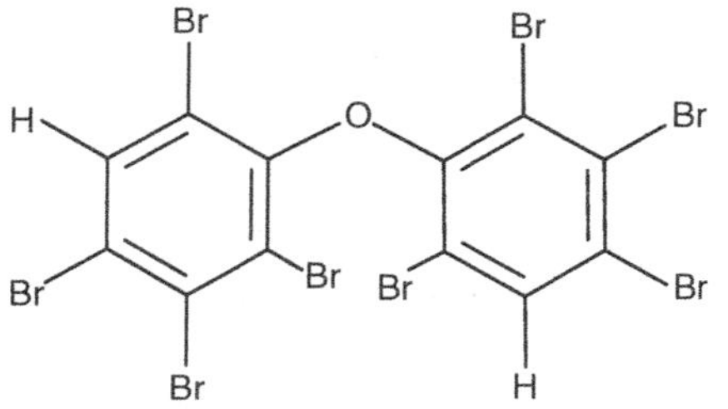

2,2′,3,4,4′,5,5′,6 - Octabromodiphenyl ether

Figure 11.8 Examples of chemical structures of polybrominated diphenyl ethers (PBDEs).

Current commercial mixtures of PBDE compounds contain the following POPs:

- Tetrabromodiphenyl ether and pentabromodiphenyl ether (main components of commercial pentabromodiphenyl ether)
- Hexabromodiphenyl ether and heptabromodiphenyl ether (main components of commercial octabromodiphenyl ether)
- Decabromodiphenyl ether – main component of commercial mixture of decabromodiphenyl ether (decaBDE). It consists primarily of the fully brominated decaBDE congener (77.4–98%), and smaller amounts of the congeners of nonaBDE (0.3–21.8%) and octaBDE (0–0.04%).

Commercial mixtures of PBDEs are highly persistent in the environment, bioaccumulative, and have a high potential for food web biomagnification and long-range environmental transport. These chemicals have been widely detected in humans (e.g. US biomonitoring program). There is also evidence of adverse and potential toxic effects in wildlife, including mammals. PBDEs can be subject to degradation through debromination, i.e. the replacement of bromine on the aromatic ring with hydrogen. As a result, higher bromodiphenyl ether congeners may be converted to lower, and possibly more toxic, congeners. The higher congeners may therefore be precursors to tetraBDE, pentaBDE, hexaBDE, or heptaBDE (UNEP 2017e).

A number of non-POP chemical alternatives is already on the market for the substitution of c-decaBDE in plastics and textiles. Furthermore, nonchemical alternatives and technical solutions such as nonflammable materials and physical barriers are also available.

Polybrominated biphenyls (PBBs) are no longer produced, but residues can still be occasionally found in the environment.

11.6 Less-Persistent and Emerging Toxic Organic Chemicals

The vast majority of organic chemicals are less persistent in the environment and bioaccumulative than PBTs and POPs, but may still exhibit a wide range of acute and chronic toxic effects on exposed organisms, including humans.

Among these, toxic organic chemicals are generally characterized by physical, chemical, environmental, health, and safety properties. Important properties are:

- Chemical structure, form, and functional groups
- Molecular weight
- Particle sizes and shape (e.g. nanoparticles, microplastics, polymer fibers)
- Polarity and water solubility (nonpolar to polar)
- Vapor pressure
- Environmental persistence
- Mobility in environmental media
- Bioaccumulation properties, e.g. K_{ow} or BCF values
- Specific toxicity characteristics: human health and ecotoxicity

TOCs vary enormously from small molecules such as gaseous formaldehyde to low boiling liquid solvents, and those with increasing structural complexity, form and size, including from nanoparticles to macromolecules such as synthetic polymers. In this section, some important examples are highlighted and discussed below.

Common organic pollutants found in air, water, and contaminated soils with low molecular weights and boiling points include basic chemical feedstocks, solvents, and intermediates indicated in Figure 11.1. Examples are low-boiling toxic aromatic, nonhalogenated, halogenated, and oxygenated hydrocarbons that are widely used in industrial, commercial, and household products (e.g. solvents) as given in Table S11.2 in Section S11.3. Generally, they are classified as volatile organic chemicals (VOCs), or air toxics in some cases, and are commonly found in indoor environments, workplaces, and urban airsheds. VOCs are organic chemicals that evaporate under normal indoor conditions of temperature and atmospheric pressure. Very volatile organics (VVOCs) have a boiling point range from <0 to 100 °C (e.g. butane, trimethyl amine, and dichloromethane), other VOCs have a range of 50–100 °C to 240–260 °C (e.g. toluene, ethanol, hexane, and xylenes). Semivolatile organic compounds (SVOCs) have a higher range from 240 to 380–400 °C (e.g. many POPs and phthalates) (US EPA 2017b).

Synthetic Polymers (Plastics)

Polymers generally consist of large complex molecules (macromolecules) made up of long chains of repeating molecular units of small molecules that are known as **monomers**. A polymer may be of natural origin, such as cellulose, amber, biopolymers, e.g. DNA, and natural rubber, or of synthetic origin such as polyethylene, Teflon, and silicones. The term, **polymer**, is commonly used to mean plastics, and vice versa, but covers a wider class of compounds including synthetic fibers, rubbers, plastics, and various biomolecules as indicated above. There are also many types of hybrid polymers, containing inorganic and organic entities such as polydimethylsiloxane.

The basic chemistry, properties and characteristics of polymers are concisely described in The Essential Chemistry Industry – Online (2013).

Monomers and Polymerization Reactions

Polymerization is the process in which **monomers** covalently bond to form polymers. There are two main types of reactions:

1. *Addition Polymerization:* Monomers simply add or join together and no other products are formed.

   An example is the formation of **polyethylene** from ethylene where n represents several thousand units of the monomer (Figure 11.9). Polyethylene is a tough flexible plastic used in high-density pipes (polypipe), bottles, insulation, packaging film, bags, and toys. Its properties can be changed by substituting different groups (methyl, phenyl, cyano, CN–) or elements such as chlorine or fluorine for hydrogen(s) in the ethylene monomer. For example, monochloroethylene or vinyl chloride monomer can produce the tough and rigid polymer, PVC (polyvinylchloride) (Figure 11.10). PVC is used in floor tiles, piping, building materials, and toys.

   Another example is the use of tetrafluoroethylene to make the polymer, Teflon (Figure 11.10). Teflon is well known for coating cooking utensils (*nonstick*), thermal insulation, and bearings. It may also contain persistent by-products or residues of perfluorooctanoic acid (PFOA), polytetrafluoroethylene (PTFE), and perfluorooctane sulfonate (PFOS).

   Polystyrene is made from the monomer, styrene, a derivative of benzene with an ethylene or ethene group attached (substituted benzene) (Figure 11.11). It is commonly rigid, practically nonbiodegradable and is used in products such as foam containers, surfboards, thermal insulation, and toys. The styrene monomer can be released during manufacture or by degradation of the polymer as an environmental residue.

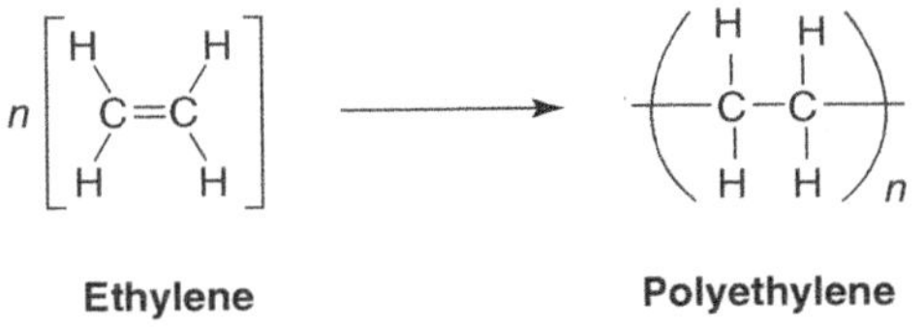

Figure 11.9 Addition polymerization of polyethylene from ethylene.

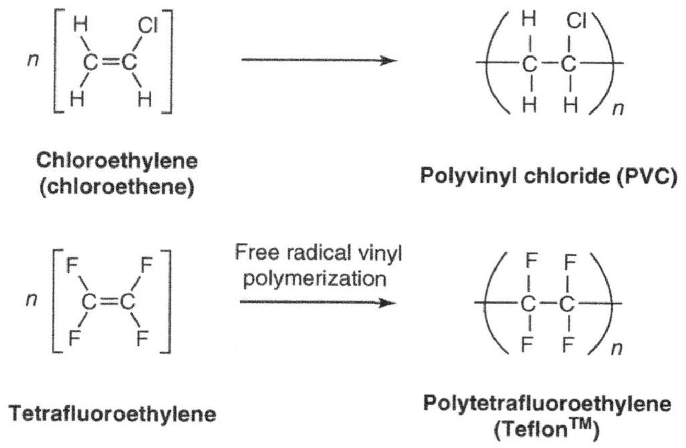

Figure 11.10 Addition polymerization of other common polymers of PVC and polytetrafluoroethylene or Teflon.

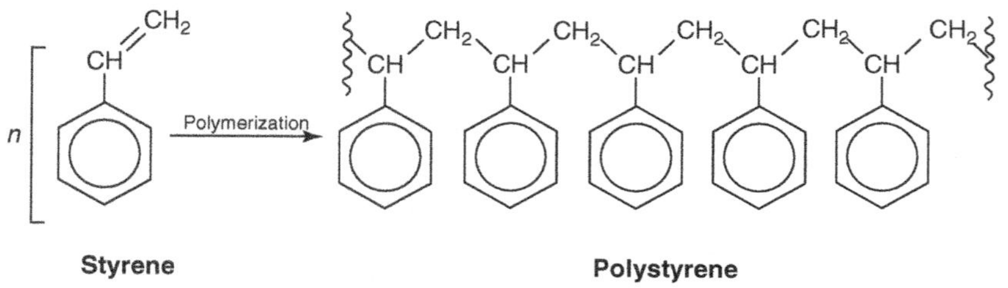

Figure 11.11 Polymerization of polystyrene from styrene monomers.

2. *Condensation Polymerization*: In this reaction, the bonding of the monomers at each step in the polymer chain releases small molecules such as water so that they may join together. Nylon is a classic example of this type of polymerization. It involves two different types of monomers joining together and is known as a **copolymer**. Other types of common polymers include polypropylene, formed in long structural units by joining propylene monomers, and polyurethane, an example of a copolymer made from urethane and ethanol monomers by addition polymerization (see Figure 11.12).

Thermoplastics and Thermosetting Polymers

Polymers can be characterized into four classes:

1. *Thermoplastics* are the plastics that do not undergo chemical change in their composition when heated and can be repeatedly warmed, softened, and remolded. Examples include polyethylene, polypropylene, PVC, and Teflon.
2. *Thermosetting plastics* readily melt and form a shape that stays solid after cooling but cannot be remolded (e.g. Bakelite, polystyrene, and polymethacrylate). They are made up of chains of many repeating molecular units, known as *repeat units*, derived from monomers. These reactions are irreversible.
3. *Elastomers* are amorphous elastic solids. They have coiled chains, which can be stretched out but readily return to their original shape when the stretching force is released. Examples include polyurethanes, poly(buta-1,3-diene), ABS (acrylonitrile butadiene styrene), and silicones.
4. Synthetic plastic *fibers* that are produced by extruding a molten polymer through small holes in a die. Examples

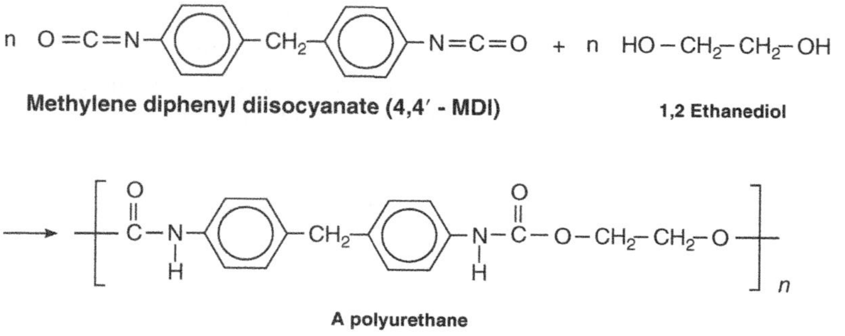

Figure 11.12 Polyurethane is an example of a copolymer made from an isocyanate (MDI) and ethanediol monomers by addition polymerization.

Table 11.4 Environmental issues and properties related to plastics.

Issues/properties	Problems
Petrochemicals–major source of plastics	Large use of non-renewable energy sources;
	Processing releases large quantities of VOCs, acid gases, and CO_2;
	Burning of plastics also
Persistence and biodegradability	Plastics generally biodegrade slowly and disintegrate into fragments, microparticles, nanoparticles, and fibers; may be subject to ingestion and bioaccumulation
Toxicity	Many plastics can cause deaths to wildlife through (i) physical obstruction, entanglement, or through ingestion of pellets, micro- and nanoparticles, and (ii) various additives and residual monomers (e.g. bisphenol A and phthalate esters) that are known to be environmental contaminants, and toxic to humans and wildlife
Waste disposal	Global pollution and life cycle issues exist over the production, use, and disposal of petrochemical-derived plastics
Incineration or burning (e.g. use of plastic wastes incl. plastic bottles for fuels and open landfill burning of wastes)	Large emissions of CO_2, fine particles, VOCs, and other toxic by-products depending upon temperature and chemical nature of plastics
Recycling	Contamination problems from additives (e.g. flame retardants). Some types of plastics are not cost effective to recycle and are disposed to landfill, incinerated, or used for fuels

include the polyamides (e.g. nylon), the polyesters, and polypropene.

The manufacture, use, and disposal of plastics, however, are associated with several important environmental pollution issues, as indicated in Table 11.4.

Monomers

Note: many plastic **monomers** (e.g. styrene and vinyl chloride) are known to be toxic to humans in workplace exposures and potentially so in environmental exposures and through residual contact.

Plasticizers

Various types of chemicals or dispersants are added to polymers to improve their physical properties such as viscosity, flexibility, extensibility, and workability. Phthalate esters (PAEs) are the dominant group of plasticizers, used mainly in PVC applications and in many other diverse plastic applications. The general structural formula of PAEs is given in Figure 11.13.

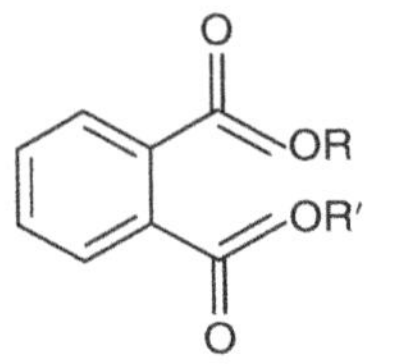

Figure 11.13 General structural formula of phthalate esters (PAEs) where R can be diethyl, dibutyl, bis(2-ethylhexyl), isodecyl, and a variety of other organic groups.

The global use of phthalates in plastics has led to ubiquitous contamination of human and natural environments. Certain low-molecular-weight ortho-phthalates have been classified as potential endocrine disruptors with some developmental toxicity reported (Halden 2010).

Examples of organic and inorganic additives used in plastics are given in Table 11.5. Some of these are well-known environmental contaminants (phthalate esters, flame retardants, and lead). Higher molecular weight phthalates are being substituted and more biodegradable plasticizers are being developed (e.g. alkyl citrates in food packaging and toys). Green plasticizers include epoxidized vegetable oils.

Surfactants

Surfactants or surface-active agents are usually organic compounds that contain both hydrophobic and hydrophilic groups on their molecules (amphiphilic agents). This means that a surfactant contains both a water insoluble group (tail) that readily dissolves in fats or oils and a water-soluble group (head).

These substances act by lowering of the surface tension in liquids, especially water. The surfactant molecules diffuse in water and assemble at interfaces between air and water or fat, and also in water where they absorb and lower the surface tension. This causes the formation of bubbles, tiny droplets, and other surface effects that allow these substances to act as cleaning or dispersing agents for a wide range of practical purposes in industry and for domestic purposes.

Table 11.5 Plastic additives: examples and functions.

Additive	Examples	Function
Plasticizer	Esters of benzene-1,2-dicarboxylic acid, phthalate esters	Acts as a lubricant for polymer chains. Large amounts give a flexible product; low quantities produce a rigid one
Stabilizer	Lead carbonate (<1%), lead phosphate or, for nontoxic requirements, mixtures of metal octadecanoates and epoxidized oil	Prevents decomposition of polymer. Without a stabilizer, poly(chloroethene), for example, decomposes on heating to give a brittle product and hydrogen chloride. Some plastics become colored (yellowing) when exposed to long periods of sunlight
Extender	Chlorinated hydrocarbons	Extends the effect of the plasticizer, but generally cannot act as plasticizers alone. They are cheaper than plasticizers, so help reduce costs
Fillers	Chalk, glass fibers	Special design requirements, or cheaper additives
Miscellaneous	Flame retardants, UV stabilizers, antistatics, processing aids, pigments	Impart specially required properties to the plastic for the manufacturing process or for end-use

Source: Data from The Essential Chemical Industry – Online (2013). Polymers-an overview. http://www.essentialchemicalindustry.org/polymers/polymers-an-overview.html (accessed 25 November 2017).

Types of Surfactants

The original surfactant is soap made from animal fats and oils and alkalis. Soap has been largely replaced in many applications by synthetic organic surfactants, because soaps can form insoluble calcium and magnesium salts with calcium and magnesium ions in hard water and insoluble scum with suspended clays or dirt in water. Commercial formulations known as detergents contain surfactants, plus a variety of other ingredients to assist the cleaning process, for example builders, bleaches, and so on (Connell and Miller 1984, pp. 278–279; Connell 2005, pp. 209–226).

Surfactants consist of many different chemical structures but are classified according to the composition of their hydrophilic group (head). There are four major groups:

- *Anionic*–contains a head that carries a negative charge. These include (i) common alkyl carboxylates, soaps such as sodium stearate, and perflurosurfactants such as PFOA and PFOS, and (ii) sulfate, sulfonate, phosphate, and alcohol esters. Sodium lauryl or dodecyl sulfate is a major anionic surfactant. Other common types of anionic surfactants are alkylbenzene sulfonates and alkyl ether sulfates.
- *Cationic*–contains a head that carries a positive charge. Quaternary ammonium salts, known as *quats*, are common types. The quaternary ammonium cation is permanently charged, independent of the pH of their solution, and has attached alkyl or aryl groups. They are used as disinfectants, fabric softeners, surfactants, and antistatics (e.g. in shampoos).
- *Nonionic*–contains a head with no ionic charge groups. The major group of nonionics is the ethoxylates made by condensing long-chain alcohols with ethylene oxide to form ethers. Their hydrophilic properties result from covalent-bonded oxygen atoms in one part of the molecule, which are capable of forming hydrogen bonds with water molecules. Examples of nonionics include fatty alcohol ethoxylates, alkylphenol ethoxylates, and fatty acid esters of glycerol. There are major environmental concerns about the alkylphenol ethoxylates and their alkylphenol precursors, because of the persistence and potential estrogenic toxicity of breakdown products such as **nonyl and octyl phenols**.
- *Zwitterionic*–contains a head with both positive and negative charge groups attached to the same molecule. These amphoteric surfactants usually have a quaternary ammonium ion or amine as the cation, and phosphate, sulfate or carboxylic ions as the anion. Amphoteric surfactants are pH balanced.

Examples of common types of surfactants are shown in Figure 11.14.

Environmental issues with surfactants, such as anionics, range from the physical and toxic effects of their surfactant properties in waters, soils, and wildlife at parts per million levels, to resistance to degradation of long branched alkyl chains such as in alkyl benzene sulfonates (ABS) and sulfates (Connell and Miller 1984, pp. 279–282; Connell 2005, pp. 209–226), and the persistence, bioaccumulation, and toxicity of certain perflurosurfactants. In particular, PFAS and PFOA are global persistent organic pollutants (POPs) (Stockholm Convention).

Emerging Organic Pollutants (CECs)

The scope and number of emerging pollutants being discovered in the environment are a major challenge for analytical

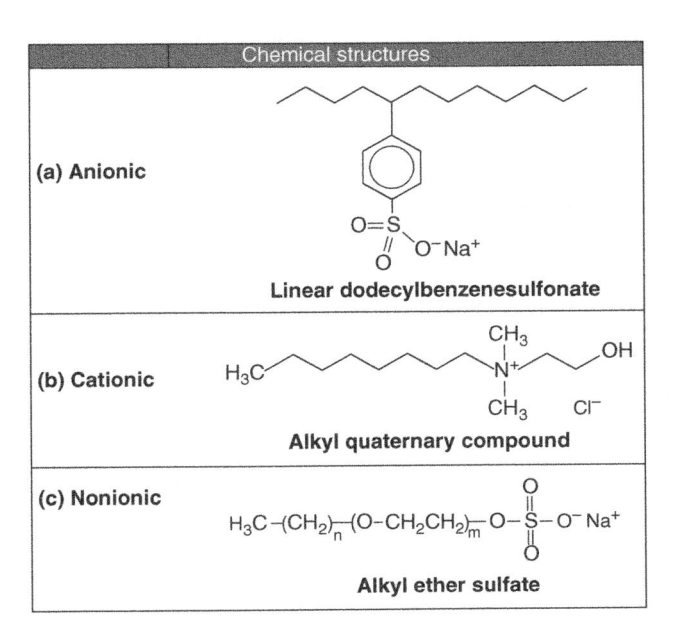

Figure 11.14 Common examples of anionic, cationic, and nonionic surfactants.

methodologies and environmental risk assessment. Over 700 emerging substances (pollutants, metabolites, and degradation products) have been identified in the European aquatic environment and categorized into 20 or more classes, according to their origin, in the NORMAN network (Geissen et al. 2015). Most of these substances are organic pollutants, both new and existing chemicals.

Emerging pollutants in the environment are somewhat complex to detect, monitor, and evaluate their environmental behavior, exposure, and potential effects on living organisms. They include (i) existing contaminants in the environment that have been detected or evaluated by advances in analytical technology and toxicology, assessment or review, and (ii) new sources of emerging contaminants that are released or disposed to the environment. The latter can refer to new or existing chemicals such as the PFAS and PFOA.

The global scale of the problem is highlighted by the presence of parts per billion levels of numerous emerging organic pollutants (e.g. PPCPs) being measured in wastewaters and biosolids from many countries in the last decade or so. The potential magnitude of this contamination is related to the estimated annual volume of treated and untreated wastewaters (2212 km^3), which is produced worldwide from sewage and industrial effluent and agricultural drainage waters (UN WWDR 2017). Over 80% of this wastewater is inadequately treated. This volume is similar to the estimated volume of freshwater in all the rivers of the world.

Furthermore, the extent of global pollution by chemicals includes contamination of the world's oceans with microplastic particles (and adsorbed pollutants), excessive exposures to airborne organic toxics and respirable particles in major urbanized airsheds, POPs, and CECs, including synthetic organic pesticides, increasingly found in soils and sediments of urban, agricultural, rural, world heritage areas, and remote lands, such as in polar regions.

Some emerging types of pollutants are introduced in Chapter 2. Many of these substances are synthetic organic pollutants as indicated in Table 11.6 that occur in chemical mixtures, along with other pollutants, and are widely dispersed with human-related wastes and emissions. They vary from new to re-emerging pollutants, from industrial chemicals to disinfection by-products, personal care products, pesticides, and also pharmaceuticals, that are designed specifically as bioactive chemicals for human and animal use.

Pharmaceuticals and Personal Care Products (PPCPs)

Worldwide, many emerging organic pollutants derived from pharmaceutical and personal care products are found among the mixtures of chemical residues increasingly discovered in our environments (e.g. Halling-Sorensen et al. 1998; Petrie et al. 2015; Ebele et al. 2017). Pharmaceuticals are described as prescription, nonprescription, and veterinary therapeutic drugs used to prevent or treat human and animal diseases. Personal care products are mainly used to improve the quality of daily life (Ebele et al. 2017). These types of products include antibiotics, analgesics, steroids, lipid regulators, antidepressants, hypertension drugs, anti-epileptics, stimulants, antimicrobials, sunscreen agents, fragrances, and cosmetics.

The large consumption of pharmaceuticals by human activities and the disposal of unused products and their wastes, such as through human and animal excretion, landfilling, incineration or reuse of contaminated biosolids and wastewaters on land commonly results in the release of their residues to aqueous and terrestrial environments from point sources (e.g. untreated or treated sewage discharges) or nonpoint sources (e.g. runoff from animal feedlots, land application of biosolids, and landfill leachate). Many scientific studies show that residues of parent compounds, metabolites, or degradation products are widely found in wastewaters, surface waters, and groundwaters at concentration levels of ng L^{-1} to low mg L^{-1} (e.g. Petrie et al. 2015). Additionally, these types of analytical investigations detect various illicit drug residues as co-extractives in wastewaters, sewage effluents, and receiving waters. Examples of various PPCPs commonly detected in wastewaters, surface waters, and groundwaters are listed in Table 11.7.

Table 11.6 Major types of organic chemicals of emerging concern (CECs).

Chemical product category	Examples
Industrial chemicals	
Solvents/intermediates	1,2,3-Trichloropropane, 1,4-dioxane
Flame retardants	Triphosphates, tribromodiphenyl ether, long-chain chloroalkanes (C18–C30), PBDEs
Plasticizers	Bisphenol A, diethyl phthalate
Biocides	Chlorophene, chloroxylenol, 2-mercapto-benzothizole, triphenyl tin, triclosan, formaldehyde, camphor
Surfactants	4-Nonylphenol ethoxylates, nonyl phenol, 4-nonylphenoxy acetic acid, perfluoroalkyl sulfates (e.g. PFOS and PFOA)
Pesticides	Permethrin, glyphosate
Plastics/monomers	Microplastic particles, plastic microbeads, styrene, formaldehyde, vinyl chloride, acrylamide
Pharmaceuticals	
Antibiotics	Amoxicillin, erythromycin, oxytetracyclin, sulfamethoxazole
Drugs (prescription and non-prescription)	Codeine, Ibuprofen, metformin, primidone, paracetamol (acetaminophen), propanolol
Illicit drugs	Methamphetamines (e.g. MDMA) and cocaine (stimulants)
Reproductive hormones	Estrone, progesterone, testosterone, Estradiol
Steroids	Coprostanol, cholesterol
Personal care products (antimicrobials, sunscreen agents, preservatives, and fragrances)	Triclosan, benzaldehyde, ethylhexyl methoxycinnamate, benzophenones, parabens, N,N-dimethytoluamide, butylated hydroxytoluene, galaxolide
Lifestyle products	Caffeine, nicotine, cotinene
Nanomaterials	Carbon nanotubes, graphene oxide, polyurethane resin
Disinfection by-products	Nitrosamines (e.g. N-nitrosodimethylamine), Dibromodichloromethane, iodoform
Algal toxins	Microcystins

Personal care products such as antimicrobials, sunscreen agents, preservatives, and fragrances are also used in high quantities resulting in similar concentrations in wastewater systems and final effluents (ng L^{-1} to µg L^{-1}) to pharmaceuticals (Petrie et al. 2015).

Emerging organic pollutants possess the potential capacity to be biologically active at low exposure doses. For example, pharmaceuticals are designed to act in this way at low levels and to target certain metabolic, enzymatic, or cell signaling mechanisms. Many PPCPs are already identified as endocrine-disrupting chemicals, along with other emerging pollutants. Antibiotic residues may also create antibiotic resistant strains in natural bacteria populations (Petrie et al. 2015).

Environmental Phenols

Alkyl phenols are major industrial chemicals that are precursors to many commercial products such as disinfectants, surfactants, fuel additives, polymers, and phenolic resins. They range from methyl and ethyl phenols (cresols and xylenols) to long-chain alkyl phenols. Other important phenols include chlorinated phenols (mono-, dichloro-, trichloro- and tetrachlor- and penta-) making up a number of different types, used commonly as pesticides and disinfectants. Examples of alkyl phenols and an ethoxylate are shown in Figure 11.15.

A number of these alkyl phenols have emerged as ECCs, mainly because of their estrogen-mimicking and hormone-like properties:

- Nonyl (and octyl) phenols used in the manufacture of emulsifiers, antioxidants, and nonionic surfactants and alkyl phenol ethoxylates.
- Bisphenol A used to make protective coatings and as a precursor for the production of the hard plastic, polycarbonate.
- Parabens are preservatives used in personal care products.
- Triclosan is an antimicrobial widely used in soaps and other personal care products.

Table 11.7 Some pharmaceutical and personal care products detected in various wastewaters, surface waters, and groundwaters.

Pharmaceutical	Therapeutic use	Pharmaceutical	Therapeutic use
Estrone	Steroid estrogen	Furosemide	Diuretic
17ß-Estradiol	Steroid estrogen	Carbamazepine	Antiepileptic
17a-Ethinylestradiol	Steroid estrogen	Gabapentin	Antiepileptic
Tamoxifen	Estrogen antagonist	Simvastatin	Lipid regulator
		Gemfibrozil	Lipid regulator
Acetaminophen (paracetamol)	NSAID	Amoxicillin	Antibiotic
Diclofenac	NSAID	Ciprofloxacin	Antibiotic
Ibuprofen	NSAID	Erythromycin	Antibiotic
Naproxen	NSAID	Ofloxacin	Antibiotic
Ketoprofen	NSAID	Oxytetracycline	Antibiotic
Tramadol	Analgesic	Sulfamethoxazole	Antibiotic
Oxymorphone	Analgesic	Diazepam	Hypnotic
Morphine	Analgesic	Sulfasalazine	Chronic bowel disorders
Venlafaxine	Antidepressant	Tamoxifen	Anticancer
Dosulepin	Antidepressant	Metformin	Antidiabetic
Amitriptyline	Antidepressant	MMDA	Hallucinogen
Sotalol	Beta-blocker	MDEA	Hallucinogen
Propranolol	Beta-blocker	Amphetamine	Stimulant
Metoprolol	Beta blocker	Methamphetamine	Stimulant
Valsartan	Hypertension	Cocaine	Stimulant
Diltiazem	Calcium-channel blocker	Benzylpiperazine	Stimulant
		Caffeine	Stimulant
Personal care products	**Chemical family/use**	**Personal care products**	**Chemical family/use**
Triclosan	Antibacterial/biocide	Methylparaben	Preservative
Bisphenol A	Plasticizer	Ethylparaben	Preservative
Benzophenones	Sunscreen agent (UV filter)	Propylparaben	Preservative
Ethylhexyl methoxycinnamate	Sunscreen agent (UV filter)	Musk galaxolide,	Fragrance
DEET	Insect repellent		

NSAID, Nonsteroidal anti-inflammatory drug.
Source: Based on Halling-Sorensen, B., Nielson, S.N., Lanzky, P.F. et al. (1998). Occurrence, fate and effects of pharmaceutical substances in the environment - a review. Chemosphere 36 (2): 357-93; Petrie, B., Barden, R., Kasprzyk-Horden, B. (2015). A review on emerging contaminants in wastewaters and the environment: Current knowledge, understudied areas and recommendations for future monitoring. Water Research 72: 3–27; Ebele, A.J., Abdallah, M.A., Harrad, S. (2017). Pharmaceuticals and personal care products (PPCPs) in the freshwater aquatic environment. Emerging Contaminants 3 (1): 1–16.

Chlorophenols and derivatives are commonly found in the environment through use in certain pesticides, disinfectants, and pharmaceuticals, as by-products, and from industrial wastes. They are persistent and difficult to biodegrade, and consequently persistent in the environment. Common chlorophenols such as pentachlorophenol (PCP), 2,3,4,6-tetrachlorophenol (2,3,4,6-TeCP), 2,4,6-trichlorophenol (2,4,6-TCP), 2,4,5-trichlorophenol (2,4,5-TCP), and 2,4-dichlorophenol (2,4-DCP) are highly toxic, mutagenic, and carcinogenic for living organisms (Olaniran and Igbinosa 2011; Igbinosa et al. 2013).

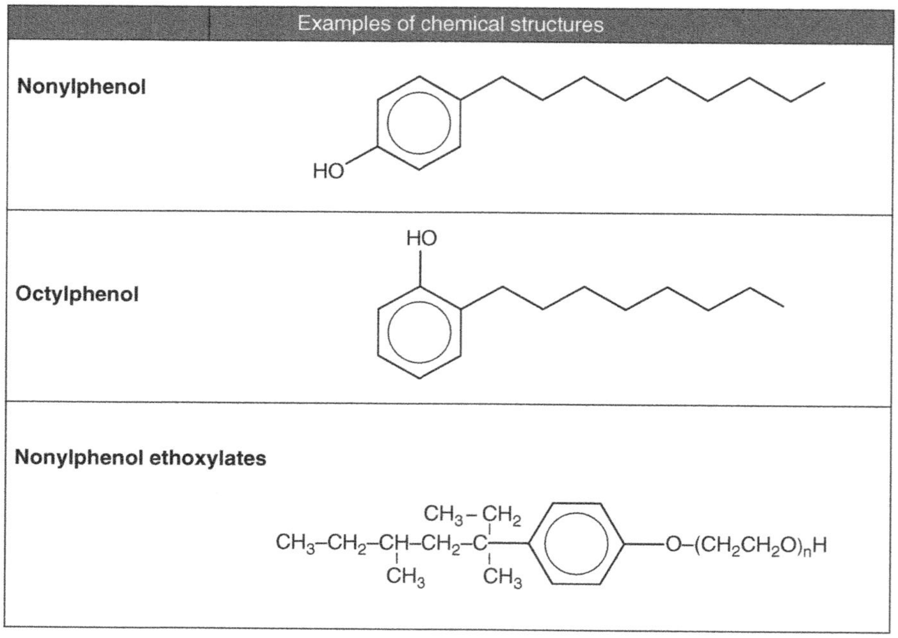

Figure 11.15 Examples of chemical structures of alkyl phenols and an ethoxylate.

11.7 Sources and Emissions of Toxic Organic Chemicals

Many thousands of organic contaminants are generated and released into the environment through human activities via multiple pathways as indicated in Figure 11.16 for PPCPs. Many organic chemicals are of environmental concern and can be toxic to wildlife, domestic animals, and humans. The synthesis of organic chemicals for human use originates primarily from basic chemical feedstocks refined from crude petroleum fuels. Major downstream chemical products produced for human uses include common industrial chemicals, such as halogenated organic compounds, plastics, pesticides, surfactants, and pharmaceuticals listed in Table 11.8.

Annual quantities produced range from thousands to tens of millions of tonnes and up to about 350 million tons for plastics. Toxic organic chemicals include millions of tonnes of synthetic pesticides and halogenated organics such as vinyl chloride and PCBs, brominated flame retardants, solvents, and surfactants such as PFOS. For example, the monomer vinyl chloride is a very toxic chemical and a human carcinogen with a large yearly production of about 47 million tons in 2010, essentially for production of PVC plastics. Persistent organic pollutants and emerging organic chemicals of concern found widely in the environment, such as pharmaceutical and personal care chemicals and microplastics, are now the focus of numerous scientific studies and monitoring programs.

As discussed in Chapter 2, the total quantity of organic (and inorganic) chemicals released to the global environment is simply unknown. Many millions of tonnes of potential organic pollutants (e.g. petroleum, industrial chemicals, plastics, and pesticides) are widely dispersed each year into air, deposited onto soils, lost to water bodies, and absorbed by biomass, mainly nontarget organisms. The proportion of hazardous organic chemicals is unavailable.

In North America (USA, Canada and Mexico), pollutant release and transfer register (PRTR) data reported in 2009 that 4.9 million tons of chemicals were released or disposed. Hazardous chemicals consisted of persistent, bioaccumulative and toxic (PBT) chemicals (almost 1.5 million tons). In 2015, the US Toxic Release Inventory (TRI) reported that industry disposed or released 3.36 billion pounds of TRI chemicals to on-site land, into the air, water, or by off-site transfers to some type of land disposal unit. Most of these chemical releases were toxic metals, nitrates, and ammonia. The top 10 organics were primarily volatile organic compounds or solvents (methanol, n-hexane, styrene, toluene, formaldehyde, xylenes, ethylene, formic acid, certain glycol ethers, and acetonitrile) that would readily redistribute to air and dissolve at low levels in waters.

Hazardous wastes, containing many types of toxic organic chemicals (e.g. PCBs, brominated flame retardants, dioxins, and phthalates) are generated in large quantities by many countries based on available data, although this is likely to be greatly underreported. Data from UNEP (2013) indicate that 64 countries generated over 250 million tons of hazardous wastes, as reported in the years 2004, 2005, or 2006, under the Basel Convention for transboundary movement of wastes. Electronic waste generation was reported to be 40 million tons in 2009 (UNEP 2013, p. 42). The proportion of hazardous

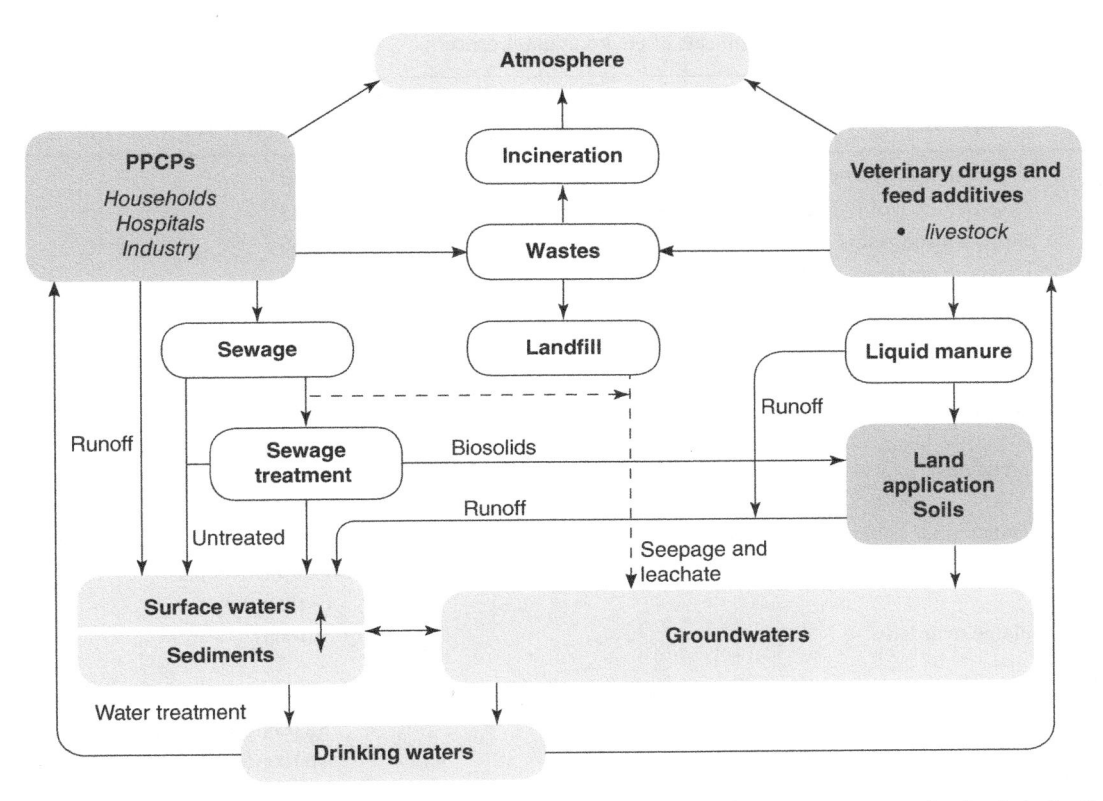

Figure 11.16 Generalized environmental pathways for release of PPCPs from human uses and animal applications.

organic chemicals in these various wastes are generally unavailable.

PPCPs and Veterinary Chemicals

Numerous sources release PPCPs from human uses and animal applications into the ecosphere via multiple environmental pathways: excretion, washings, waste disposal (e.g. manure and biosolids), discharge, runoff, infiltration, and leachate into urban, agricultural, aquatic, and terrestrial environments. Manufacturing, use, and waste disposal of pharmaceuticals and personal care products by individuals, in hospitals, community health, and for veterinary use in domestic animals, livestock, animal feed additives and aquaculture result in residues of PPCPs and their transformation products being found worldwide in surface waters, sediments, groundwater, drinking water, manure, soils, biosolids from digested sewage solids and related composts, and other environmental sources. There are now many scientific studies and reviews of the diverse types and levels of PPCPs detected in these sources and the environment.

A comprehensive literature review of human and veterinary pharmaceuticals in the global environment by aus der Beek and co-researchers identified 631 different pharmaceutical substances in environmental samples in 71 countries covering all continents. Municipal wastewaters are considered to be the main environmental pathway on a global scale although wastewater treatment removes a variable percentage of PPCP residues. Other significant sources from industrial production, hospitals, agriculture (livestock), and aquaculture are important locally (aus der Beek et al. 2016).

Environmental pathways for the distribution of residues of PPCPs and other organic pollutants are illustrated in Figure 11.16. Major sources of residues from human pharmaceutics are their excretion to sewage or wastewaters from households, hospitals, and manufacturers, and disposal of wastes to sewer, landfill, or by incineration. Dozens of pharmaceutical substances, among other organic contaminants, are commonly detected in municipal wastewaters (raw and treated), effluents, and receiving waters. Veterinary drugs and feed additives are also excreted in manure to soils, and wastes are disposed to landfill or by incineration. Urban and agricultural runoff and infiltration also transport residues to surface waters or groundwaters. Case Study 11.1 presents an example of the discharge of PPCPs from pollution sources in an Asian megacity (Beijing, China). Emerging environmental concerns, aside from ecotoxicity effects, include human health risks (e.g. children) from residues detected in drinking water sources (surface water and groundwaters), and dietary sources such as freshwater and coastal fisheries, and increasing antibiotic resistance from bacterial populations.

Table 11.8 Global production and uses of some organic chemicals of environmental concern.

Chemical	Principal uses	Global production/use (millions of metric tons year^{-1})
Petroleum	Production of hydrocarbon fuels, basic feedstocks, and petrochemicals	4362
Pesticides	Insecticides, herbicides, and fungicides	4.113[a]
Plastics (2017)	Polymers and resins	~350
Surfactants (2017)	Cleaning products, disinfectants, pesticides	16.8[b]
Fluoropolymers	Teflon, perfluorinated compounds	(100 000–180 000)
Vinyl chloride monomer (VCM)	Monomer for PVC polymer	47 (2010, Dow chemical)
Trichloroethylene and perchloroethylene	Solvents	(>200 000)
DDT, lindane, endosulfan	Pesticides	(12 800) (Endosulfan)
PCBs (obsolete)	Transformer and capacitor fluids	1–1.6[c] Cumulative total (~1930–1993)
Polybrominated diphenyl ethers (PBDEs)	Flame retardants	(50 000)
Fluorocarbons (CFCs)	Propellants, solvents, refrigerants (restricted use)	>1
Antibiotics (2010)	Antibacterial	≥63 000 tons (livestock) ~60 000 tons (humans)[d]

() tonnes.
[a]FAO (2019).
[b]HIS Markit (2019) surfactants.
[c]UNEP (2016).
[d]Crude estimate.

Case Study 11.1

Discharges of Pharmaceutical and Personal Care Products (PPCPs) from Pollution Sources in Beijing, China

The unseen problem of mass loadings of antibiotic residues and other PPCP residues in urban environments is highlighted in this case study by Zhang and co-researchers (Zhang et al. 2016).

Estimates of total annual discharges of PPCPs from pollution sources in Beijing, a megacity, are presented in Table 11.9, as summed from data in Table S1 in Zhang et al. (2016). There were 119 PPCPs from 11 categories investigated in the study of the various pollution sources listed in Table 11.9. Results from this study indicated:

- Most types of PPCP residues were found in urban sewage and untreated wastewaters. The discharge load in untreated wastewater was almost two times higher than in WWTPs.
- Antibiotics were the main types of PPCPs discharged in liquid and solid phases, with the top three being over 10 000 kg (10 tons) year^{-1}.
- Livestock farms were a serious source of the antibiotic pollution.
- Chemicals in the top 10 emission loads in order were enrofloxacin, tetracycline, and norfloxacine at over 10 000 kg (10 tons) year^{-1} for each, followed by galaxolide or HHCB (fragrance musk), ofloxacin, sulfadiazine, tonalide (fragrance musk), naproxen (nonsteroidal anti-inflammatory drug), sulfanilamide, and difloxacin.
- Discharges of caffeine, and the personal care products: galaxolide and AHTN (fragrances), and triclocarban (antibacterial agent), were between 1000 and 10 000 kg. PCPs were more prevalent in discharges from WWTPs.
- Estimated annual waste solids amounted to over 6 million tons (648.3×10^4 tons year^{-1}).

Table 11.9 Discharge inventory of total pharmaceutical and personal care products from pollution sources in Beijing, China.

Sources of PPCP emissions Beijing	Discharge loads of pollution sources	Total pharms emissions (kg year^{-1})	Total PCPs emissions (kg year^{-1})
WWTPs influent	12.64×10^8 (m^3 y^{-1})	13 006	4722
WWTPs effluent	12.64×10^8 (m^3 y^{-1})	4687	1206
Untreated urban wastewater	NA	8497	3084
Sludge	114×10^4 (tons year^{-1})	2341	3435
Hospital wastewater	1470.6×10^4 (m^3 year^{-1})	328	NA
Livestock: dry	612.3×10^4	134 640	NA
Wet dung	1874.8×10^4 (tons year^{-1})		
Aquaculture wastewater	9240×10^4 (m^3 year^{-1})	1926	NA
Landfill leachate	NA	15	NA

WWTPs, wastewater treatment plants.
NA, no data available; Notes on methods for each source (see Zhang et al. 2016).
HHCB (Galaxolide) and AHTN [1-(5,6,7,8-tetrahydro-3,5,5,6,8,8-hexamethyl-2-naphthyl) ethan-1-one].
Source: Based on Zhang, Z., Wang, B., Yuan, H. et al. (2016). Discharge inventory of pharmaceuticals and personal care products in Beijing, China. Emerging Contaminants 2: 148–156.

11.8 Behavior and Fate in the Environment

Properties and Processes

The behavior and fate of known, suspect, and new toxic organic chemicals in air, waters, sediments, soils, and biota can be evaluated from a knowledge of their physical, chemical, and environmental properties, and interactions with abiotic and biotic processes. Key physical and chemical properties can be described as their chemical forms and functional groups, molecular mass, particle sizes, melting point, boiling point, vapor pressure, water solubility, octanol/water partition coefficient, and dissociation constant(s). For detailed information, Table S11.3 contains a set of physicochemical properties for important types of TOCs discussed in this chapter.

Current techniques include the use of experimental data and software screening-tools, such as the US-EPA EPI Suite™, to estimate physical/chemical and environmental fate properties, including removal in wastewater treatment and use of a multi-media (fugacity) model (see also Chapter 4 and Case Study 11.2 in this chapter).

Important environmental properties and rate processes for transport and removal are discussed in Chapter 4. These include Henry's Law constant, soil adsorption coefficient (K_{OC}), bioconcentration factor, and processes such as diffusion, adsorption, volatilization, oxidation, photolysis, hydrolysis, biodegradation, and chemical decomposition. Persistence in environmental compartments or media are indicated by half-lives ($t_{1/2}$).

Figure 11.17 depicts the behavior and fate of organic chemicals in the environment.

Atmosphere

The atmospheric movement of pollutants such as TOCs results from transport, dispersion, diffusion, and deposition processes. Air pollutants are generally transported by downwind movement and wind currents at different atmospheric scales, including long-range transport of persistent organic pollutants such as some POPs discussed in Chapter 4. Turbulent fluctuations in velocity cause dispersion of pollutants, and diffusion occurs due to concentration gradients of pollutants. At the same time, deposition processes, including wet and dry precipitation, scavenging, and sedimentation, remove pollutants from the atmosphere to the Earth's surface. Residence time of TOCs is strongly influenced by their physical and chemical properties and interactions with prevailing atmospheric conditions.

Emissions of volatile or semi-volatile chemicals to air will exist solely in the vapor phase, or in vapor and particulate phases in the atmosphere depending largely on vapor pressure. Particulate emissions (e.g. aerosols and dry dusts) from manufacturing, spraying applications, and waste disposal (e.g. low temperature combustion and stockpiles) are other airborne sources from urban, industrial, and

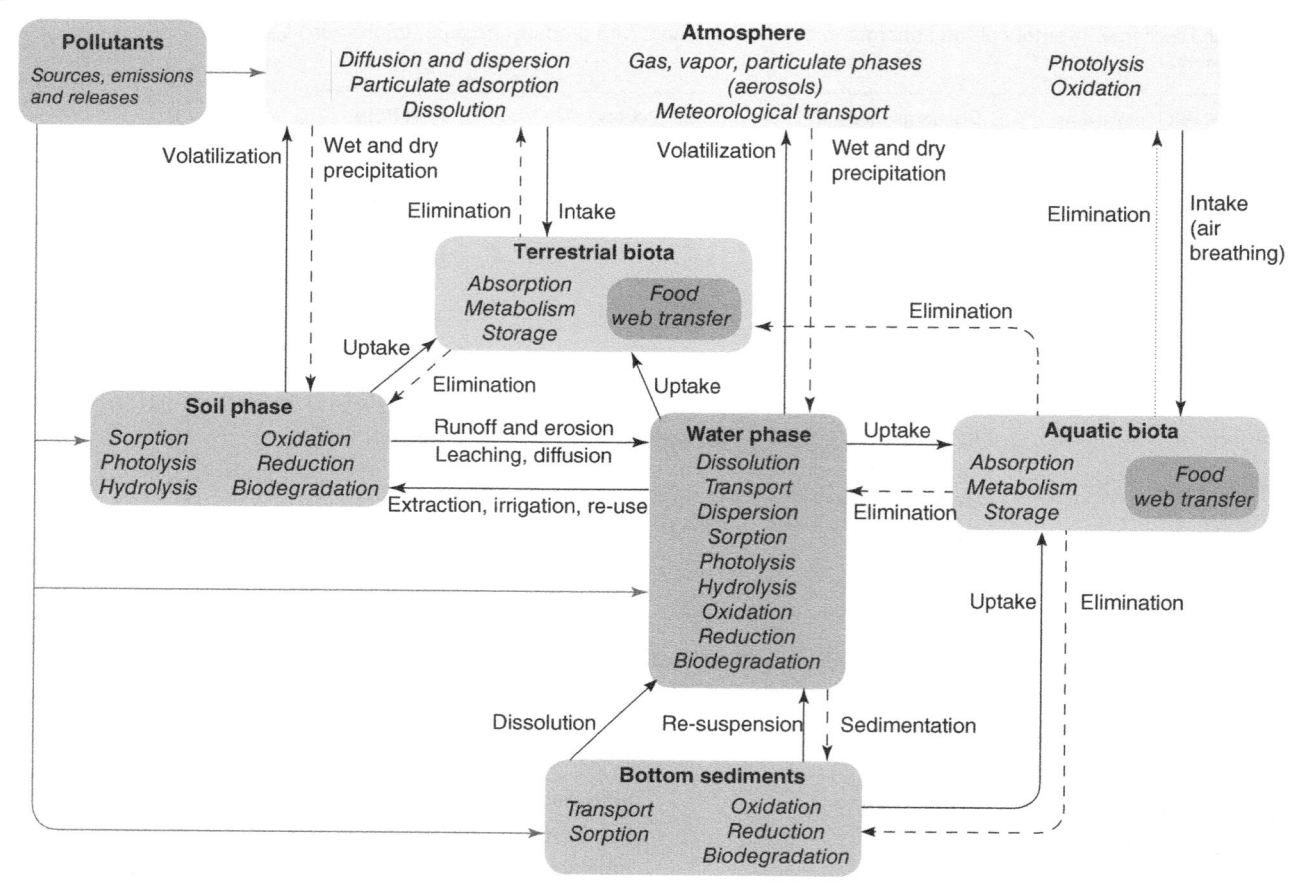

Figure 11.17 Shows a conceptual model of the dynamic environmental processes acting on organic pollutants (TOCs) that are emitted or released due to human activities into the atmosphere, terrestrial, and aquatic environments.

agricultural areas. Low-boiling petro-solvents, such as benzenes and halogenated hydrocarbons, occur in the vapor phase because of their relatively high vapor pressures, e.g. toluene (28.4 mm Hg at 25 °C) and tetrachloroethylene (18.5 mm Hg at 25 °C). Other examples with higher boiling points are the monomer, styrene, the plasticizer, diethyl phthalate, and the surfactants, PFOS and PFOA, as shown in Table S11.3.

Semi-volatile chemicals with lower vapor pressures $<10^{-4}$ mm Hg at 25 °C (e.g. DDT, TCDD, PCBs, nonyl phenol, bisphenol A, and triclosan) usually occur in both the vapor and particulate phases (see Table S11.3). The dioxin, TCDD, with a vapor pressure of 1.5×10^{-9} mm Hg at 25 °C, exists solely in the particulate phase in the air. Pharmaceuticals and some personal care products with very low vapor pressures are also expected to exist in the particulate phase. Fragrances exist to variable degrees in the vapor state. Galaxolide exists in air in the vapor and particulate phases but is degraded in the vapor phase by photochemical reaction with hydroxyl radicals (estimated half-life, 10 hours). PBDEs vary from vapor phase only to vapor-particulate phases and particulate phase only. Sunlight can degrade some PBDEs.

Vapor-phases (e.g. toluene, tetraethylene, styrene, bisphenol A) tend to be readily degraded by reaction with hydroxyl radicals that are photochemically produced, ozone, or to some extent by photolysis from direct sunlight. For example, vapor-phase styrene is degraded essentially by hydroxyl radicals and ozone, with estimated half-lives for these reactions of 7 and 16 hours. In contrast, vapor-phase DDT, PFOS, and PFOA have estimated half-lives of 4.7, 115, and 31 days for photochemical degradation with hydroxyl radicals. Particulate phases are removed from the atmosphere by wet or dry deposition.

Terrestrial Environment

The environmental fate of toxic organic chemicals in soils depends largely on their mobility in soils, capacity to volatilize from soil surfaces (wet and dry) and persistence in soils, as measured by biodegradation rates in aerobic and anaerobic soils.

Mobility is related to the adsorption property of a chemical on soil particles. The soil adsorption coefficient (K_{oc})

provides a measure of the ability of a chemical to adsorb to the organic portion of soil, sediment, and sludge. K_{OC} is calculated using the formula:

$$K_{OC} = K_D/(\% \text{ organic carbon or OC} \times 100) \text{ L kg}^{-1} \text{ soil} \quad (11.1)$$

Where K_D is the partitioning coefficient between the solid phase (soil or sediment) and the solution phase (water) at equilibrium. It is normalized by multiplying the % organic carbon content of the soil by 100 (US EPA 2019).

Chemicals with a high K_{OC} value (tightly bound to soil) are practically immobile via sorption to soil or sediment particles and show negligible migration to groundwater. Low K_{OC} values are readily mobile, can leach into soil, and may contaminate groundwater. K_{OC} classification criteria are tabulated in Table S11.4.

The mobility of TOCs in soils can be usefully predicted based on their K_{OC} values, as given for a number of organic chemical compounds in Table 11.10. For example, the K_{OC} value for the dioxin, TCDD, is 2.45×10^7 L kg^{-1}, so its log K_{OC} = 7.39, much greater than 4.5. TCDD is expected to be highly adsorbed in soil and practically immobile. From Table S11.4 criteria, DDT is also very strongly adsorbed and immobile as indicated by log K_{OC} values >5 (as estimated from Table 11.10). Many organic pollutants, including POPs such as PFOS and PBDEs, show moderate to low or negligible mobility. Nonyl phenol is expected to be immobile in most soils. Common solvents tend to have moderate to high mobility, as indicated also by their presence in groundwater contamination (e.g. benzenes and tetrachloroethylene). PFOA is also expected to have moderate to high mobility based upon its log K_{OC} values of 1.92–2.59. Its pK_a value of −0.5 to 4.2 indicates it will exist in the anionic form in the environment and will not strongly adsorb to soils containing organic carbon or clays compared to neutral chemicals. Similarly, other organic chemicals (e.g. PFOS, bisphenol A, naproxen, and triclosan) that form anions or exist partially as anions, generally do not sorb strongly to soils.

Among PPCPs found widely in the environment, sorption varies greatly. For example, high mobility is expected

Table 11.10 Soil properties relating to environmental mobility and biodegradation of organic pollutants in soils.

Organic pollutant PBTs/POPs	Soil adsorption coefficient (K_{OC}) (L kg^{-1})	Soil biodegradation half-lives ($t_{1/2}$)[a]
DDT	1.13–3.5 × 10^5	2–>15 years (aerobic)
		<1–7 days (anaerobic)
TCDD	Av 2.45 × 10^7	Slow
PBDEs	3100–276 000	>93 days[b]
PFOS	Av 1000 (250–50 100)	Resists biodegradation
PFOA	83–389	Resists biodegradation
Industrial TOCs		
Toluene	37–178	Hours – 71 days
Tetrachloroethylene	200–237	Slow
Styrene	960	16 weeks
Diethyl phthalate	69–1726	1.83/5 days (aerobic/anaerobic)
Bisphenol A	251–3886	Variable (readily degraded to nondegradable)
4-Nonylphenol	10 000–50 000	2.1–50 days
Pharmaceuticals		
Ofloxacin	44 143	Resists biodegradation
Naproxen	330	Resists biodegradation
Metformin	12–19	1–5 days (aerobic)
PCPs		
Triclosan	3550–20 000	17 days (loamy soils)
Galaxolide	20 000	4 months

Compiled from PubChem database for individual chemicals. HSDB data integrated into PubChem, National Institutes for Health.
[a]Soil biodegradation values are indicative: depend on soils, environmental, and experimental conditions.
[b]May also apply to biosolids and organic sludges.

for metformin, moderate for naproxen, and low to negligible mobility for ofloxacin (antibiotic), triclosan (biocide), and galaxolide (musk fragrant).

Major processes for removal of organic chemicals from the soil compartment involve volatilization and biodegradation. Volatilization from moist or wet soil surfaces (and surface waters) is based on Henry's Law constant. For persistent organic chemicals such as DDT, TCDD, PBDEs, and galaxolide, volatilization is expected to be an important fate process, given Henry's Law constants (Table S11.3) in the order of 10^{-6} atm-m^3 mol^{-1} (see Table 11.11). However, high adsorption to soils is expected to attenuate volatilization for DDT and other chemicals with high K_{OC} values. Volatile to moderately volatile chemicals (e.g. toluene, tetrachloroethylene, and styrene) have Henry's Law constant values $>10^{-5}$. In contrast, volatilization of bisphenol A from moist soils is not expected to be significant because of its Henry's Law constant of 4.0×10^{-11} atm-m^3 mol^{-1} and pK_a of 9.6. Chemicals that form anions do not volatilize from soils. In dry soils, volatilization is based on vapor pressure of the organic chemical (and ambient temperatures). TCDD, for instance, is not expected to volatilize from dry soil surfaces because of its vapor pressure but is expected to volatilize from moist soil surfaces, although attenuated by its high adsorption to soil.

Environmental persistence of organic chemicals in soils usually depends on biodegradation by microorganisms although volatilization, photolysis, and hydrolysis can be other important processes. Biodegradation of organic chemicals may be measured in situ (e.g. wastewater treatment), experimentally (e.g. OECD test protocols) or predicted (e.g. US EPA EPI Suite). It can vary from primary biodegradation (change in molecular structure to form new compound or degradation product) to ultimate biodegradation (complete mineralization of molecule). The probable persistence of a chemical in soil, under aerobic or anaerobic conditions, can be usefully indicated by rate of biodegradation, usually as a half-life, taking into account the importance of other removal processes. Biodegradation half-lives for some typical toxic organic chemicals in soils may vary from hours to days, weeks, months, and years. In many cases, field and experimental results for persistent chemicals are difficult to determine over long periods and, generally for TOCs, vary with soil type and prevailing environment or experimental conditions.

Environmental persistence of POPs in soils and water is indicated by half-lives of >60 days (waters), >120 days (sediments and soils), and >2 days (air). Reported half-lives for DDT in aerobic soils range from 2 to >15 years but DDT can have very short half-lives in anaerobic sediments and sludge. PFOS is reported to be resistant to biodegradation under typical environmental conditions. Biodegradation does not appear to be an important fate process for PFOA in soils or water. TCDD is regarded as recalcitrant to biodegradation but was photo-degraded in soil application tests. Less persistent chemicals may be biodegraded in hours to days or persist for weeks to months based on their half-lives for biodegradation in soils (see also Table 11.10).

Some PPCPs (e.g. naproxin and galaxolide) are known to be persistent in waters, biosolids, and soils, as opposed to pseudo-persistent because of continual discharge in wastewaters or from biosolids and sludge applications to soils. Galaxolide is reported to have a biodegradation half-life of 4 months. In contrast, water-soluble metformin demonstrated a half-life of 1–5 days when incubated in soil under aerobic conditions.

Aquatic Environment

Environmental partitioning, distribution, and fate of toxic organic chemicals in waters and sediments can be evaluated from a knowledge of key physical and chemical properties (water solubility, vapor pressure, and octanol–water partition coefficient), environmental fate properties related to environmental transport (Henry's Law constant, soil adsorption coefficient, and bioconcentration factor), and persistence (biodegradation, hydrolysis, chemical oxidation, and volatilization rates from surface waters). Removal of TOCs in wastewater treatment processes such as sewage is increasingly relevant for emerging chemicals of concern that are waterborne (e.g. PPCPs).

Environmental partitioning of organic chemicals or pollutants between water and other environmental phases and their removal from water or sediment are readily indicated by key physical, chemical, and environmental properties and processes as shown in Table 11.12.

Persistent organic pollutants such as DDT, TCDD, PCBs, PFOS, and PBDEs remain in water for >60 days and sediments for >120 days. They tend to adsorb strongly to suspended solids and sediments based on K_{OC} values

Table 11.11 Criteria for volatilization of organic chemicals from water based on Henry's law constant.

Henry's law constant (atm-m^3 mol^{-1})	Classification
>10^{-1}	Very volatile from water
10^{-1}–10^{-3}	Volatile from water
10^{-3}–10^{-5}	Moderately volatile from water
10^{-5}–10^{-7}	Slightly volatile from water
<10^{-7}	Nonvolatile

Source: Based on US EPA (2012). Sustainable Futures / P2 Framework Manual. 5. Estimating Physical/Chemical and Environmental Fate Properties with EPI SuiteTM. EPA-748-B12-001. United States Environmental Protection Agency. https://www.epa.gov/sustainable-futures/sustainable-futures-p2-framework-manual (accessed 18 April 2020).

Table 11.12 Environmental partitioning of organic chemicals between water and other phases.

Process	Property	Environmental phases
Dissolution	Water solubility	Water and solid phases
Volatilization	Henry's law constant	Air–water surface
Adsorption	Soil adsorption coefficient	Water-suspended solids/sediment
Sedimentation	Precipitation, deposition, or settling and resuspension	Suspended sediment-sediments
Hydrolysis	pH, hydronium, and hydroxide ions	Water phase
Biodegradation	Microbial activity, dissolved oxygen	Waters and sediments
Bioconcentration	Bioconcentration factor	Water-biota

and persist in the aquatic environment. The strength of sorption for PFOA is influenced by environmental factors. Water-soluble polar chemicals with low K_{oc} values such as the drug metformin and antibiotic, sulfanilamide, are not partitioned from water to solid phases. Removal from water by volatilization can be an important process for many organic chemicals including POPs as indicated by Henry's law constants but is subject to attenuation by strong adsorption processes on suspended solids and sediments. Benzene derivatives (e.g. toluene and styrene) and halogenated hydrocarbons are relatively volatile from water surfaces with estimated half-lives from hours to days. Volatilization is not expected to be an important fate process for chemicals such as PFOS and naproxen based upon their pK_a values. Hydrolysis is not expected to be important in chemicals that lack functional groups that do not hydrolyze under environmental conditions. DDT undergoes base-catalyzed hydrolysis to toxic and persistent DDE. Under acidic conditions, it has a reported half-life of 12 years. Organic fluorochemical compounds are also expected to resist hydrolysis.

POPs and other persistent chemicals have slow to recalcitrant biodegradation rates in waters. PBDEs, for instance, are not expected to biodegrade in water (or sediments). Biodegradation products from various POPs may also be toxic and persistent. Half-lives for biodegradation rates of tetrachloroethylene in aerobic and anaerobic waters were reported as 180 and 98 days, respectively, while half-lives for toluene were 4 days (aerobic water) and 56 days (anaerobic water). Biodegradation is not an important process for a number of PPCPs (e.g. galaxolide, triclosan, naproxen, and sulfanilamide) found in treated municipal effluents, in rivers, lakes, and coastal waters receiving urban discharges and agricultural drainage (see PubChem website).

The slow to recalcitrant degradation of synthetic polymers is a major environmental problem for waste disposal in landfills, aquatic environments, and plastics recycling. Polymer degradation depends largely on its structure. Environmental factors that affect degradation of polymers at variable rates include thermal degradation (organic polymers) and UV degradation (e.g. epoxies and polymer chains containing reactive aromatics). Exposure to ozone degrades polymers with unsaturated carbon bonds while hydrolysis is known to degrade polyesters.

Plastic pollution of the world's oceans has resulted from the physical breakdown of plastic litter and widespread dispersion of microplastic particles, as discussed in Chapter 16.

As mentioned earlier in this section, the potential environmental fate of organic chemicals can be evaluated by using a Fugacity Model approach discussed in Chapter 4. An example is given in Case Study 11.2.

Case Study 11.2

Predicted Environmental Fate of Organic Chemicals (Nonyl Phenol and PFOA) Using a US EPA Multi-Media Fugacity Model (Screening Tool)

The environmental fate of an organic chemical may be usefully predicted by the use of a multimedia fugacity model, such as Level III included in the EPI Suite used by the US EPA. Examples derived from model outputs for PFOA and nonyl phenol are given in Table 11.13.

Results in Table 11.13 indicate that both chemicals are partitioned mostly into the soil and sediment compartments of the environmental model, but more nonyl phenol is also distributed into the water phase compared to PFOA. The persistence of PFOA in air, water, soil, and sediments is much greater than in the case of nonyl phenol and suggests it is a potential persistent organic pollutant (POP).

Compare results in Table 11.13 with persistence criteria for PBTs and POPs in Table 11.3. How would you classify the two chemicals?

Table 11.13 Environmental fate and persistence output for 4-nonyl phenol and PFOA using level III fugacity model.

Compartments	4-Nonyl phenol		PFOA		Emissions (kg hr^{-1})
	Mass amount (%)	Half-life (hr)	Mass amount (%)	Half-life (hr)	
Air	0.286	4.97	0.581	494	1000
Water	9.47	360	1.15	4320	1000
Soil	42.5	720	45.9	8640	1000
Sediment	47.8	3,240	52.4	38 900	0
	Persistence time: 816 hr		Persistence time: 7340 hr		

Source: Data obtained from Chemspider database, The Royal Society of Chemistry, for nonyl phenol and PFOA.

11.9 Environmental Exposures to Toxic Organic Chemicals

Environmental releases of large quantities of mostly petrochemicals derived from petroleum and their derivatives (e.g. plastics, surfactants, and pharmaceuticals) have resulted in local, regional, and global exposures of human populations and wildlife to many of these substances at acute and chronically toxic levels. Persistent, bioaccumulative, and toxic chemicals have widely contaminated the abiotic environment, natural biota, ecosystem food webs, foodstuffs, and humans to an extent that includes the polar regions of the Earth. Less-persistent toxic chemicals can also cause local and regional areas of elevated contamination from industrial, urban, agricultural, and waste disposal sources.

Major sources of TOC exposures for humans and wildlife are diverse and originate from point sources (e.g. effluent or stack discharges) and nonpoint sources (e.g. urban stormwater runoff, agricultural drainage, urban and indoor emissions, and long-range transport via wind and ocean currents). The environmental fate of TOCs is variable, depending upon their diverse physical and chemical properties. Some are persistent and will concentrate in soils, sediments, or fatty tissues while others are more soluble in water and tend to be more readily degraded. As indicated by endocrine-disrupting chemicals, levels in humans and wildlife are likely to be related to how much of the chemical is used (see WHO/UNEP 2013, p. 237).

Environmental Levels

A major challenge for environmental health is arising from the rapidly growing evidence of exposures to complex mixtures of emerging organic chemicals of concern (ECs), including up to hundreds of PPCPs. These chemicals are commonly measured at chronic levels of ng L^{-1} or µg L^{-1} in raw and treated wastewaters, agricultural drainage, and environmental waters, and in the µg kg^{-1} range in sediments affected by wastewaters (e.g. Hughes et al. 2012; aus der Beek et al. 2016; Archer et al. 2017; Bradley et al. 2017; Ebele et al. 2017). Biosolids and sludges applied to agricultural lands are often contaminated with antibiotics and other PPCPs that readily adsorb onto solid particles (high K_{OC} values). Levels in biosolids range from µg kg^{-1} to mg kg^{-1} (dry weight) (e.g. Petrie et al. 2015). Antibiotic resistance is becoming a serious problem, especially where there is intensive and poorly managed use and disposal of these pharmaceuticals in wastewaters, sewage sludges, and biosolids to agricultural lands or in runoff.

Surface waters from 38 streams across the United States (2012–2014) were found to contain 406 organics that were detected at least once. Bioactive pharmaceuticals and biocides comprised 57% of these organics. Concentrations detected in this study ranged from <1 ng L^{-1} to >10 µg L^{-1} (Bradley et al. 2017).

The extent of global contamination of surface waters by persistent pharmaceutical residues is indicated in Table 11.14.

The 16 substances shown in Table 11.14 were detected in each of the five UN regional groups of the world. Globally, the water soluble, antibiotic ciprofloxacin was detected at the highest average level. In total, 631 different pharmaceutical substances, including antibiotics, nonsteroidal anti-inflammatory drugs, analgesics, lipid-lowering drugs, estrogens, and drugs from other therapeutic groups were found at measurable levels in the environment of 71 countries covering all continents (aus der Beek et al. 2016).

Groundwaters are also vulnerable to contamination with synthetic organic pollutants from wastewaters, surface waters, and leaching from waste disposal. PPCPs have been detected in groundwaters from many countries and inhabited continents. For instance, emerging organic contaminants are found in urban groundwater sources (e.g. wells) in Africa (Sorensen et al. 2015).

Again, measured levels in groundwaters generally range from ng L^{-1} to low µg L^{-1}. Common contaminants include caffeine, DEET, carbamazepine, X-ray contrast media (iopamidol and diatrozic acid), sunscreen agents,

Table 11.14 Global occurrence of common pharmaceuticals measured in surface waters.

Pharmaceutical substance	Water solubility (mg L^{-1} 25 °C)	Log K$_{OW}$	Average (µg L^{-1})	Maximum (µg L^{-1})
Diclofenac	2.37	4.51	0.032	18.74
Carbamazepine	17.7	2.45	0.187	8.05
Ibuprofen	21	3.97	0.108	303.0
Sulfamethoxazole	3942	0.89	0.095	29.0
Naproxen	15.9	3.18	0.050	32.0
Estrone	30.0	3.13	0.016	5.0
Estradiol	3.9	4.01	0.003	0.012
Ethinylestradiol	11.3	3.67	0.043	5.9
Trimethoprim	400	0.91	0.037	13.6
Paracetamol	14 000	0.46	0.161	230.0
Clofibric acid	582.5	2.57	0.022	7.91
Ciprofloxacin	11 480	0.28 (non-ionized)	18.99	6,500
Ofloxacin	2.83 × 10^4	−0.390	0.278	17.7
Estriol	27.34 (2)	2.45	0.009	0.48
Norfloxacin	1.78 × 10^5	−1.03	3.457	520.0
Acetylsalicylic acid	4600	1.19	0.922	20.96

Source: Data from aus der Beek, T., Weber, F-A., Bergmann, A. et al. (2016). Pharmaceuticals in the Environment – Global Occurrences and Perspectives. Environmental Toxicology and Chemistry 35 (4): 823–835.

beta-blockers, antibiotics, and anti-inflammatories (Sui et al. 2015).

Human Exposures

Human populations are chronically exposed to a diverse range of individual and mixtures of TOCs and other environmental chemicals, as demonstrated by national biomonitoring programs for human exposure to environmental chemicals, and thousands of scientific studies, reports, and reviews in the literature.

Major sources of exposure include (i) occupational activities such as petroleum refining and processing, manufacturing, mixing/loading, handling, use and waste treatment or disposal of organic chemicals and products (e.g. e-wastes); (ii) individuals or communities exposed to environmental emissions from adjacent industrial activities or land use accidents and spraying (e.g. solvents, surfactants, and coatings); (iii) contaminated foodstuffs (e.g. POPs residues from many different sources, including fisheries and other wildlife); (iv) drinking water contamination, from surface waters and groundwaters; (v) inhalation of contaminated particulates, aerosols, and vapors in rural and outdoor urban air; (vi) inhalation of organic contaminants in indoor air (e.g. VOCs) and contact with residues (e.g. flame retardants, BPA) in dusts in homes, dwellings, and other buildings; (vii) contact with contaminated soils, dusts, runoff, and disposal of wastewaters/sludges, such as from livestock applications (e.g. biocide or antibiotic residues) or land contamination and remediation activities.

Environmental pathways for human exposures to TOCs are primarily from airborne contamination (e.g. workplaces and households), drinking water sources, contact with soils, dusts, and residues on surfaces such as indoor environments and dietary intake of foodstuffs. Uptake of toxic chemicals through skin from contact with personal care products is another significant exposure source. There are numerous published studies and monitoring reports on environmental levels of TOCs from many countries, although key gaps remain for estimates of actual exposures or doses, especially for multiple residue exposures indicated by human biomonitoring data (see Section 11.10).

Increasing scientific reviews and chemical databases (e.g. PubChem) on levels in air, waters, soils, and dusts indicate a highly variable spatial and temporal range of exposures, depending largely on chemical properties, emission sources and usage rates, activities, and environments (e.g. workplaces, households, urban, and rural). Concentration ranges are similar to those for pesticides, from ppb to ppm. Dioxin levels in the environment are usually lower (e.g. ppt).

Wildlife Exposures

Wildlife are exposed virtually everywhere to a wide variety of TOCs including POPs in their diet, uptake from water, and soil in some cases, and through inhalation and dermal absorption. POPs, along with many other TOCs, can cause endocrine disruption in wildlife. Global exposure of wildlife to POPs have contributed, along with other chemical pollutants, to major impacts on some wildlife populations and global concern for biodiversity. In the Twenty-first Century, more recent POPs and emerging chemicals of concern such as environmental phenols and PPCPs are now accumulating in wildlife, and there is increasing evidence of transfer through food webs to wildlife in higher trophic levels (and also to humans). For example, wildlife living downstream of sewage treatment works discharges are exposed to many different EDCs including active ingredients in pharmaceuticals and additives in personal care and cleaning products (WHO/UNEP 2013, p. 209). The bioaccumulation of TOCs and their toxic effects are discussed in Sections 11.10 to 11.11.

11.10 Biological Processes Affecting Toxic Organic Chemicals in Wildlife and Humans

Wildlife and humans are exposed to a multitude of toxic organic chemicals and often, complex mixtures of these chemicals via ingestion in their diet, water intake, and dusts (e.g. household or workplace) through inhalation of contaminated particulates and vapors, and dermal absorption. Basic biological processes involved in the uptake, distribution in the body, storage, metabolism, and excretion of TOCs by wildlife and humans are shown in Figure 11.18. Food web transfer of TOCs from individuals to consumers or predators and also other routes of uptake from environmental sources, including excretion of metabolites of TOCs are also indicated.

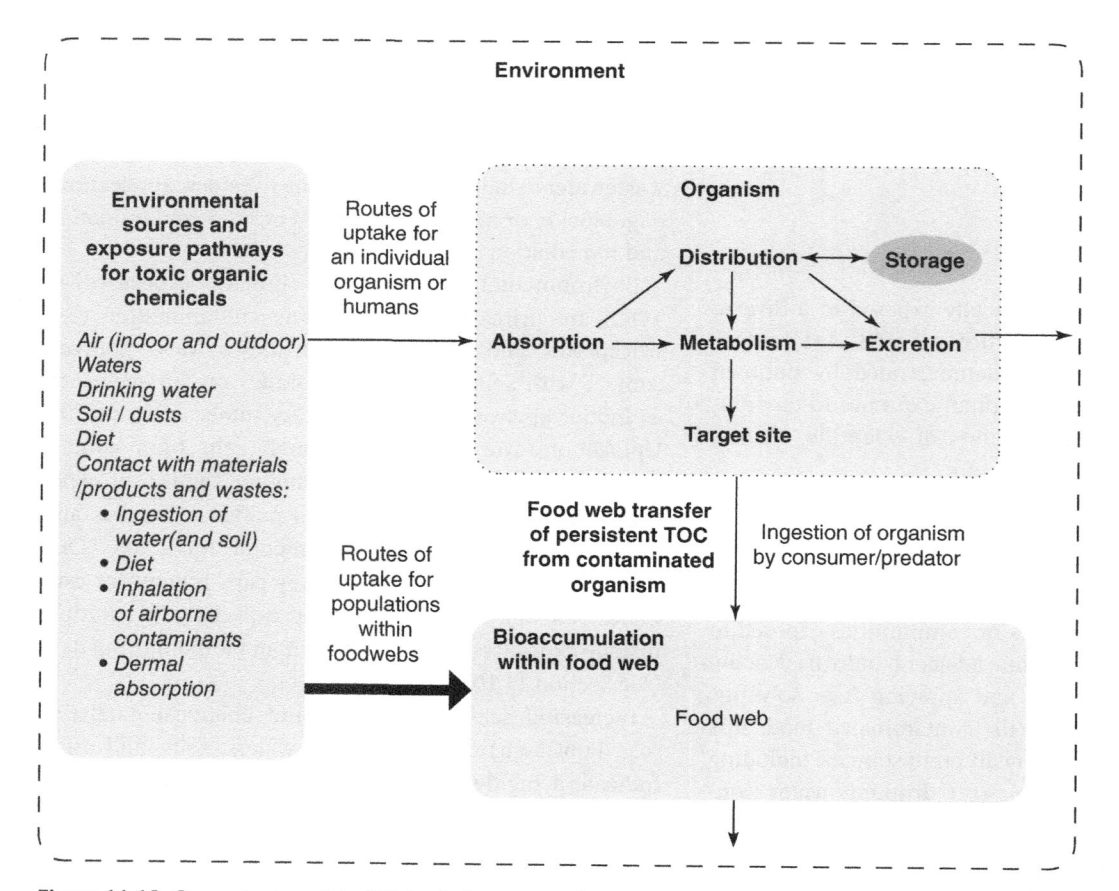

Figure 11.18 Conceptual model of biological processes involved in environmental and dietary uptake, absorption, bioaccumulation, metabolism, and excretion of toxic organic chemicals (TOCs) in (i) an organism, and (ii) populations of organisms within food webs, including transfer through ingestion of a contaminated organism by a consumer or predator. Excreted residues or metabolites of TOCs are returned to the external environment where they may be further consumed or biodegraded.

The dynamic biological processes that affect the behavior and fate of toxic organic pollutants within living organisms and food webs are described below.

Biological Processes Within Organisms

Absorption – Uptake of TOCs by wildlife and humans usually involves absorption of these chemicals from multiple sources of exposure (air, water, particulates, soils, foods, and contact with contaminated surfaces or products) via intake routes such as inhalation, ingestion, skin, or surface absorption. Pollutants are absorbed by partitioning or transport processes through internal or external surfaces, cell membranes, organs, and cell barriers via aqueous flow systems (e.g. water, blood, and sap), until returned to the external environment unchanged or transformed by metabolic processes. Food chain transfer of pollutants (e.g. POPs) accumulated within or adsorbed to an organism can also occur during consumption by a higher trophic organism (see Figure 11.18).

Factors influencing the uptake and distribution of TOCs in biological systems are related to (i) inherent physical and chemical properties of the chemical (e.g. volatility, water solubility, sorption, K_{OW}, and bioconcentration factors), and (ii) physiological characteristics (e.g. feeding behavior, routes of uptake, and habitat) and ecosystem specific properties (e.g. temperature, pH, organic matter, and food web structure).

Distribution and Storage – After internal intake into the body of an invertebrate, fish, bird or mammal, or plants, a toxic organic chemical can be transported via blood or other fluid systems (e.g. lymph) to different tissues and target sites, where it can be metabolized, excreted, or stored through bioaccumulation (see below). Storage of lipophilic TOCs can occur in fatty tissues (e.g. liver, brain, adipose) or proteins (e.g. liver, muscle). Several types of TOCs (and organometallics) are proteinphilic (e.g. perfluorinated chemicals and methylmercury) and accumulate in protein-rich tissues (liver) and fluids (blood).

Target Site – Toxicants such as toxic organic chemicals usually interact or bind with endogenous target molecules (e.g. receptors, enzymes, or proteins) in particular cells of an organism, to trigger and interfere with cellular functions and/or structures, leading to various toxic effects (e.g. neurotoxic and endocrine disruption), depending on the roles of the target molecules and delivery of sufficient doses of the toxicant at the target sites. Endocrine disruptors, for example, can interfere with hormone actions by acting directly on hormone receptors or various types of proteins that control the delivery of a hormone to its normal target cell or tissues (WHO/UNEP 2013, pp. 6–12).

Metabolism – Several enzymes (e.g. mixed function oxidases) in wildlife and humans can transform TOCs to metabolites. For vertebrates, the liver is the main site of metabolism for removal of many TOCs, including EDCs, as more polar metabolites, via feces and urine. However, metabolism may also produce more harmful or persistent metabolites (e.g. hydroxylated PCBs interfere with the transport of thyroid hormones in blood) (e.g. Letcher et al. 2000; WHO/UNEP 2013, pp. 89–99). Polyfluorinated telomer alcohols can also be metabolized in organisms to perfluorinated carboxylic acids with half-lives of several years (Lau et al. 2007).

Excretion – Toxic organic chemicals such as EDCs are excreted from the body via multiple ways, such as in feces, urine, sweat, and mothers' milk at various rates, due to their chemical properties and persistence. Nonpersistent TOCs are rapidly metabolized, mainly via the liver, and excreted through urine or feces. More persistent TOCs are slowly excreted because they tend to accumulate in different parts of the body, such as fat, and are slowly released as they are eliminated from blood to maintain equilibrium (Lee et al. 2017). The half-lives of POPs in the body range from several months to years, while less lipophilic chemicals (e.g. PAHs and BPA) have much shorter half-lives (Lee et al. 2017).Excretion via breast milk is particularly critical because significant to excessive transfer of persistent, lipophilic EDCs (e.g. POPs) can occur in breast-fed infants from residues in mothers' milk (WHO/UNEP 2013, pp. 223–224).

Bioaccumulation

In the process of bioaccumulation, a chemical (e.g. TOC) is absorbed in an organism by various routes of exposure from dietary and environmental sources. In aquatic organisms, bioaccumulation is the net result of the competing processes of chemical uptake into the organism. These processes involve respiratory and dietary intake and chemical elimination from the organism, including respiratory exchange, fecal egestion, metabolic biotransformation of the parent compound, and growth dilation (Arnot and Gobas 2006).

The bioaccumulation of persistent TOCs in wildlife and humans is well known in body fluids (blood), tissues (adipose and muscle), and organs (adrenal and brain), and depends upon physicochemical properties of individual chemicals and biological characteristics of the exposed organism. Lipophilic TOCs, such as most POPs, are readily stored in fatty tissues of wildlife and humans. However, higher concentrations of these chemicals can be released into blood and other circulating fluids in an organism when fats are mobilized and used by a fish or bird to produce eggs, by mammals to produce milk for offspring, or as a source of energy during periods of low food or starvation. Higher internal exposure to these chemicals can lead to adverse effects on the organism. Proteinphilic chemicals (e.g. PFOA and PFOS) and methylmercury can accumulate in protein-rich tissues (liver) and fluids (blood) rather than fatty tissues.

Human and wild mammal exposure to TOCs prior to and during pregnancy (prenatal exposure) is a potential problem for the developing fetus because persistent TOCs (e.g. PBDEs, organochlorine pesticides, and plasticizers) can cross the placenta to the fetus, leading to endocrine disruption effects on the reproductive system of the offspring (see review by Hamlin and Guillette 2011). Several of the halogenated phenols, POPs, or their metabolites can accumulate in blood due to their proteinphilic properties. Both the POPs and the halogenated phenols are also efficiently transferred through the placenta (Park et al. 2008). Evidence of postnatal transfer of persistent organic chemicals in mothers' milk exists from many studies (see Needham et al. 2011) and is also of similar concern for breast-fed offspring. In some populations, estimated dietary intake for a nursing infant can be similar to that of an adult or exceed tolerable doses for individual chemicals (WHO/UNEP 2013, pp. 223–224).

Various PPCPs are being found in a number of aquatic studies in the tissues of wildlife living near or downstream of sewage treatment plant discharges in highly populated areas (WHO/UNEP 2013, pp. 200–202). Pharmaceutical residues include the antiepileptic carbamazepine and the active ingredients of several antidepressants (fluoxetine, sertraline, venlafaxine, citalopram, norfluoxetine, diphenhydramine, diltiazem) in muscle or liver of wild fish or fish caged downstream of these discharges. Human contraceptives have also been found in fish muscle (EE2) and plasma (levonorgestrel) at the low parts per billion levels or so. Chemicals from personal care products found in tissues of aquatic species include benzotriazole, UV stabilizers, parabens, triclosan, and organophosphorus compounds. Triclosan has been found in a range of aquatic organisms including algae, invertebrates, fish, and dolphins (Dann and Hontela 2011). Earthworms can also accumulate triclosan from soils treated with solids from municipal wastewater treatment plants or biosolids (Kinney et al. 2008; WHO/UNEP 2013, p. 210).

Food Chain Transfer of Toxic Organic Chemicals

Chemicals (e.g. POPs) with a high persistence and bioaccumulation potential tend to be found at the highest concentrations in animals at the top of the food web (e.g. humans, seals, polar bears, birds of prey, and some large amphibians) and in tissues and body fluids that are high in

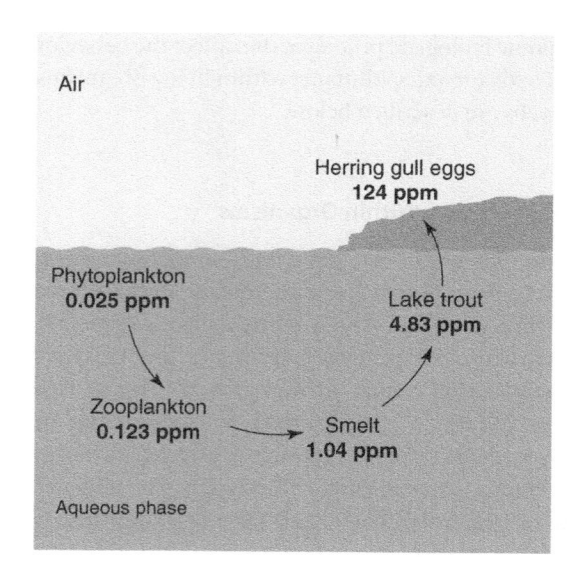

Figure 11.19 Bioaccumulation and biomagnification of PCBs in an aquatic food chain of The Great Lakes, North America. Persistent organic chemicals such as PCBs bioaccumulate. This diagram shows the degree of concentration of PCBs in each trophic level of a simplified Great Lakes aquatic food chain (in parts per million, ppm). The highest levels are reached in the eggs of fish-eating birds such as herring gulls Source: Modified from Environment Canada/US EPA (1995). The Great Lakes: An Environmental Atlas and Resource Book. Third Edition. Chapter 4, The Great Lakes Today-Concerns. Environment Canada/United States Environmental Protection Agency. https://www.nrc.gov/docs/ML0419/ML041970161.pdf (accessed 18 April 2020).

fat (e.g. blubber, mothers' milk, and egg yolk). Some PCB congeners, particularly those with substitution at the 2,4 and 2,4,5 positions on the rings, accumulate through food webs to high concentrations in humans and wildlife, while other PCB congeners are easily metabolized (WHO/UNEP 2013, pp. 191–192). Examples of biomagnification of some persistent halogenated organic compounds in aquatic and terrestrial food chains are reported for air breathing top predators. Figure 11.19 shows an example of PCB accumulation in a simplified aquatic food chain (The Great Lakes) in which the highest levels are reached in the eggs of the fish-eating herring gull.

Bioconcentration

The bioaccumulation of organic pollutants from aqueous solution can be related to a simple partitioning or sorption process between an organism and water. As discussed in Chapters 4 and 9, this relationship can be expressed by a two-phase partition coefficient of an organic chemical between water and biota, K_B, known as the bioconcentration factor (BF or BCF), which is usually measured by aquatic bioassays under equilibrium conditions: That is, BCF is the ratio of the concentration of the chemical

Table 11.15 Bioconcentration factors (BCF) for common toxic organic chemicals in aquatic organisms.

Chemical	BCF values	Hazard classification
DDT	600–1 × 10^5 (fish, bivalves)	High to very high
TCDD	1585–5.1 × 10^6	Very high
PFOS (K$^+$ salt)	200–5000 (carp)	High to very high
	830–2600 (catfish and largemouth bass) field-based study	
PFOA	<5.1–9.4 (carp)	Low to moderate
	100–230 (mullet)	
PBDE isomers	1560–>10000 (di- to penta and octa-)	Moderate to high
	42–850 (mono-, hexa-, hepta, and nona-)	
Bisphenol A	5.1–73.4 (est)	Low to moderate
4-Nonyl phenol	543.5 (est)	Moderate
Styrene	37.3 (est)	Low
Tetrachloroethylene	26–115 (fish)	Low to moderate
Toluene	13–90 (fish)	Low to moderate
Galaxolide	1548	High
Benzophenone 3	13–160 (fish)	Low to very high
	1585–5.1 × 10^6 (fish)	
Triclosan	2.7–90 (orange-red killifish)	Low to moderate
Ofloxacin	3 (est)	Low
Sulfanilamide	0.2 (est)	Low
Naproxen	3 (est)	Low
Metformin	3 (est)	Low

Sources: Data from PubChem (HSDB) website, National Institutes of Health, US, and Chemspider, The Royal Society of Chemistry. Predicted data is generated using the US EPA EPI Suite™.

in the organism (C_B) to its concentration in water at steady state:

$$BCF = K_B = C_B/C_W \quad (11.2)$$

Some BCF values for common TOCs are compiled in Table 11.15 from experimental, field data, or predicted data taken from chemical databases. The predicted or estimated BCF values were derived from the US EPA Suite. Regulatory criteria for bioaccumulation of organic chemicals refers to BCF values of 1000–5000 (United States) or ≥2000 (EU) for bioaccumulative, and ≥5000 for very bioaccumulative (EU and United States). POPs criteria include BCF values of >5000 for bioaccumulative in aquatic species.

Biomonitoring of Toxic Organic Chemicals

Monitoring of biological fluids (blood, urine, and mothers' milk) and tissues (adipose) from human and wildlife populations for internal exposures to toxic organic chemicals, including POPs, EDCs, pesticides, toxic metals, and organometallics (e.g. methylmercury), has emerged since the use of DDT and other organochlorine pesticides. For example, persistent organic chemicals are commonly measured in blood and mothers' milk, while urine is used to measure less-persistent and less-bioaccumulative chemicals (and their metabolites), due to the short half-lives in humans.

Biomonitoring has become a major strategy to measure and assess the types of contaminants and their metabolites or biomarkers, exposure pathways, concentrations, sub-populations, geographic, and time trends in the mixtures of environmental chemicals now found in biological specimens taken from target or baseline populations of humans (e.g. children), and specific species of wildlife populations (see Box 11.1 also).

Data on internal human and wildlife exposures can then be linked to human or ecological health studies and epidemiological surveys to help develop (i) exposure-response relationships in humans or exposed wildlife populations, (ii) health and ecological risk assessments, and (iii) policies, priorities, and practices for risk intervention and management of exposures to environmental chemicals.

Human biomonitoring for TOCs has progressed widely since Laug et al. (1951) detected DDT in mothers' milk from nonoccupationally exposed mothers. A concise history of human biomonitoring programs in the United States and internationally (e.g. Europe and WHO) has been reported by the United States (National Research Council 2006).

> **Box 11.1 Human Biomonitoring (HBM)**
>
> **Human Biomonitoring (HBM)** *is a science-based approach for assessing human exposures to natural and synthetic compounds from environment, occupation, and lifestyle.*
>
> *It relies on the measurement of particular substances or biological breakdown products, known as metabolites, in human tissues and/or fluids, and also includes the study of their effects and the possible influence of individual susceptibility as response modulators.*
>
> *HBM is a growing area of knowledge used for exposure and risk assessment in environmental and occupational health, and its importance has been increasing as a result of advancements in the ability to measure greater numbers of chemicals in the human body and tissues. In order to achieve this purpose, HBM focuses on the use of biomarkers as measurable indicators of early changes in biological systems* (Laderia and Viegas 2016).

> **Box 11.2 Some Industrial Chemicals Found in Human Tissues**
>
> United States biomonitoring indicates widespread human exposure to some commonly used industrial chemicals. These include:
>
> - Polybrominated diphenyl ethers are fire retardants used in certain manufactured products. These accumulate in the environment and in human fat tissue. One type of polybrominated diphenyl ether, BDE-47, was found in the serum of nearly all of the participants.
> - Bisphenol A (BPA), a component of epoxy resins and polycarbonates, may have potential reproductive toxicity. General population exposure to BPA may occur through ingestion of foods in contact with BPA-containing materials. CDC scientists found bisphenol A in more than 90% of the urine samples representative of the US population.
> - Perfluorinated chemicals are another example of widespread human exposure to persistent industrial chemicals. One of these chemicals, perfluorooctanoic acid (PFOA), was a by-product of the synthesis of other perfluorinated chemicals and was a synthesis aid in the manufacture of a commonly used polymer, polytetrafluoroethylene, which is used to create heat-resistant nonstick coatings in cookware. Most participants had measurable levels of this environmental contaminant.
>
> *Fourth Report on Human Exposure to Environmental Chemicals* – Centers for Disease Control and Prevention (2017).

Biomonitoring of populations is rapidly developing in the United States, Canada, and Europe with comparable types and numbers of analytes being measured (National Research Council 2006). From the 1970s and onwards, much effort was focused on biomonitoring of persistent pesticides, PCBs, and other POPs. More recently, human biomonitoring has expanded to include other EDCs and more easily metabolized compounds, such as phthalates, bisphenol A, and PAHs.

In the United States, The *Fourth Report on Human Exposure to Environmental Chemicals, Updated Tables, January 2017* (abbreviated as *Updated Tables, January 2017*) provides national biomonitoring data for 308 chemicals (Centers for Disease Control and Prevention 2017). Many of them are POPs and other toxic organic chemicals. Extensive statistical evidence of human internal exposure to and excretion of multiple toxic organic chemicals and other chemicals (e.g. pesticides, metals, and other biomarkers) from all age groups, and representative populations in the United States is presented. Examples of TOCs include VOCs, disinfection by-products, dioxins, PCBs, PAHs, octyl and nonyl phenol, bisphenol A (BPA), chlorinated phenols, parabens, triclosan, benzophenone-3, PFOS and PFOA, and polybrominated diphenyl ethers. Recent results indicate widespread exposure to some common industrial chemicals that are widely used in the United States and elsewhere (Box 11.2).

11.11 Environmental Toxicity of Organic Chemicals

Toxic organic chemicals can exert a wide range of physiological and behavioral effects on wildlife species and humans as demonstrated by numerous laboratory bioassay, field, clinical, and epidemiological studies. Despite this, large gaps in our knowledge of ecotoxicity and human health effects from environmental exposures remain and also apply to emerging organic chemicals.

Toxic effects for TOCs are usually complex at the molecular level with a variety of toxicity mechanisms suggested for different species. For example, dioxins and dioxin-like compounds are major examples of persistent organic pollutants that induce toxicity in vertebrates. These chemicals can bind with high, but variable affinities, to the aryl hydrocarbon receptor (AhR) in vertebrates, which can trigger or mediate a cascade of potential adverse effects that include enzyme induction responses, immunotoxicity, oxidative stress, cytokine production, hormonal, and growth factor disruptions (Ross and Birnbaum 2003). AhR-mediated effects may also affect developmental endpoints, including neurologic, immunologic, and reproductive parameters, in individual vertebrates exposed in early life stages, as shown in studies of laboratory and wildlife species (White and Birnbaum 2009).

The DDT metabolite, p,p'-DDE, is considered to inhibit calcium ATPase in the membrane of the shell gland and reduces the transport of calcium carbonate from blood into the eggshell gland. This results in a dose-dependent reduction in thickness or *thin eggshell effect* (Connell and Miller 1984, pp. 209–211).

The biological effects of pollutants usually differ greatly within and between species reflecting differences in the pattern of exposure (individual or mixtures of chemicals; levels of exposure), routes of uptake, rates of uptake, toxicity mechanisms, and sensitivity of target sites or organs, metabolism, and elimination. Organic pollutants can induce a broad spectrum of selective and nonselective toxic effects on humans and wildlife due to the variety of their chemical forms, functional groups, and properties. Toxic effects can impact upon almost all taxonomic groups.

Many environmental toxicological studies show that considerable variation is observed (i) between species in sensitivity and tolerance to exposure from a toxic organic chemical, and (ii) in the toxicity of different TOCs to a particular species. Other factors such as sex, age, nutritional state, stress, and habitat or microenvironment vary with individual sensitivity. Acute and chronic toxic effects may originate from several different exposure pathways such as from direct contact with airborne aerosols, residues on particles or surfaces (e.g. leaf), or through ingestion of foods (e.g. contaminated prey), uptake of residues from water, soils, and/or via skin absorption. Alternatively, indirect toxic effects may result from habitat modification or loss of prey or food through selective toxicity to sensitive species in food chains or webs.

Acute and chronic or sublethal toxicities of environmental chemicals to organisms (e.g. LD_{50}, LC_{50}, and EC_{50}) are derived from field observations and a vast array of bioassay data on experimental test species, particularly rodents and other mammals, aquatic organisms such as fish, crustaceans (e.g. mysid shrimp and daphnids), duckweed and algae, and more recently, beneficial insects, pollinators (e.g. honey bees), amphibians (e.g. frogs, toads, or their tadpoles), soil invertebrates (e.g. earthworms), and terrestrial plants.

Toxicity data on laboratory and field studies of lethal and various sublethal effects (e.g. physiological, biochemical, neurotoxic, reproductive, or behavioral) of many TOCs on numerous taxa are now compiled in various databases, although large data gaps remain for many key species and other chemicals in commercial use. Broadly, available data and findings suggest terrestrial animals and plants at different life stages may suffer adverse toxic effects from exposures to different TOCs, over a very wide concentration range of several orders of magnitude, from <1 to 1000 or so $mg\,kg^{-1}$ (body or plant tissue weight range), for common test exposure periods of 1–21 days. Very high toxicity occurs in the $<1\,mg\,kg^{-1}$ or $\mu g\,kg^{-1}$ range for sensitive species. For instance, single toxic doses of the dioxin, TCDD, at very low levels of $\mu g\,kg^{-1}$ body weight can cause delayed mortality in rodents and some other common terrestrial vertebrates (see Table 11.16).

Aquatic organisms generally experience acute and chronic toxicities in the range of $\mu g\,L^{-1}$ for highly toxic organic chemicals to $mg\,L^{-1}$ or more for moderate to slightly toxic organic chemicals. Early life stages of sensitive species tend to be more vulnerable.

Some acute toxicities of various mammals, avian, and aquatic species exposed to selected toxic organic chemicals in experimental studies are given in Table S11.5 in Section S11.6. Data originally taken from the TOXNET hazardous substances databank (HSBD). The persistent toxic pollutant, TCDD, is very toxic to common test species of mammals, birds, and fish but readily available data on invertebrates and plants are lacking. PCB 1254 exhibits acute toxicity to sensitive species of fish and aquatic invertebrates, and the tadpoles of the American toad. The perfluorinated surfactant, PFOS, is acutely toxic (oral) to the rat and bird species. Toxic effects include adverse development (gut) and defects (swim bladder) in embryos

Table 11.16 Single toxic doses of 2,3,7,8-TCDD for some vertebrates.

Species	LD_{50} ($\mu g\,kg^{-1}$)	Mean time to death (days)
Rabbit	10 and 115	Unknown and 6–39
Rat	22 to <100	9–42
Guinea pig	0.6 and 2	5–34
Mouse	114–284	20–25
Monkey	<70	28–47
Dog	>30 to >300	9–15
Chicken	25–50	12–21

Source: Based on Moore, J.A. (1978). Toxicity of 2,3,7,8-tetrachlorodibenzo-p-dioxin. In: *Chlorinated Phenoxy Acids and Their Dioxins*. (ed. C. Ramel), 134. Stockholm, Sweden: Swedish Natural Science Research Council.

of the zebrafish and also in growth of wheat seedlings. The volatile solvent, trichloroethane, was tested in aquatic species and plants without significant toxic effects. Fathead minnows (220 mg), *Daphnia pulex* and rainbow trout (embryos and juveniles) were the most sensitive organisms to nonyl phenol in laboratory bioassays. Low BPA exposure (158 µg L^{-1}) over two weeks increased vitellogenin in male fathead minnows. Diethyl phthalate showed low ecotoxicity to most test species. Lethal concentrations for species tested ranged from about 1 mg L^{-1} to >10 mg L^{-1}.

The toxicity of pharmaceuticals is of major concern for unintentional exposure to nontarget organisms because they were specifically designed to maximize their bioactivity at low doses and target certain metabolic, enzymatic, or cell-signaling mechanisms (Ebele et al. 2017). The evolutionary conservation of these molecular targets in a given species increases the risk of related bioactivity in nontarget organisms. Known or suspect endocrine-disrupting chemicals already include pharmaceuticals such as sex hormones, glucocorticoids, and nonsteroidal pharmaceuticals (see Ebele et al. 2017).

Petrie et al. (2015) have collated acute ecotoxicity data (EC$_{50}$ values) on aquatic species, taken from the literature for a set of 12 pharmaceuticals exposed to different test species in bioassays using a variety of toxicological endpoints. A median EC$_{50}$ value and statistical descriptions for each pharmaceutical were then derived for each pharmaceutical from collated test results to provide an indicative classification of its ecotoxicity. The EC$_{50}$ values were classified according to the Commission of the European Communities, 1996 criteria for aquatic organisms: <1 mg L^{-1} = very toxic, 1–10 mg L^{-1} = toxic, 10–100 mg L^{-1} = harmful, >100 mg L^{-1} = not classified.

Eight of the twelve pharmaceuticals (NSADs, lipid regulators, carbamazepine, an epileptic, and trimethoprim, an antibiotic) were generally classified in the harmful range, while the other antibiotics (ofloxacin, sulfamethoxazole, oxytetracycline, and erythromycin) were mostly in the toxic range.

As with other TOCs, these authors have argued further that the toxic effects of PPCPs in the aquatic environment on nontarget organisms extends beyond those observed at acute levels to chronic or sublethal effects, including certain sensitive stages of development. The situation is also complicated at environmental levels in many cases because of exposures to mixtures of parent compounds and/or their toxic metabolites, and not simply to individual toxicants.

Another major problem for environmental risk evaluation identified by researchers is that specific sublethal effects are numerous and diverse, and relate to a broad spectrum of physiological and behavioral responses, such as alterations in enzyme production, growth rate, reproduction, behavior and activity, production of tumors, teratogenic effects, and endocrine disruptions, that are difficult to separate from natural variations or adaptive responses in individuals or populations.

Toxicity of Mixtures

There is overwhelming scientific evidence that wildlife and humans are widely exposed to multiple and variable toxic organic chemicals, pesticides, and other chemical contaminants, through their environments, via contact with air, water, soils, dusts, foods, and other sources (e.g. materials and consumer products).

How to investigate, measure, evaluate, and manage the increasing evidence of ecotoxic effects and human health impacts or risks from diverse chemical mixtures, often at low environmental levels of exposures, are current major challenges for scientific research and regulatory authorities. In essence, the combined interactions of multiple pollutants are of specific concern because they may exert effects (e.g. additive, synergistic, and antagonistic) even when each individual chemical is present at concentrations below its toxicity endpoints (see White and Birnbaum 2009). As well, many chemicals can have multiple modes of action within an organism.

The toxicology of chemical mixtures is introduced in Chapter 5. From a regulatory perspective, risk assessment on the combined effects of chemicals in mixtures found in the environment and humans is not a usual practice or regulatory requirement in many countries. Currently, it is a critical area of research and development for registration of chemicals, risk assessment, and management of environmental and human health effects. Chapter 18 also considers approaches to toxicity assessment of chemical mixtures. An example of an approach being developed in the European Commission for evaluating toxicity of chemical mixtures is given in Section S11.7 in the Companion Book Online.

11.12 Effects of Toxic Organic Chemicals on Wildlife Populations and Communities

Wildlife species and populations continue to decline worldwide, often at disturbing rates as indicated in Chapter 3. This human-related phenomenon is essentially due to a number of factors including over-exploitation, loss of habitat, climate change, and chemical contamination. From a pollution perspective, many thousands of individual organic chemicals (anthropogenic and natural) are known or suspected to be toxic to wildlife and humans. This includes hundreds of individual chemicals (EDCs) that can or may also interact with endocrine systems in wildlife and humans, leading to adverse effects on their reproduction, development, and populations. As discussed in this chapter, the types of EDCs and their sources of exposure, chemical properties, and environmental fate vary widely (WHO/UNEP 2013, Chapter 3).

Recent and historical evidence of population declines in some wildlife populations are associated with reproduction effects (e.g. infertility and feminization) and nonreproduction effects (e.g. thyroid and adrenal disorders or disease, hormone cancers, and bone disorders) related to EDC exposures, such as from mixtures of POPs. Table S11.6 (reproductive effects) and Table S11.7 (nonreproductive effects) summarize some of these reported EDC-related effects from EDC exposures.

Several key examples of population decline in wild animal species attributed to chemical impacts on endocrine systems are given below. In at least two cases, the strength of the evidence is supported by population recoveries after wildlife exposures were reduced by restrictions or bans on use of the chemicals implicated (DDT and tributyl tin) (WHO/UNEP 2013).

- *EDCs and marine mammals*: Several well-known EDCs (diverse mixtures of PCBs, DDT, other POPs, mercury, and other metals) are reported to be associated with declines in the populations of marine mammals (sea otter, northern fur seal, Baltic grey seals, Steller sea lion, Galapagos sea lion, and the endangered southern resident killer whale in the North Pacific Ocean) although other environmental factors may also play a role in effects on reproduction and survival of these populations. Significantly, Bergman (2009) found that reductions in POPs exposures were associated with improved gynecological health and numbers of pregnancies in gray Baltic seals, but observed increased prevalence of colonic ulcers in young animals, possibly due to *new* or increased amounts of unidentified toxic factors. Killer whales accumulate high concentrations of POPs, including PCBs and PBDEs in their fatty tissues. Buckman et al. (2011) have shown that PCB tissue burdens in these whales were strongly correlated with increases in the expression of aryl hydrocarbon receptor, thyroid hormone α receptor, estrogen α receptor, interleukin, and metallothionein. These findings provide supporting evidence of endocrine disruption due to PCB exposures.
- *DDT and bird populations:* The intensive use of the insecticide DDT is strongly associated with population declines of top predator bird species and observed recoveries in their populations, following banning or restriction of its use in many countries. Its breakdown product, DDE, accumulates in top predator bird species, such as osprey, falcons, and eagles, and is known to interfere with the hormones (prostaglandin signaling) that control eggshell production. As a result, high DDE exposures reduced the populations of many bird species due to its effects on eggshell thickness, causing a decrease in the survival of chicks and reproductive success of the exposed species. For example, after DDT was banned in North America in 1972, osprey populations in the United States increased from 8000 in 1981 to 14 200 in 1994 to 16 000–19 000 in 2001 (Grove et al. 2009).
- *PBDEs and birds:* PBDEs are found at elevated concentrations in wild birds, especially those that are high in the food web and living near urban centers. American kestrels exposed to environmental levels of PBDEs in diet studies exhibited delayed laying of eggs, and laid eggs with thinner shells and smaller weights. These studies suggest that reproduction may be compromised in populations of wild birds accumulating high concentrations of PBDEs (WHO/UNEP 2013; Guiqueno and Fernie 2017).
- *Estrogens and fish populations*: A 7 year, whole-lake experiment showed dramatic declines in the fathead minnow population after exposures to low $ng L^{-1}$ concentrations of the synthetic estrogen 17 alpha-ethynylestradiol (EE2) led to feminization of males, through the production of vitellogenin mRNA and protein, impacts on gonadal development as evidenced by intersex in males and altered oogenesis in females (Kidd et al. 2007).

Male fish are known to have high incidences of intersex in rivers receiving sewage treatment works effluent that contains estrogens and anti-androgens (e.g. roach in the United Kingdom) (Jobling et al. 2006), and this condition decreases their reproductive success when in competition with normal males (Harris et al. 2011).
- *Tributyltin and invertebrate populations:* Exposure to the antifoulant TBT has caused imposex (development of male sex organs) in female snails and led to reproductive failure and declines in several species found in harbors and other areas of high TBT use (WHO/UNEP 2013).

Effects on Wildlife Biodiversity

The genetic effects of pollutants such as TOCs on populations are emerging as a major issue in ecotoxicology that may have marked impacts on biodiversity and ecological health through decreases in genetic biodiversity. Chemical contaminants can reduce populations by the genetic effects of somatic and heritable mutations and also nongenetic modes of toxicity (Bickham et al. 2000). The critical outcomes for wildlife are that reduced genetic diversity in a population could threaten its survival in the environment, increase risk of extinction, and may lead to impacts on how communities and ecosystems function and their diversity.

Bickham et al. (2000) have produced a conceptual model that shows how environmental mutagens (somatic and germ-line mutations) and nonmutagenic chemical toxicants (e.g. EDCs) can affect the reproductive success of individuals by means of heritable effects and damage to somatic tissues, leading to reduced genetic biodiversity of the population and potential extinction or extirpation of the population.

> Although a direct link between EDC exposure and reduced genetic diversity in populations or communities has not yet been demonstrated, it is possible that this may occur given the effects of EDCs on the reproductive success of wildlife
>
> (WHO/UNEP 2013).

11.13 Human Health Effects of Toxic Organic Chemicals

Overview

Humans may experience chemical poisonings or adverse effects from toxic organic chemicals through (i) accidental or self-poisonings, (ii) workplace exposures (manufacturing, mixing/loading, transport, and application), (iii) environmental contamination (e.g. air, soil, dusts, drinking water intake), (iv) contact with contaminated materials, personal care, and other consumer products and wastes, or (v) dietary contamination in the case of the general public. Prenatal exposure of parents to many toxic chemicals (e.g. endocrine disrupters), similar to pesticides, is also a special risk for developmental effects in fetuses and children.

There is unequivocal evidence (e.g. biomonitoring) of the environmental uptake, storage, persistence, metabolism, and elimination of hundreds of different toxic organic chemicals or metabolites in body fluids, tissues, and organs of humans, at all life stages, and from multiple populations in various countries around the globe. Acute and chronic effects of human exposures to many toxic organic chemicals are well known to originate from dose-related absorption of these substances via inhalation, ingestion (mainly dietary), and dermal contact. Despite enormous scientific progress in our knowledge of toxic organic chemicals and their biological effects and mechanisms since the 1960s, large data gaps remain in the extent of global contamination and adverse effects from numerous individual and mixtures of TOCs released into the environment, as shown by current studies of emerging chemicals of concern.

Adverse effects involve acute poisonings and long-term systemic impacts on human body systems and organs, such as the central and peripheral nervous systems, cancers, reproduction and developmental effects, and the expanding role of many TOCs (e.g. POPs and estrogens) as disrupters of essential endocrine systems.

The persistent pharmaceuticals and personal care products in the aquatic environment are another potential exposure source of endocrine disrupters for humans and wildlife. Special concern exists for unintentional exposures to pharmaceutical residues, which were specifically designed to be biologically active at low doses in humans and to target certain metabolic, enzymatic, or cell-signaling mechanisms (e.g.Archer et al. 2017; Ebele et al. 2017).

Global Burden of Disease and Toxic Organic Chemicals

Many toxic organic chemicals are known systemic toxicants, carcinogens, and endocrine disruptors. Best available estimates of their contributions to the global burden of injury and disease at work and from environmental exposures are limited in scope, because of inadequate exposure studies and epidemiological data, and are likely to be underestimates. Nonetheless, they are strongly implicated in many types of work and environment-related diseases and deaths as discussed below.

In 2015, nearly one million workers died due to exposures to hazardous substances at work, including vapors, gases, dusts, and fumes. Most of these deaths were attributed to cancers or respiratory diseases, mainly chronic obstructive pulmonary disease (COPD) (see Hamalainen et al. 2017).

Annual work-related deaths (2.78 million) account for 5% of total global deaths (based on 2015 Global Burden of Disease Study), of which 2.4 million (86.3%) were attributed to work-related diseases: circulatory (31%), cancers (26%), respiratory (17%), injuries (14%), and communicable diseases (9%). Noncommunicable diseases are dominant, while most deaths occurred in Asia (~66.6%), followed by Africa (11.8%) and Europe (11.7%) (see Hamalainen et al. 2017).

The global burden of disease attributed to environmental risks has been discussed in Chapter 3. In 2012, environmental risks contributed 23% (95% CI = 13–34%) of global deaths or 12.6 million deaths, and 22% (95%CI = 13–32%) of global disability adjusted life years (DALYs). Significantly, an estimated 8.2 million deaths (65%) are attributed to noncommunicable diseases (Prüss-Ustün et al. 2017).The global burden of disease due to selected chemical exposures was estimated to be 4.9 million per year in 2004, mainly from air pollutants. Recent WHO estimates (2016) attribute 7 million deaths year^{-1} to exposure from air pollution (outdoor and indoor). The total burden of disease due to known chemicals and fine airborne pollution particulates also exceeds the burden of all cancers estimated worldwide, indicating the size of its public health impact (Prüss-Ustün et al. 2011). There are overlaps in this type of comparison because chemical carcinogens and related air pollution particulates induce a proportion of cancers. At this stage, the burden of disease due to toxic organic pollutants is unclear and appears to be largely hidden or unresolved by current assessment methods but is likely to be significant when the complex mixtures of environmental and occupational exposures are better resolved, somewhat like insights into the magnitude of health effects from chronic exposures to respirable particulate matter and carbon black.

Industrial Accidents

Major chemical accidents can cause severe injury and death on a significant scale for exposed workers and surrounding residents. Two of the world's worst environmental disasters occurred at Bhopal in India in 1984, when more than 40 tons of poisonous methyl isocyanate gas escaped in a plume from a pesticide plant producing the pesticide Sevin, and in 1976 at Seveso, near Milan in Italy, involving the emission of a toxic plume of industrial chemicals containing 1 kg of TCDD, during batch processing of 2,4,5-trichlorophenol from 1,2,4,5-tetrachlorobenzene.

The Bhopal accident killed at least 3800 people within hours and caused significant morbidity and premature death for many more thousands (Broughton 2005). Estimates of injuries and deaths vary considerably. About half a million persons suffered some form of injury, with thousands reported as suffering severe and permanent disability, and dying overall from gas-related poisoning or diseases.

In the case of the Seveso accident, the plume of chemicals emitted, including TCDD, settled over about $18\,km^2$. TCDD soil concentrations varied in the affected area from <5 to $>50\,\mu g\,m^{-2}$. Thousands of local animals died within days and many thousands more were slaughtered to prevent TCDD from entering the food chain. Skin lesions or chloracne were reported among some exposed residents (several hundred) and chloracne later confirmed as related to the exposure event. However, long-term health effects among thousands of exposed residents have been regarded as inconclusive, although significant evidence of excess carcinogenic effects, cardiovascular and endocrine-related effects has been reported in long-term epidemiological studies (see Bertazzi et al. 2001; Pesatori et al. 2009).

Environmental Poisoning

Acute and chronic chemical poisonings are a major problem worldwide, especially in low- to middle-income countries. Children are facing high risks from chemical poisonings according to major UN agencies (e.g. WHO/UNEP 2013). Episodes of human poisoning by PCBs, dioxins, and other toxic organic chemicals are reported from environmental and food contamination. Acute and long-term effects are indicated for TCDD contamination (e.g. Seveso, Italy) and PCBs. TCDD and dioxin-like PCBs are listed as human carcinogens and can suppress the human immune system. The consumption of PCB-contaminated rice oil in Japan in 1968 and in Taiwan in 1979 caused pigmentation of nails and mucous membranes and swelling of the eyelids, along with fatigue, nausea, and vomiting. Post-natal effects of developmental delays and behavioral problems were also observed in children born up to seven years after the Taiwan incident. These effects were attributed to the pre-natal exposures and persistence of PCBs in their mothers' bodies. Also, children of mothers who ate large amounts of contaminated fish from Lake Michigan in North America showed impaired short-term memory function (UNEP 2017d).

Chronic Effects of Toxic Organic Chemicals

Chronic exposures to toxic organic chemicals are generally similar to organic pesticides and associated with a wide range of nonspecific symptoms such as headache, dizziness, fatigue, weakness, nausea, chest tightness, difficulty breathing, insomnia, confusion and difficulty in concentration, skin irritations, and reactions (e.g. VOCs and allergenic organic chemicals). Exposures are often complex, variable, and may involve mixtures of toxic chemicals over time, and other chemicals or agents. Symptoms are difficult to evaluate at low levels of exposure, usually self-reported, and prevalence varies (e.g. hazardous waste sites). Some noncarcinogenic effects on body organs and fluid systems from abnormal exposures to common toxic organic chemicals are summarized in Table 11.17.

Neurological Effects

Many toxic organic chemicals (e.g. nervous system toxicants) and pesticides affect the nervous system. Effects can vary according to (i) acute exposures causing short-term signs and symptoms of poisoning; sometimes with delayed impairments long after the exposure episode, or (ii) chronic/long-term exposures. Neurotoxic effects may vary from deficits in neurobehavioral performance and abnormalities in nerve function to neurological disease (e.g. Parkinson's disease and Alzheimer's disease) and neurodevelopmental disorders, including learning disabilities in children. An overview of similar neurotoxic effects related to pesticide exposures of adults and children is given in Section 9.10, derived primarily from reviews of human

Table 11.17 Some non-carcinogenic effects of toxic organic chemicals on humans.

Systemic effects	Observed effects	Examples of organic chemicals (workplace and environmental)
Respiratory	Many TOCs irritate the respiratory tract (nose, throat, and lungs) and some can induce asthma-like reactions	Organic solvents (e.g. formaldehyde, toluene, xylenes, perchloroethylene, and VOCs)
Cardiovascular (heart and blood vessels)	Elevated blood pressure, hardening of arteries, abnormal heart rhythms (e.g. methylene chloride and trichloroethylene), coronary heart disease (carbon disulfide)	Halogenated solvents, toluene and dinitrotoluenes, vinyl chloride
Hematological (blood systems)	Reduction in oxygen carrying capacity of red blood cells, disrupt immunological processes of white blood cells, decrease in production of blood cells	Benzene and other volatile organic compounds (VOCs), 1,2 dichloropropane, acrylonitrile, cresols, chlorinated phenols, dinitrophenols; 1,3 butadiene, TCDD
Gastrointestinal (digestive system)	Direct injury to cell surface; effects on metabolism, the immune system, or intestinal microbial flora	Formaldehyde, PCBs, dioxin, chlorophenols, and solvents
Hepatic	Liver damage	Benzenes, halogenated solvents (e.g. methylene chloride, carbon tetrachloride)
Renal	Kidney dysfunction and damage	Benzenes, halogenated solvents, ethylene glycol
Endocrine system	Disruption of normal function of organs and glands that secrete hormones. Diseases include hypothyroidism, diabetes mellitus, hypoglycemia, reproductive disorders	Various POPs (e.g. PCBs, dioxins, and DDT), BPA, phthalate esters, and endocrine-disrupting pharmaceutical residues (e.g. certain anticonvulsants and NSAIDs)
Dermal	Skin irritation damage and inflammation; dermatitis (irritant and allergic), skin absorption of solvents; chloracne effects (dioxins)	Organic solvents, epoxy resins, biocides, and halogenated compounds
Ocular	Eye irritation by direct contact (e.g. solvent vapor or liquid). Blindness (methanol)	Organic solvents (e.g. formaldehyde, xylenes, trichloroethylene, and styrene)
Nervous system[a]	Acute and chronic effects on central (CNS) and peripheral nervous system (PNS) (e.g. acute: dizziness, intoxication, respiratory failure and death; chronic: degeneration of nerve cells, muscle weakness, memory loss, and personality changes)	Organic solvents: CNS (e.g. benzenes, alcohols, glycols, and styrene) and PNS (e.g. n-hexane, methyl n-butyl ketone)

[a] See Neurotoxicity effects below.

epidemiology studies. Many animal studies support key neurotoxic effects of pesticides observed in humans.

The specific scope and global impact of the neurological effects of environmental chemicals, including TOCs, are largely unknown and complex. Landrigan et al. (2017, p. 19) argue in The Lancet that "the Evidence is strong that widely used chemicals and pesticides have been responsible for injury to the brains of millions of children and have resulted in a global pandemic of neurodevelopmental toxicity." Toxic organic chemicals that induce neurodevelopmental toxicity include PCBs, phthalates, PAHs, and brominated flame retardants.

In the case of the autism spectrum, recent research supports findings on mechanisms of behavioral disorders, in which changes in brain architecture may be due to genetics, environmental exposures such as chemicals, or an interaction between the two, and related to critical periods of development during pregnancy (e.g. Roberts et al. 2007; Shelton et al. 2012).

Cancers

Various lifestyle, dietary, and environmental factors are known to increase our risk of developing cancer, as described by the International Agency for Cancer Research (IARC) in its Monographs on the Identification of Carcinogenic Hazards to Humans, Volume 124. These risk factors include behaviors such as smoking and excessive alcohol consumption, workplace and environmental exposures to certain chemicals, ionizing radiation and sun exposure (UV), indoor emissions of household coal combustion, diesel exhaust emissions, and some viruses and bacteria. Out of over 140 000 chemicals produced for commercial use, a relatively small number of toxic organic chemicals (individual and mixtures) are classified as human or probable human carcinogens on the basis of human and animal studies by major agencies such as the IARC, US NTP, US EPA, ACGIH, and ECHA. Even so, many of these commercial chemicals lack adequate testing for evaluation of carcinogenic properties.

Among organic chemicals, the IARC has classified some common environmental contaminants such as formaldehyde, benzene, benzo(a)pyrene, trichloroethylene, the dioxin, TCDD, and dioxin-like PCBs, vinyl chloride, and some mixtures of hydrocarbon-related substances (e.g. coal tar pitches, shale oils, soot, and untreated mineral oils), as human carcinogens (Group 1). Various chemotherapeutic drugs are also included, as well as manufacturing or workplace activities (e.g. rubber manufacturing). Toxic organic chemicals that are probable human carcinogens (Group 2A) or possibly carcinogenic to humans (Group 2B) are similar to many substances or mixtures related to those classified in Group 1 (e.g. benzene derivatives, PAHs, halogenated hydrocarbons, PCBs, nitrosamines, and refined petroleum oils).

Work-related deaths from exposures to chemical carcinogens, including organic chemicals, are likely to be significant given the high proportion of cancers among an estimated almost one million workers who died in 2015 due to exposures from hazardous substances at work. The situation may be similar for environmental exposures to chemical carcinogens, but estimates are limited by the state of available evidence. WHO estimates of global cancer deaths (4.9 million) and burden of disease (86 million DALYs) in 2004 were substantial. Yet these estimates were limited by the absence of sufficient health data on known or probable carcinogens including dioxins, chlorinated organic solvents, PCBs, and chronic pesticide exposures, as well as health impacts from exposure to local toxic waste sites (see Prüss-Ustün et al. 2011).

Environmental factors, including environmental carcinogens, are estimated to be responsible for 75–80% of cancers in the United States. Workplace and environmental exposures to specific carcinogens are directly linked to about 6% of cancer deaths per year (34 000 deaths annually) in the United States (President's Cancer Panel 2010; Physicians for Social Responsibility 2017). Conservatively, the cancer burden caused by environmental carcinogens may be larger (see Physicians for Social Responsibility 2017) or underestimated (Prüss-Ustun et al. 2011).

The observed increase in incidence of endocrine-related cancers in humans in countries studied appears to be related to genetic factors and environmental factors, including chemical exposures, but very few of these factors have been clearly shown (WHO/UNEP 2013, pp. 136–137).

Reproduction and Development
Many TOCs exhibit endocrine-disruption properties, among about 800 chemicals known or suspected to be endocrine-disrupting chemicals (EDCs). As discussed in Chapter 9 on pesticides, there have been significant increases in reproduction disorders and related diseases over the last few decades in many parts of the world linked to the endocrine system (e.g. testicular and breast cancers). The mounting evidence related to EDCs is based on three strands (WHO/UNEP 2013, p. 2):

- The high incidence and increasing trends of many endocrine disorders in humans
- Observations of endocrine-related effects in wildlife populations
- The identification of chemicals with endocrine-disrupting properties linked to disease outcomes in laboratory studies

Certain human studies have also shown stronger associations with pesticide and other EDC exposures, particularly through the use of animal model data and human evidence. Children are among the most vulnerable humans. Exposures to EDCs during fetal development and puberty are considered to play a role in increased incidences of endocrine-related diseases and dysfunctions including infertility, endocrine-related cancers, behavioral and learning problems, for instance, ADHD, infections, and asthma, and possibly obesity and diabetes in humans (WHO/UNEP 2013, p. 7). In adults, EDCs have been linked with obesity, cardiovascular disease, diabetes, and metabolic syndrome.

Endocrine Disorders and Diseases
The evidence of endocrine disruption in wildlife is substantial (as mentioned in Section 11.12) and supported by mechanisms of action of EDCs as shown in experimental animals. Our knowledge of the association of endocrine effects on humans from EDC exposures is limited (WHO/UNEP 2013), although on-going research is continually finding new insights and filling gaps.

Effects of endocrine disrupters (e.g. phthalate esters, flame retardants PBDEs, dioxins, PCBs, and chlorinated pesticides) on human (male and female) and animal reproduction systems, as well as metabolic, thyroid, adrenal, and bone disorders, and immune system dysfunction and disease have been reviewed by the WHO/UNEP (2013) among others. Table S11.8 summarizes major human-related effects of EDCs from sources reported by the WHO reviews. The findings suggest that all endocrine systems are affected by EDCs and not limited to estrogenic, androgenic, and thyroid pathways. EDCs can also interfere with metabolism, fat storage, bone development, and the immune system.

Many emerging organic pollutants of concern, including PPCPs, are endocrine disrupters in animal studies but sufficient human evidence based on epidemiological studies faces many data and methodology barriers at environmental levels as found for known POPs. Several persistent pharmaceuticals (carbamazepine, naproxen, diclofenac, and ibuprofen) in environmental waters are proposed as priority emerging chemicals because of their possible adverse health effects (e.g. EDCs) in humans and wildlife (Archer et al. 2017).

> ... There is a growing concern that maternal, fetal and childhood exposure to EDCs could play a larger role in the causation of many endocrine diseases and disorders than previously believed. This is supported by studies of wildlife populations and of laboratory animals showing associations between exposure to EDCs and adverse health effects and by the fact that the increased incidence and prevalence of several endocrine disorders cannot be explained by genetic factors alone. Epidemiological studies to date have explored quite narrow hypotheses about a few priority pollutants, without taking account of combined exposures to a broader range of pollutants ....
>
> WHO/UNEP (2013, p. X).

The state of child health impacts from exposures to EDCs is now a priority concern for global health. Many EDCs are developmental neurotoxicants and are found widely in consumer products. Exposures to EDCs, including various TOCs, are now known to cause strong and often irreversible effects on developing organs in the fetus and during early childhood compared to exposures in adults. There is sufficient evidence that male reproductive disorders that originate during fetal development are increasing in human populations that have been studied, and that this is partially related to environmental exposures. Some occupational diseases and disorders are also linked to EDC exposures (e.g. WHO/UNEP 2013; Landrigan et al. 2017, pp. 19–20).

11.14 Key Points

1. The major types of organic chemicals produced are petrochemicals and derivatives, agrochemicals (including pesticides), polymers, surfactants, many other industrial, specialty and life science chemicals, pharmaceuticals and veterinary chemicals, and personal care products.
2. Many of these organic chemicals are known or potential toxic organic chemicals. Their scientific classification depends largely on evaluation of their chemical structures, functional groups, physical, chemical and environmental properties (e.g. persistence, mobility, and bioaccumulation), and studies of their toxicity.
3. Natural or recent biogenic chemicals such as terpenes from plants, phytoestrogens (e.g. isoflavones and mycotoxins), PAHs from forest or bushfires, and algal toxins (e.g. microcystin) are also potentially toxic to humans and natural biota.
4. TOCs can generally be separated according to criteria that define their persistence and fate in the environment, capacity to bioaccumulate in organisms, and induce different types of toxic effects.
5. They may be considered as five main groups for evaluation purposes: (i) persistent bioaccumulative and toxic chemicals (PBTs), (ii) persistent halogenated and nonhalogenated organic chemicals (POPs), (iii) less-persistent and less-bioaccumulative TOCs, i.e. less to more polar, (iv) contaminants of emerging concern (CECs), and (v) pharmaceutical and personal care products (PPCPs), a special sub-group of CECs.
6. POPs are an important group of persistent and bioaccumulative chemicals controlled under the United Nations (UNEP) – Stockholm Convention.
7. Environmental persistence of POPs is indicated by half-lives of >60 days (waters), >120 days (sediments and soils), and >2 days (air). Bioaccumulation is shown by a bioconcentration factor (BCF) in aquatic species of >5000 or a $K_{OW} > 5$ (in absence of BCF data).
8. Other criteria for POPs depend on the evaluation and weighting of evidence of long-range environmental transport, monitoring and modeling data of transport, transfer and fate, ecotoxicity effects, and adverse toxic effects on human health.
9. Emerging pollutants are described "as any synthetic or natural chemical or microorganism that is not routinely monitored in the environment but has the potential to enter or is already detected in the environment, and cause suspected adverse effects on humans and wildlife."
10. Emerging TOCs include (i) existing organic contaminants in the environment that have been detected or evaluated by advances in analytical technology and toxicology, assessment or review, and (ii) new sources of emerging contaminants (new or existing substances) that are released or disposed to the environment.
11. PPCPs are a special group of TOCs of recent environmental concern that include antibiotics, analgesics, steroids, lipid regulators, antidepressants, hypertension drugs, anti-epileptics, stimulants, antimicrobials, sunscreen agents, fragrances, and cosmetics.
12. Global TOCs emitted or released into the environment are in the many millions of tonnes per year. Hazardous wastes containing TOCs are also generated in large quantities.

13. Important environmental properties and rate processes for transport and removal of TOCs include physicochemical properties, Henry's Law constant, soil adsorption coefficient (K_{OC}), bioconcentration factor, and processes such as diffusion, adsorption, volatilization, oxidation, photolysis, hydrolysis, biodegradation, and chemical decomposition. Persistence in environmental compartments or media is indicated by half-lives ($t_{1/2}$).
14. The rate of degradation of synthetic polymers (plastics) is a major environmental problem for waste disposal in landfills, aquatic environments, and plastics recycling. Polymer degradation depends largely on its structure. UV degradation affects epoxies and chains containing reactive aromatic components, while polyesters are degraded by hydrolysis.
15. Plastic pollution of the world's oceans (and land) and related deaths of vulnerable wildlife continue to result from large-scale disposal of unwanted plastic wastes, environmental breakdown of plastic litter/particles, and widespread dispersion of microplastic particles, a complex emerging form of organic pollutants, including adsorbed contaminants.
16. PBTs and POPs have widely contaminated the abiotic environment, natural biota, ecosystem food webs, foodstuffs, and humans to an extent that includes the polar regions of the Earth.
17. Less-persistent toxic chemicals can also cause local and regional areas of elevated contamination from industrial, urban, agricultural, and waste disposal sources.
18. PPCPs are increasingly found widely in the global environment. Therapeutic drug residues including increasing antibiotic contamination from human and intensive livestock sources are of major concern.
19. Measurable levels of human and wildlife exposures to TOCs are prevalent and pervasive from multiple sources and environmental pathways.
20. In the Twenty-first Century, more recent POPs and emerging chemicals of concern are now accumulating in wildlife, and there is increasing evidence of transfer through food webs to wildlife in higher trophic levels and humans.
21. The bioaccumulation of organic pollutants from aqueous solution can be related to a simple partitioning or sorption process between an organism and water.

$$BCF = K_B = C_B/C_W \quad (11.3)$$

22. Biomonitoring has become a major strategy to measure and assess the types of contaminants and their metabolites or biomarkers found in biological specimens taken from target or baseline populations of humans (e.g. children) and specific species of wildlife populations.
23. Toxic organic chemicals can exert a wide range of physiological and behavioral effects on wildlife species and humans as shown by numerous laboratory, bioassay, field, and epidemiological studies.
24. Organic pollutants can induce a broad spectrum of selective and nonselective toxic effects on humans and wildlife due to the variety of their chemical forms, functional groups, and properties.
25. The combined interactions of mixtures of pollutants such as TOCs are of specific concern because they may exert effects (e.g. additive, synergistic, and antagonistic), and potentially when individual chemicals are present at concentrations below toxicity endpoints.
26. Recent and historical evidence of population decline in some wildlife populations are associated with reproduction effects (e.g. infertility and feminization) and nonreproduction effects (e.g. thyroid and adrenal disorders or disease, hormone cancers, and bone disorders) related to endocrine-disrupting chemical (EDC) exposures.
27. Many toxic organic chemicals are known systemic toxicants (e.g. neurotoxicants, respiratory toxicants, reproductive and development toxicants), carcinogens, and endocrine disrupters.
28. Available estimates of global burden of injury and disease at work and from environmental exposures due to TOCs are limited in scope and are likely to be underestimates. Even so, they are strongly implicated in many types of work and environment-related diseases and deaths.
29. Human carcinogens include toxic organic substances such as formaldehyde, benzene, benzo(a)pyrene, trichloroethylene, the dioxin, TCDD, and dioxin-like PCBs, vinyl chloride, and some mixtures of hydrocarbon-related substances (e.g. coal tar pitches, shale oils, soot, and untreated mineral oils).
30. There is substantial and increasing evidence of major human effects related to exposures to EDCs, which is also supported by wildlife studies. Many TOCs exhibit endocrine-disruption properties, among about 800 known or potential chemical toxicants.
31. "There is a growing *[scientific]* concern that maternal, fetal and childhood exposure to EDCs could play a larger role in the causation of many endocrine diseases and disorders than previously believed."

References

Archer, E., Petrie, B., Kasprzyk-Horden, B. et al. (2017). The fate of pharmaceuticals and personal care products (PPCPs), endocrine disrupting contaminants (EDCs), metabolites and illicit drugs in a WWTW and environmental waters. *Chemosphere* 174: 437–446.

Arnot, J. and Gobas, F.A.P.C. (2006). A review of bioconcentration factor (BCF) and bioaccumulation factor (BAF) assessments for organic chemicals on aquatic organisms. *Environmental Reviews* 14 (4): 257–297.

ATSDR (1998). *Draft Update Toxicological Profile for Chlorinated Dibenzo-p-dioxins*. Atlanta, GA: Prepared by Research Triangle Institute for US Department of Health and Human Services, Agency for Toxic Substances Disease Registry.

aus der Beek, T., Weber, F.-A., Bergmann, A. et al. (2016). Pharmaceuticals in the environment – global occurrences and perspectives. *Environmental Toxicology and Chemistry* 35 (4): 823–835.

Baird, C. and Cann, M. (2012). *Environmental Chemistry*, 5e. New York: W.H. Freeman and Company.

Bergman, A. (2009). Health condition of the Baltic grey seal (*Halichoerus grypus*) during two decades. Gynaecological health improvement but increased prevalence of colonic ulcers. *Journal of Pathology, Microbiology and Immunology* 107: 270–282.

Bertazzi, P.A., Consonni, D., Bachetti, S. et al. (2001). Health effects of dioxin exposure: a 20-year mortality study. *American Journal of Epidemiology* 153 (11): 1031–1044.

Bickham, J.W., Sandhu, S., Hebert, P.D.N. et al. (2000). Effects of chemical contaminants on genetic diversity in natural populations: implications for biomonitoring and ecotoxicology. *Mutation Research* 463: 33–51.

Bosveld, A.T. and van den Berg, M. (2002). Reproductive failure and endocrine disruption by organohalogens in fish-eating birds. *Toxicology* 181–182: 155–159.

Bradley, P.M., Journey, C.A., Romanok, K.M. et al. (2017). Expanded target-chemical analysis reveals extensive mixed-organic-contaminant exposure in U.S. streams. *Environmental Science & Technology* 51: 4792–4802.

Broughton, E. (2005). The Bhopal disaster and its aftermath: a review. *Environmental Health* 4 (1): 6. https://doi.org/10.1186/1476-069X-4-6.

Buckman, A.H., Veldhoen, N., Ellis, G. et al. (2011). PCB-associated changes in mRNA expression in Killer whales (Orcinus orca) from the NE Pacific Ocean. *Environmental Science & Technology* 45 (23): 10194–10202.

Centers for Disease Control and Prevention (2017). Fourth report on human exposure to environmental chemicals. Updated Tables. Atlanta, GA: US Department of Health and Human Services, Centers for Disease Control and Prevention. https://www.cdc.gov/exposurereport/ (accessed 18 April 2020).

Cheek, A.O. (2006). Subtle sabotage: endocrine disruption in wild populations. *Revista de Biologia Tropical* 54: 1–19.

Chemsafetypro (2019). Overview and comparison of PBT and vPvB Criteria in EU and USA. ChemSafetyPro.COM. https://www.chemsafetypro.com/Topics/CRA/Overview_and_Comparison_of_PBT_and_vPvB_Criteria_in_EU_and_USA.html (accessed 13 April 2020).

Chesman, B.S. and Langston, W.J. (2006). Intersex in the clam *Scrobicularia plana*: a sign of endocrine disruption in estuaries? *Biological Letters* 2 (3): 420–422.

Connell, D.W. (2005). *Basic Concepts of Environmental Chemistry*, 2e. Boca Raton, FL: CRC Press.

Connell, D.W. and Miller, G.J. (1984). *Chemistry and Ecotoxicology of Pollution*. New York: Wiley.

Dann, A.B. and Hontela, A. (2011). Triclosan: environmental exposure, toxicity and mechanisms of action. *Applied Toxicology* 31 (4): 285–311.

Ebele, A.J., Abdallah, M.A., and Harrad, S. (2017). Pharmaceuticals and personal care products (PPCPs) in the freshwater aquatic environment. *Emerging Contaminants* 3 (1): 1–16.

Engineering ToolBox (2008). Dielectric constants of liquids. [online] https://www.engineeringtoolbox.com/liquid-dielectric-constants-d_1263.html (accessed 17 March 2020).

Environment Canada/US EPA (1995). *The Great Lakes: An Environmental Atlas and Resource Book*, 3e. Chapter 4, The great lakes today concerns. Chicago, IL: Environment Canada/United States Environmental Protection Agency https://www.nrc.gov/docs/ML0419/ML041970161.pdf (accessed 18 April 2020).

Esposito, N.P., Tiernan, T.O., and Dryden, F.E. (1980). *Dioxins*. Cincinnati, OH: Industrial Environmental Research Laboratory, Office of Research and Development, United States Environmental Protection Agency.

FAO (2019). Pesticides use. FAOSTAT. Food and Agricultural Organization, United Nations, Rome (accessed 14 April 2020).

Geissen, V., Mol, H., and Klumpp, E. (2015). Emerging pollutants in the environment: a challenge for water resource management. *International Soil and Water Conservation Research* 3 (1): 57–65.

Grove, R., Henny, C., and Kaiser, J. (2009). Osprey: worldwide sentinel species for assessing and monitoring environmental contamination in rivers, lakes, reservoirs, and estuaries. *Journal of Toxicology and Environmental Health Part B* 12 (1): 25–44.

Guillette, L.J. Jr., and Moore, B.C. (2006). Environmental contaminants, fertility, and multioocytic follicles: a lesson from wildlife? *Seminars in Reproductive Medicine* 24 (3): 134–141.

Guillette, L.J. Jr., Gross, D., Rooney, A.A. et al. (2000). Alligators and endocrine disrupting contaminants: A current perspective. *American Zoology* 40: 438–452.

Guiqueno, M.F. and Fernie, K. (2017). Birds and flame retardants: a review of the toxic effects on birds of historical and novel flame retardants. *Environmental Research* 154: 398–424.

Halden, R.U. (2010). Plastic and Health Risks. *Annual Review of Public Health* 31: 174–194.

Halling-Sorensen, B., Nielson, S.N., Lanzky, P.F. et al. (1998). Occurrence, fate and effects of pharmaceutical substances in the environment – a review. *Chemosphere* 36 (2): 357–393.

Hamalainen, P., Takala, J., Kiat, T.B. (2017). Global estimates of occupational accidents and work-related illnesses 2017. Ministry of Social Affairs, Finland and Workplace Safety and Health Institute, Singapore. http://www.icohweb.org/site/images/news/pdf/Report%20Global%20Estimates%20of%20Occupational%20Accidents%20and%20Work-related%20Illnesses%202017%20rev1.pdf (accessed 30 September 2020).

Hamlin, H.J. and Guillette, L.J. Jr., (2011). Embryos as targets of endocrine disrupting contaminants in wildlife. *Birth Defects Research, Part C: Embryo Today – Reviews* 93: 19–33.

Harris, C.A., Hamilton, P., Runnalls, T.J. et al. (2011). The consequences of feminization in breeding groups of wild fish. *Environmental Health Perspectives* 119 (3): 306–311.

HIS Markit (2019). Surfactants. https://cdn.ihs.com/www/pdf/0219/IHS%20Markit%20Surfactants%20Infographic.pdf (accessed 18 April 2020).

Hughes, S.R., Kay, P., and Brown, L.E. (2012). Global synthesis and critical evaluation of pharmaceutical data sets collected from river systems. *Environmental Science & Technology* 47: 661–677.

Igbinosa, E.O., Odjadjare, E.E., Chigor, V.N. et al. (2013). Toxicological profile of chlorophenols and their derivatives in the environment: the public health perspective. *Scientific World Journal* 2013: 460215. https://doi.org/10.1155/2013/460215.

Jobling, S., Williams, R., Johnson, A. et al. (2006). Predicted exposures to steroid estrogens in U.K. rivers correlate with widespread sexual disruption in wild fish populations. *Environmental Health Perspectives* 114 (Suppl. 1): 32–39.

Kidd, K.A., Blanchfield, P.J., Mills, K.H. et al. (2007). Collapse of a fish population after exposure to a synthetic estrogen. *Proceedings of the National Academy of Sciences of the United States of America* 104 (21): 8897–8901.

Kinney, C., Furlong, E., Kolpin, D. et al. (2008). Bioaccumulation of pharmaceuticals and other anthropogenic waste indicators in earthworms from agricultural soil amended with biosolid or swine manure. *Environmental Science & Technology* 42 (6): 1863–1870.

Ladeira, C. and Viegas, S. (2016). Human biomonitoring – an overview on biomarkers and their application in occupational and environmental health. *Biomonitoring* 3 (1): 15–24.

Landrigan, P.J., Fuller, R., Acosta, N.J.R. et al. (2017). The Lancet Commission on pollution and health. *The Lancet* 391 (10119): 462–512.

Lau, C., Anitole, K., Hodes, C. et al. (2007). Perfluoroalkyl acids: a review of monitoring and toxicological findings. *Toxicological Sciences* 99: 366–394.

Laug, E.P., Kunze, F.M., and Prickett, C.S. (1951). Occurrence of DDT in human fat and milk. *Archives of Industrial Hygiene and Occupational Medicine* 3 (3): 245–246.

Lee, Y.-M., Kim, K.-S., Jacobs, D.R. et al. (2017). Persistent organic pollutants in adipose tissue should be considered in obesity research. *Obesity Reviews* 18: 129–139.

Letcher, R.J., Klasson-Wehler, E., and Bergman, A. (2000). Methyl sulfone and hydroxylated metabolites of polychlorinated biphenyls. *The Handbook of Environmental Chemistry* 3K: 315–359.

Matthies, M., Solomon, K., Vighi, M. et al. (2016). The origin and evolution of assessment criteria for persistent, bioaccumulative and toxic (PBT) chemicals and persistent organic pollutants (POPs). *Environmental Science: Processes and Impacts* 18 (9): 1114–1128.

Moore, J.A. (1978). Toxicity of 2,3,7,8-tetrachlorodibenzo-p-dioxin. In: *Chlorinated Phenoxy Acids and Their Dioxins* (ed. C. Ramel), 134. Stockholm: Swedish Natural Science Research Council.

National Center for Biotechnology Information (2020). PubChem database. Perfluorooctanesulfonic acid, CID=74483. https://pubchem.ncbi.nlm.nih.gov/compound/Perfluorooctanesulfonic-acid (accessed 14 April 2020).

National Research Council (2006). *Human Biomonitoring for Environmental Chemicals*. Washington, DC: The National Academies Press https://doi.org/10.17226/11700.

Needham, L.L., Grandjean, P., Heinzow, B. et al. (2011). Partition of environmental chemicals between maternal and fetal blood and tissues. *Environmental Science & Technology* 45 (3): 1121–1126.

Neilson, A.H. and Allard, A.-S. (2009). Chemistry of organic pollutants. In: *Environment and Ecological Chemistry*, vol. 1 (ed. A. Sabijic) UNESCO. Oxford: Eolss.

Olaniran, A.O. and Igbinosa, E.O. (2011). Chlorophenols and other related derivatives of environmental concern:

properties, distribution and microbial degradation processes. *Chemosphere* 83 (10): 1297–1306.

Park, J.S., Bergman, A., Linderholm, L. et al. (2008). Placental transfer of polychlorinated biphenyls, their hydroxylated metabolites and pentachlorophenol in pregnant women from Eastern Slovakia. *Chemosphere* 70 (9): 1676–1684.

Pesatori, A.C., Consonni, D., Rubagotti, M. et al. (2009). Cancer incidence in the population exposed to dioxin after the "Seveso accident": twenty years of follow-up. *Environmental Health* 8: 39. https://doi.org/10.1186/1476-069X-8-39.

Petrie, B., Barden, R., and Kasprzyk-Horden, B. (2015). A review on emerging contaminants in wastewaters and the environment: current knowledge, understudied areas and recommendations for future monitoring. *Water Research* 72: 3–27.

Prüss-Ustün, A., Vickers, C., Haefliger, P. et al. (2011). Knowns and unknowns on burden of disease due to chemicals: a systematic review. *Environmental Health* 10: 9.

Prüss-Ustün, A., Wolf, J., Corvalán, C. et al. (2017). Diseases due to unhealthy environments: an updated estimate of the global burden of disease attributable to environmental determinants of health. *Journal of Public Health* 39 (3): 464–475.

Rappe, C. (1978). Chemical background of the phenoxy acids and dioxins. *Ecological Bulletins* 27: 28–30. https://www.jstor.org/stable/20112709 (accessed 30 October 2020).

Roberts, E.M., English, P.B., Grether, J.K. et al. (2007). Maternal residence near agricultural pesticide applications and autism spectrum disorders among children in the California Central Valley. *Environmental Health Perspectives* 115 (10): 1482–1489.

Ross, P.S. and Birnbaum, L.S. (2003). Integrated human and ecological risk assessment: a case study of persistent organic pollutants (POPs) in humans and wildlife. *Human and Ecological Risk Assessment: An International Journal* 9 (1): 303–324.

SCHER, SCCS, SCENIHR (2012). Opinion on the toxicity and assessment of chemical mixtures. Scientific Committee on Health and Environmental Risks. Scientific Committee on Consumer Safety. Scientific Committee on Emerging and Newly Identified Health Risks. European Commission, European Union. https://ec.europa.eu/health/scientific_committees/environmental_risks/docs/scher_o_155.pdf (accessed 18 April 2020).

Shelton, J.F., Hertz-Picciotto, I., and Pessah, I.N. (2012). Tipping the balance of autism risk: potential mechanisms linking pesticides and autism. *Environmental Health Perspectives* 120 (7): 944–951.

Sorensen, J.P.R., Lapworth, J.P., Nkhuwa, D.C.W. et al. (2015). Emerging contaminants in urban groundwater sources in Africa. *Water Research* 72: 51–63.

Stepa, R., Schmitz-Felton, E., and Brenzel, S. (2017). Carcinogenic, mutagenic, reprotoxic (CMR) substances. OSHWiki (accessed 14 February 2017).

Sui, Q., Cao, X., Lu, S. et al. (2015). Occurrence, sources and fate of pharmaceuticals and personal care products in the groundwater: a review. *Emerging Contaminants* 1 (1): 14–24.

Tarcher, A.B. (ed.) (2013). *Principles and Practice of Environmental Medicine*. New York: Springer US.

The American Cancer Society (2016). Perfluorooctanoic acid (PFOA), Teflon, and related chemicals. https://www.cancer.org/cancer/cancer-causes/teflon-and-perfluorooctanoic-acid-pfoa.html (accessed 14 February 2020).

The Essential Chemical Industry – Online (2013). Polymers – an overview. http://www.essentialchemicalindustry.org/polymers/polymers-an-overview.html (accessed 25 November 2017).

UNEP (2003). *Global Report 2003 - Regionally Based Assessment of Persistent Toxic Substances*. Geneva, Switzerland: United Nations Environment Programme https://www.researchgate.net/publication/292286858_Regionally_based_assessment_of_persistent_toxic_substances_Global_reprt (accessed 17 March 2020).

UNEP (2013). *Global Chemicals Outlook – Towards Sound Management of Chemicals*. Geneva: United Nations Environmental Programme.

UNEP (2016). Polychlorinated biphenyls (PCB) inventory guidance. PCB Elimination Network (PEN), Stockholm Convention. United Nations Environment Programme.

UNEP (2017a). Stockholm convention website. United Nations Environment Programme. http://www.pops.int (accessed 18 April 2020).

UNEP (2017b). What are POPs? Stockholm convention. United Nations Environment Programme. http://chm.pops.int/TheConvention/ThePOPs/tabid/673/Default.aspx (accessed 13 November 2017).

UNEP (2017c). The new POPs under the Stockholm convention. United Nations Environment Programme. http://chm.pops.int/TheConvention/ThePOPs/TheNewPOPs/tabid/2511/Default.aspx (accessed 18 February 2017).

UNEP (2017d). The 12 initial POPs under the Stockholm convention. United Nations Environment Programme. http://chm.pops.int/TheConvention/ThePOPs/The12InitialPOPs/tabid/296/Default.aspx (accessed 13 November 2017).

UNEP (2017e). All POPs listed in the Stockholm convention. United Nations Environment Programme.http://chm.pops

.int/TheConvention/ThePOPs/ListingofPOPs/tabid/2509/Default.aspx (accessed 13 November 2017).

UNEP (2019). All POPs listed in the Stockholm convention – United Nations Environment Programme. http://www.pops.int/TheConvention/ThePOPs/AllPOPs/tabid/2509/Default.aspx (accessed 26 July 2020).

UN WWDR (2017). *The United Nations World Water Development Report 2017: Wastewater*. Paris, UNESCO: The Untapped Resource https://www.unenvironment.org/resources/publication/2017-un-world-water-development-report-wastewater-untapped-resource (accessed 29 October 2020).

US EPA (2009a). International cooperation: persistent organic pollutants: a global issue, a global response. United States Environmental Protection Agency. https://www.epa.gov/international-cooperation/persistent-organic-pollutants-global-issue-global-response#domestic (accessed 12 November 2017).

US EPA (2009b). Emerging contaminants – polybrominated diphenyl ethers (PBDEs) and polybrominated biphenyls (PBBs). Fact Sheet, EPA 505-F-09-009. United States Environmental Protection Agency, Washington, DC.

US EPA (2009c). Emerging contaminant – N-Nitrosodimethylamine (NDMA). Fact Sheet, EPA 505-F-09-008. United States Environmental Protection Agency, Washington, DC.

US EPA (2009d). Emerging contaminant – 1,4-Dioxane. Fact Sheet, EPA 505-F-09-006. United States Environmental Protection Agency, Washington, DC.

US EPA (2009e). Emerging contaminant – 1,2,3-Trichloropropane. Fact Sheet, EPA 505-F-09-010. United States Environmental Protection Agency, Washington, DC.

US EPA (2012). Sustainable futures/P2 framework manual. 5. Estimating physical/chemical and environmental fate properties with EPI Suite™. EPA-748-B12-001. United States Environmental Protection Agency. https://www.epa.gov/sustainable-futures/sustainable-futures-p2-framework-manual (accessed 18 April 2020).

US EPA (2017a). Effluent guidelines: toxic and priority pollutants under the clean water act. United States Environmental Protection Agency. https://www.epa.gov/eg/toxic-and-priority-pollutants-under-clean-water-act/ (accessed 13 February 2017).

US EPA (2017b). Technical overview of volatile organic compounds. United States Environmental Protection Agency, Washington, DC. https://www.epa.gov/indoor-air-quality-iaq/technical-overview-volatile-organic-compounds (accessed 31 October 2020).

US EPA (2019). EPI Suite™ – estimation program interface. United States Environmental Protection Agency. https://www.epa.gov/tsca-screening-tools/epi-suitetm-estimation-program-interface (accessed 18 April 2020).

Verboven, N., Verreault, J., Letcher, R.J., and Gabrielsen, G.W. (2008). Differential investment in eggs by arctic-breeding glaucous gulls (Larus Hyperboreus) exposed to persistent organic pollutants. *The Auk* 126 (1): 123–133.

White, S.S. and Birnbaum, L.S. (2009). An overview of the effects of dioxins and dioxin-like compounds on vertebrates, as documented in human and ecological epidemiology. *Journal of Environmental Science and Health. Part C, Environmental Carcinogenesis and Ecotoxicology Reviews* 27 (4): 197–211.

WHO/UNEP (2013). *State of the Science of Endocrine Disrupting Chemicals – 2012*. Geneva: World Health Organization/United Nations Environment Programme.

Zhang, Z., Wang, B., Yuan, H. et al. (2016). Discharge inventory of pharmaceuticals and personal care products in Beijing, China. *Emerging Contaminants* 2: 148–156.

12

Metals

12.1 Introduction

Natural processes, such as chemical weathering and geochemical activities, release the various elements, including metals, from the Earth's crust into the lithosphere, atmosphere, and hydrosphere. Many of these metals and their compounds are essential for life on Earth. They undergo environmental transport and transformations through natural recycling processes which make up the Earth's biogeochemical cycles.

Although natural chemicals, many metals and their compounds can become toxic to humans, livestock, and wildlife at low doses of exposure. The metals, lead and mercury, are commonly rated in the top group of pollutants.

The influence of pollution on the global environment through activities such as mining operations, burning of fossil fuels, agriculture and urbanization, has accelerated the fluxes and accumulation of many metals in the ecosphere. The current rate of global input of some metals such as mercury, lead, zinc, and cadmium is in excess of the natural rate of biogeochemical cycling. For other metals, the rate of global contributions from human activities are less compared to natural fluxes. Even so, these metals may cause localized pollution such as with mineral extraction and metal processing, leachates from mine tailings, or urban emissions and disposal of metal contaminated wastes.

This chapter covers (i) the chemical nature, sources, releases, environmental behavior, and fate of metals, (ii) exposures to organisms, their toxicity, and ecological effects, and (iii) human health effects (noncarcinogenic and carcinogenic). Chapter 13 on air pollutants also considers metal pollutants in the atmosphere, while Chapter 15 on soil and groundwater contamination includes the behavior and effects of toxic metals.

12.2 Metals

The term, metal, typically describes a chemical element which is a good conductor of electricity and has high thermal conductivity, density, malleability, ductibility, and electropositivity. They lose electrons to form positive ions when chemically bonding. However, some elements (boron, silicon, germanium, arsenic, and tellurium), referred to as metalloids, possess one or more of these properties but are not sufficiently distinctive in their characteristics to allow a precise delineation to be called a metal or a non-metal. Arsenic (As) is an important example of a toxic metalloid. Furthermore, allotropic forms of some borderline elements may also exhibit different properties. In contrast, non-metals are elements that are poor conductors of electricity and heat, have low elasticity, and are very brittle. They have no metallic luster and do not reflect light. Most are gases; one is a liquid (bromine); and several are solids at room temperature. They tend to be electronegative except for the noble or inert gases of group 18 (He to Rn).

Figure 12.1 shows an arrangement of the periodic table of elements with atomic numbers, indicating metals, metalloids, and nonmetals. The metals consist of the alkali metals (Li 3–Fr 87), the alkaline earths (Be 4–Ra 88), the block of 38 transition metals, and two series of elements with metallic properties known as the lanthanides or *rare earth* elements (La 57–Lu 71) and the actinides (Ac 89–Lr 103). The lanthanide series contains 17 metal elements (*rare earths*), which are increasingly important in high technology industries, such as optics, electronics, and nanomaterials. Elements with atomic numbers greater than 92 (uranium) are also called transuraniums. They are unstable and decay radioactively into other elements.

The actinides series contains 15 elements, all radioactive metals, including uranium, thorium, and plutonium.

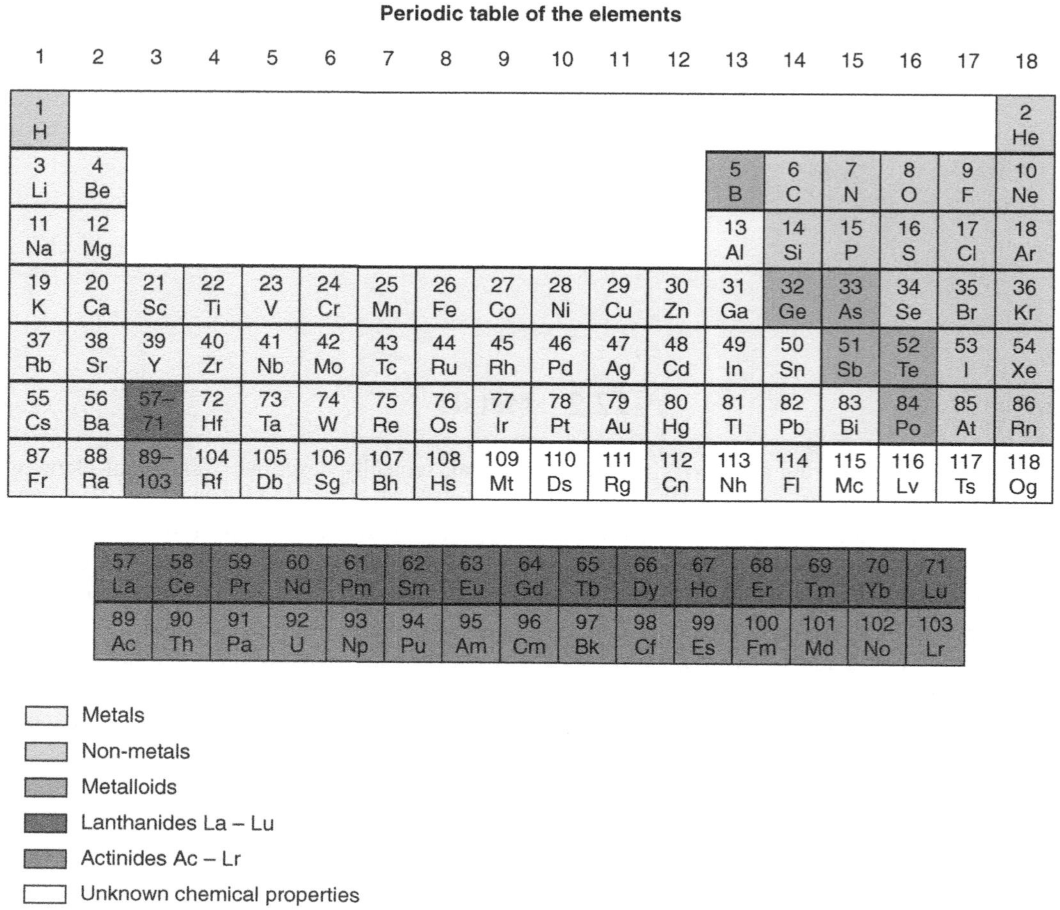

Periodic table of the elements

1	2	3	4	5	6	7	8	9	10	11	12	13	14	15	16	17	18
1 H																	2 He
3 Li	4 Be											5 B	6 C	7 N	8 O	9 F	10 Ne
11 Na	12 Mg											13 Al	14 Si	15 P	16 S	17 Cl	18 Ar
19 K	20 Ca	21 Sc	22 Ti	23 V	24 Cr	25 Mn	26 Fe	27 Co	28 Ni	29 Cu	30 Zn	31 Ga	32 Ge	33 As	34 Se	35 Br	36 Kr
37 Rb	38 Sr	39 Y	40 Zr	41 Nb	42 Mo	43 Tc	44 Ru	45 Rh	46 Pd	47 Ag	48 Cd	49 In	50 Sn	51 Sb	52 Te	53 I	54 Xe
55 Cs	56 Ba	57–71	72 Hf	73 Ta	74 W	75 Re	76 Os	77 Ir	78 Pt	79 Au	80 Hg	81 Tl	82 Pb	83 Bi	84 Po	85 At	86 Rn
87 Fr	88 Ra	89–103	104 Rf	105 Db	106 Sg	107 Bh	108 Hs	109 Mt	110 Ds	111 Rg	112 Cn	113 Nh	114 Fl	115 Mc	116 Lv	117 Ts	118 Og

57 La	58 Ce	59 Pr	60 Nd	61 Pm	62 Sm	63 Eu	64 Gd	65 Tb	66 Dy	67 Ho	68 Er	69 Tm	70 Yb	71 Lu
89 Ac	90 Th	91 Pa	92 U	93 Np	94 Pu	95 Am	96 Cm	97 Bk	98 Cf	99 Es	100 Fm	101 Md	102 No	103 Lr

- ☐ Metals
- ☐ Non-metals
- ☐ Metalloids
- ☐ Lanthanides La – Lu
- ☐ Actinides Ac – Lr
- ☐ Unknown chemical properties

Figure 12.1 Metals, metalloids, and nonmetals of the periodic table. Source: Modified from Jackson (2017).

Most are not found in nature. Most actinides are used to make larger synthetic elements, for example in nuclear reactors. Radionuclides are considered in **Chapter S22** in the Companion Book Online.

Metalloids (B, Ge, As, Sb, Te, and Po) form an interface between the metals and nonmetals, as shown in Figure 12.1, and exhibit intermediate properties between metals and nonmetals. Nonmetals, are on the right hand side of the periodic table, and include carbon, silicon, oxygen, and 17 group of the halogens (e.g. fluorine, chlorine, bromine, and iodine). Inert gases make up group 18.

Metals react as electron-pair acceptors (Lewis acids) with electron pair-donors (Lewis bases) to form various chemical groups, such as ion pairs, a metal-ligand complex, or a donor-acceptor complex. This type of equilibrium reaction can be generalized as follows:

$$M + L \leftrightarrow ML$$

$$K_{ML} = [ML]/[M][L]$$

Note: Charges on chemical species have been omitted for convenience.

In the above chemical equation, K_{ML} is the equilibrium (or stability) constant, and M represents the metal ion, L the ligand, and ML the metal-ligand complex. K_{ML} is the equilibrium (or stability) constant, which is characteristic of the metal and ligand involved. Accordingly, a metal ion has a broad general preference for either large, easily polarizable, low electronegativity ions (e.g. sulfide) or smaller, more electronegative anions (e.g. oxides) (Connell and Miller 1984; Connell 2005).

Essential and Nonessential Elements

A number of elements including metals are known as essential for life, although our knowledge is still evolving on the functions of nonessential elements in living organisms. Several low atomic weight cations, such as calcium, sodium, potassium and magnesium, have an essential role in mammalian metabolism (Goyer et al. 2004). Co, Cr (III), Cu, Fe, Mn, Mo, Se, and Zn are nutritionally essential metals for humans, but the biological roles of many elements tend to be uncertain, difficult to classify or unknown in biological species. Several elements with possible benefits at low levels of exposure are Ni, V, B and Si, and those with no known beneficial effects are reported to be Al, Sb, As, Ba, Be, Cd, Pb, Hg, Ag, Sr, and Tl (Goyer et al. 2004).

From a biological perspective, the term, *heavy metals*, is widely used in the scientific literature to describe toxic metals. Some others prefer the use of *transition metals*. The definition of heavy metals has been based primarily on (i) the specific gravity of the metals (greater than 4 or 5); (ii) location within the periodic table, for example elements with atomic numbers 22–31, 40–51, and 78–83, and the lanthanides and actinides; and (iii) specific biochemical responses in animals or plants. Generally, the following heavy metals and their compounds, including the metalloid arsenic, are of major environmental health concern: Zn, Cu, Cd, Hg, Pb, Ni, Cr, Co, Fe, Mn, Sn, Ag, and As.

Organometallic compounds form a covalent bond between the carbon of an organic compound and a metal such as an alkali, alkaline earth, or transitions metals. Organolead (e.g. tetraethyl lead), organomercury (e.g. methylmercury), and organotin (e.g. tributyltin) compounds are examples of organometallics of environmental and human health concern.

In this book, the term, *metals*, is commonly used to include metalloids. Sometimes, however, metals and metalloids of interest are identified separately.

12.3 Sources and Emissions of Metals

Metals released into the environment come from a variety of natural and pollution sources, which are outlined below from Connell and Miller (1984, pp. 289–293).

Natural Sources

In freshwater systems, chemical weathering of igneous and metamorphic rocks and soils in drainage basins is the most important source of background levels of trace metals entering surface waters (Leckie and James 1974). Considerable variation in background levels in surface water and bottom sediment is observed due to the presence of mineralized zones in the drainage basins. Decomposing plant and animal detritus also contribute small yet significant amounts of metals to surface waters and bottom sediments.

Metals released into the atmosphere from natural sources are commonly derived from (i) dusts emitted from volcanic activities, (ii) erosion and weathering of rocks and soils, (iii) smoke from forest fires, and (iv) aerosols and particulates from the surface of the oceans. Wet and dry precipitation from the air onto drainage basins and surface waters are the second most important input of metals into the world's oceans.

Overall, natural inputs of metals into the marine environment are released from (i) coastal supply sources, including river runoff from drainage basins, wave erosion of coastlines, and glaciers; (ii) deep sea supply, particularly metals released from deep sea volcanism and removed from particles or sediments by chemical processes; and (iii) atmospheric deposition and material from glacial erosion in polar regions that is transported by floating ice.

Pollution Sources

Mining Activities and Metal Production. The extraction of ore deposits on a large scale exposes fresh and waste rock surfaces and soils to accelerated weathering conditions, resulting in erosion and leaching of minerals and metals. For example, oxidation of pyrites and other sulfide minerals leads to the formation of acidic drainage waters that can leach residual heavy metals from ore stockpiles, mine tailings, and wastes (see Section 12.4). In addition, ore processing, smelting, and refining operations can cause (i) excessive atmospheric emissions, dispersion, and deposition; and (ii) wastewater discharges of toxic metals into drainage basins and human environments.

Urban Activities. Fossil fuel combustion from transport, heating, and processing (e.g. leaded fuels, coal, and diesel) emits substantial amounts of toxic metals into local and regional airsheds. Combustion of leaded fuels contributes significantly to urban lead deposition although decreasing in developed countries. Trace metals occur in domestic effluents due to metabolic wastes, corrosion of pipes, and disposal of metal-containing consumer products. Individual metal levels of up to mg L^{-1} are found, depending upon factors such as source characteristics, water flows, and level of treatment. Waste discharges and disposal of sewage sludges are significant causes of toxic metal enrichment of receiving waters or land treatment catchments (e.g. biosolids reuse). Leachates from sanitary landfills can have substantial levels of metals such as Cu, Zn, Pb, and Hg although loads are generally low.

Stormwaters. Runoff from urban catchments is a substantial source of trace metal enrichment of receiving waters. Metal composition of and levels in urban runoff are known to vary widely, influenced by factors such as urban planning, traffic, road construction, land uses in sub-catchments, accumulation of litter, physical characteristics, and climatic conditions. Peak loads of metals are usually associated with *first flush* storm events.

Industrial Activities: Emissions and Discharges. Many metals are emitted during industrial activities (e.g. cement production, coal-fired electricity generation) and are also discharged into water bodies through industrial effluents, dumping, and leaching of industrial wastes and stormwater runoff from sites. Metal levels in industrial wastewaters are often in the mg L^{-1} range.

Agricultural Runoff. Worldwide, very large quantities of sediment containing metals are eroded and discharged in downstream runoff from diverse agricultural catchments. Agricultural soils may be enriched with trace elements from animal and plant residues, phosphate-based fertilizers, specific herbicides and fungicides, and through use of biosolids (e.g. composted sewage sludges) or recycling of sewage effluent. However, trace elements tend to be stabilized through oxidation, formation of insoluble salts, and adsorption reactions, dependent on soil characteristics. Usually, the fate of metals is complex because of erosion of topsoils and potential remobilization of metals from sediments under different conditions, discussed in Section 12.4.

Tens of millions of tonnes of metals are produced yearly for human use as indicated in Table 12.1. Annual air emissions of toxic metals from anthropogenic sources are estimated to be in the thousands to tens of thousands of tonnes. Significantly, toxic metals are persistent, can be cumulative in air, water, soils, sediments, and organisms, and induce adverse effects on exposed organisms at low environmental doses.

Table 12.1 contains estimates of heavy metal emissions derived from heat and power production, nonferrous metal production, iron and steel industry, cement production, and waste disposal. These estimates indicated that stationary fossil fuel combustion was the major source of chromium (69%), nickel (90%), and selenium (89%). Nonferrous metal production was the largest source of atmospheric emissions of cadmium (73%), arsenic (69%), copper (70%), and zinc (72%) (MSC-E 2020).

The significance of metal pollution on a country scale is indicated by data for the United States. In 2015, the US Toxic Release Inventory (TRI) reported that industry disposed or released 3.36 billion pounds of TRI chemicals to on-site land, into the air, water, or by off-site transfers to some type of land disposal unit. Most of these chemical releases (70%) involved eight chemicals, mainly metals and their compounds: zinc (19%), lead (17%), manganese (7%), barium (6%), arsenics (5%), and copper (5%) (US EPA 2017, p. 34).

12.4 Behavior and Fate of Metals in the Environment

Under prevailing environmental conditions, the physical and chemical forms of metals and their properties (e.g. chemical species or valency and aqueous or lipid solubility) within air, waters, soils, sediments, and food, including prey, are vital in influencing the environmental distribution, fate, bioavailability, exposures and related biological effects within organisms, their populations and communities. These aspects are discussed below and also in other relevant chapters.

Atmosphere

Metals and their compounds can be readily emitted into the air as particulates or adsorbed onto them, and in the form of metal vapor (e.g. mercury), fumes (e.g. cadmium), or gases (arsine or nickel carbonyl), depending on the source (natural and human activities). Each year, very large quantities of toxic metals are emitted into the global atmosphere (see Table 12.1) and deposited widely on the Earth's oceans, freshwater bodies, and land, due to wet and dry precipitation processes. Atmospheric processes affecting metals and other atmospheric pollutants are described in Chapter 13.

Aquatic Systems

Natural waters and associated particulate matter are complex heterogeneous electrolyte systems containing

Table 12.1 Global emissions and releases of some major metal pollutants to the environment from production and use of chemical products and generation of hazardous wastes.

Metals	Global production/use (million tons year^{-1})	Air emissions (tons year^{-1})
Arsenic (2012)[a]	(38 655)[b]	5011[c]
Cadmium (2012)[a]	(21 861)[b]	2983[c]
Chromium (2015)[d]	30	14 730[c]
Copper (2015)[d]	18.7	25 915[c]
Lead (2010)[a]	~9.6	119 259[c]
(2015)[d]	4.7	
Mercury (2009)[a]	(3800)[b]	1960[c] (2010)
Nickel (2012)[a]	2.1	95 287[c]
Rare earth oxides (2015)[d]	(120 000)[b]	
Selenium		4601[c]
Uranium (2015)[d]	(60 500)[b]	
Vanadium (2015)[d]	(79 400)[b]	100 000–287 000[e]
Zinc (2015)[d]	13.4	57 010[c]

[a]UNEP (2013).
[b]tonnes year^{-1}.
[c]1995 emissions data unless indicated – Emissions from global modeling (MSC-E 2020).
[d]Geoscience Australia (2020).
[e]Schlesinger et al. (2017).
Sources: Based on UNEP (2013); Geoscience Australia (2020); Schlesinger et al. (2017); MSC-E (2020).

numerous inorganic and organic species distributed between aqueous phases. Trace metals entering natural waters become part of this system, and their distribution processes are controlled by a dynamic set of physico-chemical interactions and equilibria (Stumm and Morgan 1970).

The solubility of trace metals in natural waters is mainly controlled by (i) pH of the water, (ii) type and concentration of ligands and chelating agents, and (iii) oxidation state of the mineral components and the redox environment of the system (Leckie and James 1974). In addition, dynamic interactions at the solution–solid interfaces determine the transfer of metals between aqueous and solid phases. Thus, trace metals may be in a suspended, colloidal, or solid form. Generally, suspended particles are considered to be those greater than 100 μm in size; soluble particles are those less than 1 μm in size; and colloidal particles are those in the intermediate range.

Metal Species

Metal species present in suspended and colloidal particles may consist of (i) compounds or heterogeneous mixtures of metals in forms such as hydroxides, oxides, silicates, or sulfides, or (ii) clay, silica, or organic matter to which metals are bound by adsorption, ion exchange, or complexation (see Tinsley 1979). The soluble forms are usually ions, simple or complex, or unionized organometallic chelates or complexes (Stumm and Morgan 1970; Leckie and James 1974).

Table 12.2 shows examples of metal species in natural waters using a scheme proposed by Stumm and Bilinski (1972). Metals in natural waters may exist simply in the form of free metal ions surrounded by coordinated water molecules, although the concentrations of anionic species (e.g. OH^-, Cl^-, SO_4^{2-}, HCO_3^-, organic acids, and amino acids) are usually sufficient to form inorganic or organic complexes with the hydrated metal ions by replacing the coordinated water molecules. Other associations occur with colloidal and particulate material such as clays, hydrous iron and manganese oxides, and organic material (Hart and Davies 1978).

Several types of interactions occur between metal ions and other species in aqueous solutions (e.g. Stumm and Morgan 1970; Leckie and James 1974). These can be divided into two basic types of reactions.

Hydrolysis Reactions of Metal Ions

Most highly charged metal ions (e.g. Th^{4+}, Fe^{3+}, and Cr^{3+}) are strongly hydrolyzed in aqueous solution and have low pK_1 values.

$$Fe(H_2O)_6^{3+} + H_2O = Fe(H_2O)_5OH^{2+} + H_3O^+$$

Table 12.2 Forms of occurrence of metal species in natural waters.

Metal species and forms	Examples
Free metal ions	Cu^{2+}(aq), Zn^{2+}(aq), Pb^{2+}(aq), Fe^{2+}(aq), Cr^{3+}(aq), Ni^{2+}(aq)
Inorganic ion pairs	$Cu_2(OH)_2^{2+}$, $Al_2(SO_4)_3$, $Pb_2^{2+}(CO_3)_2^{2-}$, $Fe(OH)^{2+}$
Metal inorganic complexes	$[ZnHCO_3]^+$, $[Cu(CO_3)_2]^{2-}$, $[Ni(H_2O)_6]^{2+}$, $[Al(OH)_4(H_2O)_2]^-$
Organometallics	CH_3Sn^{3+} (aq), CH_3Hg^+ (aq) (CH_3HgX)
Metal organic complexes (M + L = ML)	M-SR, M-OOCR, M-amino acids, M-aminocarboxylic acids, M-phenolics, M-citrate
Metal dissolved organic matter (DOM) complexes: metal ions bound to high-molecular weight organic compounds	M-DOM: M-lipids, M-humic acids, M-fluvic acids, M-polysaccharides
Metal species dispersed in form of colloids	$Fe(OH)_3$, Mn (IV)
Metal species adsorbed onto colloids (e.g. clays), suspended inorganic and organic particles, and other matter	$M_x(OH)_y$ e.g. $Fe(OH)_3$, MCO_3, MS

Source: Modified from Stumm and Bilinski (1972).

Hydrolysis may also proceed further by the loss of one or more protons from the coordinated water.

$$Fe(H_2O)_5OH^{2+} + H_2O = Fe(H_2O)_4(OH)_2^+ + H_3O^+$$

Many divalent metals (e.g. Cu^{2+}, Pb^{2+}, Ni^{2+}, and Zn^{2+}) undergo hydrolysis also within the pH range of natural waters.

The hydrolysis of aqueous metal ions can also produce polynuclear complexes containing more than one metal ion, such as:

$$2FeOH^{2+} = Fe_2(OH)_2^{4+}$$

Polymeric hydroxo forms of metal ions may condense slowly with time to yield insoluble metal oxides or hydroxides. These types of species are important in moderate to high concentrations of metal salt solutions (Connell and Miller 1984, p. 295).

Complexation of Metal Ions

Importantly, metal ions also form complexes with inorganic and organic agents present in water from both natural and pollution sources. The main inorganic ligands that form complexes include Cl^-, SO_4^{2-}, HCO_3^-, F^-, S^{2-}, and phosphate species. These reactions are somewhat similar to the hydrolysis reactions of metal ions that may result in sequences of soluble complex ions and insoluble phases, depending on the metal and ligand concentrations and pH of water.

Inorganic ligands are usually present in natural waters at much higher concentrations than the trace metals they tend to complex. Each metal ion has a speciation pattern in simple aqueous solutions, that is dependent upon (i) the stability of the hydrolysis products and (ii) the tendency of the metal ion to form complexes with other inorganic ligands. For example, Pb (II), Zn (II), Cd (II), and Hg (II) each form a complex series when in the presence of Cl^- and/or SO_4^{2-} at concentrations similar to those in seawater. The pH at which a significant proportion of hydrolysis products are formed is dependent upon the ligand concentration, for example Cl^- competing with OH^- for the metal ion.

Metals can also bond to natural and synthetic organic substances by way of (i) carbon atoms to produce organometallic compounds, (ii) carboxylic acids to form salts of organic acids, (iii) electron-donating atoms O, N, S, P, and so on forming coordination complexes, or (iv) π-electron donating groups (e.g. alkenes or unsaturated carbon bonds, aromatic ring, etc.). Under aerobic conditions, free metal ions occur mainly at low pH. As the pH increases, the carbonate, and then the oxide, hydroxide, or even the silicate solids precipitate.

Metal speciation is also controlled by oxidation–reduction conditions. Redox environments within natural waters are usually complex, in a nonequilibrium state, and may show marked variations and gradients between air–water and water–solid, or water–sediment interfaces. Metal speciation is affected in two ways: (i) by direct change in the oxidation state of the metals ions (e.g. Fe (II) to Fe (III), and Mn (II) to Mn (IV), and (ii) by redox changes in available or competing ligands or chelates. Typical redox environments found in aquatic systems can be characterized by the use of pε–pH stability field diagrams. Stability fields for metal phases as a function of pε and pH at 25 °C and 1 atm total pressure are widely available, for example for lead, mercury, and iron (Connell and Miller 1984, p. 296).

Metal Speciation in Freshwater and Seawater

Metal speciation in freshwater and seawater differ considerably due mainly to (i) different ionic strengths, (ii) lower content of adsorbing surfaces in seawater, (iii) different concentrations of trace elements, (iv) different concentrations of major cations and anions, and (v) usually higher concentrations of organic ligands in freshwater systems. Adsorbed metals on metals mainly dominate in freshwaters and soluble metal–ligand complexes are more varied than in seawater. In mixtures of freshwater and seawater (e.g. estuaries), chloro-complexes become the dominant species for Cu, Zn, Hg, and Co, while Ni tends to remain as the free ion, and Cr forms hydroxide complexes. Adsorption is negligible for these metals, because an increase in ionic strength decreases the density of metal ions on particle surfaces due to competitive exchange in the electric double layer (see Förstner 1979a). Few metals (e.g. Cu (II) and Fe (II)) form significant amounts of organic complexes in seawater (Stumm and Brauner 1975). In most cases, abundant cations such as Mg^{2+} and Ca^{2+} compete for the organic functional groups (Förstner 1979a).

Metal Interactions Between Aqueous and Solid Phases

The behavior of metals in natural waters is strongly influenced by the interactions between the aqueous and solid phases, particularly water and sediments. Dissolved metal ions and complexes are rapidly removed from solution upon contact with surfaces of particulate matter through several different types of surface bonding phenomena (see Table 10.4, Connell and Miller 1984). The formation of insoluble metal substrates, such as sulfides due to acid-volatile sulfides (AVS), leads to the deposition and enrichment of trace elements in sedimentary environments. Enrichment and remobilization of metals in sediments are dependent on factors such as chemical composition (e.g. the amount of dissolved iron and carbonate), salinity, pH, redox values, and the hydrodynamic conditions.

Environmental factors affecting trace metal enrichment of aquatic sediments and their function as metal sinks are given as follows (see Connell and Miller 1984, pp. 297–300):

1. *Detrital Minerals.* The presence of heavy metals in silt and fine-sand fractions of sediment results in enrichment with trace elements, through adsorption processes. Alternatively, quartz, feldspar, and detrital carbonates tend to have the opposite effect.
2. *Sorption.* The general sorption capacity of solids for heavy metals follows the order; MnO_2 > humic acid > iron oxide > clay minerals. The sorption capacity of manganese oxides for trace metals is at least ten times greater than for iron oxides.
3. *Co-precipitation with Hydrous Fe/Mn Oxides and Carbonates.* Under oxidizing conditions, hydrous iron and manganese oxides act as a highly effective sink for metals. Co-precipitation of Zn and Cd with carbonate is indicated to be a depositional mechanism when other substrates, such as hydrous iron oxides, are less abundant.
4. *Complexation and Flocculation with Organic Matter.* In systems rich in organic matter, the role of Fe/Mn oxides is much less important because of competition

from more reactive humic acids, organo-clays, and oxides coated with organic matter. Organic coatings greatly affect the adsorption capacities of sediment and suspended matter. Metals complexed by humic acids become unavailable to form sulfides, hydroxides, and carbonates, which prevent the formation of insoluble salts. Chemical and electrostatic processes result in flocculation of Fe, Al, and humates, particularly in marine estuaries. An observed outcome is the enrichment of metals in humic substances from marine reducing environments.

5. *Trace Metal Precipitates.* Direct precipitation of trace metals as hydroxides, sulfides, and carbonates also occurs as a result of exceeding their solubility products K_{SP} in water bodies. For most metals in aerobic waters, the metal ion is stable at pH values of 7–8. As pH increases, the stable phase progresses from carbonate to hydroxide. Under reduced conditions, (-Eh), the insoluble sulfide remains the stable phase over a wide pH.

Overall, the processes of flocculation of metal-organic complexes and co-precipitation with hydrous Fe/Mn oxides are considered, especially important for metals introduced from human activities.

Mobilization of Metals from Sediments

Dissolved trace metals found in pore waters or interstitial waters within surface sediments are the most mobile and bioavailable forms of these metals in sediments. However, significant metal contamination of water bodies can readily occur from the release of various metals accumulated in sediments. Under suitable conditions, some metals associated with sediments and suspended particles are returned to the overlying water following remobilization and upward diffusion.

The release of these metals is controlled by at least five (5) major processes outlined below (Förstner 1979b):

1. *Elevated Salt Concentrations.* At elevated concentrations, the alkali and alkaline earth cations can compete for adsorption sites on the solid particle resulting in displacement of sorbed trace metal ions.
2. *Changes in Redox Conditions.* A decrease in the oxygen potential in sediments can occur due to conditions such as advanced eutrophication. This can result in a change in the chemical form of the metals and their water solubility. Under reducing conditions, trace metals in the interstitial waters occur as (i) sulfide complexes for Cd, Hg, and Pb; (ii) organic complexes for Fe and Ni; (iii) chloride complexes for Mn; and (iv) hydroxide complexes for Cr. With the development of oxidizing conditions in sediments, the solubility of metal ions is influenced by a gradual change from metallic sulfides to carbonates, hydroxides, oxyhydroxides, oxides, or silicates (Lu and Chen 1977).
3. *pH Changes.* Reduction in pH leads to dissolution of carbonates and hydroxides, as well as increased desorption of metal cations due to competition with hydrogen ions.
4. *Presence of Complexing Agents.* Increased use of natural and synthetic complexing agents can form stable metal complexes with trace metals that are otherwise adsorbed to solid particles.
5. *Biochemical Transformations.* This process can lead to either transfer of metals from sediments into the aqueous phase, or their uptake by aquatic organisms and subsequent release via decomposition products.

In the case of sedimentary deposition, organic matter containing metals, undergoes diagenesis, involving an increase in molecular weight and loss of some functional groups. A relatively stable and less reactive reservoir for heavy metals in sediments is formed. However, remobilization may occur through microbial processes.

Acid Mine Drainage and Acid Sulfate Soils

The exposure of iron pyrite and other sulfide minerals to atmospheric oxygen and moisture results in the oxidation of these minerals and the formation of sulfuric acid, such as in *acid mine drainage water* and also *acid sulfate waters* in disturbed coastal soils and sediments (also known as *acid sulfate* soils). Low pH conditions (along with metallic compounds including yellow–orange deposits of ferric hydroxide), are produced by a complex series of reactions mediated substantially by *Thiobacillus* and *Ferrobacillus* bacteria (see Figure 12.2). This process readily releases acid mine drainage from active and abandoned mine sites, such as coal mines, containing pyrite mineralization. It is widely known to cause serious water pollution impacts, involving relatively high levels of metals such as Fe, Mn, Zn, Cd, Pb, Cu, Ni, and Co. (Similar processes are observed in coastal development of *acid sulfate* soils leading to acidification of estuarine waters and releases of toxic metals.)

(1) $2FeS_2 + 2H_2O + 7O_2 \longrightarrow 4H^+ + 4SO_4^{2-} + 2Fe^{2+}$

(2) $4Fe^{2+} + O_2 + 4H^+ \longrightarrow 4Fe^{3+} + 2H_2O$

(3) $FeS_2 + 14Fe^{3+} + 8H_2O \longrightarrow 15Fe^{2+} + 2SO_4^{2-} + 16H^+$

(4) $Fe^{3+} + 3H_2O \longrightarrow Fe(OH_3 + 3H^+)$

Figure 12.2 Sequence of reactions involved in the oxidation of pyrites, iron (II) sulfide (FeS_2) to form *acid mine drainage water*.

Microbial–Metal Interactions

There are three major microbial processes affecting environmental transport of metals: (i) degradation of organic matter to lower molecular weight compounds, which are more capable of forming metal ion complexes; (ii) alterations to the physiochemical properties of the environment and chemical form of metals by metabolic activities, for example the oxidation–reduction potential and pH conditions; and (iii) conversion of inorganic compounds into organometallic forms by means of oxidative and reductive processes (Förstner 1979b).

This third mechanism involves bacterial or microbial methylation of a number of elements, for example Hg, As, Pb, Se, and Sn, in which methyl cobalamin appears to be the primary biological methylating agent. Craig (1980) has extensively reviewed biological transformations of metals and their environmental cycles. Methylation of mercury in sediments forms covalent dimethylmercury, $(CH_3)_2Hg$, and methylmercury compounds, CH_3HgX, under anaerobic conditions (X = suitable anions, e.g. Cl and OH). Methylmercury predominates over the more volatile dimethylmercury in neutral and acidic waters (see Baird and Cann 2012, pp. 531–533).

Terrestrial Systems

The chemical forms, mobility, fate, and bioavailability of metals in natural, agricultural, and contaminated soils are the subjects of intensive scientific studies for environmental and public health, including sustainable food production. Metal concentrations in soils and plant uptake of metals, nutritional benefits and potential toxicity to plants, herbivores, and humans are influenced by many factors including soil properties such as water content, salinity, pH, Eh, nutrients, organic carbon, climate conditions, and other factors. These aspects are considered further in this chapter (e.g. Sections 12.5 and 12.7) and also in Chapter 15 on Soils and Groundwater Contamination.

12.5 Environmental Exposures of Metals to Organisms

Living organisms are exposed to natural and anthropogenic sources of metals in different chemical forms, essentially through multiple environmental pathways and routes of intake into bodies: inhalation of metal containing particulates, gases or vapors from ambient air; absorption through root systems and leaves (plants), external surfaces or skin from contact with metals in air, dusts, soils, sediments, and/or waters; and oral ingestion of particulates, soil, waters, and dietary foods.

Humans can suffer multiple metal exposures from environmental, occupational, urban, indoor, dietary, and lifestyle (e.g. smoking) sources and contact with toxic metals in consumer products (e.g. lead paint), and materials. Exposure levels from toxic metal contamination such as in some workplaces, living near smelters or drinking groundwaters, can vary greatly above background or natural levels by factors of 10–100.

Vulnerable population groups at increased risk of exposure include pregnant women, fetuses, infants, and children (e.g. toxic metals such as lead and mercury). For example, young children exposed to lead (and some other metals) can have increased risk of adverse effects because of factors such as (i) behavioral characteristics (e.g. outdoor activity, ingestion of lead paint residues, *pica*, poor hygiene), (ii) higher absorption of food and drink per unit body weight than adults, (iii) higher lead absorption rate (50%) via gastrointestinal tract than adults (10%), enhanced by higher prevalence of nutritional deficiencies (e.g. iron and vitamin D) in children, (iv) blood–brain barrier not fully developed in young children, and (v) susceptible to hematological and neurological effects at lower thresholds than adults (WHO 2010a).

Background levels of toxic metals are relatively low in air, waters, sediments, and soils (except in areas of metal mineralization). In ambient air, values are in the range of $ng\,m^{-3}$ from <1 to $100\,ng\,m^{-3}$. Lead is usually $<50\,ng\,m^{-3}$ and Cd, As and Hg $<10\,ng\,m^{-3}$. Urban environments may have average air levels of these metals, increasing to 5 or 10 times higher. Urban lead levels can range in $\mu g\,m^{-3}$. Workplace activities with metals may result in air levels in the $mg\,m^{-3}$ range. Toxic metals in uncontaminated surface and groundwaters are usually in the low $\mu g\,L^{-1}$ or less, including drinking waters. In the United States, mercury levels have been reported at $5–100\,ng\,L^{-1}$. Arsenic and zinc levels, for instance, can occur at levels over $1\,mg\,L^{-1}$ or more. Higher metal levels (Pb, Cu, Zn, Fe, Ni, and Cr) in drinking waters may also be due to metal corrosion in pipes and fittings or water sources.

Average toxic metal levels in background soils have been reported at <0.1 (Hg) to $100\,mg\,kg^{-1}$ (dry weight), although some levels may be $>500–1000\,mg\,kg^{-1}$ due to local mineralization (e.g. Zn). Freshwater and marine sediments are similar to soils in background metal levels. Sediment quality guidelines for protection of aquatic organisms indicate low threshold levels for metals from $<1–2\,mg\,kg^{-1}$ (dry weight) for Ag, Cd, Hg, and Sb to higher ones for Pb, Ni, and Zn at 50, 21 and $200\,mg\,kg^{-1}$ (dry weight), respectively (Section 12.7).

Cumulative exposures for toxic metals are derived mainly from environmental and dietary sources. Estimated daily intakes ($\mu g\,day^{-1}$) of several toxic metals for humans from

Table 12.3 Estimated daily metal intakes (μg day^{-1}) by humans from different routes of exposure.

Metal	Air	Drinking water	Food	Other	Total
			μg day^{-1}		
As	0.02–0.20 (R)	<20	60 (US)	Soil/dust	<65–<280
	0.4–0.6 (U)		45 (Belgium)	0.14–0.28	
			126–273 (Japan)	Smoking	
				0.2–2; 6	
Cd	0.01–0.4	0.02–2	15–50	Smoking	<15–<52.4
Cr	<1 (<0.005)	0.8–16 (<1)	<200 (<10)	—	<202–<216
Pb	1 (0.5)	2–120 (1–60)	<100–500	—	<103–<621
Ni	<0.8 (0.4) (U)	<20 (3)	<300 (45)	—	<321–<386
	<23 (12) Ambient smoker				
Hg					
Vapor	0.04–0.20	0	0	Dental amalgam	3.9–21^a
				3.8–21	
Inorganic	0	0.05	0.6 (fish)	0	
			3.6 (non-fish)		
					4.2 (0.42)
Methyl mercury	0	0	2.4 (fish)	0	2.4
			(2.3)		(2.3)

R rural; U urban. Available absorption rates of metal intakes via a route of exposure are given in brackets ( ).
^{a}0.20–0.40 μg day^{-1} in absence of dental amalgams.
Source: Data from WHO (2000).

air, drinking water, food, and other sources are presented in Table 12.3.

There is considerable uncertainty in food estimates. Lead intake has been reduced in many countries to the extent that total daily intake is likely to be <100 μg. Nickel and chromium data in foods are limited. The mercury contribution from dental amalgams is potentially significant but scientific studies showing adverse effects from this source are lacking.

Tolerable intake values for selected metals are evaluated by the Joint FAO/WHO Expert Committee on Food Additives (JECFA) (see Table 12.4). Estimated exposure levels for specific metals can be compared with relevant tolerable intake values (where available) based on the exposed person's body weight (kg).

12.6 Absorption, Distribution, Transformation, and Excretion in Organisms

The absorption, distribution, transformation, and excretion of a metal within an organism, depends on the metal, the form of the metal or metal compound, and the organism's ability to regulate and/or store the metal

– US EPA (2007).

Table 12.4 Summary of tolerable intakes of selected metals.

Metals	Tolerable intake	Evaluation year
Arsenic	(2.1 μg kg^{-1} bw day^{-1})^a	Withdrawn 2011
Cadmium	PMTI 25 μg kg^{-1} bw	2013
Lead	(25 μg kg^{-1} bw day^{-1})^a	Withdrawn 2011
Mercury	PWTI 4 μg kg^{-1} bw	2011
Methyl mercury	PWTI 1.6 μg kg^{-1} bw	2007

PMTI, Provisional monthly tolerable intake; PWTI, Provisional weekly tolerable intake.
^{a}No longer considered health protective; bw, Body weight.
Source: Data from WHO (2019a).

Uptake Processes

The routes of uptake of metals by humans and other biota are primarily by respiratory intake (e.g. inhalation), ingestion (dietary, drinking water, and soil or dust), and absorption through contact with an organism's surfaces. In humans, dermal absorption of metals is limited. The actual amount an organism is exposed to through these uptake routes is the bioaccessible fraction of the metal(s). The metal form and how much of the metal is taken up by the organism are critical factors for potential toxicity.

The concept of metal bioavailability refers to the fraction of a metal species that an organism absorbs (or adsorbs) by crossing biological membranes, with the potential for distribution, metabolism, elimination, and bioaccumulation (US EPA 2007, pp. 4–4).

Absorption of metals across biological membranes can vary greatly and can be complex. It is influenced by several key factors that include aqueous solubility, particle size, valence state, lipid solubility, and the exposure matrix. Bioavailability increases as particle size decreases. For instance, inhalation is a primary route for metal exposure, essentially as inhalable and respirable particulates (e.g. airborne urban lead), along with any exposures to metal vapors or fumes (e.g. mercury and cadmium) and arsine gas (AsH_3). Larger metal particles tend to be swallowed and ingested while fine particles (<10 μm) penetrate deep into the lungs where they are more readily absorbed. Bioavailability of metals via inhalation can be much higher than by other routes of intake (US EPA 2007, pp. 4–9). Permeability of metals through human skin is generally low. Exceptions include elemental Hg and dimethyl Hg.

Uptake of metals by terrestrial wildlife is mainly via food and incidental soil ingestion, with minor inhalation intake. Soil ingestion tends to be more important than dietary intake for herbivores than for carnivores and invertebrates. Bioavailability of metals to animals and plants that live on (e.g. ground foraging) or in soils (e.g. burrowing) can be influenced by soil parameters, such as pH, cation exchange capacity (CEC), and organic carbon (US EPA 2007, pp. 6–9). Soil invertebrates such as soft-bodied invertebrates (e.g. earthworms) obtain metals through soil pore water by dermal absorption and soil ingestion. Hard-bodied invertebrates are mainly exposed through ingestion of food and incidental soil intake.

Plants can take up metals via roots through soil pore waters and also from aerial deposition on foliar surfaces (leaves). Foliar uptake of mercury is dominant, but other toxic metals (Cd and Pb) may also be accumulated through foliage depending upon soil conditions (US EPA 2007, pp. 6–3). Phytoremediation is an example where plants and associated soil microbes are used to reduce the concentrations or toxic effects of contaminants in soils by mechanisms of metal uptake, translocation, sequestration, and tolerance in plants (Ali et al. 2013).

The initial uptake of metals by aquatic organisms involves three main processes: (i) from water through respiratory surfaces (e.g. gills); (ii) adsorption from water onto body surfaces; and (iii) from ingested food, particles, or water through the digestive system. In the case of photo- and chemoautotrophic organisms, metal uptake occurs directly from solution, or for higher plants, also via the roots. Phytoplankton, for example appear to rapidly absorb metals at the cell surface, from where they diffuse into the cell membrane and are adsorbed or bound to proteins (ion exchange sites) within the cell. Generally, Bryan (1976a) considers the uptake of metals by plants to be a passive process that can be influenced indirectly by metabolism. However, some species of chemotrophs, for example those active in acid mine drainage, can metabolize metals directly from inorganic compounds such as metal sulfides (Connell and Miller 1984, pp. 302–306).

In heterotrophic organisms, the mechanisms of metal uptake are much greater than in autotrophic organisms and vary widely between species. Absorption of metals by most animals occurs by passive diffusion, probably as a soluble metal complex, through gradients created by adsorption at the body surfaces and binding by body constituents (Bryan 1976a). Rates of absorption are influenced by metals species, environmental changes (e.g. temperature, pH, salinity), and physiological and behavioral characteristics of organisms. For many metals, rates of absorption are directly proportional to the levels of availability in the environment (Bryan 1979).

More recently, the US EPA (2007, pp. 5–15) considers that unlike hydrophobic, nonionic organic chemicals, which commonly use passive diffusion for uptake across biological membranes, metal uptake involves a number of specific transport mechanisms (e.g. binding with membrane carrier proteins, transport through hydrophilic membrane channels, and endocytosis). Lipid soluble forms of metals, such as alkyl metals and neutral inorganic metal complexes (e.g. $HgCl^0$), are thought to use passive diffusion.

Metal uptake from dietary sources such as food and particulates in comparison to direct adsorption from solution is an essential mechanism for heterotrophic aquatic organisms, particularly for large animals (e.g. fish and lobsters) (Bryan 1976a; Bryan 1979). This is evident in polluted aquatic ecosystems where dietary preferences or feeding habits influence uptake from metal-enriched sediments, particulates, and detritus (Connell and Miller 1984, p. 306).

Distribution

The transport and accumulation of metals in the body within tissues, blood or plasma, or extracellular space is somewhat complex. Metals and their complexes are often ionized, with partitioning to blood and cellular components via interactions with proteins, being important. Retention or storage of metal forms often depends on sequestering by forming inorganic complexes or metal protein complexes (e.g. Pb in bones and Cd in tissues bound to low molecular weight, metallothionein proteins).

Metabolism or Transformation

General metabolism of metals is limited to oxidation–reduction reactions or alkylation/dealkylation reactions, resulting in the formation of many new inorganic or organic metal species. Different species of the same metals may have very different toxicities (e.g. As (III) and As (V), and inorganic and organic forms of mercury). Many contaminated organisms, however, are able to tolerate levels of metals in excess of known physiological needs or toxic effects (e.g. inhibition of enzymes). Detoxification mechanisms can involve storage of metals in inactive sites within organisms on a temporary or long-term basis. Temporary storage includes binding of metals to proteins, polysaccharides, and amino acids in soft tissues or body fluids. Metallothionein, however, can effectively store cadmium in liver and kidney tissues. Other storage sites such as bone, hair, feathers, fur, or exoskeleton usefully remove some toxic metals (e.g. Pb, Cd, and Hg). Remobilization of toxic metals from storage in tissues or organs can occur at different stages of development or age (Goyer et al. 2004).

Bioaccumulation

Many different organisms have shown the capacity to accumulate essential and nonessential metals from their diet and/or environments. Generally, the relative abundance of essential metals in an organism reflects levels necessary to maintain biochemical functions, for example enzyme systems. Where uptake of essential metals exceeds these levels, homeostatic mechanisms control optimum body levels, and tissue distributions. For example, in various decapod crustaceans, total body levels of Zn and Cu appear to be regulated within definite limits, showing concentration factors of about 10^4 for both metals (Phillips 1977). However, if uptake of essential or nonessential metals is excessive, homeostatic mechanisms are inhibited and bioaccumulation proceeds as uptake rate exceeds the loss rate. Phytotoxicity studies also demonstrate the capacity of various plant species to selectively accumulate specific metals/metalloids (e.g. As, Cu, Mn, Ni, Pb, and Mn) in their tissues/leaves.

The use of bioconcentration factors (BCFs) or bioaccumulation factors (BAFs) in aquatic risk assessment commonly assumes that the greater these values are, the greater is a chemical's bioconcentration or bioaccumulation potential. In the case of metals, BCF/BAF values are considered more uncertain and limited. An inverse relationship has been frequently reported between BCFs/BAFs and exposure concentrations. Factors considered responsible include the specific transport mechanisms of metals, saturable uptake mechanisms, storage of metals in organisms, and regulation of bioaccumulated metal levels. "Specifically, the current science does not support the use of a single, generic threshold BCF or BAF value as an indicator of metal hazard" (see US EPA 2007, pp. 5–12). The use of BCF/BAF values appears to be limited to site-specific measurements where extrapolation of these values across differing exposure conditions and species is minimized.

In humans, certain toxic metal compounds (e.g. Pb, Cd, and Hg) bioaccumulate in specific tissues or organs and may result in acute or chronic toxicity.

Box 12.1 Metal Bioaccumulation

Bioaccumulation of metals is the net accumulation of a metal in the tissue of interest or the whole organism that results from all environmental exposure media (e.g. air, water, soil, dusts, and food). It represents a net mass balance between uptake and elimination of the metal.

Source: Data from US EPA (2007).

Food Chain Transfer and Biomagnification

In food webs, the trophic transfer of a chemical from prey species to a predator species via dietary exposure is an important mechanism in toxicity assessment for wildlife and human populations. Bioaccumulation occurs within a trophic level. Biomagnification is a special type of trophic transfer where chemicals increase in concentrations in organisms from a lower trophic level to a higher one within the same food web. In the case of metals, the form of metals, bioavailability, and their distribution in dietary or prey organisms are critical factors in knowing their trophic transfer potential.

Prossi (1979) concluded that metal enrichment in aquatic food chains does not occur and the biomagnification mechanism has been oversimplified. Several key factors were identified:

1. The bioavailability of metals to higher trophic animals is generally determined by the rate of transfer from water rather than from food organisms.

2. Filter-feeding organisms are known to accumulate high levels of metals in their tissues but transfer only a small proportion to their predators.
3. Sediment- and detritus-feeding animals tend to accumulate higher metal levels than animals at higher trophic levels.
4. The life span of animals at higher trophic levels is usually greater than that of organisms at lower levels. Thus, age-related enrichment may be a significant factor in metal enrichment at higher trophic levels.
5. Preferential uptake and elimination of different metals and forms occurs.

Even so, Suedel et al. (1994) found from a review of various studies that total mercury and organic forms of mercury and arsenic had the potential to biomagnify in aquatic food chains although there was considerable variability and uncertainty in degree of food-web biomagnification.

Excretion and Regulation

The ability of organisms to regulate abnormal concentrations of metals determines their metal tolerance threshold, which is critical for their survival. This process occurs more with essential and relatively abundant metals (e.g. Cu, Zn, and Fe) rather than nonessential metals (e.g. Hg and Cd). Clemens (2006) has reviewed molecular mechanisms of toxic metal accumulation in plants and algae, including responses to metal exposure, as well as metal tolerance and its evolution. All plants appear to possess a basal tolerance of toxic nonessential metals. For Cd and As, this is largely dependent on the phytochelatin pathway.

In humans, metals are mainly excreted in urine because metal compounds are usually small and hydrophilic. Various metals may be excreted via protein binding in hair and fingernails (US EPA 2007, pp. 4–19). The rate of loss is a critical factor. The experimental rate of loss of various forms of metal isotopes from marine and estuarine mammals, fish, crustaceans, molluscs, and polychaetes, generally follows a two phase loss curve (fast and slow), with different half-lives. The slow phase is usually more representative.

12.7 Toxic Effects of Metals on Organisms

Different metals can induce a diverse range of physiological, biochemical, and behavioral dysfunctions in exposed organisms, including humans, via a variety of biochemical mechanisms. The effects of many metals are well documented in the scientific literature including those on the neurological, cardiovascular, hematological, gastrointestinal, musculoskeletal, immunological, and epidermal systems.

Metals cause biological effects on individual organisms following uptake and transport across cell membranes into cells where they can bind onto a cellular target (reversibly or irreversibly) and change specific biochemical processes. These effects may (i) be beneficial, such as for essential metals at optimum levels, or (ii) induce toxicity, particularly through oxidative (and nitrative) stress leading to oxidative tissue damage. The toxic effect of a metal primarily follows from the initial interaction of the free metal ion with the cellular target, generally consisting of biomolecules in the form of macromolecules (e.g. cellular proteins), membranes, or organelles. Metal toxicity can affect multiple organs, including damage to the kidney, nervous system, respiratory system, endocrine, and reproductive systems.

Essential and nonessential metals play different roles in inducing biological effects. Adverse nutritional and other effects or syndromes from stressors can result if nutritionally essential metals (e.g. Co, Cr, Cu, Fe, Mg, Mn, Ni, and Zn) are deficient in humans and other biota. Conversely, an excess amount of essential metals can produce cellular and tissue damage leading to a variety of adverse effects and diseases. In the case of chromium and copper, for example, there is a very narrow range of concentrations between beneficial and toxic effects. Dose–response relations for essential elements are variable but follow a general biphasic or U-shaped curve as discussed in Chapter 5. Nonessential metals (e.g. Al, Sb, As, Ba, Bi, Cd, Pb, Hg, Ag, Ti, U, and V) can induce general toxicity at increasing doses (non-threshold) or above, no adverse effect thresholds, although some may have beneficial effects at low levels of exposure (Tchounwou et al. 2012).

The general toxicity mechanism for metal ions has been classified into the following three categories: (i) blocking of essential biological function groups of biomolecules (e.g. proteins and enzymes), (ii) displacing the essential metal ion in biomolecules, and (iii) modifying the active conformation of biomolecules (Ochiai 1977). Nieboer and Richardson (1980) observed there are similar patterns in toxicity sequences for metal ions in a wide range of organisms that appear to be explained by their binding preferences. The most toxic ions exhibit a broad spectrum of toxicity mechanisms (see Connell and Miller 1984, p. 313).

Recent research has shown that common mechanisms of toxicity for metals may include: (i) inhibition of enzymes, such as some involved in metabolism, detoxification, and damage repair, (ii) disruption of the structure and/or function of subcellular organelles, (iii) interaction with DNA leading to mutagenesis or carcinogenesis,

(iv) covalent modification of proteins, (v) displacement of other critical metals in various metal dependent proteins, and (vi) inhibitory or stimulatory effects on the regulation of expression of various proteins. As well, competition between metals for complexation sites in proteins, involved in electron transfer reactions, may lead to toxicity (Hollenberg 2010).

Metal-induced toxicity is known to cause oxidation of sulfhydryl groups of proteins, lipid peroxidation, depletion of proteins, and DNA damage. The key factors that contribute to toxicity for different metals involve the generation of reactive oxygen (ROS) and nitrogen (RNS) species that disturb cell redox systems. ROS include free radicals (e.g. superoxide, O_2^{*-}, hydroxyl, OH^*, peroxyl, RO_2^*, and alkoxyl, RO^*), and certain non-radicals (e.g. peroxynitrite, $ONOO^-$, and H_2O_2) that act as oxidizing agents or are readily converted to radicals (Jan et al. 2015). Proposed mechanisms for metal-induced oxidative stress are shown in Figure 12.3.

Importantly, cells possess defensive mechanisms that can overcome oxidative and nitrative stress to maintain cellular homeostasis (e.g. Jan et al. 2015). These mechanisms involve antioxidants (e.g. glutathione, α-tocopherol, and ascorbate) and certain enzymes (e.g. glutathione, GSH) that can metabolize and detoxify ROS/RNS, and also cell repair systems for damage to DNA. Modes of action by different antioxidants in controlling the toxic effects induced by metals are indicated in Figure 12.3. However, metal toxicity, in the form of oxidative damage, develops when antioxidants are depleted by the production of reactive free radicals (ROS/RNS) induced by either redox inactive or redox active metals.

Figure 12.3 Possible mechanisms for metal-induced oxidative stress. Source: Modified from Ercal et al. (2001).

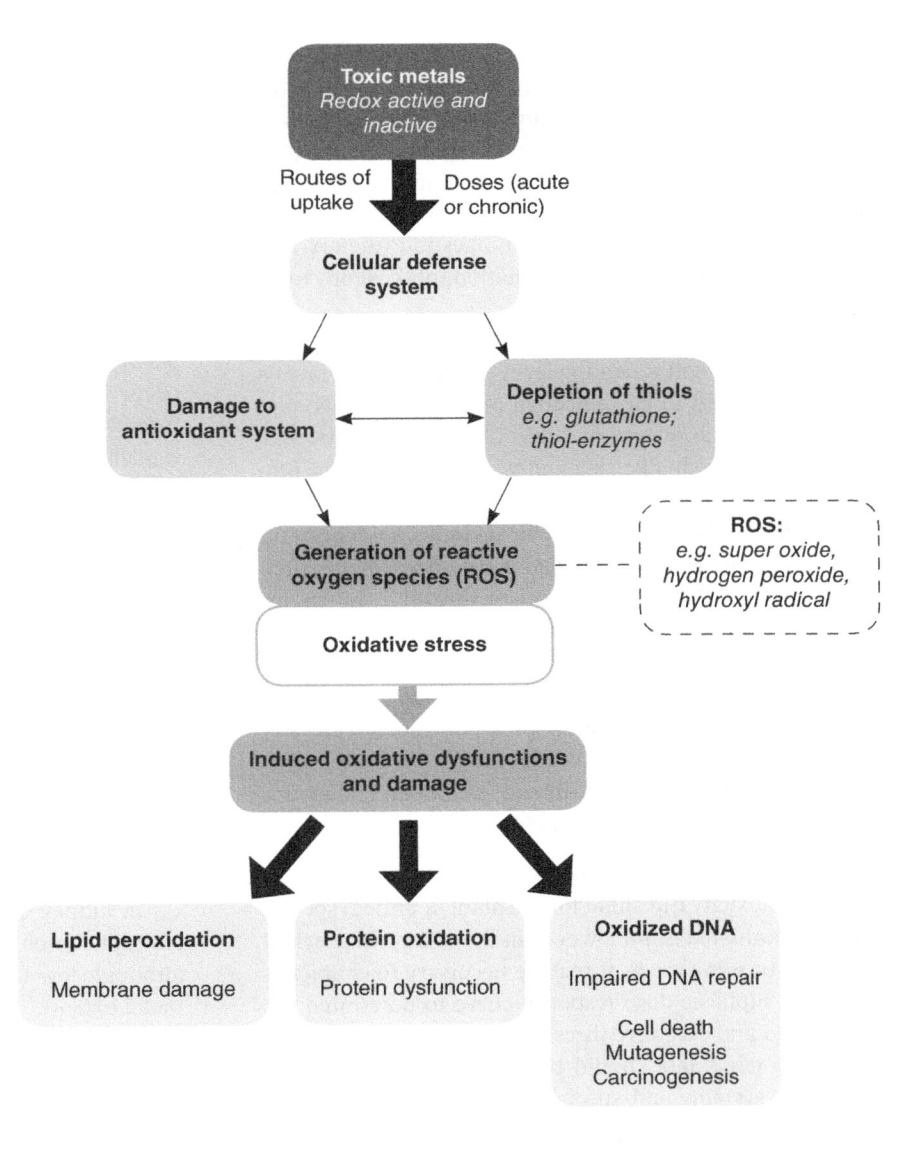

Effects on Terrestrial Organisms

Acute and Chronic Toxicity of Metals

Metal toxicity studies among terrestrial species vary greatly with a large body of literature on soil organisms, particularly plants and soil invertebrates, to limited data on wildlife vertebrates (e.g. ingestion of lead shot by waterfowl) compared to toxicity tests on laboratory species (e.g. rats, mice, domestic animals, and some bird species). Variability in soil toxicity tests is an issue due to a number of physical, chemical, and biological factors. For example, soil properties such as pH, organic matter, and cation exchange capacity (CEC) can affect bioavailability of various metals.

The evaluation of metal toxicity for exposed terrestrial organisms follows the same general principles used for ecological screening and risk assessment (see Chapter 18). Typically, receptors of interest cover soil invertebrates, plants, and wildlife (e.g. avian and mammalian species) and microorganisms involved in soil processes. Metal chemistry and bioavailability are very much associated with natural or background levels of metals in soils, soil properties, and soil-plant processes. Site-specific soil parameters can influence exposure to metal species (e.g. clay content, pH, and CEC). Exposure pathways vary from direct soil contact in the case of soil invertebrates; root and/or foliar uptake for plants; to ingestion (dietary and incidental soil) with wildlife. Inhalation intake tends to be poorly known (US EPA 2007).

In the United States, the EPA has developed a national set of ecological soil screening levels (Eco-SSLs) for metals that are designed to be protective of wildlife, plants, and soil organisms, as shown in Table 12.5. These values may be applied individually, regionally, or locally, taking into account factors such as background or site-specific levels of metals. The Eco-SSL document includes guidelines on derivation of toxicity thresholds for risk assessments such as site-specific. Although this set of threshold values for soils is a useful benchmark for metal toxicity interpretation, it does show gaps because of inadequate toxicity data for some metals (e.g. mercury, chromium, and arsenic), which reflects the current regulatory problem and uncertainty with available soil toxicity studies.

A number of key factors influence the metal toxicity for terrestrial organisms (US EPA 2007). These include:

1. *Essential metals*: it is critical to ensure that the effects level or toxicity threshold for the plant or animal species being evaluated is not lower than the estimated nutrient needs. In some cases, it may be necessary to characterize the biphasic dose response curve to determine the required and excessive thresholds.
2. *Toxicity tests*: data should be obtained from appropriate test systems and species tested that are similar to site-specific conditions. Otherwise, data from standard test species and conditions may be adjusted with uncertainty factors to obtain a final toxicity value.
3. *Metal mixtures and interactions*: mixtures of metals commonly occur in environmental media and in the presence of other chemical toxicants. Interactions between metals (e.g. Cd/Ca/Zn, Hg/Se, Cu/Mo) can occur in either their uptake or toxicity actions. These interactions are often difficult to resolve. Two common models used are the Concentration Addition model (used when metals are known or assumed to act with the same or similar MOA [modes of action] and can predict toxicity that is more than additive) and the Effect Addition model (used for metals with different modes of action). In the latter model, differing potencies are ignored, and the effect concentrations for each metal are combined to predict toxicity of the mixture.
4. *Critical body residues (CBRs)*: this approach uses the internal concentrations of chemicals to determine critical tissue or body burden loads that correlate with the onset of toxic effects. Some specific examples of tissue-specific critical loads are known for several species of vertebrate wildlife, including Pb in liver, Cd in kidney, Hg in brains, and Se in eggs. The same concept can be applied to soil invertebrates and plants although few CBRs appear to be readily available for metal toxicity.
5. *Metal toxicity thresholds:* soil toxicity tests tend to suffer with variability due to bioavailability differences in test conditions and difficulty in extrapolating toxicity

Table 12.5 US EPA ecological soil screening levels for metals and metalloids (eco-SSLs – mg kg^{-1}, dry weight in soil) for the protection of terrestrial organisms.

Metals/metalloids	Plants	Soil invertebrates	Avian	Mammalian
Antimony				
Arsenic	18	NA	43	46
Cadmium	32	140	0.77	0.36
Chromium (III)	IA	IA	26	34
Chromium (VI)	IA	IA	NA	130
Copper	70	80	28	49
Lead	120	1700	11	56
Manganese	220	450	4300	4000
Nickel	38	280	210	130
Silver	560	NA	4.2	14
Vanadium	NA	Na	7.8	280
Zinc	160	120	46	79

IA, Inadequate data to derive Eco-SSL; NA, Not available.
Note: these values may be periodically updated by US EPA.
Source: Based on US EPA (2016).

across soil types. Comparisons between field data and laboratory toxicity responses are considered to be best achieved by measuring metals in soil pore water and comparing results with those from spiked laboratory soils. However, site-specific assessments for plants, soil invertebrates, and wildlife can be derived and judged by following guidance documents developed for Ecological Soil Screening Levels (Eco-SSL) in the United States (see Table 12.5) or similar ecological screening guidelines used elsewhere for metal toxicity in contaminated land investigations.

6. *Wildlife toxicity:* the paucity of toxicity data on wildlife is a challenge for extrapolation of toxicity data to selected wildlife species from standard tests on laboratory rodents and other test species (e.g. domestic animals and waterfowl). The capacity of some species to regulate or store metals in their tissues can also complicate.

Effects on Aquatic Organisms

Acute and Chronic Toxicity of Metals

There is a large volume of available bioassay data covering the acute and chronic toxicities of metals on various aquatic organisms in water and more recently in sediments. These studies are performed on metals for a variety of marine and freshwater species at different life stages. Chronic toxicity studies may also involve a complete life cycle of a test organism. Databases containing aquatic ecotoxicity values for various metals include the US EPA Ecotoxicology database (ECOTOX), the US Library of Medicine's Hazardous Substances Databank (HSDB) (now integrated into PubChem), and the Pesticide Action Network (PAN) database.

Toxicity data for marine and freshwater test organisms exposed to metals are commonly measured as 48–96 hours LC_{50} or EC_{50} values in mg L^{-1} or µg L^{-1}. Acute and chronic toxicities can vary widely between metals, taxonomic groups, and species. For example, Table 12.6 shows some reported values for lethal toxicities of common metals to marine organisms.

Most sublethal or chronic effects appear to be biochemical in origin and often related to metabolic processes. They tend to be variable and nonlinear in dose–response and difficult to correlate with significant changes seen at the whole organism level (e.g. morphology, behavior, and reproduction). Much of the data on sublethal toxicity results from multiple and inconsistent experimental methods and lacks a mechanistic explanation for observed effects at the organism level.

Many studies show larval and juvenile life stages are critical for metal toxicity effects. For example, laboratory studies of larval and juvenile instars of the sand crab, *Portunus pelagicus*, exposed to chronic levels of Cu^{2+}, Ni^{2+}, and Cr(VI) in seawater found sublethal effects included inhibition of larval moulting, increase in the duration of the development period, and reduced size achieved by successive juvenile crab instars (Mortimer and Miller 1994).

Generally, sublethal effects have been observed as changes in: (i) morphology/histology; (ii) physiology (growth, development, swimming performance, respiration, circulation); (iii) biochemistry (blood chemistry, enzyme activity, endocrinology); (iv) behavior neurophysiology; and (v) reproduction (Bryan 1976a, 1976b; Alabaster and Lloyd 1980).

In aquatic organisms, some general observations from sublethal studies can be made as (Connell and Miller 1984, p. 320):

1. Morphological/histological changes in tissues of various fish and crustacean species following sublethal exposures to metals (e.g. Cu, Cd, and Hg) are secondary effects due to interference with enzyme processes involved in food utilization (Bryan 1976b). Many diverse changes are recorded including replacement of mucous cells in the gill epithelium by chloride cells, vertebral damage by Zn in freshwater minnow, as well as developmental and structural abnormalities, for example with sea urchin eggs and larvae (Bryan 1976b).
2. Suppression of growth and reproduction occurs widely among aquatic vertebrates and invertebrates exposed to metals at relatively low concentrations. Freshwater

Table 12.6 Lethal toxicity of metals to marine organisms (96 hours-LC_{50}, mg L^{-1}).

Classes of organisms	Metals						
	Cd	Cr	Cu	Hg	Ni	Pb	Zn
Fish	22–55	91	2.5, 3.2	0.8, 0.23	350	188	60, 60
Crustaceans	0.015–45	10	0.17–100	0.05–0.5	6, 47	—	0.4–50
Molluscs	2.2–35	14–105	0.14, 2.3	0.058–32	72, 320	—	10–50
Polychaetes	2.5–12.1	2.0–>5.0	0.16–0.5	0.02–0.09	25, 72	7.7–20	1.8–55

Source: Based on Connell and Miller (1984) and Connell (2005).

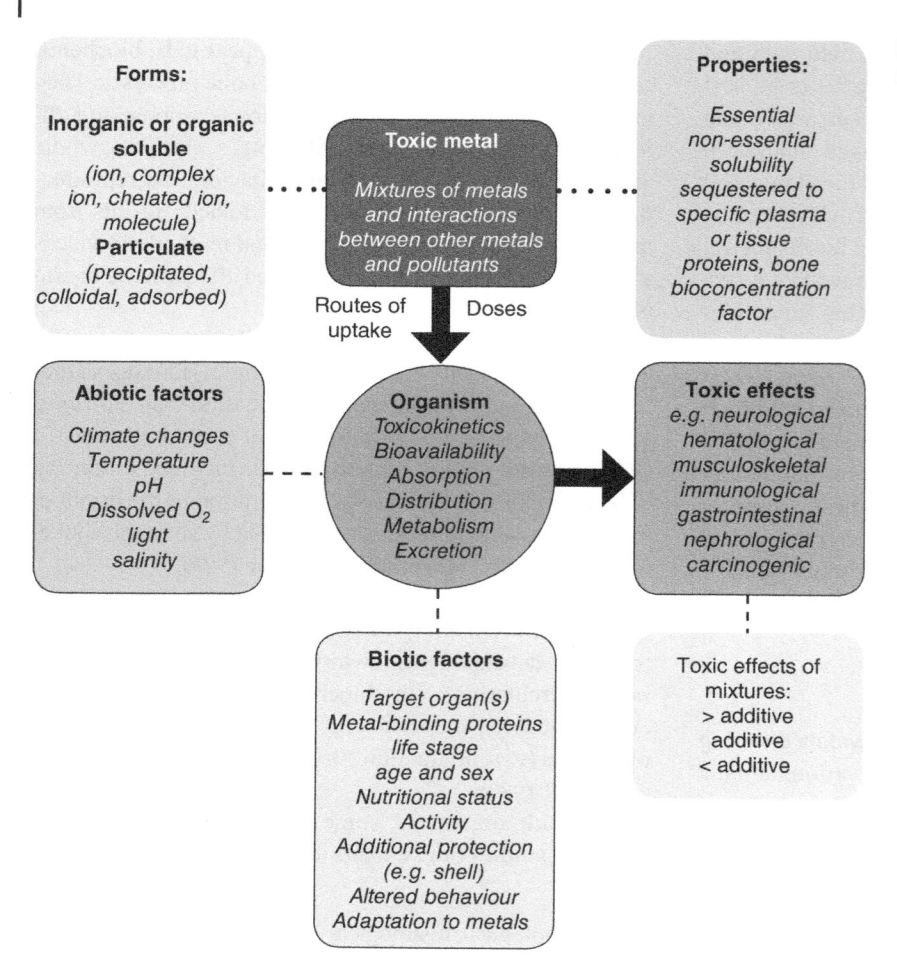

Figure 12.4 Factors influencing metal toxicity in aquatic organisms.

Forms:
Inorganic or organic soluble (ion, complex ion, chelated ion, molecule)
Particulate (precipitated, colloidal, adsorbed)

Toxic metal
Mixtures of metals and interactions between other metals and pollutants

Properties:
Essential non-essential solubility sequestered to specific plasma or tissue proteins, bone bioconcentration factor

Abiotic factors
Climate changes
Temperature
pH
Dissolved O_2
light
salinity

Routes of uptake | Doses

Organism
Toxicokinetics
Bioavailability
Absorption
Distribution
Metabolism
Excretion

Toxic effects
e.g. neurological
hematological
musculoskeletal
immunological
gastrointestinal
nephrological
carcinogenic

Biotic factors
Target organ(s)
Metal-binding proteins
life stage
age and sex
Nutritional status
Activity
Additional protection (e.g. shell)
Altered behaviour
Adaptation to metals

Toxic effects of mixtures:
> additive
additive
< additive

---

invertebrates and plants tend to be more resistant than fish although considerable variation in sensitivity exists between and within taxonomic groups (Alabaster and Lloyd 1980).

3. Reproduction in many aquatic organisms is affected in the parts per billion range for most toxic metal ions in solution.
4. Effects of metal exposure on physiological processes are complex and variable and tend to be unpredictable. Essential and nonessential metals may stimulate organisms at low concentrations and inhibit at higher concentrations.
5. Behavioral changes observed in laboratory experiments are diverse and difficult to relate to field conditions. However, metals can impair processes such as feeding, learning, swimming activity, and response to external stimuli (Bryan 1976a).

The potential for trace metal toxicity to aquatic organisms depends on the form and properties of the metal and many other physical, chemical, biological, and ecological factors or conditions as indicated in Figure 12.4.

The development of ambient water quality guidelines on national scales in many countries and ecological risk assessment has advanced our understanding of principles that apply to aquatic metal toxicity, and how to evaluate risks from metal exposures to aquatic organisms and their ecosystems. For example, the US EPA has developed national criteria for the protection of aquatic ecosystems. Numerical national water quality criteria were derived for freshwater or saltwater or both to protect aquatic organisms and their uses from unacceptable effects due to acute and/or chronic exposures of toxic chemicals, including certain metals. Table 12.7 presents a recommended set of aquatic life criteria for dissolved metals that are designed to protect aquatic organisms[1].

Protective levels covering most species range from <10 up to 1000 µg L^{-1} for different dissolved metals and exposure periods. Chronic criteria for toxic metals are <100 µg L^{-1} for

---

[1] Aquatic life criteria for toxic chemicals are the highest concentration of specific pollutants or parameters in water that are not expected to pose a significant risk to the majority of species in a given environment or a narrative description of the desired conditions of a water body being "free from" certain negative conditions.

Table 12.7 United States recommended aquatic life criteria for metals.[a]

Metal/Metalloid	Freshwater		Saltwater	
	CMC[b] acute ($\mu g\ L^{-1}$)	CCC[c] chronic ($\mu g\ L^{-1}$)	CMC[b] acute ($\mu g\ L^{-1}$)	CCC[c] chronic ($\mu g\ L^{-1}$)
Al (pH 6.5–9.0)	750	87	—	—
Arsenic	340	150	69	36
Cd[d]	1.8	0.72	33	7.9
Cr (III)[d]	570	74	—	—
Cr (VI)	16	11	1000	50
Cu	—	—	4.8	3.1
Fe	—	1000	—	—
Pb[d]	65	2.5	210	8.1
Hg	1.4	0.77	1.8	0.94
Methyl mercury				
Ni[d]	470	52	74	8.2
Se	—	—	290	71
Ag[d]	3.2	—	1.9	—
Sn tributyl (TBT)	0.46	0.072	0.42	0.0074
Zn[d]	120	120	90	81

[a] Criteria based on dissolved concentrations in water column.
[b] CMC, Criterion Maximum Concentration.
[c] CCC, Criterion Continuous Concentration.
[d] Freshwater acute and/or chronic criteria are hardness-dependent and were normalized to a hardness of 100 mg L^{-1} as $CaCO_3$ to allow the presentation of representative criteria values. See also Appendix B for parameters and calculation equations for freshwater hardness-dependent criteria.
Source: Data from US EPA (2019).

saltwater and up to 150 $\mu g\ L^{-1}$ for freshwaters. Protective levels for Hg, Cd, Ag, and Cu (saltwater), and tributyl tin are <10 $\mu g\ L^{-1}$.

Alternatively, aquatic ecotoxicity data for individual toxic metals covering major organism groups have been compiled from major sources in the scientific literature, evaluated and summarized in the PAN Pesticides Database for chemicals (PAN 2019).

Average acute toxicity ratings for selected toxic metals from this database are summarized in Table 12.8, according to toxicity ratings of 1 (very highly toxic) to 5 (not acutely toxic) for organism groups. Numerical acute toxicity values for each metal and specific organisms in each group are given in the ecotoxicity category of the PAN database.

Very highly toxic metals have average LC_{50} (values) less than 100 $\mu g\ L^{-1}$ (toxicity rating 1). Acute toxic concentrations generally varied widely, between different metals and within organism groups, from slightly toxic to very highly toxic. Average metal toxicity ranged from Hg > Cu > As > Cd > Pb > Cr > Ni > Zn. Amphibians were particularly vulnerable to most metals, followed by plankton, various invertebrates, fish, insects, nematodes and flatworms. Marine benthic communities also indicated vulnerability based on exposures to Cd, Cu, and Cr. There was a lack of data for toxicity effects of metals on aquatic plants.

As part of the trend toward developing water quality guidelines using risk-based decision schemes, acute and chronic toxicity data for metals (and other toxicants) have been used to derive trigger values, mostly from single-species toxicity data for a range of test species that meet screening criteria. A risk-based approach is shown in Table 12.9 for an Australian and New Zealand set of trigger values for metals that are designed to provide different levels of ecosystem protection (% species expected to be protected) in fresh and marine waters. For example, a 95% level of protection for trigger values should apply to a slightly moderately disturbed ecosystem and a 99% level of protection to a pristine or undisturbed ecosystem.

Trigger values indicate sensitive freshwater species are potentially vulnerable to As (V), Cd, Cr (VI), Cu, Pb, Hg, Ni, Ag, and Zn at low levels of about 10 $\mu g\ L^{-1}$ or less, and generally at levels of 100 $\mu g\ L^{-1}$ or less, except for Mn. In marine waters, Cd, Cr (VI), Cu, Pb, Hg, Ag, Zn, and to some degree Ni, are also potentially toxic at low $\mu g\ L^{-1}$ levels. Tributyl tin is exceptionally toxic. The level of risk, however, depends on the *bioavailable* metal concentration compared against an adjusted guideline value, according to a risk decision tree (e.g. correction for hardness in freshwaters, total dissolved metal level and consideration of any metal speciation and direct toxicity assessment) (ANZECC/ARMCANZ 2000). Water Quality Australia (2018) have updated these guidelines within an on-line access database platform.

Metal Toxicity in Sediments

The toxicity of metals in sediments depends greatly on factors that influence the bioavailability of metals such as sediment properties. Various studies have shown that metal levels in interstitial (pore) water that are correlated with observed biological effects are independent of total metal concentrations (dry weight) in sediments. A key controlling parameter of metal toxicity in sediments is acid-volatile sulfide (ASV), which can readily bind, on a molar basis, to toxic metal cations to form insoluble sulfide complexes. If there are sufficient AVS concentrations to limit metal ion concentrations in pore waters, then a lack of toxic effects will occur (e.g. Simpson et al. 2013).

The development of sediment quality guidelines for metals (e.g. Long et al. 1995) has advanced these principles to establish low and high range effects, and no effect or a threshold for effects (TE) level for metals in sediments.

Table 12.8 Average acute toxicity ratings of heavy metals to aquatic organisms.

Taxonomic groups	As	Cd	Cu	Cr	Pb	Hg	Ni	Zn
Fish	3	3	3	4	4	2	4	4
Amphibians	1	1	2	3	1	1	1	4
Crustaceans		3	3	4	4	2	3	4
Molluscs		3	2	3	3	3	5	3
Echinoderms			1	3				5
Cnidaria			1					3
Insects		4	3	4		3	4	4
Annelida		3	2	4	2	3	4	3
Nematodes and flatworms		4	4	3	5	2	4	4
Marine benthic community		2	2	3				
Zooplankton	3	2	2	3	3	1	3	3
Phytoplankton		5	2	2		2	3	4
Aquatic plants								
Average rating	2.33	3	2.25	3.27	3.14	2.11	3.44	3.73

Toxicity category	LC50 (µg L−1)	Toxicity rating
Very highly toxic	<100	1
Highly toxic	100–1000	2
Moderately toxic	1000–10 000	3
Slightly toxic	10 000–100 000	4
Not acutely toxic	>100 000	5

Toxicity ratings are highlighted in the table by increasing darker shades from moderate (3), highly (2), to very highly toxic (1).
Source: Data from PAN (2019).

Table 12.9 Australian and New Zealand trigger values for metals at different levels of protection for aquatic species.

Metal	Trigger values for fresh water (µg L^{-1}) Level of protection (% species)^a				Trigger values for marine water (µg L^{-1}) Level of protection (% species)^a			
	99%	95%	90%	80%	99%	95%	90%	80%
As (III)	1	24	94	360	ID	ID	ID	ID
As (V)	0.8	13	42	140	ID	ID	ID	ID
Cd H	0.06	0.2	0.4	0.8	0.7	5.5	14	36
Cr (III) H	ID	ID	ID	ID	7.7	27.4	48.6	90.6
Cr (VI)	0.01	1.0	6	40	0.14	4.4	20	85
Cu H	1.0	1.4	1.8	2.5	0.3	1.3	3	8
Pb H	1.0	3.4	5.6	9.4	2.2	4.4	6.6	12
Mn	1200	1900	2500	3600	ID	ID	ID	ID
Hg (inorganic)	0.06	0.6	1.9	5.4	0.1	0.4	0.7	1.4
Ni H	8	11	13	17	7	70	200	560
Ag	0.02	0.05	0.1	0.2	0.8	1.4	1.8	2.6
Sn (TBT)	ID	ID	ID	ID	0.0004	0.006	0.02	0.05
V	ID	ID	ID	ID	50	100	160	280
Zn H	2.4	8.0	15	31	7	15	23	43

Notes: (1) H subject to hardness correction for toxicity; ID inadequate data. (2) The primary method for deriving trigger values uses a statistical distribution approach modified from Aldenberg and Slob (1993), adopted in the Netherlands and recommended by the OECD. It is based on a probability distribution of aquatic toxicity endpoints that is used to estimate selected levels of protection for species, usually 95%.
^{a}Levels apply to different ecosystem conditions.
Source: Data taken from ANZECC/ARMCANZ (2000).

Table 12.10 Recommended sediment quality guideline values for metals, metalloids, and organometallics.

Metals/metalloids	Guideline value (mg kg^{-1}, dry weight)	SQG-high (mg kg^{-1}, dry weight)
Antimony	2.0	25
Arsenic	20	70
Cadmium	1.5	10
Chromium	80	370
Copper	65	270
Lead	50	220
Mercury	0.15	1.0
Nickel	21	52
Silver	1.0	4.0
Zinc	200	410
Organometallic		
Tributyl tin (µg Sn kg^{-1} dry weight, 1% TOC)	9.0	70

Source: Adapted from Simpson et al. (2013).

However, there is not a clearly delineated relationship between no effect levels and those that cause biological effects (Batley et al. 2005). Simpson et al. (2013) have recommended sediment quality guideline values (SQGVs) and upper guidelines (SQG-High values) for a range of metals, metalloids, and organometals as an update to Australian and New Zealand Guidelines for Fresh and Marine Water Quality released in 2000 (ANZECC/ARMCANZ 2000). These values are listed in Table 12.10 and are essentially unchanged from interim values in 2000. The guidelines for metals are primarily based on the effects range median (ERM) values (see Long et al. 1995).

Significantly, Hg, Cd, Sb, Ag, and tributyl tin have low toxicity thresholds compared to the more common contaminants of copper, chromium, lead, and zinc. There remains some degree of scientific uncertainty in toxicity between the lower and upper guideline values. In practice, the approach is to use the lower guideline as a screening level and apply a tiered decision-tree approach (e.g. chemistry-bioavailability, ecotoxicology, bioaccumulation, and benthic ecology measures) as introduced in the water quality guidelines (see Simpson et al. 2013).

12.8 Ecological Effects of Metals

Heavy metals and other pollutants such as toxic organic chemicals are widely known to cause adverse physiological effects on many animal and various plant species that have been studied in toxicity testing, and to a much lesser extent in field studies involving aquatic and terrestrial systems. Metals differ especially from organic pollutants because deficiencies in essential metal elements can also induce toxic or adverse effects in various microflora, animals, and plants. Individual and multiple metal pollutants, among other pollutants, interact with other constituents in their environments, and upon biological uptake, with physiological processes in organisms, and at the site(s) of toxic action within organisms (e.g. US EPA 2007; Rhind 2009).

Despite the need for more detailed studies on many taxa, multiple toxicity mechanisms, and a more precise understanding of toxicity endpoints and thresholds, the current state of scientific knowledge about the ecological effects of heavy metals strongly suggests that all species and ecosystems are potentially vulnerable to the interactions and effects of pollutants, such as heavy metals, because many basic physiological systems can be affected by these metals (and many organic pollutants) at very low levels, and when acting in combination. The ecological responses and longer-term risks are also exacerbated by known interactions between the effects of metal pollutants (and other persistent pollutants), and environmental stressors (e.g. nutritional deficiency and osmotic stresses), as well as by climate change events (e.g. rising air temperatures, heat stress, drought, and scarcity of food) (see Rhind 2009).

Ecological impacts of metals are derived mainly from field observations in aquatic and terrestrial systems exposed to known metal sources (e.g. smelter emissions, mining wastes, sewage, industrial discharges, or metal-contaminated solids). The direct effects of metal pollution on populations and communities are mostly characterized from studies on biota in rivers and lakes. Examples of studies on the effects of common toxic metals on freshwater fish and invertebrates are reported in Tables 10.14 and 10.15 in Connell and Miller (1984). In these types of situations, consistent patterns of change are observed and dose–response relations can be estimated from field measurements and comparisons made with laboratory toxicity data (see Alabaster and Lloyd 1980). For example, Weatherley et al. (1967) investigated mine drainage pollution (Cu, Fe, and Zn) in the Molonglo River, Australia, and found that uncontaminated areas, including two tributaries, usually contained 30–45 species of benthic invertebrate fauna. Immediately below the discharge, the number of species was reduced to about 4, and a slow recovery to a normal number of species was apparent about 50 km or so downstream.

In a marine ecosystem, the community structure (species richness, species composition, and abundance of species) of epibenthic seagrass fauna near a large lead smelter was investigated in southern Australia (Ward and Young 1982). Decreased frequencies of 20 common species, mostly fish, were correlated with the concentration of contaminant metals (Cd, Cu, Pb, Mn, and Zn) in the sediments. It was also found that frequencies of certain species, mostly crustaceans, correlated with particle size distributions. These researchers concluded that both contaminant metals and particle size distributions have substantial controlling effects on community structure.

Several general characteristics of metal toxicity effects within aquatic populations and communities have been suggested in the scientific literature and can be extended to terrestrial systems:

1. Metal ions and complexes exhibit a wide range of toxicity to natural populations and communities of organisms. Bioaccumulation of nonessential metal elements is well known in many animals and plants studied.
2. Significant to severe modifications in community structure involving reduction in the number of species, including complete absence of sensitive species, are relatively common.
3. A reduction in the number of individuals of surviving species with the amount of reduction related to the metals present and the level and period of exposure.

These characteristics are generally consistent with the classical pattern of ecological effects of toxicants on ecosystems. Toxicants are associated with an overall reduction in number of species and individuals present in relation to the severity of the toxic stress. The development of tolerance in some species has increased their capacity to accumulate relatively high concentrations of metals.

12.9 Human Health Effects of Metals

Introduction

The health effects of metals have been widely studied and reviewed in the scientific literature, and by international and many national agencies responsible for human and environmental health. Transition or heavy metals and certain metalloids are the focus of health concerns. Some transition metals can be significant toxicants and carcinogens (Kim et al. 2015). The toxicity of metals (dose–responses) depends largely on their chemical forms (e.g. atomic, ionic, or organometallic species) and properties (aqueous or lipid solubility and vapor pressure), exposure levels and routes of intake, bioavailability, bioaccumulation, and interactions with sensitive cellular and organ targets in the body. Readily soluble forms of inorganic metals are usually more toxic than insoluble forms. Increasingly, scientific evidence of toxicities at low levels of metal exposure and from mixtures of trace metals are of public health concern. The setting of safe levels for environmental exposures for some toxic metals are also being challenged by low safety margins for adopted guidelines (e.g. in drinking waters) and tolerable intakes of some toxic metals.

The main threats to human health are generally associated with exposures to the toxic group of nonessential metals: lead, cadmium, mercury, and arsenic (Järup 2003). These metals induce multi-organ effects with initial toxicity likely to depend on the most sensitive target organ. Various essential metals/metalloids such as antimony, copper, zinc, chromium (III), and selenium can exhibit well-known deficiencies or toxicities to humans, depending on doses, metal interactions, and other factors (e.g. nutritional state and immune system).

Types of Human Health Effects from Metals

Metal-induced toxicity causes a diverse range of health effects, such as biochemical, physiological, neurological, behavioral, developmental, reproductive, and carcinogenic, that are associated with known toxic metals, mainly transitional, and metalloids. Several metals such as arsenic, cadmium, mercury, and lead are associated with endocrine disruption effects. Health effects due to toxic metals can be considered as two general types:

1. **Noncarcinogenic effects:** All toxic responses other than the induction of tumors.
2. **Carcinogenic effects:** Those associated with the induction of tumors as an endpoint.

Metals and compounds of arsenic, cadmium, chromium (VI), lead, mercury, and nickel are known to be major systemic toxicants and induce multiple organ effects. Several metals are also associated with endocrine disruption at low levels of exposure. Metals listed as human carcinogens by the IARC and US National Toxicology Program (NTP) are arsenic, cadmium, and chromium (VI) and their inorganic compounds, nickel, and beryllium compounds. Inorganic lead compounds are classified as probable human carcinogens (ACS 2019) (see also IARC Group 1 classification).

Table 12.11 provides a summary of various human health effects that can be caused by several common nonessential and essential metals, depending upon chemical forms, exposure levels, conditions, and other factors (see also Table 12.12).

Biomarkers

Evaluation of toxic metal exposures and adverse human health effects or diseases commonly uses a wide range of biomarkers, mainly based on metal levels in biological specimens such as blood, urine, hair, and fingernails (see Table 12.11). For example, some biomarkers for inorganic lead are indicated in Table 12.11, but studies on blood lead

Table 12.11 Summary of human health effects of key toxic metals.

Toxic metals	Acute effects	Chronic effects	Biomarkers
Arsenic (inorganic)	Nausea, vomiting, diarrhea, fever, anorexia, hepatomegaly, melanosis, cardiac arrhythmia, upper respiratory tract symptoms, peripheral neuropathy (e.g. sensory loss), gastrointestinal, cardiovascular, hematopoietic effects, renal failure, and damage to membranes	Diabetes, hypopigmentation/hyperkeratosis, neurotoxicity (CNS and peripheral) (e.g. paresthesia and motor dysfunction), encephalopathy, liver injury and alteration of liver enzymes in blood, cardiovascular disease Carcinogenic effects: lung, bladder, skin	Urine >100 µg L^{-1} (without seafood) blood 50 µg L^{-1} Normal: hair <1 µg g^{-1} blood <1 µg L^{-1} urine <100 µg L^{-1} nails = <1 µg g^{-1} Nonoccupational:
Arsine (AsH$_3$)	Potent hemolytic agent, hemoglobinuria, renal failure	Jaundice and anemia	urine <10 µg As g^{-1} creatinine (As = inorganic As, MMA and DMA)
Cadmium	Metal fume fever (inhaled): pneumonitis (oxide fumes), pulmonary edema Nausea, vomiting, diarrhea	Proteinuria, osteomalacia. Chronic obstructive pulmonary disease and emphysema Chronic renal tubular dysfunction and disease Skeletal effects, disruption of calcium metabolism, osteomalacia and/or osteoporosis Carcinogenicity: pulmonary tumors	Blood– recent exposure and urine: proteinuria and/or ≥15 µg Cd g^{-1} creatinine Critical Cd level in renal cortex: reported levels ~200 µg g^{-1} or lower
Chromium Cr^{6+}	Corrosive reactions (nasal septum). Acute irritative dermatitis. Acute tubular and glomerular damage, renal failure (Cr^{6+} ingestion)	Toxic and carcinogenic: Chrome ulcers, pulmonary fibrosis, allergic skin reactions/ulceration. Respiratory cancers (e.g. lung cancer (chromate exposures)	Blood and urine: Cr > normal levels Cr^{6+} no safe level.
Lead	Nausea, vomiting, encephalopathy (headache, seizures, stupor, loss of consciousness)	Hematological, gastrointestinal, cardiovascular, renal, encephalopathy, peripheral neuropathy, neurobehavioral deficits, e.g. IQ deficits, cognitive and psychological effects. Reproductive effects (reduced fertility). Carcinogenicity: high level exposures associated with cancers	Blood lead levels <5 µg dL^{-1} Various effects at increasing levels Urinary and hair lead levels also
Organic lead compounds: tetraethyl lead	Nausea, vomiting, encephalopathy progressing to convulsions and coma	Encephalopathy, colorectal cancer among workers associated with industrial production	
Mercury Hg vapor	Acute corrosive bronchitis, pneumonitis, CNS effects	Neurotoxic effects (CNS), increased excitability, tremors, gingivitis, changes in personality, loss of coordination, depression, delirium/hallucination	Variable blood, urine and hair levels: subgroups include: children, pregnant and other women, indigenous populations
Inorganic Hg (mercuric) (mercurous)	Corrosive ulceration, bleeding, gastrointestinal necrosis and renal failure (less soluble and less severe effects)	Neurotoxicity: proteinuria, nephritis Hypersensitivity reaction, *pink disease*: swelling of spleen and lymph nodes, hyperkeratosis	
Methylmercury	Neurotoxic effects	Neurotoxic effects (adults), toxicity to fetuses of mothers exposed during pregnancy	
Essential metals	**Deficiency**	**Toxicity**	
Nickel	Known to be essential in some species of animals and plants (e.g. urease from Jack beans and several other species of plants is a nickel enzyme), and bacteria. Little data on deficiency effects	Acute inhalation of nickel carbonyl: severe lung damage; contact dermatitis (10–20% of general population sensitized); respiratory tract inflammation – allergic asthma, lung and nasal cancers among exposed Ni workers. Various nickel compounds classified as human carcinogens	Inhalation exposure: blood plasma and urine; levels depend on Ni species in air. Higher levels for soluble Ni chlorides and sulfates cf. to Ni oxide and sulfide

Source: Based on Järup (2003); ATSDR (2005); ATSDR (2007); ATSDR (2012a); ATSDR (2012b); WHO (2017); ATSDR (2019).

Table 12.12 Factors influencing toxicity of metals to humans.

Factors	Examples
Metal forms/properties	Vapors or fumes, particulates, inorganic or organometallic, ionic species, valency or oxidation states, complexes, solubility
Metal mixtures/interactions	Synergistic, additive, potentiating and/or antagonistic effects; interactions between metals such as molecular or ionic mimicry (e.g. Cd uptake can mimic that of zinc)
Formation of metal-protein complexes	Metallothioneins: low molecular weight proteins, rich in thiol groups, with high affinity for binding to toxic metals (e.g. Cd, Cu, Hg, Ag, and Zn) to detoxify and regulate metal homeostasis
Life stages	Age and gender: infants and children more sensitive than adults; elderly more sensitive to metals that target the kidney, and uptake of some trace elements (e.g. Cu and Zn) decreases
Lifestyle factors	Smoking and alcohol consumption; increased Cd uptake from cigarette smoke; additional metal intake from use of dietary supplements/other consumer products, and "alternative" remedies
Pregnancy and lactation	Higher demand for essential metals (e.g. Cu, Zn and Fe); bioavailability effects (e.g. lack of protein can reduce uptake); loss of Fe can increase Cd intake
Pre-existing sensitivity or disease	Allergies and diseases (e.g. brain, liver and kidney damage or disorders)
Nutritional state	Sources and effects of diet on metal absorption (e.g. lack of protein can reduce essential metal uptakes); adequate RDAs or RDIs
Immune status of host	Immature development of the immune system in children generally increases susceptibility to metals; immune reactions to metals (e.g. Au, Hg, Pt, Be, Cr, and Ni) producing hypersensitivity responses
Genetic risk factors	Heritable genetic polymorphisms (e.g. Wilson's disease – increase in Cu accumulation leading to potential damage to liver, kidney, brain, and cornea; inherited hemochromatosis characterized by excessive Fe absorption and long-term risk of liver disease (cirrhosis and cancer); genetic effects on potential sensitization reactions to different metals (e.g. Ni)

Sources: Based on Goyer and Clarkson (2001); US EPA (2007).

levels in children and adult populations show a progressive decline toward *no adverse effect* level of PbB $<5\,\mu g\,dL^{-1}$. Reported normal and biomarker levels for metal toxicity or deficiency, however, can vary to some degree between sub-population groups and countries (e.g. lead and mercury) as threshold levels remain under continual review by agencies such as WHO.

Emerging biomarkers include measures of gene expression and protein regulation to identify sensitive health endpoints. Target organs may be a more accurate measure of safe exposure doses rather than daily intake or dose ($mg\,kg^{-1}\,day^{-1}$) because metals can persist and accumulate in these organs (Goyer et al. 2004).

Factors Influencing Toxicity

Aside from route of exposure and dose absorbed, a number of chemical and host factors can influence the toxicity of a metal at a certain level of exposure, and as a result, the risk of toxicity to susceptible populations (Goyer and Clarkson 2001, p. 814). Table 12.12 lists key factors, which may increase the risk of metal toxicity in susceptible population subgroups, compared to the general population (US EPA 2007, pp. 4–5). Susceptible populations include toxic metal-exposed workers and residents (e.g. living near metal mining, smelters, or processing industries), persons drinking contaminated waters (e.g. arsenic), or eating contaminated foods (e.g. lead), pregnant women, children, the elderly, and persons with compromised immune systems.

Common Toxic Metals

A concise review of common toxic metals (arsenic, cadmium, lead, and mercury) is given in Section S12.1 in the Companion Book Online.

12.10 Key Points

1. Topics cover (i) the chemical nature, sources, releases, environmental behavior and fate, (ii) their biological toxicity and ecological effects, and (iii) human health effects (noncarcinogenic and carcinogenic).
2. The term, **metal**, typically describes a chemical element, which is a good conductor of electricity and has high thermal conductivity, density, malleability, ductibility, and electropositivity. They lose electrons to form positive ions when chemically bonding.
3. Metals also include lanthanides (rare earths) and actinides (radioactive) (see Figure 12.1).

4. Some elements (boron, silicon, germanium, arsenic, and tellurium) are known as metalloids, because they possess one or more metallic properties but are not sufficiently distinctive to be called a metal or a nonmetal.
5. Metals react as electron-pair acceptors (Lewis acids) with electron pair-donors (Lewis bases) to form various chemical groups, such as ion pairs or metal ligand complexes.
6. Certain metals such as Co, Cr (III), Cu, Fe, Mn, Mo, Se, and Zn are nutritionally essential metals for humans, but the biological roles of many metal elements tend to be uncertain or difficult to classify or unknown in biological species.
7. Generally, heavy or transition metals, including the metalloid arsenic, are of major environmental health concern: Zn, Cu, Cd, Hg, Pb, Ni, Cr, Co, Fe, Mn, Sn, Ag, and As.
8. Some metals can form organometallic compounds by covalent bonding with the carbon of an organic compound. Common toxic organometallics include organolead (e.g. tetraethyl lead), organomercury (e.g. methylmercury), and organotin (e.g. tributyl tin) compounds.
9. Sources of metals are abundant in the environment from natural geochemical processes (e.g. volcanic and erosion or weathering) and human activities (e.g. mining, industrial, urbanization, waste disposal, agricultural uses, and related stormwater runoff from these catchments).
10. Tens of millions of tonnes of metals are produced yearly for human use leading to many point sources and diffuse emissions and the release of large quantities of various toxic metals into the environment. Waterborne and atmospheric transport and deposition are major processes in the distribution of metals worldwide.
11. In 2015, the US Toxic Release Inventory (TRI) reported that industry disposed or released 3.36 billion pounds of TRI chemicals to on-site land, into the air, water, or by off-site transfers to some type of land disposal unit. Most of these chemical releases (70%) involved eight chemicals, mainly metals and their compounds: zinc (19%), lead (17%), manganese (7%), barium (6%), arsenics (5%), and copper (5%).
12. Toxic metals are persistent, can be cumulative in air, water, soils, sediments, and organisms, and induce adverse effects on exposed organisms at low environmental levels (e.g. ppb and ppm) and doses.
13. The physical and chemical forms of metals and their properties (e.g. chemical species or valency and aqueous or lipid solubility) within air, waters, soils, sediments, and food, including prey, are vital in influencing the environmental distribution, fate, bioavailability, exposures and related biological effects within organisms, their populations, and communities.
14. Living organisms are exposed to natural and anthropogenic sources of metals in different chemical forms, essentially through multiple environmental pathways and routes of intake into animal bodies or plants.
15. Human individuals and populations can suffer multiple metal exposures from environmental, occupational, urban, indoor, dietary and lifestyle (e.g. smoking) sources and contact with toxic metals in consumer products (e.g. lead paint), and materials.
16. The absorption, distribution, transformation, and excretion of a metal within an organism, depends on the metal, the form of the metal or metal compound, and the organism's ability to regulate and/or store the metal – US EPA.
17. The routes of uptake of metals by humans and other biota are primarily by respiratory intake (e.g. inhalation), ingestion (dietary, drinking water, and soil or dust) and absorption through contact with an organism's surfaces. In humans, dermal absorption of metals is limited. The metal form and how much of the metal is taken up by the organism are critical factors for potential toxicity.
18. The concept of metal bioavailability refers to the fraction of a metal species that an organism absorbs (or adsorbs), by crossing biological membranes, with the potential for distribution, metabolism, elimination, and bioaccumulation.
19. Bioaccumulation of metals is the net accumulation of a metal in the tissue of interest or the whole organism that results from all environmental exposure media. It represents a net mass balance between uptake and elimination of the metal.
20. Metals cause biological effects on individual organisms following uptake and transport across cell membranes into cells where they can bind onto a cellular target (reversibly or irreversibly) and change specific biochemical processes. These effects may (i) be beneficial, such as for essential metals at optimum levels, or (ii) induce toxicity, particularly through oxidative (and nitrative) stress leading to oxidative tissue damage.
21. Metal ions and complexes exhibit a wide range of toxicity to natural populations and communities of organisms.
22. Bioaccumulation of nonessential metal elements is well known in many animals and plants studied and also in humans (e.g. cadmium in kidneys and liver, lead in bones, and methylmercury in the fetus).

23. Depending upon metals and exposures, significant to severe modifications in community structure involving reduction in the number of species, including the complete absence of sensitive species, are relatively common.
24. Metal-induced toxicity causes a diverse range of health effects, such as biochemical, physiological, neurological, behavioral, developmental, reproductive, and carcinogenic, that are associated with known toxic metals, mainly transitional, and metalloids.
25. Vulnerable human population groups at increased risk of exposure include pregnant women, fetuses, infants, and children (e.g. toxic metals such as lead and mercury).
26. Human health effects due to toxic metals can be considered as two general types: noncarcinogenic effects and carcinogenic effects.
27. Arsenic, cadmium, chromium (VI), lead, mercury, and nickel and their compounds are known to be major systemic toxicants and induce multiple organ effects.
28. Metals listed as human carcinogens by the IARC are arsenic and its inorganic compounds; beryllium, cadmium, and hexavalent (VI) chromium and their compounds; and nickel compounds. Inorganic lead compounds are classified as probable human carcinogens.
29. Evaluation of toxic metal exposures and adverse human health effects or diseases commonly uses a wide range of biomarkers, mainly based on metal levels in biological specimens such as blood, urine, hair, and fingernails.
30. Emerging biomarkers include measures of gene expression and protein regulation to identify sensitive health endpoints. Target organs may be a more accurate measure of safe exposure doses rather than daily intake or dose ($mg\,kg^{-1}\,day^{-1}$) because metals can persist and accumulate in these organs.
31. Studies on blood lead levels (PbB) in children and adult populations show a progressive decline toward or to *no adverse effect* levels of PbB $<5\,\mu g\,dL^{-1}$.

References

Aldenberg, T and Slob, W. (1993). Cited in ANZECC/ARMCANZ (2000).

Alabaster, J.S. and Lloyd, R. (1980). *Water Quality Criteria for Freshwater Fish*. London: Butterworths.

Ali, H., Khan, E., and Sajad, A. (2013). Phytoremediation of heavy metals – concepts and applications. *Chemosphere* 91 (7): 869–881.

ACS (2019). Known and probable human carcinogens. American Chemical Society. https://www.cancer.org/cancer/cancer-causes/general-info/known-and-probable-human-carcinogens.html (accessed 18 March 2020).

ANZECC/ARMCANZ (2000). Australian and New Zealand guidelines for fresh and marine water quality. Australian and New Zealand Environment and Conservation Council/Agriculture and Resource Management Council of Australia and New Zealand. https://www.waterquality.gov.au/media/57 (accessed 28 August 2020).

ATSDR (2005). *Toxicological Profile for Nickel*. Atlanta GA: US Department of Health and Human Services, Agency for Toxic Substances and Disease Registry.

ATSDR (2007). *ToxGuide^{TM} for Arsenic*. Atlanta GA: US Department of Health and Human Services, Agency for Toxic Substances and Disease Registry.

ATSDR (2012a). *ToxGuide^{TM} for Cadmium*. Atlanta GA: US Department of Health and Human Services, Agency for Toxic Substances and Disease Registry.

ATSDR (2012b). *ToxGuide^{TM} for Chromium Cr*. Atlanta GA: US Department of Health and Human Services, Agency for Toxic Substances and Disease Registry.

ATSDR (2019). *ToxGuide^{TM} for Lead*. Atlanta GA: US Department of Health and Human Services, Agency for Toxic Substances and Disease Registry.

Baird, C. and Cann, M. (2012). *Environmental Chemistry*, 5e. New York: W.H. Freeman and Company.

Batley, G.E., Stahl, R.G., Babut, M.P. et al. (2005). Scientific underpinnings of sediment quality guidelines. In: *Use of Sediment Quality Guidelines and Related Tools for the Assessment of Contaminated Sediments* (ed. R.J. Wenning, G.E. Batley, C.G. Ingersoll and D.W. Moore), 39–119. Fairmont, MT: SETAC Press.

Bryan, G.W. (1976a). Some aspects of heavy metal tolerance in aquatic organisms. In: *Effects of Pollutants on Aquatic Organisms* (ed. A.P.M. Lockwood). Cambridge: Cambridge University Press.

Bryan, G.W. (1976b). Heavy metal contamination in the sea. In: *Marine Pollution* (ed. R. Johnston), 185–302. London: Academic Press.

Bryan, G.W. (1979). Bioaccumulation of marine pollutants. *Philosophical Transactions of the Royal Society of London Series B* 286: 483–505.

CDC (2019). Childhood lead poisoning prevention-blood lead levels in children. Centers for Disease Control and Prevention. United States National Center for

Environmental Health, Division of Environmental Health Science and Practice. https://www.cdc.gov/nceh/lead/prevention/blood-lead-levels.htm (accessed 20 March 2020).

Clemens, S. (2006). Toxic metal accumulation, responses to exposure and mechanisms of tolerance in plants. *Biochimie* 88: 1707–1719.

Connell, D.W. (2005). *Basic Concepts of Environmental Chemistry*, 2e. Boca Raton, FL: CRC Press.

Connell, D.W. and Miller, G.J. (1984). *Chemistry and Ecotoxicology of Pollution.* New York: Wiley.

Craig, P.J. (1980). Metal cycles and biological methylation. In: *The Natural Environment and the Biogeochemical Cycles* (ed. H. Hutzinger), 169. Berlin: Springer.

Ercal, N., Gurer-Orhan, H., and Aykin-Burns, N. (2001). Toxic metals and oxidative stress part I: mechanisms involved in metal-induced oxidative damage. *Current Topics in Medicinal Chemistry* 1 (6): 529–539.

Förstner, U. (1979a). Metal concentrations in river, lake and ocean waters. In: *Metal Pollution in the Aquatic Environment* (ed. U. Förstner and G.T.W. Wittmann), 71. Berlin: Springer-Verlag.

Förstner, U. (1979b). Metal transfer between solid and aqueous phases. In: *Metal Pollution in the Aquatic Environment* (ed. U. Förstner and G.T.W. Wittmann), 197–270. Berlin: Springer-Verlag.

Geoscience Australia (2020). Australia's identified mineral resources table 1 (as at December 2015). Geoscience Australia. Australian Government. http://www.ga.gov.au/scientific-topics/minerals/mineral-resources-and-advice/aimr/table1 (accessed 28 August 2020).

Goyer, R.A. and Clarkson, T.M. (2001). Toxic effects of metals. In: *Casarett & Doull's Toxicology* (ed. C.D. Klaassen), 811–868. New York: McGraw-Hill.

Goyer, R., Golub, M., Choudhury, H. et al. (2004). *Issue Paper on the Human Health Effects of Metals*, ERG (Eastern Research Group Inc.), Lexington, MA. https://pdfs.semanticscholar.org/da68/45d127c7cfa622273878a5c04d1e0628dce0.pdf (accessed 28 August 2020).

Hart, B.T. and Davies, S.H.R. (1978). *A Study of the Physico-Chemical Forms of Trace Metals in Natural Waters and Wastewaters.* Australian Water Resources Technical Paper No. 35. Canberra: Australian Government Publishing Service.

Hollenberg, P.F. (2010). Introduction: mechanisms of metal toxicity special issue. *Chemical Research in Toxicology* 23: 292–293.

Jackson, T. (2017). *The Periodic Table: A Visual Guide to the Elements.* London: Aurum Press.

Jan, A.T., Azam, M., Siddiqui, K. et al. (2015). Heavy metals and human health: mechanistic insight into toxicity and counter defense system of antioxidants. *International Journal of Molecular Science* 16 (12): 29592–29630.

Järup, L. (2003). Hazards of heavy metal contamination. *British Medical Bulletin* 68: 167–182.

Kerrie, S. and Austin, D.W. (2011). Ancestry of pink disease (infantile acrodynia) identified as a risk factor for autism spectrum disorders. *Journal of Toxicology and Environmental Health Part A* 74 (18): 1185–1194.

Kim, H.S., Kim, Y.J., and Seo, Y.R. (2015). An overview of carcinogenic heavy metal: molecular toxicity mechanism and prevention. *Journal of Cancer Prevention* 20 (4): 232–240.

Leckie, J.O. and James, R.O. (1974). Control mechanisms for trace metals in natural waters. In: *Aqueous-Environmental Chemistry of Metals* (ed. A.J. Rubin), 1. Ann Arbor, MI: Ann Arbor Science.

Long, E.R., MacDonald, D.D., Smith, S.L., and Calder, F.D. (1995). Incidence of adverse effects within ranges of chemical concentrations in marine and estuarine sediments. *Journal of Environmental Management* 19: 81–97.

Lu, C.S.J. and Chen, K.Y. (1977). Migration of trace metals in interfaces of seawater and polluted surficial sediments. *Environmental Science and Technology* 11: 174–182.

MSC-E (2020). Emissions for global modelling-emissions of heavy metals (mercury, lead, cadmium and other metals). Meteorological Synthesizing Centre-East (MSC-E) Website. Moscow, Russian Federation. http://www.msceast.org/index.php/j-stuff/content/list-layout/global (accessed 19 March 2020).

Mortimer, M.R. and Miller, G.J. (1994). Susceptibility of larval and juvenile instars of the sand crab, *Portunus pelagicus* (L.), to sea water contaminated by chromium, nickel or copper. *Australian Journal of Marine and Freshwater Research* 45: 1107–1121.

Nieboer, E. and Richardson, D.H.S. (1980). The replacement of the nondescript term "heavy metals" by a biologically and chemically significant classification of metal ions. *Environmental Pollution Series B* 1 (1): 3–26.

Ochiai, E. (1977). *Bioinorganic Chemistry. An Introduction.* Boston, MA: Allyn and Bacon.

PAN (2019). *PAN Pesticides Database – Chemicals.* North America, Berkley, California: Pesticide Action Network http://www.pesticideinfo.org/Search_Chemicals.jsp#ChemSearch (accessed 16 March 2020).

Phillips, D.J.H. (1977). The use of biological indicator organisms to monitor trace metal pollution in marine and estuarine environments – a review. *Environmental Pollution* 13: 281–317.

Prossi, F. (1979). Heavy metals in aquatic organisms. In: *Metal Pollution in the Aquatic Environment* (ed. U. Förstner and G.T.W. Wittmann), 271–323. Berlin: Springer-Verlag.

Rhind, S.M. (2009). Anthropogenic pollutants: a threat to ecosystem sustainability? *Philosophical Transactions of the Royal Society B Biological Sciences* 364 (1534): 3391–3401.

Schlesinger, W.H., Klein, E.M., and Vengosh, A. (2017). Global biogeochemical cycle of vanadium. *Proceedings of the National Academy of Sciences United States of America* 114 (52): E11092–E11100.

Simpson, S.L., Batley, G.E., Chariton, A.A. (2013). Revision of the ANZECC/ARMCANZ sediment quality guidelines. CSIRO. https://doi.org/10.4225/08/5894c6184320c.

Stumm, W. and Bilinski, H. (1972). Chemical speciation. Trace metals in natural waters: difficulties in interpretation arising from our ignorance on their speciation. In: *Advances on Water Pollution Research. Proceedings of the Sixth International Conference, Jerusalem* (ed. S.H. Jenkins), 39–52. New York: Pergamon Press.

Stumm, W. and Brauner, P.A. (1975). Chemical speciation. In: *Chemical Oceanography* (ed. J.P. Riley and G. Skirrow), 173. New York: Academic Press.

Stumm, W. and Morgan, J.J. (1970). *Aquatic Chemistry*. New York: Wiley.

Suedel, B.C., Boraczek, J.A., Peddicord, R.K. et al. (1994). Trophic transfer and biomagnification potential of contaminants in aquatic ecosystems. *Reviews of Environmental Contamination and Toxicology* 136: 21–89.

Tinsley, I.J. (1979). *Chemical Concepts in Pollutant Behavior*. New York: Wiley.

Tchounwou, P.B., Yedjou, C.G., Patlolla, A.K., and Sutton, D.J. (2012). Heavy metal toxicity and the environment. In: *Molecular, Clinical and Environmental Toxicology. Experientia Supplementum*, vol. 101 (ed. A. Luch), 133–164. Basel: Springer.

UNEP (2013). *Global Chemicals Outlook – Towards Sound Management of Chemicals*. Geneva: United Nations Environment Programme.

US EPA (2007). Framework for metals risk assessment. Report EPA 120/R-07/00. Office of the Science Advisor, Risk Assessment Forum, United States Environmental Protection Agency. Washington, DC.

US EPA (2016). Interim ecological soil screening level documents. United States Environmental Protection Agency. https://www.epa.gov/chemical-research/interim-ecological-soil-screening-level-documents (accessed 19 March 2020).

US EPA (2017). TRI national analysis 2015: releases of chemicals. United States Environmental Protection Agency. www.epa.gov/trinationalanalysis/ (accessed 28 August 2020).

US EPA (2019). National recommended water quality criteria – aquatic life criteria table. United States Environmental Protection Agency. https://www.epa.gov/wqc/national-recommended-water-quality-criteria-aquatic-life-criteria-table/ (accessed 21 May 2019).

Ward, T.J. and Young, P.C. (1982). Effects of sediment trace metals and particle size on the community structure of epibenthic seagrass fauna near a lead smelter, South Australia. *Marine Ecology Progress Series* 9: 137–146.

Water Quality Australia (2018). Australian and New Zealand guidelines for fresh and marine water quality. On-line-platform. Water Quality Australia, Commonwealth of Australia. Canberra. https://www.waterquality.gov.au/anz-guidelines (accessed 28 August 2020).

Weatherley, A.H., Beevers, J.R., and Lake, P.S. (1967). The ecology of a zinc polluted river. In: *Australian Inland Waters and Their Fauna* (ed. A.H. Weatherley). Canberra: Australian National University Press.

WHO (2000). *Air Quality Guidelines for Europe*. 2e. WHO Regional Publications, European Series, No. 91. World Health Organization, Regional Office for Europe. Copenhagen, Denmark. http://www.euro.who.int/__data/assets/pdf_file/0005/74732/E71922.pdf (accessed 19 March 2020).

WHO (2008). *Guidance for Identifying Populations at Risk from Mercury Exposure*. Geneva: World Health Organization/United Nation Environment Programme https://www.who.int/foodsafety/publications/risk-mercury-exposure/en/ (accessed 16 March 2020).

WHO (2009). Levels of lead in children's blood. World Health Organization. Europe. European Environment and Health Information System Fact Sheet 4.5. http://www.euro.who.int/__data/assets/pdf_file/0003/97050/4.5.-Levels-of-lead-in-childrens-blood-EDITING_layouted.pdf (accessed 18 March 2020).

WHO (2010a). *Childhood Lead Poisoning*. Geneva: World Health Organization https://apps.who.int/iris/bitstream/handle/10665/136571/9789241500333_eng.pdf?sequence=1 (accessed 20 March 2020).

WHO (2010b). *Exposure to Lead: A Major Public Health Concern*. World Health Organization, Geneva. http://www.who.int/ipcs/features/lead..pdf?ua=1 (accessed 16 March 2020).

WHO (2017). *Mercury and Health*. Geneva: World Health Organization https://www.who.int/news-room/fact-sheets/detail/mercury-and-health (accessed 16 March 2020).

WHO (2019a). *Evaluations by the Joint FAO/WHO Expert Committee on Food Additives (JECFA)*. Geneva: World Health Organization https://apps.who.int/food-additives-contaminants-jecfa-database/search.aspx?fc=47 (accessed 21 May 2019).

WHO (2019b). *International Programme on Chemical Safety-Lead*. Geneva: World Health Organization http://www.who.int/ipcs/assessment/public_health/lead/en/ (accessed 16 March 2020).

13

Air Pollutants

13.1 Introduction

Air pollution is now a major global risk factor for human disease and death. It is the main environmental cause of premature deaths in the human population based on global burden of disease estimates (WHO 2016; Health Effects Institute 2017). Ambient particulate matter and household pollution are the dominant risk factors, particularly in developing and transient countries. The International Agency for Research on Cancer (IARC) has classified outdoor air pollution and particulate matter in outdoor air pollution as carcinogenic to humans (IARC Group 1). A key message is that the substantial burden of disease from air pollution is a serious challenge for today and the future. The challenge of air pollution also includes the increasing rate of greenhouse gas emissions, global warming trend, and serious climate change impacts and risks evaluated in Chapter 14.

The atmospheric environment consists of a mixture of major gases, mainly nitrogen and oxygen, and trace gases such as water vapor and carbon dioxide, that generally extends about 10–16 km from the Earth's surface. The early atmosphere is believed to have been a mixture of carbon dioxide and nitrogen gases, water vapor, and trace amounts of hydrogen gas. This composition has evolved since the origin of life on Earth, before which the amount of carbon dioxide on Earth exceeded the oxygen content. With the evolution of green plants, carbon dioxide was converted by photosynthesis into atmospheric oxygen gas, and carbon was deposited in sedimentary layers.

The present composition of the atmosphere is introduced in Section 13.2. A heterogeneous mixture of potentially harmful substances such as dusts, fine particulates, various gases, metals, and organic compounds are known to enter the atmosphere, as part of biogeochemical cycles, from natural or biogenic sources and human activities. Since the Industrial Revolution of the Nineteenth Century, the impact of air pollutants on the atmosphere, and health of humans and ecosystems has expanded greatly on a global scale, through the relatively rapid use and synthesis of fossil fuels, extraction and processing of finite ecosystem resources, and expansion of urbanization due to population growth. In addition, increasing greenhouse gas emissions and levels, mainly carbon dioxide, in the atmosphere are strongly associated with an observed and projected global warming trend, and ocean acidification, leading to climate change.

In this chapter, the nature, key factors, and issues of human health and environmental impacts of air pollutants are evaluated, from local to global scales, following the general conceptual model of the book in Chapter 1.

A general model of the processes involved in air pollution that can lead to adverse effects on the atmosphere, humans, and other life forms, their ecosystems and resources is shown in Figure 13.1. A multitude of air pollutants are emitted into the atmosphere from natural and human sources. These are transported and transformed in air, resulting in exposures at various concentrations, rates, or frequencies, over short- or long-time scales, that may cause adverse or toxic effects on humans, wildlife, ecosystems, and modifications to the atmosphere, such as observed with climate change, ozone depletion in the stratosphere, and in expanding urban airsheds.

13.2 Earth's Atmosphere

The structure and composition of the Earth's atmosphere is fundamental to its diverse regional climates and capacity to sustain the biodiversity of life on Earth. The atmosphere plays a critical role in integrating the wide range of natural and air pollutant emissions that are continually released from over the Earth's surfaces on time scales that vary from weeks to years. Variations in the constituents and amounts of gases and aerosols affect air quality, weather, and longer-term climate.

Our atmosphere consists of a relatively thin layer of gases and aerosols, thought to have formed and evolved around the planet by the release of volatile compounds from the planet itself, and held in place by gravitational forces.

Chemistry and Toxicology of Pollution: Ecological and Human Health, Second Edition. Des W. Connell and Greg J. Miller.
© 2023 John Wiley & Sons, Inc. Published 2023 by John Wiley & Sons, Inc.
Companion website: www.wiley.com/go/toxicologyofpollution2e

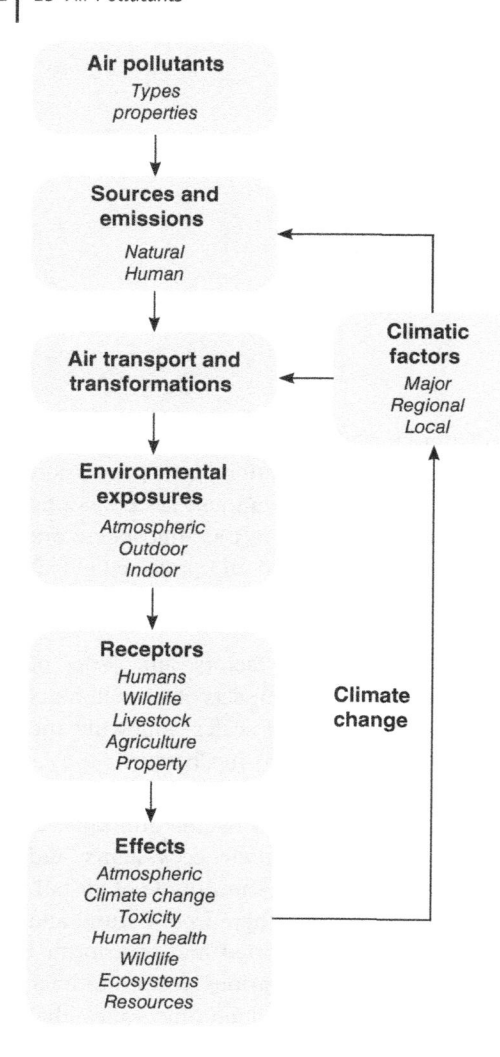

Figure 13.1 Air pollution conceptual model.

Table 13.1 Major gas concentrations in the Earth's atmosphere.

Gas	Concentration[a,b,c] (% v/v)	ppmv
Nitrogen (N_2)	78.084	780 840
Oxygen (O_2)	20.946	209 460
Argon (Ar)	0.9340	9340
Carbon dioxide	0.041	410
Neon	0.001818	18.18
Helium	0.000524	5.24
Methane	0.000179	01.79

[a] ppmv, parts per million by volume.
[b] Dry atmosphere.
[c] Water vapor (0.001–5%).

Mixtures of aerosols or airborne particulates occur near the Earth's surface and are dispersed in the troposphere at variable levels ($\mu g\,m^{-3}$ to $>10\,mg\,m^{-3}$). Most aerosols (~90% by mass) have natural origins and include thick columns of ash, sulfur dioxide gas, and formation of sulfate particulates from volcanoes, organic carbon particles, and thick smoke from forest fires, and aerosols formed from the reaction of volatile organic compounds of plant origin with other substances in the air. Abundant aerosol sources consist of sea salt derived from the wind-driven spray of ocean waves, and mineral dusts (coarse and fine) generated by dust storms in deserts or by wind erosion, usually in degraded arid and semi-arid zones. Human-derived sources of aerosols are less abundant than natural forms, making up the remaining 10% of aerosols, but can dominate the air in urban and industrial areas, and downwind.

Trace gases (e.g. carbon dioxide, methane, water vapor, and ozone) and aerosols play a crucial role in the Earth's radiative balance and in the chemical properties of the atmosphere. Current evidence shows rapid changes in the levels and atmospheric effects of trace gases over the last 200 years, mostly due to human activities such as combustion of fossil fuels (coal and oil) for energy and transportation (Seinfeld and Pandis 1998, pp. 1–4).

Atmospheric Layers

The vertical structure of the Earth's atmosphere is generally divided into five main layers that extend from the densest layer of air – the troposphere, closest to the Earth's surface, to the exosphere, where the extremely thin gases dissipate into outer space. Figure 13.2 shows these layers and their boundaries and altitudes up to the thermosphere, but omits the outer layer of the exosphere that merges into outer space. There is no clear boundary between the end of the Earth's atmosphere and where outer space begins.

The thickness of the atmosphere is only about 2% of the Earth's radius (6378 km at the equator) based on most of the atmosphere's mass being found at 100 km above sea level. The total mass of the atmosphere is given as 5.1×10^{18} kg (NASA 2017a).

The Earth's atmosphere or air is composed essentially of about 78% nitrogen gas, 21% oxygen gas, and 0.93% argon gas on a volume per volume basis. The remainder of components consists of water vapor and trace gases, generally less than 0.1%. However, water vapor is highly variable, up to 5% in the lower atmosphere, due to evaporation and precipitation processes. Table 13.1 shows dry air concentrations of major and trace gases in the Earth's atmosphere. Numerous other gaseous compounds from natural sources such as biogenics or air pollutants from human activities also occur at highly variable and trace levels (low ppm or ppb), usually in the lower atmosphere and urban airsheds.

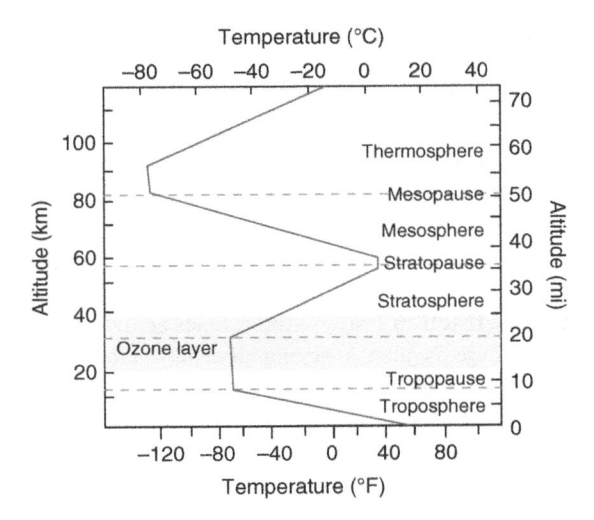

Figure 13.2 Atmospheric layers of the Earth.

The delineation is generally known as the Karmin line, 100 km (62 miles) above the Earth's surface (99.99997% of the Earth's atmosphere is below this line) (NASA 2020).

Precise description of the layers varies in the scientific literature. The main layers are described further in S13.1 (see Companion Book Online) and are characterized by their thermal properties such as temperature changes, chemical composition, movement, and density.

Atmospheric pressure and density of air decrease with altitude. At sea level, standard atmospheric pressure is 1013.25 millibars (mb) or one atmosphere (760 mmHg). One pascal equals 0.01 millibars (NWS 2020). As indicated in Figure 13.2, air temperature decreases almost linearly with altitude in the troposphere until the tropopause.

13.3 Air Pollution, Weather, and Climate

> "Climate" refers to the average weather in terms of the mean and its variability over a certain time-span and a certain area
>
> (Houghton et al. 1990, p. 87).

The Earth's global climate depends on the energy balance between how much energy the Earth receives from the sun and other flows of energy (and heat), which take place within the atmosphere and other parts of the climate system itself. Its climate system is described as an interactive system consisting of five major components: the atmosphere, the hydrosphere, the cryosphere, the land surface, and the biosphere. This system is under pressure or influenced by various external forcing mechanisms, particularly solar radiation, and the direct effect of human activities, including emissions in the troposphere. The atmosphere is described as the *most unstable and rapidly changing part of the system* (Houghton et al. 1990, p. 87).

Its role, however, is crucial to regulating the Earth's climate and temperature, as shown by the capacity of greenhouse gases in the atmosphere to trap re-emitted heat energy in the lower atmosphere and keep the Earth sufficiently warmer than otherwise to sustain life, through a natural *greenhouse effect,* as described in Chapter 14 of this book.

Earth's climate is strongly influenced by circulation of air masses in the atmosphere, together with the slower movement of water masses by ocean currents, acting on global, regional, mesoscales, and microscales. "Many physical, chemical and biological interaction processes occur among the various components of the climate system on a wide range of space and time scales, making the system extremely complex. Although the components of the climate system are very different in their composition, physical and chemical properties, structure and behaviour, they are all linked by fluxes of mass, heat and momentum: all subsystems are open" (Houghton et al. 1990, p. 89).

Types of climates can be categorized into major, regional, meso-, and microclimates (see S13.2)

13.4 Outdoor and Indoor Air Pollutants

Overview

Air pollutants are found in the atmosphere in the form of particulates or aerosols, gases or vapors, and radiation. Pollutants may partition between solid particulates and gaseous phases in the air, and partition at the air–water interface, or upon contact with, or deposition onto solid surfaces (e.g. plants and soils), depending on their sorption properties. Water-soluble pollutants can dissolve in water vapor, snow, and ice in air. Atmospheric aerosols consist of a suspension of fine solid particles or liquid droplets in air. Table 13.2 lists some other characteristics.

This section identifies the different types, sources, and some of the important properties of outdoor and indoor air pollutants that are present in human environments, including workplaces, and the environments of other living organisms, especially wildlife. In many cases, the same types of pollutants will be found in outdoor and indoor environments, although for some, their concentrations, sources, and impacts can be different. Major types, sources, and health effects of air pollutants are summarized in Table S13.1 (see Section S13.3, Chapter S13, Companion Book Online).

How these air pollutants are transported and transformed by environmental processes, and impact on the health of wildlife and humans at local to global scales, is evaluated in the following sections.

Table 13.2 Some key characteristics of air pollutants.

- Three basic types of pollutants: physical (e.g. ultraviolet radiation and particulates), chemical (e.g. carbon monoxide, sulfur dioxide, and benzene), and biological (bioaerosols).
- Emitted from human derived sources and/or natural sources, including produced from living organisms (e.g. biogenic hydrocarbons).
- Classified as **primary** and **secondary** pollutants in the atmosphere.

    A **primary** pollutant (e.g. sulfur dioxide) is an air pollutant emitted directly from a source. In contrast, a **secondary** pollutant is not directly emitted, but forms when other pollutants react in the atmosphere. An example is the formation of ozone in the atmosphere (troposphere) by the UV catalyzed reaction of hydrocarbons with nitrogen oxides during the production of photochemical smog.

Particulate Matter

Particulate matter (PM) is a complex mixture of dust (suspended solids) to ultrafine particles (>100 to <0.1 μm diameter) and liquid droplets that are emitted from human activities and natural sources and are dispersed in air by physical transport processes (see Table 13.3). Typically, solid and liquid particles found in the atmosphere cover 4–5 orders of magnitude in particle size from nanometers (<0.1 μm) to 100 μm or more (Solomon 2012).

Aside from the visibility and nuisance effects of dusts, breathing the **inhalable fraction** (up to 100 μm for larger particles) can affect the respiratory and cardiovascular systems and increase the risk of lung cancer in humans. This fraction of airborne particles enters the nose and mouth during breathing and the particles can potentially be deposited within the respiratory tract. Very small particles, however, can penetrate deep into the lung. The **thoracic fraction** PM_{10} (<10 μm) describes the mass fraction of particles penetrating beyond the larynx, and the fraction of particles reaching the deep lung is called the fine or **respirable fraction** $PM_{2.5}$ (<2.5 μm). A **submicron fraction**, PM_1, (<1 μm) is much finer than $PM_{2.5}$ and can penetrate even further into the cardiovascular system leading to higher risks of heart disease, premature births, and effects on fetal development. Within the $PM_{2.5}$ fraction is an **ultrafine fraction** $PM_{0.1}$ with particles below 0.1 μm or 100 nm. There is also a strong association between ultrafine particles (UFP) and adverse human health effects (Solomon 2012). Elevated ambient airborne particles from pollution sources are also classified as a Group 1 human carcinogen (see IARC).

Properties of Airborne Particles

The diameter of airborne particles is commonly expressed as an **aerodynamic diameter** instead of the physical diameter of the particles (see Table 13.3). The aerodynamic diameter is defined as the equivalent diameter of a spherical particle of unit density (1 g cm^{-3}), which exhibits the same aerodynamic behavior as the airborne particle.

The aerodynamic diameter (D_a) is calculated from the physical equivalent diameter (D_e) by (DeCarlo et al. 2004):

$$D_a = D_e \sqrt{(d/X)} \qquad (13.1)$$

Where D_a, aerodynamic diameter of the particle (μm); D_e, physical equivalent diameter of the particle = $(6V/\pi)^{1/3}$ (μm); V, volume of the particle (μm^3); d, density of the particle (g cm^{-3}); X, dynamic shape factor of the particle (e.g. spheroids ~1.0, cubes ~1.1–1.25).

Table 13.3 Classification of airborne particles.

Classification	Aerodynamic size range (μm)	Description
Fallout-dust	>100	Particles that readily settle out from the atmosphere
Total suspended particles (TSP)	<30	Inhalable particles that are able to remain suspended in the atmosphere for longer periods of time
PM_{10}	<10	Particulate matter (10 μm or less in diameter) that can be suspended for long periods and penetrate down to the thoracic region
Fine particles $PM_{2.5}$	<2.5	Respirable particles that can penetrate deep into the lungs and enter into the body via the blood stream of the alveoli
Coarse particles $PM_{2.5-10}$	<2.5–10	Coarse dust particles between 2.5 and 10 μm in diameter
Submicron particles PM_1	<1	PM_1 (<1 μm) is a much finer fraction of $PM_{2.5}$ and can penetrate even further into the cardiovascular system leading to higher risks of heart disease
Ultrafine particles $PM_{0.1}$	<0.1	Ultrafine particles (UFP) in the nanoparticle size range; strong association between ultrafine particles (UFP) and adverse human health effects
Nanoparticles	<0.1	Engineered nanoparticles are the same as ultrafine particles (between 1 and 100 nm in size); nanotoxicity effects
Fibrous particles, e.g. asbestos	<3	Length >5 μm, and an aspect ratio (length to width) greater than or equal to 3 : 1; asbestos is carcinogenic to humans

Sources: Compiled from Solomon (2012); Baird and Cann (2012, pp. 118–130).

The distribution of ambient particles is a function of **particle size** and is typically characterized by three main modes that reflect the main processes involved in forming particulate matter. These are the nucleation mode, the accumulation mode, and the coarse mode (Seinfeld and Pandis 1998, pp. 429–433; Solomon 2012):

(1) Nucleation mode particles derive from physical and chemical processes, such as nucleation and condensation of supersaturated vapors, produced by combustion and are short-lived (minutes to hours).
(2) Accumulation mode particles grow mainly from the nucleation mode particles by coagulation or vapor adsorption. Particles in this range can remain suspended for long periods of a few days with slow settling.
(3) Coarse mode particles are usually primary particles generated by mechanical abrasion processes in the atmosphere and in emission sources, but may contain other constituents as a result of coagulation and condensation processes.

The particle mass-size distribution of urban aerosols is often bimodal or trimodal. Generally, in urban areas, PM_{10} contributes about 40–60% of the mass of TSP, while $PM_{2.5}$ makes up about 40–60% of the mass of PM_{10}. Most of the particle number count consists of nuclei or Aitken nuclei mode (≤ 0.1–$2\,\mu m$) but usually account for only a few percent of the total mass (see Figures 13.3a and b).

Nanoparticles: In contrast to ultrafine dust particles (derived from human generated processes, or natural aerosols), engineered nanoparticles (NPs) are ultrafine particles with diameters of <100 nm, made for their specific properties (e.g. size, number of particles, surface area, shape, and optical properties). However, the comparable nature of engineered NPs to UFPs suggests that the potential for human health effects is likely to be similar to those of UFPs and precautionary measures are argued (Gwinn and Vallyathan 2006).

An overview of toxic inorganic and organic air pollutants is given below, derived primarily from the WHO (2000, 2006a, 2010a).

Inorganic Gases

Major air pollutants are the oxidized gases of carbon, sulfur, and nitrogen, also known as acid gases, and the oxidant gas, ozone. Acid gases are readily soluble in water vapor or aqueous solutions and can form acidic solutions

Figure 13.3 (a) Generalized distribution of urban atmospheric particles by size diameter and number with different modes of formation. Source: Modified from Seinfeld and Pandis 1998, pp. 429–433; Solomon 2012. (b) Generalized distribution of urban atmospheric particles by size diameter and volume with different modes of formation. Source: Modified from Seinfeld and Pandis 1998, pp. 429–433; Solomon 2012.

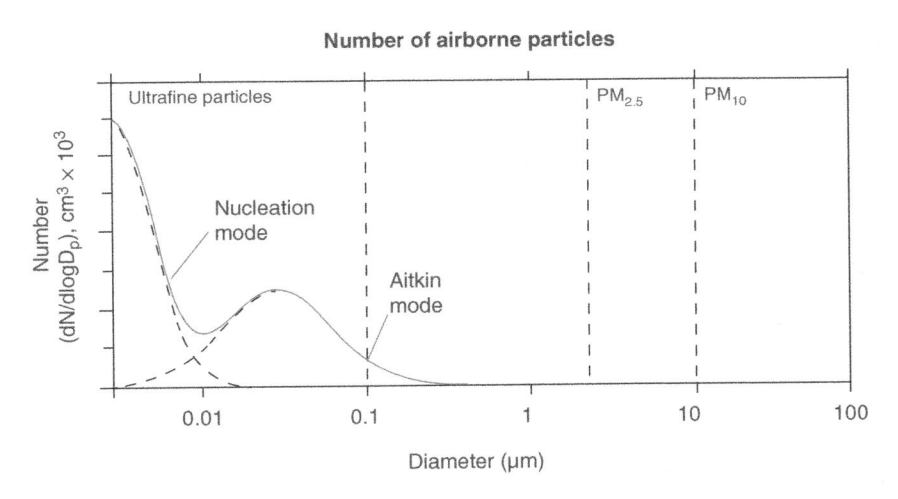

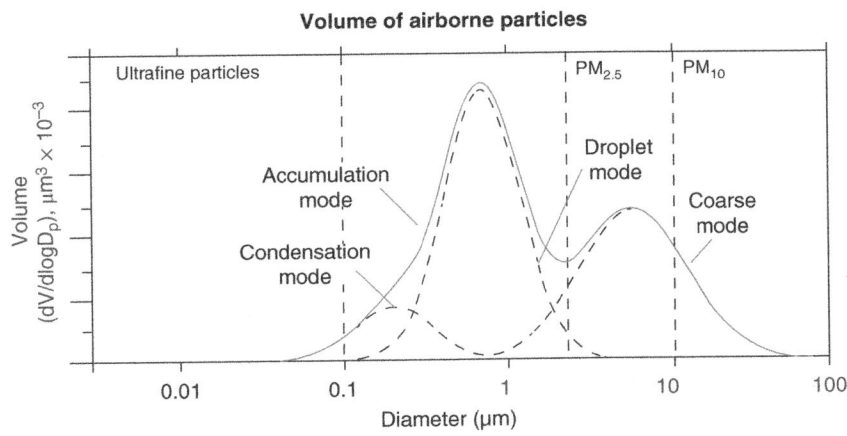

(e.g. acid rain or precipitation). Toxic halogen gasesu (chlorine, Cl_2, bromine, Br_2, fluorine F_2, and iodine I_2) also form acidic gases because they are hydrolyzed by reaction with water to form strong acidic solutions such as hydrochloric acid (HCl), hydrobromic acid (HBr), hydrofluoric acid (HF), and hydroiodic acid (HI). Hydrogen sulfide (H_2S) is another toxic acid gas that is formed from the anaerobic digestion of organic matter or naturally in crude oil, natural gas, and volcanic gases or by reaction of sulfides S^{2-} with water (e.g. heavy metal sulfide ores). Hydrogen sulfide gas reacts slowly with water and forms a weak acid.

Oxides of carbon consist of carbon monoxide (CO) and carbon dioxide (CO_2). Carbon monoxide is a colorless, odorless, tasteless, and toxic gas produced by the incomplete combustion of carbon-containing fossil fuels (e.g. motor vehicles) and biomass (e.g. wood fires), forest fires, and volcanoes. Tobacco smoke is one of the main sources of CO in indoor or enclosed environments.

Atmospheric carbon dioxide is the primary source of carbon for life on Earth and the major greenhouse gas in the atmosphere, as part of the carbon cycle. Natural sources include volcanoes, geothermal waters, and dissolution of carbonates by water and acids. Carbon dioxide gas is produced by the complete oxidation of carbon monoxide, combustion of biomass, fossil fuels, and other organic materials or compounds, through respiration by all aerobic organisms, during decay of organic matter, fermentation of sugars, and industrial oxidation processes. Alternatively, photosynthesis by green plants, algae, and cyanobacteria uses solar energy to convert carbon dioxide from the atmosphere and water into carbohydrates and oxygen gas. Carbon dioxide dissolves in water to form weak carbonic acid (H_2CO_3), involved in the formation of carbonates as a sink for global carbon, and as a cause of ocean acidification.

Sulfur compounds that occur in the atmosphere consist essentially of sulfur oxides (SO_2 and SO_3), hydrogen sulfides, sulfuric acid, sulfates, and organic sulfur compounds (e.g. dimethyl sulfide, $(CH_3)_2S$). The dominant air pollutant is sulfur dioxide, a colorless gas with a pungent odor. It comes mainly from human activities such as the burning of coal and oil at power plants, or from copper smelting. It is a precursor to sulfuric acid in acid rain. Natural sources include volcanic eruptions. Sulfur trioxide is formed from the oxidation of sulfur dioxide and readily reacts with water to produce sulfuric acid.

Nitrogen oxides: NO_X are the main air pollutants of the gaseous nitrogen compounds. Nitric oxide (NO) and nitrogen dioxide (NO_2) are colorless to reddish brown toxic gases in the atmosphere. Nitrogen oxides are emitted from the exhaust of motor vehicles, the burning of coal, petroleum, oil, or natural gas, during industrial processes such as arc welding and electroplating, producing nitric acid. Nitric oxide forms in mixtures of nitrogen gas and oxygen gas at high temperatures (e.g. internal combustion engines). A major source of indoor exposure to NO_2 is from the use of gas stoves for cooking or heating in homes. Natural sources include bacterial respiration, volcanoes, and lightning. In contrast, nitrous oxide (N_2O) is a colorless gas with a pleasant, sweetish odor, and taste, used as an anesthetic, analgesic, a propellant in food aerosols, and fuel oxidant in motor sports. It is a greenhouse gas with a strong global warming potential and can damage the ozone layer in the stratosphere (ozone depleting substance). Ammonia gas (NH_3) is an alkaline reactive gas that is pungent and irritating at low levels in air and toxic at high doses. It reacts with water to form ammonium ions (NH_4^+) and soluble salts. It is released into the air from various biological sources (e.g. nitrification and denitrification), agriculture, industrial, and combustion processes. Agriculture is the main source of ammonia emissions, including from livestock and ammonia-based fertilizers (e.g. urea and ammonium sulfate).

Gaseous hydrogen fluoride (HF) will be absorbed by atmospheric water forming an aerosol or fog of aqueous hydrofluoric acid. It will be removed by wet deposition. Particulate fluorides in the atmosphere will be removed by dry and wet deposition. Upon entering water, insoluble forms of fluorides will gravitate to the sediment. Fluorides are strongly retained by soil. In water, fluoride forms strong complexes with aluminum. Fluoride accumulates in some plants and in the skeletal system of terrestrial animals that consume fluoride-containing foliage.

Ozone (O_3) is a triatomic form of oxygen that is relatively unstable. It is a colorless to bluish gas with a characteristic odor at low levels (<2 ppm). Most ozone is formed naturally in the upper atmosphere (stratosphere) when UV solar radiation splits oxygen gas into two single atoms of oxygen, each of which recombines with another molecule of O_2 gas to form a molecule of ozone with three atoms of oxygen. Ozone gas is unstable and most converts back to oxygen gas. In the lower atmosphere (troposphere), toxic ozone gas is formed in the presence of sunlight and heat as a secondary pollutant from fossil fuel combustion (e.g. traffic emissions), through a complex series of reactions during the production of photochemical smog. Some other emissions of ozone result from its use as a powerful oxidant, for example, as a disinfectant in air and water, odor control, bleaching agent, and in chemical synteses (ozonolysis). Ozone levels in large urban environments are estimated to be a significant cause of the burden of disease and death in human populations from urban air pollution (see Section 13.9).

Organic Compounds

Organic air pollutants cover a wide range of organic substances, chemical forms, and families that often occur as mixtures in outdoor and indoor environments. Major categories include biogenic organic compounds, petroleum gases (e.g. methane, ethane, ethylene, propane, and butane), non-methane hydrocarbons, volatile organic compounds (e.g. benzenes and various low boiling solvents), and semi-volatile compounds including PAHs, POPs, industrial chemicals, and pesticides. Categories tend to be empirical and components may overlap with other categories. Definitions of VOCs also vary between organizations and countries (e.g. US EPA, European Union, and World Health Organization).

- Biogenic organic compounds – volatile biogenic organic compounds (BVOCs) are produced by plants and are released in large mass emissions from vegetation into the atmosphere. They are highly reactive chemically and include compounds such as isoprene, monoterpenes (e.g. pinene and limonene), formaldehyde, acetaldehyde, methanol, ethanol, acetone, formic acid, and dimethyl sulfide.
- Volatile organic compounds (VOCs) – organic chemicals of biogenic, petrogenic, and synthetic origins with high vapor pressures at room temperature due to their low boiling points (e.g. 50–100 °C to 240–260 °C). Human-related sources include paints, coatings, adhesives, solvents, and fossil fuels. Toxic VOCs such as benzene, methylene chloride, halogenated hydrocarbons, and formaldehyde are discussed in Chapters 2 and 11. Exposures to VOCs can lead to chronic human health effects in indoor environments, workplaces, and in polluted outdoor environments (e.g. traffic emissions and through their role in the formation of photochemical smog and toxic ozone gas).
- Semi-volatile organic compounds – usually applies to organic compounds that can be outdoor or indoor pollutants with higher molecular weights and boiling points (e.g. 240–260 °C to 380–400 °C) than VOCs such as PAHs, dioxins, POPs, and pesticides. These types of chemicals are discussed in Chapters 2, 8, 9, and 11 of this book.

Toxic Metals

Common toxic metals found in urban air include Pb, Cd, Zn, Cu, Mn, Ni, Cr, and the metalloid As. Atmospheric mercury emissions are significantly associated with coal-fired power stations and particulates. Metals and their compounds can be readily emitted into the air as particulates or adsorbed onto them, and in the form of particulates (e.g. lead) from combustion of leaded gasoline, metal vapor (e.g. mercury), fumes (e.g. cadmium), or gases (arsine or nickel carbonyl), depending on the source (natural and human activities). In recent years, lead exposure has been decreased by regulatory actions in removing lead from paint and gasoline and reducing occupational lead exposure.

As, Pb, Cd, Cr, Ni, and Hg are systemic toxicants affecting multiple organs. As, Cd, Cr (VI), and Ni are listed as human carcinogens and Pb is a probable human carcinogen (International Agency for Cancer Research, IARC). Cobalt is anticipated to be a human carcinogen (United States National Institute of Environmental Health Sciences (US-NIEHS) – National Toxicology Program).

Hazardous or Toxic Air Pollutants

Hazardous or toxic air pollutants are those air pollutants known or suspected to cause cancer or other serious health effects (e.g. reproductive or birth defects), which may include many different toxic organic chemicals (including POPs and certain pesticides), toxic metals and non-metals, inorganics, mineral acids, acid gases, and other hazardous substances (e.g. radionuclides and asbestos), usually specified for regulatory purposes. Under the Clean Air Act in the United States, the US EPA is required to regulate hazardous air pollutants (e.g. 187 pollutants listed in 2017) (US EPA 2017a).

Air Pollutants Modifying the Properties of the Atmosphere

A number of air pollutants are now well known to change critical properties of the Earth's atmosphere on regional and global scales resulting in extensive evidence of harm and damage to the ecosphere, and living organisms, including effects on the health of human and wildlife populations and ecosystem resources. These are:

- **Acid rain pollutants** – acidic gases emitted from human activities (e.g. combustion of fossil fuels) dissolve in and react with rain water and oxygen gas to form secondary pollutants – sulfuric acid, nitric acid, sulfates, and nitrates - resulting in wet and dry acid deposition of these pollutants onto soils, forests, and aquatic water bodies, causing acidification and loss of fisheries, forests, and other wildlife (e.g. Eastern USA and Europe) (see Section 13.6 for reactions).
- **Ozone-depleting chemicals** produced by human sources (e.g. refrigerants, fire extinguishers, solvents, dry cleaning agents, and fumigants) – mainly chlorofluorocarbons (CFC-11 and CFC-12) and halons – are stable and persistent in the lower atmosphere but slowly penetrate into the ozone layer of the stratosphere where they are decomposed by UV radiation. The released chlorine and bromine atoms initiate chain reactions that destroy

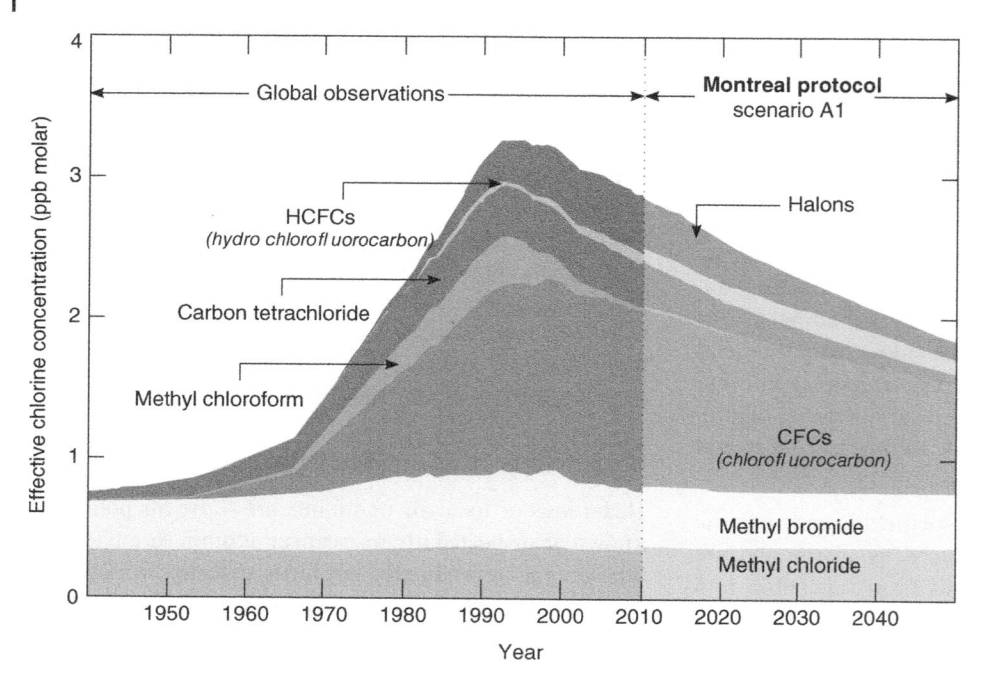

Figure 13.4 Concentrations of ozone-depleting chemicals in the atmosphere observed from before 1950 to 2010, and as projected by Montreal Protocol scenario A1 to 2050. Source: State of the Environment Committee (2011), p. 118. Published by Commonwealth Scientific Industrial and Research Organisation, Canberra, Australia.

ozone molecules and deplete the UV protective capacity of the ozone layer for life on Earth (see Section 13.6). Peak levels of chlorine and bromine in the stratosphere were reached in the mid-1990s and have declined, but ozone depletion is projected to continue for decades (see Figure 13.4).

- **Greenhouse gases** – carbon dioxide, methane, nitrous oxide, and to a lesser extent, CFCs and substitutes, sulfur hexafluoride – are strongly associated with a warming trend in the Earth's atmosphere and increasing acidification of the oceans acting as a global sink for observed excess emissions of carbon dioxide from human activities. Chapter 14 describes and evaluates greenhouse gases and global warming observations, projections, and impacts on human health and wildlife.

Ultraviolet (UVB) Radiation

Stratospheric ozone acts to limit exposure to solar UVB in the lower atmosphere. However, ozone-depleting substances (e.g. CFC aerosols) have reduced ozone concentrations and increased exposure to UVB radiation at the Earth's surface. Overexposure to ultraviolet B (UVB) light in the wavelength range of 280–315 nm can lead to harmful effects on humans and various other life forms. It can cause nonmelanoma skin cancer in humans and is a significant factor in the development of malignant melanoma and cataracts. The incident rate of nonmelanoma skin cancer is exponentially related to UV exposure, while malignant melanoma appears to be related to short periods of very high UV exposure, especially when young (Baird and Cann 2012, pp. 8–9).

Bioaerosols

Various pathogens (and allergens) are emitted to air (indoor and outdoor) and dispersed in the form of biological aerosols (bioaerosols), dusts, and liquid droplets from a wide range of environmental, body, and plant sources (e.g. infected animal or person, animal dander, fungi spores, pollen grains, endotoxins, excreta, and other biological wastes). Bioaerosols are fine suspensions of airborne particles in air that contain microbes and/or biological matter released from living organisms. They range in size from less than 1 to 100 μm or so (see Chapter 2.1 in this book).

13.5 Sources and Emissions of Air Pollutants into the Atmosphere

Outdoor Air Pollutants

Emission sources of atmospheric pollutants, primary and secondary, consist of human-related and natural sources, usually classified as point sources (e.g. industrial stacks, vents, or chimneys), line sources (e.g. vehicle traffic flows), and area sources. General types of emissions are point or stack and fugitive. The latter refer to uncontrolled or unintended pollutant releases into air from nonpoint sources, such as leaks from equipment, processes, pipelines, stockpiles, dusts from exposed surfaces, odors from wastewater ponds, and other activities (e.g. organic waste burning, clearing of vegetation, or excavation of soils).

Global emissions of major primary air pollutants (particulate matter species, CO, NO_X, SO_2, VOCs), secondary pollutants (O_3), greenhouse gases (CO_2, CH_4, and N_2O),

Table 13.4 Global air pollutant emissions (Teragrams per year, Tg yr^{-1}) from human activities based on year 2010.

Air pollutants	Annual emissions (Tg yr^{-1})
Particulate matter PM_{10}	62.5^a
Black carbon (BC)	13.5^a
	5.3^b
SO_2	95^b
CO	580^b
CO_2	37 000 (2017)^c
NO_x	70^b
O_3 (secondary pollutant)	4520 Production Troposphere^d
VOCs	1150 Biogenic^e

^{a}Klimont et al. (2017) – Supplemental material (excludes forest fire emissions but includes agricultural burning).
^{b}Granier et al. (2011) – Estimates of total global emissions for each species reported from MACCity anthropogenic emissions inventories for 2010 (specific project) (see ECCAD –The GEIA Database).
^{c}Global Carbon Project (2017).
^{d}See Figure 13.8 this chapter.
^{e}Seinfeld and Pandis (1998, pp. 82–85).
Tg (SI unit of mass) teragram = 1×10^{12} g.

Table 13.5 Estimates of anthropogenic emissions^a (Tg yr^{-1}) of particulate matter (PM) species (1995–2010).

Particulate matter species	1995	2000	2010
PM_{10}	57.8	58.4	62.5
$PM_{2.5}$	43.8	44.6	47.8
PM_1	35.9	36.7	37.8
Black carbon	6.2	6.6	7.3
Organic carbon	11.95	12.45	13.55

^{a}Estimates exclude forest fires but include agricultural burning.
Source: Data compiled from Table S8.1, ECLIPSE V5a PM estimates (Klimont et al. 2017).

toxic metals, and organic pollutants (e.g. CFCs, POPs, and PAHs) from anthropogenic sources are of critical concern for human and ecological health. Natural sources of emissions of primary pollutants on a global scale usually exceed those due to human activities, but current exposures to many human related sources of air pollutants (e.g. $PM_{2.5}$ and ozone) are recognized as the leading environmental cause of premature disease and death among exposed human populations.

Global estimates for major air pollutants from human activities (mainly from combustion of fossil fuels, and some biomass, in households, industry, transport, and waste disposal) are listed in Table 13.4. These estimates, however, do not reflect relative ambient levels and potential adverse effects of pollutants because variations exist in their physical, chemical, and toxic properties, and different atmospheric residence times.

Global anthropogenic emissions of particulate matter and primary carbonaceous aerosols including black carbon and organic carbon are shown in Table 13.5. While global emissions of PM have not changed significantly between 1990 and 2010, a strong decline in PM emissions in North America, Europe, and the Pacific was observed while Asia's contribution grew from just over 50% to nearly two thirds of the global anthropogenic total during this period. Residential combustion was reported as the most important sector (Klimont et al. 2017).

Air Emission Inventories

Estimates of air pollutant emissions on local sites, urban, regional, country, and global scales are developed from databases that compile lists of the types of air pollutants and greenhouse gases, their emission sources for various human activities and natural sources, and the amounts of air pollutants discharged from these sources, over various spatial (e.g. 1×1 km, country or region) and time resolutions (e.g. yearly or monthly).

Anthropogenic emissions include fuel production, industrial and domestic combustion (fossil fuels and biofuels), transportation (road, rail, air, and ships), industrial processes, uses of chemical products, agriculture, and waste disposal. There are a number of global, international, and national emission inventories that cover historical, current, and may also project future anthropogenic emissions of greenhouse gases and air pollutants (e.g. RAINS.GAINS, EDGAR, GEIA, ECLIPSE, REAS [Regional – South and East Asia]). Global and regional estimates for individual pollutants have developed substantially but variations in purposes, areas covered, methods used, emission factors, spatial and time resolution remain issues for analysis of trends.

Granier and co-researchers evaluated several different emission inventories of global and regional anthropogenic emissions of CO, NO_X, SO_2, and black carbon for the period 1980 to 2010. All the data sets showed a slight increase in global emissions of NO_X, and a similar pattern for CO, with an increase from 1980 to 1990, followed by a slight decrease until 2000–2005, although there was some spread between the lowest and upper estimates. For SO_2 emissions, most of the inventories indicated a significant decrease in global emissions from 1980 to 2010. Black carbon emissions showed a slight increase in most of the inventories providing these estimates, although large differences occurred in early estimates (1980–1990). Significant differences between emission estimates for the pollutants in regional inventories, however, were found for the study period, indicating a lack of consensus on the best estimates for surface emissions of air pollutants (Granier et al. 2011).

Estimation of Air Pollutant Emissions from Sources

General emission equation for air pollutants and a worked example are given in **S13.4 supplementary material**.

Indoor Air Pollutants

Emission sources of air pollutants in indoor environments have been studied intensively since the 1980s and the era of the *sick building syndrome*. The scope of the problem of indoor air pollution for human health and comfort is serious because a large proportion of the global population, including children and the elderly, can spend up to 90% or so of their time indoors exposed to multiple emission sources of common air pollutants, in relatively limited airspaces compared to outdoor environments. The range of air pollutants is extensive, from combustion products, respirable forms of particulates, asbestos fibers, second hand or environmental tobacco smoke, VOCs, solvents, toxic organic chemicals and metals such as lead, radon gas, to biological pollutants or bioaerosols (allergens, mites, bacteria, mold, and fungi) (Spengler et al. 2001).

Table 13.6 presents a summary of potential emission sources of air pollutants found within indoor environments, from dwellings to buildings and enclosed structures, offices, education facilities, shops, motor vehicles, trains, ships, and aeroplanes, among others. Potential sources cover outdoor pollutants from surrounding land uses and traffic via natural or mechanical ventilation, emissions from building materials, furnishings and fittings, equipment, and indoor activities by occupants and personal care products.

13.6 Behavior and Fate of Pollutants in the Atmospheric Environment

The distribution and fate of primary emissions of anthropogenic and natural pollutants in the atmosphere are strongly influenced by interactions between a complex set of transport, transformation, and deposition processes mediated by solar radiation, and also the presence of water vapor (troposphere).

As indicated in Figure 13.5, primary pollutants are emitted from the Earth's surface into the atmosphere, essentially the troposphere, where they are subject to atmospheric mixing, transport, transformation, and removal processes that vary on local, regional, and global scales. The troposphere acts as a chemical reservoir and reactor in which atmospheric oxidation and photochemical oxidation reactions convert carbon-, nitrogen- and sulfur-containing compounds to the oxidized states. Oxides of nitrogen along with ozone formation and removal in the presence of sunlight play central roles in the chemistry of the troposphere.

Removal of aerosol particulates, short-lived chemical pollutants, and converted products from the atmosphere occurs generally by wet and dry deposition and capture from the atmosphere by natural biological, chemical, and physical processes, and sequestration in land, oceanic, and polar sinks (e.g. carbon sequestration). Some long-lived chemical species slowly mix and move vertically from the troposphere into the stratosphere where they are involved in photochemical reactions that form and destroy ozone.

Atmospheric Transport Processes

The dispersion of air pollutants is subject to the types of emission sources (e.g. stacks, area, and traffic flows), the atmospheric physics of diffusion and turbulence, and related atmospheric stability mediated by local climatic factors, terrain characteristics, land uses, population growths, and other factors. Diffuse sources include evaporation or volatilization, and wind erosion processes.

The major mechanisms for dispersion involve horizontal (e.g. wind speed and direction), vertical (e.g. thermal buoyancy) and mechanical turbulence (e.g. canyon effect and aerodynamic downwash) rather than the much slower diffusion of chemicals from high to low concentrations. Combined effects of turbulence (e.g. from plume rise and dispersion) are associated with the phenomenon of atmospheric stability class under stable, neutral, or unstable conditions. These conditions can be determined by temperature lapse rates – the rate of temperature decrease with altitude of the surrounding air (environmental lapse rate) and the parcel of air moving upwards with the surrounding air (adiabatic lapse rate). Temperature inversions also interfere with dispersion under certain conditions when a layer of warmer air forms on top of a layer of cooler air. Other factors include urban heat island effects with temperatures elevated several degrees above surrounding areas and topographical characteristics, including larger regional basins or airsheds (LaGrega et al. 2001, pp. 213–227).

Atmospheric Transformation Processes

These types of processes in the atmosphere, although complex, are generally photochemically induced by sunlight ($h\nu$), assisted by catalytic chemical species (e.g. free radicals). Reaction steps in these processes usually consist of parallel, sequential, and competitive reactions. Physical and chemical transformations can be divided into two major categories:

1. **Aerosol particulate processes.** Particles emitted into the atmosphere can undergo transformations in size and distribution due to physical and chemical processes and meteorological conditions. Particles can grow from

Table 13.6 Potential emission sources of indoor air pollutants.

Pollutants	Outdoor sources	Building sources (materials, furnishings, and fittings)	Indoor services and equipment	Occupants and activities
Combustion products (gases (e.g. carbon monoxide and nitrogen oxides), smoke, fumes, PAHs)	Vehicle and non-road engine emissions/stacks Photochemical smog (e.g. ozone); forest fires	Toxic emissions from any fire and smoke	Contamination of HVAC systems, emissions from combustion, cooking and indoor heating sources, generators	Cooking and heating, fireplaces, smoking, use of combustion motors/generators
Gases/odors (e.g. methane, carbon monoxide, hydrogen sulfide)	Emissions from nearby vehicles, industrial, commercial, agricultural land uses, feedlots, landfills, incinerators	Sewer gas, fungal or mold growths	Contamination of HVAC systems, office equipment, (incl. ozone generating from UV sources)	Trash or rubbish, waste storage activities
Particulates (PM_{10} and $PM_{2.5}$)	Tobacco smoking, respirable dusts, including from traffic emissions, dust storms, fires, and smoke	New or renovated materials, furnishings, inadequate cleaning, construction and renovations	Emissions from office equipment	Cooking, cleaning activities, wood working, renovations, and repairs
Hazardous dusts: (asbestos fibers, synthetic vitreous fibers, and crystalline silica dusts)	Airborne asbestos, SMF, and silica dusts	Materials containing asbestos and vitreous fibers	Insulation containing hazardous fibers and toxic dusts	Mechanical disturbance or damage to materials containing hazardous fibers and/or toxic dusts
VOCs (include formaldehyde)	Traffic emissions, industrial activities, petroleum	Surface coatings, adhesives, MDF boards, emissions from new furnishings and floorings, building renovations and repairs	Emissions from office equipment using solvents and other volatile chemicals	Personal care products, cleaning activities, painting, solvent and adhesive use
Toxic and hazardous chemicals, incl. pesticides	Industrial processes, leaks and spills, pesticides	Surface coatings, adhesives, emissions/dusts from materials, furnishings, and floorings, pesticide and fire residues in dusts or on surfaces	Airborne emissions of TOCs or contaminated dusts	Cleaning activities, pesticide use, storage and handling of hazardous chemicals and fuels, artwork
Toxic metals	Lead emissions from traffic, industrial sources, contaminated soils/land	Corrosion products or particulates from toxic metal containing materials, surface coatings, lead pipes, and flashings	Toxic metal containing materials or coatings	Lead paint and dust residues, dusts from use of treated timber (e.g. CCA)
Radon gas (formed from natural ^{238}U decay series)	Soil gas via underground sources; off gassing from water wells	Outgassing from natural building materials and soils with elevated radon	—	—
Allergens/irritants	Outdoor pollen, mold spores, endotoxins, emissions, stormwaters and flood, waste handling and storage	Mold growth on or in soiled or water damaged indoor surfaces and furnishings	Inadequate fresh air ventilation, poorly maintained, and water-affected HVAC systems, contaminated equipment	Allergens derived from insects and other pests, domestic and native animals, occupants, cleaning activities
Pathogens	Toxic fungi (e.g. mold), bacteria (*legionella*), mycotoxins	Contaminated surfaces, moisture, water damage, flooding	Sewer, storm water gas/leaks, HVAC systems, water cooling towers, hot water systems and showers	Introduced pathogens due to occupants or their activities

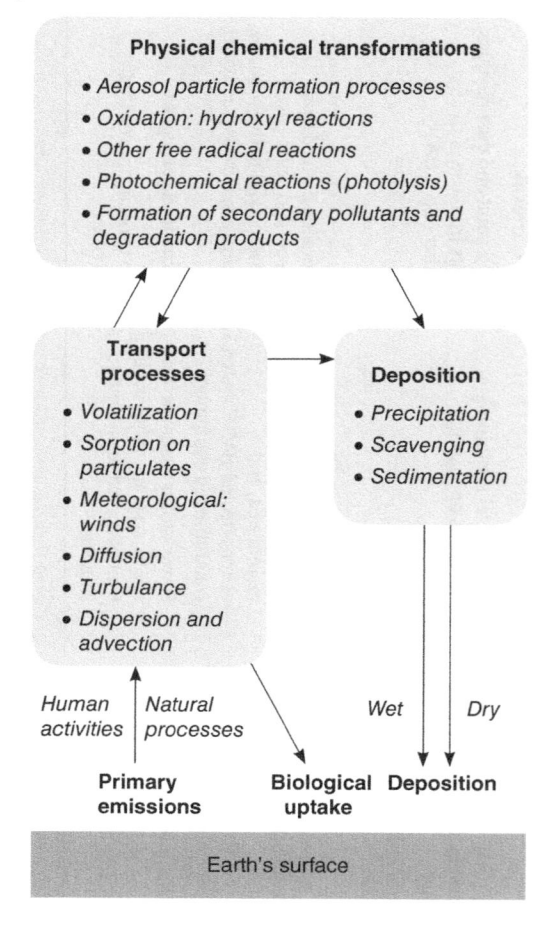

Figure 13.5 Major atmospheric and transport processes affecting air pollutants.

many of these molecular species have bond energies that are approximately equivalent to the energy of the solar cut off at sea level. At high altitudes, the total solar flux is greater, resulting in an increased amount of photolytic and excited state reactions.

Major sources of hydroxyl radicals in the lower atmosphere are given by the following reactions:

$$H_2O + O(^1D) \rightarrow 2OH$$

$$NO + HO_2 \rightarrow OH + NO_2$$

$$H_2O_2 \xrightarrow{+h\nu\,(<370\,nm)} 2OH$$

Hydroxyl reactivity with chemicals may be represented by a second-order rate mechanism. Reported rate constants average from about 10^9 to 10^{10} mol^{-1} s^{-1} for hydrocarbons and reduce to about 10^5–10^8 mol^{-1} s^{-1} for halocarbons containing labile hydrogen atoms or double bonds (Connell and Miller 1984, p. 24).

In polluted atmospheres, ozone is produced by the following reactions:

$$NO_2 \xrightarrow{+h\nu\,(<430\,nm)} NO + O^*$$

$$O^* + O_2 + M \rightarrow O_3 + M$$

M is a nonreactive third body (e.g. N_2 or H_2O). In the presence of alkenes, for example, C_3H_6,

$$O_3 + C_3H_6 \rightarrow \text{Products}$$

General reactions of organic compounds, for example arenes and alkenes, with active oxygen $O(^3P)$ and singlet molecular oxygen O_2, have been reviewed by Korte (1978). Alkenes react much faster with ozone than do most other organics, with rate constants between 10^2 and 10^5 mol^{-1} s^{-1}. Overall, the reactions of hydroxyl radicals control rates of oxidation of most chemicals. Ozone is confined to oxidation of some alkenes in the atmosphere, as in the case of photochemical smog, and possibly oxidation of some sulfur and phosphorus compounds exposed on surfaces in smog areas (Mill 1979).

Atmospheric Aerosol Particles

The general behavior and fate of aerosol particles in the atmosphere is illustrated in Figure 13.6. Primary emissions of atmospheric aerosol particles originate from a wide variety of natural and anthropogenic sources. Airborne particles undergo various physical and chemical interactions and transformations that result in changes in their particle sizes, structure, and composition such as growth from clusters of ultrafine particles, coagulation, restructuring, phase transitions, gas uptake, and chemical reactions.

the nucleation mode (e.g. ultrafine particles) by coagulation or vapor adsorption to form larger particles in the accumulation mode. Coarse particles can also change through mechanical abrasion. Alternatively, particles may also be formed by chemical reactions in the atmosphere and be changed physically and chemically as a result of such reactions. For example, sulfuric acid droplets produced by the oxidation of sulfur dioxide and hydrogen sulfide and further reaction with ammonia can form ammonium sulfate aerosols (Cadle 1972; Seinfeld and Pandis 1998, pp. 97–103).

2. **Photochemical reactions.** The photolysis of air pollutants depends on the energy of the incident solar radiation, the absorption spectrum of the molecular species, and the presence of photochemical sensitizers in the atmosphere. Direct absorption of UV-visible radiation (240–700 nm) may result in cleavage of bonds, dimerization, oxidation, hydrolysis, or rearrangement. For instance, many organic chemicals in the atmosphere that absorb UV-visible radiation are degraded because

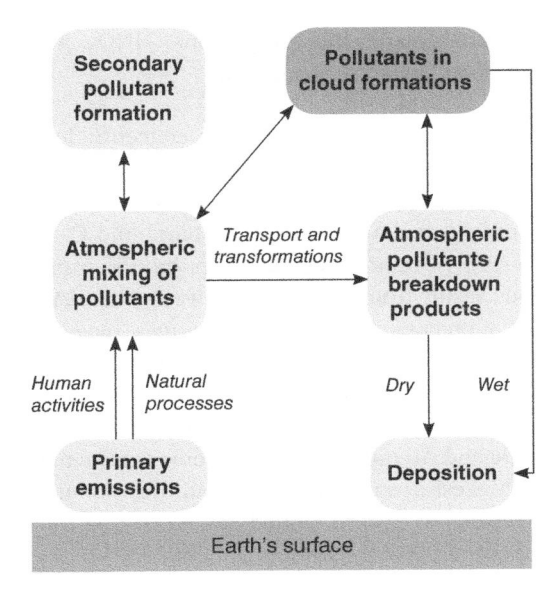

Figure 13.6 Primary emissions, secondary formation, and atmospheric processing of natural and anthropogenic aerosols. Source: Monks et al. (2009), p. 5298. Published by Elsevier, Amsterdam.

Secondary particles are formed by gas-to-particle conversion in the atmosphere and/or condensation of gaseous compounds on pre-existing aerosol particles (Monks et al. 2009, p. 5298).

When the air is saturated with water vapor, clouds are formed by condensation of water vapor on pre-existing aerosol particles under favorable meteorological conditions. Aerosol particles, however, are usually removed during their accumulation mode from the atmosphere by wet deposition, caused by precipitation and scavenging, and also dry deposition, due to convective transport, diffusion, and adhesion to the Earth's surface (Monks et al. 2009, p. 5298).

Free Radical Species

Free radicals (e.g. OH, HO_2, RO_2, and NO_3) are major reactants in the chemical processes that transform trace gases and pollutants in the atmosphere. OH represents the major atmospheric oxidant, primarily formed through photolysis of ozone. Nitrogen oxides, sunlight, and hydrocarbons produce most tropospheric ozone (see photochemical smog). Reactive halogens (e.g. Cl), on the other hand, generally destroy ozone in catalytic cycles, altering the oxidation capacity of the atmosphere. During daytime, HO_X (OH and HO_2) and RO_2 are efficiently formed by photolysis, while NO_3 is important at night because it undergoes rapid photolysis during daytime (Monks et al. 2009, pp. 5293–5295). Transformation reactions of free radicals are illustrated in the case of nitrogen oxides in Figure 13.7.

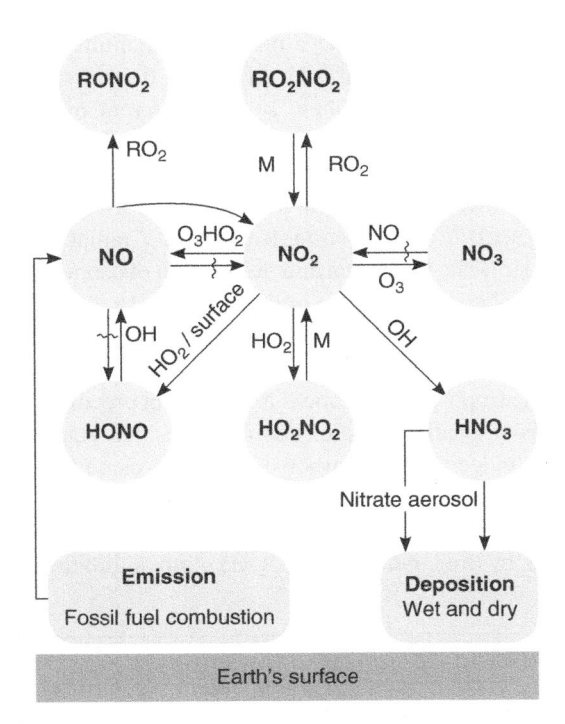

Figure 13.7 Schematic representation of the atmospheric nitrogen cycle illustrating nitrogen emissions from fossil fuel combustion. Source: Modified from Monks et al. (2009), p. 5292. Published by Elsevier, Amsterdam.

Nitrogen Oxides

Nitrogen oxides have a key role in the atmospheric chemistry of radical species that leads to the oxidation of reactive trace gases and to the photochemical formation of ozone. Different species of nitrogen compounds (e.g. NO, NO_2, NO_3, HONO, and N_2O_5) that occur in the atmosphere vary widely in their chemical and physical properties.

From Figure 13.7, it follows that the chemical processes that convert these nitrogen species into oxidized and reduced forms play a major part in their transport, atmospheric chemistry, and removal of certain nitrogen forms (HNO_3 and NO_3) via wet and dry deposition to the surface, within the nitrogen cycle. Deposition of gas-phase nitrogen compounds is also an important input of nutrients to vegetation (Monks et al. 2009, p. 5292).

Atmospheric Ozone

Ozone plays critical roles in atmospheric chemistry of the troposphere and the stratosphere. In the troposphere, it acts as (i) the main source of the hydroxyl radical, which is the key driver of atmospheric oxidation, (ii) a strong greenhouse gas, and (iii) a toxic air pollutant at low elevated levels near the Earth's surface due essentially to its photochemical formation in polluted regions of the world.

Ozone gas is formed in the atmosphere in multistep chemical processes that need solar radiation. It occurs in the stratosphere and troposphere by two different basic mechanisms:

1. In the stratosphere, the initial process involves the breaking apart of an oxygen molecule by UV radiation from the sun. The formation and destruction of ozone in the stratosphere is discussed in Case Study 13.1. There is also an influx of ozone from the stratosphere into the troposphere.
2. Within the troposphere, ozone is formed from precursor emissions of volatile organic chemicals, reactive VOCs, including biogenic hydrocarbons, and nitric oxide that are transformed, in the presence of ample sunlight, by sequences of photochemical and other reactions, mediated by free radicals (e.g. OH), into a mixture of secondary pollutants including ozone, HNO_3, and organic products (see Case Study 13.2).

In urban environments, with intense concentrations of precursor emissions, this phenomenon is known as *photochemical smog* and is characterized by relatively high concentrations of ozone at ground level (Baird and Cann 2012, pp. 76–79). In remote regions, tropospheric ozone concentrations result from a balance between chemistry and transport of precursors from natural sources (biogenic VOC emissions, biomass burning, and lightning) and polluted regions and deposition. Various researchers have reported large-scale transport of ozone and precursors over the Pacific and Atlantic Oceans and some continental regions. The highest ozone levels are found over polluted continental regions of eastern United States, Europe, and China (Monks et al. 2009, p. 5295).

Case Study 13.1

Formation and Disruption of the Ozone Layer in the Stratosphere

The absorption of solar ultraviolet radiation (UV-C photons) by oxygen in the atmosphere leads to the production of ozone molecules and warming of air within a region of 15–50 km altitude, known as the stratosphere. This region is called the ozone layer and provides a protective barrier for living organisms on the Earth's surface against exposure to harmful ultraviolet radiation (Baird and Cann 2012, pp. 3–32).

Ozone is formed by the photolytic action of high-energy solar radiation (UV-C photons) on some of the oxygen molecules in the stratosphere. Each of the oxygen molecules dissociates into two single oxygen radicals, which can then react with molecular oxygen to produce ozone and release heat or re-form oxygen molecules by reacting with ozone. As well, ozone can interact with photons to regenerate molecular oxygen and continue this natural and dynamic cyclic process, known as the *Chapman cycle*.

The following set of reaction equations show the basic mechanism of the *Chapman cycle*, which acts overall as a steady-state system within the stratosphere. It acts as an ozone screen to filter UV-C and much of the UV-B from sunlight before entering the lower atmosphere (Baird and Cann 2012, pp. 3–32; Tran et al. 2018):

$$h\upsilon + O_2 \rightarrow 2O\cdot$$
$$O_2 + O\cdot \rightarrow O_3 + heat$$
$$O_3 + O\cdot \rightarrow 2O_2$$
$$O_3 + h\upsilon \rightarrow O_2 + O\cdot$$

Catalytic Destruction of Ozone Processes

In the last 50 years or so, free radical reactions, in addition to the Chapman mechanisms, that cause catalytic destruction of ozone, have been discovered and shown to include chemical species such as HO, HO_2, NO, NO_2, Cl, and ClO (Sherwood-Roland 2006). In particular, large emissions of stable and halide-containing gases (e.g. CFCs and methyl bromide) and NO_X mainly of origin from human activities, slowly penetrate into the stratosphere from the Earth's surface. These chemicals increase the rate of ozone breakdown by various photolytic reactions (e.g. HCl into Cl and $ClONO_2$ into ClO) to create free radicals, especially Cl and NO, that act as efficient catalysts within the ozone layer of the stratosphere. For example, the following basic set of cyclic reactions indicates how a Cl atom can cause repeated catalytic destruction of thousands of ozone molecules. One chlorine atom can destroy over 100 000 ozone molecules before it is removed from the stratosphere (Baird and Cann 2012, pp. 20–43).

$$Cl + O_3 \rightarrow Cl + O_2$$
$$Cl + O\cdot \rightarrow Cl + O_2$$

Overall reaction: $O_3 + O\cdot \rightarrow 2O_2$

Significant ozone loss in the stratosphere is known to occur over temperate zone latitudes that include North America, Europe, Asia, and much of Africa, Australia, South America, and Polar regions (e.g. *Antarctica ozone hole*). As a result, there has been an increase in solar ultraviolet B radiation (290–320 nm wavelength) that reaches the surface (Sherwood-Rowland 2006; US EPA 2017b).

Case Study 13.2

Urban Air Pollution and Photochemical Smog

Urban air pollution is a major cause of adverse human health in large cities, particularly megacities (10 million persons or greater) and may influence air quality and climate on regional to global scales (e.g. Molina et al. 2007; Monks et al. 2009). The atmospheric chemistry in large cities is significantly different from rural and remote environments due to large emissions of particulates, NO_X, CO, and VOCs (non-methane hydrocarbons and oxygenated VOCs), which result in high levels of precursors and reactants of free radicals, mainly HO_X, with high chemical turnover rates (Monks et al. 2009). Major sources of HO_X are from photolysis of HONO and aldehydes, ozonolysis of alkenes, and photolysis of ozone.

A serious health outcome of urban air pollution can be the formation of photochemical smog and the production of ozone and other hazardous chemicals such as peroxyacetyl nitrate (PAN). It is a major harmful type of air pollution formed in the airsheds and downwind plumes of many cities and urbanized regions when NO_X gases and reactive VOCs (e.g. alkenes) emitted from petroleum combustion sources, especially from motor vehicle exhausts and evaporation of petroleum hydrocarbons, interact with sunlight and oxygen gas through a complex series of reactions.

At high temperatures in internal combustion engines, nitrogen and oxygen gases from air combine to form the free radical, nitric oxide (NO), which can further react with oxygen to produce nitrogen dioxide (NO_2):

$$N_2 + O_2 \rightarrow 2NO$$
$$2NO + O_2 \rightarrow 2NO_2$$

A mixture of these nitrogen oxides (NO_X) and hydrocarbons are emitted into the air from engine exhausts and react in sunlight under relatively stable conditions. Free radicals are formed when nitrogen dioxide undergoes photodissociation into nitric oxide and an oxygen atom. The oxygen atom combines with an oxygen molecule to produce harmful ozone gas, as a secondary pollutant:

$$NO_2 + h\upsilon \rightarrow NO + O$$
$$O_2 + O \rightarrow O_3$$

When the ratio of NO_2 to NO is >3, the dominant reaction is ozone formation. At a low ratio (<0.3), the nitric oxide tends to destroy the ozone:

$$NO + O_3 \rightarrow NO_2 + O_2$$

However, when unburnt hydrocarbons (VOCs) react with NO in sunlight, they can also produce more NO_2 and increase the ratio of NO_2 to NO. Critically, NO_2 can proceed to react with prevalent hydrocarbons and oxygen gas, instead of the NO, in a series of photochemical chain reactions in which hydrocarbons are oxidized to aldehydes by hydroxyl radicals, formed by disassociated oxygen atoms from NO_2, reacting with water. The aldehydes are further oxidized into secondary organic products, including aldehyde peroxides, aldehyde peroxyacids, and peroxyacetyl nitrate (PAN), a strong irritant.

$$NO_2 + O_2 + \text{hydrocarbons} + h\upsilon$$
$$\rightarrow CH_3CO - O - O - NO_2$$

Peroxyacetyl nitrate (PAN)

Overall reaction for photochemical smog can be expressed in a general form as:

$$\text{VOCs} + \text{NO} + O_2 \rightarrow \text{mixture of } O_3, HNO_3, \text{organics}$$

Primary pollutants   Secondary pollutants

Figure 13.8 presents an annual ozone budget for the troposphere. There is a mass influx of some ozone from the stratosphere (Meul et al. 2018) and a mass influx of precursor pollutants that leads to the production of ozone in the troposphere. However, most of the produced ozone is destroyed in the troposphere by chemical reactions with natural airborne chemicals or air pollutants and the remaining reactive ozone is removed by dry deposition. Ozone can also be removed by contact and reaction with surfaces (e.g. soils, plants, and buildings) (Monks et al. 2009).

Atmospheric Halogens

Reactive halogen species are potent oxidizers known to play a role in the destruction of ozone in the stratosphere (e.g. Monks et al. 2009, pp. 5307–5310). Simpson et al. (2015) have reviewed recent advances in our understanding of tropospheric halogen chemistry. Halogen species (e.g. Cl, Br, I, ClO, BrO, IO, and higher oxides) exert a strong influence on the chemical composition of the troposphere, impact on the fate of pollutants, and may affect climate. For example, they can affect methane, ozone, and particles, which are powerful climate forcing agents through

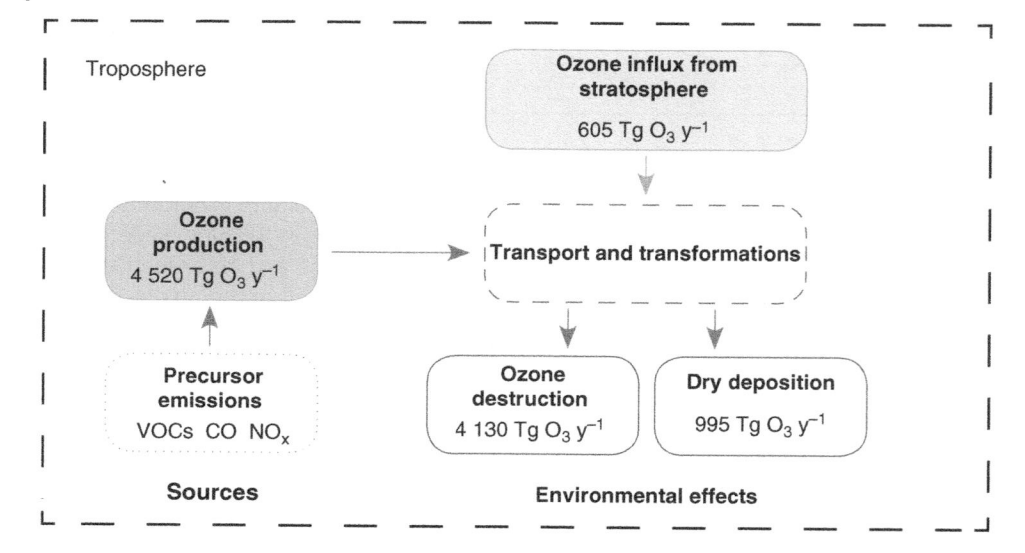

Figure 13.8 Annual tropospheric budget for ozone (annual ozone fluxes are given in Tg year^{-1}). Source: Modified from Monks et al. (2009). Published by Elsevier, Amsterdam.

direct and indirect radiative effects (Simpson et al. 2015). Significantly, these reactive species can affect the oxidation capacity of the troposphere, such as through changes in the OH/HO2 and NO/NO2 ratios, and in the oxidation of dimethyl sulfide (DMS), and also elemental mercury. Most of the reactive halogen compounds in the troposphere are of natural origin and basically related to the presence of halides in ocean water (Monks et al. 2009, p. 53010).

Deposition of Air Pollutants

Wet deposition removes aerosol particles by scavenging particles through interactions (e.g. impaction and collision) with rain and cloud drops, snow, ice, fog, etc. and removal by sedimentation, and Brownian and turbulent diffusion. Dry deposition occurs mainly through impaction, gravitational sedimentation, and Brownian diffusion and turbulence.

The rate of deposition velocity can generally be derived from flux density (F) = deposition velocity (v) × concentration of aerosols in air (C).

The aerodynamic diameter of a particle is a key factor in determining its settling characteristics and depends upon its settling velocity. A particle reaches a terminal settling velocity when the drag force and buoyancy force balance with the gravitational force (net force acting on particle is zero). The terminal settling velocity of a particle (v_{TS}) can be calculated from equation (13.2):

$$v_{TS} = \sqrt{\frac{4gd}{3C_D}\left(\frac{\rho_p - \rho}{\rho}\right)} \quad (13.2)$$

Where d is the diameter of the spherical particle; g is the gravitational acceleration; ρ is the density of the air (fluid); ρ_p is the density of particle; and C_D is the drag coefficient.

Atmospheric Residence Time of Air Pollutants

The average lifetime or residence of an air pollutant in the atmosphere depends on the rate of input from sources and the rate it is eliminated by deposition or other removal mechanisms to sinks at the Earth's surface. If the entire atmosphere is considered as the reservoir for air pollutants, then under steady-state conditions, where the mass of a substance Q in the atmosphere is not changing, i.e. all the sources of the substance are balanced by the sinks of the substance, the average atmospheric residence time (t_{avg}) can be estimated from equation (13.3) (Seinfeld and Pandis 1998, pp. 50–52):

$$t_{avg} = Q/R = Q/P \quad (13.3)$$

Where R is the rate of removal of the species and P is the rate of input of the species from sources.

Examples of atmospheric residence times and concentrations for some major gases and air pollutants are given in Table 13.7.

Residence time of a given molecule in a *reservoir* (the atmosphere) is the reservoir size divided by the total flux, or inputs or outputs of the reservoir, assuming the input is equal to output. That is, the average residence time t_{avg} of an air pollutant, under steady state conditions, can also be related to its steady state concentration C_{ss} in the atmosphere (or a reservoir) of known volume and the rate of input P into the atmosphere, as shown in equation (13.4), described in Baird and Cann (2012, pp. 216–219) for elimination or removal via a single sink. (Note: Baird and Cann (2012) use the symbol R for rate of input.)

$$t_{avg} = C_{ss}/P \quad (13.4)$$

Table 13.7 Atmospheric residence times of major gases and pollutants.

Chemical species	Atmospheric concentration	Atmospheric residence time
Nitrogen gas[a]	78.084%	1.6×10^7 y
Oxygen gas[a]	20.946%	$3 \times 10^3 - 10^4$ y
Water vapor[a]	<0.1–4%	~10 d
Carbon dioxide[b]	~410 ppm[d]	5–200 y[e]
Methane[b]	1.745 ppm	12 y
Nitrous oxide[b]	0.314 ppm	114 y
Carbon monoxide[a]	50–200 ppb	60–200 d
Nitrogen oxides[c]	0.001–50 ppb	1–3 d
Sulfur dioxide[c]	0.01–50 ppb	14 d
Ozone[b] (tropospheric)	—	0.01–0.05 y
Hydrogen sulfide[a]	<0.1–0.5 ppb	<5 d
CFC-11[b]	0.268 pbb	45 y
CFC-12[b]	0.533 ppb	100 y

[a]Railsback (2017).
[b]Table 4.1, Chapter 4 (Houghton et al. 2001).
[c]Table 1 (Trogler 1995).
[d]Concentration based on NOAA (2017c).
[e]No single lifetime can be defined for CO_2 because of the different rates of uptake by different removal processes.

For example, if the steady state concentration of a pollutant is 500 ppb, and the global rate of input P is 50 ppb year^{-1} (annual input divided by volume of atmosphere), then from equation (13.4), the average residence time would be 10 years (note: this does not apply to CO_2, which has multiple sinks of removal and sources).

Behavior and Fate of Pollutants in the Indoor Environment

There are some important differences in the transport and transformation of indoor pollutants compared to outdoor pollutants that affect levels of personal and indoor exposures and persistence. Enclosed environments (e.g. dwellings, buildings, offices, factories, and transport vehicles) can concentrate intakes or penetration of outdoor pollutants (e.g. traffic emissions) and emissions of indoor pollutants from internal sources (e.g. building materials) and activities (e.g. gas cooking, cleaning, construction, and work equipment) depending on outdoor intake, air exchange and ventilation rates, and spatial volumes of indoor environments. In many cases, fresh air exchange rates in built environments, including dwellings, are poor. As a result, the ratio of the levels of many indoor to outdoor air pollutants is often higher (Spengler et al. 2001).

Transport mechanisms of indoor pollutants such as dust and particles (PM_{10} and $PM_{2.5}$) are somewhat complex with many sources and are greatly affected by particle sizes and shapes, design of airflows, indoor activities, sorption, dry deposition and accumulation of dusts on surfaces, re-suspension of deposited dusts, and fine particles. These mechanisms also apply to the spread and deposition of dust allergens and bioaerosols. *Out-gassing* of formaldehyde and other VOCs from various materials, sometimes with long half-lives, is a key source of persistent indoor contaminants. In some situations, vapor intrusion can occur from sub-surface petroleum contamination and radon gas infiltration.

Gaseous and volatile pollutants are generally short-lived indoors, and along with airborne particulates, may be readily removed by natural and mechanical ventilation and sometimes by filtration systems. For organic contaminants (e.g. VOCs) and colloidal particles, sorption onto surfaces of indoor materials is a common sink for removal from indoor air (Spengler et al. 2001).

Major health impacts from indoor pollutants also relate to the formation of carbon monoxide, smoke particles, and numerous secondary pollutants (e.g. nitrogen oxides, PAHs, formaldehyde, and other aldehydes) from combustion of fuels (gases, coal, biomass, and petroleum products), usually for cooking, heating and light, and also tobacco smoking, in the form of environmental tobacco smoke. Similar to oxidation reactions in outdoor air, transformations of gaseous pollutants also occur with combustion products such as nitrogen oxides to form nitrous acid and nitric acid (Baird and Cann 2012, pp. 152–161).

Excess moisture, temperature, and soiling of surfaces can contribute to the indoor growth of microorganisms such as bacteria and mold, release of airborne spores and formation of potentially harmful bioaerosols, which can need remedial treatment with chemical agents and control measures (Spengler et al. 2001, pp. 45.1–45.33).

13.7 Exposures from Pollutants in the Atmospheric Environment

The rapidly growing body of evidence from international, regional, national and local monitoring, modeling programs, and scientific studies strongly indicates that humans, many plant communities, and animals are directly and indirectly exposed to mixtures of air pollutants, such as fine particulate matter and combustion products, frequently at elevated or excessive levels, compared to existing air quality guidelines for major air pollutants.

Numerous scientific studies and targeted monitoring programs continue to identify environmental contamination and impacts from old and new toxic organic chemicals and heavy metals that are emitted into the atmosphere on local to global scales. Many air pollutants occur in complex mixtures that are challenging the capacity of health and environmental risk managers to evaluate *safe* or threshold levels for regulatory controls and risk management.

WHO evaluations of the human burden of death and disease due to health, injury, and environmental risk factors have confirmed that current exposure levels to air pollutants in many parts of the world are significantly associated with premature mortality and morbidity on a global scale (e.g. WHO 2016). The IARC has classified exposure to ambient air pollution and particulate matter, as Group 1 human carcinogens, along with several other toxic organic chemicals, metals, and asbestos that are well known as air pollutants (see Section 13.9).

> **Box 13.1 Modeling Global Exposure of Air Pollution**
>
> In the State of Global Air Report (Health Effects Institute 2017), the annual median concentration of $PM_{2.5}$ is the key parameter that is highly relevant for estimating health impacts. It has been modeled for the year 2014 to provide a thorough global coverage of estimates of air quality. Modeled exposure to ambient $PM_{2.5}$ levels is designed to provide more comprehensive information for countries than measured data, which is, generally, limited to a selection of major cities and towns.

According to the State of Global Air Report, 2017, the majority of regions of the world experience annual median concentrations of air pollution that are higher than the WHO guideline level of $10\,\mu g\,m^{-3}$ for $PM_{2.5}$. As shown in Table 13.8, exposures are particularly high in the Eastern Mediterranean, South-East Asian, and Western Pacific Regions. Air pollution can also be greatly influenced by dust storms, for example in areas close to deserts, such as in the African and Eastern Mediterranean Regions defined by the WHO, where rural levels of $PM_{2.5}$ are greater than urban ones (Health Effects Institute 2017).

Table 13.8 Annual median $PM_{2.5}$ concentrations ($\mu g\,m^{-3}$) for urban and rural areas in WHO regions, as modeled for the year 2014.

WHO region	Median $PM_{2.5}$ ($\mu g\,m^{-3}$)	
	Urban	Rural
Africa	29	37
America (HIC)	8	6
America (LMIC)	15	14
EMR (HIC)	89	92
EMR (LMIC)	52	49
Europe (HIC)	13	7
Europe (LMIC)	25	19
South-East Asia	46	35
Western Pacific (HIC)	12	8
Western Pacific (LMIC)	40	30
WHO $PM_{2.5}$ AQG		10

Notes: AQG, air quality guideline; EMR, Eastern Mediterranean; HIC, high-income countries; LMIC, low- and middle-income countries.
Source: Data taken from Figure 12, WHO (2016, p. 35).

Typical air concentrations of major or priority air pollutants and other toxic air pollutants range from low $\mu g\,m^{-3}$ to $mg\,m^{-3}$ or ppb to ppm. Ambient air levels of pollutants are usually measured in the $\mu g\,m^{-3}$ (and also ppb) with the exception of carbon monoxide (and carbon dioxide) in $mg\,m^{-3}$, or expressed as ppm. Peak or excessive levels of PM_{10} and $PM_{2.5}$, for example, can reach hundreds of $\mu g\,m^{-3}$ or higher in polluted urban and industrial environments (e.g. many megacities) and from dust storm events. Toxic metals (e.g. lead) and PAHs range from $ng\,m^{-3}$ to $\mu g\,m^{-3}$.

Indoor air pollution levels are generally low in well-ventilated and clean environments but can be substantially higher in poorly ventilated ones and those with indoor combustion of fuels (e.g. WHO 2010a). Workplace exposures can be acutely or chronically toxic. Exposure levels are usually highly variable, depending largely upon physical nature of environment, personal doses, and toxicity of air pollutants. Work environments include ambient exposures, built environments, and those where mechanical ventilation and/or respiratory and other personal protection equipment are necessary. Occupational hygiene standards for atmospheric contaminants generally vary from low $\mu g\,m^{-3}$ for very toxic substances to high $mg\,m^{-3}$ or are expressed as ppm for gases and vapors. Biological pollutants are diverse and pervasive in practically all environments, with highly variable exposures that can be difficult to measure and predict. Control of exposures usually relies upon good hygiene, preventive, and intervention health practices to reduce excessive risks to human populations.

The World Health Organization has compiled some typical ranges of concentrations of ambient air indicator pollutants for different regions as shown in Table 13.9 (WHO 2006a, p. 31).

Table 13.9 Ranges of ambient concentrations (µg m^{-3}) of indicator air pollutants for different regions of the world.

Region	PM$_{10}$ annual average	NO$_2$ annual average	SO$_2$ annual average	O$_3$ 1-hour maximum average
Africa	40–150	35–165	10–100	120–300
Asia	35–220	20–75	6–65	100–250
Australia/New Zealand	28–127	11–28	3–17	120–310
Canada/United States	20–60	35–70	9–35	50–380
Europe	20–70	18–57	8–36	150–350
Latin America	30–129	30–82	40–70	200–600

Source: Compiled from WHO (2006a, p. 31) based on urban data from Table 1.

The highest concentrations of primary indicators such as PM$_{10}$ and sulfur dioxide are found in Africa, Asia, and Latin America. The highest levels of secondary pollutants such as ozone and nitrogen dioxide are measured in Latin America and in some larger cities and urban airsheds in the developed countries. Trends show a general decrease in SO$_2$ and PM$_{10}$ in developed countries (WHO 2006a, pp. 31–32).

Table S13.2 (in Section S13.5) presents a set of typical ranges and peak levels of ambient air pollutants in the United States, as representative of a major developed country. Except for carbon monoxide, it is evident that these priority and hazardous air pollutants usually occur at levels well below 1000 µg m^{-3} or 200 ppb. Higher levels for many of these pollutants that are above WHO recommended air quality guidelines are reported in low- and middle-income countries.

Critically, the human burden of death and disease falls mostly on populations living in these countries, although exposure to some air pollutants such as SO$_2$ and PM$_{2.5}$ appears to be declining in developed countries and from indoor combustion of solid fuels.

Ultrafine particulates (UFP) are another emerging air pollutant indicated and cover the size range for engineered nanoparticles.

Air Pollution Modeling

Air dispersion models are widely used to predict the concentration of pollutants in the environment or as exposure levels for receptors, following the release of these pollutants from emission sources (e.g. stacks, traffic, or diffuse or area) under different meteorological or environmental conditions (e.g. Seinfeld and Pandis 1998, pp. 1193–1203). The principles and methods of air dispersion models used for regulation and research purposes on scales from local stack or diffuse emissions, to urban and regional airsheds, and indoor environments, are not discussed here, although an example of a simple air dispersion model (Box model) suitable for ambient environments is given in **S13.6** (see Companion Book Online).

13.8 Effects of Air Pollutants on Wildlife and Natural Systems

An overview of toxic and ecological effects (acidification, eutrophication, phytotoxicity, and particulate deposition) of major atmospheric pollutants of global concern is given below. Additionally, principles and effects that apply to many toxic or hazardous air pollutants (e.g. mercury, PAHs, and dioxins) are considered in other sections and chapters of this book. An earlier textbook by Connell and Miller (1984) on Chemistry and Ecotoxicology of Pollution also includes related effects on air pollutants (e.g. SO$_2$, fluoride, PAHs, and heavy metals) in Chapter 11 of that text.

The scientific literature on ecological effects of major air pollutants is extensive and growing because of factors such as rapid urbanization, long-range transport of air pollutants (e.g. POPs, PBDEs, mercury), monitoring evidence of regional and global impacts on natural ecosystems, and international pressure for improved regulatory and management actions.

Among a large body of information, this section identifies key impacts of air pollutants at various levels of biological organization. It uses parts of a review by Industrial Economics, Incorporated (IEc) of Cambridge, Massachusetts, prepared for the Office of Air and Radiation, US EPA (IEc 2010).

Acid Deposition

Connell and Miller (1984, pp. 336–344) have described the basic chemistry of acid precipitation and its effects on terrestrial and aquatic ecosystems. Rainwater is naturally acidic due to the dissolution of carbon dioxide gas (0.04% v/v) in the atmosphere forming weak carbonic acid, H$_2$CO$_3$, which produces a hydrogen ion concentration in atmospheric precipitation of about 10^{-5} M at equilibrium (pH ~5). Acid rain or precipitation occurs when the dissolution of acidic oxides, primarily SO$_2$ and NO$_2$, and sometimes HCl, in the atmosphere produces rainwater with a pH <5 due to the formation of strong mineral acids. The chemical equations below show the overall reactions for the formation of sulfuric acid from sulfur dioxide, and nitric acid and nitrous acids from nitrogen dioxide, respectively.

$$2SO_2(g) + 2H_2O(l) + O_2(g) \rightarrow 2H_2SO_4(aq)$$

$$2NO_2(g) + H_2O(l) \rightarrow HNO_3(aq) + HNO_2(aq)$$

The major sources of sulfur and nitrogen oxides are from high temperature combustion of fossil fuels (e.g. electrical power generation and traffic emissions from internal combustion engines) although natural sources include volcanoes and geothermal springs (sulfur dioxide) and localized lightning strikes in the atmosphere at high temperatures (nitrogen dioxide). Large emissions of ammonia to air from agriculture, manure, and livestock results in the formation of ammonium (NH_4^+) ions, which are then oxidized into H^+ ions and nitrate ions by soil microbes (Baird and Cann 2012, pp. 137–139).

Terrestrial systems. Plants generally exhibit damage to epidermal layers and cells of plants from acid deposition, and changes to stomatal activity. Trees suffer increased loss of nutrients via foliar leaching and depletion of nutrient cations from soils by acidification as the neutralization capacity of soils (e.g. $CaCO_3$) is reduced (Baird and Cann 2012, pp. 139–140). The release of soluble aluminum (Al) ions can reach phytotoxic levels in various plants. Indirect effects may include increased sensitivity to other stress factors (e.g. pathogens, insects, and frost). Populations experience (i) decrease in productivity due to sensitive individuals, (ii) changes in mortality, and (iii) selective advantage of acid-resistant individuals, leading to adverse changes in community structure and competitive patterns. Decrease in species richness and diversity is common (e.g. decline in sugar maple and red spruce in Eastern US and Canadian forests) (IEc 2010). In ecosystems, nutrient cations in soils continue to be depleted; mobility of toxic Al ions increases; and leaching and runoff of sulfate, nitrate, aluminum, and calcium ions to water bodies occur. On a regional scale, changes in sulfur and nitrogen biogeochemistry have been documented (e.g. Northeastern forests in the United States) (IEc 2010). Environmental variations can occur. For instance, acidification in southern and southwestern China is more serious than in Northern China because airborne alkaline dusts are swept in from desert areas and neutralize the acidity (Baird and Cann 2012, p. 141).

Aquatic systems. Decreases in pH <5 and increases in Al ions cause pathological changes in gill tissues of fish, and increased H^+ and Al ions in the water column impair regulation of body ions. Dissolved Al ions can be toxic to many aquatic species through impairing gill regulation. Aquatic plant populations and communities decrease in biological productivity and increase in mortality of sensitive organisms and species. Selective advantage for tolerant species develops along with a decrease in species richness and diversity. At the local ecosystem level, acidification of water bodies results in decreased acid neutralizing capacity. Regional acidification of aquatic systems due to high deposition rates and nitrate saturation of terrestrial ecosystems, and increased leaching of nitrates into water bodies is observed (IEc 2010).

Overall, persistent acidification continues to be observed in various regions (United States and Europe) although evidence of reduced acid sulfate deposition exists in developed countries because of improved pollution control of sulfur dioxide emissions. In European EU-28 ecosystem areas (selected freshwater, forest, and grassland habitats), only 4% of the EU-28 areas are expected to exceed acidification critical loads by 2020 compared to 43% above critical loads in 1980. However, full recovery from past acidification may still take decades. While SO_2 emissions have decreased, relative emissions of ammonia from agriculture and nitrogen oxides from combustion processes have increased in some regions of Europe but nitrogen oxides are projected to decrease along with sulfur oxides (European Environment Agency 2014, p. 26).

Nitrogen Deposition

Increased deposition of nitrogen compounds (e.g. nitrogen oxides, nitrates, and ammonia) from the atmosphere is a critical factor that can lead to increased supply of nitrogen in terrestrial systems and eutrophication in coastal systems. This happens because most of these systems are nitrogen limited, while freshwater bodies tend to be phosphorus limited (see Chapter 8). Long-term effects of chronic nitrogen deposition are likely to result in adverse impacts on organisms, communities, biodiversity, and biogeochemical cycles of catchments and coastal waters (Connell and Miller 1984; European Environment Agency 2014). An estimated 10–45% of nitrogen produced by human activities reaches coastal waters via atmospheric deposition (US EPA 2007 cited in IEc 2010). Again in EU-28 areas, critical loads for eutrophication peaked in 1990 at 84% exceedances but are projected to decrease to 54% in 2020 assuming the amended Gothenburg Protocol is fully applied (European Environment Agency 2014, p. 26).

Terrestrial systems. Increased nitrogen uptake by plants (e.g. stomata and root systems) and microorganisms readily occurs, although chronic exposures can reduce stomatal activity and photosynthesis in some species. Direct effects on terrestrial plants tend to increase in leaf-size, foliar nitrogen levels, and cause change of carbon allocation between plant tissues. Indirect effects include decreased plant resistance to abiotic and biotic stress factors, including pathogens, pests, and frost, and disruption to symbiotic interactions between plants and mycorrhizal fungi (IEc 2010).

Some plant populations (e.g. fast growing species) exhibit selective or efficient uptake of nitrogen and increases in biological productivity and growth rates, while increases in pathogens can also result. Outcomes for communities can involve alterations in competitive patterns, losses of species adapted to nitrogen-poor or acidic environments, and increases in weed species and parasites (IEc 2010).

Impacts on local and regional ecosystems progress from nitrogen saturation to mobilization and leaching of nitrate, aluminum, calcium, and magnesium ions from soils to lakes and streams, and acidification of water bodies and soils. Increases in emissions of greenhouse gases from soils to the atmosphere are also reported. Overall, increased inputs of reactive nitrogen, loss of soils nutrients, nitrogen saturation, leaching, and acidification of surface waters have been observed throughout forests in northeastern United States and Western Europe (IEc 2010).

Aquatic systems. In estuarine and coastal waters, the assimilation of nitrogen inputs occurs by marine macro-algae, phytoplankton, and microorganisms leading to increased algal growth and macro-algal biomass, decrease and loss of seagrass beds, and de-oxygenation of water column (see also Chapter 8). As eutrophication proceeds, populations and communities suffer changes in composition of species, with excessive algal growth and loss of sensitive species of algae and macro-fauna. The nitrogen cycle of the aquatic ecosystems is disrupted by increased algal growth, shading of seagrass beds, reduced water clarity, and depletion of dissolved oxygen. Further nitrogen runoff from nitrogen-saturated catchments to estuaries and coastal waters increases the rate of eutrophication and water quality decline (Connell and Miller 1984, pp. 134–147; IEc 2010). Major pollution of coastal marine environments on regional and global scales are now widely observed due to these nitrogen-induced changes of the structure and function of estuarine and coastal waters (e.g. US estuaries and coastal waters such Chesapeake Bay) (IEc 2010).

Tropospheric Ozone
Ozone is well known as a strong oxidant and phytotoxic pollutant to sensitive plants at low environmental levels. Examples of molecular and cellular effects on plants involve oxidation of enzymes, generation of toxic reactive oxygen species (hydroxyl radicals), disruption of membrane potential, reduced photosynthesis and nitrogen fixation, and increased cell deaths (IEc 2010).

Direct physiological effects observed on forest trees include visible foliar damage, decreased chlorophyll content, accelerated leaf senescence, decreased photosynthesis, increased respiration, altered carbon allocation, water balance changes, and epicuticular wax (Karnosky et al. 2007). Indirect effects include decreased plant resistance to abiotic (e.g. temperature, humidity, wind speed, and nutrient content of soils) and biotic stress factors, including pathogens, pests, and frost, and disruption to symbiotic interactions between plants and mycorrhizal fungi (IEc 2010).

Population, community, and ecosystem effects observed on forest trees are described below (IEc 2010).

(1) Populations exhibited reduced biological productivity and reproductive success, with selection for less-sensitive individuals and potential for development of ozone resistance.
(2) Communities showed changes in structure (e.g. microbial species composition in soils), competitive patterns, and reduced biological productivity. The loss of ozone sensitive species and individuals resulted in reduced species richness and evenness.
(3) Ecosystem changes in energy flow and nutrient cycling have also been observed, as indicated by alterations in litter quantity, nutrient content, and degradation rates, and in carbon fluxes sequestration in soils.

Ozone may also act synergistically with other stress factors to induce injury, such as with acid deposition in high elevation forests of Eastern United States (IEc 2010).

Regional impacts of ozone-related damage to temperate forests and soils are known to be extensive in parts of United States and Western Europe. However, observed patterns of change in some forests with long-term ozone exposure can take >10 years or even decades (IEc 2010). Current ozone levels at or near ground levels are a major concern for vulnerable ecosystems and human health impacts in many parts of the world.

Particulate Deposition
The ecological effects of the deposition of atmospheric particulate matter (PM) appear to depend on its mass loadings, particle sizes, and other physical and chemical properties and interactions with exposed vegetation and microbial-soil environments, and their energy exchanges and nutrient cycling.

Deposition of PM on the leaves of vegetation can affect plant physiology in several ways, depending upon prevailing environmental conditions. Dust accumulation on leaf surfaces may interfere with gas diffusion between the leaf and air. Fine particles, in particular, can block or clog stomata, which causes reduced rates of transpiration and carbon assimilation, leading to reduced photosynthesis. As well, dust deposition on leaves may cause physical abrasion, increased radiative heating and leaf temperature, and inhibiting photosynthesis by reducing sunlight action on photosynthetic tissues. Acid deposition or alkaline dusts (e.g. cement or limestone) may also cause leaf surface injury (Farmer 1993; Grantz et al. 2003).

Accumulation of dust deposition in the soil environment can influence the bioavailability of inorganic soil nutrients through its effects on the rhizosphere bacteria and fungi and changes in soil chemistry (e.g. pH range). The microbial community of the rhizosphere readily decomposes litter fall and is critical for nutrient cycling like nitrogen. The deposition of sulfate and nitrate particles or alkaline dusts may result in significant pH changes depending upon soil buffer capacity.

Some localized or site-specific impacts of increased PM on ecosystems (e.g. changes in structure and composition of species in plant communities) from specific anthropogenic sources have been reported in the scientific literature. Farmer (1993) in a review of dust effects on vegetation found that most plant communities (crops, grasslands, heathlands, trees, and woodlands, arctic bryophyte, and lichen communities) are affected by dust deposition to the extent that community structure is altered. Epiphytic lichen and *Sphagnum* communities were the most sensitive of those studied. Identification of regional impacts of various sized PM on natural ecosystems is more complex in view of other interacting air pollutants on plant communities and associated soils, and the spatial and temporal scales involved in these studies. Aerosol particles are known to have direct and indirect effects on climate, for example, by altering the global radiation balance through scattering and absorptive properties in the atmosphere (e.g. Griffin 2013). There is some indication that increased PM may reduce radiation interception by plant canopies and also reduce precipitation. Further research is needed on unspecified PM interactions with other pollutants and its potential role in climate change (Grantz et al. 2003).

Toxic Air Pollutants

Airborne emissions of toxic chemicals (e.g. PAHs, POPs, fluorides, and mercury) from human activities have been found to contaminate natural and agricultural ecosystems and human environments on local to global scales. Persistent toxic chemicals can be transported over large distances, including to polar regions. Top predators in contaminated food webs are vulnerable to adverse reproductive and developmental effects (e.g. Chapters 9 and 11).

13.9 Human Health Effects of Air Pollutants

Overview

Air pollution is growing as a major global cause of human death and disease and climate change in the Twenty-first Century. The World Health Organization has described air pollution as the biggest environmental risk to health. It affects all regions, settings, socio-economic groups, and age groups (WHO 2016). In 2014, the vast majority of the world population were living in places such as urban areas where air pollution levels exceed WHO air quality guidelines.

Combined ambient and household air pollution is estimated to result in over 7 million premature deaths annually. The main causes of premature deaths are increased mortality from stroke, heart disease, chronic obstructive pulmonary disease, lung cancer, and acute respiratory infections. Most of these premature deaths occurred in low- and middle-income countries, and the greatest number in the WHO Western Pacific and South-East Asia regions (WHO 2018a).

Morbidity effects from ambient air and household air pollution exposures are also serious. Both children and adults exposed to ambient air pollution can suffer reduced lung function, respiratory infections, and aggravated asthma. Maternal exposure to ambient air pollution is associated with adverse birth outcomes, such as low birth weight, pre-term birth, and small gestational age births. There is also emerging evidence that suggests ambient air pollution may affect diabetes and neurological development in children (WHO 2018a). Estimates of the global burden of death and disease due to exposure to air pollutants are discussed below in this section.

Air pollutants with the strongest evidence for major health effects identified by the World Health Organization include particulate matter (PM), ozone (O_3), nitrogen dioxide (NO_2), and sulfur dioxide (SO_2). In particular, the First State of Global Air Report in 2017, stated: *Exposure to $PM_{2.5}$ is the leading environmental risk factor for death, accounting for about 4.2 million deaths, ranks 5th worldwide among all risks, including smoking, diet, and high blood pressure* (Health Effects Institute 2017, p. 2). The health effects of these pollutants and recommended air pollution guidelines are briefly reviewed below.

Health Effects of Major Air Pollutants

Major air pollutants considered here are respirable and fine particulate matter, and the toxic gases: carbon monoxide, ozone, nitrogen dioxide, and sulfur dioxide. Exposures to these air pollutants relate to ambient air and indoor air emissions, including within and from households or dwellings and workplaces. Detailed reviews and evaluations of reported human health studies on air pollutants are available from the World Health Organization, US EPA, and ATSDR, among other sources. Much of this information is related to the derivation and review of recommended air pollution guidelines or national standards (e.g. United States).

Particulate Matter

The following key points on the health effects of PM are derived from the updated WHO Air Pollution Guidelines of 2005 (WHO 2006a, pp. 217–305):

- Exposures to airborne particulate matter show various adverse health effects, mainly to the respiratory and cardiovascular systems, at exposure levels measured in urban populations in cities in Asia, Europe, Latin America, and North America. All populations are affected but susceptibility may vary with health and increases particularly for young children and the elderly.
- Detailed reviews of strong epidemiological evidence, including methodology re-evaluation of previous multi-city studies, confirm that short-term and long-term exposures to PM_{10} and $PM_{2.5}$ particles are closely related to adverse effects (morbidity and mortality) in human populations.
- Quantitative estimates of risk from adverse effects increase with exposure, and there is little evidence to suggest that a *no effects* threshold exists above background concentrations, estimated at 3–5 $\mu g m^{-3}$ for $PM_{2.5}$ in both the United States and Western Europe.
- Various toxicity mechanisms have been proposed for how PM may cause and/or exacerbate acute or chronic diseases. Inflammation due to reactive oxygenated species is emerging as a key mechanism but research continues on identifying specific toxic relationships between physical and chemical characteristics of PM.
- Exposure assessment studies have provided evidence supporting the use of ambient PM concentration as an indicator of population exposure to PM in epidemiological studies and the use of PM_{10} and $PM_{2.5}$ as specific indicators of PM exposures. Ultrafine particles are emerging as an indicator of health risks.

The WHO (2006a, pp. 275–280) has developed from epidemiological data a set of recommended air quality guidelines (AQGs) and interim targets for PM_{10} and $PM_{2.5}$ for short-term (24 hours mean) and long-term (annual mean) exposures. The AQGs are given in Table 13.10. The annual mean values are described in WHO (2006a, p. 278) as: "… the lowest levels at which total, cardiopulmonary and lung cancer mortality have been shown to increase with more than 95% confidence in response to $PM_{2.5}$ in the ACS study … The use of the $PM_{2.5}$ guideline is preferred."

Table 13.10 WHO air quality guideline values.[a]

Pollutant	Averaging time	Air quality guideline ($\mu g\ m^{-3}$)
Particulate matter		
$PM_{2.5}$	1 year	10
	24 h (99th percentile)	25
PM_{10}	1 year	20
	24 h (99th percentile)	50
Ozone	8 h, daily maximum	100
Nitrogen dioxide	1 year	40
	1 h	200
Sulfur dioxide	24 h	20
	10 min	500

[a]Updated AQG of WHO (2006a); modified from Krzyzanowski and Cohen (2008).

The interim targets for PM_{10} and $PM_{2.5}$ are not included in Table 13.10. However, *Countries may find these interim targets helpful in gauging progress over time in the difficult process of steadily reducing population exposures to PM* (WHO 2006a, p. 277).

Given the apparent absence of a *margin of safety* for PM exposure, the US EPA has adopted a risk-based approach, using quantitative relationships between PM exposures and selected adverse effects in exposed populations, to develop a National Ambient Air Quality Standard for PM.

Table 13.11 Blood carboxyhemoglobin levels related to adverse effects of exposure to carbon monoxide.

Observed effect	COHb[a] (%)	Exposure[b] (ppm)
Endogenous production	<0.5	0
Typical level in nonsmoker	0.5–1.5	1–8
Increased arrhythmias in coronary artery disease patients and exacerbation of asthma	0.3–2	0.5–10[b]
Enhanced myocardial ischemia and increased cardiac arrhythmias in coronary artery disease patients	2.4–6	14–40
Decreased stamina in healthy adults	5–8	30–50
Neurobehavioral/cognitive changes	5–20	30–160
Acute and delayed onset of neurological impairment	20–60	160–1000
High risk of death	>50–160	>600

[a]Reported value.
[b]Predicted from the Coburn-Forster-Kane (CFK) model with a rate of endogenous carbon monoxide production assigned a value of 0.006 mL CO kg^{-1} body weight and all other parameter values as noted in Table 3-13, ATSDR (2012, p. 160).
Source: Modified from ATSDR (2012, p. 22) Table 2-1.

The analysis is based on risk coefficients from the epidemiological studies and exposure information obtained via the US EPA's monitoring network.

The current National Ambient Air Quality Standard for PM in the United States is included in Table S13.3 (in Section S13.7). It varies from the air quality guidelines for PM developed by the WHO (2006a), particularly the 24 h mean value. These values reflect risk management differences that can be noted between international agencies and individual countries in goal setting and decision-making processes.

Carbon Monoxide

Key effects observed from carbon monoxide gas exposures:

- Inhaled carbon monoxide is readily absorbed through the alveolar membranes into blood and rapidly distributed throughout the body. In blood, it rapidly complexes with hemoglobin (Hb) to form carboxyhemoglobin (COHb), and in muscle it exists as a complex with myoglobin known as carboxymyoglobin (COMb) (ATSDR 2012).
- The primary targets for carbon monoxide poisoning are the heart, cardiovascular system, central nervous system, and the fetus and neonate (ATSDR 2012).
- At very high levels, CO can cause dizziness, confusion, unconsciousness, and death (US EPA). Symptoms of moderate poisoning may include confusion, syncope, chest pain, dyspnea, weakness, tachycardia, tachypnea, and rhabdomyolysis. Mild exposures may cause headache, nausea, vomiting, dizziness, and blurred vision (ATSDR 2012).
- Vulnerable persons include those with cardiovascular and/or respiratory diseases that reduce their ability to transport oxygenated blood to their hearts. Exercise and increased stress exacerbate these conditions and may induce angina (chest pain) (US EPA). Exercise decreases the CO elimination rate. Asthmatic children can also be vulnerable to CO exposures through associated respiratory effects. Pregnant women may suffer miscarriage if exposed to high levels of CO. Breathing low levels during pregnancy may lead to development impairment of the child (ATSDR 2012).

Long-term exposures to lower levels of carbon monoxide are considered to be of greater concern for human health than acute carbon monoxide exposures. Up to billions of people around the world are chronically exposed to carbon monoxide indoors. Epidemiological studies involving large population groups, where exposures are generally at relatively low carbon monoxide levels, have demonstrated increased incidences of low birth weight, congenital defects, infant and adult mortality, cardiovascular admissions, congestive heart failure, stroke, asthma, tuberculosis, pneumonia, etc. (ATSDR 2012). As well, epidemiological and clinical studies show that some of the adverse health effects of carbon monoxide in humans occur with increasing blood levels of COHb.

In fact, COHb is a widely used biomarker for carbon monoxide poisoning and body burden or specific target tissues where non-hypoxic modes of actions of carbon monoxide exert effects (ATSDR 2012). The Coburn-Forster-Kane (CFK) model can be used to predict steady-state blood COHb levels that correspond to a given continuous inhalation exposure to carbon monoxide in a normal adult. Measured % COHb levels in blood are related to adverse health effects in humans in Table 13.11. The CFK equation is also used to determine the levels of carbon monoxide that correspond to % COHb levels measured. AQG for ambient CO is typically $10\,mg\,m^{-3}$ (8 hours average). Recommended WHO indoor AQGs are $100\,mg\,m^{-3}$ (15 minutes average), $35\,mg\,m^{-3}$ (1 hour average), $10\,mg\,m^{-3}$ (8 hours average), and $7\,mg\,m^{-3}$ (24 hours average) (WHO 2010a p. 88).

Ozone

Key effects observed from ozone gas exposures:

- Exposure to ozone is known to constrict muscles in the respiratory airways, trapping outdoor air in the alveoli. This can lead to a range of symptoms and effects, from wheezing and shortness of breath, sore throat, pain, inflamed and damaged airways to aggravated lung diseases, such as asthma, emphysema, and chronic bronchitis, increased susceptibility to lung infections, and cause chronic obstructive pulmonary disease (COPD) (US EPA 2017c).
- Long-term exposure to ozone is linked to aggravated asthma and likely development of asthma. At higher concentrations, it may cause permanent lung damage, such as abnormal lung development in children (US EPA 2017c).
- The adverse effects of ozone on the respiratory tract, from the nasal passages to the gas-exchange areas, are reported as unequivocal based on results of animal experiments, controlled human exposures, short-term effects measured during or after outdoor activity, and morbidity and mortality studies. The evidence for cardiovascular effects, however, is less conclusive.
- Evidence for the chronic effects of ozone is supported by human and experimental information. Animal data and some autopsy studies indicate that chronic exposure to ozone induces significant changes in airways at the level of the terminal and respiratory bronchioles.

- More recent time series studies on human populations have shown a marked increase in health effects though small, positive associations between daily mortality and ozone levels, independent of the effects of particulate matter. Significantly, these time series studies have shown effects at ozone concentrations below the previous guideline of 120 μg m^{-3} without clear evidence of a threshold (WHO 2006a, pp. 307–326).
- The WHO guideline level has been reduced to 100 μg m^{-3} (daily maximum 8 hours mean) (see Table 13.10), but it is possible that health effects will occur in some sensitive individuals (WHO 2006a, pp. 307–326). Krzyzanowski and Cohen (2008) noted there is a residual risk in this guideline level as it is still within the ozone range found to increase risk of mortality by some 1–2%. The risk was considered acceptable by the WHO group because natural phenomena can also produce occasional ozone levels at this guideline level.
- As ozone concentrations increase above the guideline value, health effects at the population level become increasingly numerous and severe. For example, at 160 μg m^{-3}, ozone is associated with an estimated 3–5% increase in daily mortality (time series studies) and health effects in children and exercising healthy young adults in short-term exposures. Such effects can occur in places where concentrations are currently high due to human activities or are elevated during episodes of very hot weather (WHO 2006a, pp. 307–326).

Nitrogen Dioxide

Nitrogen dioxide gas usually occurs outdoors and indoors in association with other air pollutants (nitric oxide, fine PM, or benzene) produced by combustion sources (e.g. traffic emissions or solid fuel household stoves). Inhaling air with high levels of NO_2 can irritate airways in the human respiratory system. Short-term exposures can aggravate respiratory diseases, especially asthma, and lead to respiratory symptoms (e.g. coughing, wheezing, and breathing problems), visits to emergency rooms, and/or admissions to hospital. Acute health effects can follow exposure to 1-hour NO_2 concentrations in excess of 500 μg m^{-3}. Studies of bronchial responsiveness among asthmatics suggest an increase in responsiveness at levels upward from 200 μg m^{-3} (US EPA 2016c; WHO 2006a, pp. 331–377).

Longer-term exposures to elevated NO_2 may contribute to the development of asthma and increased susceptibility to respiratory infections. Higher-risk groups are asthmatics, as well as children and the elderly. NO_2 and other NO_x react with other chemicals in the air to form both particulates and ozone, which are also harmful when inhaled (e.g. photochemical smog reactions).

Setting a safe level of exposure for humans to NO_2 levels is somewhat complex and uncertain. Recent epidemiological studies have linked adverse health effects in children (aggravated asthma and reduced lung function growth) to existing urban ambient levels in North America and Europe. Recent indoor studies have also provided evidence of effects on respiratory symptoms among infants at NO_2 concentrations below 40 μg m^{-3}. However, it is unclear to what extent the health effects observed in epidemiological studies are attributable to the direct toxicity of NO_2 or in combination with the other primary and secondary combustion-related products (WHO 2006a, pp. 375–376). Recent scientific evidence accumulated since the WHO 2000 guidelines for NO_2 was considered insufficient to change the guideline values of 200 μg m^{-3} for 1-hour exposure and 40 μg m^{-3} for an annual mean (WHO 2006a, pp. 375–376; WHO 2010a, pp. 247–248).

Sulfur Dioxide

Sulfur dioxide is a colorless pungent gas that can readily dissolve in water. In air, it can convert to sulfur trioxide, sulfuric acid, and sulfates. Inhalation of SO_2 affects the lungs. At high levels, it can be life threatening, for example 100 ppm is immediately dangerous to life. Acute symptoms for humans from breathing high levels may consist of burning of nose and throat, breathing difficulties, and severe airway obstructions. Animal studies at high levels of SO_2 have shown decreased respiration, inflammation of the airways, and destruction of areas of the lung (ATSDR 1998).

> Controlled studies involving exercising asthmatics indicate that a proportion experience changes in pulmonary function and respiratory symptoms after periods of exposure to SO_2 as short as 10 minutes. Based on this evidence, the WHO has recommended that a SO_2 concentration of 500 μg m^{-3} should not be exceeded over averaging periods of 10 minutes duration
>
> (WHO 2006a, pp. 413–414).

Long-term exposures to persistent SO_2 are known to affect lung function such as in some workers exposed for 20 or more years to low levels of SO_2. However, the problem found with long-term studies is the interaction of confounding exposures due to other air pollutants. Asthmatics have also been shown to be sensitive to respiratory effects from low-level exposures to SO_2. The WHO (2006a, pp. 414–415) has argued "there is a basis for revising the 24-hour guideline for SO_2 downwards adopting a prudent precautionary approach to a value of 20 μg m^{-3}."

Carcinogenic Air Pollutants

Many carcinogens are found in air pollution as mixtures or individual substances. Human carcinogens can range from ultraviolet radiation, soot, asbestos fibers, crystalline silica dust, benzene, cadmium, chromium (VI), nickel, radon, diesel engine exhaust, PAHs, dioxin-like PCBs, to second-hand tobacco smoke.

Outdoor air pollution, in particular, is described as a leading cause of cancer deaths by the IARC. It consists of various mixtures of known cancer-causing substances such as fine particulate matter, benzene, soot, formaldehyde, diesel engine exhaust (diesel fumes), PAHs (e.g. benzo[a]pyrene), and TCDD, among numerous other toxic or potentially carcinogenic chemicals. In 2013, the IARC classified outdoor air pollution and particulate matter in outdoor air pollution as carcinogenic to humans (IARC 2013; Loomis et al. 2013).

Air quality guidelines (AQGs) derived by the World Health Organization for some common carcinogens are presented in Table 13.12. The AQGs adopt a risk-based approach, which postulates that there is no safe level of exposure (i.e. no threshold for effects) for genotoxic carcinogens. They are expressed as incremental unit risks for lifetime exposures.

For example, from Table 13.12, the geometric mean of the range of the estimates of the excess lifetime risk of leukemia from benzene exposure at an air concentration of $1\,\mu g\,m^{-3}$ is 6×10^{-6}. The concentrations of airborne benzene associated with an excess lifetime risk of 1/10 000, 1/100 000, and 1/1 000 000 are 17, 1.7, and $0.17\,\mu g\,m^{-3}$, respectively (WHO 2010a, p. 38).

A major on-line database for evaluation of carcinogenic risks for a number of common toxic air contaminants is the Integrated Risk Management System (IRIS) developed by the US EPA.

Health Effects of Indoor Air Pollutants

People spend a large part of their day in indoor environments (microenvironments) where they can be exposed to mixtures or individual air pollutants at levels that rival or exceed outdoor levels. Poorly designed or controlled ventilation and indoor emission sources, particularly from indoor combustion of fuels (gas, coal, kerosene, and biomass), tobacco smoking, building materials, furnishings, consumer products, occupant activities, wastes, and accumulation of excess moisture result in acute and chronic exposures or contact with numerous types of pollutants in indoor air, dusts, residues, and contaminated surfaces.

Table 13.12 WHO air quality guidelines for carcinogens.

Substance	IARC Group	Exposure	Lifetime risk estimate[a,b]	Tumor site
Arsenic[c]	1	$1\,\mu g\,m^{-3}$	2×10^{-5}	Lung
Asbestos[c]	1	$0.0005\,f\,mL^{-1}$	$10^{-6}–10^{-5\,d}$	Lung
		$0.0005\,f\,mL^{-1}$	$10^{-5}–10^{-4\,d}$	(Mesothelium)[e]
Benzene[f]	1	$1\,\mu g\,m^{-3}$	6×10^{-6}	Blood (leukemia)
Chromium VI[c]	1	$1\,\mu g\,m^{-3}$	4×10^{-2}	Lung
Formaldehyde[g]	1	$100\,\mu g\,m^{-3}$ (30 min average conc.)	1×10^{-6}	Nasal/para nasal/leukemia
Nickel compounds[c]	1	$1\,\mu g\,m^{-3}$	4×10^{-4}	Lung
PAHs (B[a]P)[c]	1	$1\,ng\,m^{-3}$	8.7×10^{-5}	Lung
Refractory ceramic fibers[c]	2B	$1\,f\,L^{-1}$	1×10^{-6}	Lung
Radon[e]	1	$1\,Bq\,m^{-3}$	$0.6 \times 10^{-5}\,NS^b$	Lung
			$15 \times 10^{-5}\,S^b$	Lung
Trichloroethylene[c]	2A	$1\,\mu g\,m^{-3}$	4.3×10^{-7}	Lung, testis
Vinyl chloride[c]	1	$1\,\mu g\,m^{-3}$	1×10^{-6}	Liver and other sites

[a] Calculated with average relative risk model based on human studies.
[b] unit risk estimates for lifetime exposure to $1\,\mu g\,m^{-3}$ or $1\,Bq\,m^{-3}$ (radon).
[c] risk estimates for lifetime exposure to 1 fiber per liter of air ($1\,f\,L^{-1}$) for ceramic fibers or 0.0005 fibers per milliliter of air ($0.0005\,f\,mL^{-1}$) for asbestos. BP, benzopyrene; NS, nonsmokers; S, smokers.
[d] Modified from WHO (2000, Chapter 3).
[e] Modified from Mesothelioma (cancer).
[f] Modified from WHO (2010a).
[g] Modified from WHO (2010b).

The health effects of air pollutants emitted from major indoor sources are compiled in Table S13.4 (in Section S13.8), mainly from Chapter 9 of WHO (2006a, pp. 189–207). Predominant air pollutants are fine particulates, combustion gases (CO, NO_X, and SO_X), free radicals and other reactive secondary pollutants, and volatile and semi-volatile toxic organic compounds (e.g. formaldehyde, benzenes, and styrene). Health effects remain the same as those for outdoor air pollutants. There are also widespread exposures and reported health effects from continual use of hazardous pesticides and biocides for pest control and public health purposes, and the use and remedial removal of toxic and/or carcinogenic metals and metalloids (lead, arsenic, nickel, chromium VI compounds), and asbestos-containing insulation. Natural radon penetration of households in parts of North America and Europe substantially increase risks of lung cancer. Biological pollutants, including bioaerosols, are major causes of skin, eye, and membrane irritations, allergic reactions, respiratory dysfunction, and diseases among occupants of indoor environments.

A critical health issue for indoor pollution is the household burning of solid fuels (e.g. coal, wood, charcoal, dung, and other agricultural residues for cooking and heating), which can result in substantial impacts on mortality and health. Use of solid fuels in open or poorly ventilated stoves for cooking and heating exposes up to billions of people, mainly in developing countries, to high concentrations of PM and gases that are many times above international air quality guidelines (e.g. WHO air quality guidelines) (WHO 2006a, pp. 189–207).

Global Burden of Human Disease and Air Pollution

Globally, ambient and household air pollution has been estimated by the WHO to cause about seven or more million premature deaths every year, mainly attributed to increased mortality from stroke, ischemic heart disease (IHD), chronic obstructive pulmonary disease (COPD), lung cancer, and acute lower respiratory infections (ALRI). Alternate estimates reported by the first State of Global Air Report (Health Effects Institute 2017) are higher for ambient air pollution ($PM_{2.5}$ and ozone) and lower for household air pollution. The latter indicated a downward trend since 1990. The total deaths (based on 2015), however, were practically similar, although different approaches were used.

Burden of disease calculations are based on combining exposure to air pollution (modeled as annual mean concentrations of $PM_{2.5}$) and its distribution in the population with exposure-risk estimates (relative risks, RR) at each level of exposure. This results in a population attributable fraction (PAF), which is multiplied by the health outcome of interest (e.g. deaths, DALYs, YLLs, etc.) to calculate the attributable burden (AB) (WHO 2016, pp. 39–40) (see also Chapters 3 and 17 of this book).

These estimates are considered by the World Health Organization to be conservative because many diseases or adverse effects (e.g. pre-term birth or low birth weight) associated with air pollution are not included in the assessment, due to insufficient robust evidence on the disease, or not including separate impacts of other air pollutants such as nitrogen oxides and ozone (WHO 2016, p. 47).

An outline of the method used for burden of disease attributable to ambient air pollution is given in WHO (2016, pp. 39–40). The basic steps are (i) population exposure assessment; (ii) exposure-response functions; and (iii) underlying health data (e.g. deaths, years of life lost [YLLs], years lived with the disability [YLDs], and disability-adjusted life years [DALYs] by disease). Epidemiological evidence is rapidly accumulating and methodologies used are evolving and under regular review (WHO 2016, p. 47). Population attributable fractions (PAF) for each country were calculated for ALRI, COPD, lung cancer, stroke, and IHD for each sex and age group.

Ambient Air Pollution

Ambient air pollution in both cities and rural areas was estimated to cause about 3 million premature deaths and 85 million DALYS worldwide in 2012 (WHO 2016, p. 47). About 87% of these deaths occur in low- and middle-income countries, which make up 82% of the world population. Western Pacific and South-East Asia regions are the most affected (WHO 2016, p. 40).

The Global Burden of Disease (GBD) Project and the first State of Global Air Report (Health Effects Institute 2017) have concluded that long-term ambient exposure to $PM_{2.5}$ contributed to the deaths of 4.2 million persons, just over 50% in China and India, and to the loss of 103 million years of healthy life (DALYs) in 2015. The absolute number of deaths increased from 3.5 million in 1990 to 4.2 million in 2015. This increase (20%) was related to increases in air pollution and also to growth and aging in the global population. However, there was an overall decrease of 12.2% in the rate of deaths due to $PM_{2.5}$ indicating that the increase in absolute deaths was related to changes in population characteristics (Health Effects Institute 2017).

Additionally, the State of Global Air Report (2017) found that ozone contributes to the global health burden through deaths and disability from COPD (33rd highest risk factor in GBD). In 2015, ambient ozone exposures contributed to an estimated 254 000 deaths and 4.1 million DALYs. Globally, there was a proportional increase of almost 60%

in ozone-related deaths, with 67% of this increase occurring in India.

Household Air Pollution

Household air pollution is one of the leading causes of disease and premature death in the developing world. More than 40% of the world's population depends on solid fuels for cooking. Exposure to smoke (and other emissions) from cooking fires causes an estimated 3.8 million premature deaths each year (2016), and 7.7% of global mortality, mostly in low- and middle-income countries. Children under five years old are particularly vulnerable to acute lower respiratory infections while adults are susceptible to ischemic heart disease, stroke, chronic obstructive pulmonary disease, and lung cancer (WHO 2018b).

The first State of Global Air Report (Health Effects Institute 2017) concluded that air pollution from household burning of solid fuels contributed to about 2.85 million deaths in 2015 from cardiovascular disease, COPD, ALRIs, and lung cancer, with a high relative risk factor for children under 5 years old for death from acute lower respiratory infections (e.g. pneumonia).

Household air pollution ranks as the second leading mortality risk factor in *low-income* countries identified by the World Bank. However, total deaths, death rates, and DALYS have decreased significantly since 1990. For example, a 39% decrease in death rates has occurred. These findings suggest that household air pollution has decreased in many places but Africa and Asia remain regions of high risk (Health Effects Institute 2017, p. 13).

13.10 Global Impacts and Risks of Air Pollutants

The World Health Organization estimates that 9 out of 10 persons breathe air containing high levels of pollutants. Air pollution is a major cause of global mortality and burden of disease for human populations. An estimated 7 million people die prematurely from outdoor and household air pollution each year (about 1 in 8 deaths globally) according to the WHO/CCAC (2018).

In 2012, the deaths of 12.6 million people were linked to the environment, representing 23% of all estimated global deaths (WHO Global Health Observatory [GHO] data, 18-4-18). Among these deaths, ambient (outdoor) exposures to fine particulate matter ($PM_{2.5}$), including black carbon, are estimated to cause about 3.7 million premature deaths annually from ambient exposures and another 4.3 million deaths are attributed to $PM_{2.5}$ exposures from household combustion of solid fuel (14% of total annual global deaths). Diseases caused by $PM_{2.5}$ exposure include stroke, ischemic heart disease, acute lower respiratory disease (e.g. pneumonia), chronic obstructive pulmonary, and lung cancers. In particular, lung cancer from $PM_{2.5}$ exposures is attributed to about 500 000 annual deaths (WHO 2018c).

Exposure to ozone is responsible for about 150 000 deaths annually from respiratory conditions (WHO/CCAC 2015, p. 32). However, The State of Global Air Report (Health Effects Institute 2017, pp. 11–12) concluded that ambient ozone exposures contributed to an estimated 254 000 deaths in 2015. There are few estimates available for other air pollutants. In 2004, chronic lead-related deaths were estimated to be 143 000, and occupational chemical exposures, including asbestos, were also linked to 581 000 deaths (Prüss-Ustün et al. 2011).

All these estimates are considered conservative and if combined are likely to be underestimates of the total risks from air pollution. It is evident that annual mortality from air pollution (ambient, household and workplace) is likely to be over 8 million per year. Children bear a significant proportion of the burden of death and disease (see also Chapter 3). Developing countries, particularly low-income countries and mainly regional areas of Africa and Asia/Western Pacific have the highest health risks, although elevated risks exist for various countries in parts of Middle East/Eastern Europe and Latin America. A key point is that almost all populations are exposed to air pollution risks.

Recently, the WHO has focused on policies for the mitigation of short-lived climate pollutants (SLCP) (black carbon, methane, ozone, and hydrofluorocarbons) to reduce global health risks (WHO/CCAC 2015). Potential health benefits indicate that 3.5–5 million premature deaths may be averted by SLCP management actions.

Global emissions of some major air pollutants from human activities (e.g. combustion of fossil fuels and biomass, mining, industrial production, agriculture, and urbanization) have resulted in modification of the atmosphere on regional and global scales leading to adverse effects on terrestrial and aquatic environments and human health.

Major ecological risks (e.g. temperate forests and freshwater lakes) are associated with the following:

(1) Acid precipitation from SO_2 and NO_2 emissions (e.g. North America, Europe, and China). Pollution control of these emissions since the 1980s has substantially reduced these risks in developed countries.
(2) Increases in nitrogen emissions and transformations (nitrogen oxides, nitrates, and ammonia) have resulted in large-scale deposition of nitrogen compounds and

increased eutrophication impacts in terrestrial, freshwater, and coastal marine environments around the world.

(3) Emissions of mainly CFCs into the stratosphere have caused catalytic reduction of ozone gas (e.g. thinning of ozone layer) and increased wildlife and human health risks (e.g. skin cancers) from solar ultraviolet B radiation exposure in many countries. Significant ozone loss in the stratosphere is known to occur over temperate zone latitudes that include North America, Europe, Asia, and much of Africa, Australia, South America, and polar regions. The Montreal Protocol has led to international reductions in emissions of ozone-depleting chemicals, although depletion due to CFCs is expected to continue for decades.

(4) Photochemical smog formation in urban environments is increasing ozone production in the lower troposphere where it acts as a strong oxidant and is phytotoxic to many natural plants and agricultural crops.

(5) Particulate deposition can affect most plant communities but identification of regional impacts is more complex because of interactions with other air pollutants. Dust storms from natural and degraded lands can impact large regions and transport PM long distances.

(6) Emissions of toxic chemicals (e.g. POPs and mercury) have been found to contaminate ecosystems and top predators over large distances, including to polar regions.

Air quality is closely linked to the Earth's climate and ecosystems globally. Many of the drivers of air pollution (e.g. combustion of fossil fuels) are also sources of high CO_2 emissions associated with global warming and climate change (Chapter 14). Some air pollutants such as ozone and black carbon are short-lived climate pollutants that greatly contribute to climate change and affect agricultural productivity.

13.11 Key Points

1. The Earth's atmosphere consists primarily of major gases: about 78% nitrogen gas, 21% oxygen gas, and 0.93% argon gas by volume. The remainder consist of water vapor and trace gases, generally at <0.1%. The vertical structure of the Earth's atmosphere can be divided into five main layers that extend from the densest layer of air – the troposphere, closest to the Earth's surface, to the exosphere.
2. Air pollution is now a major global risk factor for human disease and death, and it is the main environmental cause of premature deaths in the human population. Ambient particulate matter and household pollution are the dominant risk factors, particularly in developing and transient countries.
3. There are three basic types of air pollutants: physical (e.g. particulates), chemical (e.g. carbon monoxide, sulfur dioxide, and benzene), and biological (bioaerosols) emitted from human derived sources (e.g. combustion of fossil fuels) and/or natural sources. They are also classified as **primary** or **secondary** pollutants (formed by reactions between precursors) in the atmosphere.
4. Particulate matter (PM) is a complex mixture of dust (suspended solids) to ultrafine solid particles and liquid droplets emitted into air from human activities and natural sources covering 4–5 orders of magnitude in particle size from nanometers (<0.1 µm) up to 100 µm or more.
5. Emission sources of atmospheric pollutants are usually classified as point sources (e.g. industrial stacks, vents, or chimneys), line sources (e.g. vehicle traffic flows), and area sources (e.g. site or urban area).
6. The distribution and fate of primary emissions of anthropogenic and natural pollutants in the atmosphere are strongly influenced by interactions between a complex set of transport, transformation, and deposition processes mediated by solar radiation, and also the presence of water vapor (troposphere).
7. Types of transformation processes in the atmosphere, although complex, are generally photochemically induced by sunlight ($h\nu$), assisted by catalytic chemical species (e.g. free radicals). Physical and chemical transformations can be divided into three major categories: 1. aerosol particulate processes, 2. photochemical reactions, and 3. deposition of air pollutants.
8. There are two case studies given in this chapter: 13.1 Formation and Disruption of the Ozone Layer in the Stratosphere, and 13.2: Urban Air Pollution and Photochemical Smog.
9. The average lifetime or residence of an air pollutant in the atmosphere (t_{avg}) depends on the rate of input from sources and the rate it is eliminated by deposition or other removal mechanisms to sinks at the Earth's surface. They can range from hours, days, to many years (long-lived pollutants).
10. Indoor pollution is important because a large proportion of the global population can spend up to 90% or so of their time indoors, exposed to multiple emission sources of common air pollutants, in relatively limited airspaces compared to outdoor environments.
11. Indoor air pollutants include combustion products, respirable particulates (PM_{10} and $PM_{2.5}$), asbestos fibers, second hand or environmental tobacco smoke, VOCs, solvents, toxic organic chemicals, and metals

(e.g. lead), radon gas, to bioaerosols (allergens, mites, bacteria, mold, and fungi). Typical air concentrations of chemical pollutants range from low µg m^{-3} to mg m^{-3}. Gases and vapors can be expressed as ppb or ppm by volume.

12. Higher levels are generally found in poorly ventilated buildings or dwellings and those with indoor combustion of fuels (fossil and biomass). Personal exposure levels are usually higher than area exposure levels.

13. Workplace exposures can be acutely or chronically toxic. Occupational hygiene standards for atmospheric contaminants generally vary from low µg m^{-3} for very toxic substances to high mg m^{-3} (or converted to ppm for gases and vapors).

14. Major ecological effects and risks from air pollutants are associated with acid precipitation from SO_2 and NO_2 emissions, increased eutrophication of aquatic (mainly marine) and terrestrial systems from airborne deposition of nitrogen compounds, excessive particulate deposition, and phytotoxic effects on many natural and agricultural plants from the formation and dispersion of ozone due to photochemical smog formation in urban environments.

15. Air emissions of persistent toxic chemicals can contaminate local, regional, and global ecosystems and food webs, and cause population impacts on some top predators. Some long-lived air pollutants, mainly CFCs, can diffuse into the stratosphere and cause ozone gas depletion, resulting in increased UV exposures and adverse effects (e.g. skin cancers) among vulnerable wildlife and human populations.

16. The main air pollutants that cause major human health effects are particulate matter (PM), ozone (O_3), combustion gases (carbon dioxide, nitrogen dioxide [NO_2], sulfur dioxide [SO_2]), volatile and semi-volatile toxic organic compounds, and toxic metals (e.g. lead), including known and probable human carcinogens. Exposure to fine particulates ($PM_{2.5}$) is the leading environmental risk factor for deaths.

17. Many carcinogens are found in air pollution as individual substances or mixtures. Known human carcinogens can range from ultraviolet radiation, soot, asbestos fibers, crystalline silica dust, benzene, cadmium, chromium (VI), nickel, radon, diesel particulates/fumes, PAHs (benzo[a]pyrene), dioxin-like PCBs, to tobacco smoke.

18. Airborne biological pollutants (aerosols and dusts) are also major causes of skin, eye, and membrane irritations, allergic reactions, respiratory dysfunction, and diseases.

19. Worldwide, ambient and household air pollution has been estimated by the WHO to cause over 7 million premature deaths every year, mainly attributed to increased mortality from stroke, ischemic heart disease (IHD), chronic obstructive pulmonary disease (COPD), lung cancer, and acute lower respiratory infections (ALRI). Children and persons from developing countries bear much of the burden of death and disease from air pollution.

References

ATSDR (1998). *Toxicological Profile for Sulfur Dioxide*. Atlanta, GA: US Department of Health and Human Services, Agency for Toxic Substances and Disease Registry.

ATSDR (2007). *Toxicological Profile for Lead*. Atlanta, GA: US Department of Health and Human Services, Agency for Toxic Substances and Disease Registry.

ATSDR (2012). *Toxicological Profile for Carbon Monoxide*. Atlanta, GA: US Department of Health and Human Services, Agency for Toxic Substances and Disease Registry.

Australian National Pollutant Inventory (2012). Emission estimation technique manual for Fossil Fuel Electric Power Generation, Version 3. http://www.npi.gov.au/resource/emission-estimation-technique-manual-fossil-fuel-electric-power-generation (accessed 14 July 2018).

Baird, C. and Cann, M. (2012). *Environmental Chemistry*, 5e. New York: W.H. Freeman and Company.

Brook, R.D., Rajagopalan, S., Arden Pope, I.I.I. et al. (2010). Particulate matter air pollution and cardiovascular disease: an update to the scientific statement from the American Heart Association. *Circulation* 121: 2331–2378.

Bureau of Meteorology and CSIRO (2016). State of the climate 2016. Commonwealth of Australia. https://www.csiro.au/en/Showcase/state-of-the-climate (accessed 9 October 2018).

Cadle, R.D. (1972). Formation and chemical reactions of atmospheric particles. *Journal of Colloid and Interface Science* 39 (1): 25–31.

Cendese, C. and Gordon, A.L. (2018). Ocean current. Encyclopaedia Britannica. https://www.britannica.com/science/ocean-current (accessed 15 July 2018).

Connell, D.W. and Miller, G.J. (1984). *Chemistry and Ecotoxicology of Pollution*. New York: Wiley.

DeCarlo, P.F., Slowik, J.G., Worsnop, P. et al. (2004). Particle morphology and density characterization by combined mobility and aerodynamic diameter measurements. Part 1: theory. *Aerosol Science and Technology* 38: 1185–1205.

European Environment Agency (2014). Effects of air pollution on European ecosystems. EEA Technical report 11/2014.

Farmer, A.M. (1993). The effects of dust on vegetation – a review. *Environmental Pollution* 79 (1): 63–75.

Frumkin, H. (ed.) (2010). *Environmental Health: From Global to Local*, 2e. Hoboken, NJ: Wiley.

Global Carbon Project (2017). Global carbon budget. http://www.globalcarbonproject.org/carbonbudget/index.htm (accessed 8 March 2018).

Granier, C., Bessagnet, B., Bond, T., and D'Angiola, A. (2011). Evolution of anthropogenic and biomass burning emissions of air pollutants at global and regional scales during the 1980–2010 period climate. *Change* 109: 163–169.

Grantz, D.A., Garner, J.H., and Johnson, D.W. (2003). Ecological effects of particulate matter. *Environmental International* 29 (2–3): 213–239.

Griffin, R.J. (2013). The sources and impacts of tropospheric particulate matter. *Nature Education Knowledge* 4 (5): 1.

Gwinn, M.R. and Vallyathan, V. (2006). Nanoparticles: Health Effects - Pros and Cons. *Environmental Health Perspectives* 114 (12): 1818–1825.

Health Effects Institute (2017). State of global air 2017. Special report. Boston, MA: Health Effects Institute.

Houghton, J.T., Jenkins, G.J., and Ephraums, J.J. (1990). *Climate Change: The IPCC Scientific Assessment. Intergovernmental Panel on Climate Change*. Cambridge: Cambridge University Press.

Houghton, J.T., Ding, Y., Griggs, D.J. et al. (2001). Climate change 2001: the scientific basis. Working Group 1 (WG1), Intergovernmental Panel on Climate Change. https://www.ipcc.ch/ipccreports/tar/wg1/index.php?idp=127 (accessed 25 July 2018).

IARC (2013). Outdoor air pollution a leading environmental cause of cancer deaths. Press release. https://www.iarc.fr/news-events/iarc-outdoor-air-pollution-a-leading-environmental-cause-of-cancer-deaths/ (accessed 17 October 2013).

IEc (2010). Effects of air pollutants on ecological resources: literature review and case studies. Draft Report – February 2010. Industrial Economics, Inc., Cambridge, MA. https://yosemite.epa.gov/sab/sabproduct.nsf/0/0397DBA55D7EAC4F852576540049EF41/$File/100205+IEc+812+Pro+II+-+Effects+of+Air+Pollutants+on+Ecological+Resources+-+full+report.pdf

Karnosky, D.F., Skelly, J.M., Percy, K.E. et al. (2007). Perspectives regarding 50 years of research on effects of tropospheric ozone air pollution on US forests. *Environmental Pollution* 147: 489–506.

Klimont, Z., Kupiainen, K., Heyes, C. et al. (2017). Global anthropogenic emissions of particulate matter including black carbon. *Atmospheric Chemistry and Physics* 17: 8681–8723.

Korte, F. (1978). Abiotic processes. In: *Principles of Ecotoxicology*, SCOPE 12 (ed. G.C. Butler). New York: Wiley.

Krzyzanowski, M. and Cohen, A. (2008). Update on WHO air quality guidelines. *Air Quality and Atmospheric Health* 1: 7–13.

LaGrega, M.D., Buckingham, P.L., Evans, J.C., and Environmental Resources Management (2001). *Hazardous Waste Management*, 2e. New York: McGraw Hill.

Loomis, D., Grosse, Y., Lauby-Secretan, B. et al. (2013). The carcinogenicity of outdoor air pollution. *Lancet Oncology* 4: 1262–1263.

Meul, S., Langematz, U., and Kroger, P. (2018). Future changes in the stratosphere-to-troposphere ozone mass flux and the contribution from climate change and ozone recovery. *Atmospheric Chemistry and Physics* 18: 7721–7738.

Mill, T. (1979). Structure, reactivity conditions for environmental reactions. US-EPA report 560/11-79-012. United States Environmental Protection Agency, Washington, DC.

Molina, L.T., Kolb, C.E., de Foy, B. et al. (2007). Air quality in North America's most populous city – overview of the MCMA-2003 campaign. *Atmospheric Chemistry and Physics* 7: 2447–2473.

Monks, P.S., Grainer, C., Fuzzi, S. et al. (2009). Atmospheric composition change – global and regional air quality. *Atmospheric Environment* 43: 5268–5350.

NASA (2017a). Earth fact sheet. National Aeronautics and Space Administration, Washington, DC. https://nssdc.gsfc.nasa.gov/planetary/factsheet/earthfact.html (accessed 9 October 2018).

NASA (2017b). Earth's atmospheric layers. National Aeronautics and Space Administration, Washington, DC. https://www.nasa.gov/mission_pages/sunearth/science/atmosphere-layers2.html (accessed 8 September 2020).

NASA (2017c). Climate change: atmospheric carbon dioxide. National Ocean Service, National Aeronautics and Space Administration, Washington, DC. https://www.climate.gov/news-features/understanding-climate/climate-change-atmospheric-carbon-dioxide (accessed 7 September 2020).

NOAA (n.d.). Currents. National Ocean Service Education, National Oceanic and Atmospheric Administration, Washington, DC. https://oceanservice.noaa.gov/education/tutorial_currents/welcome.html (accessed 7 September 2020).

NWS (2020). Jetstream – an online school for weather: the atmosphere. National Weather Service. National Oceanic and Atmospheric Administration, Washington, DC.

https://www.weather.gov/jetstream/layers. (accessed 23 October 2020).

Prüss-Ustün, A., Vickers, C., Haefliger, P. et al. (2011). Knowns and unknowns on burden of disease due to chemicals: a systematic review. *Environmental Health* 10: 9.

Railsback, L.B. (2017). Some fundamentals of mineralogy and geochemistry. University of Georgia, USA. http://www.gly.uga.edu/railsback/FundamentalsIndex.html (accessed 24 July 2018).

Seinfeld, J.H. and Pandis, S.N. (1998). *Atmospheric Chemistry and Physics: From Air Pollution to Climate Change*. New York: Wiley.

Sherwood-Rowland, F. (2006). Stratospheric ozone depletion. *Philosophical Transactions of the Royal Society of London Series B, Biological Sciences* 361 (1469): 769–790.

Simpson, W.R., Brown, S.S., Saiz-Lopez, A. et al. (2015). Tropospheric halogen chemistry: sources, cycling, and impacts. *Chemical Reviews* 115 (10): 4035–4062.

Solomon, P.A. (2012). An overview of ultrafine particles in ambient air. *EM: Air and Waste Management Association's Magazine for Environmental Managers* Air & Waste Management Association, Pittsburgh, PA. 5: 18–27.

Spengler, J.D., Samet, J.M., and McCarthy, J.F. (ed.) (2001). *Indoor Air Quality Handbook*. New York: McGraw-Hill.

State of the Environment Committee (2011). Australia state of the environment 2011. Independent report to the Australian Government Minister for Sustainability, Environment, Water, Population and Communities, Canberra, Australia.

Stockholm Environment Institute (2004). GAP: global air pollution forum emission manual. Stockholm. https://www.sei.org/projects-and-tools/tools/gap-global-air-pollution-forum-emission-manual/.

Tran, C., Chong, D., Keith, K. et al. (2018). Depletion of the ozone layer. *Chemistry LibreTexts*. https://chem.libretexts.org/Textbook_Maps/Physical_and_Theoretical_Chemistry_Textbook_Maps/Supplemental_Modules_(Physical_and_Theoretical_Chemistry)/Kinetics/Case_Studies%3A_Kinetics/Depletion_of_the_Ozone_Layer (accessed 23 July 2018).

Trogler, W.C. (1995). The environmental chemistry of trace atmospheric gases. *Journal of Chemical Education* 72 (11): 973–976.

UK Met Office (2018). Global circulation patterns. Meteorological Office, London, UK Government. https://www.metoffice.gov.uk/learning/atmosphere/global-circulation-patterns (accessed 9 October 2018).

US EPA (2016a). Basic information of air emissions factors and quantification. United States Environmental Protection Agency. https://www.epa.gov/air-emissions-factors-and-quantification/basic-information-air-emissions-factors-and-quantification.

US EPA (2016b). NAAQS table. United States Environmental Protection Agency. https://www.epa.gov/criteria-air-pollutants/naaqs-table.

US EPA (2016c). Basic information about NO_2. United States Environmental Protection Agency. https://www.epa.gov/no2-pollution/basic-information-about-no2 (accessed 1 March 2018).

US EPA (2017a). Initial list of hazardous air pollutants with modifications. United States Environmental Protection Agency. https://www.epa.gov/haps/initial-list-hazardous-air-pollutants-modifications (accessed 16 October 2018).

US EPA (2017b). Health effects of ozone pollution. United States Environmental Protection Agency. https://www.epa.gov/ozone-pollution/health-effects-ozone-pollution (accessed 28 March 2018).

US EPA (2017c). Basic ozone layer science. United States Environmental Protection Agency. https://www.epa.gov/ozone-layer-protection/basic-ozone-layer-science (accessed 18 March 2018).

US EPA (2017d). Biological pollutants' impact on indoor air quality. United States Environmental Protection Agency. https://www.epa.gov/indoor-air-quality-iaq/biological-pollutants-impact-indoor-air-quality#Health_Effects (accessed 6 April 2018).

WHO (2000). *Air Quality Guidelines for Europe*, 2e. Copenhagen: World Health Organization Regional Office for Europe.

WHO (2004). Health aspects of air pollution: results from the WHO project "systematic review of health aspects". World Health Organization Regional Office for Europe, Copenhagen.

WHO (2006a). Air quality guidelines. Global update 2005. World Health Organization Regional Office for Europe, Copenhagen.

WHO (2006b). WHO Air quality guidelines for particulate matter, ozone, nitrogen dioxide and sulfur dioxide. Global update 2005. Summary of risk assessment. World Health Organization, Geneva.

WHO (2010a). WHO guidelines for indoor air quality: selected pollutants. World Health Organization Regional Office for Europe, Copenhagen.

WHO (2010b). Formaldehyde. WHO guidelines for indoor air quality: selected pollutants. Chapter 3. World Health Organization Regional Office for Europe, Copenhagen.

WHO (2013). Review of evidence on health aspects of air pollution – REVIHAAP project. Technical Report. World Health Organization Regional Office for Europe, Copenhagen.

WHO (2016). *Ambient Air Pollution: A Global Assessment of Exposure and Burden of Disease*. Geneva: World Health Organization.

WHO (2018a). Ambient (outdoor) air quality and health. World Health Organization. http://www.who.int/en/newsroom/fact-sheets/detail/ambient-(outdoor)-air-quality-and-health (accessed 12 October 2018).

WHO (2018b). Household air pollution. World Health Organization. http://www.who.int/gho/phe/indoor_air_pollution/en/ (accessed 25 March 2018).

WHO (2018c). Global Health Observatory (GHO) data – public health and environment. World Health Organization. http://www.who.int/gho/phe/en/ (accessed 16 October 2018).

WHO/CCAC (2015). Reducing global health risks through mitigation of short-lived climate pollutants. Scoping report for policy-makers. World Health Organization, Geneva.

WHO/CCAC (2018). World Health Organization releases new global air pollution data. World Health Organization and Climate & Clean Air Coalition. Geneva. http://www.ccacoalition.org/en/news/world-health-organization-releases-new-global-air-pollution-data (accessed 16 October 2018).

14

Greenhouse Gases, Global Warming, and Climate Change

14.1 Introduction

At the time of our 1984 book, *Chemistry and Ecotoxicology of Pollution*, the increasing trend in carbon dioxide levels in the atmosphere was evident. The potential climate effects of global warming caused by emissions of greenhouse gases (GHGs) were growing as a scientific concern. Woodwell (1978) had reported on observations and an estimated steady increase in CO_2 content of the atmosphere over the last 100 years, and projected relationships between fossil fuel use and atmospheric CO_2 concentrations. Available modeling predicted that doubling of the CO_2 level in the atmosphere would cause a 2.9 °C increase in the average global surface temperature (Connell and Miller 1984, pp. 364–365).

The relationship between variations in average annual surface temperatures and the levels of atmospheric CO_2 was subject to considerable scientific uncertainty. There was a lack of understanding of key factors affecting these levels, such as how increasing particulate pollution at the time affected incoming radiation (e.g. cooling effect), and also in predicting biological and ecological effects.

This chapter explores the science behind the emergence and role of GHGs, especially carbon dioxide, as major anthropogenic pollutants that have accelerated the rate of global warming and induced climate changes observed on a global scale since the pre-Industrial Revolution era. Physical, biological, ecological, and human health impacts, largely attributed to global warming due to increasing GHG emissions into the atmosphere from human activities, are increasingly observed and projected through advanced modeling in the scientific literature.

Global scientific consensus on the science of global warming, accelerating climate change, its implications, and the acute need for international policy actions has been demonstrated through the work and reporting of the Intergovernmental Panel on Climate Change (IPCC) under the United Nations. Sustainable and risk reduction policy options and actions are being addressed internationally by both decision- and policymakers in the vast majority of countries, including numerous government and nongovernment bodies, industries, and communities.

Carbon dioxide in the atmosphere plays an integral role in the regulation of the Earth's surface temperature due to what is commonly called the *greenhouse* effect. Rainfall, seasonal temperature variations, and sea levels are all affected by changes in the Earth's temperature caused by natural and human-related factors (drivers) that can alter the Earth's energy balance. This phenomenon is measured as radiative forcing (see Section 14.6). Its net outcome can lead to a warming or cooling effect of the atmosphere and the surface temperature, with observed adverse short-term and projected long-term effects on the climate and health of life on Earth. However, the patterns of changes are often subtle, variable, difficult to measure, and climate-related trends need longer-term and statistically robust observations compared to our familiarity with weather events.

Weather refers to the state of the atmosphere at any given time and place. Most of the weather that affects people, agriculture, and ecosystems takes place in the lower layer of the atmosphere. Severe or extreme weather events include hurricanes, tornadoes, blizzards, and droughts. While the weather can change in minutes or hours, a change in climate is something that develops over longer periods of up to decades to centuries. Commonly, climate is defined by (i) the long-term average of weather events (e.g. temperature and precipitation), and (ii) the type, frequency, duration, and intensity of weather events such as heat waves, cold spells, storms, floods, and droughts.

The conceptual framework used in this chapter to evaluate the contribution of GHGs to global warming and effects of climate change is shown in Figure 14.1. It generally follows the conceptual model (Figure 1.3) applied to other key pollutants in Part 3.

Carbon dioxide and other GHGs act as major pollutants because human activities (mainly burning of fossil fuels) have released sufficient amounts of these heat absorbing gases to trap additional heat in the lower atmosphere, and affect the global climate, through enhanced global warming.

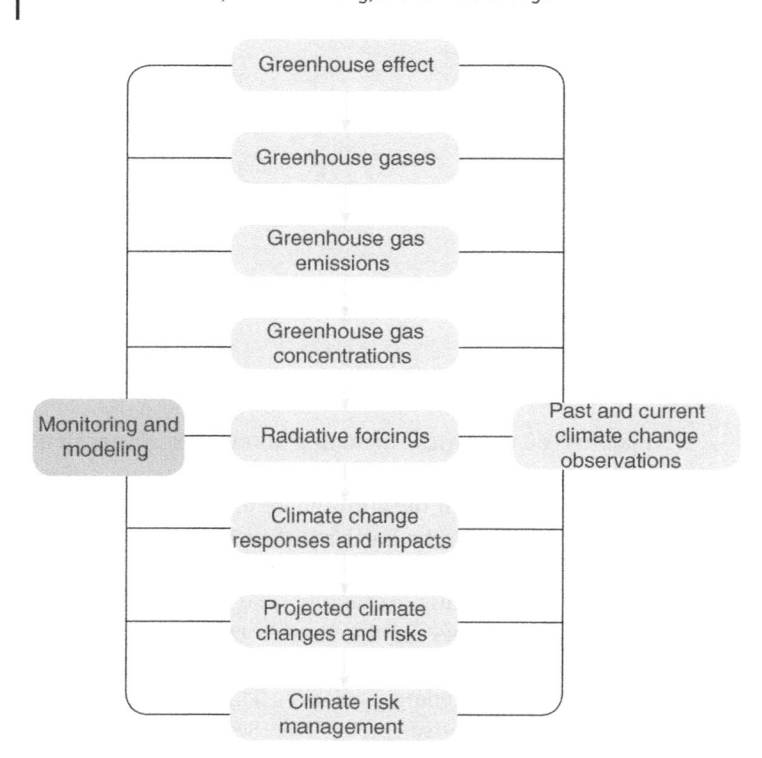

Figure 14.1 Conceptual framework for evaluation of global warming attributed to greenhouse gas emissions to the atmosphere, radiative forcings of the Earth's energy balance, and climate change responses resulting in environmental and human health effects and risks. The critical role of past and current climate change observations, monitoring, and modeling projections of future changes and risks at each step in the process is indicated.

The atmospheric process of global warming or cooling depends on the natural *greenhouse* effect and the rate of net changes in the Earth's energy balance measured by radiative forcings (rates of energy changes at the top of the atmosphere, expressed in watts per square meter, $W\,m^{-2}$) resulting from natural (e.g. solar irradiance) and anthropogenic sources (e.g. GHGs, black carbon, and other aerosols).

Global warming or cooling mechanisms can lead to climate change responses and effects (including climate feedback loops that may enhance or reduce radiative forcings), on physical environments, ecosystems, wildlife, and other biota, and human health and vital ecosystem resources.

Recent findings by the IPCC state clearly that:

> The global climate has changed relative to the pre-industrial period, and there are multiple lines of evidence that these changes have had impacts on organisms and ecosystems, as well as on human systems and well-being (high confidence)
> Hoegh-Guldberg et al. (2018).

The latest 2021 report from the IPCC, known as the "Sixth Assessment Report (AR6 WGI) warns the Earth's climate is at a critical state. It is unequivocal that human influence has warmed the atmosphere, ocean, and land. Widespread and rapid changes in the atmosphere, ocean, cryosphere, and biosphere have occurred" (IPCC 2021). The average surface temperature is projected to exceed 1.5 °C above pre-industrial levels in the next two decades.

The urgent global challenge is how to mitigate GHG emissions from human activities to prevent an increase in the average global surface temperate above a dangerous limit of 1.5 °C, as set by 175 States and the European Union in the Paris Climate Agreement (2016).

14.2 The Greenhouse Effect

The basic principle of global warming depends on the energy balance between the solar radiation of energy (UV, visible, and IR) from the Sun that warms the surface of the Earth and the thermal radiation (IR) from the Earth and the atmosphere that is emitted out into space.

Essentially, the net incoming radiation is balanced on average by what is the outgoing thermal radiation from the Earth. This radiation balance can be changed by factors such as the intensity of solar energy, reflectivity of clouds or gases, and absorption by surfaces, materials, and certain gases known as greenhouses gases. Any such change is called a radiative forcing (Houghton 2015, pp. 32–33). If the balance is disturbed by an increase in atmospheric GHGs, primarily carbon dioxide, it can be restored by a heating effect that increases the Earth's surface temperature through a natural phenomenon known as the *greenhouse effect*.

The use of the term *greenhouse effect* relates to the similar radiative properties of the Earth's atmosphere and the glass in a greenhouse. In the latter, the visible solar radiation transfers through the glass and is absorbed by soil and plants inside the greenhouse before being re-emitted as thermal radiation (longer wavelength energy) that is absorbed by the glass. Some of the absorbed radiation is re-emitted by the glass back into the greenhouse to warm it. Heat transfer in this analogy is limited to radiation and does not consider the major role of conductive heat transfer within a greenhouse or atmosphere (Houghton 2015, pp. 19–21).

The natural *greenhouse effect* is fundamental for the survival of life on Earth. Basically, it is the process by which solar radiation is absorbed by the planet's surface and is emitted into the atmosphere where some of it is absorbed by GHGs and re-emitted to warm the planet's surface to a higher temperature than would occur in the absence of an atmosphere or lack of natural GHGs. In the case of Earth, the absorbed solar radiation warms its surface to a temperature of about 255 °K (−18 °C) and is then re-radiated back into the atmosphere, in the form of longer wavelength infrared radiation (~4–100 μm, microns). However, the natural *greenhouse effect* warms the Earth's lower atmosphere to an average of about 15 °C (see Case Study 14.1).

This situation occurs because the atmosphere is transparent to the transfer of emitted thermal radiation at some wavelengths in the infrared. Radiative gases in the atmosphere (e.g. water vapor and carbon dioxide) absorb some of the thermal radiation (long wavelength) and re-radiate this energy in all directions including out to space and critically toward the Earth's surface causing it to warm.

Global warming depends largely on the atmosphere's temperature and the concentration of GHGs. The temperature of the atmosphere increases with the concentration of carbon dioxide (Houghton 2015, pp. 19–25).

The temperature of the troposphere, however, falls with increasing height due to circular convective processes of heating and cooling of large-scale air masses (average lapse rate of ~6 °C km^{-1} of height) such that the temperature at the top of the atmosphere (TOA) (~5–10 km altitude) is much colder (−30 to −50 °C). Here, the cold GHGs emit less radiation into space but act to absorb some of the radiation emitted by the Earth's surface and help keep the surface warmer than it would be (Houghton 2015, pp. 19–25).

Box 14.1

- The Earth is warmed by the absorption of visible radiation from the sun and is cooled by the re-emission of infrared radiation back into space.
- The extent of the *greenhouse effect* is determined basically by how much of the infrared radiation emitted from the Earth's surface is absorbed by the atmosphere and re-emitted, primarily by GHGs such as carbon dioxide.

Case Study 14.1

What is the extent of the greenhouse effect on the Earth's surface temperature?

This question can be answered by using the energy balance approach together with some basic physics.

The Earth's surface temperature depends on three factors:

The solar flux S available at the distance of the Earth's orbit, Earth's reflectivity or *Albedo* (A), and the amount of warming provided by the atmosphere (due to the greenhouse effect).

The average solar flux at the top of the Earth's atmosphere is about 1370 W m^{-2}. However, about 30% (or 0.3) of this incident radiation is reflected back into space.

From the **Stefan–Boltzmann law**, the energy emitted from the Earth can be calculated assuming it is acting as a true blackbody (i.e. the total radiant heat energy emitted from a blackbody is proportional to the fourth power of its absolute temperature).

Using the assumption that the energy emitted must equal the energy absorbed, the following relationship can be derived for the energy balance between incoming solar radiation and outgoing infrared radiation:

Energy emitted by Earth = energy absorbed by Earth   (14.1)

The next step is to calculate the **energy emitted by Earth**. By treating the Earth as a blackbody with an effective radiating temperature T_e, the Stefan–Boltzmann law shows

(Continued)

Case Study 14.1 (Continued)

that the energy emitted per unit area must be equal to σT_e^4. Earth radiates over its entire surface area, $4\pi R_{earth}^2$, where R_{earth} represents the Earth's radius. Thus, the total energy emitted by Earth is given by:

$$\text{Energy emitted} = 4\pi R_{earth}^2 \times \sigma T_e^4 \quad (14.2)$$

The **energy absorbed by Earth** can now be calculated. From the Sun, the Earth would look like a disk with radius R_{earth} and area πR_{earth}^2. Note this is the area of the Earth projected against the Sun's rays that enter here, not half of the surface area of Earth. (Half of Earth's surface area would be $2\pi R_{earth}^2$, but the Sun's rays do not strike all of this area perpendicularly.)

The total energy intercepted must be equal to the product of Earth's projected area and the solar flux (S), $= \pi R_{earth}^2 S$. The reflected energy is equal to this incident energy times the albedo (A) that is the reflectivity of the Earth (about 0.3). The difference between these two quantities is the energy absorbed by Earth:

$$\begin{aligned}\text{Energy absorbed} &= \text{energy intercepted} - \text{energy reflected}\\ &= \pi R_{earth}^2 S - \pi R_{earth}^2 SA \\ &= \pi R_{earth}^2 S(1-A) \end{aligned} \quad (14.3)$$

The outgoing and incoming energy (Eq. 14.1) is now equated. By substituting in Eq. (14.1), using the expressions in Eqs. (14.2) and (14.3), the energy balance then becomes:

$$4\pi R_{earth}^2 \sigma T_e^4 = \pi R_{earth}^2 S(1-A) \quad (14.4)$$

By dividing both sides by $4\pi R_{earth}^2$, a useful equation for the **Earth's energy balance** is obtained:

$$\sigma T_e^4 = S/4(1-A) \quad (14.5)$$

Rearranging Eq. (14.5), an expression to calculate the **effective temperature of the Earth** is obtained:

$$T_e = \sqrt[4]{\frac{S(1-A)}{4\sigma}} \quad (14.6)$$

If the Earth acted as a perfect blackbody, such that A = 0, then the effective temperature would be about 5.3 °C. However, the Earth reflects essentially 30% of incoming solar radiation. Thus, at A = 0.3, and substituting S = 1366 W m^{-2} and $\sigma = 5.67 \times 10^{-8}$ W m^{-2} K^{-4} in Eq. (14.6), the effective temperature of the Earth's surface would be ~255 K or −18 °C. This temperature would be too cold to sustain life on Earth.

The actual temperature of the Earth's surface is taken to be an average of 15 °C, or 33 °C higher than the effective temperature calculated above.

This temperature increase ΔT from −18 to 15 °C is due to the natural *greenhouse effect* of the atmosphere that makes life possible (see also Kump et al. 2010).

The global energy budget for the *greenhouse effect* in the recent atmosphere is indicated in Figure 14.2a. Some of the incident solar radiation (30%) is reflected by the Earth's surface and atmosphere into space. The remaining radiation (70%) is absorbed by the Earth's surface (45%) and atmosphere (25%) warming it. Some of the thermal or infrared radiation emitted from the Earth's surface passes through the atmosphere while the rest is absorbed and re-emitted in all directions by greenhouse gas molecules and clouds. The net result is to warm the Earth's surface and lower atmosphere.

Human activities since the Industrial Revolution, mainly the burning of fossil fuels and clearing of forests or groundcover, have increased the emissions of GHGs, primarily carbon dioxide, from the Earth's surface to the atmosphere, leading to an enhanced *greenhouse effect* and observations of an increased rate of global warming, compared to the pre-Industrial Revolution era.

As illustrated in Figure 14.2a, there is a net balance between the average flux of incoming solar incident radiation and the average flux of reflected solar and outgoing thermal radiation. However, increasing emissions of thermal radiation from the surface and GHGs due to human activities have caused an enhanced *greenhouse effect* as indicated by the downward flux of thermal radiation from greenhouse gas re-emissions.

Specific data on radiation fluxes (W m^{-2}) in the atmosphere taken from the Fifth IPCC report (AR5) are given in Figure 14.2b. The average incoming solar radiation is about 340 W m^{-2} at the top of the atmosphere (TOA) of which an average of 239 W m^{-2} of solar radiation is absorbed by the atmosphere and the surface. This means an average outgoing of thermal radiation of 239 W m^{-2} is needed to achieve a net energy balance. It follows that the temperature of the atmosphere and surface adjust accordingly to reach this balance (note: there would be no *greenhouse effect* on Earth unless there are colder temperatures in the higher atmosphere) (Houghton 2015, p. 24).

About half of the incoming solar energy is absorbed by the Earth's surface. The energy is transferred to the atmosphere by warming the air in contact with the surface (sensible heat), by evaporation and by thermal radiation that is absorbed by clouds and by GHGs. The atmosphere then radiates thermal radiation back to Earth and into space. The numbers state the magnitude of individual energy fluxes adjusted within their uncertainty ranges close to the energy budgets (compiled from Houghton 2015, p. 25 and data from IPCC 2013a).

Figure 14.2 (a) Global energy budget for atmosphere showing average energy fluxes normalized to 100 arbitrary units of incident radiation. Source: Modified from Schneider (1987). (b) Global energy budget for atmosphere: components of the radiation fluxes (in watts per square meter), which on average enter and leave the Earth's atmosphere.

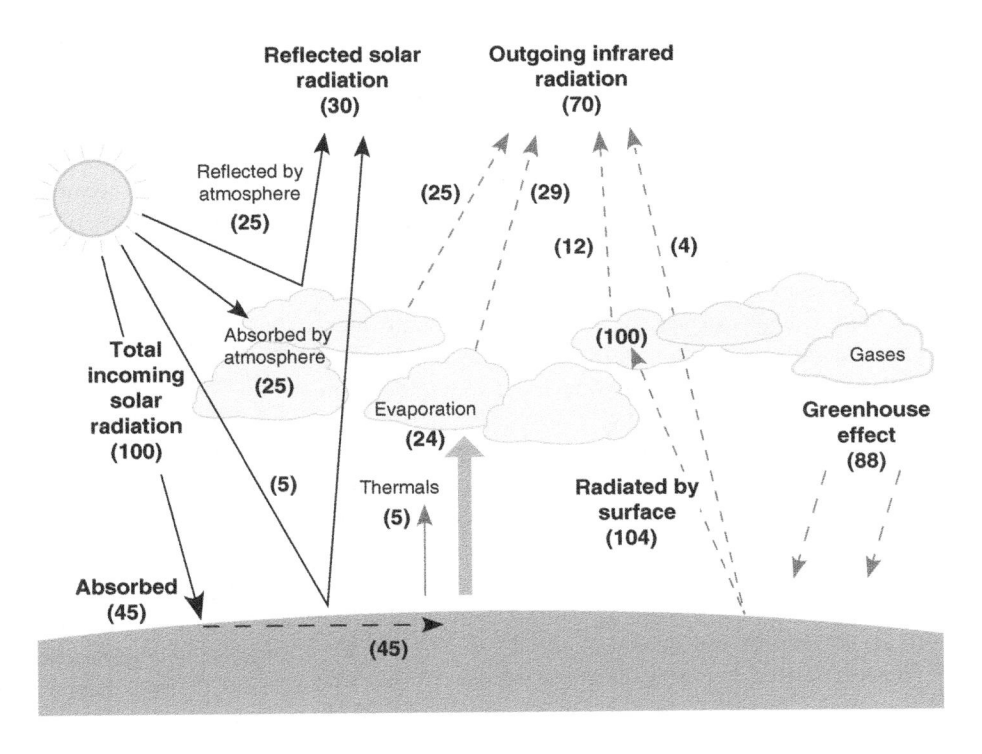

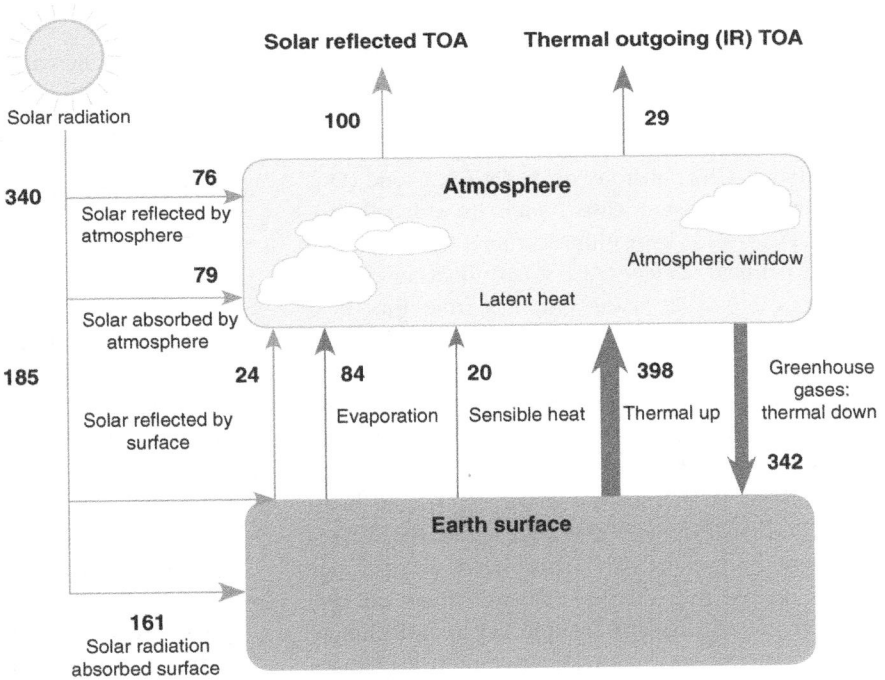

At this point, it is useful to understand the basic physics behind global warming due to the greenhouse effect that changes the Earth's surface temperature (see Case Study 14.1).

The current global concern, however, is that increasing concentrations of GHGs in the atmosphere from human activities of combustion of fossil fuels and clearing forests are accelerating the *greenhouse effect*. The average global temperature is likely to increase above 2 °C by 2100 with greater warming in the following century. Global warming is an average temperature increase near the Earth's surface and in the lowest layer of the atmosphere. It is associated with other observed climate changes in precipitation, storm intensity, sea level changes, etc. (IPCC 2014a).

14.3 Greenhouse Gases and Atmospheric Particles

Greenhouse Gases

Greenhouse gases are able to absorb thermal radiation (longer wavelength infrared radiation) emitted by the Earth's surface. Most diatomic gases with two different elements and gases with three or more atoms (e.g. CO_2, CH_4, O_3, and SF_6) are able to absorb and emit infrared radiation (IR). This energy can also be transferred through intermolecular collisions with non-IR absorbing gas molecules such as nitrogen (N_2) and oxygen (O_2) (e.g. Houghton 2015, Chapter 3; US EPA 2018).

They contribute to the greenhouse effect that traps heat in the atmosphere. In effect, GHGs influence the Earth's climate by interacting with the flows of heat energy in the atmosphere as indicated in Figure 14.2a and b. Life on Earth depends on the greenhouse effect of GHGs in the atmosphere to form a *warming blanket* over the Earth's surface to control the loss of thermal radiation to the coldness of space. Without the warming effect of GHGs, the average temperature of the Earth's surface would be about −18 °C rather than the present average of 15 °C (see Case Study 14.1).

Water vapor is a major GHG, but its variable concentration in the atmosphere is controlled more by temperature than due to human activities. The main GHGs influenced directly by human activities are carbon dioxide (CO_2), methane (CH_4), nitrous oxide (N_2O), ozone (O_3), and synthetic fluorinated gases, such as chlorofluorocarbons (CFCs) and hydrofluorocarbons (HFCs). The American Chemical Society (ACS 2020) illustrates major GHG sources and sinks using graphics from the IPCC Assessment Fourth Report. Key GHGs are outlined below.

Water Vapor – most abundant greenhouse gas in the atmosphere. It plays a key yet complex role in the climate system as a potent greenhouse gas and as a source for clouds. Even so, it is the long-lived GHGs – mainly CO_2, but also methane (CH_4), nitrous oxide (N_2O), and halocarbons (CFCs, HCFCs, HFCs) – that act as the drivers of the greenhouse effect. The short-lived water vapor and clouds readily trap thermal radiation (heat energy) emitted from the Earth's surface and act as fast climate feedbacks because water vapor responds rapidly to changes in temperature, through evaporation, condensation, and precipitation. In this way, they are measured as the largest contributors to global warming. The amount of water vapor in the atmosphere is directly related to the amount of CO_2 and other long-lived GHGs (WMO 2016). Increases in global temperature due to climate change increases water vapor concentrations and adds to the warming effect (ACS 2020).

- **Carbon Dioxide (CO_2)** – This gas is the most important of the GHGs that are increasing in atmospheric concentrations because of emissions due to human activities (Garnaut 2008, p. 33). CO_2 is mainly released to the atmosphere through burning of fossil fuels, solid waste, trees, and wood products (biomass), and also from some chemical reactions such as cement production with limestone (US EPA 2018). It is very stable in the atmosphere but is removed from the atmosphere through photosynthesis by green plants and also in exchange with ocean surfaces. It is then redistributed over hundreds to thousands of years in various forms of carbon storage (*sinks*), as part of the global carbon cycle (see Figure 14.4) (Garnaut 2008, p. 33).

  There is strong evidence from carbon isotope signatures of CO_2 that the burning of fossil fuel by human activities is the major cause of the increase in atmospheric CO_2. The CO_2 from combustion of fossil fuels has a lower $^{13}CO_2/^{12}CO_2$ ratio than the CO_2 found in the atmosphere before the Industrial Revolution. The $^{13}CO_2/^{12}CO_2$ ratio of atmospheric CO_2 has been decreasing steadily as the concentration of CO_2 has increased over the past 50 years. Combustion also uses up oxygen gas from the atmosphere and precise measurements of the O_2/N_2 ratio in the atmosphere show that the fraction of oxygen is decreasing (measured in ppm relative to a standard sample) (ACS 2020).

- **Methane (CH_4)** – reactive hydrocarbon gas that is somewhat short-lived in the atmosphere where it reacts to form water vapor and carbon dioxide. Methane is the main component in natural gas from petroleum reservoirs, shale and coal deposits. It is released during production and transport of coal, natural gases, and oil. Methane is also emitted from livestock (bioeffluents) and other agricultural practices and during decay of organic wastes in municipal solid wastes (e.g. landfills). Large amounts of methane are also stored in frozen soils and as methane hydrates in ocean sediments (Garnaut 2008, p. 33).

- **Nitrous Oxide (N_2O)** – long-lived minor greenhouse gas emitted from natural ecosystems, agriculture (e.g. fertilizer use), industrial activities, burning of biomass, and organic waste in municipal solid waste landfills. The major sink of atmospheric nitrous oxide is destruction in the stratosphere by photolysis to N_2 and O and reaction with electronically excited oxygen atoms (O) over an atmospheric lifetime of about 120 years (Houghton 2015,

p. 50). Nitric oxide (NO) is also produced, which can further enter into an ozone-depleting reaction cycle in the stratosphere (ACS 2020).

- **Ozone (O_3)** – acts as a short-lived greenhouse gas in the troposphere, absorbing some of the infrared energy emitted by the Earth. It has very strong radiative forcing effects on regional scales although its global warming contribution is difficult to estimate because of its short lifetimes and highly variable concentrations at different locations. Even so, considerable increases in tropospheric ozone have been estimated in the last 100 years. Large amounts of ozone are formed as a secondary pollutant in the troposphere by photochemical reactions between CO and CH_4 gases and also during photochemical smog formation such as from reactions between hydrocarbons and nitrogen oxides (NO_X) in sunlight. There is also a large influx of ozone from the stratosphere but much of the ozone is removed by photochemical destruction (see Chapter 13).

- **Fluorinated gases** – these long-lived halogen-containing gases in the atmosphere such as chlorofluorocarbons (CFCs), hydrochlorofluorocarbons (HCFCs), and perfluorocarbons (PFCs) are synthetic and potent GHGs that did not exist in the atmosphere before the industrial era. Only a few of these halogenated carbon types of gases such as methyl bromide, methyl chloride, and tetrafluoromethane (CF_4) occur naturally. They exhibit low reactivity and high persistence in the atmosphere and transfer into the stratosphere where CFCs are known to destroy ozone in a cyclic series of reactions. CFCs are mainly used as refrigerants and solvents. They are decreasing in atmospheric concentrations since the Montreal Protocol on Substances that Deplete the [Stratospheric] Ozone Layer (1989 and subsequent revisions) has phased out the production and use of these gases. Other fluorinated GHGs with long atmospheric lifetimes, including HCFCs and PFCs (e.g. CF_4, C_2F_6, and SF_4), are being phased out as well by other international agreements between most countries (ACS 2020). Most of these gases are slowly removed in the atmosphere but will persist for tens to many thousands of years.

- **Dimethyl Sulfide** – DMS is an organosulfur compound $(CH_3)_2S$ that is the most abundant biogenic sulfur compound present in the atmosphere. It is outgassed to the air mainly by marine phytoplankton. DMS is a significant source of cloud condensation nuclei in regions of the marine atmosphere and is implicated in potential global or regional changes to radiative forcings, including cooling effects (see Baird and Cann 2012, p. 201).

Atmospheric Particles

Aerosols consisting of very small particles or liquid drops (about 0.1–100 µm in diameter) suspended in the atmosphere play a key role in the Earth's energy balance by both absorbing solar radiation and scattering it back into space. Essentially, aerosol particles that have a light color reflect incoming sunlight. In contrast, dark particles such as black carbon (soot) can absorb energy. Natural aerosol particles (≥ 1 µm) include dust and sea salt blown into the atmosphere by strong winds and particles from volcanic eruptions.

Human-made aerosol particles (≤ 1 µm) include sulfate particles produced from sulfur dioxide (SO_2) emitted by coal combustion and black carbon particles from biomass burning and diesel engine exhaust. Aerosol particles also influence climate change by their effects on cloud formation, particle, and droplet sizes (indirect radiative forcing). These mechanisms exert an effect through changes in reflectivity and energy absorbing properties of clouds. The radiative forcing from particles can be positive (warming) or negative (cooling) depending on the nature of the particles and their interactions between particles from different sources and with clouds. Black carbon particles, for example, produce a positive radiative forcing (see Section 14.5) (Houghton 2015, pp. 53–59; ACS 2020).

Molecular Absorption of Infrared Radiation by GHGs in the Atmosphere

What Is a Greenhouse Gas?

Greenhouse gases such as water vapor and carbon dioxide are defined by their capacity to absorb and emit infrared radiation because of their particular molecular structure and properties (rotation and vibration). There are two characteristic ways (Kump et al. 2010, pp. 48–50):

1. Changing molecular rotation: quantum mechanics argues that gas molecules can only rotate at certain discrete frequencies (or wavelengths) that depend on the molecule's structure. The rotation frequency is simply the number of rotations by a molecule per second. In the case of a water molecule, an incoming IR photon at the right wave frequency can be absorbed and increase the rate of rotation. Emitting a photon causes the rotation to slow down. The water molecule absorbs almost 100% of IR radiation at wavelengths of 12 µm or longer (H_2O rotation band), including the microwave region (note: microwave ovens work by readily heating water). Carbon dioxide also absorbs in this rotation band.

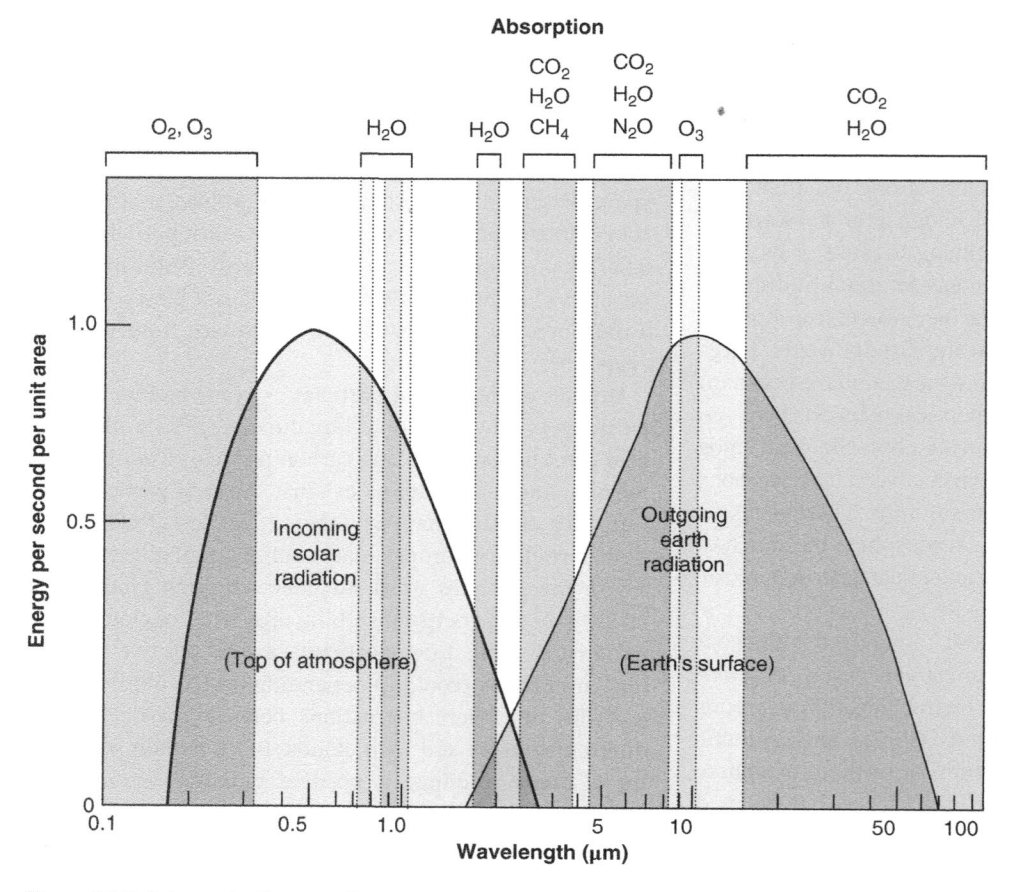

Figure 14.3 Schematic diagram of atmospheric absorption of incident solar radiation and outgoing radiation emitted from the Earth by major atmospheric gases at UV, visible, and infrared wavelengths. The shift in solar radiation wavelengths (incoming) from the visible to infrared range for outgoing radiation from the Earth's surface is evident.

2. Changing vibration of the molecule (i.e. the amplitude at which they vibrate): molecules also vibrate by moving toward and away from each other. In particular, triatomic molecules such as water and carbon dioxide can vibrate in three ways, including a bending mode.

Several other GHGs (e.g. CH_4, N_2O, O_3, CFCs) occur at very low concentrations in the atmosphere, but are effective at absorbing outgoing radiation at wavelengths different from water vapor and carbon dioxide. For example, CFCs are strong absorbers in the 8–12 μm window. Figure 14.3 illustrates the atmospheric absorption by gases of incident solar and outgoing radiation from the Earth.

Why Are Nitrogen and Oxygen Gases Not GHGs?

Common diatomic molecules found in air such as nitrogen and oxygen vibrate and rotate but are perfectly symmetrical (same atoms) and do not have a bending mode. There is no separation of positive and negative electrical charges within the molecule. The oscillating electric and magnetic fields of an electromagnetic wave do not interact with totally symmetrical molecules and so they are not absorbed, unlike water and carbon dioxide which have bending modes. Although the triatomic CO_2 forms, on average, a linear symmetrical molecule, the symmetry is disrupted when the molecule bends, and it can absorb or emit radiation in the 15 μm band.

Atmospheric Absorption of Radiation

Energy from solar radiation is the primary driver of the Earth's climate and weather patterns. Incoming solar radiation consists of an almost continuous spectrum. As it passes through the Earth's atmosphere, various gases present in the atmosphere absorb specific wavelengths of light to create characteristic absorption spectra. The incident radiation is mostly absorbed by O_2 and O_3 at short UV wavelengths and the longer outgoing infrared radiation by GHGs (H_2O, CO_2, CH_4, N_2O, and O_3) as shown in

Figure 14.3. Partial or total absorption at different wavelength regions or bands varies according to the absorption spectrum for each gas, and degree of overlap with other gases and atmospheric concentrations.

14.4 Sources, Emissions, and Sinks of Greenhouse Gases

Sources of Greenhouse Gases

Natural and major human-derived sources of greenhouse gas emissions are listed in Table 14.1. Major GHGs occur naturally, except for various synthetic fluorocarbons.

The Global Carbon Cycle

The global carbon cycle shows the exchanges or fluxes of carbon between and within major reservoirs of the Earth (the atmosphere, oceans, land, and fossil fuels) and carbon storage within these reservoirs. In Figure 14.4, a schematic example of the global carbon cycle modified from Houghton (2015) shows estimated carbon exchange masses and reservoir masses from the time prior to the industrial era (~1750) and annual anthropogenic carbon fluxes and changes in carbon masses in reservoirs due to the industrial era since 1750 (see Figure 14.4 caption for details).

Annual anthropogenic fluxes of carbon averaged over the decade from 2000 to 2009 are included to represent the perturbation of the carbon cycle during the industrial era since 1750.

Black numbers and arrows indicate reservoir mass and exchange masses estimated from the pre-industrial era (c. 1750). Bold black italic numbers and black dotted arrows indicate annual anthropogenic fluxes averaged over the decade from 2000 to 2009. These fluxes are a perturbation of the carbon cycle during the industrial era since 1750. They are indicated by (black dotted arrows) for *Fossil fuel and cement emissions of CO_2*, *Net land use change*, and *Average atmospheric increase in CO_2* in the atmosphere. The uptake of anthropogenic CO_2 by terrestrial ecosystems and by the oceans (carbon sinks) is shown by the broken arrows part of the *Net land flux* and *Net ocean flux*. The bold

Table 14.1 Sources of greenhouse gases.

Greenhouse gas	Natural sources	Major anthropogenic sources
Carbon dioxide	Respiration of living organisms and decomposition of dead organisms;	Combustion of fossil and biomass fuels and cement manufacture;
	Volcanic eruptions;	Land use changes including deforestation and agricultural changes
	Forest fires;	
	Outgassing from the oceans	
Methane	Oceans, natural wetlands, and hydrates	Fossil fuels extraction and mining, vegetation burning, waste treatment, landfills, rice cultivation and ruminant livestock
Nitrous oxide	Soil and oceanic processes;	Nitrogenous fertilizer use, biomass burning, livestock manure, fossil fuel combustion, chemical production (e.g. nylon)
	Oxidation of ammonia in atmosphere	
Fluorinated gases	No natural sources for some PFCs and all HFCs; other PFCs and SF_6 present in small amounts in Earth's crust; released to air in volcanic activity	HFCs: refrigeration, air conditioning, solvents, fire retardants, foam manufacture, aerosol propellants;
		PFCs: aluminum production;
		SF_6: electrical supply (switches and high-voltage systems)
CFCs and HCFCs	No known natural sources	Refrigeration, air conditioning, solvents, foam manufacture, and aerosol propellants
Tropospheric ozone	Chemical formation by precursor species (e.g. CO and NO_X)	Secondary pollutant formation from precursor species (CH_4, CO, NO_X and non-methane organic compounds) emitted from industry, power generation and transport
Water vapor	Water vapor in atmosphere is a function of temperature and varies considerably on short time scales	Irrigation, artificial dams and lakes, and fossil fuel production

Source: Modified from Garnaut (2008).

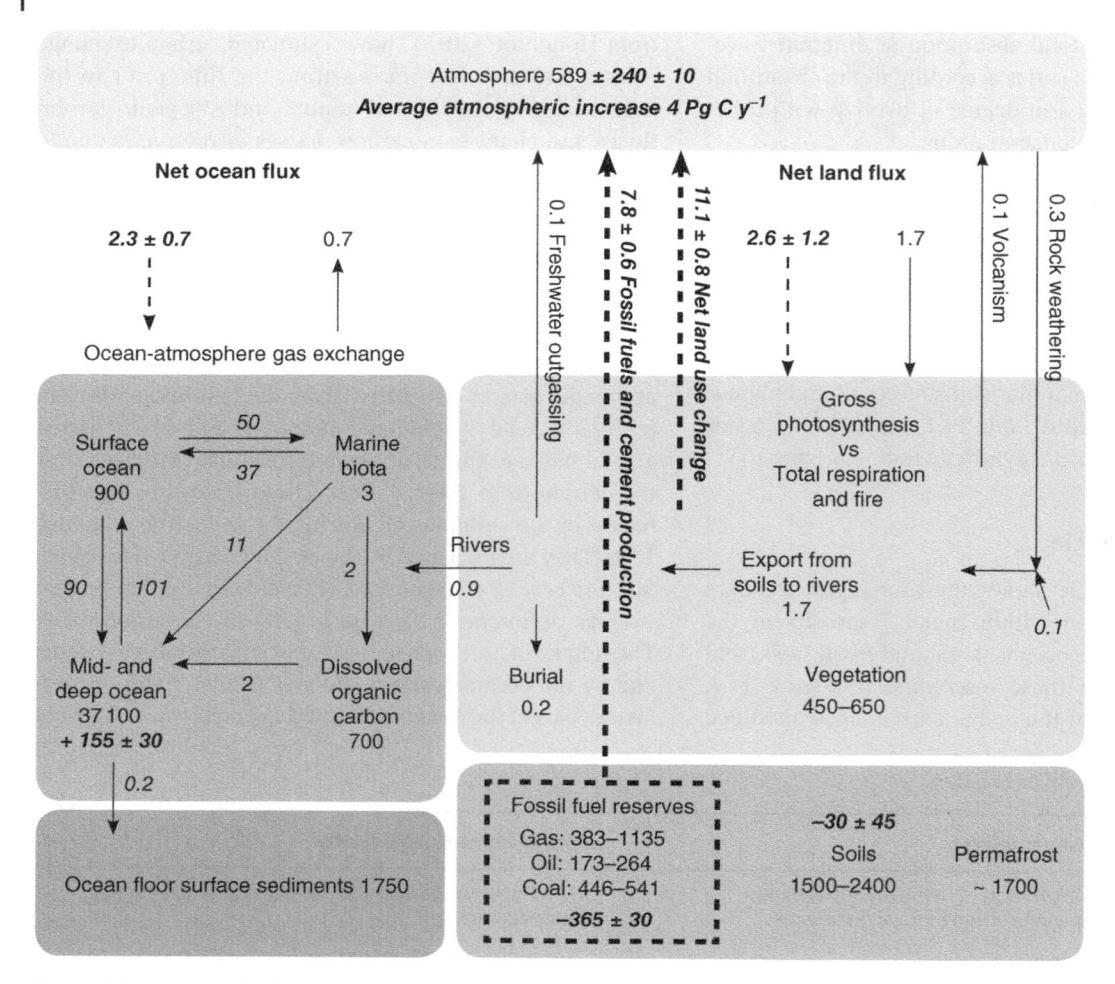

Figure 14.4 Schematic diagram of global carbon cycle. Carbon units are in Pg (10^{15} g) or gigatonnes (Gt) or thousand million tonnes, and uncertainties are given as 90% confidence intervals. Source: Modified from Houghton (2015).

black italic numbers in the reservoirs denote cumulative changes of anthropogenic carbon over the industrial period (1750–2011). By convention, a positive cumulative change means a reservoir has gained carbon since 1750. The cumulative change of anthropogenic carbon in the terrestrial reservoir is the sum of the carbon cumulatively lost through land use change and carbon accumulated since 1750 in other ecosystems. Note that the mass balance of the two ocean carbon stocks *Surface ocean* and *intermediate and deep ocean* includes a yearly accumulation of anthropogenic carbon (not shown) (Houghton 2015, p. 35).

Houghton (2015, p. 34) emphasizes the wide range of timescales over which carbon is exchanged between carbon reservoirs. For example, 50% of an increase in CO_2 in the atmosphere will be removed within 30 years while the remaining 50% will eventually take many thousands of years to be removed. The following section considers the emissions of CO_2 and other GHGs from human-related activities, mainly from fossil fuel combustion.

Global Greenhouse Gas Emissions from Human Activities

Well over 600 Gt of CO_2 are estimated to have been emitted into the atmosphere from the burning of fossil fuels since about 1700 (industrial era). During this period, CO_2 in the atmosphere has increased from about 280 ppm to over 400 ppm, an increase of about 43% (see Houghton 2015, pp. 34–36 and Figure 14.5).

Global scientific and government responses to evidence of increased emissions of CO_2 (and other GHGs) and related global warming observations have included setting up monitoring networks and detailed air emission inventories for GHGs.

Global Greenhouse Gas Inventories

GHG emissions from natural and anthropogenic sources are compiled in special types of air emission inventories known as Global Greenhouse Gas Inventories that are

Figure 14.5 Annual global emissions of carbon dioxide gas (Giga tonnes, Gt CO_2 year^{-1}) from fossil fuels and cement production, 1959–2019. Source: Based on Global Carbon Project (2019a, b, c).

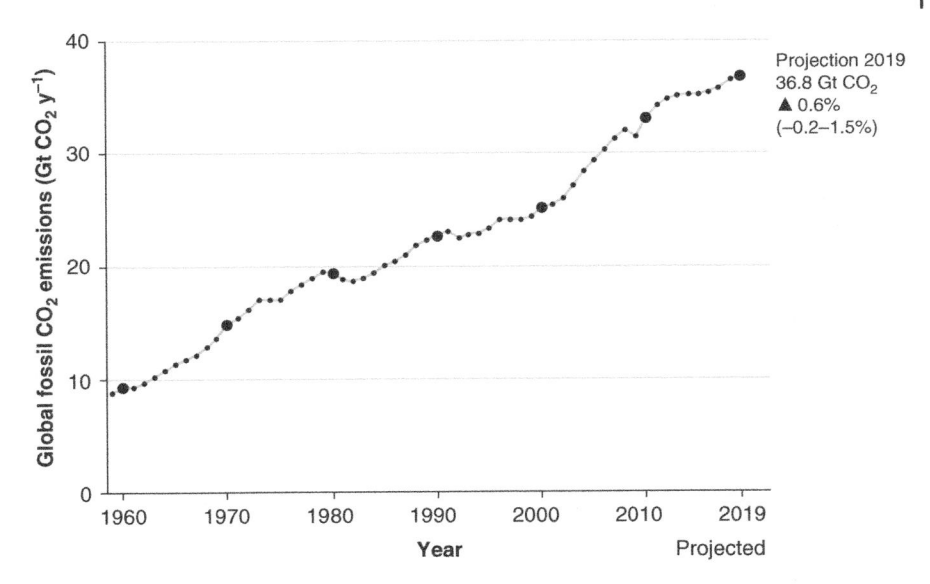

Projection 2019
36.8 Gt CO_2
▲ 0.6%
(−0.2–1.5%)

used as scientific tools for monitoring and developing atmospheric models and by policymakers to develop strategies, policies and measures for mitigation of emissions and for auditing, public reporting of performance, usually yearly, and review. National GHG inventories are usually prepared under the IPCC National Greenhouse Gas Inventories Programme using *IPCC Guidelines for National Greenhouse Gas Inventories* and managed by the IPCC's Task Force on National Greenhouse Gas Inventories (TFI) (see IPCC/TFI 2020). There is major worldwide pressure (e.g. the UN Framework Convention on Climate Change, 1992, Kyoto Protocol and the Paris Agreement) for countries, organizations, and businesses to adopt greenhouse gas accounting methods to report measured and estimated levels of major GHGs in carbon dioxide equivalents (e.g. tonnes CO_2-e) from their direct and indirect sources.

Global Emissions of Carbon Dioxide Gas

In 2018, total global emissions of carbon dioxide gas from fossil fuel, industry, and land use reached 42.1 ± 2.8 Gt CO_2. The 2019 projection is for total CO_2 emissions of 43.1 GtCO_2 (39.9 to 46.2).

> Global CO_2 emissions from fossil fuels and industry have increased every decade from an average of 11.4 GtCO_2 in the 1960s to an average of 34.7 ± 2 yr^{-1} during 2009–2018. Emissions in 2018 reached a new record high of 36.6 ± 2 GtCO_2 with a share of coal (40%), oil (34%), gas (20%), cement (4%), and flaring (1%). Global emissions in 2019 are projected to increase by an additional 0.6% (−0.2% to +1.5%), a slower growth than in the past two years.
>
> Source: Global Carbon Project (2019b).

Global Greenhouse Gas Emissions by Type of Gas, Economic Sector, and Country

Emissions of key GHGs generated by human activities on a global scale by gas, economic sector, and country (CO_2 only) are given in Figures 14.6 (by gas), 14.7 (by economic sector), and 14.8 (by countries) based on emissions data from 2010 (Figures 14.6 and 14.7) and 2014 (Figure 14.8) (US EPA 2019). Major GHGs are carbon dioxide, methane, nitrous oxide, and fluorinated gases (F-gases). The fluorinated gases include HFCs, PFCs, and SF_6. Major human-related sources of GHG emissions are outlined in Table 14.1.

The combustion of fossil fuels is the main source of GHGs, primarily CO_2 from human activities, as indicated in Figure 14.6. CO_2 emitted from forestry/land uses are 11% of total CO_2 equivalent emissions.

Within economic sectors, electricity and heat production account for 25% of GHG emissions followed closely by agriculture, forestry, and other land uses with 24% and industry emissions at 21%. Transportation (14%) and buildings (6%) are also significant sources. Building emissions arise from on-site energy generation and burning fuels for heating and cooking but do not include electricity use covered in electricity and heat production sector. Other energy emissions (10%) refer to uncovered activities in the Energy sector that include fuel extraction, refining, process, and transportation.

Global carbon emissions from fossil fuels between 1900 and 2011 have increased substantially, particularly since 1970, mostly from fossil fuel combustion and industrial processes, followed by land use activities, as shown in Figure 14.7. In 2014, six countries (China, United States,

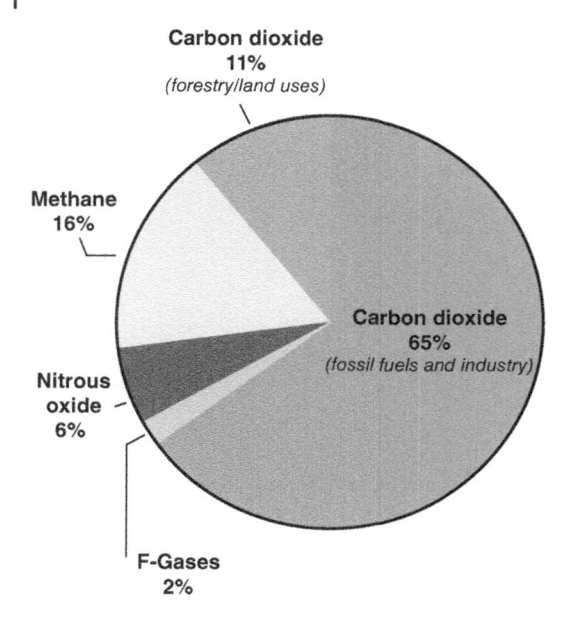

Figure 14.6 Global greenhouse gas emissions by type of gas. Source: US EPA based on global emissions data from 2010 (IPCC 2014a).

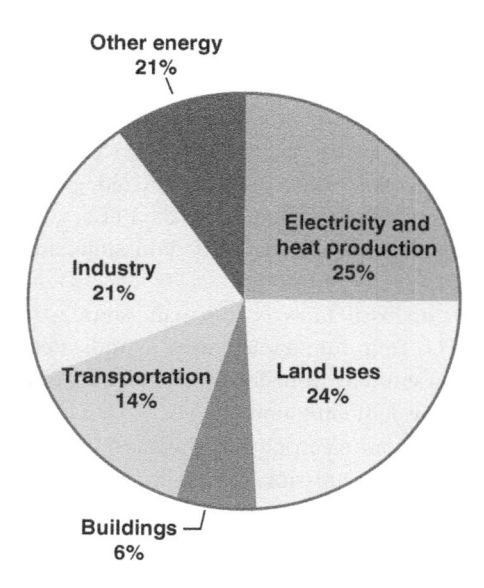

Figure 14.7 Greenhouse gas emissions (CO_2-e) by economic sector. Source: US EPA based on global emissions data from 2010 (IPCC 2014).

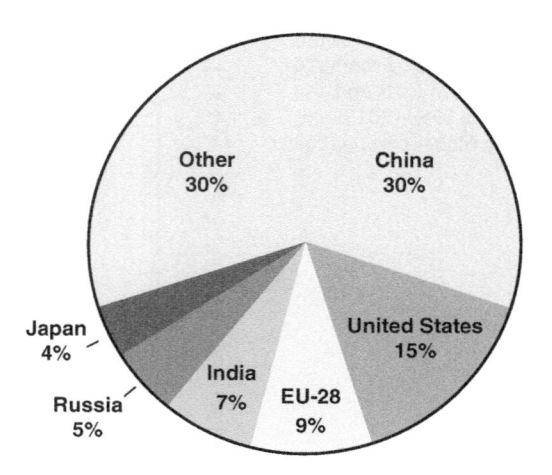

Figure 14.8 2014 Global carbon dioxide emissions from fossil fuel combustion and some industrial processes. Source: US EPA (2019).

European Union, India, Russian Federation, and Japan) generated 70% of CO_2 emissions from fossil fuel combustion and industrial sources such as cement production and gas flaring (Figure 14.8). Additionally, net global GHG emissions from agriculture, forestry, and other land uses accounted for about 24% of total GHG emissions (over 8 million tonnes of CO_2 equivalent). Some changes in land uses associated with human activities (e.g. US and Europe) have the net effect of absorbing CO_2, partially offsetting the emissions from deforestation in other regions (US EPA 2019).

Calculation of GHG Emissions – Solid Fuels

Greenhouse gas emissions from the combustion of solid fuels, where the relevant energy content factor (GJ/t) and emission factor for each gas type are available, can be estimated from Eq. (14.7) (US EPA 2016).

Note: Direct (or point-source) emission factors give the kilograms of carbon dioxide equivalent (CO_2-e) emitted per unit of activity at the point of emission release (i.e. fuel use, energy use, manufacturing process activity, mining activity, on-site waste disposal, etc.). These factors are used to calculate Scope 1 emissions (owned or controlled sources) (US EPA 2016).

$$E_{ij} = \frac{Q_i \cdot EC_i \cdot EF_{oxf}}{1000} \quad (14.7)$$

Where: E_{ij} is the emissions of gas type (j), (carbon dioxide, methane or nitrous oxide), from fuel type (i) (CO_2-e tonnes). Q_i is the quantity of fuel type (i) (tonnes). EC_i is the energy content factor of the fuel (gigajoules per tonne) according to each fuel. If Q_i is measured in gigajoules (GJ), then EC_i is 1. EF_{ijoxf} is the emission factor for each gas type (j) (It includes the effect of an oxidation factor for fuel type (i) (kilograms of CO_2-e per gigajoule) according to each fuel.)

Worked Example 14.1 shows a calculation of CO_2-equivalent emissions from combustion of a fossil fuel (black coal).

Worked Example 14.1

Calculation of CO_2-equivalent emissions* from black coal consumption

A facility consumes 10 000 tonnes of bituminous or black coal for a purpose other than for the production of electricity or to produce coke. For the type of bituminous coal, the energy content factor is 27.0 (GJ t^{-1}) and emission factors for carbon dioxide, methane, and nitrous oxide are 90, 0.03, and 0.2, respectively, in units of kg CO_2-e GJ^{-1}.

Emissions of GHGs (carbon dioxide, methane, and nitrous oxide) in tonnes of CO_2-e are estimated from Eq. (14.7) as follows:

Emissions of carbon dioxide :

$$= (10\,000 \times 27.0 \times 90)/1\,000$$
$$= 24\,300 \text{ t } CO_2 - e$$

Emissions of methane :

$$= (10\,000 \times 27.0 \times 0.03)/1\,000$$
$$= 8 \text{ t } CO_2 - e$$

Emissions of nitrous oxide :

$$= (10\,000 \times 27.0 \times 0.2)/1\,000$$
$$= 54 \text{ t } CO_2 - e$$

Total Scope 1 GHG emissions** = 24 300 + 8 + 54
$$= 24\,362 \text{ tonnes } CO_2 - e$$

* CO_2-e calculation see Box 14.2
** Scope 1 emissions – direct emissions from owned or controlled sources

14.5 Environmental Exposures from Atmospheric Greenhouse Gases

The IPCC (2014a) synthesis report states that atmospheric concentrations of GHGs have reached levels that are unprecedented in at least 800 000 years (IPCC 2014a, p. 14). Since 1750, major GHG concentrations have shown large increases: carbon dioxide (40%), methane (150%), and nitrous oxide (20%). CO_2 concentrations increased at the fastest observed decadal rate of change (2.0 ± 0.1 ppm year^{-1}) for the period 2002–2011. Atmospheric concentrations of major GHGs (2011) are given in Table 14.2; 2019 CO_2 concentrations averaged 410 ppm.

In 2019, atmospheric CO_2 concentrations were higher than at any time in at least 2 million years. CH_4 and N_2O levels were higher than at any time in at least 800 000 years (IPCC 2021).

Trends in changes of atmospheric greenhouse gas concentrations are described briefly in S14.3 of the Companion Book Online (see also Global Carbon Project 2020).

Table 14.2 Atmospheric concentrations of major greenhouse gases.

Greenhouse gas	Atmospheric concentrations (2011)
Carbon dioxide CO_2	410 (ppm)[a]
Methane CH_4	1866 (ppb)[a]
Nitrous oxide N_2O	332 (ppb)[a]
Chlorofluorocarbons (CFCs)	(ppt)
CCl_3 (CFC-11)	238 ± 0.8
CCl_2F_2 (CFC-12)	528 ± 1
CFC-113	74.3 ± 0.1
Hydrofluorocarbons (HFCs)	(ppt)
CHF_3 (HFC-23)	24.0 ± 0.3
Perfluorinated compounds (PFCs)	(ppt)
SF_6	7.28 ± 0.03
NF_3	0.9
C_2F_6 (PFC-116)	4.16 ± 0.02
Other halogenated hydrocarbons	(ppt)
Carbon tetrachloride CCl_4	85.8 ± 0.8
CH_3CCl_3	6.32 ± 0.07
Water vapor[b]	(ppm) 0.1 (South Pole) – 40 000 (tropics)

[a] AR6-WGI, IPCC (2021).
[b] Kump et al. (2010).
Source: Adapted from Myhre et al. (2013) and Kump et al. (2010).

14.6 Greenhouse Gases, Climate Change Processes, and Metrics

Radiative Forcings (RF)

The climate of the Earth is affected by variations in the balance between solar radiation coming into the atmosphere and radiation going out. Climate factors or variables that cause changes in the Earth's energy balance (equilibrium) and alter the climate system, including surface temperature, are known as radiative or climate forcing agents. These agents consist of natural and anthropogenic sources of forcings, which include GHGs, solar radiation, aerosols, and albedo. The climate forcing process is called radiative or climate forcing (see Figure 14.9). As such, radiative forcing is considered a useful tool or measure to compare the different factors (e.g. solar irradiance and GHGs) causing perturbations in the climate system (Houghton 2015, p. 33).

There are three types of climate forcings: (i) *direct radiative forcings* that act directly on the radiative budget of the Earth (e.g. added CO_2 in the atmosphere absorbs and emits infrared (IR) radiation; (ii) *indirect radiative forcings* that create an energy imbalance by first altering climate system

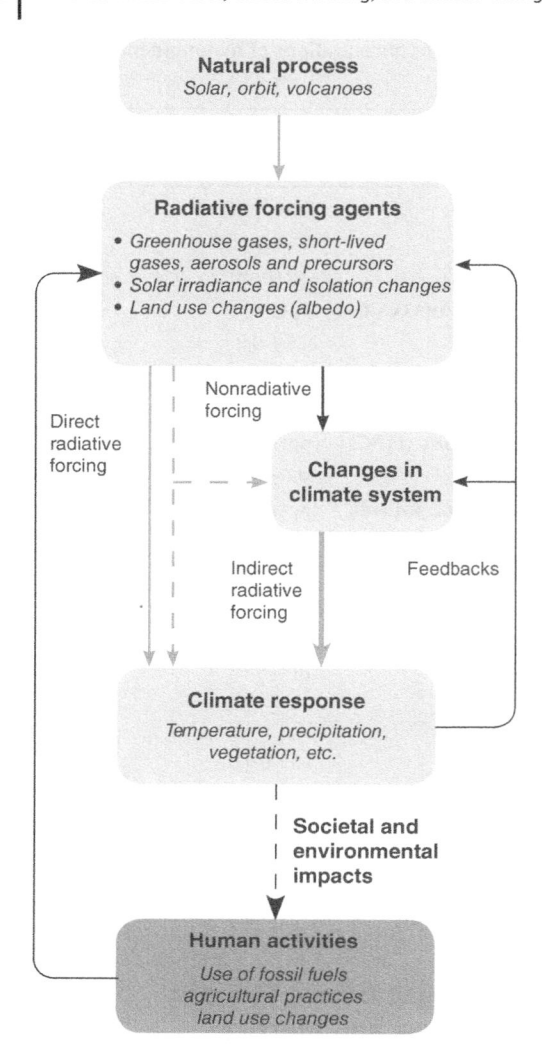

Figure 14.9 Conceptual framework of climate forcing, response, and feedbacks under current climate conditions. Examples of human activities, forcing agents, climate system components, and variables that can be involved in climate response are provided in the lists in each box. Source: Modified from NRC (2005).

components (e.g. precipitation efficiency of clouds), which then lead to changes in radiative fluxes (e.g. effect of solar variability on stratospheric ozone); and (iii) *nonradiative forcings* that create an energy imbalance that does not directly involve radiation (e.g. increasing evapotranspiration flux resulting from agricultural irrigation) (NRC 2005, pp. 12–15).

The effects of GHGs on global warming and climate change depend largely on their radiative properties, rates of emissions into the atmosphere, and resulting concentrations, influenced by environmental transport and transformation processes, and their capacity to alter rates of energy flow (flux) in the atmosphere that lead to changes in the global energy balance.

Since pre-industrial times (1750), greenhouse gas concentrations in the atmosphere have shown large increases and are *extremely likely* to be the primary drivers or forcing agents of global warming and climate change responses that are currently observed (IPCC 2014a).

It follows that the environmental behavior and fate of GHGs in the atmosphere are likely to strongly influence the global warming of the Earth.

Monitoring and understanding the role of GHGs as radiative forcing agents in global warming is crucial for managing the observed impacts and risks of the climate change process. As concluded by the IPCC (2014a):

Continued emission of greenhouse gases will cause further warming and long-lasting changes in all components of the climate system, increasing the likelihood of severe, pervasive and irreversible impacts for people and ecosystems. Limiting climate change would require substantial and sustained reductions in greenhouse gas emissions which, together with adaptation, can limit climate change risks.

Radiative forcing (RF) due to some factor or agent such as a GHG is simply defined as the difference in the Earth's energy balance (radiative equilibrium) between incoming solar radiation (sunlight) and thermal IR energy radiated back into space. A positive RF tends on average to warm the surface of the Earth (e.g. GHGs absorb infrared radiation and re-emit it back to the Earth's surface, thus increasing the Earth's energy balance). Negative RF tends on average to cool the surface by reducing the energy budget (e.g. most aerosol particles reflect solar radiation, leading to a net cooling).

The IPCC Fourth Assessment Report (AR4) defines radiative forcing as:

… a measure of the influence a factor has in altering the balance of incoming and outgoing energy in the Earth-atmosphere system and is an index of the importance of the factor as a potential climate change mechanism. In this report radiative forcing values are for changes relative to pre-industrial conditions defined at 1750 and are expressed in watts per square meter (W m^{-2}).

Essentially, RF represents the rate of energy change per unit area of the globe as measured at the top of the atmosphere (TOA).

Effective Radiative Forcing (ERF)

- Many current applications use an "adjusted" radiative forcing in which the stratosphere is allowed to relax to thermal steady state, thus focusing on the energy imbalance in the Earth and troposphere system, which is most

relevant to surface temperature change. Once the stratosphere has been allowed to adjust to a forcing, the change in energy flux at the tropopause is equivalent to that at the top of the atmosphere (TOA), which is how radiative forcings are commonly reported (NRC 2005, p. 3).
- ERF is the change in net TOA downward radiative flux after allowing for atmospheric temperatures, water vapor, and clouds to adjust, but with surface temperature or a portion of surface conditions unchanged.

For example, the IPCC AR5 report defines radiative *forcing* as the change in net downward radiative flux at the tropopause (TOA) after allowing for stratosphere temperatures to re-adjust to radiative equilibrium while holding surface and tropospheric temperatures and state variables such as water vapor and cloud cover fixed at the unperturbed values.

(The radiative forcing of a GHG is determined by its atmospheric concentration, warming capacity, residence time, and spatial distribution).

Radiative forcing (AR5, p. 126): the strength of drivers is quantified as radiative forcing (RF) in units of watts per square meter (W m^{-2}) as in previous IPCC assessments. RF is the change in energy flux caused by a driver and is calculated at the tropopause or at the top of the atmosphere (WGI).

The total anthropogenic radiative forcing over 1750–2011 is calculated to be a warming effect of 2.3 (1.1–3.3) watts per square meter (W m^{-2}), as shown in Figure 14.10a. It has increased more rapidly since 1970 than during previous decades (high confidence). Natural and anthropogenic radiative forcing estimates for 2011 are illustrated in Figure 14.10b.

Figure 14.10 (a) Total radiative forcing due to human activities (watts per square meter), which indicates the size of the energy imbalance in the atmosphere. Source: Modified from Myhre et al. (2013), IPCC (2014a) and Houghton (2015). (b) Radiative forcing (RF) estimates for natural and human-derived emissions and drivers in 2011 relative to 1750 – also known as the Radiation Forcing Bar Chart from AR5, Chapter 8. Source: Modified from IPCC (2014a, b) and Houghton (2015).

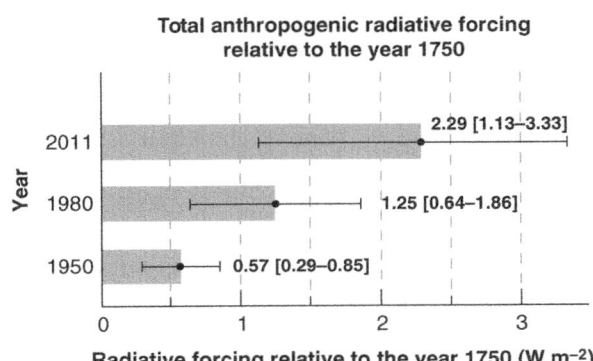

Values are global average RF partitioned according to emitted compounds or processes that result in a combination of drivers. The best estimates of the net RF are shown as black dots with corresponding uncertainty intervals: numerical values and levels of confidence in the net forcing (VH-very high, H-high, M-medium, Low-low, VL-very low). Albedo forcing due to black carbon on snow and ice is included in the black carbon aerosol bar. Volcanic forcing is not included. Small RF due to contrails (~0.05 W m^{-2} including contrail-induced cirrus) are not shown. Concentration-based RFs can be obtained by summing the like-colored bars (modified from Figure 3.13, Houghton 2015, p. 58).

Significantly, CO_2 is the largest single contributor to radiative forcing over 1750–2011 and its trend since 1970. Importantly, the total anthropogenic RF estimate for 2011 is substantially higher (43%) than the IPCC (AR4) estimate reported in 2005, because of a combination of continued growth in most GHG concentrations and an improved estimate of radiative forcing from aerosols. The radiative forcing from aerosols, which includes cloud adjustments, is better understood and indicates a weaker cooling effect than in AR4 (IPCC 2014a, b).

In the case of aerosols, radiative forcing over 1750–2011 is estimated as −0.9 (−1.9 to −0.1) W m^{-2} (medium confidence). It has two competing components: a dominant cooling effect from most aerosols and their cloud adjustments, and a partially offsetting warming contribution from black carbon absorption of solar radiation (IPCC 2014b; Houghton 2015) (see Figure 14.10b). There is high confidence that the global mean total aerosol radiative forcing has counteracted a substantial portion of radiative forcing from well-mixed GHGs.

Aerosols continue to contribute the largest uncertainty to the total radiative forcing estimate. Changes in solar irradiance and volcanic aerosols cause natural radiative forcing. The radiative forcing from stratospheric volcanic aerosols can have a large cooling effect on the climate system for some years after major volcanic eruptions. However, changes in total solar irradiance are calculated to have contributed only around 2% of the total radiative forcing in 2011, relative to 1750 (IPCC 2014b; Houghton 2015).

Solar irradiance or radiation is the source of heat for the Earth as measured by the amount of solar radiation that reaches Earth's surface. The Sun has an 11-year sunspot cycle, which causes about 0.1% of the variation in the Sun's output. Variations from this solar cycle are included in climate models (NASA 2020a).

Climate Sensitivity

How sensitive is our climate to increases in carbon dioxide gas? The term, climate sensitivity, describes the relationship between atmospheric CO_2 and warming attributed to radiative forcing. It is a key measure in climate modeling used to predict how much warming of the atmosphere is associated with increases in CO_2 levels (see enhanced greenhouse effect).

It is underpinned by the theoretical development of the radiative forcing concept (NRC 2005, pp. 19–23). Basically, the concept adopts the hypothesis that change in the global annual mean surface temperature is proportional to the imposed global annual mean forcing and is independent of the nature of the applied forcing. It also assumes that the heat transfer process in the Earth-troposphere system is in a state of radiative-convective equilibrium.

Climate sensitivity is the equilibrium temperature change in response to changes of the radiative forcing. In simple climate models, radiative forcing can be used to estimate the **equilibrium surface temperature (ΔT_S)** due to the particular forcing, as given by Eq. (14.8), for a steady state:

$$\Delta TS = \lambda \Delta F \tag{14.8}$$

Where λ is a climate sensitivity parameter, usually with units K/(W m^{-2}), and ΔF is the radiative forcing in W m^{-2}.

As a standard measure, climate sensitivity is expressed simply as how much the Earth's surface temperature would increase if pre-industrial levels of CO_2 were doubled. Estimates of climate sensitivity for different emission or climate scenarios, such as by the IPCC, are usually given as a range of values to reflect the degree of uncertainty in behavior of the climate system and some measures of model parameters.

For radiative forcing by CO_2 gas, a typical value of λ is 0.8 K per (W m^{-2}), which gives an increase in global temperature of about 1.6 K (°C) above the 1750 reference temperature due to the increase in CO_2 over that time (278–405 ppm), for a forcing value of 2.0 W m^{-2}. If the present atmospheric CO_2 level was to double the pre-industrial value, the global warming is predicted to increase the global temperature to about 3 K (°C), assuming no other forcings and linear changes in ΔT_S from the climate sensitivity λ (Houghton 2015).

Three evidence-based approaches are used to estimate climate sensitivity: historical climate records, complex mathematical models of the climate system based on physics, and paleoclimate records (e.g. ice cores) over periods of up to thousands of years.

Two common ways of estimating climate sensitivity depend on the timescales of interest and use of some form of climate models of increasing complexity, including other climate forcing factors and feedbacks: transient climate response (TCR), equilibrium climate sensitivity (ECS), and Earth System sensitivity.

TCR is defined by how much the global mean temperature would rise if atmospheric CO_2 level increases by 1% each year (compounded), until that value is doubled. It tends to follow past changes in CO_2 levels. However, if the atmospheric CO_2 level was held at the TCR point, the Earth would continue to warm over a longer time period (mainly due to a slow heating response of oceans), until an equilibrium is reached by the climate system. In this situation, ECS can then be defined as the amount of warming achieved when the entire climate system reaches an equilibrium or stable temperature in response to the doubling of CO_2.

Climate sensitivity to the effects of other GHGs and climate feedbacks also need to be considered. Calculation of TCR shows that the average surface temperature of the Earth's atmosphere would rise by 1 °C or so, if warming due to other climate forcings (e.g. other GHGs) and feedbacks are not included. By including feedbacks, the estimate of ~1 °C warming from doubling of CO_2 changes to an uncertain range of potential global warming, from about 1.5 to 4.5 °C (CarbonBrief 2018).

As indicated here, the range of climate sensitivity estimates vary considerably due largely to uncertainties in climate model parameters derived for climate feedbacks.

Note: A third way of estimating climate sensitivity is called Earth System sensitivity (ESS), which includes slow Earth system feedbacks, such as changes in water vapor, cloud cover, ice sheets, and vegetative cover. These climate feedbacks may cause larger long-term responses than current projections by IPCC.

Climate Change Feedbacks

While climate forcings (solar irradiance, GHGs, aerosols, dust, smoke, soot, or carbon black) are the initial drivers of climate, there are a number of known physiochemical or biological feedback processes occurring in the biosphere that can either amplify (positive feedback) or diminish (negative feedback) the effects of each climate forcing. The degree of amplification or reduction is a critical factor. Examples of climate feedbacks include:

1. Water vapor
2. Clouds
3. Precipitation
4. Greening of the forests
5. Increases in forest fires
6. Ice albedo
7. Release of methane
8. Plankton multiplier effect in oceans
9. Carbon dioxide fertilization

Houghton (2015, pp. 106–112) provides several examples of major climate feedbacks: temperature feedback, water vapor feedback, cloud-radiation feedback, ocean-circulation feedback, and ice albedo feedback. The most important form of feedback is related to water vapor, which is a powerful GHG. On average, the water vapor of the atmosphere will increase with a warming atmosphere leading to a positive feedback. Modeling shows the magnitude of the global average temperature would virtually double relative to a fixed water vapor situation.

Another example is the ice albedo, which is a strong positive feedback loop. Warming of the atmosphere causes the very reflective sea ice to melt, which allows the darker ocean to absorb more heat, resulting in more ice melting, and further heating of the Earth's surface (NASA 2020a).

Climate feedbacks can influence what is called climate sensitivity, and abruptly disrupt climate states at climate *tipping points*, or delay climate change response (Pittock 2005, p. 12). These phenomena are discussed below.

Climate Change Tipping Points

Beyond certain ecological thresholds, known as *tipping points*, ecosystems may collapse and change into distinctly different states (see Chapter 7 in this book). A *tipping point* in a climate system occurs when it abruptly moves from a relatively stable state to another state. This change may be irreversible. Some examples have been observed from the paleoclimate record and projected under current rates of global warming. These include:

1. Ocean circulation (e.g. disruption to Gulf stream in northern Atlantic Ocean – Atlantic Meridional Overturning Circulation [Amoc])
2. Ice loss (e.g. Greenland ice cap)
3. Rapid release of methane from wetlands and trapped in sediments as methane hydrates, mostly at high latitudes

Examples of Climate *tipping points* are given in Section S14.4.

Radiative Forcing for a Greenhouse Gas

For a GHG, such as CO_2, radiative transfer codes that examine each spectral line for atmospheric conditions can be used to calculate the change ΔF as a function of changing concentration. These calculations can be simplified into an algebraic expression that is specific to that gas. For example, the forcing change $\Delta F(t)$ for CO_2, expressed in units of watts per square meter, can be calculated from Eq. (14.9) using a specific coefficient given by the IPCC

Table 14.3 Direct global warming potential (GWP) values of major and other selected greenhouse gases relative to CO_2.

Greenhouse gas	Atmospheric concentrations (2011)	Radiative forcing (2011) (W m^{-2})	Lifetime (years)	GWP over 100 years (2011)
Carbon dioxide, CO_2	391 ± 0.2 (ppm)	1.82 ± 0.19	30–100 (avg. 60)	1
Methane, CH_4	1803 ± 2 (ppb)	0.48 ± 0.05	12	28
Nitrous oxide, N_2O	324 ± 0.1 (ppb)	0.17 ± 0.03	114	265
CFCs	(ppt)			
CCl_3 (CFC-11)	238 ± 0.8	0.062	45	4 660
CCl_2F_2 (CFC-12)	528 ± 1	0.17	100	10 200
$CBrF_3$ (Halon-1301)	—	—	—	6 290
Hydrofluorocarbons (HFCs)	(ppt)			
CHF_3 (HFC-23)	24.0 ± 0.3	0.0043	270	12 400
Perfluorinated compounds	(ppt)			
SF_6	7.28 ± 0.03	0.0041	3 200	23 500
NF_3	0.9	0.0002	740	16 100
C_2F_6 (PFC-116)	4.16 ± 0.02	0.0010	10 000	11 100
C_4F_{10} (PFC-31-10)				9 200
Other halogenated hydrocarbons	(ppt)			
Carbon tetrachloride CCl_4	85.8 ± 0.8	0.0146	26	1 730
CH_3CCl_3	6.32 ± 0.07	0.0004	—	—
Methyl bromide CH_3Br	—	0.01	0.7	2

Values compiled from the IPCC Fifth Assessment Report 2014 (AR5): Chapter 8, AR5 – Anthropogenic and Natural Radiative Forcing.
Source: Data from IPCC (2014a) and Myhre et al. (2013).

(NRC 2005, pp. 24–25):

$$f(t) = 5.35 \ln \left[\frac{CO_2(t)}{CO_2(1750)} \right] \quad (14.9)$$

where $CO_2(t)$ is the atmospheric concentration of CO_2 for year t. These types of models can relate major greenhouse gas emissions to the equilibrium global averaged temperature changes (ΔT_S) and, using transient oceanic heat uptake models, to transient temperature changes and impacts (NRC 2005). Different formulas are derived for other GHGs with coefficients that may be found, for example, in the IPCC reports.

Formulas for calculating radiative forcing from concentration are given in IPCC (2013, supplementary material for Chapter 8, p. 8 SM-7).

For short-lived species, such as aerosols, expressions of the form given in Eq. (14.9) are unavailable due to the great spatial variability in concentrations and optical properties (NRC 2005, p. 25).

Global Warming Potential

The concept of global warming potential (GWP) was developed to address the need for a simple quantitative way to compare the radiative consequences of emissions of different gases (NRC 2005, pp. 26–27). GWP is known as a relative measure of how much heat a GHG absorbs or traps in the atmosphere. It compares the amount of heat absorbed by a certain mass of a GHG to the amount of heat absorbed by a similar mass of carbon dioxide.

A GWP is calculated over a specific time interval, usually 20, 100, or 500 years. GWP is expressed as a factor of carbon dioxide (whose GWP is standardized to 1) (see NRC 2005, pp. 26–27; for a quantitative introduction to the calculation of GWP for a greenhouse gas).

The global warming potential (GWP) of a greenhouse gas is defined as its total warming effect over 100 years relative to the same amount of carbon dioxide gas. That is, the ratio of the time-integrated radiative forcing from the instantaneous release of 1 kg of the gas to that of the release of 1 kg of carbon dioxide over a specified time period (100 years) (Houghton 2015, p. 59).

Table 14.3 presents a selected set of direct GWP values for major and other selected GHGs over a period of 100 years (known as a time horizon). GWP is used as an index to convert non-carbon dioxide gases, particularly GHGs, to a carbon dioxide equivalent (CO_2-e) by multiplying the quantity of the gas by its GWP (Box 14.2).

Because the GWP of individual GHGs varies, a comparative measure, **carbon dioxide equivalents,** has been derived (see Box 14.2).

> **Box 14.2 Carbon dioxide equivalents.**
>
> It is a unit of measurement that allows the effect of different GHGs and other factors to be compared in a given mixture, using carbon dioxide as a standard unit for reference, over a period of time, usually a time scale of 100 years.
>
> $$\text{Carbon dioxide equivalent } (CO_2 - e)$$
> $$\text{of a non-carbon dioxide gas}$$
> $$= \text{quantity of the gas} \times \text{GWP of gas}$$
>
> For example, the global warming potential for **methane** over 100 years is estimated to be 25–28. This means that emissions of 1 million tonnes of methane are equivalent to the emissions of 25 million tonnes of carbon dioxide for a factor of 25. The GWP for nitrous oxide is estimated to be 265–298.
>
> (i.e. 25 million tonnes of CO_2-e = 1 million tonnes of methane × 25)

14.7 Past Climate Changes

Key Indicators and Monitoring of Climate Change

The study of past climate change or paleoclimatology relies on proxy or indirect data gathered from historical records, tree rings, lake and marine sediments, ice cores (e.g. CO_2 gas), pollen, speleothems (cave secondary mineral deposits), loess, and geomorphic features. From these natural archives, it is possible for scientists to derive past values of climate indicators (e.g. temperature, precipitation and humidity, chemical composition of air (e.g. CO_2 and water), vegetation patterns, solar activity, volcanic eruptions, geomagnetic field variations, and sea level, and to reconstruct the climate of specific periods, for example, from past proxy data networks, and reconstructions of large-scale temperature patterns over the past millennium (Pittock 2005, chapter 2; Houghton 2015, chapter 4).

Key indicators of climate change include atmospheric measures of GHG concentrations and CO_2 exchange in terrestrial biospheres, weather indicators (e.g. global mean surface temperature, number of hot days/heat index, heavy precipitation events, drought and tropical cyclones), and biophysical indicators (e.g. global mean sea level, snow cover, nonpolar glaciers, permafrost, plant and animal ranges, breeding, flowering, and migration and coral bleaching).

Over the last 160 years or so, systemic weather records, databases, and networks have become global and advanced scientific instrumentation and monitoring networks measure, observe, and analyze short-term state of the atmosphere (weather) and long-term patterns of weather conditions at different locations (e.g. US National Oceanic and Atmospheric Administration [NOAA]).

The World Meteorological Organization (WMO) measures a key set of global climate indicators that describe the changing climate for temperature and energy, atmospheric composition, ocean and water, and the cryosphere. Parameters cover: surface temperature, ocean heat, atmospheric CO_2, ocean acidification, sea level, glacier mass balance, and Arctic and Antarctic sea ice extent (see WMO Website and Chapter 17 of this book).

Past Million Years

The Earth has long experienced climate change. Our current knowledge of past climates based on records of temperature, atmospheric composition, and sea level from ice cores (Greenland and Antarctica), from ocean and lake sediment cores, and other proxy records extends over much of the past million years (Houghton 2015, p. 87).

The average temperature of the Earth has fluctuated throughout its 4.54 billion-year history. In the last 1 million years or more, long cold periods (*ice ages*), and warm periods (*interglacials*) on about 100 000 year cycles have been recorded. Naturally occurring ice ages of the past 800 000 years, ending with the early Twentieth Century, are plotted from ice core data (NOAA 2015).

The main trigger mechanisms for these ice ages over the last million years or more is attributed to variations in solar radiation reaching the Earth caused by regular orbital variations of the Earth around the Sun, known as Milankovitch cycles. Variations in GHGs have also provided a positive feedback to this forcing (Houghton 2015, p. 87). In the first half of the last interglacial period (~130 000–123 000 years ago), orbital variation caused a large increase in summer solar radiation leading to higher temperatures, 3–5 °C warmer than now, and melting in polar regions, resulting in higher sea levels, rising 4–6 m higher than today. These cycles are considered well understood and the start of the next ice age is not expected for at least 30 000 years (Houghton 2015, p. 87).

Paleoclimatic records show that abrupt changes can occur in the Earth's temperature and other climate factors.

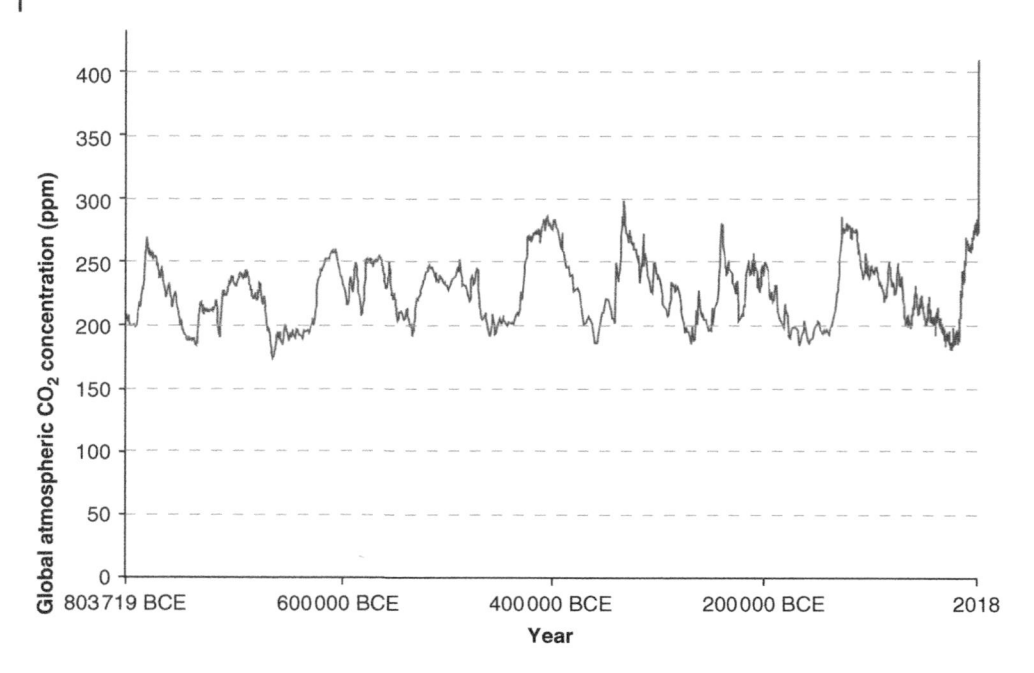

Figure 14.11 Atmospheric Carbon Dioxide Concentration: Global average carbon dioxide concentrations (parts per million) in atmosphere during ice ages and warm periods for the past 800 000 years. Source: Data from NOAA Climate.gov (2015). OurWorldinData.org/co2-and-other-greenhouse-gas-emissions CC-BY.

For example, the last interglacial period (80 000–18 000 years ago) experienced large repeated changes in climate and ocean circulation between cooler and warmer events known as Dansgaard–Oeschger cycles, raising the possibility of future abrupt climate change such as in ocean circulation (see Schmidt and Hertzberg 2011).

Past 2000 Years

Paleoclimate and monitoring data indicate that the recent global warming is abnormal compared to at least the last 2000 years. Over the past two millennia, global temperatures have warmed and cooled but no previous warming episodes appear to be as large and abrupt as recent global warming. The current increase in global average temperature also appears to be much faster than at any point since modern civilization and agriculture developed in the past 11 000 years, and probably than in any interglacial warming periods over the last million years (NOAA 2015).

In comparison to surface temperature, carbon dioxide concentrations in the atmosphere over the past 800 000 years are plotted in Figure 14.11 based on proxy climate data and modern monitoring-based records.

Figure 14.11 shows peaks and valleys of ice ages (low CO_2) and warmer interglacials (higher CO_2), during which atmospheric CO_2 levels did not exceed 300 ppm. In 2019, the CO level was 409 ± 0.1 ppm. Significantly, the annual rate of increase in atmospheric carbon dioxide over the past 60 years is about 100 times faster than previous natural increases, such as those that occurred at the end of the last ice age 11 000–17 000 years ago.

> The last time the atmospheric CO_2 amounts were this high was more than 3 million years ago, when temperature was 2–3 °C (3.6–5.4 °F) higher than during the pre-industrial era, and sea level was 15–25 m (50–80 feet) higher than today.
> Source: NOAA (2020).

In comparison, modern human civilization emerged during the Holocene Epoch, which began about 12 000 years ago.

Industrial Revolution to Present

NOAA has reported that the Earth has warmed significantly since the late Nineteenth Century, based on reviews of multiple paleoclimatic and thermometer data. The first decade of the Twenty-first Century is the warmest on record within the entire global instrumental temperature record. Recent years are also the warmest global temperatures of at least the last 1000 years.

> Global surface temperature has increased faster since 1970 than in any other 50-year period over at least the last 2000 years (high confidence).
> Source: IPCC (2021).

Significantly, the IPCC AR5 assessment (IPCC 2014, p. 11; IPCC 2014, p. 44) concluded that historical emissions have driven atmospheric concentrations of GHGs (carbon dioxide, methane, and nitrous oxide) to levels that are unprecedented in at least the last 800 000 years, leading to an uptake of energy by the climate system. Carbon dioxide levels have now increased by at least 40% since pre-industrial times, primarily due to emissions from fossil fuels and from net land use changes (IPCC 2014a, p. 44; see also IPCC 2021).

14.8 Observed and Projected Climate Change Impacts for the Environment and Human Health

Observed Climate Changes

The IPCC has evaluated the scientific evidence of observed changes in the Earth's climate system and concluded that:

> Warming of the climate system is unequivocal, and since the 1950s, many of the observed changes are unprecedented over decades to millennia. The atmosphere and ocean have warmed, the amounts of snow and ice have diminished, and sea level has risen (IPCC 2014, p. 2).

Furthermore, they have strongly emphasized the role of humans in inducing climate change in the form of global warming:

> Human influence on the climate system is clear, and recent anthropogenic emissions of greenhouse gases are the highest in history. Recent climate changes have had widespread impacts on human and natural systems.
>
> Source: IPCC (2014a, p. 2).

For example, glaciers have continued to shrink almost worldwide (Figure 14.12). Greenland and Antarctic ice sheets are observed to be losing mass at a larger rate this Century (e.g. IPCC 2014a).

Table S14.1 provides a summary of key climate change variables, observed changes and trends, and likelihood of these changes being due to global warming and human activities.

Other strong evidence is based on observations of extreme weather/climate events in different regions and latitudes of the Earth since essentially 1950, as shown by examples in Table S14.2 (see Section S14.1).

The main impacts of climate change are identified as increases in surface temperatures (land and ocean) leading to heatwaves, sea level rises, ocean acidification, and a more intense hydrological cycle associated with patterns of increasing precipitation, glacier, ice, snow and permafrost melt, and more intense and frequent floods, droughts, tropical cyclones, and storms in different regions. The degree of global and regional warming and likely rate of climate change are serious and critical factors for the impact of climate change on natural ecosystems, and degradation of ecosystem resources and services (IPCC 2014a; Houghton 2015, IPCC 2021).

Figure 14.12 Glacier – Prins Christian Sund, Greenland, August 2016. Source: Photograph by G.J. Miller (Author).

The strength of evidence in support of the role of global warming due to human activities in extreme climatic events continues to increase. The Sixth Assessment Report of the IPCC (AR6) states: "Evidence of observed changes in extremes such as heatwaves, heavy precipitation, droughts, and tropical cyclones, and, in particular, their attribution to human influence, has strengthened since AR5" (IPCC 2021). For example, in land regions, there is *high confidence* that human-induced climate change is the main driver of increased hot extremes (more frequent and more intense) since the 1950s and decreased cold extremes (less frequent and less severe).

Projections of Climate Change During the Twenty-First Century and Beyond

> If we can explain the past, we can build conceptual and then detailed computer models. These will enable us to make predictions about future climate changes, given the likelihood of future changes in those factors which drive the climate system.
>
> Pittock (2005, p. 24) – Climate Change.

Emission Scenarios and Computer Modeling

Climate models are used to construct and develop computer numerical modeling of the atmosphere and climate, and project future changes in atmospheric GHGs and climate responses to investigate the potential impacts of various anthropogenic forcings, based on GHG emissions and other human-related and natural forcings. Timescales of modeling extend up to a century or more and cover global and regional spatial scales at increasing resolutions.

Emission scenarios are a set of descriptions of likely future emissions of GHGs that incorporate many of the major driving forces – including processes, impacts (physical, ecological, and socio-economic), and potential responses that are important for informing climate change policy. According to the IPCC, the goal of scenarios is to explore the implications of climate change to better understand uncertainties and alternative futures, and then to consider how robust different decisions or options may be under a wide range of possible futures (IPCC Scenario Process) (IPCC 2014a, b).

Importantly, as described by Houghton (2015, p. 128), confidence in modeling of projections is based on validation of model simulations against (i) detailed observations of current and recent climate of both the atmosphere and oceans; (ii) detailed observations of particular climate cycles such as *El Niño* events; (iii) observations of perturbations arising from particular events such as volcanic eruptions; and (iv) paleoclimate information from past climates under different orbital forcing.

For the Fifth Assessment Report (AR5), the IPCC developed a set of emission scenarios known as representative concentration pathways (RCPs) (see Box S14.1 in Section S14.2). They take into account a number of factors and assumptions that include the amount of future greenhouse gas emissions, developments in technology, changes in energy generation and land use, global and regional economic circumstances, and population growth. Outputs from different modeling systems can then be compared by using a standard set of scenarios to provide a consistent set of starting conditions, historical data, and possible future emissions for use by climate scientists (Houghton 2015, pp. 135–136).

The RCPs include a stringent mitigation scenario (RCP2.6), two intermediate scenarios (RCP4.5 and RCP6.0) and one scenario with very high GHG emissions (RCP8.5). Scenarios without additional efforts to constrain emissions (baseline scenarios) lead to pathways ranging between RCP6.0 and RCP8.5 (Figure SPM.5a). RCP2.6 is representative of a scenario that aims to keep global warming *likely* below 2 °C above pre-industrial temperatures (IPCC 2014a, p. 8).

Cumulative emissions of CO_2 are projected to be the main drivers of global warming to the end of the Twenty-first Century and beyond, as indicated in Table 14.4, and GHGs are reported to vary over a wide range depending

Table 14.4 Cumulative CO_2 emissions for the 2012–2100 period compatible with the RCP atmospheric concentrations simulated by CHIP5 earth system models.

Scenario	Cumulative CO_2 emissions (2012–2100) Giga tonnes (Gt) CO_2	
	Mean	Range
RCP 2.6	990	510–1505
RCP 4.5	2860	2180–3690
RCP 6.0	3885	3080–4585
RCP 8.5	6180	5185–7005

Notes: 1 Gigatonne (Gt) of carbon = 10^{15} g of carbon and is equivalent to 3.667 Gt of CO_2.
Source: Data from IPCC (2014a) and Houghton (2015).

on socio-economic development and climate policy (IPCC 2014a, p. 56).

Projected changes in key climate variables (atmospheric CO_2 equivalents, global mean surface temperature, and global mean sea level rise) at the end of the Twenty-first Century for the four RCP scenarios are indicated in Table 14.5.

The change in global mean surface temperature for the period 2016–2035 relative to the 1986–2005 baseline is projected to be similar for the four RCPs and will likely be in the range 0.3–0.7 °C (medium confidence) (this range assumes no major natural or unexpected perturbations). By the mid-Twenty-first Century, the magnitude of the projected climate change is substantially affected by the choice of emissions scenarios. Climate change continues to diverge among the scenarios, due to differences in sensitivity, through to 2100 and beyond to 2300 (IPCC 2014a, pp. 58–59) (see Table 14.5; Figure 14.13).

Overall, the IPCC (2014a, p. 60) concludes that the **global mean surface temperature change** (relative to 1850–1900) for the end of the Twenty-first Century (2081–2100) "is projected to likely exceed 1.5 °C for RCP4.5, RCP6.0 and RCP8.5 (high confidence). Warming is likely to exceed 2 °C for RCP6.0 and RCP8.5 (high confidence), more likely than not to exceed 2 °C for RCP4.5 (medium confidence), but unlikely to exceed 2 °C for RCP2.6 (medium confidence)".

The **global mean sea level** rise is expected to exceed 0.4 m (RCP 4.5) or 0.7 m (RCP 8.5) by 2100, as indicated by Figure 14.14. Over the period 1901–2010, global mean sea level rose by 0.19 (0.17–0.21) m. The rate of sea-level rise since the mid-Nineteenth Century has been larger than the mean rate during the previous two millennia (high confidence).

Since the AR5 report in 2013–2014, the 2021 IPCC report, AR6-WGI, has adopted a new set of five climate change scenarios, known as Shared Socioeconomic Pathways

Table 14.5 Projected changes in atmospheric carbon dioxide equivalent concentrations, global mean surface temperature and mean sea levels using AR5 emission scenarios for climate models.

Emission scenario	Atmospheric carbon dioxide equivalent (CO_2-eq) concentrations (2100) (ppm)[a]	Global mean surface temperature change to 2081–2100[b] (°C)	Global mean sea-level rise 2081–2100[b] (m)
RCP 2.6	~490	1.0 (0.4–1.6)	0.40 (0.26–0.55)
RCP 4.5	~650	1.8 (1.1–2.6)	0.47 (0.32–0.63)
RCP 6.0	~850	2.2 (1.4–3.1)	0.48 (0.33–0.63)
RCP 8.5	~1370	3.7 (2.6–4.8)	0.63 (0.45–0.82)

Values in brackets represent likely ranges.

[a]Van Vuuren et al. (2011): each of the RCPs covers the 1850–2100 period, and extensions have been formulated for the period 2100 to 2300.

[b]Relative to 1986–2005 baseline.

Source: Data from IPCC (2014a) and Houghton (2015).

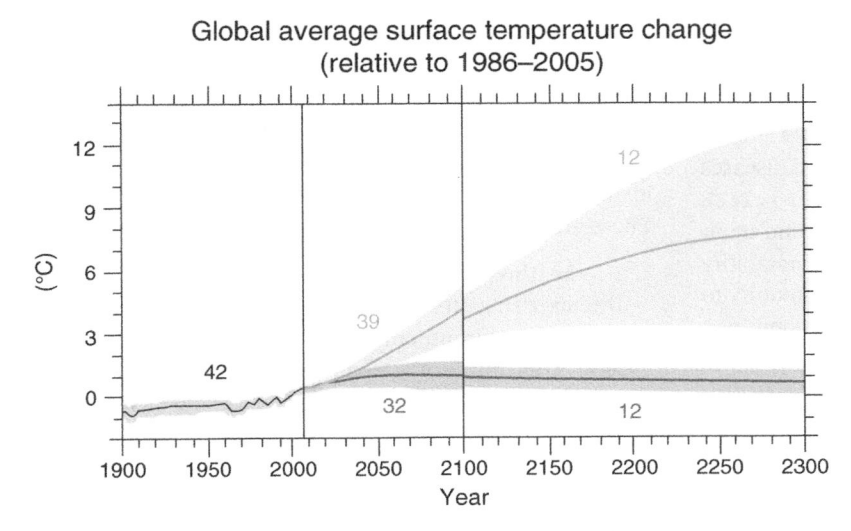

Figure 14.13 Projected global average surface temperature changes from 1900 to 2300 (relative to 1986–2005) for RCP scenarios 4.5 and 8.5. Values in bold on graphs are the numbers of climate models (CMIP5). Source: IPCC (2014a).

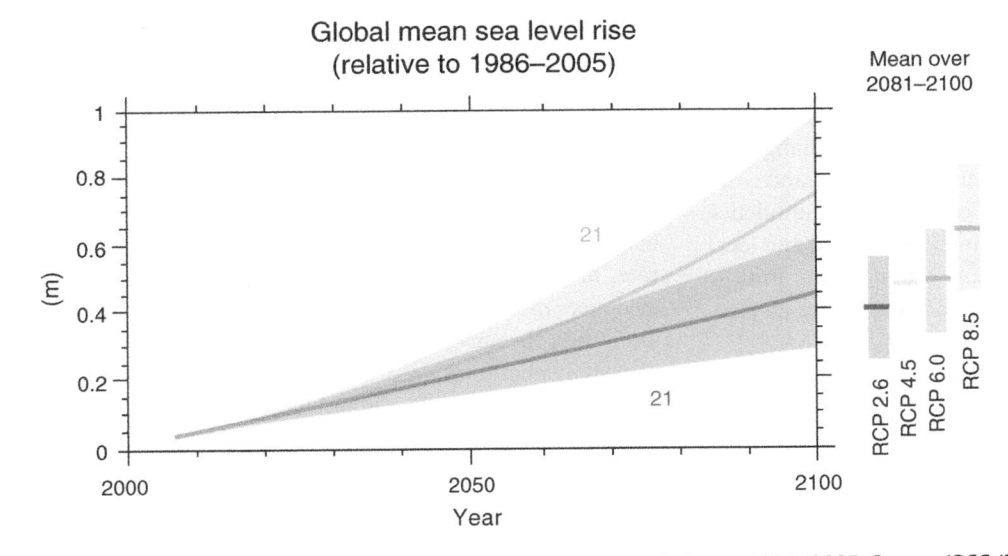

Figure 14.14 Projected global mean sea level rise (metres, m) relative to 1986–2005. Source: IPCC (2014a).

(SSPs)[1] to project global changes up to 2100, using a CMIP6 model. These scenarios are combined with mitigation targets of RCP scenarios (AR5) to evaluate different levels of climate change mitigation and the outcome of a *no climate policy* scenario (see also S14.5).

SAR6 concludes that the Earth will be 1.4–4.4 °C warmer than pre-industrial levels by the end of this Century. The projected temperature changes will depend on the rates at which emissions of GHGs are rapidly reduced to net zero or continue to rise (see IPCC 2021).

Projected changes and impacts based on climate modeling scenarios covering the Twenty-first Century indicate that it is likely there will be serious and cumulative impacts on natural ecosystems, resources, such as agriculture, forestry, water supplies, energy, fisheries, infrastructure and human settlements, and human health (deaths, injuries, and disease). For example, forests will be affected by increased climate stress, including dieback and reduced production. The health of exposed human populations will also be affected by increasing heat stress, especially later in the Twenty-first Century, and by spread of tropical diseases such as malaria to warming areas (Houghton 2015, p. 212).

The extent of climate change impacts will depend upon the response to mitigate GHG emissions, the vulnerability of exposed systems, and the global capacity of humanity to adapt. This applies to ecosystems and resources combined with risk management policies and measures implemented for selected SSP/RCP scenarios. Global warming is projected to continue causing sea level rises until the end of the Twenty-first Century (see Figure 14.14).

Complexity of Climate Change Impacts

Observed and likely climate changes from modeling show considerable variability in different regions of the world. Climate variables such as temperature, rainfall, and sea level rises vary to different amounts within and between regions and are subject to extreme weather or climate events. To compensate for such variability, changes in surface temperature and sea level rise, for instance, are estimated on a global scale as averages and confidence intervals over a longer period of time. At regional and local scales, the situation becomes more complex due to how natural and human systems respond to climate change, under the pressures of natural variability, and human activities and responses.

Here, critical concepts are the sensitivity of different systems to climate change, the adaptive capacity of a system to cope with or adjust to the degree of the changes, and the vulnerability of each system exposed to adverse effects of climate change. Vulnerability of a system can be expressed as a function of the character, magnitude, and rate of climate change, and also its degree of exposure, sensitivity, and adaptive capacity (Houghton 2015, p. 163).

Natural Ecosystems and Biodiversity

"Anthropogenic climate change is predicted to be a major cause of species extinctions in the next 100 years. It is already causing widespread local extinctions" (Cahill et al. 2013). Climate change impacts and responses on individual species, communities, and ecosystems tend to follow the basic principles and ecological responses to environmental change or stressors. However, Steffen (2009, pp. 71–72) identifies two major differences that add to the complexity of predicting responses by biota to climate change. These are:

1. … the threats to biodiversity of climate change are threats rising from changes in the basic physical and chemical environment underpinning all life – especially CO_2 concentrations, temperature, precipitation and acidity – unlike other threats such as land clearing and introduced species.
2. … the rate of current warming and other associated changes in climate are unprecedented since the last massive extinction event 60 million years ago (with the possible exception of a very rapid cooling and subsequent rapid warming of more than 5 °C in northern Europe at the start and finish of the Younger Dryas Period (from 12 000 to 11 500 years b.p.), caused by the disruption and restoration of the North Atlantic thermohaline circulation during deglaciation and freshwater runoff from North America and Greenland.

Changes in the physical environment affect the physiological processes of biota such as respiration, metabolic rate, and water use efficiency. Current climate change evidence includes observations of diverse species that alter their behavior or timing of life cycle events (e.g. dispersal, migration, and reproduction) that can lead to serious impacts on their populations, communities, and ecosystems.

Numerous examples of observed, predicted, and emerging impacts of global warming and climate change are described widely in the recent scientific literature. Many are derived from decades of careful observations, field studies, long-term biological databases, time series analyses and

---

1 The SSPs use storylines that describe broad socioeconomic trends that could shape future society. They include: a world of sustainability-focused growth and equality (SSP1); a "middle of the road" world where trends broadly follow their historical patterns (SSP2); a fragmented world of "resurgent nationalism" (SSP3); a world of ever-increasing inequality (SSP4); and a world of rapid and unconstrained growth in economic output and energy use (SSP5) (CarbonBrief 2021).

advances in ecological modeling and research, especially in the Northern Hemisphere.

The following examples indicate briefly the wide range and complexity of global warming impacts at all levels of biological organization and are largely taken from an in-depth review of the scientific literature by Chivian and Bernstein (2008) from the Center for Health and the Global Environment, Harvard Medical School (see Chivian and Bernstein 2008, pp. 63–73; 108 and Chapter 3 in this book).

- Climate change impacts on the distribution of species are already of concern. Some mobile species such as birds and butterflies will need to migrate to find and adapt to available habitats in higher altitudes and latitudes. Less mobile species will be more vulnerable. Species losses are expected and are likely to be exacerbated by increasing habitat degradation and destruction. Vascular plant species in European Alps, for example, are observed to have slowly moved their ranges up mountains toward the summits but are threatened by loss or eventual extinction by further warming.
- Shifts in the ranges of mobile marine species are similarly observed. In these cases, they move toward colder waters or greater depths, and toward the poles. The species composition and interactions in ecosystems are likely to be complex and unpredictable.
- Warming surface temperatures, at slight increases of about 1 °C above the mean summer ocean temperature, are causing coral reefs worldwide to appear *bleached* from losses of symbiotic algae, and making coral polyps susceptible to large-scale mortalities from various infectious diseases. Many coral reefs and their richly diverse species are becoming decimated by losses of coral species and their vulnerable habitats. These reefs are predicted to be severely threatened and devastated at current warming rates (see also Hoegh-Guldberg et al. 2018).
- Increasing acidification of oceans is occurring due to the absorption of carbon dioxide gas from the atmosphere by seawater and formation of carbonic acid. pH stability of the ocean after hundreds of thousands of years has been essentially reduced by 0.1 pH unit since the pre-industrial era and is predicted by the IPCC to increase by 0.3 pH units by 2100, a change in the pH level that has not occurred for more than 20 million years. Greater acidity has already been observed to interfere with the ability of some marine organisms that form calcareous shells, skeletons, and hard body parts. These include abundant and diverse organisms such as reef-forming corals, crustaceans, mollusks, and certain plankton (e.g. vulnerable photosynthetic *coccolithophores*) that are vital for marine food webs.
- Melting sea ice from ocean warming threatens sensitive marine food webs in the Arctic. Polar bears in the Arctic are faced with starvation through reduced capture and declines in seal populations (ringed seals) that are their major food source. Diminishing snow cover and retreating ice also exposes seal pups to further predation from Arctic foxes and polar bears and also stress from prolonged exposure to cold waters.
- Disruption of biological cycles by global climate change can impact on the timing of critical biological events such as the arrival and departure times of migratory birds, the breeding of amphibians, hatching of bird eggs, the flowering of plants, and the selection of organisms better suited to the new temperature conditions. These events can affect complex prey–predator relationships and reproduction of threatened populations.

Human Health Impacts from Global Warming and Climate Change

The World Health Organization has clearly and strongly emphasized the strong link between climate change, air pollution, and health. The same industry and land use sectors that produce most GHGs are also known to be the main sources of fine particulates and other key air pollutants that are attributed to the premature deaths and diseases of millions of persons each year. These pollutants include short-lived climate pollutants such as black carbon and ground-level ozone (WHO 2018) (see Chapter 13 in this book).

Exposure to climate change has the potential to affect everyone at sometime through multiple pathways (e.g. air, food, water, and weather events). People in developing countries are more vulnerable but large vulnerable sub-populations also exist in developed countries (e.g. children, older adults, pregnant women, and people on low incomes).

The 2015 *Lancet* Commission on Health and Climate evaluated the health effects of climate change on human populations. One of its key conclusions was that anthropogenic climate change threatens to undermine the past 50 years of gains in public health. The Lancet Countdown report (Watts et al. 2017) identified several pathways through which climate change primarily affects health:

- Direct effects that are diverse, being mediated, for instance, by increases in the frequency, intensity, and duration of extreme heat and by increases in average annual temperature (e.g. heat stress and related mortality). Rising incidence of other extremes of weather, such as floods and storms, increase the risk of drowning and injury, damage to human settlements, spread of waterborne disease, and mental health conditions from previous disease or injury;
- Ecosystem-mediated impacts include changes in the distribution and burden of vector-borne diseases (such as malaria and dengue) and waterborne infectious disease;

- Human institution-mediated impacts such as human nutritional deficiency from crop failure, population displacement from sea-level rise, and occupational health risks; and
- Noncommunicable diseases mediated through a variety of pathways, which can involve cardiovascular disease, acute and chronic respiratory disease from worsening air pollution and airborne allergens, or the often-unseen mental health effects of extreme weather events or of population displacement. Emerging evidence also suggests links between a rising incidence of chronic kidney disease, dehydration, and climate change.

The WHO (2018) also reports a set of findings from an assessment of human health effects due to climate change for projected increases in exposure to four climate variables: heat, droughts, floods, and heatwaves. These findings, including estimates of impacted populations, are summarized as:

- Up to 3 billion people aged over 65 years may be exposed to heatwaves by 2100, because of a combination of increasing temperatures, ageing and urbanization;
- The warmest and poorest countries of the world will be most severely affected by climate change, particularly in South Asia;
- Overall, the health impacts of climate change could force 100 million people into poverty by 2030, with strong impacts on mortality and morbidity;
- A highly conservative estimate of 250 000 additional deaths each year due to climate change has been projected between 2030 and 2050;
- Among these deaths, 38 000 are expected to result from heat exposure in the elderly people, 48 000 due to diarrhea, 60 000 due to malaria, and 95 000 due to childhood undernutrition.

These current estimates are considered to be preliminary findings, limited to only four direct climate change effects on health, and are likely underestimates of overall human impacts.

14.9 Global Risks from Climate Warming and Climate Change

The scientific relationship between atmospheric carbon dioxide and global temperatures has been understood for over 100 years (IPCC 2014a, b). Accelerated emissions of carbon dioxide and other GHGs to the atmosphere have occurred since the pre-industrial era of human civilization, leading to the phenomena of global warming and climate change.

The current findings of the IPCC concludes unequivocally that human influences have warmed the Earth (atmosphere, oceans, and land).

The observed and projected probability of adverse impacts of related global warming and climate changes on the Earth's systems, human populations, and ecosystem services have reached a critical point for global action by all countries to achieve net global emissions of carbon dioxide.

Multi-lines of evidence, scenarios, and risks evaluated by the IPCC include direct measurements of increasing atmospheric surface temperatures and sub-surface ocean temperatures, rising global mean sea levels, retreating glaciers, ocean acidification, intensifying extreme climate events, observed negative impacts or changes on natural ecosystems and wildlife (e.g. species shifts and migration patterns), and ecosystem services (crop yields, water resources, and forestry) (IPCC 2014a, b, 2021; NASA 2020b).

In recent decades, widespread changes in climate have caused strong impacts on many natural systems, and human systems on habitable continents and across the oceans. These impacts are observed widely in many regions throughout the world (IPCC 2014c, 2021). Extreme weather events were ranked as a top global risk by likelihood and impact in the Global Risks Report 2018 published by the World Economic Forum (2018).

There is a high to critical risk that many of the trends will accelerate known impacts and increase the risk of climate tipping points, leading to abrupt or irreversible climatic shifts. Major global risks from climate change based on global mean temperature projections, relative to the recent 1986–2005 baseline, and evaluation of the literature and expert judgments reported by the IPCC (2014c, p. 12, IPCC) are summarized in S14.6 in the Companion Book Online.

Impacts of climate change due to global temperature rises are already affecting the health of vulnerable populations around the world, mainly in low- and middle-income countries. For example, the evidence is clear that exposure to more frequent and intense heatwaves is increasing, with an estimated 157 million more vulnerable adults exposed to heatwaves in 2017 than in 2000. Risks from these impacts are expected to increase greatly with global warming (Watts et al. 2018a, b).

Ecosystem impacts from global warming and climate change are contributing to growing vector-borne diseases such as the transmission of dengue fever by *Aedes aegypti* (estimated 9·1% increase since 1950) and malaria, waterborne infectious diseases, reduced food security and human undernutrition. Rising incidence of extremes of weather, such as floods and storms, increase the risk of drowning and injury, damage to human settlements, spread of waterborne disease, and related impacts on mental health.

Aside from the health impacts of greenhouse gas emissions through global warming, the exposure to ambient air pollution, particularly fine particulates ($PM_{2.5}$) emitted by combustion processes, is a major cause of premature death and disease in primarily large, urbanized populations worldwide (Watts et al. 2018b). The IPCC AR6 report found that air quality could be improved more rapidly by targeting scenarios with reductions in air pollutant emissions rather than only GHG emissions. From 2040, it was projected that this could be improved further in scenarios that combine efforts to reduce both air pollutants and GHG emissions.

While there is increasing evidence of human intervention in addressing climate change symptoms, future risks, and impacts on ecological health and human health will be determined by (i) our collective responses and rates of mitigation to significantly reduce the current trend of increasing carbon emissions from combustion processes, and (ii) adaptation to global climate changes that are projected to continue to the end of the Twenty-first Century and well beyond.

14.10 Key Points

1. The scientific relationship between atmospheric carbon dioxide and global temperatures has been understood for over 100 years.
2. Greenhouse gases (GHGs), particularly carbon dioxide, as major anthropogenic pollutants that have accelerated the rate of global warming and induced climate changes observed on a global scale since the pre-Industrial Revolution era.
3. Climate is basically described as the long-term average of the weather at a given location. It is defined by (i) the average of weather events (e.g. temperature and precipitation), and by (ii) the type, frequency, duration, and intensity of weather events such as heatwaves, cold spells, storms, floods, and droughts.
4. The basic principle of global warming depends on the energy balance between the solar radiation of energy (UV, visible, and IR) from the Sun that warms the surface of the Earth and the thermal radiation (IR) from the Earth and the atmosphere that is emitted out into space.
5. This radiation balance can be changed by factors such as the intensity of solar energy, reflectivity of clouds or gases, and absorption by surfaces, materials, and certain gases known as greenhouses gases.
6. If the balance is disturbed by an increase in atmospheric GHGs, it can be restored by a heating effect that increases the Earth's surface temperature through the natural phenomenon of the *Greenhouse Effect*.
7. The *Greenhouse Effect* is the process by which solar radiation is absorbed by the planet's surface and is emitted into the atmosphere where some of it is absorbed by GHGs and re-emitted to warm the planet's surface to a higher temperature than would occur in the absence of an atmosphere or lack of natural GHGs.
8. The atmospheric process of global warming or cooling depends on the natural *greenhouse effect* and the rate of net changes in the Earth's energy balance measured by radiative forcings (rates of energy changes at the top of the atmosphere, $W\,m^{-2}$).
9. Radiative forcings result from natural (e.g. solar irradiance) and anthropogenic sources (e.g. GHGs, black carbon, and other aerosols).
10. The global energy budget for the *greenhouse effect* in the recent atmosphere is indicated in Figure 14.2a. Some of the incident solar radiation (30%) is reflected by the Earth's surface and atmosphere into space. The remaining radiation (70%) is absorbed by the Earth's surface (45%) and warming the atmosphere (25%).
11. Some of the thermal or infrared radiation emitted from the Earth's surface passes through the atmosphere while the rest is absorbed and re-emitted in all directions by greenhouse gas molecules and clouds. The result is to warm the Earth's surface and lower atmosphere.
12. Human activities since the Industrial Revolution, mainly the burning of fossil fuels and clearing of forests or groundcover, have increased the emissions of GHGs, primarily carbon dioxide, from the Earth's surface to the atmosphere, leading to an enhanced *greenhouse effect* and observations of an increased rate of global warming, compared to the pre-Industrial Revolution era.
13. The GHGs influenced directly by human activities are the long-lived carbon dioxide (CO_2), methane (CH_4), nitrous oxide (N_2O), synthetic fluorinated gases, such as chlorofluorocarbons (CFCs) and hydrofluorocarbons (HFCs). Another GHG is ozone (O_3), short-lived at ground level.
14. Water vapor is the most abundant and a potent GHG but short-lived, rapidly variable, and acts through cloud formation and fast climate feedback.
15. Atmospheric particulates also influence climate change by their warming and cooling effects through cloud formation, particle, and droplet sizes (indirect radiative forcing). Human-made aerosol particles include sulfate particles (SO_2 derived) and black carbon (soot). Aerosols can also induce a cooling effect in the atmosphere.

16. GHGs (e.g. water and carbon dioxide) absorb and emit infrared radiation because of their specific molecular properties (rotation and vibration). They are strong absorbers of emitted thermal radiation from the Earth at several wavelength windows of the IR spectrum.
17. The global carbon cycle shows the exchanges or fluxes of carbon between and within major reservoirs of the Earth (the atmosphere, oceans, land, and fossil fuels) and carbon storage within these reservoirs. In Figure 14.4, note changes and net increases in annual anthropogenic carbon fluxes and cumulative carbon masses in reservoirs (e.g. atmosphere) due to the industrial era since1750.
18. Atmospheric concentrations of GHGs have reached levels that are unprecedented in at least 800 000 years. Since 1750, GHGs have shown large increases and are *extremely likely* to be the primary drivers or forcing agents of global warming and climate change responses that are currently observed (IPCC 2014a). Average global CO_2 concentrations are at 410 ppm (2019), and now over this level.
19. In particular, carbon dioxide levels have increased by 40% since pre-industrial times, primarily due to emissions from fossil fuels and also from net land use changes.
20. In 2017, total global emissions of carbon dioxide gas from fossil fuel, industry, and land use reached 41.2 ± 2.8 Gt CO_2 (11.3 Gt C). The share of emissions from fossil fuels and industry was coal (40%), oil (35%), gas (20%), cement (4%), and flaring (1%) (Global Carbon Project 2019a).
21. In 2014, six countries (e.g. China, United States, European Union, Russian Federation, and Japan) generated 70% of CO_2 emissions from fossil fuel combustion and industrial sources such as cement production and gas flaring. Net global GHG emissions from agriculture, forestry, and other land uses accounted for about 24% of total GHG emissions.
22. Climate factors or variables that cause changes in the Earth''s energy balance (equilibrium) and alter the climate system, including surface temperature, are known as radiative or climate forcing agents.
23. These agents consist of natural and anthropogenic sources of forcings, which include GHGs, solar radiation, aerosols, and albedo. The climate forcing process is called radiative or climate forcing (see Figure 14.9). It is considered a useful tool or measure to compare the different factors (e.g. solar irradiance and GHGs) causing perturbations in the climate system.
24. Radiative forcing due to some factor or agent such as a gas is simply defined as the difference in the Earth's energy balance (radiative equilibrium) between incoming solar radiation (sunlight) and thermal IR energy radiated back into space. RF can have a positive (warming) or negative (cooling) effect on the Earth's surface.
25. Figure 14.10 shows radiative forcing (RF) estimates for natural and human-derived emissions and drivers in 2011 relative to 1750. They are rates of energy changes calculated at the top of the atmosphere, expressed in watts per square meter, $W\,m^{-2}$).
26. The total anthropogenic radiative forcing over 1750–2011 is calculated to be a warming effect of 2.3 (1.1–3.3) $W\,m^{-2}$ (see Figure 14.10), Overall, forcings have increased more rapidly since 1970 than during previous decades (high confidence) (IPCC 2014a, b).
27. While climate forcings are the initial drivers of climate, there are a number of known climate feedback processes occurring in the biosphere (e.g. water vapor feedback, and ice-albedo feedback) that can either amplify (positive feedback) or diminish (negative feedback) the effects of each climate forcing.
28. A *tipping point* in a climate system occurs when it abruptly moves from a relatively stable state to another state. This change may be irreversible. Examples projected under current rates of global warming include ocean circulation [e.g. disruption to Gulf stream in northern Atlantic Ocean), ice loss (e.g. Greenland ice cap), and rapid release of methane (e.g. warming of Tundra regions).
29. Climate sensitivity is the equilibrium temperature change in response to changes of the radiative forcing.
30. The global warming potential (GWP) of a greenhouse gas is known as a relative measure of how much heat a GHG absorbs or traps in the atmosphere. It is defined as its total warming effect over 100 years relative to the same amount of carbon dioxide gas.
31. Because the GWP of individual GHGs varies, a comparative measure has been derived, known as carbon dioxide equivalent. Carbon dioxide equivalent (CO_2-e) of a non-carbon dioxide gas = quantity of the gas x GWP of gas.
32. The main impacts of observed and projected climate change are identified as increases in surface temperatures (land and ocean) leading to heatwaves, sea-level rises, ocean acidification, and a more intense hydrological cycle associated with patterns of increasing precipitation, glacier, ice, snow and permafrost melt, and more intense and frequent floods, droughts, and storms in different regions. Climate change is also linked to increased environmental degradation by human activities (e.g. IPCC 2014a, 2021).

33. Climate models are used to construct and develop computer numerical modeling to predict future changes in atmospheric GHGs and climate responses due to the potential impacts of various anthropogenic forcings, including GHG emissions.
34. Emission scenarios are a set of descriptions of likely future emissions of GHGs that incorporate many of the major driving forces – including processes, impacts (physical, ecological, and socio-economic), and potential responses that are important for informing climate change policy (see IPCC 2014a, 2021).
35. The IPCC AR5 report used a set of emission scenarios known as representative concentration pathways (RCPs 2.6, 4.5, 6.0, and 8.5 in W m^{-2}) at different levels of radiative forcing.
36. In 2021, the AR6-WGI report adopted a set of "Shared Socioeconomic Pathways" (SSPs) to evaluate how the world might evolve without a climate policy and how different levels of climate change mitigation could be achieved when the mitigation targets of RCPs are combined with the SSPs.
37. Projected outputs from different modeling systems can then be compared by using a standard set of scenarios to provide a consistent set of starting conditions, historical data, and possible future emissions for use by climate scientists.
38. Projected changes in atmospheric carbon dioxide equivalent concentrations, global mean surface temperature, and mean sea levels using AR5 emission scenarios for climate models are summarized in Table 14.5.
39. The global mean surface temperature change (relative to 1850–1900) for the end of the Twenty-first Century (2081–2100) *is projected to likely exceed 1.5 °C for RCP4.5, RCP6.0, and RCP8.5 (high confidence). Warming is likely to exceed 2 °C for RCP6.0 and RCP8.5 (high confidence), more likely than not to exceed 2 °C for RCP4.5 (medium confidence), but unlikely to exceed 2 °C for RCP2.6 (medium confidence)* (IPCC 2014a).
40. Anthropogenic climate change is predicted to be a major cause of species extinctions in the next 100 years. It is already causing widespread local extinctions, and increasing impacts and serious threats on global biodiversity.
41. Increasing acidification of oceans is occurring due to the absorption of carbon dioxide gas from the atmosphere by seawater and formation of carbonic acid. pH stability of the ocean after hundreds of thousands of years has been reduced by 0.1 pH unit since the pre-industrial era.
42. The World Health Organization has clearly stated the strong link between climate change, air pollution, and health. A very conservative estimate of 250 000 additional deaths each year due to climate change (exposure to heat, droughts, floods, wildfires, and heatwaves) has been projected between 2030 and 2050.
43. There is a significant risk that many of the trends in climate variables will accelerate, leading to an increasing risk of abrupt or irreversible climatic shifts (see tipping points).
44. Serious to irreversible global risks exist from global warming caused climate change, as strongly indicated by global mean temperature projections, relative to the recent 1986–2005 baseline, and evaluation of the climate data, scientific literature, and expert judgments reported by the IPCC and other science-based agencies.

References

ACS (2020). *Greenhouse Gas Sources and Sinks*. American Chemical Society https://www.acs.org/content/acs/en/climatescience/greenhousegases/sourcesandsinks.html (accessed 25 May 2020).

Baird, C. and Cann, M. (2012). *Environmental Chemistry*, 5e. New York: W. H. Freeman.

Cahill, A.E., Aiello-Lammens, M.E., and Fisher-Reid, M.C. (2013). How does climate change cause extinction? *Proceedings of the Royal Society B: Biological Sciences* 280 (1750): https://doi.org/10.1098/rspb.2012.1890 (accessed 28 May 2020).

CarbonBrief (2018). Climate sensitivity. Explainer: How Scientists Estimate "Climate Sensitivity". CarbonBrief website, UK. https://www.carbonbrief.org/explainer-how-scientists-estimate-climate-sensitivity (accessed 8 June 2020).

Chivian, E. and Bernstein, A. (ed.) (2008). *Sustaining Life: How Human Health Depends on Biodiversity*. Oxford, UK: Center for Health and the Global Environment, Harvard Medical School, Oxford University Press.

Connell, D.W. and Miller, G.J. (1984). *Chemistry and Ecotoxicology of Pollution*. New York: Wiley.

Garnaut, R. (2008). *The Garnaut Climate Change Review. Final Report*. Cambridge: Cambridge University Press.

Global Carbon Project. (2019a). Supplemental Data of Global Carbon Budget 2019 (Version 1.0) [Data set]. Global Carbon Project. https://doi.org/10.18160/gcp-2019 (accessed 22 October 2020).

Global Carbon Project. (2019b). Global Carbon Budget 2019. Integrated Carbon Observation System. https://www.icos-cp.eu/science-and-impact/global-carbon-budget/2019 (accessed 22 October 2020).

Global Carbon Project (2019c). Global Carbon Budget Summary Highlights. Global Carbon Project website. http://cms2018a.globalcarbonatlas.org/en/content/global-carbon-budget (accessed 26 May 2020).

Global Carbon Project (2020). CO_2 Emissions. Global Carbon Project website. http://www.globalcarbonatlas.org/en/CO2-emissions (accessed 26 May 2020).

Hoegh-Guldberg, O., Jacob, D., Taylor, M. et al. (2018). Impacts of 1.5 °C global warming on natural and human systems. In: *Global Warming of 1.5 °C. An IPCC Special Report on the Impacts of Global Warming of 1.5 °C Above Pre-industrial Levels and Related Global Greenhouse Gas Emission Pathways, in the Context of Strengthening the Global Response to the Threat of Climate Change, Sustainable Development, and Efforts to Eradicate Poverty* (ed. V.P. Masson-Delmotte, H.-O. Zhai, D. Pörtner, et al.). Chapter 3. Geneva: IPCC.

Houghton, J. (2015). *Global Warming: The Complete Briefing*, 5e. Cambridge: Cambridge University Press.

IPCC (2013a). Climate change 2013: the physical science basis. In: *Contribution of Working Group I to the Fifth Assessment Report of the Intergovernmental Panel on Climate Change* (ed. T.F. Stocker, D. Qin, G.-K. Plattner, et al.). Cambridge, UK and New York, NY: Cambridge University Press.

IPCC (2013b). Summary for policymakers. In: *Climate Change 2013: The Physical Science Basis. Contribution of Working Group I to the Fifth Assessment Report of the Intergovernmental Panel on Climate Change* (ed. T.F. Stocker, D. Qin, G.-K. Plattner, et al.). Cambridge, UK and New York, NY: Cambridge University Press.

IPCC (2014a). Climate change 2014: synthesis report. In: *Contribution of Working Groups I, II and III to the Fifth Assessment Report of the Intergovernmental Panel on Climate Change* (ed. Core Writing Team, R.K. Pachauri and L.A. Meyer). Geneva: IPCC.

IPCC (2014b). Climate change 2014: mitigation of climate change. In: *Contribution of Working Group III to the Fifth Assessment Report of the Intergovernmental Panel on Climate Change* (ed. O. Edenhofer, R. Pichs-Madruga, Y. Sokona, et al.). Cambridge, UK and New York, NY: Cambridge University Press.

IPCC (2014c). Summary for policymakers. In: *Climate Change 2014: Impacts, Adaptation, and Vulnerability. Part A: Global and Sectoral Aspects. Contribution of Working Group II to the Fifth Assessment Report of the Intergovernmental Panel on Climate Change* (ed. C.B. Field, V.R. Barros, D.J. Dokken, et al.), 1–32. Cambridge, UK and New York, NY: Cambridge University Press.

IPCC (2021). Summary for policymakers. In: *Climate Change 2021: The Physical Science Basis. Contribution of Working Group I to the Sixth Assessment Report of the Intergovernmental Panel on Climate Change* (ed. V. Masson-Delmotte, P. Zhai, A. Pirani, et al.). Cambridge University Press. In Press.

IPCC/TFI (2020). *The Task Force on National Greenhouse Gas Inventories (TFI)*. Intergovernmental Panel on Climate Change https://www.ipcc.ch/working-group/tfi/ (accessed 26 May 2020).

Kump, L.R., Kasting, J.F., and Crane, R.G. (2010). Global energy balance: the greenhouse effect. Chapter 3. In: *The Earth System*, 3e. San Francisco, CA: Prentice Hall.

Myhre, G., Shindell, D., Bréon, F-M. et al. (2013). Anthropogenic and natural radiative forcing. In: *Climate Change 2013: The Physical Science Basis. Contribution of Working Group I to the Fifth Assessment Report of the Intergovernmental Panel on Climate Change* (ed. T.F. Stocker, D. Qin, G.-K. Plattner, M. Tignor, et al.). Cambridge, UK and New York, NY: Cambridge University Press.

NASA (2020a). *The Study of Earth as an Integrated System*. NASA Science. National Aeronautics and Space Administration https://climate.nasa.gov/nasa_science/science/ (accessed 24 May 2020).

NASA (2020b). *Scientific Consensus: Earth's Climate is Warming*. National Aeronautics and Space Administration https://climate.nasa.gov/scientific-consensus/ (accessed 29 May 2020).

NOAA Climate.gov (2015). *What's the Difference Between Global Warming and Climate Change?* National Oceanic and Atmospheric Administration https://www.climate.gov/news-features/climate-qa/whats-difference-between-global-warming-and-climate-change (accessed 17 June 2018).

NOAA Climate.gov (2020). *Climate Change: Atmospheric Carbon Dioxide*. National Oceanic and Atmospheric Administration https://www.climate.gov/news-features/understanding-climate/climate-change-atmospheric-carbon-dioxide (accessed 27 May 2020).

NRC (2005). *Radiative Forcing of Climate Change: Expanding the Concept and Addressing Uncertainties*. Washington, DC: National Research Council of the National Academies, Division on Earth and Life Studies, Board on Atmospheric Sciences and Climate, Committee on Radiative Forcing Effects on Climate, The National Academies Press.

Pittock, A. (2005). *Climate Change. Turning up the Heat*. London and VIC, Australia, EARTHSCAN, and Collingwood: CSIRO Publishing.

Schmidt, M.W. and Hertzberg, J.E. (2011). Abrupt climate change during the last ice age. *Nature Education Knowledge* 3 (10): 11.

Schneider, S. (1987). Climate modeling. *Scientific American* 256 (5): 72–80.

Steffen, W., Burbridge, A.A., Hughes, L. et al. (2009). *Australia's Biodiversity and Climate Change*. Collingwood, VIC: CSIRO Publishing.

US EPA (2016). *Greenhouse Gas Inventory Guidance – Direct Emissions from Stationary Combustion Sources*. US Environmental Protection Agency https://www.epa.gov/sites/production/files/2016-03/documents/stationaryemissions_3_2016.pdf (accessed 20 May 2018).

US EPA (2018). *Greenhouse Gas Emissions: Overview of Greenhouse Gases*. US Environmental Protection Agency (accessed 26 May 2020).

US EPA (2019). Greenhouse Gas Emissions: Global Greenhouse Gas Emissions Data. https://www.epa.gov/ghgemissions/global-greenhouse-gas-emissions-data (accessed 26 May 2020).

Van Vuuren, D., Edmonds, J., Kainuma, M. et al. (2011). The representative concentration pathways: an overview. *Climatic Change* 109: 5–31.

Watts, N., Adger, W.N., Ayeb-Karlsson, S. et al. (2017). The *Lancet* countdown; tracking progress on health and climate change. *Lancet* 389 (10074): 1151–1164.

Watts, N., Amann, M., Ayeb-Karlsson, S. et al. (2018a). The *Lancet* countdown on health and climate change: from 25 years of inaction to a global transformation for public health. *Lancet* 391 (10120): 581–630.

Watts, N., Amann, M., Arnell, N. et al. (2018b). The 2018 report of the *Lancet* countdown on health and climate change: shaping the health of nations for centuries to come. *Lancet* 392 (10163): 2479–2514.

WHO (2018). *Health & Climate Change. COP24 Special Report*. Geneva: World Health Organization https://www.who.int/globalchange/publications/COP24-report-health-climate-change/en/ (accessed 28 May 2020).

WMO (2016). *Observing Water Vapour*. World Meteorological Organization https://public.wmo.int/en/resources/bulletin/observing-water-vapour (accessed 25 May 2020).

Woodwell, G.M. (1978). The carbon dioxide question. *Scientific American* 238: 34–43.

World Economic Forum (2018). *The Global Risks Report 2018*, 13e. Geneva: World Economic Forum http://www3.weforum.org/docs/WEF_GRR18_Report.pdf (accessed 29 May 2020).

15

Soil and Goundwater Pollution

15.1 Introduction

This chapter examines soil and groundwater contaminants and related pollution as emerging and critical components of land and water resources degradation on local to global scales. Soil pollution has become a major pollution problem and acts as a long-term environmental health threat in many parts of the world, especially from the direct and indirect disposal of untreated or poorly treated liquid and solid wastes, intensive use of agricultural chemicals, and dispersion of toxic chemicals in urbanized, industrial, agricultural, or mining areas.

Soil pollution is described in the *Status of the World's Soil Resources Report* (FAO/ITPS 2015) as *an alarming issue*. This report has identified it as:

> ... the third most important threat to soil functions in Europe and Eurasia, fourth in North Africa, fifth in Asia, seventh in the Northwest Pacific, eighth in North America, and ninth in sub-Saharan Africa and Latin America
>
> (FAO/ITPS 2015).

The problems of soil pollution consist largely of the exposure of human populations, agricultural and natural ecosystems to soil and groundwater contaminants via direct contact, inhalation of dusts and vapors, and contaminated food chains, leading to potential adverse environmental and health effects for humans, crops, livestock, and wildlife. Soil and groundwater contamination may originate from human activities, and in some cases, natural biogeochemical sources.

Contaminated sites are known to exist in the thousands to tens of thousands in many developed countries and are a major environmental challenge for developing countries. Potential contaminated sites in the European Economic Area (EEA-39) are estimated to be about 3 million (EEA 2014). In some countries such as China, large areas of agricultural soils are affected by soil contamination. For instance, about 20% of China's total farmland is affected by heavy metal contamination (e.g. Cd, Ni, and As). Examples of point sources include cesium pollution from the Fukushima Dai-ichi nuclear power plant (Japan) and the Chernobyl disaster of 1986 in the Ukraine (FAO/ITPS 2015, p. 178). Regional assessments for soil contamination show that the trend for contamination is improving in Europe, North America, Australia, and New Zealand (FAO/ITPS 2015, p. 29).

The terms, *soil contamination* and *soil pollution,* are used widely in the literature and also in this chapter. The Intergovernmental Technical Panel on Soils (ITPS) under the Global Soil Partnership (GSP) has formalized definitions of the two terms as follows (Rodríguez-Eugenio et al. 2018, p. VIII):

- *Soil contamination* occurs when the concentration of a chemical or substance is higher than would occur naturally but is not necessarily causing harm.
- *Soil pollution*, on the other hand, refers to the presence of a chemical or substance out of place and/or present at a higher than normal concentration that has adverse effects on any non-targeted organism.

The diverse nature of potential and known soil and groundwater contaminants, key environmental properties, behavior and fate, toxicity, environmental and human health effects, and important contaminated land management approaches (e.g. site remediation) are described and evaluated in this chapter.

15.2 Soil and Groundwater Systems and Key Environmental Properties

Soils and groundwaters are interlinked structural and functioning components of the Earth's living land systems. Their physical nature, structure, and chemical composition are under continuous and variable change due to the Earth's climate and interactions of biogeochemical cycles. Human activities and natural contamination of

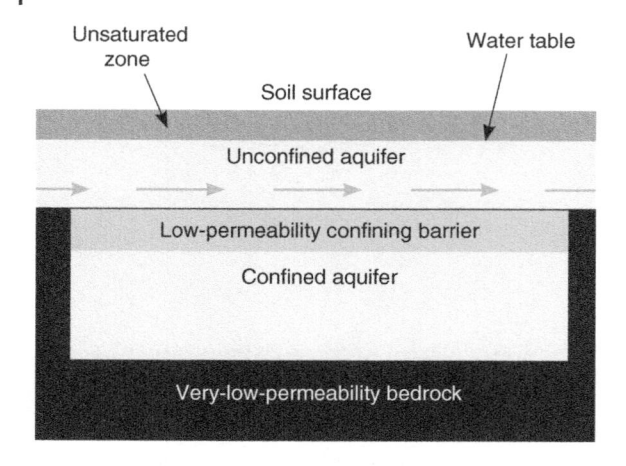

Figure 15.1 Conceptual model of a basic soil–groundwater system showing a simple soil surface and subsurface profile over bedrock with examples of an unsaturated subsurface (valdose) zone above unconfined and confined aquifers (groundwaters).

soils and groundwaters are major components of land and catchment degradation and related pollution.

A conceptual model of a basic physical soil–groundwater system is shown as Figure 15.1.

Soil is the uppermost layer of unconsolidated material or sediment covering the Earth's crust, derived from combinations of weathering and erosion of parent rock and unconsolidated sediment, followed by transport, deposition, accumulation of materials, and further changes due to various weathering processes and biological activity.

The resultant material or soil is a heterogeneous mixture of inorganic materials, organic matter, water, and living organisms that can vary widely in physical structure (soil horizons), composition, and distribution over large surface areas, and with subsurface depth to bedrock.

The soil matrix consists of a mixture of solid particles with pore or void spaces between, that make up a three-phase system: solids, water, and air. Porosity (n) is a measure of the void space between the solid particles, expressed as a percentage (0–100%) or fraction (0–1) of the void volume over the total volume. Bulk density of a soil is ρ_b = mass of solids/total volume of soil (kg m^{-3}).

The soil gas concentration (e.g. methane) is usually reported as ppmv (parts per million by volume) using the following relationship:

$$C_G \text{ (ppmv)} = (C_G \text{ (mg L}^{-1}) \times 24\,000 \text{ mL mole}^{-1}) / (\text{molecular weight, g mole}^{-1}) \quad (15.1)$$

The total contaminant concentration in soil is given by:

$$C_T = \rho_b C_S + \theta_W C_W + \theta_G C_G \quad (15.2)$$

(θ is used for the water or gas porosity of soil so that $n = \theta_W + \theta_G$)

The mineralized component of soils consists essentially of granular to fine mineral grains and colloidal particles that are subdivided according to particle sizes (see Table 15.1).

Soil texture is classified by measuring the relative proportion of gravel, sand, silt, and clay in its physical composition. The major classes of soil (sand, silt, and clay) are commonly indicated in the form of a soil texture triangle, as shown in Figure 15.2.

Soils consist largely of inorganic minerals (e.g. silicates, quartz, calcite, iron, and manganese oxides) and salts formed from major elements (Si, Fe, K, Ca, and Mg) and trace elements (e.g. B, Cl, Fe, Mn, Co, Cu, Mo, Ni, and Zn) derived from the geochemistry of parent rock materials, degraded organic matter, and wet and dry deposition from the atmosphere, as part of biogeochemical cycling. Organic matter in soils consists basically of decomposed plants (plant detritus or humus) together with microbial communities. Soil organic matter (SOM) usually ranges in content from 0.1 to 3–7% of the total soil. Common organic components include plant lignin, phenolic compounds, humic and fluvic acids. The organic matter helps stabilize soils by binding inorganic particles together to form variable aggregates of sand, silt, clay, and organic matter (LaGrega et al. 2001, p. 195).

Influence of Environmental Properties of Soils and Groundwater Systems on Contaminants

A combination of the physicochemical properties of contaminants and environmental properties of soils and

Table 15.1 Soil texture and some physical properties of soils.

Major classes of soil texture	Particle size (diameter, mm)	Porosity (%)	Permeability (hydraulic conductivity) (cm s^{-1})	Water and nutrient holding capacity
Gravel	>2 to <75	~20 to 40	1×10^5 to 1.0	Poor
Sand	>0.075 to <2	~22 to 46	1.0 to 1×10^{-3}	Poor
Silts	>0.002 to <0.075	~20 to 68	1×10^{-2} to 1×10^{-6}	Medium
Clay	<0.002	~40 to 75	1×10^{-5} to 1×10^{-9}	Good

Source: LaGrega et al. (2001, p. 172 and p. 194); Wright and Boorse (2014).

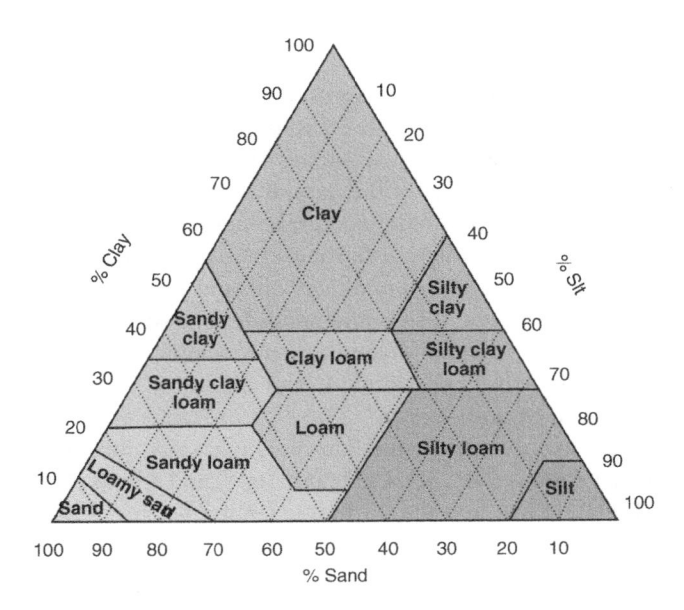

Figure 15.2 Generic soil texture triangle.

groundwaters strongly influences the concentrations, mobility, distribution, and persistence of contaminants that infiltrate these environmental media due to human activities and natural sources. Significant physical, chemical, and microbial properties of soils and groundwaters are summarized below.

Soil Systems
- Key physical properties of soils that affect the flow of groundwater and mobility of contaminants through subsurface soils are particle sizes, particle surface areas, porosity, permeability, and sorption. Very high to medium mobility can be expected through gravels and sands while silts and clays with fine to colloidal size particles, high surface area ratios, and low permeability would have slow to poor mobility for contaminants. Silts and clays also have the capacity to retain contaminants through their relatively higher porosity and related water holding capacity, as indicated in Table 15.1.
- In particular, sorption properties play an important role in soils containing clay and organic matter. Clay consists of various hydrous silicates and oxides that can be characterized by such measures as cation exchange capacity (CEC) and the specific surface area ($m^2\,g^{-1}$). These properties give measures of how clay affects the behavior of metal ions and polar molecules in soils. Metal ions such as Cu, Zn, and Ni are readily sorbed onto these particles which possess high surface areas. As such, CEC is a measure of the capacity of a soil to sorb cations with which it comes into contact, depending upon factors such as pH, clay, and organic matter content.
- Very polar cationic pesticides such as paraquat and diquat are strongly sorbed onto clay particles despite being very soluble in water. Their sorption onto clay is sufficiently strong to overcome the highly lipophilic properties of such compounds. Another widely used pesticide, glyphosate, shows similar properties. However, its molecule exists as a zwitterion, with several sites for cationic and anionic effects on its molecular structure (Connell 2005, p. 327).
- Organic matter in soils, including plant lignin (complex phenolic compound of high molecular weight) and humic substances (fulvic and humic acids), exhibits much stronger sorption than clays to both polar and lipophilic organic contaminants, such as various pesticides, POPs, and benzo(a)pyrene. The influence of organic carbon in soil is important. This property is indicated by the high log K_{OW}, log K_{OM}, or log K_{OC} values observed for these types of contaminants. The K_{OM} value, for example, is the ratio of the concentration in SOM to concentration in water at equilibrium (see Section 15.5).
- Fulvic and humic acids contain a range of phenolic groups that can form chelating complexes with metal ions that are relatively stable and resistant to breakdown. As well, they possess substantial lipophilic properties attributed to the large size of their organic molecules.
- Among other functions, soil microbial communities play an essential role in the decomposition of plant and other living matter to form the complex chemical mixtures of organic matter found in most soils. Many organic contaminants in soils and water are also degraded by the same processes but rates for individual chemicals may extend from days to years.
- Other soil properties that affect various contaminants such as metal species include pH, oxygen content, redox potential, salinity, carbonate, and sulfide content.

Groundwaters
Some of the key environmental processes affecting contaminants in groundwater are advection, sorption, and biological degradation:

- *Advection* occurs when contaminants move or migrate with the groundwater. A continuous source of dissolved contaminants usually moves through groundwater in the form of a plume consisting of diluted contaminants.
- *Sorption* occurs when contaminants adsorb to soil particles. Sorption reduces the mobility of contaminants in groundwater and increases their persistence.
- *Biological degradation* of organic contaminants such as pesticides or petroleum hydrocarbons can also occur due to the actions of microorganisms, such as bacteria and fungi, which can use carbon-containing organic

contaminants as substrates for food and energy. Rates of biodegradability are also influenced by the interactions of environmental factors such as groundwater flow, types of subsurface media in the aquifer, availability of nutrients, aerobic or anaerobic conditions, pH, and water temperature.

Section 15.5 discusses the general transport and transformation of contaminants in soil and groundwater systems.

15.3 Types and Properties of Soil and Groundwater Contaminants

The dominant chemical contaminants in soils and groundwater are generally heavy metals, hydrocarbons (mineral oils, PAHs, BTEX, and chlorinated hydrocarbons), excess macronutrients (N and P forms), POPs, and pesticide residues. Table 15.2 presents an overview of major types of contaminants, including emerging chemicals, such as antibiotics, microplastic residues, PFOS and PFOA, and asbestos, pathogens and antimicrobial-resistant bacteria and genes found in soils and/or groundwaters in variable concentrations.

From Table 15.2, it is evident that many inorganic chemicals can act as soil groundwater pollutants (e.g. heavy metals, nonmetals, and excess nutrients) under various environmental and elevated exposure conditions. Many toxic organic chemical contaminants are found in soils and groundwaters. Some common and emerging examples are compiled in Table 15.3 to help explain in this chapter how important physicochemical properties influence their behavior and fate in soil and groundwater environments. The nature of these and other chemical properties are described in other chapters of this book.

Section 15.5 discusses the main principles governing the environmental behavior and fate of inorganic and organic contaminants in soils and groundwaters.

15.4 Sources and Releases of Soil and Groundwater Contaminants

Soil and groundwater contaminants originate from natural biogeochemical and anthropogenic sources via point or local sources (e.g. sewage discharges, stack emissions, wells, or volcanoes) or diffuse emissions from dispersed (area) sources (e.g. applications of pesticides, excessive nutrients or untreated biosolids to crop lands, mine tailings, oil spills, atmospheric deposition, and flooding). Common toxic contaminants found in soils and groundwaters are heavy metals, arsenic, various petroleum hydrocarbons (BTEX and PAHs), persistent pesticides residues, and POPs.

Natural Sources

Natural sources of contaminants in soils and groundwaters are derived from mineralized soils formed from parent materials and geological substrates via groundwater flows and the water quality of the natural recharge waters. The types and concentrations of natural contaminants found in soils and groundwaters are strongly influenced by climatic conditions (e.g. arid, humid, hot, wet, and cold), environmental properties of soils and groundwaters (e.g. mineral composition, organic matter content, acidity, aerobic or anaerobic state, and salinity), and also flow rates and storage of groundwaters.

The main contaminants are inorganic salts, including major cations (Na^+, K^+, Ca^{2+}, and Mg^{2+}) and anions (Cl^-, HCO_3^-, CO_3^{2-}, and SO_4^{2-}), fluoride, nutrients, heavy metals, arsenic, selenium and other trace elements (e.g. boron and sulfur), and radionuclides. Metals are usually found at low dissolved levels in goundwaters but high levels of toxic metals (e.g. copper, zinc, and lead) can occur in soils associated with mineralized ore deposits. Natural arsenic and fluoride contamination of groundwaters cause major environmental health problems in various regions of the world. Potentially toxic and carcinogenic arsenic is derived from volcanic deposits, arsenic-containing ores and minerals, and weathering of some sulfide-bearing rocks. Harmful levels of fluoride occur in various groundwaters from dissolution of natural fluoride-containing salts and minerals such as fluorite. High natural radioactivity is common in acidic igneous rocks, mainly in feldspar-rich rocks and illite-rich rocks, and low levels may be found in groundwaters through erosion and decay. Mineral sand deposits containing thorium (monazite) emit hazardous radioactivity. Soils and rocks can also be natural sources of the radioactive gas, radon (Rn) which is hazardous when inhaled (human carcinogen) in elevated doses. Naturally occurring asbestos minerals originate from fibrous silicate minerals that occur in soils formed from ultramafic rock, especially serpentine and amphibole. Inhalation of airborne respirable fibers is also hazardous (human carcinogen). Exposures to respirable free silica may also result from deposits of fine silica sands. Organic contaminants can originate at low to trace levels from natural sources such as crude oil seeps and coal seam gas leaks (methane gas, benzenes, and PAHs), volcano eruptions, and as combustion by-products in forest fires (PAHs and dioxin-like compounds) (see FAO/ITPS 2015, p. 6).

Anthropogenic Sources

Major sources of soil pollutants from human activities are outlined in Figure 15.3. Soil (and groundwater) contamination occurs where intensive industrial activities,

Table 15.2 Major types of soil and groundwater pollutants.

Pollutant types and examples	Description
Metals/metalloids Pb, Cd, Cu, Cr (VI), Hg, Ni, Zn, V, As, Sb, Se	Heavy metals, including As, Sb, and Se, are very persistent in soils (and groundwaters) and can readily accumulate in living organisms, depending upon metal forms and bioavailability. They occur naturally at low levels in soils and many are essential micronutrients. At increased levels, they can be toxic to living organisms, including phytotoxic to plants, and also lead to food chain contamination. Metal bioavailability is influenced by soil properties (e.g. pH, ion exchange capacity, and organic matter content), environmental conditions, contact time, and biological uptake.
Nonmetals/ionic forms Fluoride, boron, cyanide, sulfur	Several nonmetallic species (fluoride, cyanide, boron, chlorine, and sulfur) are important soil and groundwater contaminants from natural and anthropogenic sources (e.g. fluoride in soil and groundwater is mainly released by weathering of fluoride-containing minerals such as fluorite and fluorapatite, while anthropogenic sources include aluminum smelters and use of phosphorus fertilizers). Key factors for these pollutants in soils are leaching, enrichment or concentration, sorption capacity of soils, and mobility in soils and waters.
Radionuclides Strontium-90 (B$^-$), Cesium-134 and 137, Plutonium-239	The most common natural and anthropogenic radionuclides found in soils are ^{40}K, ^{238}U, ^{232}Th, ^{90}Sr, and ^{137}Cs. Radionuclides are readily taken up by plants from soils. Emission of ionizing radiation during nuclear decay is the main contamination route of radionuclides.
Nitrogen and phosphorus Nitrates, phosphates, ammonia, amino acids proteins	Nitrogen and phosphorus are macronutrients that are essential for all living organisms. Excessive application of synthetic N and P fertilizers, biosolids, human or animal wastewaters to soils, or leaching to groundwaters can reduce crop yields, decompose soil organic matter, cause acidification of agricultural soils, eutrophication of waters, including deoxygenation, high nitrate levels in groundwaters, and increased salinity.
Mineral dusts Asbestos, free crystalline silica	Disturbance of friable asbestos materials, wastes, or residues disposed in soils can pose a significant risk of exposure to airborne asbestos fibers that are known to be human carcinogens. Similarly, inhalation exposures to silica dusts from occupational and nonoccupational sources can lead to silicosis.
Acid sulphate soils (ASS) Acidity and toxic metals, including Al	Potential acid sulfate soils are natural waterlogged soils and sediments containing iron sulfides (pyrite). When these soils are exposed to air (e.g. excavated or drained), the reduced iron sulfide minerals can oxidize and the soils can acidify (pH < 4) due to the formation of sulfuric acid. Upon acidification, these acid sulfate soils pose a significant environmental risk because of the low pH of the soil and leaching and transport of the acidity/toxic metals into groundwaters or surface waters.
Petroleum hydrocarbons (crude and refined oils, C_1–C_4 gases, fuel additives) Aliphatic and aromatic hydrocarbons: C_1–C_{40} range, benzenes (BTEX), polycyclic aromatic hydrocarbons (PAHs)	Petroleum hydrocarbons can readily contaminate soils, sediments, and groundwaters through discharges, spills, seepage, and leaching. Aromatic BTEX and PAH fractions in contaminated soils are potentially toxic to exposed wildlife and humans through inhalation, ingestion, and skin absorption. Low boiling BTEX and 2–3 ring PAHs are relatively volatile in soils and soluble at low levels in water, decreasing with higher carbon numbers. Their persistence may generally extend from hours to many years, increasing with increasing carbon number and boiling point.
Industrial chemicals (Halogenated and nonhalogenated chemicals)	Halogenated and nonhalogenated solvents, surfactants, phenolics, biocides, and cyanides are among numerous industrial chemicals used in industry, commerce, and/or households that are released to the soil, air, and water environments through use, disposal, spills, and leaks.
Polycyclic aromatic hydrocarbons (PAHs) Naphthalene, anthracene, pyrene, benzo(a)pyrene, benz(a)anthracene, alkyl derivatives	PAHs consist of two or more fused benzene rings by sharing a pair of carbon atoms between them. They are produced by the incomplete combustion of organic matter such as fossil fuels and plant matter. PAHs accumulate and persist in soils because of their hydrophobicity and resistance to biodegradation. Low-molecular-weight PAHs (2 or 3 rings) are volatile.
Pesticides (insecticides, herbicides, and fungicides)	Pesticides are classified on the basis of their chemical structures, their mode of action, their way of entry into the body, their target organisms, and toxic effects related to their chemical composition. Examples include DDT, dieldrin, endosulfan, dicofol, chlorpyrifos, fenthion, aldicarb, cypermethrin, imidacloprid, glyphosate, 2,4-D, paraquat, atrazine, benomyl, triphenyl tin acetate, zineb, and copper-based fungicides. As contaminants, their persistence, behavior, mobility, and bioavailability in soils and waters depend greatly on their chemical nature and interactions with environmental processes and prevailing conditions.

(Continued)

Table 15.2 (Continued)

Pollutant types and examples	Description
Persistent organic pollutants (POPs) DDT, HCB, PCBs, polychlorinated dibenzodioxins and furans, perfluorooctane sulfonic acid (PFOS), polybrominated diphenyl ethers (PBDEs)	POPs are a subgroup of toxic chemical substances released from anthropogenic sources that persist in the environment, bioaccumulate through the food chain, and have adverse effects on human health and the environment. They are hydrophobic and lipophilic compounds that show a high affinity for binding in soil organic matter. Soils are the main environmental sink for these persistent pollutants but POPs also exhibit high mobility via volatilization from soils into the atmosphere and penetration into water in its gaseous phase during warm weather.
Plastics Monomers (e.g. styrene) plasticizers (e.g. bisphenol A), microplastic residues/particles	Emerging but limited evidence of microplastic particles in agricultural soils from biosolids/sludges and mulching of plastic wastes (e.g. Ng et al. 2018). Microplastics also release additives and decomposition products: flame retardants (e.g. PBDEs) and plasticizers (e.g. bisphenol A and phthalate esters).
Pharmaceutical and personal care products (PPCPs) Drugs and metabolites, veterinary chemicals, diagnostic agents, cosmetics, fragrances, biocides	PPCPs are a major class of emerging chemical contaminants of more than 4000 pharmaceutical and chemical products, including antimicrobial agents that can promote bacterial resistance in the environment. Bioactive residues of PPCPs, including drug metabolites and antibiotics, are widely distributed in agricultural soils and leached into groundwaters, through applications of urban wastewaters, animal manures, sludges, and biosolids. Crop plants, for example, can selectively take up certain antibiotics (tetracyclines, amoxicillin, and fluoroquinolones) from soils treated with contaminated wastewaters.
Synthetic nanoparticles (inorganic and organic forms) Carbon based (fullerenes, nanotubes, carbon black), metal oxides (SiO_2, TiO_2, Fe_2O_3, Fe_3O_4, ZnO), semiconductors (e.g. cadmium-tellurite CdTe, indium phosphide InP), metals (Au, Ag, Fe, Co)	These emerging pollutants are being rapidly developed and dispersed widely in many new products and materials. They vary in physical properties (e.g. size, shape, surface charge, surface-to-volume ratio, sorption, and optical properties), chemical composition and reactivity, and applications. Knowledge of their interactions with soils, including their movement, fate, persistence, and bioavailability is currently limited and likely to be somewhat complex, given their diversity and the nanoscale or colloidal range of many soil particles. Synthetic nanoparticles are known to undergo surface modifications by humic acids, interactions with cations and dissolution, which may control their fate in the environment.
Pathogenic microorganisms	Fecal pathogens (e.g. salmonella) excreted by warm-blooded animals (e.g. manure) and humans (e.g. wastewaters, urban stormwaters, landfill leachates, and biosolids) occur widely in soils and groundwaters and can contaminate food plants and livestock food products.
Antimicrobial-resistant bacteria and genes	Bacteria developing natural or acquired resistance to antibiotics are known to be selectively favored in humans, animals, or environments in which antibiotics are used. Urban and agricultural soil environments are increasingly susceptible to disposal of antibiotic residues in urban wastes, manures, and contaminated runoff from intensive livestock farming.

Source: Compiled from FAO/ITPS (2015); Rodríguez-Eugenio et al. (2018); Ng et al. (2018); and other Chapters (e.g. 2, 9, 10, 11, and 12).

urbanization, inadequate waste disposal, mining and petroleum operations, agriculture, and military activities or accidents introduce excessive amounts of contaminants to the soil environment (FAO/ITPS 2015, p. 119). The main soil pollutants are the chemicals used in or produced as by-products of industrial activities, domestic and municipal wastes, including wastewater, agrochemicals, and petroleum-derived products. These chemicals are released to the environment accidentally, for example, from oil spills or leaching from landfills, or intentionally, as is the case with the use of fertilizers and other agricultural pesticides, irrigation with untreated wastewater, or land application of sewage sludge and other untreated biosolids (Rodríguez-Eugenio et al. 2018, p. 7).

In most developed countries, the key sources of local soil pollution are reported to be waste disposal and treatment, industrial and commercial activities, storage and transport spillages, military activities, and nuclear operations (FAO/ITPS 2015, p. 119). Common soil contaminants include inorganic compounds such as metals, metalloids, radionuclides, nonmetals (fluoride, boron, and sulfur) and cyanides, and organic compounds including petroleum hydrocarbons, PAHs, POPs, phenolics, and synthetic pesticides.

Diffuse contamination from pollution sources tends to be dominated by excessive nutrient and pesticide applications, heavy metals, arsenic, POPs, and some other organic and inorganic contaminants (FAO/ITPS 2015, p. 119). Fluorides in groundwaters from natural sources are also a major problem in many countries. Intensive application of sewage, wastewaters, and untreated biosolids to agricultural areas also increases the risk of spread of infectious diseases.

Table 15.3 Some physicochemical properties of various toxic organic chemicals.

Chemical	Molecular weight (g mole^{-1})	Water solubility mg L^{-1} 25 °C	pK$_a$	Log K$_{OW}$	Vapor pressure mm Hg or (Pa) 25 °C	Henry's Law constant atm-m^3 mol^{-1} 25 °C	Ref.
p,p′ DDT	354.5	0.025	—	6.91	1.60×10^{-7} 20 °C	8.3×10^{-6}	(1)
PCB-153	326.433	2.91×10^{-2}	—	6.98 (est)	2.22×10^{-6}	1.015×10^{-4}	(3)
TCDD	321.971	2×10^{-4}	—	6.8	1.50×10^{-9}	5.00×10^{-5}	(2) (3)
PFOS	500.126	3.2×10^{-3}	<1.0	4.49 (est)	2.0×10^{-3}	4.34×10^{-7} (6)	(2) (6)
PBDEs: Decabromo-diphenyl ether	959.2	$<1 \times 10^{-3}$	—	6.265	3.2×10^{-8}	1.62×10^{-6}; 1.93×10^{-8}; 1.2×10^{-8}; 4.4×10^{-8}	(8)
Atrazine	215.7	3.47 (26 °C)	1.60	2.61	2.89×10^{-7}	1.093×10^{-7} 20 °C	(2) (5)
Glyphosate	169.07	12 000	0.8	−3.40	9.8×10^{-8}	5.819×10^{-11} 20 °C	(2) (5)
Chlorpyrifos	350.6	1.4	—	4.96	2.02×10^{-5}	1.729×10^{-4} 20 °C	(2)
Benzene	78.114	1790	—	2.13	94.8	5.55×10^{-3}	(2) (3)
Toluene	92.141	526	—	2.73	28.4	5.104×10^{-3}	(2) (3)
Benzo(a)pyrene	252.32	4×10^{-3}	—	6.0	5.49×10^{-9} (2)	4.697×10^{-5} 20 °C	(7)
Perfluorooctanoic acid (PFOA)	414.07	3300	−0.5–4.2	4.81 (est)	3.16×10^{-2}	3.044 (3)	(2) (3)
Diethyl phthalate	222.24	1080	—	2.47	2.1×10^{-3}	1.87×10^{-5} 20 °C	(2)
Bisphenol A	228.291	120 350	9.6	3.32	4.0×10^{-8}	2.17×10^{-10} (5)	(2) (3) (5)
4-Nonylphenol	220.35	7	10.7 +/− 1.0 (4)	5.76	0.109 (Pa) 0.0+/−0.7 mmHg	3.40×10^{-5}	(2), (3) (4)
N-Nitrosodimethyl-amine (NDMA)	74.08	Miscible	—	−0.57	2.7	2.63×10^{-7}	
Triclosan	324.637	2.317×10^{-7}	—	11.64 (est)	1.74×10^{-5}	51.3–600 (est)	(2)
Ciprofloxacin	331.347	30 000	6.09 (COOH group) 8.74 (N on piper-azinyl ring)	0.28 (non-ionized)	2.85×10^{-13}	1.082×10^{-17} (est) (3)	(2) (3)
Acetaminophen (paracetamol)	151.165	14 000	9.38	0.46	6.29×10^{-5}	1.271×10^{-12} (est)	(2) (3)
Sulfadimethoxine	310.328	343	—	1.63	1.49×10^{-9}	1.405×10^{-12} (est)	(2) (3)
Benzophenone-3 (oxybenzone)	228.247	69 (est)	—	3.79	1.4×10^{-6}	2.9×10^{-8}	(2) (3)
Methyl paraben (methyl 4-hydroxy benzoate)	152.149	2500	8.5 (est)	1.96	2.37×10^{-4}	2.86×10^{-8}	(2) (3)

(1) ATSDR (2019)
(2) Pubchem website
(3) Chemspider website
(4) Mao et al. (2012); Int J Mol Sci;13 (1): 491-505
(5) GSI Environmental (2011)
(6) Witteveen+Bos and TTE consultants (2016)
(7) Connell (2005, p. 329)
(8) ATSDR (2017)

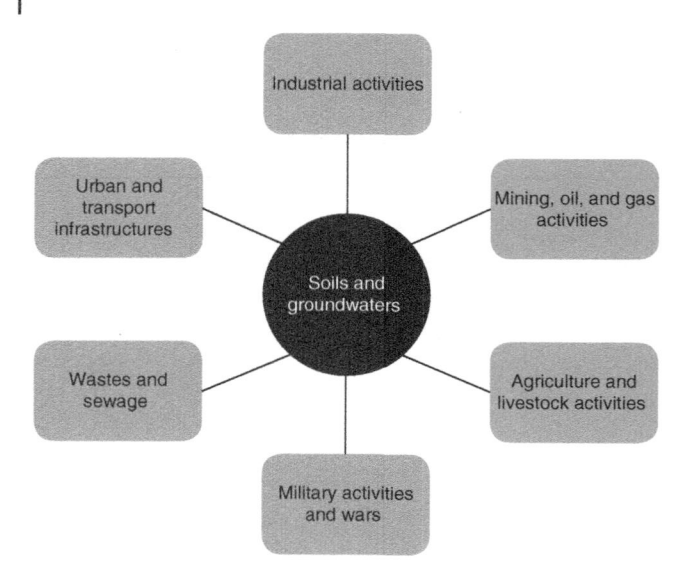

Figure 15.3 Overview of major anthropogenic sources of soil pollutants.

wastewaters and contaminated groundwaters, and stormwater runoff from urban, agriculture, and intensive livestock areas. For example, in many cases, veterinary and human therapeutic agents such as antibiotics and hormones are present in amendments added to soils, such as manures.

The global extent of soil contamination covers (i) localized areas such as in cities, industrial facilities, landfills, many mine sites, and coal-fired power generation depending upon scale of operations, and (ii) large areas such as in agriculture, on-shore oil fields, open cut mining, and water catchments affected by more insidious diffuse sources of contaminant releases. Numerous contaminated sites have been identified in developed areas and much less so in developing countries where legislation covering contaminated land management is lacking or under implementation. Many sites such as in developed countries involve multiple sources and complex mixtures of contaminants.

Table 15.4 indicates the nature and extent of contaminated sites and areas reported for some major regions of the world and countries. It is evident that site investigation, management, remediation, and monitoring are a major ongoing program in many developed countries. Large-scale remediation and site management for many contaminated

Chemicals of emerging concern (CECs) present a new challenge for soil (and water) contamination from sources such as urban wastewater and industrial discharges, land disposal of wastes, use of poorly treated

Table 15.4 Summary of contaminated sites in various regions or countries.

Region/country	Key contaminants	Contaminated sites reported	Comments
Developed countries			
Europe	Various/complex	Potential: >2.5 million, ~340 000 likely contaminated	One-third of high-risk sites positively identified; 15% of these remediated (EEA 2014)
United States	Various	>540 000 sites and 9.3 million ha of contaminated land cleaned up	—
	Complex	Superfund sites (NPL): 1322 (September, 2014)	1163 sites remediated/site managed
Canada	Metals, PHCs, and PAHs	12 723 identified	Soil-contaminated sites
Australia	As for other developed countries	Potential: 80 000 (2010)	Many investigated, remediated or site managed
Developing countries			
China	Heavy metals (e.g. Cd), arsenic, mercury	20 million ha farmland, mining and industrial sites	(~20% of total farmland)
SE Asia	Heavy metals (e.g. Cd), arsenic, mercury	Farmlands, mining, and industrial sites	Arsenic in groundwater
Latin America	Heavy metals, arsenic, mercury, fertilizers, and pesticides	Arsenic in 14 out of 20 countries; mercury in gold mining sites, including Amazon basin; other mining sites and intensive-use farmlands	—
Africa	Various: PHCs, pesticides, heavy metals, wastewaters, leaks, and spills	Mining sites, waste disposal, oil and chemical spills, use of contaminated groundwater or wastewater	Nigeria reported 7000 spills (1970–2000); Botswana and Mali: loss of >10 000 tonnes of pesticides from container leaks and contaminated land

Source: Modified from FAO/ITPS (2015, pp. 120–121). Published by Food and Agriculture Organization of the United Nations, Rome.

soil sites have been completed in the United States since the 1980s.

Similarly, many developing countries are faced with future challenges and serious threats to their environment, food production, and human health from soil pollution sources related to relatively rapid industrial growth and urbanization. Diffuse soil contamination, in particular, poses a difficult management problem for most countries because it is dispersed over large areas at variable mixtures of substances and concentrations, and its environmental behavior, bioavailability, and toxicity can be influenced by changes in soil properties.

15.5 Environmental Behavior and Fate of Soil and Groundwater Contaminants

The conceptual approach used to describe the release of contaminants into the soil environment and groundwaters and their environmental flows and distribution is illustrated in Figure 15.4.

The behavior and fate of chemical contaminants in soils, sediments, suspended solids or particulates, and groundwaters depend on (i) the inherent properties of the chemical and also mixtures of chemicals such as petroleum hydrocarbons or formulated solvents, (ii) properties of soils, other solid phases, and waters, (iii) interactions of pollutants with a set of natural transport and transformation processes, and (iv) the environmental conditions that interact with the abiotic and biotic transport and transformation processes that form part of the biogeochemical cycles. These properties, processes, and environmental factors combine to largely influence the distribution of chemical contaminants between solids such as soils, air, water, and biological phases.

In the case of subsurface soils, an example of the basic transport and fate processes acting on a chemical spill (e.g. solvent or pesticide solution) is shown in Figure 15.5, as compiled from various sources (e.g. LaGrega et al. 2001; FAO/ITPS 2015). These processes are discussed further in this section, including potential contamination of groundwaters in the saturated zone below the water table.

Critical outcomes of these environmental processes are the mobility, distribution, persistence, and biological uptake of contaminants in soil and water phases, which affect exposure concentrations (and doses) and potential toxic effects on humans, crops, livestock, and wildlife. Biological uptake and bioaccumulation of contaminants by living organisms from soils and groundwaters are discussed in Sections 15.7 and 15.8 of this chapter.

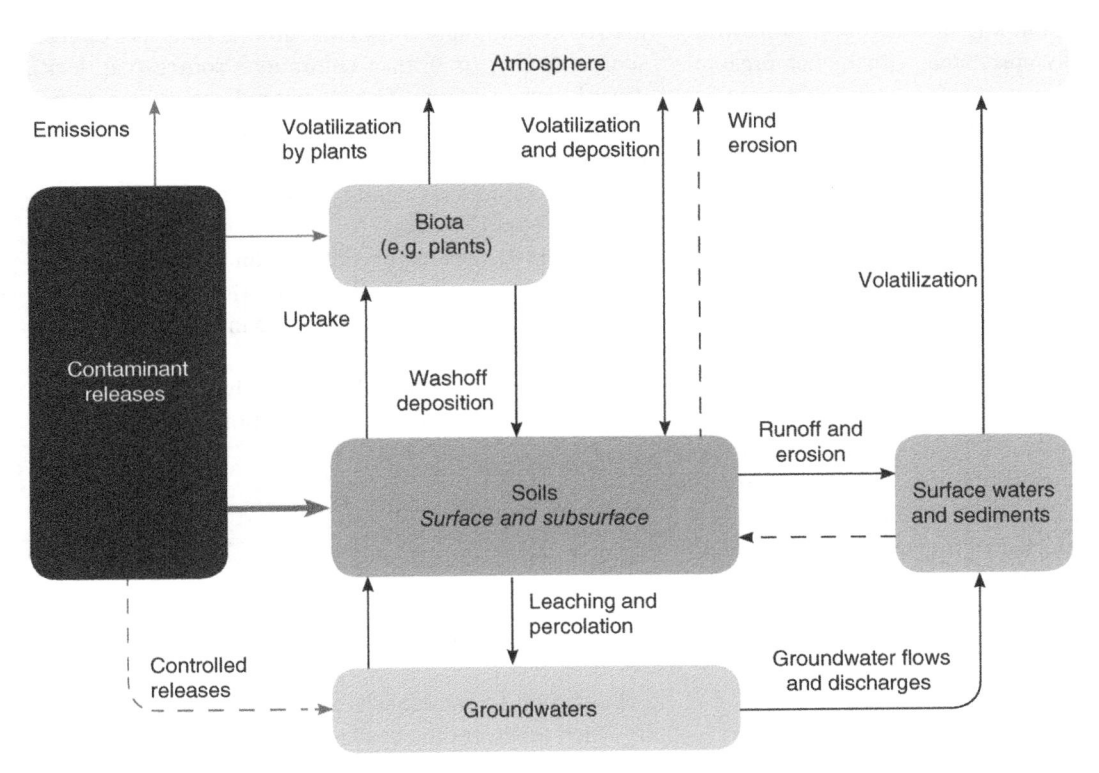

Figure 15.4 Conceptual model of contaminant releases from human and natural sources into the terrestrial environment and their environmental flows and distribution.

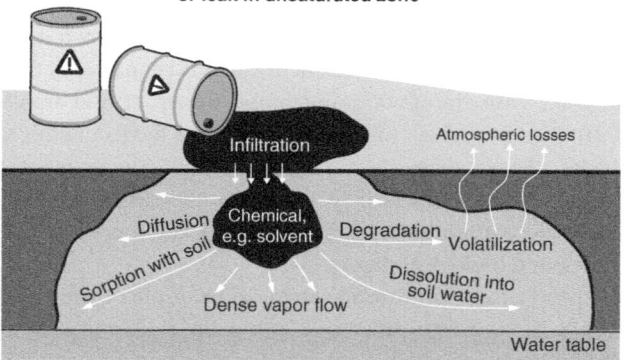

Figure 15.5 Basic fate and transport processes in the unsaturated and permeable zone of subsurface soils.

Organic compounds in soils and groundwaters are mainly degraded by the actions of microorganisms and a few chemical processes. Water leaching of organics from soils into pore water and groundwater can remove organics by volatilization into air. Generally, metals are more persistent than organic compounds. Metals and organometallic compounds are not degradable beyond the elemental state but transformation to volatile organic forms or organic complexes can result in evaporation into the atmosphere or removal by leaching, or loss in stormwater runoff. More persistent chemicals exhibit strong adsorption properties on soil particles, usually correlated with higher organic carbon content.

The fate of chemical contaminants in soils and subsurface waters is largely influenced by (i) physical transport processes (advection, hydrodynamic dispersion, molecular diffusion, dissolution, leaching, filtration, and percolation), (ii) soil erosion of surface soils by water flows and wind movement, (iii) the degree of retention or retardation from water solutions by sorption, precipitation, ion exchange, and chemical complexation, and (iv) attenuation involving removal or transfer to another medium or transformation (e.g. volatilization, chemical oxidation–reduction, hydrolysis, biodegradation, and biological uptake or transfer, such as by soil invertebrates and root systems of plants). Contaminants in groundwater tend to be removed or reduced in concentration with time and distance due to retention and attenuation processes.

These natural physical, chemical, and biological processes are discussed below.

Physical Transport Processes

Mass transport of contaminants dissolved in subsurface waters occurs through a porous medium primarily by advection, due to displacement by groundwater along streamlines. Advection is a function of groundwater velocity. Solutes are transported by the bulk groundwater flow. Nonreactive (conservative) solutes move at a rate equal to the average linear seepage flow velocity. The concentration of the dissolved contaminant changes and spreads in three directions as it migrates through the medium as a result of mechanical or hydrodynamic dispersion, and tends to be diluted. During advective transport, solutes are mechanically mixed by velocity variations due to friction when moving through pores and around solid particles. It is a function of groundwater velocity and a measure of the *dispersivity* of the aquifer medium.

Dispersivity is an important empirical property of a porous medium such as in the erosion of dispersive soils. For soils, it basically determines the characteristic dispersion of a soil by relating the components of pore velocity to the dispersion coefficient.

The dilution measured over a distance is called a concentration gradient. The movement or spread from high to lower concentration areas is known as diffusion (e.g. molecular or ionic diffusion). Mass movement by diffusion (diffusive flux) is proportional to the concentration gradient. Mechanical dispersion usually exceeds diffusion. However, molecular dispersion is the main process for contaminant transfer between fine-grained and coarse-grained sediments, and between fractures and rock matrix. Contaminants may enter groundwater as a discrete source or *slug* or from a continuous source (e.g. leak), described as a *plume*. The shape and extent of a plume is influenced by a number of factors such as sources and release variations, local geology constraints, type of aquifer, types of contaminants, and their concentrations (LaGrega et al. 2001, pp. 175–180).

Importantly, the rates of contaminant transport are determined by (i) groundwater flow rates, (ii) interactions with aquifer media (retention), and (iii) changes in water chemistry.

The porous media model for groundwater flow is based on **Darcy's law,** by an empirical relationship:

$$q = k\,i\,A \qquad (15.3)$$

where q = flow rate ($cm^3\,s^{-1}$)

k = hydraulic conductivity ($cm\,s^{-1}$)
i = hydraulic gradient ($cm\,cm^{-1}$)
A = cross-sectional area of flow measured perpendicular to the flow (cm^2).

The hydraulic gradient i is expressed as the rate of change in which the hydraulic head (or energy potential) is lost as water flows through the porous media or materials:

$$i = (h_1 - h_2)/l \qquad (15.4)$$

where h_1 = head at location 1 (cm)

h_2 = head at location 2 (cm)
l = length of flow distance (cm).

Darcy's law is considered applicable over a wide range of subsurface flow conditions under assumed laminar flows although realistic conditions may affect this assumption (see LaGrega et al. 2001, pp. 164–166).

Diffusion in water and air involves the movement of a contaminant also from areas of high to lower concentration under a concentration gradient. **Fick's law** gives the flux or amount of a contaminant flowing through a unit area during a unit time interval:

$$J = -D\,(dC/dx) \tag{15.5}$$

where J = flux (mol cm^{-2} · s)

D = diffusion coefficient (cm^2 s^{-1})
C = concentration (mol cm^{-3})
x = length in direction of movement (cm).

Contaminant transport in fractured media and heterogeneous porous media, however, is more complex to model, as well as for non-aqueous-phase liquids (NAPLs) that usually exist as a separate liquid phase in subsurface water, such as gasoline or chlorinated solvents leaking from tanks or from a spill. Basically, less dense NAPL floats on the surface of groundwater and spreads laterally, and migrates down gradient in contrast to denser NAPL, which tends to sink and migrate according largely to subsurface stratigraphy and low permeability barriers (LaGrega et al. 2001, pp. 180–183).

Particulate Matter Emissions

Fugitive emissions of particulate matter from soils such as croplands, dry and eroded pastures, landfills, hazardous waste sites, vegetation clearing, land development, unpaved roads, and stockpiling of soils are mainly influenced by wind erosion, and disturbance by human activities (e.g. excavation, grading, and vehicle movements), resulting in transport of suspended dusts offsite, including any adsorbed contaminants (and bioaerosols). The amount of soil lost from an inactive site or area depends upon factors such as soil characteristics (e.g. particulate sizes, soil moisture content, and texture), exposed surface area, and wind velocity. However, fugitive dust generated by vehicular traffic is a function of soil properties and vehicular characteristics (e.g. vehicle speed, weight, and number of wheels) (LaGrega et al. 2001, pp. 154–155). A set of air emission factors and quantification for fugitive dusts is included in document AP-42: Compilation of Air Emission Factors (US EPA 2019).

Retardation Processes

Soil–Water Distribution Process

Sorption of a chemical contaminant from water solution to a solid (sorbent) phase is a key retardation process that involves either adsorption via accumulation of chemicals at the solid–liquid surface (e.g. complexation with surface functional groups) or absorption of molecules within the solid phase (e.g. SOM).

Sorption of nonpolar or organic compounds from the aqueous phase to porous subsurface media follows a linear isotherm under equilibrium conditions when solute concentrations are low (e.g. $\leq 10^{-5}$ molar or less than half their solubility) (LaGrega et al. 2001, pp. 199–201).

$$S = K_D\,C \tag{15.6}$$

where S = mass sorbed per mass of sorbent (mg kg^{-1}); K_D = partition or distribution coefficient; and C = concentration in groundwater at equilibrium (mg L^{-1}). Here, K_D represents the slope of this relationship.

The **soil adsorption coefficient (K_D)** measures the amount of chemical substance adsorbed onto soil per amount of water. It can be expressed as:

$$K_D = C_S/C_W \text{ at equilibrium} \tag{15.7}$$

where C_S = concentration of chemical in soil and C_W = concentration of chemical substance in water. The values for K_D vary greatly because the organic content of soil is not considered in the equation. Since adsorption occurs predominantly by partition into the SOM, it is more useful to express the partition coefficient as K_{OM} or the widely used organic carbon–water partition coefficient, K_{OC}. In the case of K_{OM}:

$$K_{OM} = C_{SOM}/C_W \tag{15.8}$$

where C_{SOM} = concentration in the SOM. As shown in Chapter 4 of this book, the following relationship can be derived:

$$K_{OM} = C_S/f_{OM}\,C_W \tag{15.9}$$

where f_{OM} is the fraction of organic matter in the solid phase.

This means that

$$K_{OM} = K_D/f_{OM} \tag{15.10}$$

and

$$K_D = K_{OM}\,f_{OM} \tag{15.11}$$

where $K_{OM} > K_D$.

From equations (15.10) and (15.11), useful expressions can be derived to calculate K_{OM} and K_D values from experimental and estimated data as follows (Connell 2005, p. 336;

and Chapter 4 of this book). Octanol acts as a reasonable surrogate for SOM if a proportionality factor, x, is used to account for its lower capacity to sorb lipophilic compounds and a nonlinearity constant, a, is used to adjust for curvature of the relationship.

Thus, K_{OM} can be related to the octanol–water partition coefficient K_{OW}:

$$K_{OM} = x \cdot K_{OW}^a \quad (15.12)$$

and

$$K_D = x \cdot f_{OM} \cdot K_{OW}^a \quad (15.13)$$

For K_{OM}, a common empirical equation for this relationship is given as:

$$K_{OM} = 0.66 \cdot K_{OW}^{1.03} \quad (15.14)$$

where the constants are $x = 0.66$ and $a = 1.03$.

Equation (15.14) can be expressed as a linear relationship in the following logarithmic form to evaluate soil partitioning with water, by taking logs of both sides,

$$\log K_{OM} = 1.03 \log K_{OW} - 0.18 \quad (15.15)$$

Importantly, instead of K_{OM}, the **organic carbon–water partition coefficient**, K_{OC} is widely used as a key transport and environmental fate parameter and is defined as:

$$K_{OC} = C_S/C_W \quad (15.16)$$

or

$$K_{OC} = (K_D \cdot 100)/\% \text{ organic carbon} \quad (15.17)$$

and can be simplified to calculate the sorption coefficient, K_D, as:

$$K_D = K_{OC} \, f_{OC} \quad (15.18)$$

A more precise value of the organic carbon fraction in soil, f_{OC}, is usually measured and substituted for f_{OM} but the values for constants, a, and, x, will be different.

Similar to equation (15.14), a linear correlation between K_{OC} and K_{OW} can be developed as:

$$\log K_{OC} = a \log K_{OW} + x \quad (15.19)$$

This form of published or measured regression-based QSARs for various groups of chemical classes can be used to calculate K_{OC} values from known K_{OW} values and also solubility S (Ecetoc 2013) and Table 15.5 [1].

---

1 K_{OC} is a measure of the tendency of a chemical to bind to soils, corrected for soil organic carbon content. These values can vary substantially, depending on soil type, soil pH, the acid–base properties of the pesticide, and the type of organic matter in the soil (Weber et al. 2004).

Table 15.5 Correlations between K_{OC} and chemical properties (K_{OW} and solubility S).

Class of chemicals	Number of chemicals	Equations	Notes
Pesticides	45	$\log K_{OC} = 0.544 \log K_{OW} + 1.377$	—
Aromatics	10	$\log K_{OC} = 1.00 \log K_{OW} - 0.21$	—
Chlorinated hydrocarbons	15	$\log K_{OC} = -0.557 \log S + 4.277$	S in μmol L^{-1}
Aromatics	10	$\log K_{OC} = -0.54 \log S + 0.44$	S in mole fraction
Pesticides	106	$\log K_{OC} = -0.55 \log S + 3.64$	S in mg L^{-1}

Source: Modified from Table 3–5 (LaGrega et al. 2001, p. 115).

Worked Example 15.1 Groundwater is contaminated with 0.80 mg L^{-1} of xylene near a leaking solvent tank. Calculate the concentration of xylene likely to be sorbed from the contaminated groundwater onto subsurface soil containing 4% organic carbon at this location.

The calculated K_{OC} value for xylene is given as 240 mL g^{-1}. Assume a linear sorption model. From equation (15.18),

$$\begin{aligned} K_D &= K_{OC} \, f_{OC} \\ &= (240) \times (0.04) \text{ mL g}^{-1} \\ &= 9.6 \text{ mL g}^{-1} \end{aligned}$$

The sorbed concentration can then be calculated from equation (15.6), as follows:

$$\begin{aligned} S &= K_D \, C \\ &= (9.6 \text{ mL/g}) \times (0.8 \text{ mg}/10^3 \text{ mL}) \times (10^3 \text{ g/kg}) \\ &= \mathbf{7.68 \text{ mg/kg}} \end{aligned}$$

Examples of correlations between K_{OC} and chemical properties K_{OW} and solubility S for some sets of chemical classes are shown in Table 15.5; see also worked example (15.2) for the herbicide, atrazine, below.

K_D or K_{OC} measures the mobility of a chemical substance in soil. A very high value means it is strongly adsorbed onto soil and organic matter and does not move throughout the soil. A very low value means it is highly mobile in soil. Log K_{OC} values of some organic pollutants are given in Table 15.6.

K_{OC} is recognized as a critical input parameter for estimating the environmental distribution and environmental exposure level of a chemical substance in soils, sediments, and solid particulates. For pesticides, the higher values of K_{OC} or K_D are better because such pesticides are less likely

Table 15.6 Measured partition coefficients K_{OW} and K_{OC} for some organic pollutants in soils and sediments.

Pollutant	Log K_{OW}	Log K_{OC}
Benzene	2.11	1.78
Naphthalene	3.36	2.94
Pyrene	5.18	4.83
Dibenz(a,h)anthracene	6.50	6.22
Tetrachloroethylene	2.53	2.56
γ BHC (lindane)	3.72	3.30
p p′ DDT	6.19	5.38
PCB-153	6.72	5.62
Atrazine	2.33	2.17
Carbaryl	2.81	2.36
Malathion	2.89	3.25
Chlorpyrifos	3.31–5.11	4.13
Diuron	1.97, 2.81	2.60, 2.58

Source: Compiled from Karickhoff (1981, pp. 833–846). Published by Elsevier, Amsterdam.

to leach or occur as surface runoff to contaminate groundwater. If K_{OC} of a substance is very high (e.g. log K_{OC} > 4.5), the focus should then be on the potential adverse effects of the substance on terrestrial organisms such as earthworms and related ecotoxicity tests for terrestrial organisms.

Worked Example 15.2 The log K_{OW} value for the herbicide, atrazine, is given in Table 15.6 as 2.33. Estimate the K_{OC} value (L kg^{-1}) for atrazine, using the following linear regression-based relationship derived for triazine compounds (Ecetoc 2013, Table 9).

For triazines (e.g. atrazine):

$$\text{Log } K_{OC} = a \log K_{OW} + x \ (a = 0.30 \text{ and } x = 1.50)$$

Substituting in above equation,

$$\text{Log } K_{OC} = 0.30 \log K_{OW} + 1.50$$

(n = 16, r^2 0.32, standard error 0.38)

$$\text{Log } K_{OC} = 0.30 \times 2.33 + 1.50 = 2.20$$

$$K_{OC} = \mathbf{158} \text{ (L kg}^{-1}\text{)}$$

(Compare this predicted result with the measured atrazine value given in Table 15.6. Note: Convert Log K_{OC} value to K_{OC}).

In the case of metals, the sorption of metals to the solid matrix also results in a reduction of the dissolved metal concentration during transport in soils and water systems. The metal partition coefficient K_D is also determined by general equation (15.7) for particular metals. The K_D value in soils depends largely upon various geochemical characteristics of the soil and its pore water (pH of the system, and nature and concentration of sorbents associated with the soil or water). Allison and Allison (2005) have reported soil partition coefficient values (median and range) for various metals in soil/water systems. These K_D values include As 3.4 (0.3–4.3); Cd 2.9 (0.1–5.0); Cr (VI) 1.1 (−0.7–3.3); Cu 2.7 (0.1–3.6); Hg 3.8 (2.2–5.8); CH_3Hg 2.8 (1.3–4.8); Ni 3.1 (1.0–3.8); Pb 4.2 (0.7–5.0); and Zn 3.1 (0.7–3.6).

Surface reactions in soils include ion exchange. Clay mineral, hydroxides, and organic matter components of soil have negatively charged sites on their surfaces which adsorb and hold positively charged ions (cations) by electrostatic force. **CEC** is the total capacity of a soil to hold exchangeable cations of various metals (e.g. Ca, Mg, K, Cu, Zn, Cr, and Ni) and buffer against acidification. It generally increases with soil pH by enhancing the negative charge on clay minerals and organic matter by replacing H$^+$ ions. As pH becomes acidic, the CEC tends to increase as H$^+$ ions replace metal ions.

Dissolution/precipitation reactions are particularly important for various heavy metals. At high pH levels, most metals precipitate from solutions as hydroxides although the process can be reversed (e.g. solubility of amphoteric metals such as Ni increases as pH continues to be increased). Changes in oxidation state can also cause precipitation such as for Fe^{2+} to Fe^{3+} (LaGrega et al. 2001, pp. 203–204).

Attenuation Processes

These processes mainly involve physical dispersion, volatilization, biodegradation, and chemical processes such as oxidation and reduction, acid–base reactions, and hydrolysis to a minor degree.

Volatilization

Volatilization is a major diffusion process for the removal of many chemical contaminants from the solid phase (soils and sediments) or liquid phase (e.g. water, solvent, and petroleum) to gas phase. The loss of chemical contaminants from soils by volatilization can be expressed by a number of relationships that include K_{OM} or K_{OC}. As sorption of the chemical to soil particles increases, the rate of loss decreases and the $t_{1/2}$ value increases. Alternatively, as the vapor pressure (P) of chemical increases, the volatilization also increases and the half-life $t_{1/2}$ decreases.

Biodegradation/Biotransformations

Organic compounds are degraded by the actions of microorganisms and various abiotic processes as indicated by experimental measures of their half-lives $t_{1/2}$ in soils or groundwaters, depending on chemical and substrate properties (e.g. microbial populations and organic carbon content) and environmental conditions. Water leaching of organics from soils into pore water and also groundwater can remove organics by volatilization into air.

Generally, metals are more persistent than organic compounds. Metals and organometallic compounds are not degradable beyond the elemental state but transformation to volatile organic forms or organic complexes can result in evaporation into the atmosphere or removal by leaching or loss in stormwater runoff. More persistent chemicals exhibit strong adsorption properties on soil particles, which are usually correlated with higher organic carbon content.

Oxidation–Reduction Reactions

These reactions basically involve the loss of (oxidation) or gain of electrons (reduction) between chemicals (inorganic or organic). Redox reactions are typically a function of oxidation potential (Eh) and pH of the system. Eh is measured in volts as the electrical potential for the reaction relative to standard state of hydrogen oxidation.

Some redox reactions occur between organic chemicals and soil materials but usually are mediated by biological activity (biodegradation).

$$\text{Sulfide oxidation}: 2O_2 + HS^- = SO_4^{2-} + H^+$$

$$\text{Iron oxidation}: O_2 + 4Fe^{2+} + 4H^+ = 4Fe^{3+} + 2H_2O$$

Mobility Enhancement

Some physicochemical processes such as ionization, dissolution, complexation, and co-solvation can enhance the mobility of chemical pollutants in soils and/or groundwaters. For instance, organic solvents (e.g. petroleum fuels and alcohols) or leachates may reduce sorption sites, increase solubility, or physically alter flow paths. Metal ions may also form a stable metal–ligand complex with an inorganic ligand (e.g. Cl^-, SO_4^{2-}, CO_3^{2-}, and PO_4^{3-}) or organic ligand (e.g. humic acid) that prevents removal by precipitation or adsorption, or increases potential water solubility.

Fate of Chemical Contaminants

The persistence of soil and groundwater contaminants depends largely on how key properties of the contaminants interact with various soil properties (e.g. soil temperature, density, particle size, moisture content, porosity, organic matter or organic carbon content, pH, redox state, and sulfide content) through sorption onto particles, dissolution in water, and important removal or retardation processes (e.g. volatilization or evaporation, sorption, water leaching, molecular diffusion, chemical oxidation and reduction, complexation, and microbial-mediated transformations and degradation).

An important example is common heavy metals (Pb, Cr, As, Zn, Cd, Cu, and Hg) found at contaminated sites, which are capable of decreasing crop production, causing contamination of the food chain through bioaccumulation, and also potential groundwater contamination.

The fate and transport of heavy metals in soil (and groundwater) is influenced by the chemical form and speciation of the metals, and interactions with soil properties and related processes. In soils, heavy metals are adsorbed by initial fast reactions (minutes to hours), followed by slow adsorption reactions (days to years) which can result in different chemical forms with varying bioavailability, mobility, and toxicity. The resulting distribution and bioavailability of heavy metals in soils is largely influenced by soil properties and processes described above.

Generally, soil and groundwater contaminants are subject to removal by a combination of abiotic and biotic processes at various rates which usually follow first-order kinetics. The half-life $t_{1/2}$ is commonly used to measure the persistence of substances in soil and other phases. Field and experimental studies of half-life values of chemicals in soils can show considerable variation under different soil properties and environmental conditions. Moisture is usually an important factor for promoting the growth of microorganisms, along with temperature, pH, oxygen availability, and nutrients.

The ranges of some $t_{1/2}$ values for organic contaminants found in soil are presented in Table 15.7. Persistent organic pollutants in soils usually have $t_{1/2}$ values greater than 90 days to years, due to chemical bond types that are resistant to chemical oxidation, hydrolysis, and biotransformations.

Benzenes and PAHs, e.g. benzo(a)pyrene, can have $t_{1/2}$ values from 5 to 490 days while halogenated hydrocarbons

Table 15.7 Half-lives ($t_{1/2}$) of some organic compounds in soils.

Chemical	Aqueous solubility (mg L^{-1})	Half-life ($t_{1/2}$) (days)
Benzene	1.780	5–16
Toluene	515	4–22
Benzo(a)pyrene	0.004	5–490
DDT	0.0032	700–6000
Dieldrin	0.17	175–1100
Malathion	143	3–7
Atrazine	30	1–8
Glyphosate	1200	50–70

Source: Data taken from Table 14.3, Connell (2005, p. 329).

tend to be longer. Water-soluble organics generally have shorter $t_{1/2}$ values as they are more susceptible to biodegradation through active chemical groups or polar bonds. Synthetic contaminants such as pesticides and pharmaceuticals exhibit a wide range of persistence in soils and groundwaters depending upon structural types and related properties. The persistence of chemical fractions in mixtures of chemical contaminants, as shown by crude and refined oil spills on soils, sediments, and through leaching into groundwaters, can vary greatly from days to years, and even decades for recalcitrant high boiling components.

15.6 Exposure Assessment of Soil and Groundwater Contaminants

Human and animal exposures to soil contaminants depend upon the nature of the contaminant, soil properties, and environmental pathways of exposures due to natural sources and human activities. Primary contaminant exposures can involve several environmental pathways: ingestion of soils and swallowing of dusts, inhalation of dust particles and volatile chemicals by air-breathing animals, consumption of plants and animals that have accumulated soil contaminants, and through skin or body surface contact (dermal absorption). Secondary exposures include sources such as drinking waters and deposition of atmospheric contaminants (see Rodríguez-Eugenio et al. 2018, p. 68). Plants are exposed to contaminants in soils via root systems and leaves for airborne contaminants (pollutant gases, volatile organic compounds, and fine particulates).

Soil contamination levels vary greatly for different chemical contaminants (<1 mg kg^{-1} to 10 000 mg kg^{-1} or so, dry weight in soil) as reflected by available international soil screening levels for metals and organic contaminants such as persistent pesticides, PAHs, POPs, and aromatic benzenes. Excessive levels for toxic chemicals in soils are commonly in the range of μg kg^{-1} to mg kg^{-1} (dry weight). Naturally occurring chemicals (e.g. heavy metals) may occur at levels from hundreds to thousands of mg kg^{-1} (dry weight) in mineralized areas. Chemical spills or leaks can result in similar concentration ranges. Petroleum hydrocarbon spills can also result in TPH levels in the thousands of mg kg^{-1} (dry weight). Reference to local background levels of potential contaminants and relevant soil properties is essential in risk assessments for human and ecological health.

Contaminated groundwater levels for soluble toxic chemicals range from μg L^{-1} to mg L^{-1} as indicated by health and ecological screening levels for contaminated site investigations. Risk assessments of groundwaters usually apply water quality guidelines or standards applicable to the water use (e.g. drinking water standards for humans, irrigation of food crops, and ecological protection).

Human health risk assessments for contaminated sites commonly estimate soil and dust exposures to contaminants, combined with other sources and routes of exposure as indicated by risk screening level calculations (US EPA 2020) (see Section 15.8).

15.7 Biological Uptake and Bioaccumulation of Soil Contaminants

The transfer of contaminants in soils to organisms involves a three-phase partitioning process from solids to water, and then from pore water to biota such as soil invertebrates (e.g. earthworms) and root systems of plants. The latter pathway applies to many food plants and food chain transfer to consumers.

Connell (2005, pp. 336–339) has shown that a simple three-phase model can be used to estimate the transfer of lipophilic contaminants from soils to pore water to biota as shown in Figure 15.6.

Assuming the lipids in organisms absorb the lipophilic organic compounds, then the biotic–water partition coefficient K_B can be shown as developed from Connell (2005, pp. 336–339):

At equilibrium

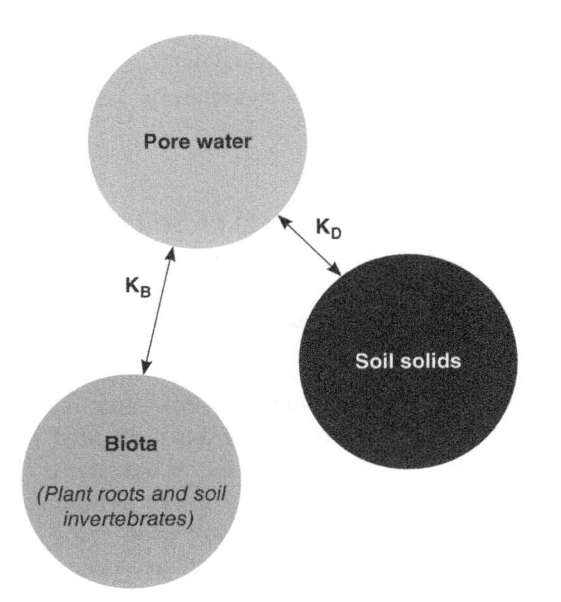

$$K_B = C_B/C_W \quad (15.20)$$

Figure 15.6 Three-phase model for transfer of lipophilic organic chemicals from soil solids to pore water to biota.

Where C_B is the concentration in biota and C_W is the concentration in water. A plot of C_B against C_W is expected to be linear at low concentrations. If n octanol is a good surrogate for biota lipid, then K_B becomes

$$K_B = f_{lipid} K_{OW}^b \quad (15.21)$$

where f_{lipid} is the lipid fraction in the biota and constant b is an empirical nonlinearity constant.

The relationship between soil solid and biota is now considered as the **bioaccumulation factor (BF)** which is the ratio between the concentration in the biota (C_B) to the concentration in the soil (C_W), i.e. BF = C_B/C_S. By including the water phase, BF becomes:

$$BF = (C_B/C_W)(C_W/C_S) = K_B/K_D \quad (15.22)$$

and K_B and K_D can be substituted by equations (15.21) and (15.13), respectively, to give:

$$BF = \left(f_{lipid} K_{OW}^b\right) / \left(x f_{OM} K_{OW}^a\right) \quad (15.23)$$

and

$$BF = (f_{lipid}/x f_{OM}) \cdot K_{OW}^{b-a} \quad (15.24)$$

In equation (15.24), the nonlinearity constants are a = 1.03 and b = 0.95, so that the difference, (a − b) = 0.08, is close to zero, such that the value of $K_{OW}^{0.08}$ is close to unity. The value of x is 0.66 as in equation (15.4).

$$BF = (f_{lipid}/0.66 \cdot f_{OM}) \cdot K_{OW}^{0.08} \quad (15.25)$$

Equation (15.25) can then be simplified to:

$$BF \approx f_{lipid}/0.66 \cdot f_{OM} \quad (15.26)$$

This means that the BF value should show little dependence on K_{OW} or other properties of the chemical. It mainly depends on the properties of the soil and the biota, particularly the ratio of the lipid in the biota to the organic carbon content of the soil (Connell 2005, p. 339).

15.8 Environmental Effects of Soil and Groundwater Contaminants

Ecotoxicity of Soil and Groundwater Contaminants

Principles of ecotoxicology that apply to major groups of pollutants found in soils and waters have been introduced and discussed in preceding chapters of this book. Studies on the toxic effects of chemical contaminants on terrestrial species, such as from acid deposition on forests in Europe, Canada, and the United States, and freshwater lakes in these regions, have expanded greatly to global concerns about the impacts of soil pollutants and other environmental stressors on the status of soil resources and ecosystem services such as biodiversity, food production, and security.

Within this context, pollutants enter the terrestrial environment, and contaminate soils and groundwaters through direct application, from diffuse sources or by long-range transport. Terrestrial organisms can be exposed to soil pollutants through dermal, oral, inhalation, and food-chain exposures due to bioaccumulation, and in certain cases, biomagnification of persistent chemicals (e.g. DDT/DDE and mercury in air-breathing, top avian predators). Terrestrial receptors include soil microbes, invertebrates, plants, amphibians, reptiles, birds, and mammals.

Toxic responses of terrestrial organisms can be assessed by acute and chronic toxicity bioassays in the laboratory or field under natural conditions. Usually, standard laboratory toxicity test protocols (e.g. OECD) are used, based on a standard set of test organisms, from which extrapolations are made to other species. Biological monitoring for some chemicals (e.g. cadmium) and related biomarkers in plant or animal tissues can be used to predict risks of adverse effects from known threshold levels.

Pollutant exposures may also result in a change in the community composition of plant or animal species. Population models can be applied to determine if toxic effects from exposure to environmental pollutants can significantly alter the ability of a population to sustain itself over time. However, plants and animals can develop tolerance over time and adapt to some environmental pollutants. Ecotoxicity data on soil pollutants can then be combined with exposure profiles, and exposure levels and effects, to estimate the likelihood of adverse effects to populations or communities in ecological risk assessments, and to predict statistical levels for harmful effects on the plants and animals of concern (Fairbrother and Hope 2005, pp. 138–142). Chapter 18 of this book discusses environmental (ecological) risk assessment.

In contaminated land investigations, remediation and management, and environmental and human health risk assessments are common components of the process and outcomes. Many countries have developed or adopted guidelines that include soil screening criteria for the investigation of soil contaminants for ecological and human health evaluation (e.g. see US EPA 2020 Regional Screening Levels). In the United States, for example, the US EPA developed ecological soil screening levels (Eco-SSLs) for a set of inorganic and organic contaminants commonly found at Superfund contaminated sites. Table 15.8 contains a compilation of Eco-SSLs for generic soil contaminants (metals and a limited group of organic contaminants) for the protection of plants, soil invertebrates, avian, and mammalian species. Insufficient data were available to develop Eco-SSLs for soil microorganisms, reptiles, and amphibians.

Table 15.8 US EPA Ecological Soil Screening Levels (Eco-SSLs) for generic soil contaminants (mg kg^{-1}, dry weight of soil).

Contaminants	Plants	Soil invertebrates	Wildlife		Year
			Avian	Mammalian	
Aluminum	a	(−)	(−)	(−)	2005
Antimony	NA	78	NA	0.27	2005
Arsenic	18	NA	43	46	2005
Barium	NA	330	NA	2000	2005
Beryllium	NA	40	NA	21	2005
Cadmium	32	140	0.77	0.36	2005
Chromium	#	#	Cr^{3+} 26	Cr^{3+} 34	2008
			Cr^{6+} NA	Cr^{6+} 130	
Cobalt	13	NA	120	230	2005
Copper	70	80	28	49	2007
Lead	120	1700	11	56	2005
Manganese	220	450	4300	4000	2007
Nickel	38	280	210	130	2007
Selenium	0.52	4.1	1.2	0.63	2007
Silver	560	NA	4.2	14	2006
Vanadium	NA	NA	7.8	280	2005
Zinc	160	120	46	70	2007
Dieldrin	NA	NA	0.022	0.0049	2007
DDT and metabolites	NA	NA	0.093	0.021	2007
PAHs LMWs	NA	29	NA	100	2005
HMWs	NA	18	NA	1.1	2005
Pentachlorophenol	5.0	31	2.1	2.8	2007

aToxicity and bioaccumulation for soluble aluminum in soils (pH < 5.5); (−) Insignificant toxicity for insoluble aluminum forms in soils; # insignificant toxicity; NA not available (incl. insufficient data).
Source: Compiled from RAIS (2019).

The risk-based Eco-SSLs are statistically designed to provide adequate protection of the above ecological receptors and their terrestrial ecosystems that commonly come into contact with and/or consume biota that live in or on soil. These screening levels are used to identify the contaminants of potential concern (COPCs) that require further evaluation in site-specific baseline ecological risk assessments according to specific US EPA guidance.

Ecosystem Effects of Soil Pollutants

Soil pollution can adversely affect soil structure, properties and functions, soil microbial, plant and animal communities in terrestrial, and also aquatic ecosystems, through runoff and groundwater discharges and atmospheric deposition of volatile soil contaminants (e.g. acid rain and ammonia). These contaminants can impact a succession of trophic levels of primary producers (plants) and consumers (herbivores), predators, and higher levels of carnivores and top predators. Known effects in soils at lower trophic levels include altered and reduced metabolism of microorganisms, inhibition of plant photosynthesis, and toxic effects on soil invertebrates. Persistent and toxic chemicals in the soil may be passed up the food chain from plants to larger animals, leading to increased mortality rates and reduced reproduction. Toxic chemical residues (e.g. heavy metals and insecticides) can also contaminate food crops and animal products used as feedstuffs for livestock and human foodstuffs.

Other important examples of soil pollution effects on ecosystems include:

- Acidification which is a major form of soil pollution in many parts of the world caused by different sources and processes: (i) the release of large quantities of nitrogen into the atmosphere through ammonia volatilization and into the soil in agriculture, combustion of fossil fuels and urban wastes, and the decomposition of organic materials in soil which can release sulfur dioxide and other sulfur compounds, leading to the formation

of acid precipitation; (ii) oxidation of sulfide minerals in air-exposed subsurface soils and mine tailings to generate sulfuric acid in acid mine drainage and in acid sulfate soils; and (iii) acidification of agricultural soils from excessive synthetic fertilizer use.
- It follows that acidification can disrupt soil chemistry (e.g. reduce availability of plant nutrients, reduce buffer capacity and pH), increase soil salinity, alter microbial activity, plant metabolism and production, and mobilize toxic metals in soils through runoff, leaching into groundwaters, and plant uptake. For example, solubilized aluminum ions from natural soils in acidified waters are highly toxic in inorganic forms to many plants and aquatic species.
- Surface runoff, erosion, and groundwater discharges from soils contaminated with nitrogen and phosphorus compounds can readily cause or enhance eutrophication of waterways and other water bodies, resulting in *algal blooms* and toxins, excessive growths of macrophytes, deoxygenation of water, sedimentation, and loss of vulnerable species. As well, atmospheric deposition of large quantities of nitrogen compounds from volatilization of ammonia and NO_x from agricultural soils and wastes also contributes to increasing eutrophication of regional catchments, freshwater bodies, and coastal waters such as in Europe, North America, and Asia.

Soil Pollutants and Agriculture

Modern agriculture practices accelerate soil pollution through the intensive use of fertilizer and pesticides designed to increase productivity and reduce crop losses. Harmful to serious adverse impacts on soil processes, crop productivity, and food chains (e.g. bioaccumulation of contaminants) can occur when pollutant concentrations in soils (and groundwaters used for irrigation) exceed threshold levels for toxicity, alter soil processes and functions, or risk levels for contamination. Table 15.9 summarizes major impacts of soil and groundwater pollutants on agriculture and food chains.

Impacts and Risks of Soil Pollution on Food Chains and Ecosystem Services

Environmental contamination of urban and nonurban soils, agriculture, ecosystems, and related food chains has severe implications for equitable and future ecosystem services including food supply and ecological health. Food production and good food quality and water are seen as major global issues to meet the quantity, nutritional, and human health needs of a projected world population of over nine billion people by 2050. The latest projections by FAO indicate that global food production will increase by 60% between 2005/07 and 2050 under its baseline scenario

Table 15.9 Impacts of soil and groundwater pollutants on agriculture and food chains.

Sources and types of pollutants	Impacts on agriculture and food chains
Synthetic fertilizers Nitrogen-containing fertilizers (e.g. ammonium sulfates and phosphates, urea); phosphorus-containing fertilizers (e.g. rock phosphates, superphosphates, ammonium phosphates, or polyphosphates)	Excess nutrients in soils, mainly nitrogen, causes toxicity and soil acidification effects on plant productivity leading to crop losses. Acidification is accelerated relative to natural processes. It can also mobilize toxic heavy metals in soils and to groundwaters. Increased nitrification microbial activity will lead to accumulation of nitrates that readily leach to and pollute groundwaters. Microbial diversity will be altered causing soil nutrient imbalances. Excess P compounds from fertilizers in soils are also known to be transported in runoff (e.g. soil erosion) to surface water bodies and cause or enhance eutrophication of aquatic ecosystems in many regions of the world.
Pesticides	While pesticides protect crops and pastures, adverse impacts and risks mainly occur for nontarget species, both on and off agricultural lands. The transfer of soil and crop residues of many persistent and toxic synthetic pesticides in ecosystems, animal, and human food chains is well established.
Manure Heavy metals, antibiotics, and soil pathogens (e.g. *Salmonella, Campylobacter,* and *Escherichia coli* bacteria and viruses)	Application of untreated manure may lead to heavy metal pollution, impacts on microbial communities, soil invertebrate populations, phytotoxicity of crops, and uptake of heavy metals and food chain transfer. Untreated manure can contain high amounts of veterinary antibiotics that can lead to a rapid increase in antibiotic resistance in soils. Common intestinal pathogens are also found in manure and may persist for months.
Urban wastes	Heterogeneous contaminants and pathogens are found in soil amendments or biosolids from untreated sewage sludge and to some degree in pretreated biosolids that may affect soil properties, inhibit plant growth, and contaminate food production.

Source: Modified from FAO/ITPS (2015); Rodríguez-Eugenio et al. (2018).

(Rodríguez-Eugenio et al. 2018, p. 47). Significantly, 95% of food production is reported to depend on soils.

The major global review on soil resources by FAO/ITPS (2015) has concluded that: *Only healthy soils can provide the needed ecosystem services and secure supplies of more food and fibre ... Soil pollution reduces food security both by reducing crop yields due to toxic levels of contaminants and by causing the produced crops to be unsafe for consumption.*

15.9 Human Health Effects of Soil and Groundwater Contaminants

Harmful Effects of Major Contaminants

Local and diffuse environmental contaminants in soils and groundwaters, mainly associated with industrial, mining, agriculture, urban, transport, waste disposal, fossil fuel, and energy activities can expose human populations to multiple health risks such as neurotoxicity, renal disease, autoimmune diseases, cancers, cardiovascular diseases, respiratory diseases, and congenital malformations. Soil-related human health risks include exposures to heavy metals (e.g. As, Cd, Pb, and Hg), organic chemical contamination (e.g. dioxins, PCBs, and other POPs), pesticides, pharmaceuticals (e.g. estrogen and antibiotics), and soil pathogens (e.g. antibiotic-resistant bacteria, prions, and anthrax).

Health impacts and risks from exposures to major soil and groundwater contaminants reported in the scientific literature are summarized in Table 15.10. Harmful human effects of exposures to these types of pollutants are well documented, as indicated in preceding chapters of this book. Their effects may be acute or emerge over much longer time scales.

For soils, evidence from case studies of multiple contaminated sites (e.g. industrial, urban, waste disposal, intensive farming, and drinking water wells) supports these findings. In many multiple exposure situations (e.g. heavy metals at a mining site), the problem or challenge is how to identify cause–effect relations between specific contaminants and harmful effects. This applies in nonurban areas where there are many different and often diffuse sources of pollution. Overall, long-term impacts of soil pollution on global food security, human health, and the environment are difficult to quantify and resolve, although the scientific principles of soil quality and pollution effects are reasonably well understood (WHO 2013, p. 24; FAO/ITPS 2015; Rodríguez-Eugenio et al. 2018, p. 55).

Human Epidemiological Studies of Contaminated Land Effects

Physical effects on human health from contaminated lands can occur from a multitude of physical hazards (e.g. shafts, tunnels, tailings dams, abandoned structures, equipment, and tanks), fire and explosion of flammable gases and liquids (solvents and fuels), and unexploded ordnances (UXOs). In the United Kingdom, the Department for Environment, Food and Rural Affairs (Defra 2009a) undertook a major scientific review of the health effects of contaminated land on human populations (nonoccupational). Previous reviews reported a lack of evidence on adverse health effects such as cancers related to chemical exposures from contaminated land, although a link between landfill sites and/or contaminated land and reproductive effects (primary congenital anomalies and low birth weight) has been suggested.

The Defra review considered epidemiological studies associated with single and multiple contaminated land sites and persons living near hazardous waste and landfill sites. A summary of the review's findings on health effects associated with exposure to contaminants from contaminated land is presented in Table 15.11.

Scale of evidence used by Defra (2009b, p. 35):

(1) **Sufficient evidence** is based on peer-reviewed reports of expert groups or authoritative reviews.
(2) **Limited evidence** includes relationships for which several epidemiological studies (including at least one case-control or cohort study) showed fairly consistent associations and evidence of exposure–risk relationship after control for potential confounders.
(3) **Inadequate evidence** is used for a relationship for which epidemiological studies were limited in number and quality (for example, small studies, ecologic studies, limited control of potential confounders), had inconsistent results, or showed little or no evidence of exposure–risk relationships.

The Defra review found that despite the methodological limitations of many studies and inconsistencies in findings, the scientific literature does contain evidence for impacts of contaminated land on specific aspects of health. Of the health effects listed in Table 15.11, sufficient evidence was evaluated for stress, anxiety, and psychiatric disorders. There was limited epidemiological evidence for renal dysfunction and bone disorders, birth defects, low birth weight, and preterm deliveries. For example, clear causality has been established for severe renal and bone disorders under conditions of high dietary exposure to cadmium (e.g. exposure via a diet dominated by locally grown rice in parts of Japan and China) but was not generally shown at much lower exposures (Defra 2009b, pp. 35–36).

In the case of cancer studies, inadequate evidence was concluded because of inconsistent findings between different sites and inadequate characterization of individual

Table 15.10 Health impacts and risks from exposures to major soil and groundwater contaminants.

Contaminants of concern	Soil and groundwater-related exposures	Health impacts and risks for exposed persons
Asbestos	Asbestos contamination in the soil and friable asbestos materials can be released to the air by the wind or by human disturbance. Asbestos has long-term health consequences if it is inhaled.	Fine inhaled asbestos fibers accumulate in lung tissue over many years and can cause: • parenchymal asbestosis • asbestos-related pleural abnormalities • lung carcinoma • pleural mesothelioma
Arsenic	Mainly through ingestion of groundwater with naturally high levels of inorganic arsenic, food prepared with this water, food crops irrigated with water high in arsenic and agricultural soil.	Intake of inorganic arsenic can lead to chronic arsenic poisoning: gastrointestinal tract, skin, heart, liver and neurological damage; diabetes; bone marrow and blood diseases; cardiovascular disease.
Cadmium	Cd in soil or water used for irrigation, some phosphate fertilizers, contaminated biosolids, and atmospheric deposition can lead to accumulation in plants, food crops (e.g. rice and leafy vegetables), or animals that enter the human food chain.	Liver and kidney damage, low bone density. Diets poor in iron and zinc vastly increase the negative health effects of cadmium. Carcinogenic (by inhalation). In Europe, correlations between cadmium and age-adjusted prostate or breast cancer rates.
Lead	Leaded fuel, lead paint residues, and mining activities are common causes for elevated lead levels in surface soils. Ingestion of Pb in groundwaters used for drinking.	Neurological damage, lowers IQ and attention, hand–eye coordination impaired, encephalopathy, bone deterioration, hypertension, kidney disease.
Mercury	Main exposure route is via eating contaminated seafood. For children, it is direct ingestion of soil.	Central nervous system (CNS) and gastric system damage; affects brain development, resulting in a lower IQ; Affects co-ordination, eyesight, and sense of touch; liver, heart, and kidney damage; teratogenic.
Fluoride	Usually associated with high levels of fluoride in drinking water (e.g. groundwaters and artesian waters).	Mottling of teeth. Skeletal fluorosis: fluoride accumulates progressively in the bone over many years. Early symptoms include stiffness and pain in the joints. Crippling skeletal fluorosis: associated with osteosclerosis, calcification of tendons and ligaments, and bone deformities.
Radionuclides: (radon Ra and decay products, iodine-131, caesium-134 and -137, and strontium-90 isotopes)	Penetration and inhalation of natural radon gas and decay products in buildings from subsurface soils and igneous rocks. Iodine-131, caesium-134 and -137, and strontium-90 isotopes in food chain contamination of animals from pasture/soil uptake (e.g. milk). Dust exposures.	Elevated incidences of lung cancer from radon gas inhalation; chronic cancer risks from exposure to iodine-131, caesium-134 and -137, and strontium-90 isotopes.
Volatile organic compounds (e.g. aromatic hydrocarbons and halogenated aliphatic hydrocarbons)	Volatilization from petroleum fuel and halogenated hydrocarbons in soils; subsurface penetration buildings and indoor air contamination.	Chronic inhalation of elevated benzene levels may cause cancer (leukemia); sensory, neurotoxic effects, liver and kidney damage.
PAHs including benzo(a)pyrene	Major exposure route for polycyclic PAHs is through contaminated food ingestion.	Potential carcinogenic risk due to the aromatic nature of PAHs, which can easily penetrate cellular membranes and bind covalently with DNA molecules, where they may cause mutations.
Dioxins and dioxin-like compounds incl. PCBs	Accumulation up food chain. Human exposure to dioxin and dioxin-like substances occurs, mainly through consumption of contaminated food (>90% of human exposure), mainly meat and dairy products, fish, and shellfish.	Dioxins are highly toxic and can cause reproductive and developmental problems, damage the immune system, interfere with hormones, and cause cancer, skin lesions, and chloracne.

Table 15.10 (Continued)

Contaminants of concern	Soil and groundwater-related exposures	Health impacts and risks for exposed persons
Pesticides	Many persistent organic pesticides accumulate in the food chain.	Exposures to organic chemicals, including various pesticides, have been linked to a wide range of health effects: acute and chronic neurotoxic, genotoxic, carcinogenic, growth, and reproductive. Few studies have been conducted on the toxicity of complex chemical mixtures in soils. Difficult to show specific cause–effect in humans from soils.
Antibiotics and antimicrobial-resistant bacteria and genes	Intake of antibiotics and AMR bacteria from food chain contamination via soil residues from farms, manure, biosolids, urban waste disposal, and wastewater discharges. Ingestion of drinking water obtained from contaminated groundwaters.	Allergic and toxic reactions or chronic toxic effects from prolonged low-level exposure to antibiotic residues. Increased mortalities and higher risks from AMR bacteria: transfer of antibiotic resistance genes from the environment to human pathogens; reduced effectiveness of antibiotics; persistent infections in the body, increasing the risk of contamination of other persons. Largely uncertain human health implications for intake of antibiotic residues and AMR bacteria present in food; increasing multidrug resistant bacteria.

Source: Compiled from Science Communication Unit, University of the West of England, Bristol (2013); Rodríguez-Eugenio et al. (2018). See also Chapters 9, 10, 11, 12, and this chapter.

Table 15.11 Summary of epidemiological evidence for health effects associated with exposure to contaminants from contaminated land.

Health effect	Level of evidence	Notes
Stress, anxiety, psychiatric disorders	Sufficient	Evidence primarily relates to hazardous waste sites, but likely to be a more generalized phenomenon.
Blood pressure	Inadequate	No evidence so far.
Renal dysfunction and bone damage	Limited	Causality has been established under conditions of high exposure to cadmium due to dietary intake; evidence for effects under lower exposure situations is contradictory and/or based on biomarker responses rather than health impacts.
Cancer	Inadequate	Evidence is inconsistent and exposure is inadequately characterized. Some specific links have been established (naturally occurring asbestos, sites significantly contaminated with dioxins).
Birth defects	Limited	Evidence from multiple site studies is relatively consistent and causality was established with proxies of exposure from hazardous waste sites.
Low birth weight/preterm deliveries	Limited	Studies are largely related to landfill sites. Evidence is relatively consistent and temporality was demonstrated on several occasions.
Fetal death/stillbirth/infant death	Inadequate	The few studies investigating this outcome do not support an effect.

Source: Compiled from Defra (2009b). Published by UK Department for Environment, Food and Rural Affairs, London.

exposures. Several studies suggested *associations between residence near landfills containing hazardous waste and cancer (for example, leukemia and cancer of the bladder). However, several other studies did not observe such a link and there is thus a general lack of consistency in the results obtained for landfills.*

Epidemiological studies of contaminated sites (e.g. industrial and hazardous waste landfills) are known to suffer with problems of characterizing exposure assessment (Martuzzi et al. 2014; Defra 2009b). Many studies rely on residential distance from a site as a surrogate measure of exposure or sometimes use exposure modeling. There is a lack of direct measurements and of individual exposures. This situation is evident in cancer studies of potentially exposed residents which are affected by long gestation periods for the disease and many other confounding factors that may contribute to the disease. Studies suffered from lack of consistency in study designs (e.g. sampling sizes and control or reference populations) and methodologies (e.g. controlling for confounding factors) (Defra 2009b, p. 34).

Contaminated soils and groundwaters are a major environmental and human health issue of current global concern. Available evidence suggests there are millions of potentially contaminated sites (e.g. Europe) and enormous areas of contaminated agricultural lands on a world scale. The magnitude of the health impacts is likely to be underestimated, particularly for historic exposures in developed countries and largely unknown in developing countries, where land contamination and hazardous wastes are emerging issues for investigation, health risk assessment, remediation, and management.

15.10 Key Points

1. Soil and groundwater pollution are major problems and challenges for global land degradation, water resources, agriculture, food supply, and the health of human populations and natural ecosystems.
2. Soil contamination occurs when the concentration of a chemical or substance is higher than would occur naturally but is not necessarily causing harm. In comparison, soil pollution refers essentially to chemical contaminants at levels in the soils that have adverse effects on any non-targeted organism.
3. Soils consist largely of inorganic minerals and salts formed from major and trace elements derived from the geochemistry of parent rock materials, degraded organic matter, and atmospheric deposition, as part of biogeochemical cycling. The organic matter helps stabilize soils by binding inorganic particles together to form variable aggregates of sand, silt, clay, and organic matter.
4. Groundwater originates from fresh water (precipitation or from lakes and rivers) that soaks or infiltrates into the subsurface soil and is stored in the pores within the soil matrix and rock fractures, forming aquifers where water can move through interconnected pores, rock fractures, and porous rock. Note: unsaturated and saturated zones.
5. Major types of soil and groundwater pollutants are inorganic (e.g. heavy metals, metalloids, radionuclides, nitrogen and phosphorus and various other non-metals, and asbestos), organic compounds (e.g. POPs, pesticides, PAHs, crude and refined oil mixtures, industrial solvents, surfactants, antibiotics, and plastics), and pathogens.
6. Soil and groundwater contaminants originate from natural and anthropogenic sources via point or local sources (e.g. sewage discharges, stack emissions, wells, or volcanoes) or diffuse emissions from dispersed (area) sources (e.g. applications of pesticides, excessive nutrients or untreated biosolids to crop lands, mine tailings, oil spills, atmospheric deposition, and flooding).
7. The fate of chemical contaminants in soils and subsurface waters is largely influenced by (1) physical transport processes (advection, hydrodynamic dispersion, molecular diffusion, dissolution, leaching, filtration, and percolation), (2) soil erosion of surface soils by water flows and wind movement, (3) the degree of retention or retardation from water solutions by sorption, precipitation, ion exchange, and chemical complexation, and (4) attenuation involving removal or transfer to another medium or transformation (e.g. volatilization, chemical oxidation–reduction, hydrolysis, biodegradation, and biological uptake or transfer such as by soil invertebrates and root systems of plants).
8. Contaminants in groundwater tend to be removed or reduced in concentration with time and distance due to retention and attenuation processes.
9. Soil and groundwater exposures depend upon the nature of the contaminant, soil properties, and environmental pathways of exposures due to natural sources and human activities.
10. Environmental pathways involve primary exposures (e.g. ingestion of soils, inhalation of dust particles, and volatile chemicals by air-breathing animals; consumption of contaminated plants and animals) and secondary exposures (e.g. drinking contaminated waters and atmospheric deposition of contaminants). Plants are exposed to contaminants in soils via root systems and leaves from airborne contaminants (pollutant gases, volatile organic compounds, and fine particulates).
11. Soil and groundwater pollutants adversely impact upon urban and natural ecosystems, agriculture, and ecosystem services (e.g. food production, nutrition, and quality). Major effects include acidification of soils and water bodies, altered and reduced metabolism of microorganisms, inhibition of plant photosynthesis, toxic effects on soil invertebrates and terrestrial plant communities, reduced crop productivity and losses, food chain contamination, and eutrophication of catchments and water bodies.

12. Toxic chemical residues (e.g. heavy metals and pesticides, and antibiotics) can also contaminate food crops and animal products used as feedstuffs for livestock and human foodstuffs.
13. Almost all human populations are exposed to multiple soil and groundwater pollutants. Exposures may include heavy metals (e.g. As, Cd, Pb, and Hg), organic chemical contamination (e.g. dioxins, PCBs, and other POPs), pesticides, pharmaceutical residues (e.g. estrogen and antibiotics), and soil pathogens (e.g. antibiotic-resistant bacteria, prions, and anthrax).
14. Known health risks include neurotoxicity, renal disease, autoimmune diseases, cancers, cardiovascular diseases, respiratory diseases, and congenital malformations.
15. Epidemiological studies of residents living on or near contaminated sites show sufficient evidence for stress, anxiety, psychiatric disorders and limited evidence for renal dysfunction and bone disorders, birth defects, low birth weight, and preterm deliveries. Evidence from cancer studies is regarded as inadequate due to methodology limitations and inconsistent findings between studies although some studies show increased incidences of cancers at higher levels of known carcinogens.
16. Contaminated soils and groundwaters are a major environmental and human health issue of current global concern. Available evidence suggests there are millions of potentially contaminated sites (e.g. Europe) and enormous areas of contaminated agricultural lands on a world scale. The magnitude of the health impacts is likely to be underestimated, particularly for historic exposures in developed countries and largely unknown in developing countries where land contamination and hazardous wastes are emerging issues for investigation, health risk assessment, remediation, and management.

References

Allison, J. D., and Allison, T. L. (2005). Partition Coefficients for Metals in Surface Water, Soil, and Waste. Report EPA/600R-05/074. United States Environmental Protection Agency, Washington, DC.

ATSDR (2017). *Toxicological Profile for Polybrominated Diphenyl Ethers (PBDEs)*. Atlanta, GA: Agency for Toxic Substances & Disease Registry. US Department of Health and Human Services, Public Health Service.

ATSDR (2019). *Toxicological Profile for DDT, DDE, and DDD (Draft for Public Comment)*. Atlanta, GA: Agency for Toxic Substances & Disease Registry. US Department of Health and Human Services, Public Health Service.

Connell, D.W. (2005). *Basic Concepts of Environmental Chemistry*, 2e. Boca Raton, FL: CRC Press.

Defra (2009a). Potential Health Effects of Contaminants in Soil – SP1. *Science and Research Reports*. UK Department for Environment, Food and Rural Affairs, London. http://sciencesearch.defra.gov.uk/Default.aspx?Module=More&Location=None&ProjectID=16185 (accessed 23 March 2020).

Defra (2009b). SP1002 Appendix 1 Objective 1: Overview of risks to human health posed by contaminated land. UK Department for Environment, Food and Rural Affairs. London. randd.defra.gov.uk › Document › Document=SP1002_9618_FRA (accessed 23 March 2020).

Ecetoc (2013). *Environmental risk assessment of ionisable compounds*. Technical Report No. 123. European Centre for Ecotoxicology and Toxicology of Chemicals, Brussels. http://www.ecetoc.org/report/estimated-partitioning-property-data/computational-methods/log-koc/ (accessed 22 March 2020).

EEA (2014). *Progress in management of contaminated sites*. European Environment Agency, Copenhagen. https://www.eea.europa.eu/data-and-maps/indicators/progress-in-management-of-contaminated-sites-3 (accessed 22 March 2020).

Fairbrother, A. and Hope, B. (2005). Terrestrial ecotoxicology. In: *Encyclopedia of Toxicology*, 2e (ed. P. Wexler), 138–142. Limerick, Ireland: Elsevier.

FAO/ITPS (2015). *Status of the World's Soil Resources (SWSR) – Main Report*. Rome, Italy: Food and Agriculture Organization of the United Nations and Intergovernmental Technical Panel on Soils.

GSI Environmental (2011). Bisphenol A. GCI Chemical Properties Database. Oakland, CA. https://gsi-net.com/en/publications/gsi-chemical-database.html (accessed 24 March 2020).

Karickhoff, S.W. (1981). Semi-empirical estimation of sorption of hydrophobic pollutants on natural sediments and soils. *Chemosphere* 10 (8): 833–846.

LaGrega, M.D., Buckingham, P.L., Evans, J.C. et al. (2001). *Hazardous Waste Management*, 2e. New York: McGraw Hill.

Mao, Z., Zheng, X., Zhang, Y. et al. (2012). Occurrence and biodegradation of nonylphenol in the environment. *International Journal of Molecular Sciences* 13 (1): 491–505.

Martuzzi, M., Pasetto, R., and Martin-Olmedo, P. (2014). Industrially contaminated sites and health. *Journal of*

Environmental and Public Health 2014: 198574. http://dx.doi.org/10.1155/2014/198574.

Ng, E., Lwanga, E.H., Eldridge, S.M. et al. (2018). An overview of microplastic and nanoplastic pollution in agroecosystems. *Science of the Total Environment* 627: 1377–1388.

RAIS (2019). *EPA Ecological Risk Assessment Guidance. The Risk Assessment Information System*. University of Tennessee, United States. https://rais.ornl.gov/guidance/epa_eco.html (accessed 23 March 2020).

Rodríguez-Eugenio, N., McLaughlin, M., and Pennock, D. (2018). *Soil Pollution: A Hidden Reality*. Rome: Food and Agricultural Organization.

Science Communication Unit, University of the West of England, Bristol (2013). Science for Environment Policy In-depth Report: Soil Contamination: Impacts on Human Health. Report produced for the European Commission D G Environment. http://ec.europa.eu/science-environment-policy (accessed 23 March 2020).

US EPA (2019). AP-42: Compilation of Air Emission Factors. United States Environmental Protection Agency. https://www.epa.gov/air-emissions-factors-and-quantification/ap-42-compilation-air-emissions-factors (accessed 22 March 2020).

US EPA (2020). Regional Screening Levels (RSLs) – Risk Assessment: Regional Screening Levels (RSLs) User's Guide. May 2020. United States Environmental Protection Agency. https://www.epa.gov/risk/regional-screening-levels-rsls-users-guide (accessed 27 August 2020).

Weber, J.B., Wilkerson, G.G., and Reinhardt, C.F. (2004). Calculating pesticide sorption coefficients (K_d) using selected soil properties. *Chemosphere* 55: 157–166.

WHO (2013). *Health and the Environment in the WHO European Region - Creating Resilient Communities and Supportive Environments*. Copenhagen, Denmark: World Health Organization European Regional Office.

Witteveen+Bos and TTE consultants (2016). Emerging Contaminants Inventory - Factsheets PFOS and PFOA: Introduction into the Properties. Netherlands. http://www.emergingcontaminants.eu/index.php/background-info/Factsheets-PFOS-intro/Factsheets-PFOS-properties (accessed 22 March 2020).

Wright, R.T. and Boorse, D.T. (2014). *Environmental Science: Toward a Sustainable Future*, 12e. Harlow, Essex, England: Pearson.

16

Solid, Liquid, and Hazardous Wastes

16.1 Introduction

Current consumption rates of resources along with waste disposal practices are considered ecologically unsustainable and harmful at all levels.

> …Wastes place great pressure on the environment and are closely linked to patterns of unsustainable consumption and production
>
> (Landon 2006, p. 70)

Managing unsustainable waste and resource uses by human populations are major global issues for developed and developing countries. Waste is a major by-product of human activities that is released into the environment in the form of materials, substances, products, and energy. Its physical states cover gases, particulates, solids, and liquids, increasingly in complex chemical, biochemical, and microbiological mixtures. Unused or waste energy is released in various forms such as heat or thermal energy from combustion or biodegradation, nonionizing and ionizing radiation, noise, and visible light. Critically, wastes have the capacity to impact on environmental and human health and quality of life, at local to global scales. As discussed in previous chapters of this book, environmental pollution, including anthropogenic climate change, is the adverse outcome from unsustainable releases of wastes into the global environment.

Wastes, including energy, are generated by human activities, from natural resource extraction (e.g. mining and forestry), energy production, agriculture, processing of materials and food sources, manufacturing of products, construction and demolition, modes of transport, to industrial, commercial and domestic use, and disposal of materials and products. From a waste management perspective, wastes are generally divided into controlled or uncontrolled, especially for collection, treatment and disposal purposes, and classified according to regulatory criteria such as nonhazardous or hazardous, or more specific types based on their source, nature and/or properties.

Over the last 1000 years, three primary drivers for waste management have emerged mainly from the experiences of developed countries with solid and liquid wastes:

- Resource value of wastes
- Public health and safety
- Environmental protection

Prior to the Industrial Revolution, the resource value of wastes became recognized through the repair and reuse of products, recycling of materials, and composting of food, animal, and plant wastes for growing crops and feeding animals. The mid-nineteenth century saw the emergence of public health practices (e.g. collection and disposal of wastes and water sanitation) following the concentration of human populations and their wastes, noxious fumes and smoke, and outbreaks of infectious disease in the new industrial cities of the world, primarily in Europe and United States. Public health developed widely up until the 1970s, followed by environmental health. Public demand for environmental protection of humans and wildlife from pollution and hazardous wastes (e.g. toxic and bioaccumulative chemicals, pesticide residues, air, water, marine, and soil pollutants) resulted in diverse legislation and regulatory controls of wastes and their management practices on local, national, then international scales for prevention and management of hazardous wastes and chemicals (e.g. POPs).

Current waste management in resource use is adopting more holistic design approaches, waste prevention, and minimization, including advanced methods of recovery of resource value and also climate change mitigation of greenhouse gas emissions. Globally, the proportion of collected municipal solid wastes is moving toward 100% in developed countries and increasing in developing countries. The collection, treatment, and disposal of solid wastes, and hazardous or special waste categories are the focus of global waste management policies and programs in many developing countries (UNEP/ISWA 2015).

Chemistry and Toxicology of Pollution: Ecological and Human Health, Second Edition. Des W. Connell and Greg J. Miller.
© 2023 John Wiley & Sons, Inc. Published 2023 by John Wiley & Sons, Inc.
Companion website: www.wiley.com/go/toxicologyofpollution2e

16.2 Wastes: Types and Hazardous Characteristics

What Are Wastes?

Waste is commonly described as a by-product of human activities in a broad sense or can be considered simply as an unwanted or unusable substance, material, or product. Legal definitions of waste vary in scope and complexity of technical classification, especially in the case of hazardous wastes. Waste legislation in the European Union defines *waste* in Article 3(1) Waste Framework Directive (WFD) as: *any substance or object which the holder discards or intends or is required to discard.*

From an economic perspective, waste has been generally defined as: *any product or substance that has no further use or value for the person or organisation that owns it, and which is, or will be, discarded. It thus excludes products or substances that are reused or sold by the organisation that owns them.* This definition can include products that are recoverable, such as through reuse, recycling, or by energy extraction (Australian Productivity Commission 2006, p. 2).

Wastes can be divided into several broad categories such as solid wastes (e.g. municipal solid wastes), liquid wastes (e.g. sewage), hazardous wastes and other special, controlled, scheduled or regulated wastes (e.g. medical or clinical wastes, toxic chemical wastes, and contaminated soils), mining and quarry wastes, and agricultural and forestry wastes.

Classifications of wastes can vary substantially between countries, provinces, states, local government areas, etc., depending largely on rate and scale of industrial development. Complex laws and regulations are increasingly needed to control how waste materials are characterized, classified, stored, collected, transported, treated, recycled, and disposed in most countries.

The main categories of wastes are outlined in Table 16.1 along with examples of wastes in each category.

Table 16.1 General types of waste categories.

Types	Examples
Solid wastes[a]	
Municipal solid wastes (household or domestic)	Waste from kerbside collection of household wastes, community wastes, and hard waste collection (mainly organic materials, such as paper, garden and kitchen waste, and also plastics and glass)
Construction and demolition	Mainly inert materials: timber, bricks, plaster and fiberboard cuts, concrete, rubble, steel, ceramics, plastic materials, and excavated earth
Commercial and industrial	Metals, plastics, timber, paper and cardboard, food wastes, discarded plant equipment and materials.
Agriculture and forestry	Biomass materials, crop wastes, manure, biosolids, forestry residues, and machinery
Mining and quarrying	Nonferrous and ferrous mining wastes, mine tailings, and overburden
Energy production	Coal, petroleum and gas extractions and processing wastes, power generation residues, discarded equipment, and materials
Water supply, sewage treatment, waste management, and land remediation	Treatment residues (e.g. water treatment, sludges, sewage sludges and biosolids, incineration ash, non-recyclables, biomass residues, excavated, and waste materials (e.g. plant, equipment and infrastructure)
Liquid wastes	
Sewage (raw or blackwater and graywater) and some combined urban stormwaters, treated and untreated effluents	Domestic, commercial, industrial, and rural sources and discharges
Graywaters or sullage	Mainly domestic
Grease trap	Commercial (e.g. retail food outlets) and food processing
Industrial wastewaters	Untreated, treated, or sewerable
Recycled wastewaters	Treated sewage, industrial, and agricultural irrigation
Contaminated stormwaters	Mainly urban and intensive agricultural runoff
Hazardous wastes	
Other wastes: special, scheduled, controlled or regulated	Pesticide wastes, metal solutions and sludges, toxic chemical wastes, and radioactive wastes
	Liquid putrescible or biodegradable (e.g. grease trap wastes), waste tires, asbestos, clinical and related wastes, industrial or trade wastes, including sludges, sewage sludges, electronic wastes, and quarantine wastes

[a] Waste categories can vary. Solid waste sources in Asia are generally described in Landon (2006, p. 71) as: residential, industrial, commercial, institutional, construction and demolition, municipal services, process, and agriculture.

Hazardous and Special Wastes

Hazardous wastes are generally wastes with properties that make them dangerous or potentially harmful to human health or the environment. Key hazardous properties or characteristics of hazardous wastes can include one or more of being flammable, explosive, toxic, or radioactive such as used by the Basel Convention on the Control of Transboundary Movements of Hazardous Wastes and their Disposal, an international agreement ratified by 166 countries. The Basel Convention also lists particular types of waste that are considered *hazardous wastes*. These include waste streams such as clinical waste, waste from specific production processes, and some constituents of waste such as zinc, mercury, lead, and asbestos. Note: potentially hazardous wastes may be found in municipal solid wastes (Australian Productivity Commission 2006, pp. 5–6).

An example of the complexity in defining and classifying wastes and their hazardous properties is described below.

The European Union Guidance Document on wastes provides clarifications in accordance with the existing EU legislation and takes into account guidelines on waste classification from various EU Member States (see Official Journal of the European Union 2018). This technical guidance may be updated as necessary in light of the experience with the implementation of the EU relevant legislation.

In particular, The Commission Notice on technical guidance on the classification of waste (2018/C 124/01) gives technical guidance on certain aspects of Directive 2008/98/EC on waste ("Waste Framework Directive" or "WFD") and Commission Decision 2000/532/EC on the list of waste ("List of Waste" or "LoW"), as revised in 2014 and 2017. It provides clarifications and guidance to national authorities, including local authorities, and businesses (e.g. for permitting issues) on the correct interpretation and application of the relevant EU legislation regarding the classification of waste, namely identification of hazardous properties, assessing if the waste has a hazardous property and, ultimately, classifying the waste as hazardous or nonhazardous.

The Waste Framework Directive (WFD) describes *what waste is and how it should be managed*. It *defines a "hazardous waste"* in its Article 3(2) as:

waste which displays one or more of the hazardous properties listed in Annex III. There are fifteen hazardous properties listed in Annex III as shown in Table 16.2.

Annex III provides principles and guidelines on how to assess the individual hazardous properties HP1 to HP15 via calculation or testing.

Table 16.2 Hazardous properties of wastes.

Code	Hazardous properties
HP1	Explosive
HP2	Oxidizing
HP3	Flammable
HP4	Irritant – skin irritation and eye damage
HP5	Specific Target Organ Toxicity (STOT)/Aspiration toxicity
HP6	Acute toxicity
HP7	Carcinogenic
HP8	Corrosive
HP9	Infectious
HP10	Toxic for reproduction
HP11	Mutagenic
HP12	Release of an acute toxic gas
HP13	Sensitizing
HP14	Ecotoxic
HP15	Waste capable of exhibiting a hazardous property listed above not directly displayed by the original waste

Waste Framework Directive (WFD) Annex III. Compiled from European Commission notice on the technical guidance on classification of waste (2018/C 124/01).

16.3 Waste Sources, Generation, and Emissions

Global waste management data from OECD countries indicate six major waste categories, of which the three major waste streams are construction and demolition (C&D), commercial and industrial (C&I), and municipal solid waste (MSW), as shown by Figure 16.1 (UNEP/ISWA 2015, p. 54). Worldwide estimates of wastes from non-OECD countries are generally variable in classifications, inadequate, or unavailable.

Global best estimates of municipal solid wastes amount to about 2 billion tonnes per year. If a broader category of *urban wastes*, including MSW, commercial and industrial (C&I) waste, and construction and demolition (C&D) is considered, the estimated waste generation is about 7–10 billion tonnes per year (UNEP/ISWA 2015, p. 52).

Mining and quarrying, and agriculture and forestry represent major waste streams that are usually managed within countries and close to sources. Globally, *order of magnitude* estimates for each of these waste categories are reported as 10–20 billion tonnes per year, exceeding *urban waste* estimates. Most agricultural and forestry wastes are reused as soil improvers, nutrients, or as biofuels, while, in contrast, mine tailings are potential sources for health and environmental impacts (UNEP/ISWA 2015, p. 52).

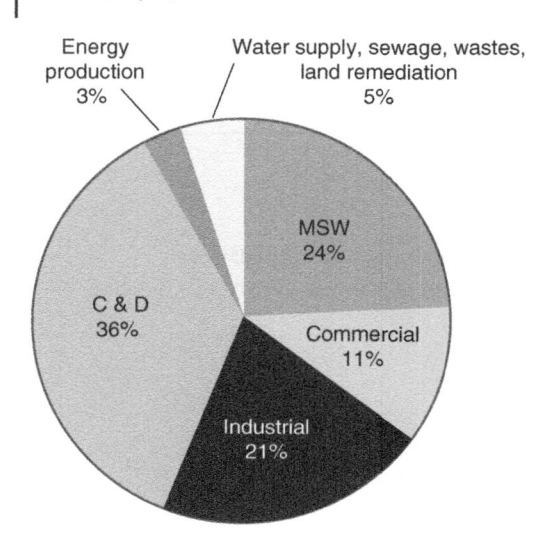

Figure 16.1 Relative quantities of waste from different sources in the material and product cycle, derived from OECD Data Figure 3.1. Source: UNEP/ISWA 2015, p. 54.

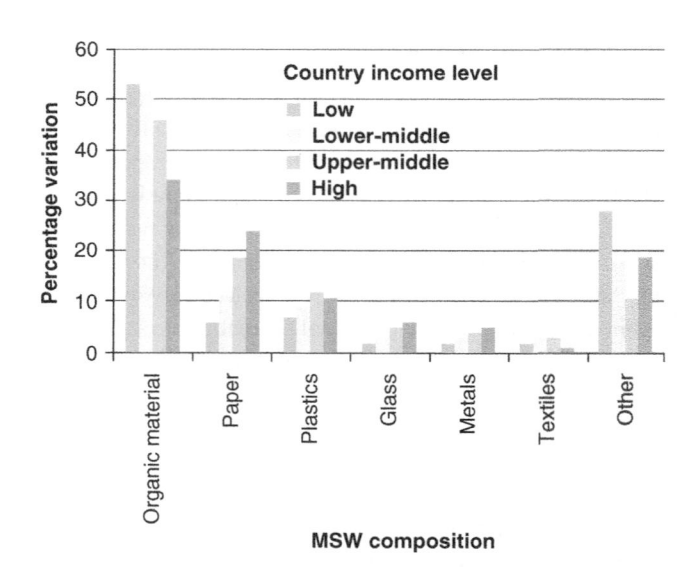

Figure 16.2 Percentage variation in MSW composition grouped by country income levels. Source: Data compiled from UNEP/ISWA 2015, p. 57.

Municipal Solid Wastes

The generation of wastes varies widely between countries. Even so, MSW generation rates depend on income, socio-cultural factors, and climatic factors. There is a strong positive correlation between waste per capita (kg year^{-1}) denoted by (y), and gross national income levels (USD) per capita (x) for 82 countries (UNEP/ISWA 2015, p. 55):

$$y = 109.6 \ln(x) - 651.45 \qquad R^2 = 0.72 \quad (16.1)$$

The trend for MSW, however, in high-income countries is that waste growth is starting to stabilize, or slightly decline while waste is steadily increasing in low- to middle-income countries where economies are growing rapidly. The predominant percentage of the composition of MSW is organic and other wastes but less so in upper middle- to high-income countries where paper, plastics, glass, and metals are a significant proportion, as shown in Figure 16.2.

Disposal of Municipal Solid Wastes

Total annual disposal of MSW on a worldwide basis is difficult to estimate because of variable methods, inconsistent reporting, and limited to no available data from many countries according to a global review of solid waste by the World Bank. Generally, landfilling and thermal treatment of waste are the most common methods of MSW disposal in high-income countries in contrast to disposal of wastes in dumps in low-income countries. Several middle-income countries have poorly operated landfills or controlled dumping.

Figure 16.3 contrasts MSW disposal practices in the world's richest (OECD) and poorest (Africa) regions, which have similar populations. The OECD region generates about 100 times the waste of Africa. In Africa, the vast

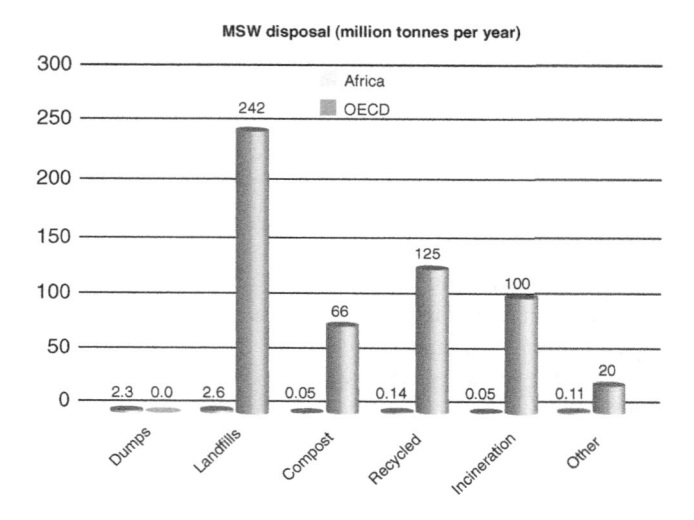

Figure 16.3 MSW disposal (million tonnes per year) by methods in OECD countries (high-income) and African countries (low-income). Source: Compiled from data derived from Hoornweg and Bhada-Tata 2012, Table 12.

majority of collected waste is dumped or sent to landfills compared to more than 60% of waste in the OECD countries being diverted from landfills, predominantly through composting, recycling, and incineration/waste to energy.

16.4 Waste Management

Waste Management Strategies

The waste hierarchy is the basic concept used as the cornerstone of most waste minimization strategies. The aim of the waste hierarchy is to extract the maximum practical benefits from products and to generate the mini-

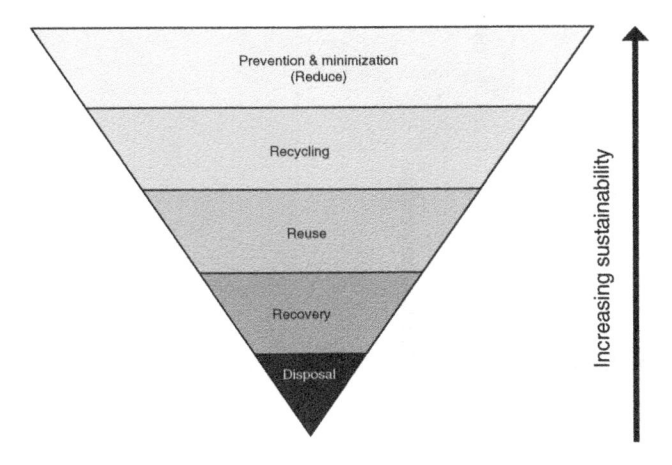

Figure 16.4 Generic waste management hierarchy showing increasing trend toward sustainability.

mum amount of waste. It may have from three to five steps, as shown in Figure 16.4.

It is an increasingly common practice to set waste minimization targets to reduce disposal by landfill and incineration techniques according to this hierarchy and disposal strategy. There are many management strategies for wastes as shown by the handling of municipal solid wastes. These depend to different degrees on the source and type of wastes involved, the capacity and financial viability of the different management methods (e.g. collection, recycling, composting, landfilling, incineration, or exporting), government policies (e.g. sustainability), legislation and other factors such as location, siting of facilities, and environment constraints (ABS 2006).

General waste management approaches used are:

1. Waste minimization including avoidance or reduction of wastes, such as reuse, recycling, and waste-to-energy recovery.
2. Waste treatment (landfilling containment, incineration and composting)
3. Disposal in controlled landfills of all non-recycled solid wastes, whether they are domestic waste collected and transported directly to a landfill site, residual materials from Materials Recovery Facilities (MRF) and composting facilities, residue from combustion of solid waste, or other substances from various solid-waste-processing facilities. These landfills are usually designed and operated to meet specific environmental guidelines and regulatory requirements.

The general conceptual approach needed to manage increasing wastes from human activities within an environmental context is illustrated in Figure 16.5. The essential objective is to achieve healthy and sustainable outcomes through waste minimization and treatment, and working toward net *zero* waste emissions, including greenhouse gases from wastes, on a global scale.

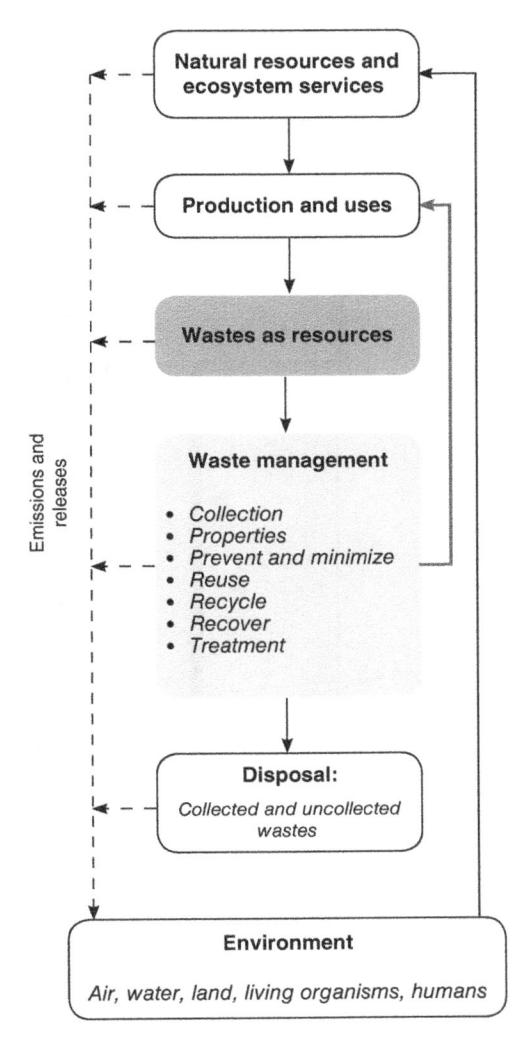

Figure 16.5 General waste management system indicating the added environmental flow of matter and energy: from the generation of wastes due to human uses of natural resources and ecosystem services, to production and use of materials, goods and services, to managing and reducing the adverse impacts of the disposal of waste materials, and their emissions, on the health and sustainability of natural and human environments.

Waste Treatment Methods

The major types of solid, liquid, and hazardous waste treatment techniques are summarized below.

Landfilling of Solid Wastes

Landfill sites are commonly used around the world for the disposal of waste materials by burial, cover with soil or alternative materials (e.g. biosolids), and capping, usually with soils. It is the oldest form of waste treatment. Modern sanitary landfills are designed to contain physical, chemical and biological wastes, control pests and vermin, hazardous wastes, and pollutants such as odors, chemical contaminants, landfill gases, and pathogens. They use leachate control systems, landfill gas collection, and environmental management systems for wastes,

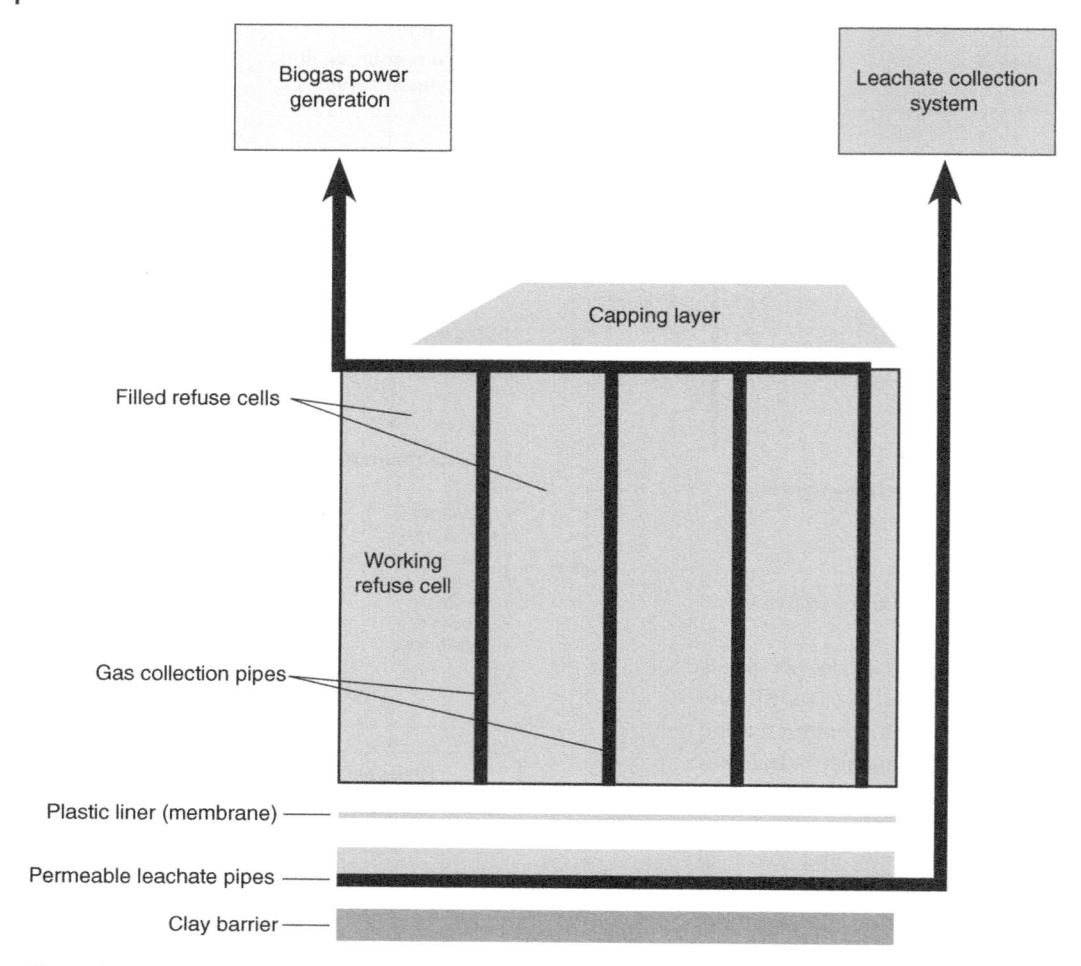

Figure 16.6 Conceptual diagram of a sanitary landfill.

based on from *cradle to burial* and rehabilitation of waste disposal sites.

Figure 16.6 presents a model of a modern sanitary landfill site containing multiple cells of refuse enclosed within an outer clay layer (walls and base) to enhance biodigestion of organic wastes and to prevent or reduce leachate escape. Landfill gases of primarily methane and carbon dioxide (GHGs) are produced by biodigestion (aerobic and/or anaerobic processes). These gases, and contaminants such as VOCs, may gradually escape as fugitive emissions, or be collected and used for energy generation. Burning of organic wastes in open or poorly managed landfills is a source of emissions of toxic combustion products. Landfilling can occupy large areas of land and contain large volumes of wastes that generate leachates and fugitive gas emissions (e.g. US EPA 2018).

This major type of landfill consists of a series of daily filled refuse cells, with compaction and soil cover, to form soil-enclosed cells in multiple layers, within an engineering designed membrane-lined outer clay barrier. A biogas collection and power generation system are indicated along with a leachate collection system (with deep monitoring wells) for potential recirculation in landfill and/or treatment, including the option of offsite discharge (e.g. US EPA 2018). Working refuse cell(s) is covered daily.

Waste Incineration

Incineration is a thermal waste treatment process that involves the combustion of organic substances, mainly composed of C and H, and sometimes combined with O, and also metals and nonmetals (e.g. halogens and nitrogen) that are contained in waste materials such as MSW and also in hazardous wastes. Incineration of waste materials converts the waste into ash (inorganics, residual carbon, and partially decomposed organics), flue gas and particulates, and waste heat.

In modern incinerators (e.g. high temperature and waste to energy conversion), flue gases may be treated to remove or reduce emissions of gaseous and particulate pollutants, including acidic gases (NO_x and SO_2), VOCs, PAHs, mercury, arsenic, and dioxins, before they are dispersed into the atmosphere via stack emissions. Treatment methods may include wet scrubbers, electrostatic precipitators, and/or baghouse particulate filters.

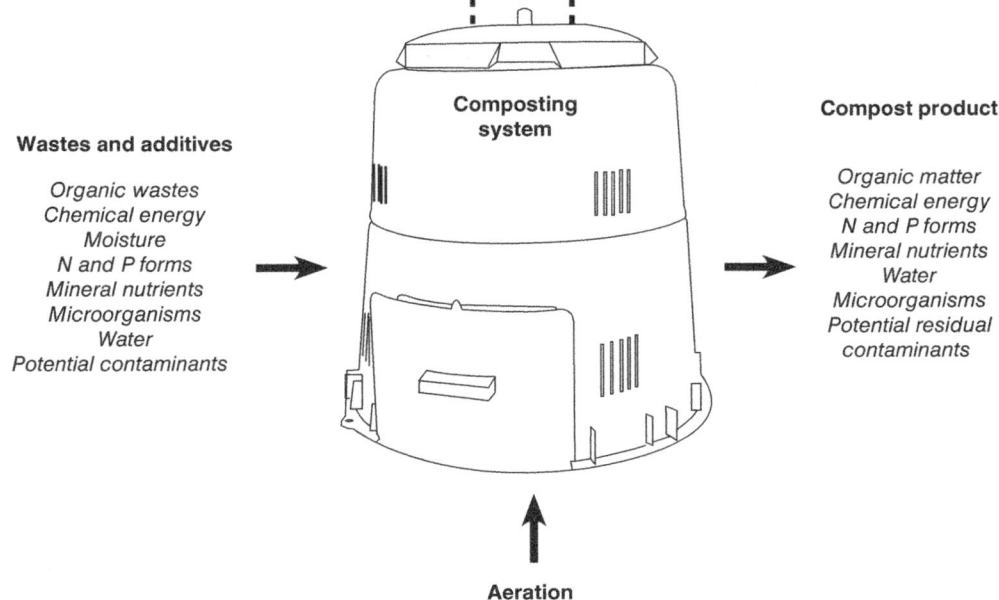

Figure 16.7 A generalized aerobic composting system indicating inputs (wastes, additives, and aeration), outputs (composting product), air emissions (and leachates), and potential residual contaminants (e.g. persistent organic chemicals, heavy metals, and pathogens). Large-scale or industrial composting systems contain air emission and leachate controls, including enclosure in cold climates or near sensitive land uses.

The operation, emissions, and monitoring of air pollutants from incinerators are usually well regulated, particularly from hazardous waste incinerators, for example, by measuring the destruction and removal efficiency (DRE) of principal organic hazardous constituents (POHC) during a trial burn. For instance, the incinerator must demonstrate a capacity to achieve a 99.99% DRE, as calculated by equation (16.2):

$$\text{DRE} = \frac{(W_{in} - W_{out})}{W_{in}} \times 100 \qquad (16.2)$$

Where DRE = destruction and removal efficiency (%); W_{in} = mass feed rate of particular POHC (kg h^{-1}); W_{out} = mass emission rate of the same POHC (kg h^{-1}).

Waste to energy (WtE) incinerators commonly recover the heat energy from combustion and convert it into electrical energy in the form of electricity and/or use the heat energy for heating such as in urban developments in cold climates (e.g. Stockholm). Modern incinerators can reduce the volume of the original waste by 95% or so, depending upon composition and degree of recovery of materials such as metals from the ash for recycling.

Waste Composting

Composting is the natural biological process that involves the decomposition of organic matter by microorganisms (fungi, actinomycetes, yeasts, and bacteria), insects, worms, and other organisms to produce reusable nutrient and microbial-rich organic matter (biosolids). It can convert complex organic materials (municipal wastes, manure, sludge, plant matter, fruit and vegetables, and food wastes) into simpler organics and humic materials under aerobic and anaerobic conditions, and typically at thermophilic temperatures (LaGrega et al. 2001, p. 604). Generally, composting processes rely on bacterial and fungal organisms to break down organic matter. In vermiculture, organic matter is decomposed primarily by worm species.

The composting process is used to treat municipal solid wastes and municipal sewage sludge, typically at higher operating temperatures (55–65 °C) to allow thermophilic bacteria to thrive and sterilize the wastes of pathogenic bacteria (see Figure 16.7).

Bioreactor Landfills

A bioreactor landfill operates to rapidly transform and degrade organic waste by increasing biodegradation and waste stabilization through addition of air and water or leachates to enhance microbial processes. There are three different basic types of bioreactor landfills:

Aerobic – In which leachate is removed from the bottom layer, piped to liquid storage tanks, and recirculated into the landfill in a controlled manner. Air is injected into the waste mass, using vertical or horizontal wells, to promote aerobic activity and accelerate waste stabilization.

Anaerobic – In which moisture is added to the waste mass in the form of re-circulated leachate and other sources to obtain optimal moisture levels. Biodegradation occurs in the absence of oxygen and produces landfill gas. Landfill gas, primarily methane, can be captured to minimize greenhouse gas emissions and for energy project.

Hybrid (Aerobic–Anaerobic) – In which waste degradation is accelerated by using a sequential aerobic–anaerobic treatment to rapidly degrade organics in the upper sections of the landfill and collect gas from lower sections. It promotes earlier onset of methanogenesis compared to aerobic landfills.

The most critical factor compared to landfills that simply recirculate leachates for liquids control is to inject moisture in the form of leachates, stormwaters, wastewaters and/or treatment plant sludges to produce and maintain optimal moisture content (35–65% of field capacity) for stimulation of natural microbial processes. Landfill gas production is also enhanced in bioreactors.

Advantages of bioreactors include (i) rapid degradation of degradable organics and sequestration of inorganics under aerobic and/or anaerobic conditions in MSW, (ii) improved quality of leachates, and (iii) potential decrease in landfill volume leading to an extended operational life of landfill. Earlier and increased generation of landfill gas and decreased fugitive emissions are also observed (e.g. US EPA 2019).

Treatment of Hazardous Waste

The general management system for hazardous wastes is illustrated in Figure 16.8. In developed countries, it typically involves waste minimization practices (including recovery and recycling, and use of selective treatment options, and disposal of some regulated wastes (e.g. asbestos and metal contaminated soils) and residual wastes (from treatment processes) to specialized landfills or cells. In less-developed countries, many hazardous wastes are often dumped or disposed to land.

Here hazardous wastes are taken to include a broader range of wastes, including other special and regulated wastes with hazardous properties (e.g. asbestos, radioactive wastes, clinical wastes, and contaminated soils).

Hazardous waste collection, handling, storage, and treatment depend very much on physical, chemical, and biological composition and properties of the wastes, and their hazardous characteristics (see also Table 16.2). As indicated in Table 16.3, treatment technologies are selective and can be complex for mixtures, toxic, radioactive, and recalcitrant substances.

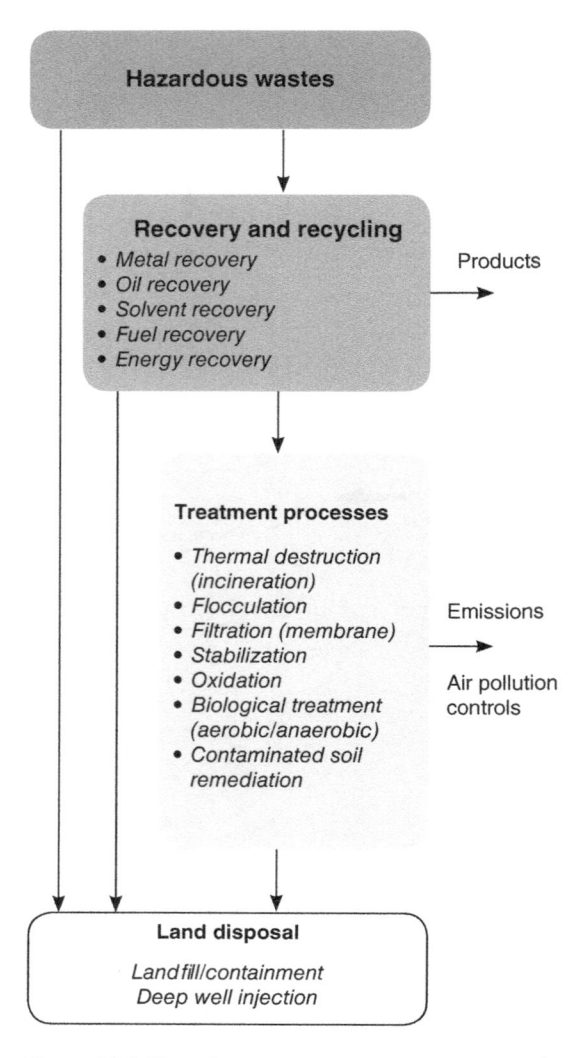

Figure 16.8 Hazardous waste recovery, treatment options, and disposal to land. Arrows show the flows of wastes and residuals from recovery and treatment.

Releases or emissions to the environment are normally regulated and require pollution control technologies, monitoring of processes (e.g. biomedical incinerators), monitoring of residues, and reporting on performance. Potential health risks are issues such as for incineration waste workers and residents living near incinerators (see Section 16.8).

Knowledge of pollutant emissions or releases from waste activities is an important part of pollution evaluation, control, and waste management. Two worked examples of emission calculations for pollutants released by waste treatment activities are given below.

Table 16.3 Examples of hazardous waste treatment technologies.

Treatment technology	Types of contaminants and wastes
Physical	
Clarification, coagulation, flocculation, sedimentation, filtration	Suspended solids, aqueous metals/inorganic solutions, industrial wastewaters
Air stripping	Volatile organic compounds
Carbon adsorption (activated carbon/granular GAC)	Aromatics, halogenated aromatics, PAHs, nonionic organic pesticides
Membrane processes (e.g. electrodialysis, reverse osmosis, ultrafiltration)	Saline and brackish solutions, inorganic ions, organics in wastewaters, halogenated hydrocarbons, phenols, organic pesticides
Dechlorination using electroreduction	Chlorinated phenols
Chemical	
Neutralization	Iron, cyanide, organics, acidic wastes
Oxidation (ozone, hydrogen peroxide, ozone/hydrogen peroxide, chlorine, ozone/UV, oxygen, potassium permanganate, catalyzed/persulfate)	Cyanide, phenols, sulfide, metals (e.g. Fe^{2+}/Fe^{3+}), acetaldehyde, benzene, halogenated aliphatics
	Degradation of PFOA/ bisphenol A
Super critical fluid extraction and supercritical water oxidation	Organic compounds, organic halogenated, nitrogen, sulfur and phosphorus compounds, PCBs
Precipitation	Metals
Stabilization/solidification (matrices of modified clays, lime, fly ash, Portland cement/silicates/pozzolans, organic polymers, *in situ* vitrification)	Heavy metals, oils/sludges, toxic organics (e.g. aromatics, PCBs) – industrial wastes, pre-treatment of liquid and solid wastes for secure landfill disposal, contaminated soils
Biological	
Aerobic (suspended, fixed film, batch, and slurry phase systems)	Volatile and non-volatile organic compounds, phenols, PAHs, solvents (e.g. PCE), petroleum oils/sludges, pesticides, contaminated ground water, leachates, contaminated soils, pharmaceutic residues
Ex situ systems (composting and land treatment); *in situ* biological treatment systems; bioreactors; soil bioremediation, biofilters	VOCs, e.g. toluene and odor chemicals
Anaerobic (e.g. dechlorination)	Halogenated solvents, phenols, dry cleaning wastes (e.g. PCE), sulfonated benzene, polysulfide rubber wastewaters
Thermal	
Liquid and vapor incinerators (thermal and catalytic)	Liquids, sludges, slurries, waste oils, halogenated solvents; VOCs
Fluidized bed	Solid (and liquid) hazardous wastes – high organics, pesticides, clinical wastes, pharmaceuticals, industrial organic wastes, POPs (e.g. PCBs); auxiliary fuels (natural gas, light petroleum)
Rotary kiln	
Fixed and multiple hearth	
Secondary combustion chambers	

Source: Compiled from LaGrega et al. (2001), Parts III and IV; Wang et al. (2019).

Worked Example 16.1 *Emissions of Carbon dioxide Released from Waste Incineration* (see Australian Department of Environment and Energy 2016)

$$E_i = Q_i \times CC_i \times FCC_i \times OF_i \times 3.664 \quad (16.3)$$

Where:

E_i is the emission of carbon dioxide released from the incineration of waste type (i) by plant during the year measured in CO_2-e tonnes.

Q_i is the quantity of waste type (i) incinerated by plant during the year measured in tonnes of wet weight value.

CC_i is the carbon content of waste type (i).

FCC_i is the proportion of carbon in waste type (i) that is of fossil origin.

OF_i is the oxidation factor for waste type (i).

Problem: Estimate the quantity of carbon dioxide released from the incineration of 3000 tonnes per year of clinical wastes.

Use default values given by IPCC in Australian Department of Environment and Energy (2016) for clinical wastes: $CC_i = 0.60$; $FCC_i = 0.40$; and $OF_i = 1.0$.

Answer:

$E_i = 3000 \times 0.6 \times 0.4 \times 1.0 \times 3.664 =$ **2638 CO_2-e tonnes**

Worked Example 16.2 *Leachate of Pollutants from Landfill (see NPI 2010, pp. 16–17)*

Problem 1: Annual estimation of total nitrogen (N) or total phosphorus (P) released in leachate from a municipal solid waste (MSW) landfill.

Step 1: Calculating the mass of total nitrogen and total phosphorus is best done using direct measurement techniques. Total nitrogen (TN) is considered to be the sum of nitrate, nitrite, ammonia, and organic nitrogen all expressed as nitrogen. Total phosphorus (TP) is the sum of all inorganic and organic forms in water, expressed as phosphorus.

Step 2: The quantity of either TN or TP emitted can be calculated:

$$Q_S = V \times C_S \tag{16.4}$$

Where: Q_S is the quantity of substances released (kg year^{-1}); V is the release rate of leachate (m^3 year^{-1}); and C_S is the average concentration of the substance in the leachate (kg m^{-3}).

Note: Only leachate that flows out of the landfill system should be measured. If leachate routinely comes from an onsite treatment plant, then the flow would be measured.

Step 3: If leachate is not collected and is released, then an estimate of annual discharge can be used:

$$V = 0.15 \times R \times A \tag{16.5}$$

Where: V is the volume of leachate discharged in a year (m^3 year^{-1}); R is the annual rainfall (m), and A is the surface area of the landfill (m^2).

Problem 2: Estimate the quantity of total phosphorus (TP) released in leachate with a TP concentration of 5 mg L^{-1} from a MSW landfill with an area of 10 ha and an annual rainfall of 1.5 m.

Step 1: Estimate annual leachate volume (discharge) from equation (16.5)

$$V = 0.15 \times 1.5\,\text{m} \times 100\,000\,\text{m}^2 = 22\,500\,\text{m}^3\,\text{year}^{-1}$$

Step 2: Estimate the annual quantity of total phosphorus released from MSW landfill using equation (16.4):

$$Q_S = 22\,500\,\text{m}^3\,\text{year}^{-1} \times 0.005\,\text{kg m}^{-3} = \mathbf{112.5\,kg\,year^{-1}}$$

(0.112 tonnes year^{-1})

16.5 Environmental Behavior and Fate of Waste Contaminants

The transport and transformation processes that influence the environmental distribution of contaminants released from waste streams generated by human activities are shown in Figure 16.9. The nature and effects of these processes on diverse type of pollutants or contaminants generated by human-related wastes are described in preceding chapters.

Waste disposal (liquids and solids from urbanization, mining, agriculture, wastewaters, biosolid and manure applications, landfilling of wastes, and soil contamination) and deposition of contaminants (e.g. nutrients, acid rain, heavy metals, pesticides, antibiotics, and pathogens) on land and in aquatic systems results in the distribution and biogeochemical cycling of these substances and their breakdown products between land and aquatic-based environments, and the atmosphere.

The fate of combustion products (particulates, black carbon, toxic acid gases, greenhouse gases, VOCs, PAHs, dioxins, and metals) emitted from burning combustible wastes and fugitive emissions from sources such as incineration, biogas, landfills, waste treatment, and other human-related waste activities (e.g. forest clearing and burning, oil spills) is largely determined in the atmosphere before wet and dry deposition on and/or sorption by soils, living biomass, and surfaces.

Increasing emissions of greenhouse gases from human activities, and to some extent by waste activities, are major contributors to disruption of atmospheric energy flows, global warming or heating, ocean acidification, and impacts due to climate changes (Chapter 14).

16.6 Exposures to Contaminants from Wastes

The principles of wildlife and human exposure to contaminants found in various environmental media (air, water, soils, sediments, dietary sources), food chains, materials, and wastes are covered in preceding chapters for different groups of pollutants (e.g. metals, petroleum hydrocarbons, pesticides, toxic organic contaminants, and PM$_{2.5}$). The release of contaminants from wastes, especially hazardous wastes of concern, results in the transfer of inorganic and organic contaminants, often as mixtures, into surrounding environments (natural, human, and workplaces) where transport and transformation processes distribute these substances in environmental media, according to their properties, interactions with these media, and prevailing environmental conditions.

It is also critical to identify who is exposed. This means we need to characterize the known, potentially exposed, and nonexposed populations or receptors, including sensitive or vulnerable sub-populations, and their communities.

The mean concentration of a contaminant that wildlife or humans may be exposed to is determined by defined environmental pathways (routes by which a contaminant

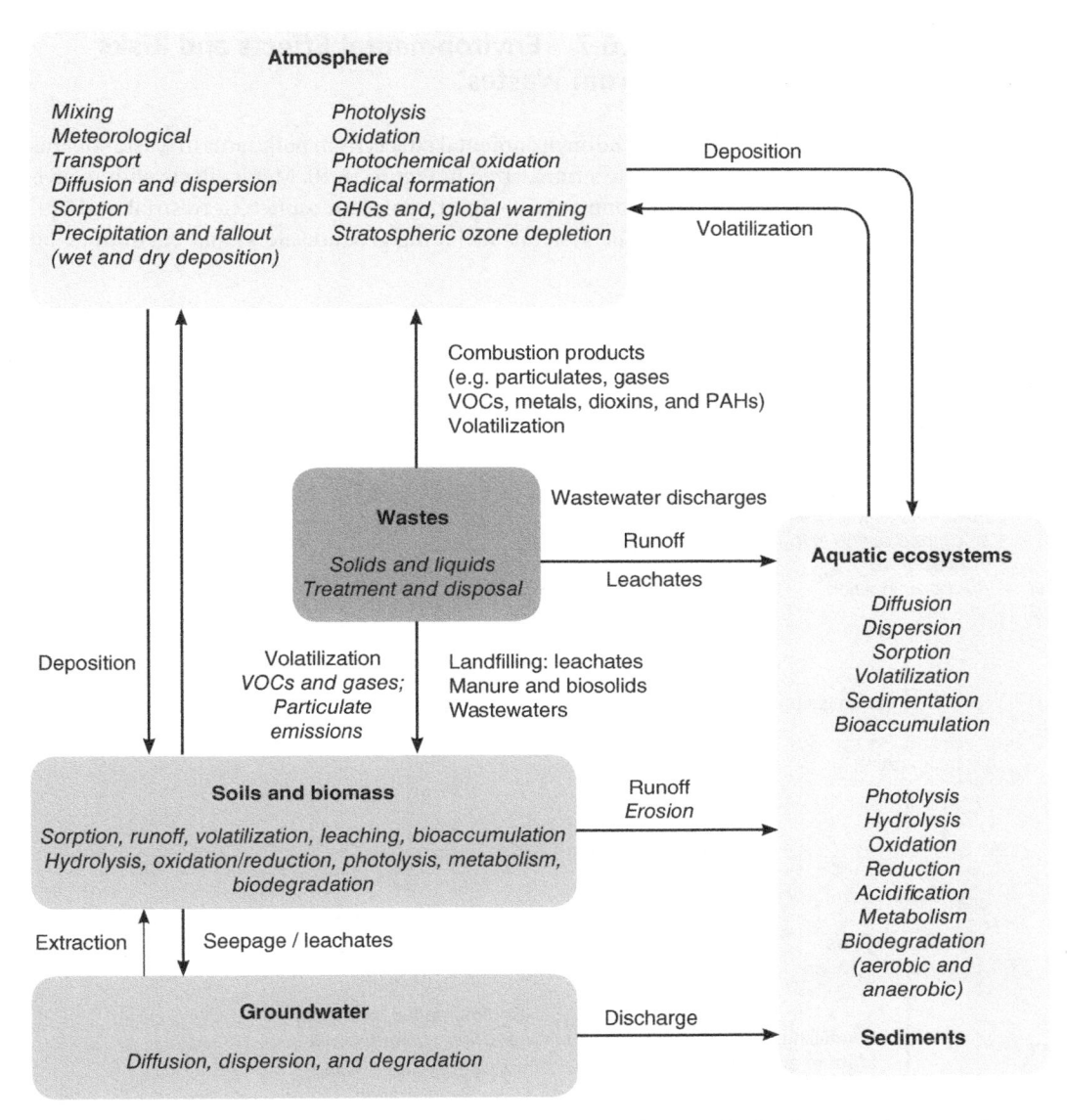

Figure 16.9 General model of behavior and fate processes affecting the distribution of environmental contaminants released from solid, liquid, and airborne waste streams. Note: pollutants in waste streams are commonly present as mixtures, from gross pollutants to particulates, chemicals, pathogens, and waste heat.

travels from the point of release to reach a receptor such as air emission source and dispersion by wind strength and direction) and exposure scenarios that are developed to describe the conditions of exposure (e.g. potential residential exposure from airborne fine particulates from a waste incinerator or a heavy metal discharge from a refinery into a freshwater lake). Exposure scenarios depend on factors such as quantities and frequency of release, volume of compartment(s), rates of transport (mobility), and transformation processes (e.g. biodegradation) that act on the contaminant, and bioavailability for uptake by receptor.

For complex mixtures of substances, such as in many waste types, potential contaminants are often difficult to detect but environmental levels of exposure to diverse contaminants can cover orders of magnitude, usually in the ppb to high ppm ranges (air, water, and tissues), $\mu g\ m^{-3}$ to $mg\ m^{-3}$ (air) or $\mu g\ kg^{-1}$ to $mg\ kg^{-1}$ (wet or dry weight (solids and tissues). Monitoring of emission and exposure levels may include direct reading instruments, sampling of environmental media or biological tissues, and advanced analytical techniques. Mathematical modeling of contaminant emission levels and rates can also be used to estimate exposure levels (e.g. air dispersion modeling of ground level concentrations of contaminants at receptors resulting from incinerator emissions).

Exposure assessment methodology, described in risk assessment (see Chapter 18), can be used to estimate exposure to specific contaminants of concern for populations that are at risk. Receptor exposures are usually reported as short-term (acute) and long-term (chronic) doses according to exposure pathways. Total doses are given by the combined doses for each exposure route from ingestion, inhalation, dermal or surface absorption, and dietary intake.

16.7 Environmental Effects and Risks from Wastes

The environmental effects from pollutants in waste streams are summarized in Figure 16.10. Major effects within environmental compartments (atmosphere, terrestrial, and aquatic systems, including groundwaters) and environmental

Atmosphere

Increasing global warming

Reducing depletion of ozone layer by CFCs

Accumulation and transport of toxic particulates and chemical pollutants (e.g. PM$_{2.5}$ and POPs)

Alteration of atmospheric composition due to human activities

Formation of toxic secondary pollutants (photochemical smog and ozone in lower troposphere)

Acid precipitation

Wastes
Solids and liquids Treatment and disposal

Incineration

Aquatic ecosystems
Deoxygenation
Salination
Sedimentation, dredging, and filling of sediments
Acid rain and oceanic acidification
Accelerated eutrophication
Bioaccumulation
Ecotoxicity from air and water chemical pollutants
Food web contamination (e.g., POPs, heavy metals, EDCs, and microplastics)
Waterborne pathogens
Impacts on fisheries, water bodies, wetlands, estuaries, oceanic ecosystems, and Polar regions
Loss of biodiversity and species
Global warming and climate change effects

Terrestrial ecosystems
Nitrogen eutrophication effects, major impacts on soil status, productivity, acidification, soil contamination

Increasing ecotoxicity, major and increasing loss of biodiversity, contamination of food chains, loss of forests, and agricultural lands

Major and increasing loss of biodiversity, global warming, and climate change

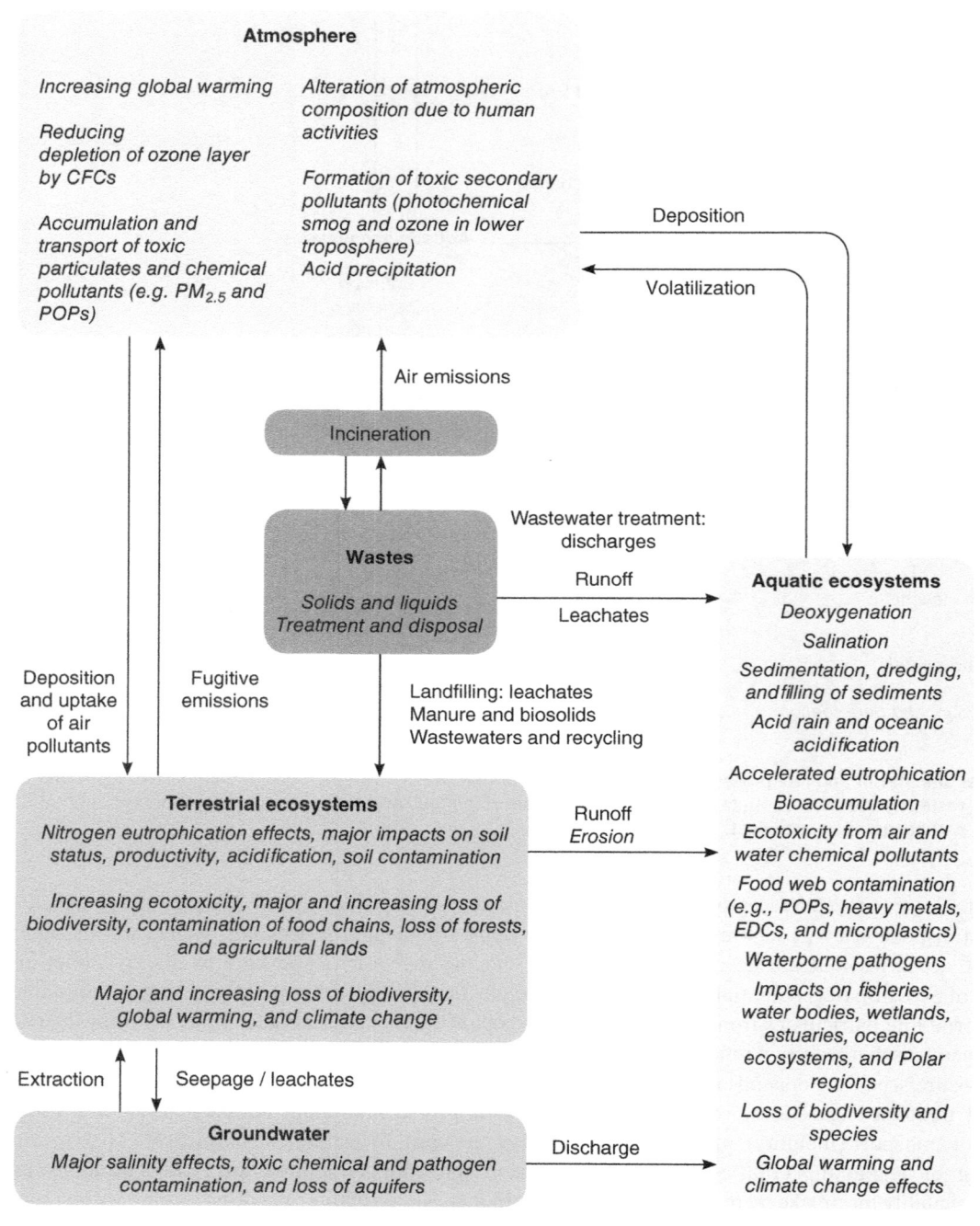

Groundwater
Major salinity effects, toxic chemical and pathogen contamination, and loss of aquifers

Figure 16.10 Generalized environmental effects of releases and emissions from pollutants in waste streams (solid, liquid, and air) via landfilling and land applications, wastewater discharge (treated and untreated), surface runoff and groundwater discharge, incineration and fugitive air emissions from urban and nonurban lands, and marine and freshwater systems.

flows between compartments are indicated. The atmosphere is experiencing global scale warming from point source and fugitive emissions of GHGs and black carbon. Resulting climate change is a major challenge for adaptation of numerous species to temperature shifts and loss of biodiversity on Earth. Ocean acidification from uptake of CO_2 from the atmosphere is affecting many marine organisms that produce calcium carbonate shells or skeletons. Increasing emissions, accumulation, and deposition of toxic air pollutants and fine particulates in urban and many regional areas are leading to large-scale ecotoxic and eutrophic effects on many terrestrial and aquatic communities (Chapter 13).

Terrestrial and aquatic pollution from growing waste disposal, land applications, catchment runoff and wastewater discharges, combined with atmospheric deposition/sorption of pollutants, is causing large numbers of *dead zones* and accelerated eutrophication in coastal and marine zones, and many freshwater lakes and wetlands. Bioaccumulation of POPs in food chains and long-range transport of these toxic chemicals are increasing evidence of food chain contamination, ecotoxicity effects (e.g. endocrine disruption), survival threats to many sensitive species, and contributing to declines in biodiversity of many species, notably mammals, fish, amphibians, reptiles, and invertebrates (e.g. insects).

Plastic pollution of the Earth's oceans is a dramatic example of emerging pollution from human waste disposal of litter/refuse (see Case Study 16.1 below).

Case Study 16.1

Plastics in the Marine Environment

Plastic pollution of the oceans occurs on a global scale. The rapidly growing influx and accumulation of synthetic organic plastic wastes (plastic debris including microplastic particles) in marine environments result from runoff in coastal catchments and the discard of plastic wastes, nets, ropes, and fishing lines at sea from human activities.

Injuries to and deaths of marine animals such as seabirds, whales, and turtles due to physical entanglement with and ingestion of plastic debris during feeding are widely observed and documented. Ingestion of plastic microparticles (and absorbed toxic pollutants) is contaminating many marine animals, from surface feeders to deep-ocean dwellers, fisheries, and their food webs (Derraik 2002; Figure 16.11).

Figure 16.11 Marine litter including plastics – Stanley Beach, Hong Kong. Source: Photograph by GJ Miller.

Annual global production of primary plastic was estimated as 270 million tonnes (2010) and plastic waste generated was 275 million tonnes (slightly higher due to some wastes from earlier years). Annual input of these wastes from coastal areas and rivers into the oceans due to mismanagement of landfills and discard of wastes was estimated to be 8 million tonnes. Marine-based inputs are ~2 million tonnes/year (Figure 16.12).

In comparison, cumulative production of plastics from 1950 to 2015 (major growth period of plastics industry) is estimated as 8300 million tonnes, with about 4900 million tonnes disposed to landfills or discarded, 800 million tonnes incinerated, and 2600 million tonnes still in use (see Figure 16.12).

The predominant source of plastic inputs into the oceans via rivers is from Asia, followed to a much less extent by Africa and South America. Other regions account for just over 1% of the world total. Inputs from land-based sources represent about 80% and marine sources about 20% (Ritchie and Roser 2018).

Eriksen et al. (2014) estimated that there was about 269 000 tonnes of plastic in surface waters of the world's oceans, well less than annual input estimates. This discrepancy is described as the "missing plastic problem." Potential sinks are marine sediments (shallow and deep sea) and ingestion by marine biota.

Insoluble plastic debris enters the marine environment in a wide range of sizes, from microscopic to the meter range. Common plastic fragments are polyethylene and polypropylene polymers. Microscopic plastics include manufactured plastic products (e.g. *scrubbers* or microbeads), industrial pellets, fragments, and fibers.

Case Study 16.1 (continued)

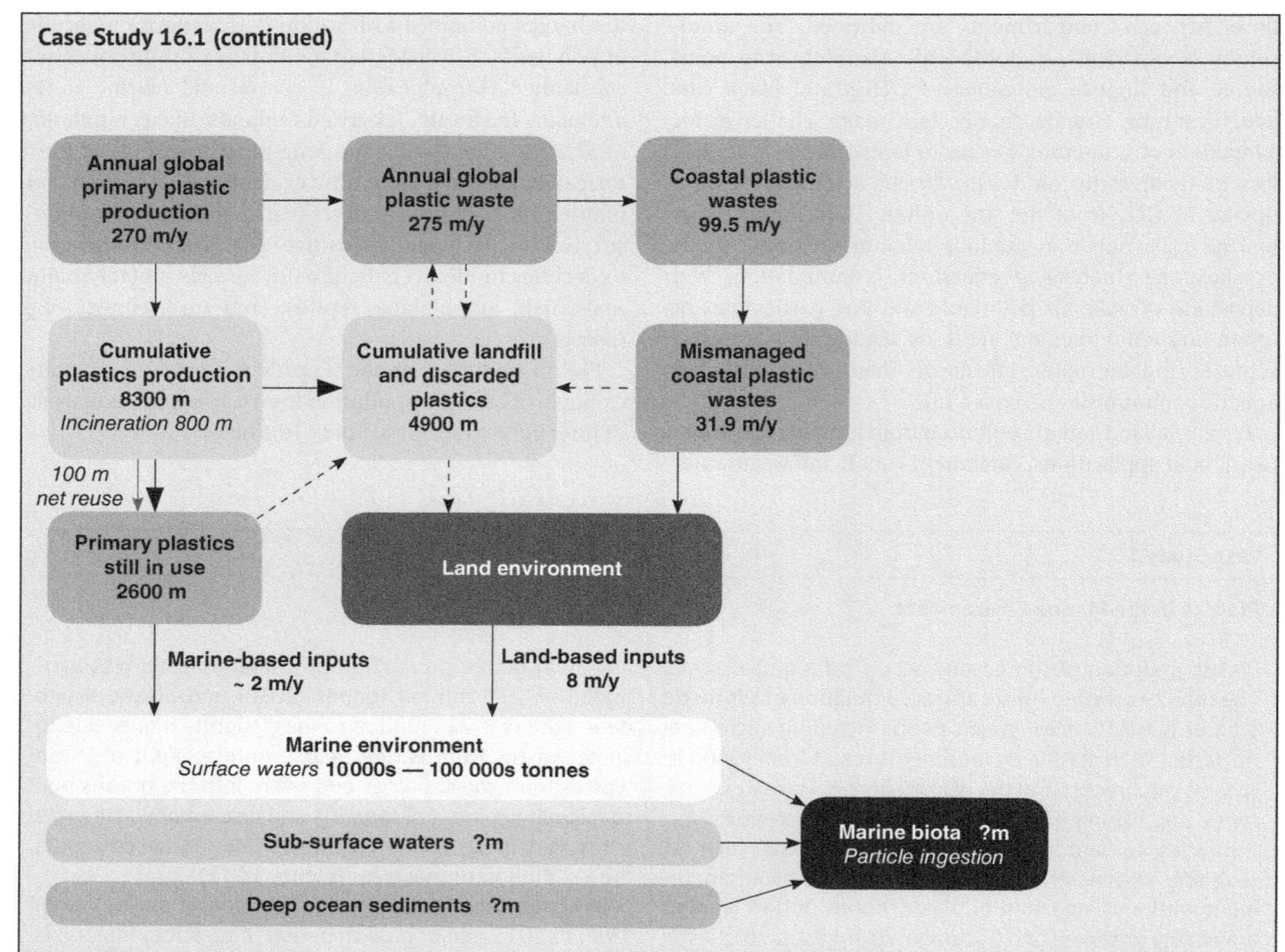

Figure 16.12 Annual global plastics production, plastic waste generation, and coastal (and marine) inputs entering the oceans based on 2010 global estimates by Jambeck et al. (2015). Estimates of global cumulative plastic production (1950–2015), plastic waste disposal and discard, and plastics *still in use* are also shown as first published by Ritchie and Roser (2018). Estimates of plastic pollution in surface waters are derived from Eriksen et al. (2014). m/y = million tonnes/year; m = million tonnes.

Generally, plastics float within the surface water layer and are transported by prevailing winds and ocean surface currents, tending to migrate and accumulate in oceanic gyres such as the *Great Plastic Cabbage Patch* in the northern Pacific Ocean (Ritchie and Roser 2018). Degradation of plastics (e.g. by mechanical abrasion from wave action, UV photo-oxidation, and biodegradation) is very slow and microplastics can be very persistent. They can also absorb persistent bioaccumulative and toxic organic pollutants and metals from seawater.

Marine plastic debris interacts with wildlife (seabirds, marine mammals, turtles, fish, and invertebrates) through three main pathways leading to physical injury, disease, death, or contamination with residues (Ritchie and Roser 2018):

(1) Entanglement, mainly with plastic ropes, netting, fishing gear, and packaging;

(2) Other physical contact or interaction with plastic debris such as collisions, obstructions, abrasions, or use as substrates. For example, damage to coral reef organisms due to impact from discarded fishing gear. Impacts may also occur on light penetration, availability of organic matter, and oxygen exchange; and

(3) Ingestion of plastics by direct uptake or indirect uptake through feeding on prey species, depending on particle sizes, feeding mechanisms, and limited by size of organism. Reported health effects on many diverse marine species vary from biochemical effects in laboratory studies (e.g. oxidative stress, metabolic disruption, reduced enzymes activity and cellular necrosis) to physiological impacts (e.g. reduced stomach capacity and potential starvation, ulcerative lesions of gut, or gastric rupture).

Case Study 16.1 (continued)

Microplastic ingestion (particles <4.75 mm) is prevalent in many organisms and can induce various potential effects at different biological levels, from sub-cellular to ecosystems. It can affect the consumption of prey, leading to energy depletion, inhibited growth, and fertility impacts, and survival in a number of species. In contrast, many organisms do not exhibit changes in feeding after ingestion of microplastic particles (Ritchie and Roser 2018).

Potential toxic effects from the ingestion of microplastics in seafoods by humans relate to ingestion of individual particles, uptake of persistent pollutants absorbed by plastic particles (e.g. PCBs), and leaching of plastic additives. There appears to be no clear evidence of any human accumulation or health risks, but further detailed research is necessary to reduce the levels of uncertainty associated with exposure to plastic pollutants and other complex mixtures of pollutants already found within marine ecosystems and human food chains.

Improving waste management, including recycling, and end-of-life management, is seen by the United Nations GEO 6 report as the most urgent short-term solution to reducing input of litter to the ocean (UNEP 2019).

16.8 Human Health Effects from Waste Management

Large numbers of health studies have investigated community and workplace concerns about potential human health effects of exposures to emissions from the collection, treatment, and disposal of solid and hazardous wastes at numerous waste sites, particularly landfills and incinerators, and other waste management activities. Relatively few studies have examined the health effects of emission exposures from large-scale recycling, materials recovery, and composting activities. A similar situation exists for human health effects from sewage collection, treatment, and discharge or reuse and spreading of biosolids or sewage sludges.

This section focuses on several comprehensive reviews of epidemiological studies of community and workplace exposures to emissions from waste management activities related to municipal solid wastes, hazardous wastes, liquid wastes (sewage), composting, and recycling. Impacts and risks from agricultural wastes and mining waste sites (e.g. tailings) are not specifically included although the same principles for epidemiological studies and monitoring would apply.

Analytical studies such as cross-sectional, cohort, and case control are widely used to test specific hypotheses on waste exposures, to determine quantitative estimates of relative risk of health outcomes, statistical significance, and to evaluate causality. People living near waste sites (e.g. landfills and incinerators) and worker exposures to emissions from waste management are the primary subjects of health related studies, essentially in developed countries. Many of these studies, especially earlier ones, suffer from inadequate exposure measures of emissions, lack of direct exposures of individuals, data on important confounders (e.g. socio-economic status, alternate sources of exposures, migration of populations, lack of specific biomarkers, and inherent latency of diseases like cancers) (Rushton 2003; Giusti 2009).

Systematic reviews of epidemiological studies have rated evidence on health outcomes, according to three grades: sufficient, limited, and inadequate[1]. Criteria used to meet these grades vary but generally follow the approach used by the International Agency for Research on Cancer (IARC) Monographs (see also Table 16.4).

Giusti (2009) describes how epidemiological studies can define the strength of the association between exposure to a potentially toxic substance and specific health effects. A common approach used in cohort studies is to calculate the relative risk (RR), i.e. the ratio of the incidence of a disease in the exposed population over the incidence of the same disease in the nonexposed population. An RR of >1 suggests an increased risk of disease or an adverse effect. Another similar measure is an odds ratio (OR), which is used to estimate the strength of association between a risk factor and health outcome or disease in a case–control study.

A simplified RR or OR model used by the World Cancer Research Fund and the American Institute for Cancer Research to define the strength of evidence of an association between exposure and disease is given by Giusti (2009). The risk level determined depends on the

---

1 *Inadequate*: available studies are of insufficient quality, consistency, or statistical power to decide the presence or absence of a causal association. *Limited*: a positive association has been observed between exposure and disease for which a causal interpretation is considered to be credible, but chance, bias, or confounding could not be ruled out with reasonable confidence (Porta et al. 2009). *Sufficient evidence*: causal relationship has been established in which chance, bias, and confounding could be ruled out with reasonable confidence (IARC).

Table 16.4 Examples of relative risks of health effects for community exposures to landfills, incinerators, and sewage contaminated waters.

Health effect	Distance from source	Relative risk (confidence interval)	Level of confidence[#]
Landfills (Elliott et al. 2001)			
Congenital malformations			
Neural tube defects	<2 km	1.06 (1.01–1.12)[a]	Moderate
Gastroschisis and exomphalos	<2 km	1.18 (1.03–1.34)[a]	Moderate
Low birth weight	<2 km	1.06 (1.052–1.062)[a]	High
Incinerators			
Congenital malformations (Cordier et al. 2004)			
Facial cleft	<10 km	1.30 (1.06–1.59)[b]	Moderate
Renal dysplasia	<10 km	1.55 (1.10–2.20)[b]	Moderate
Cancer (Elliott et al. 1996)			
All cancer	<3 km	1.035 (1.03–1.04)[b]	Moderate
Liver cancer	<3 km	1.29 (1.10–1.51)[b]	High
Non-Hodgkin's lymphoma	<3 km	1.11 (1.04–1.19)[b]	High
Soft-tissue sarcoma	<3 km	1.16 (0.96–1.41)[b]	High
Sarcoma	—	3.30 (1.24–8.76)[c]	—
(Zambon et al. 2007)			
Sewage contaminated water (Kay et al. 1994)	Sydney beaches, Australia	—	
Gastroenteritis	—	—	—
0–39 Fecal streptococci/100 mL	—	1.0	—
40–59 Fecal streptococci/100 mL	—	1.91 (1.60–2.28)	—
60–79 Fecal streptococci/100 mL	—	2.90 (1.43–5.88)	—
80+ Fecal streptococci/100 mL	—	3.17 (1.12–8.97)	—

[a](99% CI).
[b](95% CI).
[c](95 or 99% CI).
[#]Adopted scale: very high, high, moderate, low, very low.
Source: Data from Giusti (2009) and Porta et al. (2009).

combined assessment of the RR and the statistical significance found. Examples of RR values reported for a number of epidemiological studies, with RR values >1, are presented in Table 16.4 and their strength of evidence is interpreted.

Municipal Solid Wastes

Multiple studies have investigated the potential association of persons exposed to or living in the proximity of waste landfills and incinerators and the risk of adverse health outcomes (e.g. cancers, birth defects and disorders, and respiratory diseases). Recent detailed reviews of these studies (e.g. Rushton 2003; WHO 2007; Giusti 2009; Porta et al. 2009; Fazzo et al. 2017) evaluate the relevant evidence on health outcomes for community exposures in the case of multiple and individual landfill and incinerator sites:

Landfills

Human exposures from landfill sites usually involve low-level emissions of landfill gases (methane, ammonia, amines, hydrogen sulfide, carbon dioxide, and volatile organic compounds including benzene), PAHs, dioxins, toxic metals, dusts, and pathogens (animal vector borne, bioaerosols, leachates, and water runoff). The disposal of hazardous wastes in municipal landfills also increases the risk of higher exposures to toxic industrial chemicals, pharmaceutical wastes, clinical wastes, and asbestos. Other potential environmental pathways of exposure result from contact with land contamination, surface and groundwater pollution from leachates, and stormwater runoff. These exposures are influenced by factors such as the types of landfilling (e.g. sanitary, dumping, and open burning practices), mix of nontoxic and toxic wastes, and treatment methods.

Many epidemiological studies have investigated health effects among persons living in the proximity of numerous landfill (and incineration) sites in many developed countries (e.g. UK, Europe, and USA). While there is some limited evidence of adverse health risks for residents living near landfills (and incinerators), in certain cases, evidence of positive associations with adverse effects is unconfirmed by other studies. Many studies are also classified as inadequate or the strength of evidence is inconclusive. The situation is summarized largely by the WHO (2007) report on waste and health, which concluded:

Despite the methodological limitations, the scientific literature on the health effects of landfills provides some indication of the association between residing near a landfill site and adverse health effects. The evidence, somewhat stronger for reproductive outcomes than for cancer, is not sufficient to establish the causality of the association. However, in consideration of the large proportion of population potentially exposed to landfills in many European countries and of the low power of the studies to find a real risk, the potential health implications cannot be dismissed (Fazzo et al. 2017).

The main health effects evaluated in community exposure studies refer to cancers, birth defects and reproduction disorders, respiratory and skin diseases:

Cancers Porta et al. (2009) in their systematic review of studies on health effects associated with solid waste management found that there was inadequate evidence of increased risk of cancer for communities living near landfills, including major UK and European studies, and additional studies in Finland, Montreal, Canada, and Rome, Italy, that indicated slightly positive associations but were considered inconsistent (e.g. Pukkala and Pönkä 2001). In Finland, Pukkala (2014) reported that a follow-up of cancer incidence among former Finnish dump site residents found no significant excess risk of cancers.

Birth Defects and Reproductive Disorders Limited evidence of increased risk of birth defects (total, neural tube, and genitourinary) and low birth weight, however, was evaluated for national studies in the United Kingdom (Elliott et al. 2001) and from the European EUROHAZCON study (Dolk et al. 1998) for persons living near both hazardous and nonhazardous waste sites (e.g. <2 km from the sites in the United Kingdom). The risk estimates from living near hazardous sites was higher than nonhazardous sites. The main uncertainty of these studies reported by Porta et al. (2009) was (i) the completeness of data on birth defects, (ii) the use of distance from the sites for exposure classification, (iii) and the classification as toxic and nontoxic waste sites. Examples of quantitative effects reported by Elliott et al. (2001) in the UK study are given in Table 16.4.

Respiratory Diseases There appear to be few epidemiological studies covering respiratory disorders associated with municipal landfills. Other community exposure studies have focused more on incineration, hazardous waste, and composting sites. Porta et al. (2009) have identified a Finnish study (Pukkala and Pönkä 2001) that found a significantly higher prevalence of asthma among a cohort of persons living in a former dump area than in a reference cohort living nearby, but outside the landfill site area. Porta and colleagues concluded that the overall evidence of this study may be inadequate.

Incineration

Cancers Porta et al. (2009) evaluated epidemiological studies of residents living near incinerators. Unlike inadequate evidence found for landfill sites, these researchers concluded that there is limited evidence that people living near a MSW incinerator have an increased statistical risk for various cancers, stomach, colorectal, liver, and lung cancers. These findings were based on the UK studies of MSW incinerator sites by Elliott et al. (1996). Other studies on incinerators in France and Italy also suggested an increased risk for non-Hodgkin's lymphoma and soft-tissue sarcoma (Porta et al. 2009).

Birth Defects and Reproductive Disorders Despite the usual inconsistencies observed between community exposure studies of waste sites, there is some limited evidence from epidemiological studies (e.g. Cordier et al. 2004) for increased risk of congenital malformations (facial cleft and renal dysplasia) among people living close to MSW incinerators, but less so for other birth defects, and also low birth weight, compared to reviewed studies on landfill sites (see Porta et al. 2009). Examples of quantitative relative risk estimates by Cordier et al. (2004) for facial cleft and renal dysplasia are included in Table 16.4.

Respiratory Diseases or Disorders Despite known air emissions of toxic combustion particulates, gases and other chemicals from incineration of municipal solid waste and reported symptoms and diseases (e.g. asthma) among various incinerator workers, community exposure studies tend to show little evidence of increased risk of respiratory or skin disease. Shy et al. (1995), however, found some indications of an increased risk of respiratory diseases, especially in children (Porta et al. 2009).

Hazardous Wastes

Russi et al. (2008) concluded that epidemiological studies of populations living near a toxic waste site have not produced adequate evidence to show a causal link between toxic waste exposures and cancer risk. For example, Love

Canal in the US which in 1978 became a well-known chemical waste landfill (see Gensburg et al. 2009).

A recent major review by Fazzo et al. (2017) evaluated 57 papers of epidemiological investigations on the health effects of persons living near hazardous waste sites. This evaluation covered the association between residential exposure to hazardous waste sites and 95 health outcomes (diseases and disorders). Findings were:

- Sufficient evidence was found of association between exposure to oil industry waste that releases high concentrations of hydrogen sulfide and acute symptoms.
- Limited evidence of causal relationships with hazardous wastes was found for the following health outcomes:
- *Cancers:* bladder, breast and testis cancers, and non-Hodgkin lymphoma;
- *Birth defects and reproduction disorders:* congenital anomalies overall and anomalies of the neural tube, urogenital, connective and musculoskeletal systems, low birth weight and pre-term birth; and
- *Respiratory diseases:* asthma.

Inadequate evidence was defined for the other health outcomes. In particular, the association of hazardous waste with *acute infections of the respiratory system, pneumonia, and influenza* was inadequate because of the lack of consistency between studies. Even so, these researchers noted that consistent results of increased risk were reported near waste sites with air emissions of persistent organic pollutants.

The strength of evidence for various cancers and asthma was associated with resident exposures to organic pollutants (benzene, PCBs, and dioxins), heavy metals including arsenic, and EDCs for testicular and breast cancers, and congenital anomalies. Further health concerns due to toxic exposures are related to acute respiratory diseases, evidence of diabetes, and childhood neurological disorders (Fazzo et al. 2017).

Specific classifications of hazardous wastes (e.g. e-wastes, radioactive wastes, chemical wastes [Basel Convention], contaminated soils, and asbestos) are the subject of health effects and risk assessment studies. Case Study 16.2 focuses on the global environmental problem that has emerged from the generation, disposal, and recycling of e-wastes.

Investigations of clusters of leukemia in children living near nuclear power plants indicate inconclusive findings, while relative risk estimates for cancers from occupational exposures to radiation in the nuclear industry predict a small excess cancer risk for cumulative doses (e.g. Giusti 2009).

Liquid Wastes

The disposal of raw sewage and inadequately treated sewage and its sludges are major sources of fecal contamination of freshwater and marine waters worldwide, with significant to high health risks of waterborne diseases, mostly gastrointestinal disorders and diseases in exposed populations through ingestion (water and food) and contact (washing, bathing, and recreational use of contaminated waters). Urban stormwaters and agricultural runoff from animal manure, sewage sludge, compost, and land application of biosolids may contain large numbers of pathogens (e.g. Salmonella, Campylobacter, *Escherichia coli*, Giardia and Cryptosporidium, and viruses) that can pose serious environmental health risks for human and animal use of receiving waters (Gerba and Smith 2005).

Waste Management Activities

Reported health and safety studies of workers involved in waste management activities (collection, landfilling, incineration, composting, recycling, and waste remediation) have shown an increased incidence of accidents and musculoskeletal injuries (Rushton 2003; Giusti 2009). The overall evidence, however, from various epidemiological studies of disease-related effects from occupational exposures to waste emissions is generally evaluated as conflicting or inconclusive (Giusti 2009; Porta et al. 2009). Nonetheless, many waste workers are exposed to well-known hazardous substances such as fine particulates (e.g. organic dusts), VOCs, dioxins, and bioaerosols, which are confirmed by exposure monitoring (Rushton 2003; Giusti 2009). For example, several studies reviewed by Porta et al. (2009), Giusti (2009), and Fazzo et al. (2017), and others have reported exposure measurements associated with one or more increased respiratory, skin, eye, and gastrointestinal symptoms and/or effects among exposed cohorts of workers compared to control cohorts involved in activities such as waste collection, landfills, incinerators, and composting.

Serious problems are evident in the capacity and sensitivity of epidemiological studies to resolve chronic exposure and biological effect levels in exposed and control populations. The lack of direct human exposure evaluation is a critical limitation in a number of studies. The use of biomarkers in epidemiological studies is seen as a key step to help solve this type of problem. This approach would enhance the detection of excessive exposures to airborne pollutants and changes at cellular and molecular levels, prior to clinical signs or diagnosis of diseases (e.g. Giusti 2009; Porta et al. 2009; Fazzo et al. 2017). Giusti (2009) has proposed that preference should be given to prospective cohort studies of sufficient statistical power that include direct human exposure measurements and are supported by data on health effect biomarkers and susceptibility biomarkers.

Case Study 16.2

Electronic and Electrical Wastes (E-Waste): Human Health Effects

The generation, recycling, and disposal of e-wastes are a rapidly growing global pollution problem affecting human and environmental health in developed and developing countries. These wastes are also known as waste electrical and electronic equipment (WEEE).

The sources and range of e-wastes are extensive, including household appliances, electrical power tools, cell phones, TVs, computers, monitors, sensors, telecommunications equipment, electrical lighting, electrical motors, medical devices, scientific instruments, electronic components (e.g. batteries, circuit boards, cathode ray tubes, capacitors), solar panels, and inverters (Grant et al. 2013; Figure 16.13).

Figure 16.13 Electronic components for recycling. Source: Philip Laurell / Getty Images.

Generation of e-wastes is increasing rapidly and is estimated to be about 50 million tonnes each year. The top 10 countries that produce electronic wastes per person are reported as USA, Australia, Germany, Japan, UK, France, Russia, Brazil, South Africa, and China (Juniper 2018, p. 89). Most e-waste (~75–80%) is shipped, often illegally, to Asian and African countries for recycling and disposal. E-waste recycling occurs in formal (regulated facilities with pollution controls) and informal economic sectors where usually rudimentary techniques in developing countries are applied with lack of ventilation and pollution controls (Perkins et al. 2014).

Recycling processes can result in high emissions of toxic particulates and fumes from practices such as melting of electronic components, acid extraction of metals, smelting of metals, and from incomplete combustion of plastics at relatively low temperatures to recover valuable metals. Potential contaminants released during recycling of e-wastes include toxic metals (lead, mercury, cadmium, chromium, manganese, nickel, lithium, barium, and beryllium), persistent organic pollutants, e.g. brominated flame retardants PBDEs and PBBs, PCBs and combustion by-products from burning of plastics (polychlorinated dibenzodioxins, e.g. TCDD and furans, dioxin-like PCBs), and also PAHs formed from burning of plastic and other organic materials and fuels (Grant et al. 2013; Perkins et al. 2014).

Workplace and environmental exposures to individual and mixtures of e-waste contaminants result essentially from inhalation of airborne particulates, fumes and gases, ingestion of contaminated water and food supplies, skin exposures, and uptake from contact with or exposures to air, toxic dust on surfaces, and clothing due to poor hygiene. Vulnerable populations in e-waste recycling consist largely of poor and less educated workers (many are poor women and children), and persons living near recycling facilities in urban environments.

The toxic effects of individual contaminants (e.g. heavy metals, POPs, and PAHs) identified in e-wastes are largely known, but the toxicity of mixtures of these and other potential substances are more complex and much less known. Potential health effects from exposure to e-wastes may include changes in lung function, thyroid function, hormone expression, adverse reproductive health effects (increases in spontaneous abortions, stillbirths and premature births, and reduced birth weight and lengths), reduced childhood growth rates, impairment of mental health, cognitive development, reduced IQ with increasing blood lead level in children, cytotoxicity (e.g. DNA damage), genetic toxicity, potential carcinogenicity, and endocrine disrupting properties (Grant et al. 2013; WHO 2013; Perkins et al. 2014).

Grant et al. (2013) found plausible outcomes associated with exposure to e-waste in their systematic review of the literature in major electronic databases. These researchers concluded better designed epidemiological investigations of vulnerable populations (e.g. pregnant women and children) are needed to confirm these associations.

Quantitative Assessment of Some Epidemiological Studies on Community Exposures to Wastes

Selected studies with relative risk (RR > 1) for community exposures to landfills, incinerators, and also marine sewage discharges near to recreational beaches are compiled in Table 16.4. These studies represent a few studies where there is some positive association with community exposure and indicate limited evidence of a causal association (e.g. specific cancers and birth defects). Many epidemiologists consider that an increased risk of RR = 1.0–1.5 (or a decreased risk of RR = 0.7–1.0) is either a weak association or no association. An RR > 1.5 can have a moderate to strong association (Craun and Calderon 2005, p. 112). The vast majority of studies on health outcomes related to waste management activities have RR or OR values <1.5 (Giusti 2009). This reviewer concludes that ... *the evidence of adverse health outcomes for the general population living near landfill sites, incinerators, composting facilities and nuclear installations is usually insufficient and inconclusive. There is convincing evidence of a high risk of gastrointestinal problems associated with pathogens originating at sewage treatment plants.*

There is limited evidence of some adverse health effects on persons exposed to contact with contaminated solids and liquid wastes, and their emissions (gases, dusts, vapors, mists, and bioaerosols) in waste management activities (e.g. incinerator and compost workers), and for the public, through contact or ingestion of waters (and foodstuffs) contaminated by wastewaters.

Many persons in the community live in the vicinity of existing or former waste management facilities (e.g. landfills). The weight of evidence and causality of association of reported health effects from studies of residents living near waste facilities, and with exposures to potential toxic waste emissions, are usually described as inadequate or inconclusive. Limited evidence exists from a few studies for birth defects (landfills and incinerators), low birth weight (landfills), birth defects and cancers (incinerators). Verification of the strength of evidence appears to be difficult because of factors such as the weaknesses of design, lack of exposure or biomarker studies, and statistical power in many studies. It remains a key challenge for future investigations.

In 2016, the WHO Regional Office for Europe concluded that the available scientific evidence on the waste-related health effects is inconclusive, but suggests the possible occurrence of serious adverse effects, including mortality, cancer, respiratory disease, reproductive health, and milder effects affecting well-being (WHO 2016).

16.9 Key Points

1. Waste is commonly described as a by-product of human activities. Wastes can be divided into several broad categories such as solid wastes (e.g. municipal solid wastes, MSW), liquid wastes (e.g. sewage), hazardous wastes and other special, controlled, scheduled or regulated wastes (e.g. medical or clinical wastes, toxic chemical wastes and contaminated soils), mining and quarry wastes, and agricultural and forestry wastes.

2. Hazardous wastes are generally wastes with properties that make them dangerous or potentially harmful to human health or the environment. Key hazardous properties include one or more of being flammable, explosive, toxic, or radioactive.

3. Global best estimates of municipal solid wastes amount to about 2 billion tonnes per year. *Urban wastes*, including MSW, commercial and industrial (C&I) waste, and construction and demolition (C&D), are estimated to be about 7–10 billion tonnes per year. Mining and agriculture-related wastes are each reported as 10–20 billion tonnes per year.

4. The general waste management approaches used are (i) waste minimization including avoidance or reduction of wastes, e.g. reuse, recycling and waste-to-energy recovery; (ii) waste treatment (landfilling containment, incineration, and composting); and (iii) disposal in controlled landfills of all non-recycled solid wastes.

5. Waste treatment methods for solid wastes include sanitary landfills with biogas recovery and energy reuse, and leachate recovery and treatment/reuse, bioreactors and composting for solid wastes (e.g. MSW). Incineration is used for waste (MSW) to energy recovery (electricity/heating) and destruction of certain hazardous wastes (e.g. biomedical).

6. Sewage, industrial, and many hazardous wastes are treated using different combinations of physical, chemical, and biological processes to concentrate, reduce, remove or degrade gross organic and inorganic pollutant loads, toxic and hazardous substances. Non-potable reuse of treated wastewaters is also increasing.

7. The release of contaminants from wastes, especially hazardous wastes of concern, results in the transfer of contaminants, often as mixtures, into surrounding environments (natural, human, and workplace) where transport and transformation processes distribute these substances in environmental media, according to their properties, interactions with these media, and prevailing environmental conditions.

8. Waste streams are significant to major contributors of local to global pollution, as demonstrated by increases in atmospheric particulates (PM_{10} and $PM_{2.5}$), global warming from GHGs, accelerated eutrophication in coastal marine and freshwaters, and plastic pollution of the oceans.
9. There is limited evidence from many studies of adverse health effects (respiratory, birth defects, and cancers) among waste workers and persons living near existing or former waste facilities from exposures to emissions from hazardous wastes, landfills, incinerators, and to a lesser degree, composting activities.
10. The weight of evidence and causality of association of reported health effects from studies of residents living near waste facilities, and with exposures to potential toxic waste emissions, are usually described as inadequate or inconclusive. Limited evidence exists from a few studies for birth defects (landfills and incinerators), low birth weight (landfills), and birth defects cancers (incinerators).
11. Verification of the strength of evidence appears to be difficult because of factors such as the weaknesses of design, lack of exposure or biomarker studies, and statistical power in many studies, and remains a key challenge for future investigations.

References

ABS (2006). Solid waste in Australia, feature article. In: *Australia's Environment: Issues and Trends 2006* (ed. P. Harper), 3–27. Canberra: Australian Bureau of Statistics, Commonwealth of Australia.

Australian Department of Environment and Energy (2016). *National Greenhouse and Energy Reporting Scheme Measurement: Technical Guidelines for the Estimation of Emissions by Facilities in Australia*. Canberra: Department of Environment and Energy, Australian Government.

Australian Productivity Commission (2006). *Waste Management*. Productivity Commission Inquiry Report No. 38. Canberra, Commonwealth of Australia. https://www.pc.gov.au/__data/assets/pdf_file/0013/21613/overview.pdf (accessed 15 February 2020).

Cordier, S., Chevrier, C., Robert-Gnansia, E. et al. (2004). Risk of congenital anomalies in the vicinity of municipal solid waste incinerators. *Occupational and Environmental Medicine* 61: 8–15.

Craun, G. and Calderon, R.L. (2005). How to interpret epidemiological associations. In: *World Health Organization. Nutrients in Drinking Water*. 9: 108–115. Geneva: World Health Organization. https://www.who.int/water_sanitation_health/dwq/nutrientsindw.pdf?ua=1#page=117 (accessed 16 February 2020).

Derraik, J.G.B. (2002). The pollution of the marine environment by plastic debris: a review. *Marine Pollution Bulletin* 44: 842–852.

Dolk, H., Vrijheid, M., Armstrong, B. et al. (1998). Risk of congenital anomalies near hazardous-waste landfill sites in Europe: the EUROHAZCON study. *Lancet* 352: 423–427.

Elliott, P., Shaddick, G., Kleinschmidt, I. et al. (1996). Cancer incidence near municipal solid waste incinerators in Great Britain. *British Journal of Cancer* 73: 702–710.

Elliott, P., Briggs, D., Morris, S. et al. (2001). Risk of adverse birth outcomes in populations living near landfill sites. *British Medical Journal* 323: 363–368.

Eriksen, M., Lebreton, L.C.M., Carson, H.S. et al. (2014). Plastic pollution in the world's oceans: more than 5 trillion plastic pieces weighing over 250,000 tons afloat at sea. *PLoS ONE* 9 (12): e111913. doi: https://doi.org/10.1371/journal.pone.0111913.

Fazzo, L., Minichilli, F., Santoro, M. et al. (2017). Hazardous waste and health impact: a systematic review of the scientific literature. *Environmental Health* 16:107. doi: https://doi.org/10.1186/s12940-017-0311-8.

Gensburg, L.J., Pantea, C., Kielb, C. et al. (2009). Cancer incidence among former Love Canal residents. *Environmental Health Perspectives* 117 (8): 1265–1271.

Gerba, C.P. and Smith, J.E. Jr., (2005). Sources of pathogenic microorganisms and their fate during land application of wastes. *Environmental Quality* 34: 42–48.

Giusti, L. (2009). A review of waste management practices and their impact on human health. *Waste Management* 29 (8): 2227–2239.

Grant, K., Goldzien, F.C., Sly, P.D. et al. (2013). Health consequences of exposure to e-waste: a systematic review. *Lancet Global Health* 1 (6): e310–e379.

Hoornweg, D.A. and Bhada-Tata, P. (2012). *What a Waste: A Global Review of Solid Waste Management*. Urban Development Series Knowledge Papers No.15. Washington, DC: World Bank. http://documents.worldbank.org/curated/en/302341468126264791/pdf/68135-REVISED-What-a-Waste-2012-Final-updated.pdf (accessed 15 February 2020).

Jambeck, J.R., Geyer, R., and Wilcox, C. (2015). Marine pollution. Plastic waste inputs from land into the ocean. *Science* 347 (6223): 768–771.

Juniper, T. (2018). *How We're F***ing Up Our Planet (How Things Work)*. New York: DK Publishing.

Kay, D., Fleisher, J.M., Salmon, R.L. et al. (1994). Predicting likelihood of gastroenteritis from sea bathing: results from randomised exposure. *Lancet* 344 (8927): 905–909.

LaGrega, M.D., Buckingham, P.L., Evans, J.C. et al. (2001). *Hazardous Waste Management*, 2e. New York: McGraw-Hill.

Landon, M. (2006). *Environment, Health and Sustainable Development*. Maidenhead, UK: Open University Press.

NPI (2010). *Emission estimation technique manual for Municipal Solid Waste (MSW) Landfills. Version 2.0. National Pollutant Inventory*. Canberra: Commonwealth of Australia.

Official Journal of the European Union (2018). Commission notice on technical guidance on the classification of waste (2018/C 124/01). Volume 61. European Commission. https://eur-lex.europa.eu/legal-content/EN/TXT/HTML/?uri=OJ:C:2018:124:FULL&from=DA (accessed 17 November 2018).

Perkins, D.N., Brune Drisse, M-N., Nxele, T., Sly, P.D. (2014). E-Waste: a global hazard. *Annals of Global Health* 80 (4): 286–295. doi: https://doi.org/10.1016/j.aogh.2014.10.001.

Porta, D., Milani, S., Lazzarino, A.I. et al. (2009). Systematic review of epidemiological studies on health effects associated with management of solid waste. *Environmental Health* 8: 60. doi: https://doi.org/10.1186/1476-069X-8-60.

Pukkala, E. (2014). A follow-up of cancer incidence among former Finnish dump site residents: 1999–2011. *International Journal of Occupational and Environmental Health* 20 (4): 313–7. doi: https://doi.org/10.1179/2049396714Y.0000000080.

Pukkala, E. and Pönkä, A. (2001). Increased incidence of cancer and asthma in houses built on a former dump area. *Environmental Health Perspectives* 109 (11): 1121–1125. doi: https://doi.org/10.1289/ehp.011091121.

Ritchie, H., and Roser, M. (2018). Plastic pollution. *OurWorldInData.org*. https://ourworldindata.org/plastic-pollution (accessed 16 February 2020).

Rushton, L. (2003). Health hazards and waste management. *British Medical Bulletin* 68 (1): 183–197.

Russi, M.R., Jonathan, J.B., Cullen, M.R. (2008). An examination of cancer epidemiology studies among populations living close to toxic waste sites. *Environmental Health* 7: 32. doi: https://doi.org/10.1186/1476-069X-7-32.

Shy, C.M., Degnan, D., Fox, D.L. et al. (1995). Do waste incinerators induce adverse respiratory effects? An air quality and epidemiological study of six communities. *Environmental Health Perspectives* 103: 714–724.

UNEP (2019). *Global Environment Outlook GEO-6. Healthy Planet, Healthy People. United Nations Environment Programme*. Cambridge: Cambridge University Press.

UNEP/ISWA (2015). *Global Waste Management Outlook*. United Nations Environment Programme. International Solid Waste Association. https://www.unenvironment.org/resources/report/global-waste-management-outlook (accessed 14 February 2020).

US EPA (2018). *Municipal Solid Waste Landfills*. United States Environmental Protection Agency, Washington, DC. https://www.epa.gov/landfills/municipal-solid-waste-landfills (accessed 16 February 2020).

US EPA (2019). *Bioreactor Landfills*. United States Environmental Protection Agency. http://www.epa.gov/osw/nonhaz/municipal/landfill/bioreactors.htm#1 (accessed 16 February 2020).

Wang, J., Shih, Y., Wang, P.Y. et al. (2019). Hazardous waste treatment technologies. *Water Environment Research* 91: 1177–1198.

WHO (2007). *Population Health and Waste Management: Scientific Data and Policy Options*. Copenhagen: World Health Organization Regional Office for Europe. http://www.euro.who.int/__data/assets/pdf_file/0012/91101/E91021.pdf (accessed 17 February 2020).

WHO (2013). *State of the Science of Endocrine Disrupting Chemicals - 2012*. United Nations Environment Programme/World Health Organization. http://www.who.int/ceh/publications/endocrine/en/ (accessed 3 September 2020).

WHO (2016). Waste and human health: Evidence and needs. *WHO Meeting Report* 5–6 November 2015 Bonn, Germany. Copenhagen: World Health Organization Regional Office for Europe. http://www.euro.who.int/__data/assets/pdf_file/0003/317226/Waste-human-health-Evidence-needs-mtg-report.pdf (accessed 17 February 2020).

Zambon, P., Ricci, P., Bovo, E. et al. (2007). Sarcoma risk and dioxin emissions from incinerators and industrial plants: a population-based case-control study (Italy). *Environmental Health* 6: 19. doi:https://doi.org/10.1186/1476-069X-6-19.

17

Pollution Monitoring, and Assessment

17.1 Introduction

Environmental Monitoring

> Effective prediction, assessment, policy and management are built on accurate, timely and appropriate observations and monitoring programs
> (US National Science and Technology Council 1995).

Pollution monitoring has emerged as an integral part of environmental and health monitoring. Environmental monitoring and modeling are widely used to characterize, measure, and predict changes in the quality of the environment due to natural events and human activities. For example, observed global warming and climate change events are subject to environmental monitoring (and modeling) on a massive scale to evaluate, inform and advise on public policy, political, and management actions.

Basically, environmental monitoring involves the systematic sampling of air, water, soil, and biota to observe and evaluate the status of environmental quality, as indicated by variations and trends in key environmental parameters or indicators. There are a multitude of purposes, such as from baseline studies of an environment, research, and knowledge, evaluation of environmental conditions or impacts from human activities, compliance with environmental regulations, to public information, policy formulation, and decision-making for environmental management. Monitoring programs can also vary greatly in scope, spatial, and temporal scales, from local monitoring by government agencies, private corporations, contractors, and community-based groups, to global monitoring programs such as for persistent organic pollutants and climate change variables.

Today, a great deal of emphasis is placed on the quality and effectiveness of monitoring programs, including objectives, sampling and analytical plans, QA/QC, data quality analysis and management, assessment, and reporting, as discussed in this chapter.

The scope of environmental management and monitoring is growing much broader in the context of environmental health concepts that incorporate ecosystem health, human health, and sustainability. Integrated environmental monitoring of air and water pollutants, for instance, and remote sensing by aircraft and satellites have expanded rapidly. Current trends are toward integrated environment and health monitoring (see this chapter). In this way, pollution monitoring has expanded further, from target monitoring of specific pollutants in natural or human environments, or basic compliance monitoring of regulated emissions and discharges of pollutants, to increased evaluation of ecological and human health impacts and risks (see Chapter 18).

17.2 Monitoring Objectives

Previous sections of this book have been concerned with the ecological and human health relationships between pollutants and natural and human environments. The outcomes of these relationships contribute to how we monitor and understand the increasing global observations and trends of adverse impacts on many diverse ecosystems, rapid loss of biodiversity, climate change, and increasing burden of death and disease for human populations due to pollutants from human activities. Increasingly, the objectives of monitoring must be designed to meet robust scientific knowledge, protocols, methods and quality standards demanded by stakeholders, including the community, to improve the likelihood of effective monitoring outcomes.

In this chapter, the focus is on how to monitor and evaluate changes in environmental pollution and its effects on humans and natural biota and their environments, as part of sustainable approaches toward environmental management strategies, plans, and programs. In a more holistic context, effective management of pollution, on local to global scales, needs science-based knowledge of (i) adverse effects that occur in natural and human environments; (ii) pollution factors causing these effects; and (iii) how pollutants may interact and affect the health of

---

Chemistry and Toxicology of Pollution: Ecological and Human Health, Second Edition. Des W. Connell and Greg J. Miller.
© 2023 John Wiley & Sons, Inc. Published 2023 by John Wiley & Sons, Inc.
Companion website: www.wiley.com/go/toxicologyofpollution2e

humans and natural biota. Holdgate (1979) has suggested that a knowledge of the following factors is important for an effective monitoring program:

1. The pollutants entering the environment, and their quantities, sources, and distribution.
2. The effects of those substances on the environment, exposed organisms, and human health.
3. Trends in concentrations and effects, as well as sources and causes of these changes.
4. How far these inputs, concentrations, effects, and trends can be modified, and by what means and at what cost.

These factors form part of a wider environmental management program, not only involving monitoring and surveillance, but assessment of risk, what action is to be taken, and its outcomes (see also Chapters 18–20 of this book). Additionally, there are many other aspects to be considered in the total program such as legal, administrative, social, economic, and engineering. If a discharge is made to the environment without any adverse effects being observed, then this would not be considered pollution and would not need any remedial action, although in some cases, precautionary or regulated monitoring may occur depending on potential risk (e.g. very low frequency of adverse event but high consequences of an event).

As part of environmental management, environmental quality is of primary concern to all interested parties. The evaluation of quality requires subjective judgment of some aspect of the characteristics of the environment. This judgement usually involves the use to which the particular component of the environment is put. The uses of different sectors of the environment are manifold. For example, water can be used for domestic or stock consumption, industrial purposes, agricultural use, maintenance of fisheries, recreation, conservation of natural resources, and aesthetic appreciation. The water quality values and needs for these different uses are quite different and the judgment of environmental quality will vary accordingly.

In the design of an investigation or monitoring program, the environmental quality aspects of interest need to be clearly defined, such as water quality objectives and uses. For example, if the quality of domestic drinking water is of concern, monitoring would include sampling and analysis of harmful chemicals and microorganisms (e.g. fecal pathogens). In the case of protection of aquatic ecosystems, environmental pollution factors such as nutrient inputs and levels are likely to be among key targets for monitoring of any significant changes in the system due to eutrophication. Alternatively, contaminated land and groundwater investigations would design a preliminary sampling and analytical program that tests for known and/or potential contaminants based on site history and regulatory requirements for proposed land and groundwater uses.

17.3 Monitoring Strategies

Since the 1970s, environmental monitoring strategies have advanced from using a combination of physicochemical and biological parameters to integrated monitoring programs, enhanced by new sensor-based networks, data logging and acquisition technologies, and remote sensing, to obtain near real-time data for analysis and decision-making purposes. Recent approaches include integration of environmental and human health monitoring parameters on national and regional scales.

Physicochemical and Biological Monitoring

Pollution monitoring usually involves physical and/or chemical measurements in various locations or situations, for example workplace processes, occupational hygiene, emissions to the environment, and within environmental media, at the interface or within a target organism (i.e. bioindicator species), human tissues or fluids, and so on. Microbiological testing such as waterborne pathogens is widely used for human health surveillance and monitoring. Increasingly biomarkers are being used to measure adverse biological responses toward exposures to anthropogenic or environmental pollutants. Microorganisms such as fungi or mold spores are also monitored in indoor environments for human health and hygiene purposes.

These types of measurements can be designed to evaluate exposure of biota or human populations to pollutants of concern by comparison to environmental or toxicological standards, criteria or guidelines, and sometimes pollutant indices (e.g. air pollution). Ecological and human health risk assessment methods can also characterize risk levels and form the basis for risk management options.

In biological monitoring, the responses of individual organisms, populations and communities due to known or potential stresses, adverse stimuli, or pollutants can be monitored in various ways to indicate effects on an ecosystem. It also plays an essential role in ecological risk assessment (Karr and Chu 1997). Chapman (1996) and Bartram and Balance (1996), for example, have described detailed strategies and methods for biological monitoring and water quality assessment. Generally, approaches can include (i) measuring the presence and abundance of selected species at different locations within a habitat (e.g. waterway), (ii) the responses of individual organisms, such as behavioral, physiological, or morphological changes, exposed to environmental stressors or pollution, (iii) field or laboratory bioassays and toxicity testing, including whole effluent testing, (iv) monitoring the spatial and temporal accumulation of selected chemical contaminants in tissues of bioindicator organisms within ecosystems, and

(v) the use of biomarkers (e.g. physiological, biochemical, or genetic) to indicate or estimate exposure, effect, and/or susceptibility in organisms due to a stressor or pollutant, especially for long-term exposures.

Physicochemical and biological monitoring methods are essentially complementary. Both methods form an important basis for setting environmental criteria and quality evaluation. In any investigation, the techniques that should be used are those which are most appropriate to the resources and objectives of the study (Connell and Miller 1984, p. 412). The advantages and disadvantages of physicochemical and biological monitoring are compiled in Table 17.1.

In many ecosystems, bioindicator species are selected to indicate variations in contaminant exposure spatially and with time. Sessile organisms such as mollusks (e.g. mussels), benthic plants, and sea anemones are widely used to monitor local contamination (Nikinmaa 2014, pp. 147–148). For example, since 1986 NOAA's Mussel Watch program in the United States has been used to monitor many persistent toxic contaminants in bivalves and sediments in the estuarine and coastal environments of the United States.

Biomarkers are often used as biological indicators of a measurable alteration of a normal function of an organism that responds to exposure from a contaminant, or an effect induced by one. They indicate possible exposures and/or early biological responses before sublethal effects occur and should usefully evaluate potential risks to populations, communities, or ecosystems depending upon their specific properties (e.g. Nikinmaa 2014, pp. 149–153).

New biological monitoring tools are recommended by Jackson et al. (2016) from the long history of freshwater studies to focus on indicators suitable for national, regional, and global assessments of biodiversity and ecosystems. These types of tools include considering less-studied taxonomic groups, standardizing tools across regions to allow global comparisons, and measuring changes over multiple time points. New biological monitoring tools should also make use of recent advances, such as molecular tools (biomarkers), remote sensing, and local-to-global citizen science networks. Already some advances are occurring in biomonitoring of human population exposures to environmental chemical contaminants on national scales, as discussed below.

Table 17.1 Comparison between physicochemical and biological methods of environmental pollution monitoring.

Physicochemical and related microbiological methods	
Advantages	**Disadvantages**
1. Rapid assessments can be made in many cases 2. The quantitative nature of these assessments enables comparison with set standards and a rapid evaluation of the pollution in an area 3. The physiochemical evaluations usually reveal the nature of the pollution 4. The distribution of the pollutant can be related to introduced control measures	1. There is a lack of accurate data on the response of organisms to pollution factors for many areas 2. The data available on organism response tend to be related to short-term exposures with little data related to continuous long-term exposures 3. Very limited data available on sublethal effects (e.g. reproduction) 4. Very little data available on combined effects of pollution mixtures 5. Often limited data available on the chemical form or species of toxic pollutants present

Biological methods	
Advantages	**Disadvantages**
1. An evaluation of the biological consequences of pollution, which are the area of principal concern in environmental management 2. A time-averaged measure of pollution effects 3. An overall evaluation of the effects of pollutant measures 4. In some circumstances, the technique is a sensitive indicator of pollution	1. They can be expensive and time consuming 2. There is a lack of quantitative criteria 3. Variations in environmental conditions, such as season, substrate type, water depth, and so on, may cause large natural variations in populations and communities of biota irrespective of pollution 4. There is often a lack of taxonomic knowledge available to classify the organisms involved to the degree necessary 5. They do not indicate the nature of the substances causing the pollution and so possible sources cannot be identified 6. Limited use in assessing suitable environmental quality for many human uses or needs such as water quality for drinking, agricultural use, water recycling, air quality, and contamination of soils

Source: Modified from Connell and Miller (1984, pp. 411–412).

Human Biomonitoring for Environmental Chemicals

Biomonitoring is a method for measuring amounts of toxic chemicals such as heavy metals and persistent pesticides in tissues of humans and ecological receptors since the late 1950s. In the twenty-first century, human biomonitoring (HBM) has developed as a valuable monitoring tool to indicate and quantify the exposure of human populations to environmental chemicals by measuring their residues, metabolites, or reaction products in biological specimens (e.g. blood, urine, body fats, and human milk).

Advantages of this monitoring strategy include baseline population studies, confirmation of demographic exposures to hundreds of chemical contaminants found in the environment and diet of local and national populations in various countries around the world, and monitoring studies (or cycles) that inform and allow evaluation of public health measures, and assist regulatory risk assessment and management decisions on the effects of environmental chemicals (e.g. Centers for Disease Control and Prevention 2009). For example, in the United States, population biomonitoring data revealed high blood lead concentrations resulting in regulatory action to reduce lead in gasoline. Further biomonitoring data confirmed that this regulatory measure was associated with a decrease in blood lead concentrations in US populations.

New Strategies in Integrated Monitoring

There are needs for new strategies to integrate shifts in monitoring applications and allow greater access to data, statistical analysis and high-quality outputs for research, information sharing and decision-making purposes arising from advances in technology (e.g. low cost and multiple sensors), wireless networks, data acquisition and storage, real or near time, spatial and remote monitoring and data analysis, combined with local to global information systems with rapid mapping (e.g. GIS based), modeling, and visual presentations.

The significance of shifts in monitoring strategies is illustrated by changes in approaches to air pollution monitoring (e.g. ambient air and compliance monitoring) in Figure 17.1. New paradigm changes in traditional air pollution monitoring approaches are due to recent advances in (i) the development of portable, lower-cost air pollution sensors that report data in near-real time at a high-time resolution, (ii) increased computational and visualization capabilities, and (iii) wireless communication/infrastructure (Snyder et al. 2013).

In this example, advances in sensor-based technologies are opening up new opportunities for personal, community,

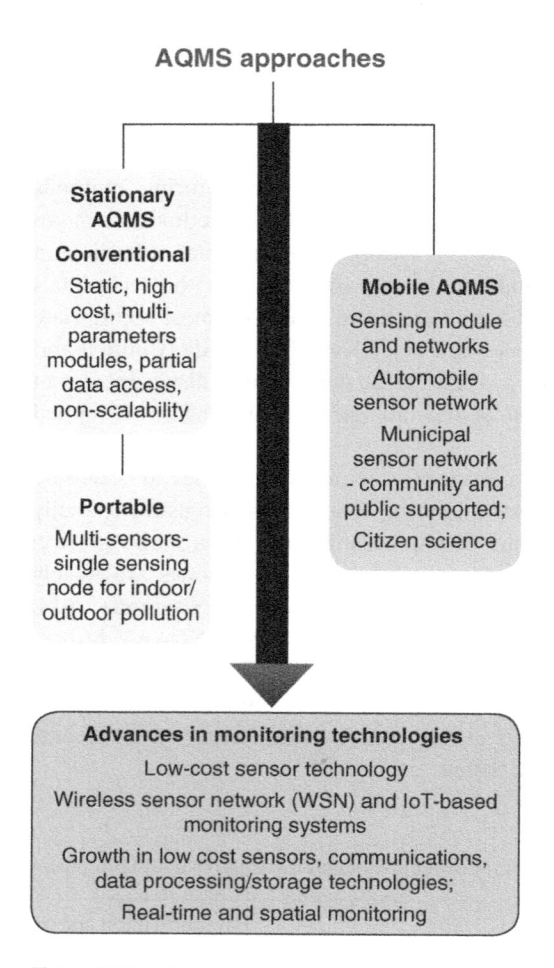

Figure 17.1 Advances in air quality monitoring systems (AQMS) from conventional stations with stationary multi-parameter and high-cost instrumentation to portable and mobile sensing nodes/modules with wireless networks and use of IoT*-based interconnected monitoring and advanced computational processing, data storage, and visualization systems, *from sensor to cloud*. Source: Snyder et al. 2013; Idrees and Zheng 2020.*
IoT – Internet of Things.

and workplace-based monitoring strategies to measure, evaluate, and reduce pollution exposures in comparison to environmental and health indicators (Snyder et al. 2013).

In pollution and interconnected monitoring of human and natural systems, research is rapidly developing along with many sensor-based and wireless network applications such as in *green building* and *smart cities* programs, climate change, and integrated urban water management.

Integrated Monitoring Programs (IMPs)

The need for integrated monitoring approaches is evident in environmental impact studies, pollution monitoring or environmental health assessments where multiple scientific disciplines, agencies, and stakeholders can be involved, such as in regional studies of air pollution or environmental management of major water bodies. In these situations, it is critical that a monitoring program coordinates and integrates the various objectives, design, monitoring approaches and protocols, data quality and management, assessments and reporting consistent with the vision, specific goals/objectives and framework of the overall environmental management plan, including the demand for rapid data and information sharing, and effective communication between participants and the public. Case Study 17.1 provides an example of Global Climate Change Monitoring.

Case Study 17.1

Global Climate Change Monitoring

Every day a global monitoring network, coordinated and provided by the World Meteorological Organization (WMO), collects weather, climate, and water-related data from observations made by tens of thousands of land, sea, and atmospheric instruments, special commercial aircraft, and Earth-observing satellites. The WMO processes these data in powerful computers using numerical models based on physical laws to produce weather, climate, and water-related forecasts, predictions, and information products and services for daily use, long-term decision-making and research (WMO 2019).

The WMO weather and climate monitoring system comprises three integrated core components as follows (ITU n.d.):

- The Global Observing System (GOS) provides observations of the atmosphere and the Earth's surface (including oceans) from the globe and from outer space. The GOS uses remote sensing equipment placed on satellites, aircraft, and radiosondes and relays data to environment control centers (see Figure 17.2).
- The Global Telecommunication System (GTS) – radio and telecommunication networks for real-time exchange of a large volume of data between meteorological centers.
- The Global Data Processing System (GDPS) - thousands of linked mini, micro, and supercomputers processes an enormous volume of meteorological data and generates warnings and forecasts.

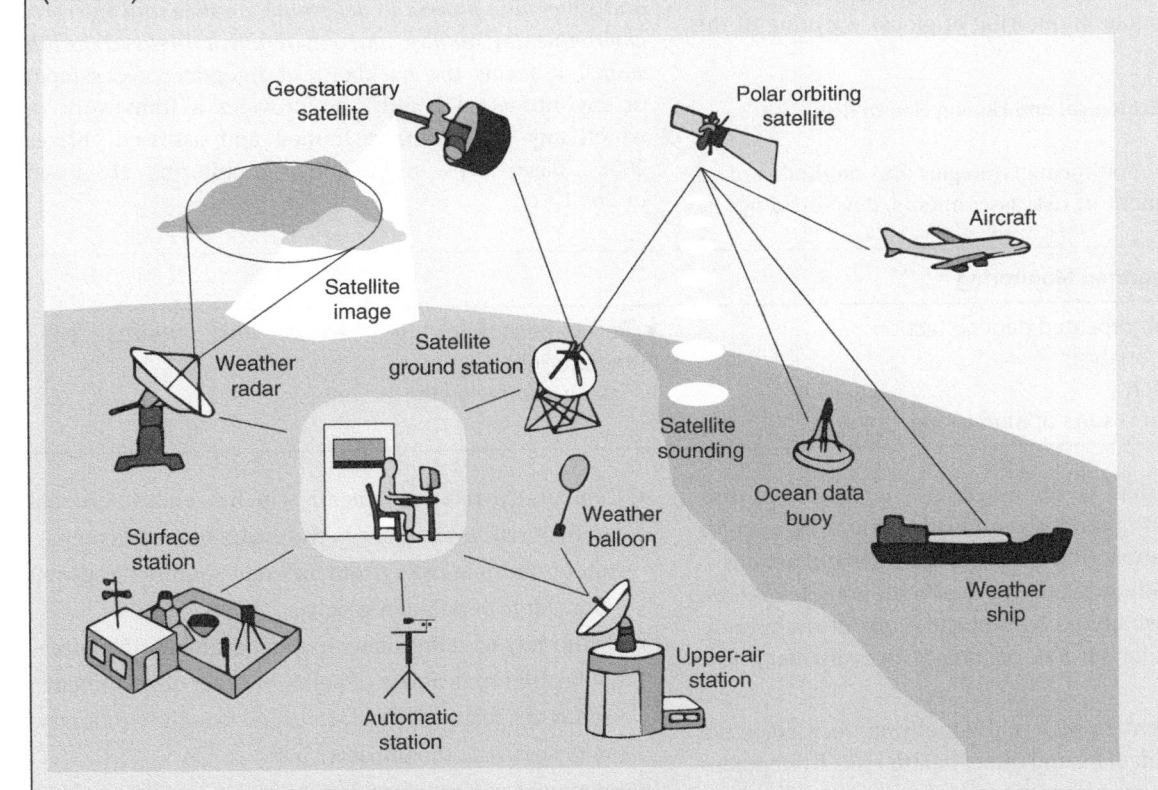

Figure 17.2 Schematic depiction of global and national networks for monitoring, communication, and data processing of weather and climate observations (modified from ITU n.d.).
Source: Published by International Telecommunication Union, United Nations, Geneva.

In practice, an IMP usually seeks to integrate various measures of environmental condition or quality such as physical, chemical, and biological indicators of ecosystem or human health to evaluate and provide an overall assessment or *weight of evidence* approach, which reflects the values, status, and health of the system or exposed populations for sustainable management purposes. IMPs offer many other potential advantages, from cost-effectiveness, research opportunities for understanding processes, and capacity to link to other key components that may include legislation, zoning, research, education, and public participation. Among many examples worldwide is the regional monitoring of Chesapeake Bay, the largest estuary in the United States (see Chesapeake Bay Program 2020).

Other variants of IMPs may include, for example, daily operational site monitoring (e.g. electric power generation or mining) for processing and compliance purposes that include using multiple monitoring locations and sensor technologies for regulated air emissions and ambient air monitoring, indoor air quality in Green Buildings and audits, stormwater and wastewater management, treatment, reuse and discharges to receiving waters, buffer zones, and revegetation programs.

Later in this chapter, Case Study 17.2 outlines the approach to remote monitoring of global warming of the oceans.

Monitoring for Ecological and Human Health Risk Assessment

Environmental monitoring strategies and methods are an integral component of risk assessments, developed by the US EPA during the 1980s and 1990s, to characterize the nature and magnitude of health risks to humans (e.g. residents, workers, recreational visitors) and ecological receptors (e.g. birds, fish, wildlife) from chemical contaminants and other stressors, which may be present in the environment. It plays a vital role in (i) measuring and modeling the frequency and magnitude of human and ecological exposures that may result from contact with environmental contaminants in air, waters, soils, sediments, biota, foodstuffs, and human environments, and (ii) dose-response studies (e.g. field studies and bioassays) for toxicity assessment of known or potential contaminants.

Chapter 18 of this book describes the scientific models, approaches, and applications of risk assessments, which use hazard exposure and toxicity assessments of environmental pollutants. In addition, Case Study 17.3 provides an example of the role of integrated monitoring in comparative risk assessment used to evaluate the global burden of disease due to ambient air pollution.

Monitoring for Integrated Environmental Health Impact Assessment

In this case, integrated monitoring is defined in integrated environmental health assessment as *an ongoing and systematic process to determine, analyse and interpret environmental quality and environmental-related health status*. It forms the backbone of integrated assessment of environmental health and provides a framework in which any issue can be framed and assessed (Briggs 2008). Basic steps in integrated monitoring are listed in Box 17.1.

Box 17.1 Integrated Monitoring

- Planned and repeated data collection
- Monitoring/analysis
- Interpretation
- Reporting of results of monitoring

- Recommendations for action (incl. results on monitoring)
- Taking and reviewing actions

Data integration is an essential component of this type of monitoring strategy and evaluation of its data for health-related impacts or risks and management actions.

It is specifically designed to deal with complex issues, which would usually be beyond the scope of more traditional forms of health risk or impact assessments. These include:

- Risk assessment – which traditionally has focused on the relatively immediate and direct health risks from potentially dangerous substances (e.g. chemicals) or practices (e.g. manufacturing processes), often in relation to cancer;

- Comparative risk assessment – which extends traditional forms of risk assessment to the evaluation and comparison of multiple risks (from different sources or agents) across large population groups;

- Health impact assessment – which evaluates the potential health implications of policy or other developments, usually at a relatively local scale (and usually without trying to aggregate the impacts); and

- Integrated environmental assessment – which assesses the overall environmental (but rarely human health) impacts of large, complex pressures or developments.

In 2008, D.J. Briggs (Department of Environmental Epidemiology and Public Health, Imperial College, London) proposed a framework for integrated environmental health impact assessment of systemic risks to address the more complex interactions between environment and health related problems. This integrated and comparative process is designed to support policy development and other interventions in the real world. As for integrated monitoring, it depends on stakeholder involvement in issue-framing, design of the assessment, and characterization of results. Some major challenges are reported to exist due to…
difficulties in ensuring effective stakeholder participation, in dealing with the multicausal and non-linear nature of many of the relationships between environment and health, and in taking account of adaptive and behavioral changes that characterize the systems concerned (Briggs 2008).

This global network enables the national climate monitoring systems of all the member states (~191 countries) of the WMO to feed data into a central Global Data Processing System that is made available to all, including developing countries that may not have access to advanced climate monitoring technology.

Global Climate Observing System

In 1992, the Global Climate Observing System (GCOS) was set up to ensure that the observations and information needed to examine climate-related issues are obtained and made available to all potential users. It is co-sponsored by the WMO, the Intergovernmental Oceanographic Commission of the United Nations Educational, Scientific and Cultural Organization (IOC-UNESCO), the United Nations Environment Programme (UN Environment), and the International Science Council (ISC). In this role, the GCOS also regularly assesses the status of global climate observations of the atmosphere, land, and ocean, and produces guidance for its improvement (GCOS 2020a).

The GCOS monitors, processes, and evaluates data on a detailed set of essential climate variables (ECVs) that are selected to measure or reflect critical changes in the climate of the Earth's atmospheric, oceanic, and terrestrial environments, as compiled in Table 17.2. Among these variables are atmospheric greenhouse gases emitted by natural processes and human activities (mainly due to combustion of fossil fuels).

Another set of seven key parameters, known as Global Climate Indicators, are used to provide a concise yet better overview of climate change impacts than available from just temperature change. These indicators describe changing climate in the most relevant domains of climate change: temperature and energy, atmospheric composition, ocean and water, and also the Cryosphere. The key indicators for each climate change domain are shown in Figure 17.3.

Outcomes

The global climate monitoring system (GCOS) provided by the WMO plays an essential role in the global collection of and access to climate data, modeling, evaluation, quality

Table 17.2 Essential Climate Variables (ECVs),[a] including greenhouse gases, air and surface temperatures, monitored by the Global Climate Observing System (GCOS).

Atmospheric	Surface: air temperature, wind speed and direction, water vapor, pressure, precipitation, surface radiation budget
	Upper-air: temperature, wind speed and direction, water vapor, cloud properties, Earth radiation budget, lightning
	Composition: carbon dioxide, methane, other long-lived greenhouse gases, ozone, aerosol, precursors for aerosol and ozone
Oceanic	Physics: temperature: sea surface and subsurface; salinity: sea surface and subsurface; currents, surface currents, sea level, sea state, sea ice, ocean surface stress, ocean surface heat flux
	Biogeochemistry: inorganic carbon, oxygen, nutrients, transient tracers, nitrous oxide, ocean color
	Biology/ecosystems: plankton, marine habitat properties
Terrestrial	Hydrology: river discharge, groundwater, lakes, soil moisture, evaporation from land
	Cryosphere: snow glaciers, ice sheets and ice shelves, permafrost
	Biosphere: albedo, land cover, fraction of absorbed photosynthetically active radiation, leaf area index, above-ground biomass, soil carbon, fire, land surface temperature
	Human use of natural resources: water use, greenhouse gas fluxes

[a] An ECV is described as a physical, chemical, or biological variable or a group of linked variables that critically contributes to the characterization of Earth's climate. ECV data sets provide the empirical evidence needed to understand and predict the evolution of climate, to guide mitigation and adaptation measures, to assess risks and enable attribution of climate events to underlying causes, and to underpin climate services. They are required to support the work of the United Nations Framework Convention on Climate Change (UNFCCC) and the IPCC.
Source: Compiled from Global Climate Observing System (GCOS 2020b). Published by the World Meteorological Organization, Geneva.

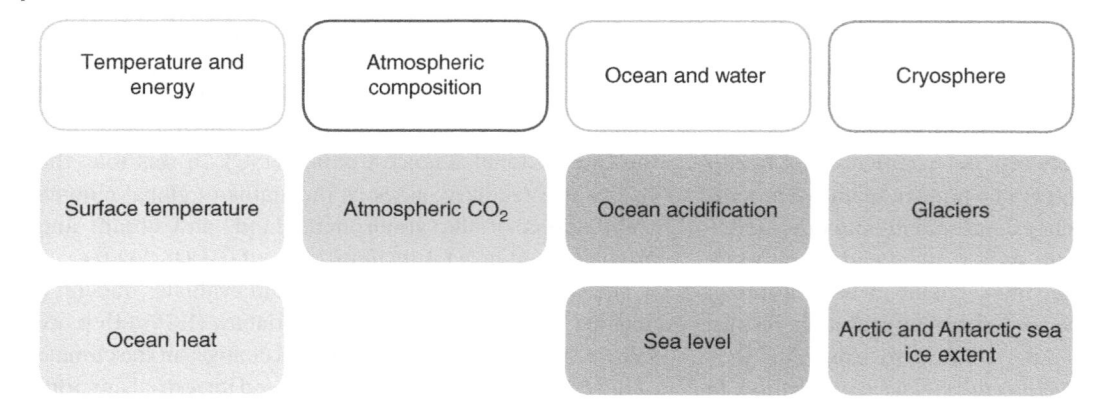

Figure 17.3 The key Global Climate Indicators for each climate change domain. Source: Global Climate Observing System (GCOS 2020c). Published by the World Meteorological Organization, Geneva.

assurance, and reporting on the status and trends of local to global climates. It regularly evaluates the status of global climate observations of the atmosphere, land, and ocean, and guides improved quality.

17.4 Environmental Metrics of Pollution Monitoring

Environmental monitoring programs are often designed to sample and analyze for physical, chemical, biological, and ecological indicators in environmental media and/or populations and communities within ecosystems or built environments. Statistical analyses of results from monitoring are commonly compared with available sets of environmental health quality indicators (e.g. standards, criteria, guidelines, and indices) that apply to the type of environment, ecosystem, individual species, populations, and communities being studied. Indicators (e.g. guidelines) are usually published by major international or national regulatory or advisory agencies. Standards generally refer to prescribed measures adopted for regulatory purposes.

Environmental standards, criteria or guidelines for environmental management, protection of human health, and regulatory purposes are widely established for common pollutants in air, waters, sediment, soils, foods, and to a much lesser degree, in natural biota. For example, water quality guidelines readily exist for many applicable physicochemical parameters, nutrients, toxicants, and microbial indicators of fecal contamination for the protection of aquatic ecosystems, human health, drinking water quality, and other water uses such as agricultural, recreational, recycling of non-potable waters, stormwaters, groundwaters, and some industrial uses. Maximum residue levels and recommended dietary intake levels are available for many contaminants (e.g. pesticides and heavy metals).

In many cases, guidelines for contaminants are unavailable or quality of data is limited. In these cases, ecological or human health risk assessment procedures are widely followed to derive and characterize a relevant guideline for exposure monitoring, toxicity or impact assessment, and risk management purposes (see Chapter 18 of this book).

17.5 Monitoring Programs and Methods

Specific environmental monitoring strategies, programs and methods are commonly available from many research and regulatory agencies, and environmental management bodies responsible for activities such as air quality, water quality, catchment management, wildlife monitoring, and contaminated land. An overview of general principles and some examples are compiled in this section from various sources to assist pollution monitoring. Detailed examples for the development of strategies and programs include Chapman (1996) and Bartram and Balance (1996) on water quality assessment.

In planning an environmental monitoring program to investigate issues such as air quality, water and sediment quality, soil or wildlife contamination, there are a number of vital scientific questions to be answered before sampling or data collection and analysis can start. These include:

- What are the issues (e.g. pollution effects; exposed populations) to be investigated by the monitoring program?
- What are the objectives of the proposed monitoring program?
- What is the scope? (e.g. potential boundaries and time frame; integration with other studies or programs).
- What living organisms and/or human populations and their environments are potentially at risk of exposure/stress and adverse effects?
- What environmental standards and guidelines apply?

- What monitoring strategies, methods and tools apply?
- Is a pilot study necessary? (e.g. statistical assessment for sampling design)
- What parameters and indicators (e.g. air quality) are to be measured, where, how, and when?
- What are their levels in the environment and their units of measurement?
- What sampling strategies, plans, and protocols apply to design, collection of samples and data, to meet analytical needs, data analysis, and objectives of study/program?
- Are analytical methods, time constraints, and costs appropriate to meet the objectives and scope of the program?
- What quality objectives, quality assurance and quality control measures are necessary?
- What data management system will be necessary to store and process data and statistical results/visual presentations?

Once these types of questions are suitably resolved, the monitoring program can be designed to meet the study objectives by proceeding in several steps as indicated by the generic framework in Figure 17.4 for pollution monitoring.

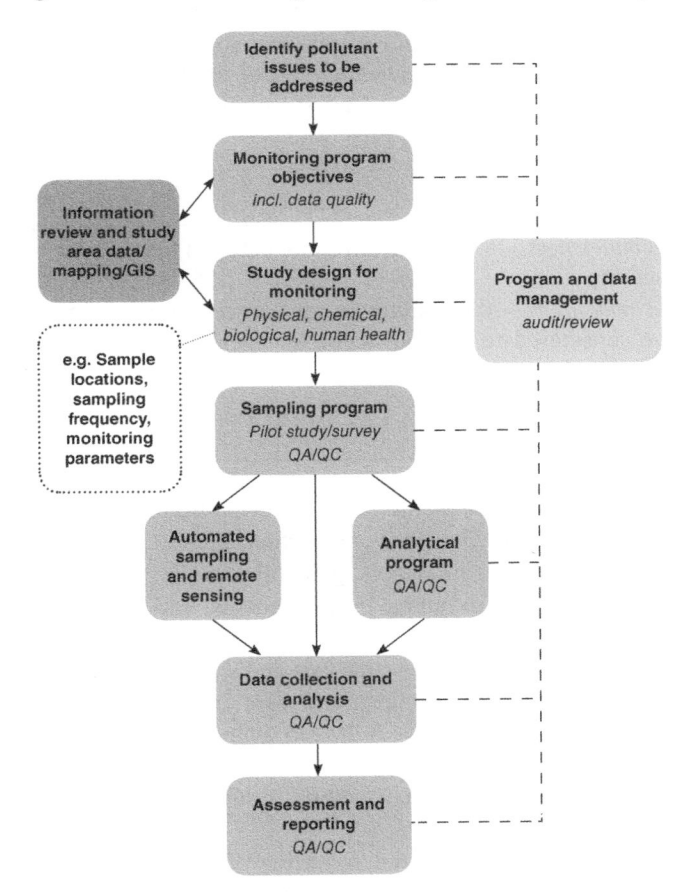

Figure 17.4 Framework for key steps in a monitoring program for pollutants.

Pollution Monitoring Programs

The key steps in Figure 17.4 apply general environmental monitoring principles. Many international, national, state, provincial and city guidelines, and protocols for environmental monitoring exist and provide detail on monitoring approaches, sampling, and analytical methods, quality assurance, data analysis and reporting, often as part of regulatory requirements for assessment and management of air, waters (surface, groundwaters, and drinking), noise quality, hazardous wastes and chemicals, contaminated lands, natural resources, and special land use zones (e.g. conservation, cultural, heritage, and multiuse parks, urban planning, and green buildings).

Defining the Issues

Why and what to monitor? The issue, problem, or question needs to be clearly defined, including extent of potential impact and risks (technical and perceived), relevant indicators and guidelines for monitoring, available background information (e.g. scientific and environmental knowledge, mapping, land use constraints, stakeholder concerns, environmental and community values), and proposed outcomes, through a consultation process between the proponents and stakeholders. The latter may extend from local to national, and sometimes, international interests. The same approaches would basically apply to research-based and compliance monitoring.

Monitoring Program Objectives and Scope

Once the monitoring issue(s) is defined, there is a further process for setting clear objectives for the program. Initially it defines and gathers the information needed, develops an understanding of the system, sources and actions of stressors, and uses this knowledge to construct a conceptual model for the system (e.g. structures of components, nature of processes, functions, flows or linkages, stressors, and receptors) to be monitored.

Objectives should describe the purpose and broad scope of the program and specific scientific and management issues to be investigated, such as why monitoring is needed, what relevant indicators are to be measured and compared with the guidelines or regulatory standards, what methodology is to be used, and how are results to be evaluated and reported (e.g. New Zealand Ministry for the Environment 2019). For example, measuring real-time levels of PM_{10} or $PM_{2.5}$ within a large urban area may involve comparing 24-hour and annual statistical measures with national guideline values to evaluate performance, potential health risks for exposed persons, and management options/actions.

Increasingly, monitoring objectives also need to address broader issues and desired outcomes for study such as within frameworks for environmental management plans, from adequate resources required to do monitoring program, community values involved, socioeconomic evaluations of management options, risk assessment, public risk communication, and political decision-making processes.

Study Design for Monitoring Program

The study design considers, determines, and defines how the monitoring objectives are to be achieved by the monitoring program that takes into account the agreed scale and complexity of the study, resources, and outcomes. The basic design process uses several steps:

- review and describe the information needed for the study,
- use the conceptual model developed for the study,
- select optimum monitoring methods (e.g. sampling and analytical) and tools to answer what, where, when, and how to monitor, together with data quality objectives, quality assurance, and quality control protocols,
- determine suitable methods to analyze, evaluate, and manage data to meet data quality objectives,
- design reporting of monitoring program and results to address study objectives and quality assurance needs, including access to data.

Sampling and Analytical Program or Plan

Sampling and analytical plans (SAP) are usefully developed together to ensure the resources of the study and sampling design (e.g. sample parameters to be measured, limits of reporting, numbers and frequency of samples to be submitted) are compatible with field and analytical equipment and procedures, analytical costs, reporting times, etc., and also data collection, analysis, and management. The entire program or plan is subjected to quality assurance and quality control (QA/QC) procedures that demonstrate the quality of the data and the right results for decision-making purposes.

Quality Assurance (QA) and Quality Control (QC)

Quality assurance is the process that involves the use of data quality objectives and indicators, planned and systematic actions, procedures, checks, and makes decisions to ensure samples and data collected in the study are representative, their integrity is maintained, and the results of analyses are accurate and reliable.

Quality control is the management system for field and laboratory procedures and related activities that control and assess results. The assessment of the reliability of sampling, analytical procedures, data and analytical results is based on data quality indicators (DQIs) of accuracy, precision, representativeness, completeness, and comparability. DQIs are defined in Box 17.2.

Box 17.2 Data Quality Indicators

Accuracy – a quantitative measure of how close reported data is to the true value.
Precision – a quantitative measure of the variability (or reproducibility) of data.
Representativeness – the qualitative confidence that data are representative of each media sampled and samples analyzed in study.

Completeness – a measure of usable data (%) from a data activity.
Comparability – the qualitative confidence that data may be considered to be equivalent for each sampling and analytical event.

Field Sampling Program

Sampling programs covering field measurements, sample, and data collection usually involve a number of essential elements within the scope of the resources available to meet the study objectives. These are:

- Review study objectives and nature of sampling (e.g. remediation of asbestos soil contamination).
- Conduct preliminary site investigation and observations of study area.
- Identify parameters (e.g. physical, chemical, biological) to be measured or observed, their properties, types of samples to be collected (e.g. air, water, sediment, tissues, blood, freshwater benthic invertebrates, and fish) and analytical requirements (e.g. specific pesticides residues, analytical method, limit of reporting, analytical costs).
- Select relevant indicators, guidelines or standards and protocols for sampling, measurements, and analytical requirements.
- Design field sampling strategies and statistical analyses for collection of samples and data measurements to meet program objectives. It is to include selected types and numbers of samples, parameters and indicators to be measured, spatial locations, timing and sampling frequencies.
- Evaluate suitability and recording of site details (e.g. location, description, access, interference from environmental factors such as adjoining land uses and flooding, workplace health and safety factors).

- Select suitable equipment based on appropriate sampling protocols (e.g. air, water, soil, sediment, and biological) including direct reading field instruments with data loggers, automated samplers, sampling devices, passive samplers, calibration, handling, preservation, transport, and storage prior to laboratory analyses and/or data processing from direct reading instruments/remote sensors.
- Define data quality objectives, quality assurance and quality control for sampling procedures, including, for example, replicates, field blanks, and certification/calibration of field and automated direct reading instruments.
- Document sampling protocols and methods used, all sampling records, and QA/QC data that apply to the study.
- Evaluate performance of the sampling program in meeting its data quality indicators.

Sampling Strategies There are a wide range of sampling methods for monitoring pollution which depend on the type of environment, the nature of the material or media being sampled, the variable levels and distributions of the contaminants, or diversity and numbers of species within a habitat/ecosystem being sampled, methods of sample analyses, and background levels and distributions or reference habitats.

Sampling design is a fundamental part of environmental data collection for decision-making based on science. The sampling design process, methods of data collection, and statistical evaluation for developing an effective quality assurance plan for a project are well described in the US EPA (2002) document: *Guidance for Choosing a Sampling Design for Environmental Data Collection (EPA QA/G-5S)*.

Four common sampling approaches used in monitoring programs are briefly described below based on a composite of various scientific literature sources (e.g. water, soil, and ecological sampling).

Judgmental Sampling In this case, the sampling approach (i.e. the number and location of samples, timing and frequency of sampling) is based on (i) knowledge of study area or investigation site characteristics, including environmental conditions, and (ii) professional judgment. In this situation, judgmental sampling is limited in application to parametric probability statistics as inferences about the target population rely on the quality of professional judgment and other factors such as sample size. Alternative methods using nonparametric approaches for environmental pollution monitoring, however, are available (e.g. Gilbert 1987).

Random Sampling Simple random sampling is effective for relatively homogeneous populations of interest (e.g. soil contaminants) such as where pollutants are not expected to be concentrated. An example is a numbered grid map over the study area and random selection of sampling grids. Advantages include ease of sample size calculations and statistically unbiased estimates of the mean, proportions, and variability. Limitations may include difficulty in random sampling of some locations due to precise identification or access (e.g. sampling of habitats).

Systematic and Grid Sampling Systematic sampling of a study area is a useful method, which divides the area into subareas that are more homogeneous than the whole area. This approach allows different sampling patterns and densities to be used in different subareas based on site history, site characteristics, and expert judgment.

A useful way to avoid bias in this type of sampling is to use systematic random sampling, in which samples are taken at regularly spaced intervals over space or time. A transect line across the study area is an example. An initial location or time is chosen at random, and then the remaining sampling locations are defined so that all locations are at regular intervals over an area such as a transect, grid, or time period. Random systematic sampling is used to search for *hot spots* and to infer means, percentiles, or other parameters and is also useful for estimating spatial patterns or trends over time. This design provides a practical and easy method for designating sample locations and ensures uniform coverage of a site, unit, or process.

Grid sampling involves the application of a regular or offset grid, or herringbone pattern, according to the size and topography of the study area. It may be based on random selection or aligned with site topography. In certain cases, grid size and sampling density can be estimated using various mathematical formulae assuming normal distributions of concentrations although caution is advised because pollutant data distributions are commonly skewed.

Stratified Sampling Stratified sampling offers an advantage over simple random sampling. In this type of sampling, the total population is divided into smaller groups (sub-populations) or strata that are judged to be more homogeneous, for example relative to the type of contaminant or environmental media. For example, stratified sampling is used to take into account different areas (or strata), which are identified within the study area (e.g. habitat such as a woodland or a lake). Sub-samples can then be taken from each group to achieve greater precision in estimates of the mean and variance, potentially reduce sample size for a given precision, and obtain reliable estimates for population subgroups of special interest.

Other sampling approaches include composite sampling, cluster sampling (e.g. adaptive cluster sampling), and ranked set (see US EPA 2002).

Analytical Laboratory Program

The analytical laboratory steps are basically designed to accurately determine the concentrations of the analytes of interest in samples or biological specimens (e.g. biomarkers) that are representative of the environment, populations of natural biota or humans being investigated, as part of a monitoring plan to evaluate an environmental matter or problem. For example, the problem may involve monitoring exposure to a known toxic analyte in the environment or presence of a biomarker in wildlife or humans.

It is evident that the statistical design of the sampling, selection of appropriate analytical instrumentation and procedures, statistical analyses of analytical results, and suitable QA/QC procedures are essential elements to meet monitoring objectives, such as accuracy and precision of results.

In the case of chemical analyses, Connell (2005, pp. 394–401) provides a concise description of techniques for common chemical analysis (e.g. wet chemistry, spectrophotometry, and chromatography) and laboratory QA/QC procedures. Major types of analytical instrumentation used for analyses of pollutants or contaminants in environmental samples are outlined in Table 17.3. Importantly, key QA/QC indicators and measures for field sampling and laboratory procedures are listed in Table 17.4.

Data Management

Data management in the context of monitoring is the process that includes the collection, validation, storage, protection, and processing of study (or project) data to ensure the accessibility, reliability, and timeliness of the data for its users. Key steps involve:

- *Data quality objectives* are set at the start of the study to ensure that the data to be acquired from monitoring will address the study objectives and the monitoring program objectives in the study design phase (e.g. hypotheses to be tested, parameters or indicators to be measured and statistically compared with guideline values), and the type, quality, and quantity of data to be collected.

Table 17.3 Major types of field monitoring and laboratory analytical instrumentation used for identifying and measuring pollutant concentrations in the environment and related samples.

Analyte	Types of major environmental analytical instrumentation
Particulates	Microscopes (phase contrast, polarizing, fluorescence), scanning electron microscopes (SEM), Energy-dispersive X-ray spectroscopy, high and low volume air samplers, continuous automatic PM_{10}, $PM_{2.5}$, TSP particulate monitors
	Aerosol particle counters, nanoparticle meters
Gases	NO_x, SO_2, CO, VOC, O_3 analyzers, portable air quality meters/dataloggers, CO_2 monitors
Water/soil/air physicochemical parameters	Wet chemistry:
Chemical species (e.g. cations and anions)	gravimetric, titrimetric,
	electrochemical meters (e.g. dissolved oxygen, pH, and specific conductivity),
	particle counters
	Visible/ultraviolet/IR/fluorescent spectrophotometers
Nutrients (e.g. total N and P), nitrate, nitrite, ammonia, and phosphates) and selected inorganic species	Manual/semiautomatic colorimeters/spectrophotometers
Waters/sediments/soils/plants	Automated continuous flow analyzers and multiple parameter discrete analyzers
	Ion chromatographs (IC)
Metals/metalloids (including alkaline and earth metals)	Atomic absorption spectrophotometers (AAS) and accessories
Metals/nonmetals	Inductively coupled plasma (ICP) spectrometers and mass spectrometers (ICP-MS)
Selected isotopes	
Volatile organic chemicals (VOCs/BTEX)	Gas chromatograph/mass spectrometers (GC/MS)
Semi-volatile and non-volatile organic chemicals	Specialized GC/MS (e.g. capillary GC/MS)
(e.g. pesticides, PCBs, PAHs, halogenated hydrocarbons, dioxins, PBDEs, PFOA, and PFOS)	High-pressure liquid chromatographs (HPLC)
	Liquid chromatographs/mass spectrometers (LC/MS)
Personal care products and pharmaceuticals (PCPPs)	Sequential mass spectroscopy (MS/MS)
Radioactivity and radionuclides (radioactive isotopes)	Field survey meters, gamma-ray spectrometers
Note: Selection of instrumentation largely related to properties of analytes and analytical capabilities of instrument	

Table 17.4 Key quality control indicators and measures for field sampling and laboratory analysis.

Quality control indicators	Measures/Comments
Field sampling/Measurements	
Sampling program	Objectives and statistical design, relevant sites, parameters and indicators, sufficient numbers of samples, frequency of sampling, timing, and adequacy of environmental media/biological samples collected, effects of climatic conditions and land uses, etc.
Calibration protocols/standards for field meters and monitoring instruments	Check and record status
Collection of field QC samples for lab analyses	Field and *rinsate* blanks, replicate samples and blind lab duplicates for inter- and intra-lab analyses. NB: sample batch size per QC sample
Certification of calibration of field meters/instruments from third party suppliers/contractors/labs	Check and record calibration certificates are valid
Chain of custody (COC) documentation	Check and record validity including sample details and analyses to be performed for each sample. Ensure COC
Sampling documentation (field sampling locations, samplers, times/dates, labeling details, methods of sampling, measurements, calibration records, samples for analysis, QC samples, and data collection logs)	Check and record adequacy of sampling documentation details
Laboratory analysis	
Internal	
Sample receipt and acceptability, specific analytes requested, sample tracking, laboratory and data analyses, QA/QC, reporting of results, corrective actions, storage of samples (e.g. waters, soils, and biological), information/data	Sign and check chain of custody forms/ analyses requested, suitability of samples for analysis. Registration of samples/tests in laboratory management system (LMS)
Validation of analytical methods used (show that the methods are acceptable for its intended purpose)	Analysis over a range of laboratory analytical standards and number of replicates to obtain acceptable measures of standard deviation, precision, limits of detection, and reporting for selected indicators
Field or trip blanks (usually deionized water)	Treated as a sample to identify errors or contamination in sample collection and analysis
Analytical blanks (background, method, reagent, and calibration blanks)	Reveals background levels of contamination from reagents used, laboratory, sampling, and so on
Calibration blank and standards	Calibration blank and a set of standards are used to calibrate field and lab meters, analytical and monitoring instruments before analyses or readings
Internal standards	Known amount of a compound different to analyte that is added to unknown sample to compare analyte and internal standard signals to calculate amount of analyte

(continued)

Table 17.4 (Continued)

Quality control indicators	Measures/Comments
Duplicates	Internal analytical precision
Matrix spikes – standard additions of known analyte	Known added levels of analyte to counter sample matrix effects.
Analytical recovery of spiked samples	Percentage analytical recovery of added standard or mix of standard concentrations of selected indicators
Laboratory control samples	Made up or previously analyzed samples
Control charts	Upper and lower control limits – used to monitor variations/exceedances in accuracy, precision, or instrument performance over a period of time
Standard reference materials	Externally certified composition of analyte(s) in reference materials used to evaluate accuracy of the analytical method
Internal inter-laboratory calibration program	Internal set of calibration samples of related laboratories conducting the same analyses
Laboratory management system	Management of all laboratory sample registration, sample tracking, procedures, data analyses, QA/QC performance, results, and reporting. Calibration of testing equipment
External	
External field duplicates and replicates	Used to measure sampling and laboratory analysis precision
Split samples	One subsample is analyzed at the project lab and the other is analyzed at an independent lab. The results are compared
External lab analysis of duplicate samples (internal or external)	Analyzed at independent lab and results compared with project lab
Known samples	QC lab sends known samples for selected indicators to project lab
Unknown samples	QC lab sends unknown samples for selected indicators to project lab
External inter-laboratory calibration program	External set of calibration samples (e.g. organized by professional association, industry, or government agency) for performance testing of laboratories conducting the same analyses

- *Data collection, organization, and storage* need to be managed using acceptable or best practices so that suitable, sufficient, and valid data are available for data analysis processing and reporting of results. For example, raw data is to be securely stored and quality checks are made for missing data, and minimizing errors and anomalies.
- *Data analysis* depends on the early planning in the monitoring strategy of the data types, quantities, methods of statistical analyses, and modeling approaches to be applied at various steps in the study. Critical outcomes for the data analysis process are that data of sufficient quality and quantity are available for data processing, selected statistical analyses (software programs), and modeling of data. Adequate QA/QC procedures are to ensure and demonstrate the integrity of data and results of data analysis.
- Many statistical methods and modeling software packages for data analysis are available.

Assessment and Reporting

The final process is to evaluate the data analyses and performance of data quality objectives according to the monitoring objectives and reporting needs of the study (e.g. objectives adopted, methods used, summary results, limitations, timing, and contents of reports), and stakeholders (e.g. funding organization, government agencies, clients, local communities, and public interest groups). Outcomes may include exposure assessments based on monitoring results, and risk characterizations which demand special attention to risk communication needs for decision-making purposes.

17.6 Remote Sensing for Pollution Monitoring

The scientific details of remote sensing are beyond the scope of this chapter. Remote sensing is the science of obtaining information about the nature of the Earth (e.g. areas, landforms, objects, oceans, and its chemical composition) from a distance, often using sensors on satellites or mounted on aircraft (and drones). These remote sensors collect data by detecting energy, usually electromagnetic radiation, that is reflected or emitted from the Earth including its atmosphere. There are two basic types of sensors: passive sensors that collect natural energy reflected or emitted from the Earth's surface, usually reflected sunlight, and active sensors that use an internal energy source, such as a laser-beam, to collect data (e.g. spectral images and distances) about the Earth, by projecting the beam onto the Earth's surface. For example, a laser-beam remote-sensing system can measure the time that it takes for the laser to travel to the Earth (or its target) and reflect back to its sensor. It can also record the spectral signatures of the reflected beam for analysis (NOAA 2019).

Environmental monitoring applications of remote sensing techniques are numerous, from land use changes, wildlife habitats under threat, flooding and drought effects, measuring ocean temperatures, erosion, and sediment transport, to analysis of the chemical composition of trace gases and fine particulates in the atmosphere. An example of remote monitoring of ocean temperatures is given below (Case Study 17.2).

Case Study 17.2

Monitoring Global Warming of the Oceans

How are global water temperatures of the oceans measured? How fast are the oceans warming?

The oceans act as a critical buffer to the effects of global warming due to heat trapped by long-lived greenhouse gases emitted into the atmosphere, mainly by human activities since the Industrial Revolution. This heat is not absorbed evenly across the Earth's surfaces. The oceans absorb the vast majority of this excess heat (~93%), but it generally varies across the ocean surfaces and decreases with depth. The heat content of the oceans is observed to be increasing since about the mid-twentieth century. As the oceans continue to warm and sea temperatures rise, effects such as heat stress, sea level increases, and related extreme climate events (e.g. intensity of hurricanes) are observed to be affecting many marine species and their ecosystems (e.g. shifts in some warm water species toward higher latitudes, and coral deaths due to *bleaching* events), marine resources (e.g. fisheries), and human use of coastal environments, as discussed in Chapter 14 in this book.

Recent evidence from oceanographic studies (Cheng et al. 2019) found that the oceans are warming at a faster rate than previously reported (e.g. IPCC 2014). A key component of this research was the use of the global array of drifting autonomous *ARGO* floats that monitor surface to sub-surface temperature (and salinity) profiles of the oceans to depths of 2000 m.

Monitoring Surface and Sub-surface Ocean Temperatures

Sea Surface Temperature (SST) is a measure of the energy due to the motion of molecules at the top layer of the ocean. Temperatures are measured from about 10 µm below the surface (infrared bands) to 1 mm (microwave bands) depths using radiometers. Since the 1980s, most global SST observations are made by spectroradiometers in satellites, unlike earlier temperature readings derived from instruments on ships, buoys and shorelines, mainly along shipping routes.

As well as satellite sensors, modern ocean temperature measurements (surface and sub-surface) are taken by a global array of high-tech autonomous floats, fitted with temperature and other types of sensors, deployed to drift on the ocean surface. Two international types of global monitoring programs have evolved using different designed floats, one known as the Global Drifter Program uses *drifters*, and the other, the Global ARGO Program, uses *floaters*. Both *drifters* and *floaters* transmit data to satellites and are used to calibrate satellite sensors (e.g. radiometers). These monitoring programs form part of the Global Ocean Observing System (GOOS) and are described below.

Global Drifter Program (GDP)

The Global Drifter Program maintains a 5 × 5 degree array of ~1300 *drifters* as part of GOOS. It is coordinated, developed, and maintained by NOAA with special scientific and technical support from Scripps Institution of Oceanography (SIO). The global drifter program was set up in the 1980s as the Surface Velocity Program (SVP) to measure surface ocean currents, sea surface temperature, and sea-level atmospheric pressure (Lumpkin and Pazos 2007).

The modern drifter consists of a surface float (or buoy) and a sub-surface drogue (sea anchor), attached by a long, thin tether. Sensors in the surface float measure ocean surface properties, such as sea surface temperature, barometric pressure, salinity, and wave height. A transmitter in the float sends data collected by the sensors to passing satellites, which then relay it to land-based centers. The drogue dominates the total area of the instrument and is centered at a depth of 15 meters beneath the sea surface (see website for Global Drifter Program - NOAA 2020).

Global ARGO Program

Argo is a twenty-first century monitoring program that is now a major component of the ocean observing system and complements other upper ocean monitoring networks. It consists of a global array of almost 4000 free-drifting profiling floats (see Figure 17.5) that continuously measures the temperature and salinity profiles of the upper 2000 m of the ocean (see Figure 17.5). High-quality data are collected by smart sensors, transmitted to satellites, and relayed to receiving stations, and made publicly available within hours (Argo 2020).

It is designed to quantitatively describe the changing state of the upper ocean and the patterns of ocean climate variability from months to decades, including heat content and freshwater storage and transport, and to enhance the interpretation of altimetric sea surface height variability measured by remote sensing.

Primary objectives of Argo-derived data include recording and evaluation of climate variability (from seasonal to decadal), use in ocean-atmospheric forecasting models, testing research models, and advancing our understanding of climate predictability.

17.7 Monitoring and Assessment of Disease Impacts From Pollution

Country, regional, and global assessment of disease attributed to pollutant-related exposures, such as air pollutants, and environmental risk factors (e.g. spread of vector-borne diseases and climate change indicators) are increasingly subject to intensive and integrated physical data monitoring and analysis in the twenty-first century. A major global example is the evaluation of the burden of disease related to the exposure of humans to ambient air pollution as outlined in Case Study 17.3.

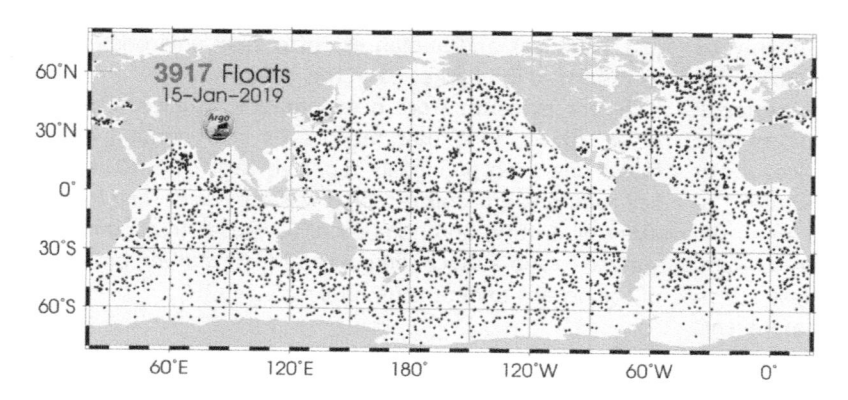

Figure 17.5 Screenshot of global distribution of Argo floats (see Argo 2020).

Case Study 17.3

Burden of Disease from Ambient Air Pollution

The Burden of Disease (BoD) is a comparative assessment of the health impact of key risk factors, including ambient air pollution, which is evaluated by measuring and estimating exposures to $PM_{2.5}$. Integrated environmental monitoring of human population exposures to $PM_{2.5}$ plays a central role in the evaluation of the global, regional, and country health effects from ambient air pollution (WHO 2018a).

The methodology is based on combining exposure data to $PM_{2.5}$ and its distribution in the population, with exposure-risk estimates over a global range of exposures using integrated exposure risk models (IERs). It results in a population attributable fraction (PAF) of disease burden, which can then be used to calculate the disease burden attributable to ambient air pollution for each health outcome (specific disease of interest).

A framework showing the role of integrated exposure monitoring and the method for burden of disease estimation is presented as Figure 17.6.

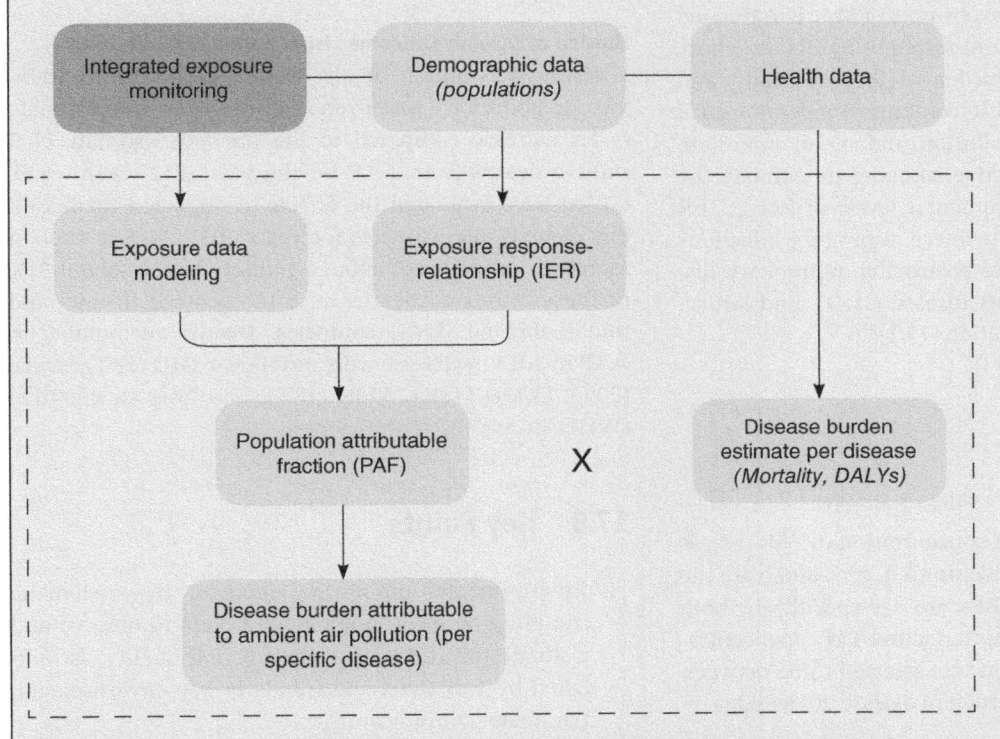

Figure 17.6 Outline of method of burden of disease estimation attributable to ambient air pollution. Source: modified from WHO 2016, 2018a. Published by the World Health Organization, Geneva.

Data Sources

Primary data sources for the estimation of disease burden due to exposure to ambient air pollution consist of (i) population data obtained from the UN Population Division, (ii) health data compiled by the World Health Organization on the total number of deaths, years of life lost (YLLs), years lived with disability (YLDs), and disability-adjusted life years (DALYs) for each country by sex and age group, for acute lower respiratory infections (ALRI), lung cancer, chronic obstructive pulmonary disease (COPD), stroke and ischemic heart disease (IHD), and (iii) integrated exposure monitoring on a global scale, mainly for $PM_{2.5}$. Monitoring approaches are briefly discussed below.

Exposure Monitoring and Assessment (World Health Organization)

Population exposures to ambient air pollution on a global scale are based primarily on extensive ground monitoring networks for $PM_{2.5}$ and PM_{10}, supplemented by other sources of information, data and modeling, particularly in regions with little or no monitoring. Data sources now include over 9000 ground level monitoring locations in some 4300 cities around the world, satellite remote sensing, population estimates, topography, chemical transport models, and information on local monitoring networks. The Data Integration Model for Air Quality (DIMAQ) combines multiple data sources and produces estimates of population exposures to $PM_{2.5}$ at high spatial resolution (0.1° × 0.1°) and air quality profiles for individual countries, regions, and globally. It also incorporates calibration of satellite data and other data sources with ground measurements. Model outputs include the percentage of the population exposed to $PM_{2.5}$ by country, in increments of $1\,\mu g\,m^{-3}$ (WHO 2016, 2018a).

Burden of Disease Attributable to Ambient Air Pollution

Exposure-Risk Relationships The relative risk (RR) for a disease caused by ambient exposure to $PM_{2.5}$ is estimated using an integrated exposure-response (IER) function, developed by the Global Burden of Disease Study, and also used by WHO. The IER combines epidemiological evidence for outdoor air pollution, indoor air pollution, second-hand smoke, and active smoking to estimate the level of disease risk at incremental levels of $PM_{2.5}$. IER functions are applied to acute lower respiratory infections (ALRI), lung cancer, chronic obstructive pulmonary disease (COPD), ischemic heart disease (IHD), and stroke, and can be described by equation (17.1):

For $z < z_{cf}$,

$$RR_{IER}(z) = 1 \quad (17.1)$$

For $z \geq z_{cf}$,

$$RR_{IER}(z) = 1 + \alpha\left\{1 - \exp\left[-\gamma(z - z_{cf})^{\delta}\right]\right\} \quad (17.2)$$

Where z is the annual mean concentration of $PM_{2.5}$; z_{cf} is the counterfactual $PM_{2.5}$ concentration; α, γ, and δ are the parameter estimates. The IERs are age-specific for both IHD and stroke (e.g. the counterfactual $PM_{2.5}$ concentration for ambient air pollution was selected to be between 2.4 and $5.9\,\mu g\,m^{-3}$ for the burden of disease 2016 update).

Population Attributable Fractions (PAF) The PAF is the proportional reduction in population disease or deaths that would occur if exposure to a risk factor (e.g. ambient air pollution) was reduced to an alternative ideal exposure scenario instead of zero (e.g. counterfactual distributions such as a theoretical minimum - exposure that results in minimum population risk).

Country PAF values for ALRI, COPD, lung cancer, stroke, and IHD are calculated from equation (17.3), for each sex and age group.

$$PAF = \frac{\sum_{i=1}^{n} P_i(RR - 1)}{\sum_{i=1}^{n} P_i(RR - 1) + 1} \quad (17.3)$$

Where i is the level of $PM_{2.5}$ in $\mu g\,m^{-3}$, and P_i is the percentage of population exposed to that level of air pollution, and RR is the relative risk derived from systematic review/meta-analysis of epidemiological studies.

Attributable Burden of Disease (AB) The attributable burden (AB) is then calculated by multiplying the population attributable fraction (PAF) value by the health outcome, for each disease of interest, sex, and age group, as shown by equation (17.4) and indicated in Figure 17.6.

$$AB = PAF \times \text{health outcome} \quad (17.4)$$

In Figure 17.6, the health outcome for each disease of interest is derived from the health data compiled by the WHO, as shown by the disease burden estimate per disease (e.g. total number of deaths and DALYs).

Burden of Disease Outcomes from Ambient Air Pollution

Globally, 4.2 million deaths were attributable to ambient air pollution (AAP) reported in 2016. This estimate is an increase compared to the previous estimate of 3 million deaths from AAP for the year 2012. Ostro et al. (2018) have evaluated the effects of upgraded methodology changes on estimates for years 2013–2015 to explain increased estimates of about 1 million deaths. About 91% of these estimated deaths in 2016 occurred in low- and middle-income (LMI) countries. Deaths attributable to AAP in 2016 by disease were as follows: IHD (38%), stroke (20%), COPD (18%), ALRI (18%), and lung cancer (6%) (WHO 2018b).

17.8 Key Points

1. Environmental monitoring includes the systematic sampling of air, water, soil, and biota to observe and evaluate the status of environmental quality, as indicated by variations and trends in key environmental parameters or indicators.
2. There are a multitude of purposes, such as from baseline and research studies to monitoring exposed populations, evaluation of environmental conditions or impacts from human activities, compliance with environmental regulations, to public information, policy formulation, and decision-making for environmental management.
3. A great deal of emphasis is placed on the quality and effectiveness of monitoring programs, including objectives, sampling and analytical plans, QA/QC, data quality analysis, management, assessment, and reporting.
4. The scope of monitoring and environmental management is growing much broader in the context of environmental health concepts that incorporate ecosystem health, human health, and sustainability. Integrated

environmental monitoring of air and water pollutants, for instance, and remote sensing by aircraft and satellites have expanded rapidly.
5. Increasingly, biomarkers are used as biological indicators of a measurable alteration of a normal function of an organism that responds to exposure from a contaminant, or an effect induced by one. They indicate possible exposures and/or early biological responses before sublethal effects occur and should usefully evaluate potential risks to populations, communities, or ecosystems depending upon their specific nature and properties.
6. Environmental monitoring programs are often designed to sample and analyze for physical, chemical, biological, and ecological indicators in environmental media and/or populations, and communities within ecosystems, including human environments. Statistical analyses of results from monitoring are commonly compared with available sets of environmental or health quality indicators (e.g. standards, criteria, guidelines, and indices), usually published by major international or national regulatory or advisory agencies.
7. The framework for a pollution-based monitoring program is given in Figure 17.4. The key steps apply general environmental monitoring principles. Many international, national, state, provincial, and city guidelines and protocols for environmental monitoring exist, and provide detail on monitoring approaches, sampling, and analytical methods, quality assurance, data analysis, and reporting, often as part of regulatory requirements for assessment and environmental management.
8. Sampling and analytical plans (SAP) are usefully developed together to ensure the resources of the study and sampling design are compatible with field and analytical equipment and procedures, analytical costs, reporting times, etc. and also data collection, analysis, and management. The entire program or plan is subjected to quality assurance and quality control (QA/QC) procedures that demonstrate the quality of the data and the right results for decision-making purposes.
9. A wide range of sampling methods are used for monitoring pollution, which depend on the characteristics of environment, populations, pollutants, and reference sites being investigated. Common sampling strategies cover judgmental, random, systematic and grid, and stratified approaches.
10. Quality assurance is the process that involves the use of data quality objectives and indicators, planned and systematic actions, procedures, checks, and makes decisions to ensure samples and data collected in the study are representative, their integrity is maintained, and the results of analyses are accurate and reliable.
11. Quality control is the management system for field and laboratory procedures and related activities that control and assess results. The assessment of the reliability of sampling, analytical procedures, data, and analytical results is based on data quality indicators (DQIs) of accuracy, precision, representativeness, completeness, and comparability.
12. Data analysis depends on the early planning in the monitoring strategy of the data types, quantities, methods of statistical analyses, and modeling approaches to be applied at various steps in the study. Critical outcomes for the data analysis process are that data of sufficient quality and quantity are available for data processing, selected statistical analyses, and modeling of data. Adequate QA/QC procedures are to ensure and demonstrate the integrity of data and results of data analysis.
13. The final process is to evaluate the data analyses and performance of data quality objectives according to the monitoring objectives and reporting needs of the study (e.g. objectives adopted, methods used, summary results, limitations, timing, and contents of reports), and stakeholders (e.g. funding organization, government agencies, clients, local communities, and public interest groups). Outcomes may include exposure assessments based on monitoring results, and risk characterizations, which demand special attention to risk communication needs for decision-making purposes.
14. Case Studies 17.1 (Global Climate Change Monitoring), 17.2 (Monitoring Global Warming of the Oceans), and 17.3 (Burden of Disease from Ambient Air Pollution) provide a global perspective on application of monitoring to measure natural and pollution-related changes and to evaluate impacts and risks discussed in chapters within this book.

References

Argo (2020). What is Argo? Argo website. Global Ocean Observing System. http://www.argo.ucsd.edu (accessed 11 February 2020).

Bartram, J. and Balance, R. (1996). *Water Quality Monitoring - A Practical Guide to the Design and Implementation of Freshwater Quality Studies and Monitoring Programmes.*

United Nations Environment Programme (UNEP) and the World Health Organization (WHO). https://www.who.int/water_sanitation_health/resourcesquality/waterqualmonitor.pdf (accessed 10 February 2020).

Briggs, D.J. (2008). A framework for integrated environmental health impact assessment of systemic risks. *Environmental Health* 7 (1) 61 doi:https://doi.org/10.1186/1476-069X-7-61.

Centers for Disease Control and Prevention (2009). *Fourth Report on Human Exposure to Environmental Chemicals, 2009*. Atlanta, GA: US Department of Health and Human Services, Centers for Disease Control and Prevention. https://www.cdc.gov/exposurereport/ (accessed 12 February 2020).

Chapman, D. ed. (1996). *Water Quality Assessments - A Guide to Use of Biota, Sediments and Water in Environmental Monitoring*. 2. United Nations Educational, Scientific and Cultural Organization/World Health Organization/United Nations Environment Programme. London: E&FN Spon. https://www.who.int/water_sanitation_health/resourcesquality/watqualassess.pdf (accessed 12 February 2020).

Cheng, L., Abraham, J., Hausfathur, Z. et al. (2019). How fast are the oceans warming? *Science* 363 (6423): 128–129.

Chesapeake Bay Program (2020). *Monitoring*. Chesapeake Bay Program. Annapolis, MD. https://www.chesapeakebay.net/what/programs/monitoring (accessed 11 February 2020).

Connell, D.W. (2005). *Basic Concepts of Environmental Chemistry*, 2e. New York: CRC Press.

Connell, D.W. and Miller, G.J. (1984). *Chemistry and Ecotoxicology of Pollution*. New York: Wiley.

GCOS (2020a). *Global Climate Observing System*. World Meteorological Organization. https://gcos.wmo.int (accessed 11 February 2020).

GCOS (2020b). Essential climate variables. Global Climate Observing System. World Meteorological Organization. https://gcos.wmo.int/en/essential-climate-variables (accessed 11 February 2020).

GCOS (2020c). *The Key Climate Indicators for Each Climate Change Domain*. Global Climate Observing System. World Meteorological Organization. https://gcos.wmo.int (accessed 11 February 2020).

Gilbert, R.O. (1987). *Statistical Methods for Environmental Pollution Monitoring*. New York: Wiley.

Holdgate, M.W. (1979). *A Perspective of Environmental Pollution*. Cambridge: Cambridge University Press.

Idrees, Z. and Zheng, L. (2020). Low cost air pollution monitoring systems: a review of protocols and enabling technologies. *Journal of Industrial Information Integration* 17: 100123.

IPCC (2014). Topic 1: observed changes and their causes. In: *Climate Change 2014: Synthesis Report. Contribution of Working Groups I, II and III to the Fifth Assessment Report of the Intergovernmental Panel on Climate Change* (ed. Core Writing Team, R.K. Pachauri and L.A. Meyer). Geneva, Switzerland: IPCC.

ITU (n.d.). *Monitoring Climate Change*. International Telecommunication Union. United Nations. http://www.itu.int/themes/climate/docs/report/06_monitoringclimatechange.html (accessed 11 February 2020).

Jackson, M. C., Weyl, O.L.F., Altermatt, F. et al. (2016). Recommendations for the next generation of global freshwater monitoring tools. *Advances in Ecological Research* 55: 616–36. https://www.altermattlab.ch/wpcontent/uploads/2018/06/Jackson.et_.al_AdvEcoRes_2016.pdf (accessed 5 February 2019).

Karr, J. and Chu, E.W. (1997). Biological monitoring: essential foundation for ecological risk assessment. *Human and Ecological Risk Assessment* 3: 993–1004.

Lumpkin, R. and Pazos, M. (2007). Measuring surface currents with Surface Velocity Program drifters: the instrument, its data, and some recent results. In: *Lagrangian Analysis and Prediction of Coastal and Ocean Dynamics* (ed. A. Griffa, A. Kirwan Jr., A. Mariano, et al.), 39–67. Cambridge, UK: Cambridge University Press.

New Zealand Ministry for the Environment (2019). Why monitor? Monitoring Programme Objectives. http://www.mfe.govt.nz/publications/air/good-practice-guide-air-quality-monitoring-and-data-management-2009/2-why-monitor (accessed 11 February 2020).

Nikinmaa, M. (2014). *An Introduction to Aquatic Toxicology*. Oxford: Academic Press. https://www.altermattlab.ch/wpcontent/uploads/2018/06/Jackson.et_.al_AdvEcoRes_2016.pdf (accessed 5 February 2019).

NOAA (2019). *What Is Remote Sensing?* National Ocean Service website. National Oceanic and Atmospheric Administration. https://oceanservice.noaa.gov/facts/remotesensing.html (accessed 20 February 2019).

NOAA (2020). *Global Drifter Program*. National Oceanic and Atmospheric Administration, Atlantic Oceanographic and Meteorological Laboratory, Physical Oceanography Division (PhOD). https://www.aoml.noaa.gov/phod/gdp/index.php (accessed 13 February 2020).

Ostro, B., Spadaro, J.V., Gumy, S. et al. (2018). Assessing the recent estimates of the global burden of disease for ambient air pollution: methodological changes and implications for low- and middle-income countries. *Environmental Research* 166: 713–725.

Snyder, E.G., Watkins, T.H., Solomon, P.A. et al. (2013). The changing paradigm of air pollution monitoring. *Environmental Science & Technology* 47 (20): 11369–11377.

US EPA (2002). Guidance on choosing a sampling design for environmental data collection for use in developing a quality assurance project plan. EPA QA/G-5S. United States Environmental Protection Agency. Washington, DC. https://www.epa.gov/sites/production/files/2015-06/documents/g5s-final.pdf (accessed 12 February 2020).

US National Science and Technology Council (1995). Setting a new course for US Coastal Ocean Science. *Final report of the Subcommittee on US Coastal Ocean Science*. National Science and Technology Council, Committee on Environment and National Resources. Washington, DC.

WHO (2016). *Ambient Air Pollution: A Global Assessment of Exposure and Burden of Disease*. World Health Organization. https://apps.who.int/iris/handle/10665/250141 (accessed 12 February 2020).

WHO (2018a). Burden of disease from ambient air pollution for 2016. Description of method. V 5. World Health Organization. https://www.who.int/airpollution/data/AAP_BoD_methods_Apr2018_final.pdf (accessed 11 February 2020).

WHO (2018b). Burden of disease from ambient air pollution for 2016. Summary of results. V2. World Health Organization. https://www.who.int/airpollution/data/AAP_BoD_results_May2018_final.pdf (accessed 11 February 2020).

WMO (2019). *What We Do*. World Meteorological Organization. https://public.wmo.int/en/our-mandate/what-we-do (accessed 1 February 2019).

18

Human Health and Ecological Risk Assessment

18.1 Introduction

The concept of **risk** has a long history inherently within the development of insurance, gambling, and mathematics as described by Bernstein (1996). The considerations of risk are necessary when there is uncertainty regarding an outcome of importance to human society which cannot be defined with precision. Some of the fascinating story of risk started with Fibonacci with his book published in 1202, *Liber Abaci* or in English *Book of the Abacus*. This book introduced the Hindu–Arabic numbering system to Europe displacing the cumbersome Roman system of numbering. It demonstrated the power of this system and it rapidly became the accepted system of numerals and calculation. It also provided the numerical basis for collecting and collating data and development of databases. The analysis of data draws on statistics leading to sophisticated and quantitative considerations of risk and uncertainty. The analysis of risk is now applied to insurance, engineering constructions, economic decisions, flood occurrence and damage, and so on. Governments and agencies throughout the world have set out guidelines to apply risk evaluation techniques in general and in specific applications, for example, OSHA (2016) and Rausand (2013). In this chapter, we are concerned with the risk to human health and the environment posed by exposure to chemicals in the natural and human environment.

Chemicals are used in an immense variety of ways for the benefit of human society. However, the use of many chemicals has a direct risk involved, for example, medicinal compounds, such as antibiotics, are consumed orally giving a risk of resultant adverse effects. On the other hand, the use of pesticides in agriculture results in the exposure of nontarget organisms to these active biological agents. In this way, human populations and organisms in the natural environment can both be exposed to biologically active chemicals. Unintentional, and often unplanned, risks can also be involved in the disposal of waste chemical products. Waste chemicals are often disposed to land, water, and air and may contain agents which can have adverse effects on human health and the natural environment.

The chemical industry provides petroleum fuels, antibiotics and other drugs, plastics, pesticides, food preservatives, agricultural fertilizers, and so on, without which our society cannot survive. About 151 000 chemicals are estimated to be in daily use and most of these substances have little or no adverse environmental effects but some may be harmful to human health or the natural environment as a direct result of usage, manufacture, or disposal of wastes. Often, these effects only become apparent after wide and prolonged usage and then control measures are introduced.

Management agencies have had difficulties in placing the risk involved in these situations into a clear perspective because of the uncertainty of the outcomes in terms of human health and the natural environment. As a result, there has been a need to develop techniques which give a systematic and transparent evaluation of the risks involved in the use of chemicals. These evaluations can then be used to gain an insight into the needs for management of the risks involved and allow the allocation of resources in the most effective way to reduce or eliminate that risk. Risk assessment gives:

1. A transparent and reproducible method to evaluate risk to human health and natural ecosystems.
2. Outcomes that are in accord with community expectations.
3. Greater confidence that efforts in environmental protection from chemicals will be successful in protecting human health and natural ecosystems.
4. Resources will be used in the most effective manner in controlling chemicals.
5. Industry and other activities will be managed in the most cost-efficient manner without wastage of resources on chemical management problems which may not be of high priority.

Risk assessment gives the benefits outlined above which provide the basis for communicating the risk to the members of the community who may be affected by the exposure to the chemicals involved. The community members who may be involved include industry workers, families, environmentalists, health workers, and so on.

Relationship of the Impact of Pollutants on the Components and Functions of Human and Natural Ecosystems to other Chapters in this Book

The living system that we are considering is dynamic and the steps involved in a chemical's interaction with the human and natural environment is illustrated in Figure 1.3 (see Chapter 1). This follows a cascading sequence of interactions of a chemical discharged or released to the environment. Initially, **sources** are considered, then through **distribution** leading to exposure of **humans and organisms in the natural environment**. These processes were considered in detail in other chapters in this book as indicated in Figure 18.1. The contents of these chapters provide a basic understanding of how chemicals interact with humans and organisms in the natural environment. However, considerations of assessing the risk to human health and organisms in the natural environment require a more specific approach utilizing existing and supplementary data and information.

Figure 18.1 Conceptual model of the impact of pollutants on the components and functions of the human and natural environment with the related chapters in this book.

Development of the Framework for the Assessment of the Risk to Human Health and the Natural Environment

The US Environmental Protection Agency (US EPA) has been a leader in the development and use of procedures for the assessment of the risks due to chemicals. These are described in their publication *Framework for Human Health Risk Assessment to Inform Decision Making* (Fitzpatrick et al. 2017). The basic framework was formulated in the 1980s and is shown in Figure 18.2. This has four basic steps – Hazard Identification, Exposure Assessment, Toxicity or Carcinogenesis – Dose–Response Relationships, and Risk Characterization. The success of this framework depends on knowing the exposure, and when the dose–response relationship is also known, then the adverse response can be derived. This framework has been retained and improved with additional factors added and procedures improved through many iterations over the years.

This information on risk assessment is used in the risk management process which consists of consideration of options, communication of risk to those involved, control decisions, and then monitoring to ensure that a correct decision has been made. All of these steps can feed back into previous steps in the process or into the initial hazard identification step and this procedure may occur several times before a satisfactory result is obtained. A learning procedure may be involved when new information is developed and fed into the system in later steps which could modify the outcome of preceding steps.

This risk assessment framework requires sets of data on dose–response relationships particularly and other factors to operate successfully. Thus, another important development was the internationally coordinated and accepted guidelines published by the Organisation for Economic Cooperation and Development (OECD 2020). This is an extensive series of guidelines for testing and evaluation to determine those chemicals which present a potential environmental human health hazard or hazard to the natural environment. The OECD describes this as *a collection of about 150 of the most relevant internationally agreed testing*

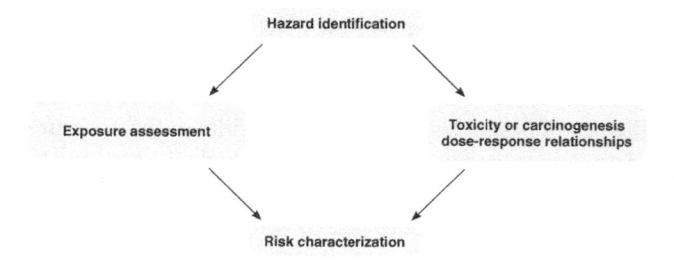

Figure 18.2 Basic framework for the assessment of risk due to exposure to chemicals.

methods used by government, industry and independent laboratories to identify and characterise potential hazards of chemicals. They are a set of tools for professionals, used primarily in regulatory safety testing and subsequent chemical and chemical product notification, chemical registration and in chemical evaluation. They can also be used for the selection and ranking of candidate chemicals during the development of new chemicals and products and in toxicology research. This group of tests covers health effects.

Utilizing the data from the OECD test guidelines mentioned above, many countries have developed risk evaluation programs for new and other chemicals providing a basis for the reduction or elimination of adverse effects. Some have indicated that this control program will be their major environmental management program for the foreseeable future.

Toxicology and Risk Assessment

Toxicology is the science of poisons and it has been successful in addressing the measurement of the toxic effects of chemicals and many related aspects. In recent times, there has been concern that chemical agents in very low concentrations over relatively long periods of time present a hazard to human health and the natural environment. However, toxicology techniques have been mainly applied in measuring short-term effects at relatively high concentrations of chemicals. But long-term adverse effects usually result from a very low exposure over lengthy periods of time and are not amenable to normal toxicology approaches. Thus, many assessments of risk due to exposure to chemicals in the environment make use of extrapolations of the known data from high exposures in laboratory experiments to the relatively low levels and extended exposure times of environmental chemicals. These extrapolated values are derived for use in the risk assessment process by division of the laboratory data by a large number described usually as a **Safety Factor or Application Factor**. These factors are considered in various sections below.

The Socioeconomic Aspects of Risk with Exposure to Chemicals

The concept of **risk** due to exposure to chemicals requires a specialized approach within the group of activities which need the evaluation of uncertain outcomes. It involves the assessment of the possible risk of adverse effects on human health and the natural environment as a result of exposure to chemicals. Here, in contrast to other parts of this book, considerations of risk involve people in the community within different age groups, occupations, family situations, attitudes to chemicals, and so on. Risk perception is an individual or societal characteristic with aversion to some hazards, indifference, or acceptance of others.

Thus, the basic concept of risk itself has many difficulties in interpretation and understanding. **Risk** can be defined as the probability of realization of a **hazard** resulting from exposure to a chemical or other agent. However, the public perception of risk involves many social and cultural factors. Some hazards are particularly dreaded and rank higher than comparable hazards as a result. For instance, hazards from nuclear power generation are rated very highly by the public but are ranked relatively low by risk experts who have studied the data involved. On the other hand, the use of alcoholic beverages is regarded by experts to be a reasonably high-risk activity but in the public mind, it is rated considerably lower. To minimize these problems, the risk is often defined in terms of probability of a particular effect resulting from exposure to a chemical. It can have the dimensions of frequency of occurrence or incidence coupled to an exposure. For example, with human health, there is a risk of 1 in 1 million (1 in 10^6) of an individual contracting cancer resulting from exposure to a chemical in air at a certain concentration breathed 24 hours/day for a lifetime, usually 70 years. Risk assessment techniques are commonly used to set acceptable levels of contaminants in the environment for both human health and the natural environment. With ecological risk assessment, the risk can be defined in terms of an adverse effect on an ecosystem, e.g. occurrence of fish kills or adverse effect on iconic or commercial species.

18.2 Risk Assesssment Processes and Principles

The concept of risk in many human activities is now widely accepted and government and regulatory agencies have formulated general principles to use in the conduct of risk assessments. Risk assessment with chemicals is fundamentally concerned with anticipating where chemicals go (exposure) and probability of harm (toxicology) when they get there. However, there are many levels of requirements for the evaluation of hazard and risk due to chemical exposure as shown in Table 18.1. In many workplaces, there is possible exposure to a chemical hazard and a preliminary evaluation is needed. An initial **Hazard Evaluation** can be carried out by **qualitative observation** on a walk-through in the workplace as to whether exposure to a hazardous chemical is possible. This can be seen as a Level 1 hazard evaluation as shown in Table 18.1. If a significant hazard is identified, further evaluations can be carried out using **Semiquantitative** techniques at Level 2. These evaluations can be carried out by basically trained personnel at an

Table 18.1 Levels and types of human health hazard evaluation and risk assessment.

Level 1 - hazard evaluation in workplaces - based on qualitative observations and evaluations without direct measurement and evaluation of exposure.
Level 2 - semiquantitative risk assessment - evaluation using a score assigned by observation and collection and collation of available data.
Level 3 - quantitative evaluation - measurement of chemical exposure and comparison with standards - human health risk assessment.
Level 4 - quantitative evaluation - use of modeling and probabilistic techniques to evaluate risk.

---

intermediary level between the qualitative hazard evaluation at Level 1 and the numerical evaluation of quantitative risk assessment using measurement of exposure levels at Levels 3 and 4. Semiquantitative techniques evaluate risks with a **score** assigned by observation as well as collection and collation of data already available. It offers a more consistent and rigorous approach to assessing and comparing risks and risk management strategies than qualitative hazard evaluation at Level 1. It avoids some of the greater ambiguities that a qualitative hazard evaluation may produce. Semiquantitative techniques can be applied widely to workplaces due to their low cost and relatively easy application. In some countries, all workplaces must be assessed for hazards due to chemicals by the Level 1 and 2 processes. The levels of Public Health Risk Assessment are not mutually exclusive and can be carried out in a sequence with each level leading to further evaluations at a higher level.

There are basically two sources of exposure to chemicals – chemicals present in the environment as a part of the normal background can constitute a **Background Risk** and those chemicals which are usually artificially added to the environment which constitute an **Incremental Risk**. The sum of the two is a measure of the **Total Risk**. For example, the presence of arsenic in drinking water can be a background risk to human health in many areas throughout the world and the addition of chlorine to disinfect the water is an incremental risk.

Hazard Evaluation in Workplaces

In many countries, it is necessary to conduct a chemical hazard evaluation in all workplaces:

- before the workplace begins operations
- when an apparent hazard is identified
- if there are changes in the workplace procedures
- in response to an incident or accident

This can involve a qualitative evaluation of the workplace at Level 1 by observations on a walk-through with such factors as the use of hazardous chemicals, the type and quantities used with timing and frequency of use, personnel exposed, and so on. This can be supplemented with information from **MSDS (Material Safety Data Sheets)** available at many sites on the Web. Using this information,

an evaluation of the potential hazard can be made and if there is a significant potential hazard, then a Semiquantitative Risk Assessment can be carried out to further evaluate the hazard.

18.3 Semiquantitative Risk Assessment

Evaluation Using a Score

There are many semiquantitative risk assessment procedures available for evaluating the risk due to exposure to chemicals. This risk assessment procedure operates by assigning a score to the two main factors which influence the level of risk – the **likelihood** of exposure and the **consequences** of exposure and presentation of these in a matrix. A discussion of the use of matrices in this way is contained in the review by Duijm (2015). These two factors correspond with exposure (likelihood) and toxicity (consequences) in the higher levels of risk assessment application – Levels 3 and 4. The chemical exposure is assigned a rating according to the scale in Table 18.2 and the probability of harm is also assigned a rating as described in Table 18.3. By observing the area in the matrix (see Figure 18.3) where the two factors, likelihood and consequences, intersect, a **Risk Score** is obtained. Alternatively,

Table 18.2 Likelihood of exposure rating.[a]

Rating	Description	Occurrence	Probability
Almost certain	Expected to occur in most circumstances	Multiple/ 12 months	>80%
Likely	Strong possibility of occurrence	Within 12 months	61–80%
Possible	May occur occasionally	Within five years	31–60%
Unlikely	Not expected to occur but may happen	Within 10 years	5–30%
Rare	May only occur in exceptional circumstances	>10 years	<5%

[a] The number of times within a specified period in which a risk may occur.

Table 18.3 Risk matrix and rating for consequences.[a]

Consequences/Likelihood	1 Insignificant *no hazard*[a]	2 Minor *slightly hazardous*[a]	3 Moderate *moderately hazardous*[a]	4 Major *highly hazardous*[a]	5 Catastrophic *extremely hazardous*[a]
5 Almost Certain	Low 5	Medium 10	High 15	High 20	Extreme 25
4 Likely	Low 4	Medium 8	Medium 12	High 16	High 20
3 Possible	Low 3	Low 6	Medium 9	Medium 12	High 15
2 Unlikely	Low 2	Low 4	Low 6	Medium 8	Medium 10
1 Rare	Very Low 1	Low 2	Low 3	Low 4	Low 5

[a]Consequence ratings indicated in the upper row of the table are based on Table 5.3.

Initiation and planning
- Review of reasons for initiation
- Identification of affected people and communities
- Community involvement
- Formulation of objectives

Hazard identification
- History of the activities causing contamination
- Possible exposure of people and communities
- Identification of sources of hazardous chemicals
- Collection of the properties of potential chemical hazards

Exposure assessment (EA)
- Conceptual models of exposure of groups in community
- Identification of pathways of exposure
- Calculation of exposure from measurement and analysis
- Modelling of exposure patterns

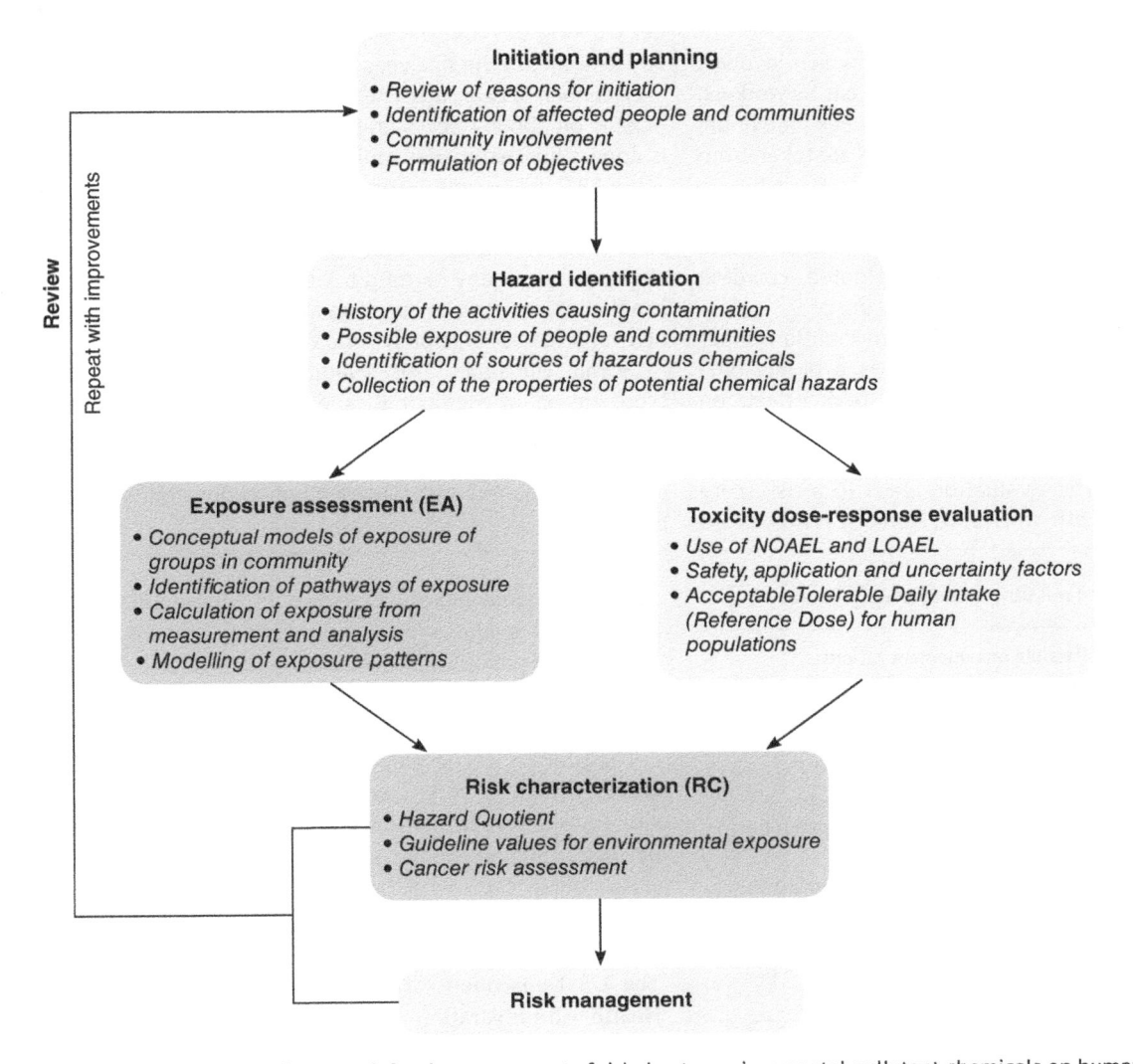

Toxicity dose-response evaluation
- Use of NOAEL and LOAEL
- Safety, application and uncertainty factors
- Acceptable Tolerable Daily Intake (Reference Dose) for human populations

Risk characterization (RC)
- Hazard Quotient
- Guideline values for environmental exposure
- Cancer risk assessment

Risk management

Review / Repeat with improvements

Figure 18.3 Conceptual framework for the assessment of risk due to environmental pollutant chemicals on human health.

the two ratings can be assigned a numerical value, for example, a value of 1 for the lowest rating and ranging up to 5 for the highest rating as shown in Figure 18.3.

These ratings can be multiplied to obtain a **Numerical Overall Risk Score**. Table 18.4 outlines possible actions which can be taken as indicated by the risk score. This system is suitable for general application but needs the likelihood, consequences, and management actions to be specific for a given situation.

Two examples are shown in Box 18.1 where this Semiquantitative Risk Assessment is applied to workplace situations in a petrochemical plant.

Box 18.1 Worked Examples of Semiquantitative Risk in the Workplace

A petrochemical plant which produces a range of organic chemical products including a range of organic solvents has workplaces where workers are subject to possible exposure to hazardous chemicals. A Hazard Evaluation at Level 1 has been carried out by a qualitative inspection as a part of a walk-through of the relevant parts of the plant. Several operations were identified as posing a hazard to workers and in particular the following situations were evaluated by Semiquantitative Risk Assessment.

Situation A. – Exposure to Benzene
The distillation of benzene is carried out which involves taking samples for analysis on a routine basis by workers. These workers operate over 40 hours per week for 8 hours per working day. In a working day, samples are taken from the still on an hourly timetable.
Likelihood of Exposure. The workers could be exposed to benzene on about 40 occasions per week when sampling for analysis. Thus, exposure would be evaluated according to Table 18.2 as *Almost Certain* with a rating of 5.
Consequences. Consulting a MSDS information sheet from the Web indicates that benzene is a highly hazardous substance having many harmful toxic effects on humans as well as being carcinogenic. Thus, benzene would be evaluated at *Highly Hazardous* with a rating of 4.
Overall Outcome. This evaluation leads to a risk score rating of High Risk with an Overall Numerical Risk Score of 20. Possible management actions in Table 18.4 suggest that senior management attention is needed to derive an action plan to reduce exposure to a low risk rating.

Situation B – Exposure to Hexane
In another part of the petrochemical plant, workers are using hexane to clean parts of the equipment used in the plant on a five yearly basis. This is carried out in a well-aerated workplace where exposure is minimized. The workers are employed on a 40 hour week and 8 hours per working day and the work involves about 1 working day (8 hours) in the five year period.
Likelihood of Exposure. The workers could be exposed to hexane on about 1 occasion per five years when cleaning is done. Thus, exposure would be evaluated according to Table 18.2 as *Possible* with a rating of 3.
Consequences. Consulting a MSDS information sheet from the Web indicates that hexane is a hazardous substance having many harmful toxic effects on humans but is not considered to be carcinogenic. Thus, hexane would be evaluated at *Moderately Hazardous* with a rating of 3.
Overall Outcome. This evaluation leads to a risk score rating of Medium Risk with an Overall Numerical Risk Score of 9. Possible management actions in Table 18.4 indicate that specific monitoring or procedures are required and management responsibility must be specified.

Table 18.4 Risk score and possible management actions.

Risk score	Possible management actions
Extreme 25	Immediate action required.
High 15–20	Action plan required, senior management attention needed.
Medium 8–12	Specific monitoring or procedures required, management responsibility must be specified.
Low 2–6	Manage through routine procedures. Unlikely to need specific application of resources.
Very low 1	Unlikely to need any application of resources.

18.4 Human Health Risk Assessment

Background

The basic principles involved with the assessment of the human health impact of a polluting chemical are described in Figure 18.2. However, this has been expanded to cover many additional factors as shown in Figure 18.3. Many government agencies and organizations responsible for management of environmental chemicals have developed frameworks for the risk assessment of environmental chemicals according to their needs.

The US EPA has developed a **Risk Assessment Framework** which is very comprehensive and covers most situations where human health can be adversely affected by environmental chemicals (Fitzpatrick et al. 2017). In the classic book "*Casarett & Doull's Toxicology - The Basic Science of Poisons*" the authors Faustman and Omenn (2013) describe **Risk Assessment** from a toxicology perspective. The US Department of Labor, Occupational Safety and Health Administration approach is concerned with recommended practices for safety and health programs with respect to occupational health (OSHA 2016). Additionally, there are programs developed by the OECD through the Inter-Organization Programme for the Sound Management of Chemicals (IOMC), WHO, and the International Program on Chemical Safety (IPCS). The framework described in Figure 18.3 outlines the overall approach to the risk assessment of chemicals in the environment from a broad perspective.

Initiation and Planning

The **Initiation** of a human health risk assessment is as a result of the identification of an adverse effect, or possible adverse effect, due to a chemical on human health. This may occur in a number of different situations as listed below.

- A chemical may be new for commercial and industrial use or evaluated for use in a new situation and would be identified as a hazard due to its observed properties in the laboratory.
- With human populations, epidemiological techniques may give indications of adverse effects on mortality, reproduction, neurotoxicity, occurrence of cancer, and so on. The adverse effects then need to be identified as being due to a chemical or some other specific agent.
- Evaluation of the potential hazards in the workplace at Level 1 and 2. This allows the identification of people and communities possibly affected by the hazardous chemical and the activities that contribute to the hazard.
- Guidelines for acceptable concentrations of contaminants in water, food, air, or soil as well as the workplace are needed.

Members of the communities involved can be identified and be involved in the formulation of the broad objectives of the risk assessment. Individuals and communities affected can be government officials conducting environmental management, private organizations involved with the environment, commercial and industrial companies, workers, private individuals, or others as appropriate. This involvement can make the risk assessment relevant from a political and socioeconomic perspective and acceptable to the community in general.

Hazard Identification

A history of the activities which resulted in possible contamination of the environment with hazardous chemicals can be prepared with quantities of the possible contaminant recorded. This should also record the exposure or possible exposure of individuals and communities including the time periods and qualitative estimates of the levels. The data collected and collated in this *Initiation and Planning* phase of the risk assessment will often allow the identity, or possible identity, of the chemical hazard or hazards to be determined. When the hazardous chemical is known, then the physicochemical and biological properties can be collected and collated using information from the Web. This information should include such properties as the K_{ow} value, K_D value, aqueous solubility, vapor pressures, and other related properties as well as such biological properties as bioaccumulation factor and persistence in biota.

Exposure Assessment

The generalized set of steps shown in Figure 18.4 allows an evaluation of the intake by the human population. Other examples of the uptake of different chemicals in different contexts are shown in Box 5.1. The identification of uptake routes is often assisted by the use of a conceptual model as illustrated in Figure 18.5. Of course, this intake will not be the same for all sectors of the population. Some, for example children, may be exposed to larger quantities of the contaminant than other sectors of the population. Contaminant releases can occur from a wide range of activities including mining, motor vehicles, contamination of soils, rural activities, and so on. So, initially sources of contamination should be identified. Often, the quantities

Quantification and evaluation of contaminant releases
↓
Identification of exposed population
↓
Potential exposure pathways
↓
Estimation or measurement of concentration of contaminants for pathways
↓
Estimation of human intake

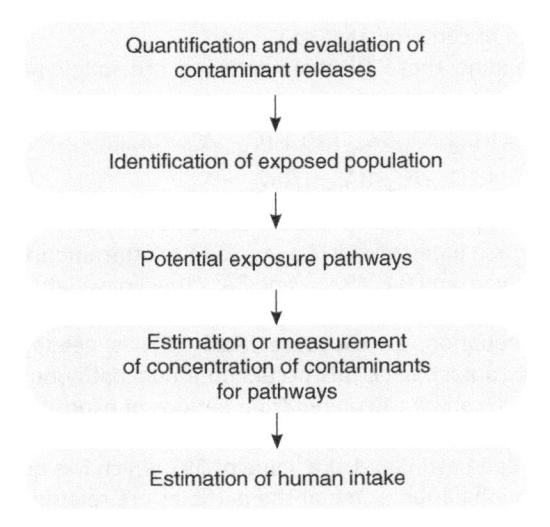

Figure 18.4 Set of generalized steps for the evaluation of the uptake of an environmental chemical by the human population.

Inhalation
- Air vapors
- Airborne dust/soil

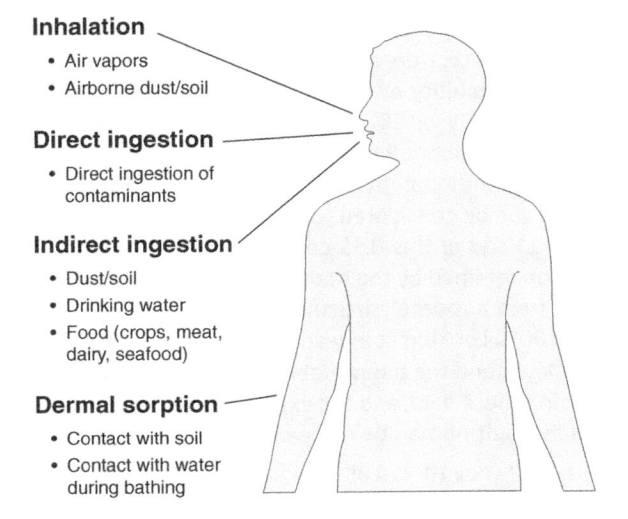

Direct ingestion
- Direct ingestion of contaminants

Indirect ingestion
- Dust/soil
- Drinking water
- Food (crops, meat, dairy, seafood)

Dermal sorption
- Contact with soil
- Contact with water during bathing

Figure 18.5 Potential exposure routes of a hazardous environmental chemical.

Box 18.2 Worked Example: Calculation of Dose with Adults Exposed to Dieldrin in Soil

The significant pathways for uptake of dieldrin from soil by human beings are:

- Dermal
- Inhalation
- Ingestion of contaminated soil

It can be assumed that drinking water, food, and bathing water will generally not be contaminated by the contaminants in soil in a particular location and will not be significant sources of contamination at a particular site. So, looking at the DI based on equation (18.1) and (18.2) then:

$$DI = DI_d + DI_{ih} + DI_{ig}$$

where DI_d is dermal absorption, DI_{ih} inhalation, and DI_{ig} is ingestion of contaminated soil.

By expanding these DI characteristics, the following equation can be obtained:

$$DI = [(C_{sd}.A_{sd}.BA_{sd}.EF) + (C_{sih}.A_{sih}.BA_{sih}.P) + (C_{sig}.A_{sig}.BA_{sig})]/bw$$

where C_{sd}, C_{sih}, and C_{sig} are the concentrations in dermal, inhalation, and ingested soil; A_{sd}, A_{sih}, and A_{sig} the amount of soil involved; and BA_{sd}, BA_{sih}, and BA_{sig} the bioavailability in the different pathways.

In this equation, an **exposure factor** (EF) is needed for dermal contact since this occurs on a noncontinuous basis. The EF can be calculated from periods of exposure obtained by observation. In the case of dieldrin, this factor has been estimated at a value of 392 which has no units. With inhalation, some of the particles are retained while others are passed out of the lungs without having any significant effect. This proportion (P) must be taken into account as indicated above to evaluate exposure from this pathway. Exposure factors are not needed with inhalation and ingestion since these occur on, or can be considered to occur on, a continuous basis.

The **bioavailability** of dieldrin in soil through the dermal route is 0.05 or 5%. The amount of soil in contact with the skin is about 8 mg, so on average A_{sd} is 8×10^{-6} kg. Turning to inhalation, the amount of particulate absorbed can be considered to be generally about 0.3 mg day^{-1} (A_{sih}) and of this 0.35 can be considered to be the proportion retained by the body (P). The bioavailability of dieldrin from airborne particulates can be considered to be 1 or 100%. Looking at ingestion, A_{sig} can be considered 0.1 mg day^{-1} and the bioavailability (BA_{sig}) as 0.1 or 10%.

By using these data, and the expression above, the total soil DI by adult human beings can be calculated as:

$$DI = [C_{sd}(8 \times 10^{-6})\, 0.05 \times 392 + C_{sih}0.3 \times 10^{-6}$$
$$\times 0.35 \times 0.1 + C_{sig}(0.1 \times 10^{-6})\, 0.1]/70\ \text{mg kg}^{-1}\text{bw}$$

In this case, adult average weight is considered to be 70 kg. Calculating this through, then:

$$DI = [(C_{sd}157 \times 10^{-6}) + (C_{si}0.02 \times 10^{-6}) + (C_{si}0.01 \times 10^{-6})]/70\ \text{mg kg}^{-1}\text{bw}$$

This indicates that the ingestion and inhalation pathways are relatively unimportant. This means that:

$$DI = C_{sd}2.24 \times 10^{-6}\ \text{mg kg}^{-1}\text{bw}$$

Thus, by substituting values for C_{sd}, which can be assumed to be the same as the concentration of contaminant in soil, the uptake from soil can be calculated.

The data produced by this exposure assessment give information on the total intake of contaminant by a general adult population group. It can be reapplied to specific groups of the population with different exposure characteristics. In this way, assessment of exposure characteristics for different groups in a population can be obtained.

Biomarkers. Another useful technique is the use of **biomarkers**. This technique utilizes the concentrations of the chemical contaminant which occur in biota already using the area as a habitat. Alternatively, specimens of biota are deliberately placed in the habitat to measure biological exposure. Examples of biota which can be used in this way are rats, mice, earthworms, and a variety of other organisms. Analysis of these biota give an indication of the concentration of the chemical likely to occur in humans and in other sectors of the environment which may be involved in pathways to biota (Figure 18.6).

Figure 18.6 An historic mercury still house where mercury used in recovery of gold was distilled but the soil in the area over 100 years later remains contaminated with mercury. Source: Photograph by Des Connell.

involved can be estimated to allow further calculations to be made on possible levels in the environment. Exposed populations could be children or adults or could result from the activities of particular groups of people on sites or in work situations. In other words, the exposure could be related to whether the main activities of an individual or group are related to industry, domestic situations, sport, and so on. When the group has been identified, the usage pattern resulting in exposure needs to be quantified. Exposure pathways can be developed for particular exposed groups of the population and it is helpful to use conceptual models. Generally, these pathways will fall into the following classes:

* Inhalation of vapors and dust
* Direct ingestion of contaminants
* Indirect ingestion from food and water contaminated by environmental processes
* Dermal sorption due to skin contact

Once the exposure pathways have been identified, it is then necessary to estimate quantitatively the amount of exposure resulting from each particular pathway. The simplest form that this can take is by the direct measurement of the amount and concentrations of contaminant, the amount of contaminant taken up by the population as well as the periods of exposure. This gives a direct measurement of the concentrations and amounts of chemicals involved in a particular pathway. Of course, these data may not be available and so alternative methods may be used. The use of a variety of models is now common for exposure assessment, some of which are outlined in Chapter 4 – Chemodynamics of Pollutants. With these techniques, calculations are made of the concentrations of chemical and amounts of chemical involved in different pathways to the population.

The general equation for intake by humans is:

$$\text{Daily Intake (DI)} = \Sigma \text{ (intake by the individual pathways)} \quad (18.1)$$

This means that $DI = \Sigma (C_i . A_i . BA_i)$, where C_i is the concentration in the medium involved (mg kg^{-1}), A_i the amount of medium involved (kg), and BA_i the bioavailability of the contaminant in the medium to the human being (unitless proportion). For Daily Intake (DI) expressed in terms of body weight (bw), then:

$$DI = [\Sigma (C_i . A_i . BA_i)/bw] \text{ mg kg}^{-1} \text{bw} \quad (18.2)$$

This may be modified to suit pathways with particular characteristics. For example, with the dermal pathway from the soil, an exposure factor is used which is estimated from the period of contact of the soil with the skin. Another example is with the particulates taken up by inhalation from the atmosphere where an evaluation of the particulates retained by the human body is needed.

Toxicity – Dose–Response Relationships

Data Available on Dose–Response Relationships

It is important to remember that generally there is very little information on the toxicity of environmental chemicals to humans. Some specific chemicals of importance have been subject to intensive investigation, such as benzene and lead, but knowledge of the adverse effects of chemicals in the environment at low levels of environmental exposure and relatively long exposure time periods is very limited. Some limited human data have been collected from the random occurrence of poisoning accidents, suicides, and so on, but this is not derived from an organized and methodological program. Also, information is available from occupational and epidemiological investigations but systematic data from organized research on Dose–Response relationships is very limited. But, on the other hand, there is a range of sound toxicological data available resulting from investigations of surrogate animals such as rats, mice, and rabbits. This is discussed in Chapter 5 – Environmental Toxicology and Ecotoxicology.

Exposure to toxic chemicals gives rise to a variety of possible responses in a population depending not only on the nature of a chemical but also on the dose and period of exposure which occurs. A generalized sequence of the effects of a chemical on populations is shown in Figure 18.7.

At high and very high doses, the exposure period is short and death is a common response. At intermediate doses, a major proportion of a population survives but the survivors exhibit severe effects in many cases. The lethal levels for 50% of the population (LD_{50} and LC_{50}) are used to record the toxic characteristics of a chemical to evaluate effects in the very high, high, and intermediate ranges of concentration. At low doses, exposure can be for months ranging into years and the LD_{50} becomes less useful since lethality occurs only with sensitive individuals. But in addition, a range of sublethal effects occurs in survivors which are not evaluated by the LD_{50} test. Other adverse biological responses occur apart from lethality. At very low doses and exposures for many years with many biota, there may be no effects that can be detected by current techniques.

In environmental toxicology and risk assessment, the concentrations generally involved are in the low, probably more often very low, dose range and long exposure periods of months to many years. This means that there is a wide difference between the *usual environmental conditions* under which humans are exposed to an environmental chemical and the *usual laboratory and test conditions* where the same chemical has been subject to toxicological

Dose	Exposure time period	Dose/concentration	Exposure period	Expected response
Increasing dose ↓	Increasing exposure period ↑	**Usual environmental conditions**		
		Very low	Very long (many years)	• No detectable effects
		Low	Long (months/years)	• Death of sensitive individuals & species • Sublethal effects in survivors
		Usual laboratory and test conditions		
		Intermediate	Intermediate (days)	• Equal numbers of deaths and survivors • Severe effects in some survivors
		High	Short (hours/days)	• Few resistant individuals and species survive
		Very high	Very short (hours)	• Death to all members of the population and ecosystem

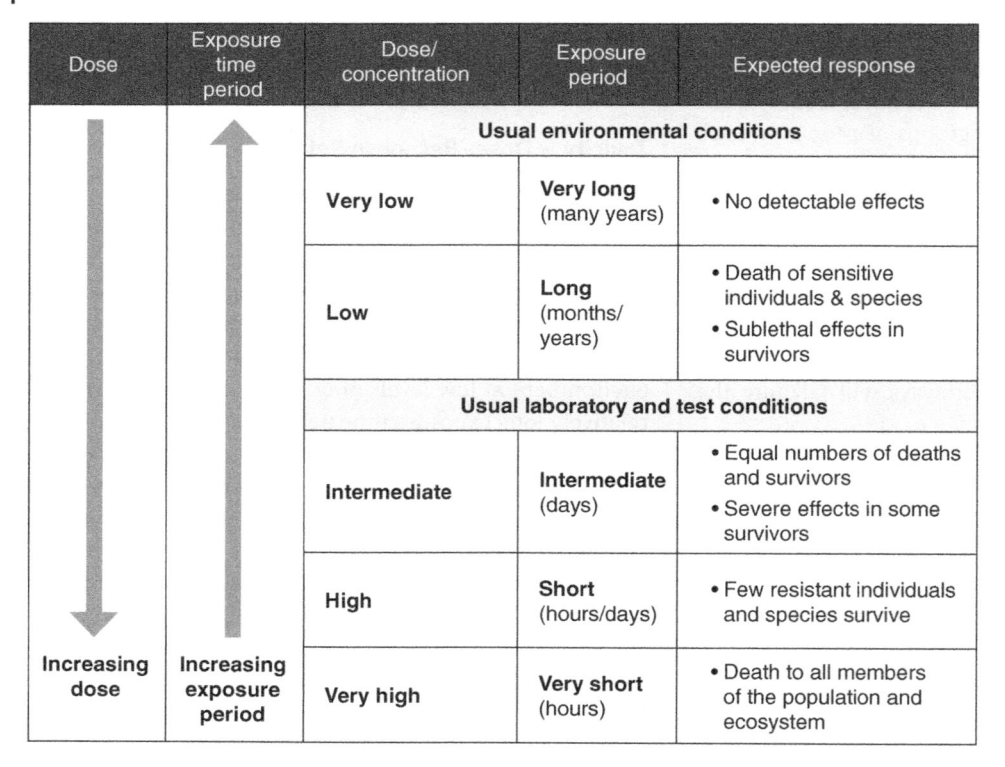

Figure 18.7 A comparison of the range of conditions in the environment and the conditions that usually exist in the laboratory and test situation with respect to hazardous environmental chemicals.

testing as shown in Figure 18.7. So, in these cases, the conventional toxicological techniques for evaluating toxicity cannot be applied. Additionally, the situation is further complicated since the test biota are not human beings but surrogate animals such as rats, mice, and rabbits.

The adverse responses of biota, specifically human beings, are needed to utilize the exposure data which can be calculated as indicated in the previous section. By combining the exposure with an evaluation of the adverse effects of that exposure, then a characterization of the risk posed by a particular chemical can be made. The dose–response data available to carry out this risk characterization fall into two broad groups:

* Epidemiological evaluations of the effects on human populations.
* Experiments conducted under controlled conditions on surrogate animals in the laboratory.

Usually, it is not a choice of which to use since both sets of data must be used according to reliability, availability, and suitability. At first hand, the data on human populations could be considered to be the most useful, since these data relate directly to the organism of interest, human beings, in the case of human health evaluations. On the other hand, the laboratory experiments on animals, such as rats, would be expected to be less useful since the two organisms, human beings and rats, are considerably different in many of their characteristics. But the use of the epidemiological data is difficult for the following reasons:

- The human population is exposed to numerous agents which may cause adverse biological effects and not only from the chemical of specific interest.
- Temporal patterns of exposure of humans are usually not known and can also be variable over long periods of time.
- The population includes many subgroups, such as children, adults, occupational groups, and so on, which may have different responses and exposures to the contaminant.
- The data available on epidemiological studies on specific chemicals are very limited.

Experiments on animals, as surrogates for humans, also have a range of limitations as outlined below:

- Experimental animals, and humans, may respond in different ways to the same chemical agent.
- Dose rates and exposure concentrations with experimental animals are much higher than with most actual or potential human environments and the exposure periods much shorter as shown in Figure 18.7.
- It is not practical to conduct experiments on surrogate animals over long periods, for example, 15–20 years with

very low dosages particularly when the life expectancy is much lower than humans.
- The exposure route in laboratory experiments may differ from that in the human exposure situation.
- A significant incidence of adverse effects in a human population could be an effect on 1 in 10^4 or 1 in 10^5 or lower. The exposure of human populations to chemicals can be in the order of millions of individuals with chemicals in air, food, water, and other media. Thus, experiments would be needed on very large numbers of animals to observe the low incidences which are of significance in the human population.

For the reasons outlined above, the LD_{50} and LC_{50} data are not commonly used in evaluating environmental effects, although these characteristics may be used in the absence of other data. The most common toxicity data used are the **Lowest Observable Adverse Effects Level (LOAEL) and the No Observable Adverse Effects Level (NOAEL)** as shown in Figure 18.8. These characteristics were described in Chapter 4 on *Environmental Toxicology and Ecotoxicology*. These values are more useful in environmental applications since they approach the low-level long-term effects usually of interest more closely than the LD_{50} and LC_{50} data. Despite the many difficulties in using the available toxicity data, both the epidemiology and animal experiment data are used according to availability and suitability in the evaluation of human health effects as outlined below.

Safety, Application, and Uncertainty Factors

Most data are available from experiments conducted on surrogate animals in the laboratory since this is a consistent and practical source of information. If data are lacking, it can be obtained by the conduct of experiments on animals with the chemical of interest. Such experiments cannot be carried out on human populations and epidemiology data may not be available and cannot be obtained when required. As well, epidemiology data usually originate by the chance exposure of a population and the design and the significance of the epidemiological results may not be optimal. Even when epidemiology or other data on human beings are available, many of the limitations mentioned previously apply to it.

This has forced environmental toxicologists to seek ways to apply laboratory experimental data despite the limitations. Of course, when epidemiological data on human populations are available and appropriate, they should be used. The most common approach in the utilization of the experimental data, and to a certain extent the epidemiology data, is to take these data and extrapolate them to the environmental exposure conditions by carrying out a division by a large number. This large number is described as a **Safety**, or **Uncertainty** or **Application Factor** which will increase the calculated toxicity as it is applied to evaluate safety. This has been a standard procedure for many years and has achieved considerable success in protecting public health.

Briefly, the purpose of the **Safety Factor** is to:

- account for the different sensitivities of individuals within the human populations,
- allow for the possible increased sensitivity of human beings as compared to animals,
- extend short-term toxicity and exposure to long-term toxicity and exposure,
- account for the use of the LOAEL as compared to the NOAEL, and
- account for the availability of a limited amount of data.

The overall Safety Factor is obtained by multiplying the individual factors which apply to the data set according to the criteria set out in Table 18.5.

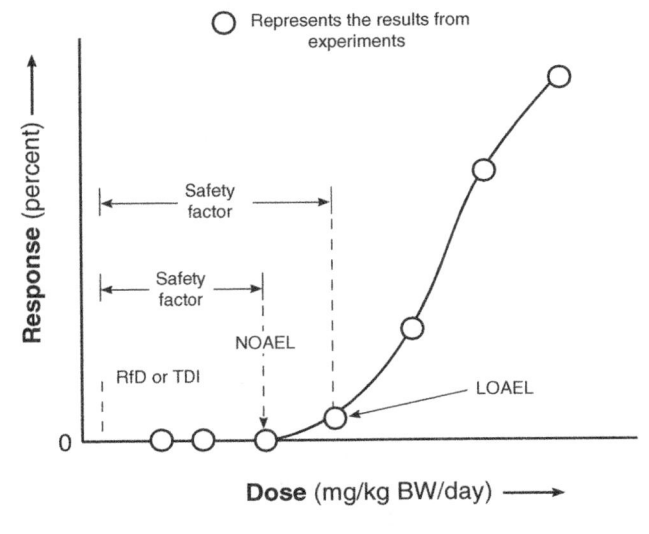

Figure 18.8 Diagrammatic illustration of the application of the Safety Factor to experimental data to obtain the TDI.

Table 18.5 Magnitude of safety factors often used in human health risk assessments.

Extrapolation	Factor
Average human to sensitive human	10
Animal to human	10
Short-term to long-term exposure	≤ 10
LOAEL used not NOAEL	≤ 10
Limited database	≤ 10

With human health assessment, the actual values of the Safety Factor range from 10 to 10 000. Values are occasionally used that are less than 10 where humans are believed to be less sensitive than the test animals. It is considered that values greater than 10 000 indicate that the information is very imprecise and may not produce reliable results. As a general rule, Safety Factors greater than a thousand indicate that more information is needed to obtain a lower Safety Factor. The lower the Safety Factor the higher the confidence in the result and if the Safety Factor was unity, our information would be precise and apply to the population being considered exactly.

Acceptable or Tolerable Daily Intake (or Reference Dose) for Human Populations

In many situations, an **Acceptable** or **Tolerable Daily Intake (ADI and TDI, respectively)** also referred to as the **Reference Dose (RfD)** is calculated as a basis for risk characterization. The TDI is the benchmark dose derived from the NOAEL by the use of Safety Factors as shown in Table 18.5. The actual values for the Safety Factors are derived reflecting the type of data available. The application of the Safety Factor to the NOAEL and LOAEL is indicated diagrammatically in Figure 18.8. The LOAEL is a higher value than the NOAEL and so a larger Safety Factor is applied with LOAEL data. The use of Safety Factors, as described here, assumes that there is a threshold dose below which no adverse effects will occur and the application of the Safety Factor will provide a measure which is below that threshold. So, the TDI (or RfD) is expressed as

$$\text{TDI} = (\text{NOAEL/SF}) \text{ mg kg}^{-1} \text{bw day}^{-1} \quad (18.3)$$

In practical terms, the TDI is a value which has a degree of uncertainty to its numerical expression. It indicates the daily exposure a human population can experience without an appreciable risk of adverse effects.

TDI can be converted to **Guideline Values** (maximum permissible levels) in drinking water or food by using:

$$\text{GV} = (\text{TDI} \times \text{bw} \times P)/C \quad (18.4)$$

where bw for adults is 60 kg, children 10 kg, and infants 5 kg. P varies with compound considered, and C is the daily drinking water or food consumption (L day^{-1}).

Risk Characterization

The final step in the risk assessment process is to characterize the risk involved in numerical terms. This requires the use of the exposure assessment to evaluate the actual amounts of chemical to which a human population is exposed, and relating this exposure to the toxicity and other information obtained from dose/response relationships. If the daily exposure is less than the TDI, then there should not be any significant adverse effects on the health of the human population. These effects relate to toxic effects rather than carcinogenicity which is a separate evaluation considered below. Sometimes the outcome of this process is quantified as the Hazard Quotient (HQ), where:

$$\text{HQ} = \text{Dose/TDI} \quad (18.5)$$

HQs above unity have the possibility of adverse effects in human populations increasing in severity with the increase in the numerical value of the HQ.

The dose of chemical can be calculated by various methods with one involving uptake from soils shown in Box 18.2. With drinking water and food, the calculation can be made as the DI using the equation below.

$$\text{DI (water)} = [(C_W \times C)/\text{bw}] \text{ mg kg}^{-1}\text{bw day}^{-1} \quad (18.6)$$

where C_W is concentration in water (mg L^{-1}); C, daily consumption of water (L); bw, body weight.

The HQ then can be calculated from equation (18.5).

A major factor in assessing the health risk with chemicals is the evaluation of the risk due to mixtures of chemicals. This was considered in *Section 5.8 Toxicity due to exposure to multiple toxicants* in Chapter 5. The toxic action with a mixture can be **synergistic** and increase the expected toxic effects, or simply **additive** or **antagonistic** and have a decreased toxic effect. Extensive research suggests that most mixtures are additive. This leads to another important concept in Risk Characterization which is the **Hazard Index (HI).** The HI is the sum of the Hazard Quotients for substances that have the same or similar toxicity mechanisms, although the research mentioned previously in this paragraph suggests that the HI may have wider application. As with the HQ, aggregate exposures with an HI below 1.0 will not likely result in adverse non-cancer health effects over a lifetime of exposure, whereas HI greater than unity results in an increased risk of adverse effects.

Cancer Risk Assessment

Environmental factors can contribute to the occurrence of cancer, also described as neoplasms and tumors, in humans. However, a clear experimentally established link between exposure to certain environmental chemicals and the occurrence of cancers has to be established. Mammals are not the only species at risk of developing cancer from exposure to chemical carcinogens since this disease occurs in all living organisms. Some of the chemicals that

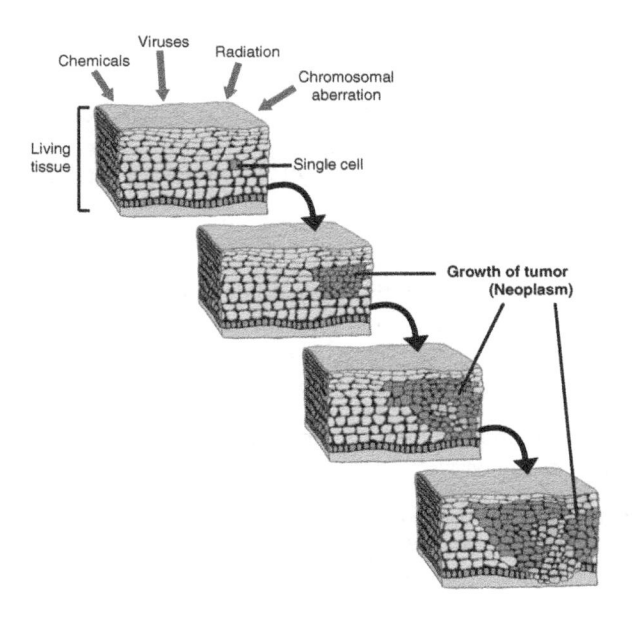

Figure 18.9 The actions of chemicals, viruses, radiation, and chromosomal aberration can, independently or in concert, initiate the formation of a cancer cell which can multiply and grow into a tumor having adverse health effects.

humans encounter in the environment are carcinogenic (cancer causing). The evaluation of cancer can be treated similarly to toxicants but with some differences since the incidence of cancer results from a complex set of chemical, biochemical, and physiological interactions as shown in Figure 18.9.

In 2017, total global deaths due to all cancers were estimated at 9.6 million (updated), making it the second leading cause of death after cardiovascular diseases (Roser and Ritchie 2015). Some estimated causes include certain physical, chemical, and biological agents (known and probable carcinogens) in food, drinking water, breathed air including occupational exposure to hazards, as well as alcohol and tobacco consumption. However, rather than a cause due to external exposure, there is chromosomal aberration which is inherent within the organism itself. With this mechanism, adverse alterations to the chromosomal structure can initiate cancer formation.

The US EPA approach to evaluation of risk as the incidence of cancer in a population has been described in a series of publications (see US EPA 2005). Essentially, this approach uses dose–response data from experiments on surrogate animals which have been carried out and recorded with the response, as **Population Incidence of Cancer (PIC, unitless proportion),** related to the lowest level doses usually in units of **ng kg^{-1} body weight day^{-1}**. This approach has also been used with epidemiological data for humans (e.g. benzene and asbestos carcinogenic risks) (see US EPA 2020).

The levels of the lowest level exposure in the experiments are usually well above the levels of actual exposure in the environment (see Figure 18.7) and thus extrapolations are necessary to obtain relevant data. In agreement with this, the plots of these lowest level data are made and extrapolations made to zero incidence and zero dose on a linear basis. The slope of this extrapolation (on a linear basis) is described as the **Slope Factor (SF)** which is Unit Tumor Incidence/Dose in units of the reciprocal of dose, e.g. 1/(ng/kg bw/day) or (ng/kg bw/day)$^{-1}$. It is usually calculated in terms of the upper 95% probability of the data plots. Thus,

$$SF(\text{ng/kg bw/day})^{-1} = PIC\ (\text{unitless proportion})/\text{Dose (ng/kg bw/day)} \quad (18.7)$$

$$PIC(\text{unitless proportion}) = \text{Slope Factor} \times \text{Dose} \quad (18.8)$$

This means to obtain the PIC, then the Slope Factor is multiplied by the Dose. This gives the incidence due to the chemical being considered alone is obtained as a unitless proportion of the number of adversely affected in a population of individuals. This is the Incremental Risk, and the Background Risk is additional to this figure. It should be kept in mind that there are a number of criticisms of this approach as exemplified by Ames and Gold (1990).

The surrogate animal bioassay provides the main source of the data on carcinogenicity. The extent of the data available (bioassay and supporting data) will vary considerably from chemical to chemical. It may be necessary to make a judgment on the significance of a positive animal finding. Factors may include such matters as consistency between studies by different laboratories, the magnitude of the finding, i.e. dose/response increase in tumor incidence, and statistically significant trends of the supporting environmental data consistent with the results of bioassays.

The **Guideline Dose** – Often, the end result of the carcinogen risk assessment process for a particular chemical agent is the Guideline Dose. This is calculated from the TDI which is obtained using the average *modified*-Bench Mark Dose (BMD) methodology for each carcinogenic endpoint. With the derivation of the TDI, it is necessary to use application or safety factors for each *modified*-BMD human equivalent dose similar to that used with toxicants as shown in Table 18.5. TDI can be converted to Guideline Values (maximum permissible levels) by using equation (18.4).

> **Box 18.3 Worked Example - Calculations for Chlordane in Drinking Water**
>
> 1. Calculation and comparison of the guideline value (GV) for chlordane in drinking water with adults, children, and infants.
> 2. Calculation of the HI for consumption of drinking water.
> 3. Calculation of the incidence of cancer at the TDI.
>
> 1. **Calculation and comparison of the guideline value (GV) for chlordane in drinking water with adults, children, and infants**
>
> *Environmental Characteristics of Chlordane*
>
> - pesticide
> - persistent
> - NOAEL in well-validated long-term studies on rats of 0.05 mg/kg bw/day
> - UF = 100 (inter and intra species variation)
> - contribution of drinking water to total uptake of chlordane, P = 1% = 0.01
> - C, daily drinking water consumption is 2 L per day for adult, 1 L for children, and 0.75 L for infants, and bw for adults is 60 kg, children 10 kg, and infants 5 kg
> - Chlordane Slope Factor is 1.61 $(mg/kg/day)^{-1}$ for adults
>
> *Step 1. Calculation of the tolerable daily intake (TDI)*
> Using equation (18.3),
> TDI = NOAEL/UF = 0.05/100 = 0.5 µg/kg/day (0.0005 mg/kg/day)
> For children, GV = (0.5 × 10 × 0.01)/1 = 0.05 µg/L (0.00005 mg/kg/day)
>
> *Step 2. Calculation of the guideline values (GV)*
> Using equation (18.4),
> GV = (TDI × bw × P)/C
> For adults, GV = (0.5 × 60 × 0.01)/2 = 0.15 µg/L (0.00015 mg/kg/day)
> For children, GV = (0.5 × 10 × 0.01)/1 = 0.05 µg/L (0.00005 mg/kg/day)
> For infants, GV = (0.5 × 5 × 0.01)/0.75 = 0.033 µg/L (0.000033 mg/kg/day)
>
> *Comparisons*
> There is a variation in the permissible maximum levels allowable in drinking water with infants being the most sensitive to chlordane. This is due to the higher consumption of drinking water with infants per body weight.
>
> 2. **Calculation of the hazard index (HI) for consumption of drinking water**
>
> Using equation (18.6) to calculate the DI at a concentration of chlordane of 10 µg/L (0.01 mg/L):
> DI (water) = [(C_W × C)/bw] mg/kg bw/day
> where C_W is concentration in water (mg/L); C, daily consumption of water (L); and bw, body weight.
> DI (water) = (0.01 × 2)/60 = 0.00033 mg/kg bw/day
> Hazard Index (HI) = DI/TDI = 0.00033/0.00015 = 2.2
>
> 3. **Calculation of the population incidence of cancer (PIC) at the TDI**
>
> It should be kept in mind that this calculated PIC is the Incremental Risk due to the exposure to the chlordane alone and does not include the Background Risk due to the presence of hazards normally present in the environment.
>
> Using equation (18.8) with the SF for adults since there are no factors for children and infants,
>
> $$\text{PIC (unitless proportion)} = \text{SF} \times \text{Dose}$$
> $$= 1.61 \, (mg/kg\,bw/day)^{-1} \times 0.0005 \, mg/kg/day = 0.0008$$
>
> This means that the Incremental Incidence of Cancer at the TDI would be about 8 individuals in 10 000.

Weight of Evidence Approach (WoE) in Risk Assessment

Weight of evidence is a key concept used for scientific and regulatory purposes in health and ecological risk assessments to make decisions about hazardous substances, including chemicals. It can be applied at various stages of the assessment, such as in hazard assessment and risk characterization to evaluate the relative strengths of lines of evidence in terms of consistency and coherence (Government of Canada 2017; Martin et al. 2018). Increasingly, WoE analysis methods are also used in setting environmental standards or criteria by regulatory agencies (e.g. lead standards or guidelines in water). For example, the Canadian Environmental Protection Act, 1999, places greater weight on stronger and more relevant lines of evidence such as direct causality, specificity to Canadian situations, and consideration of uncertainties that have been identified (Government of Canada 2017). Other important factors include making realistic assumptions that account for uncertainties and identifying limitations of the study (e.g. objectives, scope, boundaries of study, parameters measured, and data quality).

18.5 Ecological Risk Assessment

Background

Natural ecosystems present much more complex systems than human systems since there is an involvement of many species in complex inter-relationships with one another and the physical environment, whereas human health involves a single species and less-complicated relationships to the physical environment. Also, the natural systems are in a constant of change with responses to daily, seasonal, and climatic changes. These factors create difficulties in predicting ecosystem responses to contaminants compared to changes in response to natural events such as changes in temperature, rainfall, season, etc. In accord with this, there are major differences between human health and ecological risk assessment while the fundamental principles remain the same. Also, each ecosystem on the globe is essentially unique and can be different in many respects from related ecosystems which means that there is no benchmark system from which we can evaluate changes.

Techniques to evaluate aspects of the biological effects of chemicals have been described previously in Chemodynamics of Pollutants (Chapter 4), Environmental Toxicology and Ecotoxicology (Chapter 5), Genetic Toxicology and Endocrine Disruption (Chapter 6), and Pollution Ecology (Chapter 7). These are all concerned with the various aspects of the behavior and effects of chemical pollutants in the environment. However, previously many of these aspects were considered as separate factors and the biological effects related principally to individual species. There is a need to draw these aspects together into an overall approach which addresses the effects on whole ecosystems. Our capability to evaluate the effects on whole ecosystems is not particularly strong, but approaches based on **ecotoxicology** have provided a systematic approach and are described in Chapters 5 and 7.

The Ecotoxicology Concept (Also Considered in Chapters 5 and 7)

Scientific investigation of the effects of pollutants in the natural environment has focused on the effects on individual biological species. Such investigations can only give suggestions as to the likely effects on the complex ecosystems existing in the natural environment. Thus, there has been an increase in interest in the effects of pollutants on whole ecosystems since these are of primary management concern. This has led to the development of **ecotoxicology**.

Ecotoxicology is concerned with the fate and toxicity, and related effects, of chemicals in natural ecosystems. The ecotoxicology of a chemical is based on a sequence of interactions and effects controlled by the physical, physicochemical, and biological properties of a chemical as indicated in Figure 1.3. Initially, the physical state of the contaminant, whether it is a solid, liquid, or gas has a major influence on its physical dispersal from the source. Following that, physicochemical properties influencing movement into different environmental phases become important. A chemical discharged to the environment can then be subject to distribution in the atmosphere, water or soils, and sediment depending on its physical chemical properties as described in Chapter 4. At the same time, it can be chemically modified and transformed by abiotic processes or more often by microorganisms in the environment. The organisms present are then exposed to the toxicant in its original form and in its degraded or transformed state and at concentrations resulting from its dispersal. Uptake of the chemical and its degradation products occurs, and organisms can exhibit a variety of different reactions from negligible to sublethal effects such as reduced growth, reproduction decline and behavioral effects, or ultimately death. The complex natural ecosystem of which the organisms are an integral part can react in a variety of ways to the effects on the component organisms. Food chain relationships, energy flows, and so on, may be altered. Thus, the ecotoxicology of a chemical can be considered as a sequence of steps starting with a source and following through to the ecosystem response with different properties of a chemical being involved at each step as illustrated in Figure 1.3.

The Ecological Risk Assessment Framework

The diagrammatic illustration of the **ecotoxicology concept** in Figure 1.3, which is discussed above, describes the interaction of chemicals with ecosystems and is based on our current scientific understanding of the processes involved. However, this framework needs to be developed into an assessment and management framework which takes into account other additional factors such as socioeconomic and management aspects.

An ecological risk assessment framework was formulated by the US EPA in 1992 (US EPA 2016) and this process has been followed, in general terms, by many government agencies throughout the world. It generally follows and fits within the overall concept of human health risk assessment shown in Figure 18.3 but has clear specific differences.

The characterization of risk in natural ecosystems can be directed to the integrity of the whole ecosystem or toward the effects on specific biotic components. With whole ecosystems, we are concerned with the maintenance

18 Human Health and Ecological Risk Assessment

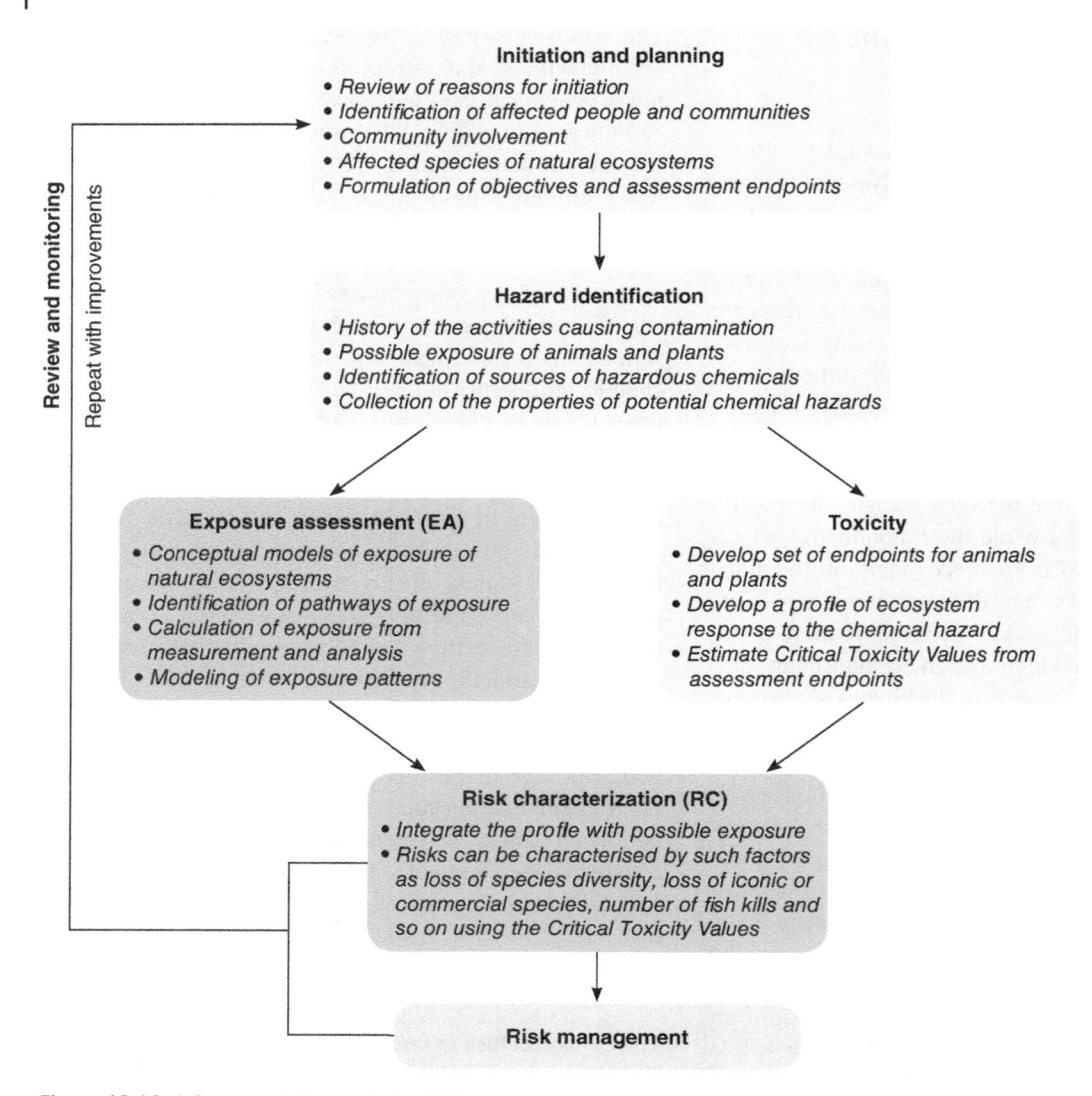

Figure 18.10 A framework for ecological risk assessment.

```
Initiation and planning
• Review of reasons for initiation
• Identification of affected people and communities
• Community involvement
• Affected species of natural ecosystems
• Formulation of objectives and assessment endpoints
            ↓
Hazard identification
• History of the activities causing contamination
• Possible exposure of animals and plants
• Identification of sources of hazardous chemicals
• Collection of the properties of potential chemical hazards
      ↙                              ↘
Exposure assessment (EA)            Toxicity
• Conceptual models of exposure of   • Develop set of endpoints for animals
  natural ecosystems                   and plants
• Identification of pathways of      • Develop a profile of ecosystem
  exposure                             response to the chemical hazard
• Calculation of exposure from       • Estimate Critical Toxicity Values from
  measurement and analysis             assessment endpoints
• Modeling of exposure patterns
      ↘                              ↙
Risk characterization (RC)
• Integrate the profile with possible exposure
• Risks can be characterised by such factors
  as loss of species diversity, loss of iconic or
  commercial species, number of fish kills and
  so on using the Critical Toxicity Values
            ↓
Risk management
```

(Left side: Review and monitoring / Repeat with improvements)

of the species present, whereas with specific biotic components, we are concerned with rare and endangered species, important game species, and so on.

The framework for ecological risk assessment is shown in Figure 18.10 which follows a similar structure to that for human health risk assessment shown in Figure 18.3. The steps are in a sequence with each step building on the knowledge acquired in the preceding step leading to a more informed understanding. The steps are considered below.

Initiation and Planning

The **Initiation** of an ecological risk assessment is a result of the identification of an adverse effect, or possible adverse effect due to an environmental stressor such as a chemical on an ecosystem. This may occur in a number of different situations as listed below for a chemical hazard.

- A chemical may be new for commercial and industrial use or evaluated for use in a new situation and would be identified as a hazard due to its observed properties in the laboratory.
- Adverse biological effects may have been detected in an ecosystem which may have been associated with a chemical hazard. Adverse effects such as declines in the populations of iconic or commercial species, occurrence of *fish kills,* and so on, may be involved.
- Guidelines for acceptable concentrations of contaminants in water, food, air or soil, as well as levels in discharges, are needed.

The first step in human health risk assessment is the identification of the hazard and similarly with ecological risk assessment, there is the identification of the problem. With human health risk assessment, the objective is to evaluate the risk to human health but there is no similar relatively simple answer with ecological risk assessment. The objective could be to protect endangered species, protect species diversity of fish, prevent fish kills, and so on, which permits the definition of **Assessment Endpoints**. It needs the available information to be collected and collated to provide a sound basis on which risk assessors, risk managers, state agencies, voluntary groups, etc., can formulate the problem in quantitative terms, e.g. no more than 5% loss of invertebrate species, no more than one fish kill in five years. This involvement can make the risk assessment relevant from a political and socioeconomic perspective and acceptable to the community in general.

Hazard Identification

A history of the activities which resulted in possible contamination of the environment with hazardous chemicals can be prepared with quantities of the possible contaminant discharge recorded. This should also record the exposure or possible exposure of ecosystems including important endangered, iconic, and commercial species. This information should include the time periods and qualitative estimates of the levels. The data collected and collated in this *Initiation and Planning* phase of the risk assessment will often allow the identity, or possible identity, of the chemical hazard or hazards to be determined. When the hazardous chemical is known, then the physicochemical and biological properties can be collected and collated using information from the Web. This information should include such properties as the K_{ow} value, K_D value, aqueous solubility, vapor pressures, and other related properties as well as such biological properties as bioaccumulation factor and persistence in biota.

Exposure Assessment

Exposure pathways to the ecosystem and the species of interest should be identified and conceptual models developed. Conceptual models can be of considerable value in providing a visual presentation of the ecosystem and its component animals and plants. Conceptual models can also integrate all the information allowing clear understanding and communication between the human parties involved as identified in the initiation and planning stage. Measurements can be made of the potential chemical hazards allowing calculation of the exposure levels to be determined. Quantitative modeling may be possible to provide more definitive data on the levels of exposure.

Toxicity

With ecological risk assessment, the toxicity evaluations using dose–response relationships from laboratory experiments are of less value since there are a large number of possible species involved. Also, the particular species in the ecosystem or ecosystems of interest may be untested with the specific chemical hazard. The species of particular interest, endangered, iconic and commercial species, probably have not been tested. It is also important to remember that the species of particular interest are unlikely to have data available for all their life stages from eggs to adults. Nevertheless, a set of possible assessment endpoints, from mortality to reduced fertility, should be developed for as many species as possible in the ecosystem and from this a profile of ecosystem responses to the chemical developed.

Toxicity relationships can be developed by methods somewhat similar to those used with human health risk assessment. The NOAEL, LOAEL (see Figure 18.8), LC_{50}, EC_{50}, and LD_{50} can be divided by an appropriate Uncertainty Factor, considering the differences discussed above, to yield a concentration related to an **Assessment Endpoint**, sometimes described as the **Critical Toxicity Value (CTV)**.

Ecological Risk Characterization

The ecological objectives of the risk assessment are defined in the **Problem Formulation** stage as **Assessment Endpoints**. These provide a focus for development of the response profile of the particular ecosystem and quantification of the related **Critical Toxicity Values** in the **Toxicity** step. This information can now be integrated with exposure from the **Exposure Assessment** step. In this way, the risk to the ecological values can be evaluated and characterized. However, every ecosystem is unique and has different characteristics from others. This means that in applying this information to specific systems, differences will occur and need to be considered. These differences include:

- Between taxa in the natural system and in the scientific information available.
- Between the assessment endpoints for the risk assessment and the test endpoints available in the scientific information, e.g. mortality as compared to growth reduction.
- Between the conditions and other factors in the laboratory and natural system particularly between short exposure in the laboratory data and long-term exposure in the natural environment.

The characterization of ecological effects can occur at the level of the individual organisms, such as the occurrence of a contaminant, but the effects are often measured at the level of a population, community, or ecosystem. Individuals of a single species constitute a *population* occupying a defined space and time. The individuals in a population may exhibit toxic effects such as reduced life expectancy which leads to a modification of the population profile and subsequent effects on the community and ecosystem as a whole. Changes may occur due to chemical stress on an ecosystem. These are:

- Reduced species diversity
- Altered community structure
- Changes to predator/prey relationships
- Changes to respiration/photosynthesis ratio
- Changes to nutrient flows

One characteristic of natural ecosystems is that they exhibit a varying degree of change in response to normally encountered conditions (temperature, rainfall, etc.) and nutrient availability (varying runoff). A problem may occur when an observed change is due to a particular pollutant or due to fluctuations inherent in the ecosystem itself.

18.6 Risk Assessment Using Probabilistic Techniques

The probabilistic methods of evaluating the risk resulting from exposure to a chemical can be used with human health and ecological risk assessment and represent a more accurate but more complex approach. The exposure step in the risk assessment procedure can result in a set of values (in mg kg^{-1} bw day^{-1} or another measure) accurately expressing the range of exposure circumstances in a real situation. These chemical exposure profile data can be expressed as a **Cumulative Frequency Distribution (CFD)** as shown in Figure 18.11, Graph A, if sufficient data are available. These plots take the same form as the Graph C. The plot in Figure 5.11 also takes the same form and records the exposure to chlorine by-products in drinking water as a **Cumulative Probability Distribution (CPD)** (Hamidin et al. 2008).

These profiles are usually approximately linear normal distributions and have considerable utility. For example, they can be used to estimate the possible frequency of occurrence of different exposure values at other times, presuming the same conditions prevail. This method has been used to estimate the exposure to benzene in service stations by Edokpolo et al. (2019) as well as the exposure of farmers to pesticide by Phung et al. (2015) and Atabila et al. (2017). Additionally, the biological response from exposure to the chemical can also be plotted as a **CFD** as shown in Figure 5.11, Graph B. In fact, these distributions are commonly used to present toxicity data; for example; the toxicity data in Figure 18.8 are presented as a CFD where the percentage of the biotic system responding is plotted against the exposure dose.

A generalized diagram of this presentation is shown in Figure 18.11, Graph B. This profile can be used to estimate the fraction of a biotic population which will respond to a specific exposure dose or concentration. The biotic system being tested could be an individual species, a population, a community, or an ecosystem. Also, when a line is statistically fitted, then extrapolations and interpolations can be made to estimate the frequency of other doses including very high and very low doses and concentrations. The two CFD profiles can be presented together as in Figure 18.11, Graph C and interpreted by combining the two distributions to characterize the risk. One of the techniques available to do this is illustrated in Box 18.4 but it should be remembered that there are many other techniques available.

Box 18.4 Example - Risk Assessment Using Probabilistic Techniques

A relatively large amount of data on factors, which can be difficult to measure, is usually required for the plots of CPDs. In this illustrative example, it is assumed that CFD curves for the Exposure and Biotic Response to Chemical Profiles have been developed as in Figure 18.11, Graph A and Graph B, respectively - now a **Probabilistic Risk Assessment** can be made.

These two profiles, exposure and biotic system responses to chemical exposure, can be integrated into one figure since the two distributions are on the same scales as shown in Figure 18.11, Graph C. Utilizing this combined graph, many entirely new interpretations can be made.

For example, the risk associated with exposure at many possible different levels of exposure dose of the toxicant in the system can be made by extrapolation and interpolation. For example, the exposure dose of 1.6 units (log 1.6 = +0.2) occurs at a cumulative frequency of 0.90 (90%) read on the left hand axis, *Cumulative Frequency of Exposure*, and about 0.03 (3%) of the biotic system would be expected to respond as read from the right hand *Cumulative Fraction of Biotic System Responding*. At a lower exposure dose of 0.4 units (log 0.4 = −0.4) which occurs at a cumulative frequency of about 50%, then <0.02 (2%) of the population would be expected to respond. This method allows different exposure scenarios to be evaluated according to appropriate assessment endpoints.

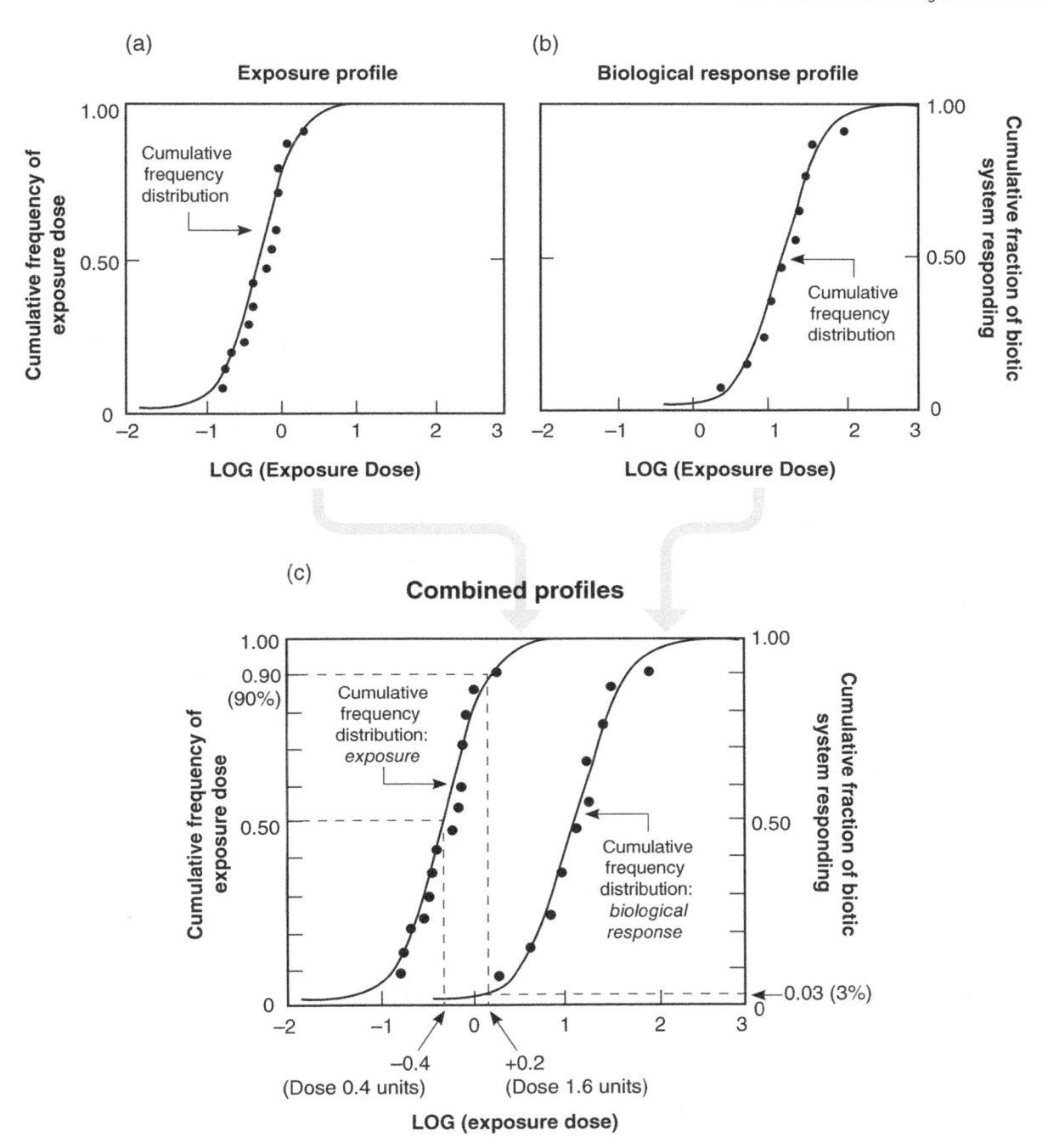

Figure 18.11 Generalized expression of the exposure profile (as cumulative frequency, Graph A) and biotic system response (Graph B) to exposure to a chemical (as cumulative frequency) and as a combined graph (Graph C) since the same scales are used in both frequency distribution plots.

There are a variety of other methods to characterize risk to human health and ecosystems from exposure to chemicals. Monte Carlo simulations of the distributions and integrating the two distributions have also been used with some success. However, some researchers have used the actual distributions. For example, Cao et al. (2011) have used a graphical method to integrate the observed exposure dose and response plots, and Edokpolo et al. (2019) have used Toxicant Sensitivity Distributions (TSD) to evaluate the risk to human health.

The techniques described above do not take into account secondary effects such as bioaccumulation, changes in predator–prey relationships, and so on. With ecological risk assessment and with human health risk assessment,

18.7 Key Points

1. **Risk Assessment** is a procedure for identifying hazards and quantifying the risks presented by chemical contaminants to human health (**human health risk assessment**) and natural ecosystems (**ecological risk assessment**).
2. The Risk Assessment process for both human health and natural ecosystems consists of five interconnected steps as outlined below. While the same general steps are followed, there are distinctive differences between the human health and ecological risk assessment. Usually, the Initiation and Planning and the Hazard Identification are similar but the remaining steps (Exposure Assessment, Toxicity: Dose–Response and Risk Characterization) are similar in principle but distinctively different.

    INITIATION AND PLANNING
    ↓
    HAZARD IDENTIFICATION
    ↙   ↘
    EXPOSURE ASSESSMENT   TOXICITY: DOSE–RESPONSE
    ↘   ↙
    RISK CHARACTERIZATION

3. In the overall risk assessment framework, the first step is initiating and planning the project as well as formulating the objectives. Then, the chemical hazard can be identified and the exposure dose to human populations and natural ecosystems quantified and modeled. The next step is evaluating dose/response relationships available for the chemical, and finally characterizing the risk involved by integrating the exposure and the dose/response information.
4. The **Initiation and Planning** of a risk assessment involves such matters as identifying the populations affected both human and natural as well as involvement of human individuals and organizations in the formulating of the objectives.
5. The **Hazard Identification** step requires the evaluation of the contamination history of the study area, evaluation of possible exposure of human and natural communities, identification of the chemical hazard or hazards, and collection and collation of the physicochemical and biological properties of the potential chemical.
6. **Exposure Dose** can be evaluated by measurement of the amounts of contamination in the various pathways and calculations of quantities available made using a range of models for chemical distribution.
7. Environmental contaminants present particular problems in the evaluation of **Toxicity: Dose/Response Relationships** with human health and natural ecosystems. Exposure to chemicals with these systems is at low concentrations for long periods and the data available are generally for relatively high concentrations and relatively short periods. So, extrapolations are needed to obtain suitable data for risk assessment.
8. With human health risk assessment, the **Toxicity: Dose/Response** data are obtained principally from experimental data on animals. To utilize these data, differences in biological response between the test animals and the human population being evaluated need to be taken into account. A **Safety Factor** is applied in toxicity evaluations to increase the apparent toxicity and thereby account for these factors.
9. **Safety Factors** are used in human health evaluations to account for the different sensitivities of individuals in a population, the differences between the test animals and the human population, temporal and other factors, and can range from ten to ten thousand. Safety Factors greater than a thousand indicate considerable uncertainty in the data available and the need for more definitive information.
10. The **TDI**, also termed ADI and RfD, can be calculated from the NOAEL by division by the Safety Factor. Thus,

    TDI = NOAEL/SF mg/kg.bw/day

11. **Human health risk** can be characterized in terms of the TDI as a HQ, where:
    HQ = DOSE/TDI. Levels of HQ <1 indicate no significant adverse effect in a human population, whereas HQ > 1 indicate possible adverse effects.
12. The **Slope Factor (SF)** (see Box 18.3) is Unit Tumor Incidence/Dose in units of the reciprocal of dose, e.g. 1/(ng/kg bw/day) or $(ng/kg\ bw/day)^{-1}$. Thus,

    SF$(ng/kg\ bw/day)^{-1}$ = PIC (unitless proportion)/ Dose (ng/kg bw/day)

    PIC(unitless proportion) = Slope Factor × Dose

    This allows the calculation of the incidence of cancer as a unitless proportion of the population.
13. **Ecological risk assessment** is based on the ecotoxicology concept and follows the same sequence of steps as human health risk assessment outlined above. It differs from human health risk assessment in that the objectives are not formulated in terms of human health outcomes but the effects of toxic chemicals on

ecosystems. Objectives include ecological characteristics such as reduced species diversity, reduced biomass, detrimental change in iconic or commercial species, a change in the types of biota present, and changes in the energy and nutrient flows in ecosystems.
14. The **Hazard Identification** with ecological risk assessment can be carried out using such information as the contamination history of the site and system, and the physicochemical and biological characteristics of the identified chemical hazard can be then collected and collated. With this information, the **Exposure Assessment** can be carried out by identification of the pathways of exposure and use of models.
15. Using the objectives identified, as above in Point 12, a set of **Assessment Endpoints** can be identified by using the available data from the Web. This allows a set of **Critical Toxicity Values** to be developed. With this procedure, there remains the problem of extrapolation from high exposure dose/short exposure times to low exposure dose/long exposure times. Use of safety factors in a somewhat similar manner to the use of these factors with human health is required.
16. **Ecological Risk Characterization** can be carried out in the terms of the original objectives by integrating the Exposure Dose with the Critical Toxicity Values.
17. **Risk Assessment using Probabilistic Techniques** utilizes the various values recorded in the risk assessment process as probabilistic distributions. This means that the exposure dose and toxicity: dose–response data are plotted as distributions rather than discrete numerical values. The risk can be characterized by various techniques, such as the Monte Carlo method, as the proportion of the population affected. These techniques provide a powerful interpretation of the risk but require a large amount of data from systematic observations.

References

Ames, B.N. and Gold, L.S. (1990). Too many rodent carcinogens: mitogenesis increases mutagenesis. *Science* 249: 970–971.

Atabila, A., Phung, D.T., Hogarh, J.N. et al. (2017). Dermal exposure of applicators to chlorpyrifos on rice farms in ghana. *Chemosphere* 178: 350–358.

Bernstein, P.L. (1996). *Against the Gods – The Remarkable Story of Risk*. New York: Wiley.

Cao, Q.M., Yu, Q.M., and Connell, D.W. (2011). Health risk characterisation for environmental pollutants with a new concept of overall risk probability. *Journal of Hazardous Materials* 187: 480–487.

Duijm, N.J. (2015). Recommendations on the use and design of risk matrices. *Safety Science* 76: 21–31. https://doi.org/10.1016/j.ssci.2015.02.014 (accessed 29 September 2020).

Edokpolo, B., Yu, Q.J., and Connell, D.W. (2019). Use of toxicant sensitivity distributions (TSD) for development of exposure guidelines for risk to human health from benzene. *Environmental Pollution* 250: 386–396.

Faustman, E.M. and Omenn, G.S. (2013). Risk assessment. In: *Casarett & Doull's Toxicology - The Basic Science of Poisons*, 8e (ed. C.D. Klaassen), 123–149. New York: McGraw-Hill.

Fitzpatrick, J., Schoeny, R., Gallagher, K. et al. (2017). US Environmental Protection Agency's framework for human health risk assessment to inform decision making. *International Journal of Risk Assessment and Management* 20 (1/2/3): 3–20.

Government of Canada (2017). Application of weight of evidence and precaution in risk assessment. Risk assessment of chemical substances, Government of Canada. https://www.canada.ca/en/health-canada/services/chemical-substances/fact-sheets/application-weight-of-evidence-precaution-risk-assessments.html (accessed 30 September 2020).

Hamidin, N., Yu, Q.J., and Connell, D.W. (2008). Human health risk assessment of chlorinated disinfection byproducts in drinking water using a probabilistic approach. *Water Research* 42: 3263–3274. (reprinted in Japanese in Journal of the Japanese Water Works Association 2009, 78: 70-72).

Martin, P., Bladier, C., Meek, B. et al. (2018). Weight of evidence for hazard identification: A critical review of the literature. *Environmental Health Perspectives* 126 (7): 076001. https://doi.org/10.1289/EHP3067.

OECD (2020). *OECD Guidelines for the Testing of Chemicals website*. Organisation for Economic Co-operation and Development. https://www.oecd.org/chemicalsafety/testing/oecdguidelinesforthetestingofchemicals.htm (accessed 29 September 2020).

OSHA (2016). *Recommended Practices for Safety and Health Programs. United States*. Department of Labor, Occupational Safety and Health Administration, Washington, DC. www.osha.gov/safetymanagement (accessed 10 October 2020).

Phung, D.T., Connell, D.W., and Chu, C. (2015). New method to set guidelines to protect human health from agricultural exposure using chlorpyrifos as an example. *Annals of Agricultural and Environmental Medicine* 22: 275–280.

Rausand, M. (2013). *Risk Assessment: Theory, Methods, and Applications*. Hoboken, NJ: Wiley.

Roser, M., and Ritchie, H. (2015). Cancer. Published online at OurWorldInData.org. https://ourworldindata.org/cancer [Online Resource] updated 2019 (accessed 29 September 2020).

US EPA (2005). Guidelines for Carcinogen Risk Assessment. EPA/630/P-03/001B, Risk Assessment Forum, United States Environmental Protection Agency, Washington, DC. https://www3.epa.gov/airtoxics/cancer_guidelines_final_3-25-05.pdf (accessed 29 September 2020).

US EPA (2016). Uses of Ecological Risk Assessments, Risk Assessment. United States Environmental Protection Agency, Washington, DC. https://www.epa.gov/risk/ecological-risk-assessment (accessed 29 September 2020).

US EPA (2020). Integrated Risk Information System (IRIS). United States Environmental Protection Agency, Washington, DC. https://www.epa.gov/iris (accessed 4 October 2020).

19

Management of Hazardous Chemicals

19.1 Introduction

Chemicals are a vital and integral part of modern society and life. According to UNEP's second Global Chemicals Outlook (GBO II) released in 2019, global trends such as population dynamics, urbanization, and economic growth are rapidly increasing chemical use. In 2017, the value of the chemical industry was more than US$5 trillion. It is projected to double by 2030. While the global production and use of chemicals has continued to grow, there is considerable cause for concern about the current impacts and threats posed for human health and the environment from exposures to hazardous chemicals and other pollutants (e.g. microplastics and pharmaceutical products) (UNEP 2019, pp. viii–ix).

As highlighted by GBO II,

> … Whether this growth becomes a net positive or a net negative for humanity depends on how we manage the chemicals challenge. What is clear is that we must do much more
>
> (UNEP 2019, p. vi).

How to identify, test, evaluate, and manage the environmental and human health risks of new and existing chemicals and their multiple products are now a major global challenge for this decade to 2030. Already well over 100 000 chemicals are estimated to be in commercial use, with about 6 000 or more of these being produced in large quantities (or high production volumes, HPVs). Many of these chemicals are known or suspected to be toxic to humans and wildlife and are being released into the global environment in large quantities. Preceding chapters in this book evaluate the serious extent of environmental and human health impacts and risks caused by or associated with exposure to chemical pollution.

Initially, the United States introduced regulation of both new and existing industrial chemicals through the Toxic Substances Control Act (TSCA) of 1976. Since the early 1990s, the United Nations and its Agencies have coordinated and promoted an international strategy and plan for the sound and sustainable management of chemicals, with support from the OECD and other organizations.

Many countries (e.g. the United States, other members of the OECD, members of the European Union, and China) have now taken various actions to control the use and dispersal of chemicals in the environment. This usually requires the regulation, registration of chemical substances, assessment of chemical properties, uses, and safety measures, and management of new and existing industrial and agricultural chemicals, including chemical inventories and central databases. Nevertheless, how to manage sustainable production, use, and disposal of hazardous chemicals to protect the environment and human health is a worldwide global problem, since chemicals are international commodities with manufacture being undertaken in one country to supply many others. Without coordination, a complex system of different national needs of countries would develop. This was seen as likely to limit the effectiveness of moves to control chemicals, including high economic costs.

International and national approaches to chemical assessment and management of hazardous chemicals are covered in this chapter. The worldwide challenge is how to accelerate actions and effectively reduce the known and emerging pollution impacts and risks of hazardous chemicals as we move toward 2030 and beyond. To do this, we also need a detailed understanding of the scientific principles of pollution as emphasized in this book.

19.2 Goals and Strategies for Managing Chemicals

International goals and targets for the sound and sustainable management of chemicals are developed and promoted worldwide by the United Nations and its agencies. In 1992, at the Earth Summit (UN Conference on Environment and Development, UNCED) held in Rio de

Janeiro, Brazil, Heads of State and Government adopted the Rio Declaration on fundamental principles for the environment and sustainable development (e.g. the polluter pays principle, the right-to-know, and the precautionary approach), and Agenda 21, a global action plan aimed at sustainable development, which also included actions on safer use of toxic chemicals, managing hazardous wastes, and reducing pollution.

The 1992 UN Conference also led to the formation of the Inter-Organization Programme for the Sound Management of Chemicals (IOMC) which was established in 1995 to promote sound management of chemicals worldwide.

In 2002, the World Summit on Sustainable Development (WSSD) adopted the goal for chemicals and wastes to achieve.

> ... the sound management of chemicals throughout their life cycle and of hazardous waste in ways that lead to minimization of significant adverse effects on human health and the environment ... by 2020
>
> (UNEP 2019, p. 5).

In February 2006, at the first UN International Conference on Chemicals Management (ICCM1) in Dubai, a Strategic Approach to International Chemicals Management (SAICM) was adopted by Ministers, Heads of State, and stakeholders (SAICM 2020). It consisted of an Overarching Policy Strategy and a Global Plan of Action. The scope of the Strategic Approach included:

- Environmental, economic, social, health, and labor aspects of chemical safety.
- Agricultural and industrial chemicals, with a view to promoting sustainable development and covering chemicals at all stages of their life cycle, including products.

The objectives of the Policy Strategy focused on five themes: (i) risk reduction, (ii) knowledge and information, (iii) governance, (iv) capacity-building and technical cooperation, and (v) illegal international traffic. Six core activity areas were identified to implement the objectives to achieve the overall goal by 2020:

1. Enhance the responsibility of stakeholders.
2. Establish and strengthen national legislative and regulatory frameworks for chemicals and waste.
3. Mainstream the sound management of chemicals and waste in the sustainable development agenda.
4. Increase risk reduction and information-sharing efforts on emerging policy issues.
5. Promote information access.
6. Assess progress toward the 2020 goal of minimizing the adverse effects of chemicals on human health and the environment.

Importantly, 11 basic elements have also been recognized as critical at the national and regional levels to achieve the goal of sound chemicals and waste management. These elements are listed in Section 19.7.

The 2020 timeline for minimizing the adverse effects of chemicals and also wastes was reiterated at the Rio plus 20 Summit in 2012 (UNEP 2019, p. 5). In 2013, the first Global Chemical Outlook report (GCO-I) prepared by UNEP found increasing global consumption and production of chemicals, and associated impacts and risks for human health and the environment, at local to global scales (UNEP 2013).

The challenge to achieve the 2020 goal through sound management of chemicals and wastes has proven to be formidable. For instance, "…By 2025, the World's cities will produce 2.2 billion tonnes of waste every year, more than three times the amount produced in 2009." Despite the efforts of many countries to manage hazardous chemicals under multilateral environmental agreements, much remains to be achieved (see Box 19.1).

> From 2010 to 2014, only 57 per cent of the parties to the Basel Convention on the Control of Transboundary Movements of Hazardous Wastes and Their Disposal had provided the requested data and information. The figure was 71 per cent for the Rotterdam Convention on the Prior Informed Consent Procedure for Certain Hazardous Chemicals and Pesticides in International Trade, and 51 per cent for the Stockholm Convention on Persistent Organic Pollutants
>
> (UNEP 2020).

International efforts to manage chemicals were advanced in 2015 by the adoption of the 2030 Agenda for Sustainable Development and its seventeen (17) Sustainable Development Goals (SDGs), which included specific goals related to chemicals and waste management:

- SDG 3: Good Health and Well-Being – Ensure healthy lives and promote well-being for all at all ages.
  Target 3.9: By 2030, substantially reduce the number of deaths and illnesses from hazardous chemicals and air, water, and soil pollution and contamination.
- SDG 12: Responsible Consumption and Production – Ensure sustainable consumption and production patterns.
  Target 12.4: By 2020, achieve the environmentally sound management of chemicals and all wastes throughout their life cycle, in accordance with agreed international frameworks, and significantly reduce their release to air, water, and soil in order to minimize their adverse impacts on human health and the environment.

A number of other SDGs (e.g. biodiversity loss, clean water and sanitation, climate action, and facilitate access to clean energy) also apply to chemicals and waste management.

Since 2015, the International Conference on Chemicals Management (ICCM), the governing body of the SAICM, has followed up on SDGs and initiated a process to prepare by 2020, recommendations regarding a Strategic Approach "for the sound management of chemicals and waste beyond 2020." In response, UNEP's GCO-II report has envisaged a sustainable future and identified a range of actions for consideration by policy-makers around the world and informing chemicals and waste management beyond 2020 (UNEP 2019).

Future management approaches to chemicals and wastes beyond 2020 are discussed in Section 19.7 of this chapter, including a proposed action plan by UNEP.

19.3 Collection of Data on Chemicals

Information and Data Sources

In practice, sound management of chemicals requires evaluation of information and data on the properties of a chemical substance, so that predictions can be made as to how the chemical substance (or mixtures containing it) will behave in natural and human environments, and on its possible effects on the environment and human health.

Some of these properties are gathered during the normal process of developing a chemical to enable its effective synthesis or commercial use. For example, the chemical structure, water solubility, boiling point, and melting point are usually derived for all chemicals to ensure effective manufacture and use (Connell and Miller 1984, p. 419).

Toxicological assessment needs additional data specific to the behavior and fate of a chemical in the natural or human environment (e.g. partitioning between phases, biodegradation, and bioaccumulation). Knowledge of the ecotoxicity of a chemical substance to wildlife, and its acute and chronic toxicity to humans has become essential worldwide for the development, registration, evaluation of chemical hazards, risks, and safety, and also marketing of chemical substances (e.g. hazard classification and labeling, and safety data sheets).

Basic data requirements for evaluation of chemicals are outlined below. This applies to new chemicals and also the many existing chemicals that have been inadequately assessed for their properties, toxicities, and environmental and health risks.

1. Chemical identification data including impurities
2. Patterns of production, use, disposal, and release
3. Physicochemical properties
4. Environmental behavior and fate properties
5. Ecotoxicity data
6. Human toxicity data
7. Applicable analytical methods

In particular, the OECD, US EPA, and EU have developed and advanced a variety of laboratory and other tests for the properties of chemicals, environmental behavior and fate, ecotoxicity, and human toxicity that are discussed in Section 19.3.

An example of detailed information sources that need to be gathered for management of chemicals is given in the European Union program for registration and evaluation of chemicals (REACH)[1] which is discussed in Case Study 19.1. Here, registrants need to collect and submit all relevant and available information on the intrinsic properties of a substance, regardless of the quantity manufactured or imported. These properties include:

- substance identity
- physicochemical properties
- chemical categories
- environmental fate, including chemical and biotic degradation
- exposure/uses/occurrence and applications
- mammalian toxicity
- toxicokinetics
- ecotoxicity

In the case of REACH, the critical first step is to collect and assess all of the available information on a substance and any relevant information that may clarify the intrinsic properties of the substance. There are a large number of available sources as indicated by REACH:

- in-house company and trade association files (including test data);
- databanks and databases of compiled data;
- agreed data sets such as the OECD HPV Chemicals Program;
- published technical and scientific literature;
- internet search engines and relevant websites;
- quantitative structure–activity relationships or QSAR models; and
- data sharing in the substance information exchange forum (SIEF).

---

1 REACH is a regulation of the European Union, adopted to improve the protection of human health and the environment from the risks that can be posed by chemicals, while enhancing the competitiveness of the EU chemicals industry. ... REACH stands for Registration, Evaluation, Authorisation and Restriction of Chemicals. It is managed by the European Chemicals Agency known as ECHA (see ECHA 2020a).

Types of existing data to be collected and assessed can include:

- *in vivo* animal studies;
- *ex vivo* studies (e.g. tissues from animals);
- *in vitro* studies (for example, using bacteria or cultured cells);
- information from human exposure;
- predictions based on information from structurally related substances (read-across and chemical categories);
- predictions from computational prediction methods, quantitative structure–activity relationships (QSAR); or
- from other literature sources.

Assessment of information (including data) may also include a weight-of-evidence approach to improve the level of certainty on the information.

Registrants of chemicals then have to consider whether the existing data are of sufficient quality to fill in the gaps in the information needed. Depending upon the annual tonnes of a chemical substance involved, any studies using vertebrate animals that need to be conducted are to occur only as a last resort.

Data sharing is very important to minimize animal testing of chemicals where existing and comparable toxicity data are available for a chemical from other sources including Registrants. Alternative methods and approaches to avoid (or reduce) testing of vertebrates are also discussed in Section 19.4.

Some important database sources for chemical information on properties, hazards, toxicity, risk assessment, and management are listed in Table 19.1.

19.4 Laboratory and Field Testing of Chemicals

Testing of chemicals for the assessment of their human health and environmental hazards and risks has evolved

Table 19.1 Some important chemical databases for assessment and management of chemical substances.

Database	Information	Source
Toxnet (Including HSDB Hazardous substances databank and Toxline)	Most information for Toxnet will continue to be collected and reviewed; selected information will be through PubChem, Pubmed, and Bookshelf. HSDB – peer reviewed toxicology data for over 5000 hazardous chemicals.	US National Library of Medicine (NIH) (See PubChem also HSDB data)
eChemportal	Simultaneous searching of reports and data sets by chemical name and number, by chemical property, and by GHS classification. • properties of chemicals • physical–chemical properties • ecotoxicity • environmental fate and behavior • toxicity	OECD
PubChem	Chemical information repository consisting of three primary databases: substance, compound, and bioassay.	US National Center for Biotechnology Information(NCBI)
ChemSpider	Chemical structure database providing fast access to over 34 million structures, properties, and associated information.	Royal Society of Chemistry
IUCLID	Essential tool for any organization or individual that needs to record, store, submit, and exchange data on chemical substances in the format of the OECD Harmonised Templates (e.g. EU-OECD).	ECHA/OECD
INCHEM Key source of information on chemical safety and the sound management of chemicals	Rapid access to internationally peer reviewed information on chemicals published through IPCS. All types of chemicals from the full range of exposure situations (environment, food, and occupational) are included. Searching across all collections or within individual collections is available.	International Programme on Chemical Safety (IPCS) Canadian Centre for Occupational Health and Safety (CCOHS)
SIDS Screening information data set	Summarizes the literature on high production chemicals and provides an initial assessment for decision-makers.	Organisation for Economic Co-operation and Development (OECD)

greatly since the 1950s because of growing public, industry, national, and international concerns about how to prevent, control, or manage chemical hazards, known impacts, and risks from exposures to the numerous chemicals produced and used commercially. Many government agencies such as the US EPA and ECHA in the EU, nongovernment research, academic, and industry bodies, together with major international organizations such as the OECD have advanced the development of a multitude of methods for testing the physical, chemical, and environmental properties of chemicals, and their toxicity to wildlife and humans (e.g. OECD 2020a).

For example, important physicochemical properties of chemical substances that need to be generally considered for testing, registration, and assessment purposes are listed in Table 19.2.

The OECD has been the leading international organization in developing and coordinating standardized testing of chemicals.

> The current OECD Guidelines for the Testing of Chemicals is a collection of about 150 of the most relevant internationally agreed testing methods used by government, industry and independent laboratories to identify and characterise potential hazards of chemicals. They are a set of tools for professionals, used primarily in regulatory safety testing and subsequent chemical and chemical product notification, chemical registration and in chemical evaluation. They can also be used for the selection and ranking of candidate chemicals during the development of new chemicals and products and in toxicology research
> (see OECD Guidelines for the Testing of Chemicals, Section 1, OECD 2020a).

OECD Test Guidelines for Chemicals

Tables 19.3–19.6 list selected test guidelines for chemical properties published by the OECD (2020a). Other agencies (e.g. the EU and US EPA) or some countries (e.g. China) also develop other specific test methods for registration of chemicals.

Physical and Chemical Properties

Table 19.3 Selected OECD testing guidelines – physical–chemical properties.

Density of liquids and solids	OECD 109 (02-10-2012)
Water solubility	OECD 105 (27-07-1995)
Boiling point	OECD 103 (27-07-1995)
Adsorption–desorption using a batch equilibrium method	OECD 106 (21-01-2000)
Partition coefficient (n-octanol/water): Shake Flask method	OECD 107 (27-07-1995)
Partition coefficient (n-octanol/water): HPLC method	OECD 117 (23-11-2004)
Vapor pressure	OECD 104 (11-7-2006)
Hydrolysis as a function of pH	OECD 111 (23-11-2004)

Source: OECD (2020a).

Effects on Biotic Systems (Ecotoxicity)

Table 19.4 Selected OECD testing guidelines – effects on biotic systems.

Avian acute oral toxicity test	OECD 223 (29-07-2016)
Fish, acute toxicity test	OECD 203 (18-06-2019)
Fish, early-life stage toxicity	OECD 210 (26-06-2013)
21-day fish assay	OECD 230 (08-09-2009)
Amphibian (tadpoles) thyroid assay (XETA)	OECD 248 (18-06-2019)
Bumblebee, acute oral toxicity test	OECD 247 (09-10-2019)
Bumblebee, acute contact toxicity test	OECD 247 (09-10-2019)
Honey bee (*Apis Mellifera* L.), chronic oral toxicity test (10-day feeding)	OECD 245 (09-10-2017)
Earthworm reproduction test (*Eisenia fetida/Eisenia andrei*)	OECD 222 (29-07-2016)

Source: OECD (2020a).

Table 19.2 Selected physicochemical properties of chemicals.

Physical state, appearance, and odor	n-Octanol/water partition coefficient K_{ow}
Melting and boiling points	Flammability
Relative density	Oxidizing properties
Flash point	Explosive properties
Autoignition temperature	Granulometry
Vapor pressure	Explosive limits
Henry's Law constant	Viscosity
Surface tension	Dissociation constant
Water solubility	

Environmental Fate and Behavior

Table 19.5 Selected OECD testing guidelines – environmental fate.

Degradation: biotic	Aerobic and anaerobic transformation in aquatic sediment OECD 307
Abiotic	Phototransformation of chemicals in water – direct photolysis OECD 316
Bioaccumulation	Bioaccumulation in fish: aqueous and dietary exposure OECD 305
	Bioaccumulation in terrestrial oligochaetes OECD 317

Source: OECD (2020a).

Health Effects (Toxicology)

Table 19.6 Selected OECD testing guidelines – health effects (toxicology).

Acute toxicity (oral, dermal, inhalation)	OECD 401, 402, 403
Skin and eye irritation	OECD 437, 439, 460
Sensitization (skin)	OECD 406, 442C, 442D, 442E
Repeated oral dose toxicity (28 days and 90 days)	OECD 407, 408
Mutagenicity and genotoxicity	OECD 471, 473, 474, 475, 487, 490
Reproductive and developmental toxicity	OECD 414, 421, 422, 426
Carcinogenicity	OECD 451, 453
Toxicokinetics	OECD 417

Source: OECD (2020a).

The other OECD Test Guidelines category includes tests such as 501 Metabolism in Crops, 505 Residues in Livestock, and 503 Metabolism in Livestock.

EU-REACH Test Guidelines for Chemicals

Usually the studies for REACH information requirements on ecotoxicity, human toxicity, and physical–chemical properties are generated using test guidelines. A set of official test guidelines approved by the OECD and EU is given in Tables 19.7–19.9 (OECD 2020b).

Physical and Chemical Properties

No prioritized test guidelines for physical and chemical properties are published on the ECHA website.

Nonanimal Testing Guidelines

Under EU-REACH, ECHA develops, promotes, and supports alternative methods and approaches for registrants to avoid unnecessary animal testing of chemicals as outlined below. Under these guidelines,

> Registrants may only carry out new tests when they have exhausted all other relevant and available data sources
>
> (ECHA 2020b).

Examples of the main alternative approaches are:

- The read-across approach which allows the use of relevant information from analogous substances to predict properties of the target substances. ECHA has developed and published a read-across assessment framework (RAAF) together with an example to illustrate its use.
- Data sharing where companies are encouraged to share any available data on their substance if requested by a registrant of an analogue substance.
- The QSAR Toolbox is a software application which supports companies in identifying data that may be relevant for assessing the hazards of chemicals. ECHA develops and manages the tool in cooperation with the OECD.

Usefully, ECHA publishes a large proportion of the data it receives from registrations on its website.

Table 19.7 Human health test guidelines – EU REACH.

Human health	Test guidelines
Serious eye damage/eye irritation	Bovine Corneal Opacity and Permeability Test Method (BCOP), EU B.47, OECD 437 (adopted in 2009, revised in 2013 and 2017 by OECD)
	Isolated Chicken Eye Test Method (ICE), EU B.48, OECD 438 (adopted in 2009 and revised in 2013 and 2017 by OECD)
Skin corrosion/eye irritation	Corrosion
	Transcutaneous Electrical Resistance Test Method (TER), EU B.40, OECD 430, (2000, revised in 2013 and 2015 by OECD)
	Irritation
	– Reconstructed Human Epidermis Test Method (RHE) (includes more than one protocol), EU B.46, OECD 439 (in EU 2009 and in OECD 2010, revised in 2013 and 2015 by OECD)
Skin sensitization	In Chemico Skin Sensitization: Direct Peptide Reactivity Assay (DPRA), OECD 442C (adopted 2015)
Genotoxicity	In vitro mammalian cell gene mutation tests using the thymidine kinase Gene, OECD: 490 (adopted 2015, updated 2016), EU: none at present

Source: OECD (2020b).

Table 19.8 Environmental fate test guidelines – EU REACH.

Degradation	OECD 314: Simulation Tests to Assess the Biodegradability of Chemicals Discharged in Wastewater, 2008
Bioaccumulation	OECD 317: Bioaccumulation in Terrestrial Oligochaetes, 2010
	OECD 305: Bioaccumulation in Fish: Aqueous and Dietary Exposure, 2012

Note: Last updated 19-03-2014.
Source: OECD (2020b).

Table 19.9 Ecotoxicity test guidelines – EU REACH.

Aquatic (Last updated 19-09-2016)	OECD 236: Fish Embryo Acute Toxicity (FET) Test, July 2013
	OECD 209: Activated Sludge, Respiration Inhibition Test (Carbon and Ammonium Oxidation), 23 July 2010
	OECD 210: Fish, Early-life Stage Toxicity Test, 26 July 2013 (replaces OECD 210: Fish, Early-life Stage Toxicity Test, 17 July 1992)
	OECD 211: Daphnia magna Reproduction Test, 2 Oct 2012 (replaces OECD 211: Daphnia magna Reproduction Test, 16 Oct 2008)
Terrestrial (Last updated 19-03-2014)	OECD 226: Predatory mite (Hypoaspis [Geolaelaps] aculeifer), reproduction test in soil, 2008
	OECD 232: Collembolan Reproduction Test in Soil, 2009
Sediment (last updated 19-03-2014)	OECD 233: Sediment-Water Chironomid Life Cycle Toxicity Test Using Spiked Water or Spiked Sediment, 2010

Source: OECD (2020b).

19.5 Regulation and Assessment of Chemicals

Many countries and regions have developed their own policies and regulations to manage the health and environmental hazards of industrial chemicals and related chemical products. Detailed support is readily available for all countries through international organizations such as the IOMC, UNEP, OECD, WHO, ILO, and FAO (see Section 19.7). For example, the WHO Regional Office for Europe has described approaches to help set up national chemicals registers and how they can improve the strategy of sound chemicals management, including useful information sources and examples (WHO 2018).

A central component of chemicals management under such policies and regulations is the setting up of chemical registers and inventories that compile lists of chemical substances that are produced or imported into a country or region. Increasingly, registration, evaluation, and recording of new and existing chemicals are common goals and practices to improve sound chemicals management worldwide.

As regulation of chemicals in a management context is a complex and detailed topic, the scope of this section provides a brief overview of the US, European, Chinese, and Indian approaches to regulation of chemicals, and focuses on a Case Study 19.1 of the European REACH model and key elements of its chemical assessment and management.

Examples of International Regulation of Chemicals

United States of America

The TSCA of 1976 was the first legislation to regulate both new and existing industrial chemicals. It was passed by the US Congress in response to a lack of US legislation for the control of industrial chemicals, such as toxic and persistent PCBs found in the environment during the 1970s. The recently formed Environmental Protection Agency (EPA) in 1970 was empowered to enforce this Act and regulations (Silbergeld et al. 2015). Although this regulation was a leading model for many years, in practice, it suffered from a number of limitations such as in the control of existing and emerging hazardous chemicals (see Silbergeld et al. 2015 for a critical review of legal and scientific limitations of the TSCA prior to the 2016 amendment).

In 2016, the Act was amended and updated by the Frank R. Lautenberg Chemical Safety for the 21st Century Act. Essentially, it provides the "EPA with authority to require reporting, record-keeping, and testing requirements, and restrictions relating to chemical substances and/or mixtures. Certain substances are generally excluded from TSCA, including, among others, food, drugs, cosmetics, and pesticides" (US EPA 2019). New chemical substances require premanufacture notification and evaluation by the EPA. Existing commercial chemicals are listed in a TSCA chemicals inventory (>83 000 chemicals). Once new chemicals are commercially manufactured or imported, they are placed on the list.

The EPA can issue regulations designed to require manufacturers, importers, or processors to test chemical substances and mixtures. This testing is required to develop data about health or environmental effects when there is insufficient data for EPA risk assessors to determine whether a chemical substance or mixture presents an unreasonable risk to health or the environment (US EPA 2019).

European Union

Prior to the 2016 amendment of TSCA in the United States, comprehensive chemical regulation in the European community was enacted in 2007 under the REACH legislation (see Case Study 19.1), to advance registration, testing, assessment, and authorization of new and all existing industrial chemicals. Producers and importers of chemicals have the burden of proof to demonstrate safety (health and environmental) prior to authorization to market their chemical-related products. The European Chemicals Agency (ECHA) is responsible for registration, evaluation, verification, control, and management of chemicals (Silbergeld et al. 2015).

REACH has influenced the policies for the control of chemicals in other countries, such as China, Turkey, Japan, Taiwan, and South Korea which have adopted similar regulations. China is now the leading worldwide producer and seller of chemicals and chemical-containing products (Silbergeld et al. 2015).

China

Regulations on Safe Management of Hazardous Chemicals, known as Decree 591, was entered into force on 1 December 2011. Decree 591 is the highest chemical control law in China. It regulates hazardous chemicals through the entire supply chain, from manufacture and importation, to distribution and storage, transportation, and use.

Domestic manufacturers and importers of hazardous chemicals in China have to apply for licenses to operate and also submit detailed HazChem registrations to the State Administration of Work Safety (SAWS) (renamed the Ministry of Emergency Management, MEM, in 2018), prior to manufacturing or importation. Decree 591 also implements the Globally Harmonised System (GHS) in China which requires companies to provide SDSs and labels prepared in accordance with relevant national standards (Chemsafetypro 2019a). This regulatory version of the REACH model, sometimes known as China REACH, lacked requirements for registration of chemicals with data, in the case of tens of thousands of existing chemicals in the Chinese chemicals inventory (Silbergeld et al. 2015).

Recently, the Ministry of Ecology and Environment (MEE) introduced new environmental regulations to strengthen environmental risk assessment and control of chemical substances, and also regulatory requirements. The proposed regulatory framework is seen to integrate, and expand on China's existing regulatory programs. Significantly, it would establish an environmental risk control and management system for chemical substances. As well, it would authorize governmental authorities to promulgate use restrictions, material restrictions, or even bans on certain chemical substances in the future, and impose hefty penalties on violators (Luo 2019).

India

India is reported to be the third largest manufacturer of chemicals in Asia. Regulation of chemicals through Acts and Rules is diverse. Key chemical regulations in India are described as the Manufacture, Storage, and Import of Hazardous Chemical (Amendment) Rules (1989, 1994, 2000), enacted by the Ministry of Environment & Forests (MoEF) in 1989. These Rules regulate the manufacture, storage, and import of hazardous chemicals in India, while the transport of hazardous chemicals is controlled under the Motor Vehicles Act (1988). More specifically, the Ozone Depleting Substance (R&C) Rules (2000) enforce the control of the production, import, and use of ozone-depleting substances (ODCs), of which most are banned in India.

Currently, there are no REACH-like or TSCA-like regulations for chemicals in India. There is a lack of chemical registration and also no national chemical inventory. This situation is expected to change in the 2020s. The Ministry of Environment, Forest, and Climate Change has formed a National Coordination Committee which has prepared a draft National Action Plan for Chemicals for India which includes draft plans for inventory and registration (Chemsafetypro India 2020).

Case Study 19.1

Model of Regulation of Chemicals in the European Union (EU-REACH)

For many years, a large number of chemical substances have been manufactured and marketed in the European Union. Well over 100 000 commercial chemicals are known to exist, often with inadequate knowledge of their human health and environmental hazards. The regulation of these chemicals and evaluation of the hazards and risks posed by exposure to hazardous chemicals is a major and complex challenge for governments, industry, and consumers to ensure their safe management.

The current regulation of chemicals in the EU, known as REACH, was established under EU regulation (EC No 1907/2006) and enacted in 2007. Its name is derived from the scope of its four processes: Registration, Evaluation, Authorization, and Restriction of Chemicals. The fundamental aims of REACH are to improve the protection of human health and the environment from risks that are posed by chemicals, while enhancing innovation and competitiveness of the EU chemicals industry. The regulation also seeks the progressive substitution of chemical *substances of very high concern* with suitable alternatives as they become available.

The REACH regulation applies to all the chemicals that are produced or imported into the EU, including those used in industrial processes and everyday consumer products. Registration of chemicals is required by all companies that produce or import these chemical substances

Case Study 19.1 (Continued)

in quantities of 1 tonne or more per year. Importantly, the burden of proof is placed on manufacturers and importers to gather information on the properties of their chemicals which will allow their safe handling and use. This information is then documented in a register and stored in a central database.

The ECHA based in Helsinki, Finland, manages the technical, scientific, and administrative aspects of REACH, including public databases on hazard information for consumers and professionals.

The ECHA website covers the legislation, information on chemicals, and detailed support that applies to the regulation of chemicals in the EU. Information on chemicals includes a critical set of Guidance documents for stakeholders to understand and meet requirements of REACH and associated regulations for registration, assessment, and management of chemicals (European Commission 2019).

The scope of other legislation for chemical management covers:

- The Classification, Labelling, and Packaging (CLP) Regulation ((EC) No 1272/2008) is based on the United Nations' Globally Harmonised System (GHS) and its purpose is to ensure a high level of protection of health and the environment, as well as the free movement of substances, mixtures, and articles.
- The Biocidal Products Regulation (BPR, Regulation (EU) 528/2012) concerns the placing on the market and use of biocidal products, which are used to protect humans, animals, materials, or articles against harmful organisms like pests or bacteria, by the action of the active substances contained in the biocidal product.
- The Prior Informed Consent Regulation (PIC, Regulation (EU) 649/2012) administers the import and export of certain hazardous chemicals and places obligations

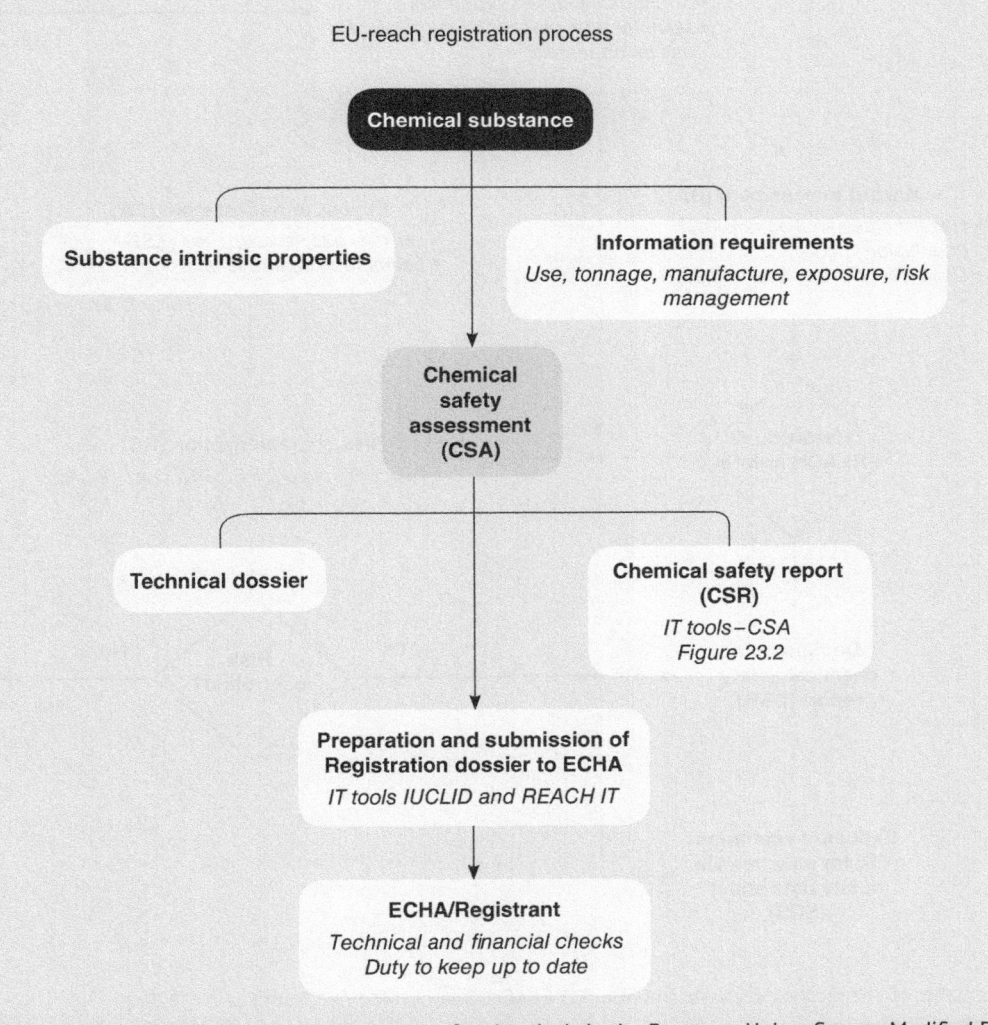

Figure 19.1 Basic outline of EU-REACH registration process for chemicals in the European Union. Source: Modified ECHA (2020d).

(Continued)

Case Study 19.1 (Continued)

on companies who wish to export these chemicals to non-EU countries;
- The Chemical Agents Directive (Directive 98/24/EC) sets out the minimum requirements for the protection of workers from risks to their safety and health – arising or likely to arise – from the effects of chemical agents in the workplace or the use of chemical agents at work. It lays down indicative and binding OELs, as well as biological limit value.
- The Waste Framework Directive sets out measures addressing the adverse impacts of the generation and management of waste on the environment and human health, and for improving efficient use of resources which are crucial for the transition to a circular economy.
- POPs Regulation (EU) No 2019/1021 of the European Parliament and of the Council of 20 June 2019 on persistent organic pollutants. The POPs Regulation bans or severely restricts the production and use of persistent organic pollutants in the European Union.

Readers are advised to consult the ECHA website (ECHA 2020c) and Guidance information for current requirements and guidance documents (e.g. technical and scientific details and methods) as these are subject to progressive updates, amendments, and regulatory changes.

Registration of Chemicals

The basic registration process for chemicals under REACH legislation is outlined in Figure 19.1. The process incorporates a risk assessment framework known as a chemical

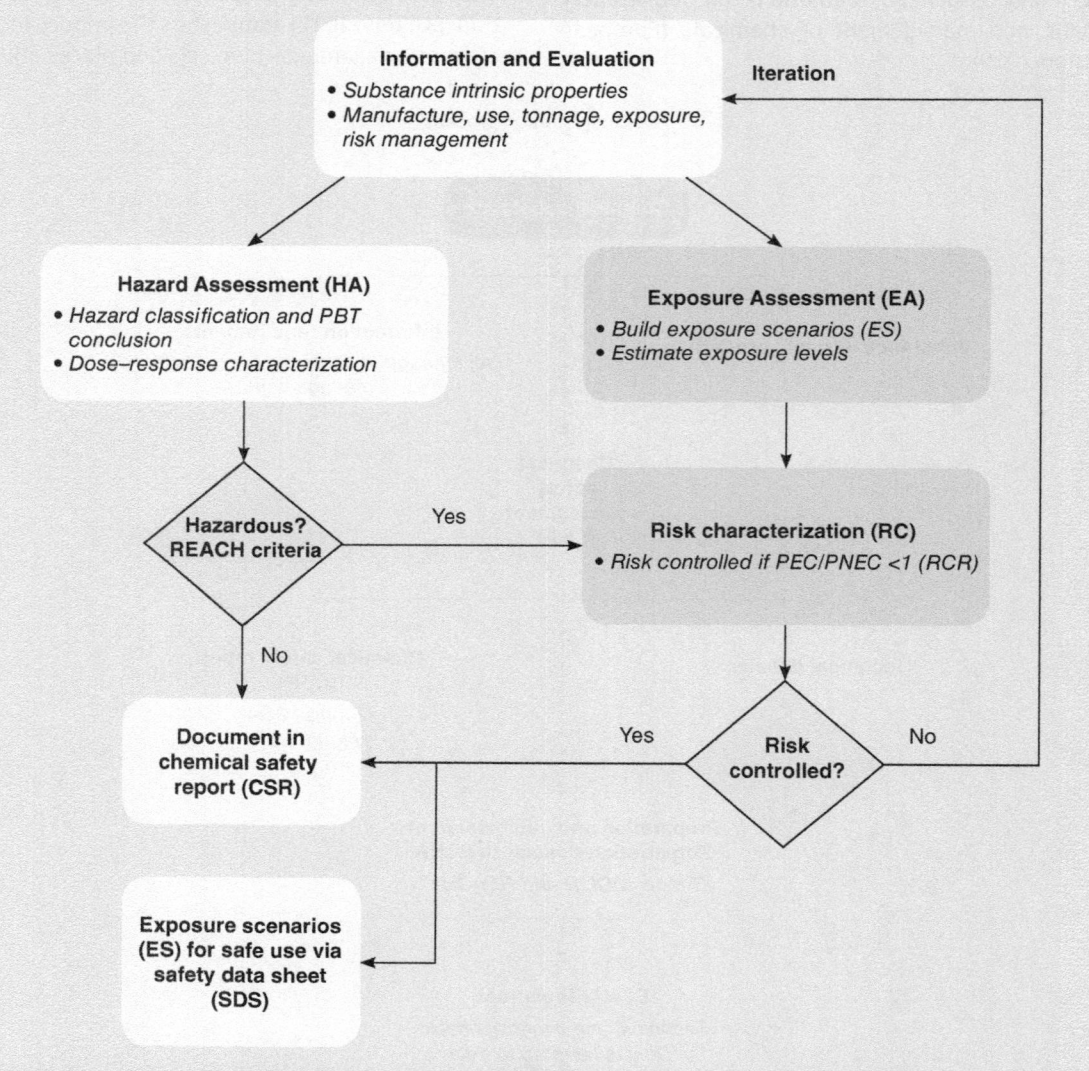

Figure 19.2 Elements of chemical safety assessment in EU-REACH. Source: Modified from ECHA (2020d).

Case Study 19.1 (Continued)

safety assessment (CSA) for new chemicals, and is also applicable to existing chemicals. Figure 19.2 presents a decision-making flow chart of the key steps and components of the CSA used by a registrant to evaluate, demonstrate, classify, and communicate the safety of a chemical under exposure scenarios that apply to its use.

As shown in Figure 19.2, the CSA is based on the fundamental principles of the risk assessment framework for chemical substances (see Chapter 18). It underpins regulatory guidance on the information gathering, including intrinsic properties and relevant data, and evaluation steps for hazard assessment, exposure scenarios, and risk characterization, according to REACH criteria, followed by the decision-making and reporting process for registration. The latter includes preparation of a registration dossier and submission to ECHA (see Figure 19.1).

Where risk control of the substance for exposure scenario(s) is insufficient or unsafe, iteration of the CSA is necessary to identify and show how sufficient risk control or reduction can be achieved for registration.

Chemical Safety Report (CSR)
- Results of the CSA are documented in the chemical safety report (CSR) (ECHA 2020e). If no hazards are identified for the chemical substance, then exposure assessment and risk characterization are not required.
- Where exposure assessment and risk characterization are required, ECHA has developed a CSA and reporting tool, known as Chesar. This tool can be used to create the exposure assessment and the relevant part of the CSR, as well as the corresponding exposure scenarios for communication.

Table 19.10 describes the key steps involved in a full CSA, i.e. where exposure assessment and risk characterization are also needed (see Figure 19.2).

Note: A series of REACH guidance documents is given in the ECHA website. It describes the information requirements and CSA, including theoretical aspects, information, and guidelines on methodology and criteria.

Some additional information on guidance and IT Tools used in CSA steps is given below.

Hazard Assessment Step

Qualitative and Quantitative Structure–Activity Relationships (Q)SARs

The use of QSARs in chemical toxicology is introduced in Chapter 5 of this book. QSAR models are increasingly important in the registration and risk assessment of hazardous chemicals, especially as an alternative screening method for chemical properties and animal toxicity testing. They are theoretical models that can be used to predict the physicochemical, environmental fate, and ecotoxicology properties of chemical compounds from the knowledge of their chemical structure.

Generally, QSAR results are used as part of a weight-of-evidence (WoE) approach or as supporting information. The ECHA has produced a practical guide on how to use and report QSARs for registration purposes (ECHA 2016).

The most commonly used and free QSAR models for physiochemical properties are the US EPA EPI Suite™ and US EPA TEST. For aquatic toxicity, the Danish QSAR database and US EPA ECOSAR Toolbox are available for use. ECOSAR can only be used to evaluate organic chemical substances with discrete structures and does not apply to inorganic or organometallic chemicals, polymers, and chemicals with MW > 1000, mixtures, and nanomaterials. Basically, it is used for screening purposes, checking the need for further aquatic testing or more sensitive species, and estimated values are limited to qualitative risk assessment (Chemsafetypro 2019b).

In particular, the US EPA EPI Suite™ of QSAR models can be used to predict physicochemical properties, aquatic toxicity, and environmental fate properties Chemsafetypro (2019c). It consists of a suite of software programs as shown in Table 19.11.

An advantage of the EPA Suite program is that it can be run simply by the input of the chemical name, CAS No., and Smiles code[2]. For example,

Chemical name: Ethanol, 2-(2-nitrophenyl) amino)-
CAS no: 4926-55-0
Smiles: O = [N+]([O-])c1c(NCCO)cccc1

How to Derive Quantitative Dose–Response Relationships

The ECHA website provides detailed guidance on calculation of the following (see also Table 19.10 below):

PNECs – Environment

How to derive quantitative **thresholds** for the environment (PNECs) (threshold property).

- Worked examples are also given in chemsafetypro.com website.
- Section 19.6 also contains a PNEC calculation for acetone in an aquatic ecosystem (freshwater), modified from Chemsafetypro (2019d).

How to assess the type and extent of hazard related to a substance, including PBT properties, and decide on hazard classification and labeling of substance.

DNEL/DMELs – Human Health

How to calculate derived no-effect level (DNEL) and derived minimum effect level (DMEL) for exposed humans (workers and general populations) in EU-REACH.

Methodology: ECHA (2010) – DNEL/DMEL Derivation from Human Data

Specific or various exposure patterns may be required for exposure scenarios, e.g. acute and/or long-term

(Continued)

---

2 Smiles (Simplified Molecular Input Line Entry System) is a molecular text code of the structure of a chemical.

Case Study 19.1 (Continued)

Table 19.10 Outline of chemical safety assessment steps for EU-REACH.

CSA steps	Key information/actions/outcomes
Information requirements and evaluation	• Substance intrinsic properties. • Manufacture, use, tonnage, exposure, conditions, risk management.
Hazard assessment	
Type and extent of hazards of the substance	Determine the potentially harmful **properties** of the substance: • Physicochemical properties that may be harmful (incl. explosivity, flammability, and oxidizing potential). • Environmental properties (incl. assessment for multiple environmental compartments[a] and PBT/vPvB properties). • Human health properties. • Dose-response relationships. Derive quantitative **thresholds** for human health (DNELs/DMELs) and the environment (PNECs) (threshold of property). Assess type and extent of hazard related to substance, including PBT properties, and decide on hazard classification of substance.
Exposure assessment	Qualitative exposure assessment (property with no threshold) or quantitative exposure estimate (property with threshold).
Information needed	Substance properties, identified uses, and known existing conditions of use over life cycle of substance.
Distinguish situations to be assessed	Differentiate: manufacture, formulation, industrial uses, professional uses, consumer uses, and articles uses.
Conditions of use	Gather information on operational conditions.
Exposure scenarios (ES)[b]	Generate exposure scenarios for all identified uses and stages of life cycles. Use ES to estimate exposures for environment/human health scenarios. Examples are given further in this section.
Physicochemical hazards, e.g. flammability	Qualitative assessment – determine that the conditions of use, including risk management measures, are sufficient to prevent accidents such as at workplace.
Environmental	Assess emissions of the substance from the process, and the fate and distribution, together with environmental conditions, of the substance in the environment to determine environmental concentrations. Emissions and concentrations in the environment can be either measured and/or modeled. Conduct various exposure assessments for several environmental compartments[a]. Assess exposure separately around the local point sources, and for regional exposure from several sources in a given region.
Human health	Several exposure assessments for each identified use, such as for different routes and timeframes are usually needed. The types of exposure assessment are related to the properties and uses of the substance.
Exposure assessments and limitations	Ensure exposure assessment methods and tools are applicable to the property profile of your substance and the use conditions.
Risk characterization	Final step in chemical safety assessment where it should be determined whether risks arising from the manufacture, import, and uses of substances are controlled. It is carried out for each exposure scenario.[b] It involves comparing the toxicological or effect thresholds with the estimated exposure concentrations to humans and the environment.
Methodology	Balance the information on the hazards of substances with the information on the exposure to the substances (for human health, the environment, and, where relevant, for physicochemical properties).

Case Study 19.1 (Continued)

Table 19.10 (Continued)

CSA steps	Key information/actions/outcomes
Assessment	Compare the quantitative exposure estimates with the relevant toxicological thresholds (derived no-effect levels (DNELs) for human health or predicted no-effect concentration (PNEC) for the environment).
	Divide the relevant estimated exposure level (or concentration) by the relevant effect level (DNEL or PNEC) to obtain the risk characterization ratio (RCR).
	Ensure for each relevant use and separate assessment that each RCR is below 1 (i.e. the level of exposure is lower than the threshold level).
	If not, iteration of the CSA is applied until risks can be demonstrated to be under control (e.g. refining properties of substance, improving operational conditions, and/or risk management measures).
	Risks for hazardous physicochemical properties assess the likelihood and severity of an adverse effect.
	If a property has no threshold, a qualitative exposure assessment is made to determine if risk is sufficiently controlled or not.
Risk outcomes	If risk is sufficiently controlled, describe emission scenario(s) and proceed to chemical safety report (CSR).

The elements to be included in the CSR are listed in Annex I, section 7 of REACH.
Notes:
PNEC (Predicted no-effect concentration – see Worked Example 19.1 for definition. It is an effect threshold of a chemical property).
DNEL – Derived No-Effect Level is defined as the level of chemical exposure above which humans should not be exposed. In human health risk assessment, the exposure level of each human population known to be or likely to be exposed is compared with the appropriate derived no-effect level, DNEL.
DMEL – Derived minimal effect level is defined as a level of exposure below which the risk levels of cancer become tolerable. Exposure levels below a DMEL are judged to be of very low concern. For non-threshold carcinogens, only DMELs may be obtained because even a very low level of non-threshold carcinogens may lead to cancer. No DNEL can be derived.
[a]Environmental compartments may consist of fresh surface water, soil, sediment, STP microorganism, air, and predator.
[b]An exposure scenario (ES) describes, in a structured format, the operational conditions and risk management measures leading to safe use. Exposure scenarios apply to the whole life cycle for a substance with harmful properties for humans, environment, or harmful physicochemical properties.
Source: Modified from ECHA (2020d).

Table 19.11 List of programs in US EPA EPI Suite™ for estimation of various properties of a chemical.

Programs	Predicted properties
Physicochemical	
KOWWIN™	Log octanol–water partition coefficient, Log KOW
HENRYWIN™	Henry's Law constant
MPBPWIN™	Melting point, boiling point, and vapor pressure of organic chemical substances
WATERNT™	Water solubility
Environmental fate	
AOPWIN™	Gas-phase reaction rate for the reaction between the most prevalent atmospheric oxidant, hydroxyl radicals, and a chemical
BIOWIN™	Aerobic and anaerobic biodegradability of organic chemical substances
BioHCwin™	Biodegradation rate of chemical containing only carbon and hydrogen (i.e. hydrocarbons)
BCFBAF™	Fish concentration factors

(Continued)

Case Study 19.1 (continued)

Table 19.11 (Continued)

Programs	Predicted properties
HYDROWIN™	Aqueous hydrolysis rate constants and half-lives
STPWIN™	Predicts removal of a chemical in a typical activated sludge-based sewage treatment plant. Values are given for total removal and three processes that may contribute to removal: biodegradation, sorption to sludge, and air stripping
Fugacity model LEV3EPI™	Uses a Level III multimedia fugacity model and predicts partitioning of chemicals among air, soil, sediment, and water under steady-state conditions for a default model "environment"
Aquatic toxicity ECOSAR™	Short-term and long-term aquatic toxicity

Source: Chemsafetypro (2019c).

inhalation, systemic effects; long-term oral, systemic effects, etc.

For example, "the DNEL (oral route, systematic effects) of methanol for general population is **0.088 g/kg bw/day**. It means that if an adult person (assuming **60 kg** body weight) intakes less than **5.28 g** (0.088*60) methanol per day by oral route, the risk of methanol causing ocular toxicity/vision damage can be controlled."

- *Detailed worked example(s) is available from Chemsafetypro (2019e).*

Note: For many local effects (irritation), DNELs cannot be derived. It may also apply to, for example, a mutagen/carcinogen where no safe threshold level can be obtained. In these cases, a semiquantitative value, known as the **DMEL** or **Derived Minimal Effect Level** may be developed.

Exposure Assessment Step

"Essential guidance on exposure assessment, including exposure scenarios, theory, and methodology," is contained in the ECHA website (ECHA 2020c).

Computer-based software tools for predicted exposure concentrations (PECs) include the European Union System for the Evaluation of Substances (EUSES) which is a free tool developed by the European Commission to assist authorities, research institutes, and companies to estimate **environmental** exposure levels of industrial chemicals and biocides. PECs (or measured ECs) are used in the Risk Characterization step below.

EUSES is reported as easy to use and needs few data on substance properties to calculate PECs for tier 1 assessment. If the use of default exposure estimates and tier 1 assessment do not lead to a PEC/PNEC < 1, a refined assessment is possible in EUSES by including more specific information on releases.

Risk Characterization Step

As described in Table 19.10, this is the final step in CSA used to determine whether risks arising from the manufacture, import, and uses of substances are controlled. It is carried out for each exposure scenario. It involves comparing the toxicological or effect thresholds with the estimated exposure concentrations to humans and the environment.

Worked Example 19.1 *Basic Quantitative Chemical Safety Assessment of a Chemical in the Aquatic Environment (freshwater)*
Problem: Characterize the environmental risk of acetone (neutral organic solvent) at a predicted environmental concentration (PEC) in an exposure scenario for an aquatic environment (freshwater) (see Section 19.6 for worked example 19.1).

19.6 Risk Characterization Example (19.1)

Measured physicochemical properties of acetone are readily available from the literature and supported by predicted values from QSAR modeling (e.g. USEPA EPI Suite™).

Hazard Assessment Step

- Physicochemical properties in this scenario are acceptable and not harmful (e.g. explosivity, flammability, and oxidizing potential).
- Acetone does not have PBT/vPvB properties.

Derive the PNEC of acetone in freshwater environment scenario using ecotoxicology data and relevant assessment factors (AFs) given in Table 19.13 obtained from Chemsafetypro (2019d).

Methodology

Predicted No-Effect Concentration (PNEC) is the concentration of a substance in any environment below which adverse effects are most unlikely to occur during long-term or short-term exposure. In environmental risk assessment, PNECs are derived in the Hazard Assessment step and then will be compared to actual or PEC derived in the Exposure Assessment to determine if the risk of a substance is acceptable or not. If PEC/PNECs are <1, the risk is acceptable.

The PNECs are usually calculated by dividing toxicological dose descriptors by an assessment factor. Common toxicity endpoints used for deriving PNECs are mortality (LC50), growth (ECx or NOEC), and reproduction (ECx or NOEC). These endpoints are defined below.

- **LC50/EC50** (Median Lethal Concentration/Median Effective Concentration) are the concentrations at which 50% mortality or inhibition of a function (e.g. growth or growth rate) was observed. They are usually obtained from short-term ecotoxicology studies.
- **NOEC (**No Observed Effect Concentration**)** is the highest tested concentration for which there is no statistical significant difference of effect when compared to the control group. It is usually obtained from long-term ecotoxicology studies. In some studies, only **LOEC (lowest observed effect concentration)** can be obtained, in which case NOEC can be calculated as the LOEC value divided by 2 (LOEC/2).
- **ECx** is the concentration at which x% (10% for EC10) effect was observed or derived statistically when compared to the control group. It is usually obtained from long-term ecotoxicology studies.

Assessment Factors (AFs)

These factors are used to address the differences between laboratory data and natural conditions, taking into account interspecies differences and intraspecies differences. AFs applied for long-term tests are smaller because the uncertainty of the extrapolation from laboratory results to the natural environment is reduced. Uncertainties can also be reduced further when more data on more species in the same environmental compartment are available, and used in the statistical analysis (see Table 19.12 for common AFs).

Calculation of PNEC Values for Acetone in Different Environmental Compartments

Derived PNECs for acetone are calculated by dividing dose descriptors for each compartment with the assessment factors, as indicated in Table 19.13.

Table 19.12 Common assessment factors used to calculate PNEC from ecotoxicity studies.

PNEC type	Available data	AFs
PNEC-water or PNEC-soil	At least one short-term L(E)C 50 from each of three trophic levels	1000
	One long-term EC10 or NOEC from one trophic level	100
	Two long-term results (e.g. EC10 or NOECs) from species representing three trophic levels	50
	Long-term results (e.g. EC10 or NOECs) from at least three species representing two trophic levels	10
	Species sensitivity distribution (SSD) method	1–5
	Field data or model ecosystems	Case by case
PNEC-STP microorganism	Short-term EC50 from activated sludge respiratory inhibition	100
	Long-term NOEC from activated sludge respiratory inhibition or biodegradability test	10
	Long-term NOEC from inhibition of nitrification bacteria	1
	Short-term EC50 from activated sludge respiratory inhibition	100
PNEC-sediment	One long-term test (NOEC or EC10) on one sediment living organism	100
	Two long-term tests (NOEC or EC10) with two species of sediment living organism	50
	Three long-term tests (NOEC or EC10) with three species of sediment living organism	10

Source: Chemsafetypro (2019d).

Table 19.13 Calculation of derived predicted no-effect concentrations (PNECs) for different environmental compartments.

Compartment	Ecotoxicology dose descriptors	Assessment factor	PNEC value
Freshwater	NOEC Algae growth inhibition 100 mg L^{-1}	10	1 mg L^{-1}
	NOEC Daphnia reproduction 10 mg L^{-1}		
	NOEC Freshwater Fish 20 mg L^{-1}		
STP microorganism	Three hours – NOEC >1 000 mg L^{-1}	10	100 mg kg^{-1}
	Activated sludge inhibition test		
Soil	LC50 Earthworm acute toxicity >1000 mg kg^{-1}	1000	1 mg kg^{-1}

Source: Chemsafetypro (2019d).

In the above example, **PNEC**-water is calculated as **1 mg L^{-1}**. For a freshwater environment, toxicological data **(10 mg L^{-1})** from the most sensitive species (*Daphnia*) is used for the PNEC-freshwater calculation. An assessment factor of **10** based on NOECs is used to take into account the differences between laboratory conditions and natural conditions (see Table 19.12).

Exposure Assessment Step

In this example, the PEC of acetone in the freshwater environment scenario is 5 mg L^{-1} derived from environmental modeling (e.g. EUSES tool) of the freshwater exposure scenario for use of acetone. Various emission models, for example, are given in the REACH Guidance on exposure assessment.

Risk Characterization Step

Generally, in the **risk characterization** step, each PNEC is then compared to measured or PEC to determine if the risk of a substance is acceptable or not.

The risk is judged to be acceptable if PEC/PNEC <1 (risk characterization ratio).

In this scenario, the PEC value of acetone in the aquatic environment (freshwater) is taken to be **5 mg L^{-1}**. Therefore, acetone is likely to cause an **unacceptable risk** to the aquatic environment (freshwater) in this risk scenario when compared to its PNEC value.

$$\text{I.e. PEC/PNEC} = \left(5 \text{ mg L}^{-1}\right) / \left(1 \text{ mg L}^{-1}\right)$$
$$= 5 \text{ (risk characterization ratio)}$$

Specialized risk assessment may be required for more complex scenarios and life cycle evaluations (e.g. new materials such as manufactured nanometals).

19.7 Managing Chemicals to 2020 and Beyond

International Coordination and Support for Sound Chemical Management

The IOMC initiates, facilitates, and coordinates international action to achieve the WSSD goal for sound management of chemicals. The participating organizations are the Organisation for Economic Co-operation and Development (OECD), the Food and Agriculture Organization of the United Nations (FAO), the International Labour Organization (ILO), the United Nations Environment Programme (UNEP), the United Nations Industrial Development Organization (UNIDO), the United Nations Institute for Training and Research (UNITAR), and the World Health Organization (WHO).

The IOMC, individual member organizations, various other intergovernmental and governmental agencies such as ECHA provide a wide range of supporting information on chemicals, policies, risk management, and technical tools to assist countries, and regions, to develop and promote (i) national chemical management programs as promoted by SAICM and also (ii) multilateral treaties and various voluntary agreements on management of chemicals and wastes (e.g. Box 19.1).

For example, the Rotterdam Convention on the Prior Informed Consent Procedure bans or severely restricts certain hazardous chemicals and pesticides in international trade, as listed in Annex III of this Convention. Over 150 countries are reported to be parties to this multilateral treaty.

Among members of IOMC, the OECD is an example of a major organization of developed countries that implements SAICM objectives, elements of its plans, and supports member and nonmember countries. For example,

> **Box 19.1 Key Multilateral Treaties on Management of Chemicals and Wastes**
>
> - Montreal Protocol on Substances that Deplete the Ozone Layer (entry into force in 1989)
> - Basel Convention on the Control of Transboundary Movements of Hazardous Wastes and their Disposal (entry into force in 1992)
> - International Labour Organization (ILO) Conventions: C170 – Chemicals Convention (entry into force in 1993) and C174 – Prevention of Major Industrial Accidents Convention (entry into force in 1997)
> - Rotterdam Convention on the Prior Informed Consent Procedure for Certain Hazardous Chemicals and Pesticides in International Trade (entry into force in 2004)
> - Stockholm Convention on Persistent Organic Pollutants (POPs) (entry into force in 2004)
> - World Health Organization (WHO) International Health Regulations (IHR) (2005) (entry into force in 2007)
> - Minamata Convention on Mercury (entry into force in 2017)
>
> Source: Modified from UNEP (2019).

the scope of key risk reduction, knowledge and information, governance, and capacity-building outputs from its chemical program are outlined below (OECD 2020c):

- Risk reduction outputs: sharing experience on policies and tools to assist countries to manage and reduce the risks of chemicals, (e.g. risk management tools and experiences, sustainable chemistry, pesticide risk reduction, and chemical accidents).
- Knowledge and information outputs: providing information, methods, and other tools for testing and assessing chemicals; providing frameworks for work sharing and generating information on the hazards and risks of chemicals [e.g. biocides, new and existing chemicals, (Quantitative) Structure–Activity Relationships project, hazard and exposure assessment methods, pollutant releases and transfer registers, hazard classification and labeling, testing and assessment of nanomaterials, templates, pesticide registration work sharing].
- Governance outputs: assisting countries to implement systems and agreements set out in OECD Council Acts (e.g. Mutual Acceptance of Data [MAD] system, Good Laboratory Practice [GLP], Principles and Test Guidelines, including toxicogenomics, endocrine disrupters, and nonanimal methods).
- Capacity-building outputs: disseminating OECD products and making them more accessible, relevant, and useful for members and nonmembers (e.g. MAD, e-ChemPortal for safety data, food safety cooperation; dissemination of products of all EHS work areas).

National and Regional Management of Hazardous Chemicals and Wastes

As discussed in Section 19.1, SAICM has addressed the need for comprehensive national management of chemicals and wastes by countries, through its six core activity areas, and provided a set of 11 basic elements, considered as critical at national and regional levels to achieve sound chemicals and waste management toward 2020.

These elements are listed below:

1. Legal frameworks that address the life cycle of chemicals and waste.
2. Relevant enforcement and compliance mechanisms.
3. Implementation of chemicals and waste-related multilateral environmental agreements, as well as health, labor, and other relevant conventions and voluntary mechanisms.
4. Strong institutional frameworks and coordination mechanisms among relevant stakeholders.
5. Collection and systems for the transparent sharing of relevant data and information among all relevant stakeholders using a life cycle approach, such as the implementation of the Globally Harmonised System (GHS) of Classification and Labeling of Chemicals.
6. Industry participation and defined responsibility across the life cycle, including cost recovery policies and systems as well as the incorporation of sound chemicals management into corporate policies and practices.
7. Inclusion of the sound management of chemicals and waste in national health, labor, social, environment, and economic budgeting processes and development plans.
8. Chemicals risk assessment and risk reduction through the use of best practices.
9. Strengthened capacity to deal with chemicals accidents, including institutional-strengthening for poison centers.
10. Monitoring and assessing the impacts of chemicals on health and the environment.
11. Development and promotion of environmentally sound and safer alternative.

Minimizing Global Chemical Pollution Beyond 2020

Since 2006, significant gains in implementing the SAICM objectives and plan of action have been made toward achieving the 2020 goal as detailed in the IOMC website (UNEP 2019). Regulation, assessment, and management of industrial and agricultural chemicals have generally advanced well in the OECD countries and various non-OECD countries. Despite this, the IOMC, and the GCO-II report produced by UNEP, recognized that sufficient progress toward the 2020 goal will not be achieved and there is a critical need to understand the various challenges of implementation among all countries, and to revise and advance SAICM objectives and action plans for well beyond 2020.

Looking beyond 2020 has now become more critical because, as the 2019 GBO-II report found, the global goal to minimize adverse impacts of chemicals and waste will not be achieved by 2020. It warned: business as usual will not be an option. Even so, this report recognized that it is possible to achieve this goal in the context of the 2030 Agenda under a sustainability scenario. It concluded that solutions exist, but more ambitious worldwide action by all stakeholders is urgently required (UNEP 2019, p. IX).

As an outcome, UNEP's GCO-II report presents a set of 10 options to implement actions for an international approach to chemicals and waste management by policy- and decision-makers, and stakeholders, to achieve relevant SDGs and targets beyond 2020. These are outlined in Table 19.14. They are designed to meet a sustainability scenario,

> … where legacy problems are addressed and future legacies are avoided, including through green and sustainable chemistry innovation and sustainable consumption and production.

The 10 areas of action are further described in the Annex of the UNEP's Synthesis Report [Part V, Ch. 4] (UNEP 2019).

In the final chapter, we look at the future implications of global pollution and the role of sustainability as a solution.

19.8 Key Points

1. The international and national management of chemicals is a major global challenge for all countries on how to identify, test, evaluate, and manage the environmental and human health risks of new and existing chemicals and their multiple products. Worldwide production, distribution, and consumption of chemicals continue to grow markedly.

2. International goals and targets for the sound and sustainable management of chemicals are developed, coordinated, and promoted worldwide by the United Nations and its agencies, primarily through the Inter-Organisation Programme for the Sound Management of Chemicals (IOMC), established in 1995.

3. The 1992 Earth Summit in Rio de Janeiro introduced fundamental principles for the environment and sustainable development (e.g. the polluter pays principle, the right-to-know, and the precautionary approach), and Agenda 21, a global action plan aimed at sustainable development, which also included actions on safer use of toxic chemicals, managing hazardous wastes, and reducing pollution.

4. In 2002, the WSSD in Johannesburg, South Africa, adopted a goal (WSSD 2020) for chemicals and wastes to achieve "…the sound management of chemicals throughout their life cycle and of hazardous waste in ways that lead to minimization of significant adverse effects on human health and the environment…" by 2020.

5. A SAICM was adopted at the ICCM1 in Dubai (2006). It consisted of an Overarching Policy Strategy and a Global Plan of Action to achieve the 2020 Goal.

6. The objectives of the Policy Strategy focused on (i) risk reduction, (ii) knowledge and information, (iii) governance, (iv) capacity-building and technical cooperation, and (v) illegal international traffic. Six core activities and 11 basic elements were identified to implement international and national measures for chemical and hazardous waste management to minimize significant adverse effects on human health and environment.

7. International efforts to manage chemicals were further advanced in 2015 by the adoption of the 2030 Agenda for Sustainable Development and its 17 SDGs, which included specific goals related to chemicals and waste management.

8. In practice, sound management of chemicals requires evaluation of information and data on the properties of a chemical substance, so that predictions can be made as to how the chemical substance (or mixtures containing it) will behave in natural and human environments, and on its possible effects on the environment and human health.

9. Regulatory authorities for registration, assessment, and management of new and existing chemicals in commercial use increasingly need manufacturers and importers of chemicals to supply details on chemical properties.

Table 19.14 Proposed policy actions by UNEP GCO II to achieve SDG goals and targets for management of chemicals beyond 2020.

Develop effective management systems	Address prevailing capacity gaps across countries, strengthen national and regional legislation using a life cycle approach, and further strengthen institutions and programs.
Assess and manage risks	Refine and share chemical risk assessment and risk management approaches globally, in order to promote safe and sustainable use of chemicals and address emerging issues throughout the life cycle.
Mobilize resources	Scale-up adequate resources and innovative financing for effective legislation, implementation, and enforcement, particularly in developing countries and economies in transition.
Use life cycle approaches	Advance widespread implementation of sustainable supply chain management, full material disclosure, transparency, and sustainable product design.
Assess and communicate hazards	Fill global data and knowledge gaps, and enhance international collaboration to advance chemical hazard assessments, classifications, and communication.
Strengthen corporate governance	Enable and strengthen the chemicals and waste management aspects of corporate sustainability policies, sustainable business models, and reporting.
Educate and innovate	Integrate green and sustainable chemistry in education, research, and innovation policies and programs.
Foster transparency	Empower workers, consumers, and citizens to protect themselves and the environment.
Bring knowledge to decision-makers	Strengthen the science–policy interface and the use of science in monitoring progress, priority-setting (e.g. for emerging issues), and policy-making throughout the life cycle of chemicals and waste.
Enhance global commitment	Establish an ambitious and comprehensive global framework for chemicals and waste beyond 2020, scale-up collaborative action, and track progress.

Source: Modified from UNEP (2019).

10. The scope of information and properties can cover: chemical identification data, including impurities, patterns of production, use, disposal, and release, physicochemical properties, environmental behavior and fate properties, ecotoxicity data, human toxicity data, and applicable analytical methods.
11. Trends for the assessment of chemicals include data from standardized test methods, development of important IT tools and database sources for chemical information on properties, hazards, toxicity, risk assessment and management, and alternative methods to minimize animal testing where existing and comparable toxicity data are available for a chemical from other sources, including *in silico* testing and data sharing.
12. Standardized testing of chemical properties has been promoted by the OECD, US EPA, and EU. Available test guidelines cover physicochemical, environmental fate (e.g. biodegradability and bioaccumulation), ecotoxicity, and human health toxicity requirements for assessment and registration of chemicals.
13. Variations in test guidelines used, such as in toxicity testing, occur between countries but there is a trend toward harmonization of test methods, and also use of nonanimal test methods.
14. Alternative approaches to animal testing include (i) the read-across approach which allows the use of relevant information from analogous substances to predict properties of the target substances, (ii) data sharing where companies are encouraged to share any available data on their substance, and (iii) QSAR methods.
15. Regulation and management of chemicals have advanced since the 1976 TSCA was passed in the United States to regulate both new and existing industrial chemicals, and internationally under the United Nations agencies (e.g. IOMC) and the OECD.
16. A central component of chemical management under national or regional policies and regulations is the setting up of chemical registers and inventories that compile lists of chemical substances that are produced or imported into a country or region.
17. Increasingly, registration, evaluation, and recording of new and existing chemicals are common goals and practices to improve sound chemicals management worldwide.
18. The European community (now the EU) enacted REACH legislation in 2007, described in Case Study 19.1 of this chapter, to advance registration, testing, assessment, and authorization of new and all existing industrial chemicals.
19. Under REACH, producers and importers of chemicals have the burden of proof to demonstrate safety (health and environmental) prior to authorization to market their chemical related products. The ECHA is responsible for registration, evaluation, verification, control,

and management of chemicals, including a central database of chemicals, legislation, and information (e.g. data and guidelines for registrants and data).

Case Study 19.1 EU-REACH is given as a model of current regulatory approaches. It outlines and describes the basic approach, regulatory requirements, registration, and assessment framework. The registration is based on REACH guidance documents/software to help the registrant provide a detailed CSA to demonstrate safe use of the chemical (environment and human health) under exposure scenarios. It is then followed by preparation of a registration dossier (CSR and technical dossier) for submission and evaluation by ECHA.

20. The CSA steps for REACH are outlined in Case Study 19.1 as (i) information requirements/evaluation for chemical, (ii) hazard assessment, (iii) exposure assessment, and (iv) risk characterization. Outcomes for the CSA are included in the CSR.
21. The IOMC, its members, various other intergovernmental (e.g. OECD) and governmental agencies provide a wide range of supporting information on chemicals, policies, risk management, and technical tools to assist countries, and regions, to develop and promote (i) national chemical management programs as promoted by SAICM, and also (ii) multilateral treaties and various voluntary agreements on management of chemicals and wastes (e.g. Box 19.1).
22. Since 2006, significant gains in implementing the SAICM objectives and plan of action have been made toward achieving the 2020 goal, as detailed in the IOMC website, generally for industrial and agricultural chemicals in OECD countries and some non-OECD countries.
23. Despite this, the IOMC, and the GCO-II report produced by UNEP, recognized that sufficient progress toward the 2020 goal will not be achieved and there is a critical need to understand the various challenges of making this happen among all countries, and to revise and advance SAICM objectives and action plans for well beyond 2020.
24. Looking beyond 2020, however, has become more critical because as the 2019 GBO-II report warned: business as usual will not be an option. Even so, this report recognized that it is possible to achieve this goal in the context of the 2030 Agenda under a sustainability scenario. It concluded that solutions exist, but more ambitious worldwide action by all stakeholders is urgently required (UNEP 2019, p. IX).
25. As an outcome, UNEP's GCO-II report presents a set of 10 options to implement actions for an international approach to chemicals and waste management by policy- and decision-makers, and stakeholders, to reach relevant SDGs and targets, designed to meet a sustainability scenario, "… where legacy problems are addressed and future legacies are avoided, including through green and sustainable chemistry innovation and sustainable consumption and production."

References

Chemsafetypro (2019a). Overview of Chemical Regulations in China https://www.chemsafetypro.com/Topics/China/Overview_of_Chemical_Regulations_in_China.html (accessed 7 March 2020).

Chemsafetypro (2019b). How to Use ECOSAR to Predict Aquatic Toxicity. https://www.chemsafetypro.com/Topics/CRA/How_to_Use_US_EPA_ECOSAR_to_Predict_Aquatic_Toxicity.html (accessed 5 March 2020).

Chemsafetypro (2019c). How to Use US EPA EPI Suite to Predict Chemical Substance Properties. https://www.chemsafetypro.com/Topics/CRA/How_to_Use_US_EPA_EPI_Suite_to_Predict_Chemical_Substance_Properties.html (accessed 5 March 2020).

Chemsafetypro (2019d). How to Calculate Predicted No-Effect Concentration (PNEC). https://www.chemsafetypro.com/Topics/CRA/How_to_Calculate_Predicted_No-Effect_Concentration_(PNEC).html (accessed 7 March 2020).

Chemsafetypro (2019e). How to Derive Derived No-Effect Level (DNEL). https://www.chemsafetypro.com/Topics/CRA/How_to_Derive_Derived_No-Effect_Level_(DNEL).html (accessed 17 October 2019).

Chemsafetypro India (2020). Overview of Chemical Regulations in India and Latest Developments. https://www.chemsafetypro.com/Topics/India/Overview_of_Chemical_Regulations_in_India.html (accessed 5 September 2020).

Connell, D.W. and Miller, G.J. (1984). *Chemistry and Ecotoxicology of Pollution*. New York: Wiley.

ECHA (2010). DNEL/DMEL Derivation from Human Data. https://echa.europa.eu/documents/10162/23047722/draft_r8_dnel_hd_20100211_en.pdf/a407f1ba-6817-4b1b-a7b4-aeaa1fe965fe (accessed 18 October 2019).

ECHA (2016). *Guide – How to Use and Report (Q)SARs 3.1. European Chemicals Agency*. Helsinki, Finland: European

Union https://echa.europa.eu/documents/10162/13655/pg_report_qsars_en.pdf (accessed 5 March 2020).

ECHA (2020a). *Understanding REACH*. European Chemicals Agency, European Union https://echa.europa.eu/regulations/reach/understanding-reach (accessed 4 March 2020).

ECHA (2020b). *Animal Testing Under REACH*. European Chemicals Agency, European Union https://echa.europa.eu/animal-testing-under-reach (accessed 7 March 2020).

ECHA (2020c). European Chemicals Agency. https://echa.europa.eu (accessed 7 March 2020).

ECHA (2020d). *Guidance on Information Requirements and Chemical Safety Assessment*. European Chemicals Agency, European Union https://echa.europa.eu/guidance-documents/guidance-on-information-requirements-and-chemical-safety-assessment (accessed 7 March 2020).

ECHA (2020e). *Chemical Safety Report*. European Chemicals Agency, European Union https://echa.europa.eu/regulations/reach/registration/information-requirements/chemical-safety-report (accessed 5 March 2020).

European Commission (2019). Environment-REACH. https://ec.europa.eu/environment/chemicals/reach/reach_en.htm (accessed 22 September 2019).

Luo, W. (2019). *China Proposes Regulatory Overhaul Targeting Environmental Risks from Chemicals*. Beveridge & Diamond https://www.bdlaw.com/publications/china-proposes-regulatory-overhaul-targeting-environmental-risks-from-chemicals/ (accessed 18 January 2019).

OECD (2020a). *OECD Test Guidelines for the Chemicals*. Organisation for Economic Co-operation and Development www.oecd.org/chemicalsafety/testing/oecdguidelinesforthetestingofchemicals.htm (accessed 4 March 2019).

OECD (2020b). *OECD and EU Test Guidelines*. Organisation for Economic Co-operation and Development https://echa.europa.eu/support/oecd-eu-teguidelines (accessed 4 March 2020).

OECD (2020c). *Chemical safety and biosafety*. Organisation for Economic Co-operation and Development http://www.oecd.org/chemicalsafety/ (accessed 7 March 2020).

SAICM (2020). *Strategic Approach to International Chemicals Management Website*. United Nations Environment Programme http://www.saicm.org (accessed 3 March 2020).

Silbergeld, E.K., Mandrioli, D., and Cranor, C.F. (2015). Science and the Unbearable Burdens of Regulation. *Annual Review of Public Health* 36: 175–191.

UNEP (2013). *Global Chemicals Outlook – Towards Sound Management of Chemicals*. Geneva: United Nations Environment Programme.

UNEP (2019). *Global Chemicals Outlook II. From Legacies to Innovative Solutions: Implementing the 2030 Agenda for Sustainable Development – Synthesis Report*. Geneva: United Nations Environment Programme, UNEP.

UNEP (2020). *Why Do Chemicals and Wastes Matter?* United Nations Environment Programme https://www.unenvironment.org/explore-topics/chemicals-waste/why-do-chemicals-and-waste-matter (accessed 4 February 2020).

US EPA (2019). *Summary of the Toxic Substances Control Act*. United States Environmental Protection Agency https://www.epa.gov/laws-regulations/summary-toxic-substances-control-act (accessed 1 October 2019).

WHO (2018). *National Chemicals Registers and Inventories: Benefits and Approaches to Development*. World Health Organization, Regional Office for Europe: Copenhagen. https://www.euro.who.int/__data/assets/pdf_file/0018/361701/9789289052948-eng.pdf?ua=1 (accessed 5 March 2020).

20

Pollution: Moving Toward a Healthy and Sustainable Future

20.1 Introduction

Pollution is a massive, overlooked cause of disease, death and environmental degradation
(The Lancet Commission on Pollution and Health, Landrigan et al. 2017).

As described by The Lancet Commission, pollution endangers the health of billions of persons, degrades the Earth's ecosystems, undermines the economic security of nations, and is responsible for an enormous burden of disease, disability, and premature death (Landrigan et al. 2017, p. 4).

In this final chapter, we elucidate the role of pollution in environmental changes that are having impact on our ecological and human health. The observed evidence, scale of events, multiple effects on living organisms of exposures to toxic substances, and increasing rates of changes due to pollution have been underestimated, despite our existing knowledge of serious impacts and growing threats to our future planetary health.[1] These pollution-related threats already include accelerated global warming and climate changes, unprecedented in human history. A crucial and urgent challenge for society is how to achieve sustainable solutions for pollution problems that will protect our future planetary health.

Our book has focused on a scientific understanding of key principles of pollution and assessment of toxic and indirect effects of pollutants on ecosystems and human health. Yet, it is essential in this Chapter to introduce the role of the sustainability concepts, principles, and systematic approaches that can solve pollution and other environmental problems. An integral focus of these science-based solutions is to prevent and minimize pollution and to mitigate global warming to help achieve sustainable development goals and targets for planetary health[1] (2030 Agenda of UN).

[1] Planetary health – see Chapter 3 or Section 20.2.

20.2 Sustainability

Concepts and Principles of Sustainability

Sustainability is widely seen from a systems perspective. It can be briefly described as the capability of a system to endure and maintain itself. It is generally used in a multidisciplinary sense, which essentially depends on the context of its use (e.g. ecological, economic, or behavioral). In ecology, for example, the word sustainability commonly refers to the ability of biological systems to remain healthy, maintain biodiversity, and be productive over a long period of time. In this sense, natural ecosystems can act as models for sustainable living systems. For future human development, the systems approach is readily applicable to the design, analysis, and evaluation of sustainability in practice (see subsection on Sustainable Systems and Processes below).

There are many definitions of sustainability, mainly focused on the capacity of human behavior to sustain the Earth. Among these, Bender and co-authors (Bender et al. 2012, chapter 14) considered key concepts underpinning the use of sustainability from several perspectives, including complexity, space, time, and societal values. These authors derived a set of nine working definitions for sustainability. The following key definitions indicate the scope considered, from an individual bounded system to the global system, and one with an ethical perspective, as a guide to human behavior for living in a sustainable way:

Sustainability is the quality a system has if its relationship with its surrounding environment is able to continue over a specified area and time frame (Definition 7).

Sustainability is the quality the global system has if the relationships between and within its subsystems are able to persist and nourish each other (Definition 8).

Sustainability is ethical human behavior that is aware and nourishing of every interaction, therefore

Chemistry and Toxicology of Pollution: Ecological and Human Health, Second Edition. Des W. Connell and Greg J. Miller.
© 2023 John Wiley & Sons, Inc. Published 2023 by John Wiley & Sons, Inc.
Companion website: www.wiley.com/go/toxicologyofpollution2e

contributing to the persistence of the global environment (Definition 9).

Another definition that relates sustainability to the decisions that society makes about its health and long-term survival is given in the Editorial of the Planetary Health Journal of The Lancet:

> …it is useful to think of civilisational health in terms of a more familiar and somewhat less grandiose term: sustainability. Analogous to human health, but on a much greater scale, sustainability is the ability of a society to make choices that are beneficial to its long-term survival
> (The Lancet Planetary Health Journal, Editorial 2019).

Principles of sustainability are not so easily defined because its meaning is often interchangeable with the concept of sustainable development and related principles that emerged in the 1980s. The economist Herman Daly had a major impact on the worldview of sustainability by proposing a set of principles or indicators that are essential for a sustainable human society. They focus on the need for sustainable rates of resource use and replenishment, and also the rate at which pollution and wastes can be absorbed or removed by natural systems. They are also known as Herman Daly's Three Rules. These have been expressed essentially as:

1. Sustainable use of renewable resources means that consumption of these resources should not be greater than the rate at which resources regenerate.
2. The sustainable use of nonrenewable resources requires that the rate of consumption is not greater than the pace at which renewable substitutes can be put into place.
3. The sustainable pace of pollution and wastes requires that production not be greater than the pace at which natural systems can absorb, recycle, or neutralize them.

From Sustainability to Sustainable Development

In practice, the concept of sustainability is transformed into a decision-making process designed to achieve sustainable development of human civilization, that is based on the sustainable capacity of the Earth's natural resources (materials and energy) and its natural living systems to meet the needs of human populations, their healthy lifestyles, and use of sustainable technologies to remove pollution and wastes from human activities. The process assumes human values such as ethics, equity, efficiency, and effectiveness.

A conceptual framework for the transition of sustainability into sustainable development is given in Figure 20.1. An overview of this framework is introduced in the first

Sustainability
- Concepts and principles

Sustainable development
Sustainable development goals

Sustainability systems and processes

Sustainable design
- Green chemistry
- Green engineering

⇔ Metrics and assessment

Sustainable technology
- Materials
- Renewable energy
- Processes and products
- Waste recycling

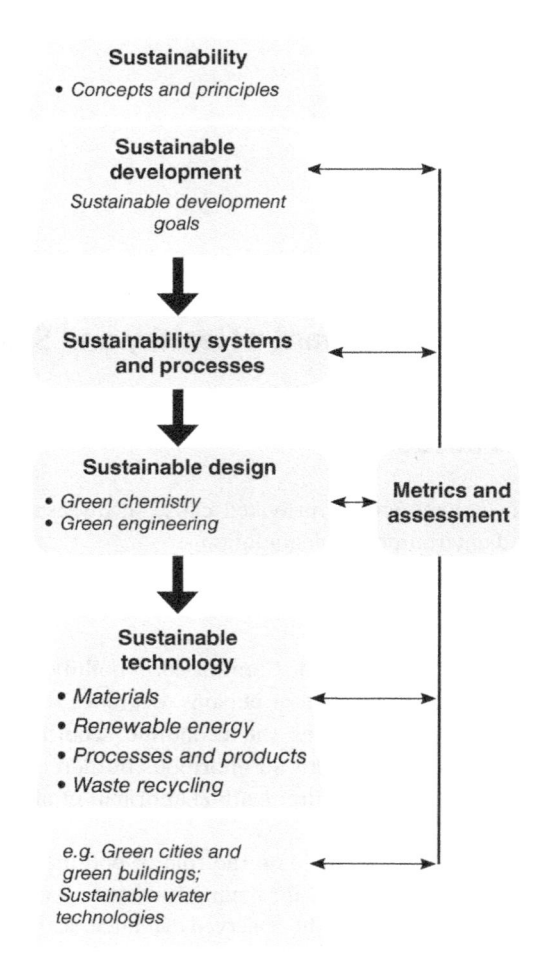

e.g. Green cities and green buildings; Sustainable water technologies

Figure 20.1 Overview of sustainability from a systems and process perspective that can be applied to sustainable development, design, and technologies to achieve sustainable use of resources, processes, and products at local to global scales. Source: World Commission on Environment and Development (1987).

part of this chapter to allow a final assessment of (i) the overall state of environmental change and human health effects due to pollution, such as global warming from greenhouse gases, and exposures to toxic chemicals, and (ii) sustainable approaches and actions to help urgently solve the causes and effects of pollution on planetary health. These approaches take into account many scientific principles and findings related to pollution and types of pollutants identified in the previous chapters of this book, and those of sustainability, from a variety of sources, such as The Lancet Commission on Pollution and Health and the Rockefeller Foundation – The Lancet Commission on Planetary Health.

The most common definition of sustainable development arises from the Brundtland Commission of the United Nations (20 March 1987), as published in the book

Our Common Future (World Commission on Environment and Development 1987).

> Sustainable development is development that meets the needs of the present without compromising the ability of future generations to meet their own needs
> (World Commission on Environment and Development 1987).

Principles of sustainable development are widely discussed in the literature and continue to evolve (e.g. equity and equality). Some common principles address the following aspects:

- Conservation of our ecosystem
- Maintaining ecosystem services
- Conservation of biodiversity
- Protecting human resources
- Development of society
- Conservation of cultural heritage
- Living within the capacity of Earth

Generally, sustainable development looks at how to meet human development goals within an interactive framework of sustaining the capacity of natural systems, their resources, and the ecosystem services that support human society, its economy, human health and well-being.

A refined set of 17 sustainable development goals (2030 Agenda SDGs), as approved under the UN, is also defined, as discussed in Chapter 19.

Elements of Sustainability

The definition of sustainable development originated in the 1980s based on three dimensions or pillars: economic growth, social inclusion, and environmental balance.

The 2005 World Summit on Social Development identified sustainable development goals, such as economic development, social development, and environmental protection.

Three key dimensions or pillars of sustainability are commonly referred to as:

- Environmental sustainability
- Social sustainability
- Economic sustainability

These three elements are seen as interdependent. As such, a basic model of sustainability can be shown by using three overlapping ellipses to indicate the three dimensions. Recently, assessment of cultural sustainability, such as in large cities, has been adopted and promoted as a fourth dimension of sustainability, within the scope of some UN-related agencies. Figure 20.2 is a basic conceptual model of sustainability showing the four dimensions or *pillars* of sustainability, which need to interact to achieve sustainable development.

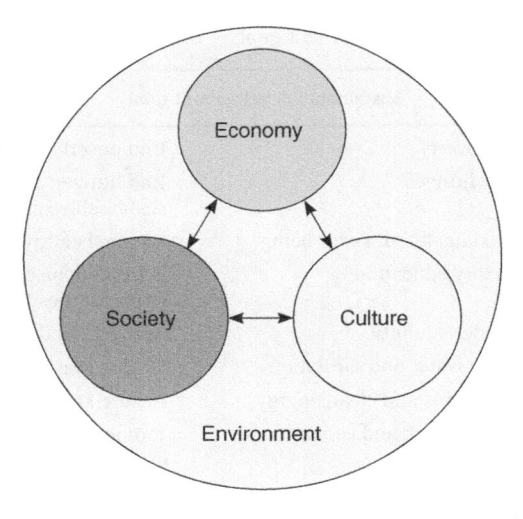

Figure 20.2 Conceptual model of sustainability showing interdependence of four dimensions of sustainability.

Sustainable Development Goals

The United Nations established a set of sustainable development goals known as Agenda 21. The sustainable development goals (SDG Agenda 2030) cover a broad spectrum of environmental goals such as clean energy and climate action, as well as goals for economic growth, hunger, poverty, health, education, equality, peace, and justice (Table 20.1).

The current SDG Agenda 2030, moving forwards from Agenda 21, includes 17 Sustainable Development Goals (SDGs) and 169 targets, as adopted on 25 September 2015 by Heads of State and Government at a special UN summit. The 2030 Agenda integrates what is seen as three dimensions of sustainable development – economic, social, and environmental – in a balanced way (UN 2015).

Sustainable Systems and Processes

The use of a systems approach is generally seen as central to the development of sustainability within modern society.

> The systems approach provides a simple and consistent basis for investigating sustainability at all levels of society, from the global scale down to the individual
> (Dandy et al. 2008, p. 226).

The challenge for sustainable development is how to translate from linear-based systems (e.g. exploitation of nonrenewable resources, production, consumption,

Table 20.1 Sustainable development goals – United Nations Agenda 2030.

SDG no.	Sustainable development goal	
1	No poverty	End poverty in all its forms everywhere
2	Zero hunger	End hunger, achieve food security and improved food nutrition and promote sustainable agriculture
3	Good health and well-being	Ensure healthy lives and promote well-being for all at all ages
4	Quality education	Ensure inclusive and equitable quality education and promote lifelong learning opportunities for all
5	Gender equality	Achieve gender equality and empower all women and girls
6	Clean water and sanitation	Ensure availability and sustainable management of water and sanitation for all
7	Affordable and clean energy	Ensure access to affordable, reliable, sustainable, and modern energy for all
8	Decent work and economic growth	Promote sustained, inclusive, and sustainable economic growth, full and productive employment and decent work for all
9	Industry, innovation, and infrastructure	Build resilient infrastructure, promote inclusive and sustainable industrialization, and foster innovation
10	Reduced inequalities	Reduce inequality within and among countries
11	Sustainable cities and communities	Make cities and human settlements inclusive, safe, resilient, and sustainable
12	Responsible consumption	Ensure sustainable consumption and production patterns
13	Climate action	Take urgent action to combat climate change and its impacts
14	Life below water	Conserve and sustainably use the oceans, seas, and marine resources for sustainable development
15	Life on land	Protect, restore, and promote sustainable use of terrestrial ecosystems, sustainably manage forests, combat desertification, and halt and reverse land degradation, and halt biodiversity loss
16	Peace and justice	Promote peaceful and inclusive societies for sustainable development, provide access to justice for all and build effective, accountable and inclusive institutions at all levels
17	Partnerships for the goals	Strengthen the means of implementation and revitalize the Global Partnership for Sustainable Development

Source: Based on UN (2015).

wastes, and their disposal) to more complex and nonlinear systems of ecology and interactions with human society such as economic models, information systems, engineering processes, and value-based decision-making processes.

A conceptual approach by Ashby (2013, pp. 320–321) to sustainable development for a society, as illustrated in Figure 20.3, uses a nested plot of ellipses expanding in time scales, from the life of a product to the lifetime of a civilization, and spatial scales, ranging from a product system to industrial systems, and then a social system as a whole. For each system, at different conceptual scales of time and space, an ellipse represents an approach to thinking about the degree of interaction between industrialization and the environment, as indicated in Figure 20.3.

- Product system – intervention is used to limit and control pollution (and wastes) during the lifetime of the product. Increasingly trends are toward quality assurance systems and product stewardship.
- Environmental design of products and processes – sustainable techniques are used for the entire design process to optimize efficiency, safety, minimize potential pollution emissions, wastes, and environmental impacts, and meet environmental quality practices.
- Industrial ecology – studies the interactions between industrial systems and natural ecosystems or the biosphere to understand how biophysical and ecological systems can be used to create, design, model, and build sustainable industrial systems that optimize their use of natural resources and minimize the impacts of their processes, products, and wastes on the environment and society, within an equitable and ethical framework.
- Sustainable development – concerned from a society and future civilization perspective about how natural systems and human development can be managed and balanced to sustain natural resources, human well-being, and the broader concept of planetary health.

Industrial ecology has emerged as part of sustainable development as a multi-discipline systems approach to design sustainable industrial processes, and create industrial systems that mimic material and energy flows

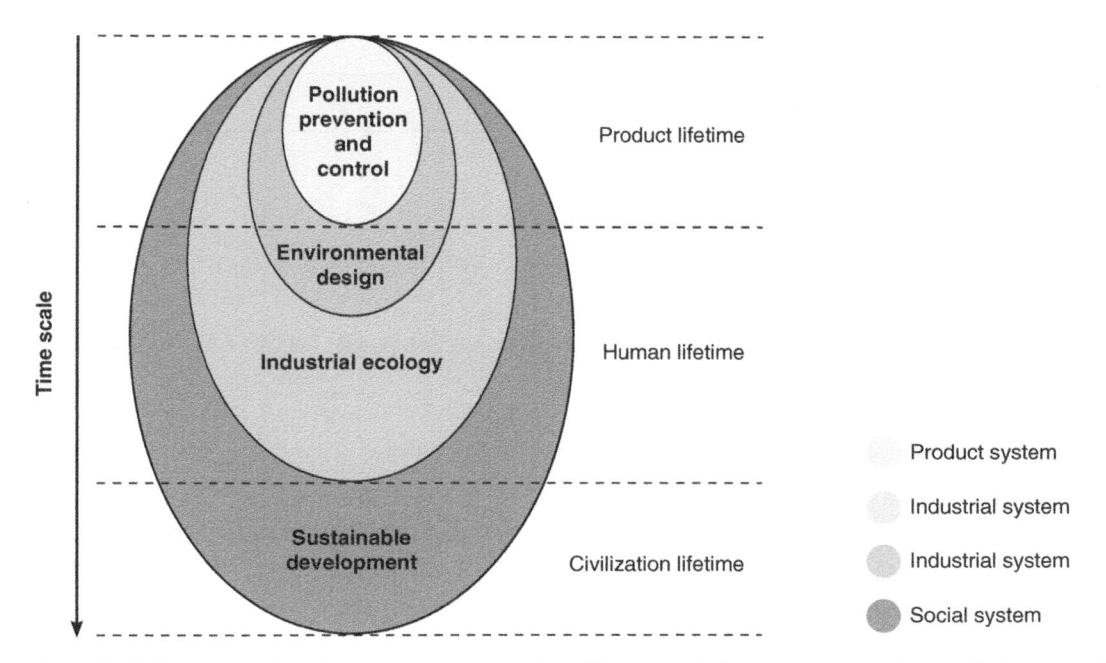

Figure 20.3 Conceptual-based systems approach, using different spatial and temporal scales, to think about relationship between industrialization of society and natural ecosystems in the context of sustainable development. Source: Modified from Ashby (2013).

through natural systems, including minimizing any wastes (Ashby 2013, pp. 320–321).

Sustainability and Design

Sustainable design is a fundamental approach to sustainable development (e.g. conservation of ecosystem services, processes and systems, materials, products, planning, and buildings) that acts to reduce negative environmental and health impacts on natural systems and humans. It focuses on (i) basic objectives of sustainability such as to reduce consumption of nonrenewable resources, minimize waste, and to create healthy and productive environments, and (ii) how human society (e.g. economy, technology, and culture) can adapt to concepts of sustainable and healthy environments, at local to global scales.

Many different creative, theoretical, and practical techniques are applied, for example, from biomimicry[2], designing for reuse, recycling, and energy efficiency, to using green chemistry and green engineering principles in developing low-impact and nontoxic chemicals, materials, or products. Life cycle assessment for materials and products is increasingly used in the design process to provide a complete evaluation of their impacts from the sources of their raw materials or inputs, transport, processing, refining, manufacturing, maintenance, use, reuse, recycle, and any disposal, as well as energy sources and uses.

---
2 **Biomimicry** is learning from and then imitating nature's forms, processes, and ecosystems to create more sustainable designs.

Green Chemistry (Sustainability Chemistry)

In the early 1990s, the term **green chemistry** was introduced by the United States Environmental Protection Agency as part of a measure to prevent growing environmental pollution from chemical production and use of products. It introduced a set of 12 principles to guide the chemical industry. Table 20.2 presents a set of these principles attributed to Paul Anastas and John Warner in 1998. Green chemistry is defined by the US EPA as "the design of chemical products and processes that reduce or eliminate the use or generation of hazardous substances. Green chemistry applies across the life cycle of a chemical product, including its design, manufacture, use, and ultimate disposal. Green chemistry is also known as sustainable chemistry" (US EPA 2017a).

Figure 20.4 shows how these principles refer to five main categories: reaction efficiency, safety and control, use of renewable resources, design and production of low toxicity substances, and waste prevention.

Some examples of green chemistry technologies are given in Table 20.5. Specific explanations and examples for various green chemistry principles are available in the website: The Essential Chemical Industry-online (2018) produced by the Centre for Industry Education Collaboration, in the Department of Chemistry, University of York in York, England.

In 2003, Anastas and Zimmerman (2003) published a set of 12 green engineering principles, which are listed in Table 20.3.

Table 20.2 Principles of green chemistry.

The 12 principles of green chemistry
1 **Prevention** It is better to prevent waste than to treat or clean up waste after it is formed.
2 **Atom economy** Synthetic methods should be designed to maximize the incorporation of all materials used in the process or product.
3 **Less hazardous chemical syntheses** Wherever practicable synthetic methods should be designed to use and generate substances that possess little or no toxicity to human health and the environment.
4 **Designing safer chemicals** Chemical products should be designed to preserve efficacy of function while reducing their toxicity.
5 **Safer solvents and auxiliaries** The use of auxiliary substances (e.g. solvents, separation agents, etc.) should be made unnecessary whenever possible and innocuous when used.
6 **Design for energy efficiency** Energy requirements should be recognized for their environmental and economic impacts and should be minimized. Synthetic methods should be conducted at ambient temperature and pressure.
7 **Use of renewable feedstocks** A raw material feedstock should be renewable rather than depleting whenever technically and economically practical.
8 **Reduce derivatives** Unnecessary derivatization (blocking group, protection/deprotection, temporary modification of physical/chemical processes) should be avoided whenever possible.
9 **Catalysis** Catalytic reagents (as selective as possible) are superior to stoichiometric reagents.
10 **Design for degradation** Chemical products should be designed so that at the end of their function they do not persist in the environment and instead break down into innocuous degradation products.
11 **Real-time analysis for pollution prevention** Analytical methodologies need to be further developed to allow for real-time, in-process monitoring and control prior to the formation of hazardous substances.
12 **Inherently safer chemistry for accident prevention** Substances and the form of substances used in a chemical process should be chosen so as to minimize the potential for chemical accidents, including releases, explosions, and fires.

Source: Anastas and Warner (1998).

Green Engineering

Green engineering focuses on product design, materials, processes and systems, and energy efficiency. Its four basic goals are to achieve waste reduction, materials management, pollution prevention, and product enhancement. These principles also complement those of green chemistry.

Sustainability Principles and Metrics

Sustainability is measured by assessing performance of social, environmental, and economic principles such as the triple bottom line. However, specific measures or indicators of sustainability, also known as metrics, are commonly used to evaluate performance. Table 20.4 lists some key principles and metrics used in sustainability.

Strategies and Technologies for Sustainable Pollution Prevention and Reduction

Several important and diverse examples of sustainable strategies and technologies are described below, which vary from life cycle assessment of products, and green

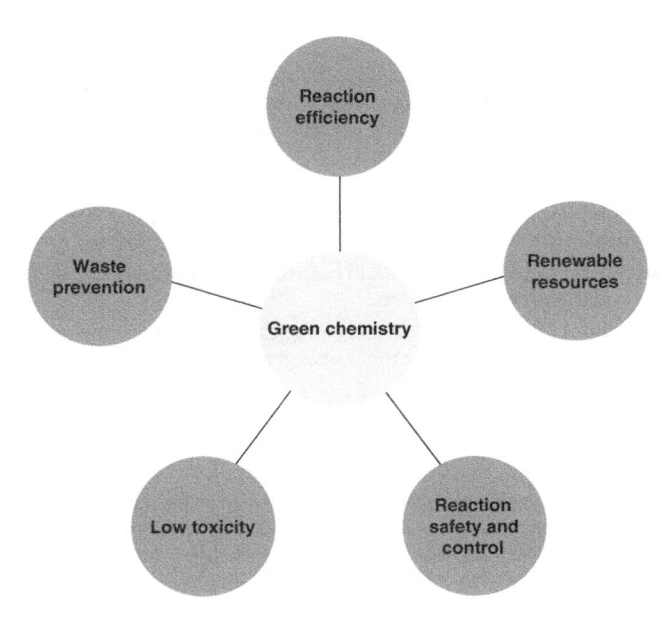

Figure 20.4 Green chemistry principles clustered in five categories. Source: Modified from Whitmee et al. (2015).

chemistry technologies, to development of green cities and green buildings.

Some examples of green chemistry technologies used to prevent or reduce environmental impacts or reduce harmful properties of chemicals are given in Table 20.5.

In human environments facing the negative impacts of urbanization, the design and development of green buildings, healthy and green cities are advancing in many cities worldwide, with the support of various sustainable city programs (e.g. UNEP and C40 Cities) and through green building certification and assessment programs (e.g. US Green Building Council's LEED, DGNB in Germany, CASBEE in Japan, LEED Canada, and Green Star in Australia).

Green Cities and Buildings

The need for sustainable and healthy built environments is highlighted by the World Green Building Council in their 2020–2022 Strategy. Buildings are responsible for 39% of global carbon emissions and 50% of global material use. From a health perspective, 91% of people live where air pollution levels exceed WHO limits. People are also more likely to have asthma due to living in a home with damp or mold.

In the Twenty-first Century, sustainable planning for new cities and urban renewable programs is being adopted in many countries. There is growing trend toward integrating green, healthy, and smart cities and communities using IoT sensors, devices, and communication technologies connected to the Internet. IoT applications increasingly include environmental pollution control, energy, climate, and water management systems on scales from green and

Table 20.3 Principles of green engineering.

The 12 principles of green engineering	
1. Inherent rather than circumstantial	Designers need to strive to ensure that all material and energy inputs and outputs are as inherently nonhazardous as possible
2. Prevention instead of treatment	It is better to prevent waste rather than to treat or clean-up waste after it is formed
3. Design for separation	Separation and purification operations should be designed to minimize energy consumption and materials use
4. Maximize efficiency	Products, processes, and systems should be designed to maximize mass, energy, space, and time efficiency
5. Output-pulled versus input pushed	Products, processes and systems should be *output pulled* rather than *input pushed* through the use of energy and materials
6. Conserve complexity	Embedded entropy and complexity must be viewed as an investment when making design choices on recycle, reuse, or beneficial disposition
7. Durability rather than immortality	Targeted durability; not immortality, should be a design goal
8. Meet need, minimize excess	Design for unnecessary capacity and capability (e.g. *one size fits all*) solutions should be seen as a design flaw
9. Minimize material diversity	Material diversity in multicomponent products should be minimized to promote disassembly and value retention
10. Integrate local material and energy flows	Design of products, processes, and systems must include integration and interconnectivity with available energy and material flows
11. Design for commercial *afterlife*	Products, processes, and systems should be designed for performance in a commercial *afterlife*
12. Renewable rather than depleting	Material and energy inputs should be renewable rather than depleting

Source: Adapted from Anastas and Zimmerman (2003) and US EPA (2017b).

Table 20.4 Some common sustainability principles and metrics.

Tragedy of the commons	An economic theory that applies to a situation in a common or shared resource system (e.g. air, land, water, or forest) where individual users act independently according to their own self-interest, resulting in degradation or loss of that resource through their collective action.
Precautionary principle	It considers potential risks in the process of policy-making. For example, it requires policy makers to not only consider the environmental risk of emissions from an industrial process but also take preventive measures to eliminate or reduce the risk to acceptable levels for human health or the environment.
Polluter pay's principle	Commonly accepted practice that those who produce pollution should bear the costs of preventing, managing, or causing damage to human health or the environment.
Ecological footprint	Used to measure the impact of human activities in terms of the area of biologically productive land and water required to produce the goods consumed and to assimilate the wastes generated (usually as global hectares per person, gha/person).
Carbon footprint	The amount of greenhouse gases (primarily carbon dioxide) released into the atmosphere by a particular human activity, process, product, etc. It is usually measured as tonnes of CO_2 equivalent gases emitted per year.
Water footprint	It measures the amount of water used to produce single or multiple goods and services for human use, or consumed by individuals, countries, and globally, similar to carbon footprints. The water footprint measures both direct and indirect water use of a process, product, company or sector and includes water consumption and pollution throughout the full production cycle from the supply chain to the end-user (see Water Footprint Network website).
Life cycle assessment	Life cycle assessment is the complete assessment of materials from their extraction, transport, processing, refining, manufacturing, maintenance, use, disposal, reuse, and recycle stages (see Ashby 2013).
Eco-audit	Eco-audit is a tool to find the environmental impact of a product across all life cycle stages and to identify the problems in all aspects of a supply chain, from extraction of raw materials to manufacturing, distribution, use, and disposal. The purpose of an analysis of a product is to establish the embodied energy, water usage, annual CO_2 to atmosphere, carbon footprint, recycle fraction in current supply, toxicity, approximate processing energy and sustainability criteria. Knowledge to guide design decisions is needed to minimize or eliminate adverse eco-impacts (see Ashby 2013).
Zero waste	The conservation of all resources by means of responsible production, consumption, reuse, and recovery of products, packaging, and materials without burning and with no discharges to land, water, or air that threaten the environment or human health (see Zero Waste International Alliance website).
External costs	These costs (also known as externalities) are important in sustainability. They refer to the economic concept of uncompensated social or environmental effects. For example, when people buy fuel for a car, they pay for the production of that fuel (an internal cost), but not for the costs of burning that fuel, such as air pollution and chronic human health effects among exposed persons. Externalities: A benefit or cost that affects an individual or group of people who did not choose to incur that benefit or cost (Whitmee et al. 2015).
Pollution offsets	The attempt to offset the results of pollution from some activity of process by improving the environment in an equal benefit. Carbon trading, for example, allows carbon polluters to offset the effect of excess carbon in the environment by trading credits with those whose activities reduce an equal amount of carbon. Pollution offsets can exist for any kind of polluting materials as long as an equal and direct benefit can be established.

Source: Adapted from Whitmee et al. (2015), Ashby (2013) and Baird and Cann (2012).

smart buildings to urban, commercial, industrial, and agriculture activities. They offer a pathway to interact with and develop sustainability in complex systems to achieve sustainable development goals at a faster rate.

Green building principles, design, technologies, and practices are being implemented widely, mainly in developed countries. Sustainable designs are often characterized by innovation, energy efficiency, multiple functions, and trends toward healthy environments and lifestyles.

Common principles of green building practices are generally applied to each phase of building development over the life cycle of green buildings, from sustainable design for purpose, construction, commissioning, operation, maintenance, refurbishment, and decommissioning.

These involve:

- Sustainable siting and design (e.g. green building codes and ISO Standards)
- Energy efficiency and reduced carbon emissions (e.g. renewable energy)
- Water efficiency (including wastewater reuse)
- Sustainable and nonhazardous building materials

Table 20.5 Some examples of green chemistry technologies.

Prevention/reduction of pollution	Examples of green chemistry technologies
Biodegradable chemicals	• Detergents based on sodium salts of alkylbenzene sulfonates were poorly biodegraded during sewage treatment and naturally in receiving waters because the alkyl group was branched. These compounds have been replaced with sodium salts of linear alkylbenzene sulfonic acids, which are readily degraded.
Replacement of harmful chemicals	• Detergents contain compounds known as builders to remove magnesium and calcium ions from hard water. Sodium phosphates were used for this purpose, but these caused considerable problems leading to eutrophication of water bodies. Zeolites (aluminosilicates) are now used to replace soluble phosphates. • Supercritical (liquid) carbon dioxide is widely used as a solvent to replace toxic chlorinated solvents such as perchloroethylene in dry cleaning of clothes. • Waterborne paints are replacing paints that use volatile organic compounds such as aromatic and other hydrocarbons which are harmful to the atmosphere and humans.
Energy efficiency	• Many organic wastes from processing/manufacture can be converted into biofuels (e.g. plant and vegetable oil wastes), liquid fuels (e.g. solvent wastes) shredded packaging, furniture, wood chips, or solid fuels to avoid landfill or stockpiling. • Waste to energy conversion, however, needs to ensure compliance with emission standards.
Catalysts to increase energy efficiency	• Catalysts are being developed so that a process can be run at lower temperatures and pressures to reduce energy use at high temperatures and pressures (e.g. the use of molecular sieves means that processes such as the purification of ethanol can occur at ambient temperatures instead of by distillation).
Reaction efficiency	• Calculation of yield and atom economy for chemical reactions/processes to determine the efficiency of the reaction.

Source: Based on The Essential Chemical Industry-online (2018).

- Healthy indoor environmental quality (including ventilation, thermal comfort, lighting, noise control, moisture control, and air quality)
- Integration of sustainable waste management strategies and practices

Integrated Urban Water Management

Integrated urban water management (IUWM) is another major strategy that is fundamental for urban development and basin management to achieve sustainable economic, social, and environmental goals. It combines water supply, sanitation, stormwater, and wastewater management (e.g. non-potable reuse) with land use planning and economic development. Benefits include enhancing potable and non-potable water supply and its security, and reducing pollution impacts from wastewaters and urban stormwaters, and maintaining aquatic biodiversity and wildlife habitats. There are many case studies. An example of an IUWM planning manual, which refers to six case studies from North America and Australia, is provided by CSIRO/Water Research Foundation (2010), including six concise case studies from North America and Australia.

Sustainable strategies, better designs, and new technologies for managing water catchments are advancing worldwide. These are needed to confront the challenge for managing water resources from increasing water demand, loss of access, and pollution effects from population growth, urbanization, and other human activities, combined with current climate change impacts and those projected for this Century.

20.3 Pollution and Planetary Health

The scientific effects and risks from pollution on planetary health and their implications for the future are at the heart of this chapter. From Chapter 3 and Whitmee et al. (2015), planetary health is described basically as "the health of human civilization and the state of the natural systems on which it depends."

This section examines the impact of pollution on the state and health of the two fundamental elements (human civilization and natural systems) that interact to form the concept of planetary health. Critically, it is the available capacity of natural systems to supply essential ecosystem services (see Figure 20.5) that underpins future sustainability of human development and society.

In the Twenty-first Century, pollution and other forms of human disruption of essential ecosystem services are resulting in significant to unsustainable impacts and risks to planetary health.

A situational assessment is provided here under the subsections on Pollution and Environmental Change and Pollution and Human Health, taking into account the key

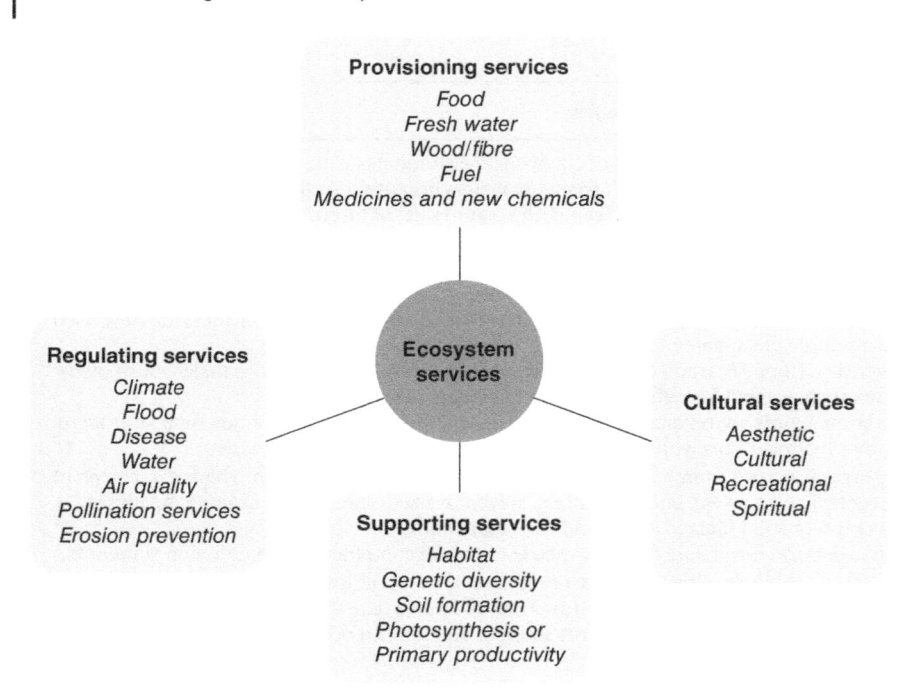

Figure 20.5 Ecosystems services provided by natural systems. Source: Modified from Whitmee et al. (2015).

Provisioning services
Food
Fresh water
Wood/fibre
Fuel
Medicines and new chemicals

Regulating services
Climate
Flood
Disease
Water
Air quality
Pollination services
Erosion prevention

Ecosystem services

Cultural services
Aesthetic
Cultural
Recreational
Spiritual

Supporting services
Habitat
Genetic diversity
Soil formation
Photosynthesis or
Primary productivity

findings of the Lancet Commissions on Planetary Health (Whitmee et al. 2015); Pollution and Health (Landrigan et al. 2017), and the evaluation of pollutants described in chapters of this book.

To address the challenges of environmental and health outcomes from human-derived pollution, Section 20.4 presents an overview of current and proposed options for sustainable solutions to help urgently resolve the impacts of pollution on planetary health as we move toward 2030 and beyond.

Pollution and Environmental Change

Pollution from all sources is inextricably linked with the scale of environmental impacts from human activities on the Earth's biosphere and its ecosystem services. It plays a major role in the disruption and degradation of the biophysical processes and ecosystems, from local to global scales. Increasing emissions of greenhouse gases and black carbon particles, primarily from global combustion of fossil fuels, are highly probable causes of recent observations of accelerated global warming and climate changes. The nature of the scientific mechanisms associated with GHGs and global warming is well known (see Chapter 14). Pathways for future trends in global heating and temperature increase depend on the strength and rate of human responses to mitigation of emissions (see IPCC scenarios). At the same time, there is an urgent global challenge for human adaptation to meet prevailing and future pollution, environmental, and societal changes and impacts.

The Rockefeller – Lancet Commission on Planetary Health described the extent of human alteration of natural systems as (Whitmee et al. 2015, pp. 1975–1976):

> Human beings have converted about a third of the ice-free and desert-free land surface of the planet to cropland or pasture and annually roughly half of all accessible freshwater is appropriated for human use. Since 2000, human beings have cut down more than 2.3 million km^2 of primary forest. About 90% of monitored fisheries are harvested at, or beyond, maximum sustainable yield limits. In the quest for energy and control over water resources, humanity has dammed more than 60% of the world's rivers, affecting in excess of 0.5 million km of river. Humanity is driving species to extinction at a rate that is more than 100 times that observed in the fossil record and many remaining species are decreasing in number.

The environmental changes (and risks) from human-related pollution and other activities can be evaluated by considering the key drivers of human society that induce human activities to place pressures on the environment, leading to environmental changes in life support systems, loss of biodiversity and ecosystem services that affect human health and society, and overall sustainability of planetary health.

The linked process between key societal drivers, pressures, environmental changes, and effects on humans and society is illustrated in Figure 20.6. Human-related

Major environmental risk factors such as chronic pollution and related climate change that apply to changes in biodiversity and the state of ecosystems include the different types of threats, the pressures, ecosystem processes involved, shifts in species distribution, and the rates of change over time (e.g. species extinction) and space. With increasing rates of change, the tipping points or thresholds at which different ecosystems or their important functions under stress may rapidly change or even collapse are also of great scientific concern (see Chapters 3 and 7).

The Living Planet Index (WWF 2018, p. 18), which measures biodiversity abundance levels based on 16 704 populations of 4 005 vertebrate species across the world, shows an overall decline of 60% in species population sizes (1970–2014), while current rates of species extinctions are 100 to 1 000 times higher than the background extinction rate. Similarly, wetland-dependent species are also in serious decline. Since 1970, population declines are reported to have affected 81% of inland wetland species and 36% of coastal and marine species (Ramsar Convention on Wetlands 2018). Human impacts through habitat loss and fragmentation, pollution, invasive species, climate change, and overharvesting are seriously reducing insect and other invertebrate abundance, diversity, and biomass, on a global scale (Harvey et al. 2020).

Since the start of the Industrial Revolution, the growth of modern society has accelerated the effects of many environmental changes to global scales, from what were once considered sustainable impacts on the Earth's natural resources, to recent unsustainable levels of natural resource uses and consumption. The key factors or drivers of these environmental changes for human and environmental health are related to human society and its development, as introduced in Chapter 1 of this book:

- Population (P) and urbanization growth,
- Consumption (related strongly to affluence, (A)), and
- Available technology and scientific development (T).

The impact (I) of the interactions of these factors can be generally expressed by the relationship: I = PAT (Eq. 1.1).

Other human driving factors to consider for sustainable development are poverty and inequity, political and economic systems, and cultural values (e.g. Landon 2006, pp. 27–28).

The projected global population growth by 2050 of another 2–3 billion persons combined with rapid economic development and urbanization in developing countries are expected to continue to drive human pressures that cause unsustainable or negative environmental changes (e.g. pollution effects, climate change effects, biodiversity losses, freshwater, and food insecurity), and increases in global burdens of disease, particularly noncommunicable

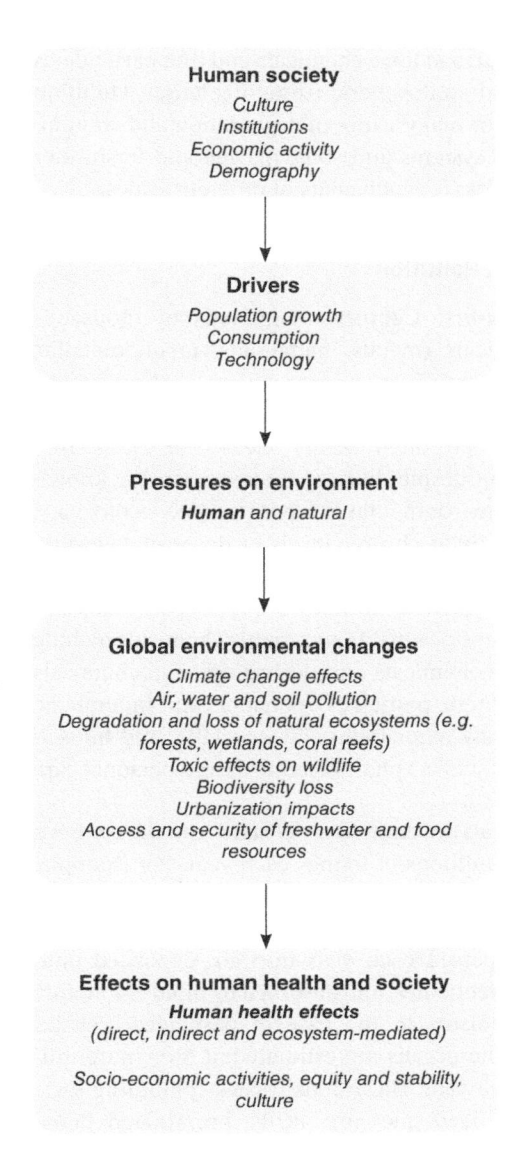

Figure 20.6 The cause–effect process between key human societal drivers and pressures on environmental changes such as pollution and effects on human health and society. Source: Modified from Whitmee et al. (2015).

drivers have placed pressures, including pollutant emissions and exposures, on the global environment that lead to a highly diverse range of environmental changes and impacts on the biodiversity of living organisms and ecosystems. Furthermore, multiple direct, indirect, and ecosystem-mediated health effects on humans are occurring at local to global scales due to these environmental changes and impacts. It also needs to be recognized that the human-induced pressures are generally superimposed upon existing interactions of natural factors, stressors or background pollutants in the environment, and on humans and other living organisms.

diseases, and continual threats, epidemics or pandemics from emerging and rapid vector-borne diseases (e.g. SARS–CoV-2 and the pandemic COVID 19). The impact of technology is determined by the multiplier effect of available technology, which is likely to be influenced by socio-economic responses to adverse effects from environmental changes (and health effects) and rates of change to a low carbon economy, conversion to use of renewable resources, and sustainable management of hazardous chemicals and wastes.

Opportunities for increased resilience and to advance sustainable solutions that can prevent or reduce negative environmental changes caused by pollution and other human-related impacts are outlined and discussed in Section 20.4.

Planetary Boundaries

Pressures on the Earth's biophysical systems can be evaluated by considering the planetary boundaries framework developed by Rockström et al. (2009) (see Chapter 1). This framework identifies and defines essential biophysical processes and systems that sustain the Earth's functions for humans and other life forms, within a conceptual boundary known as the *safe operating space*. The framework relates to and allows evaluation of nine global or regional pressures acting on the Earth's biophysical systems (Whitmee et al. 2015, pp. 1979–1980).

Human-induced changes in these systems are already at global or regional scales (e.g. climate change, biodiversity, and pollution) that add up to serious global issues for sustainability. Substantial changes in these systems have the capacity to produce rapid, nonlinear, and potentially irreversible changes in the Earth's environment that could adversely affect human development and health.

The Earth's global systems are changing for seven out of nine of these planet boundaries as described by Rockström and co-researchers. Specifically, biosphere integrity (as measured by extinction rates), biogeochemical flows (as measured by nitrogen and phosphorus flow rates), and land-system change (as measured by area of remaining forests) are estimated to be exceeding the boundary for the defined safe operating space. Additionally, ocean acidification is estimated to be approaching the identified threshold value, and freshwater use shows high spatial variation, exceeding regional thresholds in areas of low water availability or high consumption (Whitmee et al. 2015, pp. 1979–1980).

It is clearly evident that chemical pollution from human activities is a dominant factor in environmental changes caused by global warming and related climate change, ocean acidification, nitrogen and phosphorus pollution, multiple exposures to toxic chemicals and fine particulates or aerosols, and stratospheric ozone depletion. Pollution is a significant or major cause of degradation and contamination of land-systems and soils, marine and freshwater resources, and loss of biodiversity at different scales.

Toxic Chemical Pollution

In the Twenty-first Century, many tens of thousands of toxic chemicals (metals, metalloids, organometallic, inorganics, radionuclides, and organics) continue to be produced and used commercially worldwide. Many of these chemicals are intentionally released or escape into the environment despite little or limited scientific knowledge of their environmental behavior or fate, ecotoxicity, or human toxicity at chronic levels or doses of exposure. Major types of toxic elements include arsenic, lead, mercury, cadmium, nickel, chromium VI, copper, zinc, and fluorine (e.g. fluoride ion). Toxic organic chemicals include numerous petrochemicals and derivatives, agrochemicals (especially synthetic pesticides), polymers and monomers, surfactants, many other industrial, specialty, and life science chemicals such as pharmaceuticals and personal care products (PPCPs).

Toxic chemicals are emitted or released into the environment in many millions of tonnes each year. For example, well over 10 millions of tonnes of potential organic pollutants (e.g. petroleum, industrial chemicals, and pesticides) are widely dispersed each year into air, deposited onto soils, lost to water bodies, and absorbed by biomass, mainly nontarget organisms in the case of pesticides. Plastics disposed into the oceans are estimated at 8 or more million tonnes each year. Hazardous wastes containing toxic chemicals (e.g. lead, mercury, PCBs, brominated flame retardants, dioxins, and phthalates) are also generated in large quantities by many countries based on available data, although this is likely to be greatly underreported. Data from UNEP (2013) indicate that 64 countries generated over 250 million tonnes of hazardous wastes, as reported in the years 2004, 2005, or 2006, under the Basel Convention for transboundary movement of wastes. Electronic waste generation was reported to be 40 million tonnes in 2009 (UNEP 2013, p. 42). The proportion of hazardous organic chemicals in these various wastes is generally unavailable.

Environmental releases of such large quantities of mostly heavy metals, petroleum hydrocarbons, petrochemicals derived from petroleum (e.g. plastics, surfactants, synthetic pesticides, and pharmaceuticals) have resulted in local, regional, and global exposures of wildlife and human populations to many of these substances at acute and chronically toxic levels. Exposures of wildlife or non-target organisms to toxic pesticides from human use of pesticides are associated with major ecological impacts and global concern for biodiversity, driven largely by agriculture that affects about 40% of the Earth's land surface.

Persistent, bioaccumulative, and toxic chemicals, including POPs, have widely contaminated the Earth's environment, wildlife, ecosystem food webs, foodstuffs, and human populations to an extent that includes the remote polar regions of the Earth. Less persistent toxic chemicals have also caused local and regional areas of elevated contamination from industrial, urban, agricultural, mining, and waste disposal sources.

PPCPs are a special group of toxic chemicals of recent environmental concern that include antibiotics, analgesics, steroids, lipid regulators, antidepressants, hypertension drugs, antiepileptics, stimulants, antimicrobials, sunscreen agents, fragrances, and cosmetics. Multiple residues of these compounds are increasingly found in sewage discharges and stormwater runoff and receiving water bodies such as rivers, estuaries, and coastal waters. Antibiotic residue contamination from human and intensive livestock sources is of major environmental concern for spread of antibiotic resistant bacteria.

Plastic pollution of the world's oceans (and land) has resulted from the physical breakdown of plastic litter and widespread dispersion of microplastic particles, a complex emerging form of organic pollutants that are increasingly found in waters, soils, sediments, and the tissues of wildlife and humans. Marine plastic pollution has increased tenfold since 1980, affecting at least 267 species (e.g. marine turtles, seabirds, and marine mammals) (IPBES 2019, p. 13).

Toxic chemicals can exert a wide range of physiological and behavioral effects on wildlife species and humans as shown by numerous laboratory, bioassay, field, and epidemiological studies. Organic pollutants can induce a broad spectrum of selective and nonselective toxic effects on humans and wildlife due to the variety of their chemical forms, functional groups, and properties. The combined interactions of mixtures of pollutants such as TOC are of specific concern because they may exert effects (e.g. additive, synergistic, and antagonistic), and potentially when individual chemicals are present at concentrations below toxicity endpoints.

At the ecosystem level, toxic chemicals can reduce ecosystem structure and functions through species and abundance losses, reduced productivity, changes to energy flows and land uses, which can indirectly affect ecosystem services and human health. Major effects of pesticides on populations and communities of animals and plants in ecosystems relate to (i) their selective or broad spectrum of toxic effects on target and nontarget species, and (ii) their capacity to cause significant changes in species abundance and associated shifts in population dynamics that may severely impact upon communities and related terrestrial and aquatic ecosystems. These effects also apply to agricultural ecosystems. Overall, adverse ecological effects of different pesticides affect all levels of biological organization, from local to global scales, over the short-term to the long-term.

Recent and historical evidence of population declines in some wildlife populations is associated with reproduction effects (e.g. infertility and feminization) and nonreproduction effects (e.g. thyroid and adrenal disorders or disease, hormone cancers, and bone disorders) related to endocrine disrupting chemical (EDC) exposures.

Detailed evidence exists for the ecotoxicity of some chemicals within both aquatic and terrestrial ecosystems although available ecotoxicity data is largely lacking for many chemicals despite advances in testing methods. Examples include feminization of fish and developmental delays and malformations in amphibians. While levels of some pollutants such as DDT, PCBs, and dioxins in wildlife have decreased, other pollutants, such as brominated fire retardants and perfluorinated compounds, have increased (Whitmee et al. 2015) (see also Chapters 6 and 11).

Pollution and Global Change Threats to Ecosystem Services

Some of the major threats to ecosystem services involve massive impacts from human activities on a global scale from climate change, deforestation, desertification, urbanization, wetland drainage, pollution, dams, and water diversion, as reviewed by Melillo and Sala (2008, chapter 3). Pollution impacts are usually more diffuse and difficult to evaluate compared to the physical impacts of habitat destruction, urbanization, and infrastructure projects on ecosystems. Some known pollution-related threats, including climate change, on ecosystem services are described by Persson et al. (2010) in a report by the Stockholm Environment Institute.

Pollution and Human Health

The Lancet Commission on Pollution and Health stated that pollution is the largest environmental cause of disease and premature death in the world. The victims are commonly the poor, vulnerable, and marginalized persons. Children are at high risk of pollution-related disease, disability, and death during exposures to pollutants across life stages, from *in utero* and early infancy, to later childhood and beyond. The worldwide challenge remains that

> ... Despite its substantial effects on human health, the economy, and the environment, pollution has

been neglected, especially in low-income and middle-income countries, and the health effects of pollution are underestimated in calculations of the global burden of disease

(Landrigan et al. 2017).

Burden of Disease

The global burden of premature deaths attributed to pollution was estimated to be 9 million persons in 2015, mainly due to ambient and household air pollution (see Chapter 3). Premature deaths caused by pollution were responsible in 2015 for 16% of all deaths worldwide. As well, premature deaths (12.6 million) from living in unhealthy environments, including effects from pollution, accounted for almost 23% of total deaths in 2012 (Landrigan et al. 2017, p. 9).

Health Effects of Pollution

Major human health effects and environmental risks related to pollution exposures occur through air, water, soil, or land pollution, food contamination, and contact with hazardous substances or materials such as in workplaces and homes. The impacts of environmental changes due to pollution, such as from global warming and climate change, are known to cause loss and degradation of natural systems, as well as to atmospheric, land, forestry, freshwater resources, and food and nutritional resources. These types of human health effects caused directly or indirectly by pollution or specific pollutants such as toxic chemicals, are characterized below.

Air Pollution

As discussed in Chapter 13, the World Health Organization estimates that 9 out of 10 persons breathe air containing high levels of pollutants. Air pollution is a major cause of global mortality and burden of disease for human populations. An estimated 7 million people die prematurely from outdoor and household air pollution each year (about one in eight deaths globally) (WHO/CCAC 2018). More premature deaths caused by air pollutants are likely due to other workplace and environmental exposures. Combined estimates of premature deaths may exceed 8 million persons, even allowing for overlaps, as underestimates are likely to be significant.

Deaths from fine particulate exposures are mainly attributed to increased mortality from stroke, ischemic heart disease (IHD), chronic obstructive pulmonary disease (COPD), lung cancer, and acute lower respiratory infections (ALRI). Children and persons from developing and transition countries (e.g. India and China) bear much of the burden of death and disease from air pollution.

The main airborne pollutants causing health effects are fine particulates ($PM_{2.5}$), ground-level ozone, combustion gases (carbon dioxide, nitrogen dioxide, and sulfur dioxide), volatile and semi-volatile toxic organic compounds, and toxic metals (e.g. lead), including known and probable human carcinogens.

Many carcinogens are found in air pollution as individual substances or mixtures. Known human carcinogens can range from ultraviolet radiation, soot, asbestos fibers, crystalline silica dust, benzene, cadmium, chromium(VI), nickel, radon, diesel particulates/fumes, PAHs (benzo[a]pyrene), dioxin-like PCBs, dioxins, to tobacco smoke. Airborne biological pollutants (aerosols and dusts) are also major causes of skin, eye, and membrane irritations, allergic reactions, respiratory dysfunction, and diseases.

Worldwide exposure to fine particulates ($PM_{2.5}$) is the leading environmental risk factor for deaths. Combustion of fossil fuels, particularly coal and diesel, produces large amounts of fine particulates, among other air pollutants, including nitrogen oxides, carbon monoxide, and carbon dioxide gases. Uncontrolled burning of biomass is another serious source of air pollution affecting millions of persons in local to regional areas in dwellings, human settlements, and cities (e.g. household biosolid fuels, clearing and burning of forests for agriculture, open agricultural burning, and wildfires, bushfires, or landscape fires).

For example, landscape fires also have important effects on health and the environment through emissions of fine particulate matter and ozone. Smoke from landscape fires, mainly related to deforestation and land clearing, is estimated to cause more than 300 000 premature deaths worldwide per year. Short-term respiratory effects are increasingly reported in the literature (e.g. South-East Asia and Brazil) (Landrigan et al. 2017).

The health-related impacts of global air pollution from GHG emissions and climate change are considered below.

Global Warming and Climate Change

The role of greenhouse gases (and particulates) in global warming and the related-effects of climate change are discussed in Chapter 14. Figure 20.7 shows a conceptual diagram of the relationship between human development and climate-related risks and impacts.

Reviews and evaluation of human health impacts and projected risks from global warming and climate change are available from the IPCC Reports, The *Lancet* Commission on Health and Climate in 2015, The Lancet Countdown 2017 Review, and WHO Updates. For instance, direct effects from heat-related climate change is already occurring. Watts et al. (2018) reported that exposure to more frequent and intense heatwaves is increasing, with an estimated 125 million additional vulnerable adults

Figure 20.7 Simplified schematic diagram showing how risk from climate-related impacts results from the interaction of climate-related hazards with the exposure and vulnerability of humans and natural systems. Human responses to reduce climate-related risks involve changes in socio-economic processes, including (i) mitigation of climate change sources, such as GHG emissions from human development, and (ii) adaptation to address impacts of climate change on the vulnerability of humans and natural systems. Mitigation and adaptation responses are indicated in the figure.

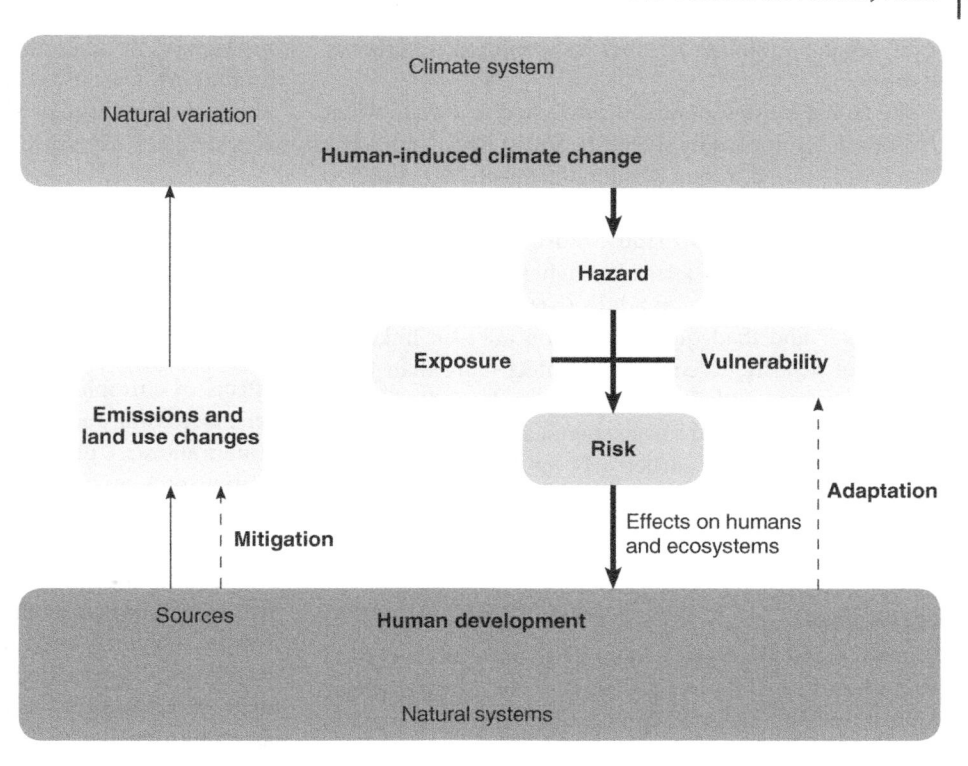

exposed to heatwaves between 2000 and 2016. Risks from these impacts are expected to increase greatly with global warming.

Ecosystem-related impacts from global warming and climate change are contributing to growing vector-borne diseases such as the transmission of dengue fever by *Aedes aegypti* (estimated 9.4% increase since 1950) and malaria, waterborne infectious diseases, reduced food security, and human undernutrition. Rising incidence of extremes of weather, such as floods and storms, increases the risk of drowning and injury, damage to human settlements, spread of waterborne disease, and related impacts on mental health (Watts et al. 2018).

Extreme Events Intensity/Frequency

The WHO (2018) reports a set of findings from an assessment of human health effects due to climate change for projected increases in exposure to four climate variables: heat, droughts, floods, and heatwaves. The warmest and poorest countries of the world will be most severely affected by climate change, particularly in South Asia. Exposed numbers of vulnerable people are very likely to be high. For example, up to 3 billion people (>65 years of age) may be exposed to heatwaves by 2100, because of a combination of increasing temperatures, ageing, and urbanization. Severe impacts are expected beyond 2020.

Overall, the WHO projects that the health impacts of climate change could force 100 million people into poverty by 2030, with strong impacts on mortality and morbidity.

A highly conservative estimate of 250 000 additional deaths each year due to climate change has been projected between 2030 and 2050. Deaths are expected to result primarily from heat exposure in elderly people (38 000), diarrhea (48 000), malaria (60 000), and childhood undernutrition (95 000).

The potential magnitude of physical and mental health effects on many populations (illness, disability, and geographical displacement) is likely to be very high and presents as a massive challenge for climate change risk management and global sustainability (see Section 20.4).

Water Pollution

The major health risk for humans from water pollution is the contamination of water sources by infectious agents (pathogens) of fecal origin (human and warm-blooded animal wastes), and toxic chemicals (e.g. heavy metals, arsenic, industrial chemicals, pesticides, disinfection byproducts, personal care, and pharmaceutical residues). Water pollution causes widespread illnesses and premature deaths in humans, primarily following contact with or ingestion of waterborne pathogens in unsafe water sources and those with inadequate sanitation (treatment to control infectious agents and also personal hygiene such as hand washing).

The most common diseases linked to water pollution are acute and chronic gastrointestinal diseases, mainly diarrheal (e.g. 70% of deaths attributed to water pollution), followed to a lesser degree by typhoid fever and paratyphoid fever, among a number of other less common waterborne diseases.

The global burden of deaths attributed to unsafe water and unsafe sanitation by the GBD study (2015) and WHO data (2012) varied from 1.8 million deaths (GBD study) to 0.8 million deaths (WHO data). Unsafe sanitation was a major cause in the GBD study. Most deaths occurred in children (<5 years of age) although increased numbers of deaths were observed in adults (>60 years of age). Polluted water and inadequate sanitation are also linked to a range of parasitic infections that affect more than 1 billion people, especially in low-income and middle-income countries. Data on deaths from chemical pollution of water are generally lacking, particularly for levels of chemical contamination of drinking water in most low-income and middle-income countries (see Landrigan et al. 2017, pp. 14–16).

Soil Pollution

Human populations are exposed to multiple soil (and groundwater) pollutants on a global scale. Anthropogenic sources include sewage, animal and hazardous wastes, stack emissions, petroleum wells, or diffuse emissions from applications of pesticides, excessive nutrients or untreated biosolids to crop lands, mine tailings, oil spills, atmospheric deposition, and flooding. Toxic exposures may include heavy metals (e.g. As, Cd, Pb, Hg), organic chemical contamination (e.g. dioxins, PCBs and other POPs), pesticides, and pharmaceuticals (e.g. estrogen and antibiotics). Infectious exposures of soil pathogens (e.g. antibiotic resistant bacteria, prions, and anthrax) also occur.

Contaminated soils and groundwaters are a major environmental and human health issue of current global concern. Available evidence suggests there are millions of potentially contaminated sites (e.g. Europe) and enormous areas of contaminated agricultural lands on a world scale. Toxic chemical residues (e.g. heavy metals, pesticides, and antibiotics) can also contaminate food crops and animal products used as feedstuffs for livestock and human foodstuffs. Known health risks include neurotoxicity, renal disease, autoimmune diseases, cancers, cardiovascular diseases, respiratory diseases, and congenital malformations.

The magnitude of the health impacts is likely to be underestimated, particularly for historic exposures in developed countries and largely unknown in developing countries where land contamination and hazardous wastes are emerging issues for investigation, health risk assessment, remediation, and management. Epidemiological studies of residents living on or near contaminated sites show sufficient evidence for stress, anxiety, psychiatric disorders, and limited evidence for renal dysfunction, bone disorders, birth defects, low birth weight, and preterm deliveries. Generally, findings from cancer studies are evaluated as inadequate by scientific reviewers, due to methodology limitations and inconsistent findings between studies, although some studies show increased incidences of cancers at higher levels of known carcinogens (see Chapter 15).

Wastes

Human activities generate waste streams that are significant to major contributors of local to global pollution. Examples include rapidly growing urbanization and emissions of atmospheric particulates (PM_{10} and $PM_{2.5}$), accelerated global warming from GHGs, unsustainable effects of eutrophication in coastal marine and freshwaters, increasing deoxygenated *dead zones* in various coastal waters and seas, plastic pollution of the oceans, global generation of E-wastes, hazardous waste landfills, chemical contamination of agricultural lands, and industrial sites.

Each year billions of tonnes of wastes are generated by human activities. Global best estimates of municipal solid wastes amount to about 2 billion tonnes per year. Urban wastes, including MSW, commercial and industrial (C&I) waste, and construction and demolition (C&D), are estimated to be about 7–10 billion tonnes per year. Mining and agriculture-related wastes are each reported as 10–20 billion tonnes per year. Plastic wastes disposed to the oceans are estimated to be about 9 million tonnes per year. Global estimates of hazardous wastes are unavailable although hundreds of millions of tonnes are known to be generated.

There is limited evidence from many studies of adverse health effects (respiratory, birth defects, and cancers) among waste workers and persons living near existing or former waste facilities from exposures to emissions from hazardous wastes, landfills, incinerators, and to a lesser degree, composting activities. The weight of evidence and causality of association of reported health effects from studies of residents living near waste facilities, and with exposures to potential toxic waste emissions, are usually described as inadequate or inconclusive. Limited evidence exists from a few studies for birth defects (landfills and incinerators), low birth weight (landfills), and birth defects and cancers (incinerators). Verification of the strength of evidence is considered difficult because of factors such as the weaknesses of design, lack of exposure or biomarker studies, and statistical power in many studies, and remains a key challenge for future investigations and health risk management (see Chapter 16).

20.4 Pollution, Health, and Sustainable Solutions

The concept of the sustainability of planetary health, together with human health, is illustrated in Figure 20.8 by plotting the timeline of human development to the present, and projected into the future, against measures of negative

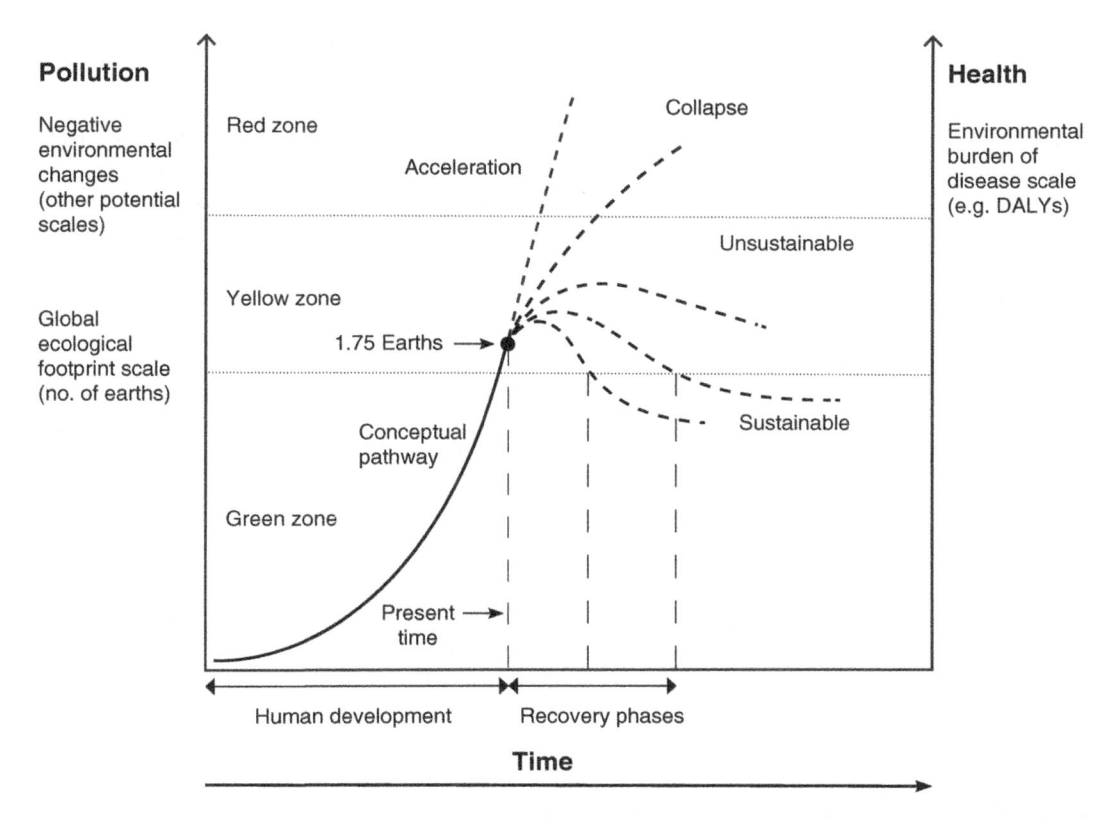

Figure 20.8 A global concept of the effects of pollution and other environmental factors due to human development on the sustainability of planetary health and human health. The nature of potential sustainable scenarios, pathways, and recovery phases is indicated.

environmental change (e.g. increasing global ecological footprint, global warming, and chemical pollution) and global burden of disease (e.g. number of premature deaths due to pollution or environmental factors, millions/year).

In Figure 20.8, the rate of negative environmental change and impact on public health accelerates from the period of the Industrial Revolution to the present time. Measures of the sustainability of the Earth decrease from the green zone to the unsustainable yellow zone of the present time, as indicated by the point of the global ecological footprint of 1.7 gha/person in Figure 20.8 (see also Chapter 1). Future pathways for development scenarios that move toward increasing unsustainability, and potentially irreversible states (red zone), or alternatively, achieve recovery through global transformative actions, are graphically depicted. Outcomes depend on rates of change, the vulnerability of exposed populations and natural systems, tipping points, and capacity for societies to mitigate, sustain ecological resilience, and to adapt to sustainable development. Global burden of disease as estimated in the Twenty-first Century indicates at least 9 million persons die prematurely due to pollution (16% of total deaths in 2015) each year. If total negative environmental impacts and pollution are considered, about 23% of total deaths in 2012 were attributed to these factors.

At present, rates of premature deaths, disease, and disability for human populations and the next generations appear to remain complex, and uncertain. They confront the notions of sustainable human populations and societies. However, the current global rate can be compared to some key measures of planetary boundaries and the global ecological footprint, which are exceeding or moving toward threshold values or planetary boundaries of uncertainty and unsustainability (see also Figure 1.4).

Moving Toward Planetary Health

How can the challenge of pollution, its impacts, and threats on planetary health be solved? An overview of progressive answers to this urgent and transformative global challenge is presented in this section. Key messages from The 2015 Lancet Commission on Planetary Health concisely state the threats to planetary health and related human health, inadequate societal responses, and nature of solutions that need to be applied. These are given below (Whitmee et al. 2015).

- The concept of planetary health is based on the understanding that human health and human civilisation depend on flourishing natural systems and the wise stewardship of those natural systems. However, natural

systems are being degraded to an extent unprecedented in human history.
- Environmental threats to human health and human civilisation will be characterised by surprise and uncertainty. Our societies face clear and potent dangers that require urgent and transformative actions to protect present and future generations.
- The present systems of governance and organisation of human knowledge are inadequate to address the threats to planetary health. We call for improved governance to aid the integration of social, economic, and environmental policies and for the creation, synthesis, and application of interdisciplinary knowledge to strengthen planetary health.
- Solutions lie within reach and should be based on the redefinition of prosperity to focus on the enhancement of quality of life and delivery of improved health for all, together with respect for the integrity of natural systems. This endeavour will necessitate that societies address the drivers of environmental change by promoting sustainable and equitable patterns of consumption, reducing population growth, and harnessing the power of technology for change.

In 2017, The Lancet Commission on Pollution and Health found:

> ...despite the great magnitude of pollution and current gaps in knowledge about its effects on human health and the environment, pollution can be prevented. Pollution is not the inevitable consequence of economic development.

Importantly, it argued that many effective strategies, policies, laws, and regulations for prevention and pollution control (e.g. clean air, clean water, and chemical safety) have been adopted in high-income and some middle-income countries, and are now available to be transferred and adapted for use by cities and countries on a global scale. The regulatory approaches are based on the polluter-pays principle, and in many cases, underpinned by environmental and health risk assessment. Examples of the prevention or control of hazardous pollutants include the phasing out or banning of lead, asbestos, and DDT, and clean-up of serious hazardous waste sites (see Landrigan et al. 2017).

The Lancet Commission on Pollution and Health made six key recommendations to advance effectively toward human and planetary health. These are listed below:

1. Make pollution prevention a high priority, nationally and internationally, and integrate it into country and city planning processes.
2. Mobilise, increase, and focus the funding and the international technical support dedicated to pollution control.
3. Establish systems to monitor pollution and its effects on health at national and local levels, evaluating the success of interventions, guiding enforcement, informing civil society and the public, and assessing progress toward goals. New technologies, such as satellite imaging and data mining can increase efficiency, expand geographic range, and lower costs. Open access to these data and consultation is essential.
4. Build multi-sectorial partnerships to advance pollution control and accelerate the development of clean energy sources and clean technologies that will ultimately prevent pollution at source.
5. Integrate pollution mitigation into planning processes for non-communicable diseases. Interventions against pollution need to be a core component of the Global Action Plan for the Prevention and Control of Non-Communicable Diseases.
6. Research is needed to understand and control pollution and to drive change in pollution policy.

Available scientific evidence shows that the effects of pollution on planetary health and human health are underestimated. Furthermore, the scientific capacity to conduct systematic health, fate, exposure, and toxicity testing of a large proportion of existing and new chemicals in commercial use that are released into the environment is inadequate and too slow. In toxicology and risk assessment, nonanimal testing is actively advancing for evaluation of chemical safety, including *in silico* or computational methods (interpretation of computing and information technology with molecular biology) (Rauno 2011). Note: >165 million unique chemical substances known to exist in 2020 (Chemical Abstract Services, CAS Registry).

Advances in pollution control measures need to be better integrated into a sustainability framework designed to meet, monitor, evaluate, and progress toward sustainable development goals and targets beyond 2020. Large-scale use of new technologies of the digital age to enhance and communicate research, monitoring, and assessment of pollutants is essential to achieve objectives such as sustainable management of chemicals by 2030, and mitigation of GHG emissions to limit global warming to an average rise in atmospheric temperature of 1.5 °C or less than 2 °C at the Earth's surface compared to the pre-industrial age average level (1750).

Conceptual models, strategies, and interventions to obtain sustainable societal, economic, and environmental solutions for global pollution and health are outlined below. A basic conceptual framework to achieve sustainable planetary health with an emphasis on pollution prevention and reduction is shown in Figure 20.9.

20.4 Pollution, Health, and Sustainable Solutions | 527

Human society
*Culture
Institutions
Economic activity
Demography*

↓

Drivers
*Population growth
Consumption
Technology*

↓

Reduced pressure → **Pressures on environment**
Human and natural

↓

Global environmental changes
*Climate change effects
Air, water and soil pollution
Degradation and loss of natural ecosystems
(e.g. forests, wetlands, coral reefs)
Toxic effects on wildlife
Biodiversity loss
Urbanization impacts
Access and security of freshwater and food resources*

↓

Effects on human health and society
Human health effects
(direct, indirect and ecosystem-mediated)

Socio-economic activities, equity and stability, culture

↓

Mitigation
*Prevent/reduce pressures, emissions and wastes
Sustainable technologies*
← **Responses**
*Interventions
Risk management
Sustainability solutions (environmental, socio-economic, cultural)*
→ **Adaptation**
*Sustainability
Apply risk management*

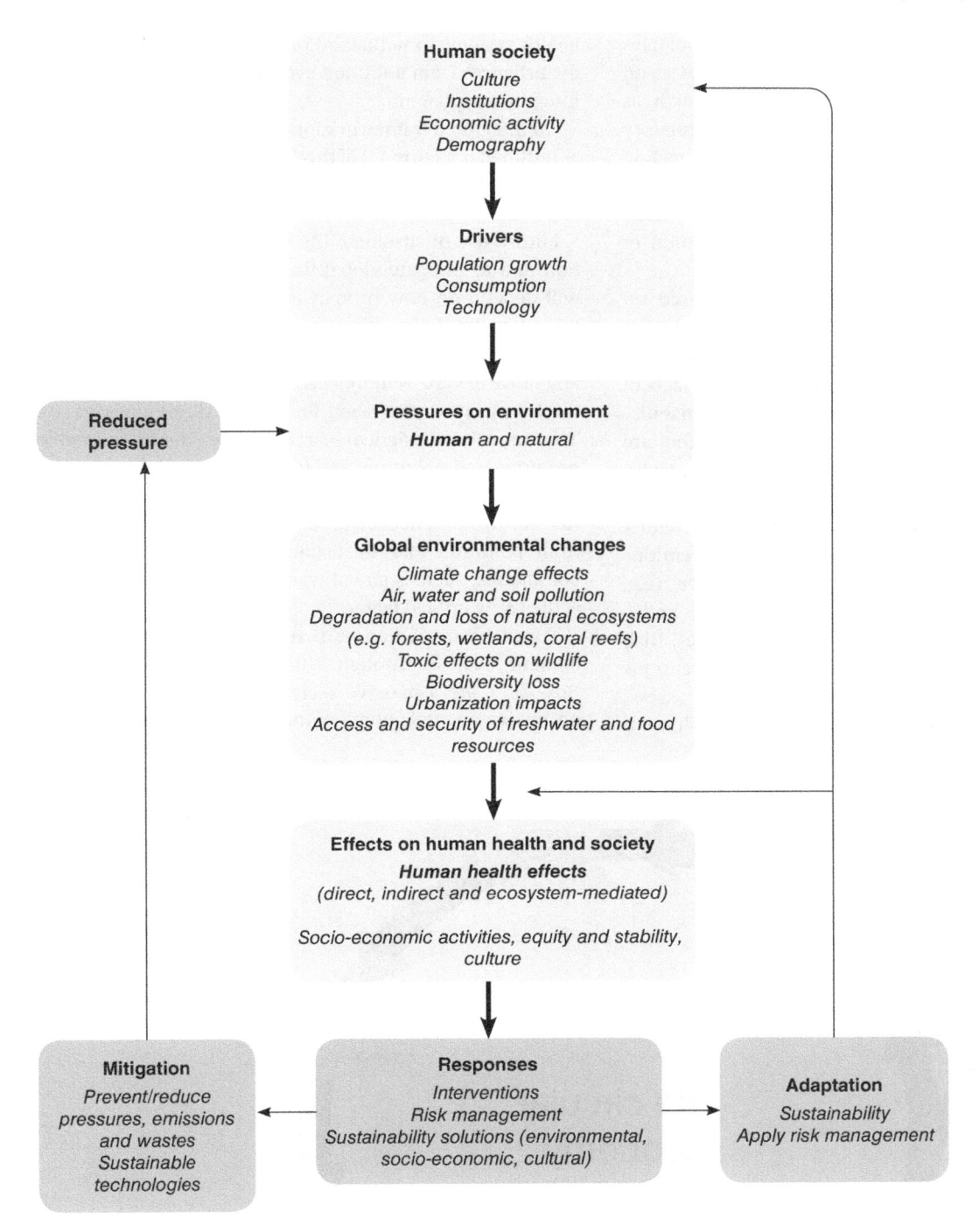

Figure 20.9 Conceptual framework for moving toward sustainable planetary health with an emphasis on pollution prevention and reduction strategies and interventions, involving societal responses of combined mitigation measures, adaptation changes, and risk management actions.

In Figure 20.9, the core process showing the links between the main societal drivers (e.g. population growth and urbanization), pressures (e.g. toxic pesticide applications in agriculture and greenhouse gas emissions), environmental changes, including climate change, and effects on human health and society (e.g. premature deaths and diseases) is illustrated, as derived from Figure 20.6. Pollution control responses involve evaluation of options, decision-making, and measures to prevent and reduce pollution, in the form of sustainability concepts, principles, planning and design, use of sustainable technologies (e.g. green chemistry and green and healthy cities), and monitoring of performance metrics or criteria such as sustainable development goals (SDGs), targets, regulatory standards, or other criteria. Progress toward meeting timetable targets is likely to become more critical for issues, such as priority pollutants and health effects, depending on the extent and rate of implementation of sustainable measures.

As indicated in Figure 20.9, responses are based on (i) mitigation to prevent or reduce pressures on the drivers of environmental change, including emissions of pollutants (e.g. from combustion of fossil fuels) and impacts of nonrenewable wastes, and (ii) adaptation of sustainability policies, strategies, and actions by human society that are targeted at key drivers and pressures that affect planetary health, biodiversity, and pollution-related diseases (PRDs) among human populations. In this context, risk reduction or risk management actions such as interventions also target direct or indirect human health effects (e.g. PRDs) due to environmental changes caused by pollution or in combination with other human activities. Risk management actions such as health-based interventions are derived from health or environmental risk assessments. Enhancing the resilience of ecosystems and human health systems to withstand or to recover from ecological disturbances from pollution events or climate change is an integral component.

An integrated framework for the evaluation of pollutants as outlined in Figure 1.3 of this book is essential to underpin risk assessment and risk management actions as part of sustainable outcomes.

Future use of sustainability measures to prevent pollution and to mitigate global warming and climate change will depend on how human societies manage large-scale transition from the dominant linear economic model of development and consumption of nonrenewable resources and move toward a circular economy model as promoted by the European Commission for Europe (see Figure 20.10). This model is designed to reduce use of nonrenewable resources and pollution, and to minimize pollution-related diseases and enhance health, as described in Box 20.1. The rate of transition needed to reduce global burden of disease from pollution and unsustainable trends for planetary boundaries, such as global warming and biodiversity loss, is the likely critical factor.

Transformative changes that lead to sustainable development, such as through a more circular economy (see Box 20.1), are a massive societal challenge that depends on environmental, population, socio-economic, cultural,

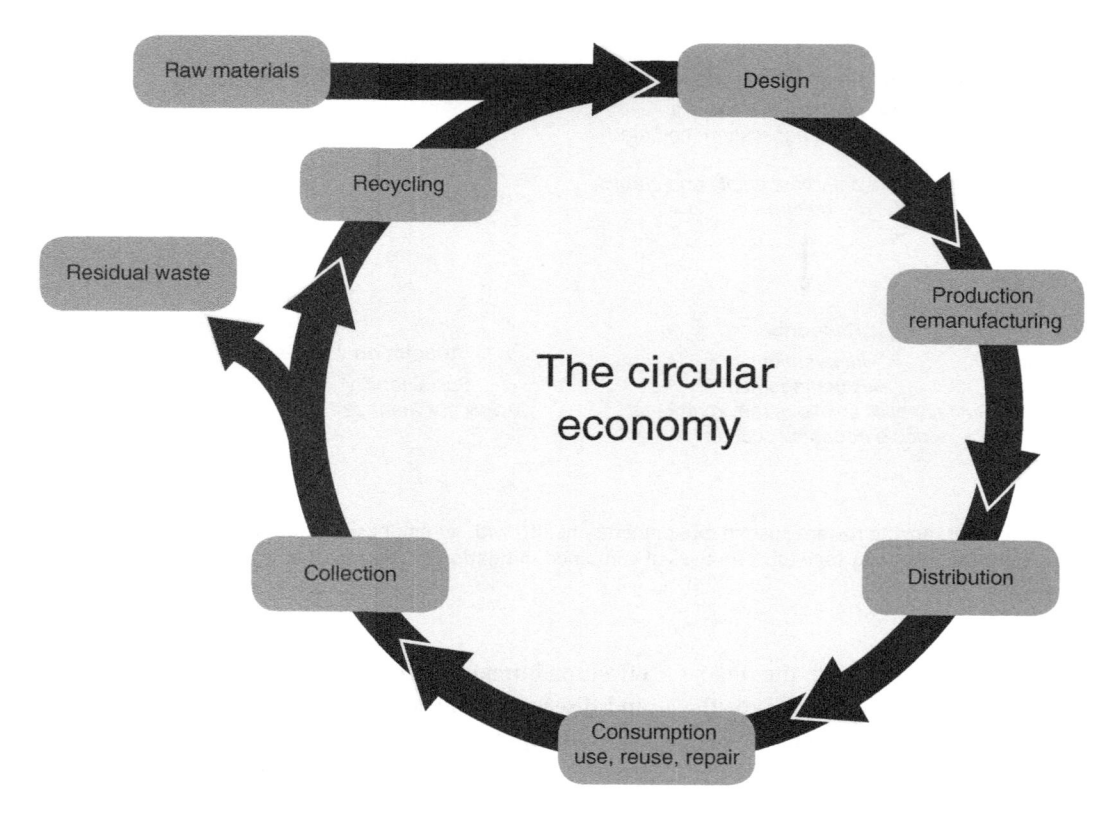

Figure 20.10 The circular economy. Source: Adapted from Whitmee et al. (2015) and European Commission (2014).

> **Box 20.1 The Circular Economy**
>
> *What is a circular economy?*
> It is an economic model that decouples development from the consumption of nonrenewable resources and minimizes the generation of pollution and other forms of waste by recycling and reuse.
>
> In a fully circular economy, the only new inputs are renewable materials, and all nonrenewable materials are recycled. The underlying assumption is that waste is an inherent inefficiency, a loss of materials from the system, and thus a cost.
>
> *What are key benefits?*
> Transition toward a circular economy will reduce pollution-related disease and improve health.
>
> *What are the core principles of the circular economy?*
> There are three:
>
> - preservation of natural capital by reducing use of nonrenewable resources and use of ecosystem management
> - optimization of resource yields by circulating products and materials so that they are shared, and their life cycles extended; and
> - fostering system effectiveness by designing out pollution, greenhouse gas emissions, and toxic materials that damage health.
>
> *How to transition toward a circular economy?*
> The steps include large-scale transition to nonpolluting sources of energy (wind, solar, and tidal), the production of durable products that require lower quantities of materials and less energy to manufacture than those being produced at present; incentivization of recycling, re-use, and repair; and replacement of hazardous materials with safer alternatives.
>
> Source: Modified from Landrigan et al. (2017).

technological, and political changes to current human societies, with open governance. Ideally, it presumes equitable and ethical-based values. A sustainable working framework is necessary to build upon and facilitate local to global health and to reduce the adverse effects of pollution to sustainable levels, such as urgently reducing the rate of GHG emissions and slowing global warming by 2050 to meet a desired 1.5 °C target increase.

Strategies, Interventions, and Controls for Pollution Solutions

The Lancet Commission on Planetary Health and Pollution has reviewed effective strategies and examples of interventions for pollution control primarily developed and proven in high-income countries. Applied within a sustainability framework, they offer both models and opportunities for developing countries to adapt and move rapidly toward sustainable development goals in areas such as pollution prevention and reduction, waste minimization, and chemical safety. Similarly, developed countries need to integrate sustainable development principles and practices into their societies as a priority across economic sectors, to overcome existing barriers to sustainable solutions for planetary health, as clearly demonstrated by the rate of GHG emissions and rising effects of global climate change, increasing loss of biodiversity (including local and global extinctions), and the estimates of premature deaths due to pollution and other environmental factors.

Essential outcomes for all countries are to reduce impacts and risks caused by human drivers and pressures of environmental change on planetary health. Table 20.6 summarizes some of the strategies and interventions used to reduce effects of pollutants or pollution on planetary health within the context of sustainability. Many other examples of strategies or interventions are described, for instance, in Whitmee et al. (2015) and Landrigan et al. (2017), among others in the rapidly emerging literature.

The concept of planetary stewardship has re-emerged and is being strongly advocated for individuals, communities, and nations to act collectively, live within planetary boundaries, effectively manage our common resources at local to global scales, and shift to a common goal that is sustainable for planetary health, human health, and well-being.

> ...Successful control strategies deployed by high-income countries include reducing exposure at source (such as removing lead from gasoline), banning asbestos, and crafting policies to reduce water and air pollution. Such strategies have proven incredibly cost-effective. Removal of lead from gasoline has returned approximately $200 billion to the US economy each year since 1980
>
> (The Lancet Planetary Health's editor-in-chief Raffaella Bosurgi, Cemma 2017).

The following example indicates the potential benefits that can be achieved by using systematic strategies,

Table 20.6 Summary of some strategies or interventions for prevention or reduction of global pollutants.

Pollutant/pollution	Strategies/interventions	Examples
Urban air pollutants (CO, NO_X, PM_{10}/$PM_{2.5}$, SO_2, O_3, VOCs, Pb, benzenes)	Urban air quality strategies	Develop sustainable control strategies and planning to prevent, reduce, and eliminate pollutants at source (point and nonpoint). Use sustainable technologies for downstream control of pollutants. Design, monitor, model, and evaluate air quality to meet health and environmental emission and ambient air quality goals for airshed from point and nonpoint sources.
	Pollution control at the source	Banning of CFCs that deplete stratospheric ozone; phasing out of tetraethyl lead from gasoline to remove toxic lead emissions; use low sulfur fuels.
	Downstream pollution control of emissions	Removal of acid gases by acid scrubbers; adsorption of VOC emissions by activated carbon filters.
Global warming pollutants and climate change	Intervention to reduce the sources or enhance the sinks of GHG emissions	Replace fossil fuel combustion for heating, electricity, and transport; avoid burning agricultural residues, protect natural vegetation, large-scale reforestation and net soil organic carbon; substitute alternative and biofuels.
Toxic and hazardous chemicals and wastes	International cooperation and sound management of chemicals; National registration, inventories, assessment and management of chemicals produced, imported, and used	International goals and targets for the sound and sustainable chemical management of United Nations and its agencies, and Inter-organization Programme for the Sound Management of Chemicals (IOMC). See also EU-REACH.
Genotoxics and carcinogens	Increased toxicity screening, testing/evaluation, and restrictions on environmental and occupational carcinogens	Prevention of exposure by banning highly hazardous and carcinogenic chemicals such as asbestos, benzene, PCBs, and DDT.
Endocrine disrupting chemicals (EDCs)	Prevention of use and exposures to EDCs (e.g. POPs, various toxic metals and chemicals	Banning and phasing out of POPs (e.g. PFAS and PFOA); Reduced use of phthalate esters and bisphenol A (BPA) in consumer products.
Pesticides	Banning or phasing out of persistent, toxic, bioaccumulative pesticides; Use of low toxicity and less persistent pesticides; Use of integrated pest management (IPM)	Neurotoxic organochlorine and organophosphate insecticides; Restrictions on uses of high-volume toxic pesticides (e.g. atrazine, glyphosate, and neonicotinoid insecticides); Reduced pesticide applications and pest resistance.
Land/groundwater contamination, land use safety assessment	Site investigations for hazardous pollutants/materials, remediation and validation	Toxic metals/arsenic, BTEX, petroleum hydrocarbons, PAHs, phenolic, pesticides, PCBs, PFAS/PFOA, asbestos residues.
Water quality pollutants (gross organics, suspended solids, salinity, nutrients, toxic chemicals, pathogens)	Point and nonpoint control of pollutant sources; Integrated catchment management; Downstream sustainable treatment of drinking waters, wastewaters and stormwaters; Drinking water interventions for specific ions, and contaminants	Urban and rural catchment management of loading and export of pollutants (e.g. organic wastes, nutrients, sediments, and pesticides). Advanced treatment systems, disinfection or removal of pathogens, and reuse of treated waters (non-potable), and treated biosolids. Removal of salinity (desalination); Na related hypertensions, Pb and As contamination.

Source: Adapted from Whitmee et al. (2015) and Landrigan et al. (2017).

interventions, and control measures to reduce the known impacts, risks, and costs to planetary health, providing international agreement and inter-sectorial collaboration can be achieved on a global scale.

In 2016, Prüss-Ustün from the WHO, and co-researchers, published an important example of how environmental interventions, using a systematic approach, have the potential to modify and significantly reduce risks to global human health from environmental factors, including exposures from pollutants. Using an update of the global burden of disease (BoD) attributable to the environment, 23% (95% CI: 13–34%) of global deaths and 22% (95% CI:

13–32%) of global disability adjusted life years (DALYs) were linked to environmental risks in 2012. Most of the global BoD was due to noncommunicable diseases, with children (<5 years of age) and adults between 50 and 75 years particularly susceptible.

The analysis showed that through eliminating hazards and reducing modifiable environmental risks, almost 25% of the global BoD could be prevented (Prüss-Ustün et al. 2016).

20.5 Key Points

1. Pollution can be seen as a silent pandemic.
2. In this final chapter, we highlight the extent of pollution impacts and threats to planetary health and human health and introduce the role that sustainability concepts, principles, and systematic approaches need to play to solve urgent and expanding pollution problems from a societal perspective.
3. There are many definitions of sustainability, mainly focused on the capacity of human behavior to sustain the Earth.
4. Sustainability depends on the capacity of the Earth's natural resources (materials and energy) and its natural living systems to meet the needs of human populations, their healthy lifestyles, and use of sustainable technologies to remove pollution and wastes from human activities.
5. Three key and interdependent dimensions or pillars of sustainability are commonly referred to as: environmental sustainability, social sustainability, and economic sustainability. A fourth dimension of cultural sustainability may also be applied to some assessments.
6. In practice, the concept of sustainability is transformed into a decision-making process designed to achieve sustainable development of human civilization, ideally through equitable and ethical values.
7. "Sustainable development is development that meets the needs of the present without compromising the ability of future generations to meet their own needs." World Commission on Environment and Development. *Our Common Future;* Oxford University Press: Oxford, 1987.
8. A schematic overview of sustainability from a systems and process perspective is given in Figure 20.1. It can be applied to sustainable development, design, and technologies to achieve sustainable use of resources, processes, and products at local to global scales.
9. Sustainable design is a fundamental approach to sustainable development (e.g. processes and systems, materials, products, planning, buildings, and transport) that acts to reduce negative environmental and health impacts on natural systems and humans.
10. Many different creative, theoretical, and practical design techniques are applied (e.g. from biomimicry, designing for reuse, recycling, and energy efficiency, to using green chemistry and green engineering principles in developing low-impact and nontoxic chemicals, materials, or products). Life cycle assessment for materials and products is increasingly used in the design process.
11. Green chemistry involves the use of "a set of principles that reduces or eliminates the use or generation of hazardous substances in the design, manufacture and application of chemical products."
12. Strategies and technologies for sustainable pollution prevention and reduction are evolving rapidly. Examples include green chemistry technologies, green buildings, and green cities, and global monitoring systems using satellites and other forms of remote sensing.
13. Planetary heath is under great environmental stress and threats, from human related pollution, among other human impacts. Planetary health is described basically as "the health of human civilization and the state of the natural systems on which it depends."
14. The environmental changes (and risks) from human-related pollution and other activities can be evaluated by considering the key drivers of human society that induce human activities. These place pressures on the environment, leading to environmental changes in life support systems, loss of biodiversity and ecosystem services that affect human health and society, and overall sustainability of planetary health (see Figure 20.6).
15. Increasing emissions of greenhouse gases and black carbon particles, primarily from global combustion of fossil fuels, are highly probable causes of accelerated global warming and climate change events (see Chapter 14). Pathways for future trends in global heating and temperature increases depend on the strength and rate of human responses to mitigation of emissions (see IPCC scenarios).
16. The Earth's global systems are changing for seven out of nine of these planet boundaries. Crucially, biosphere integrity (as measured by extinction rates), biogeochemical flows (as measured by nitrogen and phosphorus flow rates), and land-system change (as measured by area of remaining forests) are estimated to be exceeding the boundary for the defined safe operating space.
17. Ocean acidification is also estimated to be approaching the identified threshold value and freshwater

use shows high spatial variation, exceeding regional thresholds in areas of low water availability or high consumption.
18. Chemical pollution from human activities is a dominant factor in environmental changes caused by global warming and related climate change, ocean acidification, nitrogen and phosphorus pollution, toxic chemicals and exposures, and atmospheric pollution by fine particulates or aerosols, toxic gases, and stratospheric ozone depletion.
19. Pollution is a significant or major cause of degradation and contamination of land-systems and soils, marine and freshwater resources, and loss of biodiversity at different scales.
20. The biodiversity and population sizes of numerous species are estimated to be under rapid and serious threat. For example, The Living Planet Index (WWF 2018), which measures biodiversity abundance levels based on 16 704 populations of 4 005 vertebrate species across the world, shows an overall decline of 60% in species population sizes since 1970 while current rates of species extinctions are 100 to 1 000 times higher than the background extinction rate.
21. For humans, The Lancet Commission on Pollution and Health stated that pollution is the largest environmental cause of human disease and premature death in the world. The victims are commonly the poor, vulnerable, and marginalized persons. Children are at high risk of pollution-related disease, disability, and death during exposures to pollutants across life stages.
22. The global burden of premature deaths attributed to pollution was estimated to be at 9 million persons in 2015, mainly due to air pollution, and estimated to be 16% of all global deaths.
23. Overall, the WHO predicts that the health impacts of climate change could force 100 million people into poverty by 2030, with strong impacts on mortality and morbidity. A highly conservative estimate of 250 000 additional deaths each year due to climate change has been projected between 2030 and 2050.
24. The global concept of the effects of pollution and other environmental factors due to human development on the state of sustainability of planetary health, including human health, and degree of potential recovery responses by human society is illustrated in Figure 20.8.
25. The present state is indicated as unsustainable, based on the global ecological footprint. Global burden of premature deaths and diseases is significant and likely to continue growing without major coordinated interventions.
26. Figure 20.9 shows a conceptual framework for sustainable planetary health with an emphasis on pollution prevention and reduction.
27. Pollution control responses, involving evaluation of options, decision-making, and measures to prevent and reduce pollution, take the form of sustainability concepts, principles, planning, and design, use of sustainable technologies (e.g. green chemistry and green and healthy cities), and monitoring of performance metrics, indicators, or criteria used for measuring sustainability (e.g. environmental, economic, or social).
28. Sustainable solutions to prevent and minimize pollution are available but depend on how human society can effectively interact with complex environmental and human systems, at local to global scales, and make sufficient rates of change to current societal, economic, and political systems to achieve measurable and sustainable goals for planetary health.
29. However, the future challenge for sustainable development is how to translate from linear-based systems (e.g. exploitation of nonrenewable resources, production, consumption, wastes, and disposal) to more complex and nonlinear systems of ecology and interactions with human society such as socio-economic models, information systems, engineering processes, and value-based decision-making processes.

References

Anastas, P.T. and Warner, J.C. (1998). *Green Chemistry: Theory and Practice*. New York: Oxford University Press.

Anastas, P.T. and Zimmerman, J.B. (2003). Design through the twelve principles of green engineering. *Environmental Science & Technology* 37 (5): 94A–101A.

Ashby, M.F. (2013). *Materials and the Environment*, 2e. New York: Elsevier.

Baird, C. and Cann, M. (2012). *Environmental Chemistry*, 5e. New York: W.H. Freeman and Company.

Bender, H., Judith, K., and Beilin, R. (2012). Sustainability: a model for the future. In: *Reshaping Environments* (ed. H. Bender), 305–334. Cambridge: Cambridge University Press.

Cemma, M. (2017). *What's the Difference? Planetary Health Explained. Global Health Now Webpage.* John Hopkins Bloomberg School of Public Health https://www.globalhealthnow.org/2017-09/whats-difference-planetary-health-explained (accessed 5 September 2020).

CSIRO/Water Research Foundation (2010). *Integrated Urban Water Management Planning Manual*. Water Research Foundation/Commonwealth Scientific and Industrial Research Organisation https://publications.csiro.au/rpr/download?pid=csiro:EP10449&dsid=DS1 (accessed 4 September 2020).

Dandy, G., Walker, D., Daniell, T., and Warner, R. (2008). *Planning and Design of Engineering Systems*, 2e. New York: Taylor & Francis.

Editorial (2019). The bigger picture of planetary health. *The Lancet Planetary. The Lancet Planetary Health* 3 (1): PE1. https://www.thelancet.com/journals/lanplh/article/PIIS2542-5196(19)30001-4/fulltext (accessed 26 February 2020).

European Commission (2014). *Towards A Circular Economy: A Zero Waste Programme for Europe*. Minsk: DG Environment 8 October 2014. https://www.oecd.org/env/outreach/EC-Circular-econonomy.pdf. (accessed 12 October 2020).

Harvey, J.A., Heinen, R., Armbrecht, I. et al. (2020). International scientists formulate a roadmap for insect conservation and recovery. *Nature Ecology & Evolution* 4: 174–176.

IPBES (2019). *Summary for Policymakers of the Global Assessment Report on Biodiversity and Ecosystem Services of the Intergovernmental Science-Policy Platform on Biodiversity and Ecosystem Services*. Bonn: IPBES Secretariat.

Landon, M. (2006). *Environment, Health and Sustainable Development*. Maidenhead: Open University Press.

Landrigan, P.J., Fuller, R., Acosta, N.J.R. et al. (2017). The Lancet Commission on Pollution and Health. *The Lancet* 391 (10119): 462–512.

Melillo, J. and Sala, O. (2008). Ecosystem services. In: *Sustaining Life: How Human Health Depends on Biodiversity* (ed. E. Chivian and A. Bernstein). Oxford: Center for Health and the Global Environment, Harvard Medical School, Oxford University Press.

Persson, L., Arvidson, A., Lannerstad, M. et al. (2010). *Impacts of Pollution on Ecosystem Services for the Millennium Development Goals. SEI Project Report*. Stockholm Environment Institute https://www.sei.org/publications/impacts-pollution-ecosystem-services-millennium-development-goals/ (accessed 27 February 2020).

Prüss-Ustün, A., Wolf, J., Corvalán, C. et al. (2016). Diseases due to unhealthy environments: an updated estimate of the global burden of disease attributable to environmental determinants of health. *Journal of Public Health* 39 (3): 464–475.

Ramsar Convention on Wetlands (2018). *Global Wetland Outlook: State of the World's Wetlands and their Services to People*. Gland: Ramsar Convention Secretaria.

Rauno, H. (2011). *Frontiers in pharmacology. In Silico Toxicology – Non-Testing Methods* https://doi.org/10.3389/fphar.2011.00033.

Rockström, J., Steffen, W., Noone, K. et al. (2009). A safe operating space for humanity. *Nature* 461: 472–475.

The Essential Chemical Industry-online (2018). *Green Chemistry*. York, UK: Centre for Industry Education Collaboration, York University https://www.essentialchemicalindustry.org/processes/green-chemistry.html (accessed 26 February 2020).

UN (2015). *Transforming Our World: The 2030 Agenda For Sustainable Development. A/RES/70/1*. United Nations https://sustainabledevelopment.un.org/content/documents/21252030%20Agenda%20for%20Sustainable%20Development%20web.pdf (accessed 28 February 2020).

UNEP (2013). *Global Chemicals Outlook – Towards Sound Management of Chemicals*. Geneva: United Nations Environment Programme.

US EPA (2017a). *Basics of Green Chemistry*. Washington, DC: United States Environmental Protection Agency https://www.epa.gov/greenchemistry/basics-green-chemistry#definition (accessed 28 February 2020).

US EPA (2017b). *Green Engineering*. Washington, DC: United States Environmental Protection Agency https://www.epa.gov/green-engineering (accessed 19 October 2020).

Watts, N., Amann, M., Ayeb-Karlsson, S. et al. (2018). The Lancet Countdown on health and climate change: from 25 years of inaction to a global transformation for public health. *The Lancet* 391 (10120): 581–630.

Whitmee, S., Haines, A., Beyrer, C. et al. (2015). Safeguarding human health in the Anthropocene epoch: report of The Rockefeller Foundation – Lancet Commission on planetary health. *The Lancet* 386 (10007): 1973–2028.

WHO (2018). Climate change and health. World Health Organization. COP24 special report (2018). In: *Health and Climate Change*. Geneva: World Health Organization.

WHO/CCAC (2018). *World Health Organization Releases New Global Air Pollution Data*. World Health Organization and Climate & Clean Air Coalition http://www.ccacoalition.org/en/news/world-health-organization-releases-new-global-air-pollution-data (accessed 16 October 2018).

World Commission on Environment and Development (1987). *Our Common Future*. Oxford: Oxford University Press.

WWF (2018). *Living Planet Report – 2018: Aiming Higher. Summary. World Wildlife Fund* (ed. M. Grooten and R.E.A. Almond). Gland: World Wildlife Fund.

Supplementary Chapters S21–S24 listed in Table of Contents will be available in the online website www.wiley.com/go/toxicologyofpollution2e

Index

a

Acclimation periods and temperature 538–539
Acetaldehyde 30, 38, 250, 337
Acidification 15, 85–87, 309, 401, 409, 413–414, 520
Acid mine drainage 85–86, 309, 312, 414
Acid precipitation or rain 87, 337, 349–350
Acid sulfate soils 309, 401, 414
Actinides 303–304
Advection 80–81, 342, 399, 406
Aerosols 27, 333–335, 342–343, 346, 366, 371, 377, 379–380, 382
Aflatoxins 116, 260
Agrichemicals 4, 35, 236
Airborne particles 334–335, 342–343
Air emission factors, US EPA 407
Air pollutants 32, 333
  ambient air indicator levels for global regions 348–349
  atmospheric residence time 346–347
  behavior and fate in atmosphere 340–347
  carcinogenic 356
  characteristics of 27, 333–338
  deposition 349–352
  ecological effects and risks 349–352, 358–359
  effects on human health 1, 352–358
  effects on the atmosphere 357–358
  exposures and typical levels 347–349
  global emissions and inventories 339
  health effects of particulate matter 352–354
  indoor air sources and health effects 340-1, 347, 356–357
  primary and secondary 334
  sources and emissions of 338–341

Air pollution 17, 333, 522
  advances in monitoring 446
  conceptual model 331–332
  global burden of disease 352, 357, 522
  global impacts and risks 331, 352, 358–359
  household 341, 358
  modeling 349
Air toxics 32, 242, 267
Air-water partitioning 69
Albedo 367–368, 377–381
Allergens 339–342, 347, 390
Aluminum 336, 350–351, 401, 413–414
Ambient air pollution, WHO global study 459–460
Ames test 114–115
Ammonia, various 6–8, 29–38, 146, 161–168, 336–358, 413–414
Amphibians 54–55, 209, 212, 214, 289, 389, 521
Anoxia 145, 152, 154
Antibiotics 25, 30, 271, 282–283, 400–465, 521
Antiknock agents 237
Aquatic algae 29, 131–130, 136, 144, 149, 158, 160–162, 171
Aqueous solubility 182, 193, 199, 202
Aquifers 29, 143, 235, 398, 432
Arctic ecosystems 1, 54, 242, 247, 249, 352, 389
Argon gas 332
Aromatics 232, 234, 241, 243
Arsenic 180, 183, 310–311, 322–324
Asbestos 400–401, 416–417, 423, 428, 438
Asphaltenes 232–233
Asthma, introduction 11–13, 31, 122, 221
Atmosphere, structure and composition of 331–333

Atmospheric aerosol particles 370–371
Atmospheric concentrations, pollutants 348–349
Atmospheric particles, effects on atmosphere 371
Atmospheric pressure 30, 267, 333, 458
Atmospheric processes, of pollution 340–347
Atrazine 32, 55, 85, 181, 183, 187–188, 192–195, 202, 214–215
Attributable burden of disease (AB) 460
Attributed mortality (AM) 25, 459–460
Average daily dose or intake (ADD or ADI) 93
Azoxystrobin 183, 188–189

b

Basel convention 37–38, 274, 423, 438, 488, 503, 520
Bees 186, 206, 208–209, 211–212
Benzene(s) 233, 237–239, 242–244, 250, 470, 477, 482
Benzoylurea insecticides 180, 186
Bhopal 9, 293
Bifenthrin 181, 183, 186
Bioaccumulation 70, 202–203, 243–244, 313, 410–414
Bioaerosols 24, 27, 31, 334, 338, 340, 347, 357
Bioalcohols 230, 239, 245
Bioassays 11, 13, 15, 102–114, 202, 205–209, 317, 412, 448, 477
Bioavailability 401–402, 410, 472
Bioconcentration 70, 202–203, 243
Bioconcentration factor KB 182, 202–203, 243, 313, 411
Biodegradation 2, 71–72, 79, 105, 182, 203, 239–279, 409–411
Biodiesel 30, 231, 235–236, 239, 240, 245

Biodiversity   1, 51–56, 213, 291–292
Bioenergy   235
Biofuels   235–237
Biogases   235–237
Biogeochemical cycles   10, 15, 29, 34, 53, 303, 331, 350, 405
Biological pollutants   30–31
Biomagnification   71–72, 313–314
Biomarker(s)   109–100, 287–288, 322–324, 444–445, 454, 461, 472
Biomass   109, 115, 235
Biomimicry   513
Biomonitoring   11, 13, 59, 199, 220, 257, 266, 287–292, 445–446
Biopesticides   179–183, 188
Bioreactors, landfill   427–429
Biosolids   400–402, 414, 416
Biotic exposure   63, 77
Biotic indices   136–139, 156–157
Biotransformation, pesticides example   203–204
Birth defects   11, 33, 211–221, 264, 337, 415–419, 436–440, 524
Bisphenol A   12, 33, 116–120, 256–260, 269–278, 402–403
Blood, whole, serum or cord   200, 220, 322–323
Blue-green algae (cyanobacteria)   162, 171
BOD (biochemical oxygen demand)   146–147
Boiling point   30, 63, 65, 237, 267, 277–278, 337, 489, 491
Bradford Hill's criteria   43
Breast milk   220, 222
Brominated flame retardants   11, 34–38, 265–266, 274–288, 520
Brundtland commission, sustainable development   510
BTEX   400–401
Bulk density, of soils   194, 398
Burden of disease (BoD)   10–18, 44–51, 292–293, 331, 357, 448, 509, 522–530
  ambient air pollution (PM2.5)   458–460
  chemical exposures   49–51
Butadiene   5, 32, 35, 116, 250, 257, 260, 268, 294

C

Cadmium   303–316, 319, 321–323, 416
Cancer   108, 217, 219–222, 415–418, 436–440, 476–478, 499
Cancer, basic stages and mechanism of   115, 117, 122
Carbamate insecticides, types   179–180, 183, 185
Carbaryl   185
Carbohydrate metabolism, interference with   96
Carbonates   144, 308–309, 336
Carbon, black   358–359
Carbon cycle, global   374
Carbon dioxide   370
  atmospheric levels for past 800,000 years   384
  equivalents   383
  global emissions   374–377
  incineration emissions   429
  radiative forcing   381–382
Carbon footprint   516
Carbonic acid   86–87, 336, 349, 389
Carbon monoxide   354
Carbon, transformation of   145
Carcinogens   16, 33–35, 38–55, 108, 219, 322–323, 356, 522
  genotoxic and nongenotoxic   115–118
  metals   337
Cation exchange capacity (CEC), soils   399, 409
Cell communication   114
Cell death (apoptosis)   114, 118
Cells, germ and somatic   114
Cellular membrane function, disruption of   96
Cellular organelles and enzymes   91
CFCs (chlorofluorocarbons)   337–338
Chemical databases   490
Chemical equilibria   66
Chemical industry, history of development and value   4–10, 487
Chemical management, information and data   489–492, 502–503
Chemical mixtures   13, 259, 271, 290, 299, 417
Chemical oxygen demand (COD)   147
Chemical pollutants   29–30
Chemical production   5, 9, 35–38, 232, 258, 373, 513
Chemical properties, prediction of (US EPA EPI Suite™)   499–500
Chemical regulation, EU-REACH (Case study 19.1)   494–502
Chemicals
  data, registration, and evaluation   12, 489–490
  health and environmental effects   10–13
  regulation and assessment by countries   493–494
  risk characterization methods-REACH   500–502
  sound management   502
Chemical safety report, ECHA   495–500
Chemicals management
  international organizations   12, 487, 502–503
  multilateral treaties   502–503
  strategy, and global plan of action   487–489
  towards 2030   502–504
Chemical tests
  EU-REACH guidelines   492–493
  OECD guidelines   491–492
Chemical toxicity, non-animal testing guidelines   490, 492
Chemical transformation/degradation   82–83
Children, pesticide exposures and effects   216–220, 222
Chlorinated hydrocarbons (CHC), pesticides   183–184
Chlorophyll-a   55, 130–131, 170–173, 351
Chlorothalonil   181, 183, 189
Chlorpyrifos   94, 100–102, 181, 183, 185, 403
Chromium   306, 311–316, 321–323
Chromosome damage, clastogens and aneugens   117
Chromosomes   113
Circular economy model   528
Clay   398–399
Climate
  computer modeling   386–388
  extreme events   365, 388–390, 523
  feedbacks, types of climate change processes   381
  forcings   377–378, 381
  sensitivity   380–381
  system   333, 377–378, 380–381, 385
Climate change
  effects on ecosystems and biodiversity   56, 388–389
  global effects and risks   390–391, 523
  impacts on human populations   389–390, 523
  industrial revolution to present   384–385
  key indicators of past change   383
  observed impacts   383, 385

Index | 573

past million years   383–384
preliminary estimates of annual global deaths   390
tipping points   381
twenty-first century and beyond   385–388
Coal   6–8, 34, 36, 231–235, 239
Coal seam gas   234–235
Combustion   24–25, 30, 36, 82, 108
Commercial chemicals   4, 6–8, 10, 29, 257–258, 294, 493–494
Comparative risk assessment   46, 50, 448
Compartment models   10–11, 73
Conceptual model, for pollutant impact   466
Consequences, rating of   468–470
Contaminants of emerging concern   255–256, 258, 270–272, 404
Contaminant transport, fractured and porous media   406–407
Contaminated land, epidemiological studies   415–418
Contaminated sites   397, 404, 410–18, 524
Cooling systems   535–537
COPD   49, 292, 354, 357–358, 459–460, 522
Copper   306, 314, 316, 321–322
Coral, effects on   1, 51, 54–56, 86, 215, 247, 249, 389, 434
Cumulative frequency distributions   482–483
Cumulative probability scale   99
Cypermethrin   57, 65, 186, 217, 401
Cytological testing   11

d

Daily intake   199–220, 311, 324, 469, 473, 476, 478
DALYs   12, 44–50, 292, 295, 357–358, 459–460, 525, 531
Dams and reservoirs   46, 70, 86, 150, 160, 415, 521
Darcy's law   406–407
Dead zones (DZ)   16, 55, 145, 160, 433, 524
Deaeration   145, 150
Deforestation   15, 53, 56, 373, 376, 521–522
Degradation process   195–196, 198, 206
Deltamethrin   186, 217
Dengue   31, 389–390, 523
Deoxygenating substances   143–144

Deoxygenation, ecological effects   129–132, 158–159
Deoxyribonucleic acid (DNA) *see* DNA
Dermal sorption   93, 471, 473
Detritus   28, 70, 144–145, 154–168, 244, 305, 312, 314, 398
Diabetes   11–13, 44, 120–122, 221, 294–295, 323
Dichlorodiphenyltrichloroethane (DDT)   10, 291, 352, 416, 438
Diesel fumes   24, 27, 242, 250, 294, 356, 371, 522
Dietary intake, of pesticides   199–220
Diffusion   406–407
Digestion, biological   15, 149, 236, 426
Dimethyl sulfide   336–337, 346, 371
Dioxins and furans, selected   10–12, 30–37, 260–265, 288–289, 293–295, 415–419, 520–524
Disability-adjusted life years (DALYs)   44–46, 48, 50
Discharge, effects of organic matter   2, 25, 130–132, 136–138
Disease
types and theories of causation   1, 13, 31, 42–44
vector-borne   41, 46–47
Disinfectants   30, 181, 255–256, 270, 272–273, 276, 336
Disinfection by-products   25, 30, 102, 256, 271–272, 288
Dissolved oxygen (DO)   145–146
criteria   153–154
DO "sag"   148–149
effects of reduction   152–156
environmental variations of   150–152, 159
DNA
genetic toxicology   113–119
repair mechanisms   114, 117
Dose-response
cumulative probability curves   98–103, 482
relations   95–100, 103–104, 469, 473–474, 477, 481
Doses, calculations   12–13, 15, 93–104
Drivers, of pollution   2–3, 519
Dyestuffs   4, 30, 255–256

e

Earth, state of   1, 3, 9, 13, 15–18
Earth system feedbacks   381
Eco-audit   516
Ecological cascade   56
Ecological footprint   1, 516, 525

Ecological health   41–42
Ecological risk assessment   479–482
Ecological soil screening levels, US EPA   411–413
Ecosystem services   1, 51–53, 56–57, 414–415, 425, 517–518, 521
Ecosystems, impacts   1, 10, 13–16, 22–23, 30–32, 41–42, 51–57
Ecotoxicology   2, 11, 13, 63, 91, 125, 132–134, 479, 497, 501–502
Electromagnetic radiation   22, 26, 28, 457
Elements, essential and nonessential   304–305
Emerging pollutants, organics   33–34, 267–274
Emissions and releases, of pollutants   34–38
Endocrine disruption   113, 118–122, 213–215, 291–295, 433, 479, 503, 521, 530
Endocrine system   119
Endosulfan   181, 183–184, 217
Energy balance, of Earth   24, 333, 365–368, 371, 377–378
Energy budget, of Earth's atmosphere   368–369, 378
Energy, global primary fuel consumption   232
Energy pollutants   28
Environmental compartment   15, 63–69, 80, 87, 277, 432, 498–501
Environmental health and risks   41–42
Environmental phases, distribution of chemicals   71–72
Environmental toxicants, classes of   108–109
Enzyme action, modification   96
Epidemiological evaluations   417–418
Epigenetic mechanisms of action   122
Epigenome   122
Epilimnion   131–132, 166–168
Epoxides   95, 114
Erosion   24–25, 51–52, 57, 190–191, 305–306, 398, 400, 405–406
Estrogen receptors (ERs)   118–121
Estrogens, natural and synthetic   55, 291
Estuaries   149–150, 160, 167, 247–248, 308–309, 351, 521
Ethanol   30, 68–69, 99, 231, 235–241, 250, 267, 337, 497, 517
Ethylene/oxide   5, 30, 35, 116, 219, 256–257, 267, 270, 274, 337
European chemicals agency (ECHA)   491–500

European environment agency 21, 38, 350
Eutrophication, natural and accelerated 160–162
Exponential growth 162
Exposome 13, 122
Exposomics 13
Exposure and human health 357, 411, 417, 471–473, 496, 498–500
Exposure duration 93–94
Exposure factor (EF) 472–473
Exposure frequency 93–94
Exposure pathways 92–94, 460, 471, 473, 481
Exposure-risks relationship 460
Exposure science 13
Extinctions, global and local 54, 519

f

Fertilizers, introduction 3, 7–9
Fick's law 407
Fipronil 180–181, 183, 186
Fires 25, 27, 239, 242, 305, 336, 373–374, 415, 522
Fish, effects of estrogens 55, 291
Flame retardants 11, 30, 33–34, 36, 38
Fluorides 336, 352, 402
Fluorocarbons 371
Fluoropolymers 30, 36, 58, 256, 276
Food additives 34, 311
Food chain, introduction 10–11
Formaldehyde 12, 30, 35, 38, 250, 260, 272, 294–295, 347, 356
Fossil fuels
  emissions and health effects 235, 239, 242, 249–250, 303
  environmental behavior and fate 239–242
  types of 231–235
Fracking chemicals 235
Free radical species 94, 343
Freshwater systems 143, 305, 308
Freundlich isotherm 67–68, 87, 194
Fuel additives 237–240
Fuels, physicochemical properties 237
Fugacity 66–67, 78–81, 240, 281–282, 500
Fugacity model, multi-media example 281–282
Fugitive emissions 190, 196, 338, 407, 426, 428–433
Fumigants 179–181, 190, 192, 216, 218, 220
Fungicides 180–181, 183, 187–189

g

Gas, coal seam and shale 234–235
Gases 335–336
Gasoline 50, 232–233, 236–237
  phasing out of lead 237
Genes 44, 51, 53, 100, 113–114, 117, 121–122, 402, 417
Genetic diversity 51, 292, 518
Genome 113
Genotoxicity 113–117
GHG emissions
  mitigation of 54, 358, 375, 386, 388, 391, 421, 449, 518, 523, 526–528
  scenarios 385–388
Global burden of disease 44, 49–51, 357, 448, 460
Global climate indicators 450
Global climate observing system (GCOS) 449–450
Global cooling, mechanism 366–367, 371
Global deaths, pollution risk factors 1, 51
Global distillation, of pollutants 75
Global energy budget 369
Global harmonization system (GHS), pesticides 216
Global nitrogen fixation 166
Global warming 365, 522–523
  conceptual model 366
  human health impacts 389–391, 522–523
  potential, greenhouse gases 382
  projected climate changes (AR6 2021) 386
  some historical observations 383–385
Glyphosate 179–183, 187, 192–193, 196, 206–207, 219
Grasshopper effect 75
Green chemistry 513–515
Green chemistry, technologies 517
Green cities and buildings 515–517
Green engineering, goals and principles 514–515
Greenhouse effect 365–369
  basic mathematical model of 367–368
  human induced warming 368–369
Greenhouse gases (GHGs) 365
  accelerated rate of global warming 365
  anthropogenic emissions 374–376
  atmospheric absorption of IR 370–373
  atmospheric concentrations 377
  calculation of emissions from solid fuels 377
  emissions, and sinks 373–376
  environmental exposures 377
  global inventories 374–375
  nature and sources of 373–374
Groundwater contaminants 397
Guideline dose 477
Guideline value (GV) 476, 478

h

Haber-Bosch process 7–8, 161, 166
Habers Rule 102–103
Habitat destruction 52–53, 56
Half-life 73, 261–262
  pesticides 85, 182, 184, 187, 190–193, 196, 202–203, 410
  radionuclides 543, 546
Halogenated organics, examples of 35–36, 274
Halogens 304, 345–346
Hazard evaluation 467
Hazard identification 471
Hazard index (HI) and hazard quotient (HQ) 476, 478
Hazardous air pollutants (HAPs) 337
Hazardous chemicals
  international management 487–489, 504
  predicted no-effect concentrations 494–502
  regulation and testing 491–494
Hazardous substance 22
Hazards, environmental 3, 75, 490, 493–494
Health, WHO definition 4, 41
Heated effluents 535–537, 540
Henry's law constant 69–70, 182–183, 190, 198
Herbicides, types 180–181, 183, 187–189
Hexachlorobenzene (HCB) 32, 69, 188, 261–262
Hexachlorocyclohexanes (HCH) 183, 199, 200, 204
Holocene epoch 385
Hormones, nature of 30, 33–34, 55, 113–116, 118–121
Human health evaluations 92, 474
Human poisonings 48–49, 198, 216, 292–293
Human sewage 143

Hydraulic gradient 406
Hydrocarbons
    aliphatic and aromatic 232–235
    biogenic 234
Hydrochloric acid 12, 29, 38, 235, 336
Hydrofluoric acid 36, 336
Hydrofluorocarbons (HFCs) 358, 370, 377, 382
Hydrogen sulfide, various 15, 24, 29, 38, 82, 152, 154, 157, 168
Hydrolysis process 195–196
Hydrophilicity 68
Hypolimnion 131, 166–168, 171
Hypoxia 145, 152–153, 155

i

Imidacloprid, insecticide 181, 183, 186, 202
Immune system 10, 15, 55, 59, 221, 260, 263, 293–295, 322, 324
Impact equation, I = PAT 2–3, 519
Indicator and monitoring species 136, 139
Industrial ecology 512–513
Industrial revolution 3–4, 21, 55, 331, 365, 370, 385, 421, 457, 519
Ingestion, process 15, 30, 43, 58, 93
Inhalation, process 15, 26, 30–31, 58, 93
Injuries 1, 43–50, 52, 292–293, 388, 438
Inorganic chemicals 4, 6, 30, 38, 180, 274, 400
Inorganic gases 29, 335
Insecticides, introduction 179–182
Insects, various 5, 31, 56–57, 181, 209, 213, 289, 433
In silico methods, non-animal testing 505, 526
Institute of health metrics and evaluation (IHME) 46, 50
Integrated urban water management 517
Intergovernmental panel on climate change (IPCC) 56
International agency for cancer research (IARC) 43, 56
Interventions, pollution reduction 530–531
Ionizing radiation
    health effects of 543, 549–553
    recommended dose limits 553
    sources and exposures of 546–548

k

K_D values, sorption of metals 409
Kerogens 234
Kidneys (renal) 11, 58, 294, 323, 415, 417, 436–437, 524
Kinetic rate processes 84–85, 410
Kinetics, environmental transformation, and degradation 84–85
K_{OW} (or P) 65, 68–72, 74, 76, 79

l

Lakes, OECD key quantitative characteristics of 170–173, 175
Landfill gas 237
Langmuir isotherm 67, 194
Lanthanides (rare earths) 403–404
LC_{50} 99–103
LC_{50}, incipient or threshold 98, 102–104
LD_{50} 99–101, 501
Leachates 426–428, 431–432, 436
Leaching, pollutants 87, 190–197, 401, 406, 410–411, 414, 430
Lead (see Case study 3.1) 49–51, 303, 305–324, 400, 413, 416
Lead, in blood (PbB) 50, 322–324
Lethal effects 106–108
Life cycle assessment 516
Lifetime average daily dose (LADD) 93–94
Lifetime dermal average daily dose (LDADD) 94
Likelihood of exposure rating 468
Lipid metabolism, disruption 96, 121
Lipids 22, 58, 68, 70–72, 74, 96, 106, 182
Lipophilic compounds 68, 70–71, 74, 78–79, 83, 106, 201, 402, 408
Lipophilicity of chemical pollutants 68, 182, 185, 203
Liquid fuels 4, 232, 235, 517
Liver 58–59, 203–204, 285–286, 323–324, 416–417, 436–437
Liver enzyme induction, pesticides 211
Living planet index, impacts on biodiversity and extinctions 519
Log-normal plot 102–103
Long-range transport (POPs) 74–75
Lowest observable adverse effects level (LOAEL) 100–101, 475
LT50 (lethality) 102–104
Lungs 27, 31, 95, 109, 312, 334, 472

m

Malaria 7, 10, 31, 179, 183, 338, 389–390, 523
Manganese 165, 306–308, 316, 398, 413, 439
Megacities 3, 345, 348
Melting point 489, 491, 499
Mental retardation 50
Mercury, methyl-and dimethyl 285–287, 310–311, 322–324
Mercury, various 5, 10, 12, 15, 32–37, 49, 55, 303, 305–324
Metabolism, effects on 11, 24, 72, 94, 96, 106, 116–121, 294–295
Metalloids 29, 76–77, 303, 305, 321–322, 401–402, 454, 520
Metals
    aquatic ecotoxicity 317–318
    in aquatic systems 306–309
    bioaccumulation 76–77, 313
    bioavailability 312–313, 316, 318–319, 322–324
    biological processes in organisms 311–314
    biomarkers 322–324
    carcinogens 322–323
    daily metal intake estimates 310–311
    ecological effects of 321–322
    environmental behavior and fate 306–310
    environmental exposures and levels 310–311
    factors influencing human toxicity 324
    factors influencing toxicity in aquatic organisms 318
    food transfer and biomagnification 313–314
    human health effects of 322–324, 337
    properties of 303–305
    sediments and toxicity 319–321
    sources and emissions of 305–306
    toxic effects of common metals 323
    toxic effects on organisms 314–321
    transition or heavy 305
Metal toxicity
    dose-response relations 314, 317, 321–322
    mechanisms 314–315, 321
Methane 370
Methane hydrate 231, 234
Methanol 235–236, 246
Microorganisms 2, 11, 15, 23–24, 30–31, 43, 51, 84–85

Microplastics, in environment 28, 241, 402, 432, 434–435
Microsomal enzyme system, modification 96
Millennium development goals 52
Mining 2, 3, 16, 24–25, 85–86, 324, 376, 397, 404, 415–416
Mixed function oxidase (MFO). 82, 204, 209, 211, 243-244, 285
Mobility of chemical pollutants 399, 401–402, 408, 410
Model, chemical pollutant evaluation (*see* Figure 1.3) 13–17, 57, 126
Models, bioaccumulation 72–73, 203
Mode of action (MOA) 56, 121–122, 185, 187, 210
Molds 24, 181
Molecular weight 22–23, 26, 63–64, 82, 105, 182–183, 186
Molecules, characteristics and effects 105
Monitoring
  air quality systems 446
  data management 454, 457
  data quality indicators 452
  global climate change 447
  global warming of oceans 457–458
  health impact and risk assessment 448–449, 459–460
  integrated-basic steps (*see* Box 17.1) 448
  integrated environmental 446–448
  objectives, programs, and methods 443–444, 450–457
  physicochemical and biological 444–445
  QA/QC 452, 454–457
  sampling and analytical plan and programs 452–457
Monitoring strategies 444–445
  human biomonitoring 446
Monitoring strategies, *see* Case study 17.1 447
Monomers 267, 269
Monomictic change, lakes and reservoirs 167
Mortality, introduction 16, 31, 45–49, 98–99
mRNA 113, 119, 121, 291
MSDS (Material safety data sheets) 468, 470
MTBE 237–240, 243, 257
Multiple antibiotic resistant bacteria 34, 272, 400–402, 415, 521, 524
Municipal solid wastes 370, 421–425, 427, 435–436, 524
Mutagenicity 114–117

Mutagens 114–118, 292
Mycotoxins 114, 256, 341

n

Nanomaterials 5, 8, 29, 34, 272, 303, 497, 503
Nanoparticles 22, 27–29, 66, 108–109, 258, 267, 269, 334–335, 349, 402
Nanoparticles, engineered 28
Natural gas 7–8, 35, 231–236, 242, 256–257, 336, 370
Natural oil seeps 29, 238–239, 242–243, 400
Natural pesticides 179–180
Neonatal effects 47
Neonicotinoids, types 180, 186, 210
Nervous system 55, 118, 185, 210–260, 292–294, 314, 354, 416
Neurotoxicity 218, 323
Nickel 306, 311, 316, 321–323
Nitric oxide 336, 344–345, 355, 371
Nitrification 145–146, 164, 336, 414
Nitrite 82, 146, 164, 167, 196, 315, 430
Nitrogen
  cycle 164
  deposition 350–351
  fixation of 164
  and phosphorus, in aquatic systems 413–414
  transformations of 164–166
Nitrogen dioxide 342–343, 345, 355
Nitrogen oxides 24, 237, 334–350, 358, 371, 522
Nitrosamines 195–196
Nitroso dimethylamine 116, 272, 403
Nitrous oxide 342–343, 345, 370–371
N-nitroso-dimethylamine 34
Nonlinear processes 53, 449
Non-metals 403–404
Nonpoint sources 10, 25, 128, 143, 271, 282, 338, 530
No observable adverse effect level (NOAEL) 100–101, 475–476
Novel entities, chemicals 18
Nuclear energy 543–549
Nuclear fission and fusion 543–548
Nuclear fuel cycle 547–548
Nuclear power accidents 548–549, 551–552
Nutrients 162
  assimilation and enrichment 162–163
  ecosystem changes 171–172
  enrichment factors 166–169
  loadings and trophic state 169–171

o

Occupational exposures 93–94, 100, 198–201, 218–221
Ocean acidification 86–87, 520
Octanol-water partition coefficient 68–72, 182–184, 202–203, 408
OECD 489–491, 502–504
  screening tool 75
  test guidelines 91, 205, 491–493
Oil and tar sands 234
Oil spills 238–239 *see also* Petroleum
  ecological impacts 247–249
  effects on human health 249
  weathering effects 239–241, 243
Oligotrophic status, of water body 131–132, 160–161, 167–171, 173
Organic carbon-water partition co-efficient 194–195, 407–409
Organic chemical pollutants, types of 257–258, 337–338
Organic chemicals, production 256–258, 276
Organic matter 155
Organic pollutants
  chemicals 256
  toxicity classifications 259–260
Organochlorines, types 179–180, 183–184
Organometallics 305
Organophosphate insecticides, types 179, 180, 182–185
Organotin (tributyl tin) 181, 214
Oxidation process 10, 15, 24, 82–83
Oxygen dynamics and balance, aquatic systems 150
Oxygen gas 147, 331–332, 336–337, 345, 347, 370, 372
Oxygen, solubility in water 145–150, 152–154
Ozone 336, 343–346, 351, 354–355, 371
Ozone-depleting chemicals 337–338
Ozone layer, stratosphere 344

p

PAHs (polyaromatic hydrocarbons) 232–233, 337
Pandemic 1, 125, 294, 520
Parasites 23, 31, 44, 179, 212, 216, 351
Particles
  aerodynamic diameter of 27, 346
  air distribution of 334–335
  modes of formation 335
  types and sizes 27, 334–335, 342–343

Index | 577

Particulate matter (PM)   27, 334
  anthropogenic emissions of   339, 385
Particulate pollutants   27, 334, 351–352
Partition coefficient (K), definition of   66
Partitioning behavior, fundamentals   63–64, 66–72
Partition, phases   70
Pathogens   3, 22–25, 28–31, 34, 43–46, 135, 212, 338, 351, 400
PBTs (persistent, bioaccumulative, and/or toxic chemicals)   258–261, 267, 274, 279, 281, 496–498
  classification and criteria   258–261
PCBs *see* Polychlorinated biphenyls (PCBs)
Pentachlorophenol (PCP)   188, 273
Perchlorate   12, 33–34, 116, 118
Perchloroethylene   32, 36, 276, 294, 517
Perfluorinated alkyl substances (PFAS)   57-59 264–267
Perfluorooctane sulfonic acid (PFOS)   58–59 264–266
Perfluorooctane sulfonyl fluoride (PFOS-F) 264–265
Perfluorooctanoic acid (PFOA)   58–59, 264–265
Periodic table, of chemical elements   304
Permethrin   181, 186, 206–208, 217, 219–220
Persistence, of pesticides   179, 182, 185–187, 191–193, 195–198
Persistent organic pollutants (POPs)   32, 261–266
Persistent toxic substances (PTS)   32
Pesticides   179, 182
  absorption, biotransformation, and excretion   200–204
  acute toxicity to mammals (LD 50)   216–217
  aqueous solubility   182, 193, 199, 202–203
  in atmosphere   191, 193–194, 196–197
  bioaccumulation and bioconcentration   186-7, 191, 201–203
  chronic toxic effects on humans   217–222
  classification and types (*see* Table 9.1)   179–182
  ecosystem effects   211–216
  effects on children   220–222
  effects on plant communities   205, 214–216
  effects on pollinators   186, 205–206, 209, 211–212
  environmental behavior and fate   190–198
  environmental exposures   198–200
  GHS hazard classification   217–218
  global decline of wildlife   199, 209, 213
  groups or families of   183–189
  human cancers   219–221
  human poisonings   216
  interactions   193, 210–214
  key properties   191
  mobility   182, 187, 191, 193–194
  nature and properties   182–183
  neurological effects on humans   218–219
  reproduction and development effects   221
  in soils   184–185, 187, 191–199, 203
  sources, emissions, and losses   189–190
  toxic action (MoA) of   210–211
  toxicity (acute and chronic)   204–209
  typical environmental levels   199
  uses (world and USA)   189–190
  in water   190–192, 197–200, 215
Petrochemicals   4–5, 7–8, 34, 127, 235, 256–259, 269, 282, 520
Petroleum   4, 6–8, 30, 35, 231–234, 237
  bioaccumulation and food webs   243–245
  biodegradation of hydrocarbons   239–241
  ecosystem effects   246–249
  effects on health and ecosystem services   249–250
  environmental fate   239–242
  exposures and uptake   242–244
  gases   231–232
  oil spills, recovery and remediation   246–249
  refined   232–234
  sources and inputs to oceans   237–239
  toxicities in aquatic organisms   244–246
PFCs, sulfur hexafluoride   338
PFOS   11, 57–59, 256, 264–267, 278–282, 286–289, 400, 402–403
Pharmaceuticals and personal care products, (PPCPs)   271–277
  discharges in Beijing, China (*see* Case study 11.1)   276–277
  environmental pathways   275
  in waters   271–274
Phases, exchanges and equilibria   63, 67, 77–81
Phenols, environmental   272–274
Phenoxyacetic acids (2,4-D, and 2,4, 5-T)   181, 183, 187, 189
Phosphate   29, 35–37, 164–165, 272, 306–307, 401, 414–416, 517
Phosphoric acid   29, 35, 184
Phosphorus cycle   164
Phosphorus, transformations of   164–165
Photochemical reactions, air pollutants   28, 83, 197, 340, 342, 371
Photochemical smog   344–345
Photodecomposition, of pesticides   193, 195, 198
Photolysis   10, 58, 239, 241–245, 277–278, 431, 491
Photosynthesis   144–145, 149–150, 155, 164, 166, 171
Phototransformation.   83, 491
Phthalates   12, 30, 32–33, 256–269, 288, 520
Physicochemical properties   63–66
Physiological and behavioural properties   14–15
Phytoplankton   1–2, 55, 130–132, 166–173, 320, 351, 371
Planetary boundaries, effects of chemical pollutants   18, 520
Planetary health   17–18, 517–518, 525
Planetary stewardship   525–526, 529
Plastic additives   270
Plasticizers   269
Plastic pollution
  global inputs into oceans   433–435, 520
  impacts on marine animals   433–435
Plastic pollution (*see* Case study 16.1)   433–435
Plastics   269
PM2.5   27, 334–335, 459–460, 522
PM1, and PM10   27, 334-335
Point sources   10, 143, 197, 271, 282, 338, 397, 530
Polarity   63, 65
Pollen   27, 31, 338, 341, 383
Pollinators   186, 205–206, 209, 211–212
Pollutants
  nature, properties, and sources   21–25, 29, 63–66

Pollutants (contd.)
  priority 33
  types and general effects 26–31, 128
Pollution 2, 13–17, 56–57, 143, 509
  control 1, 3, 38, 166, 350, 358, 428, 439, 515–530
  ecology, nature of 125–128
  effects on planetary health 1, 517–525
  energy 28
  genetic variability 128, 213
  human health effects 522–524
  human-induced exchanges 1–2, 53–56, 518–519
  key drivers of change and risks 2–3, 519
  The Lancet Commission 17, 21, 44, 50, 509
  modified communities 14–15, 125–126, 139, 156–159, 211
  offsets 516
  organic matter 129–132, 136
  plant nutrients and eutrophication 129–132
  premature global deaths related to 522
  prevention, reduction, and sustainability 509, 525–529
  strategies, interventions, and controls 529–531
  suspended solids 24, 28, 66, 80, 129, 557–561
  tolerance 128
  toxic chemicals 520
  underestimation of impacts 1, 509
  variations with temperature and latitudes 135–136
Pollution monitoring 443
  environmental metrics and criteria 450
  remote sensing (see Case study 17.2) 457–458
Polybrominated diphenyl ethers (PBDEs). 265-266, 291
Polychlorinated biphenyls (PCBs) 10–12, 202, 263, 291
Polychlorinated dibenzofurans (PCDFs) 263–264
Polychlorinated dibenzo-p-dioxins (PCDDs) 263–265
Polymerization reactions 267–268
Polymers 267–269
Polystyrene 268
Polyurethane 268
Polyvinyl chloride (PVC) 267–268

POPs and PBTs, criteria for defining 261–262, 287
POPs, listings under Stockholm Convention 261–262
Population attributable fraction (PAF) 46–48, 459–460
Population, growth 2–3, 54, 519
Porosity, soils 66, 398–399, 401
Precautionary principle 516
Pressures
  population 519
  socioeconomic 3, 29, 44, 386, 467
Probit, transformation 98–99
Properties, of pollutants 10, 13–15, 21–29, 63–65
Propylene 5, 35, 257, 268, 433
Protein biosynthesis, interference of 96
Proteinphilic bioaccumulation, chemicals 58, 285–286
Protozoa 31, 45, 155–158, 180, 240
PRTR data, emissions of pollutants 38, 274
Pyrethroids insecticides, types 180, 183, 185–186

q

Q10 Law 166
Quantitative structure activity relationships (QSAR) 75–76, 104–106, 481, 489–490, 492, 497, 505

r

Radiation
  doses, types, and effects on humans 450–452
  fluxes in atmosphere 368–369, 378–379
  nonionizing 24–25
  solar 24, 83, 129, 132, 150, 333–344, 366–372, 377–383
  types of ionizing 544, 551–552
  ultraviolet (UVB) 338, 551–552
  units, ionizing 543–545
Radiative forcing (RF) 377–382
  conceptual model 365–369, 371, 377–380
  definitions, and effective or adjusted 377–379
  due to human activities 378–380
  natural changes 377–380
Radioactive decay, rate of 543, 546
Radionuclides (radioactive isotopes) 304
  ecological effects of 550

  environmental behavior and fate 548–549
Radon gas 29, 33, 116, 340–341, 347, 356–357, 400, 416, 522, 544, 546–547, 550–552
Rare earths 303 see also Actinides
REACH, European Union 489, 492–497
Receptors 95, 97, 446
Recovery, of impacted ecosystems 16, 53, 133, 156, 159–160, 213–214, 246–249, 321, 350, 525
Recycling 421–422, 424–425, 427–428, 433–435, 438
Red list on species extinction, IUCN 54
Redox conditions 83–84
Reduced life expectancy (RLE) model 103–104
Reference dose (RFD) 476
Remote sensing 46, 139, 443–461
Reproductive and developmental toxicity 118, 291–292
Reproductive toxic substances (CMR) 33
Residential exposure 93, 220–221, 417, 431, 438
Resins 30, 34–36, 232–235, 256, 258, 272, 288, 294
Resistance, pesticide 179, 204, 210, 213
Respiration 72, 95–96, 129–156, 168, 171
Respiratory pathways and system 93, 96, 107
Ribonucleic acid (RNA) 91, 113
Risk
  characterization, human health 476
  concept of 465, 467
  evaluation techniques 465–466
  management 466, 469
  qualitative observation 467–468
  total 468
Risk assessment 465–466
  cancer 476–478
  chemical exposure 468, 482
  ecological risk 479–482
  framework and processes 466–467, 469
  hazard concept 467–468
  human health 470–478
  probabilistic techniques 482–484
  semi-quantitative 467–470
  socioeconomic aspects 467
  toxicology 467
River impoundments and sediment 588–589

s

Safety factor (application factor) 475–476, 478
SAICM (Strategic Approach to International Chemicals Management) 488–489, 502–504
Salinity
   dryland 565–566
   effects of 563, 567–568
   irrigation-induced 564–568
   primary and secondary 564–565, 568
   and salinization, nature and composition 563–564
Salinization, global impacts 564–565, 568
Sampling strategies, contamination 451–453
Saprobic systems 129–130, 157–156
Sea grasses 55, 245
Secondary treatment 143, 149, 166
Second World War 4, 8, 179, 182, 260
Sedimentation 557–559
Sediment flux, estimates of 558–559, 561
Sediment partitioning with water. 70
Sediments
   ecological effects of 559–561
   transport and deposition 559
   types, sources and releases 557–558
Sewage treatment 145–146, 149
Shared socioeconomic pathways (SSPs), IPCC AR6 386, 388
Signaling, cell 114
Silent Spring, by Rachel Carson 74
Silica, health effects of 27, 250, 341, 356, 400–401, 522
Silt 398–399
Silver 316, 321
Slope factor 477–478
Smoke 24, 27, 49, 324–358, 381, 395, 460, 522
Soil
   acidification 401, 413–414
   composition 398–399
   erosion 406, 414
   gas concentration, relationships 398
   salinity 563–564
   types and properties 397–399
Soil and groundwater contaminants 400–405, 410–413, 415
   environmental effects 412–413
   human health effects and risks 415–418
Soil and groundwater pollution, agriculture and food 414–415
Soil contaminants 397
   environmental behavior and fate 405–411
   exposures and bioaccumulation 411–412
   half-lives of some organics 410–411
   management 397, 404–405, 412, 414, 418
Soil-groundwater systems 397–400
Soil organic matter 398–399
Soil pollutants, ecosystem effects 413–414
Soil pollution 397, 524
   effects on ecosystem services 414–415
Soil-pore water, three phase model for pollutants 411–412
Soils, oxidation-reduction reactions 410
Solar energy, radiation 366, 368
Solubility, of chemicals 22, 65
Solvents 30, 32, 35–36, 65
Sorption coefficient (K_D) 408
Sorption of metals, solid matrix 307–309, 333, 409
Sorption, organic contaminants 193–196, 407–410
Specialty chemicals 4–5, 7, 255–256, 258
Species diversity 51, 57, 130–137, 156–160, 481–482
Species sensitivity distribution (SSD) 100–101
Specific ion toxicity, salinity 567
Spinosyns 183, 210
Steroids 30, 119, 121, 156, 209, 211, 272–273
Stockholm convention (POPs) 261–262
Stokes' law 558–559
Stormwaters 34, 305, 402, 422, 438, 517, 530
Stratification 131–132, 150–159, 166–167, 172
Strobilurins, fungicides 180, 188–189
Styrene 267–269
Sub-lethal effects 106–109, 245–246, 317
Sulfides 304
Sulfur dioxide or sulfur trioxide 355, 371
Sulfuric acid 6–38, 309, 336–342, 350, 355, 410, 414
Surface coating chemicals 28, 30, 58, 256, 341
Surfactants 5, 7, 30, 34, 57–58, 108, 269–271
Suspended solids 133, 557–560
Sustainability, concepts, principles, and metrics 509–511, 514, 516
Sustainable design 513
Sustainable development 42, 510–511
Sustainable development goals (SDGs), United Nations 511–512
Sustainable solutions, planetary and human health 1–2, 524–529
Sustainable systems, processes and models 510–513

t

Target organisms, of pesticides, Table 9.1 179, 181-182
TCDD (tetrachlorodibenzodioxin) 289
Teflon (PTFE) 267–268
Temperature
   of the atmosphere 367
   incipient lethal 537
Teratogens 33, 116–118, 260
Tertiary treatment 144, 166
Testing, in vitro and in vivo 114–115, 490
Tetraethyl lead 237, 249, 305, 323
Thermal death point 537–539
Thermal discharges 535–536
Thermal ecology 135
Thermal pollution, effects on organisms and communities 537–541
Thermal stratification 150–151, 156, 159, 166, 172
Thermal tolerance zones 537–538
Thin eggshell effect 209, 211–212, 215
Thorium 543–544, 546, 548, 552
Thresholds 52–53, 103–104
Tipping points, ecological 52–53, 56, 125, 519
TOA, top of atmosphere (tropopause) 367–369, 378–379
TOCs (toxic organic chemicals)
   aqueous properties and fate 280–281
   behavior and fate processes 277–282
   bioaccumulation / food chain transfer 286–287
   bioconcentration (BCF) values 287
   biological processes in humans and wildlife 284–286
   carcinogens and cancer deaths 294–295
   effects on reproduction and development 295
   endocrine disruption, effects of 291–292, 295–296

TOCs (toxic organic chemicals) (contd.)
  environmental exposures and levels 282–284
  environmental fate, multi-media fugacity model 281–282
  environmental poisoning 293
  environmental toxicity 288–290
  exposures and effects on wildlife and biodiversity 290, 292
  global burden of disease and deaths 292–293
  global production 274
  human biomonitoring 287–288
  human exposures and health effects 292–296
  important properties of 267
  industrial accidents 293
  neurological effects 293–294
  non-carcinogenic effects 294
  soil properties and fate 278–280
  sources and emissions 274–277
  types and classification of 256–259, 267–274
Toluene, various 5, 26, 30, 35, 233, 238, 429
Total daily dose or intake (TDD or TDI) 93
Total organic carbon (TOC) 147
Toxic algal blooms 55, 57, 133, 162, 171–173, 414
Toxicant behavior 94–95
Toxicant-dose relationships 109, 473–476
Toxicant sensitivity distribution (TSD) 100–102, 108
Toxicants, types and classifications of 108–109
Toxic chemical pollution, overview 520–521
Toxicity
  chemicals mixtures 13, 104, 259, 271, 290, 417
  endpoints 205–206, 480–483
  mechanisms 91, 95, 104, 121–122, 204, 209, 214
  of pesticide mixtures 206, 209, 213–214, 217

Toxicological reactions 95
Toxicology 16, 91
  genetic 113
Toxic organic chemicals see TOCs (toxic organic chemicals)
Toxic substances control act, USA (TSCA) 487, 493
Transformative changes, sustainability 525–526, 528–529
Trans-generational effects 122
Transport processes, environmental 81–82
Triazines, types 180, 188, 192
Tributyl tin 29, 181, 214, 291, 319, 321
Trichloromethane (THM) 94
Trophic states or levels 172–173, 215–216
Turbidity, effects 559–560

u

Uncertainty, in risk assessment 469, 475–476, 481
Units of measurement, chemical pollutants 25–26
Uranium 303, 306, 543–548
Urbanization, introduction 1–3, 16
Urea 4, 7, 29, 165, 336, 414
US EPA (United States Environmental Protection Agency) 493

v

Vinyl chloride 36, 38, 116, 256, 260, 267–269
Viruses, introduction 23, 27, 31, 33, 43, 113–118
Volatile organic chemicals (VOCs) 24, 30, 256, 267, 293–294, 337, 341, 347
Volatilization, examples 194–195, 280–281

w

Waste contaminants 429–430
  behavior and fate processes 430–431
  exposure pathways and scenarios 430–432

Waste generation
  and emissions 423–424, 524
  global estimates 423–424
Waste management 421, 424–425
  evaluation of health and safety 438
Wastes 1–4, 421, 524
  bioreactor 427–428
  composting 427
  electronic and electrical (e-Wastes) 439
  environmental effects from waste streams 432–435, 524
  hazardous 422–423
  hazardous treatment 428–429
  hierarchy 425
  human health effects 435–440, 524
  incineration 426–427
  liquid 422
  minimization methods 424–425
  solid landfills 425–426
  types and properties of 422–423
Water pollution 2, 24, 33, 523–524
Water vapor 370
Weather 333, 365
Weight-of-evidence 18, 43, 214, 440, 448, 478, 490, 524
Well-being concept 41–42
Wetlands 46, 55
Wildlife, population declines or losses 54
World Health Organization (WHO) 1
World Meteorological Organization (WMO) 447, 449

x

Xylenes, various 5, 30, 35, 233

z

Zero waste 425, 516
Zinc, various 77, 303, 306, 310, 316, 322, 400, 423, 520
Zooplankton 132, 172, 215, 248, 286, 320